W9-CUK-510

Springer Series in

CHEMICAL PHYSICS 87

Springer Series in

CHEMICAL PHYSICS

The purpose of this series is to provide comprehensive up-to-date monographs in both well established disciplines and emerging research areas within the broad fields of chemical physics and physical chemistry. The books deal with both fundamental science and applications, and may have either a theoretical or an experimental emphasis. They are aimed primarily at researchers and graduate students in chemical physics and related fields.

71 **Ultrafast Phenomena XIII**
Editors: D. Miller, M.M. Murnane, N.R. Scherer, and A.M. Weiner

72 **Physical Chemistry of Polymer Rheology**
By J. Furukawa

73 **Organometallic Conjugation**
Structures, Reactions and Functions of d–d and d–π Conjugated Systems
Editors: A. Nakamura, N. Ueyama, and K. Yamaguchi

74 **Surface and Interface Analysis**
An Electrochmists Toolbox
By R. Holze

75 **Basic Principles in Applied Catalysis**
By M. Baerns

76 **The Chemical Bond**
A Fundamental Quantum-Mechanical Picture
By T. Shida

77 **Heterogeneous Kinetics**
Theory of Ziegler-Natta-Kaminsky Polymerization
By T. Keii

78 **Nuclear Fusion Research**
Understanding Plasma-Surface Interactions
Editors: R.E.H. Clark and D.H. Reiter

79 **Ultrafast Phenomena XIV**
Editors: T. Kobayashi, T. Okada, T. Kobayashi, K.A. Nelson, S. De Silvestri

80 **X-Ray Diffraction by Macromolecules**
By N. Kasai and M. Kakudo

81 **Advanced Time-Correlated Single Photon Counting Techniques**
By W. Becker

82 **Transport Coefficients of Fluids**
By B.C. Eu

83 **Quantum Dynamics of Complex Molecular Systems**
Editors: D.A. Micha and I. Burghardt

84 **Progress in Ultrafast Intense Laser Science I**
Editors: K. Yamanouchi, S.L. Chin, P. Agostini, and G. Ferrante

85 **Quantum Dynamics Intense Laser Science II**
Editors: K. Yamanouchi, S.L. Chin, P. Agostini, and G. Ferrante

86 **Free Energy Calculations**
Theory and Applications in Chemistry and Biology
Editors: Ch. Chipot and A. Pohorille

87 **Analysis and Control of Ultrafast Photoinduced Reactions**
Editors: O. Kühn and L. Wöste

O. Kühn L. Wöste (Eds.)

Analysis and Control of Ultrafast Photoinduced Reactions

With 374 Figures, 239 in color

Dr. Oliver Kühn
Freie Universität Berlin
Institut f. Chemie und Biochemie
Takustr. 3
D-14195 Berlin, Germany
E-Mail: ok@chemie.fu-berlin.de

Professor Ludger Wöste
Freie Universität Berlin
Institut für Experimentalphysik
Arnimallee 14
D-14195 Berlin, Germany
E-Mail: woeste@physik.fu-berlin.de

Series Editors:
Professor A. W. Castleman, Jr.
Department of Chemistry, The Pennsylvania State University
152 Davey Laboratory, University Park, PA 16802, USA

Professor J.P. Toennies
Max-Planck-Institut für Strömungsforschung, Bunsenstrasse 10
37073 Göttingen, Germany

Professor K. Yamanouchi
University of Tokyo, Department of Chemistry
Hongo 7-3-1, 113-0033 Tokyo, Japan

Professor W. Zinth
Universität München, Institut für Medizinische Optik
Öttingerstr. 67, 80538 München, Germany

The cover picture shows wave packet interferences in excited Bromine dimers
(courtesy of M. Héjjas and N. Schwentner)

ISSN 0172-6218

ISBN-10 3-540-68037-3 Springer Berlin Heidelberg New York
ISBN-13 978-3-540-68037-6 Springer Berlin Heidelberg New York

Library of Congress Control Number: 2006938281

Springer is a part of Springer Science+Business Media.

springer.com

Typesetting: by the authors and techbooks using a Springer LaTeX macro package
Cover design: eStudio Calamar Steinen

Printed on acid-free paper SPIN: 11847595 54/3100/techbooks - 5 4 3 2 1 0

Preface

Dynamical processes in molecules like bond shaking, breaking or making commonly take place on a time scale from the pico- down to the femtosecond range. The advent of equally fast laser sources and real-time observation schemes like pump-probe spectroscopy has facilitated the direct insight into such processes when initiated by light. In parallel the development of advanced computational methods treating the dynamics of photoexcited molecular systems allowed a convergence between theoretical description and experimental observation of such ultrafast dynamical processes. Consequently, the idea emerged, not only to analyze, but also to control molecular dynamics in real time by adequately designed light fields. Stimulated by theoretical concepts for influencing the motion of molecular wave packets by means of simple few-parameter electromagnetic field sequences, experiments were driven toward a practical realization of arbitrarily shaped laser pulses. This development culminated in the active feedback control of even complex systems. In addition this offers the unique possibility not only to determine the outcome of chemical reactions, but also to retrieve specific information about the chosen dynamical pathways, that is, to perform analysis by control.

This book illustrates a vital research field by covering a broad spectrum of molecular systems with growing complexity while demonstrating at the same time the convergence of experimental and theoretical approaches. After a general introduction in Chapter 1, Chapter 2 starts with small isolated molecules in the unperturbed environment of the gas phase and Chapter 3 proceeds to more complex systems, but still in vacuum. A higher level of complexity is then reached in Chapter 4 where small molecules in a rare gas matrices are discussed serving as prototype examples for condensed phase dynamics. This establishes the links toward applications which first focus in Chapter 5 on ultrafast dynamics at surfaces and interfaces. Then larger molecules and clusters in intense laser fields are treated in Chapter 6. Chapter 7 interrogates the vibrational, Chapter 8 the electronic spectroscopy of solvated systems. Finally, in Chapter 9 the ladder of complexity reaches out to real-time analysis of photoinduced processes in biological systems. The structure of this outline reflects the coordinated activities of the Collaborative Research Center Sfb450 named - like this book - “Analysis and Control of Ultrafast Photoinduced Processes”. This Sfb comprises currently seventeen experimental and theoretical research groups in the larger Berlin area.

On a Sfb workshop, held here in Spring 2006 together with internationally renowned researchers, the idea emerged, to write a book which should achieve two goals: The first one is to provide a systematic and didactic introduction into the field suitable for graduate students from Physics, Chemistry, and Biology. The second one, is to present the state-of-affairs of the field such that it may serve as a useful reference for the experts. The complexity-oriented treatment of the theme was an essential element for realizing this concept. This was assured by the assignment of coordinators to the chapters, who harmonized the individual contributions.

Most of the results documented in the following would not have been possible without the generous financial support from the Deutsche Forschungsgemeinschaft through the Collaborative Research Center Sfb450 “Analysis and Control of Ultrafast Photoinduced Processes” during the last eight years. We are also most grateful to the participating institutions, in particular, the Freie Universität Berlin, which hosts our administration. In this regard we are indebted to the professional and enthusiastic support of Peter Abt and Sigrid Apelt.

Berlin,
October 2006

Oliver Kühn
Ludger Wöste

Contents

List of Contributors

M. Bargheer
Max-Born-Institut für Nichtlineare Optik und Kurzzeitspektroskopie
Max-Born-Str. 2a
D-12489 Berlin, Germany
bargheer@mbi-berlin.de

T. M. Bernhardt
Institut für Oberflächenchemie und Katalyse
Universität Ulm
D-89069 Ulm, Germany
thorsten.bernhardt@uni-ulm.de

V. Bonačić-Koutecký
Institut für Chemie
Humboldt-Universität zu Berlin
D-12489 Berlin, Germany
vbk@chemie.hu-berlin.de

A. Borowski
Institut für Chemie und Biochemie
Freie Universität Berlin
Takustr. 3
D-14195 Berlin, Germany
borowski@chemie.fu-berlin.de

B. Brauer
The Fritz Haber Research Center and Department of Physical Chemistry
The Hebrew University of Jerusalem
Jerusalem, Israel
bbrauer@fh.huji.ac.il

J. Bredenbeck
Physikalisch Chemisches Institut
Universität Zürich
Winterthurer Str. 190
CH-8057 Zürich, Switzerland
jbredenb@pci.unizh.ch

C. Bressler
Laboratoire de Spectroscopie Ultrarapide
École Polytechnique Fédérale de Lausanne
Institut des Sciences et Ingénierie Chimiques
Faculté des Sciences de Base, BSP
CH-1015 Lausanne-Dorigny, Switzerland
christian.bressler@epfl.ch

B. Brüggemann
Chemical Physics
Lund University
P.O. Box 124
SE-22100 Lund, Sweden
Ben.Bruggemann@chemphys.lu.se

F. Burmeister
Bessy GmbH
Albert-Einstein-Str. 15
D-12489 Berlin, Germany
florian.burmeister@icm.uu.se

T. Burnus
II.Physikalisches Institut
Universität zu Köln
Zülpicher Str. 77
D-50937 Köln, Germany
burnus@ph2.uni-koeln.de

A. Castro
Institut für Theoretische Physik
Fachbereich Physik
Freie Universität Berlin
Arnimallee 14
D-14195 Berlin, Germany
alberto@physik.fu-berlin.de

A. Cohen
The Fritz Haber Research Center and Department of Physical Chemistry
The Hebrew University of Jerusalem
Jerusalem, Israel
arikco@fh.huji.ac.il

I. Corral
Institut für Chemie und Biochemie
Freie Universität Berlin
Takustr. 3
D-14195 Berlin, Germany
ines@chemie.fu-berlin.de

M. Chergui
Laboratoire de Spectroscopie Ultrarapide
École Polytechnique Fédérale de Lausanne
Institut des Sciences et Ingénierie Chimiques
Faculté des Sciences de Base, BSP
CH-1015 Lausanne-Dorigny, Switzerland
majed.chergui@epfl.ch

A. L. Dobryakov
Humboldt Universität zu Berlin
Department of Chemistry
Brook-Taylor-Str. 2
D-12489 Berlin, Germany
dobr@chemie.hu-berlin.de

J. Dreyer
Max-Born-Institut für Nichtlineare Optik und Kurzzeitspektroskopie
Max-Born-Str. 2a
D-12489 Berlin
dreyer@mbi-berlin.de

W. R. Duncan
Department of Chemistry
University of Washington
Seattle, WA 98195-1700, USA
wat@u.washington.edu

W. Eberhardt
Bessy GmbH
Albert-Einstein-Str. 15
D-12489 Berlin, Germany
eberhardt@bessy.de

T. Elsaesser
Max-Born-Institut für Nichtlineare Optik und Kurzzeitspektroskopie
Max-Born-Str. 2a
D-12489 Berlin, Germany
elsasser@mbi-berlin.de

V. Engel
Institut für Physikalische Chemie
Universität Würzburg
Am Hubland
D-97074 Würzburg, Germany
voen@phys-chemie.uni-wuerzburg.de

N. P. Ernsting
Humboldt Universität zu Berlin
Department of Chemistry
Brook-Taylor-Str. 2
D-12489 Berlin, Germany
nernst@chemie.hu-berlin.de

R. Ernstorfer
Hahn-Meitner-Institut
Abt. Dynamik von Grenzflächenreaktionen
Glienicker Str. 100
D-14109 Berlin, Germany
ernstorfer@hmi.de

H. Fidder
Institut für Experimentalphysik
Freie Universität Berlin
Arnimallee 14
D-14195 Berlin, Germany
fidder@physik.fu-berlin.de

C. Frischkorn
Institut für Experimentalphysik
Fachbereich Physik
Freie Universität Berlin
Arnimallee 14
D-14195 Berlin, Germany
christian.frischkorn@physik.fu-berlin.de

M. Fushitani
Institute for Molecular Scinece
Okazaki, Aichi, 444-8585, Japan
fusitani@ims.ac.jp

W. Gawelda
Laboratoire de Spectroscopie Ultrarapide
École Polytechnique Fédérale de Lausanne
Institut des Sciences et Ingénierie Chimiques
Faculté des Sciences de Base, BSP
CH-1015 Lausanne-Dorigny, Switzerland
Wojciech.gawelda@epfl.ch

R.B. Gerber
The Fritz Haber Research Center and Department of Physical Chemistry
The Hebrew University of Jerusalem
Jerusalem, Israel
benny@fh.huji.ac.il

L. González
Institut für Chemie und Biochemie
Freie Universität Berlin
Takustr. 3
D-14195 Berlin, Germany
leti@chemie.fu-berlin.de

J. Güdde
Fachbereich Physik
und Zentrum für Materialwissenschaften
Philipps-Universität
D-35032 Marburg, Germany
jens.guedde@physik.uni-marburg.de

M. Gühr
Stanford PULSE Center, SLAC
2575 Sand Hill Road,
Menlo Park, CA 94075, USA
mguehr@stanford.edu

R. van Grondelle
Faculty of Sciences
Vrije Universiteit Amsterdam
De Boelelaan 1081
1081 HV Amsterdam, The Netherlands
rienk@nat.vu.nl

E. K. U. Gross
Institut für Theoretische Physik
Fachbereich Physik
Freie Universität Berlin
Arnimallee 14
D-14195 Berlin, Germany
hardy@physik.fu-berlin.de

L. Gundlach
Hahn-Meitner-Institut
Abt. Dynamik von Grenzflächenreaktionen
Glienicker Str. 100
D-14109 Berlin, Germany
gundlach@hmi.de

S. Haacke
École Polytechnique Fédérale de Lausanne
Lab. of Ultrafast Spectroscopy
ISIC, FSB - BSP
CH-1015 Lausanne-Dorigny, Switzerland
Stefan.Haacke@ipcms.u-strasbg.fr

P. Hamm
Physikalisch Chemisches Institut
Universität Zürich
Winterthurer Str. 190
CH-8057 Zürich, Switzerland
phamm@pci.unizh.ch

A. Heidenreich
School of Chemistry
Tel Aviv University
Ramat Aviv
69978 Tel Aviv, Israel
andreas@chemsg1.tau.ac.il

G. von Helden
Fritz-Haber-Institut
der Max-Planck-Gesellschaft
Faradayweg 4-6
D-14195 Berlin, Germany
helden@fhi-berlin.mpg.de

I. V. Hertel
Max-Born-Institut für Nichtlineare Optik und Kurzzeitspektroskopie
Max-Born-Str. 2a
D-12489 Berlin, Germany
hertel@mbi-berlin.de

K. Heyne
Institut für Experimentalphysik
Fachbereich Physik
Freie Universität Berlin
Arnimallee 14
D-14195 Berlin, Germany
heyne@physik.fu-berlin.de

U. Höfer
Fachbereich Physik
und Zentrum für Materialwissenschaften
Philipps-Universität
D-35032 Marburg, Germany
hoefer@physik.uni-marburg.de

I. Horenko
Institute of Mathematics II
Freie Universität Berlin
Arnimallee 6
D-14195 Berlin, Germany
horenko@math.fu-berlin.de

H. Ibrahim
Institut für Experimentalphysik
Fachbereich Physik
Freie Universität Berlin
Arnimallee 14
D-14195 Berlin, Germany,
heide.ibrahim@physik.fu-berlin.de

J. Jortner
School of Chemistry
Tel Aviv University
Ramat Aviv
699878 Tel Aviv, Israel
jortner@chemsg1.tau.ac.il

A. Kammrath
Department of Chemistry
University of California
Berkeley, CA 94720, USA
aster@berkeley.edu

J.T.M. Kennis
Faculty of Sciences
Vrije Universiteit Amsterdam
De Boelelaan 1081
1081 HV Amsterdam, The Netherlands
john@nat.vu.nl

T. Kiljunen
University of Jyväskylä
Department of Chemistry
P.O. Box 35
FIN-40014 Jyväskylä, Finland
toni.kiljunen@jyu.fi

S. K. Kim
School of Chemistry
College of Natural Sciences
Seoul National University
Seoul 151-747, Korea
seongkim@snu.ac.kr

T. Klamroth
Institut für Chemie
Universität Potsdam
Karl-Liebknecht-Str. 25
D-14476 Potsdam, Germany
klamroth@rz.uni-potsdam.de

C. P. Koch
Institut für Theoretische Physik
Fachbereich Physik
Freie Universität Berlin
Arnimallee 14
D-14195 Berlin, Germany
ckoch@physik.fu-berlin.de

M. V. Korolkov
Institut für Chemie und Biochemie
Freie Universität Berlin
Takustr. 3
D-14195 Berlin, Germany
and B.I. Stepanov Institute of Physics
National Academy of Sciences
Skaryna AVe. 70
220602 Minsk, Belarus
korolkov@chemie.fu-berlin.de

O. Kühn
Institut für Chemie und Biochemie
Freie Universität Berlin
Takustr. 3
D-14195 Berlin, Germany
ok@chemie.fu-berlin.de

T. Laarmann
Max-Born-Institut für Nichtlineare Optik und Kurzzeitspektroskopie
Max-Born-Str. 2a
D-12489 Berlin, Germany
laarmann@mbi-berlin.de

D.S. Larsen
Department of Chemistry
University of California, Davis
One Shields Avenue
Davis, CA 95616, USA
dlarsen@ucdavis.edu

I. Last
School of Chemistry
Tel Aviv University
Ramat Aviv
69978 Tel Aviv, Israel
isidore@post.tau.ac.il

A. Lauer
Institut für Experimentalphysik
Fachbereich Physik
Freie Universität Berlin
Arnimallee 14,
D-14195 Berlin, Germany
Alexandra.Lauer
@physik.fu-berlin.de

R. J. Levis
Temple University
Philadelphia, PA 19122, USA
rjlevis@temple.edu

A. Lindinger
Institut für Experimentalphysik
Fachbereich Physik
Freie Universität Berlin
Arnimallee 14
D-14195 Berlin, Germany
lindin@physik.fu-berlin.de

J. Manz
Institut für Chemie und Biochemie
Freie Universität Berlin
Takustrasse 3
D-14195 Berlin, Germany
manz@chemie.fu-berlin.de

M. A. L. Marques
Departamento de Física
Universidade de Coimbra
Rua Larga
3004-516 Coimbra, Portugal
marques@teor.fis.uc.pt

V. May
Institut für Physik
Humboldt Universität zu Berlin
Newtonstr. 15
D-12489 Berlin, Germany
may@physik.hu-berlin.de

E. Meerbach
Institute of Mathematics II
Freie Universität Berlin
Arnimallee 6
D-14195 Berlin, Germany
meerbach@math.fu-berlin.de

G. Meijer
Fritz-Haber-Institut
der Max-Planck-Gesellschaft
Faradayweg 4-6
D-14195 Berlin, Germany
meijer@fhi-berlin.mpg.de

A. Mirabal
Institut für Experimentalphysik
Fachbereich Physik
Freie Universität Berlin
Arnimallee 14, D-14195 Berlin, Germany
Aldo.Mirabal@physik.fu-berlin.de

R. Mitrić
Institut für Chemie
Humboldt-Universität zu Berlin
D-12489 Berlin, Germany
mitric@chemie.hu-berlin.de

F. van Mourik
École Polytechnique Fédérale de Lausanne
Lab. of Ultrafast Spectroscopy
ISIC, FSB - BSP
CH-1015 Lausanne-Dorigny, Switzerland
frank.vanmourik@epfl.ch

M. Neeb
Bessy GmbH
Albert-Einstein-Str. 15
D-12489 Berlin, Germany
neeb@bessy.de

M. Nest
Institut für Chemie
Universität Potsdam
Karl-Liebknecht-Str. 25
D-14476 Potsdam, Germany
mnest@rz.uni-potsdam.de

D. M. Neumark
Department of Chemistry
University of California
Berkeley, CA 94720, USA
dneumark@berkeley.edu

E. T. J. Nibbering
Max-Born-Institut für Nichtlineare Optik und Kurzzeitspektroskopie
Max-Born-Str. 2a
D-12489 Berlin, Germany
nibberin@mbi-berlin.de

E. Papagiannakis
Faculty of Sciences
Vrije Universiteit Amsterdam
De Boelelaan 1081
1081 HV Amsterdam, The Netherlands
E.Papagiannakis@few.vu.nl

O. V. Prezhdo
University of Washington
Department of Chemistry
Seattle, WA 98195-1700, USA
prezhdo@u.washington.edu

D. A. Romanov
Temple University
Philadelphia, PA 19122, USA
daroman@temple.edu

P. Saalfrank
Institut für Chemie
Universität Potsdam
Karl-Liebknecht-Str. 25
D-14476 Potsdam, Germany
peter.saalfrank
@rz.uni-potsdam.de

A. Saenz
Institut für Physik
Humboldt-Universität zu Berlin
Hausvogteiplatz 5-7
D-10117 Berlin, Germany
saenz@physik.hu-berlin.de

S. Schenkl
École Polytechnique Fédérale de Lausanne
Lab. of Ultrafast Spectroscopy
ISIC, FSB - BSP
CH-1015 Lausanne-Dorigny, Switzerland
selma.schenkl
@ibmt.fraunhofer.de

B. Schmidt
Institute of Mathematics II
Freie Universität Berlin
Arnimallee 6
D-14195 Berlin, Germany
burkhard@math.fu-berlin.de

M. Schröder
Institut für Chemie und Biochemie
Freie Universität Berlin
Takustr. 3
D-14195 Berlin, Germany
maikesch@chemie.fu-berlin.de

C. Schütte
Institute of Mathematics II
Freie Universität Berlin
Arnimallee 6
D-14195 Berlin, Germany
schuette@math.fu-berlin.de

T. Schultz
Max-Born-Institut für Nichtlineare Optik und Kurzzeitspektroskopie
Max-Born-Str. 2a
D-12489 Berlin, Germany
schultz@mbi-berlin.de

C. P. Schulz
Max-Born-Institut für Nichtlineare Optik und Kurzzeitspektroskopie
Max-Born-Str. 2a
D-12489 Berlin, Germany
cps@mbi-berlin.de

N. Schwentner
Institut für Experimentalphysik
Fachbereich Physik
Freie Universität Berlin
Arnimallee 14
D-14195 Berlin, Germany,
schwentner@physik.fu-berlin.de

J. Stanzel
Bessy GmbH
Albert-Einstein-Str. 15
D-12489 Berlin, Germany
stanzel@bessy.de

D. Tannor
Department of Chemical Physics
Weizmann Institute of Science
76100 Rehovot, Israel
David.Tannor@weizmann.ac.il

M. Vengris
Vilnius University
Faculty of Physics Laser Research Center
Sauletekio 10
LT-10223 Vilnius Lithuania
Mikas.Vengris@ff.vu.lt

L. Wang
Institut für Physik
Humboldt Universität zu Berlin
Newtonstr. 15
D-12489 Berlin, Germany
luxia@physik.hu-berlin.de

F. Willig
Hahn-Meitner-Institut
Abt. Dynamik von Grenzflächenreaktionen
Glienicker Str. 100
D-14109 Berlin, Germany
willig@hmi.de

M. Wolf
Fachbereich Physik
Freie Universität Berlin
Arnimallee 14
D-14195 Berlin, Germany
wolf@physik.fu-berlin.de

J. P. Wolf
GAP-Biophotonics
University of Geneva
20, rue de l'Ecole de Médecine
CH-1211 Geneva 4, Switzerland
and LASIM (UMR 5579)
Université Claude Bernard Lyon 1
43, Bd du 11 Novembre 1918
F-69622 Villeurbanne, France
jean-pierre.wolf@physics.unige.ch

L. Wöste
Fachbereich Physik
Freie Universität Berlin
Arnimallee 14
D-14195 Berlin, Germany
woeste@physik.fu-berlin.de

G. van der Zwan
Faculty of Sciences
Vrije Universiteit Amsterdam
De Boelelaan 1081
1081 HV Amsterdam, The Netherlands
zwan@few.vu.nl

1

Introduction

Joshua Jortner

School of Chemistry, Tel Aviv University, Israel

On dynamics

Remarkable progress has been made in the elucidation of ultrafast dynamics and its control driven by femtosecond laser pulses in small molecules, large-scale molecular systems, clusters, nanostructures, surfaces, condensed phase and biomolecules. The exploration of photoinduced ultrafast response, dynamics, reactivity and function in ubiquitous molecular, nanoscale, macroscopic and biological systems pertains to the interrogation and control of the phenomena of energy acquisition, storage and disposal, as explored from the microscopic point of view. Photoinduced ultrafast processes in chemistry, physics, material science, nanoscience, and biology constitute a broad, interdisciplinary, novel and fascinating research area, blending theoretical concepts and experimental techniques in a wide range of scientific disciplines. The foundations for the analysis and control of ultrafast photoinduced processes were laid during the last eighty years with the development of nonradiative dynamics from small molecules to biomolecules [1–6], while during the last twenty years remarkable progress was made with the advent of femtosecond dynamics and control at the temporal resolution of nuclear motion [1, 7–12]. This scientific historical development can be artistically described by ascending the 'magic mountain' of molecular, cluster, condensed phase and biological dynamics by several paths (Fig. 1.1), all of which go heavenwards toward a unified and complete description of structure-electronic level structure-spectroscopy-dynamics-function relations and correlations.

The genesis of intramolecular nonradiative dynamics dates back to the origins of quantum mechanics, when the 1926 groundbreaking work of Schrödinger and Heisenberg laid the foundations for the description of time-dependent phenomena in the quantum world. In 1928 Bonhoeffer and Farkas [13] observed that predissociation in the electronically excited ammonia molecule, which involves the decay of a metastable state to a dissociative continuum, i.e., $NH_3 \overset{h\nu}{\rightarrow} NH_3^* \overset{1/\tau}{\rightarrow} NH_2 + H$, is manifested by spectral line broadening, with

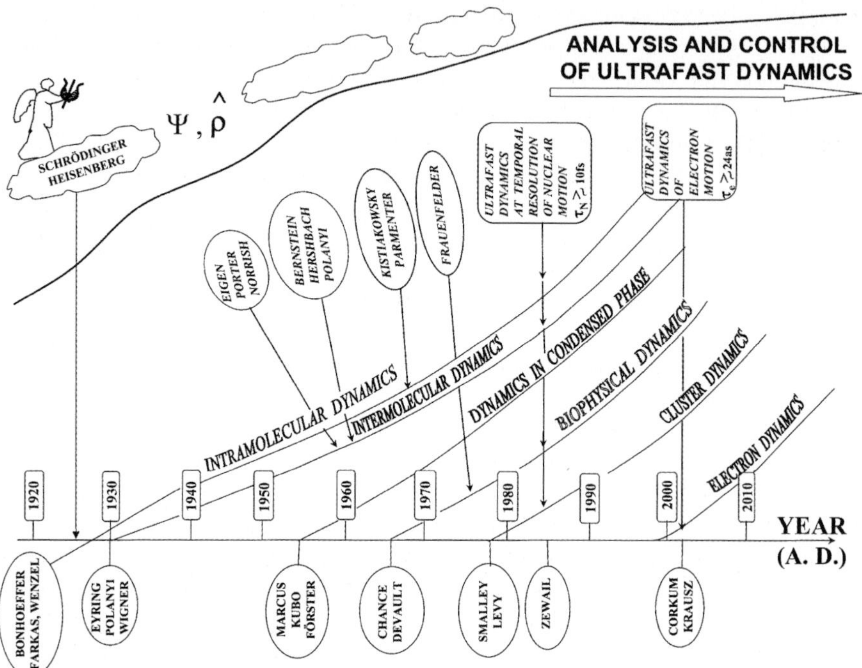

Fig. 1.1. An artist's view of the 'magic mountain' of the evolution of dynamics. The names of some of the pioneers who initiated each scientific area are marked on the paths.

a spectral linewidth Γ that considerably exceeds the radiative linewidth. This seminal work established the first spectroscopic-dynamic relation, providing experimental verification of the Heisenberg energy-time uncertainty relation, and pioneering the field of intramolecular dynamics. At about the same time, Wenzel [14] worked on another facet of nonradiative dynamics for the theory of atomic autoionization, establishing the basic unified theory of nonradiative processes. For a metastable (predissociating or autoionizing) state into a (dissociative or ionization) continuum, the decay lifetime τ was quantified in terms of the Golden Rule $\tau^{-1} = (2\pi/\hbar)|V|^2(dn/dE)$, where V is the matrix element of the Hamiltonian inducing the nonradiative transition, and dn/dE is the density of states. The Golden Rule played a central role in providing a conceptual basis for intramolecular dynamics from reactive processes in diatomics to radiationless transitions in 'isolated' large molecules. In 1931 Eyring and Polanyi [15] constructed the first potential energy surface for chemical reactions, a concept with continuous impact on the field. In the context of the present book it is noteworthy to point out that both studies of Bonhoeffer and Farkas [13] and of Eyring and Polanyi [15] were conducted in Berlin-Dahlem, the location of the Sfb 450 research center, (see below). Further important developments in the realm of intermolecular dynamics were pioneered in the 1930s by Eyring, Polanyi, and Wigner, with remarkable evolution in the 1970s for collision dynamics in molecular beams [16]. Subsequently, experiment and theory moved toward the realm of large, complex systems. A distinct field of dynamics in the condensed phase was pioneered in the 1950s by Marcus [17], Förster [18], and Kubo [19]. Marcus [17] advanced the cornerstone of the electron transfer theory in solution in terms of the Gaussian free energy (ΔE) relation for the rate $k \propto \exp[-(\Delta E + \lambda)^2/4\lambda k_B T]$ (where λ is the medium reorganization energy), leading to central kinetic and spectroscopic results for correlation rules, free-energy relations, uniqueness of the inverted region and charge transfer spectroscopy. Conceptually and physically isomorphic classes of condensed phase dynamics pertain to the Förster theory of electronic energy transfer [18]. At the same time, Kubo and Toyozawa [19] developed the theory of electron-hole recombination in semiconductors, which bears analogy to electron transfer, although at that time the interrelationship between their work and the Marcus theory was not realized. The extension of dynamics to the protein medium emerged in the 1960s with the development of biophysical electron transfer dynamics, with experimental and theoretical studies of charge separation in photosynthesis [20]. In 1975 further progress by Fraunfelder [21] in biophysical dynamics led to the description of the energy landscapes of proteins. Concurrently, progress was made in the dynamics of large-scale chemical systems. In 1965 Kistiakowsky and Parmenter [22] observed intersystem crossing within the benzene molecule in the low-pressure gas phase, stating that 'a strictly intermolecular nonradiative transition is difficult to reconcile with concepts of quantum mechanics' [22]. Three years later the theory of intramolecular radiationless transitions in 'isolated' molecules was developed [23]. In 1969, the theory of time-dependent coherence phenomena in large molecules was advanced [24] providing the conceptual basis for molecular wave packet dynamics. The conceptual framework for intramolecular radiationless transitions and coherence effects encompassed both interstate dynamics involving internal conversion, a well as intersystem crossing, and intrastate dynamics involving vibrational energy redistribution. In the 1970s, the practice and concepts of dynamics moved toward large finite systems. Cluster dynamics, which constituted the border line between molecular and condensed phase phenomena, emerged with the work of Smalley, Wharton and Levy [25] on the vibrational predissociation of HeI_2 clusters. Cluster dynamics built bridges to the response, dynamics, reactivity and structure of large finite systems [26–28], i.e., size-selected clusters [29], superfluid quantum clusters and nanodroplets [30], finite ultracold gases (temperatures of 100μK–10nK) [31] and nanostructures [32]. Some fascinating dynamic processes involve resonant and dissipative vibrational energy flow and intramolecular vibrational energy distribution in clusters [33], selective, size dependent reactivity and microscopic catalysis on metal clusters [27,28], Coulomb instability leading to fission or Coulomb explosion of multicharged elemental, molecular and metal clusters [34, 36], the expansion of optical molasses that are isomorphic to cluster Coulomb explosion [31], as well as transport of elementary excitations in nanostructures [35] that opens avenues for molecular- and nano-electronics [35].

Since 1985 the exploration of ultrafast chemical and biological dynamics stemmed from concurrent progress in experiment and theory. The advent of femtosecond dynamics by Zewail in 1987 [37–39] allowed for the exploration of dynamics in molecules, clusters, condensed phase, surfaces and biomolecules on time scales for intramolecular motion (10-100fs) and for intermolecular motion (100fs-1ps) [1,7–9,11]. Notable novel phenomena pertained to radiationless transition, wave packet dynamics, coherence effects, transition-state spectroscopy, cluster dynamic size effects, nonadiabatic condensed phase dynamics, ultrafast electron and proton transfer, charge separation in photosynthesis, and nonlinear optical interactions [1,7–9,11,40–45]. Recent advances in the elucidation of structure-dynamics-function relations in molecules, clusters, nanostructures, surfaces, and biomolecules are described in this book.

The interrogation and analysis of dynamics at the temporal resolution of nuclear motion raised important issues regarding the manipulation of the optical properties, response, reactivity and functionalism by the use of shaped laser pulses. Since the middle 1980s, the advances in the realm of nuclear dynamics driven by femtosecond laser pulses underling the theoretical proposals for different optical control schemes that rested on the dynamic response of a molecular target to the temporal shape, phase and intensity of a laser pulse. Early considerations of optical control based on the coherence properties of infrared (IR) laser radiation were advanced by Paramonov and Savva [46] and subsequently by Joseph and Manz [47]. General control concepts and schemes [5,48,49] rest on the Tannor-Rice theory of pump-dump control [48–50], the Shapiro-Brumer theory of coherent control [51–53], and the theory of Rabitz and his colleagues for control by the use of tailored laser fields produced by pulse shaping [54–57]. The foundations of optimal control theory by pulse shaping [5] were laid by Tannor and Rice [48] on the basis of a variational formulation, where the optimized pulses are obtained from the (radiation field dependent) functional $J = \langle \Psi(t)|\hat{P}|\Psi(t)\rangle$, where $\hat{P}$ is the projection operator selecting the desired target. The general concept was extended by Rabitz and his colleagues [54] from the perturbative domain of weak laser fields to arbitrarily strong optimal laser pulses. Another important progress was achieved by Judson and Rabitz [56] with the adaptation of algorithms for closed loop learning for pulse shaping. In the realm of control of nuclear dynamics, these significant developments were quite unique, as theory preceded experiment. These theoretical schemes stimulated significant control experiments. These were first carried out on a variety of systems with increasing size from metallic dimers and trimers to clusters [58–80], and later on complex systems of large molecules even in solution [81–93], confirming theoretically proposed concepts. This book describes progress in this important field in both experiment and theory, and strives toward the creation of a conceptual framework and experimental methodology of optical control.

To strike the last cord in this historical overview, we focus on most recent developments in the area of dynamics and control, which pertain to 'pure' electron dynamics in chemical and physical systems [94–97]. Electron dynamics involves changes in electronic states, with the nuclear motion being frozen. Characteristic temporal limits of ~ 24 attoseconds for electron dynamics, correspond to one atomic unit of time. Ultrafast dynamics and its control are currently moving from 'femtosecond chemistry' with the time-resolution of nuclear motion toward 'attosecond chemistry' with the temporal resolution of electronic motion.

The Sfb 450 research program

During the last decade the research area of ultrafast, femtosecond dynamics on the time scale of nuclear motion moved from the realm of analysis of ultrafast processes toward the new horizons of analysis and control of nuclear dynamics. These directions were advanced by the research groups participating in the Collaborative Research Center "Analysis and control of ultrafast photoinduced reactions" (Sfb 450) research program supported by the Deutsche Forschungsgemeinschaft. The central goals of the Sfb 450 program pertain to the following three interconnected elements: (i) To couple suitably designed laser fields with the electronic and nuclear level structure in a variety of molecular, cluster, surface, condensed-phase and biological systems, from small molecules with a few number of degrees of freedom to large systems with increasing complexity. (ii) To characterize the resulting nuclear-electronic motion of the system by the real-time interrogation of its dynamics. (iii) To direct the nuclear motion for the attainment of a stable product state that was not selectively obtained by conventional or by photochemical methods. To achieve these goals, it was imperative to develop methods of analysis for the interrogation and specification of the dynamics, together with the characterization of the temporal structure, amplitude, phase and intensity of the laser field that will allow for the control of the dynamics. Analysis and control of electron-nuclear dynamics driven by suitably shaped laser fields constitutes the two cornerstones of this research program. Such controlled nuclear dynamic processes in a nuclear-electronic level structure, coupled to a suitably shaped laser field, supplements and complements photoselective chemistry with additional elements of manipulation of functionality.

The first stage of this research program (1998-2001) focused on the analysis of ultrafast reactions in suitable model systems. The experimental pump-probe methodology involved excitation by an ultrashort transform-limited laser pulse, followed by the interrogation by a second, time-delayed laser pulse. The systems studied ranged from diatomics and triatomics to small metal clusters, with a small number of degrees of freedom, to complex systems, i.e., polyatomic molecules, biomolecules and large clusters in the gas phase, in liquids, in solids and on surfaces. The dynamics of the system was then reflected in the time dependence of the signal. The experimental and theoretical work focused on dynamics in the time domain. At that early stage, the control of dynamics [46–57] was still a vision, being in the initial developmental steps from 'theoreticians' dreams' toward experimental reality. During the second period of this research program (2001-2004) control of dynamics based on the Tannor-Rice and the Rabitz schemes was experimentally realized for small systems. Suitable modulators were used in conjunction with closed-loop genetic algorithms for the shaping of the laser fields, while the theory of optimal control was applied and extended. The close interrelationship, strong interactions and complementarity between experiment and theory were instrumental for progress in this exciting research field. The complex laser fields for the attainment of optimal control, which were generated by the closed-loop learning algorithms, provided spectral components with a temporal distribution and frequency (coherently superimposed) distribution that can be mapped on the dynamics of the nuclei. These experimental and theoretical interrelationships established a novel and significant link between the two central objectives of this research program, pertaining to analysis by control. The third period of this research program (2004-2007) relies on the experimental capabilities for the generation of shaped, complex laser fields together with the theoretical concepts for the exploration of analysis and control of dynamics in complex systems, establishing the relations between controlled dynamics, reactivity and function. The chromophores were spatially and structurally enlarged, with increasing the number of active intramolecular degrees of freedom, while the number of the intermolecular degrees of freedom (of the 'bath') was increased by microscopic solvation of chromophores and by the increase of the number of ligands surrounding them. The methods for the control of dynamics, reactivity and function were extended with increasing the spectral range of the lasers from the IR to the UV spectral range over a wide intensity domain, and with the combination with static strong electric fields. The 'bottom-up' experimental and theoretical approaches adopted in this research program make significant contributions toward the establishment of an integrated approach for the understanding and operations of control of nuclear dynamics in systems of increasing complexity. The scientific information underlines the development of exquisite experimental probes and theoretical methods for nuclei 'in motion' in the course of dynamics and its control. This research will open avenues toward the future exploration of novel processes and their applications. Fundamental problems pertain to the optical control of dynamics of complex systems, to the exploration of dynamics and control in finite ultracold systems (quantum clusters and gases), as well as to the extension in the realm of dynamics and control toward 'pure' electron dynamics, with the nuclear motion being frozen. Some notable applications, which rest on basic experimental and theoretical research, involve the optical interrogation and manipulation of biological systems, as well as to the development of sensing methods in remote environments, and to the advent of optically driven molecular memory devices for quantum computing.

Perspectives

Experimental horizons

Future developments in ultrafast dynamics and its control will continue to emerge from concurrent experimental and theoretical efforts, some of which will be based on the broad scope of techniques, concepts, theories and simulations advanced in the present volume. Progress in the establishment of the conceptual framework for the field will drive toward new scientific-technological developments.

On the experimental front, the current availability of Ti:sapphire lasers, with chirped pulse and amplification methods, provide pulses (wavelengths 700-1000nm, repetition rates 50-100MHz), with pulse lengths in the range of 100fs-5fs. The shaping of fs pulses is traditionally conducted by liquid crystal modulators, with feedback control being driven by the use of genetic algorithms. The attainable temporal pulse width of 10fs is sufficient for the interrogation of ultrafast dynamics in molecules, clusters, nanostructures, condensed phase, biomolecules and biological systems on the time scale of nuclear motion. New techniques are currently advanced to transcend the fs time domain with the production of attosecond pulses [95–100], which will be of central importance for the interrogation of dynamics with the temporal resolution of electronic motion, e.g., inner-shell Auger processes in atoms and molecules, and some other processes of electron dynamics [98–100]. One central possibility for control of these 'pure' electronic processes will be achieved by changing the phase of a single laser cycle [101–104]. For dynamics with the temporal resolution of nuclear motion, the extension of the wavelength domain of fs lasers will be of considerable interest. Subfemtosecond X-ray pulses were generated [99, 105] and utilized for the interrogation of ultrafast structural dynamics, which will be alluded to below. UV and XUV ultrashort pulses by high-harmonic generation from the output of high-power near-IR lasers [94, 106] and from free-electron lasers [107] are pertinent for the interrogation of dynamics and control of electronic excitations and of ionization in solids, liquids, clusters and molecules, e.g., large gap insulators such as rare gases [108] and in highly excited molecular states. The production of ultrashort IR pulses from free-electron lasers will be significant for the intramolecular vibrational excitations and for the control of IR-induced conformational isomerization in molecules and biomolecules. The development of intense far-IR lasers in the terahertz regime is in the planning stage and will be useful for the interrogation of low-frequency intermolecular vibrational motion in large molecular scale systems and in biomolecules.

Some of the most important novel experimental developments in chemistry, physics and biology pertain to structural dynamics that involves the interrogation of time-resolved structures. The utilization of synchrotron radiation and X-ray pulses explored time-resolved dynamics on the ns-ps time scale [109–121]. This time domain is relevant for condensed phase and biological structural dynamics. The most significant advancement and development of ultrafast femtosecond time-resolved electron diffraction, crystallography and microscopy [122–124] led to joint atomic-scale spatial and temporal resolutions [124]. Prime examples involved structural changes in 'isolated' molecules in beams, interfaces, surfaces, two-dimensional layers, nanostructures and self-assembled systems and nano-to-micro structures in materials and biological systems [124]. Time-resolved structural interrogation opens avenues for the exploration of complex transient structures and assemblies in material science, nanoscience and biology [124]. Time-resolved ultrafast X-ray diffraction methods [105, 117–120] show great promise in molecular and material science. Ultrafast X-ray pulses are currently produced from laser plasma generation (pulse widths 100-500fs). Prime phenomena that were already explored involve dynamics of melting and of phonon coherence effects.

Table-top ultraintense ultrafast lasers in the near-IR are characterized by a maximal intensity of 10^{20}-10^{22} Wcm^{-2}, which constitutes the currently available highest light intensity on earth [125–128]. Concepts were introduced for the attainment of pulses with a peak intensity as high as 10^{29} Wcm^{-2} [127, 128]. The ultraintense lasers that are currently operated in the near-IR domain span the intensity range of 10^{14}-10^{21} Wcm^{-2}, with a pulse duration of 10-100fs. Intense VUV free-electron lasers with pulse lengths of 100fs became recently operative in the intensity domain of 10^{13}-10^{14} Wcm^{-2} [107, 108]. The coupling of macroscopic dense matter with ultraintense laser fields is blurred by the effects of inhomogeneous dense plasma formations, isochoric heating, beam self-focusing and radiative continuum production [129]. To circumvent the debris problem from macroscopic solid targets, it is imperative to explore efficient laser energy acquisition and disposal in clusters, which constitute large, finite systems, with a density comparable to that of the solid or liquid condensed phase and with a size that is considerably smaller than the laser wavelength. The physics of the response to near-IR and VUV ultraintense lasers is distinct, as in the former case a quasistatic description of the laser field is applicable, while the latter case marks the failure of the quasistatic approximation for the field, as implied by the large value of the Keldysh parameter [130, 131].

The traditional control methods of fs pulses from Ti:sapphire lasers are based on the shaping of the pulse train, amplitude and phase. A significant extension of this technique to include the (linear and circular) polarization shaping of the pulse was already accomplished [132, 133], which results in 'fully shaped' near-IR pulses. Some interesting proposals for the use of linearly polarized fs pulses involve selective electron transfer [134], while circularly polarized IR π laser pulses can induce nuclear torsional motion for the preparation of pure enantiomers from an oriented racemate [135]. Regarding control in different spectral domains outside the near-IR, the newly available VUV, UV and IR ultrafast pulses cannot be shaped by the conventional devices that use liquid crystals, and new techniques will be necessary. Shaping of XUV pulses is under way [136] by phase-only shaping of the fundamental near-IR, 800nm driver pulses for high-harmonic generation. Also, the use of conventional shaping devices for the tailoring of intense near-IR laser pulses is limited to the intensity range below $\sim 5{\cdot}10^{14}$ Wcm^{-2} due to damage to the shapers [137,138]. The control of reaction products in ultraintense laser fields (peak intensities $\geqslant 10^{15}$ Wcm^{-2}) is technically and conceptually different from the exploration of control in ordinary fields. Ultraintense field control can be achieved by using different laser parameters, i.e., pulse intensity, temporal length, shape, phase and train, in different experiments. As pulse shaping via learning algorithms is inapplicable under these experimental conditions, the changing of the laser parameters is called for. Simulation methods recently developed for multielectron ionization and electron dynamics of clusters in ultraintense laser fields [130,131] will provide guides for the experimental choice of laser pulses for optimal control. Two scenarios were recently advanced for control in ultraintense laser fields, i.e., the control of extreme multielectron ionization levels in elemental and molecular clusters [137–139], and the control of the branching ratios in nucleosynthesis driven by Coulomb explosion of completely ionized large clusters (nanodroplets) of methane, ammonia, and water [140].

Conceptual framework

In what follows we shall allude to analysis and control of dynamics of systems of increasing complexity from manipulation of functionality of clusters toward biosystems, and then address some basic open questions in the realm of control. Next, we proceed to the new world of response of clusters and plasmas to ultraintense laser fields, where nonperturbative effects are fundamental and new phenomena of multielectron ionization and electron dynamics are exhibited. These issues will bring us to progress in attosecond electron dynamics. We will conclude this presentation with the dynamics and control of matter under extreme conditions in finite, ultracold systems that involve superfluid boson, e.g., $(^4\mathrm{He})_n$ and $(\mathrm{p\text{-}H_2})_n$ clusters (temperature 2.2-0.1K) [30,141], and optical molasses (temperatures 10μK-100μK) [31,141], together with the perspectives for the production of molecular and cluster species for Bose-Einstein condensation in the temperature range of 10μK-10nK [141–144].

The exploration of the control of ultrafast processes driven by tailored laser pulses allowed for the determination of how the optical properties, response and reactivity will be determined by the interplay between spatial structure, size (in the case of finite clusters and nanostructures systems), the system's energy landscapes, its electronic and vibrational level structure, and the nature of the laser field. Laser-selective chemistry is combined with the functionality, which is size-selective with manifestations of specific effects in finite systems. The extension of the concepts and techniques of analysis control to systems of increasing complexity, from large clusters, to large-scale chemical systems and to biological systems, will be of considerable importance and significance. It is often common to refer to the increase of the system size as a benchmark for increasing their complexity, without alluding to more rigorous specifications. Complexity can be characterized by spatial, energetic or temporal structure with nonperiodic variations [145]. On the basis of such a definition, the control of dynamics pertains to the manipulation of complexity. An example that comes to mind is the 'transition' between fission and Coulomb explosion of multicharged, large finite systems that can be induced by laser control of the ionization level of a large molecule, of a covalent cluster [35] or of a protein in the gas phase [146]. Theoretical studies of optimal control of nuclear dynamics in complex systems in the gas phase were recently pursued [43, 147] by the Rice-Tannor-Kosloff pump-dump scheme, searching for the connective pathway between the initial wave packet and the objective. The methodology was based on molecular dynamics in conjunction with quantum computations for the transient structures across the pathway ('on the fly') [43,147]. The maximization of the yield resulted in coupled equations for the optimal pump and dump pulses that cannot be solved for complex systems. A new strategy for pump-dump control was based on the concept of the intermediate target that involves a localized wave packet in the excited potential surface at an optimal time delay which guarantees maximal overlap for damping into the ground state objective [43,148]. This extra condition allowed for the decoupling of the equations for the pump and the dump pulses, was tested for the isomerization of moderately large Na_3F_2 clusters, and shows promise for larger complex systems [43,147,148]. Another promising approach for large systems is IR control of configurational changes. Theoretical studies of the IR control of isomerization of glycine (with 24 vibrational degrees of freedom) were conducted, being based on the propagation of the ensemble of trajectories obtained from quantum chemistry computations coupled to IR fields whose parameters were optimized by genetic algorithms [149]. This approach will be relevant for conformational dynamics in building blocks for biomolecules. Two major obstacles in the development of control methods for complex large systems should be addressed. First, for large molecular scale systems and biosystems, vibrational sequence congestion implies that laser excitation carries the congested thermal vibrational population of the ground electronic-vibrational state to the excited state, blurring the excited state wave packet. In the early stages of laser photoselective chemistry [150] this difficulty was overcome by supersonic beam cooling of large molecules and of building blocks for biomolecules. Second, of considerable interest will be the control of dynamics in such complex systems in the condensed phase, e.g., in water. The implications of energetic inhomogeneous spectral shifts induced by the solvent, together with the role of the solvent as a 'heat bath' for relaxation and dephasing, require close scrutiny in the context of control.

Current progress in the realm of optimal control points toward further extensions of the conceptual framework. Under favorable conditions it should be possible to infer on the intramolecular or intracluster nuclear dynamics from the shape of the optimized pulses. This inversion problem [151, 152] constitutes the 'holy grail' that will allow for analysis by control. Since tailored laser pulses have the ability to select pathways that optimally lead to a chosen target, the analysis of these (temporal and frequency) pulse shapes should enable to obtain information on these selected pathways. More theoretical work is required, which will allow for the design of interpretable optimal pulses for the driving of complex systems by invoking concepts for the solution of the inversion problem.

The area of laser-matter interactions is currently transcended by moving toward attosecond-femtosecond electron and nuclear dynamics in ultraintense laser fields (pulse peak intensity 10^{15}-10^{20} Wcm^{-2}). Of considerable interest is cluster electron and Coulomb explosion dynamics [130, 131, 153–178]. Extreme cluster multielectron ionization in ultraintense laser fields is distinct from electron dynamic response in ordinary fields, where perturbative quantum electrodynamics is applicable, and from the response of a single atomic and molecular species in terms of mechanisms, the ionization level and the time scales for electron and nuclear dynamics. Extreme multielectron cluster ionization involves three sequential-parallel processes of inner ionization, of nanoplasma formation and response, and of outer ionization [130, 131, 155, 158–160, 166, 172]. Cluster electron dynamics triggers nuclear dynamics, with the outer ionization being accompanied by Coulomb explosion [131,155,161,164,165,167–169,171,173–178], which produced high-energy (1keV-30MeV) ions and nuclei in the energy domain of nuclear physics. A realistic endeavor pertains to table-top dd nuclear fusion driven by Coulomb explosion (NFDCE) of deuterium containing clusters [163–165,167,168,173–178], for which compelling experimental and theoretical evidence was advanced. Predictions [164,165] that Coulomb explosion of deuterium containing heteroclusters (e.g., $(CD_4)_n$, $(D_2O)_n$) will result in considerably higher deuteron energies and dd fusion yields due to energy boosting effects were experimentally confirmed in Saclay [176], in the Lawrence-Livermore Laboratory [174, 175], and in the Max-Born Institute [177]. A theoretical-computational demonstration was recently provided for a seven-orders-of-magnitude enhancement in the neutron yield from NFDCE of light-heavy heteroclusters, e.g., $(DI)_n$, as compared to the yield from deuterium clusters of the same size [178]. The eighty years quest for table-top nuclear fusion driven by chemical reactions was achieved by 'cold-hot' fusion in the chemical physics laboratory, opening avenues for experimental and technological progress. The realm of nuclear reactions driven by cluster Coulomb explosion was extended from dd fusion to nucleosynthesis involving heavy nuclei, which is of interest in the context of nuclear astrophysics [140]. Further progress in this field will involve the experimental and theoretical studies of multielectron ionization and Coulomb explosion of nanodroplets [140, 177]. Under cluster vertical ionization conditions the energetics of the nuclei is considerably enhanced (in the energy range of 100keV-100MeV) for Coulomb exploding nanodroplets. The constraints for complete inner ionization of nanodroplets have to be established. Concurrently, incomplete outer ionization and laser attenuation effects in these large systems will limit the energetic domain for the Coulomb explosion of the bare nuclei. Interesting conceptual and technical developments are expected to emerge when cluster dynamics is transcended toward nuclear reactions.

We alluded to ultrafast adiabatic and nonadiabatic nuclear dynamics and control. Have we reached the temporal borders of the fundamental processes in chemistry and biology [179]? Indeed, the time scales for nuclear motion provide the relevant temporal limit for biophysical and biological dynamics. On the other hand, and most significantly for chemical transformations and for the response and function of nanostructures, even shorter time scales - from attoseconds to femtoseconds - can be unveiled for electron dynamics [94–97]. 'Pure' electron dynamics pertains to changes in the electronic states, without the involvement of nuclear motion, bypassing the constraints imposed by the Franck-Condon principle. In this new world, electron dynamics may prevail on the attosecond temporal resolution. An interesting development in the area of attosecond-femtosecond electron dynamics constitutes a 'spin off' of ultraintense laser-cluster interactions (discussed above) which drive phenomena of nanoplasma response and dynamics. Two notable and related developments in the realm of electron dynamics in intense fs laser fields recently emerged. First, the advent of nonsequential double ionization, involving (e,2e) recollision processes [94,180–185], provides significant information (from the electron momentum correlation function) on the electronic wavefunction of the target molecule from which the electron departed [184]. From the practical point of view, the electron can diffract from the molecular ion core, determining the spatial structure of the molecule [180,186]. The (e,2e) processes in atoms result in nonsequential ionization from the same core, while for molecules or clusters these processes can occur from different cores. Work on (e,2e) processes in diatomics [184] and in the C_{60} molecule [180] was already conducted. It will be interesting to extend these aspects of (e,2e) dynamics to elemental and molecular clusters. Second, single- (or sub-) optical cycle lasers driving atoms provides novel dynamic information on cycle and phase dependent electric field induced ionization rates and electron recollision times [101–104]. Coherent control experiments of electron dynamics demonstrated the possibility of directing fast electron emission from Xe atoms to the right or to the left with changing the light phase [102]. Of considerable interest will be the extension of these studies of electron dynamics driven by single (or few) optical cycle lasers in molecules and in elemental and molecular clusters.

The theory of electron dynamics in small molecules, driven by attosecond laser pulses, was advanced by Bandrauk [187–189]. Recent theoretical studies and quantum mechanical calculations [190, 191] addressed optimal ultrafast (6fs) lasers driving electron dynamics in molecules, establishing the scheme for state selective electronic excitation involving dipole switching in lithium cyanide [190] and the formation of a 'giant dipole' in N-methyl-6-quinolone [191]. A new mechanism was advanced for the induction of a selective, unidirectional electron ring current in oriented molecules driven by electronic excitation with a circularly polarized ultrashort (3.5fs) laser pulse [192, 193]. The implications of this proposal were examined for $X \rightarrow E_{+}$ population transfer, described by electron wave packet dynamics, in Mg-porphyrine. The ring current generated by the laser pulse is stronger by about two-orders-of-magnitude than that induced in this system by the available permanent magnetic field [192]. It was suggested that these types of specific electronic currents may in turn induce magnetic fields with characteristic effects on superconducting quantum interference devices [192]. These studies provide clues for the extension of electron dynamics to multielectron dynamics in large molecules.

The exploration of 'pure' electron dynamics without the involvement of nuclear motion is not limited to the attosecond-femtosecond time scale and can be realized on longer time scales, when the electron motion is slow. This is the case for the dynamics of wave packets of electronic high n Rydberg states of atoms [194–201], which circulate along classical Kepler paths with diameters of thousands of Bohr radii on the microsecond time scale. While such electronic wave packets driven by ps pulses lead to Rydberg state ionization near the turning point of the Kepler orbit, subpicosecond, half-cycle pulses can ionize a radially localized Rydberg wave packet over its entire trajectory [200]. New avenues for the exploration of 'slow' electron dynamics open up. The dynamics of Rydberg wave packets in molecules [201], e.g., NO, is also of considerable interest. For high n molecular Rydbergs the electron motion is slow on the time scale of nuclear motion, and the inverse Born-Oppenheimer separability has to be invoked. Rydberg electronic wave packets exhibit nonadiabatic coupling with other degrees of freedom, and are amenable to control by interference effects [201].

Significant developments encompass the realm of dynamics of ultracold finite systems [141], involving molecules, clusters, optical molasses and finite Bose-Einstein condensates in the temperature domain of T = 2.7K-10^{-8}K [141]. For ultracold systems, the upper temperature limit (T = 2.7K) is arbitrarily taken as the current temperature of the expanding universe, while the lowest temperature is chosen as that of low-density atomic or molecular Bose-Einstein condensates [141]. The higher temperature domain of the ultracold world for large molecules, e.g., aniline and anthracene (with rotational temperatures of 0.3K-2.7K), was reached by cooling in supersonic expansions in He from high-pressure pulsed supersonic nozzles [202], allowing for the study of kinetic energy and permutation symmetry effects in anthracene$(^4\mathrm{He})_n$ clusters [202]. Small molecules were cooled to the mK temperature range [203–206] by deceleration and electrostatic trapping of OH radicals at (rotational) temperatures of 50-500mK [205], and of the $^{15}ND_3$ molecule at a temperature of 1mK [206]. The relatively deep and spatially large traps for ground state, neutral, ultracold molecules show promise [203–206] for high-resolution spectroscopic and dynamic applications in large molecules and clusters. Exotic ultracold systems encompasses quantum clusters $(^4\mathrm{He})_n$, $(^3\mathrm{He})_n$, or (para-H_2)$_n$ (at T = 0.1-2.2K) [30,141,207,208], optical molasses of irradiated Rb atoms (T = 10^{-4}-10^{-6}K) [31], finite atomic clouds of Bose-Einstein atomic condensates of ^{7}Li, ^{23}Na, and ^{87}Rb (T = 10^{-7}-10^{-8}K) [209–211], and finite Bose-Einstein molecular condensates of clouds of diatomics, e.g., $^6Li_2$, $^{23}Na_2$, or $^{87}Cs_2$ (at T = 10^{-8}-10^{-7}K) [144–146,212–222]. Some notable example for dynamics in the ultracold world are: (i) The expansion of optical molasses, which is analogous to cluster Coulomb explosion, thus building a bridge between the ultraslow (ms) dynamics of ultracold finite gaseous samples and ultrafast (ps-fs) cluster dynamics [31]. (ii) The tunneling of an excess electron from a bubble in $(^4He)_n$ clusters, as a probe for superfluidity in finite boson systems [223]. The unique properties and features of ultracold quantum clusters, optical molasses and atomic and molecular gases, can be traced to quantum effects of zero-point energy and kinetic energy of the 'light' constituents in clusters and permutation symmetry effects in all systems. Outstanding problems in this field involve size effects on the superfluid transition in helium-4 clusters [141], energetics of excess electron bubbles in large helium clusters [223], electron tunneling dynamics from such bubbles that constitute a 'pure' electron dynamic process on the ms time scale [223], finite size effects on Bose-Einstein condensation in confined systems [141], probing superfluidity in finite boson systems, and a molecular description of Bose-Einstein condensation [141]. Interesting further developments in this field will focus on collective excitations, as well as nuclear and electron dynamics in large finite quantum systems. These will involve the attempt for the production of finite ultracold clusters. Two distinct classes of such ultracold clusters will be considered, involving either highly vibrationally excited 'floppy' clusters (produced via Feshbach resonances) [212–222] or rigid clusters in low vibrational states produced by photoassociation [144–146]. It will be interesting to explore the possibility of Bose-Einstein condensation in ultracold assemblies of such clusters. Another interesting problem pertains to the minimal cluster size for the attainment of Bose-Einstein condensation within a single cluster [141]. The threshold size for the superfluid transition in a boson cluster is expected to be property dependent [141]. Other interesting problems in this area pertain to the theoretical investigation of optically induced tunneling of electrons from bubbles in helium clusters [223]. This process can be controlled by the competition between electron tunneling from the bubble and ultrafast radiationless (nonadiabatic or adiabatic) relaxations of the bubble excited electronic states to lower electronic states. The exploration of electron tunneling from electronically excited states of electron bubbles in ultracold quantum clusters brings us back to the realm of laser control of electron dynamics.

Scientific-technological applications

The research directions and developments discussed herein provide perspectives for new scientific-technological developments. A number of research directions within the framework of the Sfb 450 program reach a stage when one can begin to consider technological spin-offs. Examples involve remote laser manipulation, analytic and sensing methods [224–226]. Recent accomplishments involve the use of half cycle laser pulses for the chemical analysis [226], and remote sensing by multiple filamentation of ultrashort Terawatt laser pulses in air [224, 225]. Optical manipulation of complex systems shows promise for applications to biological systems. Primary examples in this field are photodynamic therapy, based on optical manipulation of molecules with endoperoxide groups [227], and the detection of biological molecules in tissues [228]. The analytical methods have potential for probing biosystems, e.g., bacteria, while sensing methods and controlled dynamics of atmospheric processes is of current interest. Although the primary thrust of the research program is based on the integration of experiment and theory, it is imperative to mention some theoretical developments of considerable interest in the context of future technology transfer. The first is molecular motors, due to their important role as functional molecular devices [229–235]. Chiral molecular rotors were described, being driven by a linearly polarized laser pulse [229,231,233–235], with the application of control methods for the preselected directions [231]. Unidirectional molecular torsional motion can also be induced by circularly polarized π laser pulses for the driving of such molecular rotors. Potential applications in the field of nanotechnology will be of interest. The second is optically pumped and probed logic machines for quantum computing. Elaborate molecular machines for information storage and disposal can take advantage of the self evident, but most useful, fact that the optical response of photophysical systems depends on their present state [231]. It was proposed [236, 237] that the stimulated Raman adiabatic passage spectroscopy (STIRAP) [238] can be used for information storage and retrieval on the molecular level. The utilization of the STIRAP pulse sequence provides a strategy for complete and robust population transfer in a multilevel system with sequential coupling [237, 239]. In fact, the use of the STIRAP pulse sequence for this problem emerged automatically from the local optimization procedure [239]. STIRAP spectroscopy is of considerable importance for quantum computation via local control [237], to build finite-state molecular machines that can be programmed [236, 237]. An alternative approach is based on optimal control for quantum computing [240, 241].

It is apparent that more experimental and theoretical developments are expected in this fascinating research area, some of which should emerge from this overview. It is expected that the scientific quality, vitality and impact of this research field of analysis and control of ultrafast photoinduced reactions will continue well into the future.

References

1. V. Sundström (ed.), *Nobel Symposium 101. Femtochemistry and Femtobiology: Ultrafast reaction dynamics at atomic scale resolution*, (World Scientific, Imperial College Press, London, 1997)
2. J. Jortner, Spiers Memorial Lecture. Faraday Discuss. **108**, 1 (1997)
3. V. May, O. Kühn, *Charge and energy transfer dynamics in molecular systems*, (Wiley-VCH, Weinheim, 2004)
4. A. Nitzan, *Chemical dynamics in condensed phases*, (Oxford University Press, 2006)

5. D.J. Tannor, *Introduction to quantum mechanics: A time-dependent perspective*, (University Science Books, Sausalito, 2006)
6. R.D. Levine, *Molecular reaction dynamics*, (University Press, Cambridge, 2005)
7. A.H. Zewail, *Femtosecond Chemistry*, (World Scientific, Singapore, 1994)
8. J. Manz, L. Wöste (eds.), *Femtosecond Chemistry*, Vol. 1 and 2 (VCH Verlag, Weinheim, 1995)
9. M. Chergui (ed.), *Femtochemistry: Ultrafast chemical and physical processes in molecular systems*, (World Scientific, Singapore, 1996)
10. A. Douhal, J. Santamaria (eds), *Femtochemistry and Femtobiology*, (World Scientific Publishing, Singapore, 2002)
11. M.M. Martin, J.T. Hynes (eds.), *Femtochemistry and Femtobiology: Ultrafast events in molecular science*, (Elsevier, Amsterdam, 2004)
12. W.A. Castleman, Jr. (ed), *Femtochemistry VII : Fundamental ultrafast processes in Chemistry, Physics, and Biology*, (Elsevier, Amsterdam, 2006)
13. K.F. Bonhoeffer, L. Farkas, Z. Phys. Chem. **134**, 337 (1928)
14. G. Wenzel, Z. Phys. **29**, 321 (1928)
15. H. Eyring, M. Polanyi, Z. Phys. Chem B **12**, 279 (1931)
16. J.D. McDonald, P.R. Le Berton, Y.T. Lee, D.R. Herschbach, J. Chem. Phys. **56**, 769 (1972)
17. R.A. Marcus, J. Chem. Phys. **24**, 979 (1956); ibid. p. 679.
18. Th. Förster, Discuss. Farad. Soc. **27**, 7 (1959)
19. R. Kubo, Y. Toyozawa, Prog. Theoret. Phys. **13**, 160 (1955)
20. D. DeVault, B. Chance, Biophys. J. **6**, 825 (1966)
21. R.H. Austin and H. Fraunfelder, Biochemistry **14**, 5355 (1975)
22. G.B. Kistiakowski, C.S. Parmenter, J. Chem. Phys. **42**, 2942 (1965)
23. M. Bixon, J. Jortner, J. Chem. Phys. **48**, 715 (1968)
24. J. Jortner, R.S. Berry, J. Chem. Phys. **48**, 2757 (1968)
25. R.E. Smalley, L. Wharton, D.H. Levy, J. Chem. Phys. **64**, 3266 (1976)
26. J. Jortner, Z. Phys. D **24**, 247 (1992)
27. A. Kaldor, Z. Phys. D **19**, 353 (1991)
28. A. Schnachez, S. Abbet, U. Heiz, W.D. Schneider, H. Hakkinen, R.N. Barnett, U. Landman, J. Phys. Chem. A **103**, 9573 (1999)
29. M. Kappes, R. Kunz, E. Schumacher, Chem. Phys. Lett. **91**, 413 (1982)
30. J.P. Toennies, A.F. Vilosov, Ann. Rev. Phys. Chem. **49**, 1 (1998)
31. L. Pruvost, T. Serrre, H.T. Duong, J. Jortner, Phys. Rev. A **61**, 053408 (2000)
32. S.N. Kahanna, A.W. Castleman (eds.), *Quantum phenomena in clusters and nanostructures*, (Springer Berlag, Berlin, 2003)
33. A. Hartmann, J. Pittner, V. Bonačić-Koutecký, A. Heidenreich, J. Jortner, J. Chem. Phys. **108**, 3096 (1998)
34. J. Purnell, E.M. Snyder, S. Wei, A.W. Castleman, Jr., Chem. Phys. Lett. **229**, 333 (1994)
35. I. Last, Y. Levy, J. Jortner, Proc. Natl. Acad. Sci. USA **99**, 107 (2002)
36. J. Jortner, A. Nitzan, M.A. Ratner, in *Introducing molecular electronics*, ed. by C. Cuniberti, G. Fagas, K. Richter (Springer Verlag, Berlin, 2005), p. 13
37. R. Bersohn, A.H. Zewail, Ber. Bunsenges. Phys. Chem. **92**, 373 (1988)
38. A.H. Zewail, R.B. Bernstein, Chemical Engineering News **66**, 24 (1988)
39. M. Dantus, M.J. Rosker, A.H. Zewail, J. Chem. Phys. **87**, 2395 (1987)
40. J. Jortner, M. Bixon, p. 349 in [1]
41. J. Manz, p. 80 in [1]

42. W. Domcke, D.R. Yarkony, H. Köppel (eds.), *Conical intersections: Electronic structure, dynamics, and spectroscopy*, (World Scientific, Singapore, 2004)
43. V. Bonačić-Koutecký, R. Mitrić, Chem. Rev. 105, 11 (2005)
44. V. Bonačić-Koutecký, R. Mitrić, T.M. Bernhards, L. Wöste, J. Jortner, Adv. Chem. Phys. **132**, 179 (2005)
45. S. Mukamel, *Principles of nonlinear optical spectroscopy*, (Oxford University Press, Oxford, 1995)
46. G.K. Paramonov, V.A. Savva, Phys. Lett. A **97**, 340 (1983)
47. T. Joseph, J. Manz, Mol. Phys. **58**, 1149 (1986)
48. D.J. Tannor, S.A. Rice, J. Chem. Phys. **83**, 5013 (1985)
49. D.J. Tannor, S.A. Rice, Adv. Chem. Phys. **70**, 441 (1988)
50. S.A. Rice, M. Zhao, *Optical control of molecular dynamics*, (Wiley, New York, 2000)
51. P. Brumer, M. Shapiro, Faraday Discuss. Chem. Soc. **82**, 177 (1986)
52. M. Shapiro, P. Brumer, J. Chem. Phys. **84**, 4103 (1986)
53. M. Shapiro, P. Brumer, *Principles of the quantum control of molecular processes*, (Wiley, Hoboken, N.J., 2003)
54. A.P. Peirce, M.A. Dahleh, H. Rabitz, Phys. Rev. A **37**, 4950 (1988)
55. S. Shi, H. Rabitz, Chem. Phys. **139**, 185 (1989)
56. R.S. Judson, H. Rabitz, Phys. Rev. Lett. **68**, 1500 (1992)
57. W.S. Warren, H. Rabitz, M. Dahleh, Science **259**, 1581 (1993)
58. T. Baumert, B. Buhler, M. Grosser, R. Thalweiser, V. Weiss, E. Wiedenmann, G. Gerber, J. Phys. Chem. **95**, 8103 (1991)
59. T. Baumert, G. Gerber, Isr. J. Chem. **34**, 103 (1994)
60. J.L. Herek, A. Materny, A.H. Zewail, Chem. Phys. Lett. **228**, 15 (1994)
61. A. Shnitman, I. Sofer, I. Golub, A. Yogev, M. Shapiro, Z. Chen, P. Brumer, Phys. Rev. Lett. **76**, 2886 (1996)
62. H. Schwoerer, R. Pausch, M. Heid, V. Engel, W. Kiefer, J. Chem. Phys. **107**, 9749 (1997)
63. C. Nicole, M.A. Bouchene, C. Meier, S. Magnier, E. Schreiber, B. Girard, J. Chem. Phys. **111**, 7857 (1999)
64. L. Pesce, Z. Amitay, R. Uberna, S.R. Leone, R. Kosloff, J. Chem. Phys. **114**, 1259 (2001)
65. Z.W. Shen, T. Chen, M. Heid, W. Kiefer, V. Engel, Eur. Phys. J. D **14**, 167 (2001)
66. G. Grégoire, M. Mons, I. Dimicoli, F. Piuzzi, E. Charron, C. Dedonder-Lardeux, C. Jouvet, S. Matenchard, D. Solgadi, A. Suzor-Weiner, Eur. Phys. J. D **1**, 187 (1998)
67. T. Hornung, M. Motzkus, R. de Vivie-Riedle, J. Chem. Phys. **115**, 3105 (2001)
68. G. Rodriguez, J.G. Eden, Chem. Phys. Lett. **205**, 371 (1993)
69. G. Rodriguez, J.C. John, J.G. Eden, J. Chem. Phys. **103**, 10473 (1995)
70. R. Pausch, M. Heid, T. Chen, W. Kiefer, H. Schwoerer, J. Chem. Phys. **110**, 9560 (1999)
71. R. Pausch, M. Heid, T. Chen, W. Kiefer, H. Schwoerer, J. Raman Spectrosc. **31**, 7 (2000)
72. R. Uberna, Z. Amitay, R.A. Loomis, S.R. Leone, Faraday Discuss. **113**, 385 (1999)
73. S. Vajda, A. Bartelt, E.C. Kaposta, T. Leisner, C. Lupulescu, S. Minemoto, P. Rosenda-Francisco, L. Wöste, Chem. Phys. **267**, 231 (2001)

74. A. Bartelt, S. Minemoto, C. Lupulescu, S. Vajda, L. Wöste, Eur. Phys. J. D **16**, 127 (2001)
75. S. Vajda, C. Lupulescu, A. Bartelt, F. Budzyn, P. Rosendo-Francisco, L. Wöste, p. 472 in [10]
76. A. Bartelt, A. Lindinger, C. Lupulescu, S. Vajda, L. Wöste, Phys. Chem. Chem. Phys. **5**, 3610 (2003)
77. C. Lupulescu, A. Lindinger, M. Plewicky, A. Merli, S.M. Weber, L. Wöste, Chem. Phys. **269**, 63 (2004)
78. J.B. Ballard, H.U. Stauffer, Z. Amitay, S.R. Leone, J. Chem. Phys. **116**, 1350 (2002)
79. B. Schäfer-Bung, R. Mitrić, V. Bonačić-Koutecký, A. Bartelt, C. Lupulescu, A. Lindinger, S. Vajda, S.M. Weber, L. Wöste, J. Phys. Chem. A **108**, 4175 (2004)
80. T. Brixner, G. Krampete, T. Pfeifer, R. Selle, G. Gerber, M. Wollenhaupt, O. Graefe, C. Horn, D. Liese, J. Baumert, Phys. Rev. Lett. **92**, 208301 (2004)
81. C.J. Bardeen, V.V. Yakovlev, K.R. Wilson, S.D. Carpenter, P.M. Weber, W.S. Warren, Chem. Phys. Lett. **280**, 151 (1997)
82. A. Assion, T. Baumert, M. Bergt, T. Brixner, B. Kiefer, V. Seyfried, M. Strehle, G. Gerber, Science **282**, 919 (1998)
83. T. Hornung, R. Meier, M. Motzkus, Chem. Phys. Lett. **326**, 445 (2000)
84. S. Vajda, P. Rosendo-Francisco, C. Kaposta, M. Krenz, L. Lupulescu, L. Wöste, Eur. Phys. J. D **16**, 161 (2001)
85. T. Brixner, N.H. Damrauer, P. Niklaus, G. Gerber, Nature **414**, 57 (2001)
86. C. Daniel, J. Full, L. González, E.C. Kaposta, M. Krenz, C. Lupulescu, J. Manz, S. Minemoto, M. Oppel, P. Rosenda-Francisco, S. Vajda, L. Wöste, Chem. Phys. **267**, 247 (2001)
87. N.H. Damrauer, C. Dietl, G. Krampert, S.H. Lee, K.H. Jung, and G. Gerber, Eur. Phys. J. D **20**, 71 (2002)
88. J.L. Herek, W. Wohlleben, R.J. Cogdell, D. Zeidler, M. Motzkus, Nature **417**, 553 (2002)
89. C. Daniel, J. Full, L. González, C. Lupulescu, J. Manz, A. Merli, S. Vajda, L. Wöste, Science **299**, 536 (2003)
90. H. Rabitz, R. de Vivie-Riedle, M. Motzkus, K.L. Kompa, Science **288**, 824 (2000)
91. T. Brixner, N.H. Damrauer, G. Krampert, P. Niklaus, G. Gerber, J. Mol. Opt. **50**, 539 (2003)
92. T. Brixner, N.H. Damrauer, B. Kiefer, G. Gerber, J. Chem. Phys. **118**, 3692 (2003)
93. G. Krampert, P. Niklaus, G. Vogt, G. Gerber, Phys. Rev. Lett. **94**, 068305 (2005)
94. P.B. Corkum, Phys.Rev. Lett. **71**, 1994 (1993)
95. M. Hentschel, R. Kienberger, Ch. Spielmann, R.A. Reider, N. Milosevic, T. Brabec, P. Corkum, U. Heinzmann, M. Drescher, F. Krausz, Nature **414**, 509 (2001)
96. V. Strelkov, A. Zaïr, O. Tcherbakoff, R. López-Martens, E. Cormier, E. Mével, E. Constant, Appl. Phys. B **78**, 879 (2004)
97. M. Kitzler, K. O'Keefee, M. Lezius, J. Mod. Opt. **53**, 57 (2005)
98. M. Drescher, F. Krausz, J. Phys. B **38**, S727 (2005)
99. M. Drescher, M. Hentschel, R. Kienberger, G. Tempea, C. Spielmann, G.A. Reider, P.B. Corkum, F. Krausz, Science **291**, 1923 (2001)

100. A. Föhlisch, P. Feulner, F. Hennies, A. Fink, D. Menzel, D. Sanchez-Portral, P.M. Echenique, W. Wurth, Nature **436**, 373 (2005)
101. T. Berbee, F. Krausz, Rev. Mod. Phys. **72**, 545 (2000)
102. G.G. Paulus, F. Lindner, H. Walther, A. Baltuska, E. Goulielmakis, M. Lezius, F. Krausz, Phys. Rev. Lett. **91**, 253004 (2003)
103. X. Liu H. Rottke, E. Eremina, W. Sandner, E. Goulielmakis, K. O. Keeffe, M. Lezius, F. Krausz, F. Lindner, M. G. Schätzel, G. G. Paulus, H. Walther, Phys. Rev. Lett. **93**, 263001 (2004)
104. C.C. Chirilau, R.M. Porvliege, Phys. Rev. A **71**, 021402 (2005)
105. L.X. Chen, W.J.H. Jäger, G. Jennings, D.J. Gasztola, A. Mundholm, J.P. Hessler, Science **292**, 262 (2001)
106. J.L. Krause, K. Schafer, K.C. Kulander, Phys. Rev. Lett. **68**, 3535 (1992)
107. J. Andruszkov et al., Phys. Rev. Lett. **85**, 3825 (2000)
108. H. Wabnitz et al., Nature **420**, 482 (2002)
109. A. Rousse, C. Rischel, J.-C. Gauthier, Rev. Mod. Phys. **73**, 17 (2001)
110. C.H. Chin, R.W. Schoenlein, T.E. Glover, P. Balling, W.P. Leemans, C.V. Shank, Phys. Rev. Lett. **83**, 336 (1999)
111. A.M. Lindenberg, I. Kang, S.L. Johnson, T. Missala, P.A. Heimann, Z. Chang, J. Larsson, P. H. Bucksbaum, H. C. Kapteyn, H. A. Padmore, R. W. Lee, J. S. Wark, R. W. Falcone, Phys. Rev. Lett. **84**, 111 (2000)
112. D. von der Linde, Science **302**, 1345 (2003)
113. K. Sokolowski-Tinten, C. Blome, J. Blums, A. Cavalleri, C. Dietrich, A. Tarasevitch, I. Uschmann, E. Forster, M. Kammler, M. Horn-von-Hoegen, D. von der Linde, Nature **422**, 287 (2003)
114. D.A. Oulianov, I.V. Tomov, A.S. Dvornikov, P.M. Rentzepis, Proc. Natl. Acad. Sci. USA **99**, 12556 (2002)
115. F. Ráksi, K.R. Wilson, Z. Jiang, A. Ikhlef, C.Y. Côté, J.-C. Kieffer, J. Chem. Phys. **104**, 6066 (1996)
116. J.R. Helliwell, P.M. Rentzepis (eds.), *Time resolved diffraction*, (Oxford University Press, Oxford, 1997)
117. L.X. Chen, Annu. Rev. Phys. Chem. **56**, 221 (2005)
118. K. Moffatt, Faraday Discuss. **122**, 65 (2003)
119. A. Plech, M. Wulff, S. Bratos, F. Mirloup, R. Vuilleumier, F. Schotte, P. A. Anfinrud, Phys. Rev. Lett. **92**, 125505 (2004)
120. L. Nugent-Glandorf, M. Scheer, D.A. Samuels, V. Bierbaum, S.R. Leone, Rev. Sci. Instrum. **73**, 1875 (2002)
121. C. Bressler, M. Chergui, Chem. Rev. **104**, 1781 (2004)
122. J.C. Williamson, M. Dantus, S.B. Kim, A.H. Zewail, Chem. Phys. Lett. **196**, 529 (1992)
123. H. Ihee, V.A. Lobatsov, U.M. Gomez, B.M. Goodman, R. Srinivasan, C. Ruan, A.H. Zewail, Science **291**, 458 (2001)
124. A.H. Zewail, Ann. Rev. Phys. Chem. **57**, 65 (2006)
125. M.D. Perry, G.A. Mourou, Science **264**, 917 (1994)
126. G.A. Mourou, C.P.J. Barty, M.D. Perry, Phys. Today **51**, 22 (1998)
127. J. Nees, N. Naumova, E. Power, V. Yanovsky, I. Sokolov, A: Maksimchuk, S.-W. Bahk, V. Chvykov, G. Kalintchenko, B. Hou, G. Mourou, J. Mod. Opt. **52**, 305 (2005)
128. G. A. Mourou, T. Tajima, S. V. Bulanov, Rev. Mod. Phys. **78**, 309 (2006)
129. U. Andiel, K. Eldmann, K. Witte, I. Uschmann, E. Förster, App. Phys. Lett. **80**, 198 (2002)

130. I. Last, J. Jortner, J. Chem. Phys. **120**, 1336 (2004)
131. U. Saalmann, Ch. Siedschlag, J.M. Rost, J. Phys. B **39**, R39 (2006)
132. T. Brixner, G. Gerber, Opt. Lett. **26**, 557 (2001)
133. T. Brixner, G. Krampert, P. Niklaus, G. Gerber, Appl. Phys. B **74**, S133 (2002)
134. P. Krause, T. Klamroth, P. Saalfrank, J. Chem. Phys. **123**, 075105 (2005)
135. K. Hoki, D. Kröner, J. Manz, J. Chem. Phys. **267**, 59 (2001)
136. T. Pfeifer, D. Walter, C. Winterfeldt, C. Spilman, G. Gerber, Appl. Phys. B **80**, 277 (2005)
137. S. Zamith, T. Martchenko, Y. Ni, S.A. Aseyev, H.G. Muller, H.J.J. Vrakking, Phys. Rev. A **70**, 011201(R) (2004)
138. T. Martchenko, Ch. Siedschlag, S. Zamith, H.G. Muller, H.J.J. Vrakking, Phys. Rev. A **72**, 053202 (2005)
139. A. Heidenreich, I. Last, J. Jortner (submitted, 2006)
140. I. Last, J. Jortner, Phys. Rev. Lett. **97**, 173401 (2006)
141. M. Rosenblit, J. Jortner, Adv. Chem. Phys. **132**, 247 (2005)
142. F. Masnou-Seeuws, P. Pillet, Adv. At. Mol. Opt. Phys. **47**, 013402 (2004)
143. C.P. Koch, J.P. Palao, R. Kosloff, F. Masnou-Seeuws, Phys. Rev. A **70**, 013402 (2004)
144. V. Bonačić-Koutecký, J. Phys. Chem. B (in press, 2006)
145. L.P. Kadanoff, *From order to chaos*, (World Scientific, Singapore, 2003)
146. I. Last, Y. Levy, J. Jortner, Intl. J. Mass Spectrometry **249/250**, 184 (2006)
147. M. Hartmann, J. Pittner, V. Bonačić-Koutecký, J. Chem. Phys. **114**, 2123 (2001)
148. R. Mitrić, M. Hartmann, V. Bonačić-Koutecký, J. Phys. Chem. A **106**, 10477 (2002)
149. Bonačić-Koutecký, R. Mitrić, to be published (2006)
150. J. Jortner, R.D. Levine, *Photoselective Chemistry*, Adv. Chem. Phys. **47**, 1 (1981)
151. Z.M. Lu, H. Rabitz, Phys. Rev. A **52**, 1961 (1995); J. Phys. Chem. **99**, 13731 (1995)
152. J. Full, L. González, J. Manz, Chem. Phys. **239**, 126 (2006)
153. T. Ditmire, J.W.G. Tisch, E. Springat, M.B. Mason, N. Hay, R.A. Smith, J. Marangos, M.H.R. Hutchinson, Nature **386**, 54 (1997)
154. T. Ditmire, Phys. Rev. A **57**, R4094 (1998)
155. V.P. Krainov, M.B. Smirnov, Phys. Rep. **370**, 237 (2002)
156. U. Saalmann, J.M. Rost, Phys. Rev. Lett. **89**, 132401 (2002)
157. M. Lezius, V. Blanchet, M.Yu. Ivanov, A. Stolow, J. Chem. Phys. **117**, 1575 (2002)
158. Ch. Siedschlag, J.-M. Rost, Phys. Rev. Lett. **89**, 173401 (2002)
159. Ch. Siedschlag, J.-M. Rost, Phys. Rev. A **67**, 13404 (2003)
160. Ch. Siedschlag, J.-M. Rost, Phys. Rev. Lett. **93**, 043402 (2004)
161. Ch. Siedschlag, J.-M. Rost, Phys. Rev. A **71**, 031401(R) (2005)
162. I. Last, J. Jortner, Phys. Rev. A **62**, 013201 (2000)
163. I. Last, J. Jortner, Phys. Rev. A **64**, 063201 (2001)
164. I. Last, J. Jortner, Phys. Rev. Lett. **87**, 033401 (2001)
165. I. Last, J. Jortner, J. Phys. Chem. A **106**, 10877 (2002)
166. I. Last, J. Jortner, J. Chem. Phys. **120**, 1348 (2004)
167. I. Last, J. Jortner, J. Chem. Phys. **121**, 3030 (2004)
168. I. Last, J. Jortner, J. Chem. Phys. **121**, 8329 (2004)

169. I. Last, K. Levy, J. Jortner, J. Chem. Phys. **123**, 154301 (2005)
170. I. Last, J. Jortner, Phys. Rev. A **71**, 063204 (2005)
171. I. Last, J. Jortner, Proc. Natl. Acad. Sci. USA **102**, 1291 (2005)
172. I. Last, J. Jortner, Phys. Rev. A **71**, 063201 (2006)
173. J. Zweiback, T.E. Cowan, R.A. Smith, J.H. Hurtlay, R. Howell, C.A. Steinke, G. Hays, K.B. Wharton, J.K. Krane, T. Ditmire, Phys. Rev. Lett. **85**, 3640 (2000)
174. J. Zweiback, T.E. Cowan, J.M. Hartley, R. Howell, K.B. Wharton, J.K. Crane, V.P. Yanovski, G. Hays, R.A. Smith, T. Ditmire, Phys. Plasma **9**, 3108 (2002)
175. K.W. Madison, P.K. Patel, M. Allen, D. Price, R. Fitzpatrick, T. Ditmire, Phys. Rev. A **70**, 053201 (2004)
176. G. Grillon, Ph. Balcou, J.-P. Chambaret, D. Hulin, J. Martino, S. Moustaizis, L. Notebaert, M. Pittman, Th. Pussieux, A. Rousse, J.-Ph. Rousseau, S. Sebban, O. Sublemontier, M. Schmidt, Phys. Rev. Lett. **89**, 065005 (2002)
177. S. Ter-Avertisyan, M. Schnürer, D. Hilscher, U. Jahnke, S. Bush, P.V. Nickles, W. Sandner, Phys. Plasmas **12**, 012702 (2005)
178. A. Heidenreich, I. Last, J. Jortner, Proc. Natl. Acad. Scis. USA **103**, 10589 (2006)
179. J. Jortner, Proc. Roy. Soc. London, Series A **356**, 477 (1998)
180. V.E. Bahrdway, P.B. Corkum, D.M. Rymer, Phys. Rev. Lett. **93**, 5400 (2005)
181. Th. Weber, M. Weckenbrock, A. Staudte, L. Spielberger, O. Jagutzki, V. Mergel, F. Afaneh, G. Urbasch, M. Vollmer, H. Giessen, R. Dörner, Phys. Rev. Lett. **84**, 443 (2000)
182. R. Moshammer, B. Feuerstein, W. Schmitt, A. Dorn, C.D. Schröter, J. Ullrich, H. Rottke, C. Trump, M. Wittmann, G. Korn, K. Hoffmann, W. Sandner, Phys. Rev. Lett. **84**, 447 (2000)
183. C. Figueira de Morisson Faria, X. Liu, W. Becker, H. Schomerus, Phys. Rev. A **69**, 0201402(R) (2004)
184. E. Eremina, X. Liu, H. Rottke, W. Sandner, M.G. Schätzel, A. Dreischuh, G.G. Paulus, H. Walther, R. Moshammer, J. Ullrich, Phys. Rev. Lett. **92**, 173001 (2004)
185. R. Dörner, Th. Weber, M. Weckenbrock, A. Staudte, M. Hattass, R. Moshammer, J. Ullrich, H. Schmidt-Böcking, Adv. in At., Mol., and Opt. Phys. **48**, 1 (2002)
186. S. Niikura, F. Légaré, R. Hasbani, A.D. Bandrauk, M.Y. Ivanov, D.M. Villeneuve, P.B. Corkum, Nature **417**, 917 (2002)
187. A.D. Bandrauk, N.H. Shon, Phys. Rev. A **66**, 031401 (2002)
188. S. Chelkowski, A.D. Bandrauk, A. Apolonski, Phys Rev. A **70**, 013815 (2004)
189. G. L. Kamta, A.D. Bandrauk, Phys. Rev. A **74**, 033415 (2006)
190. P. Krause, T. Klamroth, P. Saalfrank, J. Chem. Phys. **123**, 074105 (2005)
191. T. Klamroth, J. Chem. Phys. **124**, 144310 (2006)
192. I. Barth, J. Manz, Y. Shigeta, K. Yagi, J. Am. Chem. Soc. **128**, 7043 (2006)
193. I. Barth, J. Manz, Angew. Chem. Int. Ed. **45**, 2962 (2006)
194. J.A. Yaezell, C.R. Stroud, Jr., Phys. Rev. A **35**, 2806 (1987)
195. J.A. Yeazell, C.R. Stroud, Jr., Phys. Rev. Lett. **60**, 1494 (1988)
196. L.D. Noordam, R.R. Jones, J. Mod. Opt. **44**, 2515 (1997)
197. T.C. Weinacht, J. Ahn, P.H. Bucksbaum, Phys. Rev. Lett. **80**, 5508 (1998)
198. T.C. Weinacht, J. Ahn, P.H. Bucksbaum, Nature **397**, 233 (1999)
199. H. Maeda, D.V.L. Norum, T.F. Gallagher, Science **307**, 1757 (2005)

200. C. Raman, C.W.S. Conover, C.I. Sikenik, P.H. Bucksbaum, Phys. Rev. Lett. **76**, 2436 (1996)
201. H.H. Fielding, Ann. Rev. Phys. Chem. **56**, 91 (2006)
202. U. Even, I. El-Hroub, J. Jortner, J. Chem. Phys. **115**, 2069 (2001)
203. H.L. Bethlem, G. Berden, F.M.H. Crompvoets, R.T. Jongma, A.J.A. van Roij, G. Meijer, Nature **406**, 491 (2000)
204. F.M.H. Crompvoets, H.L. Bethlem, R.T. Jongma, G. Meijer, Nature **411**, 174 (2001)
205. S.Y.T. van de Meerakker, P.H.M. Smeets, N. Vanhaecke, Phys. Rev. Lett. **94**, 023004 (2005)
206. J. Van Veldhoven, H.L. Bethlem, G. Meijer, Phys. Rev. Lett. **94**, 083001 (2005)
207. J.P. Toennies, A.F. Vilesov, Angew. Chem. Int. Ed. **43**, 2622 (2004)
208. K.B. Whaley, R.E. Miller (eds.), *Helium nanodroplets: A novel medium for Chemistry and Physics*, Special Issue, J. Chem. Phys. **15**, 22 (2001)
209. M. Anderson, J. Ensher, M. Matthews, C. Wieman, E. Cornell, Science **269**, 198 (1995)
210. K. B. Davis, M.-O. Mewes, M. R. Andrews, N. J. van Druten, D. S. Durfee, D. M. Kurn, W. Ketterle, Phys. Rev. Lett. **75**, 3969 (1995)
211. C. Bradley, C. Sackett, J. Tollet, R. Hulet, Phys. Rev. Lett. **78**, 985 (1997)
212. E. Timmermans, P. Tommasini, M. Hussein, A. Kerman, Phys. Rep. **315**, 199 (1999)
213. E.A. Donley, N.R. Claussen, S.T. Thompson, C.E. Wieman, Nature **417**, 529 (2002)
214. C.A. Regal, C. Thicknor, J.L. Bohn, D.S. Jin, Nautre **424**, 47 (2003)
215. J. Herbig, T. Kraemer, M. Mark, T. Weber, C. Chin, H.-C. Nägerl, R. Grimm, Science **301**, 1510 (2003)
216. S. Drr, T. Volz, A. Marte, G. Rempe, Phys. Rev. Lett. **92**, 020406 (2004)
217. K. Xu, T. Mukaiyama, J.R. Abo-Shaeer, J.K. Chin, D.E. Miller, W. Ketterle, Phys. Rev. Lett. **91**, 210402 (2003)
218. S. Jochim, M. Bartemstein. A. Altmeyer, G. Hendl, S. Riedl, C. Chin, J. Hecker Denschlag, R. Grimm, Science **302**, 2101 (2003)
219. M. Greiner, C.A. Regal, S.D. Jin, Nature **426**, 537 (2003)
220. M.W. Zwierlein, C.A. Stan, C.H. Schunck, S.M.F. Raupach, S. Gupta, Z. Hadzibabic, W. Ketterle, Phys. Rev. Lett. **91**, 250401 (2003)
221. N. Vanhaecke, W. De Souza Melo, B.L. Tolra, D. Comparat, P. Pillet, Phys. Rev. Lett. **89**, 063001 (2002)
222. C. Chin, A.J. Kerman, V. Vuletić, S. Chu, Phys. Rev. Lett. **90**, 033201 (2003)
223. M. Rosenblit, J. Jortner, J. Chem. Phys. **124**, 194506 (2006)
224. S. Skupin, L. Bergé, U. Peschel, F. Lederer, G. Mjean, J. Yu, J. Kasparian, E. Salmon, J.-P. Wolf, M. Rodriguez, L. Wöste, R. Bourayou, R. Sauerbrey, Phys. Rev. E **70**, 046602 (2004)
225. L. Bergé, S. Skupin, F. Lederer, G. Méjean, J. Yu, J. Kasparian, E. Salmon, J.-P. Wolf, M. Rodriguez, L. Wöste, R. Bourayou, R. Sauerbrey, Phys. Rev. Lett. **92**, 225002 (2004)
226. M. Fischer, M. Hoffmann, H. Helm, R. Wilk, F. Rutz, T. Kleine-Ostmann, M. Koch, P. U. Jepsen, Optics Express **13**, 5205 (2005)
227. I. Corral, L. González, A. Leuer, K. Heyne, to be published (2006)
228. J.M. Dela Cruz, V. V. Lozovoy, M. Dantus, J. Photochem. Photobiol. A **180**, 307 (2006)

229. K. Hoki, M. Yamaki, S. Koseki, Y. Fujimura, J. Chem. Phys. **118**, 497 (2003)
230. K. Hoki, M. Yamaki, Y. Fujimura, Angew. Chem. Int. Ed. **42**, 3084 (2003)
231. K. Hoki, M. Yamaki, S. Koseki, Y. Fujimura, J. Chem. Phys. **119**, 12393 (2003)
232. Y. Fujimura, L. Gonzalez, D. Kröner, J. Manz, I. Mehdaoui, B. Schmidt, Chem. Phys. Lett. **386**, 248 (2004)
233. K. Hoki, M. Sato, M. Yamaki, R. Sahnoun, L. González, S. Koseki, Y. Fujimura, J. Phys. Chem. B **108**, 4916 (2004)
234. M. Yamaki, K. Hoki, Y. Ohtsuki, H. Kono, Y. Fujimura, Phys. Chem. Chem. Phys. **7**, 1900 (2005)
235. M. Yamaki, K. Hoki, Y. Ohtsuki, H. Kono, Y. Fujimura, J. Am. Chem. Soc. Commu. **127**, 7300 (2005)
236. F. Remacle, R.D. Levine, J. Chem. Phys. **114**, 10239 (2001)
237. S. Sklarz, D.J. Tannor, Chem. Phys. **322**, 87 (2006)
238. K. Bergmann, H. Theuer, B.W. Shore, Rev. Mod. Phys. **70**, 1003 (1998)
239. D.J. Tannor, R. Kosloff, A. Bartana, Faraday Discuss. **113**, 365 (1999)
240. C.M. Tesch, R. de Vivie-Riedle, Phys. Rev. Lett. **89**, 157901 (2002)
241. U. Troppmann, C. M. Tesch, R. de Vivie-Riedle, in *Quantum information processing*, 2nd edition, ed. by G. Leuchsand, T. Beth, (Wiley-VCH, Weinheim, 2005)

2

Analysis and control of small isolated molecular systems

Albrecht Lindinger[1], Vlasta Bonačić-Koutecký[2], Roland Mitrić[2], David Tannor[3], Christiane P. Koch[4], Volker Engel[5], Thorsten M. Bernhardt[6], Joshua Jortner[7], Aldo Mirabal[1], and Ludger Wöste[1]

[1] Institut für Experimentalphysik, Freie Universität Berlin, Germany
[2] Institut für Chemie, Humboldt-Universität zu Berlin, Germany
[3] Department of Chemical Physics, Weizmann Institute of Science, Israel
[4] Institut für Theoretische Physik, Freie Universität Berlin, Germany
[5] Institut für Physikalische Chemie, Universität Würzburg, Germany
[6] Institut für Oberflächenchemie und Katalyse, Universität Ulm, Germany
[7] School of Chemistry, Tel Aviv University, Israel

Coordinated by: Albrecht Lindinger

2.1 Motivation and outline

Elementary dynamical processes in molecules like bond shaking, breaking or making commonly occur on the femtosecond time scale. With the advent of ultrafast laser sources, such as Ti:sapphire, adequately short light pulses of just a few optical cycles can be delivered. This allows the direct, time-resolved observation of dynamical molecular processes, as convincingly shown by Zewail et al. [1–5].

A powerful tool in this regard is pump-probe spectroscopy. It employs a first ultrafast laser pulse to excite the molecular system to a transient intermediate state (pump). This broadband excitation creates a coherent superposition of vibrational eigenstates; the resulting nonequilibrium configuration causes the propagation of a vibrational wave packet on the potential energy surface of the corresponding electronic state. The evolution of the induced wave packet motion is subsequently interrogated with a second, time-delayed ultrashort laser pulse (probe), which induces easily observable processes, such as fluorescence, photoelectron emission or multiphoton ionization. Oscillatory features in the observed signal reflect vibrating wave packet motions, which correspond to the vibrational structure of the photoexcited system. Nonoscillatory signal structures may arise from radiative and nonradiative decay channels; or they reflect photoinduced chemical reactions like dissociations, fragmentation cascades or the formation of new bonds. Such reaction processes can be distinguished by observing the individual signal channels of all involved molecules. This is achieved by employing photoionizing pump-probe schemes, which permit the interrogated particles being monitored size-selectively by means of mass spectrometry.

Time-dependent dynamics calculations allow to simulate the observed dynamical behavior. When performed on sufficiently small and isolated molecules, this interplay of theory and experiment yields the pathways and transition mechanisms of the photon-induced processes enabling their detailed analysis [6, 7]. For larger molecules the task becomes significantly more difficult; this is due to the rapidly growing amount of internal vibrational degrees of freedom that can couple with each other, leading to internal vibrational energy redistribution (IVR). Dephasing and loss of coherence occurs and finally the system is thermalized. If the system is embedded in a liquid or a solid, or if it is deposited on a surface, the thermalization process commonly occurs still much faster. In order to avoid this complication while entering into the subject, the book follows this growing degree of complexity and focuses in the beginning of Chap. 2 on the ultrafast reaction dynamics of small isolated molecular systems. The following chapter will treat larger molecules; then later systems coupled to a condensed environment will be treated.

Pump-probe analysis leads conveniently across an electronically excited transition state, which is easily prepared by a pump pulse. The approach is described in detail in the next Sect. 2.2. However, for understanding the dynamical behavior of the entire system, the comprehension of its electronic ground state, especially when vibrationally excited, is crucial. Such systems can be prepared by using the pump pulse for vertical photodetachment of a stable negative ion (Ne), which commonly leads it to the vibrationally hot electronic ground state of its corresponding neutral (Ne). The resulting wave packet dynamics is then probed by a time-delayed probe pulse, which reionizes the system to a positive ion (Po). As described in Sect. 2.3, this charge reversal pump-probe spectroscopy (NeNePo) allows, when combined with time-dependent quantum calculations, to understand the occurring isomerization dynamics and reactivity of the system, and the onset of IVR-processes, which progressively gain importance at growing particle sizes [8,9]. Small clusters are well suited systems to probe the evolution of dynamical processes in a range, where each atom counts.

The successful analysis of a dynamical system raises hope, that also control of the photoinduced processes can be achieved on a real time basis. Long before an experimental realization far-reaching theoretical concepts [10–21] were developed that use tailor-made pulse sequences or coherent superpositions of states to influence the dynamical pathway of a molecular system. The main goal of these control schemes is to guide the system into a distinct reaction channel. So, guided by theory new techniques were developed to shape the employed femtosecond laser pulses such, that successful control experiments could be performed. Experimentally this requires the synthesis of optimally composed pulse shapes; they can be found by using adaptive feedback loops, as suggested by Judson and Rabitz [22]. Theoretically this is accomplished by employing optimal control theory [11, 16, 20, 21, 23, 24]. As presented in Sect. 2.4, a much deeper insight into the reaction and relaxation dynamics of ultrafast photoinduced processes can be gained from this beautiful convergence between theory and experiment.

Photoassociation experiments of ultra-cold atoms in magneto optical traps (MOT) and in Bose-Einstein condensates (BEC) have raised the question, whether the formed molecules, which are vibrationally still extremely hot, can internally be cooled down. Here again, coherent control scenarios offer exciting perspectives to prepare ultracold molecules and to gain insight into their chemistry. Section 2.5 presents new concepts, which employ shaped ultrafast laser pulses to optimize photoassociation yields followed by radiative cooling cycles to pump the photo-associated molecules in their ultracold environment down to the lowest vibrational state.

2.2 Probing the dynamics of electronically excited states

V. Bonačić-Koutecký, R. Mitrić, A. Lindinger, L. Wöste

The principle of a pump-probe observation scheme via an electronically excited state is presented in Fig. 2.1. The originally rather cold molecules are excited from a low vibrational level of the electronic ground state. Due to the spectral width of the employed fs-pump pulse, a coherent superposition of several vibrational states is then created in the electronically excited state, which leads to the formation of a vibrational wave packet. If a bound electronic state is excited (Fig. 2.1a), this wave packet will oscillate between the inner and outer turning point of the potential energy curve (or surface), reflecting the vibrational motion of the excited molecule. The temporal evolution of this wave packet can be monitored by the probe pulse, which - at a variable delay - excites the particle into a size-selectively detectable ion state. Ionization can either be achieved directly by one photon or it is accomplished by a multiphotonic sequence, which preferably leads across higher excited electronic states. Since the efficiency of the ionization step depends critically upon the position of the wave packet along the reaction coordinate, the obtained ionization efficiency (time dependent Franck-Condon factor) changes significantly as a function of the delay between excitation (pump) and ionization (probe). By tuning this delay time, the temporal evolution of the oscillating wave packet appears as an intensity modulation on the corresponding ion channel, as shown later, for example, in Fig. 2.3.

If, however, a bound-free transition into a dissociative electronic state is excited (see Fig. 2.1b), no oscillatory behavior occurs, but the ion signal will

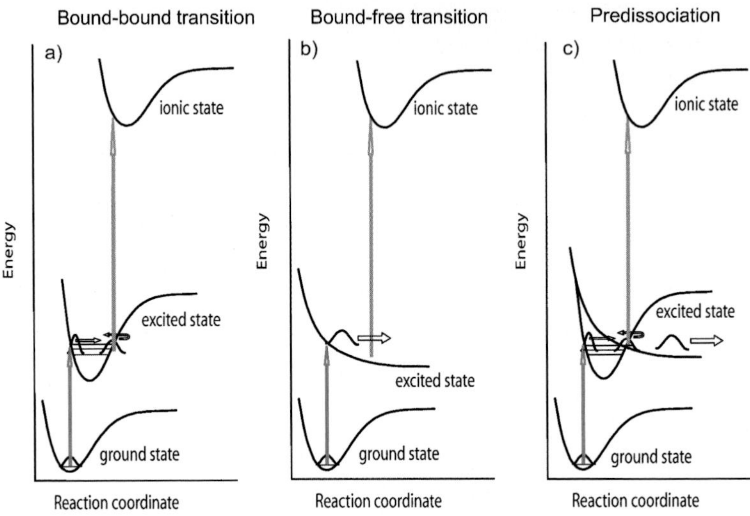

Fig. 2.1. The principle of pump-probe spectroscopy by means of transient two-photon ionization: A first fs-laser pulse electronically excites the particle into an ensemble of vibrational states creating a wave packet. Its temporal evolution is probed by a second probe pulse, which ionizes the excited particle as a function of the time-dependent Franck-Condon window; a) shows the principle for a bound-bound transition, where the oscillatory behavior of the wave packet will appear; b) shows a bound-free transition exhibiting the decay of the fragmenting particle, and c) shows the process across a predissociated state, where the oscillating particle progressively leads into a fragmentation channel.

show the temporal behavior of an exponential decay of the photofragmenting system. In parallel, correlating signals of the produced photofragments will emerge on the corresponding ion channels. In case of fragmentation cascades, multi exponential decay curves become observable, as shown later in Fig. 2.5. Predissociation occurs, when bound and dissociative electronic states are connected by curve crossings or conical intersections. Pump-probe signals taken from their excitation will, therefore, show both: wave packet oscillations of the vibrating molecule superposed by the exponential decay curve of the progressively dissociating system (see Fig. 2.6). The case is most interesting with regard to coherent control scenarios of photoinduced unimolecular dissociation processes. In Sect. 2.2 these three cases will be discussed and experimental examples be given. Then, two examples will be treated theoretically and compared with the experiment: The nonadiabatic fragmentation dynamics of electronically excited Na_2K and the geometrical rearrangement of electronically excited Na_2F.

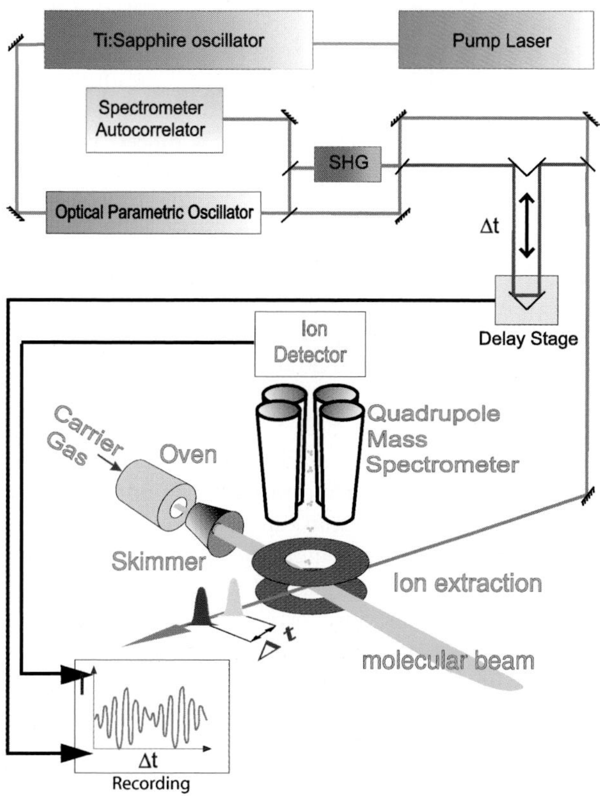

Fig. 2.2. Experimental setup of the pump-probe experiment showing the molecular beam irradiated by laser pulse sequences coming from a Ti:sapphire fs-laser system. The resulting photo-ions are detected across a quadrupole mass spectrometer.

2.2.1 Experimental set-up of the pump-probe experiment

Supersonic molecular beams are an ideal environment to prepare high densities of cold and isolated molecules or clusters and to allow their size-selective observation by means of photoionization mass spectrometry, where pump-probe concepts are easily integrated. The experiments presented here mainly focus on molecules and clusters containing metal atoms. This is due to their pronounced dynamical properties, their highly chromophoric character, and their rather low ionization energies, which are easily reached with the available laser sources. Such metal particle beams are commonly formed by expanding a metal vapor together with an inert or a reactive carrier gas from an oven cartridge across a small nozzle of some microns in diameter into the vacuum. As a result, a collision zone is created in front of the nozzle, in which the molecules and clusters are formed by adiabatic cooling and subsequent nucleation. The typical realization of such an experimental setup is shown in Fig. 2.2. The supersonic molecular beam is extracted from the expansion zone by a skimmer, which leads into a differentially pumped detection chamber. There, an ion extraction system injects the created photo-ions into a quadrupole mass spectrometer (QMS), which is placed perpendicular to the particle beam. Window ports at the detection chamber allow to irradiate the interaction zone perpendicular to both, the neutral particle beam and the extracted ion beam.

For the experiments presented here a commercial Ti:sapphire laser oscillator was used; it was pumped by a frequency-doubled Nd:YLF-, or by an argon-ion laser. The laser operates at a repetition rate of 80 MHz; it is tunable in a wavelength range between 730 and 850 nm producing pulses of 80 fs duration at a total power of 1.6 W. These conditions allow to operate the experiment at a duty cycle of 100%, since each molecule of the continuous molecular beam is irradiated several times by the pulsed laser. Furthermore, the correspondingly low laser peak power prevents undesired multiphotonic transitions, which would camouflage the sought information. The wavelength range of the laser can significantly be extended by using a second harmonic generator (SHG) and/or an optical parametric oscillator (OPO). The employed laser pulses are analyzed by a spectrometer, autocorrelator, and spectrally-resolved cross-correlation (XFROG) . For performing the pump-probe experiment, the laser pulses were split up and recombined in a Michelson interferometer system, allowing to generate pump-probe sequences of a variable delay.

2.2.2 Pump-probe spectra of bound electronically excited states

The result of a pump-probe measurement obtained from $^{39,39}K_2$ is shown in Fig. 2.3. The spectrum was recorded at a pump wavelength of about 834 nm; so the electronic K_2 $\mathrm{A}^1\Sigma_u^+$-state is excited. The probe step was achieved by a delayed two-photon transition of the same wavelength (one-color experiment). The signal exhibits a quite distinct oscillatory behavior with vibrational periods of 250 fs, which correspond well to the eigenfrequencies of the photoexcited K_2 A-state. These vibrations, however, are almost harmonically modulated every 10 ps. An explanation for this can be extracted from the corresponding Fourier-transform (FFT) of the recorded signal, which is presented in the insert of Fig. 2.3: Expected is a curve presenting the anharmonic progression of those vibrational states, which were coherently excited within the bandwidth of the pump pulse. The obtained FFT-curve shows spectral resonances around 65 cm^{-1}; this corresponds well to the known vibrational spacing of the A-state of K_2. The obtained intensity distribution, however, shows quite a surprising behavior: Two largely spaced peaks at the wings of the progression dominate the spectrum; they obviously cause the 10 ps large-amplitude modulation of the distinct beat structure. The progression is perturbed [25]. This is caused by a spectrally coinciding spin-orbit coupled "dark" $\mathrm{b}^3\Pi_u$-state of $^{39,39}\mathrm{K}_2$, which significantly alters the lifetimes and Franck-Condon factors

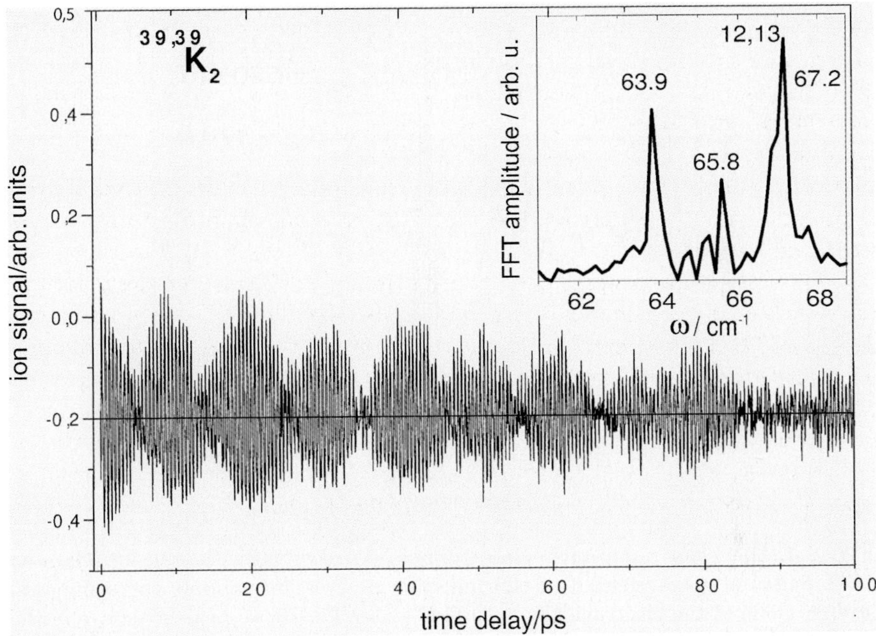

Fig. 2.3. Pump-probe spectrum of electronically excited $^{39,39}K_2$. The insert in the Figure gives the corresponding Fourier-transform of the signal (FFT), which shows the spectral positions of the anharmonic vibrational progression of the excited electronic state. The relative intensities indicate that the sequence is perturbed (see text) [25].

of the observed spectral transition. The phenomenon does not occur for the $^{39,41}K_2$ isotopomer. More details about the differently perturbed systems are given in Sect. 2.2.4.

Transient two-photon ionization experiments on trimer systems were motivated by the need for a time-resolved verification of the pseudo-rotation motion, which can be considered as a superposition of the asymmetric stretch (Q_x) and the bending vibration (Q_y) [26]. In this respect the situation of a triatomic molecule with its three modes is quite different from an isolated oscillating dimer, which vibrates in its single mode until it eventually radiates back to the electronic ground state or predissociates. The coupling of vibrational modes in a trimer system can, therefore, be considered as the onset of internal vibrational redistribution (IVR) [8]. The aspect will be treated later in Sect. 2.2.3 for the example Ag_2Au in great detail.

A typical result for a bound-bound transition in trimers, which was obtained for the electronic Na_3 (B←X)-transition with transform-limited pulses of about 100 fs duration, is shown in Fig. 2.4. The progression shows a pronounced molecular vibration, indicating only one vibrational mode of 320 fs

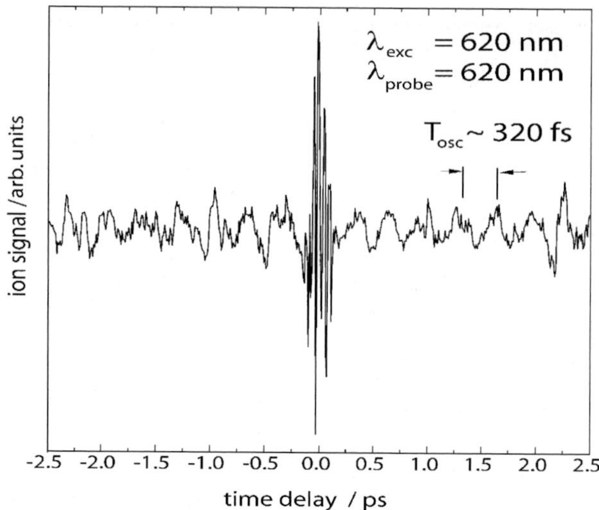

Fig. 2.4. One-color pump-probe spectrum of Na_3 recorded with transform-limited 80 fs pulses at a wavelength of 620 nm. The progression exhibits the symmetric stretch mode of the electronically excited B-state. The dense peak structure in the center is caused by the temporal overlap of the pump and probe pulse, hence it provides an autocorrelation of the employed laser pulse. In the negatively counted time range "pump" becomes "probe", and "probe" becomes "pump", so an almost symmetrical spectrum appears [27].

duration, which corresponds to the symmetric stretch (Q_s) mode of Na_3 in the B-state.

2.2.3 Pump-probe spectroscopy of dissociated and predissociated electronic states

Fragmentation becomes more important as the number of internal degrees of freedom inside the molecule or cluster increases. Here again, the fs-pump-probe observation scheme provides a deep insight into the dynamics of photoinduced cluster fragmentation. The principle of such an experiment is indicated in Fig. 2.1b and c. The particles are electronically excited with fs-laser pulses (pump) into a dissociated or predissociated electronic state. There they dissociate, or they oscillate a few times and then dissociate. The temporal behavior of the sequence is monitored with the probe pulse, which interrogates the system by ionizing the excited particles after a variable time delay Δt.

Typical results of time-resolved pump-probe photodissociation experiments across a bound-free transition are presented for sodium clusters of different sizes in Fig. 2.5. The result reveals the rapidly growing number of different dissociation channels for larger aggregates. The two-color pump-probe

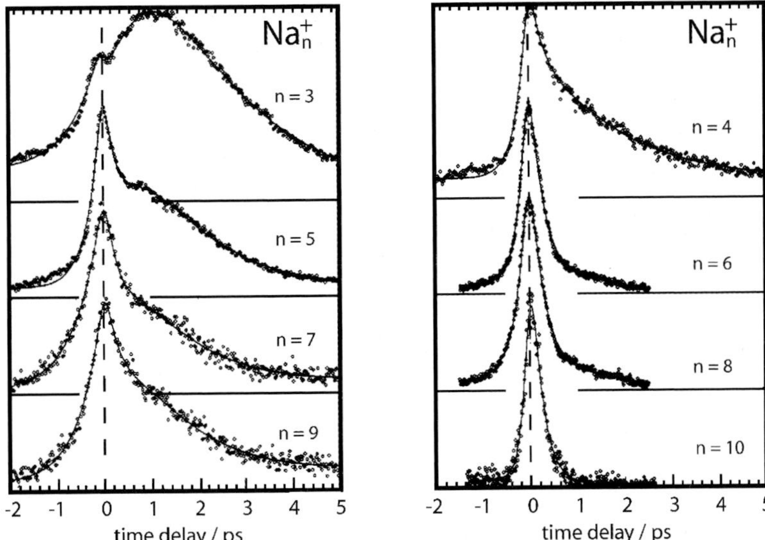

Fig. 2.5. Pump-probe spectra of a two-color experiment probing the bound-free transitions in Na_n $(3 \leq n \leq 10)$. For $\Delta t > 0$: $E_{pump} = 1.47$ eV and $E_{probe} = 2.94$ eV. For $\Delta t < 0$: $E_{pump} = 2.94$ eV and $E_{probe} = 1.47$ eV. [28]

experiments are performed on Na_n with $3 \leq n \leq 10$. For $\Delta t > 0$, the energy E_{pump} was 1.47 eV, whereas E_{probe}, the energy of the probe pulse, was 2.94 eV. Time decays with $\Delta t < 0$ inverted this sequence to $E_{pump} = 2.94$ eV and $E_{probe} = 1.47$ eV. The general trend shows a faster decay with growing cluster sizes. In addition, there is a strong dependence on the excitation wavelength [29]. The size-related increase of fragmentation speed can qualitatively be explained by the growing amount of internal degrees of freedom, which allow -via increasing IVR- to populate more dissociative channels. In order to describe the features, which appear in Fig. 2.5, in more detail, several processes -besides IVR- must be taken into account, as there are the direct fragmentation of the examined cluster size by the pump pulse, and those fragmentation processes that occur to particles, which have populated the observation channel temporarily with fragments of larger clusters, before they fragment again [28].

This fragmentation behavior could be explained in a simple energy level model of the different Na_n cluster sizes [28]. As shown in Fig. 2.5, the temporal evolution of the ion signals exhibits an interesting odd-even alternation as a function of the cluster size, which is associated with two different dissociation channels. Na_n clusters with even n dissociate preferably into an odd numbered cluster and one Na-atom, while odd-numbered clusters dissociate into an odd-numbered cluster and one dimer. However, the cluster size-dependence of the underlying differing decay times is yet not fully understood.

In the case of predissociation one observes both: a wave packet oscillation and an overlaid exponential decay. The result of such an experiment performed on K_3 is shown in Fig. 2.6a. Around $\Delta t = 0$ the signal is at maximum; the occurring signal modulation represents the cross-correlation between the pump and probe pulse. For the following delay time, first a pronounced oscillation occurs, which reflects the wave packet oscillation in the excited state. A magnified segment of this oscillation is shown in Fig. 2.6b, whereas its Fourier-transform is presented in Fig. 2.6c. Three vibrational modes appear; they correspond to the K_3 normal vibrations with $Q_s = 109\ \mathrm{cm}^{-1}$, $Q_x = 82\ \mathrm{cm}^{-1}$, and $Q_y = 66\ \mathrm{cm}^{-1}$. Superposed to these oscillations is an ultrafast unimolecular decay with a lifetime of about 6 ps, which indicates that the observed state is predissociative [30]. This fast fragmentation prevented so far the observation of this excited state by means of stationary resonant multiphoton ionization.

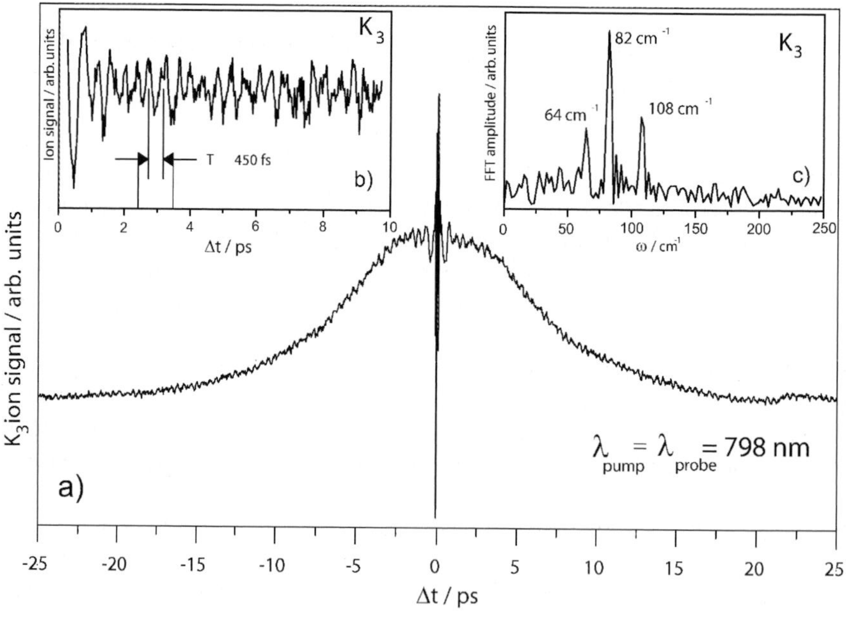

Fig. 2.6. (a) Transient two-photon ionization spectrum of K_3. The spectrum is superposed by a fast unimolecular decay of approx. 5 ps. The progression shows a pronounced oscillation (b). The corresponding Fourier-transform indicates the three normal modes (c). [30]

2.2.4 Experimental and theoretical treatment of the nonadiabatic fragmentation of Na_2K

Trimers are usually considered as simple systems from the theoretical point of view because their potential energy surfaces can be precalculated at a high level of accuracy using different quantum chemical methods. Nevertheless, metallic trimers can have complex electronic structure with a manifold of low lying electronic states violating Born-Oppenheimer approximation as in the case of Na_2K. Here, the Wigner–Moyal representation of the vibronic density matrix for the simulation of pump–probe spectra based on ensembles of classical trajectories is used [31]. According to the experimental conditions, the analytical expression for the signals has been derived under the assumption of weak fields and short pulses, as will be briefly outlined below. The signals are simulated by an ensemble of independent classical trajectories which can be propagated either at precalculated energy surfaces or combined with molecular dynamics (MD) "on the fly". In the case of Na_2K precalculated energy surfaces will be used.

First i) the electronic structure and then ii) the semiclassical dynamics and signals are treated.

i) As already indicated, the mixed alkali trimer Na_2K is characterized by a manifold of electronically excited states. This can be seen from Fig. 2.7, which represents the one dimensional cut along the fragmentation coordinate R_2 for NaK+Na at fixed bond length R_1 (NaK)=3.658 Å and bond angle α=63.79^o. The illustration shows that for the pump–excitation of 1.61 eV, the fragmentation channel NaK (1 $^3\Pi$) and Na (1 ^2S) is not accessible. Therefore, after one photon excitation the fragmentation channel NaK (1 $^3\Sigma^+$) + Na (1 ^2S) can be reached over adiabatic and nonadiabatic transitions involving an avoided crossing between the 5 $^2A'$ and the 6 $^2A'$ states in the former case and the crossings among 5 $^2A'$, 1 $^2A''$ and 4 $^2A'$ states in the latter case. Notice that the bending angle α is an important coordinate for the crossings and therefore all possible transitions cannot be clearly seen from Fig. 2.7.

ii) Semiclassical methods for nuclear dynamics, which make use of classical trajectories with quantized initial conditions, are particularly suitable for exploring ultrafast processes in complex systems. Moreover, the ability to include all degrees of freedom opens a perspective to large systems, where the separation into the chromophore unit and bath is not possible. This approach has theoretically been extended to include quantum effects such as coherence and tunneling in the framework of semiclassical methods [32–35]. It is particularly suitable to use classical trajectories for simple systems, since it allows the comparison with a full quantum dynamical treatment. Therefore, the conceptual aspects are briefly outlined:

Classical MD in different forms, including ab initio MD "on the fly", are now applicable to relatively large systems, and classical trajectories can be used as inputs in semiclassical approaches for simulations of observables. The approach here bases on classical ab initio molecular dynamics (AIMD).

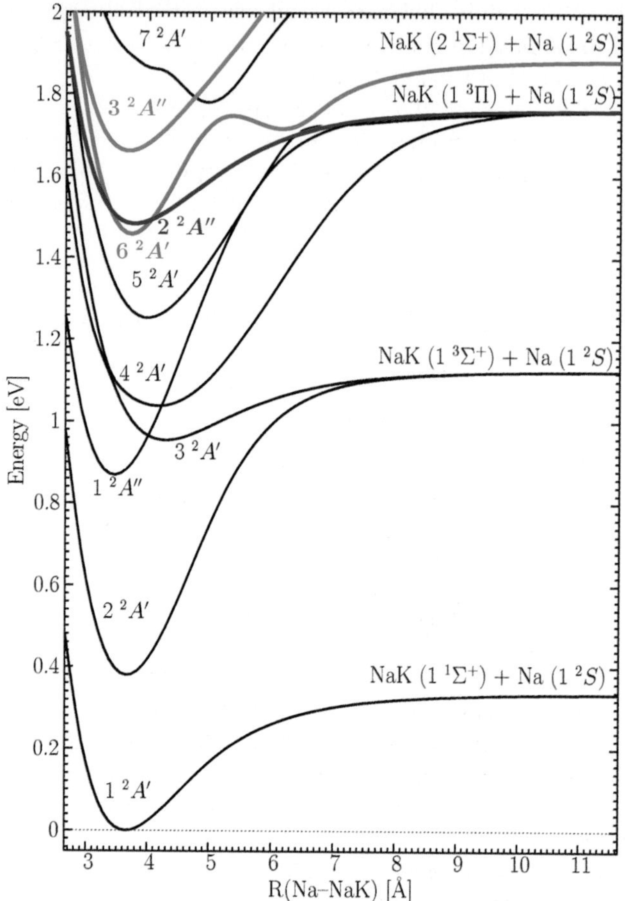

Fig. 2.7. One–dimensional cut of the ground and of nine excited states of Na_2K taken along the coordinate $R_2 = R$(Na2–K) for fixed bond length $R_1 = R$(Na1–K) = 3.658 Å and a bond angle α(Na1–K–Na2) = 63.79°.

The time evolution of the density operator $\hat{\varrho}(t)$ is described by the quantum mechanical Liouville equation

$$i\hbar\frac{\partial\hat{\varrho}}{\partial t} = \left[\hat{H}, \hat{\varrho}\right] \tag{2.1}$$

where $\hat{H}$ is the Hamiltonian of a molecular system involving several electronic states which are coupled with electromagnetic field $\varepsilon(t)$,

$$\hat{H} = \hat{H}_{mol} + \hat{H}_{int} \equiv \sum_a |a\rangle\hat{h}_a(\mathbf{Q})\langle a| - E(t)(\sum_{a,b}|a\rangle\hat{\mu}_{ab}(\mathbf{Q})\langle b| + h.c.), \tag{2.2}$$

with the vibrational Hamiltonian $\hat{h}_a(\mathbf{Q})$ of the adiabatic electronic state a, the collection of vibrational coordinates $\mathbf{Q}$, and the dipole approximation for interaction with the electromagnetic field.

The density matrix formulation offers an appropriate starting point for establishing semiclassical approaches, because the quantum Liouville equation has a well defined classical limit in the Wigner representation, which is given by the classical Liouville equation of nonequilibrium statistical mechanics:

$$\frac{\partial \varrho}{\partial t} = \{H, \varrho\} \tag{2.3}$$

Here $\varrho = \varrho(\mathbf{q}, \mathbf{p}, t)$ and $H = H(\mathbf{q}, \mathbf{p}, t)$ are functions of classical phase space variables $(\mathbf{q}, \mathbf{p})$, and

$$\{H, \varrho\} = \frac{\partial H}{\partial \mathbf{q}} \frac{\partial \varrho}{\partial \mathbf{p}} - \frac{\partial \varrho}{\partial \mathbf{q}} \frac{\partial H}{\partial \mathbf{p}} \tag{2.4}$$

is the classical Poisson bracket. This classical limit can be derived from (2.1) by performing a Wigner transformation [36–38] and expansion in terms of $\hbar$. If this expansion is terminated at the lowest order of $\hbar$ the commutator can be replaced by the classical Poisson bracket:

$$\left[\hat{A}, \hat{B}\right] \rightarrow i\hbar\{A, B\} + O(\hbar^3) \tag{2.5}$$

This leads to the classical equation (2.3). Higher order terms in $\hbar$, are responsible for the introduction of quantum effects in the dynamics. This semiclassical limit of the density matrix formulation of quantum mechanics, based on the Wigner-Moyal representation of the vibronic density matrix offers a methodological approach, which is suited for an accurate treatment of ultrafast multistate molecular dynamics and pump-probe spectroscopy using classical trajectory simulations [31, 39–42].

Keeping the conceptual simplicity of classical mechanics this approach allows an approximate description of quantum phenomena such as optical transitions by averaging over the ensemble of classical trajectories. Moreover, the introduction of quantum corrections can be made in a systematic manner. The method requires drastically less computational effort than full quantum mechanical calculations, and it provides a physical insight into ultrafast processes in complex systems. Additionally, it can directly be combined with quantum chemistry methods for electronic structure. So, the multistate dynamics can be carried out at different levels of accuracy including precalculated energy surfaces as well as the direct ab initio MD "on the fly", in which the forces are calculated, when they are needed in the course of the simulation. The approach is related to the Liouville space theory of nonlinear spectroscopy in the density matrix representation developed by Mukamel et al. (cf. [43]). Following the proposal by Li et al. [39] and formulation given by the authors [31, 40–42], the method is briefly outlined in connection with its application to simulations of the time resolved pump-probe signals, involving both adiabatic and nonadiabatic dynamics.

Assuming that the pump and probe process are both first order in the fields (weak field limit), and assuming that interference and non-Condon effects are negligible, an analytic expression for the pump-probe signal can be derived in the framework of the Wigner distribution approach [31, 40–42, 44]:

$$\begin{aligned} S[t_d] &= \lim_{t\to\infty} P_{22}^{(2)}(t) \\ &\approx \int d\mathbf{q}_0 d\mathbf{p}_0 \int_0^\infty d\tau_1 \exp\left\{-\frac{(\tau_1 - t_d)^2}{\sigma_{pu}^2 + \sigma_{pr}^2}\right\} \times \\ &\quad \exp\left\{-\frac{\sigma_{pr}^2}{\hbar^2}[\hbar\omega_{pr} - V_{21}(\mathbf{q}_1(\tau_1;\mathbf{q}_0,\mathbf{p}_0))]^2\right\} \times \\ &\quad \exp\left\{-\frac{\sigma_{pu}^2}{\hbar^2}[\hbar\omega_{pu} - V_{10}(\mathbf{q}_0,\mathbf{p}_0)]^2\right\} P_{00}(\mathbf{q}_0,\mathbf{p}_0). \end{aligned} \tag{2.6}$$

This expression is valid for adiabatic dynamics and can be interpreted in the following way: At the beginning, the system is prepared in the electronic ground state (0) where the corresponding Wigner distribution $P_{00}(\mathbf{q_0},\mathbf{p_0})$ is assumed to be known (initial condition). This initial phase space density is spectrally filtered during the pump process to a state 1 by the third Gaussian of (2.6). Subsequently, the filtered ensemble propagates on state 1 and is spectrally filtered again during the delayed probe pulse into state 2. This is expressed by the second Gaussian in (2.6). It is important to notice that the Gaussian form of spectra during the pump and probe process is a direct consequence of both, the classical approximation and the short time limit (cf. [44]). The final time resolution of the signal is determined by the pump-probe autocorrelation function given by the first Gaussian in (2.6).

As can be seen from (2.6) the simulation of pump–probe signals involves averaging over an ensemble of initial conditions. It can naturally be determined from the initial vibronic Wigner distribution $P_{00}^{(0)}$ in the electronic ground state. For this purpose the Wigner distribution of a canonical ensemble in each of the normal modes is computed according to (2.7):

$$P(q,p) = \frac{\alpha}{\pi\hbar} exp\left[-\frac{2\alpha}{\hbar\omega}(p^2 + \omega^2 p^2)\right], \tag{2.7}$$

with $\alpha = tanh(\hbar\omega/2k_bT)$ and the normal-mode frequency ω, corresponding to the full quantum mechanical density distributions. The ensemble of initial conditions needed for the MD on the neutral ground state energies emerges from sampling the phase space distribution given by expression (2.7). This allows one to include temperature effects corresponding with the experimental situations. It permits also to take into account the quantum effects of the initial ensemble at low temperatures.

Expression (2.6) can be generalized to include nonadiabatic effects [42] and simultaneous excitation of several electronic states, which gives the following general analytic expression for the pump–probe signal:

$$S[t_d] \sim \sum_n |\langle\psi_n(0)|\boldsymbol{\mu}|\psi_g(0)\rangle|^2 \cdot \int d\boldsymbol{q}_0 d\boldsymbol{p}_0 \int_0^\infty d\tau_1 \exp\left\{-\frac{(\tau_1 - t_d)^2}{\sigma_{pu}^2 + \sigma_{pr}^2}\right\}$$
$$\times \frac{1}{N_{Rand}} \sum_\nu \int_{E_{min}}^{E_{max}} dE \exp\left\{-\frac{\sigma_{pr}^2}{\hbar^2}[E - V_{cat,n(\tau_1)}\{\boldsymbol{q}_{n,\nu}(\tau_1;\boldsymbol{q}_0)\}]^2\right\}$$
$$\times \exp\left\{-\frac{\sigma_{pu}^2}{\hbar^2}[E_{pu} - V_{n(0),g}(\boldsymbol{q}_0)]^2\right\} P_{00}(\boldsymbol{q}_0,\boldsymbol{p}_0) \tag{2.8}$$

which is a modification of (2.6). Equation (2.8) includes:

1. Initial thermal distribution $P_{00}(\boldsymbol{q}_0, \boldsymbol{p}_0)$ for the electronic ground state with the initial coordinates $\boldsymbol{q}_0$ and momenta $\boldsymbol{p}_0$. In the simulations an initial state Wigner distribution is used. The temperature is T=50 K with 1000 starting points for the trajectories running over at least 5 ps. The average over trajectories, which were obtained from different randomizations ν due to the hopping algorithm is expressed by the normalization factor N_{Rand} and by a corresponding summation over ν.
2. Pump pulse window (third exponential in (2.8)) with a pump pulse duration σ_{pu} and a difference between the excitation energy for the pump step E_{pu} and the energy gap $V_{n(0),g}$ between the ground state g and the excited state n of the neutral molecule at the time $\tau_0 = 0$.
3. Probe pulse window (exponential in the second row of (2.8)) with a probe pulse duration σ_{pr}, an energy gap $V_{cat,n(\tau_1)}$ between the current propagating state $n(\tau_1)$ at the time τ_1 and the cationic state at the coordinate $\boldsymbol{q}_{n,\nu}(\tau_1)$. An inclusion of the continuum of the kinetic energy of the detached electron eKE is expressed by the integration over
$$E = E_{pr} - eKE \tag{2.9}$$
for the energy interval $[E_{min}, E_{max}]$.
4. Time window (exponential in the first row of (2.8)) with the pump–probe correlation function located around the time delay t_d, which determines the time resolution of the signal.
5. Weighting factor of the electronic states which is introduced by the square of the norm of the transition dipole moment $\boldsymbol{\mu}$ between the electronic ground state $|\psi_g\rangle$ and the desired excited electronic state $|\psi_n\rangle$ for the ground state equilibrium geometry.

In order to avoid quantitative calculations of the nonadiabatic couplings among the states involved and to use Tully's fewest switches surface hopping procedure [45] for treating the nonadiabaticity, the approach was drastically simplified. The generalization of the Landau-Zener formula has been introduced for the surface hopping probability $p_{n \to i}$ from the current surface labeled by n to the other surface $i \neq n$:

$$p_{n \to i} = \exp\left(\frac{-2\pi(E_i - E_n)^2}{|\boldsymbol{v} \cdot \boldsymbol{\nabla}(E_i - E_n)|}\right), \tag{2.10}$$

with the energy E_n of the nth surface, the gradient of the energy difference $\boldsymbol{\nabla}(E_i - E_n)$ between the nth and i^{th} surface, and the velocity $\boldsymbol{v}$ at the time τ (note that the atomic units ($\hbar$=1) are used). The probability to remain on the n–th surface is given by:

$$p_{n \to n} = 1 - \sum_{i \neq n} p_{n \to i}. \tag{2.11}$$

A uniform random number between zero and one is used to decide, to which surface to hop. If there is not enough kinetic energy, a hop is rejected.

Simulated and recorded pump–probe spectra

In order to start the simulation of signals the histograms of the corresponding energy gaps between the ground state and the excited states $V_{n(0),g}$ have been calculated, corresponding to Franck–Condon profiles. The results are given in Fig. 2.8. They show that the experimental laser bandwidth covers the upper edge of the energy gap to the 6 $^2A'$ state, although the contribution of the 3

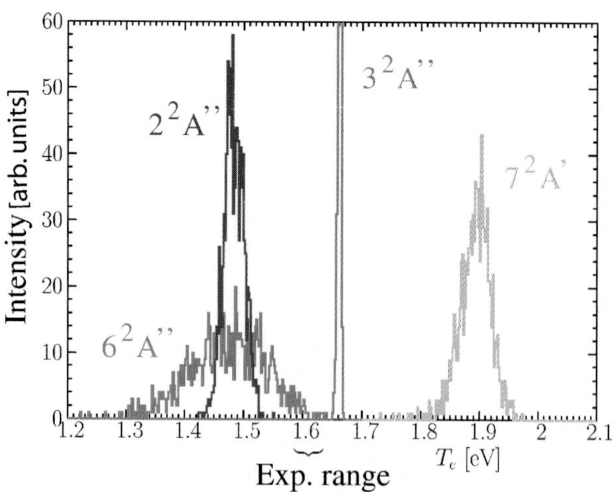

Fig. 2.8. Histograms of energy gaps (transition energies) T_e between the ground and excited states of Na_2K for a 50 K initial ensemble obtained from 1000 sampled phase points of a canonical Wigner distribution.

$^2A''$ state, which is separated by 0.026 eV from the experimental range cannot be excluded. Also the 2 ^{2}A" state is close enough and must be considered. In contrast, the mean value of the energy gap for the 7 ^{2}A' state is more than 0.2 eV separated from the experimental range. It is, therefore, excluded from further considerations.

First the energy gaps between the cationic ground state and excited electronic states of the neutral Na_2K are presented in Fig. 2.9; they reveal dynamical processes, which will be mirrored in the pump–probe signals. The energy interval of the energy gaps differs drastically for the different excited states. Due to the similarity between the electronic ground state and the 3 $^2A''$ state, the energy difference is nearly constant over a wide range, which results in a small energy interval. In contrast, the substantial change in the energy difference between the 6 $^2A'$ state and the electronic ground state around the equilibrium geometry of the latter induces the broad energy interval of the energy gap. The 2 $^2A''$ state is located in between. Therefore, it gives rise to a medium size energy interval. So, by choosing different probe pulse energies, the dynamical features, which correspond to different electronic states, can selectively be probed.

The horizontal lines in Fig. 2.9 indicate important energies: E_{gap}=3.22 eV corresponds to the zero kinetic energy of the detached electron (ZEKE) in a two photon ionization process. E_{gap}=2.62 eV and E_{gap}=2.12 eV represent the restricted and the full continuum, respectively. In Fig. 2.9a the bunch of trajectories shows an oscillation period of 290 fs, which is slightly more than half of the oscillation period of the bending vibration of the 3 $^2A''$ state. No hopping to other excited states occurs with the exception of one trajectory.

If the molecular dynamics is started in the 2 $^2A''$ state (Fig. 2.9b) a fragmentation can be observed after 1 ps. The oscillatory feature, which shows a periodicity of 870 fs, is assigned to the bending vibration of the 2 $^2A''$ state. This will be discussed later. The E_{gap}=2.45 eV indicates MD on the 6 $^2A'$ state. Trajectories in the energy range between 2.45 and 3.35 eV reveal the molecular dynamics on the 2 $^2A''$, 1 $^2A''$, 5 $^2A'$ and 4 $^2A'$ states without fragmentation.

Figure 2.9c shows that 80% of the trajectories diverge within the first 5 ps, which indicates that fragmentation takes place. Thus, fragmentation is a dominant process in this case. The low energy features reveal again dynamics occurring on the 6 $^2A'$ state. Contrary to Fig. 2.9b, however, no oscillations occur, and therefore no stable dynamics can be initiated in the 2 $^2A''$ state. The remaining trajectories reveal dynamics in lower lying states.

It is important to realize that the characteristic features of the pump–probe signals depend significantly on the kinetic energy taken into account. Similar to the cases of NeNePo spectroscopy (cf. Sect. 2.3), the ZEKE conditions provide better resolution of underlying processes in contrast to those cases with excess kinetic energy for which the integration over the energy (continuum) must be considered. Therefore, two sets of results are presented here. First, continuum with restricted values was taken into account (Fig. 2.9d and

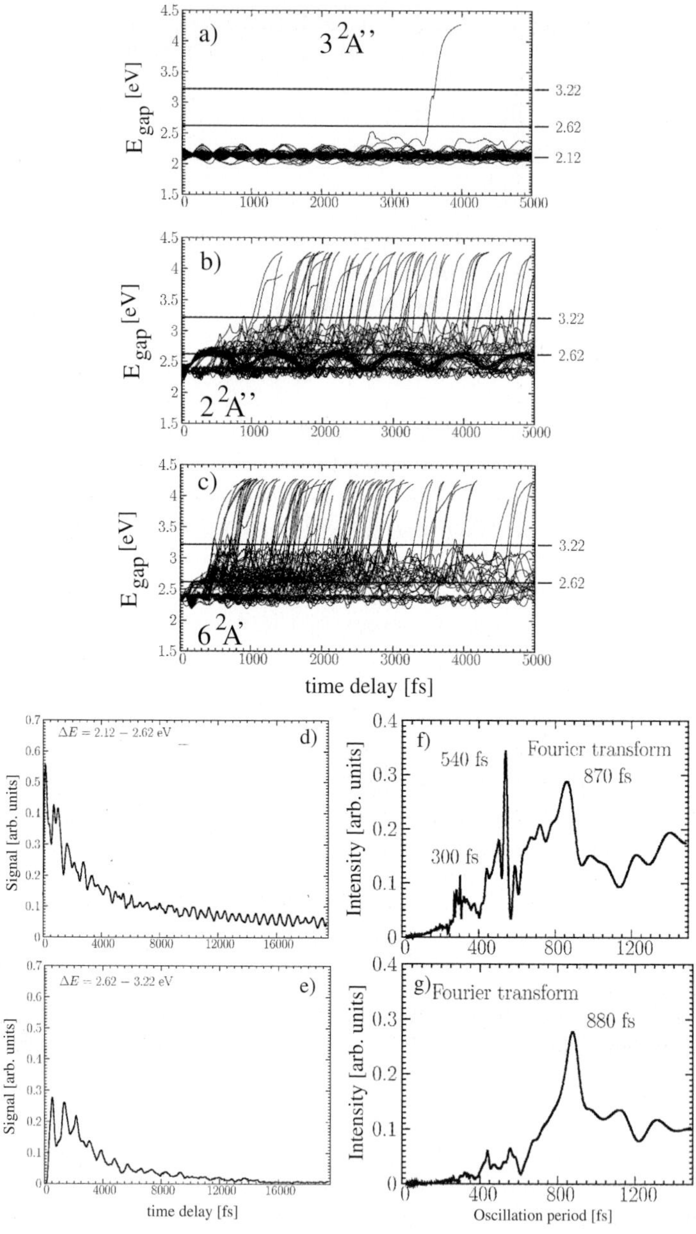

Fig. 2.9. Bunches of energy gaps between the neutral excited states and the cationic ground state of Na_2K (E_{gap}) for 100 representative trajectories which started on a) 3 ^{2}A", b) 2 ^{2}A", and c) 6 ^{2}A' states. An initial temperature of 50 K was assumed. The values of 3.22, 2.62, and 2.12 eV correspond, respectively, to ZEKE (see below), to the lower limit of the restricted, and to the full continuum. Simulated pump–probe signals account (d) $\Delta E = E_{pr} - e\mathrm{KE} = 2.12 - 2.62$ eV and (e) $\Delta E = 2.62 - 3.22$ eV. The corresponding Fourier transforms are given in f) and g).

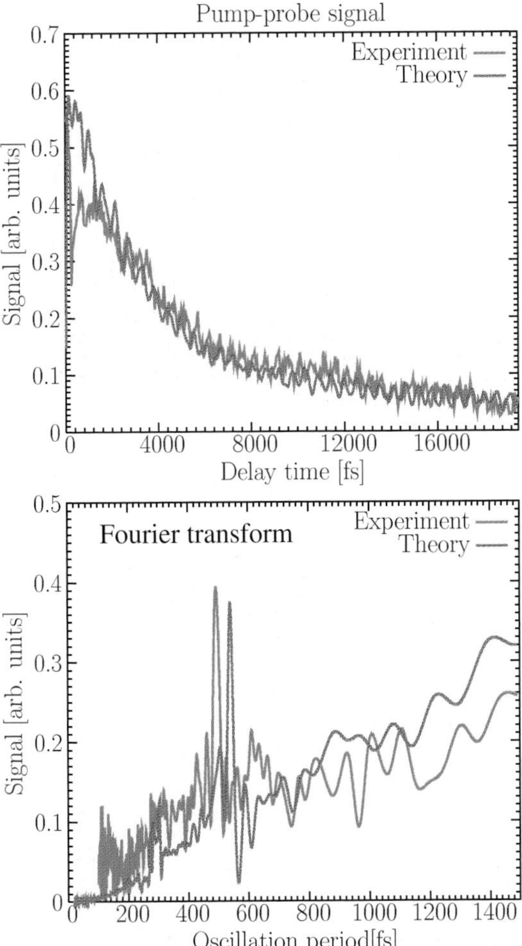

Fig. 2.10. Comparison of theory and experiment for the pump-probe signal of the nonadiabatic fragmentation of electronically excited Na_2K and the relating Fourier transform signal [46].

e) and then the complete continuum (Fig. 2.10), which corresponds better to the experimental conditions [46]. In the first case, different oscillations can be revealed in the pump–probe signals. From the Fourier transforms (cf. Fig. 2.9f and g) it is possible to identify the period of oscillations and to assign them to the particular vibrational mode within the given electronic state. In the case of complete continuum some characteristic features such as oscillations of a particular vibronic mode can easily be smeared out. This can be seen from a comparison of Figs. 2.9 and 2.10.

For the continuum restricted to the energy intervals $I_1 = [2.12 - 2.62]$ and $I_2 = [2.62 - 3.22]$ eV (cf. energy gaps of Fig. 2.9) the pump–probe signals calculated using Eq. 2.8 are given in Fig. 2.9d and e. The corresponding Fourier transforms are given in Figs. 2.9f and 2.9g, respectively. The pump–probe signals decay, but the oscillatory behavior exhibits characteristic features of the 3 $^2A''$ state (Fig. 2.9a) and of the 2 $^2A''$ state (Fig. 2.9b). The oscillation period of 540 fs can be assigned to the bending vibration mode and the oscillation period around 300 fs corresponds to the stretching vibration, both characterizing the 3 $^2A''$ state, as it can be seen from Fourier transforms of Fig. 2.9f. The decay of the signal with an oscillatory feature with the period of 870 fs stems from the bending mode on the 2 $^2A''$ energy surface. The pump–probe signal shown in Fig. 2.9d exhibits mainly features of the 2 $^2A''$ state alone. This is due to the negligible contribution of the 3 $^2A''$ state and the signature of the 6 $^2A'$ state only during earlier times. In the Fourier transform, the peak corresponding to an oscillation period of 870 fs represents the main contribution; it can be assigned to the bending vibration of the 2 $^2A''$ state.

After including the full continuum, the oscillatory behavior with a period of $\sim$ 870 fs is completely suppressed, as shown in Fig. 2.10a. At early times the signal is characterized by an oscillation period of 300 fs for the symmetric stretch and 520 fs corresponding to the bending vibrational modes. At later times the oscillation of the bending mode dominates. These features are confirmed by the Fourier transforms shown in Fig. 2.10b.

The results in Fig. 2.10b clearly indicate that the pump–probe signal obtained from (2.8), which accounts for full continuum is in excellent agreement with experimental findings [46]. This allows to conclude that the experimental results are strongly influenced by the excess kinetic energy of the ionization step. The only fingerprint of an individual electronic state which remains present in the experimental and theoretical signal is the oscillatory structure with a period of 540 fs, which corresponds to the 3 $^2A''$ state.

Altogether, the analysis shows that nonadiabatic dynamics over several excited states of Na_2K leads to a decay of the pump–probe signal which corresponds to the fragmentation process of the photoexcited system. Time scales and mechanism of the underlying processes can easily be determined, if the kinetic energy excess is restricted is in accordance with the experimental results.

2.2.5 Experimental and theoretical treatment of the geometrical rearrangement in electronically excited Na_2F

In comparison with metallic trimers the replacement of one metal atom by a fluorine atom, which leads to the Na_2F trimer with polar Na–F bonds, simplifies the situation drastically, since the first excited state is well separated from other excited states, and it has relatively low transition energies [41, 47, 48]. Therefore, the time resolved theoretical and experimental studies of this system offer the opportunity to identify the time scales of adiabatic processes in the pump-probe spectra [41, 49]. Again, the Wigner–Moyal representation of the vibronic density matrix for the simulation of pump–probe spectra based on ensembles of classical trajectories is used [31]. The analytical expression for the signals was again derived under the assumption of weak fields and short pulses (cf. (2.6)). The signals are simulated by an ensemble of independent classical trajectories, which are obtained for Na_2F from ab initio molecular dynamics "on the fly" involving excited electronic states.

Due to one excess electron in Na_2F, an accurate description of the excited states is particularly simple; it is even possible in the framework of the one–electron "frozen ionic bonds" approximation [41]. In this method the optical response of a single excess electron can explicitly be considered, in the field of the other (n-1) valence electrons, which are involved in the strongly polar ionic Na–F bonding (cf. [41]). In the framework of this approach, the fast computation of the adiabatic MD "on the fly" is particularly favorable and can be combined with (2.6) derived in the framework of the Wigner distribution approach. The only difference is that now the ground state and the first excited state of the neutral Na_2F are involved, as well as the ground state of the cationic Na_2F for probing.

After vertical excitation, the conformational change in the first excited state of Na_2F starts from the triangular ground state geometry and leads then to the linear structure. Considerable lowering of the energy in the excited state occurs as shown in Fig. 2.11a. A relatively small energy gap between the ground and the first excited state is a consequence of an avoided crossing, obtained from breaking the Na-Na bond and overshooting the linear geometry connecting two equivalent triangular geometries. In spite of the avoided crossing, the nonadiabatic coupling is weak and therefore adiabatic dynamics can be performed.

As a starting point the generation of a temperature dependent initial Wigner phase space distribution on the neutral ground state is needed. A canonical thermal ensemble at a given temperature is suitable for low temperatures at which a harmonic approximation holds. This initial ensemble has to be brought to the first excited state with a Franck-Condon transition probability, which involves the corresponding excitation energies shown in Fig. 2.11C. The propagation of this ensemble on the first excited state involves classical trajectory simulations. The probe window includes the time-dependent energy gaps between the cationic and the neutral excited states, taken at the propagated coordinate (cf. Fig. 2.11 B-a). The calculation of the signal requires a summation over the entire phase space (cf. Fig. 2.11 B-b).

The results presented in Fig. 2.11B are based on simulations using an ensemble of 300 classical trajectories at an initial temperature of 300 K, which corresponds to the experimental conditions [49]. The theoretically and experimentally obtained results are given in Fig. 2.11B. Pulse duration and laser wavelength used in the experiment and in the simulations are shown in Fig. 2.11B. The probe wavelength of 3.06 eV corresponds to the minimum of the

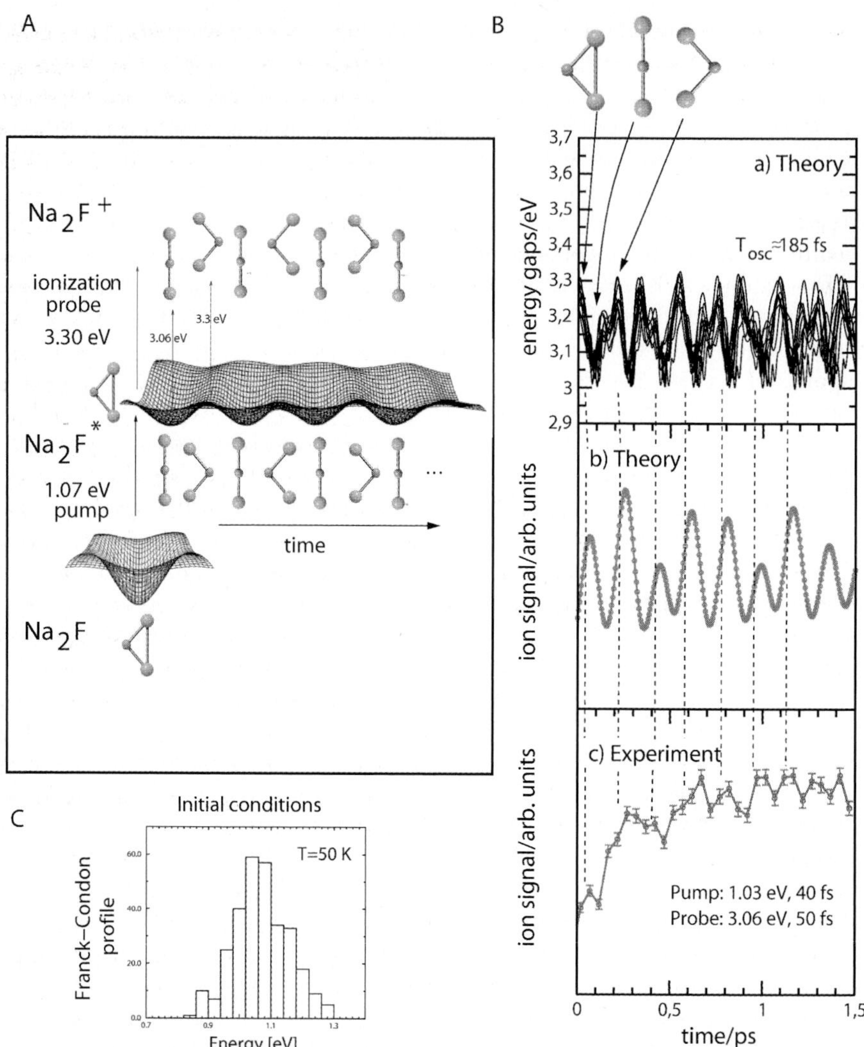

Fig. 2.11. A) Pump-probe NeExPo scheme for Na_2F involving the neutral ground state (Ne), the electronic excited state (Ex), and the cationic ground state (Po). B) Temperature dependent (T=50 K) initial conditions; histogram of transition energies between first excited and ground state. C) Comparison of theoretical with experimental results [49]: (a) bunch of energy gaps between the first excited states of Na_2F and the cationic ground state reflecting the dynamics in the first excited state; (b) simulated pump-probe signal; (c) experimentally observed transient. The involved structures are shown in the top panel. [49]

first excited state with a linear geometry, as can also be seen from Fig. 2.11A. From the bunch of the time dependent energy gaps shown in Fig. 2.11B-a the periodic relaxation dynamics with a period of $\approx$ 185 fs can be identified. The maxima correspond to the bent and the minima to the linear structures. At longer times, anharmonicities in the bending mode introduce aperiodicity. Because the energy gaps are essential for the determination of fs pump-probe signals, the oscillations in the simulated signal have the same period of 185 fs, as can be seen from Fig. 2.11B-b. Since the linear geometries are probed for an ionization energy of 3.06 eV, they give rise to maxima in the signal. The periodic feature of the signal allows the identification of structural rearrangements from triangular to linear geometry during the butterfly type relaxation dynamics in the first excited state of Na_2F. This occurs due to the breaking of the Na-Na metallic bond. The strong Na-F ionic bonds on the other hand remain almost intact, which hinders fragmentation to take place. The period of 185 fs corresponds to half of the normal bending mode frequency in the first excited state. Therefore, the observed oscillations can be assigned primarily to the bending mode. Consequently the IVR is very small, because the stretching modes do not significantly contribute. This means that the time scale for the metallic bond breaking is $\sim$ 90 fs. A recorded transient Na_2F ion signal is shown in Fig. 2.11B-c. The comparison of the theoretical pump-probe signal (Fig. 2.11B-b) with the experimentally obtained transient (Fig. 2.11B-c), shows very good agreement. Both curves exhibit oscillations with a period of 185 fs corresponding to a periodic butterfly type rearrangement between bent and linear geometries [49]. It is important to notice that theoretical predictions [41] have initiated the experimental work, which confirmed predicted findings and consequently the proposed simple theoretical approach. In fact, theoretical results on Na_2F based on quantum mechanical dynamics and precalculated energy surfaces [50] fully support the theoretical results presented in Fig. 2.11. Therefore the conclusion can be drawn that excited state dynamics in the framework of the "frozen ionic bond" approximation, combined with the Wigner distribution approach is capable to describe accurately processes on the femtosecond time scale.

2.3 Probing the dynamics of the transition state

V. Bonačić-Koutecký, T. M. Bernhardt, R. Mitrić, L. Wöste, and J. Jortner

As shown before, pump-probe spectroscopy allows the investigation of ultrafast nuclear dynamics in molecules and clusters during the geometric transformation along the reaction coordinates. This involves the preparation of a transition state by optical excitation of a stable species into a nonequilibrium nuclear configuration in the pump step, and the probing of its time evolution by laser induced processes. Neumark et al. and Lineberger et al. have demonstrated that such nonequilibrium can also be produced by vertical photodetachment of stable negative ions. The state of the neutral species can be close to the stable geometry of anions [51–60], or it can provide the starting point for an isomerization process in the neutral ground state [61–63]. The vertical one-photon detachment spectroscopy was advanced by introducing the charge reversal (NeNePo) pump-probe technique [8, 9]. The approach employs, after preparing the system by vertical photodetachment from its anion, a time delayed probe pulse to ionize the vibrationally excited neutral molecule to a positive ion. This allows to probe structural processes and isomerization relaxation and internal vibrational energy redistribution (IVR) in neutral molecules and clusters as a function of the cluster size and composition. An extension of the NeNePo technique by two-color pump- excitations has been also proposed [64]. As shown in Chap. 3, besides the NeNePo approach, also time resolved photoelectron spectroscopy [51] is a powerful technique, which can be applied to study dynamics of molecules and clusters [51].

In theory, the conceptual framework and scope of ultrafast spectroscopy is provided by simulations which allow to determine the time scales and the nature of configurational changes as well as IVR in vertically excited or ionized states [31,40–42,65–68]. The separation of the time-scales of different processes is essential for identifying them in the experimentally observed features. Moreover, the distinction between resonant and dissipative IVR in finite systems can be addressed as a function of the molecular size and its atomic composition. For the investigation of the dynamics in fs-spectroscopy, the generation of the initial conditions and multistate dynamics for the time-evolution of the system itself and for the probe or the dump step are needed. For this purpose, two basic requirements have to be fulfilled. First, accurate determination of electronic structure is mandatory. In the case that the electronic states involved are well separated, the Born-Oppenheimer approximation is valid and the adiabatic dynamics is appropriate, as it is usually the case for the ground states involved in the NeNePo spectroscopy. The second basic requirement is the accurate simulation of ultrafast observables such as pump-probe signals. This involves an appropriate treatment of optical transitions such as ultrafast creation and detection of the evolving wave packet or classical ensemble. In the latter case, the dynamics is described by classical mechanics, and the average over a sufficiently large number of trajectories has to be made in order to simulate the spectroscopic observables.

Since atomic clusters, in particular with metallic atoms [69], usually do not contain a "chromophore type" subunit and do not obey regular growth patterns [70], it is mandatory to include all degrees of freedom in the simulation of dynamical processes. Consequently, first principle (ab initio) molecular dynamics "on the fly" (AIMD), without precalculation of the energy surfaces, represents an appropriate choice to study ultrafast processes in elemental clusters with heavy atoms for which in the first approximation, the classical description of nuclear motion, is acceptable. The basic idea is to compute forces acting on nuclei from the electronic structure calculations which are carried out "on the fly" [71]. Related AIMD methods with plane wave basis sets have significantly contributed to the success of the method applied to clusters [72].

2.3.1 Multistate adiabatic nuclear dynamics and simulation of NeNePo signals

The goal of theoretical fs-NeNePo spectroscopy is to provide conditions under which different processes and their time scales can be observed and to establish the scope of this experimental technique [31, 67, 68, 73].

To illustrate these goals, first the electronic and structural properties of noble metal clusters will be addressed. Then attention will be paid to MD "on the fly" and to the theoretical approach for the simulation of signals. Furthermore, the analysis of the signals and the comparison with the experimental findings will be presented allowing for the identification of processes and conditions under which they can be observed. Finally, reactivity aspects and the scope of NeNePo spectroscopy will be addressed.

2.3.1.1 Electronic structure

Noble metal molecules are good candidates for probing multistate adiabatic nuclear dynamics because of their relatively simple electronic structure in comparison with transition metals, and their similarity to s-shell alkali metals. Their structural, reactive, and optical properties have attracted numerous theoretical [66–68, 74–87] and experimental studies [86–100] over the years. This is particularly the case for silver clusters with a large s-d gap in contrast to gold clusters. In the latter case, the s-d gap is considerably smaller, due to the relativistic effects which strongly lower the energy of the s-orbital. These differences in the electronic structure are also reflected in different structural properties of small silver and gold clusters [78, 79, 86, 87]. Increasing interest in gold and silver clusters is due to their newly discovered size-selective reactivity toward molecular oxygen and carbon monoxide [80–83, 101–104]. All together, the noble metal clusters represent an attractive research direction for fs-chemistry.

Relativistic effective core potentials (RECP) are mandatory for accurate theoretical description of these species. Gradient corrected density functional theory (GDFT) is presently the method of choice for the ground state properties of metallic clusters provided that the exchange and correlation functionals used, allow for the accurate determination of binding energies and structural properties, which is not always the case [81]. This is particularly important for a reliable calculation of the energy ordering of different isomers, which assume related or very different structures with close lying energies.

In the early work on ground state structural properties of neutral and charged silver clusters one-electron relativistic effective core potentials (1e-RECP) with corresponding AO basis sets were developed [74–76], which have been later revisited in connection with the DFT method [67] employing Becke and Lee, Yang, Parr (BLYP) functionals [105, 106] for exchange and correlation. The justification for the use of 1e-RECP is that the d-electrons are localized at the nuclei of the silver atoms, and almost do not participate therefore in bonding. Recent DFT calculations on structural properties using 19e-RECP, and ion mobility experiments carried out on Ag_n^+ [87] clusters, have confirmed the early findings [74, 75].

In contrast, the use of 1e-RECP for gold clusters might be useful only if the results agree with those obtained from 19e-RECP, due to the fact that the former one is computationally less demanding (for details cf. reference [79]). Moreover, for reactivity studies involving oxidized clusters, 19e-RECP is mandatory also for silver clusters which is due to the activation of d-electrons by p-electrons of the oxygen atom [81–83].

2.3.1.2 NeNePo pump–probe signals

Semiclassical molecular dynamics using classical trajectories with quantized initial conditions in the frame of Wigner distribution approaches outlined in Sect. 2.2.4 will be used for the simulation of NeNePo signals. For this purpose (2.6) is modified as follows:

$$
\begin{aligned}
S[t_d] &= \lim_{t\to\infty} P_{22}^{(2)}(t) \\
&\approx \int d\mathbf{q}_0 d\mathbf{p}_0 \int_0^\infty d\tau_1 \exp\left\{-\frac{(\tau_1 - t_d)^2}{\sigma_{pu}^2 + \sigma_{pr}^2}\right\} \times \\
&\exp\left\{-\frac{\sigma_{pr}^2}{\hbar^2}\left[\hbar\omega_{pr} - V_{neu,po}(\mathbf{q}_1(\tau_1; \mathbf{q}_0, \mathbf{p}_0))\right]^2\right\} \times \\
&\exp\left\{-\frac{\sigma_{pu}^2}{\hbar^2}\left[\hbar\omega_{pu} - V_{neu,neg}(\mathbf{q}_0, \mathbf{p}_0)\right]^2\right\} P_{00}(\mathbf{q}_0, \mathbf{p}_0). \qquad (2.12)
\end{aligned}
$$

where $V_{neu,neg}$ and $V_{neu,po}$ are the energy gaps between the neutral and anionic state and between cationic and neutral states, respectively. For the simulation of the NeNePo signals the initial conditions are obtained using (2.7).

Notice that (2.12) corresponds to zero electron kinetic energy conditions (ZEKE). Usually in NeNePo experiments these conditions are not satisfied and therefore the integration over the excess kinetic energy of the photoelectrons has to be carried out and expression (2.12) adequately modified.

2.3.1.3 Experimental setup for NeNePo spectroscopy

The accurate temperature control of the initial molecular cluster ensemble in the NeNePo experiment is an important issue and has already been emphasized [68]. Only through the experimental knowledge of the temperature parameter, a detailed comparison with theoretically obtained NeNePo signals

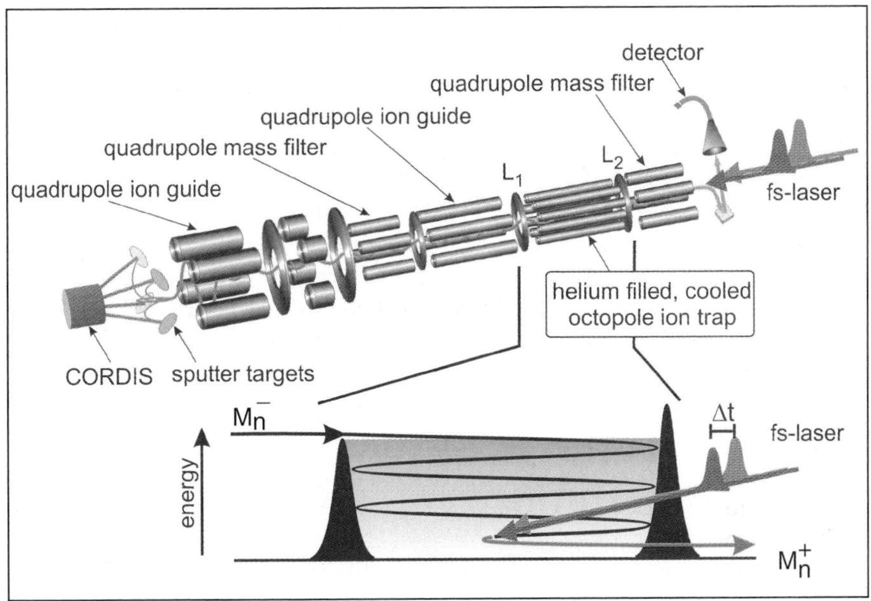

Fig. 2.12. Schematic representation of the setup for the temperature-controlled NeNePo pump-probe experiment. The upper part illustrates the arrangement of the particle source, the quadrupole mass filter and ion guides, and of the octupole ion trap with entrance and exit lenses L_1 and L_2. The process of trapping and cooling molecules inside the octupole ion trap is schematically depicted in the lower part [68].

becomes possible and different contributions to the observed nuclear dynamics can be distinguished. Therefore, the original experimental setup [8, 107] had been extended to enable the control of cluster temperature in the range between 20 and 300 K [108]. The NeNePo experiment was carried out in a helium filled, temperature variable radio frequency (rf)-octopole ion trap. The complete experimental setup is depicted in Fig. 2.12 [68]. The strong effect of the temperature on the observed nuclear dynamics is apparent from the NeNePo signals depicted in Fig. 2.14a. The molecular ions are generated by sputtering metal targets with accelerated xenon ion beams (CORDIS-source [109]). The emerging anions are subsequently mass-filtered and guided into the octupole ion trap (cf. upper part of Fig. 2.12). Inside the ion trap the cluster ions rapidly loose energy by collisions with the helium buffer gas (cf. lower part of Fig. 2.12) and perfect thermalization is reached within a few milliseconds. The ions are spatially confined by the rf-field and the electrostatic potential of octupole entrance and exit lenses. The average residence time of the molecular anions in the octupole ion trap before interaction with a laser pulse is on the order of a few hundred milliseconds. The femtosecond laser beams enter the rf-ion trap collinearly with the axis of the apparatus from the opposite side

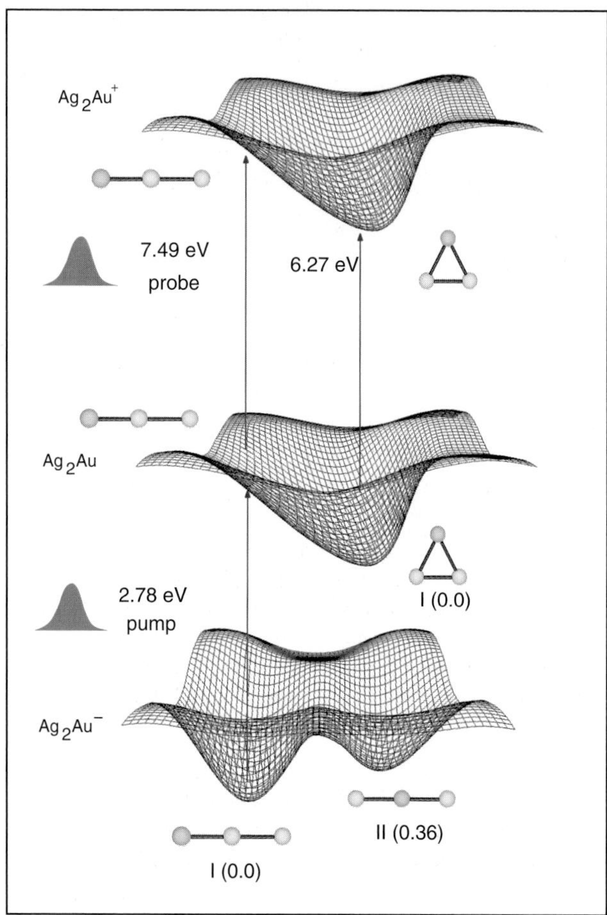

Fig. 2.13. Scheme of the multistate femtosecond dynamics for NeNePo pump-probe spectroscopy of Ag_2Au with structures in different charge states and energy intervals of the pump and probe steps.

as the injected ions. The first ultrafast laser pulse (pump) then detaches the excess electron of the anion resulting in a neutral cluster in the geometry of the anion. This leads to nuclear relaxation dynamics, which can be probed in real-time by femtosecond time-delayed ionization of the cluster to the cationic state (probe). As soon as cations are prepared inside the ion trap, they will be extracted by the electrostatic field of the octupole exit lens and can be mass analyzed with the final quadrupole mass filter. The recorded ion current at the detector as a function of the pump-probe delay time Δt gives rise to the transient NeNePo signal that reflects the time dependent ionization prob-

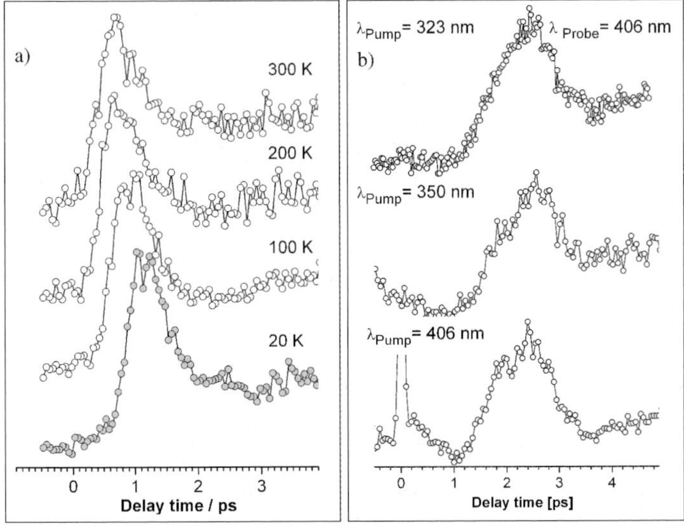

Fig. 2.14. a) NeNePo signal for Ag_2Au obtained at different ion trap temperatures. Note the change of the signal with increasing temperature. For the interpretation of the signal at T=20 K see text and Fig. 2.16. b) Ag_2Au NeNePo signal obtained at three different pump (electron detachment) energies. Note that the dynamics at 406 nm probe wavelength is independent of the pump photon energy. The peak at t=0 fs for the one-color experiment is due to pump-probe interferences (lower trace).

ability of the neutral clusters due to the nuclear dynamics initiated by the initial photodetachment. However, only by comparison of transient NeNePo signals with theoretically simulated signals the conditions can be identified under which a ZEKE-like situation can be achieved, as illustrated below.

2.3.1.4 Geometry relaxation and onset of IVR in the Ag_2Au trimer

The mixed silver-gold trimer $Ag_2Au^-/Ag_2Au/Ag_2Au^+$ has been chosen as an example to demonstrate the ability of the ab initio Wigner distribution approach to accurately predict the NeNePo signals, to interpret them, and to identify conditions under which the separation of time scales of processes, such as geometric relaxation and IVR can be achieved. This has been realized by using the experimental setup at low temperature and close to the zero-kinetic energy electron (ZEKE) conditions as described above. Furthermore, the aim was to study the influence of the heavy atom on the time scale of fs-processes since a comparison with the "light" Ag_3 trimer [31] can be made. The simulations of the NeNePo pump-probe spectra have been performed by using the ab initio Wigner distribution approach combined with the ab initio MD "on the fly" in the framework of the density functional theory [67,68] as described above. The mixed Ag_2Au trimer has the following structural properties: the anionic Ag_2Au^- trimer assumes a linear structure with one Au-Ag heterobond. The symmetric linear isomer with two hetero Ag-Au bonds lies 0.36 eV higher in energy. In the neutral state of Ag_2Au both linear structures are transition states between the two equivalent triangular geometries which correspond to the most stable structure. In the cationic state the obtuse triangle is the minimum. It is important to notice that the structural properties of these mixed trimers are sensitive to the details of the methodological treatment like the choice of RECP and of the functional in the DFT procedure. Therefore, the explicit treatment of d-electrons is necessary for quantitative considerations. The energetic scheme relevant for NeNePo together with the structural properties of the neutral and the charged Ag_2Au is shown in Fig. 2.13.

Since for the simulations the initial temperature of 20 K has been chosen in correspondence with the experimental conditions, it can be assumed that only the most stable structure is populated in the anionic state. Under these conditions, the harmonic approximation is valid and therefore the initial conditions for the MD simulations have been obtained by sampling from the canonical Wigner distribution given in (2.7). Due to the low temperature the Franck-Condon transition probabilities to the neutral state assume an almost Gaussian shape centered around 2.78 eV. The experimentally determined adiabatic detachment energy of Ag_2Au^- amounts also to 2.78 eV [100]. The first excited neutral state is separated from the anion by about 4 eV [100]. Pump photon energies of 2.78 eV and 4.00 eV should, therefore, be suitable to prepare the neutral Ag_2Au in the electronic ground state. The experimental transient NeNePo signals do indeed show the same temporal evolution independently from the pump wavelength (Fig. 2.14b).

In order to simulate the NeNePo signals, an ensemble of trajectories (e.g. $\sim$ 500) has to be propagated in the neutral state, and the time-dependent energy gaps to the cationic state need to be calculated along the trajectories. The energy gaps are presented in Fig. 2.15. They provide visual information about the time evolution of individual processes such as the onset of geometrical changes and of IVR. Within the first 2 ps after the photodetachment, the swarm of energy gaps decreases from 7.5 eV to 6.5 eV, and subsequently all energy gaps exhibit oscillations in the energy interval between 6.1 and 6.5 eV. This allows to distinguish two different types of processes: i) the geometric relaxation from the linear toward the triangular structure, taking place within the first 2 ps and ii) subsequent IVR process within the triangular structure. The minimum energy gap value of $\sim$ 6.1 eV corresponds to the structure with the closest approach of the terminal silver and gold atoms, which is referred to as an internal collision within the cluster. Therefore the adjustment of the pump-probe energies experimentally allows to probe these processes. The highest value of the IP is 7.5 eV, choosing higher probe pulse energy will lead to the signal which is rising very rapidly and which subsequently remains constant due to the contribution of the continuum. Pulse energies between 6.5 eV

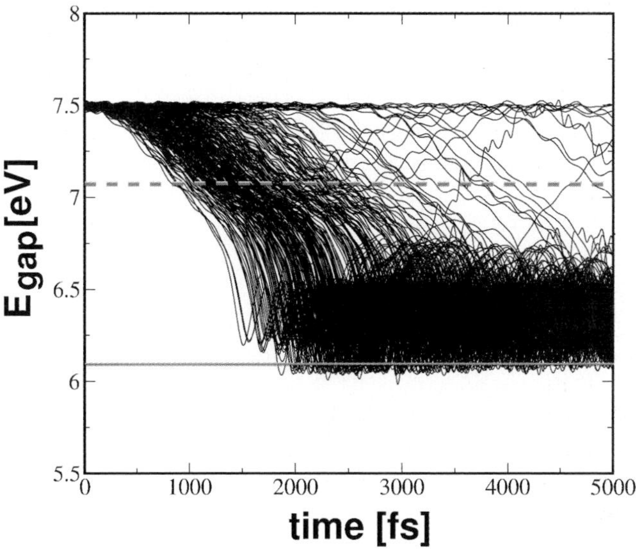

Fig. 2.15. Bunch of the cation-neutral energy gaps of Ag_2Au (right side). Energies of 7.09 eV (dashed line) and of 6.1 eV (full line) indicate the proximity of the Franck-Condon region and of the minimum of neutral species, respectively. They are used for simulating the signals.

and 7.5 eV probe the onset of the geometric relaxation processes. Therefore, it is expected that the signals exhibit maxima at delay times when the probe energy is resonant with energy gaps and decrease to zero at later times. Pulse energies below 6.5 eV probe the arrival and the dynamics at the triangular structure. It is to be expected that at $\sim$ 6.1 eV pulse energies, the signal will rise after $\sim$ 2 ps and remain constant at later times. This is illustrated also in Fig. 2.16a in which the theoretical NeNePo and NeNePo-ZEKE signals are compared with the experimental results at low temperature (T$\sim$20K) for three energies: E_{pr} = 7.7 eV; probing above the ionization threshold of the linear structure, E_{pr} = 7.10 eV; probing the Franck-Condon region, and E_{pr} = 6.10 eV; probing the triangular geometry region corresponding to the minimum of the neutral Ag_2Au [68].

The NeNePo signal in the upper trace of Fig. 2.16 has been obtained with the probe energy E_{pr2ph} = 7.7 eV. The signal rises fast between about 700 fs and 1.5 ps and remains constant afterwards due to the contribution of the continuum. The middle trace of Fig. 2.16 shows the signal obtained with the probe pulse energy E_{pr2ph}= 7.1 eV. The Ag_2Au^+ ion intensity is minimal around zero time, but starts to rise already after about 500 fs with

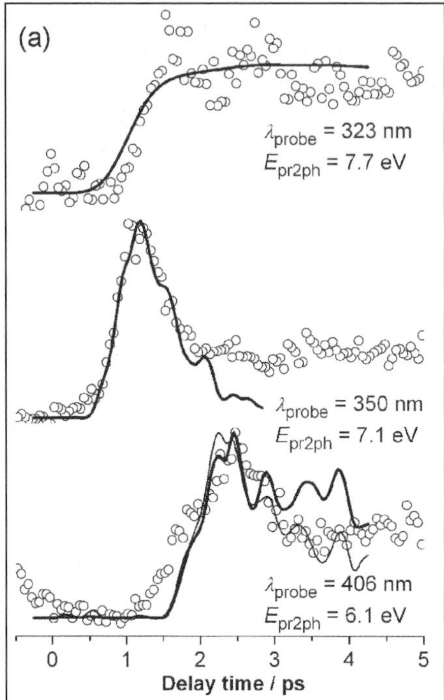

Fig. 2.16. Experimental NeNePo signals (open circles) obtained for three different probe pulse wavelengths in comparison with the simulated time dependent signals (solid lines) for the different probe pulse energies. The ionization probe step is two photonic as confirmed by power dependent measurements. The signals are normalized in intensity: lower graph: The experimental data obtained at $\lambda_{pr} = 406$ nm ($E_{pr2ph} = 6.1$ eV) are overlaid by the simulated NeNePo-ZEKE (bold line) and NeNePo (thin line) signals at $E_{pr} = 6.1$ eV, middle: experimental data obtained at $\lambda_{pr} = 350$ nm ($E_{pr2ph} = 7.1$ eV) are overlaid by the simulated NeNePo-ZEKE signal at $E_{pr} = 7.1$ eV, upper graph: the experimental data obtained at $\lambda_{pr} = 323$ nm ($E_{pr2ph} = 7.7$ eV) are overlaid by the simulated NeNePo signal at $E_{pr} = 7.41$ eV. A common time zero between experiment and theory has been chosen for all probe energies. The deviation in the time origin corresponds to less than 0.1 eV in the probe energy.

a considerably steeper slope than in Fig. 2.16 to reach a maximum already at 1.1 ps. The signal decreases again comparably fast and stays at a constant low level after 2 ps. The lower trace of Fig. 2.16 displays the NeNePo signal for 6.1 eV reached with two photon probe energy (E_{pr2ph}) measured at 20 K. The signal stays at the same low level for 1.1 ps, then rises gradually until it reaches its maximum value around 2.5 ps. It subsequently decreases again and remains almost constant at about half of its maximal intensity from 3.5 ps on.

The comparison between the theoretically obtained NeNePo signals (solid lines in Fig. 2.16) and the measured time dependent NeNePo signals enables the assignment of the observed pronounced probe energy dependence to the fundamental processes of nuclear dynamics. At E_{pr2ph} = 6.1 eV (Fig. 2.16 lower trace) the onset of IVR and the dynamics of Ag_2Au initiated by the collision of the terminal Au and Ag atoms can be probed exclusively. The good agreement between the experimental (open circles) and the simulated NeNePo-ZEKE signals (bold line) in Fig. 2.16 is apparent, indicating that the experimental signal starts to rise when the system approaches the triangular potential well. The signal maximum can be assigned to the time of intracluster collision at around 2.4 ps followed by IVR in the potential minimum of the neutral triangular geometry. The experimental signal offset at longer delay times is somewhat lower with respect to the maximum than expected from the simulated NeNePo-ZEKE signal. This might be attributed to contributions from the rather similar NeNePo type signal (thin line in the lower trace of Fig. 2.16). This shows explicitly in which regime ZEKE conditions hold in the experiment. The middle trace of Fig. 2.16 presents the comparison of simulated and experimental transient ion signals at 7.1 eV probe energy. Because the initial peak of the experimental transient is perfectly matched by the simulated NeNePo-ZEKE signal (solid line) at the corresponding wavelength the experimental conditions in this case allow for direct exclusive probing of the geometrical relaxation of Ag_2Au. The trimer passes through bending angles of $\phi = 166^o$ at the signal onset around 500 fs to $\phi = 13^o$ at the signal maximum and finally up to $\phi = 96^o$ at 2 ps, where the terminal atoms already interact and the intracluster collision is closely ahead (cf. upper trace of Fig. 2.17). The experimental signal offset at times later than 2 ps can again be attributed to the imperfect NeNePo-ZEKE conditions. The possible reason for the good agreement of the experimental transient signal with the simulated NeNePo-ZEKE transient signal is most likely due to a particularly favorable Franck-Condon overlap in the case of 7.1 eV two photon probe energy. Finally, at high ionization energy E_{pr2ph} = 7.7 eV, a comparably weak experimental transient signal is detected. This signal is in agreement with the simulated NeNePo transient (solid line) at a probe energy of 7.41 eV just below the highest theoretically predicted ionization energy which corresponds to the linear transition state structure (see top trace of Fig. 2.16). Thus, the experiment at E_{pr2ph} = 7.7 eV apparently monitors the system when it leaves this transition state region. Still there is a considerable signal onset time of about 700 fs which reflects the very shallow slope of the PES around the linear transition state geometry.

In summary, experiment and theory are in excellent agreement. The simulated signal at E_{pr}=7.70 eV rises after 1 ps and remains constant subsequently without allowing to identify the dynamical processes which take place due to the contribution of the continuum. The signals at E_{pr} = 7.1 eV reflect geometric relaxation from linear to triangular geometry of the neutral Ag_2Au. The signals at E_{pr} = 6.10 eV are due to IVR.

For an analysis of the vibrational energy redistribution, the kinetic energy was decomposed into normal mode contributions. Figure 2.17 shows a single-trajectory example together with the two distinct Ag-Ag-Au bond angles ϕ. From this representation, valuable insight into the IVR in this model system can be gained. The bond angle ϕ in Fig. 2.17 decreases from an initial value of about 180° at t = 0 to a minimum value of 54° at t = 2.36 ps. However, the kinetic energy begins to increase notably only at t$\sim$ 2.0 ps, when ϕ falls short of 90°. Accordingly, within the next 360 fs ϕ decreases much more rapidly and the kinetic energy in the bending mode passes a pronounced maximum.

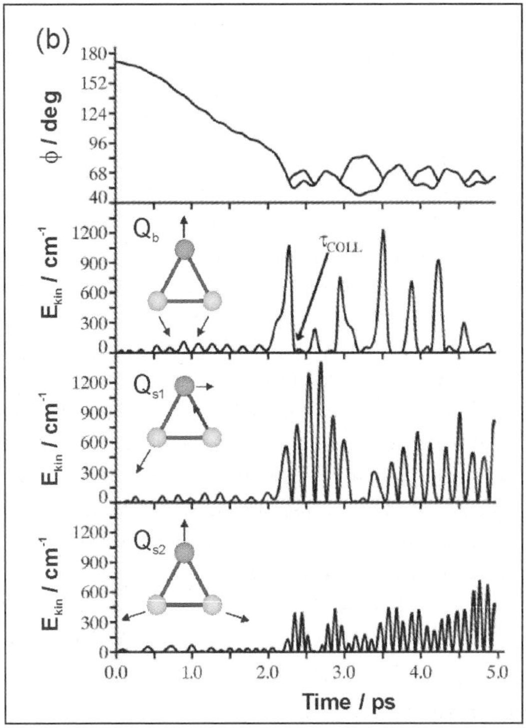

Fig. 2.17. A single-trajectory example of the evolution of the Ag-Ag-Au bond angle ϕ (upper panel) and of the kinetic energy in the three vibrational normal modes Q_b, Q_{s1}, and Q_{s2} (lower panels). In the upper panel two functions are given for ϕ, since for triangular geometries two Ag-Ag-Au bond angles can be defined. The lower curve for ϕ reflects the atom connectivities of the initial linear geometry and therefore monitors the geometrical relaxation until the closest approach of the terminal atoms (internal collision). The upper curve is the larger of the two Ag-Ag-Au bond angles and indicates partial escapes from the potential well of the triangular geometry. In the second panel from top, the collision time τ_{COLL} = 2.36 ps, determined from the first pronounced rise and the subsequent sharp minimum of the kinetic energy in the bending mode, is marked in the diagram.

Shortly afterwards, the kinetic energy decreases to zero, as the system passes the potential minimum and the terminal atoms subsequently further approach each other ("internal collision"), until the kinetic energy is consumed by running against the repulsive part of the potential. In parallel to the increase of the kinetic energy in the bending mode Q_b, intense oscillations are triggered in the first, antisymmetric stretching mode Q_{s1} and to a smaller extent only in the symmetric stretching mode Q_{s2}. Apparently, the simultaneous gain of kinetic energy of the Q_b and of the Q_{s1} mode is a consequence of the fact that both normal coordinates together make up the major components of the linear-to-triangular geometric relaxation coordinate.

Intense kinetic energy oscillations of the stretching modes appear between two pronounced kinetic energy maxima of the bending mode, when the system is in the deep region of the potential, so that enough energy is available for the stretching modes. This relation is particularly apparent for the antisymmetric stretching mode Q_{s1}, manifesting an extensive energy exchange and a close coupling of these modes. Shortly after the bending mode has passed its maximum kinetic energy, its kinetic energy drops to zero (manifesting internal collision), which for the single trajectory of Fig. 2.17 occurs at 2.36 ps. Since the other mode energies increase at the same time, IVR is manifested; the drop of the kinetic energy in the bending mode cannot be solely explained by a conversion of kinetic to potential energy in the bending mode. This behavior is also found for other trajectories, whereupon one can generally state that notable IVR sets in at the instant of internal collision.

These results imply that the nature of IVR in Ag_2Au is related to the one found for Ag_3 (cf. [31, 40]). However, two important aspects should be emphasized. First, time scales are much longer than in the case of Ag_3, due to the heavy atom effect. Second, importantly, in contrast to the Ag_3, the experimental results for Ag_2Au reveal for the first time geometric relaxation separated from an IVR process, indicating that the experimental signals are close to the ZEKE-like conditions, which has been proposed by theory as a necessary condition for the separation of time scales of these processes [68].

2.3.1.5 Vibrating rhombus of Ag_4

In order to further illustrate the ability of the theoretical approach to treat systems with more degrees of freedom, one example, Ag_4, has been chosen for the presentation. In the case of the silver tetramer, the global minima of the anion and of the neutral cluster assume related rhombic structures. Therefore, after photodetachment at low temperatures ($T \approx 50$ K), which assures that only the rhombic isomer is populated, the nonequilibrium rhombic configuration close to the global minimum of the neutral species is reached, as shown on the left hand side of Fig. 2.18.

The probe in the Franck-Condon region with $E_{pr} = 6.41$ eV reveals the vibrational structure of the rhombic configuration after photodetachment. For the probe with, e. g. $E_{pr} = 6.46$ eV, the dynamics in the vicinity of the neutral

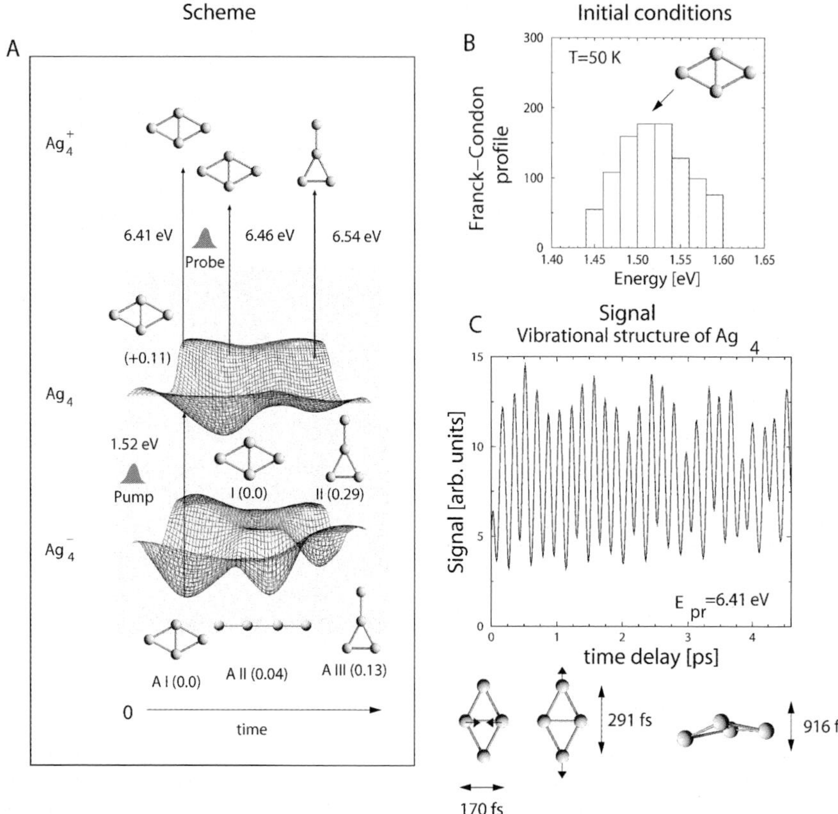

Fig. 2.18. Scheme of multistate fs-dynamics for NeNePo pump-probe spectroscopy of $Ag_4^-/Ag_4/Ag_4^+$ with structures and energy intervals for the pump and probe steps (A). Simulated NeNePo-ZEKE signals for the 50 K initial condition ensemble (B) at the probe energy of 6.41 eV and a pulse duration of 50 fs (C). Normal modes responsible for relaxation leading to oscillatory behavior of the signal are also shown [67].

rhombic structure can be monitored. The simulated NeNePo-ZEKE signal at 6.41 eV for a probe duration of 50 fs shown in Fig. 2.18 exhibits oscillations with a vibrational period of $\sim$ 175 fs which is close to the frequency of the short diagonal stretching mode, indicating the occurrence of the geometric relaxation along this mode toward the global minimum. The analysis of the signal also reveals contributions from two other modes shown in Fig. 2.18. In summary, this example illustrates that an identification of the structure of a gas-phase neutral cluster in experimental NeNePo signals is possible due to its vibronic resolution [67]. The theoretically predicted main features of the pump-probe signals for Ag_4 have been also found experimentally.

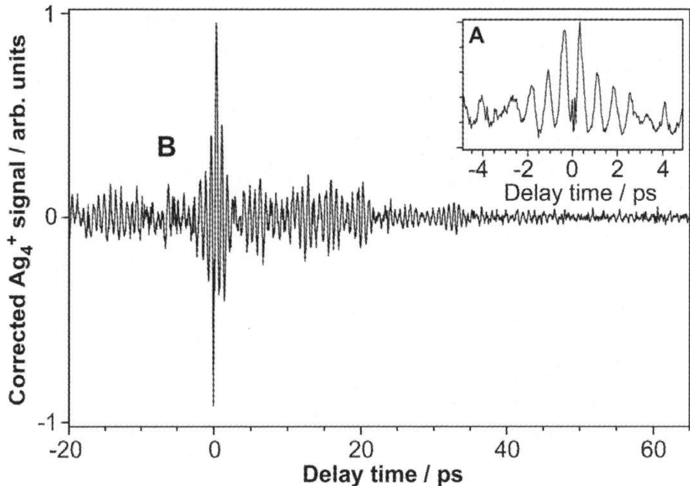

Fig. 2.19. NeNePo spectrum of Ag_4 recorded in a one-color experiment (385 nm) at 20 K ion temperature. Trace A (insert) shows the uncorrected, mass selected Ag_4^+ yield. Trace B is a composite of two measurements and the signal has been corrected for the FFT analysis. Figure taken from [108].

One-color NeNePo-spectra of Ag_4^- measured at 385 nm and an anion temperature of 20 K are shown in Fig. 2.19. [108] Trace A (top right) shows the (uncorrected) mass selected Ag_4^+ yield as function of the delay time between the pump and probe pulses from -4.9 ps to +4.9 ps in steps of 20 fs and exhibits a pronounced oscillatory structure, characterized by a period of about 740 fs. The intensity of the maxima decreases for larger delay times and additional, weaker structures are observed at delay times > 2.8 ps overlapping the 740 fs beat structure.

Trace B in Fig. 2.19 is a composite of two measurements covering delay times from -20 ps to +20 ps and +20 ps to +65 ps. The spectra have been baseline corrected and transformed for the following fast Fourier transform (FFT). Trace B shows that the oscillations in the Ag_4^+ signal intensity extend up to 60 ps, with decreasing intensity. Pronounced partial recurrences are observed. The FFT analysis of this time-resolved signal reveals as dominant feature several peaks centered around 45 cm^{-1} [108]. Comparison to photoelectron data of Ag_4^- [62] leads to the conclusion that the oscillations observed in the NeNePo spectra of Ag_4 are due to vibrational wave packet dynamics in the $2a_g$ mode of either the $^3B_{1g}$ or $^1B_{1g}$ "dark" electronically excited state of rhombic Ag_4 which is probed by a two photon ionization step to Ag_4^+. Ab initio calculations of the harmonic frequencies of the low-lying electronic states of rhombic Ag_4 support this assignment and confirm the observed pronounced anharmonicity of this vibrational mode of $2\nu_0\chi_0 = 2.65 \pm 0.05$ cm^{-1}. The $2a_g$ mode was not resolved in the previous anion photoelectron spectroscopy studies, due to its low frequency of 45 cm^{-1} which lies below the resolution of conventional anion photoelectron spectrometers. The results on Ag_4 demonstrate the successful application of femtosecond NeNePo spectroscopy to study the wave packet dynamics in real time. This is manifested by a beat structure in the cation yield, of a "purely" bound potential, in contrast to the transition-state experiments on the noble metal trimers which connect linear with triangular structures. The spectra of Ag_4 enable the characterization of a selected vibrational mode with a resolution, which is superior to that of conventional frequency domain techniques.

2.3.1.6 Reactivity aspects elucidated by the fragmentation of Ag_2O_2 and isomerization in Ag_3O_2

Noble metal clusters exhibit fascinating reactive properties such as strongly size and charge dependent reactivity toward small molecules, e. g. O_2 [110]. One particularly appealing example in this respect is the reactive behavior of the silver dimer toward dioxygen in the gas phase which has been investigated in detail by different groups [83, 94, 111–116]. Under the conditions of an rf-ion trap experiment, the anionic dimer adsorbs one O_2 molecule in a straightforward association reaction mechanism [83]. Photoelectron spectroscopic studies confirm that the oxygen is molecularly bound to Ag_2^- [116]. In contrast to Ag_2^-, the positively charged silver dimer shows a strongly temperature dependent O_2 adsorption behavior: O_2 is first adsorbed molecularly on Ag_2^+, but in an activated reaction step, the O-O bond can dissociate leading to the adsorption of atomic oxygen at temperatures above 90 K [112]. A NeNePo experiment starting from the stable $Ag_2O_2^-$ complex is thus expected to probe the real-time nuclear dynamics associated with the change in the reactive O_2 adsorption behavior initiated by the pump-photodetachment step. Figure 2.20a displays the experimental NeNePo spectrum of the bare silver dimer without adsorbed oxygen obtained in a one-color pump-probe experiment (406 nm) at 100 K anion temperature. The NeNePo trace exhibits two remarkable features: (i) A pronounced maximum in the recorded Ag_2^+ signal at 190 fs pump-probe delay time and (ii) distinct vibrational dynamics at longer delay times (>400 fs). The amplitude of the vibrational structure at delay times >400 fs is about ten times smaller than the maximum signal. The vibrational period of the observed signal oscillation was determined by FFT analysis to be 180 ± 1 fs ($\nu = 185 \pm 1$ cm^{-1}). The femtosecond NeNePo dynamics detected in the Ag_2^+ signal in Fig. 2.20a can be understood on the basis of the known spectroscopic properties of Ag_2^- and Ag_2 [97] (cf. Fig. 2.21 a and b). Through photodetachment with 406 nm (3 eV) photons, the electronic ground state of Ag_2 (X-$1^1\Sigma_g$) is populated, but also the lowest excited triplet state $1^3\Sigma_u$. This latter triplet state is, however, only very weakly bound (Fig. 2.21a). The system is thus populated in the repulsive part of the potential energy curve in the dissociation continuum of this state. During the propagation along the triplet potential curve, the wave packet might

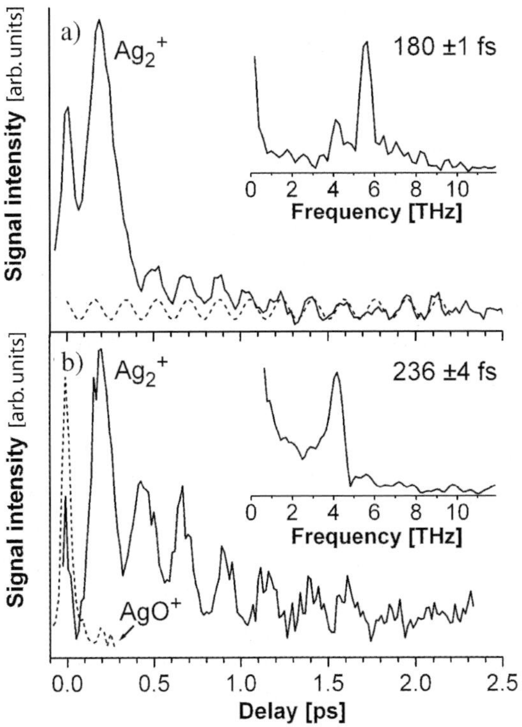

Fig. 2.20. (a) NeNePo spectrum of Ag_2 recorded in a one-color experiment (406 nm) at 100 K ion temperature. (b) Pump-probe spectra of the NeNePo fragment signals Ag_2^+ (solid line) and AgO^+ (dashed line, magnified by a factor of 10) resulting from neutral Ag_2O_2 dissociation after photodetachment of $Ag_2O_2^-$ (406 nm, 100 K).

be transferred via resonant two photon transitions at two close lying locations to the cationic state for detection (via C-$2^3\Pi_u$ and B-$1^1\Pi_u$-states of Ag_2). The existence of these resonances explains the strong enhancement of the Ag_2^+ signal at 190 fs delay time. The lack of a periodical revival with comparable amplitude confirms the assumption that the wave packet further propagates freely to the dissociation on the triplet state potential. The wave packet at the same time prepared on the X-$1^1\Sigma_g$ ground state of Ag_2 oscillates between the inner and outer turning point of the potential leading to the observed periodic signal at delay times >400 fs with 180 fs oscillation period (2.21b). The localization of the wave packet is highest at the turning points of the potential where it can be efficiently transferred into the ground state of Ag_2^+ by irradiation with the probe pulse. The ionization step requires three 406 nm photons, where a resonant transition via the A-$1^1\Sigma_u$-state is possible [117].

If a small partial pressure of O_2 is added to the helium buffer gas inside the rf-ion trap, the complex $Ag_2O_2^-$ is immediately formed, before the silver dimer

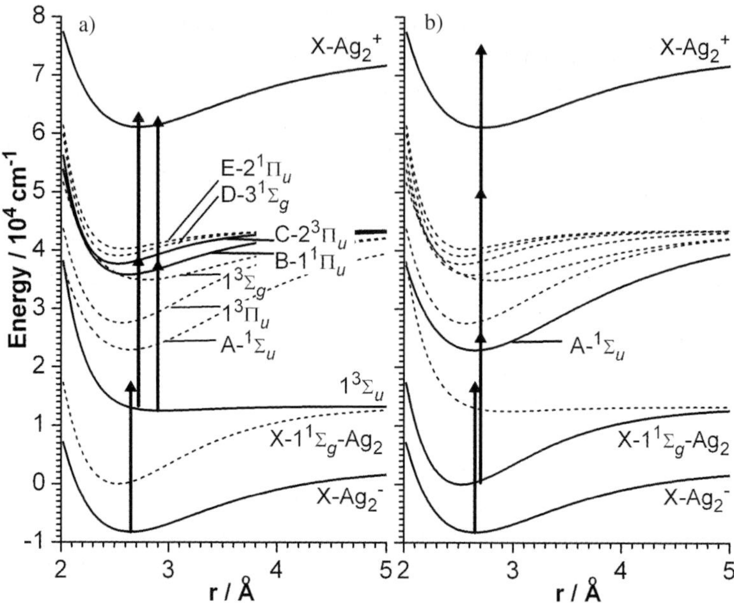

Fig. 2.21. Potential energy curves for the NeNePo process of Ag_2 in the triplet state (a) and the neutral singlet ground state (b).

can interact with the femtosecond laser pulses [83]. Thus, under these conditions, the NeNePo experiment exclusively probes the dynamics launched by photodetachment from the $Ag_2O_2^-$ cluster complex. Figure 2.20b shows the result of the NeNePo experiment starting form $Ag_2O_2^-$ performed under otherwise identical conditions as in the case of Ag_2^- (Fig. 2.19a). It is first interesting to note that at zero delay a signal of $Ag_2O_2^+$ was detected resulting from the NeNePo process. This indicates that the neutral Ag_2O_2 formed by photodetachment is unstable and rapidly dissociates. The observation is in accordance with gas-phase reactivity measurements which show that the neutral complex Ag_2O_2 is not bound [111]. Surprisingly, two fragmentation paths which lead to the formation of the product ions Ag_2^+ and AgO^+ (solid and dashed lines in Fig. 7b, respectively) seem to exist. The Ag_2^+ signal is a factor of ten larger than the AgO^+ signal. The AgO^+ signal exhibits only a peak at zero delay time with a width corresponding to the cross correlation of the laser pulses (80 fs). This suggests that Ag_2O_2 decays on a short time scale of less than 80 fs to fragments. Most likely, AgO^+ arises from fragmentation in the cationic state, i.e. by decay of $Ag_2O_2^+$, which is generated by vertical multi photon transition from $Ag_2O_2^-$ in the interference (cross correlation) time range of the two laser pulses of the same color. The much more intense Ag_2^+ signal shows pronounced oscillatory dynamics, which differs significantly from bare Ag_2 (cf. Fig. 2.20a). First, the amplitude of the observed vibrations is much larger than that of the long delay time dynamics of pure Ag_2. Second, the oscillation is damped, and third, the NeNePo spectra do not show the short time scale dynamics as it appears for bare Ag_2 in Fig. 2.20a. The FFT-analysis of the oscillatory dynamics in Fig. 2.20b leads to a vibrational period of 236 $\pm$ 4 fs ($\nu = 141 \pm 3$ cm^{-1}) which is significantly red-shifted in comparison to the 180 fs period of bare Ag_2. The red shift of the vibration points toward the substantial influence of the fragmentation of the oxygen ligand on the dynamics. Qualitatively, the fragmentation of the O_2 molecule apparently leaves the silver dimer fragment with higher vibrational levels populated. An estimation based on known spectroscopic constants of Ag_2 and the Morse approximation gives a vibrational excitation up to levels around v=40 for ν=141 cm^{-1}, which means a vibrational energy gain of approximately 0.7 eV compared to the bare Ag_2 prepared by the photodetachment from Ag_2^-. Due to the substantial anharmonicity of the potential in this frequency region, the wave packet experiences a rapid dephasing after only a few vibrational periods as can be seen from Fig. 2.20b. The amplitude of the vibration in Fig. 2.20b can be fitted in good approximation by an exponentially damped sine function with a lifetime of 650 $\pm$ 50 fs, which gives an approximate time scale for the dephasing of the wave packet [117]. This observation reflects the strong influence of the molecular adsorbate on the metal cluster structure. Such adsorbate induced structural changes, geometric as well as electronic, have recently been identified as the origin for the cooperative adsorption of multiple adsorbate molecules on small noble metal clusters. This cooperative action is regarded essential for the catalytic activity of gas-phase noble metal clusters in, e.g., the CO combustion reaction [83, 115]. In the particular case of negatively charged silver clusters Ag_n^- with an odd n, the joint experimental and theoretical work showed that a weakly bound first O_2 cooperatively promotes the adsorption of a second O_2 molecule, which is then differently bound with the O_2 bond elongated and thus activated for further oxidation reactions such as CO combustion [83]. The possible prospects of these intriguing catalytic properties of free noble metal clusters for real time laser spectroscopic investigations and photon-induced control of catalytic reactions will be illustrated for the prototype example of Ag_3O_2.

The scheme of the multistate dynamics for Ag_3O_2 is presented in Fig. 2.22a. Anionic isomers in which the isomer I has a linear Ag_3 subunit, and the isomer II a triangular Ag_3 subunit, are very close in energy and the exact energy sequence is difficult to predict theoretically, even when highly accurate methods are applied. However, due to the pronounced differences in vertical detachment energies (VDE) of both isomers, they can be selected by appropriate choice of the pump-pulse energy. It is assumed that only isomer I is populated in the anionic ground state and a canonical Wigner distribution function given in (2.7) at T=20 K is used to generate the initial conditions. After the photodetachment with the pump-pulse energy of 2.75 eV a transition state in neutral Ag_3O_2 is reached. Three isomers are present for Ag_3O_2 as shown in Fig. 2.22. In the structure of the global minimum (isomer I),

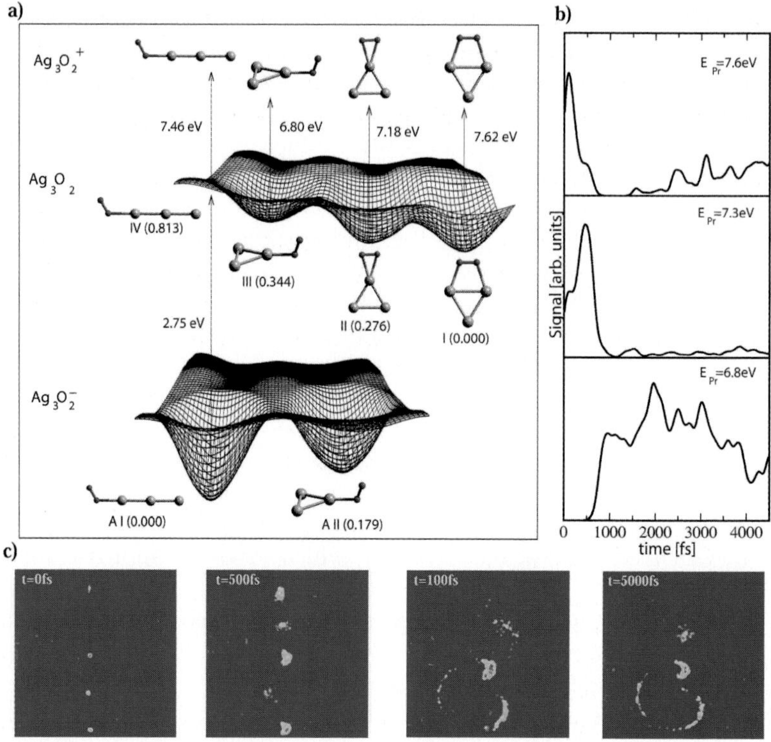

Fig. 2.22. Scheme of the multistate fs-dynamics for NeNePo pump-probe spectroscopy of $Ag_3O_2^-/Ag_3O_2/Ag_3O_2^+$ with structures and energy intervals for pump and probe steps. (b) Simulated NeNePo-ZEKE pump-probe spectra, (c) snapshots of the dynamics in the neutral Ag_3O_2 after the photodetachment.

the molecular oxygen bridges two silver atoms (cf. Fig. 2.22a). In the two higher lying isomers, molecular oxygen is bound to one silver atom, forming one and two bonds, respectively. Therefore, after the photodetachment a dynamical process dominated by the Ag_3 relaxation from the linear to the triangular structure is induced. At later times the energy gained by the cluster isomerization is partly transferred to molecular oxygen leading mainly to the mixture of two higher lying isomers (II and III). After 5 ps the global minimum (isomer I) is populated only by about 10%. Therefore, in principle the population of the chosen minimum with its corresponding reactive center can be controlled with a tailored laser field. The snapshots of the dynamics in the neutral state are presented in Fig. 2.22c; they show the onset of the geometric relaxation of the Ag_3 subunit at 500 fs, the arrival to the triangular Ag_3 subunit at 1 ps, and the final ensemble of isomers after 5 ps. The simulated NeNePO-ZEKE pump-probe signals for Ag_3O_2 are presented in Fig. 2.22b. The signal for the probe pulse energy of 7.6 eV exhibits a fast decay after 1 ps due to the escape from the initial Franck-Condon region reached upon the photodetachment. However, this signal starts to rise again after 2 ps since the global minimum with the ionization potential of 7.62 eV is being populated at later times. This signal offers an opportunity to identify the isomerization process leading to the global minimum in the ground state and also to determine its time scale. If the energy of the probe pulse is lowered to 7.3 eV the onset of isomerization of the Ag_3 subunit starting at 500 fs can be selectively probed. Further lowering of the energy to 6.8 eV allows to probe the IVR process induced after the intracluster collision within the Ag_3 subunit. It should be pointed out that in contrast to the Ag_2Au, here the IVR process is extended also to the O_2 subunit. It is responsible for the final populations of the isomers, since they mainly differ by the bonding of the O_2 subunit. In general, it can be expected that future studies of the dynamics of reactive complexes between metal clusters and small molecules may shed a new light on the influence of the dynamics and particularly IVR on their reactive (catalytic) properties.

2.3.1.7 The scope of NeNePo spectroscopy

Finally, the question can be raised: what general information can be inferred from simulated NeNePo-ZEKE signals on the multistate energy landscapes and dynamics? First, the theoretical simulations allowed to establish the connections between three objectives: the structural relation of anionic and neutral species, the influence of the nature of the nonequilibrium state reached after photodetachment, and the character of subsequent dynamics in the neutral ground state [44]. Three different situations can be encountered in which i) transition state ii) global minimum and iii) local minimum can influence the dynamics after photodetachment. Second, different types of relaxation dynamics can be identified in NeNePo-ZEKE signals. Moreover, iv) the fragmentation and signature of fragments can be also identified and v) different nature of IVR, dissipative versus resonant can be used to characterize the catalytic capability of cluster species and to use NeNePo technique for verifying these important properties.

i) In cases where the anionic structure is close to a transition state of the neutral electronic ground state (e.g. trimers), large amplitude motion toward the stable structure dominates the relaxation dynamics. In other words, the dynamics is incoherent but localized in phase space. IVR can be initiated as a consequence of the localized large amplitude motion. Large amplitude structural relaxation after the transition state is responsible for a pronounced single peak in NeNePo-ZEKE signals at a given time delay and probe excitation wavelengths. In addition, subsequent IVR processes can be identified but only under ZEKE-like conditions since the integration over the continuum of electron kinetic energies leads to the loss of the fine features in the signals.

ii) In cases where the anionic structure is close to the global minimum (i. e. the stable isomer) of the neutral electronic ground state, vibrational relaxation reflecting the structural properties of the neutral stable isomer (e.g. Ag_4) takes place. The dynamics can be dominated by a single (e.g. Ag_4) or only by few modes which are given by the geometric deviations between anionic and neutral species. Other modes and anharmonicities weakly contribute, leading to dephasing on a timescale up to several ps (longer than 2 ps for Ag_4). Vibrational relaxation gives rise to oscillations in NeNePo signals (for different pulse durations) which can be analyzed in terms of normal modes. This allows to gain indirect information about vibrational spectra of a neutral cluster and use them as a fingerprint for identification of their structure.

iii) In cases where the anionic structure (the initial state) is close to a local minimum (energetically high lying isomer) of the neutral electronic ground state, the local minimum governs the dynamics after the photodetachment. Vibrational relaxation within the local minimum is likely to dominate the ultrashort dynamics.

Moreover, the local minimum can act as a strong capture area for nuclear motion with timescales up to several ps. As a consequence, isomerization processes toward other local minima and/or toward the global minimum structure are widely spread in time. In other words, structural relaxation dynamics is characterized as being incoherent and delocalized in phase space. The signals exhibit (at different excitation wavelengths of the probe laser) fingerprints of vibrational relaxation within the local minimum, providing structural information. After the systems escape from the local minima, the beginning of structural relaxation can be identified by an onset of signals appearing at the probe wavelengths ($\approx$1 ps for $Ag_3O_2^-$).

iv) The fragmentation patterns as well as the characteristics of fragments can also be identified in NeNePo signals. So, connected with the optimal control schemes the NeNePo technique provides a promising technique to introduce the control of the chemical reactivity, such as for the example of the oxidation of CO by noble metal oxide clusters, which is of relevance for heterogeneous catalysis. Moreover, the identification of IVR in the framework of the NeNePo technique represents a powerful tool which has a significant impact on the investigation of reactivity and catalysis.

2.4 Optimal control of dynamical processes

A. Lindinger, V. Bonačić-Koutecký, R. Mitrić, V. Engel, D. Tannor, A. Mirabal, and L. Wöste

2.4.1 General control concepts

The control of the selective product formation in a chemical reaction, using ultrashort pulses by choosing proper time duration and delay between the pump and the probe (or dump) step or their phase, is based on the coherence properties of laser radiation [10,11,13,14,118–121]. First, single-parameter control-schemes were proposed and tested. The scheme, introduced by Tannor and Rice [11,118] uses the time parameter for control, taking the advantage of differences in potential energy surfaces of different electronic states. Within the Brumer-Shapiro phase-control scheme [13,14,119], constructive and destructive interference between different light induced reaction pathways is used in order to favor or to suppress different reaction channels. Single parameter control schemes like linear chirps [122,123] corresponding to a decrease or increase of the frequency as a function of time under the pulse envelope were confirmed experimentally [124–135]. They represented the first step toward shaping the pulses in the framework of the optimal control theory (OCT) which involves many parameters. Variational optimization in the weak electric field limit was first introduced by Tannor and Rice [136]. Variational optimization with applications to arbitrary i.e. weak or strong fields was introduced by Rabitz et al. [12,20,21,137], with important extensions by Kosloff et al. [17], Jakubetz et al. [121], and Rice et al. [16].

A related approach, that has grown in popularity recently, is known as 'local optimization'. [138–149]. The control method known as "tracking" is closely related [147]. In these methods, at every instant in time the control field is chosen to achieve a monotonic increase in the desired objective (see Fig. 2.28b). Typically in these methods, two conditions are used at each time step: one to determine the phase of the field and one to determine the amplitude. In contrast with OCT, which incorporates information on later time dynamics through forward-backward iteration, these methods use only information on the current state of the system. The examples of local control theory will be presented in Sect. 2.4.2.

Technological progress due to fs pulse shapers [150–154] lead to the closed loop learning control (CLL) which was introduced by Judson and Rabitz in 1992 [22] opening the possibility to apply optimal control to more complex systems. Since the potential energy surfaces (PES) of multidimensional systems are complicated and mostly not available. The idea is to combine a fs-laser system with a computer-controlled pulse shaper to produce and optimize specific fields acting on the system initiating a photochemical process. After detecting the product, the learning algorithm [150,155,156] modifies the field based on information obtained from the experiment and from the objective (the target). The optimal shape for the chosen target is then reached iteratively. Judson and Rabitz's CLL approach initiated flourishing of the field and the success has been demonstrated by numerous control experiments [151–154,157–175]. However any multiparameter optimization scheme has a manifold of local solutions which are reachable depending on initial conditions. Particularly in the case of closed-loop learning control research activities are directed toward improvements of these aspects [176,177].

Metallic dimers [177–185] and diatomic molecules [186, 187] have been studied in numerous contributions, since they were suitable model systems for verifying the scopes of different control schemes and they are accessible by experimental pulse shaping techniques [46,162,164–166,188–193]. Theoretical and experimental studies using two parameter control have been performed: In addition to the pump-probe time delay the second control parameter involved the pump [178, 188, 189] and/or probe [179, 182] wavelength, the pump-dump delay [180, 190, 191], the laser power [181], the chirp [183, 192] or temporal width of the laser pulse [184]. Optimal pump-dump control of K_2 has been carried out theoretically in order to maximize the population of certain vibrational levels of the ground electronic state using one electronic excited state as an intermediate pathway [177, 185–187].

2.4.1.1 Pump-dump scheme

In order to tackle the very elementary chemical process of branching reaction channels, the ground electronic state potential energy surface in Fig. 2.23 can be considered. This potential energy surface, corresponding to collinear ABC, has a region of stable ABC and two exit channels, one corresponding to A+BC and one to AB+C. This system is the simplest paradigm for a control of chemical product formation: a two degree of freedom system is the minimum that can display two distinct chemical products. The objective is, starting out in a well defined initial state (v=0 of the ABC molecule) to design an electric field as a function of time which will steer the wave packet out of channel 1, with no amplitude going out of channel 2, and vice versa [11, 136].

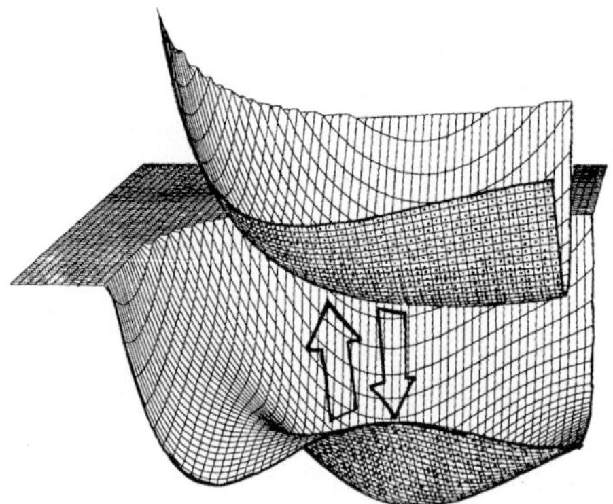

Fig. 2.23. Stereoscopic view of the ground and excited state potential energy surfaces for a model collinear ABC system with the masses of HHD. The ground state surface has a minimum, corresponding to the stable ABC molecule. This minimum is separated by saddle points from two distinct exit channels, one leading to AB+C the other to A+BC. The object is to use optical excitation and stimulated emission between the two surfaces to 'steer' the wave packet selectively out one of the exit channels.

A single excited electronic state surface is introduced at this point. The motivation is severalfold: 1) transition dipole moments are generally much stronger than permanent dipole moments. 2) the difference in functional form of the excited and ground potential energy surface will be the dynamical kernel; with a single surface one must make use of the (generally weak) coordinate dependence of the dipole. Moreover, the use of excited electronic states facilitates large changes in force on the molecule, effectively instantaneously, without necessarily using strong fields. 3) the technology for amplitude and phase control of optical pulses is significantly ahead of the corresponding technology in the infrared.

The object now will be to steer the wave function out of a specific exit channel on the ground electronic state, using the excited electronic state as an intermediate. Insofar as the control is achieved by transferring amplitude between two electronic states, all the concepts regarding the central quantity μ_{eg} will now come into play.

Consider the following intuitive scheme, in which the timing between a pair of pulses is used to control the identity of products [11]. The scheme is based on the close correspondence between the center of a wave packet in time and that of a classical trajectory (Ehrenfest's theorem). The first pulse produces an excited electronic state wave packet; the time delay between the pulses controls the time that the wave packet evolves on the excited electronic state. The second pulse stimulates emission. By the Franck-Condon principle, the second step prepares a wave packet on the ground electronic state with the same position and momentum, instantaneously, as the excited state wave packet. By controlling the position and momentum of the wave packet produced on the ground state through the second step, one can gain some measure of control over product formation on the ground state. This 'pump-dump' scheme is illustrated classically in Fig. 2.24. The trajectory originates at the ground state surface minimum (the equilibrium geometry). At $t = 0$ it is promoted to the on the excited state potential surface (a two dimensional harmonic oscillator in this model) where it originates at the Condon point, i.e. vertically above the ground state minimum. Since this position is displaced from equilibrium on the excited state, the trajectory begins to evolve, executing a two-dimensional Lissajous motion. After some time delay, the trajectory is brought down vertically to the ground state (keeping both the instantaneous position and momentum it had on the excited state) and allowed to continue to evolve on the ground state. Figure 2.24 shows that for one choice of time delay it will exit into channel 1, for a second choice of time delay it will exit into channel 2. Note how the position and momentum of the trajectory on the ground state, immediately after it comes down from the excited state, are both consistent with the values it had when it left the excited state, and at the same time are ideally suited for exiting out their respective channels.

A full quantum mechanical calculation based on these classical ideas is shown in Figs. 2.25-2.26 [136]. The dynamics of the two-electronic state model, was solved, starting in the lowest vibrational eigenstate of the ground elec-

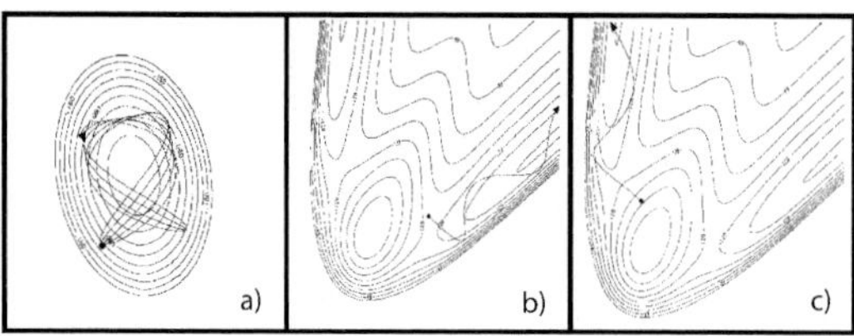

Fig. 2.24. Equipotential contour plots of the ground and the excited state potential energy surfaces (here a harmonic excited state is used because that is the way the first calculations were done). a) The classical trajectory that originates from rest on the ground state surface makes a vertical transition to the excited state, and subsequently undergoes Lissajous motion, which is shown superimposed. b) Assuming a vertical transition down at time t_1 (position and momentum conserved) the trajectory continues to evolve on the ground state surface and exits from channel 1. c) If the transition down is at time t_2 the classical trajectory exits from channel 2.

tronic state, in the presence of a pair of femtosecond pulses that couple the states. Because the pulses were taken to be much shorter than a vibrational period, the effect of the pulses is prepare a wave packet on the excited/ground state which is almost an exact replica of the instantaneous wave function on the other surface. Thus, the first pulse prepares an initial wave packet which is almost a perfect Gaussian, and which begins to evolve on the excited state surface. The second pulse transfers the instantaneous wave packet at the arrival time of the pulse back to the ground state, where it continues to evolve on the ground state surface, given its position and momentum at the time of arrival from the excited state. For one choice of time delay the exit out of channel 1 is almost completely selective (Fig. 2.25), while for a second choice of time delay the exit out of channel 2 is almost completely selective (Fig. 2.26). Note the close correspondence with the classical model: the wave packet on the excited state is executing a Lissajous motion almost identical with that of the classical trajectory (the wave packet is a nearly Gaussian wave packet on a two-dimensional harmonic oscillator). On the ground state, the wave packet becomes spatially extended but its exit channel, as well as the partitioning of energy into translation and vibration (i.e. parallel and perpendicular to the exit direction) are seen to be in close agreement with the corresponding classical trajectory.

This scheme is significant for three reasons: it shows that control is possible, it gives a starting point for the design of optimal pulse shapes; and it gives a framework for interpreting the action of two pulse and more complicated pulse sequences. Nevertheless, the approach is limited: in general with the best choice of time delay and central frequency of the pulses one may achieve only partial selectivity. Perhaps most importantly, this scheme does not exploit the phase of the light. Intuition breaks down for more complicated processes and

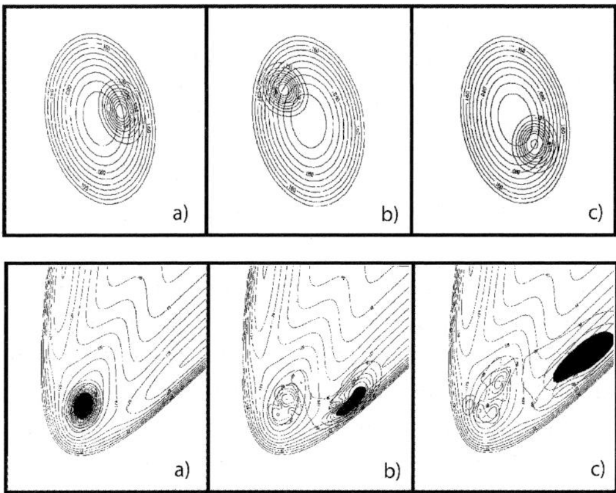

Fig. 2.25. Top: Magnitude of the excited state wave function for a pulse sequence of two Gaussians with time delay of 610 a.u.=15 fs. a) t=200 a.u., b) t=400 a.u., c) t=600 a.u. Note the close correspondence with the results obtained for the classical trajectory (Fig. 2.25). Bottom: Magnitude of the ground state wave function for the same pulse sequence, at a) t=0, b) t=800 a.u., c) t=1000 a.u. Note the close correspondence with the classical trajectory of Fig. 2.24b). Although some of the amplitude remains in the bound region, which does exit exclusively from channel 1.

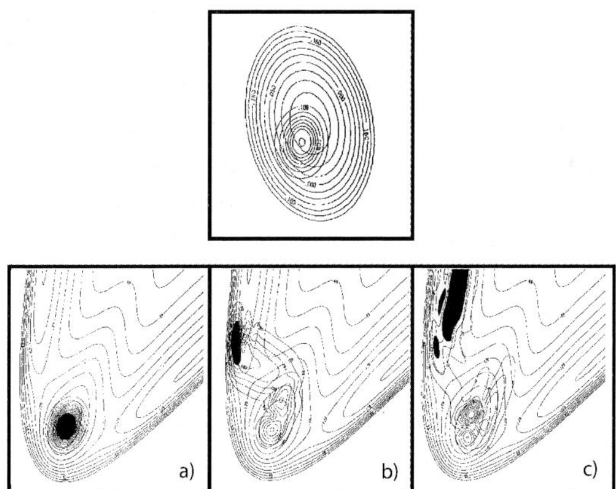

Fig. 2.26. Magnitude of the ground and excited state wave functions for a sequence of two Gaussian pulses with time delay of 810 a.u. Top: excited state wave function at 800 a.u., before the second pulse. Bottom: a) ground state wave function at 0 a.u. b) ground state wave function at 1000 a.u. c) ground state wave function at 1200 a.u. That amplitude which does exit does so exclusively from channel 2. Note the close correspondence with the classical trajectory of Fig. 2.24c.

classical pictures cannot adequately describe the role of the phase of the light and the wave function. Hence attempts were made to develop a systematic procedure for improving an initial pulse sequence.

Before turning to these more systematic procedures for designing shaped pulses, an interesting alternative perspective on pump-dump control will be pointed out. A central tenet of Feynman's approach to quantum mechanics was to think of quantum interference as arising from multiple dynamical paths that lead to the same final state. The simple example of this interference involves an initial state, two intermediate states, and a single final state, although if the objective is to control some branching ratio at the final energy then at least two final states are necessary. By controlling the phase with which each of the two intermediate states contributes to the final state, one may control constructive vs. destructive interference in the final states. This is the basis of the Brumer-Shapiro approach to coherent control [24]. It is interesting to note that pump-dump control can be viewed entirely from this perspective [24]. Now, however, instead of two intermediate states there are many, corresponding to the vibrational levels of the excited electronic state. The control of the phase which determines how each of these intermediate levels contributes to the final state, is achieved via the time delay between the excitation and the stimulated emission pulse. This "interfering pathways" interpretation of pump-dump control is shown in Fig. 2.27.

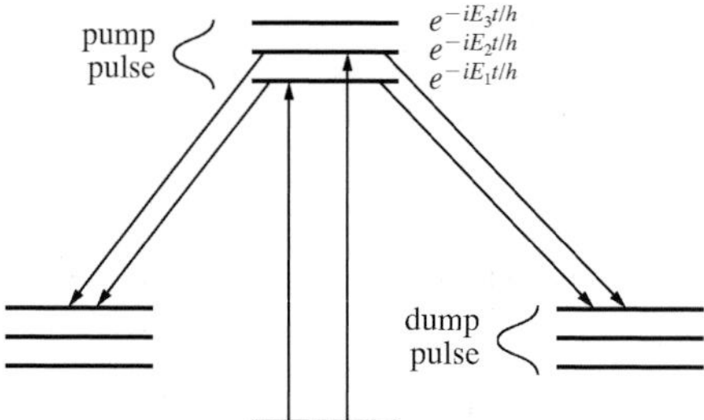

Fig. 2.27. Multiple pathway interference interpretation of pump-dump control. Since each of the pair of pulses contains many frequency components, there are an infinite number of combination frequencies which lead to the same final energy state, which generally interfere. The time delay between the pump and dump pulses controls the relative phase among these pathways, and hence determines whether the interference is constructive or destructive. The frequency domain interpretation highlights two important features of coherent control: First, if final products are to be controlled there must be degeneracy in the dissociative continuum. Second, that a single interaction with the light, no matter how it is shaped, cannot produce control of final products: at least two interactions with the field are needed to obtain interfering pathways.

2.4.1.2 Variational formulation of control

In the following a brief outline of theoretical and experimental approaches will be given. In OCT the optimized pulses are obtained from the functional in the framework of the variational method [12, 17]:

$$J = \langle \psi(T)|\hat{P}|\psi(T)\rangle. \tag{2.13}$$

J is the functional of the radiation field $E(t)$. Therefore, the maximization has to be carried out with respect to the variation of the functional form of $E(t)$ which involves temporal shape and spectral content. $\hat{P}$ is the projector operator which selects the desired target.

A constraint which the energy per pulse is given by

$$E = \int_0^T dt|E(t)|^2, \tag{2.14}$$

which, together with (2.13) implies that the following functional must be optimized:

$$J_\alpha = \langle \psi(T)|\hat{P}_\alpha|\psi(T)\rangle - \lambda \left[\int_0^T dt|E(t)|^2 - E \right], \quad \alpha = target. \tag{2.15}$$

Here λ is a Lagrange multiplier.

If one imposes the constraint that the Schrödinger equation must be satisfied, the modified objective functional, which should be optimized, takes the following form:

$$\begin{aligned} \bar{J}_\alpha = \langle \psi(T)|\hat{P}_\alpha|\psi(T)\rangle + 2Re \int_0^T dt \langle \chi(t)| i\hbar \frac{\partial}{\partial t} \\ - \hat{H}|\psi(t)\rangle - \lambda \left[\int_0^T dt|E(t)|^2 - E \right]. \end{aligned} \tag{2.16}$$

If $\psi(t)$ satisfies the time-dependent Schrödinger equation, the second term on the right hand side of (2.16) vanishes for any $\chi(t)$, and the third term is zero when $E(t)$ satisfies (2.14). Both of these terms allow variations of $\bar{J}_\alpha$ with respect to $E(t)$ and $\chi(t)$ independently to the first order in $\delta E(t)$.

The condition $\delta \bar{J}_\alpha / \delta \psi = 0$ generates a partial differential equation for the Lagrange multiplier function $\chi(t)$

$$i\hbar \frac{\partial \chi(t)}{\partial t} = \hat{H}\chi(t), \tag{2.17}$$

which is the time dependent Schrödinger equation subject to the final condition

$$\chi(T) = \hat{P}_\alpha \psi(T), \tag{2.18}$$

and a partial differential equation for $\psi(t)$

$$i\hbar \frac{\partial \psi(t)}{\partial t} = \hat{H}\psi(t), \tag{2.19}$$

subject to the initial condition

$$\psi(0) = \psi_0. \tag{2.20}$$

Finally, the optimal applied field is defined by the condition $\delta \bar{J}_\alpha / \delta E(t) = 0$ which leads to

$$E(t) = -\frac{i}{\lambda\hbar} \left[\langle \chi_g(t)|\hat{\mu}_{ge}|\psi_e(t)\rangle - \langle \psi_g(t)|\hat{\mu}_{ge}|\chi_e(t)\rangle \right] \tag{2.21}$$

with

$$\lambda^2 = \frac{1}{E} \int_0^T dt |\langle \chi_e(t)|\hat{\mu}_{eg}|\psi_g(t)\rangle - \langle \psi_e(t)|\hat{\mu}_{eg}|\chi_g(t)\rangle|^2, \tag{2.22}$$

where μ_{eg} is the dipole operator and the indices e and g label the excited and the ground state energy surfaces. The equation of motion of the coupled amplitudes on two potential energy surfaces reads

$$i\frac{\partial}{\partial t} \begin{pmatrix} \psi_e \\ \psi_g \end{pmatrix} = \begin{pmatrix} \hat{H}_e & \hat{V}_{ge} \\ \hat{V}_{eg} & \hat{H}_g \end{pmatrix} \begin{pmatrix} \psi_e \\ \psi_g \end{pmatrix}, \tag{2.23}$$

with the interaction potential $\hat{V}_{ge}$ defined as $\hat{\mu}_{ge}E(t)$.

The numerical calculations then involve an iterative procedure which includes the following:
(i) initial guess for the pulse shape $E(t)$;
(ii) integration of the Schrödinger equation forward starting from the initial condition in the ground state;
(iii) application of the projector operator which selects the target (the exit channel to $\psi(t)$) to obtain $\chi(t)$ as an initial value for backwards propagation;
(iv) propagation of $\chi(t)$ backwards in time;
(v) during the propagation, calculation of the overlap function

$$O(t) = i \left[\langle \chi_e(t)|\hat{\mu}_{eg}|\psi_g(t)\rangle - \langle \psi_e(t)|\hat{\mu}_{eg}(t)|\chi_g(t)\rangle \right]; \tag{2.24}$$

and
(vi) after completion of the backwards propagation, renormalization of the result needed to obtain the new pulse

$$E(t) = O(t) \left(\frac{1}{E} \int_0^T dt |O(t)|^2 \right)^{-1/2}. \tag{2.25}$$

(vii) iterative procedures starting from step (ii) should be repeated until convergence has been achieved (cf. [17, 122]).

2.4.1.3 Local vs. global in time optimization

As described above, the optimal control equations typically have the structure of five coupled differential equations: one for the wave function, one for the *dual* wave function, an initial condition on the wave function, a final condition on the dual, and finally, an equation for the optimal field, which in turn is expressed in terms of the wave function and its dual.

Typically, the OCT equations have to be solved via an iterative procedure, involving forwards in time propagation of the wave function, followed by backwards in time propagation of the dual, until self-consistency is achieved with respect to the equations for the wave function, the dual and the control field. Because of the structure of this forwards-backwards propagation, the optimal field 'knows' about the future, i.e. the form of the optimal field at time t takes into account the dynamics at time $t' > t$. Thus, the optimal field may be willing to tolerate a nonmonotonic increase in the objective during the action of the pulse, since, given knowledge of the future, that may be the best way to attain the highest objective at the final time (see Fig. 2.28a).

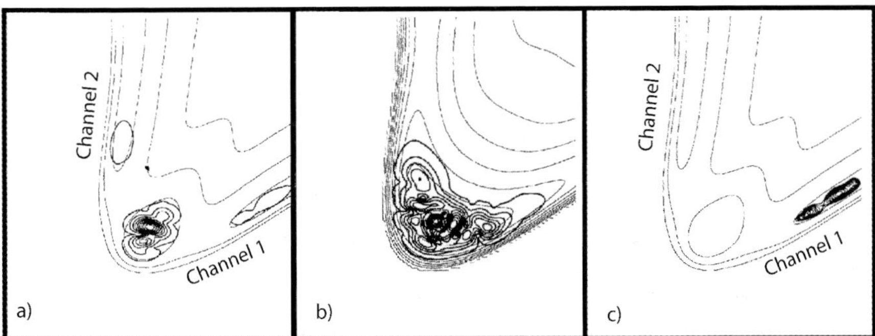

Fig. 2.28. Representative plot showing the difference between global-in-time and local-in-time optimization. (a) In global-in-time optimization, the objective may decrease at intermediate times, but is guaranteed to take on its maximum value at the final time. (b) In local-in-time optimization, the objective increases monotonically in time.

A different class of techniques that have been developed for control of atomic and molecular dynamics is called 'local optimization' [138–149]. In these methods, at every instant in time the control field is chosen to achieve monotonic increase in the desired objective see Fig. 2.28b. Typically in these methods, two conditions are used at each time step, one to determine the phase of the field and one to determine the amplitude. In contrast with OCT, which incorporates information on later time dynamics through forward-backward iteration, these methods use only information on the current state of the system.

At first glance, one would expect that the solution(s) that come out of an optimal control calculation would give a higher value of the objective than that from the local optimization: since the approach to the objective in OCT at intermediate times is unconstrained, while in local optimization it is constrained to increase monotonically in time, one expects a higher yield from OCT since the space of allowed solutions includes those from local optimization as a subset see Fig. 2.29a. However, there is one fallacy with the above argument. The optimal control equations generally have multiple solutions; these solutions are in general *local*, not global maxima in the function space (see Fig. 2.29b). Thus, the value of the objective attained with the local-in-time optimization algorithm can be larger than that obtained with the optimal control algorithm, since the latter may be stuck in the region of a poor quality local maximum!

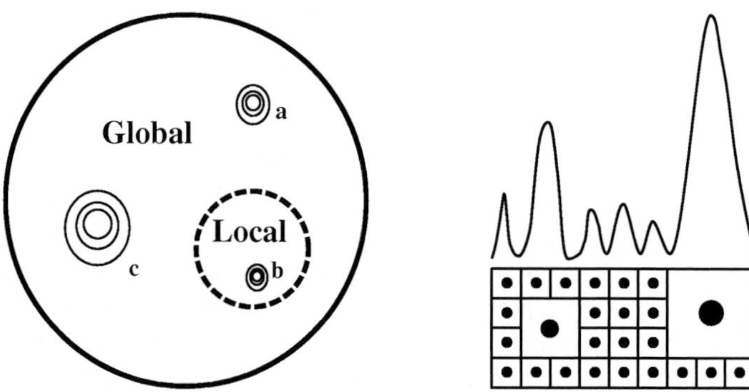

Fig. 2.29. Left: Venn diagram of the set of control fields. Fields that yield monotonically increasing solutions are a *subset* of the set of fields that lead to either an increase or a decrease at intermediate times. (Concentric circles indicate neighborhoods of fields that lead to a local maximum in the objective in the function space.) Right: The checkerboard is a schematic depiction of the space of all allowed control fields. The dots indicate the positions of the optimal fields, i.e. fields that lead to a local maximum of the objective in the function space. The squares show "basins of attraction", i.e. neighborhoods of control fields that converge to the same optimal field. The one-dimensional function above the checkerboard symbolizes the value of the objective, and it, (as well as the concentric circles in the figure on the left) are intended to convey the idea of multiple *local* maxima (see text).

There are several other attractive features to the local methods. 1) Since the increase in yield is continual, these fields are often amenable to immediate interpretation. 2) Since these methods use only information on the current state of the system, they could in principle be adapted for laboratory implementation. 3) These methods, because they differ so radically in algorithm from OCT, are capable of identifying entirely different classes of mechanisms which may be appealing because of other properties, e.g. robustness.

2.4.2 Local control theory and the analysis of control processes

When approaching the problem of the analysis of laser control from a theoretical point of view, the question to be asked is: 'How can one relate the outcome of a laser-control process to the properties of the driving electric field?' To answer this question, the approach of local control theory as introduced in Sec. 2.4.1 is discussed. For this purpose, two specific examples will be presented, which illustrate, - without going into the most general formulation - the basic ideas. They also demonstrate how the "analysis" aspect is inherent to this theory.

2.4.2.1 Infrared dissociation of a diatomic molecule

As a first example, the infrared dissociation of a diatomic molecule is treated. Therefore, only the radial motion in the single coordinate r is considered and the rotational degree of freedom is frozen. In order to initiate a fragmentation, the energy of the vibrational motion must exceed the dissociation energy, that is, the control field should be such, that an effective energy transfer from the field to the molecule is guaranteed (a process which, in what follows, is referred to as 'heating'), and a sufficient condition that this takes place is that the rate of energy transfer is positive at all times.

The Hamiltonian of a molecule with reduced mass m can be given as

$$\hat{H}_0 = \frac{\hat{p}^2}{2m} + \hat{V}(r), \tag{2.26}$$

where $\hat{p}$ is the momentum operator, and $\hat{V}(r)$ denotes the potential energy which provides a set of bound states. The internal energy rate then is calculated as the time-derivative

$$\frac{d}{dt}\langle \hat{H}_0 \rangle_t = \frac{d}{dt}\langle \psi(t)|\hat{H}_0|\psi(t)\rangle = \frac{i}{\hbar}E(t)\langle \psi(t)|[\hat{T},\hat{\mu}]\psi(t)\rangle. \tag{2.27}$$

Here, $|\psi(t)\rangle$ is the time-dependent molecular state and $[\hat{T},\hat{\mu}]$ denotes the commutator between the kinetic energy operator $\hat{T}$, and the dipole operator $\hat{\mu}$, where the latter appears in an additional term of the total Hamiltonian containing the dipole interaction of the molecule with an external field $E(t)$:

$$\hat{W}(t) = -\hat{\mu}(r)\, E(t). \tag{2.28}$$

Within local control theory, the field is constructed from the instantaneous dynamics of the system. In particular, the choice

$$E(t) = \lambda Im\langle \psi(t)|[\hat{\mu},\hat{T}]|\psi(t)\rangle, \tag{2.29}$$

leads to energy absorption if the value of λ is taken to be positive, as can be readily taken from (2.27) realizing that the expectation value of the commutator is purely imaginary [194].

Assuming now, for the purpose of illustration, a linear dipole moment ($\hat{\mu}(r) = \mu_0 + \mu_1 r$) this will lead to the rate expression

$$\frac{d}{dt}\langle \hat{H}_0 \rangle_t = \frac{\mu_1}{m} E(t) \langle \hat{p} \rangle_t . \tag{2.30}$$

It is now easy to anticipate that, if the field is in phase with the expectation value of the momentum operator, the rate is positive ('heating'), whereas for a field being out of phase, a negative rate is obtained, i.e. the system is 'cooled'. Thus, here the entanglement between control field and system dynamics is clearly visible. Returning to the problem of infrared dissociation, this means that a control field which permanently pumps energy into the system, is directly connected to the momentum. If energy is transferred into the system, the frequency of the vibrational motion will, due to the anharmonicity of the potential curve (and the decreasing spacing of the energy levels), decrease as time goes along. Thus, the driving field is "down-chirped" by construction. Here, local control theory directly leads to the conclusion that a down-chirped field induces an effective infrared dissociation [195, 196].

The interpretation of the control field in terms of expectation values can easily be carried over to the classical picture Therefore, in (2.27), the commutator is replaced by the Poisson bracket [197] as

$$[\hat{T}, \hat{\mu}] \leftrightarrow (i\,\hbar)\,\{T, \mu\} = (i\,\hbar)\left(\frac{\partial T}{\partial r}\frac{\partial \mu}{\partial p} - \frac{\partial T}{\partial p}\frac{\partial \mu}{\partial r}\right) = -(i\,\hbar)\frac{\mu_1}{m} p(t), \tag{2.31}$$

where $p(t)$ is the classical momentum. This is the well studied problem of a classically driven oscillator [198]. In particular, the induced molecular fragmentation corresponds to the case of a resonantly driven oscillator with ever increasing energy. The connection between classical and quantum treatments of control processes has been explored in detail before, see e.g. [199–201].

To illustrate what has been said above, a simplified model of the NaI molecule is regarded. This system served as one of the first gas-phase examples to demonstrate the power of femtosecond spectroscopy [1, 2] and laser control [202], for additional work see, e.g. [203–207]. Fig. 2.30, upper panel, shows an excitation scheme where a first pulse ($E_p(t)$) excites the system from its electronic ground state $|0\rangle$ to an excited state $|1\rangle$. The scenario to be discussed below is, that an additional control field $E(t)$ is applied to induce an excited state dissociation. Within the model, the nonadiabatic coupling of the two electronic states occurring around 7 Å is neglected [208], for the effect of the predissociation channel see [209]. The lower panel of the figure shows the dipole moment in the excited state [210]. At smaller distances, the function vanishes indicating a covalent bonding situation, whereas for longer bond-lengths the dipole moment is linear, illustrating the ionic character of the electronic state.

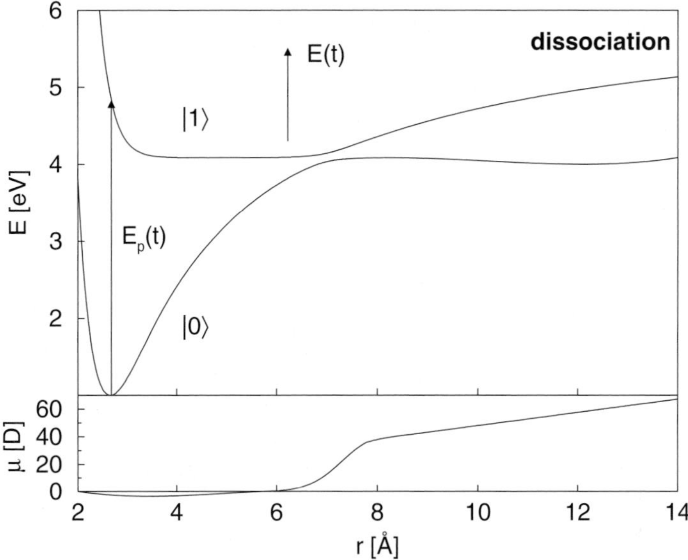

Fig. 2.30. Upper panel: excitation scheme for the NaI molecule. An initial transition ($|1\rangle \leftarrow |0\rangle$) induced by the pump field $E_p(t)$ prepares a vibrational wave packet in the excited state. An additional control field $E(t)$ then heats the system so that excited state dissociation becomes effective. The lower panel shows the permanent dipole moment in the excited state. Its functional form illustrates the change from a covalent (at distances smaller than the region where the avoided crossing occurs) to an ionic bonding situation at larger bond lengths.

The pump-pulse transition from the vibronic ground state is treated by time-dependent perturbation employing a pulse-width of 50 fs and a photon energy of 3.875 eV. This prepares a vibrational wave packet which enters into the determination of the electric field via (2.29). Results obtained from a calculation with fields of different strength parameters λ are contained in Fig. 2.31. There, the excited state populations $P(t)$ (upper panels), the bond-length expectation-values $\langle r\rangle_t$ (middle panels) and the control fields $E(t)$ (lower panels) are shown. The coordinate expectation-values reflect the vibrational wave packet dynamics which proceeds with increasing amplitude and decreasing frequency. In the case of the weaker field (right panels), three oscillations are completed before the system has absorbed sufficient energy for fragmentation to take place (at around 8.5 ps). It is then, that the fraction of still bound molecules $P(t)$, calculated from the norm of the wave packet for values $r \leq$ 20 Å) decreases, and a fragmentation yield of approximately 60% is obtained at longer times. The control field (lower right panel) follows the vibrational dynamics, exhibiting the same characteristic oscillations which demonstrates the entanglement of field and system dynamics. The same trends are present

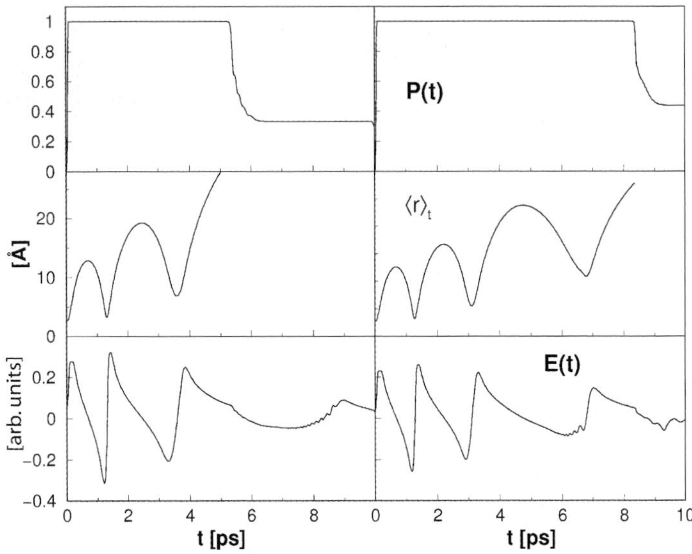

Fig. 2.31. Analysis of the laser induced excited state fragmentation of NaI. Results are displayed for different values of the field strength parameter $\lambda = 6 \cdot 10^{-6}$ a.u. (left panels) and $\lambda = 5 \cdot 10^{-6}$ (right panels). The lower panels show control fields $E(t)$ which oscillate with the instantaneous vibrational period of the wave packet motion. This can be seen in comparing the fields to the bond length expectation values (middle panel). The vibrational motion proceeds with increasing amplitude and decreasing frequency until the bound state population $P(t)$ (upper panels) starts decreasing, i.e. fragmentation takes place.

for the stronger driving field (left panels of Fig. 2.31). There, it takes only two vibrational periods until the necessary energy is absorbed so that, on the way outward a substantial part of the wave packet enters the exit channel at about 5 ps. Here, about 70% of the molecules undergo dissociation.

It is noted that at longer times, no additional fragmentation takes place. This is due to the dispersion of the wave packet which moves in an extremely anharmonic potential. As soon as the packet becomes de-localized, the expectation values entering into the construction scheme for the fields become meaningless [211]. In other words, the efficiency, but also the interpretation of the fields in terms of a classical-like motion rests on the property that the wave packets are localized in comparison with the extension of the system, for a discussion see, e.g. [212].

2.4.2.2 Population transfer between electronic states

As a second example, electronic transitions in the Na_2 molecule are regarded. In Fig. 2.32a, three electronic states are displayed. The femtosecond spec-

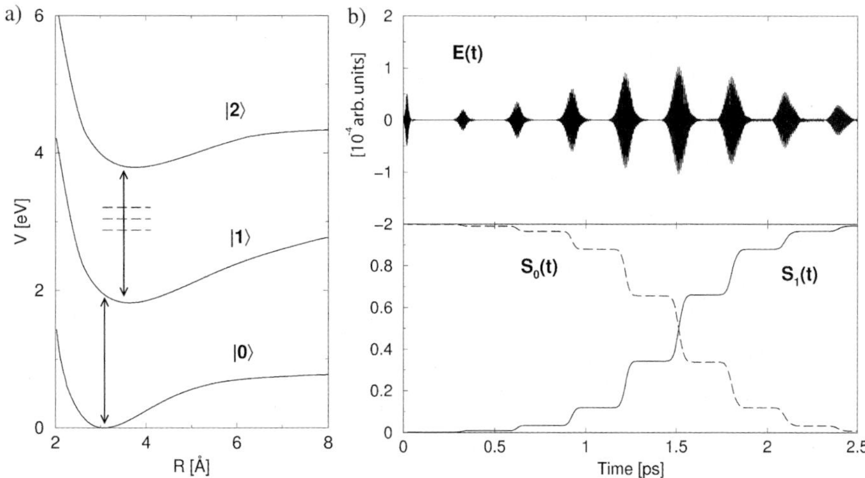

Fig. 2.32. a) Excitation scheme of the Na_2 molecule. In coupling three electronic states to a control field the objective is a complete population transfer from the ground ($|0\rangle$) to the first excited state ($|1\rangle$) without loosing population into the higher state ($|2\rangle$). b) Lower panel: time-dependence of the population in the electronic ground ($S_0(t)$) and first excited state ($S_1(t)$). The population in the second excited state ($|2\rangle$) is too small to be seen in the figure. The upper panel contains the control field and, at early times, a seed pulse which transfers a small amount of population to the state $|1\rangle$. The pulse train reflects the vibrational dynamics in the excited state.

troscopy of this dimer, employing pulses of around 620 nm wavelength is well understood [213–215]. In what follows, a complete population transfer form the ground state $|0\rangle$ to the first excited state $|1\rangle$ without a loss into the excited state $|2\rangle$ (and as a consequence to any other higher lying states) is considered. As above, local control theory is used with the hope to find a field which effectively triggers the transition and also is easily understood in terms of the system dynamics.

To put the objective into the frame of LCT, the rate of population transfer to the excite state is calculated. Therefore, the molecular state vector is represented as

$$|\psi(t)\rangle = \sum_{n=0}^{2} \psi_n(r,t)\,|n\rangle, \tag{2.32}$$

where $\psi_n(r,t)$ is the vibrational wave function in the electronic state $|n\rangle$. The population in the target state $|1\rangle$ is then calculated with the help of the projector $\hat{A}_1 = |1\rangle\langle 1|$ as

$$P_1(t) = \langle\psi(t)|\hat{A}_1|\psi(t)\rangle. \tag{2.33}$$

Upon taking the time-derivative and evaluating appearing commutators one finds [145, 148]

$$\frac{d}{dt}P_1(t) = -E(t)\frac{2}{\hbar}Im\{\langle\psi_1(r,t)|\mu_{10}|\psi_0(r,t)\rangle + \langle\psi_1(r,t)|\mu_{12}|\psi_2(r,t)\rangle\}, \tag{2.34}$$

where Im denotes the imaginary part and μ_{nm} is the transition dipole-moment between the states $|n\rangle$ and $|m\rangle$. Two things become apparent in inspecting the latter equation. First, in principle, by choosing the field proportional to the negative imaginary part of the sum of overlap integrals, the rate can be forced to assume only positive numbers. This then results in an increase of the target-state population at each increment of time. Second, if this choice is made, the field will, by construction, be determined from the ground- and excited state dynamics, thus an interpretation of the field properties should be readily obtainable. It should be noted, that the construction scheme needs a small initial population in the excited state because otherwise the dipole matrix elements (and thus the control fields) are identically zero at all times. This is usually achieved by applying a weak 'seed pulse' preceding the control pulse.

In this numerical example, a control field constructed as

$$E(t) = -\lambda Im\{\langle\psi_1(r,t)|\mu_{10}|\psi_0(r,t)\rangle + \langle\psi_1(r,t)|\mu_{12}|\psi_2(r,t)\rangle\}, \tag{2.35}$$

is employed. It has a strength parameter of $\lambda = 1.4 \cdot 10^{-4}$ a.u. This field (and also the seed pulse) is displayed in Fig. 2.32b (upper panel). It consists of a pulse train where several sub-pulses can be distinguished. Also shown is the ground- and excited-state population. The latter increases in steps every time, the field is non-zero, until a 100% population transfer is obtained. It is worth to note that the second excited state $|2\rangle$ is never populated, for an analysis see [216]. Thus LCT delivers a field being optimal in the sense that it completely populates the target state thereby minimizing the loss into other excitation channels.

Returning to the initially posed question, the appearance of the electric field is now analyzed. An analysis of a single sub-pulse shows that its frequency corresponds to the energy separation between the potentials in the Franck-Condon window of the $|1\rangle \leftarrow |0\rangle$ transition. Because the second excited state is not populated, the field oscillations are determined by the temporal variation of the first overlap integral appearing in (2.35) which can be evaluated as

$$\langle\psi_1(r,t)|\mu_{10}|\psi_0(r,t)\rangle = \sum_{e,g} a_e^*(t)\; a_g(t)\; \langle\psi_e|\mu_{10}|\psi_g\rangle e^{i(E_e-E_g)t/\hbar}. \tag{2.36}$$

Here, the wave functions are expanded in terms of the ground- and excited-state vibrational eigenfunctions $\psi_g(r)$ and $\psi_e(r)$ with energies E_g and E_e, respectively. Note, that the coefficients $a_e(t), a_g(t)$ are explicitly time-dependent.

From the latter equation it is clear, that the fast field oscillations are determined by the energy differences between vibrational states in the excited and ground electronic states. Thus, the frequency is automatically adjusted to induce resonant transitions. In this way, a vibrational wave packet is built in the excited state which, due to the repulsive force, moves toward longer bond-lengths so that, after a while, the overlap with the ground state wave packet, being still localized in the vicinity of the potential minimum, is diminished. As a result, the field amplitude decreases and approaches zero. The situation changes when the excited state wave packet, after performing a vibrational period, returns to the Franck-Condon region. Because then the overlap integral increases, the field amplitude rises again and the same scenario repeats itself. In this way, the vibrational motion is reflected in the structure of the pulse train: the temporal separation of the sub-pulses equals the well known vibrational period on Na_2 in its A $(^1\Sigma_u^+)$ electronic state [215]. Thus, the form of the field is clearly understandable. Additionally, the sub-pulse intensity increases until the two electronic states are equally populated and it decreases at later times.

In order that the population in the upper state does not decrease, no loss into other states has to occur. Concerning the coupling to the ground state, this means that the interference between the excited state and ground state wave packets has to be completely constructive. This is indeed the case and the scenario of wave packet interferometry [217] is encountered, which has been realized experimentally by Scherer et al. [218], see also more recent work [219]. There, a pair of phase-locked pulses are used to constructively interfere a wave packet (prepared by the first pulse) returning to the Franck-Condon region with another packet (prepared by the second pulse). By shifting the relative phase of the pulses by π, destructive interference can be obtained as well. In this approach it is straightforward to rationalize this phase-sensitivity: if the control field is multiplied with a phase factor of (-1), the rate (2.34) becomes negative, so that the excited state population is diminished, i.e. one encounters destructive interference.

The given example illustrates again, that local control theory delivers fields which are accessible to an interpretation. It is to be kept in mind, however, that the construction scheme is a purely theoretical and one aim of future investigations should be the comparison of fields derived from various algorithms with the ones from LCT, having the analysis aspect of laser control processes in mind.

2.4.3 Feedback control

The steering of dynamical processes requires a successive photon-molecule interaction, across a sequence of temporally open Franck-Condon windows of the evolving quantum system. To reach this goal, shaped fs-pulses can be applied which influence - as proposed by Judson and Rabitz [22] - the dynamics temporally and spectrally . These tailored pulses can be produced by a pulse shaper which can either be realized by an acousto-optic modulator [220] or a liquid crystal modulator mask [221] placed in the Fourier plane of a zero dispersion compressor [222] (see Fig. 2.33). Modulators with two arrays can be utilized to allow for independent spectral phase and amplitude modulation [223]. Additionally, the polarization can still be shaped as shown later in Sect. 2.4.6.

In the optimization experiment, a learning algorithm based on evolutionary algorithms [155,156] produces new pulse shapes regarding the experimental input and the defined objective. Then the procedure is repeated several times in a loop until the optimized laser pulses give rise to the aimed experimental yield [22]. As first demonstrated by Gerber et al., the approach has opened new roads in the field of laser control, in particular, with regard to the optimization of the yield of chosen reactivity channels [224].

The experimental setup for feedback control measurements used here (see Fig. 2.33) combines a fs-laser system with a programmable liquid crystal pulse shaper and a molecular beam apparatus as already described in Sect. 2.2.3. Each pulse shape is determined by an array of numbers called individual, representing the spectral phase and amplitude values. Individuals consisting of random numbers are created in the beginning and then modified by crossover and mutation operators. After being written on the modulator, each pulse form creates an ion signal which represents its fitness. In the last step of the

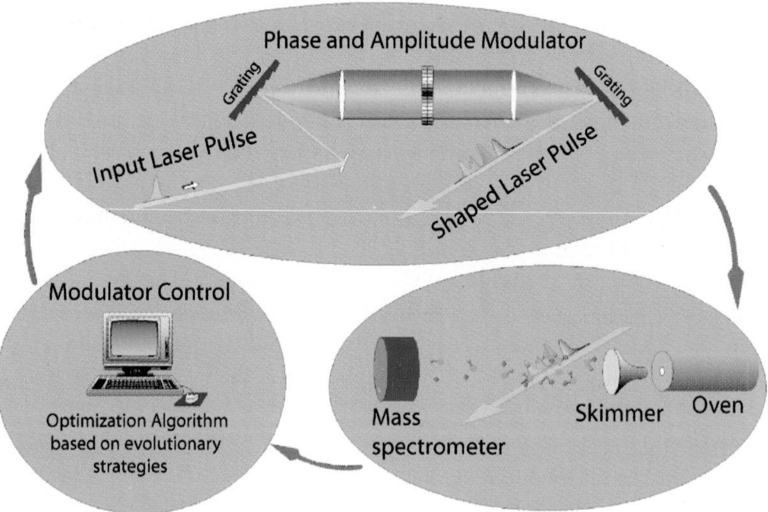

Fig. 2.33. Schematic of the closed loop experiment. Arbitrary pulse shape patterns altered by cross over and mutation operators are send to a pulse shaper. The iteratively improved pulse shapes are tested by mass selectively detecting the ion product yield. The resulting yields are a measure for the quality (fitness) of the applied pulse forms.

first cycle the best individuals are selected and serve as parents for the next iteration. For details of the applied algorithm see [162]. A Ti:sapphire laser oscillator is utilized, which provides pulses at a repetition rate of 80 MHz in the weak field regime. This allows a good signal-to-noise ratio through a very high degree of averaging, and it avoids undesired multiphotonic transitions, which would also complicate the interpretation of the obtained data.

When having obtained an optimum pulse shape the important question arises, what can be retrieved from this result with regard to the underlying dynamical processes. In order to answer this, the optimized pulse form has to be measured accurately by intensity cross correlation and spectrally resolved cross correlation (SFG-XFROG). So the cross correlation traces of the shaped test pulse and the reference pulse are recorded for different frequency components of the spectrum. When operating in the weak field regime, SFG-XFROG offers the advantage of providing an intuitive viewgraph, indicating the time-frequency character of the pulse shape. In order to fully characterize the pulse form, the amplitude and phase can be calculated from the SFG-XFROG spectrogram by an iterative phase-retrieval algorithm.

2.4.4 Controlling multiphoton ionization processes in NaK

The NaK dimer was chosen as a model system for The investigation of the multiphoton ionization process of the NaK dimer is a good model system for testing the feedback optimization procedure. The system is easily excited and ionized in a resonant three-photon process by wavelengths within an easily accessible range around $\lambda_0 = 770$ nm. Furthermore, the wave packet dynamics of low lying excited states of this system is well known from fs-pump-probe experiments; its oscillation periods are around T_{osc}^{NaK}=440 fs [225]. Also from a theoretical point of view the dynamics of the system is easily described by optimal control theory [17] as will be shown later. So the challenging question arises, whether the optimization algorithm will identify the fingerprints of these spectroscopic features.

2.4.4.1 Experimental approach

In the experiments shown here the spectral width of the laser of 9 nm was dispersed across all 128 pixels of the modulator, and pure phase optimizations were performed. So the resulting $n = 128$ spectral phase parameters $\varphi_n(\omega)$ ($\in [0 - 2\pi]$) of the shaped spectral field

$$E_{n,mod}(\omega) = E_{n,in}(\omega) \cdot A_n \; e^{-i\varphi_n(\omega)} \tag{2.37}$$

were the optimization variables, while the amplitude values were set to $A_n \equiv 1$. In Fig. 2.34a the learning curve of the optimization measurement is shown. A pronounced rise in the NaK^+ ion signal with increasing iteration (generation) number is clearly visible. The three values for each generation show the best, the worst and the mean of all values in one generation. The ion signal at the beginning is very small because the phase values of the first generation are randomly chosen. This leads to complex pulse structures spread over several ps. Such a randomly shaped pulse yields small ion signals. With successive iterations the signal exceeds the outcome of the transform-limited pulse after about 40 generations. Leveling off the signal is achieved after about 120 generations.

The mass spectrum (Fig. 2.34b) shows the distribution of NaK ions produced by a transform-limited pulse. The NaK^+ intensity is dominant. Larger Na_nK^m clusters are not detectable in the beam. The ion distribution after optimization is represented in Fig. 2.34c. The optimization factor amounts to

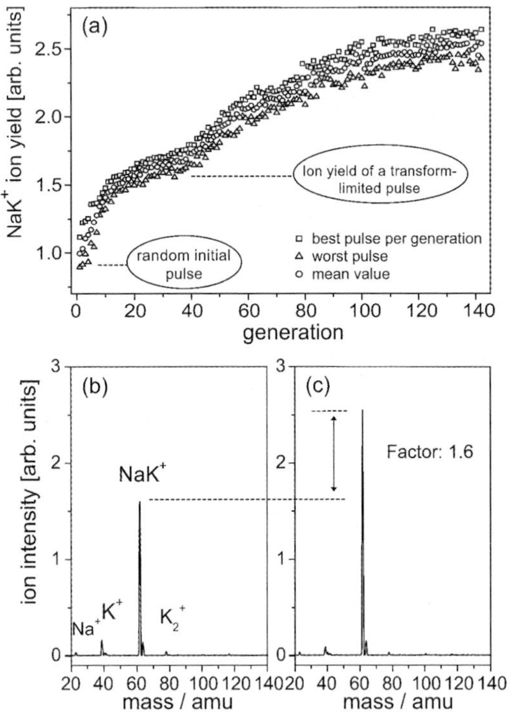

Fig. 2.34. (a) Progression of the NaK^+ ion signal during optimization. At the beginning the yield is small since the initial pulses are randomly formed. After approximately 120 generations convergence is reached. The value achieved by a transform-limited pulse is exceeded after approximately 40 generations. Shown are the values for the best, the worst and the mean for each generation. (b) The mass spectrum of the transform-limited pulse exhibits only monomers and dimers. (c) The mass spectrum recorded with an optimized pulse reveals an increased NaK^+ yield of $I_{opt}/I_{tl} = 1.6$. [226]

$I_{opt}/I_{tl} = 1.6$, where I_{opt} is the optimized ion yield and I_{tl} the yield from the transform limited pulse. Thus, the phase optimized control pulse leads to a noticeably higher ion yield than the short pulse, even though the peak intensity is reduced due to its time stretched profile. In the following the optimized pulse form will be analyzed.

The intensity cross correlation of an optimized pulse form for NaK^+ is depicted in the upper part and the SFG-XFROG trace in the lower part of the Fig. 2.35. The cross correlation shows a sequence of three main pulses. The first and second pulses are separated by $\Delta t_{1,2} \approx 650$ fs and the second and third pulses by $\Delta t_{2,3} \approx 220$ fs. The intensities of the sub pulses differ substantially: the first pulse is weak, the central pulse is strong and is followed by a weaker third pulse. The intensity ratio of these pulses is about 1:4:2.

The SFG-XFROG trace in Fig. 2.35 reveals a slight quadratic frequency chirp for the first sub pulse and a positive chirp for the middle sub pulse. The third pulse is also slightly chirped and blue shifted with respect to the second pulse. The pulse separation of $\Delta t_{1,2} \approx 220$ fs and $\Delta t_{2,3} \approx 650$ fs correspond

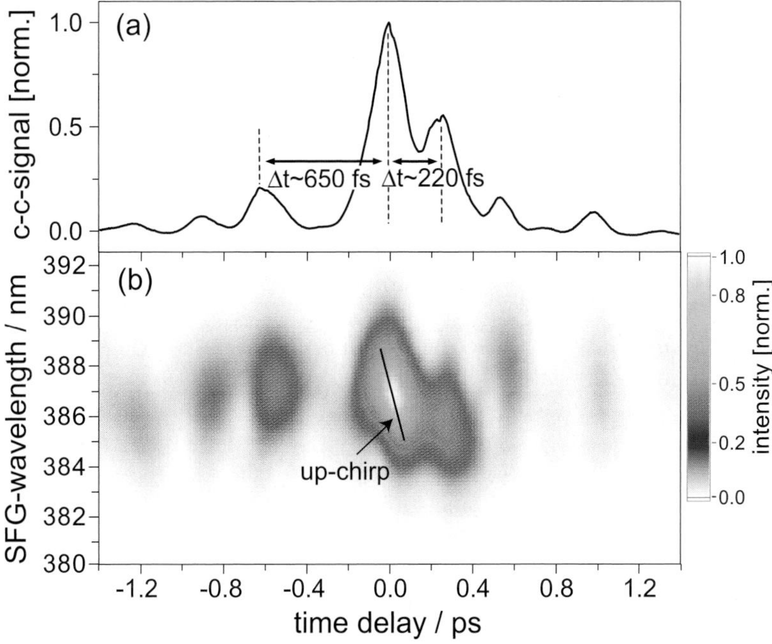

Fig. 2.35. Phase optimized control pulse for the ionization of NaK. (a) The intensity cross correlation of the pulse reveals a pulse train consisting of three main pulses. The distance between the first and the central pulse is $\Delta t_{1,2} \approx 650$ fs, whereas the central and the third pulse are separated by $\Delta t_{2,3} \approx 220$ fs. (b) The SFG-XFROG trace reveals a positively chirped central pulse and a blue shift of the third pulse. [226]

to half and one and a half oscillation periods in the excited $A^1\Sigma^+$ state, respectively, which was observed by means of fs pump-probe spectroscopy at $\lambda_{pump} = \lambda_{probe} = 770$ nm [225].

The phase profile of the pulse shape, calculated from the SFG-XFROG trace with a phase retrieval algorithm, reveals a complex structure with a phase shift of approximately $\Delta\varphi \approx 1.5 \cdot \pi$ over the whole range. For the positively chirped middle pulse a quadratic phase is found, from which the main component of the linear chirp could be calculated to $b_2 \cong +1220$ fs^2.

The result emphasizes a first simple explanation of the optimized pulse shape. In the process involved are all relevant electronic states, namely the ground state $X(1)1\Sigma^+$, the excited states $A(2)^1\Sigma^+$, $b(1)^3\Pi$, $B(3)^1\Pi$, $A(6)^1\Sigma^+$ and the ionic ground state $X(1)$ NaK^+ (see Fig. 2.36). Prior to the interaction with the laser pulse the system is located in the electronic and vibrational ground state $X(1)^1\Sigma^+$. The first subpulse creates a wave packet in the excited $A(2)^1\Sigma^+$ state (see Fig. 2.36). Contributions of wave packet motion in higher states are neglected here, since they do neither appear in the pump-probe spectra of NaK [225].

The generated wave packet subsequently evolves across the potential of the $A(2)^1\Sigma^+$ state. Due to the Franck-Condon principle the wave packet is created at the inner turning point of the $A(2)^1\Sigma^+$ state and it requires $T_{osc}^{NaK} = 440$ fs to return to the starting point [225]. The second subpulse arrives after $\Delta t_{1,2} \approx 650$ fs when the wave packet is located at the outer turning point. Also the pump-probe spectrum shows a markedly enhanced NaK^+ ion yield at the same spacings of two transform limited pulses of $\Delta t = (2n+1) \cdot 220$ fs. So , ionization occurs most efficiently, when the subpulse is applied, while the wave packet is situated at the outer turning point of the $A(2)^1\Sigma^+$ state. This position is optimally located with regard to a resonant two photonic ionization across the $B(3)^1\Pi$ state.

Since the intensity of the second subpulse is rather high it can very well also excite remaining particles from the $X(1)^1\Sigma^+$ ground state to the excited $A(2)^1\Sigma^+$ state and create another wave packet there. So half an oscillation period later also these particles are ionized by the third subpulse which appears precisely $\frac{1}{2}T_{vib} = 220$ fs later. The relative intensities of the pulses support this interpretation. The excitation into the $A(2)^1\Sigma^+$ state by the first pulse is a one photon transition. The second pulse performs a two photon ionization (and additionally a one photon excitation into the $A(2)^1\Sigma^+$ state) and is therefore more intense. The intensity of the last pulse again reflects the two photon process, as presented in Fig. 2.36.

The question can be raised whether the experimentally obtained pulse shapes can be reproduced by optimal control theory, which would allow also to gain a significantly more detailed insight into the underlying processes.

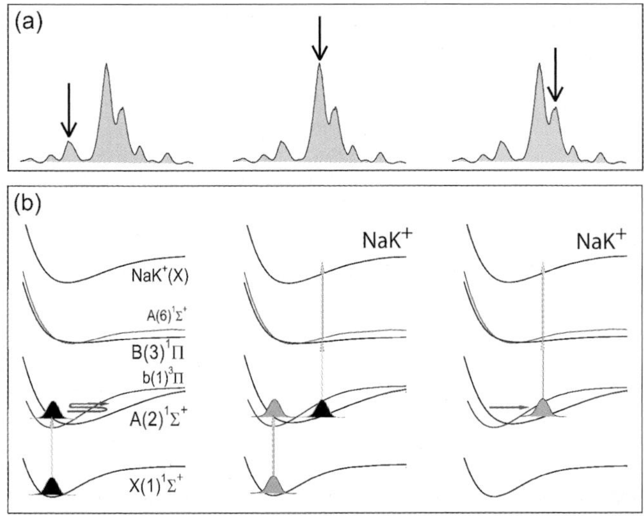

Fig. 2.36. Emphasized three step mechanism of the phase optimized photoinduced transition from the ground neutral state of NaK to the ground ionic state of NaK^+. (a): Optimized pulse train taken from Fig. 2.36a. The active subpulses are indicated by arrows. (b): The first subpulse creates a wave packet in the excited $A(2)^1\Sigma^+$ state of NaK, which is transferred into the ion state at the outer turning point of the potential by the second subpulse. This sequence is repeated with the second and third subpulse. The shown potential energy curves of NaK were taken from Magnier et al. [227]. [226]

2.4.4.2 Theoretical approach

The optimization of laser fields for controlling the photoionization in NaK has been performed using optimal control theory formulated by Kosloff et al. [17] as outlined previously . The combination of i) electronic structure, ii) quantum dynamics and iii) optimal control considering iv) experimental conditions will briefly be described. i) accurate potential energy surfaces for the ground and excited states of NaK are available in the literature [227] and have been extended by calculation of the cationic ground state necessary for the consideration of the ionization process. For this purpose the calculations using the ab initio full CI method for the valence electrons and an effective core potential with core-polarization (ECP-CPP) together with adequate AO basis sets [7s6p5d2f/5s5p4d2f] for Na and [7s5p7d2f/6s5p5d2f] for K atoms have been performed. Investigation of photoionization processes in the energy interval of 4.83 eV corresponding to three photons of 1.61 eV used in the experiments involves the three excited states $2\ ^1\Sigma^+$, $3^1\Pi$, and $6\ ^1\Sigma^+$ of the neutral NaK which are resonant with one- and two-photon energies, respectively. ii) Quantum dynamics simulations have been carried out by representing the wave function on a grid and using a nonperturbative approach based on a Chebychev polynomial expansion of the time evolution operator [228]. The interaction with the time-dependent electric field, which involves ground and three excited states of the neutral NaK as well as a manifold of cationic states, has been treated within the dipole approximation and using the rotating wave approximation (RWA) being justified in the weak field regime. The rotational motion has been neglected because of the large atomic masses and short time scales involved.

The outlined procedure involves the following steps. The Hamiltonian $\hat{H}$ is given in Born–Oppenheimer and dipole approximation

$$\hat{H}(t) = \hat{T} + \hat{V} - \hat{\mu}_{ge} E(t) , \tag{2.38}$$

where $\hat{T}$ is the operator of the nuclear kinetic energy, $\hat{V}$ stands for the potential energy curves of the considered electronic states (both operators are diagonal), and $\hat{\mu}_{ge}$ is the transition dipole moment between the considered electronic states and is off diagonal.

In rotating wave approximation e. g. a cosine-like real electric field can be replaced by

$$E_{\mathrm{RWA}}(t) = \frac{1}{2} A(t) e^{i\omega t} . \tag{2.39}$$

The time propagation follows the principle of a successive application of the time evolution operator

$$|\Psi(t_0 + n\Delta t)\rangle = \prod_{j=0}^{n-1} \hat{U}(t_0 + (j+1)\Delta t, t_0 + j\Delta t)|\Psi(t_0)\rangle, \tag{2.40}$$

with

$$\hat{U}(t + \Delta t, t) = e^{-i\hat{H}(t)\Delta t} , \tag{2.41}$$

assuming that the Hamiltonian does not change significantly within time Δt and can be replaced by piecewise constant terms.

After a normalization of the eigenvalue spectrum of $\hat{H}$ to the range [-1,1], the polynomial expansion of the normalized time evolution operator $e^{-i\hat{H}'\Delta t'}$ can be obtained by means of the Chebychev propagator [228]

$$e^{-i\hat{H}'\Delta t'} \approx \sum_{n=0}^{N} (2 - \delta_{0,n})(-i)^n J_n(\Delta t') T_n(\hat{H}') , \tag{2.42}$$

where T_n represents the n–th Chebychev polynomial and $J_n(\Delta t')$ are Bessel functions of the first kind of order n. Since the Bessel function decreases exponentially if the order n becomes larger than the argument $\Delta t'$, an exponential convergence of the expansion coefficients of (2.42) can be achieved.

iii) The objective of the optimal control is the maximization of the photoionization yield and the target operator corresponds to the total occupation of the cationic states. For the transition dipole moments between the excited electronic states of the neutral species and the ground state cation [178,179,181–184] the constant value of 5 Debye was chosen. It is in the range of transition dipole moments between electronic states of the neutral NaK and is sufficiently large to provide the robustness of the optimized pulses according to [229] and the experience. The influence of the nuclear distance-dependent transition dipole moments (cf. [184]) has been tested and found to be negligible. However, an explicit treatment of the electronic continuum for the cationic ground state dramatically influences the optimization of the ionization process and therefore it is mandatory for the appropriate treatment. For this purpose, the electronic continuum was discretized by introducing fourteen replica of the cationic ground state with energy differences of 95 cm^{-1} in the range from 1075 cm^{-1} to 2310 cm^{-1} for the electron kinetic energies . This energy range covers both, the direct and the sequential photoionization from the outer turning points of the involved electronic states. For the optimization of the pulses the Krotov algorithm [230] has been employed with additional penalty factors which allows to take into account the experimental conditions. The experimental parameters were described in the last Section where the optimization experiment was presented. The experiments were carried out in the weak field regime which is comprised in the theoretical treatment and the magnitude of the simulated laser field was adjusted to the experimental values according to the method given in [183].

2.4.4.3 Optimized pulses: Comparison between theory and experiment

Theoretically optimized pulses in the framework of OCT, obtained according to the procedure outlined above using experimentally optimized pulses as an initial guess, are shown in Fig. 2.37 [231]. They are compared with the experimentally optimized pulse using the CLL technique described above, which provided an increase of the ion yield by 60% with respect to that generated by a transform limited pulse. The leading features of both phase modulated pulses obtained from OCT and CLL are in a good agreement, as shown in Fig. 2.37a. The snapshots of the wave packet propagation under the influence of the theoretically optimized pulse (Fig. 2.37b) serve to assign the subpulses to underlying processes and to reveal the mechanism responsible for the population of the cationic state.

The role of the P_1 subpulse is to transfer a part of the population from the ground electronic state to the first excited $2\ ^1\Sigma^+$ state. This creates a wave packet in the $2\ ^1\Sigma^+$ state which propagates almost to the outer turning point within 180 fs. Subsequently, at the outer turning point the dominant P_2 subpulse simultaneously transfers the population to the $3\ ^1\Pi$ state by a one-photon process as well as to the cationic ground state by a resonant

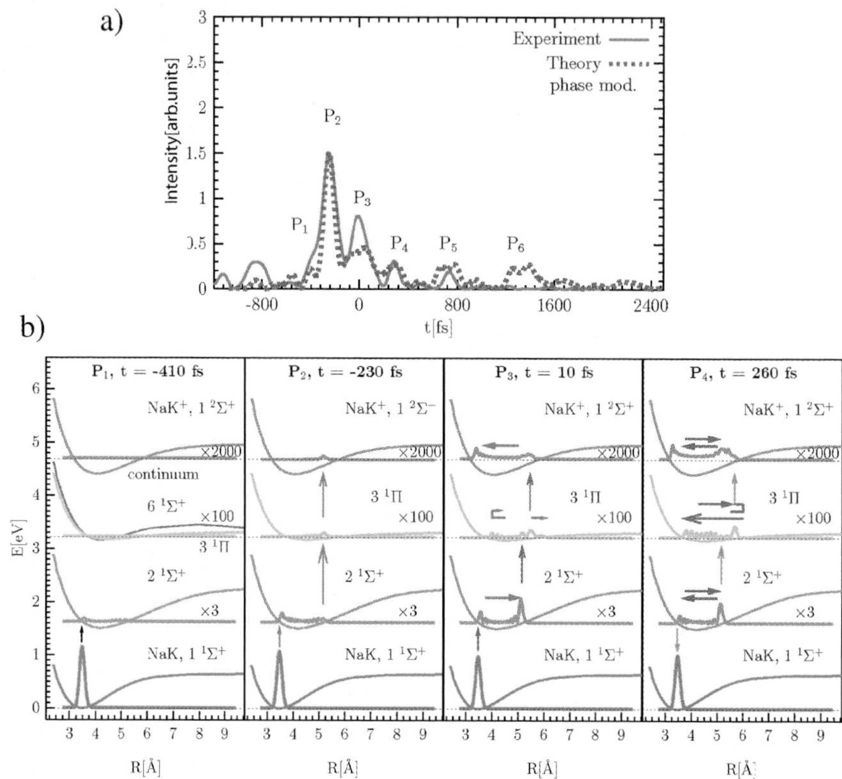

Fig. 2.37. (a) Comparison of the theoretically (dotted line) optimized phase-modulated pulse (starting with the experimentally optimized pulse) with the experimentally (solid line) optimized pulse, using CLL procedure; (b) snapshots of the wave packet propagation corresponding to P_1 (-410 fs), P_2 (-230 fs), P_3 (10 fs), and P_4 (260 fs) [231].

two-photon process as can be seen in Fig. 2.37b. In addition, the P_2 subpulse increases the population of the $2\ ^1\Sigma^+$ state at the inner turning point. Subsequently, the P_3 subpulse brings the wave packet to the $3\ ^1\Pi$ state after the outer turning point has been reached. In contrast to the dominant subpulse P_2, the P_3 subpulse also transfers population to the cationic state by the one-photon sequential processes since the split part of the wave packet, before transferred by P_2, propagates on the $3\ ^1\Pi$ state as well. At later times, e.g. at P_4, the superposition of the wave packets complicates the propagation by interference, as can be seen from the corresponding snapshot.

The separation of the early subpulses up to P_4 reflects the motion on the $2\ ^1\Sigma^+$ state with a periodicity of $\sim$ 440 fs (oscillation period in the $2\ ^1\Sigma^+$ state), while after the P_4 subpulse the periodicity is disturbed by the influence

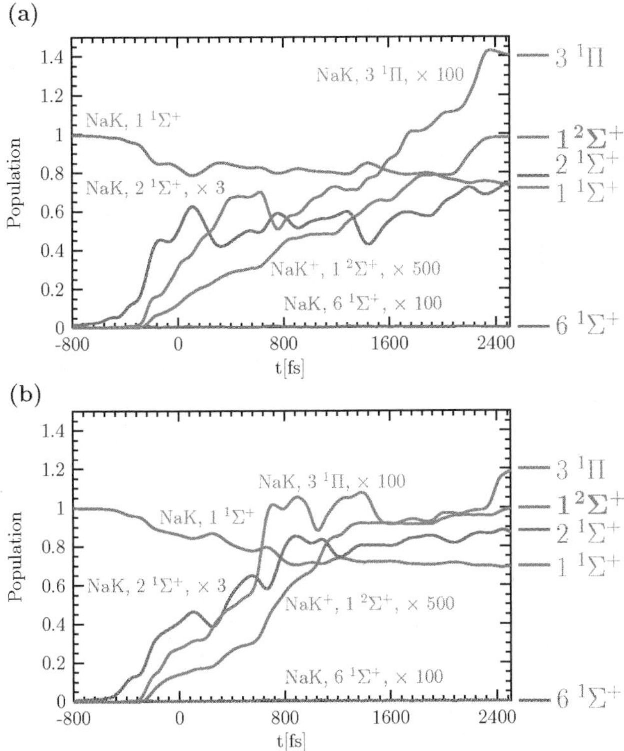

Fig. 2.38. Time-dependent population of participating electronic states of the neutral and cationic NaK, obtained from simulations with an initial guess using (a) the experimentally optimized pulse [231] (b) two Gaussian pulses [231].

of the $3\ ^1\Pi$ state. The described steps leading to the desired population of the cationic state can be also identified from the analysis of the state populations displayed in Fig. 2.38a. (Notice that, besides excitation, also dump processes appear (see Fig. 2.38a). Due to the increased population of the $2\ ^1\Sigma^+$ and $3\ ^1\Pi$ states, the dump processes appear at 200 fs, 800 fs and 1360 fs from the $2\ ^1\Sigma^+$ to the ground state and at 700 fs from the $3\ ^1\Pi$ to the $2\ ^1\Sigma^+$ state. Consequently, a staircase-like behavior in populations of these states is present.) Moreover, the later subpulses cause substantial increase in the population of the cationic states, showing the important contributions of the later subpulses with low intensities.

Based on the above analysis of the underlying dynamics driven by the optimized pulse, the following mechanism for the optimal ionization process of NaK can be proposed. It involves an electronic transition followed by a direct two-photon ionization from the outer turning point of the $2\ ^1\Sigma^+$ state. This behavior supports the proposed explanation of the experimental optimal pulse shape given in experimental publications [165,166]. However, according to the analysis of the theoretically optimized pulses described above, the sequential one-photon ionization process mediated by the $3\ ^1\Pi$ state takes over the important role at later times.

Additional insight into the energetic and temporal structure of the optimal pulse can be gained from the Wigner-Ville representation shown in Fig. 2.39b. The dominant feature is the increase of the photon energy with time. This up-chirp in the energy regime of 1.59-1.63 eV can be qualitatively explained by an overlap between the propagating wave packet on the $3\ ^1\Pi$ state and the successively higher lying vibronic levels of the cationic state. For an identification of the quantitative features, amplitude and phase modulations would be more adequate. However, the X-FROG trace obtained from the experimental result [166] also shows a pronounced up-chirp in full agreement with the features displayed in Fig. 2.39b. Moreover, the up-chirp was found to enhance the NaK ion signal according to recent chirp dependent experiments [164].

In order to verify the robustness of theoretically optimized pulses, the results obtained using two Gaussian pulses separated by 660 fs as an initial guess, are compared with experimental results in Fig. 2.39. The experimental pulse is again roughly reproduced, and the leading features of the theoretical pulse remain unchanged with respect to those obtained by an experimentally optimized pulse as an initial guess (cf. Fig. 2.37). The main differences between the optimized pulses obtained with distinct initial guesses concern relative intensities of the weaker subpulses which lead only to very small relative

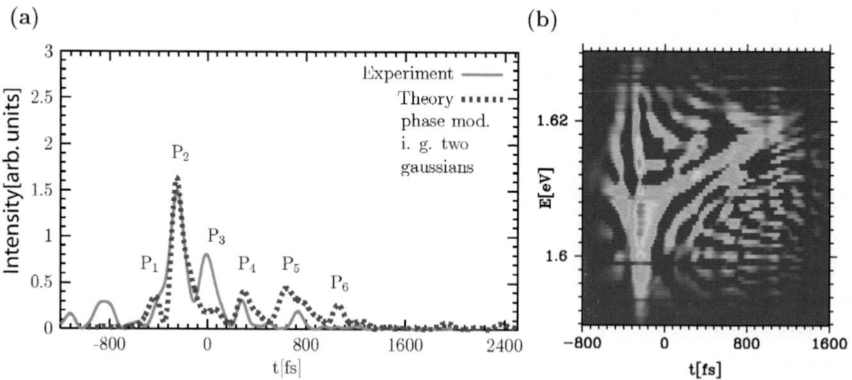

Fig. 2.39. (a) Comparison of the theoretically (dotted line) optimized phase-modulated pulse (starting with two Gaussian pulses) with the experimentally (solid line) optimized pulse, using CLL procedure; (b) Wigner-Ville distribution of the theoretically optimized pulse [231].

changes in the time dependent populations (cf. Fig. 2.38b). The Wigner-Ville representations of both theoretically optimized pulses are almost identical, verifying the robustness of the derived pulses and therefore the validity of the proposed mechanism. These findings were obtained only if the continuum of the cationic state was taken into account as described above.

In summary, the agreement between experimentally and theoretically optimized pulses, which is independent from the initial guess, shows that the shapes of the pulses can be used to deduce the mechanism of the processes underlying the optimal control. In the case of optimization of the ionization process in NaK, this involves a direct two-photon resonant process followed by a sequential one-photon processes at later times. These findings obtained from a simple system are promising for using the shapes of tailored pulses to reveal the nature of processes involved in the optimal control. Therefore, the studies were extended on isotope selective photoionization processes in NaK.

2.4.5 Isotope selectivity

The selection of particular isotopes is a topic of great interest in several fields of science and technology. Previously employed laser isotope separation schemes utilize minor isotope shifts of spectral lines [232, 233]. Yet, narrowband tunable cw-lasers and detailed knowledge of the system is required for effective isotope selection and allows only to separate the fraction present in a single quantum state. In a more recent treatment femtosecond laser pulses were employed to obtain isotope selective molecular dynamics [25,234] and to perform isotope separation by generating spatially localized vibrational wave packets due to differences in the free evolution of the different isotopes [235]. Here an approach is presented, where feedback control by means of shaped fs-laser pulses is applied for optimizing specific isotopomer yields.

2.4.5.1 Feedback optimization of isotopomer selective ionization in K_2

As presented in Sect. 2.2, the wave packet dynamics of several electronically excited states of K_2 has systematically been investigated by means of fs-pump-probe spectroscopy. For the $A^1\Sigma_u^+$ state the results revealed oscillation periods of $T_{osc}^{K_2}$= 500 fs [25, 234]. While for the $^{39,41}K_2$ isotope dephasing occurs, a recurrence can be observed in the $^{39,39}K_2$ isotope starting at 5 ps. Thereby a π phase shift appears which leads to an almost anti-phased behavior of the oscillations between the two isotopes in a range between 5 and 15 ps. In the first approach to optimize the ratio of the two isotopes it was tried to take advantage of this long-time recurrence phenomenon. A shaped pump-probe scheme was developed where first the K_2 isotopes where irradiated with a phase and amplitude modulated pump pulse and 9 ps later (where the highest optimization ratios are expected from pump-probe results) with the probe pulse that ionizes both isotopes. The implemented closed loop experiment combines ion detection with a programmable phase and amplitude pulse shaper that is driven by a self-learning optimization algorithm based on evolution strategies. The goal of the evolutionary algorithm is to find a laser field (at 833 nm center wavelength) which maximizes or minimizes the ratio $\mathrm{R} = \mathrm{I}(^{39,39}\mathrm{K}_2)/\mathrm{I}(^{39,41}\mathrm{K}_2)$ for the above described process. This preliminary experiment yielded a gain of the isotopomer ratio by a factor of about $\mathrm{S}_{max} = 1.4$ and a decrease by approximately $\mathrm{S}_{min} = 0.75$, measured with respect to the regular isotope ratio of $\mathrm{R}_n = 6.9$ [236]. The obtained maximization and minimization ratios are even better than expected from the pump-probe spectra ($\mathrm{S}_{max} = 1.25$ and $\mathrm{S}_{min} = 0.8$) which already demonstrates the potential of the closed loop approach.

In order to explore this phenomenon to its full extent single shaped laser pulses were used, whereby each pulse performs the entire ionization process (see Fig. 2.40). In this case, long-time wave packet recurrences are not ex-

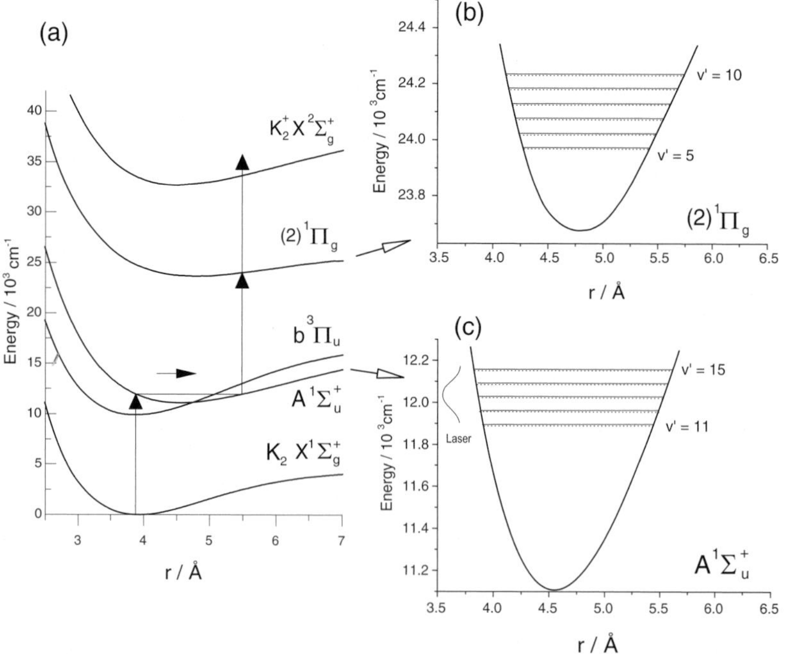

Fig. 2.40. (a) Ionization path of K_2 proposed in [237,238]. On the right hand side are the potential energy curves and vibrational levels of the $2^1\Pi_g$ (b) and $\mathrm{A}^1\Sigma_u^+$ state (c). The solid lines indicate the vibrational levels of the $^{39,39}\mathrm{K}_2$ isotope and the dotted lines the levels of the $^{39,41}\mathrm{K}_2$ isotope. The states $v'_A = 12, 13$ of the lighter isotope are disturbed by spin-orbit coupling with the $b^3\Pi_u$ state and therefore shifted by $+1.2\ \mathrm{cm}^{-1}$ and $+2.1\ \mathrm{cm}^{-1}$, respectively [25,239]. (c) also shows the spectrum of the unshaped laser light at 833 nm central wavelength. [240]

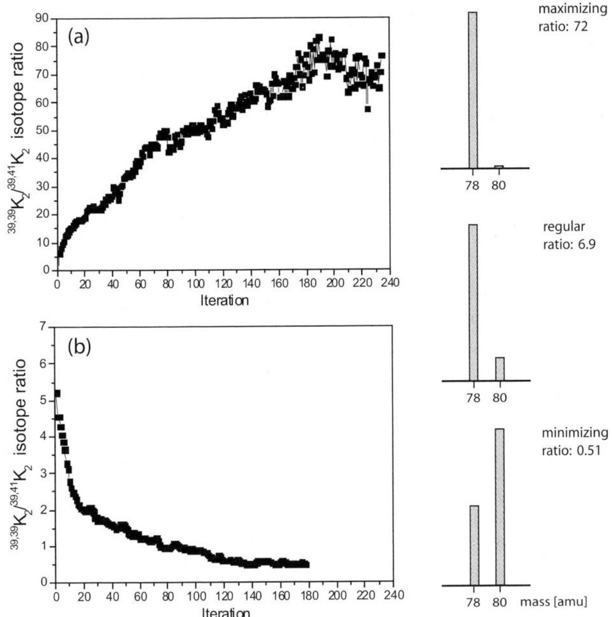

Fig. 2.41. Progressions of the mean $^{39,39}K_2/^{39,41}K_2$ ion ratio during single pulse optimization for maximization (a) and minimization (b). The considerable alteration of the isotope ratio by a factor of about 140 is also illustrated in the corresponding mass spectra on the right. [240]

pected to contribute, since the pulse shaper limits the shaped pulse duration to about 5 ps. The learning curves and the corresponding mass spectra (see Fig. 2.41) show a considerable alteration of the ion ratio by a factor of about $R_{max}/R_{min} = 140$ between minimization and maximization which provides noticeably better isotope selection results than the shaped pump-probe case ($R_{max}/R_{min} \approx 2$). This may be surprising since the shaped pump-probe experimental conditions were particularly chosen to ensure an optimal wave packet separation. Apparently, other effects are much more crucial for isotope selection than wave packet separation by recurrence phenomena. The slight enhancement of the shaped pump-probe results compared to the expected pump-probe isotope ratios already indicate this. The also performed exclusive phase optimizations provide a factor of only about 2 between maximization and minimization. Thus, the spectral amplitudes are most decisive for the observed effect which is in concordance with the results from the pure amplitude measurements (factor of about 40). This demonstrates the efficiency of combined phase and amplitude modulation.

To decipher the underlying isotope selection process, information is gained about the generated pulse shapes by recording the pulse spectra and the XFROG traces. The pulse spectrum for maximization of the $^{39,39}K_2/\ ^{39,41}K_2$

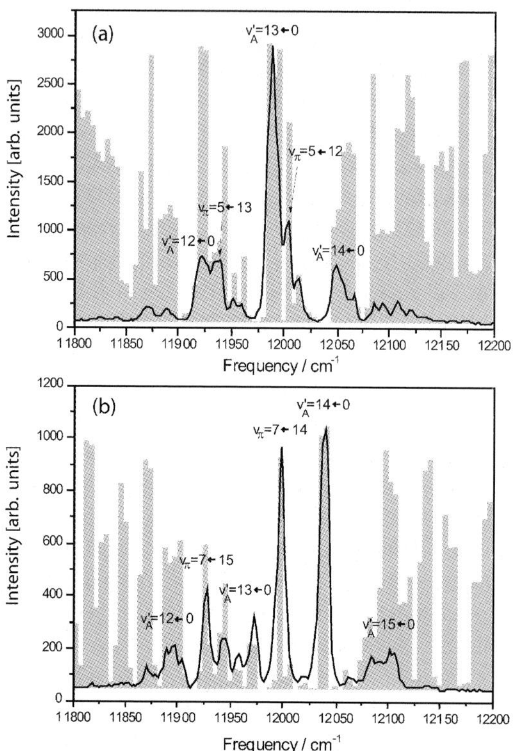

Fig. 2.42. Laser pulse spectra for maximization (a) and minimization (b) of the isotope ratio $^{39,39}K_2/^{39,41}K_2$. The gray bars denote the corresponding transmission pixel pattern of the optimized pulses. The highest peaks are assigned to transitions between particular vibrational energy levels of the different electronic states $X^1\Sigma_g^+$, $A^1\Sigma_u^+$, and $2^1\Pi_g$ of the $^{39,39}K_2$ isotope in (a) and the $^{39,41}K_2$ isotope in (b). Even indications of the mentioned distortion of the v'=13 vibrational level in the $A^1\Sigma_u^+$ state may be observable as a small peak shift, yet this cannot be firmly claimed since it is at the limit of the shaper resolution. [240]

isotope ratio at a center wavelength of 833 nm (see Fig. 2.42a) displays several distinct sharp peaks with differing intensities and intensity attenuation to almost zero between the peaks. The sharp peaks in the spectrum for minimization of the isotope ratio (see Fig. 2.42b) are located at different frequencies compared to the maximization case, shifted on average by about $-12\ \text{cm}^{-1}$. The peaks are at positions where the spectral intensity in the maximization case is low and vice versa which predominantly accounts for the high isotope selectivity. Some peaks in the presented pulse spectra can be assigned to transitions to the estimated vibrational levels in the $2^1\Pi_g$ state (see Fig. 240b) calculated by solving the time-independent Schrödinger equation numerically for the potential curve [238]. These levels exhibit isotope shifts themselves which may enhance the achievable selectivity. This demonstrates the potential of the presented method since it is not restricted to a single transition.

In the pure amplitude optimizations it can be noticed that some of the lower frequency peaks are missing. This may be due to the strongly hindered time modulation in this case resulting in a short pulse which can be understood with the superposition of the spectral components (phase-locked). The short interaction time does not allow the wave packet to propagate far in the excited state, therefore all transitions are close to the initial position of the wave packet in the ground state. Thus, higher frequencies are required to resonantly reach the $2^1\Pi_g$ transition state within the ionization step from the $\mathrm{A}^1\Sigma_u^+$ state. Moreover, some peaks belonging to the $2^1\Pi_g \leftarrow \mathrm{A}^1\Sigma_u^+$ transition are missing. This demonstrates the significance of pulse time elongation and phase modulation for molecular wave packet dynamics leading to an enhanced variability in the ionization path. Hence, another advantage of isotope selection by phase *and* amplitude modulation compared to cw-isotope separation methods is apparent.

Without theory an accurate explanation of the complex pulse shapes visible in the XFROG traces (Fig. 2.43) cannot be given, but it is possible to extract some general aspects. The frequently obtained subpulse distance of 250 fs can be explained by a stepwise excitation whereby the excitation to the inner turning point of the $A^1\Sigma_u^+$ potential is followed by an excitation 250 fs later at the outer turning point. This interpretation is backed by the fact that the Franck-Condon window via the resonant $2^1\Pi_g^+$ state is favorable at this distance [238]. Additionally, isotope dependent constructive and destructive interferences of the generated wave packets may occur within the ionization paths, respectively, which could further improve the isotope specific ionization efficiency [241]. It would as well explain the successful optimizations observed in the phase only experiments. Moreover, oscillation periods on other excited states may play a role as well.

In order to get more insight into the involved processes the vibrationally resolved excitation transitions were calculated (assuming no time evolution) by taking the spectral pulse intensities at the determined vibrational transitions into account. This has been done by calculating the vibrational state specific populations $N_{m,n} \propto f_{m,n} \cdot I(\nu_{m,n})$, where $f_{n,m}$ are the Franck-Condon factors and $I(\nu_{m,n})$ the relative spectral intensities at each involved transition for both isotopes. For the electronic transitions the selection factors are estimated to $\mathrm{S}_{trans} = \sum_{m,n} N_{m,n}(^{39,39}\mathrm{K}_2)/\sum_{m,n} N_{m,n}(^{39,41}\mathrm{K}_2)$. This calculation provides excitation step specific information. This can be calculated to $\mathrm{R}(\mathrm{A}^1\Sigma_u^+ \leftarrow \mathrm{X}^1\Sigma_g^+) = \mathrm{S}_{max}(\mathrm{A}^1\Sigma_u^+ \leftarrow \mathrm{X}^1\Sigma_g^+)/\ \mathrm{S}_{min}(\mathrm{A}^1\Sigma_u^+ \leftarrow \mathrm{X}^1\Sigma_g^+) = 31.3$ for the first electronic excitation and $\mathrm{R}(2^1\Pi_g^+ \leftarrow \mathrm{A}^1\Sigma_u^+) = \mathrm{S}_{max}(2^1\Pi_g^+ \leftarrow \mathrm{A}^1\Sigma_u^+)/\ \mathrm{S}_{min}(2^1\Pi_g^+ \leftarrow \mathrm{A}^1\Sigma_u^+) = 1.33$ for the second. Thus it is evident that both electronic excitation steps are utilized to achieve the high selectivity, whereby the first step predominantly contributes. The ionization step from the $2^1\Pi_g$ state can be neglected for determining the optimal iso-

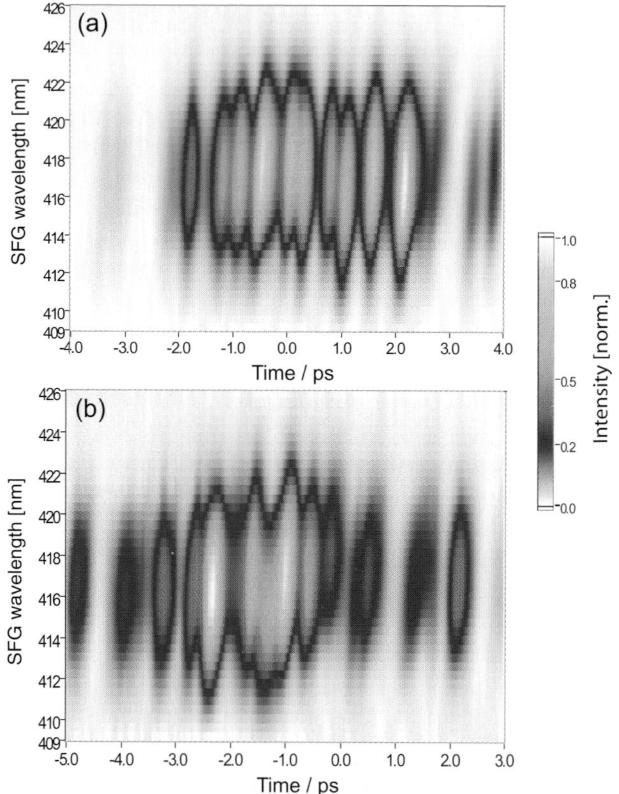

Fig. 2.43. XFROG traces for maximization (a) and minimization (b) of the isotope ratio $^{39,39}K_2/^{39,41}K_2$. The often observed subpulse distance of 250 fs ($\frac{1}{2}T_{osc}$ of the $A^1\Sigma_u^+$ state) and the wavelength shift of up to 1 nm between two adjacent subpulses indicate a successive excitation of the electronic states on the optimal ionization path. Particularly for minimization the higher frequency subpulses assigned to the $A^1\Sigma_u^+ \leftarrow X^1\Sigma_g^+$ transition are followed by the lower frequency ones assigned to the $2^1\Pi_g^+ \leftarrow A^1\Sigma_u^+$ transition (see Fig. 2.42b). [240]

tope selection since its transition probability is almost equal for the different isotopes. Thus, the total isotope selection calculated by $R = R(A^1\Sigma_u^+ \leftarrow X^1\Sigma_g^+) \cdot R(2^1\Pi_g^+ \leftarrow A^1\Sigma_u^+)$ amounts to 41.7 assuming time-independence. According to this calculation, the obtained frequency pattern accounts for a major contribution to the isotope selection. The substantial remaining difference to the experimental result can be attributed to the time evolution of the optimal pulses. The stepwise excitation process $2^1\Pi_g \leftarrow A^1\Sigma_u^+ \leftarrow X^1\Sigma_g^+$ with half oscillation periods in between, observed in the XFROG traces (Fig. 2.43), may account for the enhancement of the optimization factors. This can be simulated by removing spectral lines assigned to the $A^1\Sigma_u^+ \leftarrow X^1\Sigma_g^+$ transition (e.g. the peak v'$_A$ = 14 ← 0 in Fig. 2.42b) to determine the $2^1\Pi_g \leftarrow A^1\Sigma_u^+$ transition factor and vice versa, which leads to an about two times larger factor in the variation of the isotope ratio. Yet, the fact that even this value is lower than the experimental result demonstrates the potential of the optimization method since all involved processes are inherently utilized to achieve the optimal yields, i.e. the above mentioned interference effects described in [241] may additionally contribute.

Isotope selective optimizations were also performed at differing center wavelengths (810 nm and 820 nm) in order to learn about the molecular system (i. e. its potential energy curves and vibrational states) and the chosen optimized ionization paths [242]. Large optimization factors between minimization and maximization of about R_{max}/R_{min} = 14 and 50 were also obtained for the center wavelengths 810 nm and 820 nm, respectively. The pulse spectra for maximization of the $^{39,39}K_2/\ ^{39,41}K_2$ isotope ratio at the employed center wavelengths 810 nm, 820 nm, and 833 nm reveal several distinct sharp peaks with differing intensities and intensity attenuation to almost zero between the peaks (see Fig. 2.44). The most pronounced ones can be assigned to transitions from v"=0 in the electronic ground state to the vibrational states v'=12-19 of the $A^1\Sigma_u^+$ state in the $^{39,39}K_2$ isotopomer.

The pulse spectra for minimization of the isotopomer ratio (see Fig. 2.44b) reveal several sharp peaks located at different frequencies compared to the maximization case. They are shifted by about -11 to -16 cm^{-1} with rising shift at rising frequency. The peaks are located at positions where the spectral intensity in the maximization case is low and vice versa which can be regarded as a major reason for the high isotope selectivity. The observed peaks can be assigned to transitions from v"=0 to v'=12-19 of the $^{39,41}K_2$ isotope. The peak series at the different center wavelengths can be viewed as an extension of the first isotope optimization experiment performed at 833 nm.

Other peaks are also visible in the pulse spectra in Fig. 2.44 and may be attributed to vibrational transitions to other electronic states during the ionization path. In particular, the transition from the $A^1\Sigma_u^+$ to the $2^1\Pi_g$ state is present in the optimized pulse spectra also for 810 nm and 820 nm. The $2^1\Pi_g$ state thereby serves as a resonant transition state for the ionization step [238]. This transition occurs at the favorable Franck-Condon window at the outer turning point of the wave packet in the $A^1\Sigma_u^+$ state.

In the presented K_2 experiments the employed parameters laser bandwidth, pulse duration, and liquid crystal array pixel resolution match to fulfill both, the required frequency resolution and the sufficient freedom for time evolution. For this reason K_2 is preeminently suited as a model system, but the described selection method will work for larger molecules or clusters as well. It may even be fast enough to prevent disturbance by IVR, contrary to the above mentioned technique using wave packet recurrence phenomena. When comparing the approach with conventional laser isotope separation methods, the greatest advantage is, that the broad bandwidths allow the total isotope selection of a system with all initially populated quantum states instead of

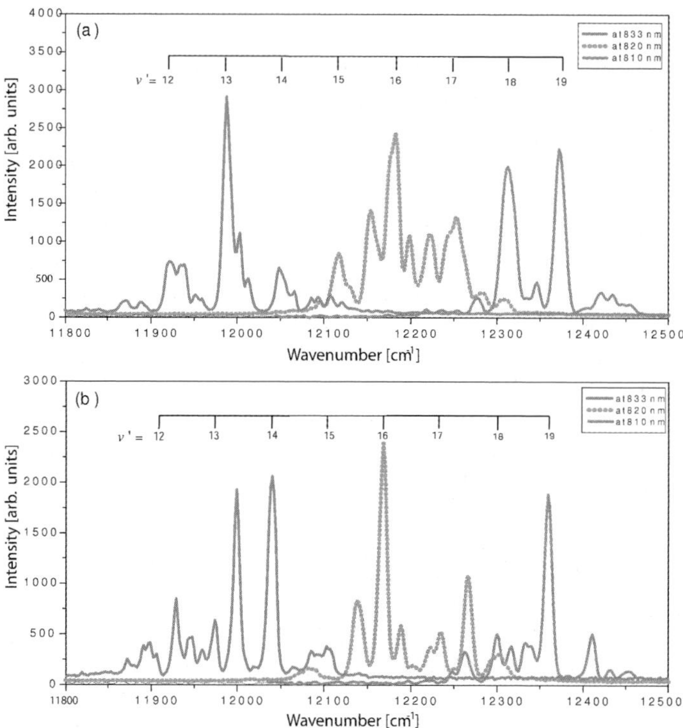

Fig. 2.44. Laser pulse spectra for maximization (a) and minimization (b) of the isotope ratio $^{39,39}K_2/^{39,41}K_2$ at different center wavelengths of the initial pulse. The highest peaks are assigned to transitions between particular vibrational energy levels of the electronic transition $A^1\Sigma_u^+ \leftarrow X^1\Sigma_g^+$ of the $^{39,39}K_2$ isotope in (a) and the $^{39,41}K_2$ isotope in (b). The vibrational series v"=0 to v'=12-19 over all measured optimized pulses is clearly visible in both spectra. [242]

only exciting a fractional single state selection by a narrow laser. Ionization is not explicitly necessary since one could as well populate selected electronic states isotope specific, which would allow isotope selection of neutral molecules/clusters.

The achieved fs-time dependent dynamics of a highly resolved frequency pattern due to superposition of the spectral components open the perspective to alternative spectroscopical treatment which combines time- and energy-resolved observations adapted to the dynamical evolution of the photoexcited system.

2.4.5.2 Theoretical aspects of the isotopomer selective ionization

in NaK

The spectroscopically rather weak differences of different isotopic species represent an excellent probe for the demonstration of selectivity power and efficiency of optimization processes. Since such experiments allow to obtain high optimization yields on selective isotopomers, this calls for deep understanding of underlying processes based on theoretical considerations.

The conceptual basis for optimal control theory to achieve the isotope selection relies on the differences of the wave packet propagation of different isotopomers and on addressing the isotope specific vibrational levels differently with a femtosecond laser field [243], as it will be illustrated below. Theoretical results for phase-only modulation will be compared with experimental pulses obtained from CLL approach and therefore the direct assignment of the processes to experimental features will be made [244].

In order to achieve an optimization of the employed laser fields with regard to controlling the isotope selective ionization in NaK, the previously outlined OCT procedure [17] has to be extended for the treatment of isotopomers. As a target function the simultaneous maximization of $(^{23}\mathrm{Na}^{39}\mathrm{K})^+$ and minimization of $(^{23}\mathrm{Na}^{41}\mathrm{K})^+$ has been considered (simulating the optimization of the isotopomer ratio) in the optimization procedure. Again, the combination of (i) electronic structure, (ii) dynamics, and (iii) optimal control considering (iv) experimental conditions will briefly be described.

(i) Investigation of photoionization processes in the energy interval of 4.83 eV corresponding to three photons of 1.61 eV used in the experiments involves the three excited states $2\,^1\Sigma^+$, $3\,^1\Pi$, and $6\,^1\Sigma^+$ of neutral NaK which are resonant with one– and two–photon energies, respectively. The accurate potential energy surfaces for the ground and excited states of NaK as well as for the cationic ground state necessary for consideration of the ionization processes have been used [231] as already described for optimization of photoionization in NaK.
(ii) Quantum–dynamics simulations have been carried out by representing the wave function on a grid and using nonperturbative approach based on a Chebychev polynomial expansion of the time evolution operator as described previously. In analogy the photoionization of the single isotope of NaK, the interaction with the time dependent field involving the ground, three excited states of the neutral species and a manifold of cationic states (14) imitating the continuum, altogether 18 states for each isotopomer have been treated within the dipole approximation and using the rotating wave approximation. This is justified in the weak–field regime. The rotational motion has been neglected because of the larger atomic masses and short time scales involved.

Now, inclusion of different isotopomers involves the block diagonal extension of the vibrational Hamilton matrix in which each isotopomer is represented by a separate block including all considered electronic states.

(iii) The objective of the optimal control is maximization of the photoionization yield of one isotopomer (here $^{23}Na^{39}K$) while the photoionization yield of the other isotopomer (in this case $^{23}Na^{41}K$) is minimized at the same time. Therefore, one should introduce in the target functional in the time range $[t_i, t_f]$:

$$J = \langle \psi(t_f) | \hat{\boldsymbol{P}}_{\text{Target}} | \psi(t_f) \rangle + \int_{t_i}^{t_f} \left[2 \cdot \text{Re} \left\{ \langle \boldsymbol{\chi}(t) | - \frac{\partial}{\partial t} - \frac{i}{\hbar} \hat{\boldsymbol{H}}(t) | \psi(t) \rangle \right\} - \lambda \frac{|E(t)|^2}{s(t)} \right] dt \tag{2.43}$$

with the target operator

$$\hat{\boldsymbol{P}}_{\text{Target}} = \begin{pmatrix} \overbrace{\begin{matrix} 0 \; 0 & \\ \ddots & \\ 0 & \sum_{i=0}^{19} |\nu_i^{(1)}\rangle\langle\nu_i^{(1)}| \end{matrix}}^{\text{Isotope (1)}} & 0 \\ 0 & \underbrace{\begin{matrix} 0 \; 0 & \\ \ddots & \\ 0 & \hat{\mathrm{I}} - \sum_{j=0}^{19} |\nu_j^{(2)}\rangle\langle\nu_j^{(2)}| \end{matrix}}_{\text{Isotope (2)}} \end{pmatrix} \tag{2.44}$$

The functional in (2.43) includes three terms. The first term represents the expectation value of the target operator at time t_f. The second and the third terms represent the constraints which insure that the wave function satisfies the time-dependent Schrödinger equation and limit the energy of the laser field $E(t)$, respectively. Associated with these two constraints are two Lagrange multipliers $\boldsymbol{\chi}(t)$ and λ [148]. Additionally, a sine squared shape function $s(t)$ is used. This enables smooth on and off switching of the laser field at borders of the time interval (cf. [184]).

The target operator defined in (2.44) is used to simultaneously maximize photoionization of the $^{23}Na^{41}K$ isotopomer and to minimize photoionization yield of $^{23}Na^{41}K$ isotopomer. ($\hat{I}$ labels the identity operator). It consists of two diagonal blocks (one for each isotopomer). Each block represents a 18×18 matrix corresponding to the considered electronic states of a given isotopomer. Its nonzero elements project onto the lowest 20 vibrational eigenstates of the cationic state. Upon variation with respect to the real and imaginary components of the laser field $E(t)$ the iterative scheme for the pulse optimization is obtained in analogy to the procedure described previously.

2.4.5.3 Comparison between theoretically and experimentally obtained optimal pulses

Experimentally, the optimization procedure is similar as outlined in Sect. 2.4.3. The ^{23}Na^{39}K/^{23}Na^{41}K ion ratio is used as fitness by sequentially recording the ion yields in time steps of less than half a second. Convergence is reached when the feedback signal does not change significantly within the following generations. The temporal intensity of the generated optimized pulse form is retrieved from the pulse spectra and the sum frequency cross correlations by using PICASO [245]. During the experiment the obtained optimization factors stay almost unchanged for repeated optimization runs [246], whereas the optimized pulse shapes may slightly differ. Therefore the best pulse under the experimental conditions was chosen for comparison with theory.

In the following, theoretically and experimentally optimized phase–modulated laser fields for the simultaneous maximization of the ^{23}Na^{39}K isotopomer and minimization of the ^{23}Na^{41}K isotopomer will be compared and analyzed. The phase–modulated pulses obtained from OCT and from experimental CLL exhibit relatively simple characteristic features and are compared in the upper part of Fig. 2.45. The experimentally shaped pulse is characterized by two dominant subpulses and theoretical results exhibit an additional dominant subpulse at later times. This is independent from the initial conditions which involve either Gaussian pulse (cf. left, upper part of Fig. 2.45) or experimentally optimized pulse (right, upper part of Fig. 2.45). For the former case also the snapshots of the wave packet dynamics for both isotopomers are presented (cf. middle, part of Fig. 2.45) and time dependent populations of the involved states (cf. bottom part of Fig. 2.45). They serve to illustrate the dynamics induced by the pulse and to deduce the mechanism responsible for the isotope selective ionization.

First the role of the dominant subpulses P_3, P_4, and P_5 for the mechanism of the photoionization is addressed. The P$_3$ subpulse populates the cationic state of ^{23}Na^{39}K in a direct resonant two-photon process. The ionization occurs at the outer turning point because the transition dipole moment between the first and the second excited state is much larger than at the inner turning point. At this time step, the other isotopomer ^{23}Na^{41}K is less efficient for photoionization, as can be seen from the snapshots of Fig. 2.45. The contribution of ^{23}Na^{39}K is larger because the wave packet corresponding to this isotopomer is located at the outer turning point.

Moreover, the snapshots in Fig. 2.45 show that the difference in propagation of the wave function on $2^1\Sigma^+$ state of different isotopomers is caused by dephasing of the wave packets. For example, in the case of P_3 subpulse, during propagation of the wave packet on the $2^1\Sigma^+$ state for the isotopomer ^{23}Na^{41}K toward the inner turning point the population transfer from the electronic ground state accelerates this motion. Therefore the population of the $3^1\Pi$ state and of the cationic state is reduced (cf. snapshot of the isotopomer

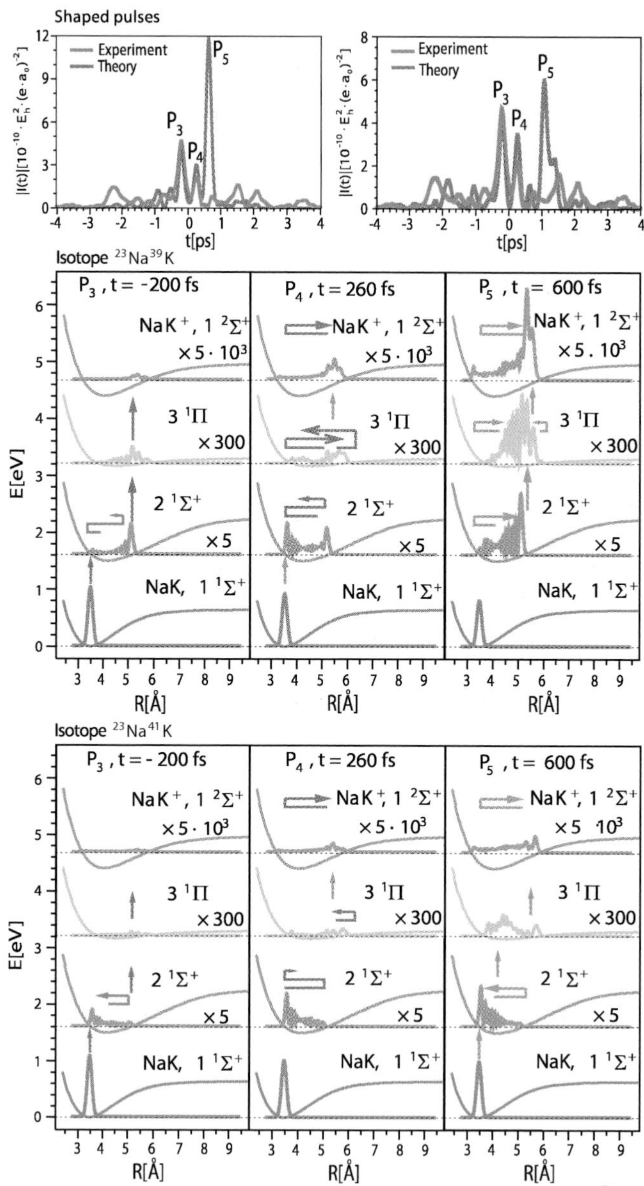

Fig. 2.45. Top: Comparison of pure phase modulated pulses for isotopomer ratio maximization obtained from OCT and from experimental CLL. The initial conditions for the theoretical optimizations involve either a Gaussian pulse (left side) or the experimentally optimized pulse shape (right side). Center and bottom: Snapshots for wave packet propagation of isotopomers $^{23}Na^{39}K$ and $^{23}Na^{41}K$ corresponding to the subpulses P_3 (-200 fs), P_4 (260 fs), and P_5 (600 fs) for a simulation started from a Gaussian pulse.

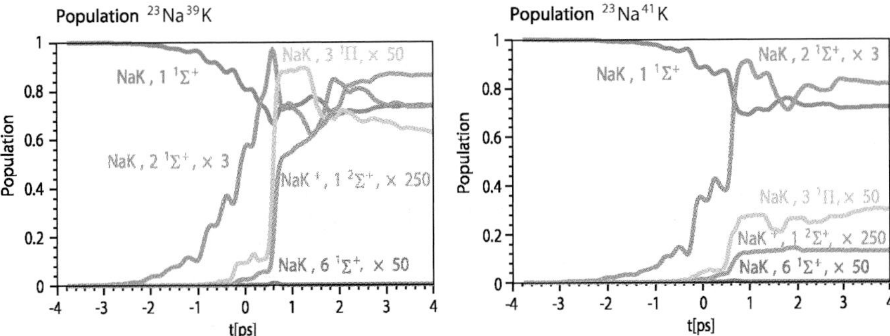

Fig. 2.46. Time-dependent populations of the participating electronic states of neutral and cationic $^{23}Na^{39}K$ (left) and $^{23}Na^{41}K$ (right) for the same initial condition.

$^{23}Na^{41}K$ for subpulse P_3 in Fig. 2.45) in contrast to the situation occurring for the isotopomer $^{23}Na^{39}K$ at the time of P_3. After 460 fs which corresponds to one full oscillation period on the first excited state, the subpulse P_4 transfers population from the ground state to the first excited state for the isotopomer $^{23}Na^{39}K$. The transfer of the population to the cationic state occurs by a sequential one photon ionization process over the second excited state which is populated by the P_3 subpulse. The subpulse P_4 introduces a sequential process for the ionization but does not increase substantially the population of the cationic state of the isotopomers (cf. Fig. 2.46).

The dominant subpulse P_5 strongly enhances the population of the cationic state of the isotopomer $^{23}Na^{39}K$ due to a direct two–photon process, which occurs very efficiently. The wave packet motion is particularly slowed down at the outer turning point. The reason for this can be depicted from the snapshots: Population as well as depopulation of the $2^1\Sigma^+$ state from the electronic ground state occurs at the inner turning point at the beginning of P_5 (cf. Fig. 2.45). Therefore, a longer residence time of the wave packet at the outer turning point occurs and this provides efficient transfer to the $3^1\Pi$ state and to the cationic state by the later part of P_5 subpulse. The role of the individual subpulses is also reflected by the variation of the population of cationic states as shown in Fig. 2.46. As mentioned above, P_3 and P_4 do not contribute substantially to the population of the cationic state. In addition to the important role of subpulse P_5 for the ionization of the isotopomer $^{23}Na^{39}K$ the later subpulses contribute with a fraction of 39% via sequential one–photon processes to the cationic population.

The maximization of the isotopomer $^{23}Na^{39}K$ is experimentally less efficient than in the case of OCT ($\sim$ 2 versus $\sim$ 8). This is independent from the initial conditions. Similar results and mechanism are obtained from OCT approach if the optimization has been started using experimentally shaped pulses in the frame of CLL. The reason for this can be connected with the presence of the dominant P_5 subpulse in the OCT which is responsible for efficient two–photon process. Experimentally optimized pulse has reached the solution which is characterized by only two dominant subpulses. However, their spacing is similar to those obtained by OCT showing the agreement between the experimental solution and the parts of theoretically optimized pulses.

In order to examine the conditions under which the pulse features obtained by the experimental CLL approach can be obtained, the theoretical optimization procedure was interrupted after three iterations starting from the experimental pulse shape. The obtained pulse shape is almost identical to the experimental one as shown in Fig. 2.47 which is characterized by two dominant subpulses. The time dependent populations of the involved states are shown as well. Moreover, an optimization factor of ~ 3 for the theoretical simulation is close to the experimental efficiency of ~ 2. This confirms that the parameters of the simulation were well adapted to the experimental conditions. By analyzing the snapshots, a difference in the propagation of the wave packets on the $2^1\Sigma^+$ state of different isotopomers was found, again due to photoinduced dephasing. The ionization occurs at the outer turning point due to larger transition dipole moment between the first and the second excited state.

The gained insight reveals that photoinduced isotopomer specific acceleration of the wave packets followed by ionization at the outer turning point of the $2^1\Sigma^+$ state is responsible for the isotopomer selection observed experimentally.

In summary, it should be pointed out that selective phase-optimization of ionization of $^{23}Na^{39}K$ isotopomer is in general due to the different isotopomer specific wave packet dynamics in the first excited state. The optimal pulse induces dephasing of initially formed wave packet leading to different spatial localization of the two wave packets at the same time corresponding to two isotopomers. This enables isotope selective ionization since only the portion of the wave packet at the outer turning point can be efficiently ionized due to the larger transition dipole moment between the first and second excited states than at the inner turning point. The pulses obtained from the OCT with different initial conditions have common leading features. This means that theoretically optimized pulses are sufficiently robust and independent from the initial conditions. Consequently the analysis of the underlying processes provides equivalent mechanisms. The subpulses which are responsible for the described mechanism are also present in the experimental findings. In this context it is interesting to note that two qualitatively different search strategies were utilized for OCT and experimental CLL, respectively. Yet, the nondeterministic evolution search strategy in CLL and the deterministic iterative search strategy in OCT provide similar results.

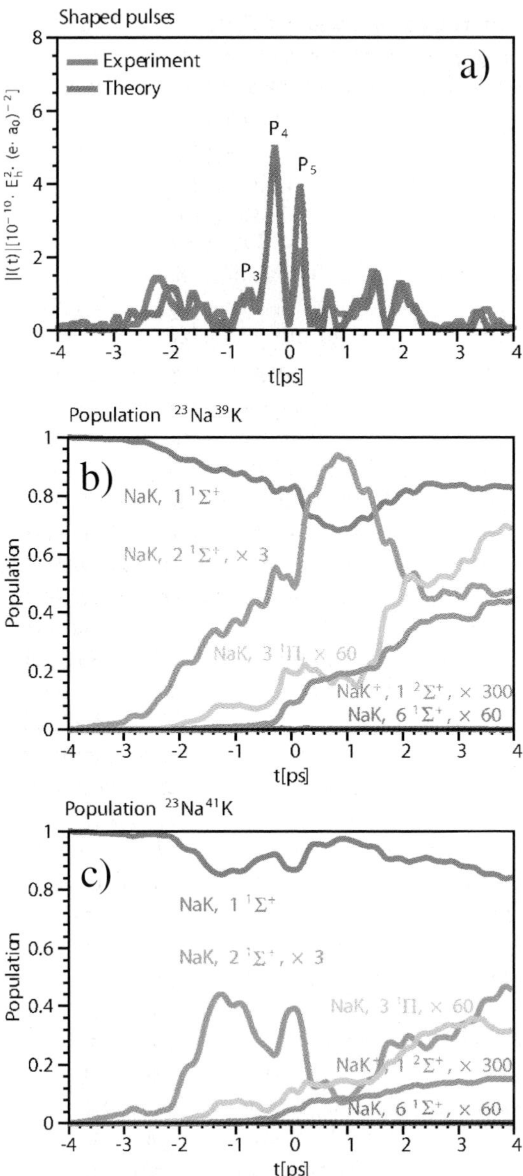

Fig. 2.47. Left: Pulse shape (blue line) obtained from a optimization started from an experimentally optimized pulse (red line) and interrupted after three iterations. Center and right: Time dependent populations of the electronic states of neutral and cationic NaK for both isotopomers.

2.4.6 Optimization procedures and their improvement

The development of pulse shapers, which allow to apply optimally shaped laser pulses in the amplitude and phase domain has opened new perspectives for driving molecular reaction dynamics in real time. An important aspect of the approach is the retrieval of the information coded in the optimized laser pulse shape which can lead to new perspectives regarding the investigation of molecular dynamics [46, 226]. This calls for systematic improvements of the method, which will briefly be presented here. One improvement of the approach examines the implementation of genetic pressure within the algorithm for performing control pulse cleaning [247]. The aim is to remove extraneous pulse features in order to expose the most relevant structures and thus reveal mechanistic insights. A second improvement on the method addresses parametric optimization [177,185,248,249] were the search space is narrowed down by introducing a few pulse parameter optimization. The suggested parameters in the time- and frequency domain are time distances, intensities, zero order spectral phases, chirps of different subpulses, and spectral peak patterns. In this manner, the relevance of certain structural features of optimal pulses can be investigated. The third approach involves the addition of the polarization as the third parameter besides phase and amplitude shaping in order to have full control over all degrees of the light field. This allows to attain new optimization paths which adds a significant degree of controllability.

2.4.6.1 Control pulse cleaning and conflicting objectives

The procedure of control pulse cleaning is targeted to remove extraneous control field features in a closed-loop quantum dynamics optimization experiment. This is accomplished by applying genetic pressure during the optimization run in order to simplify the optimized pulses and to reveal the most relevant features. Thereto, a new fitness function has to be formed where the target goal f is optimized and simultaneously genetic pressure is implemented on certain pulse components. This is solved by dividing the target f by the average spectral transmissions raised to a positive power γ, expressed as the fitness function

$$F = \frac{f}{(\frac{1}{N}\sum_{l=1}^{N} T_l)^{\gamma}} \tag{2.45}$$

with the transmissions T_l with $l = 1, 2, ... N$ for all $N = 128$ pixels. The weighting exponent γ can be freely chosen to set the desired degree of genetic pressure. The phase values were thereby allowed to be modulated freely without genetic pressure. This new fitness function was applied for the maximization of the ionization of NaK at a center wavelength of 770 nm. The influence of different levels of genetic pressures reveals the enhanced efficiency of the optimized pulse at strong control pulse cleaning. The comparison to a short pulse at equal energy demonstrates the successful removal of unimportant components and the enhancement of those that count [250]. Apparently,

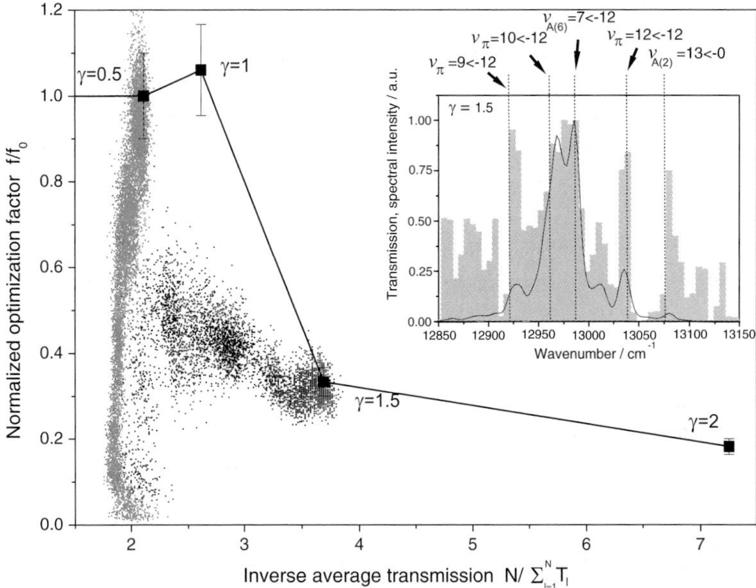

Fig. 2.48. Correlation plot of the normalized optimization factor f/f_0 versus the inverse average transmission. The solid line depicts the estimated Pareto-optimal front drawn through the optimal solutions for different γ. The optimization courses for $\gamma = 0.5$ (gray points) and $\gamma = 1.5$ (black points) are shown by plotting the values of every individual. The inset shows the spectrum (solid line), and the pixel transmissions (gray bars) of the optimized pulse for $\gamma = 1.5$. Vibronic transitions to the first and second excited states are denoted. The vibrational energy levels associated with these transitions were calculated by numerically solving the time-independent Schrödinger equation for the potential curves from [227]. [250]

this effect is due to the allocation of the available laser energy to the most relevant spectral features in the case of strong control pulse cleaning.

At increasing genetic pressure, features arise from the transmission spectra and become more distinct. These features can be assigned to transitions associated with vibrational levels of different electronic states of the $^{23}Na^{39}K$ isotopomer which underlines the multistep character of the selected ionization paths. The insert of Fig. 2.48 shows the spectrum (solid line) and the pixel transmissions (gray bars) of the optimized pulse with $\gamma = 1.5$ for the center wavelength of 770 nm. It exposes otherwise not visible vibronic transitions to the first and second excited states which come out best at this weighting factor. This demonstrates the feasibility of control pulse cleaning and shows that otherwise hidden information can be unraveled.

Fig. 2.48 depicts a correlation plot of the normalized optimization factor f/f_0 (with f_0 being the ion yield without genetic pressure) and the inverse average transmission $N/\Sigma_{l=1}^{N}T_l$. Both objectives are functions of the same 256 pixel settings in the modulator. Plotted is the so called Pareto-optimal front (solid line) drawn through the optimal solutions for different γ. This presentation reveals the correlation between both conflicting objectives and provides a tool to distinguish between optimal and inferior solutions. As examples, the optimization courses for γ =0.5 (gray points) and γ =1.5 (black points) are shown by plotting the corresponding objective values of every individual. Both optimizations evolve from the lower left to the upper right. While the ion yield increases at almost constant transmission for the optimization with γ =0.5, the optimization for γ=1.5 first tends to increase f/f_0 and then decreases by proceeding to lower average transition which indicates the enhanced efficiency of cleaning at the end of the optimization. This concept may be generalized to multi-criterion optimizations with several physical relevant objectives and new algorithms could be utilized where one receives the entire Pareto-optimal front within one run.

2.4.6.2 Parametric optimization

In the following, a different approach to receive further information will be applied by performing a few parameter optimization, where the search space is reduced to permit only simple pulse trains or spectral patterns. In this manner the relevance of certain structural features of the optimized pulses can be investigated. A further advantage of parametric optimization, besides search space reduction and application of physically relevant and intuitive parameters, is the possibility to define the framework of an experiment. One can for example choose a parameter set for an intended purpose and will receive the optimized result within this parameter set. Here, the question is raised what will be the result if different numbers of subpulses are allowed, which will tell the relevance of the additional subpulses. Thereto, the complex electrical field of a pulse train after passing the shaper was constructed by

$$\tilde{E}_{out}(t) = e^{i\omega_0 t}\sum_n \varepsilon_n(t-t_n)e^{i\varphi_0 t} = e^{i\omega_0 t}\sum_n \tilde{\varepsilon}_n(t-t_n) \tag{2.46}$$

with ε and $\tilde{\varepsilon}$ being the field envelope and complex field envelope of the subpulses, respectively, and t_n being the time distances between the subpulses. The resulting temporal field is then Fourier-transformed to receive an electrical field in the frequency description. The desired output for the electrical field is hence produced from the modulator by writing the filter function $\tilde{H}(\omega) = \tilde{E}_{out}(\omega)/\tilde{E}_{in}(\omega)$. The possible waveforms are limited by physical setup issues [221] like pixel quantization, laser bandwidth, gap-to-stripe ratio, and phase resolution, leading to a maximum complexity of the pulse, to pulse replica and other side effects [251].

The experiments were performed by pure phase modulation such that $|\tilde{H}(\omega)| = 1$ in order to keep the total pulse energy constant. Due to this

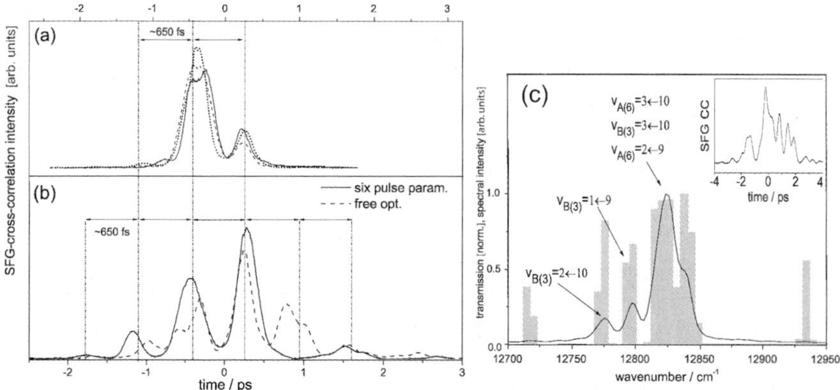

Fig. 2.49. The SFG cross-correlation results are shown for optimizing three (a) or six (b) subpulses (solid line). The results of three different runs are shown in (a). Subpulse distances of 650 fs corresponding to $1.5T_{osc}$ are visible in all cases. The optimized pulse for free optimization is shown in (b) as a dashed line. The spectrum of the optimized pulse for parameterization in the frequency domain employing 11 narrow Gaussian distributions is shown in (c) (the inset depicts the cross-correlation trace). The spectral peaks are assigned to vibronic transitions. [249]

constraint, no exact transfer function can be determined for those transformations which require a change of spectral amplitudes. A reliable algorithm is applied that approximates a desired temporal pulse shape by using a fast iterative routine [252] that implements the Gerchberg-Saxon-Algorithm [253]. The full search space was reduced to an 11 or 23 parameter space for three or six subpulses, respectively, where the subpulse distances, intensities, constant phase differences, and linear chirps between were subject to optimization. Fig. 2.49 shows the SFG cross-correlation results for optimizing three (a) or six (b) subpulses for the ionization of NaK at 780 nm center wavelength [249]. The acquired pulse forms of three different runs are presented in (a) in order to test the robustness of the optimized pulse shape. Particular subpulse distances corresponding to $1.5T_{osc}$ in the excited $A(2)^1\Sigma^+$ state can be observed for all runs. However, the optimization factor increased from approximately 1.2 for three subpulses to about 1.5 for six subpulses, which comes closer to the value of 1.6 for free optimization. This indicates that apparently more than three subpulses are required for an efficient transfer to the ionic state, which is in concordance to the above given explanation in cooperation with theory for a free phase-only optimization at 770 nm center wavelength [231]. Hence, a repeated, stepwise excitation is likely to yield better results. Allowing linear chirps to the subpulses may further enhance the ion yields. The unrestricted optimization produced a comparable waveform except for one more subpulse at around 0.8 ps. The remaining difference between the parametric and the free optimization may further originate from the search space restriction since some pulse components were still not modulated in the parametric case. Yet, this example shows the potential of the introduced method to gradually gain knowledge about the molecular system.

Another possibility is to perform spectral parameterization, where narrow Gaussian amplitude distributions are subject to modulation in order to find certain vibronic transitions. Parameters are the distances and intensities of these distributions. The phase can thereby be modulated freely to allow unrestricted temporal pulse modulation, as realized here. For the NaK model system some involved transitions could be identified as to the first and second excited state (see Fig. 2.49c). This so called transition finder can generally be applied to search for those transitions in molecular systems which are utilized by the optimized path.

2.4.6.3 Combined phase, amplitude, and polarization shaping

In the pulse shaping experiments so far only the scalar properties of the electric field were optimized and the vectorial character, namely the polarization has been neglected. Since wave functions of quantum mechanical systems are three-dimensional objects, including the polarization should increase the degree of control tremendously. At first, the group of G. Gerber created laser pulses formed in phase and polarization by simply removing the polarizer of a double array pulse shaper that is commercially available [254]. This setup allows to change the major axes ratio of the ellipse of the electric field, but for a given frequency it is restricted in the orientation of the major axes, which are fixed along the horizontal and vertical axes.

Example pulses have shown the advantages of this setup and experiments like the ionization of K_2 [255], and the ionization of aligned I_2 [256] has demonstrated the importance of the polarization in these molecular excitation processes. Recently, the limitation in the polarization modulation was overcome by passing a double array modulator and a single array consecutively [257]. This setup presented by Y. Silberberg et al., is capable to change the major axes ratio and the orientation of the major axis independently so that it features full control over phase and polarization [257].

Both setups do not yet include the amplitude as the last parameter to modulate all properties of the electric field. Recently, two setups were introduced, which allow to modulate all three parameters phase, amplitude, and polarization independently [258, 259]. In this Section, the serial setup that is shown in Fig. 2.50 is used. It presents the first implementation of all three parameters phase, amplitude, and polarization in a closed loop experiment. The main idea of the setup is the use of four liquid crystal arrays one by one, which is accomplished within a symmetrical 4f-setup that the laser pulse passes twice. Then the arrangement is combined with a feedback loop and applied on a dimer system which offers the advantage that the dipole moments are only parallel or perpendicular. The NaK system is well suited for

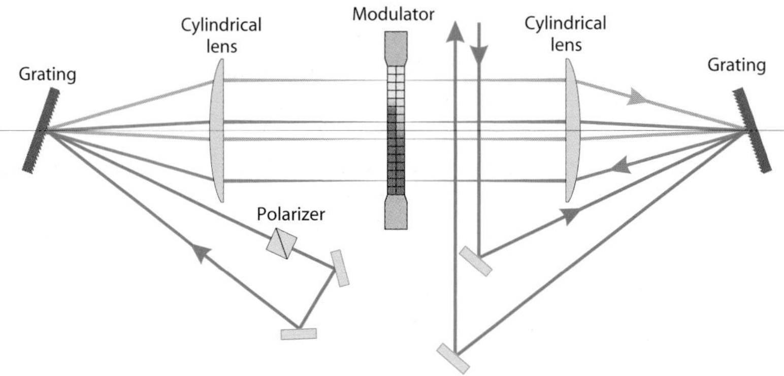

Fig. 2.50. The serial pulse shaper setup consists of a zero dispersion compressor which is passed two times. The incident laser pulse hits the first grating at a smaller incident angle, then the dispersed spectral components are focused by a cylindrical lens onto the lower half of the liquid crystal array by a combination of 800 grooves per mm of the grating and 250 mm as the focal length of the cylindrical lenses. After refocusing with the second cylindrical lens and recollimation by the second grating, this part of the modulator acts together with the polarizer as an amplitude filter. The laser pulse is redirected into the 4f setup under a larger incident angle, passes the grating and lenses and gets through the other part of the modulator. In this part of the double liquid crystal array, the phase and polarization are set and the outgoing pulse is modulated in phase, amplitude, and polarization.

this investigation since the energy potential curves are known and several optimizations were carried out, like the enhancement of multiphoton ionization [231] and isotope separation [260]. The optimizations were carried out with a central wavelength of 780 nm.

The employed pulse shaper is a zero dispersion compressor consisting of two gratings with 800 lines per mm and a pair of cylindric lenses of 250 mm focal length. The modulator has two layers of liquid crystal arrays with 640 pixels each. The setup is passed twice under different incident angles corresponding to different sectors of the modulator. In the first passage, the spectral amplitude is set by the cooperation of the polarizer and the retardances ϕ_a and ϕ_b of the first half of the double array. In the second pass, the phase and the polarization are modulated by the interplay of the retardances ϕ_c and ϕ_d of the other half of the modulator. The resulting electrical field can be written as:

$$E_{out}(\omega) = E_{in}(\omega) e^{\frac{i}{2}(\phi_a+\phi_b+\phi_c+\phi_d)} \cos\left(\frac{1}{2}(\phi_a - \phi_b)\right) \begin{pmatrix} \cos\left(\frac{1}{2}(\phi_c - \phi_d)\right) \\ i\sin\left(\frac{1}{2}(\phi_c - \phi_d)\right) \end{pmatrix} \tag{2.47}$$

As one can see, the control of the four retardances leads to a simultaneous and independent manipulation of the phase, the amplitude, and the polarization. For a difference retardance $\phi_c - \phi_d$ equal to zero the polarization stays linear in horizontal orientation, for $\pi/2$ it changes to circular polarization and for a difference of π, it is linear again but with vertical polarization. For intermediate values the electric field is elliptically polarized, whereas the helicity is determined by the sign of the difference retardances. The outcoming electrical field of one wavelength can generally be described as an ellipse with the major axes fixed in horizontal and vertical orientation.

In order to characterize the obtained pulses, they are measured with a modified SFG cross-correlation: A pivoted half wave plate turns the desired polarization component of the shaped pulse into the direction of the following horizontal polarizer and, thereby, the polarization component of interest is cut out. This takes the favored direction of the BBO-crystal into account and enables to measure every polarization direction equitable. By measuring the two orthogonal cross-correlations the temporal structure of the pulse can be measured. The corresponding spectral intensities are recorded behind a horizontal and vertical polarizer.

An evolutionary algorithm controls the parameter phase, amplitude, and polarization by setting the retardances for every pixel of the modulator during the optimizations. In this setup, the parameters are not coupled, so each component can be optimized independently. After having proven the functional capability of the serial setup within the closed loop experiment, optimizations on the $^{23}Na^{39}K^{+}$ ion yield were performed by measuring the current of the emerging ions. A phase and amplitude optimization as well as a phase, amplitude, and polarization optimization were performed. This allowed to compare the ionization efficiency and to reveal the effect of the additional polarization modulation. The learning curves of both optimizations are shown in Fig. 2.51a and depict a large enhancement compared with the ionization, in which an unshaped short pulse is used. Here one can observe the influence of the reduced search space in the higher convergence speed of the phase and amplitude optimization. The ratio obtained by the phase and amplitude optimization is 1.2 related to a short pulse. The optimization of phase, amplitude, and polarization are more efficient with a ratio of 1.8. Therefore, the additional change of the polarization within one pulse structure enhanced the ratio by another 50%.

The cross-correlation traces of the phase, amplitude, and polarization shaped pulse in Fig. 2.51b clearly shows the appearance of different polarization states within one pulse structure. The cross-correlation shows a subpulse structure with changing polarization states. As described in more detail before (see Sec. 2) the induced three-photon ionization process starts at the $A(2)^1\Sigma^+$ ground state. The population is then transferred to the first exited state which is also a $A(2)^1\Sigma^+$ state. Here the wave packet propagates until it is ionized by a resonant two photon process across the intermediate state $B(3)^1\Pi$ state. Since the dipole moment of the transition between two Σ states is orthogonal to the one of a Σ-Π transition, alternating polarization directions occur.

This demonstrates the vectorial properties of the scalar product $\boldsymbol{\mu}\cdot\boldsymbol{E}$ in the transition dipole moment operator and, thus, the orientation of the electric

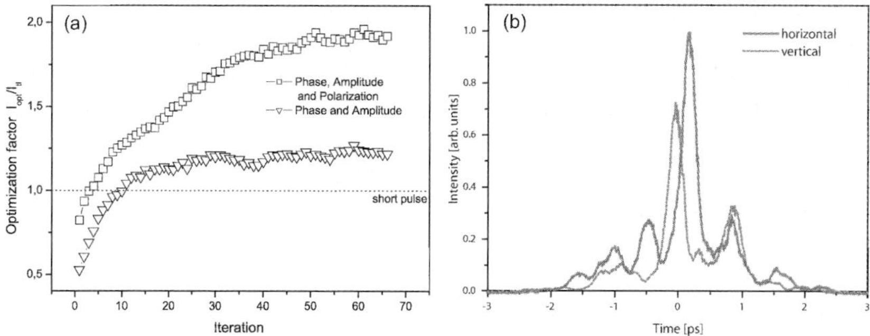

Fig. 2.51. (a) Evolution of the NaK^+ yield of each generation during the optimization, normalized to the one obtained with a transform limited short pulse (dashed line). The triangles represent the optimization of phase and amplitude, which reaches a optimization factor of 1.2. The boxes exhibits the optimization of all three parameters phase, amplitude, and polarization, ending up with a factor of 1.8. (b) Cross correlation traces of the optimized pulse shapes in horizontal and vertical orientation. The structure shows clear sub-pulses with alternating polarization states.

field respective to the dipole moment is very important. The conventional phase and amplitude shaped pulses neglect this vectorial character and only consider the Franck-Condon factors for the transition probability.

Further information arising from the cross-correlations is the distance of 660 fs between the first sub pulses. The value is consistent with the well-known oscillation time of the $A(2)^1\Sigma^+$ state achieved from pump-probe spectra [261] and the data obtained by optimizations in phase and amplitude [231]. So the algorithm enhances the ion yield by turning the polarization of each sub-pulse to the orientation parallel to the dipole moment of the transition. The distance remains constant and, therefore, the ionization process is accomplished by taking a similar path.

In conclusion, the addition of the third parameter, the polarization, improves the efficiency of the ionization compared to the phase and amplitude optimization. The analysis of the cross-correlations and the obtained spectra of the phase, amplitude, and polarization shaped pulse explains the higher optimization factors.

2.4.7 Controlling photodissociation processes

The first successful feedback control experiment of G. Gerber et al. [158] reported the optimization of distinct fragmentation channels vs. the yield of the mother ion signal of the unimolecular decay of $CpFe(CO)_2Cl$. The experimental conditions did not yet allow to decipher the underlying molecular dynamics process from the obtained optimum pulse shapes. To obtain this, experiments on simple systems were needed, which operated at significantly lower laser powers in the weak field regime, so undesired multiphotonic transitions and deformations of the potential energy surfaces could be avoided. Pump-probe experiments on Na_2K indicated this system as a suitable candidate for an optimum control experiment. As shown in Fig. 2.10 the pump-probe spectrum exhibits - superimposed on an exponential decay with a time constant of 3.28 ps - an oscillatory behavior with a period of roughly 500 fs. The exponential rise time of 3.25 ps of the fragment NaK indicates its origin from the equally fast decaying Na_2K. In the control experiment the one color pump-probe scheme is replaced by tailor-made pulse shapes optimized in suitable feedback loops. Very important in this regard is the information content, which can be acquired from this optimization process. In the case of optimizing the mother ion Na_2K^+, a double-pulse sequence is retrieved from the analysis of the autocorrelation trace [27]. The time difference between the first and second pulse is approximately 1240 fs which corresponds to 2.5 oscillation periods of the electronically excited trimer. It is known from pump-probe spectroscopy, that the photoionization of the electronically excited Na_2K molecule occurs most efficiently at this time, when the wave packet is at the outer turning point.

The channel for optimizing the NaK^+ fragment of Na_2K reaches an optimization factor of $I_{opt}/I_{tl} = 1.8$. In Fig. 2.52a and b the optimized pulse is shown in its cross correlation and SFG-XFROG representation, respectively. The pulse structure is a pulse train consisting of three main pulses with a most intense middle pulse. The time delays between the first and second pulse and between the second and third pulse are $\Delta t_{1,2} \approx 660$ fs and $\Delta t_{2,3} \approx 440$ fs,

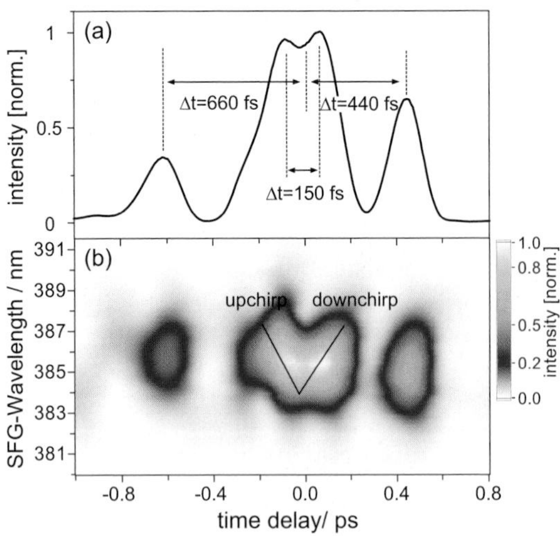

Fig. 2.52. Phase optimized control pulse optimizing the NaK^+ production in the presence of alkali trimers. (a): Cross correlation trace. The pulse train consists of three main pulses with time distances of $\Delta t_{1,2} \approx 660$ fs and $\Delta t_{2,3} \approx 440$ fs. (b) The SFG-XFROG trace reveals a substructure of the central pulse. [226]

respectively. The intensity of the central pulse is almost three times as high as that of the first pulse and twice as high as the third one leading to a sub pulse intensity ratio of 1:2.9:1.8.

With the SFG-XFROG trace in Fig. 2.52b the time evolution of the frequency components of the pulses is displayed. The central pulse reveals a pronounced positive and negative chirped twin substructure. On the red side of the spectrum the pulse is divided into two sub pulses ($\Delta t_{2a,2b} = 150$ fs) whose intensity maxima converge in a V-type shape toward shorter wavelengths. Thus, a positive chirp is followed by a negative one. The main intensity on the short wavelength side of the middle pulse marks the center of the pulse. The above mentioned pulse distances refer to this maximum. The structure of the optimized pulse form (Fig. 2.52) exhibits features of the NaK dynamics which are similar to the already discussed control pulse (Fig. 2.35) for the maximization of the NaK ionization. This is evident, because the supersonic molecular beam contains both: NaK dimers and Na_2K trimers, which cannot be distinguished mass-spectroscopically, when the trimer fragmentizes to NaK+Na. Therefore both types of NaK particles contribute to the signal: Photoionized NaK and photodissociated Na_2K. The central, most intense pulse reveals therefore a substructure which does not only correspond to the NaK dynamics, it also indicates a contribution of the fragmentation of Na_2K. This substructure can be explained by a process that circumvents the predissociative curve crossing and therefore finds a direct exit channel (see Fig. 2.53b and c): The leading red shifted part stimulates the population of the trimer down to a vibrationally excited state of the electronic ground state where due to conservation of momentum the wave packet moves in the same direction (Fig. 2.53c). The red shifted trailing part of the central pulse then excites the ground state wave packet, when it is just located beyond the curve crossing onto the repulsive potential curve of the trimer. Afterwards the third pulse ionizes the NaK fragments produced from the photodissociated Na_2K. Since this last pulse arrives after another full round trip, it can also effectively ionize the still remaining wave packet in the non-fragmented NaK, which were not yet ionized - as shown in Fig. 2.53b - by the central part of the second subpulse. Thus, also in this case the third pulse improves the ionization efficiency.

The rise in the NaK^+ yield of about 20% compared to the optimization of the pure ionization of NaK corresponds to the gain in NaK^+ determined in the pump-probe spectrum. The process delivers a convincing example for - sometimes - counterintuitive, but more efficient dynamical pathways, which can be identified by analyzing optimized pulse shapes.

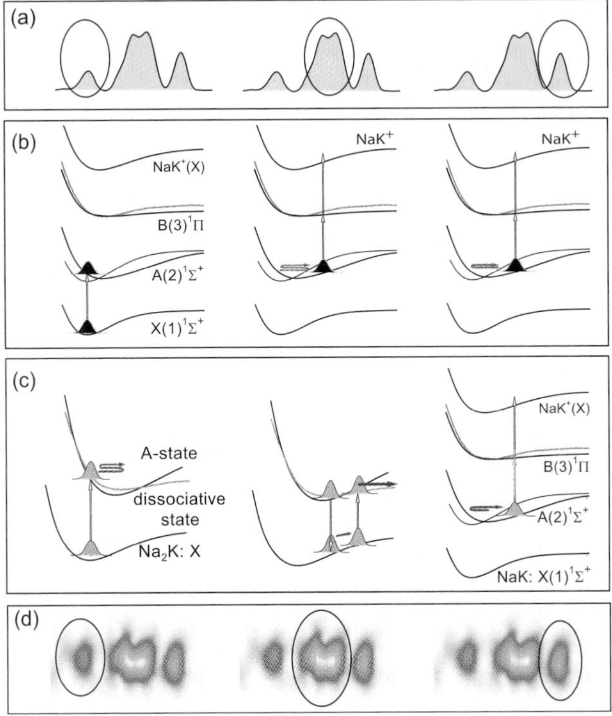

Fig. 2.53. Optimized multiphoton induced ionization and fragmentation pathways into NaK. (a): Temporal profile of the control pulse. The active subpulses are indicated. (b): Three step multi photon ionization scheme of NaK. (c): Schematic of the multi photon fragmentation of the Na_2K trimer into NaK fragments by circumventing the predissociative curve crossing. (d): SFG-XFROG traces. The active subpulses are indicated. [226]

2.5 Optimal control on ultracold molecules

C. P. Koch, A. Lindinger, V. Bonačić-Koutecký, R. Mitrić, and L. Wöste

The field of cold molecules has emerged over the last few years with many research groups working at the production of dense samples of cold and ultracold molecules [262]. Translational temperatures in these gases are below 1 Kelvin or 1 micro-Kelvin, respectively. This intense activity is motivated by many possible applications of (ultra)cold molecules which range from high precision measurements, allowing for example for the determination of a possible dipole moment of the electron as a check of the standard model of elementary particles, to the realization of a molecule laser by Bose-Einstein condensed molecular gases, or to the emergence of a new ultracold chemistry [262]. The lowest temperatures to date have not been achieved by direct cooling of the translational degrees of freedom of the molecules [263], but by first cooling atoms and then assembling these atoms into molecules [264,265]. To this end, an external field, either magnetic or optical, is applied.

Photoassociation offers the advantage of relying on optical transitions which are generally abundant. It is defined as the absorption of a photon red-detuned from the atomic line by a pair of colliding ground state atoms, creating a weakly-bound molecule in an electronically excited state, cf. Fig. 2.54 [266,267]. Series of resonances have been observed in experiments using continuous-wave (cw) lasers [266,267] which occur when the laser detuning matches the binding energy of an excited state vibrational level. In particular, excited state potentials of homonuclear dimers provide for long range molecular levels due to their scaling as $1/R^3$ at large internuclear distances R. These highly excited levels can be efficiently populated in free-bound transitions.

In order to obtain molecules in the singlet ground or lowest triplet state, the photoassociation step must be followed by a stabilization step, i.e. by a bound-bound transition. In cw photoassociation, the excited state molecules are short-lived and decay via spontaneous emission, most often through a vertical transition at large distances R, giving back a pair of atoms. However, specific mechanisms have been identified which favor radiative decay at shorter distances and allow for the creation of ground state molecules [267,268]. Stable molecules formed by spontaneous emission have been detected in a magneto-optical trap (MOT) for a number of species (e.g. [264,269,270]). These molecules are translationally ultracold, but internally highly excited.

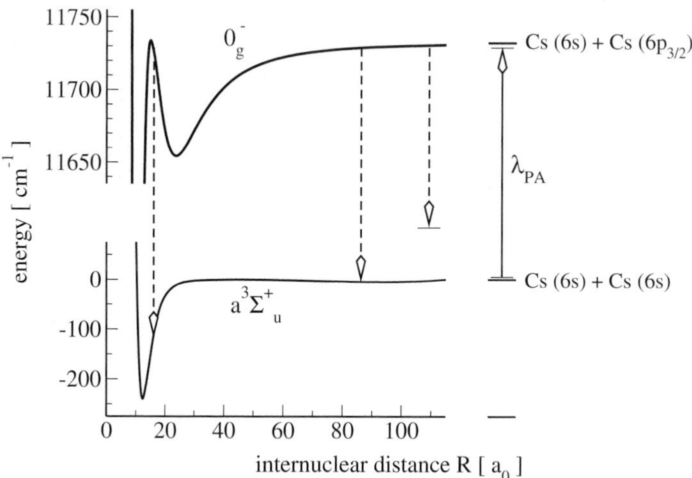

Fig. 2.54. Photoassociation: Example of two Cs atoms colliding over the lowest triplet state excited to the 0_g^- state by a cw laser λ_{PA} and ground state molecule formation via spontaneous emission (the two dashed arrows at short distances).

At this point, two possible routes for control are conceivable: (i) optimization of a *coherent* formation scheme of ground state molecules avoiding spontaneous emission, and (ii) stabilization or cooling of the internal degrees of freedom after ground state molecule formation. Both routes have been explored theoretically (cf. [271, 272] and [273]), and experimental effort toward optimal photoassociation is under way [274, 275].

2.5.1 Experimental realization of optimized molecular depletion in a magneto-optical trap (MOT)

Here, first experiments on the optimal control of excitation processes of translationally ultracold but vibrationally still very hot rubidium dimers inside a magneto-optical are presented. In order to optimize the photoexcitation of the Rb_2 molecules the employed shaped femtosecond laser pulses were optimized in feedback loops by genetic algorithms based on evolutionary strategies. As a result the irradiated molecules undergo ionization or fragmentation. The approach represents a first experimental steps toward pulsed photoassociation.

The experimental setup is shown in Fig. 2.55. About 10^7 ^{85}Rb atoms are captured in the MOT at a density of 10^{10} atoms/cm^3 and temperatures of 100 μK. In this environment diatomic rubidium molecules continuously form due to either three body collisions or light assisted two-body

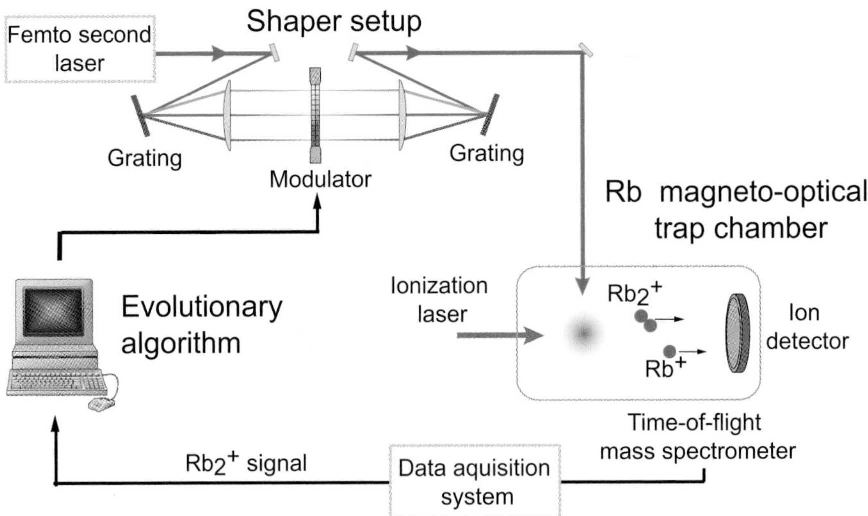

Fig. 2.55. Experimental setup for iterative closed-loop maximization of ultracold Rb_2 excitation from the ground electronic singlet or triplet states by shaped femtosecond laser pulses. The ultracold molecules are formed in a magneto-optically trapped gas of rubidium atoms. Their abundance is evaluated by monitoring the RTPI-signal of Rb_2^+.

collisions of trapped Rb atoms. They populate predominantly the highest vibrational states below the dissociation limit in the triplet electronic ground state [276, 277]. These Rb_2 molecules, which are no longer trapped by the MOT, are then detected via resonant two photon ionization (RTPI) and time-of-flight mass analysis. The RTPI laser operates at 15 Hz between 600 and 610 nm and at a pulse energy of 20 mJ. In the steady state of molecule formation and loss a maximum count rate of 0.5 Rb_2^+ molecular ions per laser pulse is observed.

For the experiments the fs-laser is tuned in the range between 780 and 820 nm. To study the laser pulse interaction with rubidium molecules, the atomic resonance components were removed from the pulse spectrum by a notch filter, realized by a physical block in the shaper's Fourier plane. In this way atomic losses from the MOT could be reduced below the detection threshold.

The rubidium dimers interact with the fs-laser pulses over the entire accessible range of central wavelengths from 780 nm to 820 nm. As shown in Fig. 2.56, the molecular signal decreases rapidly at small pulse energies and levels off to 25% at a pulse energy of 0.6 nJ. As only molecules in the electronic ground state are detected, the signal reduction can be attributed to excitation by the fs-pulses. The process can be modeled by a simple rate equation for the number of ground state molecules:

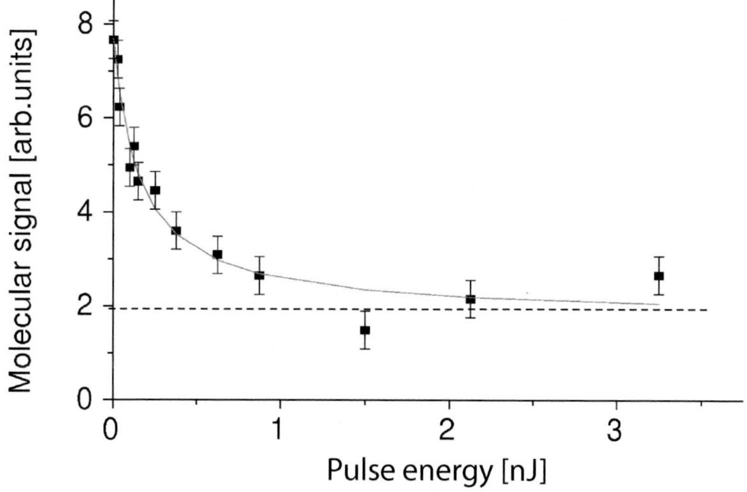

Fig. 2.56. Reduction of the Rb_2^+ molecular ion signal as function of transform limited femtosecond pulse energy. The pulses have a central wavelength of 800 nm and 10 nm FWHM. The D_1 atomic resonance at 795 nm is filtered out of the pulse. The RTPI laser is set to 602.6 nm. [274]

$$\frac{dN_{Mol}}{dt} = -R_{fs}N_{Mol} - R_{loss}N_{Mol} + R_{PA}N_{At}^2 \tag{2.48}$$

where R_{fs} is the excitation rate by the fs-laser, R_{loss} the molecular loss rate from the detection volume and R_{PA} the production rate of molecules from trapped atoms. In the steady state where $\frac{dN_{Mol}}{dt} = 0$ and assuming that R_{fs} is proportional to the fs-laser intensity the dependence of N_{Mol} on this intensity is of anti-proportional character:

$$N_{Mol} \sim \frac{R_{PA}N_{At}^2}{R_{loss} + \alpha I_{fs}} \tag{2.49}$$

where α is a proportionality constant. The curve in Fig. 2.56 represents a fit to the data based on this model. A model assuming a quadratic or higher order dependence of the excitation rate on intensity could not fit the data. This indicates that, in this regime of pulse energies, the interaction with the molecules has the character of an effective one-photon excitation [278].

According to [276], the molecules in the MOT initially populate the highest levels in the $a^3\Sigma_u^+$ state. Due to selection rules and Franck-Condon factors, they are preferably excited to the 0_g^- and 1_g $5s5p_{1/2}$ states (see Fig. 2.57). Emission back to the electronic ground state could only lead to signal reduction if the vibrational level population moves out of the excitation window of the ionization laser. However, scans of the detection laser with and without fs-beam show reduced but qualitatively similar spectra which should not

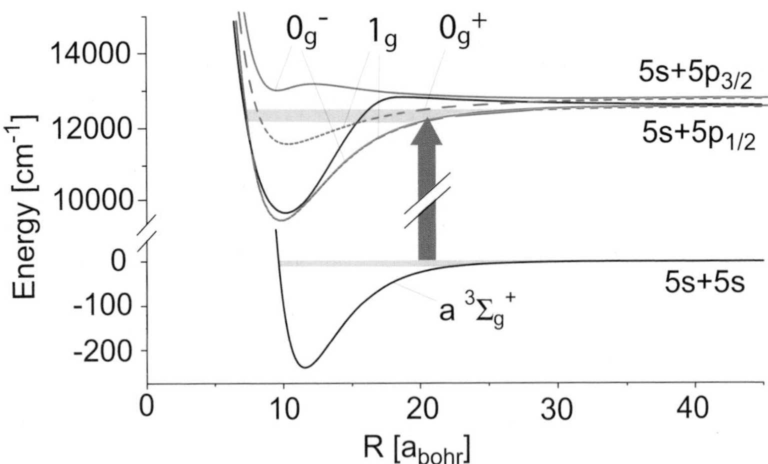

Fig. 2.57. Potential curves of the rubidium dimer including spin-orbit interaction. Initially, the molecules are expected to populate the highest levels in the $a^3\Sigma_u^+$ state [276, 277]. The molecular excitation by the femtosecond pulses is indicated by the arrow. The shaded areas show the initial and final distribution of molecular vibrational states. [274]

be the case for a vibrational redistribution. Instead it can be expected that excited molecules absorb further photons, so the whole process of molecular loss can be regarded as a resonance enhanced multiphoton excitation, followed by dissociation, predissociation or ionization. This happens either within one pulse, or, as the laser repetition rate is comparable to the lifetime of the first excited state, it occurs in the subsequent pulse. At high energies all molecules in the laser focus are excited or dissociated and the residual signal in Fig. 2.56 is due to molecules which did not interact with the femtosecond laser, indicated by the dashed line. This shows that most of the molecules are produced within a small volume inside the MOT which is consistent with the picture that they form at the MOT center where the atom number density is at its maximum [279].

In order to demonstrate the practical applicability of coherent control concepts to ultracold molecules, the Rb_2^+ signal acts as an input for the self-learning optimization algorithm which autonomously programs the pulse shaper in a closed loop experiment. The algorithm is based on evolutionary strategies, it is described in detail in [280]. Due to the small molecular ion count rate the signal is averaged over 128 RTPI laser pulses for each individual of the algorithm. In order to reduce the search space for the learning algorithm a mixed scheme of parametric amplitude and free phase optimization was chosen. So the algorithm tries to find the optimal pulse shape under the restriction that only a few sharp spectral peaks contribute to the pulse shape. During an optimization the parameters of these peaks, their spectral positions and amplitudes, are altered to find the best fitting excitation pulse. Moreover, the phase was optimized freely in order to allow a temporal modulation of the pulse.

The goal of the adaptive algorithm was to minimize the molecular RTPI signal. For each iteration the ion signals corresponding to the best and worst individuals are recorded together with the mean fitness of the whole generation. As depicted in Fig. 2.58a, all three signals decrease during the particulate optimization to about 70% of the initial value after 20 iterations. The spectra of the final best individuals of two successive runs shown in Fig. 2.58b display several peaks which coincide in some but not all spectral positions. The frequency span of the fs-pulse supports an assignment of the excitation to the 0_g^- and 1_g $5s5p_{1/2}$ states (see Fig. 2.57). By comparing the excitation yield of the best individuals with transform-limited pulses of the same energy it is observed that the optimized pulse excites the molecules on average 25% more efficiently, which demonstrates the feasibility and potential of adaptive control.

The observed excitation enhancement can be attributed to an increased spectral intensity at particular molecular resonances found by the evolutionary algorithm. Starting from a narrow band in the $a^3\Sigma_u^+$ state [276,277], molecules are excited into bound states below the D_1 resonance. By shifting the peak positions, the algorithm finds transition frequencies from this band to certain vibrational states, thereby sharing the pulse energy more efficiently than a

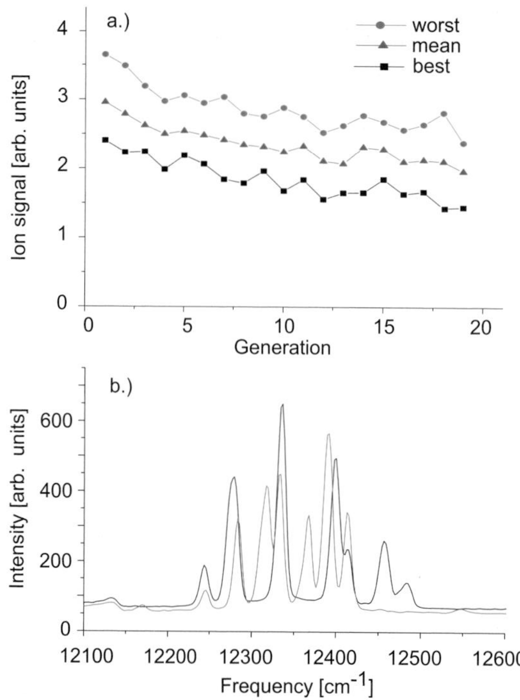

Fig. 2.58. (a) Molecular ion signal resulting from the best, the worst and the mean individual of the population for each generation during a closed loop experiment. (b) Femtosecond laser pulse spectrum of the final best individuals of two successive optimization runs under equal conditions with similar final optimization result. [274]

broad Gaussian pulse. The algorithm therefore has a large number of possible solutions to choose from and so the final pulse shapes after the optimization are not identical. In the spectral region between 12000 and 12500 cm^{-1}, the vibrational level separation is about 10 cm^{-1} in the 0_g^- and 1_g $5s5p_{1/2}$ states, respectively. The high density of states also explains the limited potential of the optimization because the optimization factor depends on the chosen peak-width which is limited by the shaper resolution. The Franck-Condon factors may also be relevant for the excitation process since they differ for different vibronic transitions and favor particular frequencies which are enhanced in the experimentally acquired spectra. Yet, as the initial ground state population distribution in the vibrational states is not accurately known, no quantitative treatment or assignment can be made yet.

2.5.2 Theoretical proposals for ground state molecule formation via short-pulse photoassociation

Photoassociation with short laser pulses has first been discussed by Machholm et al. [281] for Na_2. Considering the timescales of the ultracold collisions, the excited state vibrational dynamics and spontaneous emission, the use of pulses of picosecond duration was suggested. Such pulses allow for small detunings, i.e. for excitation at long range where the free-bound transition matrix elements become large. Due to the high density of excited state vibrational levels at small detuning, the spectral width of a picosecond pulse comprises several vibrational levels, such that a spatially localized wave packet at large R is formed. Subsequentely, Korolkov et al. [282, 283] and Backhaus et al. [284] investigated photoassociation of OH and HCl, respectively.

Vala et al. considered photoassociation of ultracold atoms with *chirped* picosecond pulses in order to achieve adiabatic transfer, and hence complete population inversion, for a range of internuclear distances R [285]. This has led to the notion of the photoassociation window [286]: The resonance condition for the carrier frequency determines the Condon point R_C, and the finite spectral width of the pulse is transferred into a finite spatial range around R_C. Within the photoassociation window, adiabatic population transfer is possible. Luc-Koenig et al. realized that in addition to improving the excitation rate, chirping a pulse offers the possibility of wave packet shaping since the sign of the chirp controls the wave packet dispersion [286,287]. For a positive chirp, small frequencies precede larger ones. Therefore vibrational levels with larger binding energy and smaller vibrational period are excited before those with small binding energy and long vibrational period, i.e. the wave packet dispersion is increased as compared to a transform-limited pulse. For a negative chirp, the opposite is the case, and the wave packet dispersion can be minimized. In particular, the strength of the chirp can be chosen such that the wave packet becomes spatially focused at the inner turning point of its vibrational motion. The optimal chirp is easily obtained in terms of the vibrational energies of the levels which are resonantly excited within the bandwidth of the pulse [287].

The spatially focused excited state wave packet which is obtained by a negative chirp represents an ideal intermediate in a coherent two-color pump-dump process to form ground state molecules [271]. The scheme is depicted for photoassociating two cesium atoms colliding via the lowest triplet potential to the $0_g^-(P_{3/2})$ excited state in Fig. 2.59. In this example, the central frequency is detuned by $2.7\,\mathrm{cm}^{-1}$ from the atomic resonance leading to a Condon point of $R_C \sim 90\,a_0$. The spectral bandwidth of this pulse is about one wavenumber corresponding to a transform-limited full-width at half-maximum (FWHM) of 15 ps. The photoassociation pulse, depicted in blue, depletes the ground state population within the photoassociation window, cf. the initial scattering wave function (red solid line) and the ground state wave function after the pulse (blue dashed line). It creates a time-dependent wave packet in the excited

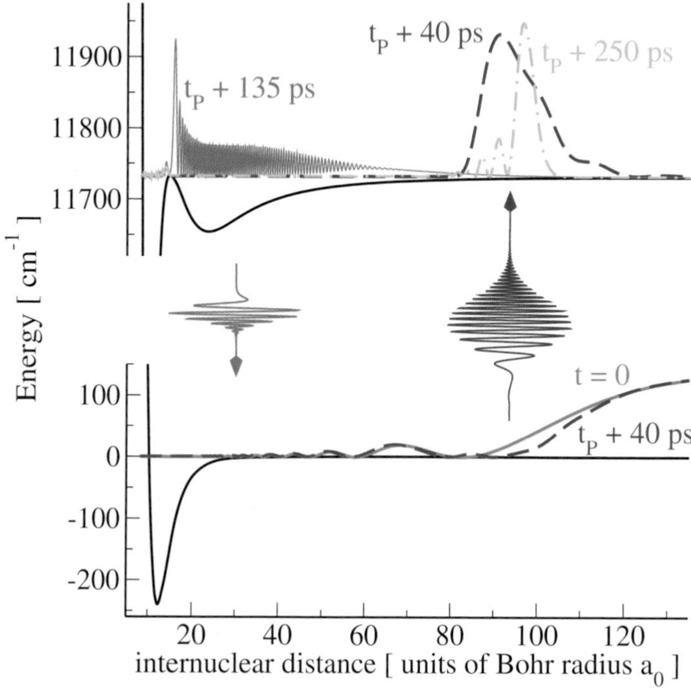

Fig. 2.59. Ground state molecule formation in a coherent pump-dump scheme for the example of photoassociation to the Cs_2 $0_g^-(P_{3/2})$ state, t_P denotes the time at which the photoassociation pulse has maximum amplitude (reprinted from [271]).

state which consists of about 15 vibrational levels and which moves under the influence of the excited state potential. For optimally chosen negative chirp, a spatially focused wave packet at the inner turning point (thin green line) is obtained after half a vibrational period. After a full vibrational period, the wave packet is found at its outer turning point, the position where it was created (yellow dot-dashed line). The dynamics of the excited state wave packet can be employed to optimize ground state molecule formation by a dump pulse, which is applied delayed in time after the photoassociation pulse (green pulse in Fig. 2.59): The transition matrix elements $\langle\varphi_v^g|\mu|\Psi(t)\rangle$ between the excited state wave packet $|\Psi(t)\rangle$ and all bound ground state levels $|\varphi_v^g\rangle$ show a pronounced maximum for a pump-dump delay of half a vibrational period, i.e. when the wave packet is focused at its inner turning point [271].

The $0_g^-(P_{3/2})$ state chosen for this example is particularly favorable. It results from a mixing of the $^3\Pi_g$ and $^3\Sigma_g^+$ states due to spin-orbit coupling. At long range, the $^3\Pi_g$ ($^3\Sigma_g^+$) potential goes as $+C_3/R^3$ ($-2C_3/R^3$). Due to the combination of different signs and different weights, a purely long-range well appears in the $0_g^-(P_{3/2})$ potential. For Cs_2 this leads to a double-well structure, while for the other homonuclear alkali dimers the inner well occurs at energies far above the dissociation limit and only the long-range outer well is bound. The inner part of this long-range well is softly repulsive ($-1/R^3$) as compared to the hard repulsive walls of all other potentials. Therefore the wave packet motion is slowed down such that the wave packet spends a considerable amount of time at the inner turning point (about 5 ps in the example of Cs_2). This time window determines the resolution of the pump-dump delay which is required in order to hit the wave packet at the optimal moment, when it is focused at the inner turning point.

The specific case of cesium offers an additional advantage: The inner part of the $0_g^-(P_{3/2})$ long-range well is located at fairly short distances ($R \sim 15\,a_0$). The dump pulse therefore induces a vertical transition into levels with considerable binding energies ($\sim 100\,\mathrm{cm}^{-1}$, see also Fig. 2.54) [271]. In the case of Rb_2 $0_g^-(P_{3/2})$, much smaller binding energies of about $5\,\mathrm{cm}^{-1}$ can be expected.

An alternative route toward formation of ground state molecules is provided by the mechanism of resonant spin-orbit coupling [268, 288] which occurs for example in the $0_u^+(P_{1/2})$ state of rubidium and cesium but also in heteronuclear alkali dimers [270]. The $0_u^+(P_{1/2})$ state results from the mixing of the $A^1\Sigma_u^+$ and $b^3\Pi_u$ states and shows an avoided crossing with the $0_u^+(P_{3/2})$ state at short internuclear distance ($R \sim 10\,a_0$ in the case of Rb_2), i.e. the spin-orbit coupling modifies the potentials, and hence the vibrational spectrum over a large range of energies and not only close to the dissociation limit. Consequently, resonantly perturbed vibrational wave functions are observed which exhibit large amplitude at the outer turning points of both diabatic potentials. They are compared to almost regular vibrational wave functions which also occur in the spectrum shown in Fig. 2.60. The outer peak of the resonantly perturbed wave functions allows for large free-bound transition matrix elements and hence an efficient photoassociation step while the inner peak facilitates stabilization to bound levels of the ground state. It is therefore expected to provide a suitable route for a pump-dump scheme [272]. In such a scenario, the photoassociation (pump) pulse should be red-detuned with respect to the D_1 line, i.e. the $nS + nP_{1/2}$ dissociation limit. The detuning should be as small as possible to take advantage of the large free-bound transition dipole matrix elements close to the atomic resonance. On the other hand, the pulse should not contain frequencies exciting the atomic resonance. For photoassociation of Rb_2 and transform-limited pulses with FWHM of a few picoseconds, the optimum detuning was identified to be 4-10 cm^{-1}. The dump pulse will then yield *singlet* ground state molecules with binding energies on the order of $10\,\mathrm{cm}^{-1}$ [272] providing an excellent starting point for a further stabilization to obtain molecules in their absolute rovibronic ground state.

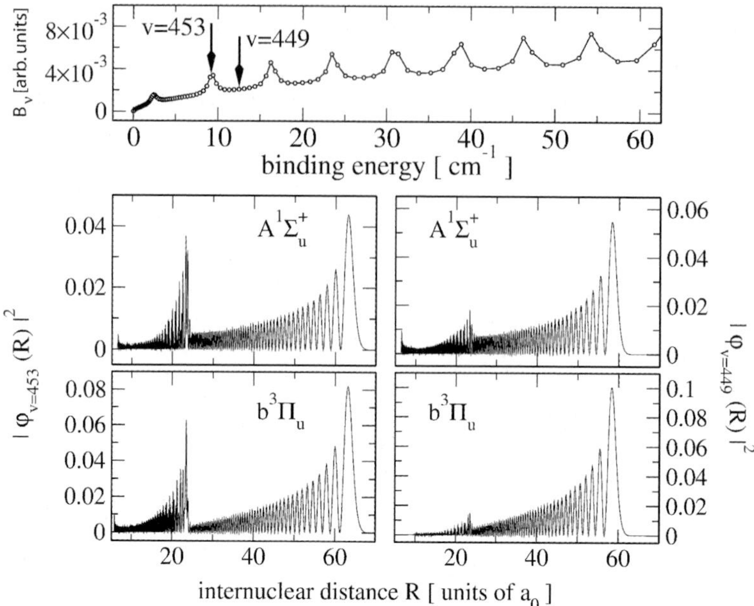

Fig. 2.60. Vibrational wave functions of the 0_u^+ states: resonantly coupled (left) and regular (right).

2.5.2.1 Molecular wave packet localization by pump-probe experiments in a MOT

A further step toward the synthesis of translationally and vibrationally ultracold molecules was achieved by pump-probe experiments in which pump wavelengths of transform-limited fs-pulses around 800 nm were used, while the system was probed with corresponding pulses around 500 nm. The results show clear indications of wave packet oscillations probably in the first excited state of rubidium dimers, with oscillation periods depending on the red detuning from the atomic D_1-line, as it will be published soon. These wave packet oscillations occur in particular excited state potential energy curves and are associated with certain nuclear distances where the initial excitation takes place. This first evidence for wave packet localization in a MOT raises hopes for the control of these wave packets with shaped laser pulses, namely to cool the system down. By using linear chirps for the pump pulse a modified response of the system was observed. The next evident step will be the application of optimally shaped laser pulse sequences aiming toward the maximum production of Rb_2 molecules by photoassociation. Afterwards the same pulse sequence will cause internal vibrational cooling cycles by photoinduced population transfer via an intermediate state. This requires high atom densities, which can be achieved by a DarkSPOT. Further relatively long laser pulse sequences in the ps range combined with a high spectral resolution will be required to allow the efficient excitation of long range potentials.

2.5.3 Theoretical aspects of vibrational stabilization and internal cooling

Vibrational cooling starts from an ensemble of molecules in different vibrational states. It requires a dissipation mechanism to dispose of energy *and* entropy. If the molecules have been formed via a magnetic Feshbach resonance or by photoassociation pulses with narrow bandwidth, a single vibrational level is populated. In this case, it is not necessary to *cool* the internal degree of freedom, but a coherent state-to-state transfer to the vibrational ground state is sufficient to stabilize the molecules. The energy of the molecule is carried away by the light, while the entropy remains unchanged. This is not just a technical point: Coherent stabilization is much easier to achieve than true cooling, since there are only a few dissipation mechanisms available in a dilute gas, and the overall cooling rate is usually limited by the time scale of dissipation [289–291].

The same consideration holds if the molecules are produced by photoassociation pulses with broad bandwidth, provided the formation is followed quickly by the stabilization step. In this case, the molecule can still be described by a coherent wave packet which serves as initial condition for the stabilization.

Vibrational stabilization of ultracold molecules using optimal control theory was first investigated for the sodium dimer, cf. Fig. 2.61 [273]. The starting point was one of the last bound levels of the Na_2 singlet ground state, and

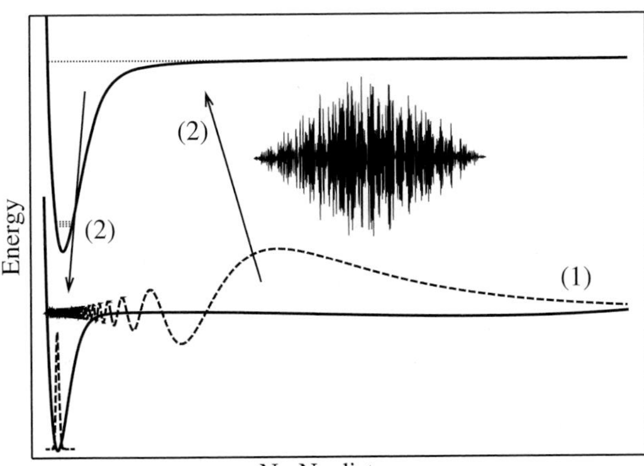

Fig. 2.61. Vibrational stabilization to transfer translationally cold, but internally highly excited molecules to their vibronic ground state using short, shaped laser pulses (reprinted from [273]).

the vibrational ground state was reached by many Raman-type transitions through the $A^1\Sigma_u^+$ first excited state. No direct Franck-Condon route exists since the overlap matrix elements of the initial and the target state with the vibrational levels of the first excited state occur in clearly disjoint spectral regions (the position of these excited state levels is indicated by lines in Fig. 2.61). It is therefore not possible to 'guess' pulses yielding a satisfactory transfer efficiency, and optimal control theory becomes imperative.

The Krotov method with the constraint to minimize the pulse energy was utilized [273]. Solutions with transfer efficiency close to one were found for the whole range of the vibrational spectrum. However, for very highly excited vibrational levels fairly intense pulses were required, and the spectrum of the optimal pulses turned out to be extremely broad. A restriction of the spectral bandwidth could not be found within the currently available algorithms. Both a constraint in time-domain to restrict the bandwidth to that of a reference field as well as filtering in frequency domain similar in spirit to [292] failed to efficiently confine the spectrum [273]. It was concluded that a new, joint time-frequency approach is necessary to impose constraints on the spectrum.

2.5.3.1 Photostabilization of ultracold Rb_2 molecule by optimal control theory

The aim of this Section is to establish conditions under which the efficient population of the ground state vibrational level $v = 0$ can be achieved and conditions under which this can be experimentally realized. So the application of full quantum optimal control to the photostabilization of the Rb_2 molecule is presented, which includes the ground electronic state ($1\ 0_g^+(I)(^1\Sigma_g^+)$ state) and two excited states ($1\ 0_u^+(I)(^1\Sigma_u^+)$ and $0_u^+(II)$ $(1\ ^3\Pi_u)$) based on accurate potential energy curves and accounting for the spin–orbit coupling.

Since primarily the stabilization and not the photoassociation process itself is addressed, a highly vibrationally excited bound state slightly below the dissociation limit in the ground electronic $1\ 0_g^+(I)(1\ ^1\Sigma_g^+)$ state is chosen as a starting point. The photoassociation starting from the unbound scattering states at large distances has already been studied theoretically [272].

It is demonstrated that the fully optimized laser pulse drives the system into the ground vibrational state of Rb_2 with nearly 100 percent efficiency. However, the spectral range of the optimized pulse is rather large. Therefore, different constraints are introduced and their efficiency in the context of possible experimental realization is examined.

Potential energy curves for Rb_2 have been previously reported on different levels of theory [293–295]. Both all-electron relativistic studies [295] as well as effective core potential calculations in the framework of the multireference configuration interaction (MRCI) methods [293, 294] were able to produce accurate potential energy curves and spectroscopic constants which are in good agreement with the experimental values.

Since the aim is to carry out quantum dynamical simulations and optimal control of photostablization, in addition to potential energy curves, also the transition dipole matrix elements between involved electronic states are needed. Therefore, a new set of data was produced employing the multireference configuration interaction method (MRCI) with two active valence electrons. This approach is computationally less demanding than those reported in [293–295] and provides sufficiently accurate results. The lowest electronic states of the Rb_2 molecule were considered which dissociate into the 5^2S+5^2S or 5^2S+5^2P atomic states. Due to considerably large spin–orbit splitting between the $5^2P_{1/2}$ and $5^2P_{3/2}$ states of the Rb atom ($\Delta E_{SO} = 237$ cm^{-1}) it is mandatory to include spin–orbit effects in the calculations. Therefore, a [9s9p6d1f] uncontracted atomic basis set [293] was employed together with a nine electron relativistic effective core potential including the spin–orbit operator ($9e^-$–RECP–SO) [296] and the Stuttgart group core polarization potential (CPP). The eigenfunctions of the total electronic Hamiltonian including spin–orbit coupling were obtained by diagonalizing the matrix representation of the $\hat{H}_{el} + \hat{H}_{SO}$ operator. For this purpose, all singlet and triplet electronic states which dissociate into the 5s + 5s and 5s + 5p atomic states are taken into account. These eigenfunctions have been subsequently used to calculate the transition dipole moment matrix elements between the involved states.

The wave packet propagation has been carried out by numerical solution of the time-dependent Schrödinger equation using the Chebychev expansion of the time evolution operator [297]. Optimal control of photostabilization has been performed in the framework of the Kosloff–Rice–Tannor optimal control scheme [17]. Both have been outlined previously.

The aim of the optimal control is to maximize the population of the vibrational ground state ν=0 and therefore to define the target functional at a specified final time t_f as the projector operator for the desired vibrational state.

Quantum dynamical simulations were performed in the manifold of three electronic states which are eigenstates of the spin–orbit Hamiltonian. The are labeled by $0_g^+(I)$, $0_u^+(I)$ and $0_u^+(II)$ and are shown in Fig. 2.62. Transition dipole moments between the ground state ($0_g^+(I)$) and two excited states ($0_u^+(I)$ and $0_u^+(II)$) are shown in the insert of Fig. 2.62. For large distances transition dipole moments converge to the values for the isolated Rb atom. However, at the distance of about ~ 5 Å abrupt change of the transition dipole moments occurs due to the change of the spin character of involved states at the avoided crossing. For smaller distances the $0_u^+(I)$ and $0_u^+(II)$ states have dominantly triplet and singlet character, respectively. At the avoided crossing, the character of the states is reversed which is then reflected in the reversal of transition dipole moments values.

The optimal control of photostabilization has been performed starting from a bound vibrational state (ν=130) which is about 5 cm^{-1} below the dissociation limit of the ground state ($0_g^+(I)$). This state has been selected since it has an outer maximum at 20 Å which lies already in a flat part of the

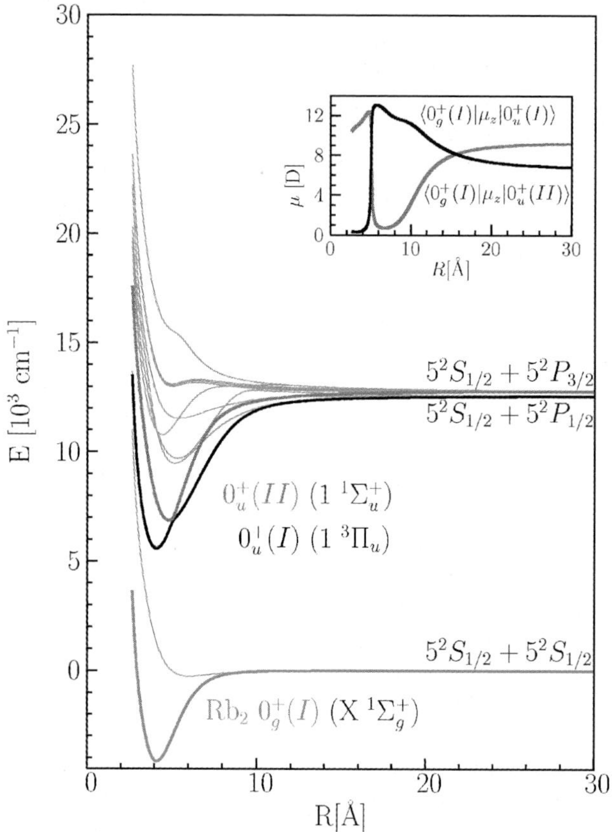

Fig. 2.62. Calculated potential energy curves for electronic Ω states of Rb_2. The states which are included in the quantum dynamics simulation are shown in red, blue and black lines. The insert shows the variation of the transition dipole moment between the ground state $0_g^+(I)$ and the $(0_u^+(I)$ and $0_u^+(II))$ states. The region around the avoided crossing between the $(0_u^+(I)$ and $0_u^+(II))$ at which the spin character of the states changes is marked by a lens.

potential and thus mimics the state created by the photoassociation process. Since the aim is to populate the (ν=0) state, the target operator is defined as the projection operator on the vibrational ground state. The time interval for the optimization is 15 ps since further reduction of the optimization time leads to significantly lower populations of the target state. Therefore it can be expected that the experimental realization will also require shaped pulses in a similar temporal range.

It should be pointed out that for successful optimal control, using electric fields of moderate intensity as it is the case in the experimental setups, the initial state and the target state must have non-vanishing Franck-Condon factors with the same group of vibrational states in the electronically excited states. If this is fulfilled the "simple" single cycle pump-dump control of photostabilization can be sufficient. This problem of the connectivity between the initial and the target state will be generally addressed for the pump–dump control in complex systems where the concept of the intermediate target [298] was introduced. The basic idea of this approach is to find an optimal excited state ensemble which insures the connectivity with both the initial and the target state and use it for the pulse optimization. Since here one starts from a highly excited vibrational state, the Franck–Condon overlap with the (ν=0) is unfavorable and does not allow the direct pulse optimization using $\nu = 0$ state as a target. Therefore, sequential stepwise optimization was carried out, in which first an optimal set of intermediate excited state vibrational levels is populated. The pulse obtained in this way is subsequently used as an initial guess for maximizing further the population of the (ν=0) level in the ground state.

The optimal pulse for the photostabilization is presented in Fig. 2.63. The insert serves to illustrate subpulse richness. The spectral analysis of the pulse (cf. Fig. 2.63b) shows two main broad features centered around 10200 cm^{-1} and 13000 cm^{-1}. The higher energy part of the pulse is responsible for the sequence of pump-dump processes involving highly excited vibrational states of all three involved electronic states. These processes populate sequentially lower and lower vibrational states, thus increasing Franck-Condon overlap with target state ν=0. The second lower energy part of the pulse at 10200 cm^{-1} is mainly responsible for subsequent optimal cooling in the $0_g^+(I)$ electronic state and leads to a cascade of transitions steering the population into the ν=0 state. However, it should be pointed out that both parts of the pulse are spectrally very broad and that their tails are also responsible for stepping down the vibrational ladder. This is due to the fact that efficient transitions can take place only between the states with large Franck-Condon factors. The latter occurs either at the inner or at the outer turning point of the ground state potential and involve both low lying (inner) and high lying (outer) vibrational states of the electronically excited states which is the main reason for the spectral broadness of the optimal pulse.

The time dependent populations of the electronic states are presented in Fig. 2.63c. They reflect the mechanism of the photostabilization process which can be followed from the snapshots of the wave packet dynamics shown in Fig. 2.64. The initial excitation populates the $0_u^+(II)$ state and creates a wave packet which is propagating toward the inner turning point (cf. Fig. 2.64). This motion of the wave packet is accompanied by mutual exchange of excited states populations but involving also the ground state, as can be seen from snapshots corresponding to 3.75 and 7.55 ps. At 9.9 ps the excited state wave packet is located in the bound region of the excited state potentials. This is achieved by successive inward propagation in the ground state and subsequent excitation to the $0_u^+(II)$ on the left side of the avoided crossing. Within the

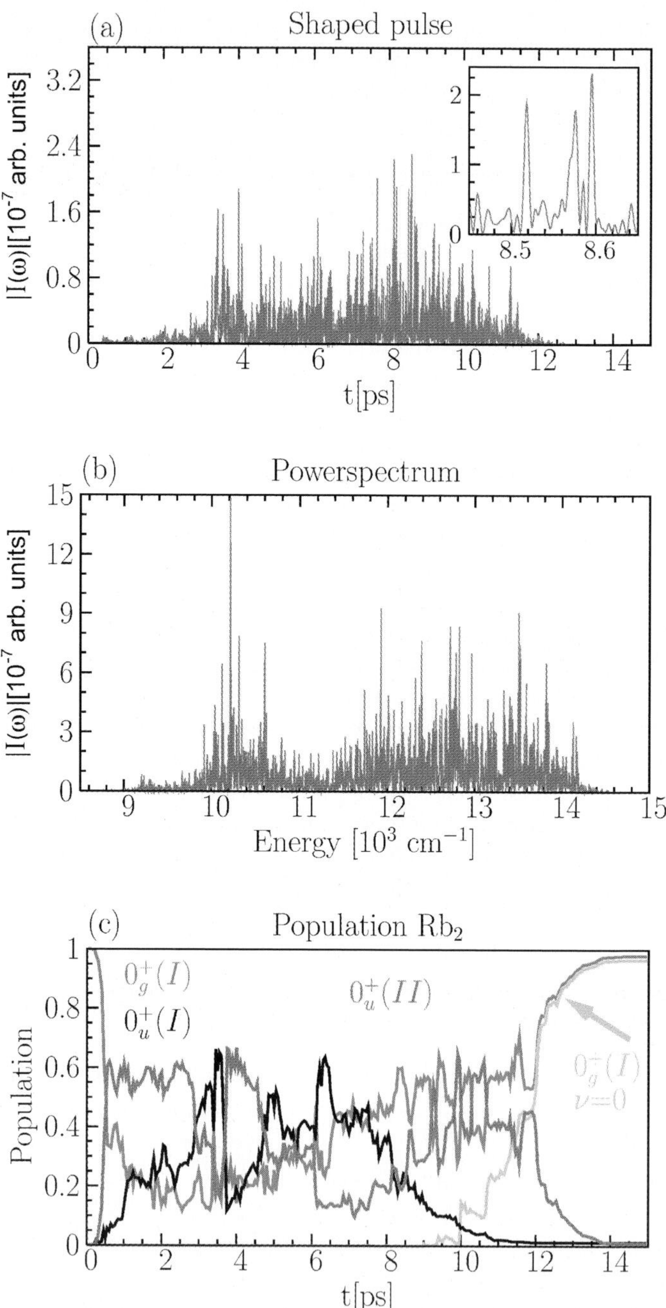

Fig. 2.63. (a) Intensity $|I(t)|$ of the optimal laser field $\epsilon(t)$ driving the photostabilization of Rb_2. The insert marked by a lens shows a details of the pulse between 8.45 and 8.65 ps. (b) Power spectrum of the optimal pulse. (c) Time-dependent populations of the electronic states. Population of the target level ν=0 is shown in turquoise color.

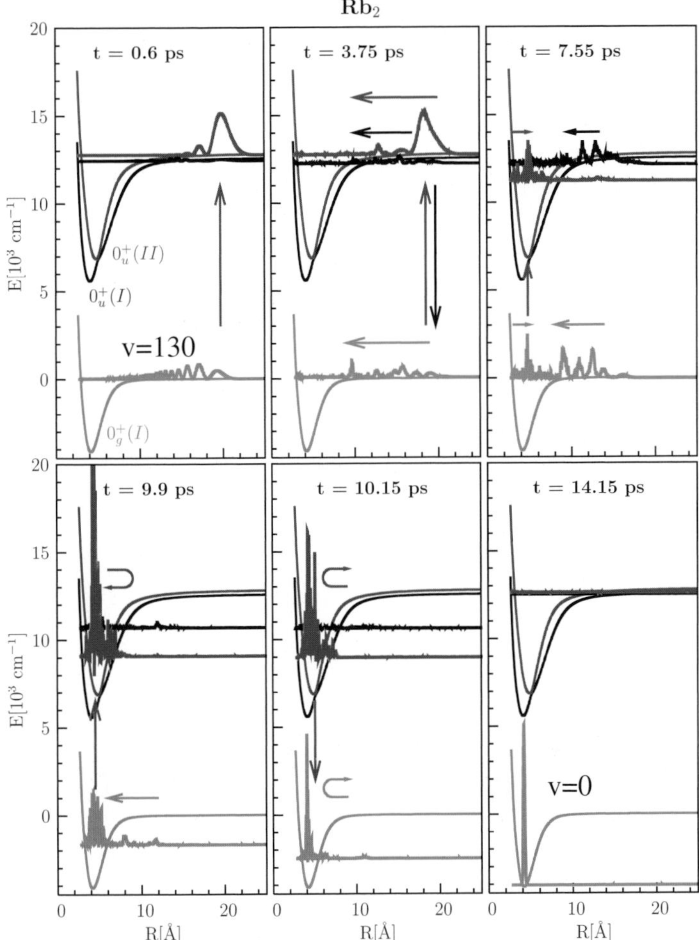

Fig. 2.64. Snapshots of the wave packet dynamics under the influence of the optimal pulse shown in Fig. 2.63, achieving 100 percent population of the vibrational ground state of Rb_2.

next 5 ps the stepping down the vibrational ladder is completed leading to almost 100 percent population of the $\nu = 0$ vibrational state at 14.15 ps.

The simulations show that efficient cooling and photostabilization can be achieved if the high lying but bound vibrational states are initially populated. The pulses obtained from the optimal control algorithm are spectrally very broad and present a challenge for the experimental realization due to various constraints imposed by the experimental setup. In order to investigate how the restrictions on the pulse shape influences the efficiency of the optimization, additional simulations were carried out in which (i) the total optimization time was restricted to 9 ps instead of 15 ps and (ii) the spectral range of the pulse is restricted to the experimentally accessible range from 11770 cm^{-1} to 13000 cm^{-1}. The results are shown in Fig. 2.65.

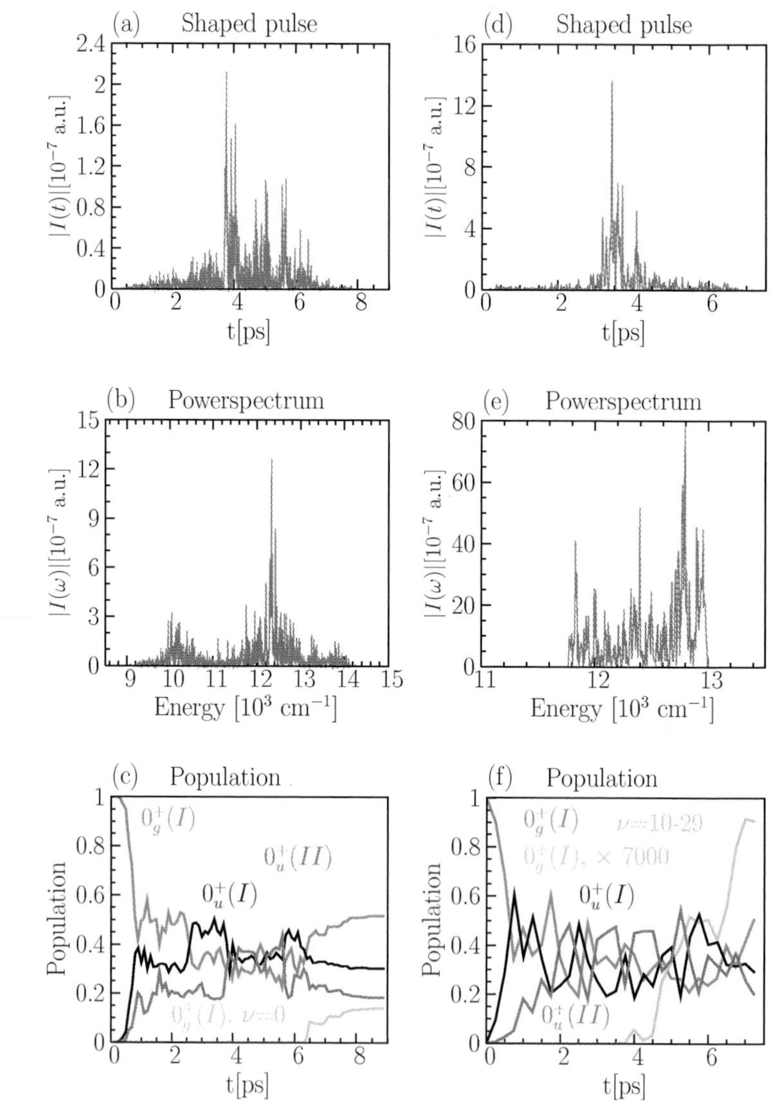

Fig. 2.65. Optimal pulse, powerspectrum and time-dependent state populations for the constrained pulse optimization: (left hand side) The pulse temporal range has been restricted to 9 ps and (right hand side) the pulse spectral range has been restricted to the range from 11770 cm^{-1} to 13000 cm^{-1}.

Restricting the time interval for the optimization leads to a pulse shape (cf. left side of Fig. 2.65) which still has two spectral components centered around 10000 cm^{-1} and 12000 cm^{-1} as in the case of 15 ps optimization (cf. Fig. 2.63). As can be seen from the time dependent state populations this leads to only 20% population of the ground state vibrational level. In contrast to the restriction of the temporal range, constraining the spectral range of the optimal pulse leads also the formation of bound ground state Rb_2 molecule but results in the maximal occupation of higher vibrational levels (ν=20-30) while the occupation of the ν=0 state becomes very low (cf. right hand side of Fig. 2.65). This means that although the stable molecule is formed under these conditions it is vibrationally relatively hot.

For the experimental realization, the two color scheme in which both pulses are shaped might provide an alternative route to the vibrational ground state.

It was demonstrated that photostabilization of ultracold Rb_2 molecule in the lowest ground state vibrational level can be achieved using tailored laser pulse obtained from the full quantum mechanical optimal control theory. This allowed to determine: (i) the mechanism leading to efficient vibrational cooling involving the ground and excited electronic states and (ii) the temporal and spectral range which are necessary in order to realize the photostabilization experimentally. Moreover, the results show that limiting the temporal and spectral range of the optimal pulse can lead to significant decrease in the yield of ultracold molecules.

In summary, it was shown that optimal control with tailored laser fields represents a promising approach for the photostabilization of ultracold molecules. The presented theoretical simulations provide a new insight into the mechanism of the cooling process itself, and may lead to the proposal of new strategies for formation of cold molecules taking into account constraints imposed by the experiment. One such possible strategy could involve a separate two–color pump–dump experiment in which first, the photoassociation is induced by a shaped pump pulse and subsequently a second shaped dump pulse is used to steer the molecule into the lowest ground state vibrational level. The second possibility involves the modeling of potential curves which allow to obtain the optimal pulse with a suitable temporal and spectral range and to find the adequate molecular system with related electronic properties on which experimental realization is more convenient.

2.6 Future perspectives

Up to now, most optimization experiments on molecular systems were performed for one polarization direction and free phase and/or amplitude modulation of the laser pulse. For future applications it is now intriguing to investigate the influence of the polarization. This new dimension will open different optimization paths and may be utilized for finding more efficient excitation paths, for molecular rearrangement into particular isomers, or change of the chirality [299] of biologically relevant units. Moreover, electric ring currents in molecules associated with magnetic fields may be induced by circularly polarized ultrashort laser pulses [300,301]. These studies may also be performed by parametrically shaped pulses in phase, amplitude, and polarization, where the subpulse features like subpulse spacings, intensities, chirps, and polarization are subject to modulation. Here, the experimenter can choose the limitations under which the algorithm has to solve the problem and receives, moreover, a solution of physically relevant parameters which aids the interpretation. The results can be compared with full quantum mechanical optimization simulations and allow moreover a fast online interaction between theoretical modeling and measurement. All these investigations will be extended to a broader spectral range, shorter pulses, and higher shaping resolution in order to have more flexibility in investigating molecular processes and to achieve control of electron dynamics. White light filaments offer an exciting perspective in this regard (see Chap. 8).

To date the formation of ultracold ground state molecules by short laser pulses has been limited by the small excitation rates for the photoassociation step. On the other hand, the stabilization step to the electronic ground state is expected to be fairly efficient [272]. The photoassociation efficiency could be improved in two different ways: (i) statically, by changing the initial probability distribution, or (ii) dynamically, by accelerating the atoms toward each other.

Once the obstacle of the low photoassociation rate is overcome, the formation of ultracold molecules in their vibronic ground state can be tackled. Experimentally, an incoherent 4-photon process was shown to produce $v = 0$ RbCs molecules [302] in a MOT. Due to stabilization via spontaneous emission, the transfer efficiency was extremely small. Employing a coherent process using short, shaped laser pulses is expected to improve the efficiency of the process. This will pave the way to many applications of ultracold molecules, in particular to reaching a stable molecular Bose-Einstein condensate (BEC).

The combination of ultracold and ultrafast holds the promise of achieving *absolute* molecular quantum state control: Ultracold matter can be prepared in a single quantum state, and shaped ultrafast laser pulses are predicted to reach close to one hundred percent efficiency for state-to-state transfer processes. This paves the way to a novel ultrafast-ultracold chemistry which is based on complete control over the reactants. There the shaped laser pulses play the role of engineering the required potential energy surfaces.

Applying pulse shaping techniques to ultracold matter offers furthermore the possibility to study the interplay of control and many-body quantum correlations as present for example in a BEC. In that case, a direct comparison between control of the many-body and the two-body quantum system could be facilitated by switching from a BEC to a Mott insulator state in an optical lattice and vice versa.

References

1. T.S. Rose, M.J. Rosker, A.H. Zewail, J. Chem. Phys. **88**, 6672 (1988)
2. T.S. Rose, M.J. Rosker, A.H. Zewail, J. Chem. Phys. **91**, 7415 (1989)
3. A.H. Zewail, Faraday Discuss. Chem. Soc. **91**, 207 (1991)
4. A. Mokhtari, P. Cong, J.L. Herek, A.H. Zewail, Nature **348**, 225 (1990)
5. M. Dantus, R.M. Bowman, M. Gruebele, A.H. Zewail, J. Chem. Phys. **91**, 7489 (1989)
6. T. Baumert, B. Buhler, M. Grosser, V. Weiss, G. Gerber, J. Phys. Chem. **95**, 8103 (1991)
7. V. Engel, H. Metiu, J. Chem. Phys. **93**, 5693 (1990)
8. S. Wolf, G. Sommerer, S. Rutz, E. Schreiber, T. Leisner, L. Wöste, R.S. Berry, Phys. Rev. Lett. **74**, 4177 (1995)
9. R.S. Berry, V. Bonačić-Koutecký, J. Gaus, T. Leisner, J. Manz, B. Reischl-Lenz, H. Ruppe, S. Rutz, E. Schreiber, Š. Vajda, R. de Vivie-Riedle, S. Wolf, L. Wöste, Adv. Chem. Phys. **101**, 101 (1997)
10. G.K. Paramonov, V.A. Savva, Phys. Lett. **97A**, 340 (1983)
11. D.J. Tannor, S.A. Rice, J. Chem. Phys. **83**, 5013 (1985)
12. A.P. Peirce, M.A. Dahleh, H. Rabitz, Phys. Rev. A **37**, 4950 (1988)
13. P. Brumer, M. Shapiro, Faraday Discuss. Chem. Soc. **82**, 177 (1986)
14. M. Shapiro, P. Brumer, J. Chem. Phys. **84**, 4103 (1986)
15. T. Joseph, J. Manz, Molec. Phys. **58**, 1149 (1986)
16. S.H. Tersigni, P. Gaspard, S.A. Rice, J. Chem. Phys. **93**, 1670 (1990)
17. R. Kosloff, S.A. Rice, P. Gaspard, S. Tersigni, D.J. Tannor, Chem. Phys. **139**, 201 (1989)
18. W. Jakubetz, J. Manz, H.J. Schreier, Chem. Phys. Lett. **165**, 100 (1990)
19. J.E. Combariza, B. Just, J. Manz, G.K. Paramonov, J. Phys. Chem. **95**, 10351 (1991)
20. H. Rabitz, S. Shi, Adv. Mol. Vibr. Collision Dyn. **1A**, 187 (1991)
21. W.S. Warren, H. Rabitz, M. Dahleh, Science **259**, 1581 (1993)
22. R.S. Judson, H. Rabitz, Phys. Rev. Lett. **68**, 1500 (1992)
23. S.A. Rice, M. Zhao, *Optical control of molecular dynamics* (John Wiley & Sons, Inc., New York, 2000)
24. M. Shapiro, P. Brumer, *Principles of the quantum control of molecular processes* (Wiley-Interscience, New York, 2003)
25. S. Rutz, R. de Vivie-Riedle, E. Schreiber, Phys. Rev. A **54**, 306 (1996)
26. G. Delacrétaz, E. Grant, R. Whetten, L. Wöste, J. Zwanziger, Phys. Rev. Lett. **56**, 1598 (1986)
27. A. Lindinger, C. Lupulescu, A. Bartelt, Š. Vajda, L. Wöste, Spectrochimica Acta Part B **58**, 1109 (2003)
28. H. Kühling, K. Kobe, S. Rutz, E. Schreiber, L. Wöste, J. Phys. Chem. **98**, 6679 (1994)
29. E. Schreiber, K. Kobe, A. Ruff, S. Rutz, G. Sommerer, L. Wöste, Chem. Phys. Lett. **37**, 175 (1995)
30. H. Ruppe, S. Rutz, E. Schreiber, L. Wöste, Chem. Phys. Lett. **257**, 356 (1996)
31. M. Hartmann, J. Pittner, V. Bonačić-Koutecký, A. Heidenreich, J. Jortner, J. Chem. Phys. **108**, 3096 (1998)
32. A. Donoso, C. Martens, J. Phys. Chem. **102**, 4291 (1998)
33. A. Donoso, Y. Zheng, C. Martens, J. Chem. Phys. **119**, 5010 (2003)

34. A. Donoso, C. Martens, Phys. Rev. Let. **87**, 223202 (2001)
35. S. Hammes-Schiffer, J. Phys. Chem. A **102**, 10443 (1998)
36. E. Wigner, Phys. Rev. **40**, 749 (1932)
37. M. Hillary, R.F. O'Connel, M.O. Scully, E.P. Wigner, Phys. Rep. **106**, 1984 (1984)
38. J.E. Moyal, Camb. Phil. Soc. **45**, 99 (1949)
39. Z. Li, J.Y. Fang, C.C. Martens, J. Chem. Phys. **104**, 6919 (1996)
40. M. Hartmann, J. Pittner, V. Bonačić-Koutecký, A. Heidenreich, J. Jortner, J. Phys. Chem. **102**, 4069 (1998)
41. M. Hartmann, J. Pittner, V. Bonačić-Koutecký, J. Chem. Phys. **114**, 2106 (2001)
42. M. Hartmann, J. Pittner, V. Bonačić-Koutecký, J. Chem. Phys. **114**, 2123 (2001)
43. S. Mukamel, *Principles of nonlinear optical spectroscopy* (Oxford University Press, Oxford, 1995)
44. V. Bonačić-Koutecký, R. Mitrić, Chem. Rev. **105**, 11 (2005)
45. J.C. Tully, J. Chem. Phys. **93**, 1061 (1990)
46. Š. Vajda, A. Bartelt, E.C. Kaposta, T. Leisner, C. Lupulescu, S. Minemoto, P. Rosendo-Francisco, L. Wöste, Chem. Phys. **267**, 231 (2001)
47. V. Bonačić-Koutecký, J. Pittner, J. Koutecký, Chem. Phys. **210**, 313 (1996)
48. V. Bonačić-Koutecký, J. Pittner, J. Koutecký, Z. Phys. D **40**, 441 (1997)
49. Š. Vajda, C. Lupulescu, A. Merli, F. Budzyn, L. Wöste, M. Hartmann, J. Pittner, V. Bonačić-Koutecký, Phys. Rev. Lett. **89**, 213404 (2002)
50. M.C. Heitz, G. Durand, F. Spiegelman, C. Meier, J. Chem. Phys. **118**, 1282 (2003)
51. A. Stolow, A.E. Bragg, D.M. Neumark, Chem. Rev. **104**, 1719 (2004)
52. A. Weaver, R.B. Metz, S.E. Bradforth, D.M. Neumark, J. Chem. Phys. **93**, 5352 (1990)
53. R.B. Metz, D.M. Neumark, J. Chem. Phys. **97**, 962 (1992)
54. D.M. Neumark, Acc. Chem. Res. **26**, 33 (1993)
55. B.J. Greenblatt, M.T. Zanni, D.M. Neumark, J. Chem. Phys. **111**, 10566 (1999)
56. M.T. Zanni, B.J. Greenblatt, A.V. Davis, D.M. Neumark, J. Chem. Phys. **111**, 2991 (1999)
57. D.M. Neumark, Annu. Rev. Phys. Chem. **52**, 255 (2001)
58. B.J. Greenblatt, M.T. Zanni, D.M. Neumark, J. Chem. Phys. **112**, 601 (2000)
59. R. Wester, A.V. Davis, A.E. Bragg, D.M. Neumark, Phys. Rev. A **65**, 051201 (2002)
60. C. Frischkorn, A.E. Bragg, A.V. Davis, R. Wester, D.M. Neumark, J. Chem. Phys. **115**, 11185 (2001)
61. S.M. Burnett, A.E. Stevens, C.S. Feigerle, W.C. Lineberger, Chem. Phys. Lett. **100**, 124 (1983)
62. K.M. Ervin, J. Ho, W.C. Lineberger, J. Chem. Phys. **91**, 5974 (1989)
63. P.G. Wenthold, D. Hrovat, W.T. Borden, W.C. Lineberger, Science **272**, 1456 (1996)
64. D.W. Boo, Y. Ozaki, L.H. Andersen, W.C. Lineberger, J. Phys. Chem. A **101**, 6688 (1997)
65. I. Andrianov, V. Bonačić-Koutecký, M. Hartmann, J. Manz, J. Pittner, K. Sundermann, Chem. Phys. Lett. **318**, 256 (2000)

66. M. Hartmann, R. Mitrić, B. Stanca, V. Bonačić-Koutecký, Eur. Phys. J. D **16**, 151 (2001)
67. R. Mitrić, M. Hartmann, B. Stanca, V. Bonačić-Koutecký, P. Fantucci, J. Phys. Chem. A **105**, 8892 (2001)
68. T.M. Bernhardt, J. Hagen, L.D. Socaciu, J. Le Roux, D. Popolan, M. Vaida, L. Wöste, R. Mitrić, V. Bonačić-Koutecký, A. Heidenreich, J. Jortner, ChemPhysChem **6**, 243 (2005)
69. V. Bonačić-Koutecký, P. Fantucci, J. Koutecký, Chem. Rev. **91**, 1035 (1991)
70. D.J. Wales, *Energy landscapes* (Cambridge University Press, Cambridge, 2003)
71. C. Leforestier, J. Chem. Phys. **68**, 4406 (1978)
72. R.N. Barnett, U. Landman, Phys. Rev. B **48**, 2081 (1993)
73. O. Rubner, C. Meier, V. Engel, J. Chem. Phys. **107**, 1066 (1997)
74. V. Bonačić-Koutecký, L. Češpiva, P. Fantucci, J. Koutecký, J. Chem. Phys. **98**, 7981 (1993)
75. V. Bonačić-Koutecký, L. Češpiva, P. Fantucci, J. Koutecký, J. Chem. Phys. **100**, 490 (1994)
76. V. Bonačić-Koutecký, J. Pittner, M. Boiron, P. Fantucci, J. Chem. Phys. **110**, 3876 (1999)
77. V. Bonačić-Koutecký, V. Veyret, R. Mitrić, J. Chem. Phys. **115**, 10450 (2001)
78. H. Hakkinen, M. Moseler, U. Landman, Phys. Rev. Lett. **89**, 033401 (2002)
79. V. Bonačić-Koutecký, J. Burda, M. Ge, R. Mitrić, G. Zampella, R. Fantucci, J. Chem. Phys. **117**, 3120 (2002)
80. R. Mitrić, M. Hartmann, J. Pittner, V. Bonačić-Koutecký, Eur. Phys. J. **24**, 45 (2003)
81. W.T. Wallace, R.B. Wyrwas, R.L. Whetten, R. Mitrić, V. Bonačić-Koutecký, J. Am. Chem. Soc. **125**, 8408 (2003)
82. M.L. Kimble, A.W.J. Castleman, R. Mitrić, C. Bürgel, V. Bonačić-Koutecký, J. Am. Chem. Soc. **126**, 2526 (2004)
83. J. Hagen, L.D. Socaciu, J. Le Roux, D. Popolan, T.M. Bernhardt, L. Wöste, R. Mitrić, V. Bonačić-Koutecký, J. Am. Chem. Soc. **126**, 3442 (2004)
84. C. Sieber, J. Buttet, W. Harbich, C. Félix, R. Mitrić, V. Bonačić-Koutecký, Phys. Rev. A **70** (2004)
85. P. Ballone, W. Andreoni, *Metal clusters* (Wiley Inc., Chichester, 1999), p. 71
86. F. Furche, R. Ahlrichs, P. Weis, C. Jacob, S. Gilb, T. Bierweiler, M.M. Kappes, J. Chem. Phys **117**, 6982 (2002)
87. P. Weis, T. Bierweiler, S. Gilb, M.M. Kappes, Chem. Phys. Lett. **355**, 355 (2002)
88. L. König, I. Rabin, W. Schulze, G. Ertl, Science **274**, 1353 (1996)
89. C. Felix, C. Sieber, W. Harbich, J. Buttet, I. Rabin, W. Schulze, G. Ertl, Chem. Phys. Lett. **313**, 105 (1998)
90. I. Rabin, W. Schulze, G. Ertl, C. Felix, C. Sieber, W. Harbich, J. Buttet, Chem. Phys. Lett. **320**, 59 (2000)
91. C. Felix, S. Sieber, W. Harbich, J. Buttet, I. Rabin, W. Schulze, G. Ertl, Phys. Rev. Lett. **86**, 2992 (2001)
92. W. Harbich, C. Felix, C. R. Physique **3**, 289 (2002)
93. M. Haruta, Catal. Today **36**, 153 (1997)
94. T.H. Lee, K.M. Ervin, J. Phys. Chem. **98**, 10023 (1994)
95. B.E. Salisbury, W.T. Wallace, R.L. Whetten, Chem. Phys. **262**, 131 (2000)
96. K.J. Taylor, C.L. Pettiette-Hall, O. Cheshnovsky, R.E. Smalley, J. Chem. Phys. **96**, 3319 (1992)

97. J. Ho, K.M. Ervin, W.C. Lineberger, J. Chem. Phys. **93**, 6987 (1990)
98. H. Handschuh, G. Ganteför, P.S. Bechtold, W. Eberhardt, J. Chem. Phys. **100**, 7093 (1994)
99. G. Lüttgens, N. Pontius, P.S. Bechtold, M. Neeb, W. Eberhardt, Phys. Rev. Lett. **88**, 076102 (2002)
100. Y. Negishi, Y. Nakajima, K. Kaya, J. Chem. Phys. **115**, 3657 (2001)
101. A. Sanchez, S. Abbet, U. Heiz, W.D. Schneider, H. Hakkinen, R.N. Barnett, U. Landman, J. Phys. Chem. A **103**, 9573 (1999)
102. L.D. Socaciu, J. Hagen, T.M. Bernhardt, L. Wöste, U. Heiz, H. Hakkinen, U. Landman, J. Am. Chem. Soc. **125**, 10437 (2003)
103. S.A. Varganov, R.M. Olson, M.S. Gordon, H. Metiu, J. Chem. Phys. **119**, 2531 (2003)
104. G. Mills, M.S. Gordon, H. Metiu, J. Chem. Phys. **118**, 4198 (2003)
105. A.D. Becke, Phys. Rev. A **98**, 3098 (1988)
106. C. Lee, W. Yang, R.G. Parr, Phys. Rev. B **37**, 785 (1985)
107. T. Leisner, Š. Vajda, S. Wolf, L. Wöste, R.S. Berry, J. Chem. Phys. **111**, 1017 (1999)
108. H. Hess, K.R. Asmis, T. Leisner, L. Wöste, Eur. Phys. J. D **16**, 145 (2001)
109. R. Keller, F. Nöhmeier, P. Spädtke, H. Schönenberg, Vacuum **34**, 31 (1984)
110. T.M. Bernhardt, Int. J. Mass. Spectrom. **243**, 1 (2005)
111. L. Lian, P.A. Hackett, D.M. Rayner, J. Chem. Phys. **99**, 2583 (1993)
112. L.D. Socaciu, J. Hagen, U. Heiz, T.M. Bernhardt, T. Leisner, L. Wöste, Chem. Phys. Lett. **340**, 282 (2001)
113. M. Schmidt, P. Cahuzac, C. Brechignac, H.P. Cheng, J. Chem. Phys. **118** (2003)
114. M. Schmidt, A. Masson, C. Brechignac, Phys. Rev. Lett. **91**, 243401 (2003)
115. L.D. Socaciu, J. Hagen, J. Le Roux, D. Popolan, T.M. Bernhardt, L. Wöste, Š. Vajda, Chem. Phys. **120**, 2078 (2004)
116. Y.D. Kim, G. Ganteför, Chem. Phys. Lett. **383**, 80 (2004)
117. L.D. Socaciu-Siebert, J. Hagen, J. Le Roux, D. Popolan, Š. Vajda, T.M. Bernhardt, L. Wöste, Phys. Chem. Chem. Phys. **7**, 2706 (2005)
118. D.J. Tannor, S.A. Rice, Adv. Chem. Phys. **70**, 441 (1988)
119. M. Shapiro, P. Brumer, Int. Rev. Phys. Chem. **13**, 187 (1994)
120. T. Joseph, J. Manz, Mol. Phys. **58**, 1149 (1986)
121. W. Jakubetz, B. Just, J. Manz, H.J. Schreier, J. Phys. Chem. **94**, 2294 (1990)
122. J. Somlói, V.A. Kazakov, D.J. Tannor, Chem. Phys. **172**, 85 (1993)
123. B. Amstrup, J.D. Doll, R.A. Sauerbrey, Szabó, A. Lörincz, Phys. Rev. A **48**, 3830 (1993)
124. R.G. Gordon, S.A. Rice, Annu. Rev. Phys. Chem. **48**, 601 (1997)
125. L.C. Zhu, V. Kleiman, X.N. Li, S.P. Lu, K. Trentelman, R.J. Gordon, Science **270**, 77 (1995)
126. C. Chen, D.S. Elliott, Phys. Rev. Lett. **65**, 1737 (1990)
127. S.M. Park, R.J. Lu, R.J. Gordon, J. Chem. Phys. **94**, 8622 (1991)
128. G.Q. Xing, X.B. Wang, X. Huang, R. Bersohn, J. Chem. Phys. **104**, 826 (1996)
129. A. Shnitman, I. Sofer, I. Golub, A. Yogev, M. Shapiro, Z. Chen, P. Brumer, Phys. Rev. Lett. **76**, 2886 (1996)
130. T. Baumert, B. Buhler, M. Grosser, R. Thalweiser, V. Weiss, E. Wiedenmann, G. Gerber, J. Phys. Chem. **95**, 8103 (1991)
131. T. Baumert, G. Gerber, Isr. J. Chem. **34**, 103 (1994)

132. E.D. Potter, J.L. Herek, S. Pedersen, Q. Liu, A.H. Zewail, Nature **355**, 66 (1992)
133. J.L. Herek, A. Materny, A.H. Zewail, Chem. Phys. Lett. **228**, 15 (1994)
134. A. Assion, T. Baumert, V. Seyfried, V. Weiss, E. Wiedenmann, G. Gerber, Z. Phys. D **36**, 265 (1996)
135. T. Baumert, J. Helbing, G. Gerber, Adv. Chem. Phys. **101**, 47 (1997)
136. D.J. Tannor, S.A. Rice, J. Chem. Phys. **85**, 5805 (1986)
137. S. Shi, H. Rabitz, Chem. Phys. **139**, 185 (1989)
138. R. Kosloff, A.D. Hammerich, D.J. Tannor, Phys. Rev. Lett. **69**, 2172 (1992)
139. A. Bartana, R. Kosloff, D.J. Tannor, J. Chem. Phys. **99**, 196 (1993)
140. V. Malinovsky, C. Meier, D.J. Tannor, Chem. Phys. **221**, 67 (1997)
141. V. Malinovsky, D.J. Tannor, Phys. Rev. A **56**, 4929 (1997)
142. J. Vala, R. Kosloff, Opt. Express **8**, 238 (2001)
143. P. Marquetand, C. Meier, V. Engel, J. Chem. Phys. **123**, 204320 (2005)
144. C. Meier, M.C. Heitz, J. Chem. Phys. **123**, 044504 (2005)
145. S. Gräfe, C. Meier, V. Engel, J. Chem. Phys. **122**, 184103 (2005)
146. S. Sklarz, D.J. Tannor, Chem. Phys. **322**, 87 (2006)
147. Y. Chen, P. Gross, V. Ramakrishna, H. Rabitz, K. Mease, J. Chem. Phys. **102**, 8001 (1995)
148. D.J. Tannor, *Molecules in laser fields* (M. Dekker Inc., 1999), p. 403
149. D.J. Tannor, R. Kosloff, A. Bartana, Faraday Discuss. **113**, 365 (1999)
150. T. Baumert, T. Brixner, V. Seyfried, M. Strehle, G. Gerber, Appl. Phys. B **65**, 779 (1997)
151. D. Yelin, D. Meshulach, Y. Silberberg, Opt. Lett. **22**, 1793 (1997)
152. A. Efimov, M.D. Moores, N.M. Beach, J.L. Krause, D.H. Reitze, Opt. Lett. **23**, 1915 (1998)
153. E. Zeek, K. Maginnis, S. Backus, U. Russek, M.M. Murnane, G. Mourou, H.C. Kapteyn, G. Vdovin, Opt. Lett. **24**, 493 (1999)
154. E. Zeek, R. Bartels, M.M. Murnane, H.C. Kapteyn, S. Backus, G. Vdovin, Opt. Lett. **25**, 587 (2000)
155. D.E. Goldberg, *Genetic algorithms in search, optimization, and machine learning* (Addison-Wesley, Reading, 1993)
156. H.P. Schwefel, *Evolution and optimum seeking* (Wiley, New York, 1995)
157. A. Assion, T. Baumert, M. Bergt, T. Brixner, B. Kiefer, V. Seyfried, M. Strehle, G. Gerber, *Springer Series in Chemical Physics, vol. 63* (Springer, Berlin, 1998), p. 471
158. A. Assion, T. Baumert, M. Bergt, T. Brixner, B. Kiefer, V. Seyfried, M. Strehle, G. Gerber, Science **282**, 919 (1998)
159. T.C. Weinacht, J.L. White, P.H. Bucksbaum, J. Phys. Chem. A **103**, 10166 (1999)
160. T. Hornung, R. Meier, M. Motzkus, Chem. Phys. Lett. **326**, 445 (2000)
161. Š. Vajda, P. Rosendo-Francisco, C. Kaposta, M. Krenz, L. Lupulescu, L. Wöste, Eur. Phys. J. D **16**, 161 (2001)
162. A. Bartelt, S. Minemoto, C. Lupulescu, Š. Vajda, L. Wöste, Eur. Phys. J. D **16**, 127 (2001)
163. Š. Vajda, C. Lupulescu, A. Bartelt, F. Budzyn, P. Rosendo-Francisco, L. Wöste, *Femtochemistry and Femtobiology* (World Scientific Publishing: Singapore, 2002), p. 472
164. A. Bartelt, A. Lindinger, C. Lupulescu, Š. Vajda, L. Wöste, Phys. Chem. Chem. Phys. **5**, 3610 (2003)

165. C. Lupulescu, A. Lindinger, M. Plewicky, A. Merli, S.M. Weber, L. Wöste, Chem. Phys. **296**, 63 (2004)
166. A. Bartelt, Steuerung der Wellenpaketdynamik in kleinen Alkaliclustern mit optimierten Femtosekundenpulsen. Ph.D. thesis, Freie Universität Berlin (2002)
167. R.J. Levis, G.M. Menkir, H. Rabitz, Science **292**, 709 (2001)
168. T. Brixner, M. Strehle, G. Gerber, Appl. Phys. B **68**, 281 (1999)
169. A. Efimov, M.D. Moores, B. Mei, J.L. Krause, C.W. Siders, D.H. Reitze, Appl. Phys. B **70**, 133 (2000)
170. D. Zeidler, T. Hornung, D. Proch, M. Motzkus, Appl. Phys. B **70**, 125 (2000)
171. D. Meshulach, Y. Silberberg, Nature **396**, 239 (1998)
172. T. Hornung, R. Meier, D. Zeidler, K.L. Kompa, D. Proch, M. Motzkus, Appl. Phys. B **71**, 277 (2000)
173. T.C. Weinacht, J. Ahn, P.H. Bucksbaum, Nature **397**, 233 (1999)
174. R. Bartels, S. Backus, E. Zeek, L. Misoguti, G. Vdovin, I.P. Christov, M.M. Murnane, H.C. Kapteyn, Nature **164** (2000)
175. J. Kunde, B. Baumann, S. Arlt, F. Morier-Genoud, U. Siegner, U. Keller, Appl. Phys. Lett. **77**, 924 (2000)
176. H. Rabitz, R. de Vivie-Riedle, M. Motzkus, K. Kompa, Science **288**, 824 (2000)
177. T. Hornung, M. Motzkus, R. de Vivie-Riedle, Phys. Rev. A **65**, 021403 (2002)
178. M. Braun, V. Engel, Z. Phys. D **39**, 301 (1997)
179. H. Schwoerer, R. Pausch, M. Heid, V. Engel, W. Kiefer, J. Chem. Phys. **107**, 9749 (1997)
180. Z.W. Shen, T. Chen, M. Heid, W. Kiefer, V. Engel, Eur. Phys. J. D **14**, 167 (2001)
181. R. de Vivie-Riedle, K. Kobe, J. Manz, W. Meyer, B. Reischl, S. Rutz, E. Schreiber, L. Wöste, J. Phys. Chem. **100**, 7789 (1996)
182. C. Nicole, M.A. Bouchene, C. Meier, S. Magnier, E. Schreiber, B. Girard, J. Chem. Phys. **111**, 7857 (1999)
183. L. Pesce, Z. Amitay, R. Uberna, S.R. Leone, M. Ratner, R. Kosloff, J. Chem. Phys. **114**, 1259 (2001)
184. K. Sundermann, R. de Vivie-Riedle, J. Chem. Phys. **110**, 1896 (1999)
185. T. Hornung, M. Motzkus, de Vivie-Riedle, J. Chem. Phys. **115**, 3105 (2001)
186. G. Grégoir, M. Mons, I. Dimicoli, C. Dedonder-Lardeux, S. Jouvet, S. Martrenchard, D. Solgadi, J. Chem. Phys. **112**, 8794 (2000)
187. Z. Shen, V. Engel, R. Xu, J. Cheng, Y. Yan, J. Chem. Phys. **117**, 6142 (2002)
188. G. Rodriguez, J.G. Eden, Chem. Phys. Lett. **205**, 371 (1993)
189. G. Rodriguez, P.C. John, J.G. Eden, J. Chem. Phys. **103**, 10473 (1995)
190. R. Pausch, M. Heid, T. Chen, W. Kiefer, H. Schwoerer, J. Chem. Phys. **110**, 9560 (1999)
191. R. Pausch, M. Heid, T. Chen, W. Kiefer, H. Schwoerer, J. Raman Spectrosc. **31**, 7 (2000)
192. R. Uberna, Z. Amitay, R.A. Loomis, S.R. Leone, Faraday Discuss. **113**, 385 (1999)
193. J.B. Ballard, H.U. Stauffer, Z. Amitay, S.R. Leone, J. Chem. Phys. **116**, 1350 (2002)
194. J.J. Sakurai, Modern Quantum Mechanics, Benjamin Cummings Publishing, Menlo Park (1985)
195. S. Chelkowski, A.D. Bandrauk, P.B. Corkum, Phys. Rev. Lett. **65**, 2355 (1990)

196. T. Witte, T. Hornung, L. Windhorn, D. Proch, R. de Vivie-Riedle, M. Motzkus, K.L. Kompa, J. Chem. Phys. **118**, 2021 (2003)
197. P.A.M. Dirac, *The Principles of quantum mechanics*, vol. 4th Edition (Oxford Science Publications, Oxford, 1958)
198. J.V. Josè, E.J. Saletan, *Classical dynamics* (Cambridge University Press, Cambridge, 1998)
199. C.D. Schwieters, H. Rabitz, Phys. Rev. A **44**, 5224 (1991)
200. C.D. Schwieters, H. Rabitz, Phys. Rev. A **48**, 2549 (1993)
201. Y. Chen, P. Gross, V. Ramakrishna, H. Rabitz, K. Mease, H. Singh, Automatica **33**, 1617 (1997)
202. J.L. Herek, A. Materny, A.H. Zewail, Chem. Phys. Lett. **228**, 15 (1994)
203. T. Taneichi, T. Kobayashi, Y. Ohtsuki, Y. Fujimura, Chem. Phys. Lett. **231**, 50 (1994)
204. C.J. Bardeen, J. Che, K.R. Wilson, V.V. Yakovlevi, P. Cong, B. Kohler, J.L. Krause, M. Messina, J. Phys. Chem. A **101**, 3815 (1997)
205. E. Charron, A. Giusti-Suzor, J. Chem. Phys. **108**, 3922 (1998)
206. M. Grønager, N.E. Henriksen, J. Chem. Phys. **109**, 4335 (1998)
207. B.H. Hosseini, H.R. Sadeghpour, N. Balakrishnan, Phys. Rev. A **71**, 023402 (2005)
208. S. Gräfe, P. Marquetand, N.E. Henriksen, K.B. Møller, V. Engel, Chem. Phys. Lett. **398**, 180 (2004)
209. P. Marquetand, V. Engel, Chem. Phys. Lett. **407**, 471 (2005)
210. G.H. Peslherbe, R. Bianco, J.T. Hynes, B.M. Ladanyi, J. Chem. Soc., Faraday Trans. **93**, 977 (1997)
211. C. Cohen-Tannoudji, B. Diu, R. Laloe, *Quantum Mechanics*, Vol. I, Wiley, New York (1977)
212. Y. Zhao, O. Kühn, J. Phys. Chem. A **104**, 4882 (2000)
213. T. Baumert, M. Grosser, R. Thalweiser, G. Gerber, Phys. Rev. Lett. **67**, 3753 (1991)
214. V. Engel, T. Baumert, C. Meier, G. Gerber, Z. Phys. D-Atoms, Molecules and Clusters **28**, 37 (1993)
215. T. Baumert, G. Gerber, Adv. At. Molec. Opt. Phys. **35**, 163 (1995)
216. S. Gräfe, M. Erdmann, V. Engel, Phys. Rev. A **72**, 013404 (2005)
217. H. Metiu, V. Engel, J. Opt. Soc. Am. **7**, 1709 (1990)
218. N.F. Scherer, R.J. Carlson, A. Matro, M. Du, A.J. Ruggiero, V. Romero-Rochin, J.A. Cina, G.R. Fleming, S.A. Rice, J. Chem. Phys. **95**, 1487 (1991)
219. V. Blanchet, C. Nicole, M. Bouchene, B. Girard, Phys. Rev. Lett. **78**, 2716 (1997)
220. C.W. Hillegas, J.X. Tull, D. Goswami, D. Strickland, W.S. Warren, Opt. Lett. **19**, 737 (1994)
221. A.M. Weiner, D.E. Leaird, J.S. Patel, J.R. Wullert, IEEE J. Quant. Elect. **28**, 908 (1992)
222. O.E. Martinez, IEEE J. Quant. Elect. **23**, 59 (1987)
223. M. Wefers, K.J. Nelson, J. Opt. Soc. Am B **12**, 1343 (1995)
224. T. Brixner, G. Gerber, ChemPhysChem **4**, 418 (2003)
225. J. Heufelder, H. Ruppe, S. Rutz, E. Schreiber, L. Wöste, Chem. Phys. Lett. **269**, 1 (1997)
226. A. Bartelt, A. Lindinger, Š. Vajda, C. Lupulescu, L. Wöste, Phys. Chem. Chem. Phys. **6**, 1679 (2004)

227. S. Magnier, M. Aubert-Frécon, P.J. Millié, J. Mol. Spetrosc. **200**, 96 (2000)
228. H. Tal Ezer, R. Kosloff, J. Chem. Phys. **81**, 3967 (1984)
229. M. Demiralp, H. Rabitz, Phys. Rev. A **57**, 2420 (1998)
230. V.F. Krotov, Control and Cybernetics **17**, 115 (1988)
231. B. Schäfer-Bung, R. Mitrić, V. Bonačić-Koutecký, A. Bartelt, C. Lupulescu, A. Lindinger, Š. Vajda, S.M. Weber, L. Wöste, J. Phys. Chem. A **108**, 4175 (2004)
232. P.T. Greenland, Contemp. Phys. **30**, 405 (1990)
233. W.H. King, *Isotope shifts in atomic spectra* (Plenum, New York, 1984)
234. J. Heufelder, H. Ruppe, S. Rutz, E. Schreiber, L. Wöste, Chem. Phys. Lett. **269**, 1 (1997)
235. I.S. Averbukh, M.J.J. Vrakking, D.M. Villeneuve, A. Stolow, Phys. Rev. Lett. **77**, 3518 (1996)
236. S. Rutz, E. Schreiber, Eur. Phys. J. D **4**, 151 (1998)
237. G. Jong, L. Li, T. J. Whang, W. C. Stwalley, J. A Coxon, M. Li, A. M. Lyyra, J. Mol. Spect. **155**, 115 (1992)
238. R. de Vivie-Riedle, B. Reischl, S. Rutz, E. Schreiber, J. Phys. Chem. **99**, 16829 (1995)
239. A.J. Ross, P. Crozet, C. Effantin, J. d'Incan, R.F. Barrow, J. Phys. B **20**, 6225 (1987)
240. A. Lindinger, C. Lupulescu, M. Plewicki, F. Vetter, A. Merli, S.M. Weber, L. Wöste, Phys. Rev. Lett. **93**, 033001 (2004)
241. M. Leibscher, I.S. Averbukh, Phys. Rev. A **2001**, 043407 (2001)
242. A. Lindinger, C. Lupulescu, F. Vetter, M. Plewicki, S.M. Weber, A. Merli, L. Wöste, J Chem. Phys. **122**, 024312 (2005)
243. A. Lindinger, C. Lupulescu, M. Plewicki, F. Vetter, A. Merli, S.M. Weber, L. Wöste, Phys. Rev. Lett. **93**, 033001 (2004)
244. B. Schäfer-Bung, V. Bonačić-Koutecký, F. Sauer, S.M. Weber, L. Wöste, A. Lindinger, J. Chem. Phys., in press (2006)
245. J.W. Nicolson, J. Jaspara, W. Rudolph, Opt. Lett. **24**, 1774 (1999)
246. F. Vetter, M. Plewicki, A. Lindinger, A. Merli, S.M. Weber, L. Wöste, Phys. Chem. Chem. Phys. **7**, 1151 (2005)
247. J.M. Geremia, W. Zhu, H. Rabitz, J. Chem. Phys. **113**, 10841 (2000)
248. T. Hornung, R. Meier, M. Motzkus, Chem. Phys. Lett. **326**, 445 (2000)
249. S.M. Weber, A. Lindinger, F. Vetter, M. Plewicki, A. Merli, L. Wöste, Eur. Phys. J. D **33**, 39 (2005)
250. A. Lindinger, S.M. Weber, C. Lupulescu, F. Vetter, M. Plewicki, A. Merli, L. Wöste, A.F. Bartelt, H. Rabitz, Phys. Rev. A **71**, 013419 (2005)
251. A.M. Weiner, Prog. Quant. Electr. **19**, 161 (1995)
252. M. Hacker, G. Stobrawa, T. Feurer, Opt. Expr. **9**, 191 (2001)
253. R. Gerchberg, W. Saxon, Optik **35**, 237 (1971)
254. T. Brixner, G. Gerber, Optics Letters **26**, 557 (2001)
255. T. Brixner, G. Krampert, T. Pfeifer, R. Selle, G.G.M. Wollenhaupt, O. Graefe, C. Horn, D. Liese, T. Baumert, Phys. Rev. Lett. **92**, 208301 (2004)
256. T. Suzuki, S. Minemoto, T. Kanai, H. Sakai, Phys. Rev. Lett. **92**, 13 (2004)
257. L. Polachek, D. Oron, Y. Silberberg, Optics Letters **31**, 631 (2006)
258. M. Plewicki, F. Weise, S.M. Weber, A. Lindinger, Appl. Opt., Appl. Opt. **45**, 8354 (2006)
259. M. Plewicki, S.M. Weber, F. Weise, A. Lindinger, Appl. Phys. B, in press (2006)

260. F. Vetter, M. Plewicki, A. Lindinger, A. Merli, S.M. Weber, L. Wöste, Phys. Chem. Chem. Phys. **7**, 1151 (2005)
261. L.E. Berg, M. Beutter, T. Hansson, Chem. Phys. Chem **253**, 327 (1996)
262. J. Doyle, B. Friedrich, R.V. Krems, F. Masnou-Seeuws, Eur. Phys. J D **31**, 149 (2004)
263. G. Meijer, ChemPhysChem **3**, 495 (2002)
264. A. Fioretti, D. Comparat, A. Crubellier, O. Dulieu, F. Masnou-Seeuws, P. Pillet, Phys. Rev. Lett. **80**, 4402 (1998)
265. E.A. Donley, N.R. Claussen, S.T. Thompson, C.E. Wieman, Nature **417**, 529 (2002)
266. J. Weiner, V.S. Bagnato, S. Zilio, P.S. Julienne, Rev. Mod. Phys. **71**, 1 (1998)
267. F. Masnou-Seeuws, P. Pillet, Adv. in At., Mol. and Opt. Phys. **47**, 53 (2001)
268. O. Dulieu, F. Masnou-Seeuws, J. Opt. Soc. Am. B **20**, 1083 (2003)
269. C. Gabbanini, A. Fioretti, A. Lucchesini, S. Gozzini, M. Mazzoni, Phys. Rev. Lett. **84**, 2814 (2000)
270. A.J. Kerman, J.M. Sage, S. Sainis, T. Bergeman, D. DeMille, Phys. Rev. Lett. **92**, 153001 (2004)
271. C.P. Koch, E. Luc-Koenig, F. Masnou-Seeuws, Phys. Rev. A **73**, 033408 (2006)
272. C.P. Koch, R. Kosloff, F. Masnou-Seeuws, Phys. Rev. A **73**, 043409 (2006)
273. C.P. Koch, J.P. Palao, R. Kosloff, F. Masnou-Seeuws, Phys. Rev. A **70**, 013402 (2004)
274. W. Salzmann, U. Poschinger, R. Wester, M. Weidemüller, A. Merli, S.M. Weber, F. Sauer, M. Plewicki, F. Weise, A.M. Esparza, L. Wöste, A. Lindinger, Phys. Rev. A **73**, 023414 (2006)
275. B.L. Brown, A.J. Dicks, I.A. Walmsley, Phys. Rev. Lett. **96**, 173002 (2006)
276. C. Gabbanini, A. Fioretti, A. Luchesini, S. Gozzini, M. Mazzoni, Phys. Rev. Lett. **84**, 2814 (2000)
277. M. Kemmann, I. Mistric, S. Nussmann, H. Helm, Phys. Rev. A **69**, 022715 (2004)
278. T. Ban, D. Aumiler, G. Pichler, Phys. Rev. A **71**, 022711 (2005)
279. C.G. Townsend, N.H. Edwards, C.J. Cooper, K.P. Zetie, C.J. Foot, Phys. Rev. A **52**, 1423 (1995)
280. A. Bartelt, S. Minemoto, C. Lupulescu, Š. Vajda, L. Wöste, Eur. Phys. J. D **16**, 127 (2001)
281. M. Machholm, A. Giusti-Suzor, F.H. Mies, Phys. Rev. A **50**, 5025 (1994)
282. M.V. Korolkov, J. Manz, G.K. Paramonov, B. Schmidt, Chem. Phys. Lett. **260**, 604 (1996)
283. M.V. Korolkov, J. Manz, G.K. Paramonov, Chem. Phys. **217**, 341 (1997)
284. P. Backhaus, J. Manz, B. Schmidt, Phys. Chem. A **102**, 4118 (1998)
285. J. Vala, O. Dulieu, F. Masnou-Seeuws, P. Pillet, R. Kosloff, Phys. Rev. A **63**, 013412 (2000)
286. E. Luc-Koenig, R. Kosloff, F. Masnou-Seeuws, M. Vatasescu, Phys. Rev. A **70**, 033414 (2004)
287. E. Luc-Koenig, F. Masnou-Seeuws, M. Vatasescu, Eur. Phys. J D **31**, 239 (2004)
288. C.M. Dion, C. Drag, O. Dulieu, B. Laburthe Tolra, F. Masnou-Seeuws, P. Pillet, Phys. Rev. Lett. **86**, 2253 (2001)
289. A. Bartana, R. Kosloff, D.J. Tannor, J. Chem. Phys. **99**, 196 (1993)
290. A. Bartana, R. Kosloff, D.J. Tannor, J. Chem. Phys. **106**, 1435 (1997)

291. A. Bartana, R. Kosloff, D.J. Tannor, Chem. Phys. **267**, 195 (2001)
292. J. Werschnik, E. Gross, Journal of Optics B **7**, S300 (2005)
293. S.J. Park, Y.J. Choi, Y.S. Lee, G. Jeung, Chem. Phys. **257**, 135 (2000)
294. S.J. Park, S.W. Suh, Y.S. Lee, G. Jeung, J. Mol. Spec. **207**, 129 (2001)
295. D. Edvardsson, S. Lunell, C.M. Marian, Mol. Phys. **101**, 2381 (2003)
296. L.A. LaJohn, P.A. Christiansen, R.B. Ross, T. Atashroo, W.C. Ermler, J. Chem. Phys. **87**, 2812 (1987)
297. R. Kosloff, Ann. Rev. Phys. Chem. **45**, 145 (1994)
298. R. Mitrić, M. Hartmann, J. Pittner, V. Bonačić-Koutecký, J. Phys. Chem. A **106**, 10477 (2002)
299. K. Hoki, D. Kröner, J. Manz, Chem. Phys. **267**, 59 (2001)
300. I. Barth, J. Manz, Angew. Chem. Intern. Ed. **45**, 2962 (2006)
301. A.B. Alekseyev, M.V. Korolkov, O. Kühn, J. Manz, M. Schröder, J. Photochem. & Photobiol. A **180**, 262 (2006)
302. J.M. Sage, S. Sainis, T. Bergeman, D. DeMille, Phys. Rev. Lett. **94**, 203001 (2005)

3

Complex systems in the gas phase

Vlasta Bonačić-Koutecký[1], Brina Brauer[2], Florian Burmeister[3], Wolfgang Eberhardt[3], R. Benny Gerber[2], Leticia González[4], Gert von Helden[5], Aster Kammrath[6], Seong K. Kim[7], Jörn Manz[4], Gerard Meijer[5], Roland Mitrić[1], Matthias Neeb[3], Daniel M. Neumark[6], Thomas Schultz[8], and Jörg Stanzel[3]

[1] Institut für Chemie, Humboldt-Universität zu Berlin, Germany
[2] The Fritz Haber Research Center and Department of Physical Chemistry, The Hebrew University of Jerusalem, Israel
[3] BESSY, Berlin, Germany
[4] Institut für Chemie und Biochemie, Freie Universität Berlin, Germany
[5] Fritz-Haber-Institut der Max-Planck-Gesellschaft, Berlin, Germany
[6] Department of Chemistry, University of California Berkeley, USA
[7] School of Chemistry, College of Natural Sciences, Seoul National University, Korea
[8] Max Born Institut Berlin, Germany

Coordinated by: Gert von Helden

3.1 Motivation and outline

Experimental and theoretical investigations of ultrafast processes in molecules or clusters can lead to a detailed understanding of the dynamical processes involved. Their control by tailored laser pulses can be employed to determine how the interplay of size, structures and light fields can be used to manipulate optical properties and chemical reactivity of these systems. Moreover, laser-selective femtochemistry [1–11] can be combined with the functionalism of nanostructures, providing new perspectives for the basic research as well as for technological applications.

Several examples of dynamical behavior and control of simple systems in the gas phase have been presented in the previous chapter. Here, more "complex" systems in the gas phase will be treated. A question that comes up in that context is "What is complex?". Of course, there is no clear definition of "complex" and a system that might appear large and complex to an atomic physicist might appear small and manageable to a biochemist. However, it is clear that the dynamics of systems containing more than a few atoms can be demanding for our intuitive understanding as well as pose a challenge to theory. The examples that will be presented in this chapter show indeed that they are complex enough, so that not all aspects of their behavior can be understood. Choices have thus to be made on what are open questions to be answered about a particular system as well as on what can or needs to be learned.

In this chapter, several specific examples are given. The examples include the ultrafast dynamics of solvated electrons, dynamics on electronically excited state surfaces of DNA base pairs, electronic and nuclear dynamics in metal cluster anions, vibrations and conformations in gas-phase amino acids, the dissociation dynamics of metal carbonyls as well as work on exploring and controlling the dynamics of free clusters and biomolecules.

All systems are investigated as isolated systems in the gas phase. For the clusters considered in this chapter the properties can have a strong size dependence and therefore mass selection is important. Understanding dynamical properties of clusters in the gas phase is of crucial importance before starting to investigate the influence of different environments. For biological molecules the situation is even more demanding. When commenting about experiments on biological molecules in the gas phase, biochemists frequently criticize that the properties of an isolated gas-phase biomolecule are completely irrelevant to any problem they face. However, to understand the structure and dynamics of biological molecules in their native environments, it is often imperative to first understand these properties for the isolated systems in the gas phase. There, one can measure the intrinsic properties and it is indeed frequently observed that they differ considerably from those observed when the molecule is in solution, embedded in a solid or deposited on surfaces. The reasons for those differences are found in the various bonding, electrostatic and dispersion interactions to the environment, which can - and will - affect the properties of the molecules. Before investigating large peptides and proteins, it is useful first to start with model systems, as the knowledge of the gas-phase structures and dynamics of such model systems can provide important insight into fundamental intramolecular interactions and can serve as calibration points for theoretical models.

In the following, a short introduction to the specific topics considered in this chapter is given, followed by a discussion of theoretical and experimental results.

3.1.1 Solvated electron in clusters

The solvated electron has long attracted interest both for its chemical importance and as the simplest quantum solute, providing a testing-ground for theoretical methods. The hydrated electron, in particular, plays a significant role in many biological processes. Its dynamics have been studied experimentally by a number of groups, whose results are in good agreement with each other [12–16]. The electron, excited from the ground 's-like' state to the excited 'p-like' state, undergoes decay on three timescales, with lifetimes of 50 fs, ~300 fs, and ~1 ps. These dynamics have been interpreted and the different timescales are assigned to various combinations of excited state solvation dynamics, internal conversion from the excited to the ground state, localized solvent dynamics on the ground state and extended solvent dynamics on the ground state. In an effort to understand the systems, numerous theoretical studies have been made.

In another approach, a number of studies were undertaken on the cluster analog of the hydrated electron. Ayotte and Johnson [17] traced the absorption spectra in small and medium sized $(\text{water})_n^-$ clusters (n=6-50) and found the excitation energy to increase with increasing cluster size, allowing for a smooth extrapolation to the bulk absorption spectrum. In work foreshadowing the results presented in Sect. 3.4.1, the same group showed that resonant 2-photon detachment can, in fact, be observed at 1.5 eV for large anionic water clusters [18].

Relatively few studies have been made of electron dynamics in the related system of methanol. A few theoretical studies have been done, largely by Rossky and co-workers. The groups of Barbara [19] and Laubereau [20] have carried out time-resolved studies of the formation and equilibration of the solvated electron in bulk methanol, and of the dynamics following excitation of the equilibrated electron, with slightly varying results, possibly attributable to different temporal resolution of the two experiments. Thaller [20] observes dynamics on three time-scales, analogous to water but somewhat slower, with the first having a lifetime of ~105-170 fs, the second ~0.67-3.5 ps and the longest several picoseconds. Studies of the corresponding clusters have been limited [21, 22].

3.1.2 DNA base clusters

In biological systems, nature employs a limited set of amino acids, DNA and RNA bases to construct a molecular machinery performing the myriad of functions which ultimately add up to living organisms. The complexity required to allow functions such as molecular recognition, energy transformation or reaction catalysis is not simply encoded in the building blocks, but is rather a product of the well defined secondary structure of the assembled biomolecules (proteins, DNA and RNA). To understand biological function, it is therefore necessary to go beyond the mere characterization of the biological building blocks and it is desirable to explore the effect of intermolecular interactions which control crucial molecular properties in a structured environment. Those interactions are the key to understand the structure-function relationship in biological systems. Biological function also implies a dynamical process, e.g. the energy flow along a reaction coordinate. Femtosecond lasers allow us to observe reaction processes in real time and to resolve the intricate details usually depicted by a simple reaction arrow.

However, if it is desired to follow the flow of energy, and to predict resulting changes in nuclear or electronic structure, one must deal with an enormous complexity given by the large number density of nuclear and electronic degrees of freedom. In addition, for large systems, weak and noncovalent interactions can be particular important and consequently, such systems present a challenge to theory. The investigation of photochemical and photophysical processes in the building blocks of DNA, are an accessible starting point for the investigation of charge and energy flow in biomolecules. By the study of DNA base pairs and microhydrated bases, it might be possible to identify the relevant excited states and reaction / relaxation pathways in DNA and the investigation of molecular clusters may offer a first step to bridge the gap between our detailed understanding of processes in the gas phase and the experimentally less-accessible systems in the condensed phase. However, whether the investigation of such small building blocks is sufficient for an extrapolation to real, biologically relevant systems is an open question. The wealth of existing experimental and theoretical data in this field was recently summarized in an extensive review [23]. The experiments shown in Sect. 3.4.2 are build on these data and try to gauge the respective importance of several excited state reaction coordinates in different cluster environments.

3.1.3 Vibrational and conformational dynamics in gas-phase aminoacids and peptides

For most biomolecules, not only the connectivity of atoms (the primary structure) is important but also the three-dimensional folding arrangement of atoms, the secondary and tertiary structures. In fact, in biological molecules, these higher order structures largely determine their physiological function. For larger molecules, a vast amount of possible conformations exists and it is surprising that nature almost always finds the same conformation that has the desired function. It is thus very important to investigate the structural dynamics of peptides and proteins at a basic level and much experimental and theoretical work is devoted to that.

It is frequently observed that even small systems, such as isolated aminoacids in the gas phase, have complex potential energy surfaces with many low lying conformers that often are simultaneously populated. In the 1980s, UV hole burning techniques coupled to molecular beam methods have demonstrated that different conformers of simple amino acids can be selectively excited and investigated [24]. When coupled to a tunable infrared (IR) laser, this hole-burning spectroscopy allows to record conformer-specific IR spectra in a heterogeneous mixture of conformations [25]. Especially the IR absorption spectrum is a unique identifier for the structure: line intensities and frequencies give direct information on the forces that hold the molecule together. Conformer selective IR spectroscopy in the gas phase has been applied to various interesting systems, such as isolated or paired nucleobases [26], to the amino acids phenylalanine [25] and tryptophan [27, 28], to beta-sheet model systems [29] and other small peptides [30]. Most of these experiments have been performed using laser systems that cover the near- and mid-IR and are limited to wavelengths shorter than about 7 μm. In this range X–H (X=C,O,N) and C=O stretching vibrations are probed and the experiments provide insight into the conformational arrangement, as those vibrational transitions exhibit shifts in absorption frequency and changes in intensity in the presence of, for instance, intramolecular hydrogen bonds or solvating molecules. However, most vibrational modes have characteristic frequencies that lie further in the IR and are not accessible using standard laser systems. Those low frequency modes are frequently floppy, large amplitude motions and delocalized vibrations which can be of particular importance in the dynamics of conformational change. Recently, specialized tabletop systems [31] and free electron lasers [32] have shown to be suitable tools for experiments in the low frequency range on gas-phase biomolecules.

In many biomolecules, a conformational change can be induced thermally and the barriers that are crossed in such processes have thus heights that are comparable to the energy of IR photons. In most experiments, the local minima are investigated and only little information is available on the barriers that separate those minima. In recent elegant experiments, Zwier and coworkers showed that conformational change in gas-phase biomolecules can be induced by the absorption of 3 μm photons and that this change can be monitored by UV excitation methods [33]. The barriers, however, that separate the different conformations are much lower in energy than the 3 μm excitation energy. Information on their height can be obtained by populating the relevant levels via stimulated emission pumping (SEP) [34].

The aim of the experiments presented here is to induce conformational change via direct IR excitation to energy levels that are close to the barriers. As a prerequisite, however, information on the energy levels for the different conformers needs to be collected. This can be conveniently done via IR spectroscopy. Next, the IR induced conformation change can be used to locate the various barriers and to investigate which minima they separate. Then, the conformational dynamics can be controlled via excitation with phase and amplitude shaped ultrafast pulses or via the timed excitation with picosecond IR pulses of suitable wavelengths [35]. The experiments will be accompanied by theoretical investigations where first, the potential energy surface will be mapped by quantum chemical methods, initially employing harmonic theory, next to include anharmonic corrections and then to include dynamics. In Sect. 3.4.4 we report on the first steps on the avenue where the conformational structures of the amino-acid phenylalanine are investigated via IR spectroscopy in the range from 300 – 1900 cm^{-1}.

3.1.4 Fragmentation dynamics of organometallic compounds

Organometallic compounds containing metal and carbonyl chromophores have been the object of profound interest because fragmentation processes occur in electronically excited states . A large number of studies have concentrated on pump–probe experiments which allow the excited states of transition metal complexes to be investigated (see e.g. [36–39]). This has, of course, also prompted numerous theoretical studies (a few examples can be found in [40–46]) which aim to support experiments and shed some light on the ultrafast photodissociation dynamics of such complexes.

In brief, the principle of a pump–probe experiment is the following: an ultrashort laser pulse—the pump—excites the molecule from the vibrational ground state of the electronic ground state to some electronic excited state, which can be bound, dissociative, or even more interesting, predissociative (see Chapter 2). In any case, the system evolves in time showing vibrations or dissociation or both. Such behavior can be monitored e.g. by multiphoton ionization with the probe pulse which arrives after a well-defined time delay, interrogating the system at a particular location. Interestingly, since the detection of the products is done after ionization, the path the system follows after excitation is not trivial. Indeed, fragmentation may occur either on neutral or on ionic surfaces without any difference in the detected fragmented ions. This problem has raised considerable attention. For instance, Trushin and coworkers have studied dissociative ionization of different metal carbonyls, $Ni(CO)_4$, $Fe(CO)_5$ and $Cr(CO)_6$ at intensities between 10^{12}–10^{14} W cm^{-2} finding that neutral dissociation can be practically neglected; fragmentation is instead rationalized by resonances in ions [47]. The example shown below (cf. Sect. 3.4.5) shows, on the other hand, that dissociation can also start on neutral surfaces.

Transition metal carbonyl complexes have additional interest because they were the first candidates in which adaptive photodissociation control was pioneered. Using $CpClFe(CO)_2$ ($Cp{=}\eta^5C_5{-}H_5$) the group of G. Gerber showed that it is possible to control the breaking of different fragmentation channels [39]. As introduced in Chapter 1, in adaptive optimal control, genetic algorithms search the best pulse shapes to prepare specific products based on fitness information, such as product yields. Also called *closed-loop* control this experimental approach prepares the desired target solving the Schrödinger equation exactly in real time through solving a many-parameter problem [11]. In most cases, however, the significance of the obtained optimal electric field is obscured by the complexity of the pulses.

An interesting question is how to use the experimental pulse shape in order to elucidate the physical mechanism which makes the control possible. This demanding issue has been addressed in different ways. Experimentally, one could repeat the optimizations with a reduced number of parameters and compare the results [11, 48]. Theoretically, it has been suggested that the learning algorithm could also be programmed to find out the control variables needed in the experiment [49]. Other strategies to learn about the system's dynamics incorporate spectral pressure during the optimization loops [50], try to extract information from the experimental data with different inversion algorithms [51–56], or even include the liquid crystal spatial light modulators in optimal control theory to establish a realistic link between experimental and theoretical optimal pulses [57]. Other examples were also explained in Chapter 2, Sect. 2.3.

3.1.5 Analysis and control of ultrafast processes in clusters

The size-selective optical and reactivity properties of clusters in the nonscalable size regime in which each atom counts are determined by the number of atoms and the corresponding structures [58–61]. The investigation of the dynamics of these systems with finite density of states is especially attractive since the separation of time scales of different processes is possible [60]. Joint theoretical and experimental time-resolved ultrafast studies carried out on clusters provided findings on the nature and the time scales of processes, such as geometrical relaxation, internal vibrational energy distribution (IVR), charge separation, and Coulomb explosion [60, 62–67]. For example, time resolved photoelectron spectroscopy offers a unique opportunity to follow structural changes or ultrafast electron relaxation in photoexcited particles with high precision. This will be illustrated by joint experimental and theoretical study of size-dependent excited state relaxation dynamics in small anionic gold clusters . Furthermore, due to advances in laser technology, tailored laser fields can be produced by pulse shapers which can control molecular dynamics guiding it to a chosen target, such as a given fragmentation channel, a particular isomer or a desired reaction product [11,40,65,68,69] (cf. also references in Chapter 2).

The role of theory has been essential from conceptual as well as from a predictive point of view. Time-resolved observations are strongly dependent on the experimental conditions, such as laser wave lengths, duration of pulses and their shapes, competition between one- and many-photon processes, strength of the electric field etc. Here, theory has the task not only to provide insight into the nature of time dependent processes, but also to identify the conditions under which they can be experimentally observed [40, 65–67, 70–81]. Consequently, theory can be directly involved in conceptual planning of time-resolved experiments.

Other prominent examples are theoretical proposals for different optical control schemes using a laser field parameters for the manipulation of ultrafast process pioneered by Rice and Tannor, Shapiro and Brumer and Peirce, Dahleh and Rabitz [2,82–87]. They stimulated control experiments which were carried out first on simple systems such a metallic dimers and trimers [88–110], and later on more complex systems [39,40,48,69,111–116], confirming theoretically proposed concepts. Since tailored laser pulses have the ability to select pathways which optimally lead to the chosen target, their analysis should allow one to determine the mechanism of the processes and to provide the information about the selected pathways (inversion problem) [56]. Therefore, theoretical approaches are needed, which are capable to design interpretable optimal laser pulses for complex systems (e.g. clusters, organic, organometallic, or bio-molecules) by establishing the connection between the underlying dynamical processes and their shapes. In this case, the optimal control can be used as a tool for the analysis.

In Section 3.4.6 of this chapter, an overview of ultrafast time-resolved pump-probe spectroscopy and optimal control of moderately complex systems is presented which is structured as following: The nonadiabatic dynamics involving electronic ground and excited states will be addressed in Sect. 3.4.6.1 based on our theoretical approach which combines ab initio molecular dynamics (MD) "on the fly" and the Wigner distribution approach (cf. Sect. 3.2.3). This approach will be used to study the photoisomerization processes through the conical intersections on the prototype examples of nonstoichiometric Na_3F and Na_3F_2 clusters with two and one excess electrons (cf. Sects. 3.4.6.2, 3.4.6.3 and 3.4.6.4).

In Sects. 3.4.6.6-3.4.6.12, also new strategies for optimal control in complex systems will be outlined. First, tailored pump-dump pulses will be used to drive photoisomerization process in the Na_3F_2 cluster, avoiding the pathway with high excess energy involving the conical intersection, and populating only one selected isomer [65, 77]. This is achieved by introducing a new strategy for optimal control based on an intermediate target in the excited state corresponding to a localized ensemble which provides a connective pathway between the initial step and the target in the ground state. The connection between the shapes of the optimized pulses and the underlying processes will be explored.

Secondly, the new strategy for infrared control of conformational changes based on combination of quantum chemical MD "on the fly" including IR field and genetic algorithm for pulse optimization will be presented and illustrated on two prototype examples: Na_3F cluster and the glycine molecule. Summary and outlook for the future joint experimental and theoretical work, which are based on our new strategies for optimal control will be given in Sect. 3.4.6.12.

3.2 Theoretical basis

3.2.1 Methods to determine steady state properties

L. González

Steady state properties of a molecular system can be obtained very accurately within the Born-Oppenheimer approximation, solving the time-independent electronic Schrödinger Equation [117]

$$H\Phi_i(q) = E_i(q)\Phi_i(q). \tag{3.1}$$

Here $E_i(q) \equiv V_i(q)$ represents the electronic energy, or the adiabatic potential energy $V_i(q)$ for the nuclear motion along the nuclear coordinate q for the electronic state i, and $\Phi_i(q)$ is the time-independent wavefunction describing the molecular system. Solving (3.1) from first principles is the spirit behind the so–called ab initio methods [118], which start from the Hartree-Fock approximation [117]. Here, the electronic wavefunction Ψ is described in terms of a single spin-adapted Slater determinant $D_i(q)$

$$D_i(q) = ||\chi_1^{\uparrow}(q)\chi_1^{\downarrow}(q)\dots\chi_k^{\uparrow}(q)\chi_k^{\downarrow}(q)\dots\chi_n^{\uparrow}(q)\chi_n^{\downarrow}(q)||, \tag{3.2}$$

where $\chi_1(q),\dots,\chi_k(q),\dots\chi_n(q)$ denote the doubly ($\uparrow\downarrow$) occupied orbitals containing $2n$ electrons. This assumption is often adequate for describing properties in the electronic ground state. A better description at little additional computational costs can be obtained using density functional theory [119]. Further refinements to the energy require the inclusion of dynamical correlation [117, 118], either using perturbation theory (Møller-Plesset MPn series) [120] or via truncated configuration interaction (CI) (Quadratic CI or Coupled Cluster series) [121]. When a radiation field is present, however, the molecules are electronically excited and the assumption of a single Slater determinant is not accurate any more. The realistic description of electronically excited states requires a multiconfigurational wavefunction. The Multiconfiguration Self-Consistent Field method (MCSCF) is a truncated CI expansion where not only the CI coefficients (c_{ji} in (3.3)), but also the molecular orbitals coefficients are variationally determined [122].

$$\Phi_i(q) = \sum_j c_{ji} D_j(q) \tag{3.3}$$

In order to select the important configurations which are relevant for the system to study, a popular method is the Complete Active Space SCF (CASSCF) method [123]. In this method, the configurations are chosen by selecting the electrons and orbitals which play a role in the electronic structure of the molecule. In cases where several states are near degenerate or crossings between electronic states are present, it is necessary to perform a state-average CASSCF (SA-CASSCF) calculation for each symmetry and spin; i.e. the electronic states of interest are optimized simultaneously in an averaged way, minimizing the weighted sum of their energies. To get more accurate energies dynamical correlation must be included, either perturbationally –with CASPT2 approaches [124, 125]– or variationally, e.g. using multireference CI (MRCI) methods. In the latter method the critical step is the selection of the reference configurations, which is not a trivial task and which is often based on chemical intuition. For instance, when calculating potential energy surfaces (PES) a reasonable understanding of the entire region is required; it is not enough to know the dominant configurations of the reactants and products, but the configurations describing each of the intermediates the molecule passes through are also needed. MRCI is probably one of the most accurate methods to describe electronically excited states and to compute spectroscopic properties, however it is computationally very demanding and can only be employed for small systems or one has to choose a small number of references and a small number of excited configurations relative to each reference.

3.2.2 Solutions of the time dependent Schrödinger equation for complex systems in reduced dimensionality

L. González and J. Manz

The dynamics of a molecular system can be simulated by solving the time-dependent Schrödinger equation,

$$i\hbar\frac{\partial}{\partial t}\begin{pmatrix}\psi_0\\ \vdots\\ \psi_n\end{pmatrix} = \begin{pmatrix}H_{00} & \cdots & H_{nn}\\ \vdots & & \vdots\\ H_{n0} & \cdots & H_{nn}\end{pmatrix}\cdot\begin{pmatrix}\psi_0\\ \vdots\\ \psi_n\end{pmatrix} \tag{3.4}$$

where the $\psi_i(t)$ are the nuclear wave functions for the electronic states i. The matrix elements H_{ij} of the Hamiltonian consist of the the kinetic energy T, the potential energy V_i, the kinetic or nonadiabatic couplings, $T_{ij}^{(1)}$ and $T_{ij}^{(2)}$, between the electronic states i and j, and the laser coupling $W_{ij}(t)$. For a single coordinate q,

$$H_{ij} = \delta_{ij}\,(T + V_i) - \frac{\hbar^2}{2\,m}\,(\,2T_{ij}^{(1)}\frac{\partial}{\partial q} + T_{ij}^{(2)}\,) + W_{ij}(t)\ . \tag{3.5}$$

The coordinate dependent kinetic couplings are defined as

$$T_{ij}^{(1)}(q) = \langle \Phi_i(q)|\frac{\partial}{\partial q}|\Phi_j(q)\rangle \tag{3.6a}$$

and

$$T_{ij}^{(2)}(q) = \langle \Phi_i(q)|\frac{\partial^2}{\partial q^2}|\Phi_j(q)\rangle. \tag{3.6b}$$

where $\Phi_j(q)$ are the time-independent electronic wave functions. Analogous expressions hold for multidimensional systems. The laser-matter interaction can be described in terms of the semi-classical dipole approximation, in which the molecule is treated quantum mechanically and the field is treated classically,

$$W_{ij}(q,t) = -\boldsymbol{E}(t)\langle \Phi_i(q)|\boldsymbol{M}|\Phi_j(q)\rangle. \tag{3.7}$$

Here $\langle \Phi_i(q)|\boldsymbol{M}|\Phi_j(q)\rangle$ is the transition dipole moment vector between electronic states i and j, and it plays an important role determining whether or not a transition is allowed and determining its strength. $\boldsymbol{E}(t)$ is the time-dependent electromagnetic laser field, often given by the following expression:

$$\boldsymbol{E}(t) = \boldsymbol{e}E^0\cos(\omega t - \eta)\cdot s(t), \tag{3.8}$$

where $\boldsymbol{e}$ is the polarization direction of the field, E^0 is the amplitude of the field, ω the carrier frequency, η is the phase, and $s(t)$ defines the shape of the laser pulse. In the present application, the amplitudes E^0 are always kept under the "weak field limit" to avoid intrapulse pump-dump processes where this would be unphysical, undesired multiphoton processes, or Stark shifts. In theoretical simulations, $s(t)$ is typically replaced by a function of the form of (3.9) [84], in the experiments it can have different forms, see Chapter 2.

$$s(t) = \sin^2\left(\frac{\pi(t-t_i)}{t_p}\right) \quad \text{for} \quad t_i \leq t \leq t_i + t_p. \tag{3.9}$$

In (3.9), t_p is the pulse duration and t_i is the initial time of the pulse. In the case of pump probe spectroscopy $\boldsymbol{E}(t) = \boldsymbol{E}_{pump}(t) + \boldsymbol{E}_{probe}(t)$ where $\boldsymbol{E}_{pump}(t)$ and $\boldsymbol{E}_{probe}(t)$ denote the respective pump and probe pulses with time delay t_d; then we have, for example, $t_i = 0$ and $t_i = t_d$ for the pump and probe pulses respectively.

Before solving (3.4), an initial wavefunction $\Psi(t_0)$ is required. Unless the system has been preexcited, this will be the vibrational ground state of the electronic ground state. The eigenfunctions and corresponding eigenvalues of a bound potential can be calculated numerically with the Fourier Grid Hamiltonian (FGH) method [126, 127], which uses a matrix-discretization of the Hamiltonian operator in the coordinate space. Equation (3.4) is then solved numerically propagating in time the initial wavefunction $\Psi(t_0)$ represented in the grid until $t = t_f$. The final $\Psi_i(t_f)$ is then the final wavefunction which can be analyzed to obtain state populations $P_i(t_f) = \int |\Psi_i(t_f)|^2 dq$, quantum yields and other properties. The time propagation of the wavefunction can be solved using different schemes [128], for example using the second order differencing method (SOD) [129] or the split operator (SPO) method [130–132]. In cases of nondiagonal parts of the Hamiltonian (e.g. nonadiabatic potential coupling or coupling with laser field, like in (3.4)) it is convenient if the wavefunctions are transformed back and forth into proper representations in which the operators are diagonal. In cases without kinetic couplings, the widely-used SPO method is based on a symmetric splitting of kinetic T and potential energy plus the laser dipole coupling $V + W(t)$ terms as

$$e^{-\frac{i}{\hbar}\Delta t H} \simeq e^{-\frac{i}{2\hbar}\Delta t T} e^{-\frac{i}{\hbar}\Delta t(V+W(t))} e^{-\frac{i}{2\hbar}\Delta t T}. \tag{3.10}$$

The error is of third order in Δt because T and $V + W(t)$ do not commute. Since T is a multiplication in the momentum space, the corresponding exponential is evaluated by transforming the wavefunction from the coordinate space to the momentum space using a fast Fourier transformation (FFT) [129], multiplying by $exp(-ik^2\hbar^2\Delta t/2m\hbar)$ (where k is the wave vector) and transforming back to the coordinate space. The SOD method may be applied in more general cases including kinetic couplings, see e.g. [40].

3.2.3 Nonadiabaticity and femtosecond signals

V. Bonačić-Koutecký and R. Mitrić

In simulations of nonadiabatic dynamics, one can employ Tully's molecular dynamics with quantum transitions (MDQT) together with the Tully's fewest-switches surface hopping method [133]. The basic feature of the surface-hopping methods is that the propagation is carried out on one of the pure adiabatic states, which is selected according to its population, and the average over the ensemble of trajectories is performed. This is based on the assumption that the fraction of trajectories on each surface is equivalent to the corresponding average quantum probability determined by coherent propagation of quantum amplitude. Moreover, a choice between adiabatic and diabatic representation has to be made. In the former case the nonadiabatic couplings have to be calculated, and in the latter case the overlap between the wavefunctions of two states is needed in the framework of the method used for calculations of the electronic structure. For example, the MD as well as nonadiabatic couplings calculated "on the fly" can be directly connected with MDQT and then used to simulate fs-signals. Therefore, we briefly outline the concept involving the adiabatic representation.

The time dependent wavefunction $\Psi(t, \boldsymbol{r}, \boldsymbol{R})$, which describes the electronic state at time t, is expanded in terms of the adiabatic electronic basis functions ψ_j of the Hamiltonian with complex valued time dependent coefficients

$$\Psi(t, \boldsymbol{r}, \boldsymbol{R}) = \sum_{j=0}^{M} c_j(t)\psi_j(\boldsymbol{r}; \boldsymbol{R}(t)). \tag{3.11}$$

The adiabatic states are also time dependent through the classical trajectory $\boldsymbol{R}(t)$. Substitution of this expansion into the time dependent Schrödinger equation, multiplication by ψ_k from the left, and integration over $\boldsymbol{r}$ yields a set of linear differential equations of the first order for the expansion coefficients which are equations of motion for the quantum amplitudes:

$$\mathrm{i}\dot{c}_k(t) = \sum_j \left[\epsilon_j \delta_{kj} - \mathrm{i}\dot{\boldsymbol{R}}(t) \cdot \langle \psi_k | \nabla_{\boldsymbol{R}} | \psi_j \rangle \right] c_j(t). \tag{3.12}$$

Here ϵ_j are the eigenvalues of the electronic Hamiltonian, and $\langle \psi_k | \nabla_{\boldsymbol{R}} | \psi_j \rangle$ are nonadiabatic coupling vectors.

The system of equations (3.12) has to be solved simultaneously with the classical equations of motion for the nuclei

$$M\ddot{\boldsymbol{R}} = -\nabla_{\boldsymbol{R}} E_m(\boldsymbol{R}), \tag{3.13}$$

where the force is the negative gradient of the potential energy of the "current" m-th adiabatic state. The hopping probabilities g_{ij} between the states are determined by

$$g_{ij} = 2\frac{\Delta t}{c_i c_i^*}[Im(c_i^* c_j \varepsilon_i \delta_{ij}) - Re(c_i^* c_j \dot{\mathbf{R}} \langle \psi_i | \nabla_{\boldsymbol{R}} | \psi_j \rangle] \tag{3.14}$$

and can occur randomly according to the fewest-switches surface hopping approach introduced by Tully [133]. This approach has been designed to satisfy the statistical distribution of state populations at each time according to the quantum probabilities $|c_i|^2$ using the minimal number of "hops" necessary to achieve this condition.

The energy conservation in MDQT is achieved by rescaling the classical velocities in the direction of the nonadiabatic coupling vector [133]. The transition is classically forbidden if there is not enough kinetic energy in this direction. In this case, two alternatives are commonly used. Either this component of velocity is inverted or it remains unchanged.

For the determination of the pump-probe signals in the framework of the ab initio Wigner distribution approach accounting for passage through the conical intersection the expression for the cationic occupation $P_{22}^{(2)}$ given by (2.6) in the Chapter 2, Sect. 2.2 on NeNePo spectroscopy has to be modified. This is due to the necessity of considering that the propagation of the ensemble starts in the excited state but can hop to the different electronic state according to the fewest-switches hopping algorithm. Therefore, not only the common averaging over the whole ensemble of the initial conditions due to the Wigner approach is required, but also, for a given initial condition, an averaging over trajectories obtained from different random numbers according to the hopping algorithm must be carried out [72]. Consequently, the coordinates and momenta of the propagated state can be labeled by $\mathbf{q}_x^\nu$ and $\mathbf{p}_x^\nu$, where x is either the excited or the ground state, as determined by the hopping procedure. The quantities ν numerate the set of random numbers used in the hopping algorithm, satisfying the same initial condition. Therefore, the average over the number of hoppings N_{hop} has to be performed and for the cationic population the following expression yields

$$\begin{aligned}P_{22}^{(2)}(t) &= \int d\mathbf{q}d\mathbf{p}\, P_{22}^{(2)}(\mathbf{q},\mathbf{p},t)\\ &\sim \int d\mathbf{q}_0 d\mathbf{p}_0 \int_0^t d\tau_2 \int_0^{t-\tau_2} d\tau_1\\ &\quad \frac{1}{N_{hop}}\sum_\nu \exp\left[-\sigma_{pr}^2 \frac{[\hbar\omega_{pr} - V_{+1,x}(\mathbf{q}_x^\nu(\tau_1;\mathbf{q}_0,\mathbf{p}_0))]^2}{\hbar^2}\right] \times\\ &\quad \exp\left[-\sigma_{pu}^2 \frac{[\hbar\omega_{pu} - V_{10}(\mathbf{q}_0,\mathbf{p}_0)]^2}{\hbar^2}\right] I_{pu}(t-\tau_1-\tau_2) I_{pr}(t-\tau_2-t_d) \times\\ &\quad P_{00}^{(0)}(\mathbf{q}_0,\mathbf{p}_0), \end{aligned} \tag{3.15}$$

which is a modification of (2.6) in Chapter 2, Sect. 2.2 on NeNePo spectroscopy valid for the adiabatic case. The quantity $V_{+1,x}$ labels the energy gap between the propagating state and the cationic state at the instant of time. From this expression, the pump-probe signal can be calculated, after the integration over the pump–probe correlation function $\int_0^\infty d\tau_2 I_{pu}(t-\tau_1-\tau_2)I_{pr}(t-\tau_2-t_d)$ is performed explicitly:

$$
\begin{aligned}
S[t_d] &= \lim_{t\to\infty} P_{22}^{(2)}(t) \\
&\sim \int d\mathbf{q}_0 d\mathbf{p}_0 \int_0^\infty d\tau_1 \exp\left\{-\frac{(\tau_1 - t_d)^2}{\sigma_{pu}^2 + \sigma_{pr}^2}\right\} \times \\
&\frac{1}{N_{hop}} \sum_\nu \exp\left\{-\frac{\sigma_{pr}^2}{\hbar^2}[\hbar\omega_{pr} - V_{+1,x}(\mathbf{q}_x^\nu(\tau_1;\mathbf{q}_0,\mathbf{p}_0))]^2\right\} \times \\
&\exp\left\{-\frac{\sigma_{pu}^2}{\hbar^2}[\hbar\omega_{pu} - V_{10}(\mathbf{q}_0,\mathbf{p}_0)]^2\right\} P_{00}^{(0)}(\mathbf{q}_0,\mathbf{p}_0). \qquad (3.16)
\end{aligned}
$$

This expression is applicable if interference and non-Condon effects are negligible. According to expression (3.16), the initial ground state density $P_{00}^{(0)}$ is promoted to the first excited state with the Franck-Condon transition probability given by the last exponential of (3.16). The propagation, the passing through the conical intersection, and the probe transition to the cationic state are described by the second exponential. This expression can be generalized for more than two states by introducing in (3.16) the sum of weighting factors corresponding to transition moments between the electronic states involved, for which also time dependent energy gaps have to be calculated. The probe pulse window, being located around the time delay t_d between the pump and the probe pulse and the resolution of the signal determined by the square of the pulse durations, are given by the first exponential. As it is required in the Wigner distribution approach, an ensemble average over the initial conditions has to be performed. The latter can be obtained from a sampling of the initial vibronic Wigner distribution $P_{00}^{(0)}$ of the ground electronic state.

3.3 Experimental approaches

3.3.1 Time-resolved photoelectron spectroscopy

Time-resolved photoelectron spectroscopy (TRPES) of clusters or molecules is an unique tool to track both nuclear dynamics as well as electron dynamics in photoexcited particles with temporal resolution. In femtosecond TRPES, a femtosecond laser pump pulse promotes the species of interest to an electronically excited state, transfers an electron from a chromophore to the solvent, or induces a reaction. After a variable delay the system is characterized by detaching an electron with a femtosecond probe pulse.

Such experiments can be performed with anionic or neutral species and examples for both are shown in this chapter. Anions have some significant advantages over neutrals. Firstly, they are easily mass separated and it is possible to ensure that only a single species interacts with the laser. This is particularly useful in cluster studies. For neutral clusters, mass selection is only possible in the cationic species after electron detachment and corresponding measurements must use electron-ion coincidence techniques. Secondly, much lower photon energies are required to detach an electron from an anion than from a neutral cluster. Therefore experiments can generally be carried out with single-photon excitation and detachment and the dynamics can be readily followed down to the ground state.

On the other hand, the relatively low density in mass-selected anion beams is a major disadvantage. This becomes especially problematic in time-resolved studies, where spectra must be collected at a large number of pump-probe delays for each species. To trace population dynamics of different states as a function of the pump-probe delay, it is essential that signal levels remain fairly constant over long periods of time.

In TRPES experiments, the experimental observable is the photoelectron kinetic energy distribution as a function of pump-probe delay. This energy can either be measured directly by time-of-flight (TOF) methods or in conjunction with velocity map imaging. In the latter case, the photoelectron angular distribution is also obtained as a function of pump-probe delay. These two pieces of information are powerful tools for deciphering the dynamics of relaxation or reaction following the initial excitation.

The energy difference between the cluster A^- before electron detachment and the corresponding species X after detachment is supplied by the photon energy $h\nu$. The photon energy is known and the kinetic energy of the detached electron is directly correlated with the energy of the remaining system:

$$E_{el} = h\nu + (E_{A^-} + \mathrm{KE}_{A^-}) - (E_X + \mathrm{KE}_X) , \tag{3.17}$$

where E_{el} corresponds to the electron kinetic energy, E_{A^-} and E_X to the internal energy of A^- and X, and KE to the kinetic energy of the cluster. This equation holds true for both, single- and multiphoton (pump-probe) ionization. Conservation of momentum must apply between the light electron and the heavier cluster core, hence electron detachment will deposit only a minuscule amount of kinetic energy into the cluster and we can set (K_{EA^-}) - $(\mathrm{K}_{EX}) = 0$. Therefore, (3.17) simplifies into:

$$E_{el} = h\nu - (E_X - E_{A^-}), \text{or rather} \tag{3.18}$$

$$E_{el} = h\nu - BE. \tag{3.19}$$

Thus the measured electron binding energy BE is equal to the energy difference between the initial state A^- and the final state X. Selection rules for the electron detachment process determine the nuclear and electronic character of the final state X. Usually, Koopmans' picture of single-electron detachment without rearrangement of the remaining electronic core is sufficient to obtain intuitive insight into the electronic selection rules. The vibrational transition can be described by Franck-Condon factors, based on the tacit assumption that nuclear positions and momenta are preserved during the electron detachment process. If the initial and final state have similar structure and potential energy surfaces, the Franck-Condon factors resemble an identity matrix and the vibrational character is preserved throughout the ionization process.

In pump-probe experiments the characterization of binding energies can be used to observe the energy flow between the electronic and nuclear system in real time. This offers the unique possibility to follow geometry changes in photoexcited particles, e.g. wave packet motion [134], isomerization [135, 136] and melting [137]. This is straightforward when the excited electronic state of interest corresponds to an energetically isolated singly excited state, i.e. where a coupling to other excited electronic states can be excluded.

In clusters with a high electronic density of states, electronic relaxation pathways are dominating. In open d-shell metal clusters a relatively dense set of electronic states and a small HOMO-LUMO gap ensures a high probability of internal conversion from a singly excited electronic state into multi-electron excited states, i.e. the cluster analogue to inelastic electron-electron-scattering. Such processes can take place on a time scale on the order of 100 fs, a value comparable to bulk those found for metals [138, 139]. In such systems it is hardly possible to follow nuclear dynamics on the potential energy surface of an individual electronic state. In large bandgap clusters, like e.g. small Hg-clusters [140, 141], the creation of secondary electrons by inelastic electron scattering is energetically excluded at excitation energies less than twice the band gap. In these systems relaxation takes place via internal conversion from a singly excited electronic state into another singly excited state, whereas the excess energy flows into the vibrations. These processes can therefore be regarded as the cluster analogue to electron-phonon-scattering in solids.

TRPES thus provides an additional level of information with respect to time resolved ion yield experiments as the energy content and charge redistribution of a photoexcited species can be measured as a function of time.

Two-color TRPES has the advantage over one-color TRPES that pump-probe signals are easily distinguished from pump-only and probe-only signals. I.e., the time-dependent pump-probe information is revealed in a spectral region where no interfering single-color photodetachment signals occur. It is also advantageous to use pump-photon energies below the electron binding energy of the unexcited cluster to exclude the competing channel of direct detachment. As the cohesive energy of metal clusters is usually quite high (2-4 eV) fragmentation is very unlikely upon excitation with the fundamental of a Ti:Sa-laser [142]. Fragmentation does, however, play a large role in the study of weakly bound molecular clusters and is discussed in the corresponding sections.

Thus the three requirements for a TRPES experiment are: A stable source for the molecules or clusters of interest, an efficient method for detecting photodetached electrons, and a laser with sufficient tunability to pump a variety of transitions in different systems. To reduce the load on vacuum pumps it is also desirable that the ion source should be pulsed, rather than continuous, while, from the perspective of minimizing the time required to collect the necessary amount of data, both laser and ion source should be able to run at a high repetition rates.

3.3.2 Time resolved photoelectron spectroscopy of anionic molecular clusters

A. Kammrath and D. M. Neumark

The $(\text{water})_n^-$ and $(\text{methanol})_n^-$ clusters used in Sect. 3.4.1 are produced by expanding a water/argon or a methanol/argon mixture via a pulsed valve into vacuum [143–145]. For making water clusters, a drop of water is placed in-line of a gas manifold. Argon is passed over it and expanded through the valve. For methanol clusters the argon is bubbled through a sample of methanol placed in a u-tube in-line. Anions are formed by secondary electron detachment, crossing the expansion with an electron beam at several hundred to over a thousand electron volts. The electron beam may be pulsed, triggered to intercept the expansion, or continuous. The anions are extracted perpendicularly into a Wiley-MacLaren type mass spectrometer, depicted in Fig. 3.1a, and mass-selected using an electro-static gate before interacting with the laser pulse.

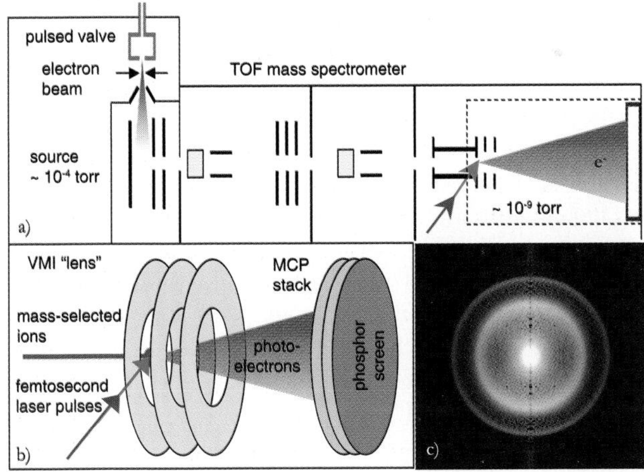

Fig. 3.1. Overview of the experimental set-up for fTRPES by Velocity Map Imaging of anionic water and methanol clusters. b) Detail of the Velocity Map Imaging apparatus. c) Transformed image taken from $(H_2O)_{15}^-$ at the overlap of pump (1650nm) and probe (790 nm) pulses. The outer ring is the transient signal due to pump+probe detachment through the excited state.

Velocity map imaging (VMI), was first demonstrated for neutrals by Eppink and Parker [146] and for anions by Bordas [147] and later Surber [148]. The VMI stack, as shown in Fig. 3.1b, consists of three plates forming an electro-static lens, which focuses electrons of the same velocity to the same radius on a detector (typically an microchannel plate (MCP) stack coupled to a phosphor screen) projecting the electron cloud onto a two-dimensional array. If the electrons are detached with cylindrical symmetry around an axis parallel to the plane of the detector, then the photoelectron angular distribution, as well as the photoelectron kinetic energy distribution can be readily extracted by a variety of transform methods [149–151], which recreate the original three-dimensional electron distribution from the two-dimensional projection. Fig. 3.1c is a transformed image taken of $(H_2O)_{15}^-$ at the overlap of pump (1650 nm) and probe (790 nm) pulses. The rings observed in the image correspond to the peaks in the photoelectron spectrum, which is obtained by angular integration of the transformed image. Detection by VMI has nearly 100% efficiency, and, comparing to the magnetic bottle detector, has the additional advantages of providing photoelectron angular distributions and the property that even very low energy electrons may be detected. When the VMI lens extracts the electrons collinear to the parent ion beam, then there is also no need to decelerate the ion beam to obtain good energy-resolution. The primary disadvantage is that large amounts of signal are required to obtain a high-quality transform, and thus a high-quality photoelectron spectrum.

The femtosecond laser pulses are obtained from a diode-pumped Ti:Sapphire laser which outputs ∼80 fs pulses at 2 nJ and repetition rate of 100MHz, tunable from about 770-820 nm. These pulses are stretched and regeneratively amplified at 500 Hz, then compressed once more to 80 fs, yielding an output of ∼1 mJ, tunable from about 780-810 nm. This fundamental can then be doubled, tripled or quadrupled in BBO crystals, or used to pump an optical parametric amplifier, which (with doubling and quadrupling of the various outputs) allows limited tunability over a range from ∼240-2500 nm.

The output of the laser is split to provide pump and probe pulses. One branch is aligned onto a computer-controlled translation stage with micrometer accuracy to generate the pump-probe delays. The beams are then collinearly recombined and focused into the chamber at the interaction region using a lens of focal length 50-100 cm.

3.3.3 Experimental setup for the investigation of DNA base clusters

T. Schultz and S. K. Kim

For the investigation of neutral DNA bases and clusters, presented in Sect. 3.4.2, samples were heated to 60-220 °C and expanded with ≈ 1 bar helium into vacuum. To obtain clusters with water, a small water reservoir was introduced into the helium line. The expansion occurred through the conical nozzle of a pulsed valve (General Valve, series 9) operating at 100-150 Hz. The cluster distribution was controlled through the helium pressure, pulsed valve opening time and time-delay between the gas- and the laser-pulse. The molecular beam was skimmed and crossed the co-propagating pump and probe laser beams in the interaction region of an electron spectrometer [152], a mass spectrometer, or an electron-ion coincidence spectrometer (Fig. 3.2) [153]. Electron energies were analyzed in a TOF spectrometer using a magnetic bottle setup for improved collection efficiencies. Ion masses were determined in a linear TOF mass spectrometer. For femtosecond electron-ion coincidence experiments (FEICO), extraction fields for the mass spectrometer were pulsed and sufficiently delayed to avoid interference with the electron trajectories.

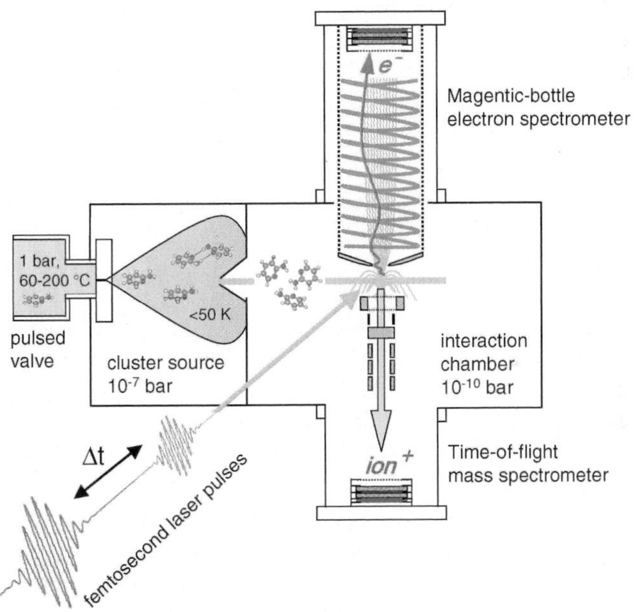

Fig. 3.2. Experimental setup for the characterization of excited state dynamics. Femtosecond pump and probe pulses excite and ionize isolated molecules and clusters. Electron and ion signals are monitored as a function of the pump-probe delay Δt.

Pump and probe laser pulses were generated by commercial and regeneratively amplified Ti:Sa lasers (Spectra Physics, Tsunami, and Spitfire for 80 fs pulses at 800 nm or CLARK MXR for 140 fs pulses at 780-820 nm). Molecules and clusters were excited by pump pulses in the ultraviolet, obtained via optical parametric amplifiers (Light Conversion, TOPAS) and/or nonlinear crystals. A probe pulse at 800, 400 or 200 nm ionized the photoexcited species via one- or multiphoton absorption. The intensities in the focused pump and probe beams were attenuated to optimize the pump-probe contrast and to avoid single-color signals. Typical count rates corresponded to 0.1-10 ionization events per laser shot. The time-delay between pump and probe pulses was adjusted with a motorized delay stage. Excited state dynamics were investigated by collecting mass spectra as a function of the pump-probe delay: when the excited state populations decay, the corresponding ion signals drop proportionally. Signals were accumulated for hundreds or thousands of laser pulses at each delay and 10-20 delay scans with alternating scan direction were averaged for a time resolved spectrum. Excited state lifetimes were determined by mono- or bi-exponential fits and error bars were estimated from multiple measurements. Where applicable, calibration compounds (e.g., NO, butadiene, indole or toluene) were used to determine the laser cross-correlation between pump- and probe beams and to calibrate the mass- and electron spectrometers [152, 154].

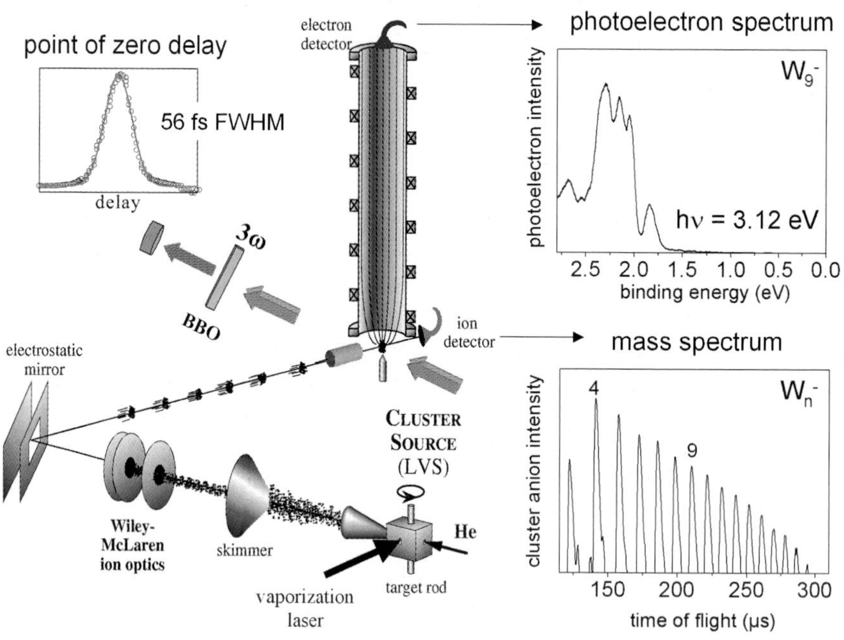

Fig. 3.3. Experimental setup for time-resolved photoelectron spectroscopy (TR-PES) on metal cluster anions. Details see text.

3.3.4 Time resolved photoelectron spectroscopy of metal cluster anions

J. Stanzel, F. Burmeister, M. Neeb, and W. Eberhardt

Clusters are produced in a laser vaporization source operating at repetition rates between 10-100 Hz. The material of interest is vaporized by a focused laser pulse (532 nm, Nd:YAG). Using a supersonic expansion of pulsed He buffer gas the vaporized atoms are condensed to form clusters. After expansion into the vacuum and forming a molecular beam cluster anions are accelerated by a pulsed, two-stage Wiley-McLaren optics to ~1000 eV kinetic energy. An electrostatic mirror behind the acceleration optics deflects the anions by 90 degrees to separate them from the neutral particles of the molecular beam. On their way into the electron spectrometer the anions are mass selected by their respective TOF (see mass spectrum in Fig. 3.3). The length of the drift tube is ~1.7 m, which results in typical drift times between 100 and 500 μs. Electrons are excited and detached in a magnetic bottle TOF spectrometer using two subsequent laser pulses. Prior to laser detachment the anions are focused and decelerated by a pulsed electric field to minimize the Doppler broadening in the electron spectra.

The electrons are detected by a channeltron at the end of a 2 m long TOF electron spectrometer. To reach maximal collection efficiency and a high transmission rate, magnetic fields are used to guide the electrons through the spectrometer. A strong magnet near the interaction zone and opposite the electron detector introduces a strong inhomogenous magnetic field to guide the diverging electrons from all directions toward the detector. The inhomogenous magnetic field from the permanent magnet is superimposed by a homogenous magnetic field shortly above the interaction zone to guide the electrons toward the detector. The kinetic energy of the electrons is given by their time of flight to the electron detector. The ion kinetic energy is minimized prior to electron detachment to minimize the Doppler broadening of the electron energy. The kinetic energy resolution ranges between ~ 10 meV at E_{el}=0.5 eV and ~40 meV at $E_{el} = 3$ eV.

A commercial Ti:Sa laser system is used which consists of a mirror-dispersion controlled fs-oscillator, a multipass amplifier and a prism compressor. The spectral bandwidth amounts to 40 nm FWHM. The system runs at 1 kHz while only each tenth to twentieth pulse is synchronized with our cluster source. A pulse energy of ~0.8 mJ and a temporal width of ~35 fs is routinely delivered after compression. In order to perform two-color pump-probe experiments, the fundamental is converted into the second harmonic by a nonlinear crystal (BBO). Pulse energies of around ~100 μJ (second harmonic, 400 nm) and ~300 μJ (fundamental, 800 nm) are delivered in the interaction region. The zero point of time delay is determined using a nonlinear crystal which is installed inside the electron spectrometer and closely behind the interaction zone on a rotational feedthrough. Using two collinear pump and probe pulses, i.e. the fundamental and second harmonic, the THG is formed inside the crystal and measured by a diode after separating the THG from the initial wavelengths by a prism behind the spectrometer. A cross correlation curve (56 fs) is shown in Fig. 3.3. This ensures a well defined time basis particularly important for measuring electronic relaxation processes that take place on a time scale of 10 to 100 fs. The pulses are only marginally focused in front of the spectrometer to avoid any Stark shifts. A beam diameter of 2-3 mm has been used for pump-probe spectroscopy.

3.3.5 IR spectroscopy on gas-phase biomolecules

G. von Helden and G. Meijer

The experimental setup for performing experiments on gas-phase aminoacids (see Sect. 3.4.4) consists of a pulsed molecular beam machine equipped with a laser desorption source coupled to a TOF mass spectrometer [28,155]. The sample is mixed with graphite powder, rubbed onto a graphite bar and placed directly under the nozzle of the pulsed valve, through which neon or argon with a backing pressure of 3 bars is expanded as a buffer gas. The molecules are desorbed from the sample using a ~1 mJ pulse at 1064 nm which is synchronized with the opening of the pulsed valve in order to achieve an efficient internal cooling in the supersonic expansion of the rare-gas beam. The neutral molecular beam is skimmed and enters the extraction region of the TOF. The molecules are ionized with a tunable UV laser beam (frequency doubled output of a dye laser pumped with the third harmonic of a Nd:YAG laser) crossing the molecular beam perpendicularly. The ions are detected on a microchannel plate detector and the TOF transient is recorded and averaged using a digital oscilloscope. Vibrational spectroscopy is performed using the infrared radiation produced by the free electron laser for Infrared experiments (FELIX) [32], described below. The IR laser beam is aligned perpendicularly to the molecular beam and counter propagating to the UV beam. A scheme of the setup is shown in Fig. 3.4.

Ionization of the molecules is possible via a resonant 2-photon process (R2PI) around 260 nm, exciting the aromatic chromophore in the molecule. To obtain the R2PI spectrum, the intensity of the ion signal on the mass channel corresponding to the parent ion is monitored as a function of the UV excitation wavelength. This spectrum is a superposition of the R2PI spectra of all conformers present in the molecular beam. The UV-excitation laser is then parked on the S_1 - S_0 transition of the conformer of interest, and the IR laser, which is fired just before the UV laser, is scanned over the fingerprint region of the molecule (300-1900 cm^{-1}). When the IR laser is resonant with an infrared transition of the molecule, the vibrational ground state of the S_0 state is depopulated, resulting in a dip of the ion signal. At each IR wavelength, the signal obtained when FELIX is irradiating the beam is divided by the signal obtained without FELIX and the natural logarithm is taken. The FELIX

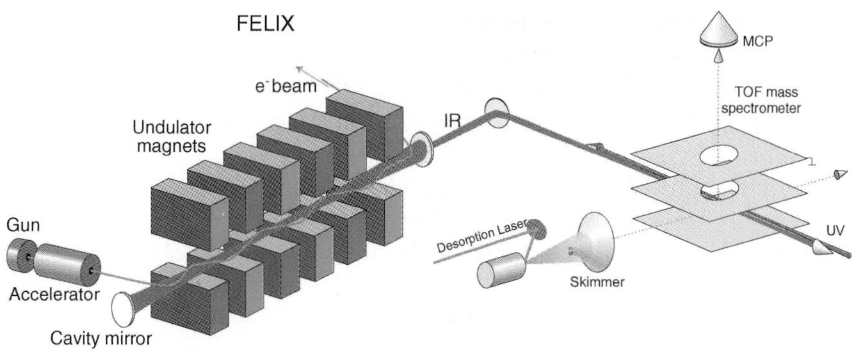

Fig. 3.4. Experimental setup to perform IR spectroscopy on gas-phase biomolecules using the free electron laser FELIX. See text for details.

fluence is not constant over the range scanned in this experiment, and the IR spectra are linearly corrected for this.

An essential ingredient in the experiments is the free electron laser FELIX [32]. In a free-electron laser (FEL) unbound, free electrons moving at relativistic speeds through a magnetic field structure serve as the gain medium and in principle, such lasers can produce photons of any wavelength. In practice, technical limitations restrict the wavelength range and most FELs operate in the IR wavelength range. The mechanism of working is briefly as follows: a relativistic beam of electrons, coming from an accelerator, is injected into an assembly of alternating permanent magnets, an undulator, which is inside an optical resonator, consisting of two high-reflectivity mirrors at either side. The magnetic field in the undulator is perpendicular to the direction of the electron beam and periodically changes polarity a (large) number of times along its length. This causes a periodic deflection, a 'wiggling' motion, of the electrons while traversing the undulator. This transverse motion is quite analogous to the oscillatory motion of electrons in a stationary dipole antenna and hence will result in the emission of radiation with a frequency equal to the oscillation frequency. The high electron velocity results in a strong Doppler-shift and the energy of the radiation emitted by the relativistic particles is concentrated in a narrow cone around the forward direction, sometimes referred to as the 'head-light' effect. However, the electrons are typically spread out over an interval that is much larger than the radiation wavelength and the initial spontaneous emission is thus not coherent and usually very weak. The interaction of the radiation field that builds up in the optical resonator with fresh electrons that enter the undulator can cause a spatial modulation of the electrons and cause them to emit coherently until saturation sets in at a power level that is typically $10^7 - 10^8$ times higher than that of the spontaneous emission.

The time structure of the light in such a FEL closely mimics the time structure of the electron beam. In FELs that use (room-temperature) rf-accelerator technology the light output consists of a microsecond duration 'macro-pulse', composed of a train of equidistant femto to pico-seconds duration 'micro-pulses', typically spaced by 1–100 nano-seconds. The macro-pulse repetition frequency is up to several tens of Hz. In FELs that use either superconducting rf-accelerators or electrostatic acceleration, continuous trains of pico-second duration micro-pulses or quasi-cw light pulses can be produced, respectively.

Due to its characteristics, FELIX is uniquely suited to be used in gas-phase experiments. It is continuously tunable over the 5 – 250 μm range. At a given setting of the beam energy, however, the tuning range is limited to about a factor of three in wavelength. In practice, that means that, at a given electron beam setting, for example the range from 5 to 15 μm can be continuously scanned within a few minutes. The macropulse length is 5 μs and the repetition rate 10 Hz. The micropulse length can be adjusted and ranges from 300 fs to several ps. The bandwidth is transform limited and can range from 0.5 % FWHM of the central wavelength to several percent. The micropulse repetition rate can be selected to be either 25 MHz or 1 GHz, resulting in a micropulse spacing of 40 or 1 ns respectively. This corresponds to 1 or 40 optical pulses circulating in the 6 m long cavity of FELIX. In the 1 GHz mode, the output energy can be up to 100 mJ / macropulse. The range between 2 and 5 μm can be covered as well by optimizing FELIX to lase on the 3rd harmonic. This is accomplished by replacing the metal cavity mirrors by dielectric mirrors that have a high reflectivity in the desired wavelength range. Lasing on the fundamental is thus suppressed and gain is present at the 3rd harmonic. Using this approach, up to 20 mJ in a 5 μs long macropulse is obtained. In all the experiments presented here, FELIX is, however, only used at its fundamental frequency.

3.4 Exemplary results

In the next subsections, results from six experimental as well as theoretical investigations are presented: In the first example, investigations of the ultrafast dynamics of an electron in water and methanol clusters are presented. In the next example, experiments and theory on the electronic and nuclear dynamics of DNA bases is presented. The third example is concerned with the electronic and nuclear dynamics in metal cluster anions. This is followed by a section on vibrations of small gas-phase biomolecules, a section on the dissociation dynamics of metal carbonyls and finally by a section on theoretical work on of the dynamics of clusters and biomolecules in the gas phase.

3.4.1 Dynamics of the solvated electron in clusters:$(H_2O)_n^-$ and $(CH_3OH)_n^-$

A. Kammrath and D. M. Neumark

The nature and dynamics of the solvated electron, which was first observed in liquid ammonia in 1864, is still far from being understood. Key issues are the timescales that are involved in the dynamics, which can range from very fast electronic dynamics to slow nuclear dynamics. Here, an investigation of electrons interacting with water and methanol clusters is presented.

3.4.1.1 Photoelectron spectroscopy of $(H_2O)_n^-$ and $(CH_3OH)_n^-$

Photoelectron spectroscopy of anionic water clusters has revealed the presence of three distinct structural isomers. Fig. 3.5a shows the vertical detachment energies (VDEs) of the three experimentally observed isomers of $(\text{water})_n^-$ as a function of cluster radius ($n^{-1/3}$), with the points taken from the work of Verlet, et al. [156] and the dotted line from Coe, et al. [157]. Isomer III, quite weakly bound, is observed only for small clusters (below n∼35) and is not well characterized. Isomer II is observed for all cluster sizes studied (n=11-200), while isomer I, previously reported by Coe, et al. [157] is observed for a similar range (n∼11-170). Both isomer I and large (n>50) isomer II clusters show a linear increase in VDE with $n^{-1/3}$ in qualitative agreement with the continuum dielectric theory of Makov and Nitzan [158] Isomer I, more tightly bound, is favored under warmer source conditions. Isomer II, with lower VDE is favored under colder source conditions, indicating that in these clusters the excess electron may be trapped in a metastable state. Based on theoretical work of Barnett, et al. [159] predicting that an internally solvated electron will exhibit a higher VDE, isomer II was assigned as binding the excess electron to the surface of the cluster. Isomer I was correspondingly assigned to an internally bound electron, approximating the bulk solvated electron in water which resides in a roughly circular cavity formed by surrounding water molecules. There are, however, complications–particularly in the assignment of isomer I to an internally solvated system. Both early calculations by Barnett, et al. [159, 159] and more recent theoretical studies by Turi, et al. [160] predict that the internally solvated electron should be more tightly bound than isomer I clusters by as much as an eV. Photoelectron spectra of $(\text{water})_n^-$ with $n = 50\text{-}200$ taken at 4.7 eV have revealed no more tightly bound isomers than isomer I [161] however there is always the possibility that such isomers simply are not made in the expansion in sufficient quantity to be observed. Studies by Johnson and co-workers have shown that small isomer I clusters bind the excess electron with both hydrogen atoms of a single water molecule in a surface motif [162]. This double-acceptor motif can be traced to clusters as large as n=21, [163] making it difficult to draw any definite conclusion as to the nature of the excess electron binding in large isomer I clusters.

Figure 3.5b shows the VDEs of the two isomers observed [145] for $(CH_3OH)_n^-$. For methanol as well as water, isomer II is favored over isomer I at higher backing pressures (cooler source conditions), however, isomer II of methanol is so loosely bound as to more nearly resemble isomer III of water than isomer II. Care must also be taken in drawing any parallel between isomer I of methanol and water, as the double-acceptor motif is unavailable in methanol clusters. It is possible that the binding motif required by the methanols isomer I is unavailable at smaller cluster sizes, thus explaining why methanol clusters are only observed at much larger sizes than water clusters.

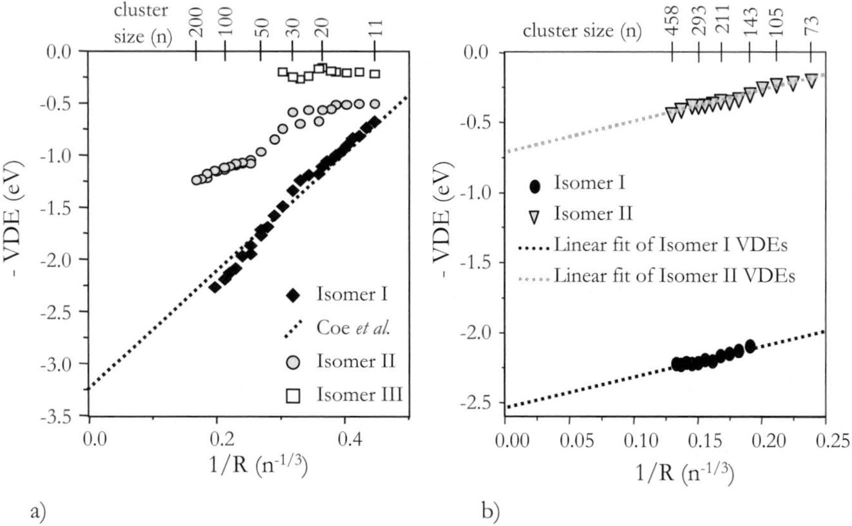

Fig. 3.5. a) VDEs shown as a function of cluster radius for the three different isomers observed in anionic water clusters. b)VDEs shown as a function of cluster radius for the two different isomers observed in anionic methanol clusters, taken at a backing pressure of 20 psi for isomer I, and 30 psi for isomer II.

3.4.1.2 Excited state of $(CH_3OH)_n^-$

Figure 3.6 shows the photoelectron spectrum of $(CH_3OH)_{190}^-$ (isomer I) taken at 20 psi backing pressure with the laser at 1.55 eV and focused (black) to give a power of ~$4x10^{10}$ W/cm^2, superimposed with the spectrum obtained using unfocused laser at 1.55 eV (solid gray line, ~$7x10^7$ W/cm^2) and that taken with the laser at 3.1 eV. At low power of 1.55 eV, only a small feature at very low energy is observed. In the spectrum using the higher power at 1.55 eV, a strong feature is observed overlapping the feature from observed with the laser at 3.1 eV, indicative of resonant two-photon detachment (R2PD) through an excited state accessible at 1.55 eV.

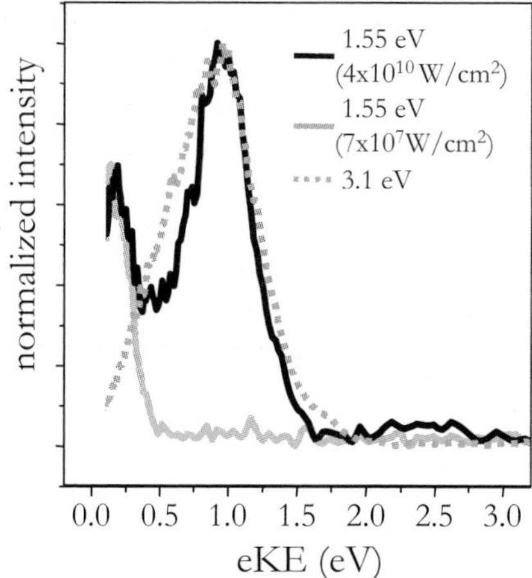

Fig. 3.6. Photoelectron spectra taken at low power 1.55 eV (solid gray line) and high power 1.55 eV (black line) compared to photoelectron spectrum taken at 3.1 eV (dotted gray line) for $(CH_3OH)^-_{190}$.

It can be seen that the R2PD feature is narrowed relative to the direct detachment feature, and shifted slightly to the high electron kinetic energy (eKE) side. This is analogous to the case observed by Weber, et al. [18] for $(H_2O)^-_n$, which they interpret as being due to a selection of vibrational states in the first excitation step which is preserved upon detachment to the neutral. The presence of an excited state in methanol's isomer I accessible at 1.55 eV is consistent with the bulk absorption spectrum in methanol, which has a broad maximum centered ~1.9 eV, and suggests that the binding motif in this isomer of the cluster is comparable to the binding of the excess electron in bulk methanol.

3.4.1.3 TRPES of $(H_2O)^-_n$ and $(CH_3OH)^-_n$ isomer I

Figure 3.7a and 3.8a show time-resolved photoelectron spectra for $(H_2O)^-_{45}$ and [4] $(CH_3OH)^-_{265}$ respectively, when the pump laser is at a wavelength of 790 nm (1.55 eV) and probed with the laser at 395 nm (3.1 eV). Intensity is plotted as a function of eKE on the horizontal axis, while increasing pump-probe delay is shown from back to front. For both clusters, feature A is due to pump plus probe, two-photon detachment, feature B is due to one-photon detachment at 395 nm, with some R2PD at 790 nm, and feature C is due to one-photon detachment at 790 nm.

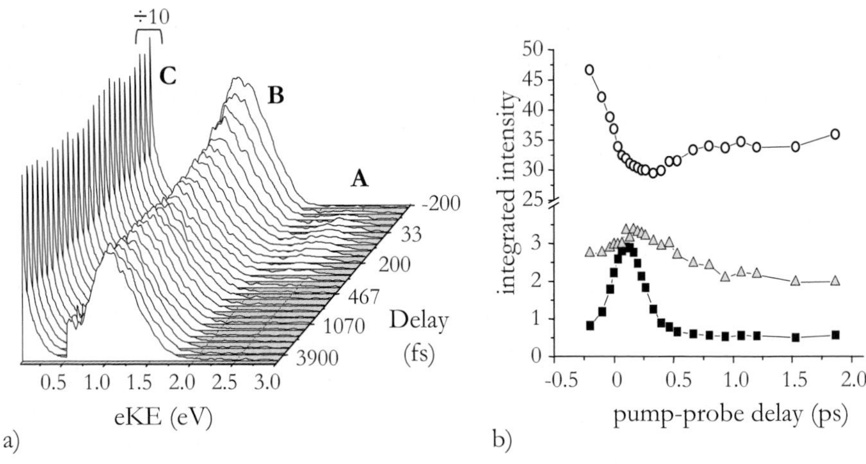

Fig. 3.7. a) Time-resolved photoelectron spectrum of $(H_2O)^-_{45}$ taken with 1.55 eV pump, 3.1 eV probe, with pump-probe delay increasing back to front. b) Integrated intensity as a function of pump-probe delay over eKE ranges corresponding to feature A (black squares), the high-energy edge of feature B (gray triangles) and the center of feature B (open circles).

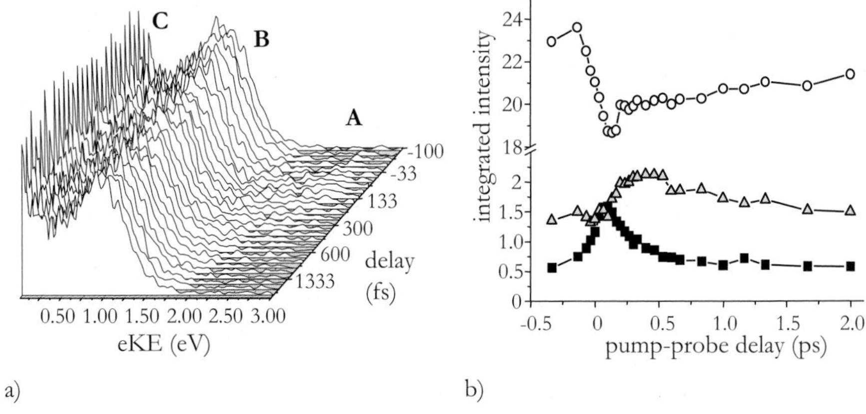

Fig. 3.8. a) Time resolved photoelectron spectrum of $(CH_3OH)^-_{265}$ taken with 1.55 eV pump, 3.1 eV probe, with pump-probe delay increasing back to front. b) Integrated intensity as a function of pump-probe delay over eKE ranges corresponding to feature A (black squares), the high energy edge of feature B (gray triangles) and the center of feature B (open circles).

Figure 3.7b and 3.8b show as a function of pump-probe delay the integrated intensity over feature A (black squares), the high energy shoulder of feature B (gray triangles) and a slice through the center of feature B (open circles). Both clusters show a drop in the intensity of feature B as the pump-pulse comes before the probe pulse (after time-zero), with only a partial recovery in the first few picoseconds. Feature A decays on an ultrafast time-scale, and this decay is accompanied by accumulation of intensity on the high eKE shoulder of feature B. The intensity observed on the high eKE shoulder of feature B begins immediately to be depleted on a timescale of hundreds of femtoseconds to ~1 ps, accompanied by a similarly slow recovery of intensity in the central part of feature B.

In Fig. 3.7a it can be clearly seen that feature A decays without undergoing any shift in eKE. Furthermore, for $n>25$, no significant increase is observed in the emission of low energy electrons as feature A decays. Thus the decay of the excited state is not likely due to either solvation dynamics or autodetachment, and is attributed instead to internal conversion from the excited to the ground state. In the case of methanol, there is again no evidence for autodetachment. Due to the broadness of feature A and relatively weak signal-to-noise ratio, a shift in eKE of feature A cannot be altogether ruled out. However, examining the decay rate over different energy slices through feature A yields the same time-zero as well as the same decay lifetime (within error), which makes it unlikely that any shift in eKE is taking place. The initial decay of the excited state in methanol is therefore also attributed to internal conversion.

For both water and methanol clusters, a decay of feature A, is accompanied by the appearance of a high energy shoulder on feature B. The decay of intensity in this shoulder corresponds to "sliding" of intensity to lower eKE over the course of a few picoseconds, so that it is finally lost in the mass of feature B. This shift in eKE is typical of solvation dynamics, and the shoulder is attributed to the excited ground state population, relaxing via ground state solvent dynamics.

3.4.1.4 Size dependent trends and comparison to bulk studies

Timescales for the decay of feature A are obtained, for both water and methanol, by fitting the integrated intensity of feature A to the convolution of a Gaussian (representing the cross-correlation of the pump and probe pulses) with an exponential decay. For $(H_2O)_{45}^-$ the decay time-scale obtained is $\tau =$ 130 fs. The lifetimes of the excited state in water's isomer I are observed to depend linearly on the inverse of cluster size $(1/n)$ in the range of $25<n<100$, extrapolating to 54 ± 30 fs (72 ± 22 fs for D_2O clusters) in the bulk [143,164]. This is in close agreement with the fastest timescales of the ultrafast dynamics experimentally observed for the excess electron in bulk water (50 fs in H_2O, 70 fs in D_2O) [12,13,15].

This may be taken to support the "nonadiabatic" interpretation of the bulk dynamics proposed by Pshenichnikov [13], according to which excitation of the equilibrated electron is followed by internal conversion with a life-time of 50 fs (70 fs in D_2O), then localized and extended ground state solvent relaxation on timescales of ~300fs and ~1ps respectively [15]. Further support for this interpretation is offered by the work of Paik, et al. [165], who observe ground-state solvation in clusters on two time-scales similar to the two longer time-scales observed in the bulk. Recent theoretical work by Scherer and Fischer [166] also recovers a 50 fs timescale for the internal conversion step in the bulk.

For $(CH_3OH)_{265}^-$ the lifetime obtained for the decay of feature A is $\tau \sim$ 210 fs. Preliminary results indicate that the lifetimes in methanol clusters do, indeed, also decrease linearly with 1/n, over the range of cluster sizes studied (n=145 – 535), extrapolating to ∼150 fs in the bulk [167]. This is in agreement with the fastest timescale observed in the bulk by Thaller [20]. Moreover, theoretical work by Zharikov and Fischer [168] has predicted that the lifetime for internal conversion of the excess electron in bulk methanol should be about three times longer than in water, or 150 fs, in agreement with the extrapolation of the experimental results for clusters. This would, perhaps, suggest a revision of the interpretation of the dynamics of the solvated electron in bulk methanol, following the nonadiabatic mechanism proposed for water.

The origin of the linear dependence of τ with 1/n both of these two cases is unclear. However, it is intriguing that the same dependence should hold for both water and methanol. It would be interesting to see whether the same dependence is observed in other polar molecules, such as acetonitrile, or whether it is a property of solvents with the general formula R-OH, of which water and methanol are the simplest examples.

3.4.2 DNA bases in the gas phase

T. Schultz and S. K. Kim

3.4.2.1 Structure of DNA bases and DNA

The structure of DNA is determined by hydrogen-bonding and stacking interactions between the bases. In single and double stranded DNA, the sugar-phosphate backbone forms a scaffold, holding adjacent bases in a stacked conformation [169]. This allows the formation of the DNA double-helix for two complementary strands of DNA, famously described by Watson and Crick [170]. In this helix, the complementary DNA bases adenine-thymine or guanine-cytosine form two or three specific Watson-Crick H-bonds. In the later discussion of photochemical properties of DNA, it will be of particular interest to gauge the relative importance of stacking and H-bonding interactions affecting the excited state properties. Free DNA bases can adopt several tautomeric structures, complicating the interpretation of experimental data in cluster experiments. E.g., for adenine, microwave spectroscopy showed the 9H tautomer as the dominant species in molecular beams [171], but IR-UV hole-burning experiments [172] also found minor contributions from the 7H tautomer. For base-pairs in vacuum, theoretical investigations predicted hydrogen bound structures [173]. But the formation energy of the Watson-Crick base pairs was similar to that of other hydrogen bound structures and a multitude of conformers were predicted. IR-UV hole-burning experiments identified no Watson-Crick base pairs in a molecular beam [174–176]. In water, H-bonding occurs to adjacent water molecules and DNA bases adopt a stacked structure. Ab initio calculations indicate that < 4 water molecules are sufficient to induce the transition from hydrogen bound to planar structures in adenine and AT base pairs [177].

3.4.2.2 Ultrafast processes in individual bases

Time-resolved photoelectron spectroscopy of adenine reproduced the excited state dynamics observed by time-resolved mass spectroscopy (TRMS) [179, 180] and allowed a firm assignment of the electronic states populated by UV photoexcitation [178]. Figure 3.9 shows the two-dimensional trace of photoelectron energy as a function of the pump-probe delay for 250 nm excitation. The measured electron-kinetic-energies can be directly converted to electron binding energies (also: ionization potentials, IP) with the equation: $IP + electron\ energy = \sum(h\nu)$. A short-lived trace with lifetime t < 100

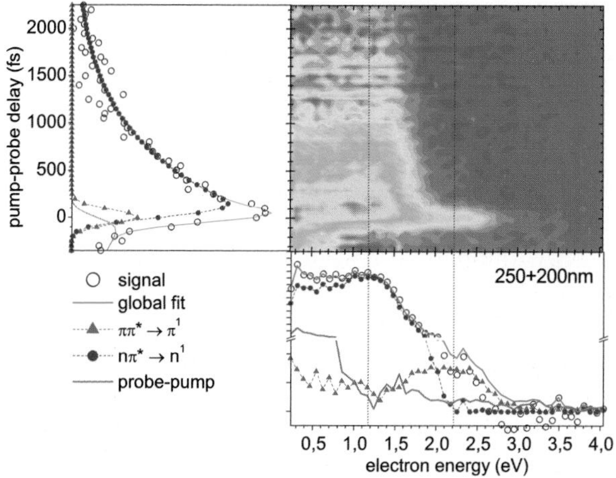

Fig. 3.9. Time-resolved photoelectron spectrum for adenine (250 + 200 nm) and projections onto the integrated photoelectron spectrum (bottom) and decay trace (left). A global fit revealed spectra and dynamics for $\pi\pi^* \rightarrow \pi^1$ ionization, $n\pi^* \rightarrow n^1$ ionization and a probe-pump process. Dotted lines indicate electron kinetic energies corresponding to the vertical π^1 ($IP_0 = 8.95$ eV) and n^1 ($IP_1 = 10.05$ eV) ionization potentials from the ground state. Figure adapted from [178].

fs and a vertical ionization potential $IP_V \approx 9.0$ eV was assigned to $\pi\pi^* \rightarrow \pi^1$ ionization. The discrepancy to the respective ground state ionization potential of 8.48 eV [181] is due to the 0.48 eV vibrational excess energy in the $\pi\pi^*$ excited state and correspondingly shifted Franck-Condon factors. A longer-lived trace with $\tau \approx 1$ ps and $IP_V = 10.1$ eV was assigned to $n\pi^* \rightarrow n^1$ ionization. Franck-Condon factors for this transitions showed a near-identical shift to that observed for ionization of the $\pi\pi^*$ state, when compared to the respective ground state ionization potential of 9.58 eV. Similar electron spectra were observed for thymine, uracil and cytosine [182] and a fast $\tau < 100$ fs process was always assigned to the corresponding $\pi\pi^* \rightarrow n\pi^*$ internal conversion. The subsequent dynamics of the $n\pi^*$ state, however, differ: for the pyrimidine bases adenine and guanine this state decays to the ground state within picoseconds, but for the purine bases thymine, uracil and cytosine the excited state population persists up to nanoseconds [183, 184]. Ab initio investigations of adenine suggested three possible relaxation mechanisms for internal conversion to the ground state: (i) Broo suggested a direct internal conversion pathway from the $n\pi^*$ to the ground state [185]. (ii) Domcke and Sobolewski proposed a pathway via conical intersection with a $\pi\sigma^*$ state on the azine group, which would have quasi-dissociative character along the N-H bond [186]. A second $\pi\sigma^*$ state on the amino group may play a similar role [154] (iii) Marian identified a conical intersection between $\pi\pi^*$ and ground state, involving an out-of-plane twist of a ring carbon [187]. The emission of atomic hydrogen from electronically excited adenine supplied experimental evidence for the existence of channel (ii) at excitation energies well above the S_1 origin [188, 189], but cannot quantify the relative importance of this channel. Experiments using femtosecond upconversion [190] and time-resolved photoelectron spectroscopy [191] also invoked the $\pi\sigma^*$ state. Conversely, the lack of isotope effects upon deuteration and similar internal conversion rates after methylation were interpreted as evidence against the $\pi\sigma^*$ state [192].

3.4.2.3 Excited state dynamics of base pairs

For the hydrogen bound Watson-Crick base pairs, theory predicted an excited state electron-proton transfer reaction (or proton coupled electron transfer, PCET), which could efficiently quench the excited state lifetime [193, 194]. Compared to the dissociative $\pi\sigma^*$ state discussed above, the electron and proton are stabilized by the adjacent acceptor base and the resulting barriers to internal conversion may be greatly reduced. This relaxation mechanism was experimentally confirmed in the model base pair aminopyridine dimer (Fig. 3.10) [195]. As a result, the excited state lifetime in the near-planar, hydrogen bound dimer was reduced by a factor > 20 compared to the lifetime of the monomer. A TD/TH isotope effect of almost 7 confirmed the rate-determining role of the amine proton transfer coordinate. Close to the $\pi\pi^*$ origin (< 1000 cm^{-1} excess energy), the relaxation rate increased with the excitation energy as expected for a rate limiting barrier on the potential

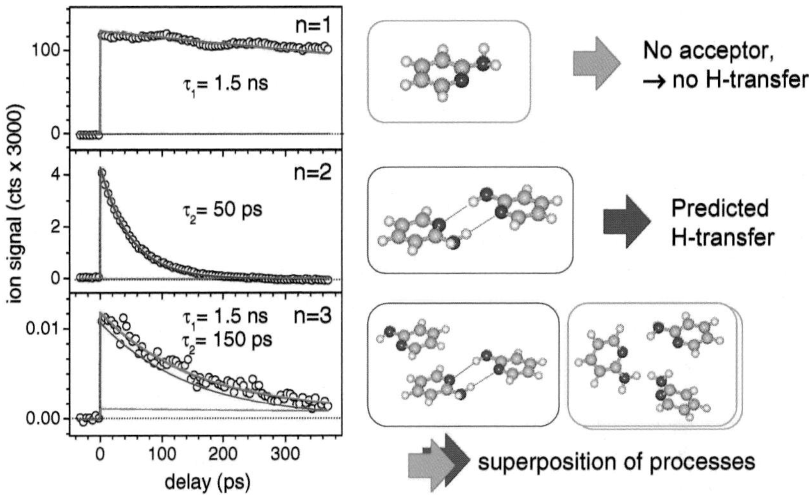

Fig. 3.10. Decay traces and PM3 semiempirical structures for aminopyridine clusters. Excited state relaxation in the hydrogen bound dimer is accelerated by an electron-proton transfer. Slower rates for the trimer indicate a strong geometry dependence of this process. Very narrow cluster distributions (cf. ordinates) were necessary to avoid fragmentation from larger clusters.

energy surface. But for larger energies (> 2000 cm^{-1} excess energy) the rate decreased. We explain this unexpected effect with a change in the equilibrium geometry of the cluster: excitation of the strongly anharmonic intermolecular vibration will increase the chromophore distance and lead to higher barriers for the proton transfer, thus slowing the observed relaxation rate.

For DNA base pairs containing adenine and thymine, we found no indication for intermolecular relaxation channels [196]. This does not disprove the existence of PCET predicted for the Watson-Crick base pair, because the observed gas phase clusters are expected to adopt non-Watson-Crick geometries. The relative signal contributions from $\pi\pi^*$ and nπ^* state ionization in the TRMS of adenine monomer were identified by their respective lifetimes (Fig. 3.10, left) and agreed well with those measured by photoelectron spectroscopy (see Fig. 3.9, above). The lifetimes in the dimer are identical to those in the monomer, but signals from the nπ^* state were partially quenched. We assigned this to the presence of competing relaxation channels with $\pi\sigma^*$ character, located on the amine and azine moiety [154]. Direct photoelectron spectroscopy is not possible for the base pairs, because the monomer signal always dominates the integrated signal. We therefore resorted to femtosecond electron-ion coincidence spectroscopy [153] to directly characterize the electronic character in the dimer (Fig. 3.11, right). This time-consuming experiment confirmed the assignment of $\pi\pi^*$ and nπ^* excited states. Please note, that ionization with 266 + 2x400 nm photons can deposit up to 2.4 eV ex-

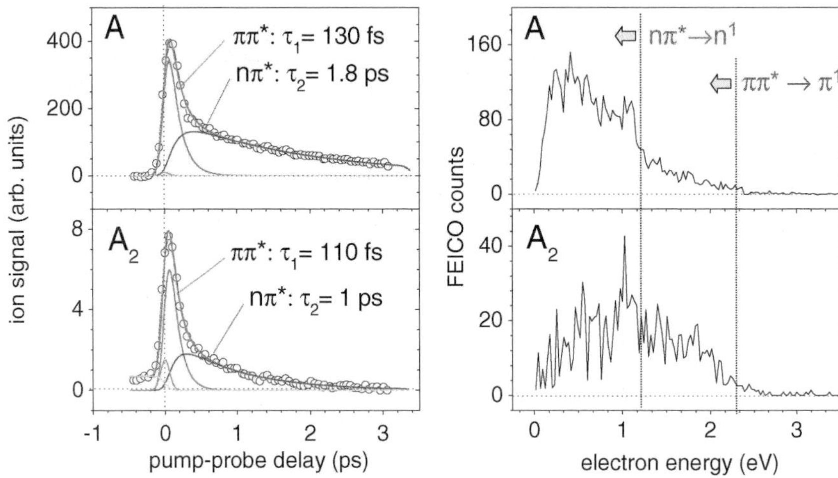

Fig. 3.11. Time-resolved mass spectra (266 + 3x800 nm, left) and photoelectron spectra (266 + 2x400 nm, at pump-probe delay Δ t = 0, right) for adenine and adenine dimer. Contributions from the nπ^* state are quenched in the dimer.

cess energy in the ion, enough for subsequent fragmentation of the cluster. In particular, the slow electrons from n$\pi^* \rightarrow$ n^1 ionization are correlated to hot ions if we assume fast internal conversion from the n^1 to the π^1 ionic ground state. We believe that the signal drop for electron energies < 0.8 eV in the dimer spectrum (Fig. 3.11, bottom right) is due to this fragmentation process. For three-photon, 800 nm ionization, we observe a scrambling of the ionization correlations, but the quenching effect persists. Therefore, fragmentation effects alone are not sufficient to explain the quenching of n^1 signals.

Fig. 3.12 illustrates the $\pi\sigma^*$ relaxation channel, which offers an alternative relaxation pathway to the $\pi\pi^* \rightarrow$ nπ^* internal conversion. The polarizability of the second chromophore in the dimer can stabilize the large dipole moment of the $\pi\sigma^*$ state. Ab initio calculations predicted a stabilization energy of 0.1 -0.15 eV for different cluster isomers [154]. To our surprise, the strongest stabilization was always predicted for the remote σ^* orbital, i.e. the $\pi\sigma^*$ state does not resemble a charge transfer state to the second chromophore. It is not obvious whether such free σ^* states can exist in the condensed phase (water). Data for the adenine-thymine base pair resembled that presented for the adenine dimer and can be found in the literature [196]. First FEICO data (Fig. 3.13) indicate that the electronically excited state of the AT cluster is similar to that of adenine dimer, but not thymine dimer. In particular, the strong $\pi\pi^* \rightarrow \pi^1$ band observed in thymine dimer is absent in the AT base pair. It remains to be explored why, and how, the excitation might become localized on the adenine chromophore. As discussed for the adenine dimer above, we expect significant cluster-ion fragmentation for hot ions in the 266

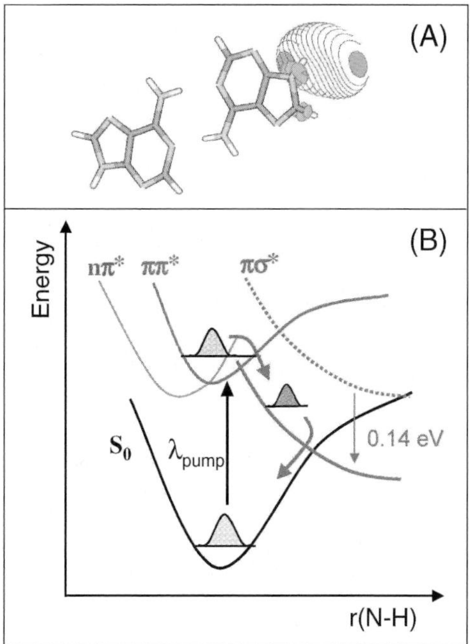

Fig. 3.12. (A) Shape of the lowest σ^* orbital in a doubly hydrogen bound isomer of adenine dimer. (B) Schematic depiction of potential energy surfaces based on [187, 193]: the $\pi\sigma^*$ state energy is reduced in the dimer and offers an IC pathway competing with the nπ^* state.

+ 2x400 nm ionization process and the corresponding signals for small electron energies approach zero. fragmentation in the cluster cation was also observed in TRMS spectra of thymine base pairs with 266 nm excitation, 2x400 nm ionization: The $\pi\pi^*$ state with a t $<$ 100 fs lifetime was observed in the dimer mass channel, but a longer lived state (probably nπ^*) with $\approx$ 35 fs lifetime fragmented asymmetrically and was observed in the mass channel of protonated thymine cation.

3.4.2.4 Solvated bases and base pairs

In the last chapter, we presented a detailed account of photophysical processes in isolated DNA bases and base pairs. We now explore the role of microsolvation, e.g. the effect of a few water molecules on the photophysical properties. This may not be sufficient to extrapolate to fully solvated systems, but it is a first step toward a polarizable and possibly acidic or basic environment. It should be noted, that the polarizability inside of proteins and other biological materials is rather small (e.g., $\varepsilon = 4 - 20$ in proteins compared to $\varepsilon \approx 80$ in water). The interaction with several water molecules in such an environ-

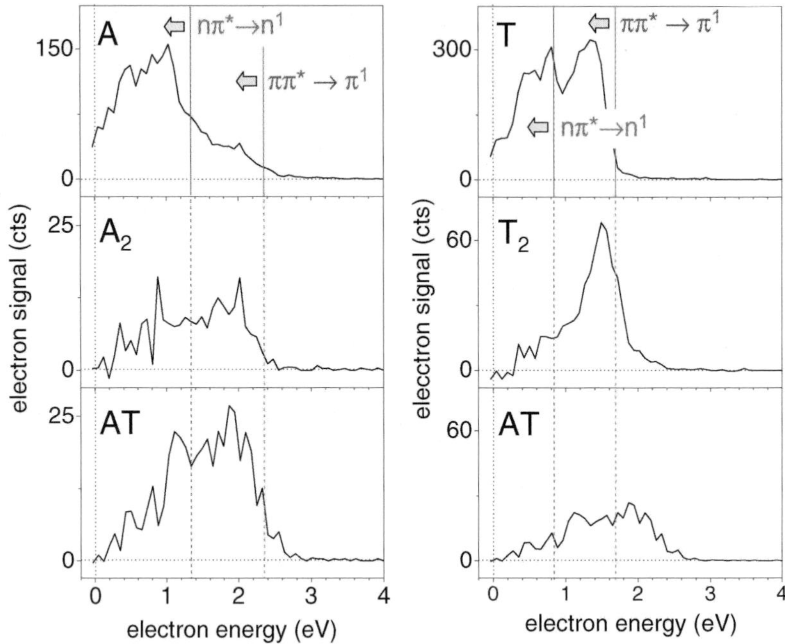

Fig. 3.13. Coincidence photoelectron spectra of adenine and thymine base pairs with 266 nm pump and 2x400 nm probe at time-zero. The n^1 and π^1 ionization potentials (dotted lines) for adenine and thymine are indicated for reference.

ment may perhaps be better approximated by isolated clusters than by fully solvated systems.

TRMS spectra of adenine-water clusters showed a complete quenching of signals from the $n\pi^* \rightarrow n^1$ ionization channel (Fig. 3.14). As for the DNA base pairs above, we assigned this to the energetic stabilization of $\pi\sigma^*$ states, which offer a competing relaxation pathway for the $\pi\pi^*$ state population [154]. Ab initio calculations for several isomers of adenine-$(H_2O)_1$ and adenine-$(H_2O)_3$ (Fig. 3.14, right) confirmed, that the stabilization of the σ^* orbitals on the amino and azine group in adenine-water is considerably stronger than in adenine dimer. Again, we find the strongest stabilization to occur when the occupied σ^* orbital does not overlap with orbitals of the solvent molecule, indicating that such states might not be energetically accessible in liquid water. TRMS data for thymine-$(H_2O)_n$ (n = 1-3) clusters resembled those for adenine: only $\pi\pi^* \rightarrow \pi^1$ signals were observed and contributions from $n\pi^* \rightarrow n^1$ were quenched. For cytosine-$(H_2O)_n$, only partial quenching was observed. A detailed theoretical treatment and of the excited state relaxation pathways in thymine and cytosine water clusters has not yet been attempted.

For the adenine$_2$-$(H_2O)_n$ clusters with $n \geq 3$, we observed an additional long lived state with > 500 ps lifetime. Due to fragmentation, this long-lived

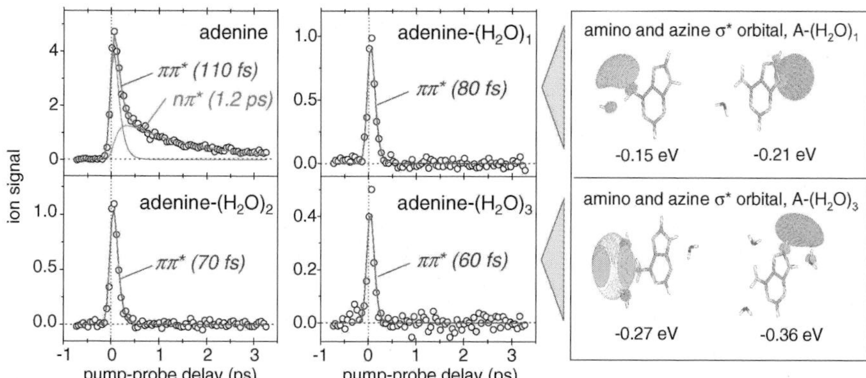

Fig. 3.14. Excited state dynamics of adenine-$(H_2O)_n$ clusters with 266 nm excitation, 800 nm ionization. The longer-lived component due to the $n\pi^*$ state vanished in the water clusters, indicating a competing relaxation pathway.

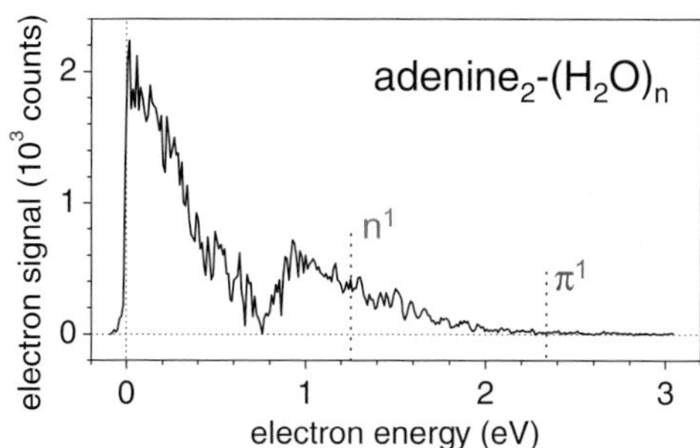

Fig. 3.15. Photoelectron spectrum of adenine$_2$-$(H_2O)_n$ with 266 nm excitation, 2x400 nm ionization with a pump-probe delay of 50 ps. Measured ionization potentials strongly deviate from the adenine π^1 and n^1 ionic states (dotted lines).

state was observed in all adenine$_2$-$(H_2O)_n$ clusters n = 3-5 and in the adenine monomer, but vanished when we reduced the cluster distribution to n = 2. We therefore concluded, that the long-lived state occurs only in larger clusters with n > 2, where a p-stacked geometry was predicted by theory [177]. A similar long lifetime was observed for stacked adenine in single-stranded poly(dA) strands, and double stranded poly(dA)-poly(dT) strands in water [23,197]. To characterize the nature of this long-lived state, we investigated the electron spectra at a pump-probe delay time $\Delta t = 50$ ps (Fig. 3.15). The electron bands are strongly shifted relative to the π^1 and n^1 states of the adenine monomer and indicate a very different electronic structure. Ab initio calculations for stacked adenine2-$(H_2O)_3$ clusters identified excimer states with $\pi\pi^*$ character and ≈ 0.5 eV reduced excitation energies, which may act as population sink and trap the excited state population for the observed ps-ns duration.

3.4.3 Metal cluster anions

J. Stanzel, F. Burmeister, M. Neeb, and W. Eberhardt

3.4.3.1 Nuclear dynamics in excited gold cluster anions

As discussed in Sect. 3.3.1, the relaxation pathway of a photoexcited cluster strongly depends on the number of electronic states that are accessible at a certain excitation energy. In our first experimental example, we present measurements on a small series of photoexcited gold cluster anions (Au_5^-, Au_6^-, Au_7^- and Au_8^-). In these clusters, excitation with an 1.56 eV pump pulse leads to the excitation of the first electronically excited state [198,199]. This state is energetically well isolated with respect to other electronically excited states. A conical intersection with other electronically excited states can be excluded [199]. This enabled us to follow geometry changes of a defined excited electronic state as a function of time.

Time-resolved photoelectron detachment spectra of Au_5^-, Au_6^-, Au_7^-, and Au_8^- are displayed in Fig. 3.16. The time-dependent photoelectron peak in these spectra is due to photodetachment of the first electronically excited state into the electronic ground state of the neutral cluster. At early times, i.e. when there is temporal overlap between pump- and probe pulse, the time-dependent photoelectron peak appears at a binding energy given by VDE−1.56 eV, where VDE is the vertical detachment energy of the corresponding cluster which is well known from the literature (see e.g. [200]).

The time-dependent photoelectron feature of the pentamer anion initially appears at a binding energy of ∼1.5 eV (Fig. 3.16). Within 1 ps an oscillation of the binding energy is observed which is fitted by a sine in Fig. 3.17a. The oscillation period amounts to 315 fs. Fig. 3.17a shows the periodic change in the energy of the photoelectrons in terms of the potential energy difference of the final and intermediate state in which dynamics occurs. The energy difference between X and A^- increases as a function of the internuclear distance. The change in the potential energy difference corresponds to a change in the kinetic energy of the photoelectron: The smaller the potential energy difference the higher the kinetic energy of the detached electron. Upon wave packet motion, which is triggered at t=0 by the absorbed photon $h\nu_{\mathrm{pump}}$, the nuclei start to move on the intermediate potential energy surface. Probing the system after a particular time, i.e. at increasing internuclear distances, the measured photoelectron energy decreases until reaching the outer turning point at $t = \frac{1}{2} \cdot T$. At this point, the photoelectron energy is minimal and starts to increase on the way back to the inner turning point where it reaches its maximal value again. With increasing pump-probe delay the wave packet or ensemble of trajectories broadens due to population of different vibrational

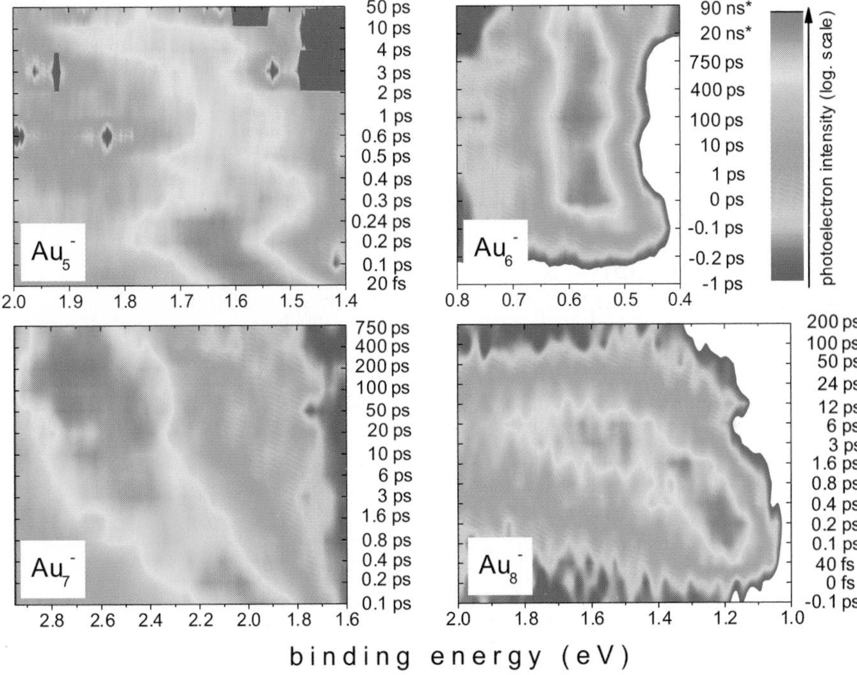

Fig. 3.16. Contour plot of time-resolved photoelectron detachment spectra of Au_5^-, Au_6^-, Au_7^- and Au_8^-. The time delay is indicated within the plot. The binding energy refers to the vacuum level of the anionic ground state. (*) The 20 and 90 ns spectra (Au_6^-) have been recorded using an Nd:YAG probe laser (details see text). Note the different time scales.

modes. This might be an explanation for the fading photoelectron intensity which is observed after ~5 ps.

Fig. 3.18 shows the calculated photoelectron spectrum of photoexcited Au_5^- as a function of time. The nuclear dynamics of the pentamer has been simulated by time-dependent DFT calculations "on the fly" conducted by the group of V. Bonačić-Koutecký [137]. Snapshots of the nuclear arrangement are shown in the lower panel. Within the first picoseconds atomic rearrangements around the trapezoidal structure at t=0 can be seen while at later times broadening around the equilibrium positions occurs. Note, however, that the initial trapezoidal structure is still maintained at 5 ps. Upon projecting the excited anionic state onto the final neutral state the kinetic energy of the photoelectrons have additionally been deduced. The oscillation period is reproduced as well as the relative energy fluctuation of the oscillating wave

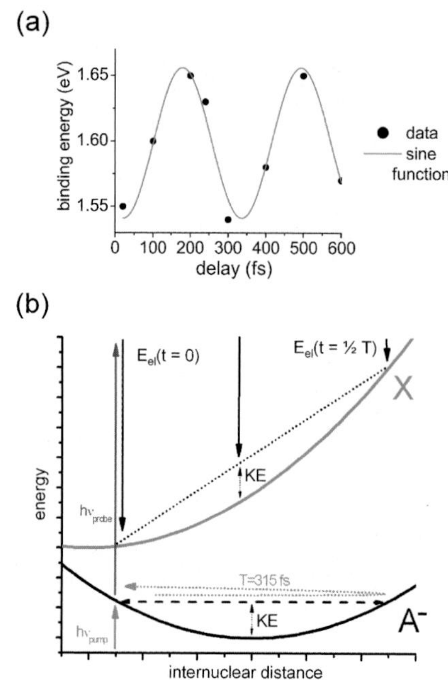

Fig. 3.17. Au_5^-. (a) Integrated photoelectron yield as a function of the pump-probe delay. The data have been fitted by a sine function with an oscillation period of 315 fs. (b) Oscillatory wave packet motion and the periodic change of the energy of the photodetached electron as schematically shown by the involved potential curves of the photoexcited state A^- and final state X. T is the vibrational period, E_{el} is the electron energy and KE is the kinetic energy of the nuclei. According to the Franck principle the kinetic energy of the nuclei as well as their actual position is not altered during the photodetachment process. This is indicated by the dotted line within the potential curve of the X state.

packet within the first ps. At later times the oscillation blurs more and more due to the increasing spreading of the atoms around the equilibrium position.

Adding one atom, the observed dynamics changes completely. The photoexcited state of the hexamer anion is extremely long-lived. Its lifetime significantly exceeds the delay that we are able to reach in our experiment by varying the optical path between pump and probe pulse ($\approx$ 1 ns). Therefore we measured additional data points for longer delays of up to 90 ns using the third harmonic of an Nd:YAG laser ($h\nu_{\text{probe}}^{\text{Nd:YAG}} = 3.5$ eV, temporal width $\approx$ 10ns) as the probe. As can be seen in Fig. 3.16, the time-dependent spectra of Au_6^- do not show any change in energy of the two-photon photodetachment peak at 0.55 eV. A sharp peak is observed for at least 90 ns. Note that this is also the time window by which the clusters remain within the interaction region. For negative delays, i.e. the probe pulse precedes the pump pulse, no

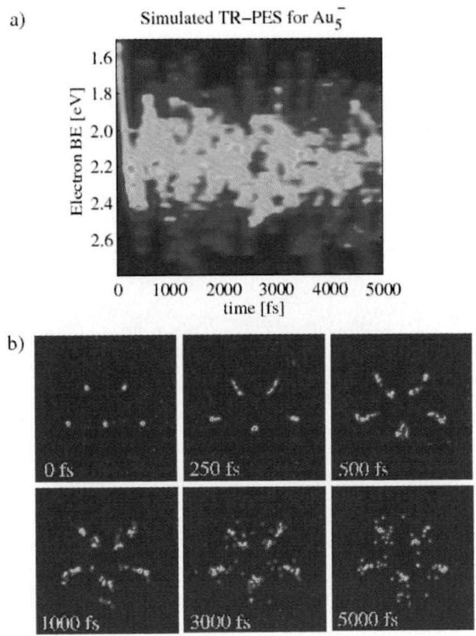

Fig. 3.18. Au_5^-. Molecular dynamics simulations by Bonačić-Koutecký et al. [137] (a) Contour plot of the simulated time-dependent photoelectron spectra. (b) Geometry snapshots for different time delays.

photodetachment feature is observed. After 100 fs the peak intensity reaches its maximum and persists for at least 90 ns. From the overall constant peak shape, intensity and energy we deduce the existence of an extremely long-lived excited state of $\tau > 90$ ns in Au_6^- which might decay by fluorescence later.

Such a long lifetime can be explained by the excitation into a local minimum of the potential energy surface, where the available energy is not sufficient to overcome the potential energy barrier into another geometry.

Photoexcited Au_7^- and Au_8^- reveal a completely different scenario. First, we concentrate on the dynamics of Au_7^- which is displayed in Fig. 3.16. The maximum of the photodetachment peak at 2 eV moves continuously to a higher binding energy. Within 10 ps it moves by 0.5 eV. Thereafter the shift significantly slows down and after ~200 ps the maximum has shifted to ~2.8 eV where it stays constant up to 750 ps. The continuous shift to higher binding energy suggests an unidirectional energy transfer into the nuclear system. The relativistic time-dependent DFT simulations "on the fly" of the nuclear relaxation are shown in Fig.3.19. Using for example a single trajectory, the potential energy of the electronic ground X^-, intermediate A^- and neutral final state X is shown. The binding energy of the detached electron is equal to the energy difference between X and A^- at a certain delay along the particular trajectory (see introduction 3.2) . Immediately after excitation, the energy of A^- decreases due to a geometry change of the cluster. The corresponding

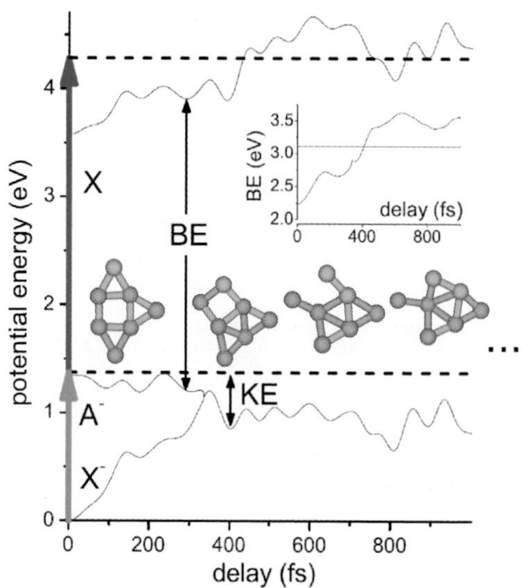

Fig. 3.19. TDDFT simulations by Bonačić-Koutecký et al. [137]. The relaxation dynamics of Au_7^- involving initial X^-, intermediate A^- and the neutral final state X are shown. The energy scale is given with respect to the ground state of X^-. The horizontal lines (dashed) correspond to the energy of the pump and the probe pulse, respectively. KE is the kinetic energy of the nuclei, BE the binding energy of the detached electrons. The inset shows BE as a function of time. Some of the evolving isomers and intermediate structures are indicated in the plot.

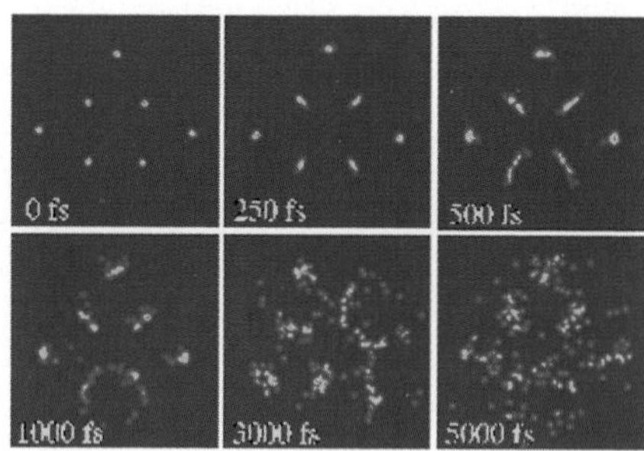

Fig. 3.20. Geometry snapshots of photoexcited Au_7^- calculated by Bonačić-Koutecký et al. [137].

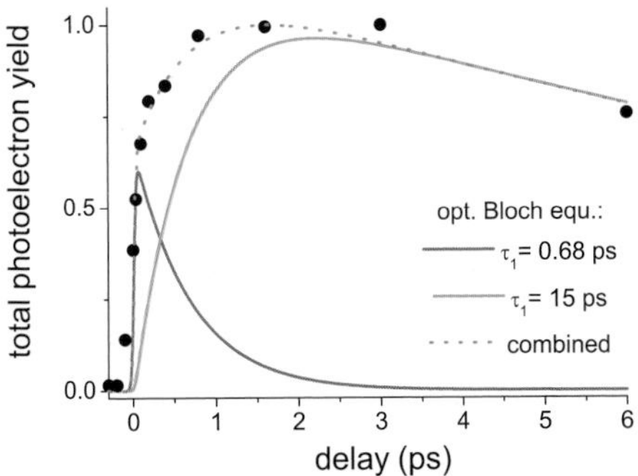

Fig. 3.21. Time-dependent photoelectron signal of Au_8^-. The intensity has been fitted by a model based on the optical Bloch equations for a two-level system with an additionally coupled rate equation yielding two time constants τ_1 and τ_2. τ_1 is interpreted as the time constant for the transfer from the initially populated first excited electronic state into the electronic ground state. The blue line corresponds to the population of the initially excited state A^-, whereas the green line reflects the population dynamics of X^-.

fraction of potential energy is transferred into kinetic energy of the nuclei. Simultaneously the final state energy increases, even more strongly. As the overall energy gap between X and A^- increases, consequently the observed binding energy of the photoelectrons raises with time as shown in the inset of Fig. 3.19. After hundreds of femtoseconds an internal conversion between A^- and X^- occurs which leads to a dramatic change in geometry. As a local potential energy maximum is reached at the crossing point a fast energy flow into the vibrational system follows. This results in a vibrationally hot cluster in the electronic ground state. The calculation reveals strong geometry fluctuations suggesting a melting-like behavior in finite system. This is clearly demonstrated in Fig. 3.20 where geometry snapshots of 30 trajectories are shown. At 5 ps the original structure is fully distorted as compared to the equilibrium structure at $t = 0$ ps. The fluctuations in the electronic ground state cluster result from a considerable excess energy due to internal conversion. As indicated by the characteristics of the nuclear kinetic energy KE in Fig.3.19, the excess energy amounts up to $\sim$0.6 eV for the calculated trajectory . Considering 3N -6 vibrational degrees of freedom, this corresponds to a vibrational temperature $T = 1200$ K, which is of the same order of magnitude as the bulk melting temperature $T_{bulk}^{melt} = 1338$ K. Therefore, at 20 ps the main peak intensity broadens substantially from 2.2 to 3 eV.

The dynamics of Au_8^- is also dominated by geometry relaxation as indicated by an overall energy shift of the photodetachment peak at 1.2 eV. After 200 fs a shift of the time-dependent feature starts toward higher binding energies. At 6 ps the peak extends from 1.0 eV, the original onset, to almost 2.3

eV. By the broad peak structure a multitude of coexisting geometrical structures and isomers are indicated which is interpreted as a melting of Au_8^-. A crossing between A^- and X^-, from which the energy for isomerization and melting is obtained, is experimentally associated with a significant increase of the total photoelectron intensity. The intensity rise takes place on a much longer time scale than expected from the pulse duration of the pump laser. Taking the pulse duration into account, an intensity rise up to only 80 fs would be expected. However, an intensity rise up to several hundreds of femtoseconds is experimentally observed. This suggests a significant change in transition dipole moment which comes along with the coupling between the initially excited A^- state and the electronic ground state of the anion. Fitting the intensity increase by a model based on the optical Bloch equations for a two-level system, an IC time of 680 fs is revealed [137]. This model consist of two contributions: The first one corresponds to the lifetime of the initially populated A^- state. As this state couples to the electronic ground state X via IC, the electronic transition moment rises and so does the electron yield. Upon IC an energy transfer into the nuclear system takes place. The significant broadening of the photodetachment peak suggests a wealth of structures after 3 ps as seen by Fig. 3.16. In contrast to Au_7^-, where thermal equilibrium is reached after around 100 picoseconds, this can not be stated for Au_8^- as the intensity drops exponentially after 3 ps with a time constant of 15 ps (second contribution of the Bloch model). This means that the potential energy difference exceeds the probe photon energy already after 15 ps. All above examples show nuclear dynamics which can be followed on a fairly extended time scale, because no radiationless decays of electronically excited states take place that correlate to a different neutral final state upon photodetachment.

3.4.3.2 Electronic relaxation processes in Au_7^-

Using an excitation energy of 3.12 eV the dynamics in Au_7^- changes significantly. In this case the excited state is close in energy to other excited states [198, 199] which opens up the possibility of electronic relaxation pathways.

Fig. 3.22a shows TRPE spectra of Au_7^-. Two photodetachment peaks appear at E_I=0.8 (peak I) and E_{II} = 1.3 eV (peak II). While the intensity of peak I continuously decreases, peak II increases at early times and then starts to decrease after 800 fs. Note that the dynamics happens on a much shorter time scale as compared to the 1.56 eV-photoexcitation of Au_7^- (Sect. 3.4.3.1). The initial intensity rise of peak II suggests a depletion of state I and a simultaneous filling of state II via internal conversion between two excited electronic states. The involved states are schematically depicted in Fig. 3.22b and c. Photoelectrons contributing to peak I are due to detachment from the excited state Z^- which results from photoexcitation of an electron of the HOMO to an empty orbital at 3.12 eV above. Upon photodetachment with the probe laser, the electronic ground state of the neutral cluster (X) is reached. Pho-

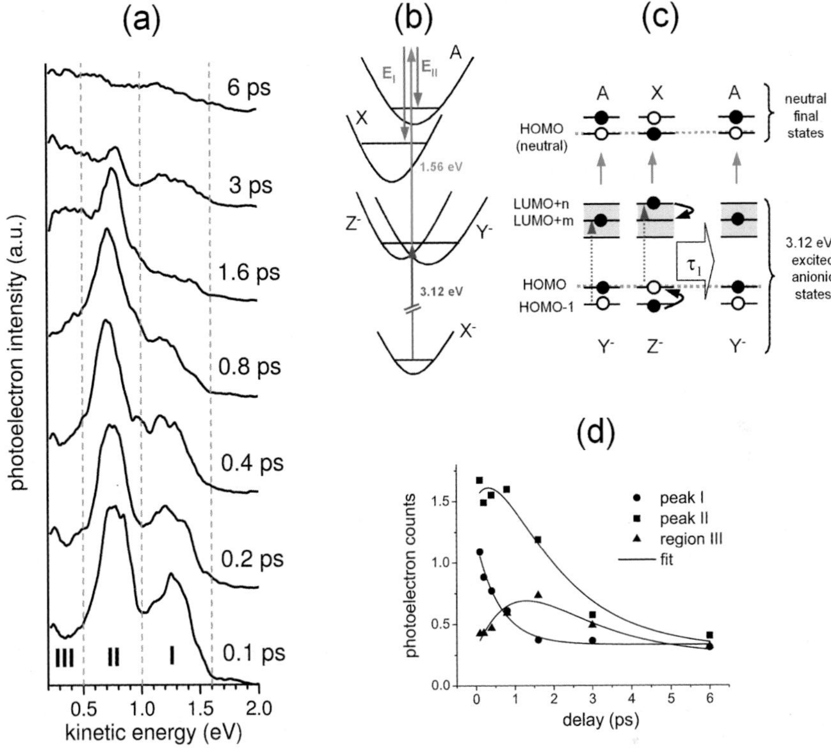

Fig. 3.22. (a) Time resolved photoelectron spectra of Au_7^- with $h\nu_{pump} = 3.12$ eV and $h\nu_{probe} = 1.56$ eV. The pump-probe delay is indicated next to each spectrum.(b)Potential energy scheme. Upon absorption of a 3.12 eV photon the two vibrationally degenerate intermediate states Z^- and Y^- are populated. Subsequent photodetachment with the 1.56 eV probe laser yields two different neutral final states resulting in photoelectrons with E_I and E_{II}. (c) Electronic configurations of the involved states: Note that Y^- is also filled from Z^- via internal conversion within $\tau_1 = 0.6$ ps. (d) Integrated photoelectron intensity for the three energy intervals indicated in (a). The three data sets have been fitted simultaneously yielding the time scale of internal conversion $\tau_1 = 0.6$ ps.

toelectrons contributing to peak II are due to detachment from the excited state Y^-. Detachment with the probe pulse transfers the system into the final state A which corresponds to the first excited state of neutral Au_7. Y^- can either be populated upon photoabsorption of the pump pulse or upon internal conversion from Z^-. As photodetachment from Z^- and Y^- results in different neutral final states, the kinetic energy of the photoelectrons changes abruptly upon internal conversion.

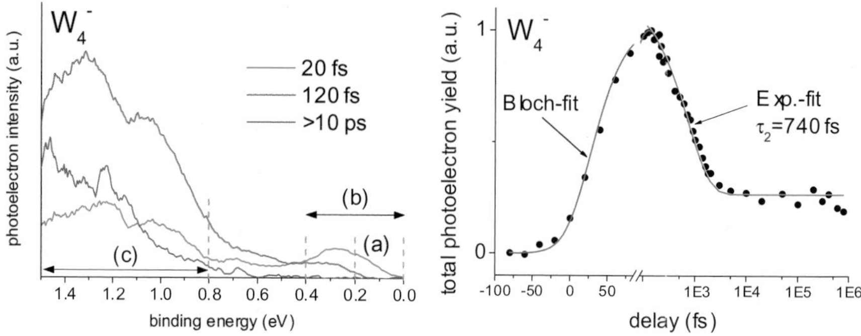

Fig. 3.23. Left: TRPE-spectra of W_4^-. The temporal evolution of the binding energy regions (a), (b) and (c) are fitted in Fig. 3.24. Right: Integrated photoelectron yield at $0 < \mathrm{BE} < 1.5\ eV$ as a function of time. The rising edge is fitted by a Bloch model described in the text, the falling edge is fitted by an exponential function yielding $\tau_2 = 740\ fs$. Note the logarithmic scale at a delay exceeding 100 fs.

Fitting the intensity evolution of peak I and II as well as the intensity in region III (intensity < 0.5 eV kinetic energy) by a rate model based on three coupled differential equations [201] the corresponding time constants have been deduced. The fit is shown in Fig 3.22d. Z^- is transferred into Y^- within τ_1=600 fs (peak I). Y^- decays with a time constant τ_2=1.6 ps (peak II). The remaining population decays with τ_3=700 fs.

This example shows that electronic relaxation processes can happen on a much faster time scale than nuclear dynamics. Therefore the latter cannot be monitored accurately by TRPES as soon as electronic relaxation processes prevail. Taking Au_7^- as example this is the case for an excitation energy of 3.12 eV, where at least five close lying electronic states exist [198, 199], but not if Au_7^- is excited to the lowest excited state by a 1.56 eV photon where coupling to other excited states is nonexistent. In Au_7^- excited by 3.12 eV photons internal conversion from a singly excited state configuration into other vibrationally degenerate electron-hole configurations dominates the relaxation mechanism. In case of metal clusters with a high electronic density of states in the valence region, a multitude of electronically excited states can be reached that correlate to different neutral final states upon photodetachment with the probe laser. As a consequence the probability of electronic relaxation processes is furthermore increased. This will be shown in the following section.

3.4.3.3 Bulk-like electronic relaxation process in W_4^-

In our final experimental example we present measurements of the tungsten tetramer anion W_4^-. The electronic structure of tungsten clusters is completely different compared to the previous examples. The tungsten atom is characterized by an open d-shell electronic configuration ($5d^4 6s^2/5d^5 6s^1$). As

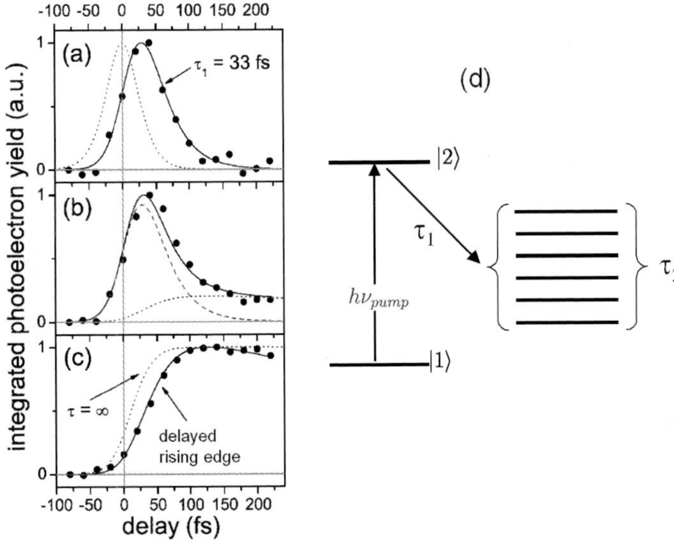

Fig. 3.24. W_4^-: Simultaneous fit of the three different binding energy regions indicated in Fig. 3.23. Region (a) ($0 < \text{BE} < 0.2\ eV$), region (b) ($0 < \text{BE} < 0.4\ eV$), and region (c) ($0.8 < \text{BE} < 1.5\ eV$). (a) is fitted by a set of two-level optical Bloch equations yielding a time constant $\tau_1 = 33\ fs$. The time-dependent photoelectron signal of region (c) is characterized by a delayed rise time as compared to region (a) which is attributed to the generation of "secondary" electrons that are created following the decay of the electronic states resulting in the emission at energies of region (a). For region (b) we applied a fit model consisting of two contributions, namely the population dynamics according to the Bloch model describing region (a), and the model describing the development of region (c).

a consequence the energy separation of occupied and unoccupied orbitals is low and the electronic density of states in the valence region is accordingly high. This has been shown by photoelectron spectroscopy [202]. Density functional theory (DFT) calculations clearly show a high level density for excitations near 1.56 eV, i.e. the fundamental of our Ti:Sa-laser system [203]. Tungsten clusters are therefore well suited for study of the relaxation behavior of systems where the number of energetically accessible excited electronic states is increased as compared to the preceding sections. In W_4^- we were able to monitor the internal conversion from a singly excited electronic state into a wealth of multiple excited states by TRPES.

TRPE-spectra of W_4^- are shown in Fig. 3.23 where $h\nu_{pump} = 1.56$ eV and $h\nu_{probe} = 3.12$ eV has been used. The whole dynamics takes place on a much faster time scale than in gold cluster s. At temporal overlap between pump- and probe laser a peak appears at BE ~ 0.3 eV. This feature vanishes almost completely after 140 fs. The vanishing of this low binding energy feature is accompanied by a significant rise in intensity in the binding energy interval 0.8 eV $<$ BE $<$ 1.5 eV. The latter is decreasing with an exponential time constant of $\tau_2 = 740$ fs. Note that the measured integrated intensity of the peak at BE = 0.3 eV is much smaller than the integral of the broad band at higher binding energy (region (c)).

Looking more carefully to the increase of the photodetachment intensity, i.e. at early times up to 220 fs, three characteristic regions can be identified in Fig. 3.23. Region (a) corresponds to 0-0.2 eV, i.e. the low binding energy edge of the 0.3 eV peak, region (b) ranges from 0-0.4 eV, and region (c) ranges from 0.8 - 1.5 eV. As we will show in the following, the initially populated excited state is decaying into a wealth of multiexcited electronic configurations which appear as the broad photoelectron band at higher binding energies, i.e. region (c). The three time-dependent spectral regions (a), (b), and (c) are fitted simultaneously.

As shown in Fig. 3.24, the intensity of region (a) rises and decays extremely quickly and a lifetime of 33 fs can be deduced from a normal Bloch fit [204]. Note that τ_1 is on the same order of magnitude as the pulse duration of the Ti:Sa-laser (Δt = 35 fs). Such short events can be analyzed using the optical Bloch equations when the exact point of zero delay is known. Our experimental setup allows the determination of the point of zero delay directly inside the interaction zone. The corresponding cross correlation curve is indicated by the dotted line in Fig.3.24a. Although the shape of the time-dependent photoelectron signal is hardly distinguishable from the cross correlation curve, the time constant can nevertheless be determined by the significant shift with respect to the point of zero time delay. The intensity regions (b) and (c) cannot be fitted by a simple Bloch fit. The initial dynamics of the photoelectron signal in region (b), i.e. for t $\leq$ 20 fs, is almost the same as for (a). But for t $\geq$ 20 fs, there is a background signal appearing, which is due to secondary processes. The contribution of the background signal can be taken into account by fitting the data applying a model containing two contributions. The first one, n_1, is populated directly upon absorption of the pump laser and exhibits the same temporal behavior as the intensity of region (a). The second contribution n_2 is due to a secondary process and is taken into account by adding the rate equation $\frac{dn_2}{dt} = \frac{n_1(t)}{\tau_1} - \frac{n_2(t)}{\tau_2}$ to the optical Bloch equations for a two-level system. In Fig. 3.24b, these two contributions are plotted as the dashed ($n_1(t)$) and dotted ($n_2(t)$) lines. Note that $n_1(t)$ and $n_2(t)$ have been convoluted with the probe laser envelope.

The photoelectron intensity of region (c) is mainly due to secondary processes and can be fitted by the second contribution $n_2(t)$ alone. We note that the time-dependent photoelectron signal cannot be described by a normal Bloch or rate model here, since the rising edge is significantly delayed with respect to a Bloch fit, even if an infinite lifetime is used (dotted line in Fig. 3.24c). As the time-dependent photoelectron intensity in region (c) has reached its maximum after t = 120 fs, it exhibits an exponential decay with a time constant $\tau_2 = 740$ fs (Fig. 3.23).

The delayed rise time of the intensity region (c) clearly demonstrates that the involved electronic states are populated not directly via absorption of the pump laser, but due to a secondary process. The delayed rise time is consistent with the decay of the intensity in region (a), as shown by the agreement with the applied fit model. We note that the population $n_2(t)$, i.e. region (c), cannot be attributed to a defined electronic state, but to an ensemble of multielectron excited states with an average decay time of $\tau_2 = 740$ fs (see schematics in Fig. 3.24(d)) . The observed internal conversion from a singly excited state into an ensemble of multielectron excited states is interpreted as the cluster analogue to inelastic electron-electron scattering. The initially excited "hot" electron scatters with an electron below the HOMO. As a consequence additional electrons are transferred above the HOMO. The striking intensity difference between the initially populated peak at 0.3 eV and the intensity region (c) confirms this interpretation.

In summary it has been shown that the characteristic electronic structure as well as the chosen excitation energy are decisive parameters for the relaxation of optically excited electronic states in metal clusters. In small gold clusters, population of an isolated electronic state allows to map the ongoing nuclear dynamics, such as oscillatory wave packet motion and melting. By increasing the excitation energy, the phase space can be significantly enlarged, which opens the door for electronic relaxation processes. Electronic relaxations take place via internal conversion between different electronic states and generally occur on a significantly shorter time scale. This has been demonstrated by choosing Au_7^- as an example. In tungsten clusters the electronic density of states is significantly higher than in gold. As a consequence, the probability of electronic relaxation processes is furthermore increased. The significantly smaller HOMO-LUMO bandgap in tungsten clusters facilitates the electron transfer above the HOMO due to secondary processes like inelastic electron-electron-scattering. In the tungsten tetramer W_4^- this process takes place on timescales as short as 33 fs.

3.4.4 IR spectroscopy of gas-phase phenylalanine

G. von Helden and G. Meijer

3.4.4.1 UV spectra

Phenylalanine can be brought into the gas phase via laser desorption and can be interrogated using UV and IR lasers. Adiabatic cooling with a neon buffer gas in the molecular beam expansion causes the molecule to vibrationally and rotationally cool to temperatures between 2 and 20 Kelvin. When only using the UV laser, an R2PI spectrum as shown in Fig. 3.25 can be obtained. This spectrum is very similar to spectra obtained when bringing phenylalanine into the gas phase via heating, followed by cooling in a molecular beam, as performed by Levy et al. [205] and Simons, Snoek and coworkers [25]. Via hole-burning experiments, it has been shown that the peaks in the UV spectrum stem from a distribution of conformers. In [25] and [205], a labeling scheme is used which we will adopt here as well. In the wavelength range shown here, all peaks in the UV spectrum stem from different conformers, which are labeled A, B, C, D, X, E, and F [25]. Outside this range, toward higher wavenumbers, peaks resulting from three other conformers are observed [25]. As has been recognized by Simons et al., conformer "X" is only observable in R2PI spectra and is not observed in laser induced fluorescence (LIF) spectra [205].

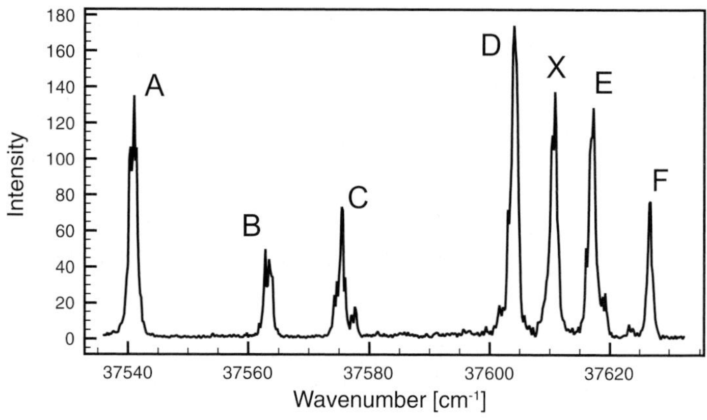

Fig. 3.25. UV-R2PI spectrum of jet-cooled Phenylalanine. All peaks shown in this spectrum stem from different conformers. The peaks are labeled as in [25].

A striking difference of the here presented spectrum to the R2PI spectrum of Simons et al., [25] is however the large relative intensity of conformer "E". In the spectra measured by Simons, "E" is only present in very minute quantities. Here, the peak corresponding to "E" is one of the stronger peaks. An experimental difference is the nature of the carrier gas used. Simons et al. use argon while here, neon is used. When we use argon in the present experiments, "E" becomes much depleted as well. The reason for this carrier-gas dependence can be found in the interaction of carrier gas atoms with the phenylalanine and will be discussed elsewhere [206].

It is difficult to extract structural information from the UV spectra. Some information can be obtained from, for example, the conformer specific measurement of the dispersed fluorescence [205] or the ionization potential [207]. In addition, the comparison of calculated and measured high resolution UV spectra can facilitate a structural assignment [208]. However, the most versatile method to deduce structural information on the conformers of phenylalanine and similar systems remains conformer specific IR spectroscopy, either in the X–H stretch (X=C,N,O) region around 3 μm [25] or in the fingerprint region beyond 5 μm [28].

3.4.4.2 IR spectra

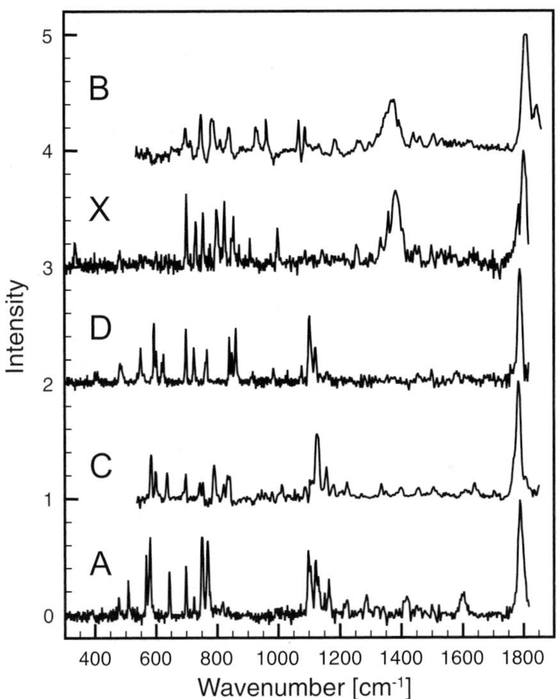

Fig. 3.26. IR spectra of five conformers of Phenylalanine. They are obtained by measuring the IR-UV ion dip spectra with the UV laser parked on the corresponding resonance in the UV-R2PI spectrum (Fig. 3.25).

When the UV excitation and ionization laser is parked on one of the UV transitions in Fig. 3.25, the IR spectrum from the corresponding conformer can be measured via ion dip spectroscopy. In Fig. 3.26, the IR spectra of five conformers are shown. The spectra for conformers X, D and A were measured in the range from 300 – 1900 cm^{-1} while the spectra of B and C were only measured in the 500 – 1900 cm^{-1} range. Clearly, in all spectra, distinct peaks can be recognized having widths ranging from 3–5 cm^{-1} in the lower wavenumber region to up to 20 cm^{-1} around 1800 cm^{-1}. Those widths are only slightly larger than the corresponding width of the infrared laser.

The spectra of the five conformers clearly differ from each other. Nonetheless, they fall into two groups where the spectra of conformers B and X resemble each other and are qualitatively different from those of conformers D, C, and A. Near 1800 cm^{-1}, each spectrum shows a peak that results from C=O stretching motion. For B and X, those peaks are found at 1808 cm^{-1} and 1798 cm^{-1}, respectively while for D, C, and A, those peaks are shifted to the red and can be found at 1787, 1781 and 1787 cm^{-1}, respectively. Hydrogen bonding can shift the C=O stretching frequency to the red and it seems possible that in D, C and A, the C=O group is involved in H-bonding while this group is free in B and X. This is in agreement with previous structural assignments [25,208]. Further structural information can be deduced from the broad peaks near 1350 cm^{-1} in the spectra of B and X and the group of peaks between 1100 and 1150 cm^{-1} in the spectra of D, C, and A. Those peaks can stem from C–O–H bending motion (coupled with C–OH stretching motion). When the H-atom of the COH group is involved in H-bonding, the C–O–H bending frequency usually shifts to the blue. It thus seems likely, that the H of the COH group in B and X is involved in a "strong" H-bond. The observed spectra are thus compatible with structures as those shown in Fig. 3.27 where for X and B, the H-atom of the COOH group is H-bonded to the N-atom of the NH_2 group and the C=O group is free, while for A, C, and D, the H-atoms of the NH_2 group have a H-bond to the C=O group and the C–O–H is not involved in H-bonding.

The structures shown in Fig. 3.27 are calculated at the B3LYP level using a 6-311+G(2d,p) basis set. They are essentially the same as those derived by others [209]. Indicated above and below the arrows are the zero-point corrected heights of some low lying barriers for interconversion between those conformers. All transition states are calculated using the smaller 6-31+G(d) basis set. The relative energies of the conformers can be calculated by taking the difference between barrier heights for forward and reverse reactions. However, at this modest level of theory, relative energies are only of qualitative value. The corresponding vibrational spectra (calculated with the 6-311+G(2d,p) basis set.) are shown in Fig. 3.28 where they are compared to the experimental spectra. All peak positions in the calculated spectra are scaled with a uniform factor of 0.98. The agreement between the calculated and experimental spectra is very good, giving confidence in the structures that the calculations are based on (Fig. 3.27). Peaks corresponding to C=O stretching motion around 1800 cm^{-1} are predicted a few cm^{-1} to the red from the experiment, irrespective of whether the C=O group is involved in H-bonding (B and X) or not (D, C, and A). Vibrations originating from C–O–H bending motion around 1350 cm^{-1} (B and X) or around 1100 cm^{-1} (D, C, and A) are even better reproduced.

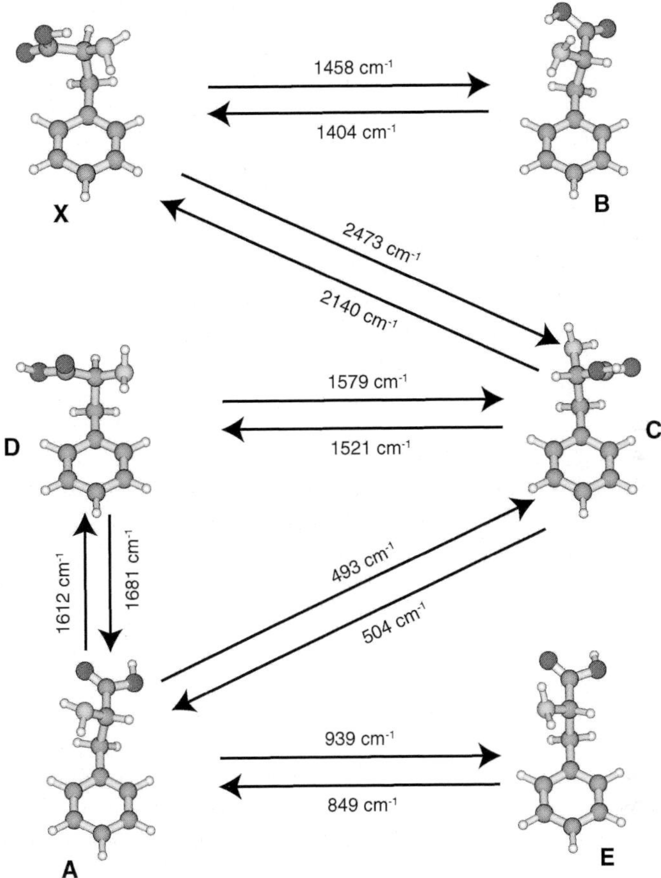

Fig. 3.27. Structures of six phenylalanine conformers, calculated at the B3LYP/6-311+(2d,p) level. The arrows indicate low lying transition states, which are calculated at the B3LYP/6-31+(d) level. The heights of the transition states (B3LYP/6-31+(d) level) are zero point corrected and are shown next to the arrows.

3.4.4.3 Anharmonic calculations

B. Brauer and R. B. Gerber

The situation is more difficult below 1000 cm^{-1} . In that range, large amplitude motion as well as delocalized vibrations where many atoms move are dominant. When comparing theory to the experiment, one can notice a good qualitative agreement. However, in order to obtain the agreement, the calculations were scaled by a factor 0.98, as it is common practice for this type of calculations [210]. Scaling is usually performed to correct for two shortcomings of the calculations. The first is due to the harmonic approximation used.

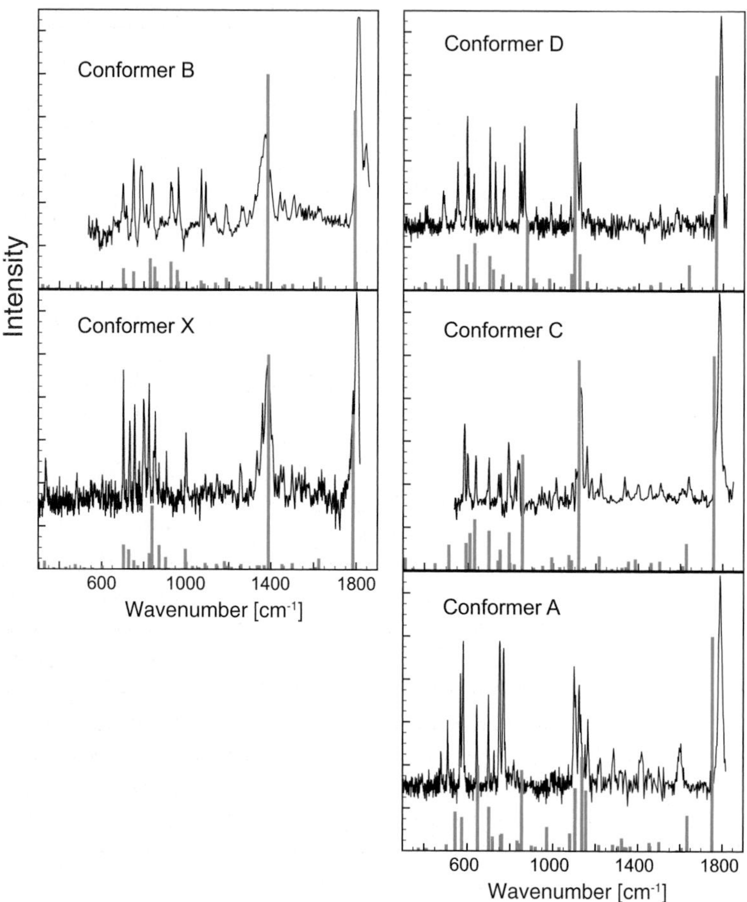

Fig. 3.28. Experimental IR spectra of five Pheylalanine conformers, compared to theoretical stick spectra calculated at the B3LYP/6-311+(2d,p) level.

It is assumed that the vibrations can be described using a purely quadratic potential and that different modes do not couple. In reality, however, the potential is "softer" than a harmonic potential and modes do couple with each other. This will let the calculated frequencies deviate from the experiment, even if the underlying potential energy surface was the "true" one. A second reason is that one might want to correct for deficiencies in the underlying potential. It is for example known that in Hartree-Fock calculations, potentials are calculated much too steep and for that reason, such calculations are usually scaled more and factors as small as 0.85 are common.

To investigate such effects, anharmonic calculations for phenylalanine conformer X, using the Correlation-Corrected Vibrational Self-Consistent Field (CC-VSCF) method [211–213] were performed by B. Brauer and R.B. Gerber.

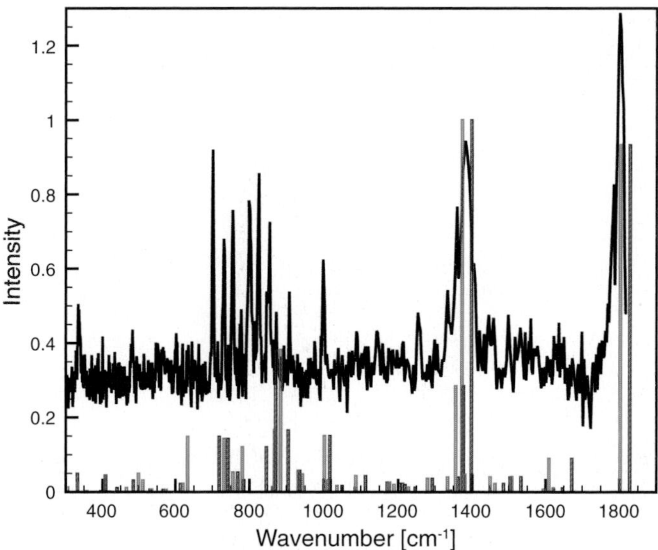

Fig. 3.29. IR spectrum of phenylalanine conformer X compared to harmonic calculations (blue sticks) and calculations where anharmonic corrections are included (red sticks).

The VSCF method assumes separability of the total vibrational wavefunction into a product of single mode wavefunctions. The potential of the system is approximated by considering a sum of single mode and pair-coupling terms. This reduces the computational effort in the VSCF . The extended CC-VSCF method uses perturbation theory to correct for correlation effects in VSCF. The potential surface used in the calculation was adapted from the PM3 semi-empirical method so that the equilibrium harmonic frequencies would be equivalent to those of a DFT calculation. The resulting harmonic (blue) and anharmonic (red) stick spectra can be seen in Fig. 3.29. From the comparison, it appears that, in the here considered frequency range, anharmonicities are comparatively small effects that shift the harmonic frequencies about 1–2 % to the red. The usage of a scale factor of 0.98 is therefore well justified and the shape of the potentials near the minima are thus well described at the B3LYP/6-311+(2d,p) level, which is used to calculate the spectra in Fig. 3.28.

However, the effect of anharmonicities can be larger for N-H or O-H stretching modes. The harmonic and anharmonic frequency values for selected modes of conformer X can be found in Table 3.1.

Some modes have an anharmonicity that is much larger than 2 %. For example, the NH_2 bending mode has an anharmonicity of 3.9 %. Of this, 1.4 % are due to (diagonal) single mode anharmonicity and 2.5 % due to coupling to

Table 3.1. Harmonic and anharmonic frequencies (wavenumbers in cm^{-1}) for selected modes of conformer X. The harmonic frequencies are calculated at the B3LYP/TZP level and the anharmonic frequencies using the CC-VSCF method. The single mode and coupling contributions (in %) of the anharmonicity are also given.

		NH_2 asym. str.	NH_2 sym. str.	O-H str.	C=O str.	NH_2 bend
	Harmonic	3605	3506	3433	1829	1673
	Anharmonic	3329	3235	3261	1803	1610
Contribution	single-mode	0.2	3.2	4.7	0.7	1.4
	coupling	8.5	5.1	0.6	0.8	2.5

other modes. Because of the low intensity of this mode, however, this discrepancy is not apparent in Fig. 3.28. The N-H and O-H stretching modes have particular high anharmonicities that can be well above 8%. Additionally, the contribution to the anharmonic frequency due to anharmonicity along a single mode can be positive or negative. In fact, in the region ca. 1000-1400 cm^{-1}, the single mode anharmonic contribution tends to be positive, while the contribution of coupling tends to be negative. This gives a cancellation of effects. Details on the mid-IR spectra of different phenylalanine conformers and their computation can be found elsewhere [214].

3.4.4.4 Conformational dynamics of biomolecules: an outlook

The barriers between different conformers of biomolecules can be very low, enabling the thermal conformational redistribution in native environments. When studying gas-phase biomolecules, this opens the possibility to investigate conformational changes on the electronic ground state surface using IR lasers. Such experiments have been performed by Zwier and coworkers using 3 μm laser light [33]. However, this is far above the barriers for isomerization (see Fig. 3.27) and the usage of a free electron laser gives the opportunity to perform such experiments closer to the barrier for isomerization, thereby allowing mapping of the potential energy surface and investigation of the dynamics over a large energy range. Phase and amplitude shaping of the FEL should then allow attempts to *control* the conformational distribution by creating vibrational wave packets that direct the reaction path in a desired direction. First steps in this direction have been taken by mapping the potential energy surface of relevant molecules near several minima in conformational space and are presented here.

Next, one may design IR pump dump laser pulses which drive the wave function from a reactant configuration via an intermediate target state to the selective product [35, 215]. The individual pump and dump steps can be achieved using IR laser pulses with analytical shapes [84], or using optimal control. [216] The approach allows various choices of intermediate target states e.g. a delocalized vibrational eigenstate, with energy close to the potential barrier which separates the reactants and products [35, 215] (Here, the time delay between the IR pump and dump pulses is irrelevant - different from the UV pump-dump approach of Tannor and Rice [82, 217]). Alternatively, the intermediate eigenstate of the system may be coupled to a dissipative environment- this is advantageous in cases of preferential deactivation to the product states [218, 219]. Still another possibility is an intermediate target state which represents tunneling from the reactant to the product domain [220, 221]. Various applications of IR laser pulse control to model systems are encouraging, see e.g. [222–225], and also the novel mechanisms of IR laser control which are presented in Sect. 3.4.6.10. Another extension is IR+UV laser pulse control, pioneered by N. Henriksen [226, 227]. This allows new applications in many domains, e.g., laser separation of different products [228], or control of molecular functions e.g. ignition of unidirectional intramolecular rotations [229], see also [230]. Optimal laser driven transitions between molecular vibrational states may even be used for quantum computing [305].

3.4.5 Organometallic complexes: Control of excited state dynamics of $CpMn(CO)_3$

L. González and J. Manz

The organometallic complex $CpMn(CO)_3$ ($Cp{=}\eta^5C_5H_5$) resembles the system which Gerber and coworkers [39] used, for the first time, to illustrate the applicability of adaptive optimal control schemes. With the goal of understanding the reaction dynamics occurring under experimental optimal pulses, pump-probe experiments and optimal control experiments on $CpMn(CO)_3$ were performed by Wöste and coworkers [40]. Fig. 3.30 shows schematically the competing dissociation and ionization processes of $CpMn(CO)_3$ under light irradiation which results in carbonyl dissociation.

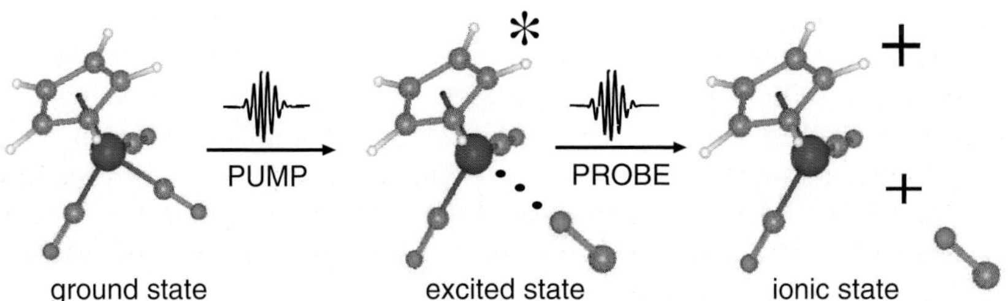

Fig. 3.30. Schematic pump-probe processes leading to dissociation in $CpMn(CO)_3$

Exemplarily, optimal control was set up to maximize the yield of the parent ion $CpMn(CO)_3^+$ while hindering competing fragmentation channels that result in $CpMn(CO)_2^+ + CO$ and other fragments. The obtained experimental optimal control pulse, with its intensity and phase profiles, is shown in Fig. 3.31. The pulse shape shows two main peaks with intensities in a 2:3 ratio. Details about pump-probe and control experiment set ups can be found in Chapter 2, Sect. 2.2. Here we just give the parameters used in this particular experiment. The wavelength used in the control experiment was 800 nm (1.55 eV); however, the optimized first subpulse is blue shifted to 798.7 nm while the frequency of the second subpulse is mostly unchanged (800.1 nm) with respect to the initial central wavelength. The initial transform-limited pulse had 87 fs of duration but the resulting optimized subpulses are separated by ca. 85 fs and have durations of about 40 fs.

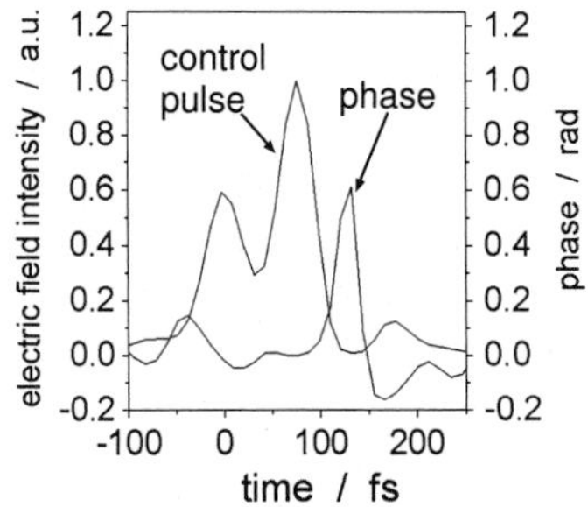

Fig. 3.31. Optimal control pulse which maximizes $CpMn(CO)_3^+$ with intensity and phase profile [40]

Our challenge was the decoding of the optimal pulse in an attempt to unravel how the structure of the optimal pulse determines the dynamics of the system toward the desired chemical channel: in this case, the maximization of ionization and minimization of dissociation. The time profile of the pulse suggests that the initial transform-limited pulse has been divided into two subpulses designed to perform an optimal pump-probe type experiment, where optimal frequencies and optimal time delays are employed. In order to confirm this hypothesis, pump-probe experiments were performed using the time profile suggested by the optimal pulse, that is, pulse durations of ca. 40 fs. Specifically, the pump-probe spectra shown in dots in Fig. 3.32 were recorded using a blue pulse of λ=402.5 nm (3.08 eV) and t_p=45 fs for the pump and a red pulse of λ=805 nm (1.54 eV) and t_p=35 fs for the probe. As it can be seen, superimposed on a 170 fs decay, oscillations with a period of ca. 80–85 fs appear in both the parent and the fragment ion signals.

Even if at first sight there appears to be some connection between the time scales of the processes occurring in the pump-probe experiment and those of the optimal control field, as it will be shown later, the mechanisms behind

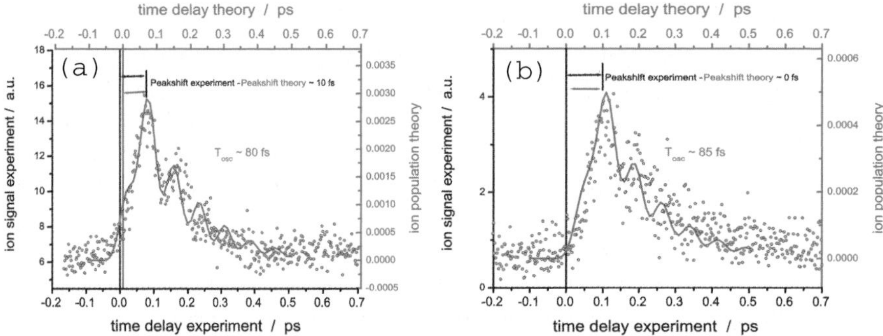

Fig. 3.32. Experimental (dots) and theoretical (lines) pump-probe spectra performed using optimal pulse durations. Adapted from [40]

both processes do not need to be the same. The challenge for theory is to interpret the origin of the vibrational and decay processes in the pump-probe transients signals (cf. Fig. 3.32), as well as to correlate these dynamics with the one taking place within the optimal pulse (cf. Fig. 3.31). The disentangling of the reaction dynamics underlying this complex system involves a detailed analysis of the processes that take place under irradiation (Section 3.2.2). Furthermore, the dynamics reflects the complexity of the photochemistry, which then needs to be preanalyzed with quantum chemical calculations (Sect. 3.2.1). Therefore, our analysis starts with an accurate MRCI/CASSCF calculation of the low-lying excited and ionic states of $CpMn(CO)_3$ and corresponding PES along the decisive metal-carbonyl fragmentation coordinate $q = R_{Mn-CO}$, see [231, 232]. The a^1A' electronic ground state conforms to a close shell electronic configuration of $\cdots(20a')^2(21a')^2\ (22a')^2(12a'')^2(13a'')^2$ assigned to $\cdots(2\pi_{Cp})^2(3d_{z^2})^2(3d_{x^2-y^2})^2(3\pi_{Cp})^2(3d_{xy})^2$. The low-lying virtual orbitals correspond to $3d_{yz}(23a')$, $3d_{xz}(14a'')\cdots$ and $\pi^*_{CO}(24a',\ 25a',\ 15a''$ and $16a'')\cdots$. According to eq. 3.3, the electronic states $\Phi_i(q)$ can be described as a linear combination of different state configurations or Slater determinants $D_j(q)$,

$$\Phi_i(q) = \sum_j c_{ij}(q) D_j(q), \tag{3.20}$$

with the corresponding CI expansion coefficients $c_{ij}(q)$. Likewise, the ionic states can also be described by a linear expansion of configurations:

$$\Phi_i^{ion}(q) = \sum_j c_{ij}^{ion}(q) D_j^{ion}(q), \tag{3.21}$$

where the D_l^{ion} are the state configurations representing the ionic state and c_{ij}^{ion} are the CI expansion coefficients.

Figure 3.33 shows the relevant PES along q [40, 232]. They are characterized by several avoided crossings. In particular, two crossings around the

Franck-Condon geometry between the b^1A' and c^1A' states and between the a^1A'' and b^1A'', are the most relevant for the forthcoming simulations. Interestingly, according to the excitation energies from the electronic ground state a^1A' to the neutral excited states, and from the excited states to the ionic ones, it is reasonable to assume that the first experimental optimal subpulse requires two red photons to excite $CpMn(CO)_3$ to an electronic excited state and that the delayed second subpulse excites population to an ionic state with three red photons, see Fig. 3.33.

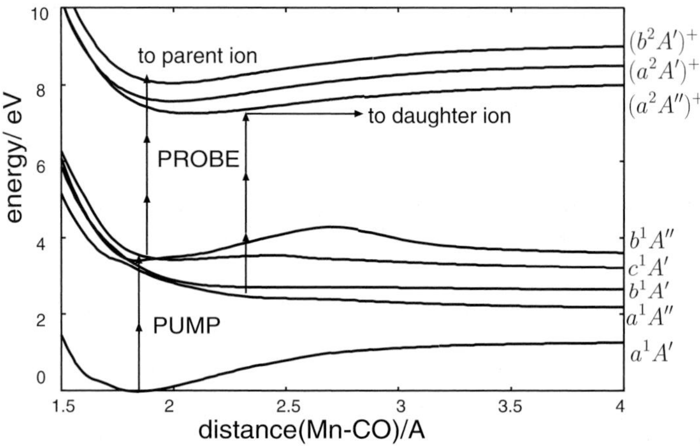

Fig. 3.33. Ab initio multiconfigurational MRCI/CASSCF potential energy curves along the metal-carbonyl dissociation degree of freedom [40, 232]. Arrows indicate multiphoton processes which lead to the parent or daughter ions.

In order to get a dynamical insight into the possible processes which occur after irradiation, the time-dependent Schrödinger equation 3.4 has to be solved. Since we will be propagating on electronic states (see Fig. 3.33), (3.4) is specified as:

$$i\hbar\frac{\partial}{\partial t}\begin{pmatrix}\psi_0\\ \vdots\\ \psi_7\end{pmatrix}=\begin{pmatrix}H_{00}&\cdots&H_{07}\\ \vdots&&\vdots\\ H_{70}&\cdots&H_{77}\end{pmatrix}\cdot\begin{pmatrix}\psi_0\\ \vdots\\ \psi_7\end{pmatrix}\tag{3.22}$$

where the labels i or j = 0, 1-4 and 5-7 denote the ground (a^1A') state, the excited states a^1A'', b^1A', c^1A', and b^1A'' of the neutral system and the lowest states $a^2A''^+$, $a^2A'^+$, and $b^2A'^+$, of the ionic system, respectively. The diabatic couplings were computed by means of the procedure suggested by Peyerimhoff et al. [233] and of Baer [234], for the details see [235]. The results indicate that the coupling between the states b^1A' and c^1A' is very weak, while the coupling between the states a^1A'' and b^1A'' is very strong. As a consequence, we can anticipate that nonadiabatic losses are only expected to occur between the a^1A'' and b^1A'' states [232, 235].

According to (3.7), transition dipole couplings involving the transition from the ground to excited neutral states, and from the excited neutral to the ionic states, are also needed. While the former can be calculated by quantum chemical approaches (see Sect. 3.2.1) in a standard fashion, the latter is a difficult task. Instead of the usual Condon approximation made in theoretical simulations, that is, setting the neutral-to-ionic dipole coupling to a constant equal to one, we have derived a procedure to estimate non-Condon couplings using multiconfigurational wavefunctions given by (3.20) and (3.21). As worked out in [236] the resulting dipole couplings contain three terms: (i) the CI coefficients corresponding to determinants of the neutral molecule which differ in only one spin orbital from the ionic determinant (indicated by $\sum'$), (ii) the overlap between the non-frozen orbitals χ_m of the initial neutral molecule and those of the final ion, and (iii) the single electron matrix element between the initial and final orbital which represents the photodetached electron, in zero kinetic energy (ZEKE) approximation,

$$\langle \Phi_i | \boldsymbol{M} | \Phi_j^{ion} \rangle \approx \sum\nolimits_k' \sum\nolimits_l' c_{ik} c_{jl}^{ion} \cdot \langle \chi_m | \tilde{\chi}_m \rangle \cdot \langle \chi_k | \boldsymbol{\mu} | \chi_{ZEKE} \rangle \tag{3.23}$$

where $\boldsymbol{M} = \sum_n \mu_n$ is the sum of all electronic dipole operators, $\mu_n = -e \cdot r_n$. In the present case, the ionic states are composed essentially of a single configuration [235]. Furthermore, $\langle \chi_m | \tilde{\chi}_m \rangle \approx 1$, and the last term in (3.23) is approximated by a constant [56, 236]. Then the final expression takes the form,

$$\langle \Phi_i | \boldsymbol{M} | \Phi_j^{ion} \rangle \approx const \sum\nolimits_k' c_{ik} \tag{3.24}$$

where c_{ik} are the CI expansion coefficients defined in (3.20).

By inserting in (3.22), a) the accurate ab initio PES for the neutral and ionic states, b) the nonadiabatic coupling between electronic excited states, and c) the neutral-to-neutral and neutral-to-ionic transition dipole couplings with the laser, and d) assuming that the two- and three-photon excitations shown in Fig. 3.33 may be represented by single photon transitions, see Fig. 3.34, allows to perform systematic investigations of pump-probe and control simulations. Here we just summarize the findings recollected in [40,56,236] and shown schematically in Fig. 3.34. The oscillations and decay of the molecular system (cf. Fig. 3.32) can be attributed to the dynamics occurring on the predissociative b^1A'' state, see Fig. 3.34a. After excitation by the pump pulse, the wave packet in the b^1A'' state decays within 170 fs nonadiabatically into the a^1A'' state which is strongly repulsive; simultaneously it vibrates in the b^1A'' state. The formation and periodic leakage of the wave packet by a diabatic transition to the a^1A'' state (see curved arrow in Fig. 3.34) is then monitored by the probe pulse which prepares parent and daughter ions on corresponding ionic surfaces. Our simulations then indicate that fragmentation starts in the neutral a^1A'' state and ends in the ionic ground state. At the same time, the excited neutral $CpMn(CO)_3^*$ also dissociates to the neutral fragments $CpMn(CO)_2$ + CO. Inversion of the pump probe signals yields a refined quantum chemistry based Hamiltonian with scalings of selected parameters within quantum chemical accuracy [56].

The final goal of our investigations is to understand the optimal pulse shape shown in Fig. 3.31 which is tailored to maximize the production of parent ions, $CpMn(CO)_3^+$. Fig. 3.34b indicates the essential steps of the dynamics hidden in the optimal field. The subpulses contained in the optimal pulse can indeed be interpreted as a pump and probe type sequence with optimal frequencies and optimal time delays. The predominant mechanism can be explained as an excitation to the bound c^1A' state where the system vibrates, and —because of the very weak nonadiabatic coupling with the dissociative counterpart b^1A' state— it leads exclusively to parent ions.

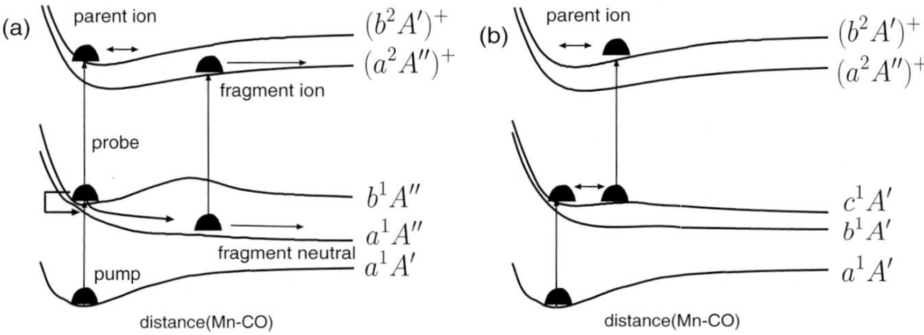

Fig. 3.34. Schematic representation of the mechanisms behind (a) the pump-probe signals shown in Fig. 3.32 which produce parent and fragment ions, and (b) the optimal control field shown in Fig. 3.31 which produces predominantly parent ions.

In summary, the present wave packet dynamics simulations based on high level ab initio calculations for the model system $CpMn(CO)_3$ achieve two goals: They allow to analyze the important initial processes in the complex polyatomic system induced by laser pulses, and to explain the mechanism of the experimental optimal laser pulse which yields preferential formation of the parent ions while suppressing competing fission of ligands [40]. In this case, the rather simple form of the optimal laser pulse, together with the intuitive hypothesis of an optimal pump-probe type process, helped to discover the laser driven dynamics, as follows: Essentially, the present system, like many other organometallic molecules, has many close lying excited electronic states. The experimental optimal laser pulse with pump-probe type patterns exploits this property of the model system such that the frequencies of the sequential "pump" and "probe" parts select a well-timed, optimal excitation path way via adequate excited states to the desired product channel, while avoiding alternative paths to competing dissociative channels. An important aspect of this mechanism is the discovery that the dissociation toward ionic products may be induced already in dissociative excited states of the neutral molecule, i.e. the "pump" and "probe" components of the optimal laser pulse have to suppress both dissociations of the excited neutral as well as of the ionized parent molecule.

For the present system, the assumed one-dimensional model appears to be appropriate, at least for describing the initial dynamics induced by the laser pulses. Subsequently, couplings between the chosen dissociative metal-ligand bond and other degrees of freedom will induce additional competing processes, such as intramolecular vibrational redistribution IVR, or intersystem crossings IC, e.g., via conical intersections; see also the discussion of the models in reduced dimensionality depending on the hierarchy of time scales for sequential processes, in Chapter 4, Sect. 4.10. Competing processes such as IVR or IC may , indeed, contribute to the experimental pump probe spectra, causing deviations from the quantum simulations based on the simple one-dimensional model, at later times ($t \gtrsim 150$ fs), cf. Fig. 3.32. In other systems, "multi-dimensional" processes such as IVR, IC etc may compete against the target channel e.g. against ionization, or bond selective dissociation, even earlier, thus calling for even more ambitious control scenarios. As a consequence, the resulting optimal laser pulses will appear more complex. This conjecture is in accord with rather complex optimal laser pulses which have been designed experimentally for other systems, see e.g. [39].

The demonstrated approach relies on several approximations. First, we have employed the one-dimensional model which neglects competing processes which involve additional degrees of freedom, as discussed above. Second, all simulations consider resonant one-photon transitions even if the PES are consistent with the fact that the first sub-pulse excites the system in a two-photon transition without any intermediate state supporting it (see Fig. 3.33). This is what we call *nonresonant* multiphoton transitions (NMT) . Likewise, the second subpulse probes the system in a three-photon nonresonant excitation. These NMT are ubiquitous in optical experiments in the strong field regime and seem to be common in closed-loop feedback experiments. However, the theory for NMT is not very widespread in the field of femtosecond spectroscopy. In [237] we suggest a formalism for fs NMT. An effective time-dependent Schrödinger equation is derived with effective couplings to the radiation field including powers of the field strength and effective transition dipole operators between the initial and final state. NMT transitions can also be implemented in the optimal control theory as we demonstrate in [238].

The third approximation is that our simulations use a zero kinetic energy (ZEKE) approach . With this is meant that the photodetachment process is not simulated with a discretization of the continuum, as described for example in [239] or [240]; rather, we assume that ionization occurs through quasiresonant transitions between neutral and ionic states [241]. However, the account of the continuum states or even a time-dependent description of the photodetached electron may uncover important details of the ionization process. Work along this line is in progress.

3.4.6 Theoretical approach for ultrafast dynamics and optimal control in complex systems based on the MD "on the fly": Applications to atomic clusters and molecules

V. Bonačić-Koutecký and R. Mitrić

3.4.6.1 Multistate nonadiabatic nuclear dynamics in electronically excited and ground states

Study of ultrafast photochemical processes involving nonadiabatic radiationless decay over one or several conical intersections among electronic excited states in multidimensional systems represents still a challenge for theorists. The importance of conical intersections in organic photochemistry was recognized long time ago ([242–245] and references therein). Identification of conical intersections in numerous molecules and clusters and their relevance for ultrafast spectroscopy attracted attention of many researchers [243]. Since most of organic photochemistry takes place in solution, the influence of environment is an important but complicated issue. Therefore, the study of ultrafast processes in the gas phase such as in elemental clusters stimulates the development and use of accurate computational methods, in particular if a direct comparison with experimental observables is accessible [65].

In order to address nonadiabatic transitions in complex systems involving avoided crossings and conical intersections between electronic states, semiclassical methods based on ab initio multistate nonadiabatic dynamics are needed for the simulation of fs pump-probe signals, in particular if all degrees of freedom have to be considered. For this purpose, in addition to the calculation of forces in the electronic ground and excited states, the computation of nonadiabatic couplings between electronic states states "on the fly" is required. The nonadiabatic dynamics can then be combined with the Wigner-Moyal representation of the vibronic density matrix allowing one to determine the fs-signals in analogy outlined for adiabatic dynamics.

The situation is still very different for ab initio adiabatic and nonadiabatic MD "on the fly" involving excited electronic states than in the case of ground states which has been outlined in chapter on the NeNePo spectroscopy. In spite of recent efforts and successes [71, 72, 246–248], further development of such theoretical methods which combine accurate quantum chemistry methods for electronic structure with MD adiabatic and nonadiabatic simulations "on the fly" has promise to open many new possibilities for the successful investigation of fs-processes. This research area will essentially remove borders between quantum chemistry and molecular dynamics communities. In this context, very intense research is presently going on along two main directions. One is to achieve fast calculations of forces in excited states, as well as of nonadiabatic couplings, at the level of theory accounting for electron correlation effects with controllable accuracy which are suitable for implementation in different adiabatic and nonadiabatic MD schemes "on the fly" [248]. The second is to introduce quantum effects for the motion of nuclei, particularly in the case of nonadiabatic dynamics [249–254], in systems with a considerable number of degrees of freedom, allowing for their identification in spectroscopic observables such as fs-signals [248,250,252,255,256]. In our work we deal with systems containing heavy atoms, and therefore the quantum dynamical effects do not play important role as it will be shown on prototype examples.

As already pointed out, ab initio nonadiabatic MD "on the fly" involving excited states and simulation of observables is demanding from theoretical and computational point of view and still needs further developments. Therefore, in order to meet the requirements on high accuracy and realistic computational demand, the choice of the systems has to be made for which the description of electronic excited states and nonadiabatic coupling is relatively simple. This is the case for nonstoichiometric alkali-halide clusters with two and one excess electron (e. g. Na_nF_{n-2} and Na_nF_{n-1}). Their structural and optical properties have attracted the attention of many theoretical and experimental researches [80, 81, 257–268] due to the localization of the excess electrons, which are not involved in ionic bonding. We have selected two prototypes Na_3F and Na_3F_2 for presentation because of addressing multiple passage versus single passage through conical intersection, respectively, and because of necessity to use different approaches for the calculations of the electronic structure.

First, we illustrate our approach for the nonadiabatic relaxation on the example of Na_3F cluster for the following reasons: i) Due to two excess electrons which are not involved in ionic NaF bonding, the treatment of electron correlation is mandatory for description of the geometric relaxation in excited states. Moreover, appearance of conical intersections among excited states is connected with multireference character of the wavefunctions and violation of Born-Oppenheimer approximation. ii) The experimental pump-probe spectrum is available [268]. iii) Our initial simulations of pump-probe spectra of Na_3F involving adiabatic dynamics "on the fly" combined with Wigner distribution approach [81] provided valuable information about relaxation dynamics for short times before nonadiabaticity starts to govern ultrafast processes. This was shown explicitly by comparing the results with experimental findings [81]. However, at later times the simulated pump-probe signals remained constant while the experimental one decreases indicating that other processes such as nonadiabatic transitions became important. All three aspects stimulated us to adapt and combine methods for treating nonadiabatic dynamics over several electronic states at high level of accuracy which will be presented here. This allowed us to calculate nonradiative lifetime quantitatively taking into account all degrees of freedom and to identify the processes involved in the radiationless decay. It will be shown that inclusion of nonadiabaticity is mandatory to achieve a complete agreement with experimental findings.

Second, nonstoichiometric sodium fluoride clusters with a single excess electron (e.g., Na_3F_2) represent a particularly simple situation concerning the description of excited states. In this case, a strong absorption in the visible-infrared energy interval occurs due to the excitations of the one excess electron placed in a large energy gap between occupied (HOMO) and unoccupied (LUMO) one-electron levels which resemble the "valence" and the "conductance" bands in infinite systems. Therefore, these clusters offer the opportunity to explore the optical properties of finite systems with some bulk characteristics such as F-color centers. Moreover, a simple but accurate description of the excited states is possible to achieve in the framework of the one-electron "frozen ionic bonds" approximation. In this method, the optical response of the single excess electron can be explicitely considered in the field of other (n-1) valence electrons which are involved in strongly polar ionic Na-F bonding [71].

The calculation of excited state energies and of gradients based on the "frozen ionic bonds" approximation (as outlined in [71]) is from a computational point of view considerably less demanding in comparison with other approaches such as RPA, CASSCF or CI, and provides comparable accuracy. Therefore, this approach allows to carry out adiabatic and nonadiabatic molecular dynamics in the excited state, by calculating the forces "on the fly" and nonadiabatic couplings (cf. [71,72]) applicable to relatively large systems. This is particularly convenient for the simulation of time-dependent transitions for which an ensemble of trajectories is needed. Of course, the application is limited to systems for which the "frozen ionic bonds" approximation offers an adequate description.

In addition to methodological aspects, the study of the dynamics in the first excited state of Na_3F_2 and the radiationless transition to the ground state allow for the prediction and verification of the consequences of conical intersections (between the first excited state and the ground state) in fs pump-probe signals in the gas-phase without the necessity to consider the environment. The latter external medium effects complicate the issue, as for example in the case of photochemistry in solution or in the case of the cis-trans photoisomerization of the visual pigment due to the influence of the protein cavity [269, 270].

Therefore, the photoisomerization in the Na_3F_2 cluster through a conical intersection will be addressed first and then in Sect. 3.4.6.3 the new strategy for optimal control will be applied in order to suppress the passage through the conical intersection and to selectively populate one of the chosen isomers.

3.4.6.2 Passage through multiple conical intersections in Na_3F cluster

We first describe our approach for electronic structure and dynamics "on the fly" and then for the simulation of the signals. For the ground and three lowest lying excited states of the Na_3F cluster configuration interaction calculations with single and double excitations (CISD) have been carried out. In accord with the previously employed "frozen ionic bonds" approximations [71, 72],

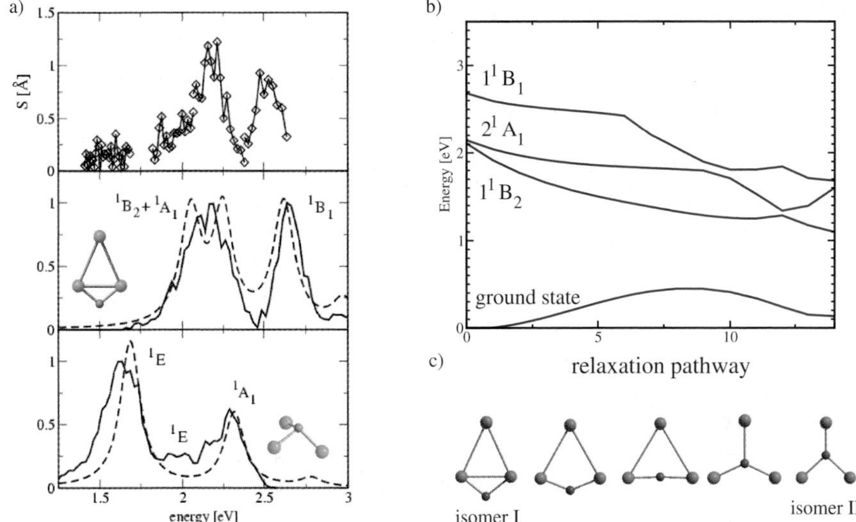

Fig. 3.35. (a) Comparison between experimental [267] and theoretical absorption spectra of Na_3F. Finite temperature absorption spectrum obtained from Monte-Carlo simulation performed at 300 K (full line) and the TDDFT absorption spectrum (dashed line) obtained from a Lorentzian convolution of the vertical lines for the rhombic isomer and for pyramidal isomer. (b) Sections from the potential energy surfaces for the ground and three excited state of Na_3F, connecting the two ground state isomers. The relaxation pathway corresponding to the initial adiabatic relaxation in the $1\ ^1B_1$ excited state has been constructed by linear interpolation of internal coordinates between the structure of isomer I and isomer II keeping the C_{2v} symmetry. (c) Structures along the relaxation pathway

only the two excess (nonstoichiometric) electrons have been correlated, effectively thus employing a full CI for these electrons. The analytic energy gradients and the nonadiabatic couplings of the first order

$$D^x_{JI} = \left\langle \Psi_J(r,R) \middle| \frac{\partial \Psi_I(r,R)}{\partial R^x} \right\rangle \tag{3.25}$$

have been calculated for the resulting CI wavefunctions employing the Columbus package for CI method. The details of the computational procedure are given in [271, 272].

We have employed the effective core potentials (ECP) for sodium and fluorine atoms with adequate basis sets [263, 264, 273] which have previously been verified to yield reasonably accurate transition energies [71].

For the time integration of the classical trajectories the velocity Verlet algorithm has been used, with a 0.5 fs time step. The accuracy of the integration has been verified by the conservation of the total energy, which was achieved with an error smaller than ±0.0005 a.u. in all trajectories.

In the ground electronic state, the most stable isomer of Na_3F assumes a rhombic structure with the C_{2v} symmetry (cf. Fig. 3.35a). The second isomer with a three dimensional structure and C_{3v} symmetry lies only 0.12 eV higher in energy but is separated from the lowest isomer by relatively high isomerization barrier of ∼ 0.22 eV. As demonstrated in Fig. 3.35a by comparing simulated and experimental absorption spectra at T=300 K, the rhombic isomer is dominantly populated and no interconversion between two isomers occurs. The stationary absorption spectrum of the most stable isomer exhibits a dominant transition to the 1^1B_2 and 2^1A_1 states which are located at 2.2 eV and second intense transition at 2.7 eV corresponding to the 1^1B_1 state (cf. Fig. 3.35a). These absorption features are characteristic for a structure with delocalized two excess electrons [263, 264, 273].

Here we wish to emphasize the role of nonadiabatic transitions between the electronic states and their influence on the pump-probe signals. For this purpose we select the third excited state (1^1B_1) due to available experimental pump-probe spectrum involving this state [268].

We first briefly describe the results based on the adiabatic simulation of pump-probe signals (cf. Fig. 3.36) and then introduce nonadiabaticity (cf. Fig. 3.37). In fact, as can be seen from the sections of the potential energy surfaces (Fig. 3.35b) along the steepest descent pathway in the third 1^1B_1 excited state, initially, adiabatic relaxation in this state should occur. This is due to the fact that the 1^1B_1 state is well separated from other excited states, as long as fluorine atom is outside the Na_3 subunit. The fingerprint of the initial dynamics in the 1^1B_1 state which involves a periodic motion of the fluorine atom along the C_{2v} symmetry axis is reflected in the adiabatic pump-probe signal given in Fig. 3.36. The comparison between the simulated pump-probe signals obtained from the adiabatic classical and quantum dynamics carried out only in the 1^1B_1 state with the experimental signal is shown in Fig. 3.36. Classical dynamics is based on our ab initio Wigner distribution approach accounting for all degrees of freedom while quantum dynamics is restricted to three degrees of freedom by introducing C_{2v} symmetry constraint. In the case of classical dynamics the initial conditions can be generated by sampling the canonical Wigner distribution at the given temperature, for example at T=50 K. This is justified by the fact, that at T=50 K only the lowest energy isomer is present, and the harmonic approximation is valid at such a low temperature. Both theoretical signals exhibit oscillations whose amplitude is decreasing due to the spreading of the phase space ensemble or the wave packet but after 2 ps they remain constant due to the fact that the 1^1B_1 state cannot be depopulated. The fact that the signals are constant after 2 ps is in contrast with experimental signal which decreases almost to zero within 3 ps. Therefore, the adiabatic dynamics cannot describe the pump-probe signal adequately throughout the whole measured time interval. The decrease of the experimental signal is an indication that radiationless relaxation takes place which leads to the depopulation of the 1^1B_1 state. Notice analogous oscillatory behavior of both theoretically simulated signals illustrating that our ab initio Wigner approach based on semiclassical dynamics "on the fly" provides sufficiently accurate results and that quantum dynamics effects do not play any role.

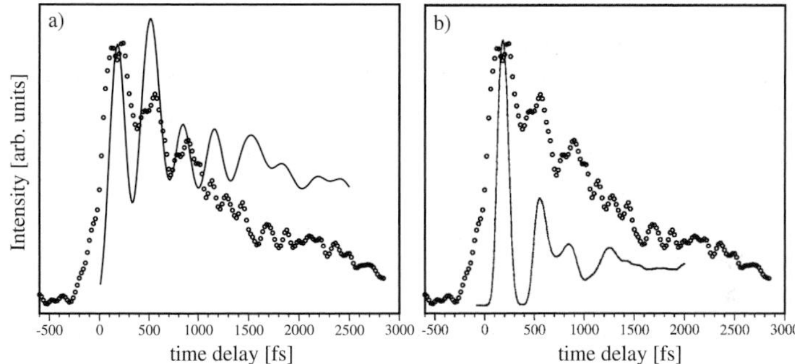

Fig. 3.36. Comparison between the simulated (full line) pump-probe spectrum obtained from the (a) adiabatic dynamics in the $1\,^1B_1$ in the framework of the Wigner distribution approach (T=50 K) and (b) the full quantum mechanical simulations (black line) for reduced dimensionality [81], with the experimental pump probe spectrum (black circles). The experimental data are taken from [268]

An indication that nonadiabatic transitions involving conical intersections should be expected can be seen from the section of the potential energy surface along the initial relaxation pathway shown in Fig. 3.35b. As the fluorine atom enters the Na_3 trimer, the crossings between the states occur. The presence of these crossings is responsible for nonadiabatic transitions between the electronic states and provides the channel for the depopulation of the initially populated 1^1B_1 state. In the following, we concentrate on the new simulations which include nonadiabatic transitions and take into account all degrees of freedom allowing us to obtain the full agreement with the experimental pump-probe signal. Furthermore, this approach allows to determine the nonradiative lifetime of an excited electronic state in a multidimensional system.

In the case of nonadiabatic simulations the initial conditions for the classical trajectory propagation are generated also by sampling the canonical Wigner distribution function at T=50 K. The trajectories were propagated "on the fly" in the framework of the full-CI method and the nonadiabatic transitions were determined by the Tully's fewest switches surface hopping algorithm. Nonadiabatic couplings were also calculated "on the fly" as described above. Since for the simulations of the pump-probe spectra according to (3.16) the averaging over an ensemble of trajectories is required, an ensemble consisting of 200 trajectories has been excited to the third 1^1B_1 excited state and subsequently propagated. The nonadiabatic transitions in the manifold consisting of the ground and three lowest excited electronic states were included.

The snapshots of the nonadiabatic dynamics are presented in Fig. 3.37a. They show that at early stages of the dynamics (until 500 fs) the oscillatory motion of the fluorine atom is the dominant process, but at later times the resulting phase space ensemble becomes very broad indicating that nonadiabatic transitions start to take place. Due to the large excess of energy gained the accessible phase space region increases enormously and eventually the cluster is fragmented into a Na_2F subunit and a single sodium atom.

The time dependent energy gaps between the state in which the dynamics currently takes place and the cationic states are presented in Fig. 3.37b. They are important ingredients used for the simulation of the pump-probe signal according to Eq. (3.16). At early times (less than 500 fs) the energy gaps exhibit oscillations in energy between 2.3 and 1.9 eV which reflect the initial oscillatory motion along the C_2 axis (cf. Fig. 3.37a). This is also the dominant process in the case of the adiabatic dynamics (cf. Fig. 3.36). At later times, however, hoppings to other electronic states become more frequent leading to discontinuities in the energy gaps (cf. Fig. 3.37b), and the values of the gaps become substantially larger, between ~ 2 and 6 eV, due to the transitions to lower lying energy states.

In order to obtain a direct comparison of the simulated pump–probe spectrum with an experimental observable, the contribution of the kinetic energy of the photoelectrons created upon ionization has been included in (3.16) by integrating over the excess of kinetic energy. For the pump pulse we assume zero pulse duration leading to the instantaneous excitation of the whole ensemble to the 1^1B_1 state. The energy of the probe pulse used in the simulation was E_{pr}=2.05 eV, with the pulse duration σ_{pu}=100 fs corresponding to the experimental conditions. The simulated pump-probe signal based on nonadiabatic dynamics exhibits oscillations at early times, with the period of ~ 380 fs as shown in Fig. 3.37c. At later times the signal decays due to the depopulation of excited states. The reason for the decrease of the signal is the fact that due to nonadiabatic transitions the ensemble of trajectories effectively leaves the time dependent Franck-Condon window of the probe pulse. The simulated signal is nearly in a perfect agreement with the experimental one (cf. Fig. 3.37c) demonstrating the ability of our approach to accurately describe radiationless decay processes which are ubiquitous in the photochemical processes.

In order to determine the nonradiative lifetime of the 1^1B_1 state the populations of the involved states were monitored by counting the number of trajectories propagating in each state at the given time and are presented in Fig. 3.37d. The population of the third excited state decays exponentially with a time constant of 900 ps. The first nonadiabatic transitions occur at ~ 200 fs. The transfer of the population occurs simultaneously to both second and the first excited state, so that the population of the second excited state

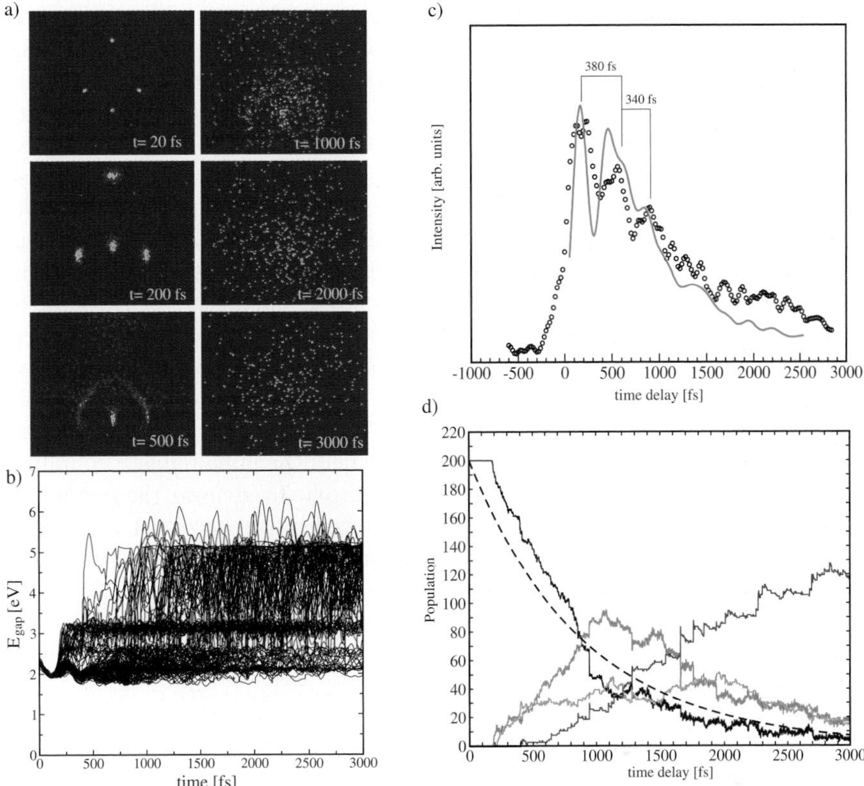

Fig. 3.37. (a) Snapshots of the nonadiabatic dynamics starting initially in the $1\ ^1B_1$ excited state. In order to obtain a continuous distribution each trajectory from the ensemble has been folded with a gaussian (width=0.05 a. u.) and the contributions of all trajectories have been summed. (b) Time dependent energy gaps between the electronic state in which the propagation takes place and the cationic ground state for the ensemble of 200 trajectories during the nonadiabatic dynamics simulations. (c) Comparison between the simulated (red line) pump-probe spectrum obtained from the nonadiabatic dynamics involving three excited states in the framework of the Wigner distribution approach state with the experimental pump-probe spectrum (black circles) [268]. (d) Time dependent populations of three excited states and the ground state. The dashed line represents the exponential fit on the population of the $1\ ^1B_1$ state, allowing to determine the nonradiative lifetime of 900 fs.

reaches its maximum at ~ 1.2 ps. The maximum in the population of the first excited state follows later at ~ 2 ps. After 500 fs the ground state starts to be populated and after 3 ps the population of the excited electronic states has almost completely decayed and the system resides in the ground electronic state with large excess of energy which partly leads to the fragmentation of the cluster.

In summary, we have demonstrated that our approach provides accurate simulation of the pump-probe signals and the ultrafast dynamics in systems where the radiationless transitions play a dominant role. We determine the nonradiative lifetime for the 1^1B_1 electronically excited state of Na_3F to be ~ 900 fs [274]. Our approach offers the opportunity to study photochemical processes in excited states of complex systems. This is particularly important for determination of emissive properties and fluorescence which are directly dependent on the rate of radiationless relaxation processes. In the future, we plan to extend our approach to the systems where the nonadiabatic couplings are not available analytically but can be approximated from the overlap of wavefunctions similarly to the method introduced by Persico et al. [275]. This should open a broad range to applications leading to better understanding of nonradiative processes in larger systems.

3.4.6.3 Photoisomerization through a conical intersection in Na_3F_2 cluster

By adding one F atom to Na_3F, giving rise to the Na_3F_2 cluster with one excess electron, structural, optical and dynamical properties change dramatically illustrating nonscalable properties in the size regime in which each atom counts. The goal is to show that in the case of Na_3F_2 the breaking of bonds in the first excited state leads to the conical intersection and to the direct return to the ground state which can be then identified in observables such as fs pump-probe signals. Therefore, first the optical properties and the characterization of the conical intersection of this cluster will be given, and subsequently the analysis of the nonadiabatic dynamics and of the signals will provide the information about the time scales of the different processes such as bond-breaking and the passage through conical intersection.

3.4.6.4 Optical response properties and conical intersection

The absorption spectra obtained for both isomers of Na_3F_2 using the "frozen ionic bond" approximation are shown in Fig. 3.38 and compare well with those calculated by taking into account all valence electrons [263]. The lowest energy isomer I, with the ionic Na_2F_2 subunit to which the Na atom is bound (forming Na–Na and Na–F bonds), gives rise to the low energy intense transition in the infrared. This is a common feature found for Na_nF_{n-1} clusters due to the localized excitation of the one-excess electron, as mentioned above. In contrast, the transition to the first excited state of isomer II (C_{2v}) with the Na_3 subunit, which is bridged by two F atoms, has a higher energy close to the energies of transitions usually arising from excitations in metallic subunits. After the vertical transition at the geometry of isomer I, the geometric relaxation in the first excited state takes place, involving a breaking of the Na-Na bond which leads to the first local minimum of the excited state (cf. Fig. 3.38) with a moderate lowering of the energy. Afterwards, the relaxation process proceeds to the absolute minimum with the linear geometry corresponding to the conical intersection for which a further considerable decrease of energy takes place. The linear geometry of the conical intersection is also reached after vertical transition to the first excited state at the geometry of the second isomer with C_{2v} structure. Accordingly, the investigation of the dynamics in the first excited state involves the breaking of metallic and ionic bonds starting from isomer I, and just metallic bonds starting from isomer II, as well as the passage through the conical intersection.

Consequently, one expects strong thermal motions within the ensemble, leading to phase space spreading and IVR. All processes can be monitored by a second ionizing probe pulse with excitation energies between $\sim$ 2.9 eV and $\sim$ 4.8 eV, as shown by the scheme given in Fig. 3.38. The first value is close to the initial Franck-Condon transition region and probes the relaxation dynamics on the potential surface of the first excited electronic state before the branching process, which is due to the conical intersection, does occur. The ground state dynamics after passage through the conical intersection allows to monitor processes involved on the ground state potential surface.

For characterization of conical intersections we use the algorithm introduced by Robb et al. [276]. It is useful for the determination of the lowest structure and the energy at the intersection seam as well as for analyzing the topology of the intersection in the space spanned by the internal degrees of freedom. The intersection seam is (N-2) dimensional as it is characteristic for *conical intersections*. The analysis of the wavefunctions of the ground and the first excited state in the close neighborhood of the conical intersection yields positive and negative linear combinations of two "valence bond like" ("VB") structures $Na^+–F^-–Na^+–F^-–\dot{Na} \pm \dot{Na}–F^-–Na^+–F^-–Na^+$. One of them contributes dominantly to the ground, and the other one contributes to the first excited state, thus giving rise to two states with different symmetries. The location of the excess electron is indicated by the dot above the sodium atom. Of course, at the point of the conical intersection, the arbitrary linear combination of the above "valence bonds" structures is possibly due to degeneracy. The two "VB" structures differ in the translocation of the single excess electron or of the charge from one to the other end of the linear system. In other words, the length of the linear chain is sufficiently long to allow for an energy gap closing, in analogy to the dissociation limit of the H_2^+ molecule for which

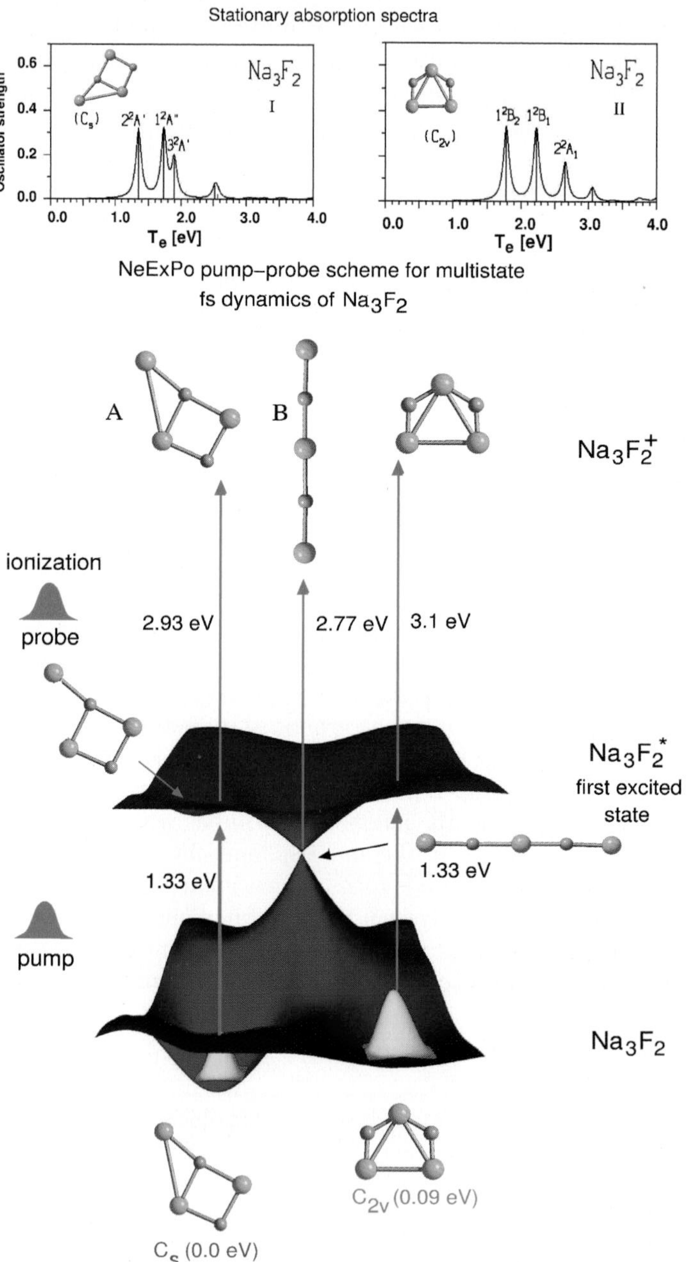

Fig. 3.38. Absorption spectra for two isomers I and II of Na_3F_2 obtained from one electron "frozen ionic bonds" approximation [72] (upper part). Scheme of the multistate fs-dynamics for NeExPo pump-probe spectroscopy of Na_3F_2 including conical intersection with structures and energy intervals for the pump and probe steps [72].

the degeneracy of the ground and excited state occurs due to equal energies of $\dot{\text{H}}$–H$^+$ and H$^+$–$\dot{\text{H}}$ structures. We conclude that the presence of the conical intersection in Na_3F_2 through which the isomerization process can take place is the consequence of the electronic structure properties. Therefore, due to general characteristics, it can be found for other systems by designing the analogous electronic situation.

In fact, the analogy can be drawn to conical intersections found in organic photochemistry involving biradicaloid species, which are generated by partial breaking of double hetero bonds due to geometric relaxation in the singlet excited states. The condition for the occurrence of conical intersections in so called "critical biradicals" has been formulated in the framework of the two-orbital two-electron model and can be fulfilled in the case that the electronegativity difference between the two centers is sufficient to minimize the repulsion between the ground and the excited states [242]. In fact, it has been confirmed experimentally that the conical intersection is responsible for the cis-trans isomerization of the retinal chromophore in the vision process [269, 270].

Moreover, investigation of the nonadiabatic dynamics through the conical intersection of the Na_3F_2 cluster has advantages. The system has 10 degrees of freedom and permits the calculation of an ensemble of trajectories based on the accurate ab initio description of the excited and ground electronic states and on corresponding MD. Thus it provides the conceptual framework for fs-observables such as fs pump-probe signals which will be addressed below.

3.4.6.5 Analysis of the nonadiabatic dynamics and of the signals

Important aspects of the analysis of the nuclear dynamics will be first addressed. The procedure used for calculation of nonadiabatic dynamics "on the fly" has been described in the previous section. For analytical expressions for gradients and nonadiabatic couplings in the frame of "frozen ionic bold" approximation cf. [72]

The simulation of the classical trajectory ensemble, consisting of a large number of sampled phase space points, can be started on the first excited electronic state using initial conditions described above. The geometric relaxation (over the local minimum) toward the linear structure corresponding to the conical intersection and its passage through the conical intersection as well as the subsequent relaxation dynamics on the electronic ground state can be visualized by considering the phase space density of the cluster ensemble shown in Fig. 3.39 for different propagation times. Initially at t=0 fs, the phase space density is localized corresponding to the C_s structure (cf. Fig. 3.39). During the subsequent $\sim$ 90 fs, the bond breaking between both sodium atoms occurs corresponding to a local minimum on the first excited state (cf. Fig. 3.39). Consecutive ionic bond breaking between the Na and the F atoms of the Na_2F_2 subunit can be observed after 220 fs (cf. Fig. 3.39) of the phase space density. After $\sim$ 400 fs, the region of the conical intersection

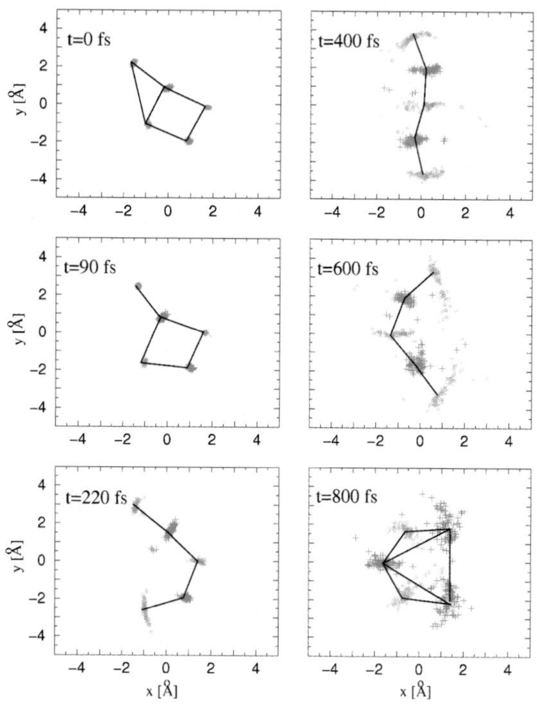

Fig. 3.39. Snapshots of the phase space distribution (PSD) obtained from classical trajectory simulations based on the fewest-switches surface-hopping algorithm of a 50 K initial canonical ensemble [72]. Na atoms are indicated by green circles and F atoms by red crosses. Dynamics on the first excited state starting at the C_s structure (t=0 fs) over the structure with broken Na-Na bond (t=90 fs) and subsequently over broken ionic Na-F bond (t=220 fs) toward the conical intersection region (t=400 fs), Dynamics on the ground state after branching of the PSD from the first excited state leads to strong spatial delocalization (t=600 fs). The C_{2v} isomer can be identified at ∼800 fs in the center-of-mass distribution.

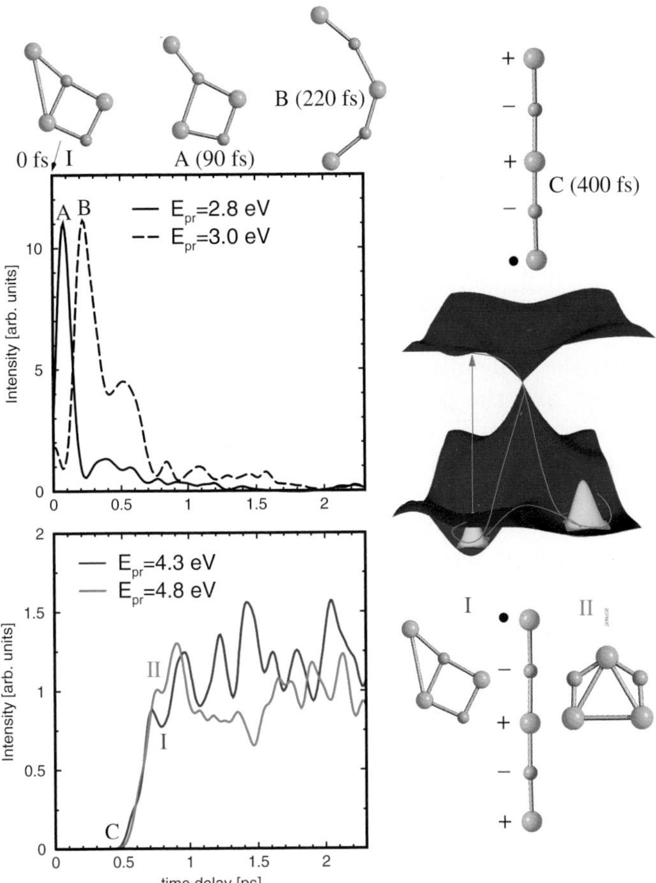

Fig. 3.40. Simulated NeExPo pump-probe signals for the 50 K initial temperature Na_3F_2 ensemble at different excitation energies of the probe laser monitoring the geometric relaxation on the first excited state involving bond-breaking processes and passage through the conical intersection as well as geometric relaxation and IVR processes on the ground state after the passage (left side). The isomerization through the conical intersection is schematically illustrated on the right hand side [72].

corresponding to the linear structure is reached (cf. Fig. 3.39). This triggers the branching of the phase space density from the excited electronic state to the ground state. At this stage, the system gains an additional kinetic energy of $\sim$ 0.67 eV. Due to this large vibrational excess energy, strong anharmonicities between the vibrational modes are present, which are responsible for the phase space spreading. The subsequent relaxation dynamics on the electronic ground state is characterized by an even larger phase space spreading, particularly after 800 fs. This is due to the fact that the vibrational excess energy rose to $\sim$ 1.3 eV which corresponds to an equilibrium temperature of $\sim$ 3400 K (cf. Fig. 3.39). However, in spite of increasing phase space spreading, structural information of the cluster ensemble can be gained up to a propagation time of $\sim$ 800 fs by considering the center of mass positions of the atomic phase space distributions in Fig. 3.39. As shown below, one can obtain detailed information about the branching ratio between these structures as well as energetic distributions in the cluster ensemble from the pump-probe signals. For times beyond 1 ps, no structures can be identified in the phase space distribution. The ensemble is geometrically completely delocalized at least up to the propagation time of 2.5 ps, which is understandable due to the large vibrational excess energy.

In summary, the dynamics through the conical intersection between the first excited state and the ground state represents an elementary physical event for the cluster ensemble in the sense that it initiates the transition from structurally and energetically localized pattern involving consecutive metallic and ionic bond breaking processes to energy delocalized pattern. Thus, the molecular dynamics might be divided into a reversible and an irreversible part separated by the passage through the conical intersection.

Simulations of signals are based on (3.16), with energy gaps obtained from the classical trajectory simulations using the fewest switching surface hopping algorithm (3.11–3.14) for the ensemble at an initial temperature of 50 K. In order to obtain comprehensive information on the dynamical processes of Na_3F_2, a zero pump pulse duration (σ_{pu}=0) is suitable, which involves a complete excitation of the ground state ensemble prepared at the initial temperature. The ultrafast structural relaxation processes involving the bond breaking can be resolved using a probe pulse duration of 50 fs. The simulated signals are shown for four different excitation energies (wavelengths) of the probe pulse in Fig. 3.40, which allow to analyze the underlying processes:

i) E_{pr}=2.8 eV and E_{pr}=3.0 eV correspond to transition energy values between the first excited and the cationic state at the time of the Na-Na metallic and the Na-F ionic bond breaking, respectively (cf. Fig. 3.40). Thus the signals for those transition energies provide information on the structural relaxation involving the bond breaking processes in the first excited state of Na_3F_2 before the conical intersection is reached. In fact, they exhibit maxima at $\sim$ 90 fs and $\sim$ 220 fs (cf. Fig. 3.40), in agreement with the timescales for the metallic and ionic bond breaking obtained from the analysis of the phase space distribution shown in Fig. 3.39. Both signal intensities decrease rapidly after 0.4-0.5 ps, indicating the branching of the phase space density from the first excited electronic state to the ground state due to the conical intersection.

ii) E_{pr}=4.3 eV and E_{pr}=4.8 eV (cf. Fig. 3.40) correspond to transition energies between the ground state and the cationic state at the C_s geometry and the C_{2v} geometry, respectively. In such a way, the signals shown in Fig. 3.40 monitor the ratio of both isomers in the phase space distribution after the passage through the conical intersection up to a time delay between pump and probe of $\sim$ 1 ps. This time represents the limit up to which structural information can be resolved in the phase space distribution (cf. Fig. 3.39). For larger time delays, the signals provide only information about the energetic redistribution, thus IVR. In fact, both signals start to increase after an incubation time of $\sim$ 0.4 ps since the ground state becomes populated, providing the time scale for the passage through the conical intersection (cf. Fig. 3.40). Furthermore, the signal at E_{pr}=4.8 eV exhibits a maximum at 0.8-0.9 ps, indicating the larger ratio of the C_{2v} structure in correspondence with the results obtained from the phase space distribution (cf. Fig. 3.39). This signal drops rapidly after 0.9 ps and the signal at E_{pr}=4.3 eV increases indicating that the population of the C_s structure is larger at 0.9-1.0 ps (cf. Fig. 3.40). The latter time dependence also exhibits oscillatory features beyond 1 ps, i.e. corresponding to the IVR regime. This leads to the conclusion that a somewhat periodic energy flow is present in the cluster ensemble. However, in view of the high vibrational excess energy, these oscillations cannot be attributed to particular normal modes.

In summary, these results provide information about the dynamics of the Na_3F_2 system in full complexity. They show that distinct ultrafast processes, which are initiated by the Franck-Condon pump pulse transition to the first excited electronic state, are involved in the dynamics of the Na_3F_2 cluster. These include geometric relaxation, consecutive bond breaking of metallic and ionic bonds, passage through the conical intersection, and IVR processes [72]. Moreover, the timescales of these processes can be identified in the pump-probe signals, and each of them can be selectively monitored by tuning the probe excitation energy. However, in order to populate only one of the isomers, the pathway has to be found which avoids a large excess of energy disposed through the conical intersection. This offers the opportunity to tailor laser pulses that will drive the system into the desired target, and will be addressed in Sect. 3.4.6.9. Similar situations can be expected in considerably larger systems providing that the characteristic electronic aspects remain preserved.

3.4.6.6 Control of ultrafast processes in multidimensional systems

The conceptual framework underlying the control of the selectivity of product formation in a chemical reaction, using ultrashort pulses rests on the proper choice of the time duration and the delay between the pump and the probe (or dump) step or/and their phase. This is based on the exploitation of the coherence properties of the laser radiation due to quantum mechanical interference effects [82, 83, 85, 86, 277]. The Genesis of the optimal control theory (OCT) and close loop learning algorithm (CLL) is given in Chapter 2 on small systems, together with corresponding references.

Concerning complex systems, until recently, the limitation in the theory was imposed by difficulties in precalculating multidimensional potential surfaces of molecules and clusters. Therefore, usually control of complex systems is treated using low dimensional models [278]. In order to bypass this obstacle, ab initio adiabatic and nonadiabatic MD "on the fly" without precalculation of the ground and excited state energy surfaces is particularly suitable provided that an accurate description of the electronic structure is feasible and practicable. Moreover, this approach offers the following advantages. The classical-quantum mechanical correspondence between trajectory and a wave packet is valid for short pulses and short time propagation. MD "on the fly" can be applied to relatively complex systems, and moreover, it can be implemented directly in the procedures for optimal control. This allows to identify properties which are necessary for assuring the controllability of complex systems and to detect mechanisms responsible for the obtained pulse shapes. In that context, the Liouville space formulation of optimal control theory developed by Yan, Wilson, Mukamel and their colleagues [279–291] , in particular its semiclassical limit in the Wigner representation [77, 281], is very suitable in spite of its intrinsic limitations. For example, quantum effects such as interference phenomena or tunneling and zero point vibrational energy are not accounted for. The study of clusters with varying size offers an ideal opportunity to test these concepts and methods as well as to investigate conditions under which different processes can be experimentally controlled and observed.

The ultimate goal of optimal control is not only to advance maximum yield of the desired process but also to use the shapes of the tailored pulses to understand the processes which are responsible for driving a complex system to the chosen target. Moreover, applications of optimal control for driving isomerization processes not only in clusters but also in biomolecules and their complexes represents an attractive perspective. In this context, we have developed strategies for two schemes: i) pump-dump control involving excited and ground states and ii) infrared control involving ground states only, which will be outlined below.

3.4.6.7 Optimal control of pump-dump processes and analysis of dynamic processes in complex systems

It is still an open, central issue if and under which conditions optimal control involving more than one electronic state can be achieved for systems with increasing complexity. For these systems, energy landscapes of the ground and excited states can substantially differ from each other or they can exhibit very complicated features. In this context there are several open basic questions, which should be addressed. An important question concerns the existence of a connective pathway between the initial state and the region of the energy landscape (objective) which is reached via a different electronic state. In addition, even if such connective pathway does exist, the optimal path must be found and the method used for nuclear dynamics and for tailoring laser pulses should involve the realistic computational demand. Therefore, the development of new strategies for optimal control is required. An attractive possibility offers the concept of the intermediate target [77, 292] in the excited state. It is defined as a localized ensemble (wave packet) corresponding to the maximum overlap between the forward propagating ensemble on the electronic excited state (starting from the initial state) and the backwards propagated ensemble from the objective in the ground state at optimal time delay between both pulses.

The classical nuclear dynamics is the only realistic approach in the case that separation of active from passive degrees of freedom cannot be made for complex systems and therefore a large number of them have to be treated explicitly. Furthermore, quantum corrections can be also introduced under the given circumstances. As will be shown below, the classical MD "on the fly" can be extremely useful for realization of new strategies for optimal control such as the construction of the intermediate target. The role of the intermediate target is to guarantee the connective pathway between the initial state and the objective and to select the appropriate parts of both energy surfaces involved. This issue is directly related to the inversion problem [293–297].

In the case of the pump-dump control for two-phase unlocked ultrafast fields in the weak response regime, we have shown that the intermediate target serves first to optimize the pump-pulse. This leads to the decoupled optimization of the pump and the dump pulses, which is very advantageous from the computational point of view. An appropriate formalism for the realization of the strategy for optimal control of complex systems based on the concept of the intermediate target is the density matrix formulation of the OCT. It combines the Wigner-Moyal representation of the vibronic density matrix with ab initio molecular dynamics (MD) "on the fly" in the electronic excited and the ground states without precalculation of both energy surfaces. This method, labeled as ab initio Wigner distribution approach, was outlined in previous sections in connection with simulations of pump–probe spectra and now will be used for different strategies to optimize laser fields. In the framework of this approach, due to available analysis based on the MD, the shapes of the optimized pulses can be directly interpreted and connected with the underlying ultrashort processes. After the outline of the theoretical basis for our new optimal control strategy using intermediate target the application will be presented. The pump and dump pulses will be optimized driving the isomerization process in the nonstoichiometric Na_3F_2 cluster, avoiding conical intersection between the ground and the first excited state and maximizing the yield in the second isomer.

3.4.6.8 Intermediate target as a new strategy for pump-dump optimal control in complex systems

The goal of the optimal control strategy described here is to optimize temporal shapes of phase-unlocked pump and dump pulses (i.e. pump and dump pulses) and the time delay between them, which drive the system starting from the lowest energy isomer over the first excited state to the second isomer , i.e. the objective.

The analytic form of the pump and dump pulses in the optimal phase-unlocked pump-dump control is

$$\epsilon_{P(D)}(t) = E_{P(D)}(t)\exp\left(-i\omega_{eg}t\right) + E^*_{P(D)}(t)\exp\left(i\omega_{eg}t\right), \tag{3.26}$$

where $E_{P(D)}$ is a slowly varying envelope of the fields, and ω_{eg} is the energy difference between the minima of the excited and the ground states. The objective in the ground state is represented in the Wigner formulation by an operator $\hat{\mathrm{A}} = \mathrm{A}(\Gamma)|g><g|$. $\mathrm{A}(\Gamma)$ is the Wigner transform of the objective in the phase space $\Gamma = \{q_i, p_i\}$ of coordinates and momenta, and $|g><g|$ is the ground electronic state projection operator. $A(\Gamma)$ can be defined, for example, as

$$A(p,q) = \prod_{i=1}^{N} \frac{1}{\sqrt{2\pi}\Delta q_i} e^{-\frac{(q_i-\bar{q}_i)^2}{2(\Delta q_i)^2}} \Theta(E_{min} - \sum_{i=1}^{N} \frac{p_i^2}{2m_i}), \tag{3.27}$$

where $\bar{q}_i$ labels Cartesian coordinates of the second isomer, and Δq_i represent the corresponding deviations. The role of the step function Θ is to insure that the kinetic energy is below the lowest isomerization barrier E_{min}. This corresponds to the spatial localization of the phase space density and arbitrary distribution of momenta. The optimized pulses can be obtained from the functional

$$J(t_f) = A(t_f) - \lambda_P \int_0^{t_f} |E_P(t)|^2\, dt, -\lambda_D \int_0^{t_f} |E_D(t)|^2\, dt, \tag{3.28}$$

where $A(t_f)$ is the yield at the time t_f, which for weak fields can be calculated in second-order perturbation theory [279, 287, 298]. It involves the propagated excited and ground state ensembles induced by the pump and dump pulses, the time dependent energy gaps between the two states, and the initial distribution of the phase space in the Wigner representation. Optimal field envelopes can be obtained by calculating the extrema from the control functional (3.28) by using the variation procedure [279–282, 284–288, 298–301]. This leads to the pair of coupled integral equations for the field envelopes:

$$\int_0^{t_f} d\tau' M_P(\tau, \tau'; E_D) E_P(\tau') = \lambda_P E_P(\tau) \tag{3.29}$$

$$\int_0^{t_f} d\tau' M_D(\tau, \tau'; E_P) E_D(\tau') = \lambda_D E_D(\tau). \tag{3.30}$$

The integral kernels corresponding to response functions are given by

$$M_P(\tau,\tau';E_D) = \int\!\!\int d^2\Gamma_0 \int_0^{t_f} d\tau'' \int_0^{\tau''} d\tau''' A(\Gamma_g(t_f-\tau'';\Gamma_e(\tau'''-\tau;\Gamma_0)))$$
$$e^{i(\omega_{eg}-U_{eg}(\Gamma_e(\tau'''-\tau;\Gamma_0)))(\tau''-\tau''')}e^{i(\omega_{eg}-U_{eg}(\Gamma_0))(\tau-\tau')}$$
$$\rho_{gg}(\Gamma_0)E_D(\tau''')E_D^*(\tau'') \quad \tau \geq \tau', \tag{3.31}$$

$$M_D(\tau,\tau';E_P) = \int\!\!\int d^2\Gamma_0 \int_0^{\tau'} d\tau'' \int_0^{\tau''} d\tau''' A(\Gamma_g(t_f-\tau;\Gamma_e(\tau'-\tau'';\Gamma_0)))$$
$$e^{i(\omega_{eg}-U_{eg}(\Gamma_e(\tau'-\tau'';\Gamma_0)))(\tau-\tau')}e^{i(\omega_{eg}-U_{eg}(\Gamma_0))(\tau''-\tau''')}$$
$$\rho_{gg}(\Gamma_0)E_P(\tau''')E_P^*(\tau'') \quad \tau \geq \tau'. \tag{3.32}$$

Γ_e and Γ_g correspond to propagated excited and ground state ensembles, and U_{eg} is the time dependent energy gap between the excited and the ground state. Since both equations depend on the pump and dump pulses, they are coupled and can, in principle, be solved iteratively yielding optimized pump and dump pulses. However, this is computationally unrealistic even for systems of moderate complexity because the coupled classical simulations on the ground and excited states have to be performed. The calculation of objective A in (3.31) and (3.32) requires the propagation of the ensemble on the ground state Γ_g, starting at different initial conditions. These conditions are obtained from the propagated ensemble Γ_e of the excited state at each time step. Therefore, the strategy involves decoupling of (3.31) and (3.32) which is possible only in the short pulse regime on the fs time scale and the necessary steps are outlined below.
(i) In the zero order approximation of an iterative procedure and in the ultrafast regime it is justified to calculate the kernel functions M_P and M_D with strongly temporally localized pulse envelopes $E_P \approx \delta(t)$ and $E_D \approx \delta(t-t_d)$. Then the zero order response functions take the following forms:

$$M_P^{(0)}(\tau,\tau') = \int\!\!\int d^2\Gamma_0 A(\Gamma_g(t_f-t_d;\Gamma_e(t_d-\tau;\Gamma_0)))$$
$$e^{i(\omega_{eg}-U_{eg}(\Gamma_0))(\tau-\tau')}\rho_{gg}(\Gamma_0) \quad \tau \geq \tau' \tag{3.33}$$

$$M_D^{(0)}(\tau,\tau') = \int\!\!\int d^2\Gamma_0 A(\Gamma_g(t_f-\tau;\Gamma_e(\tau';\Gamma_0)))$$
$$e^{i(\omega_{eg}-U_{eg}(\Gamma_e(\tau';\Gamma_0)))(\tau-\tau')}\rho_{gg}(\Gamma_0) \quad \tau \geq \tau'. \tag{3.34}$$

The equations for pump and dump pulses now become decoupled. Consequently, the pump pulse optimization involves the propagation on the excited state $\Gamma_e(t_d-\tau;\Gamma_0)$ from $\tau=0$ until $\tau=t_d$ starting with Γ_0 (initial ensemble) in (3.33). For the dump optimization, according to (3.34), the dynamics on the ground state has to be carried out $\Gamma_g(t_f-\tau;\Gamma_e(t_d))$ for $\tau'=t_d$ until t_f with the initial conditions given by the ensemble of the excited state $\Gamma_e(t_d)$ at t_d

which corresponds to the intermediate target. $\Gamma_e(t_d)$ at t_d can be determined from the maximal overlap between a forward propagated ensemble from the first isomer on the excited state and a backwards propagated ensemble on the ground state from the second isomer.
(ii) Equations (3.29) and (3.33) yield an optimal pump pulse which localizes phase space density at the intermediate target.
(iii) The optimized dump pulse projects the intermediate target to the ground state and optimally localizes the phase space density into the objective (second isomer) at a final time t_f. This means that the connective pathway between the initial state and the objective is guaranteed by the intermediate target at a time t_d. For this purpose, the function $A(\Gamma_g(t_f - t_d); \Gamma_e(t_d))$ must have nonvanishing contributions as follows from (3.33) and (3.34). This procedure can be continued iteratively, but it is most likely that the zeroth and first order iterations lead to sufficient accuracy. In summary, the concept of the intermediate target represents a new strategy which insures the connective pathway between the initial state and the objective, and moreover allows to reach the objective with maximal yield optimizing pump and dump pulses independently. This allows the application of the optimal pump-dump control to complex systems without restricting the number of degrees of freedom and insures controllability, provided that the intermediate target can be found, which is illustrated in the next section.

3.4.6.9 Application of the intermediate target concept for optimal control of photoisomerization in Na_3F_2

The isomerization in Na_3F_2 through a conical intersection between the first excited and the ground state is a nonselective process due to the high internal energy (~ 0.65 eV) which populates almost equally both isomers in the ground state and does not allow for selective population of the second isomer [72]. Therefore, the optimal control strategy described above, which is based on the concept of the intermediate target, represents an adequate tool to find the optimal pathway allowing one to populate isomer II with maximal yield and to suppress of the pathway through the conical intersection [77].

For this purpose several steps are needed. First, the initial ensemble of isomer I has to be generated. Then the intermediate target involving excited and ground state dynamics has to be determined. Finally, the pump and dump pulses have to be optimized. For the initial ensemble a 50 K canonical ensemble in the ground state of isomer I in the Wigner representation can be constructed using, for example, a set of $\sim$1000 randomly sampled coordinates and momenta. In the pump step (photon energy of 1.33 eV), the ensemble is first propagated on the excited state (e.g., for 300 fs). In order to determine the intermediate target and the optimal time delay t_d, the ensemble has to be dumped to the ground state (in steps of e.g. 25 fs) and subsequently propagated (for e.g., 1 ps). It can be shown that isomer II is reached by the ensemble at $t_d = 250$ fs and the residence time of 500 fs at least, can be

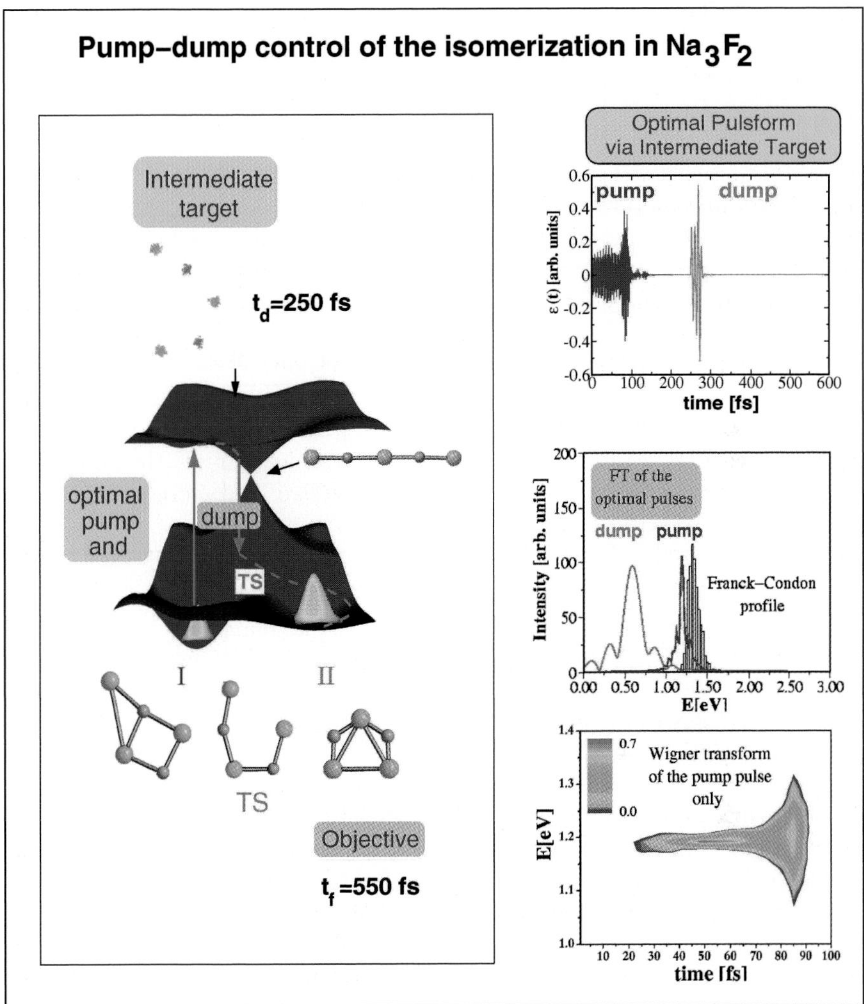

Fig. 3.41. Scheme for pump-dump optimal control in the Na_3F_2 cluster with geometries of the two ground state isomers and of the transition state separating them, the conical intersection, and the intermediate target (left side). The optimal electric field corresponding to the pump and dump pulses [77] (upper panel right side). The mean energy of the pump pulse is 1.20 eV and the mean energy of the dump pulse is 0.6 eV. Fourier transforms of the optimal pump and dump pulses and the Franck-Condon profile for the first excited state corresponding to the excitation energy T_e=1.33 eV (middle panel right side). Wigner transform of the optimal pump pulse (bottom panel right side).

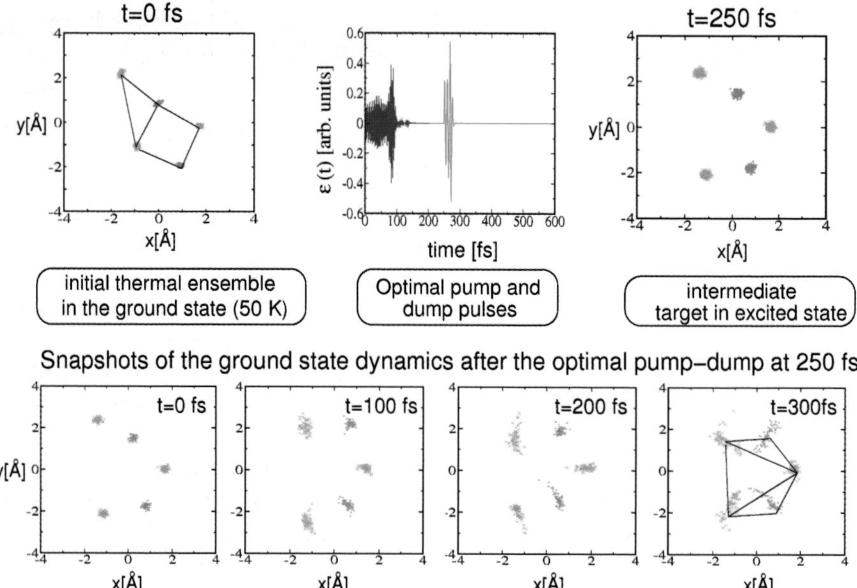

Fig. 3.42. Initial thermal ensemble, optimal pump and dump pulses, intermediate target (upper panels). Snapshots of the dynamics obtained by propagating the ensemble corresponding to the intermediate target after the optimized pump-dump at 250 fs on the ground state showing the localization of the phase space density in the basin corresponding to isomer II [77] (low panels).

achieved. The ensemble averaged geometry which determines the coordinates of the intermediate target is shown in Fig. 3.41 and 3.42. Note that the "geometry" of the intermediate target is closely related to that of the transition state separating the two isomers on the ground state. The role of the intermediate target to ensure the connective pathway from the initial state to the objective over the excited state is evidenced by its relation to the transition state which separates both ground state isomers. The average kinetic energy of the intermediate target corresponds to ~ 75 % of the isomerization barrier in the ground state. This guarantees that, after the dump, the ensemble will remain localized in the basin of isomer II.

The optimization of the pump pulse leads to a localization of the phase space density around the intermediate target. The intermediate target operator can be represented in the Wigner representation (3.33), by a minimum uncertainty wave packet:

$$A(p_i, q_i) = \prod_{i=1}^{3N=15} \frac{1}{2\pi \Delta p_i \Delta q_i} e^{-\frac{(q_i-\bar{q}_i)^2}{2(\Delta q_i)^2}} e^{-\frac{(p_i-\bar{p}_i)^2}{2(\Delta p_i)^2}} . \tag{3.35}$$

The response function $M(\tau, \tau')$ for the pump pulse (3.33) can be calculated, for example, on a time grid of 1 fs and can be symmetrized and diagonalized according to (3.29). In this case, the largest eigenvalue was obtained to be 0.82 corresponding to the globally optimized pulse which has 82% efficiency to localize the ensemble in the intermediate target.

The optimized pump pulse, shown in Fig. 3.41, consist of two portions with durations of $\sim$ 70 fs and $\sim$ 10 fs, respectively. Fourier and Wigner-Ville transforms of the pump pulse, shown also in Fig. 3.41, provide physical insight. Comparison of Fourier transform with the Franck-Condon profile of isomer I shows that the excitation of the low lying vibrational modes at $\sim$ 1.2 eV of the initial ensemble is dominantly responsible for reaching the intermediate target. This spectral region corresponds to lower lying vibrational modes which open the C_s structure of isomer I by breaking the Na-Na and one of the Na-F bonds. The Wigner-Ville transform shows that this energetically sharp transition corresponds to the first temporal portion of $\sim$ 70 fs of the pump pulse. In contrast, a very short second portion after 80 - 90 fs of $\sim$ 10 fs is energetically much wider. It is related to tails of the Fourier transform, which are symmetric with respect to the 1.2 eV transition, reflecting equally distributed velocities in the initial ensemble.

The dump pulse optimization leads to a spatial localization of the phase space density in the objective (isomer II). For this purpose, the intermediate target operator (3.35) can be propagated on the ground state, and the dump pulse is obtained from (3.30) and (3.34). The largest eigenvalue, e.g. of 0.78, can be obtained. This corresponds to 78% efficiency of localization of isomer II. The optimized dump pulse is very short $\sim$ 20 fs (cf. the part of the signal after $t_d = 250$ fs in Fig. 3.41). This implies that the time window around t_d for depopulation of the excited state is very short. Otherwise the system would gain a large amount of energy in the excited state (leading to the conical intersection). The Fourier transform of the dump pulse is centered around 0.6 eV corresponding to the Franck-Condon transition at t_d as shown on the right side of Fig. 3.41. Finally, in order to illustrate the efficiency of optimized pulses, snapshots of the ground state ensemble propagated after the dump process are shown in Fig. 3.42. It can clearly be seen that the phase space density is localized in isomer II (the objective) after $t_d + 200$ fs $= 450$ fs.

Using the strategy for optimal pump-dump control based on the intermediate target, we have shown that the isomerization pathway through the conical intersection can be suppressed and that optimized pulses can drive the isomerization process to the desired objective (isomer II). This means that the complex systems are amenable to control, provided that the intermediate target exists. Furthermore, the analysis of the MD and of the tailored pulses allows for the identification of the mechanism responsible for the selection of appropriate vibronic modes necessary for the optimal control.

In summary, optimal pump-dump control of fs-processes, involving two electronic states, requires the identification of the connective pathway between the initial state and the objective. This is possible if the intermediate target in the excited state can be found, which selects the appropriate parts of energy surfaces for the control. This was illustrated for the example of Na_3F_2 for which the optimal pump and dump pulses populate the objective (isomer II) with maximal yield, taking the optimal pathway and avoiding the conical intersection. Note that other mechanisms for optimal control through conical intersections have been proposed [302, 303]. The identification of the mechanism responsible for the shape of the pulses serves as a guide toward the understanding the theoretically and experimentally obtained tailored fields which still represents a challenging task for future work. In this way, the control is used not only to achieve desired goal but also to identify and to analyze the underlying ultrafast processes responsible for favoring one pathway and for suppressing the others. Control as a tool for analysis of the dynamics complements the closed loop learning (CLL) control technique and sheds light on the nature of the "black box".

3.4.6.10 Infrared control of conformational changes in complex systems

An important goal of laser selective femtochemistry is to design ultrashort laser pulses in the infrared spectral region which drive the system in the ground electronic state to the desired isomer or reaction product (cf. [222]). In order to achieve this it is mandatory to use IR pulses on a subpicosecond time scale in order to "beat" the IVR process which would otherwise prevent efficient localization of energy in selected vibrational modes. Infrared IR laser pulse control has been suggested first by Savva and Paramonov for selective vibrational excitations [84] and by Joseph and Manz for vibrationally state selective dissociation [304]. The two approaches have been combined in [215] for selective isomerization of complex molecules considering double well potential.

We here present a strategy to control the dynamics of complex systems in their ground electronic states based on the Wigner distribution approach combined with ab initio MD "on the fly" taking into account all degrees of freedom. In contrast to previous approaches based on low dimensionality [222], we include all degrees of freedom and show on examples that this is mandatory for systems with low frequency normal modes. We apply our strategy on two prototype examples, the Na_3F cluster and on the glycin molecule, and show that selective isomerization driven by the ultrashort laser fields can be achieved. We point out that prototype specific processes can be identified in optimized pulse shapes. The aim is to learn from the above prototype examples about the scope of the IR control which should provide corresponding information necessary for experimental realization.

Our approach for the ground state optimal control is based on the classical Liouville equation in which the dipole interaction with the laser field is explicitly taken into account:

$$\frac{\partial \varrho}{\partial t} = \frac{\partial H}{\partial \mathbf{q}}\frac{\partial \varrho}{\partial \mathbf{p}} - \frac{\partial \varrho}{\partial \mathbf{q}}\frac{\partial H}{\partial \mathbf{p}} - E(t)\frac{\partial \mu}{\partial \mathbf{q}}\frac{\partial \varrho}{\partial \mathbf{p}} \tag{3.36}$$

This equation is solved by propagating an ensemble of classical trajectories with quantum initial conditions and employing ab initio or semiempirical MD in which both forces and the dipole derivatives are calculated "on the fly". For the laser field $E(t)$ in (3.36) the following analytic expression has been chosen:

$$E(t) = \sum_{i=1}^{n} E_i \exp\left(-\frac{(t-\alpha_i)^2}{2\beta_i^2}\right) cos\left(\gamma_i t + \delta_i t^2 + \epsilon_i\right) \qquad (3.37)$$

Expression (3.37) represents combination of gaussian pulses modulated by the linearly chirped oscillatory contribution. Of course, different forms can be chosen as possible pulse parameterizations. The aim of the optimal control is to maximize the expectation value of the quantum mechanical operator $\hat{A}$ at a given final time t_f which is in the Wigner representation given by:

$$J(t_f) = \int\int d\mathbf{q}d\mathbf{p}\mathbf{A}(\mathbf{q},\mathbf{p})\varrho(\mathbf{q},\mathbf{p},\mathbf{t}). \qquad (3.38)$$

where $A(\mathbf{q},\mathbf{p})$ is a target operator which is localized around the desired isomer. The parameters of the pulse given by (3.37) are iteratively optimized using the genetic algorithm. It should be pointed out that in the optimal control approach the energy of the pulse has to be restricted and this is usually done by the Lagrange multiplier method. In the case that the pulse is optimized directly using the genetic algorithm the restriction of the pulse energy is automatically satisfied in the optimization. This is because the genetic algorithm requires to restrict the range of the optimized parameters.

Alternative way for the pulse optimization would be to use the standard optimal control theory. This requires forward and backward propagation of the phase space ensemble and the calculation of the overlap between the two phase space ensembles. However, since we represent the ensemble by the finite number of discrete trajectories, evaluation of the overlap would require to propagate extremely large number of trajectories and use the fitting procedure to obtain a smooth phase space distribution. Therefore, our strategy is more practicable in the context of controlling isomerization of complex systems accounting for all degrees of freedom.

3.4.6.11 Optimal control of isomerization in clusters and molecules; prototype examples Na_3F and glycine

The first example on which we illustrate our approach is the Na_3F clusters with two isomers in the ground state which differ in energy the by 0.09 eV. The isomerization barrier between the isomers is ~0.12 eV. The scheme of the IR control together with the structures of both isomers is presented in Fig. 3.43a. The aim is to selectively excite the isomer I and transfer optimally the population to the isomer II at final time t_f=400 fs. For this purpose we propagate an ensemble consisting of 32 trajectories sampled from the canonical Wigner distribution at T=50 K. For the propagation we use ab initio

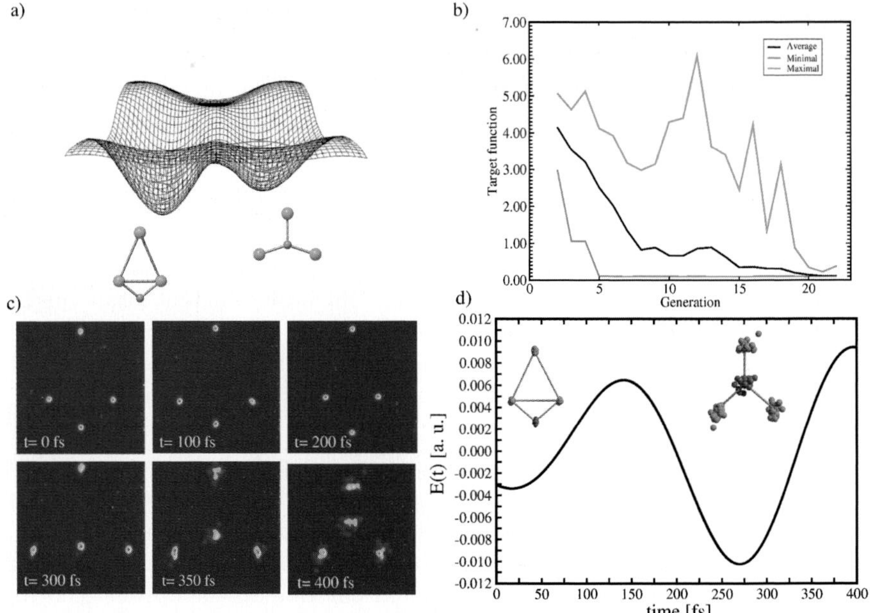

Fig. 3.43. (a) Scheme for the IR control in Na_3F (b) optimization efficiency showing average, maximal and minimal deviation from the target (isomer II), c) snapshots of the ground state dynamics driven by the optimal pulse, (d) optimal pulse (inserts show initial and final ensemble).

MD "on the fly" in the framework of the Hartree-Fock method which is the least computationally demanding approach and provides acceptable accuracy for the ground state of the considered system. The pulse optimization has been performed iteratively using the genetic algorithm and starting from the random pulse generated according to (3.37).

The efficiency of the optimization is presented in Fig. 3.43b showing that after 22 generation an uniform population containing the optimal pulse has been obtained. The snapshots of the dynamics driven by the optimal pulse and the optimal pulse itself are shown in Fig. 3.43c and d, respectively. The pulse has almost a sinusoidal shape with a period of $\sim$ 250 fs. This period corresponds very closely to the vibrational period of one of the normal modes of the isomer I which involves the motion of the fluorine atom along the C_2 axis. Fast and selective resonant excitation of this normal mode is ultimately responsible for the population transfer from the isomer I to the isomer II. In summary, we show that in a simple, highly symmetric system a selective isomerization process can be achieved by selectively exciting only one normal mode by a short infrared pulse with a periodic structure.

In order to illustrate the scope of our method with possible future applications on a complex systems we have chosen as prototypes the glycine molecule

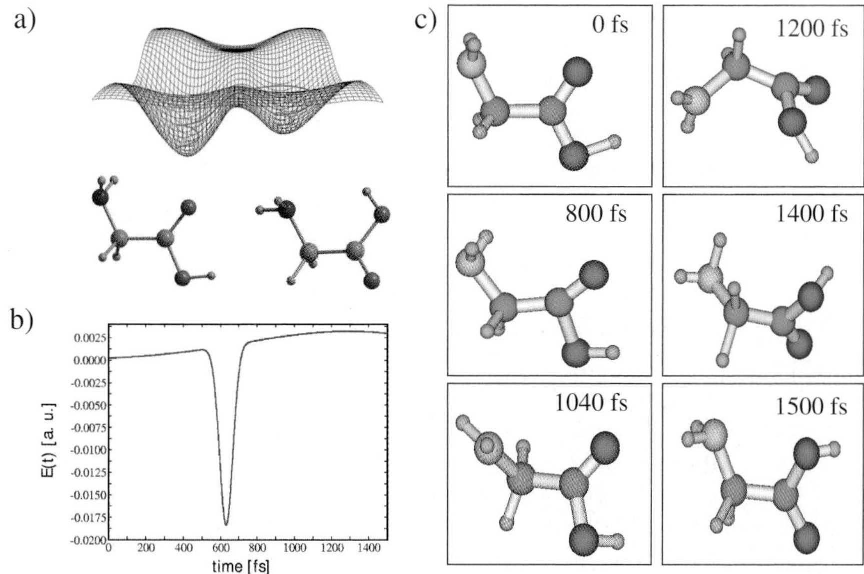

Fig. 3.44. (a) Scheme for the IR control in glycin, (b) the optimal pulse driving the isomerization and (c) snapshots of the laser driven dynamics.

which has two ground state isomers differing in energy by 0.33 eV (cf. Fig. 3.44). An ensemble of 32 trajectories has been propagated in the framework of the semiempirical AM1 method. This choice has been made because of the future applications to larger biomolecules such as peptides. Since the conformational change from the isomer I to the isomer II in glycine involves low frequency torsional modes, a longer time interval of t_f=1500 fs had to be selected for the control.

The optimal pulse driving the isomerization process and the snapshots of the laser driven dynamics are presented in Fig. 3.44. The pulse has two temporal components, one weak, slowly varying subpulse with a half period of $\sim$ 1500 fs and a strong gaussian shaped subpulse centered at $\sim$ 600 fs with a width of $\sim$ 100 fs. It should be pointed out that this pulse shape is very different from the one obtained for Na_3F. The reason for this is twofold. First, the number of degrees of freedom is larger for glycine than for Na_3F, and secondly the energy separation of the isomers is also much larger (0.33 eV vs. 0.09 eV). Due to this difference, the IVR process is expected to be much faster for glycine. Thus, this prevents the simple conformational change by a periodic excitation of the torsion normal mode because of fast dissipation of energy which would lead to heating of the molecule. Instead of that, the conformational change is achieved by an ultrashort impulsive and not periodic excitation with a gaussian subpulse with the width of $\sim$ 100 fs. These results represent the prototype information, valuable for control of larger biomolecules for which similar situation is expected.

In summary, we demonstrated that our theoretical approach is suitable to control ground state isomerization in complex systems by shaped IR fields. We have also shown that the mechanism underlying the optimal control can be significantly different in small symmetric systems and in larger molecular systems with floppy low frequency modes. This will allow us in the future to study systematically to which extent the conformational changes in functional biomolecules or their complexes with metallic clusters can be controlled by shaped IR pulses.

3.4.6.12 Perspectives for "on the fly" MD

Analysis and control of ultrafast processes in atomic clusters in the size regime in which "each atom counts" are of particular importance from a conceptual point of view and for opening new perspectives for many applications in the future. Simultaneously this research area challenges the development of theoretical and computational methods from different directions, including quantum chemistry, molecular dynamics, and optimal control theory, removing borders among them. Moreover, it provides stimulation for new experiments.

By changing cluster size, and therefore structural and optical properties, different ultrafast processes can be monitored, and their time scales can be determined. They include bond breaking, geometric relaxation of different nature, IVR, isomerization, as well as reactions channels. These processes can be identified by analyzing adiabatic or nonadiabatic dynamics and simulated fs-signals. Therefore, the conditions under which the identified processes are experimentally observed can be precisely determined. This predictive power of theory can be then directly used for conceptual planning of experiments. Moreover, the tailored fields obtained in the framework of optimal control theory can drive selected processes, such as isomerization toward one of the isomers, or the chosen reaction channel for which particular bond breaking or new bonding rearrangements promote the emanation of the reaction products.

Theoretical methods which combine ab initio MD "on the fly" with the Wigner distribution approach, which is based on classical treatment of nuclei and on quantum chemical treatment of electronic structure, represent an important theoretical tool for analysis and control of ultrashort processes in complex systems. Moreover, the possibility to include, in principle, quantum effects for nuclei by introducing corrections makes this approach attractive for further developments. However, for this purpose, new proposals for improving the efficient inclusion of quantum effects for nuclei and fast but accurate calculations of MD "on the fly" in the electronic excited states are mandatory. Both aspects represent attractive and important theoretical research areas for the future.

The strategies based on localization of the wave packet or ensemble (e.g. an intermediate target), insuring the connective pathway between the initial state and the target in complex systems involving at least two different electronic states, are attractive for following reasons. They allow simplification of optimization of pump- and dump- pulses for complex systems. They also permit selection of important parts of energy surfaces which makes the inversion problem accessible. Finally, the analysis of the underlying dynamics makes it possible to assign the shapes of optimized pulses to processes, allowing one to unravel the mechanisms responsible for optimal control. This also allows the use of optimal control schemes as analysis tools for complex systems, which is an important conceptual issue with a promising perspective for applications in biomolecules, clusters, or even their complexes.

Due to the structure-reactivity relationships of clusters, the reactive centers can be identified. Furthermore, their size selectivity can be exploited for invoking reactions toward organic and inorganic molecules or for finding the cooperative effects needed for promoting these reactions. This research direction opens new roads for using tailored laser fields to drive the laser induced selective chemical reactions involving clusters. It also takes advantage of their functional properties, which might invoke a large impact in different application areas.

3.5 Outlook

Before ending this chapter, we would like to look in the crystal ball to have a glimpse of what is ahead. Clearly, future developments are driven by a) technological advances, b) conceptual advances and c) the need to solve "practical" problems.

Some 20 years ago, the introduction of the Ti:sapphire laser and chirped pulse amplification revolutionized ultrafast studies of molecules. Nowadays, with such light sources, pulse durations as short as 10 fs can be reached. For substantially shorter pulses, other technologies are already on the horizon. However, the extension of the light sources to other wavelength regimes maybe more important for molecular physics than ever shorter laser pulses. For example, free electron lasers as well as high harmonics generation of the output of standard high power lasers provide ultrashort UV and XUV pulses of ever shorter wavelength and proposed free electron lasers might eventually provide tunable ultrashort and ultraintense X-Ray pulses to, for example, allow time resolved diffraction experiments on single molecules. There is also an intense research effort to produce ultrashort pulses in the far-infrared, the terahertz regime. Such light pulses can, for example, be used to follow low frequency, large amplitude vibrational motion in biomolecules. However, many of the above mentioned light sources make the application of liquid crystal modulators difficult, if not impossible and new concepts for shaping those light pulses need to be explored. Over the last decade, mainly phase and amplitude of the light pulse were used as parameters in control experiments. The addition of the polarization of the pulse adds a whole new dimension to the game, and the usage of such "fully shaped" pulses in control experiments of complex systems only started recently.

Advances in the general understanding of the dynamics and control of small systems allow us to apply those concepts for ever larger and ever more complex systems. Again, the gas phase is a very suitable medium for those studies, as it allows to test fundamental concepts and theoretical predictions for the isolated, unperturbed system. In most experiments, the parameters that are controlled are dissociation or ionization. Another not so widely explored parameter is the function of molecular devices. Here, coherent control might be used to trigger molecular switches, or eventually control molecular nanomachines by inducing motion such as some internal rotation. Possibly, coherent control of vibrational states might be even of use in quantum computing [305].

While in most experiments, the "control" of the dynamics is the primary goal, the optimal pulse can also teach us about the molecule. This inversion of the pulse and deciphering of the dynamics is a rich playground for the interaction of experiment and theory and we expect this field to stimulate further theoretical developments and inspire future experiments.

There are, however, also obstacles that might limit the possibilities of controlling complex systems. For large systems, the vibrational density of states can be extremely high, even at modest excitation levels. In that case, anharmonicities that couple different states become extremely important as they mediate IVR and can cause dephasing. Such processes are difficult to describe theoretically and can limit the amount of control that can be imposed on a systems with the fields available to us. It is therefore of fundamental importance to get a detailed understanding of the vibrational modes and their couplings in complex systems.

Acknowledgments

DMN and AK thank the National Science Foundation for support under Grant No. CHE-0350585. JM thanks the Fonds der Chemischen Industrie for continuous support. B.B. and R.B.G. thank the US-Israel Binational Science Foundation for support of this work (BSF-2004009) and Dr.I.Compagnon for helpful discussions.

References

1. A.H. Zewail, *Femtochemistry* (World Scientific, Singapore, 1994)
2. J. Manz, L. Wöste (eds.), *Femtosecond Chemistry Vol. 1 and 2* (VCH Verlagsgesellschaft mbH, Weinheim, Germany,, 1995)
3. M. Chergui (ed.), *Femtochemistry* (World Scientific, Singapore, 1996)

4. V. Sundström (ed.), *Femtochemistry and Femtobiology: Ultrafast reaction dynamics at atomic scale resolution* (World Scientific: Imperial College Press, London, 1997)
5. A.H. Zewail, J. Phys. Chem. A **104**, 5660 (2000)
6. R.N. Zare, Science **279**, 1875 (1998)
7. C.V. Shank, Opt. Lett. **12**, 483 (1987)
8. G.R. Fleming, *Chemical applications of ultrafast spectroscopy* (Oxford University Press, 1986)
9. G.R. Fleming, T. Joo, M. Cho, A.H. Zewail, V.S. Lehotkov, R.A. Marcus, E. Pollak, D.J. Tannor, S. Mukamel, Adv. Chem. Phys. **101**, 141 (1986)
10. K. Wynne, R.M. Hochstrasser, Adv. Chem. Phys. **107**, 263 (1999)
11. T. Brixner, G. Gerber, ChemPhysChem **4**, 418 (2003)
12. K. Yokoyama, C. Silva, D. Son, P. Walhout, P. Barbara, J. Phys. Chem. A **102**, 6957 (1998)
13. M. Psenichnikov, A. Baltuska, D. Wiersma, Chem. Phys. Lett. **389**, 171 (2004)
14. C. Silva, P. Walhout, K. Yokoyama, P. Barbara, Phys. Rev. Lett. **80**, 1086 (1998)
15. A. Thaller, R. Laenen, A. Laubereau, Chem. Phys. Lett. **398**, 459 (2004)
16. M. Assel, R. Laenen, A. Laubereau, J. Chem. Phys. **111**, 6869 (1999)
17. P. Ayotte, M. Johnson, J. Chem. Phys. **106**, 811 (1997)
18. J. Weber, J. Kim, E. Woronowicz, G. Weddle, I. Becker, O. Cheshnovsky, M. Johnson, Chem. Phys. Lett. **339**, 337 (2001)
19. C. Silva, P. Walhout, P. Reid, P. Barbara, J. Phys. Chem. A **102**, 5701 (1998)
20. A. Thaller, R. Laenen, A. Laubereau, J. Chem. Phys. **124**, 024515 (2006)
21. C. Desfrancois, H. Abdoul-Carime, N. Khefila, J. Schermann, V. Brenner, P. Millie, J. Chem. Phys. **102** (1995)
22. L. Turi, J. Chem. Phys. **110**, 10364 (1999)
23. C.E. Crespo-Hernandez, B. Kohler, J. Phys. Chem. B **108**, 11182 (2004)
24. T. Rizzo, Y. Park, L. Peteanu, D. Levy, J. Chem. Phys. **84**, 2534 (1986)
25. L. Snoek, E. Robertson, R. Kroemer, J. Simons, Chem. Phys. Lett. **321**, 49 (2000)
26. E. Nir, C. Janzen, P. Imhof, K. Kleinermanns, M. de Vries, Phys. Chem. Chem. Phys. **4**, 732 (2002)
27. L. Snoek, R. Kroemer, M. Hockridge, J. Simons, Phys. Chem. Chem. Phys. **3**, 1819 (2001)
28. J. Bakker, L. Aleese, G. Meijer, G. von Helden, Phys. Rev. Lett. **91**, 203003 (2003)
29. M. Gerhards, C. Unterberg, Phys. Chem. Chem. Phys. **4**, 1760 (2002)
30. J. Bakker, C. Plützer, I. Hunig, T. Haber, I. Compagnon, G. von Helden, G. Meijer, K. Kleinermanns, ChemPhysChem **6**, 120 (2005)
31. H. Fricke, A. Gerlach, M. Gerhards, Phys. Chem. Chem. Phys. **8**, 1660 (2006)
32. D. Oepts, A. Vandermeer, P. Vanamersfoort, Infrared Phys. & Tech. **36**, 297 (1995)
33. B. Dian, A. Longarte, S. Mercier, D. Evans, D. Wales, T. Zwier, J. Chem. Phys. **117**, 10688 (2002)
34. B. Dian, J. Clarkson, T. Zwier, Science **303**, 1169 (2004)
35. M. Dohle, J. Manz, G.K. Paramonov, Ber. Bunsenges. Phys. Chem. **99**, 478 (1995)
36. L. Banares, T. Baumert, M. Bergt, B. Kiefer, G. Gerber, Chem. Phys. Lett. **267**, 141 (1997)

37. S.A. Trushin, W. Fuss, W.E. Schmid, L. Kompa, J. Phys. Chem. A **102**, 4129 (1998)
38. S.A. Trushin, W. Fuss, L. Kompa, W. Schmid, Chem. Phys. **259**, 313 (2000)
39. A. Assion, T. Baumert, M. Bergt, T. Brixner, B. Kiefer, V. Seyfried, M. Strehle, G. Gerber, Science **282**, 919 (1998)
40. C. Daniel, J. Full, L. González, C. Lupulescu, J. Manz, A. Merli, S. Vajda, L. Wöste, Science **299**, 536 (2003)
41. T. Matsubara, C. Daniel, A. Veillard, Organometallics **13**, 4905 (1994)
42. C. Daniel, E. Kolba, L. Lehr, J. Manz, T. Schröder, J. Phys. Chem **98**, 9823 (1994)
43. K. Finger, C. Daniel, P. Saalfrank, B. Schmidt, J. Phys. Chem **100**, 3368 (1996)
44. M. Erdman, O. Rubner, Z. Shen, V. Engel, Chem. Phys. Lett. **341**, 338 (2001)
45. O. Rubner, V. Engel, J. Chem. Phys **115**, 2936 (2001)
46. M.J. Paterson, P.A. Hunt, M.A. Robb, O. Takahashi, J. Phys. Chem. A **106**, 10494 (2002)
47. S.A. Trushin, W. Fuss, W. Schmid, J. Phys. B **37**, 3987 (2004)
48. J.L. Herek, W. Wohlleben, R.J. Cogdell, D. Zeidler, M. Motzkus, Nature **417**, 553 (2002)
49. J.M. Geremia, W.S. Zhu, H. Rabitz, J. Chem. Phys. **113**, 10841 (2000)
50. T. Hornung, M. Motzkus, R. de Vivie-Riedle, J. Chem. Phys. **115**, 3105 (2001)
51. T. Hornung, M. Motzkus, R. de Vivie-Riedle, Phys. Rev. A **65**, 021403 (2002)
52. L. Kurtz, H. Rabitz, R. de Vivie-Riedle, Phys. Rev. A **65**, 032514 (2002)
53. W. Zhu, H. Rabitz, J. Chem. Phys. **111**, 472 (1999)
54. A. Mitra, H. Rabitz, Phys. Rev. A **67**, 033407 (2003)
55. J.L. White, B.J. Pearson, P.H. Bucksbaum, J. Phys. B **37**, L399 (2004)
56. J. Full, L. González, J. Manz, Chem. Phys. **329**, 126 (2006)
57. T. Mancal, V. May, Chem. Phys. Lett. **362**, 407 (2002)
58. V. Bonačić-Koutecký, P. Fantucci, J. Koutecký, Chem. Rev. **91**, 1035 (1991)
59. A.W. Castleman Jr., K.H.a. Bowen Jr., J. Phys. Chem. **100**, 12911 (1996)
60. J. Jortner, Faraday Discuss. **108**, 1 (1997)
61. U. Landman, Int. J. Mod. Phys. **B 6**, 3623 (1992)
62. Q. Zhong, A.W.a. Castleman Jr., Chem. Rev. **100**, 4039 (2000)
63. A. Stolow, A.E. Bragg, D.M. Neumark, Chem. Rev. **104**, 1719 (2004)
64. T.E. Dermota, Q. Zhong, A.W.J. Castleman, Chem. Rev. **104**, 1861 (2004)
65. V. Bonačić-Koutecký, R. Mitrić, Chem. Rev. **105**, 11 (2005)
66. M. Hartmann, J. Pittner, V. Bonačić-Koutecký, A. Heidenreich, J. Jortner, J. Chem. Phys. **108**, 3096 (1998)
67. M. Hartmann, J. Pittner, V. Bonačić-Koutecký, A. Heidenreich, J. Jortner, J. Phys. Chem. **102**, 4069 (1998)
68. T. Baumert, T. Brixner, V. Seyfried, M. Strehle, G. Gerber, Appl. Phys. B **65**, 779 (1997)
69. S. Vajda, P. Rosendo-Francisco, C. Kaposta, M. Krenz, L. Lupulescu, L. Wöste, Eur. Phys. J. D **16**, 161 (2001)
70. I. Andrianov, V. Bonačić-Koutecký, M. Hartmann, J. Manz, J. Pittner, K. Sundermann, Chem. Phys. Lett. **318**, 256 (2000)
71. M. Hartmann, J. Pittner, V. Bonačić-Koutecký, J. Chem. Phys. **114**, 2106 (2001)
72. M. Hartmann, J. Pittner, V. Bonačić-Koutecký, J. Chem. Phys. **114**, 2123 (2001)

73. M. Hartmann, R. Mitrić, B. Stanca, V. Bonačić-Koutecký, Eur. Phys. J. D **16**, 151 (2001)
74. V. Bonačić-Koutecký, M. Hartmann, J. Pittner, Eur. Phys. J. D **16**, 133 (2001)
75. R. Mitrić, M. Hartmann, B. Stanca, V. Bonačić-Koutecký, P. Fantucci, J. Phys. Chem. A **105**, 8892 (2001)
76. S. Vajda, C. Lupulescu, A. Merli, F. Budzyn, L. Wöste, M. Hartmann, J. Pittner, V. Bonačić-Koutecký, Phys. Rev. Lett. **89**, 213404 (2002)
77. R. Mitrić, M. Hartmann, J. Pittner, V. Bonačić-Koutecký, J. Phys. Chem. A **106**, 10477 (2002)
78. V. Bonačić-Koutecký, R. Mitrić, M. Hartmann, J. Pittner, Int. J. Quant. Chem. (2004)
79. N.E. Henriksen, V. Engel, Int. Rev. Phys. Chem. **20**, 93 (2001)
80. M.C. Heitz, G. Durand, F. Spiegelman, C. Meier, J. Chem. Phys. **118**, 1282 (2003)
81. M.C. Heitz, G. Durand, F. Spiegelman, C. Meier, R. Mitrić, V. Bonačić-Koutecký, J. Chem. Phys. **121**, 9906 (2004)
82. D.J. Tannor, S.A. Rice, J. Chem. Phys. **83**, 5013 (1985)
83. D.J. Tannor, S.A. Rice, Adv. Chem. Phys. **70**, 441 (1988)
84. G.K. Paramonov, V.A. Savva, Phys. Lett. **97A**, 340 (1983)
85. P. Brumer, M. Shapiro, Faraday Discuss. Chem. Soc. **82**, 177 (1986)
86. M. Shapiro, P. Brumer, J. Chem. Phys. **84**, 4103 (1986)
87. A.P. Peirce, M.A. Dahleh, H. Rabitz, Phys. Rev. A **37**, 4950 (1988)
88. T. Baumert, B. Buhler, M. Grosser, R. Thalweiser, V. Weiss, E. Wiedenmann, G. Gerber, J. Phys. Chem. **95**, 8103 (1991)
89. T. Baumert, G. Gerber, Isr. J. Chem. **34**, 103 (1994)
90. J.L. Herek, A. Materny, A.H. Zewail, Chem. Phys. Lett. **228**, 15 (1994)
91. A. Shnitman, I. Sofer, I. Golub, A. Yogev, M. Shapiro, Z. Chen, P. Brumer, Phys. Rev. Lett. **76**, 2886 (1996)
92. H. Schwoerer, R. Pausch, M. Heid, V. Engel, W. Kiefer, J. Chem. Phys. **107**, 9749 (1997)
93. C. Nicole, M.A. Bouchene, C. Meier, S. Magnier, E. Schreiber, B. Girard, J. Chem. Phys. **111**, 7857 (1999)
94. L. Pesce, Z. Amitay, R. Uberna, S.R. Leone, R. Kosloff, J. Chem. Phys. **114**, 1259 (2001)
95. Z.W. Shen, T. Chen, M. Heid, W. Kiefer, V. Engel, Eur. Phys. J. D **14**, 167 (2001)
96. G. Grègoire, M. Mons, I. Dimicoli, F. Piuzzi, E. Charron, C. Dedonder-Lardeux, C. Jouvet, S. Matenchard, D. Solgadi, A. Suzor-Weiner, Eur. Phys. J. D **1**, 187 (1998)
97. T. Hornung, M. Motzkus, de Vivie-Riedle, J. Chem. Phys. **115**, 3105 (2001)
98. G. Rodriguez, J.G. Eden, Chem. Phys. Lett. **205**, 371 (1993)
99. G. Rodriguez, P.C. John, J.G. Eden, J. Chem. Phys. **103**, 10473 (1995)
100. R. Pausch, M. Heid, T. Chen, W. Kiefer, H. Schwoerer, J. Chem. Phys. **110**, 9560 (1999)
101. R. Pausch, M. Heid, T. Chen, W. Kiefer, H. Schwoerer, J. Raman Spectrosc. **31**, 7 (2000)
102. R. Uberna, Z. Amitay, R.A. Loomis, S.R. Leone, Faraday Discuss. **113**, 385 (1999)
103. S. Vajda, A. Bartelt, E.C. Kaposta, T. Leisner, C. Lupulescu, S. Minemoto, P. Rosenda-Francisco, L. Wöste, Chem. Phys. **267**, 231 (2001)

104. A. Bartelt, S. Minemoto, C. Lupulescu, S. Vajda, L. Wöste, Eur. Phys. J. D **16**, 127 (2001)
105. S. Vajda, C. Lupulescu, A. Bartelt, F. Budzyn, P. Rosendo-Francisco, L. Wöste, *Femtochemistry and Femtobiology* (World Scientific Publishing, Singapore, 2002), vol. World Scientific Publishing: Singapore, pp. 472–480
106. A. Bartelt, A. Lindinger, C. Lupulescu, S. Vajda, L. Wöste, Phys. Chem. Chem. Phys. **5**, 3610 (2003)
107. C. Lupulescu, A. Lindinger, M. Plewicky, A. Merli, S.M. Weber, L. Wöste, Chem. Phys. **296**, 63 (2004)
108. A. Bartelt, Steuerung der Wellenpaketdynamik in kleinen Alkaliclustern mit optimierten Femtosekundenpulsen. Ph.D. thesis, Freie Universität Berlin (2002)
109. J.B. Ballard, H.U. Stauffer, Z. Amitay, S.R. Leone, J. Chem. Phys. **116**, 1350 (2002)
110. B. Schäfer-Bung, R. Mitrić, V. Bonačić-Koutecký, A. Bartelt, C. Lupulescu, A. Lindinger, S. Vajda, S.M. Weber, L. Wöste, J. Phys. Chem. A **108**, 4175 (2004)
111. C.J. Bardeen, V.V. Yakovlev, K.R. Wilson, S.D. Carpenter, P.M. Weber, W.S. Warren, Chem. Phys. Lett. **280**, 151 (1997)
112. T.C. Weinacht, J.L. White, P.H. Bucksbaum, J. Phys. Chem. A **103**, 10166 (1999)
113. T. Hornung, R. Meier, M. Motzkus, Chem. Phys. Lett. **326**, 445 (2000)
114. T. Brixner, N.H. Damrauer, P. Niklaus, G. Gerber, Nature **414**, 57 (2001)
115. C. Daniel, J. Full, L. González, E.C. Kaposta, M. Krenz, C. Lupulescu, J. Manz, S. Minemoto, M. Oppel, P. Rosenda-Francisco, S. Vajda, L. Wöste, Chem. Phys. **267**, 247 (2001)
116. N.H. Damrauer, C. Dietl, G. Krampert, S.H. Lee, K.H. Jung, G. Gerber, Eur. Phys. J. D **20**, 71 (2002)
117. A.Szabo und N. S. Ostlund, *Modern Quantum Chemistry: Introduction to advanced electronic structure theory* (Macmillan Publishing Co., Inc., New York, 1982)
118. F. Jensen, *Computational Chemistry* (Wiley, Chichester, 1999)
119. W. Koch, M.C. Holthausen, *A chemist's guide to Density Functional Theory* (Wiley-VCH, 2001)
120. C. Møller, M.S. Plesset, Phys. Rev. **46**, 618 (1934)
121. P.E.M. Siegbahn, in *Lecture notes in Quantum Chemistry*, ed. by B.O. Roos (Springer, Berlin, 1992)
122. H.J. Werner, in *Ab initio methods in Quantum Chemistry II*, ed. by K.P. Lawley (Wiley, New York, 1987)
123. P.E.M. Siegbahn, J. Almöf, A. Heiberg, B. Roos, J. Chem. Phys **74**, 2384 (1981)
124. K. Andersson, P.A. Malmqvist, B.O. Roos, A.J. Sadlej, K. Wolinski, J. Phys. Chem **94**, 5483 (1990)
125. K. Andersson, P.A. Malmqvist, B.O. Roos, J. Chem. Phys **96**, 1218 (1992)
126. R. Meyer, J. Chem. Phys. **52**, 2053 (1970)
127. C.C. Marston, G.G. Balint-Kurti, J. Chem. Phys **91**, 3571 (1989)
128. N. Balakrishnan, C. Kalyanaraman, N. Sathyamurthy, Phys. Rep. **280**, 79 (1997)
129. D. Kosloff, R. Kosloff, J. Comput. Phys. **52**, 35 (1983)

130. J.A. Fleck, J.R. Morris, M.D. Feit, Appl. Phys. **10**, 1929 (1976)
131. M.D. Feit, J.A. Fleck, A. Steiger, J. Comput. Phys. **47**, 412 (1982)
132. M.D. Feit, J.A. Fleck, J. Chem. Phys **78**, 301 (1983)
133. J.C. Tully, J. Chem. Phys. **93**, 1061 (1990)
134. A. Assion, M. Geisler, J. Helbing, V. Seyfried, T. Baumert, Phys. Rev. A **54**, R4605 (1996)
135. V. Blanchet, M.Z. Zgierski, T. Seideman, A. Stolow, Nature **401**, 52 (1999)
136. T. Schultz, J. Quenneville, B. Levine, A. Toniolo, T.J. Martinez, S. Lochbrunner, M. Schmitt, J.P. Schaffer, M.Z. Zgierski, A. Stolow, J. Am. Chem. Soc. **125**, 8098 (2003)
137. J. Stanzel, F. Burmeister, M. Neeb, W. Eberhardt, R. Mitric, C. Burgel, V. Bonacic-Koutecky, to be published (2006)
138. N. Pontius, P.S. Bechthold, M. Neeb, W. Eberhardt, Phys. Rev. Lett. **84**, 1132 (2000)
139. N. Pontius, M. Neeb, W. Eberhardt, G. Luttgens, P.S. Bechthold, Phys. Rev. B **67**, 035425 (2003)
140. J.R.R. Verlet, A.E. Bragg, A. Kammrath, O. Cheshnovsky, D.M. Neumark, J. Chem. Phys. **121**, 10015 (2004)
141. A.E. Bragg, J.R.R. Verlet, A. Kammrath, O. Cheshnovsky, D.M. Neumark, J. Chem. Phys. **122** (2005)
142. V.A. Spasov, Y. Shi, K.M. Ervin, Chem. Phys. **262**, 75 (2000)
143. A. Bragg, J. Verlet, A. Kammrath, O. Cheshnovsky, D. Neumark, J. Am. Chem. Soc. **127**, 15283 (2005)
144. A. Kammrath, J. Verlet, A. Bragg, G. Griffin, D. Neumark, J. Phys. Chem. A **109**, 11475 (2005)
145. A. Kammrath, J. Verlet, G. Griffin, D. Neumark, (in preparation)
146. A. Eppink, D. Parker, Rev. Sci. Inst. **68**, 3477 (1997)
147. C. Bordas, F. Paulig, H. Helm, D. Huestis, Rev. Sci. Inst. **67**, 2257 (1996)
148. E. Surber, A. Sanov, J. Chem. Phys. **116**, 5921 (2002)
149. V. Dribinski, A. Ossadtchi, V. Mandelshtam, H. Reissler, Rev. Sci. Inst. **73**, 2634 (2002)
150. G. Garcia, L. Nahon, I. Powis, Rev. Sci. Inst. **75**, 4989 (2004)
151. M. Vrakking, Rev. Sci. Inst. **72**, 4084 (2001)
152. S. Lochbrunner, J.J. Larsen, J.P. Shaffer, M. Schmitt, T. Schultz, J.G. Underwood, A. Stolow, J. Electron. Spectrosc. Relat. Phenom. **112**, 183 (2000)
153. V. Stert, W. Radloff, C.P. Schulz, I.V. Hertel, The Eur. Phys. J. D **5**, 97 (1999)
154. H.H. Ritze, H. Lippert, E. Samoylova, V.R. Smith, I.V. Hertel, W. Radloff, T. Schultz, J. Chem. Phys. **122** (2005)
155. M.G.H. Boogaarts, G. von Helden, G. Meijer, J. Chem. Phys. **105**, 8556 (1996)
156. J. Verlet, A. Bragg, A. Kammrath, O. Cheshnovsky, D. Neumark, Science **307**, 93 (2005)
157. J. Coe, G. Lee, J. Eaton, S. Arnold, H. Sarkas, K. Bowen, C. Ludewigt, H. Haberland, D. Worsnop, J. Chem. Phys. **92**, 3980 (1990)
158. G. Makov, A. Nitzan, J. Chem. Phys **98** (1994)
159. R. Barnett, U. Landman, C. Cleveland, J. Jortner, J. Chem. Phys **88**, 4429 (1988)
160. L. Turi, W.S. Sheu, P. Rossky, Science **309**, 914 (2005)
161. A. Kammrath, J. Verlet, G. Griffin, D. Neumark, J. Chem. Phys **125**, 076101 (2006)

162. N. Hammer, J.W. Shin, J. Headrick, E. Diken, J. Roscioli, G. Weddle, M. Johnson, Science **306**, 675 (2004)
163. N. Hammer, J. Roscioli, J. Bopp, J. Headrick, M. Johnson, J. Chem. Phys. **123**, 244311 (2005)
164. A. Bragg, J. Verlet, A. Kammrath, O. Cheshnovsky, D. Neumark, Science **306**, 669 (2004)
165. D. Paik, I.R. Lee, D.S. Yang, J. Baskin, A. Zewail, Science **306**, 672 (2004)
166. P. Scherer, S. Fischer, Chem. Phys. Lett. **421**, 427 (2006)
167. A. Kammrath, J. Verlet, G. Griffin, D. Neumark, (in preparation)
168. A. Zharikov, S. Fischer, J. Chem. Phys. **124**, 054506 (2006)
169. P.Y. Turpin, L. Chinsky, A. Laigle, B. Jolles, J. Mol. Struct. **214**, 43 (1989)
170. J.D. Watson, F.H.C. Crick, Nature **171**, 737 (1953)
171. R.D. Brown, P.D. Godfrey, D. McNaughton, A.P. Pierlot, Chem. Phys. Lett. **156**, 61 (1989)
172. C. Plützer, K. Kleinermanns, Phys. Chem. Chem. Phys. **4**, 4877 (2002)
173. M. Kabelac, P. Hobza, J. Phys. Chem. B **105**, 5804 (2001)
174. E. Nir, C. Plützer, K. Kleinermanns, M. de Vries, Eur. Phys. J. D **20**, 317 (2002)
175. C. Plützer, I. Hunig, K. Kleinermanns, E. Nir, M.S. de Vries, ChemPhysChem **4**, 838 (2003)
176. A. Abo-Riziq, L. Grace, E. Nir, M. Kabelac, P. Hobza, M.S. de Vries, Proc. Nat. Acad. Sci. USA **102**, 20 (2005)
177. M. Kabelac, P. Hobza, Chem. A **7**, 2067 (2001)
178. S. Ullrich, T. Schultz, M.Z. Zgierski, A. Stolow, J. Am. Chem. Soc. **126**, 2262 (2004)
179. D.C. Luhrs, J. Viallon, I. Fischer, Phys. Chem. Chem. Phys. **3**, 1827 (2001)
180. H. Kang, K.T. Lee, B. Jung, Y.J. Ko, S.K. Kim, J. Am. Chem. Soc. **124**, 12958 (2002)
181. S. Peng, A. Padva, P.R. Lebreton, Proc. Nat. Acad. Sci. USA **73**, 2966 (1976)
182. S. Ullrich, T. Schultz, M.Z. Zgierski, A. Stolow, Phys. Chem. Chem. Phys. **6**, 2796 (2004)
183. Y.G. He, C.Y. Wu, W. Kong, J. Phys. Chem. A **107**, 5145 (2003)
184. Y.G. He, C.Y. Wu, W. Kong, J. Phys. Chem. A **108**, 943 (2004)
185. A. Broo, J. Phys. Chem. A **102**, 526 (1998)
186. A.L. Sobolewski, W. Domcke, Eur. Phys. J. D **20**, 369 (2002)
187. C.M. Marian, J. Chem. Phys. **122** (2005)
188. M. Zierhut, W. Roth, I. Fischer, Phys. Chem. Chem. Phys. **6**, 5178 (2004)
189. I. Hunig, C. Plutzer, K.A. Seefeld, D. Lowenich, M. Nispel, K. Kleinermanns, ChemPhysChem **5**, 1427 (2004)
190. T. Pancur, N.K. Schwalb, F. Renth, F. Temps, Chem. Phys. **313**, 199 (2005)
191. H. Satzger, D. Townsend, M.Z. Zgierski, S. Patchkovskii, S. Ullrich, S. Stolow, Proc. Nat. Acad. Sci. USA **103**, 10196 (2006)
192. H.Y. Kang, B.Y. Jung, S.K. Kim, J. Chem. Phys. **118**, 11336 (2003)
193. A.L. Sobolewski, W. Domcke, C. Hättig, Proc. Nat. Acad. Sci. USA **102**, 17903 (2005)
194. A.L. Sobolewski, A theoretical treatment of PCET in the AT base pair was presented in the Femtochemistry VII conference (2005).
195. T. Schultz, E. Samoylova, W. Radloff, I.V. Hertel, A.L. Sobolewski, W. Domcke, Science **306**, 1765 (2004)

196. E. Samoylova, H. Lippert, S. Ullrich, I.V. Hertel, W. Radloff, T. Schultz, J. Am. Chem. Soc. **127**, 1782 (2005)
197. C.E. Crespo-Hernandez, B. Cohen, B. Kohler, Nature **436**, 1141 (2005)
198. S. Gilb, K. Jacobsen, D. Schooss, F. Furche, R. Ahlrichs, M.M. Kappes, J. Chem. Phys. **121**, 4619 (2004)
199. R. Mitrić, C. Burgel, V. Bonačić-Koutecký, Private communication (2006)
200. H. Hakkinen, B. Yoon, U. Landman, X. Li, H.J. Zhai, L.S. Wang, J. Phys. Chem. A **107**, 6168 (2003)
201. J. Stanzel, F. Burmeister, M. Neeb, W. Eberhardt, to be published (2006)
202. H. Weidele, S. Becker, H.J. Kluge, M. Lindinger, L. Schweikhard, C. Walther, J. Ziegler, D. Kreisle, Surf. Rev. Lett. **3**, 541 (1996)
203. K. Balasubramanian, D.G. Dai, Chem. Phys. Lett. **265**, 538 (1997)
204. T. Hertel, E. Knoesel, M. Wolf, G. Ertl, Phys. Rev. Lett. **76**, 535 (1996)
205. S. Martinez, J. Alfano, D. Levy, J. Mol. Spec. **156**, 421 (1992)
206. U. Erlekam, M. Frankowski, G. von Helden, G. Meijer, (to be published)
207. K. Lee, J. Sung, K. Lee, Y. Park, S. Kim, Angew. Chem. - Int. Ed. **41**, 4114 (2002)
208. Y. Lee, J. Jung, B. Kim, P. Butz, L. Snoek, R. Kroemer, J. Simons, J. Phys. Chem. A **108**, 69 (2004)
209. A. Kaczor, I. Reva, L. Proniewicz, R. Fausto, J. Phys. Chem. A **110**, 2360 (2006)
210. A. Scott, L. Radom, J. Phys. Chem. **100**, 16502 (1996)
211. J.O. Jung, R.B. Gerber, J. Chem. Phys **105**, 10332 (1996)
212. G.M. Chaban, J.O. Jung, R.B. Gerber, J. Chem. Phys. **111**, 1823 (1999)
213. B. Brauer, G.M. Chaban, R.B. Gerber, Phys. Chem. Chem. Phys. **6**, 2543 (2004)
214. M. Blom, I. Compagnon, G. von Helden, G. Meijer, B. Brauer, R.B. Gerber, (to be published)
215. J.E. Combariza, B. Just, J. Manz, G.K. Paramonov, J. Phys. Chem. **95**, 10351 (1991)
216. W. Jakubetz, J. Manz, H.J. Schreier, Chem. Phys. Lett. **165**, 100 (1990)
217. D.J. Tannor, S.A. Rice, J. Chem. Phys. **85**, 5805 (1986)
218. M.V. Korolkov, J. Manz, G.K. Paramonov, J. Chem. Phys. **105**, 10874 (1996)
219. N. Došlić, K. Sundermann, L. González, O. Mo, J. Giraud-Girard, O. Kühn, Phys. Chem. Chem. Phys. **1**, 1249 (1999)
220. N. Došlić, O. Kühn, J. Manz, K. Sundermann, J. Phys. Chem. A **102**, 9645 (1998)
221. H. Naundorf, K. Sundermann, O. Kühn, Chem. Phys. **240**, 163 (1999)
222. M.V. Korolkov, J. Manz, G.K. Paramonov, Adv. Chem. Phys. **101**, 327 (1997)
223. M.V. Korolkov, J. Manz, G.K. Paramonov, Chem. Phys. **217**, 341 (1997)
224. Y. Fujimura, L. González, K. Hoki, D. Kröner, J. Manz, Y. Ohtsuki, Angew. Chem. - Int. Ed. **39**, 4586 (2000)
225. K. Hoki, D. Kröner, J. Manz, Chem. Phys. **267**, 59 (2001)
226. B. Amstrup, N.E. Henriksen, J. Chem. Phys. **97**, 8285 (1992)
227. N.E. Henriksen, Adv. Chem. Phys. **91**, 433 (2005)
228. N. Elghobashi, J. Manz, Isr. J. Chem. **43**, 293 (2003)
229. Y. Fujimura, L. González, D. Kröner, J. Manz, I. Mehdaoui, B. Schmidt, Chem. Phys. Lett. **386**, 248 (2004).
230. K. Hoki, M.Yamaki, Y. Fujimura, Angew. Chem. Int Ed. **42**, 2976 (2003)

231. J. Full, L. Gonz'alez and C. Daniel, J. Phys. Chem. A **105**, 184 (2001)
232. J. Full, C. Daniel, L. González, Phys. Chem. Chem. Phys. **5**, 87 (2003)
233. G. Hirsch, P.J. Bruna, R.J. Buenker, S.D. Peyerimhoff, Chem. Phys. **45**, 335 (1980)
234. M. Baer, Chem. Phys. Lett. **35**, 112 (1975)
235. L. González, J. Full, Theor. Chem. Acc. **116**, 148 (2006)
236. J. Full, L. González, J. Manz, Chem. Phys. **314**, 143 (2005)
237. D. Ambrosek, M. Oppel, L. González, V. May, Chem. Phys. Lett. **380**, 541 (2003)
238. V. May, D. Ambrosek, M. Oppel, L. González, Opt. Comm. **264**, 502 (2006)
239. M. Seel, W. Domcke, J. Chem. Phys **95**, 7806 (1991)
240. C. Meier, V. Engel, U. Manthe, J. Chem. Phys **109**, 36 (1998)
241. J. Schön, H. Köppel, J. Phys. Chem **103**, 8579 (1999)
242. J. Michl, V. Bonačić-Koutecký, *Electronic aspects of organic photochemistry* (John Wiley & Sons Inc., New York, 1990)
243. W. Domcke, D.R. Yarkony, H. Köppel (eds.), *Conical intersections, Advanced series in Physical Chemistry, Vol. 15* (World Scientific, Singapore, 2004)
244. F. Bernardi, M. Olivucci, M.A. Robb, Chem. Soc. Rev. **25**, 321 (1996)
245. M.A. Robb, M. Garavelli, M. Olivucci, F. Bernardi, Rev. Comp. Chem. **15**, 87 (2000)
246. C. Van Caillie, R.D. Amos, Chem. Phys. Lett. **308**, 249 (1999)
247. F. Furche, R. Ahlrichs, J. Chem. Phys. **117**, 7433 (2002)
248. N.L. Doltsinis, D. Marx, J. Theor and Comp. Chem. **1**, 319 (2002)
249. A. Donoso, C. Martens, J. Phys. Chem. **102**, 4291 (1998)
250. M.D. Hack, D.G. Truhlar, J. Phys. Chem. A **104**, 7917 (2000)
251. A. Donoso, C.C. Martens, J. Chem. Phys. **112**, 3980 (2000)
252. W.H. Miller, J. Phys. Chem. **105**, 2942 (2001)
253. G. Stock, M. Thoss, Phys. Rev. Lett. **78**, 578 (1997)
254. M. Ben-Nun, J. Quenneville, T.J. Martínez, J. Phys. Chem. A **104**, 5161 (2000)
255. J.C. Tully, *Classical and quantum dynamics in condensed phase simulations* (World Scientific, Singapore, 1998)
256. S. Hammes-Schiffer, J. Phys. Chem. A **102**, 10443 (1998)
257. U. Landman, D. Scharf, J. Jortner, Phys. Rev. Lett. **54**, 1860 (1985)
258. G. Galli, W. Andreoni, M.P. Tosi, Phys. Rev. A **34**, 3580 (1986)
259. F. Rajagopal, R.N. Barnett, A. Nitzan, U. Landman, E. Honea, P. Labastie, M.L. Homer, R.L. Whetten, Phys. Rev. Lett. **64**, 2933 (1990)
260. V. Bonačić-Koutecký, C. Fuchs, J. Gaus, J. Pittner, Koutecký, Z. Physik D **26**, 192 (1993)
261. E.C. Honea, M.L. Homer, R.L. Whetten, Phys. Rev. B **63**, 394 (1989)
262. S. Pollack, C.R.C. Wang, M.M. Kappes, Chem. Phys. Lett. **175**, 209 (1990)
263. V. Bonačić-Koutecký, J. Pittner, J. Koutecký, Chem. Phys. **210**, 313 (1996)
264. V. Bonačić-Koutecký, J. Pittner, J. Koutecký, Z. Phys. D **40**, 441 (1997)
265. G. Durand, F. Spiegelman, P. Poncharal, P. Labastie, J.M. L'Hermite, M. Sence, J. Chem. Phys. **110**, 7884 (1999)
266. D. Rayane, I. Compagnon, R. Antoine, M. Broyer, P. Dugourd, P. Labastie, J.M. L'Hermite, A. Le Padallec, G. Durand, F. Calvo, F. Spiegelman, A.R. Allouche, J. Chem. Phys. **116**, 10730 (2002)
267. G. Durand, M.C. Heitz, F. Spiegelman, C. Meier, R. Mitrić, V. Bonačić-Koutecký, J. Pittner, J. Chem. Phys. **121**, 9898 (2004)

268. J.M. L'Hermite, V. Blanchet, A. Le Padellec, B. Lamory, P. Labastie, Eur. Phys. J. D **28**, 361 (2004)
269. R.W. Schoenlein, L.A. Peteanu, R.A. Mathies, C.V. Shank, Science **412**, 1991 (254)
270. M. Ottolenghi, M. Sheves, Isr. J. Chem. **35** (1995)
271. H. Lischka, M. Dallos, R. Shepard, Mol. Phys. **100**, 1647 (2002)
272. M. Dallos, H. Lischka, R. Shepard, D.R. Yarkony, P.G. Szalay, J. Chem. Phys. **120**, 7330 (2004)
273. V. Bonačić-Koutecký, J. Pittner, Chem. Phys. **225**, 173 (1997)
274. R. Mitrić, V. Bonačić-Koutecký, J. Pittner, L. H., J. Chem. Phys. **125**, 024303 (2006)
275. G. Granucci, M. Persico, A. Toniolo, J. Chem. Phys. **114**, 10608 (2001)
276. M.J. Bearpark, M.A. Robb, H.B. Schlegel, Chem. Phys. Lett. **223**, 269 (1994)
277. M. Shapiro, P. Brumer, Int. Rev. Phys. Chem. **13**, 187 (1994)
278. M. Abe, Y. Ohtsuki, Y. Fujimura, W. Domcke, J. Chem. Phys. **123**, 144508 (2005)
279. Y.J. Yan, R.E. Gillilan, R.M. Whitnell, K.R. Wilson, S. Mukamel, J. Phys. Chem. **97**, 2320 (1993)
280. J.L. Krause, R.M. Whitnell, K.R. Wilson, Y.L. Yan, S. Mukamel, J. Chem. Phys. **99**, 6562 (1993)
281. J.L. Krause, R.M. Whitnell, K.R. Wilson, Y.L. Yan, *Femtosecond Chemistry* (VCH, Weinheim, 1995), p. 743
282. B. Kohler, V.V. Yakovlev, J. Che, J.L. Krause, M. Messina, K.R. Wilson, N. Schwentner, R.M. Whitnell, Y.L. Yan, Phys. Rev. Lett. **74**, 3360 (1995)
283. J. Che, M. Messina, K.R. Wilson, V.A. Apkarian, Z. Li, C.C. Martens, R. Zadoyan, Y. Yan, J. Phys. Chem. **100**, 7873 (1996)
284. C.J. Bardeen, J. Che, K.R. Wilson, V.V. Yakovlev, P. Cong, B. Kohler, J.L. Krause, M. Messina, J. Phys. Chem. A **101**, 3815 (1997)
285. Y.J. Yan, Annu. Rep. Prog. Chem. Sect. C: Phys. Chem. **94**, 397 (1998)
286. Y.J. Yan, Z.W. Shen, Y. Zhao, Chem. Phys. **233**, 191 (1998)
287. Y.J. Yan, J. Che, J.L. Krause, Chem. Phys. **217**, 297 (1997)
288. Y.J. Yan, J.S. Cao, Z.W. Shen, J. Chem. Phys. **107**, 3471 (1997)
289. R. Xu, J. Cheng, Y. Yan, J. Phys. Chem. A **103**, 10611 (1999)
290. Z. Shen, Y. Yan, J. Cheng, F. Shuang, Z. Y., H. G., J. Chem. Phys. **110**, 7192 (1999)
291. Z. Shen, V. Engel, R. Xu, J. Cheng, Y. Yan, J. Chem. Phys. **117**, 6142 (2002)
292. R. Mitrić, C. Bürgel, J. Burda, V. Bonačić-Koutecký, Eur. Phys. J. D **24**, 41 (2003)
293. M. Gruebele, G. Roberts, M. Dantus, R.M. Bowman, A.H. Zewail, Chem. Phys. Lett. **166**, 459 (1990)
294. H. Metiu, V. Engel, J. Chem. Phys. **93**, 5693 (1990)
295. Z.M. Lu, H. Rabitz, Phys. Rev. A **52**, 1961 (1995)
296. Z.M. Lu, H. Rabitz, J. Phys. Chem. **99**, 13731 (1995)
297. B.A. Armstrup, G.J. Toth, H. Rabitz, A. Lörinz, Chem. Phys. **201**, 95 (1995)
298. B. Kohler, J.L. Krause, F. Raksi, C. Rose-Petruck, R.M. Whitnell, K.R. Wilson, V.V. Yakovlev, Y. Yan, S. Mukamel, J. Phys. Chem. **97**, 12602 (1993)
299. B. Kohler, J. Krause, F. Raksi, K.R. Wilson, R.M. Whitnell, V.V. Yakovlev, Y.L. Yan, Acc. Chem. Res. **28**, 133 (1995)
300. C.J. Bardeen, J. Che, K.R. Wilson, V.V. Yakovlev, V.A. Apkarian, C.C. Martens, R. Zadoyan, B. Kohler, M. Messina, J. Chem. Phys. **106**, 8486 (1997)

301. J.X. Cheng, Z.W. Shen, Y.J. Yan, J. Chem. Phys. **109**, 1654 (1998)
302. R. de Vivie-Riedle, K. Sundermann, M. Motzkus, Faraday Discuss. **113**, 303 (1999)
303. R. de Vivie-Riedle, A. Hofmann, in *Conical intersections: Electronic structure, dynamics and spectroscopy* (World Scientific, Singapore, 2004), p. 803
304. T. Joseph, J. Manz, Mol. Phys. **58**, 1149 (1986)
305. C.M. Tesch, R. de Vivie-Riedle, Phys. Rev. Lett. **89**, 157901 (2002)

4

Coherence and control of molecular dynamics in rare gas matrices

Matias Bargheer[1], Alexander Borowski[2], Arik Cohen[3], Mizuho Fushitani[4], R. Benny Gerber[3], Markus Gühr[4,5], Peter Hamm[6], Heide Ibrahim[4], Toni Kiljunen[7], Mikhail V. Korolkov[2,8], Oliver Kühn[2], Jörn Manz[2], Burkhard Schmidt[9], Maike Schröder[2], and Nikolaus Schwentner[4]

[1] Max Born Institut Berlin, Germany
[2] Institut für Chemie und Biochemie, Freie Universität Berlin, Germany
[3] Department of Physical Chemistry, Hebrew University Jerusalem, Israel
[4] Institut für Experimentalphysik, Freie Universität Berlin, Germany
[5] Stanford PULSE Center, SLAC, Menlo Park, USA
[6] Institut für Physikalische Chemie, Universität Zürich, Switzerland
[7] Department of Chemistry, University of Jyväskylä, Finland
[8] Institute of Physics, National Academy of Sciences, Minsk, Belarus
[9] Institute of Mathematics II, Freie Universität Berlin, Germany

Coordinated by: Nikolaus Schwentner

4.1 Introduction

N. Schwentner

A bridge is established in this chapter connecting the subject of isolated molecules from Chaps. 2 and 3 with the multidimensional and more complex systems treated in the following chapters. Explicitly this chapter considers the dynamics of a chromophore coupled to a conceptionally infinite crystalline environment of rare gas matrices. The emphasis lies on an in-depth experimental and theoretical characterization of the energetics, nonradiative transitions, and decoherence dynamics resulting from the interaction with the environment. In view of this task a restriction in the chromophore degrees of freedom is helpful. Throughout this chapter the focus lies on dihalogen chromophores and only in one exception (Sect. 4.5) the implications of more internal degrees of freedom are illustrated for the still moderate sized nitrous acid molecule. The free dihalogens have been spectroscopically investigated in great detail, just reminding on the I_2 spectral atlas for calibration purposes. Thus a sound and complete set of spectroscopic constants is available which can be exploited for the matrix applications. In fact the dihalogens provide a convenient set of bound, repulsive and crossing electronic states which is a prerequisite for prototypical photochemical studies. Therefore, a manifold of pioneering ultrafast experiments was based on the perturbation of electronically and vibrationally excited dihalogens by collisions, especially, with rare gas atoms covering perturber densities from the dilute gas up to the liquid phase. This background on available data provides a link in analogy to Chaps. 2 and 3 as already mentioned. (The citations will be given in the respective sections in order not to overburden these introductory remarks.) Based on this history it is obvious to employ rare gas matrices as a prototypical solid state environment for the dihalogens. It is worth mentioning already at this point that an intuitively appealing monotonic increase in relaxation, nonradiative transitions and decoherence rates with perturber density is documented from the gas up to the liquid phase. It turns out that this trend is broken in important examples for the further increase in density to the crystalline environment and the aspect of order in the environment deserves special consideration.

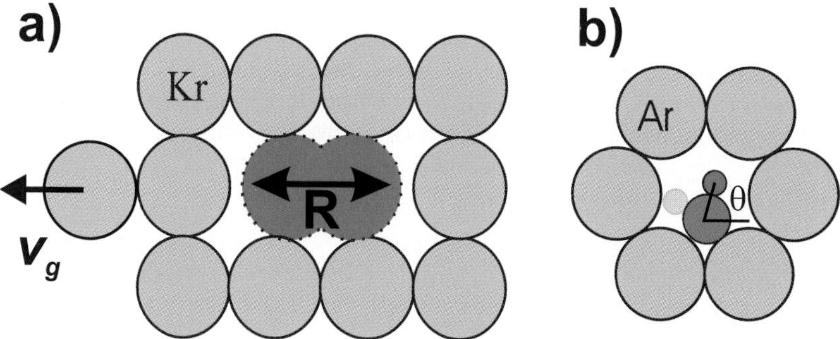

Fig. 4.1. a) Scheme for a double substitutional site (I_2 molecule in solid Kr) with internuclear coordinate R. The energy flow with phonon group velocity is indicated by v_g. b) Single substitutional site (ClF molecule in solid Ar) with angle θ of the molecular axis. After photodissociation the F fragment can scatter into a new orientation with respect to Cl.

The rare gases crystallize in a face-center-cubic (fcc) structure. The structural and dynamical properties have been investigated experimentally extensively and a tremendous theoretical effort was devoted to the derivation of crystal properties from the gas phase data. The inertness of the matrix justifies to conceptually start with the rare gas crystal structure and the dihalogens electronic and vibrational states. An art has evoked from the joint efforts of several theoretical groups to calculate the energetics of the dihalogens in the matrix from pair potentials with high accuracy. The Diatomics-in-Molecules (DIM) method to be discussed in Sect. 4.6 is perhaps the most successful approach in this respect. Starting in this conceptual spirit it is obvious that the coordinates connecting the dihalogen atoms with the atoms in the first shell of surrounding matrix atoms (the cage coordinates) play a pivotal role in the cage modified dynamics of the chromophore and also in the cage dynamics induced by the chromophore. While it will be shown that specific positions of cage atoms count, it is nevertheless tutorial to distinguish two limiting cage structures which are sketched in Fig. 4.1. The chromophore replaces two matrix atoms in a so called double substitutional site in Fig. 4.1a. The crystal lattice blocks a rotation of the molecular axis in this case which is optimal to study vibrational dynamics in the internuclear coordinate R of the chromophore and its coupling to the cage. For a single substitutional site and especially for an asymmetric chromophore displayed in Fig. 4.1b the hindering of rotation is expected to be less pronounced. Therefore, the angular coordinate θ, for example, in a librational motion becomes relevant in addition to R. Indeed these two cases lead to qualitatively different dynamics, as will be shown, emphasizing the importance of these two conceptual coordinates. The flow of energy from the first shell of cage atoms into the crystal and away from the chromophore is crucial for the cage dynamics. The flow will be related to the crystal properties i.e. phonon spectrum with group velocities v_g and coherences in the cage dynamics will be documented. Here, DIM-based classical trajectory simulations are pivotal for unraveling underlying mechanisms on an atomic level.

The discussion concerning advantages or even a preference of frequency resolved versus time resolved types of spectroscopy was heated for some time. It lost its impetus mostly due to the increasing accumulation of results and success of recent ultrafast spectroscopy. From a mathematical point of view both families are tied together by a Fourier transformation and thus in the theoretical community the principal equivalence of both methods is well appreciated. In fact the breakthrough of the time-dependent view has been stimulated largely by the relative ease of performing multidimensional time propagations as compared to the solution of eigenvalue problems. Of course, this comes at the expense of spectral resolution, which is, however, not a crucial issue in the condensed phase. Moreover, the inherent time evolution based on Heller's approach delivers a rather tutorial introduction into dissipative systems. Thereby the sequence of interactions with several degrees of freedom, often subsumed in the phrase of a bath, generates a time arrow with phase and energy relaxation. Dissipative systems and especially cases in which a sequence of coherent pulses interferes with the evolving ensemble at variable time delays, are the play grounds where time resolved spectroscopies develop their special merits. In practice both methods are by no means antipodes and for the experimentalist a clever synthesis of the two spectroscopies delivers the ultimate information as will be demonstrated for example in Sect. 4.2. It becomes evident, that even in the very same system an increasing interaction with the bath shifts the emphasis from frequency resolved to time resolved investigations. Section 4.3 illustrates that the chromophore-bath interaction can be steered by an appropriate combination of coherent pulses while the

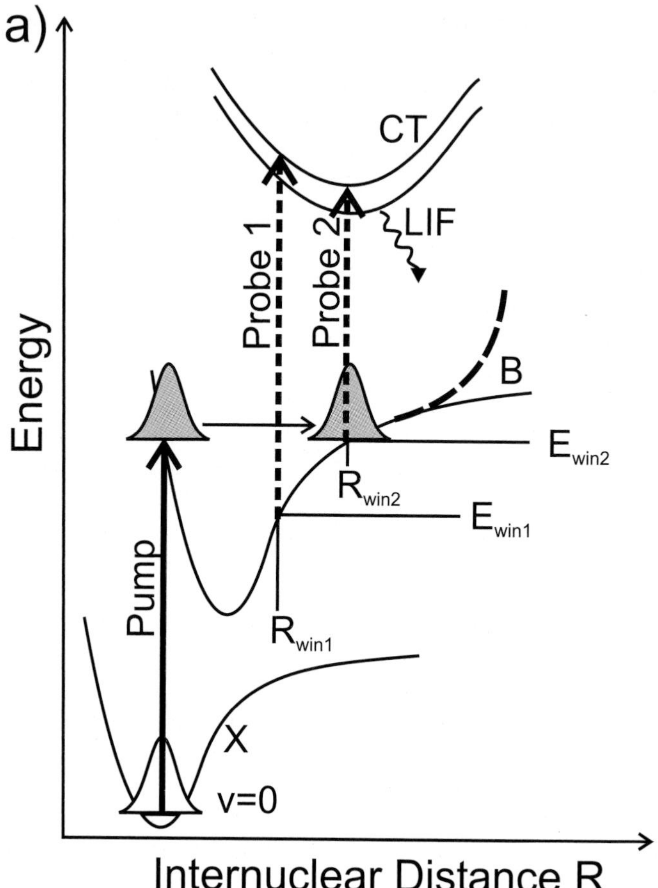

Fig. 4.2. Pump-probe scheme for dihalogens in rare gas matrices. The pump pulse transfers population from the electronic ground state v=0 level to the excited electronic B state (or any other covalent state) and excites several vibrational levels coherently. The thereby created wave packet moves quasi-periodically in the potential which is bent up due to the cage effect (dashed potential). The fs probe pulse transfers parts of the wave packet situated at a Franck-Condon window located at R_{win} and E_{win} to a charge transfer (CT) state after a time delay of Δt. The population on the CT states relaxes in the CT manifold, causing laser induced fluorescence (LIF). The fluorescence intensity is detected versus the pump-probe delay Δt in the pump-probe spectra.

selection and characterization of the pulse sequence, on the other hand, becomes intuitive in a spectral comb picture.

In this chapter, like in others, pump-probe spectroscopy is applied to a large extent and is developed systematically to deliver detailed information on all aspects of population evolution. The basic scheme is illustrated in Fig. 4.2

with an ultrashort pump pulse which prepares a coherent superposition of vibrational states in the Franck-Condon region of an excited electronic state, determined by the photon energy as well as the structure of the coupled potential energy surfaces (PES) in the matrix. The coherent superposition results in a propagating vibrational wave packet which can be interrogated with a second pulse lifting it at a probe window close to the internuclear distance R_{win} with a preferred energy E_{win} to higher lying states which in the present examples will be charge transfer states (CT). The fluorescence intensity from these CT states reflects this probability to lift the wave packet. With a variable time delay, the time course of the evolution of the wave packet in the internuclear coordinate R_{win} and energy E_{win} is displayed. The ultrashort laser pulses are obtained, using a Ti:Sapphire pumped NOPA (Noncollinear Optical Parametric Amplifier) setup. The pulses are tunable between 470 and 750 nm, with a duration around 30 fs at energies up to 10 μJ per pulse [1].

Section 4.2 demonstrates that a systematic variation of pump and probe wavelength allows to derive an effective potential of the chromophore including the dissipative interaction with the bath. The wave packet motion can be transformed to a mean trajectory reflecting deceleration processes due to transfer of kinetic energy to the bath.

While the pump-probe spectrum follows the evolving population, it also reflects phase information in the wave packet by displaying the resulting interference patterns in the population distribution. These aspects are worked out in Sect. 4.3 starting with the obvious broadening of a wave packet by dispersion in an anharmonic oscillator. Focusing is a means to discriminate between the still coherent components which follow the phase code in a, for example, chirped pulse while the incoherent part continues to broaden by its statistically scrambled phases. A more intriguing and in the application more valuable feature are revivals which display the time evolution of interference patterns in the wave packet which is spread out by dispersion. It will be shown that with fractional revivals coherence relations among subgroups of eigenstates can be characterized.

Till now the focus was on disturbances of wave packets in the intermediate electronic state by the bath. Coherences (phonons) in the bath, however, can modulate the relative electronic energy separations of the chromophore in a coherent way and thus transfer coherence back into the chromophore. The solvation energies of ionic and charge transfer states are especially sensitive to variations in density due to a modulation of dielectric screening. Section 4.4 demonstrates that coherent phonons in the matrix (i.e. bath) can be generated by the chromophore's electronic excitation and in addition recorded by the variation of E_{win} and R_{win} in pump-probe spectra. The theoretical backing is provided by DIM simulations which capture adequately the anisotropic dihalogen-matrix interaction in the different electronic states as shown in Sect. 4.7.

Linear polarization in the pump-pulse selects a subensemble in the isotropic distribution of molecules with either parallel or perpendicular alignment of the molecular axis relative to the polarization direction depending on $\Delta\Omega$ (change in angular momentum) in the electronic transition. A variation of the probe pulse polarization between the two directions is exploited in Sect. 4.2 to separate the dynamics of wave packets propagating simultaneously on different electronic states. This enables one, for instance, to unravel the subpicosecond timescale of intersystem crossing.

An obvious extension are ultrafast depolarization studies in order to follow the redistribution of the aligned ensemble back to an isotropic one. They show in Sect. 4.2 indeed time scales in dissociation-recombination events of the order of the vibrational frequency. The ultimate application uses the torque which is imposed on the molecular axis of a molecule by the electric field vector in a nonresonant pump-pulse. It originates from the differential polarizability parallel and perpendicular to the molecular axis. The kick induced by an ultrashort pulse is exploited in the gas phase to dynamical align molecules in the revivals of rotational/librational wave packets. For molecules in a crystal the trend to align along the electric field vector has to compete with the crystal field imposed by the environment and the conditions to transfer the method to the solid phase are treated extensively in Sect. 4.9.

The scheme in Fig. 4.2 can be developed further by applying a second delayed pump pulse with a well defined phase relation of the electric field with respect to the first one. The interference among the two wave packets generated by the two pulses establishes a wide variety of coherence studies introduced in Sect. 4.3. Such phase locked pulse sequences open up even the possibility to actively control the strength of the interaction of the chromophore with the bath as will be demonstrated in Sect. 4.3. Phase-locked pulse pairs were prepared in the present experiments in a Michelson interferometer. It turns out that the spectral fringes have to be adapted to the anharmonic molecular potential. The anharmonicity can be conveniently mimicked by a glass slab in one arm of the interferometer leading to the so called unbalanced interferometer. Pulse shapers based on LC (liquid crystal) displays are ready to generate flexible phase locked pulses in a computer controlled way.

Proceeding toward more involved coherent interactions of several pulses the coherent anti-Stokes Raman spectroscopy (CARS) in the box car geometry comes into play. It characterizes coherence in the coupling between ground and first excited as well as between first excited and CT states in Fig. 4.2. The power in this method has been demonstrated recently and its further application is very promising [2]. It is, however, not included in this survey. CARS provides one way to follow dynamics on the ground electronic surface, however, also pump-probe spectroscopy can be applied if the weak transition dipoles in the infrared spectral region can be overcome. This problem has been solved recently and a study on the influence of specific excited vibrational modes on the cis-trans isomerization of matrix-isolated nitrous acid is presented in Sect. 4.5. The study delivers a perspective for dealing with more internal degrees of freedom in the chromophore connected to the matrix.

The detailed look at the molecular dynamics provided by the experiment quite naturally stimulates theoretical efforts. As emphasized before the theoretical workhorse is the DIM method. Despite its semiempirical nature, it has proven to be rather reliable for the considered dihalogen-rare gas situations. Essentially it delivers full-dimensional potential energy surfaces for all valence states, including nonadiabatic couplings as will be outlined in Sect. 4.6. Besides the information contained in these surfaces, such as matrix-induced level shifts or caging barriers, they are the necessary ingredient for dynamics simulations. At first glance, a diatomic in a low-temperature rare gas matrix appears to be the simplest condensed phase system one can imagine. Recalling the experimental results, however, it becomes clear that one faces a high-dimensional problem of coupled electronic and nuclear degrees of freedom. Thus it is not surprising that trajectory-based classical and semiclassical methods enjoy great popularity. Here, it is of decisive advantage that the DIM method efficiently provides forces and nonadiabatic couplings "on-the-fly". The surface hopping approach outlined in Sect. 4.8 is the method of choice for combining classical nuclear dynamics with nonadiabatic transitions between different electronic states. In the present context this gives access, for instance, to the electronic cage effects highlighting the ultrafast intersystem crossing processes or the propensities of electronic state populations in the matrix, see Sect. 4.8. However, being based on classical trajectories, the experimentally observed effects of quantum coherence in the evolution of nuclear wave packets cannot be accounted for. Here, it has proven valuable that the DIM-based trajectories can give information on the effective dimensionality of the dynamics within a certain time window. This led to the development of reduced one- and two-dimensional models for the description of the early dynamics within the first hundreds of femtoseconds after excitation. These reduced models allow for quantum simulations and thus providing an important test for the semiclassical methods. Furthermore, reduced dimensionality quantum models are ideally suited for the development of strategies for coherent laser control. Stimulated by the observation of ultrafast spin flip effort has especially been devoted to design laser pulses which trigger the selective preparation of singlet or triplet state in a narrow time- and spatial window as will be discussed in Sect. 4.10.

4.2 Chromophore-matrix interaction

M. Bargheer, M. Fushitani, M. Gühr, H. Ibrahim, and N. Schwentner

4.2.1 An effective experimental intramolecular potential

To handle a multidimensional system it is in many cases appropriate to break it down into a functional unit (e.g. a chromophore) in which the dynamics of interest takes place and the dissipative surrounding introduced as "bath" in Sect. 4.1. In the present model system it is obvious that the dihalogen plays the role of the chromophore and for its intramolecular potential one may use the gas phase potential as a starting point. The observed dynamics will be influenced by the matrix interaction and it is convenient to derive an effective potential which includes these effects. The potential thus reflects the

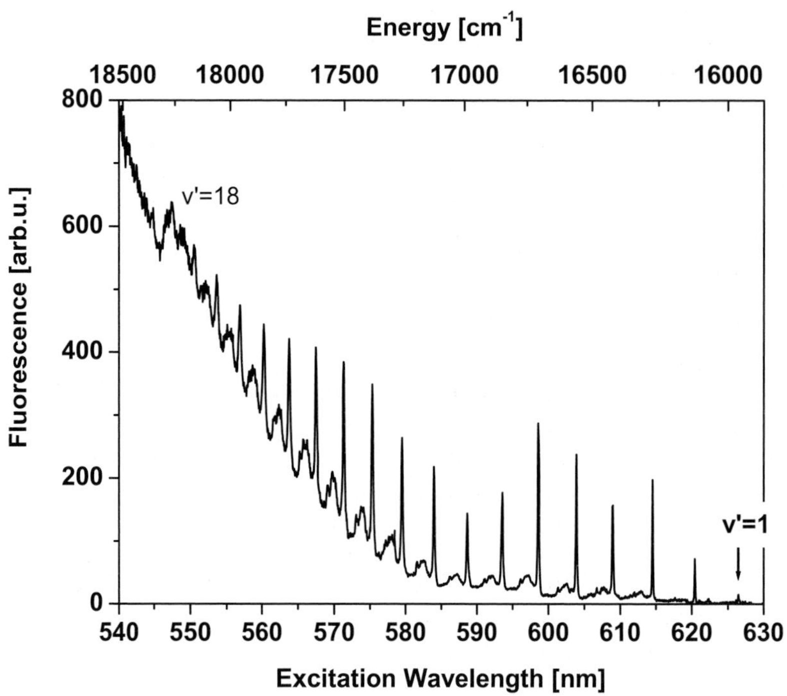

Fig. 4.3. Dye laser excitation spectrum of Br_2 in Ar matrix, taken at 6K. The sharp zero phonon lines (ZPL) indicate the vibrational levels of the B state v'. In between the ZPL the broad phonon side bands (PSB) are visible.

multidimensionality of the systems in the projection on the chromophore internuclear separation. Near the potential minimum of a bound state one deals with small vibrational elongations and the interaction with the matrix can be treated in many cases as a rather weak perturbation as will be shown. Near and above a dissociation limit one expects large vibrational amplitudes, however, the system is stabilized by the caging effect of the matrix. Substantial and fast energy dissipation into matrix coordinates will be demonstrated which in the picture of an effective potential would correspond to a time dependent potential. The potential is no longer unique in a one dimensional representation and a specification of the evolution in the cage coordinates is required for the derivation and application. As prototypical examples those diatomic molecules are used which take double substitutional sites in the host lattice (e.g. $I_2 : Kr$, $Br_2 : Ar$, $Cl_2 : Ar$). In this case the relevant number of degrees of freedom is reduced to one vibrational coordinate R representing the internuclear distance of the diatomic. Rotations and even librations are strongly hindered and also the center-of-mass coordinate of the diatomic plays a minor role, since one can assume that the molecules are symmetrically embedded in the matrix. Deep in the potential well of the B state of Cl_2 and Br_2 well resolved vibrational progressions have been observed [3]. Fig. 4.3 shows a recent excitation spectrum of the B state fluorescence of Br_2 in an Ar matrix, with improved spectral resolution compared to literature data [3]. The spectrum displays a series of sharp lines which are called Zero Phonon Lines (ZPL) since they represent an exclusive intramolecular vibration without participation of matrix oscillations. On the high energy side they are accompanied by a structured band which originates from an additional excitation of a matrix vibration and is therefore called Phonon Side Band (PSB) . The structure in the PSB is correlated with the phonon density of states of the matrix. The progression starts with v'=1 in the B state and a well isolated ZPL. The ratio of the PSB to ZPL areas increases with v' and the related larger vibrational amplitudes, as expected from the extensively developed perturbative treatment of the chromophore matrix - interaction [4]. The ratio can be cast, for example, in the Huang Rhys coupling constant S [4,5]. The theory predicts a progression of PSB which extends to higher members with increasing S. The superposition of all these contributions causes a background rising with v' and in addition a broadening of the ZPL. Beyond v'=18 the contributions merge to a continuum and deliver no further spectroscopic evidence for a derivation of the potential. We are, however, still far below the dissociation limit. To extend the potential to higher energies time resolved data are required.

The electronic states involved in a pump-probe spectrum can be characterized in a threefold manner by selecting the pump, the probe and, the fluorescence wavelength, respectively [6–11]. Starting from the ground state $X(^1\Sigma_0)$, the pump-pulse prepares a well localized wave packet in an excited electronic (valence) state of the molecule. Ideally, the electronic states in the Franck-Condon region are either energetically well separated and one can selectively excite a specific state, or only one transition of the energetically accessible states is allowed. In the generic example of the B-state $(^3\Pi_0)$ of the diatomics, which will be examined in detail in the following, the spectroscopic arguments are different for I_2 and Cl_2, respectively. In Cl_2 the A $(^3\Pi_1)$ and $\mathrm{A}'(^3\Pi_2)$ states are energetically very close to the B state. All states are spin-forbidden, but the transition dipole moment to the A and A' states is much weaker than that to B, because of the angular momentum selection rules .

In the heavier I_2, the selection rules are much more relaxed but the spin-orbit coupling energetically separates the different electronic states. However, it is not always possible to select exclusively a single electronic state and thus two or more wave packets are prepared coherently.

The second spectroscopic signature that can be used is the probe transition, which in these experiments is typically to the ion-pair manifold (I^+I^-). Again one can use the information on the energetic positions (obtained from calculations or stationary spectroscopy). Transitions to the ion-pair manifold have the additional advantage to obey a very strict selection rule $\Delta\Omega = 0$, because the transition dipole involves the transfer of an electron parallel to the internuclear bond. In many cases, the probe transition allows for spectroscopical discrimination between different wave packets in the valence state prepared either by an nonselective pump or by nonadiabatic transitions. The third spectroscopic information that can be used to identify the involved electronic states is the fluorescence wavelength, which contains information on the electronic states, to which the wave packet was probed.

The pump-probe data, i.e. the intensity of the LIF as a function of the time delay carries the information on the vibrational "states" in the electronic potential of the valence state. To be more precise, one can determine the round-trip time of the wave packet, which changes in the evolution because the wave packet loses energy by collisions with the surroundings and thus explores different vibrational "states" of the molecule. Therefore, the vibrational states are lifetime-broadened compared to the gas-phase and contain PSB in wavelength resolved studies. In the case of very strong coupling the vibrational structure is entirely lost like for v' > 18 (see Fig.4.3). For $I_2 : Kr$ the absorption spectrum shows no sharp lines at all, because of rapid electronic dephasing. Femtosecond pump-probe spectroscopy can provide the information on the vibrational properties of the molecule even in the presence of very strong dissipation [10, 11]. In the following this is exemplified for $I_2 : Kr$ [11] and it is shown how the round trip times can be used to construct an effective one-dimensional potential of the B state and of the ion-pair state E, by recording a series of pump-probe spectra with the sequence $X \overset{h\nu_{pump}}{\rightarrow} B \overset{h\nu_{probe}}{\rightarrow} E$, where both $h\nu_{pump}$ and $h\nu_{probe}$ are tuned. We start from the gas-phase potential of the B state, and assume that the solvent cage will mainly change the outer limb of the potential. The electronic state can be approximated by a Morse potential with the classical frequencies:

$$2\pi\nu(E) = \omega_e\sqrt{1 - E/D_e}. \tag{4.1}$$

$D_e = \omega_e^2/(4\omega_e x_e)$ corresponds to the dissociation energy and $\omega_e x_e$ is the anharmonicity. In the quantum analog, the frequency $\nu(E)$ is the spacing of vibrational levels. Fig. 4.4a shows a series of pump-probe spectra for fixed probe wavelength λ_{probe}. The increase of the first round trip time $T_1(E) = 1/\nu(E)$ of the wave packet with E is a direct measure of the anharmonicity $\omega_e x_e$. The change of the frequency, induced by the cage can be directly viewed in the pump-probe spectra by comparison to the gas-phase periods indicated in two transients by vertical bars. The cage decreases the anharmonicity with

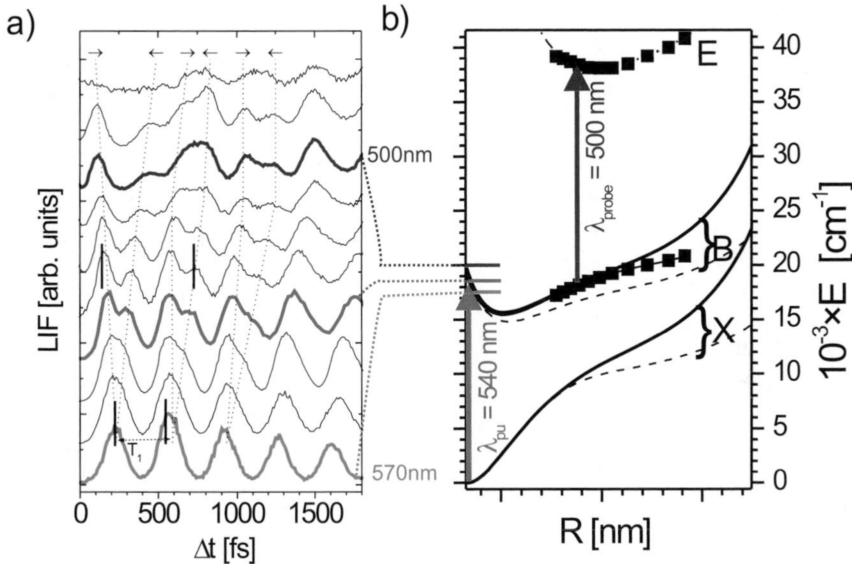

Fig. 4.4. a) Pump-probe spectra for $I_2 : Kr$ for fixed $\lambda_{probe} = 500$ nm and λ_{pump} varied in 10 nm steps . The vertical black bars indicate the first round trip time in the gas-phase, for arrows see text. b) Potential energy diagram of I_2 in Kr including the ground state X, the valence B state and the ionic E state relevant for the pump-probe spectra. The arrows depict pump and probe transitions. The horizontal lines indicate pump energies corresponding to $\lambda_{pump} = 570$, 540 and 500 nm, respectively. The squares indicate the shape of the potentials as derived from the experiment. Dashed lines correspond to DIM calculations in the fixed Kr fcc lattice and solid lines to a relaxed cage geometry. (Adapted from [11])

respect to the free molecule. The first round trip time of a wave packet excited with energy $E = h\nu_{pump}$ already includes the energy loss experienced in the first collision with the cage. From the systematics of the spectra it is evident that the measured time T_1 is a slight overestimation of the frequency $\nu(E)$, since the wave packets returning from the collision with lower energy $E - \Delta E$ oscillate faster. For $\lambda_{probe} = 500$ nm the probe window is located at the internuclear distance $R_{win} = 0.37$ nm. The arrows on the maxima in Fig. 4.4a indicate in which direction the wave packet moves through the window. For $\lambda_{probe} = 560$ nm (not shown) the inward-outward splitting disappears, implying that the wave packet is probed at its turning point. A detailed analysis of the systematics of tuning both pump and probe reveals that for each energy E_{pump} one has to choose the probe wavelength such that the first excursion of the wave packet is probed at its turning point, because other wavelengths bias the probe sensitivity either for the high- or low-energy part of the superposition of excited vibrational levels. The potential $U(R)$ can be determined by numeric integration according to [12] of the experimentally determined function $T_1(E)$ (extrapolated toward the minimum),

$$\Delta R(U) = \frac{1}{\pi\sqrt{2\mu}} \int_0^U \frac{T_1(E)}{\sqrt{U-E}} dE. \tag{4.2}$$

μ corresponds to the reduced mass. The resulting potential $U(R)$ is plotted in Fig. 4.4b as solid squares. Theoretical potentials are treated extensively in Sect. 4.6. For comparison the results of "diatomics in molecules" (DIM) calculations for a matrix cage in the undistorted face-centered cubic (fcc) Kr solid (dashed line) and for a cage which is relaxed to the tighter minimum energy configuration around the ground-state I_2 molecule (solid line) are induced. The experimental result displays an outer limb of the potential which is intermediate between these two extremes, because the I_2 wave packet pushes the cage outward during the oscillation. Indeed, the change of the potential energy due to the induced matrix motion can be semi-quantitatively reproduced in molecular dynamics simulations [13]. From the same set of systematic measurements one can also construct the ionic E state potential by using the resonance condition of the probe pulse ($h\nu_{pr} = U_E(R_{win}) - U_B(R_{win})$). For this purpose one has to identify all wavelength combinations which probe wave packets with the energy $E_{win} = U(R_{win})$ at the outer turning point (squares in Fig. 4.4b).

Thus, for the construction of the potential, the tuning of the pump wavelength has been used mainly - the probe variation is merely used for a refinement of the vibrational frequencies $\nu(E)$ and to derive the final state (E) of the pump-probe process. The B state potentials of $I_2 : Kr$ and $Br_2 : Ar$ were constructed in this way [11, 14], using the wave packet dynamics . These observed signatures can also be used (e.g. in the following section) to separate wave packets on different Potential Energy Surfaces (PES).

4.2.2 Vibrational energy dissipation

In the regime of weak coupling with resolved ZPL and PSB the perturbative treatment delivers a consistent description of the spectroscopic features and the nonradiative dissipation of chromophore vibrational energy into matrix phonons. The extensively applied "energy gap law" [4, 5] connects the rate constant for relaxations with the coupling constant S and the energy gap i.e. the number of phonons required, to dissipate one vibrational quantum. While this perturbative treatment predicts rate constants strongly rising with v' in accordance with the experiments (Fig. 4.3) it nevertheless is inappropriate in the regime of strong coupling near the dissociation limit. The energy gap law rests on a stepwise relaxation from v' to v'-1 while the trajectory to be derived in Fig. 4.5 displays in the first cage collision a loss of kinetic energy which corresponds to many vibrational quanta of the chromophore. In addition near the dissociation limit the assumed energy separation of large vibrational quanta and small phonon energies is no longer present. In this regime molecular dynamics calculations [13] are successful. We treat now the experimental determination of energy relaxation rates in this interesting strong coupling limit.

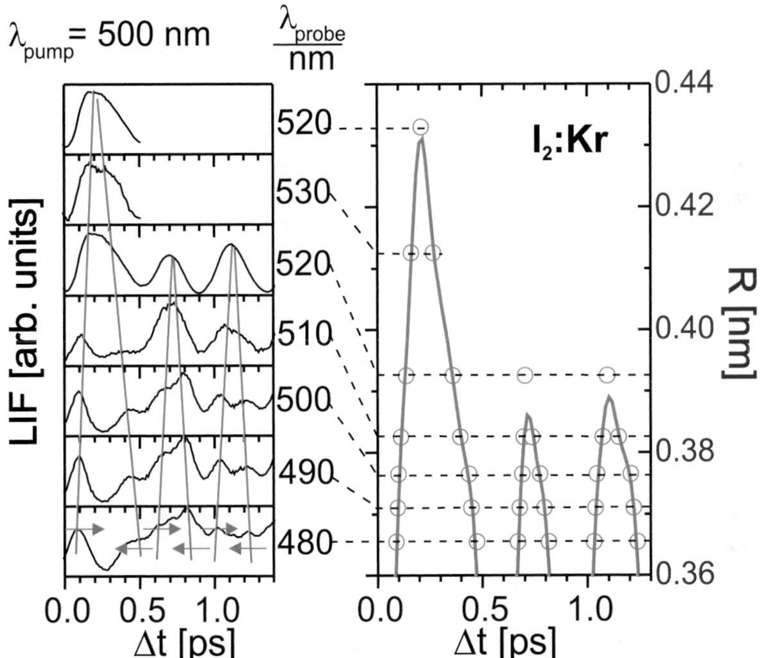

Fig. 4.5. a) Pump-probe spectra of $I_2 : Kr$ with $\lambda_{pump} = 500$ nm. Each probe wavelength λ_{probe} corresponds to a specific internuclear distance R (see text). b) Trajectory constructed from spectra in a) [15].

Concentrating on the tuning of λ_{probe} one can derive a trajectory of the $I_2 : Kr$ vibration by the following procedure. Consider the series of spectra plotted in Fig. 4.5a for fixed $\lambda_{pump} = 500$ nm, corresponding to a wave packet with energy $E = 20000$ cm^{-1}. Each of the spectra with different λ_{probe} samples the wave packet at a different internuclear distance R_{win} given by the difference potential $h\nu_{probe} = U_E(R_{win}) - U_B(R_{win})$. Therefore, the peak positions $t(\lambda_{probe})$ mark the transit times of the wave packet for a series of R_{win} and can be inverted to yield a trajectory of the wave packet $R(t)$. The result is plotted in Fig. 4.5b. We have probed only the evolution on the outer wing, which displays the interesting dynamics during the molecule-cage collisions.

The asymmetry of the trajectory for outward and inward motion demonstrates, that the molecule-cage interaction is especially strong near the outer turning point. A comparison with molecular dynamics simulations [8, 16] reveals that in the first collision energy is lost from the molecular coordinate to a breathing mode of the cage. More importantly the large elongation of the molecular bond allows four Kr atoms to close a "belt" around the molecular axis. A compression of the molecular bond now consumes energy and requires time in order to push the four Kr "belt" atoms outwards again. This process gives rise to the pronounced asymmetry of the trajectory, and the kinetic energy loss is apparent in the slope of R(t), which represents the relative velocity of the two iodine atoms. Indeed, by differentiation $\frac{d}{dt}R(t)$ =v one can calculate the velocity and thus the kinetic energy as a function of time. This method to determine the kinetic energy in the molecular coordinate during a round trip relies on data with excellent modulation in the pump-probe spectra. Apart from $I_2 : Kr$ [11] it has been applied to $Br_2 : Ar$ [14]. A more general approach to measure the dissipation of energy uses the envelope of the pump-probe spectra. For all investigated diatomics in rare gases (ClF, Cl_2, Br_2, I_2 in Ar or Kr), the envelopes show a single broad maximum in the 1 to 100 ps range, shifting in time with pump and probe wavelengths [1, 17, 18]. When the wave packet with actual energy $E(t)$ is probed at its turning point, it spends a maximum time in the window ($E(t_{\max}) = E_{win}$) and is recorded with highest sensitivity leading to the maximum at this $t_{\max}$. If it is started with λ_{probe} at or slightly below the probe window energy ($E_0 = E_{win}(\lambda_{pr})$), one only observes a decay of the intensity. Now lowering the energy of the probe window E_{win} by changing the probe wavelength to λ'_{pr}, keeping λ_{pump} fixed leads to a delay of the maximum to $t_{\max}(\lambda'_{pr})$. In this way one can determine the energy relaxation rate

$$\frac{dE}{dt} = -\frac{E_{win}(\lambda'_{probe}) - E_{win}(\lambda_{pr})}{t_{\max}(\lambda'_{probe}) - t_{\max}(\lambda_{probe})} \tag{4.3}$$

For a known electronic potential one can also use the relation

$$\frac{dE}{dt} = -\frac{E_{win}(\lambda_{probe}) - E_{pu}}{t_{\max}(\lambda_{probe})} \tag{4.4}$$

where $E_{pu} = h\nu_{pu}$ is the starting energy of the wave packet and only one probe window is needed. Repeating this comparison for various pump-probe combinations we can derive the energy dependent rates presented in Fig. 4.6. The results for four different molecules in a rare gas environment show, that the energy loss rises exponentially with the internal energy of the molecular vibration, until near the gas-phase dissociation limit a considerable fraction of the initial energy is lost within the first vibrational round trip time - typically approaching 50% or approximately 20 vibrational quanta. Clearly, this is a regime of very strong coupling of the molecule to the bath, which cannot be treated in a perturbative way. On the other hand a simple classical estimate, where the colliding atoms are taken as hard spheres with energy- and momentum conservation, already yields the right order of magnitude for the energy loss in the collision.

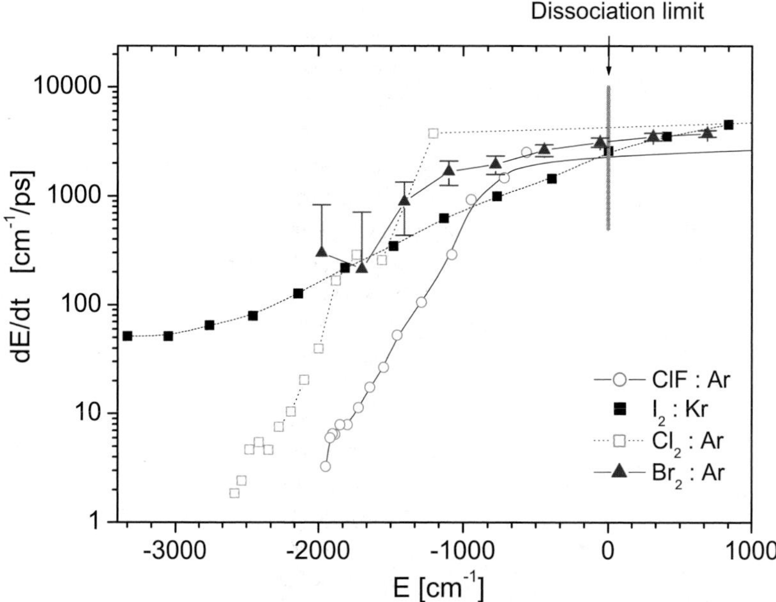

Fig. 4.6. Comparison of the energy relaxation rates dE/dt as a function of the excess energy E of the wave packet above the B state dissociation limit derived from pump-probe spectra using the "maximum criterion" [15].

This self-contained method to construct an effective one-dimensional potential and a dissipative trajectory from a systematic series of pump-probe spectra was also successfully applied to the $Br_2 : Ar$ case [14]. Although not universally applicable, it yields very useful information for the verification of molecular dynamics simulations. In addition the scheme includes the spectroscopic properties of the probing in the experimental analysis and one does not have to rely on the correct modeling of the probe-process which includes the difficult task of handling ion-pair potentials in a polarizable medium. Now one can compare experimentally derived trajectories with results from theoretical modeling [6–9, 13, 16, 19–23]. We find that simulations and experiment are in excellent agreement concerning the following central aspects: i) steepened outer wing of PES due to the cage effect, ii)vibrational periods and iii) very large energy dissipation rates.

Classical molecular dynamics simulations are the backbone of the simulations which already reproduce these features with good quantitative agreement. The vibrational coherence which is prepared by the pump pulse indeed survives even in case of very strong interactions of the molecular "chromophore" with the rare-gas "bath" Sect. 4.5.

4.2.3 Ultrafast reorientation in dissociation recombination sequences

Sect. 4.2.1 was restricted to halogens which occupy double substitutional sites in the chosen matrix in order to reduce the complexity and to suppress the possibility of a reorientation of the molecular axis with respect to the cage. In this section this additional degree of freedom is added and the tilting of the molecular axis in a sequence of dissociation and recombination events is treated. This tilt can originate from a variation of scattering angles in the recombination process. It is relevant for cage exit processes [23, 24] and it depends on the cage geometry.

Therefore, double and single substitutional site geometries of I_2 and ClF are compared in panels (a) and (b) of Fig. 4.1, respectively. Generally, the larger dihalogens I_2, Br_2, and Cl_2 occupy double substitutional sites in Ar and Kr matrices, and undergo a lattice-guided motion schematically shown in panel (a) of Fig. 4.1. The molecules vibrate rather freely along $\langle 110 \rangle$, i.e., along the bond axis direction, but rotations are hindered. While thermal rotation of ClF is suppressed by the crystal fields, the excitation to the covalent states can induce ultrafast tilting of the molecular axis by scattering off the cage atoms. The small F projectile, carrying a high kinetic energy, can scatter asymmetrically from the cage atoms leading to a new molecular orientation upon recombination. The pump-probe scheme is similar in both cases, i.e., the oscillating B-state wave packet is probed to a charge-transfer manifold, and the subsequent emission is monitored with respect to the time delay between the pulses (see Fig. 4.2).

Starting from a sample of randomly oriented molecules (a powder-like sample), a linearly polarized pump pulse selects an anisotropic ensemble of excited state molecules and dissociates them. The anisotropy is probed by a second linearly polarized pulse during the recombination process. For the initial parallel transitions with $\Delta\Omega = 0$ angular selection rule B $^3\Pi_0 \leftarrow$ X $^1\Sigma_0$, a $\cos^2\theta$ distribution of excited molecules is obtained, where θ is the angle between the directions of the linearly polarized field and the bond axis. Probing this ensemble with a parallel-polarized pulse creates the emission intensity nominated as $I_{\|}$, whereas a perpendicular probe pulse produces the $I_{\perp}$ emission with three times lower intensity for a $\Delta\Omega = 0$ transition, provided the bond direction is conserved [25, 26]. Depolarization means randomization of

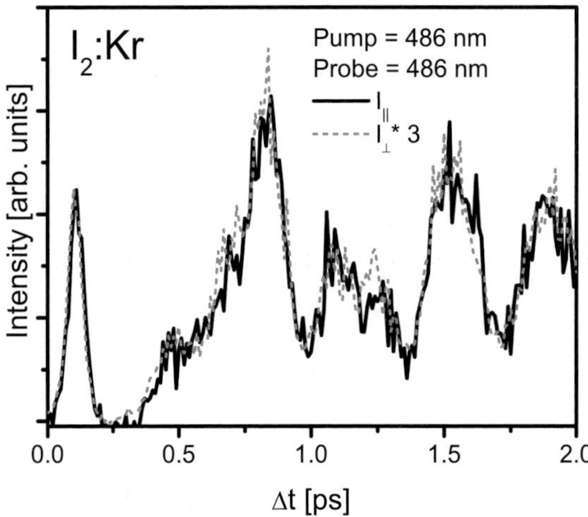

Fig. 4.7. Polarization-dependent pump-probe spectra for I_2 in Kr demonstrate negligible angular reorientation of the molecule. The spectrum for parallel pump-probe excitation has been divided by a factor of 3 and agrees perfectly with perpendicular excitation. The ratio of $I_\perp/I_\parallel = 1/3$ is maintained throughout the entire spectrum. [15]

the molecular axis direction after the pump pulse. Polarization-sensitive femtosecond pump-probe spectroscopy is exploited to measure this reorientation dynamics in [26]. Moreover, depolarization is reflected in an isotropic and nonpolarized fluorescence for excitation with the polarized beams. A characteristic "piston-in-a-cylinder" effect of the environment is reproduced in Fig. 4.7, where the double substitutional I_2 molecule exhibits perfectly fixed bond direction in Kr. The fluorescence intensities $I_\parallel$ and $I_\perp$ are plotted for pump and probe linear polarizations parallel or perpendicular to each other, respectively. The persistent 3:1 ratio of the signals confirms that the photoselection is preserved on the timescale of 2 ps. Actually, the polarization analysis showed that the ratio remains constant even on a nanosecond timescale [26]. It is important to note that this happens despite the large amount of energy deposited initially by the 486 nm pulses, i.e., by excitation significantly above the gas-phase dissociation limit of 500 nm similar to the case of ClF. All the peaks in the spectra in Fig. 4.7 are due to vibrational wave-packet motion, and the molecule can be described as a one-dimensional quantum system in a condensed environment. This is an important property for the previous sections, where vibrational coherences were discussed.

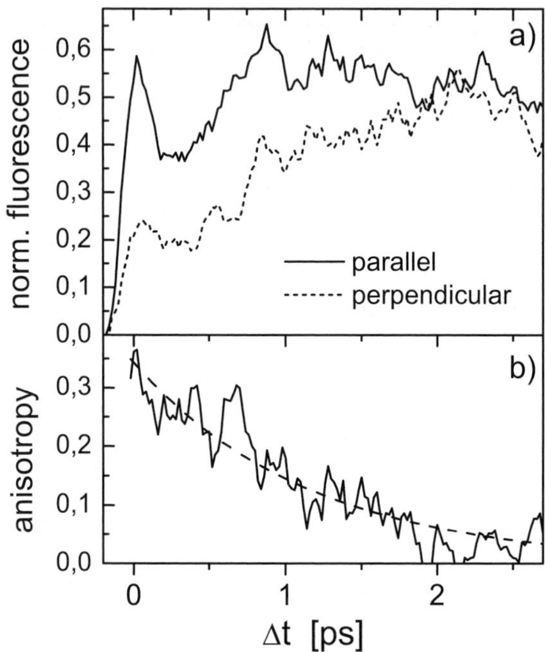

Fig. 4.8. Pump-probe spectra for *ClF* in *Ar* demonstrate ultrafast angular reorientation. a) The spectra for parallel and perpendicular pump-probe excitations converge after 2 ps but show a ratio of nearly $I_\perp/I_\parallel = 1/3$ directly after excitation. b) The anisotropy $(I_\parallel - I_\perp)/(I_\parallel + 2I_\perp)$ is shown together with an exponential decay with the time constant of 1.2 ps. The depolarization proves the ultrafast angular reorientation of the *ClF* molecule. [15]

For *ClF* in *Ar*, and for other single substitutional cases as well, the initial orientation in the cage is not as clear as it is for double substitutional cases. Computational estimates (Sect. 4.9) yield either $\langle 111 \rangle$ or $\langle 100 \rangle$ directions as the minima [17,24]. More importantly, the trapping site is nearly isotropic and the potential barriers for reorientations are relatively low and closer to rotational constants. Therefore, the depolarization measurements are of crucial importance in shedding light on the problem of preferred directionality. The depolarization can be seen in the pump-probe spectra with 387 nm (pump) and 317 nm (probe) pulses presented in Fig. 4.8. Initially, the intensity ratio is 3:1; however, the perpendicular and parallel intensities converge, indicating an ultrafast reorientation. The decay curve of the signal anisotropy yields a depolarization time constant of 1.2 ps, which is a few oscillation periods of the molecule.

The scattering character of the molecular angular redistribution is important. The anisotropy decay is much faster than the rotation period implies (32 ps for the ground state X $^1\Sigma_0$, $J = 0$). The initial anisotropy value (at t=0) of 0.35 is close to the expected one for $I_\perp/I_\parallel = 1/3$ and reduces to 0.25 during the first 0.5 ps in Fig. 4.8. Together with the trajectory calculations showing motion predominantly in $\langle 100 \rangle$ directions [23], this suggests an initial angle-conserving motion with the cage atoms tetra-atomic window acting as a guiding channel for the F fragment, followed by a later turn-over induced by the cage atoms set to motion. Apart from the first peak of the out-stretching wave packet, the weak signal-oscillation features exhibit rather noncoherent vibrational dynamics of the confined molecule due to the inhomogeneous scattering events.

The possibility to manipulate the naturally occurring alignment of a single substitutional molecule with respect to the lattice axes has been examined theoretically by applying an intense linearly polarized laser field, nonresonant to molecular transitions [27–29]. This field aligns the molecule with respect to the polarization direction by virtue of the interaction with the molecular anisotropic polarizability (see Sect. 4.9). These studies served to provide means for controlling the subsequent photodynamics explained above. In particular, the aim was to force the molecule axis toward holes or walls of the surrounding cage thus affecting the dissociation/recombination yields and suppressing/enhancing the depolarization. For example, forcing the ClF bond along the $\langle 111 \rangle$ direction enhances the possibility of direct dissociation [24]. Depending on the relative directions of the alignment field and the crystallographic axes, cooperative and competitive effects occur. In the former case, the laser field polarization is set to coincide with a minimum energy orientation of the molecule in the matrix, and high alignment degree can be achieved even for low field strengths. Otherwise, high efficiency of this mechanism is restricted to high fields and low temperatures [27, 28]. The laser intensity needed to overcome the inherent obstacle of angular potential energy barrier in a $ClF : Ar$ system already approaches the damage threshold of the crystal, which is 10^{12} W cm^{-2} for 150 fs pulses at 775 nm [23]. This is in contrast to gas phase molecules, since there the scheme is limited only by the onset of molecular ionization.

4.2.4 Nonradiative transitions and ultrafast spin-flip

Ultrafast nonradiative transitions are predicted theoretically and are treated in Sects. 4.8 and 4.10. The experiments here aim in following these processes and ClF is selected as an appropriate molecule. F_2 would be even closer to most of the calculations, it has, however, unfavorable spectroscopic properties. Nonradiative transitions are induced in these ultra fast pump-probe experiments by exciting the initial state selectively with a short pump pulse thus generating a propagating wave packet. The feeding dynamics from the excited state into the target state is monitored by a time delayed probe pulse

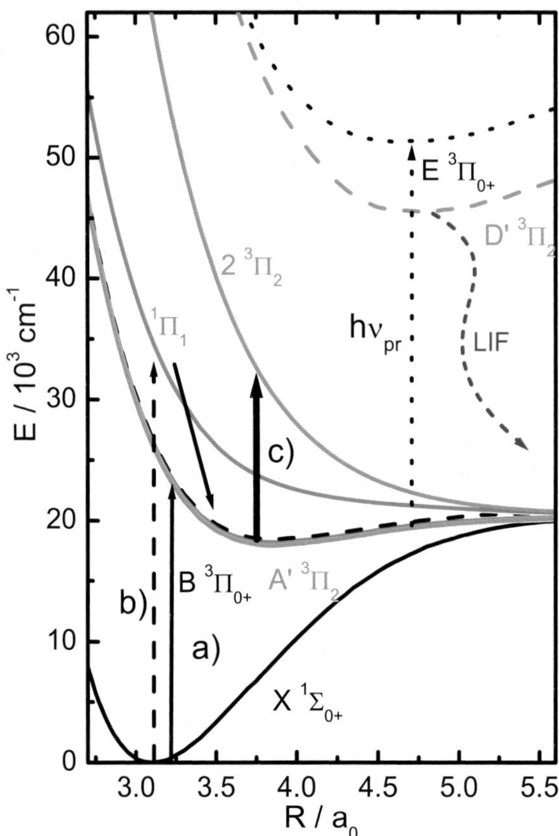

Fig. 4.9. Potential energy surfaces of ClF in the gas phase and solvated D' state. Transitions a), b) and c) are described in the text and in Fig. 4.10 [30].

which carries the wave packet in the target state at a well defined internuclear distance to the charge transfer states of ClF and by finally recording the fluorescence intensity from these charge transfer states versus the time delay. The potential energy surfaces of ClF in the gas phase are depicted in Fig. 4.9, the charge transfer states are red-shifted with respect to the gas phase due to the influence of the matrix. The target states are the bound triplet states which are in energetic order $\mathrm{A}'(^3\Pi_2)$, $\mathrm{A}\ (^3\Pi_1)$ and $\mathrm{B}(^3\Pi_0)$. The internuclear distance R of the probe window position is determined by the difference potential $\Delta\mathrm{V(R)}$ and the probe photon energy $\mathrm{h}\nu_{pr}$ via $\Delta\mathrm{V(R)} = \mathrm{h}\nu_{pr}$. For the $\mathrm{B}(^3\Pi_0)$ target state a strong probe transition to the $\mathrm{E}(^3\Pi_0)$ charge transfer state is observed as expected from the dipole selection rules.

A frequency doubled probe pulse corresponds to a probe window position R_p around 4.7 a_0 according to Fig. 4.9 assuming that the repulsive contributions from the cage wall (which are not included in Fig. 4.9) are similar for the states involved [18, 31].

All ${}^3\Pi_0 \leftrightarrow X({}^1\Sigma_0)$ transitions violate the spin selection rule and in addition, the $A'({}^3\Pi_2) \leftrightarrow X({}^1\Sigma_0)$ transition violates the Ω selection rule, having $\Delta\Omega$=2. Therefore, for all interhalogens a trend of increasing radiative lifetimes corresponding to a decrease in transition moments from the μs range for the B state to ms for the A' state is well known. Population of the B state of ClF leads to a radiative decay with a lifetime of 141 ms exclusively from the A' state [32] indicating nonradiative relaxation from B to A' within the B state radiative lifetime. The B, A, A' states are nested with small energy separations. In the charge transfer state similar close lying bunches of ${}^3\Pi$ states with $\Delta\Omega$ = 0, 1 and 2 are expected. Therefore, the probe transition shown in Fig. 4.9 may not only lift wave packets from the B state but energetically resonant ones also from the A and A' states to the charge transfer states.

Next consider the selection of initial states. The strongest absorption of ClF corresponds to the dipole allowed $X({}^1\Sigma_0) \rightarrow C({}^1\Pi_1)$ transition with its Franck-Condon maximum around 280 nm. The $X({}^1\Sigma_0) \rightarrow B({}^3\Pi_0)$ transition has its Franck-Condon maximum around 400 nm and it is the strongest one in this spectral region. Thus by spectral selection using the frequency doubled wavelength of 387 nm from the Ti:Sa fundamental an exclusive initial population of the B state was achieved in [18, 31]. The resulting pump-probe spectrum serves as a reference for the wave packet dynamics in the bound ${}^3\Pi$ manifold for initially exciting its B state component. It is reproduced in Fig. 4.10a for parallel polarization of pump and probe [31]. The $C({}^1\Pi_1)$ state is selected with a pump wavelength of 280 nm from a frequency doubled NOPA. The resulting pump-probe spectrum reflects now the dynamics of the spin-flip from ${}^1\Pi$ to ${}^3\Pi$ and is shown in Fig. 4.10b once more for a parallel polarization of pump and probe [31].

To choose another initial state among the large manifold of repulsive states, a steady state population in the A' state is generated which is vibrationally relaxed and isotropic according to the fast depolarization of the internuclear axis after excitation [26]. Delayed excitation of this transient population with a wavelength of 774 nm allows now to pick out a new repulsive state in the Franck-Condon range of the A' state. Since one starts from a triplet state another triplet state will be selected. From energy resonance and selection rule considerations a preferential population of the $2^3\Pi_2$ state is expected as indicated in Fig. 4.9. The resulting pump-probe spectrum is displayed in Fig. 4.10c.

The spectrum of Fig. 4.10a corresponds to a direct excitation of the $B({}^3\Pi_0)$ target state, and it serves as a reference of the detection sensitivity. The arrows indicate the passage times and directions of the wave packet in the probe window (Fig. 4.9) derived empirically in [31] from plausibility arguments and the observed structures. The molecular dynamics simulations (see Sect. 4.8)

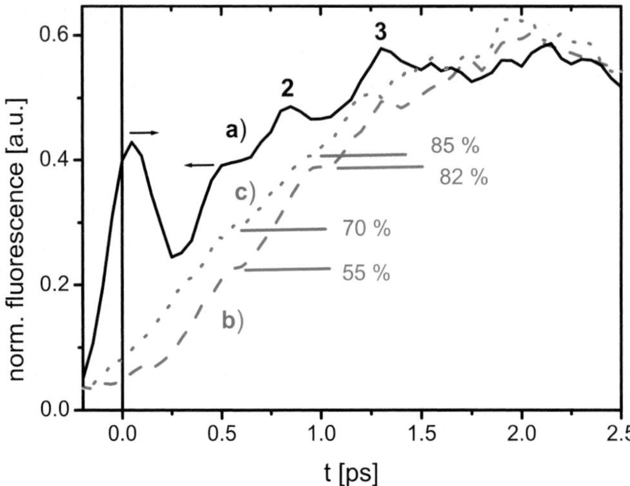

Fig. 4.10. Fluorescence intensity of *ClF* from the D' state by probing $B^3\Pi$ population with 318nm after a time delay t. a) pumping the $B^2\Pi$ state directly with 387nm, b) recording spin-flip by pumping the $^1\Pi_1$ state with 280nm c) recording $2^3\Pi$ relaxation by pumping A' population (from 280nm) with an additional pulse (774nm) to the $2^3\Pi$ state.

in the B-state of *ClF* in an *Ar* matrix neglecting nonradiative electronic transitions [24] are consistent with this interpretation and thus confirm the approximate position of the probe window internuclear distance. Based on the simulation in [24] the structures in Fig. 4.10a can be explained in the following way: The wave packet is excited in the $B(^3\Pi_0)$ state above the gas phase dissociation limit and passes in the first 100 fs the probe window leading to the strong first maximum designated by the rightward arrow. The *ClF* molecule is oriented predominantly toward the center of a fourfold window in the *Ar* cage (Fig. 1 in [24]). The wave packet has sufficient kinetic energy to cross the window, and it pushes the cage atoms outwards. The wave packet is reflected in a head-on collision from an *Ar* atom behind the window, crosses on its way back once more the cage window and is detected after about 500 fs in the probe window (leftwards arrow). The expected strong maximum is suppressed by the induced expansion of the cage window which reduces the solvation energy of the *ClF* charge transfer state and thus decreases the detection efficiency for the probe wavelength of 318 nm by shifting the probe window away from the wave packet. A shorter probe wavelength weakens the influence of the solvation energy and indeed for a probe wavelength of 302 nm the minimum in Fig. 4.10a around 300 fs is flattened out [31]. The collision with the cage is connected with a large energy loss and the following wave packet oscillations take place within the cage. The wave packet is recorded close to the turning point leading to the maxima indicated by 2 and 3. A slight increase in intensity is observed up to 2 ps and attributed to vibrational energy relaxation in the ClF intramolecular potential. In this way the kinetic energy during passage of the probe window decreases and the detection sensitivity increases [31]. Finally the central energy of the wave packet falls below the probe window energy leading to the decrease in intensity later on.

Here, the emphasis is on the nonradiative transition form the $^1\Pi_1$ and 2 $^3\Pi_2$ states toward the target state and this short summary of the B $^3\Pi_0$ dynamics is only required to relate the nonradiative transitions to the internal clock of the oscillating wave packet in the cage potential. Excitation of the $^1\Pi_1$ and the 2 $^3\Pi_2$ states leads to a population transfer to the target state within 1.5 ps according to Figs. 4.10b and c. The curves a to c are normalized to equal intensity for 2 ps and later on all three curves show an identical decay within the noise limit. This behavior indicates, that a complete population transfer takes place in the rising part. The rising parts are rather similar in Fig. 4.10b and c and even the same (soft) steps appear in both spectra. The only apparent difference is a systematically faster rise for excitation of the 2 $^3\Pi_2$ state.

This general observation is already a major result with respect to the simulations for the F_2 molecule in the cage in Sect. 4.8. It confirms the very fast and efficient nonradiative transitions from singlet to triplet states and also among the families of singlet and triplet states.

In a more detailed interpretation one can relate the soft steps with the wave packet round trip history. The first passage through the window (rightwards arrow) with the strong maximum for B $^3\Pi_0$ excitation leads only to a weak rise for $^1\Pi_1$ and 2 $^3\Pi_2$ excitation, which is of the order of 10 %. This difference to the B $^3\Pi_0$ excitation indicates that the probe window is indeed most sensitive for population in the target state and that only a weak population transfer occurs from the inner turning point up to the probe window position. After the interaction with the cage and on the passage of the window on the way back (leftward arrow) a steep rise in the target state population up to 55 % and 70 % for the $^1\Pi_1$ and 2 $^3\Pi_2$ state respectively (compared to the B $^3\Pi_0$ excitation) is observed. This delayed rise displays indeed the wave packet dynamics and goes beyond a kinetic “rate constant” description. Obviously the combined effect of improving energy resonance among the bound and repulsive states and increasing energy loss in the cage interaction funnel the wave packet toward the target state at large internuclear separation. The following rather flat region correlates with the well separated states and weak cage interaction during the inner turn. However, a second steep increase to 82 % and 85 % for $^1\Pi_1$ and 2 $^3\Pi_2$ state excitation is detected when the wave packet returns from a second collision with the cage (maximum 2). The population transfer is completed within the sensitivity of the experiment after the third cage interaction (maximum 3). Turning back to the comparison with the F_2 simulation of Sect. 4.8, one finds in the more detailed inspection once more a qualitative agreement in the sense that the transitions are governed by the condition of energy resonance combined with cage interaction. The transition can be related to the internal wave packet clock because here a difference in time scales occurs for ClF experiment and F_2 simulation. The F_2 simulated dynamics is faster due to scattering restricted within the first shell of cage atoms. Therefore, a comparison with oscillation cycles seems to be more appropriate. The simulations also contain oscillatory components in the population transfer. They do not yet show up in the experiment, which displays a stepwise rise and it is an interesting question where this difference originates from. Another aspect is transfer during the first elongation. The experiment shows some transfer up to the first probe window passage but less than predicted in the F_2 simulation. It is still open if in the ClF and F_2 comparison the discrepancy displays a specific property of the two systems or if a more distant probe window in the experiment would be the appropriate choice for ClF.

4.2.5 Curve crossings and matrix induced predissociation

The nonradiative transitions among nested potential energy surfaces have already been treated in the previous Sect. 4.2.4. Here, curve crossing and predissociation in the context of different environments with different symmetries is discussed, focusing on the predissociation of the B $^3\Pi$ state of I_2 via states, that cross the B state in its potential minimum (see potential energy sketch in Fig. 4.11a).

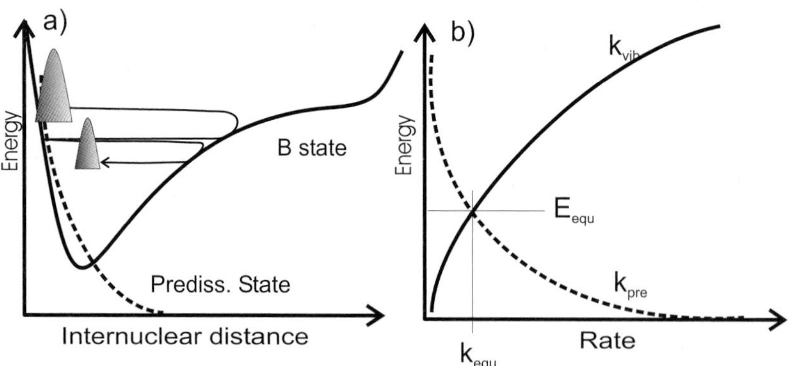

Fig. 4.11. Predissociation of the B state of I_2 in solid Kr. a) The potentials of the electronic B state (solid) and the predissociative state (dashed) cross near the minimum of the B state. A population in the B state relaxes from the originally prepared vibrational levels to lower ones and gets closer to the curve crossing. Thereby, predissociation gets more effective. b) Rate picture of the process: the vibrational relaxation with rate k_{vib} funnels the population to the curve crossing. The predissociation with rate k_{pre} is getting more effective closer to the crossing. Most of the population gets lost when the population has reached E_{equ}, the vibrational energy at which the rates are equal $k_{pre} = k_{vib} = k_{equ}$.

The predissociation of the B state is induced by the environment. Zewail and coworkers investigated the I_2 B state predissociation in rare gases from the dilute gas phase up to densities of the liquid phase [33–35] and they derived a unifying relation for a predissociation rate which was confirmed for all four rare gases [33]. It predicts a linear increase of the predissociation rate in the B state with buffer gas density. Rates of about 1 ps^{-1} are obtained for Ar at the supercritical density and somewhat larger rates are predicted for Kr. The trend of this model is confirmed by experiments on I_2 in liquid solvents n-hexane [36] and CCl_4 [8]. Extrapolating the density dependence of the predissociation rate to the rare gas solids leads to a very short lifetime of the B state, even shorter than the subpicosecond lifetime in liquids. In stark contrast, the observed B state lifetime of I_2 in Kr and Ar matrices is much longer than expected [7, 8].

The I_2 chromophore sits in a highly symmetric cage in rare gas solids in contrast to the disordered high pressure buffer gases and the liquids. Model calculations predict a cancelation effect in the summation of the angular terms of the nonradiative transition matrix elements due to the higher symmetry of the crystalline surrounding. Treating caging and predissociation in van der Waals complexes Roncero, Halberstadt, and Beswick [37, 38] derived the angular dependence of the electrostatic coupling between the B state and the crossing a$^3\Pi_g$ state used in [8]. While coupling to other crossing states may be present it is this B-a coupling which seems to dominate. Batista and Coker use a semiempirical DIM Hamiltonian and propagate trajectories on 23 nonadiabatically coupled states for liquid Xe [39] and also for solid Ar and Xe [22]. A similar Hamiltonian was applied to I_2-Ar by Buchachenko and Stepanev [40]. These simulations lead to a strong coupling of the B state to a variety of crossing states and to very rapid predissociation in accordance with the liquid phase results. However, they do not reproduce the significantly slower predissociation in the solid phase. Therefore, the surface hopping approach (see Sect. 4.8.1) was reconsidered and improved in [13]. Now, the calculated B state pump-probe signal indicated a slower decay in solid Ar compared to liquid Xe.

The symmetry dependence of the coupling between the B state and the crossing predissociative state was investigated experimentally in [41] by two different strategies: a temperature change of the Kr matrix induces a local asymmetry and a co-doping of the krypton with argon deforms the cage around the I_2 molecule. Most important, there is a strong influence of the local symmetry on the vibrational energy relaxation rate of the population in the B state. The vibrational energy relaxation follows an exponential form (sketched as the dashed line in Fig. 4.11b) that has been discussed in Sect. 4.2.2. A temperature variation from 10 to 40 K increases the vibrational energy relaxation rate by a factor of two. The co-doping with 20 % argon increases the rate by a factor of five. The predissociation probability is given as a function of energy E above the state crossing by the Landau-Zener formula: $p_{LZ} = 1 - \exp(-\frac{A}{\sqrt{E}})$, A being proportional to the coupling element between the two states [12]. Therefore, the corresponding predissociation rate k_{pre} increases with decreasing energy of the B state population. The energy of this population, however, is a dynamic quantity due to energy relaxation with k_{vib}.

The predissociated population via pump-probe spectroscopy on the lower lying A state of I_2 was observed. The predissociation rate increased with Ar co-doping concentration and increasing temperature. The observed predissociation rate increased from 0.06 ps^{-1} at 10 K to 0.11 ps^{-1} at 40 K, similar to the behavior of the vibrational relaxation rate. This is in accordance with a rate picture (Fig. 4.11b), which predicts, that the observed predissociation rate should lie close to the crossing point of energy relaxation curve k_{vib} and predissociation rate curve k_{pre}, which is calculated according to Landau-Zener. Thus, the observed predissociation rate is attributed to k_{equ} in Fig. 4.11b with the population having a vibrational energy of E_{equ} above the crossing [41]. With rising temperature and co-doping, k_{vib} rises and shifts k_{equ} upwards, when k_{pre} is kept constant. We could essentially follow the observed predissociation rate with this model, without changing the coupling matrix element of the crossing states. Furthermore, the coupling element was with 15 cm^{-1} significantly lower than in former studies [7] reports 65 cm^{-1}). Thus, the cage symmetry effect on predissociation was mainly attributed to its influence on the vibrational energy relaxation.

4.2.6 Cage exit dynamics

Permanent molecular dissociation is the route for chemical reactivity in solids, either directly via complexation with surrounding rare gas (RG) atoms or via migration of the photoproducts, and is in this sense an ultimate event in various photoprocesses. Following an electronic excitation to a dissociative state, the molecule expands as in the gas phase until the fragments experience the repulsion of the lattice potential. The chance for dissociation is governed by the possibility of the fragment atoms to cross the crystal-field barrier and to penetrate into voids outside the initial trapping site of the molecule. Quite clearly, the structure of the trapping site is of fundamental importance in determining the outcome of the photophysical experiment. While the researcher can exert control optically in choosing the wavelengths, intensities, polarizations, and temporal properties of the laser pulses, it depends on the mechanical constraints whether a sudden, transient, or delayed cage exit can occur. The orientation of the molecule and its translational confinement with respect to the surrounding cage is the predetermining factor for the initial photodynamics. Indeed, one challenge for theory to reproduce experimental observations lies in a proper description of the potential energy surfaces (see Sect. 4.6) in the many-body system, which to a large extent dictate the dynamics of the wave packets created [23, 24]. A second hotspot for theory (see Sect. 4.8) lies in treating the quantum nature of the wave packets and the nonadiabatic dynamics in particular [42].

Important trapping sites for molecules appear as single, double, or multi - substitution vacancies, i.e., one, two, or more lattice atom positions are occupied by the impurity molecule. The atomic (van der Waals) radii of the molecular constituents and the lattice constants of the commonly used Ar, Kr, and Xe solid hosts guide the prediction of the vacancy size needed for accommodating the impurity. Obviously, matching the vacancy size with the molecular van der Waals radii gives the preferred site that will be occupied by the molecule. Double substitution, as mentioned already, is the prevailing condition in the following cases: $Cl_2 : Ar$, $Br_2 : (Ar, Kr)$, and $I_2 : Kr$ [43]. Upon molecular dissociation, those head-on atoms force the molecular fragments to recombine under an extensive loss of kinetic energy, which is distributed among the cage atoms. The perfect caging has been overcome only by an extensive irradiation with photon energies beyond the covalent manifold of states and cage exit is attributed to destruction of the local cage for systems such as $Cl_2 : Ar$ [44–46] and $I_2 : Ar$ [47,48]. As long as the dissociation channels are those introduced by electronic excitations within the covalent cage-bound states, large molecules such as Br_2 and I_2 exhibit no bleaching (dissociation) of the pump-probe signal.

The probability for cage exit is rather high for molecules which are small enough to fit in a single-substitutional trapping site of the rare gas solid [43]. The atomic structure of the cage can provide accessible windows for a sudden exit of a fragment. This is the case for ClF molecules in Ar and Kr hosts [32], as well as for $Cl_2 : Xe$ [49,50] and F_2:(Ar, Kr) [51–53]. The trapping-site symmetry resembles that of an octahedron, with some distortion caused by the molecule – lattice interaction. Consequently, the pathways for dissociation are very different depending on how the molecule is oriented relative to the cage at the moment of photoexcitation. The rotation of the confined ground-state molecule is hindered to a variable extent depending on the temperature. It is bound to a small-amplitude libration about preferred crystallographic axes as long as the potential well exceeds in magnitude the thermal energy. Subsequent to a dissociative electronic excitation of the molecule, a collision of the fragments with cage atoms in the $\langle 110 \rangle$ direction (first shell) leads to vigorous angular scattering. Instead, only moderate tilting of the molecular axis is accompanied with a penetration through the tetra-atomic window and a collision with the second shell in the $\langle 100 \rangle$ direction. According to the $ClF : RG$ simulations [23, 24], the cage structure suppresses the direct dissociation in these directions (see Sect. 4.9). In the $\langle 111 \rangle$ direction, the fragment encounters a triatomic window of cage atoms, which is in a way to an interstitial octahedral site. Together with the atomic size and the excess kinetic energy, the ability of the fragments to penetrate the windows depends on the electronic state of the molecule. This has been studied in atomic fragment calculations [43] and nonadiabatic dynamics simulations for F_2 molecule [21, 54, 55](see Sect. 4.8).

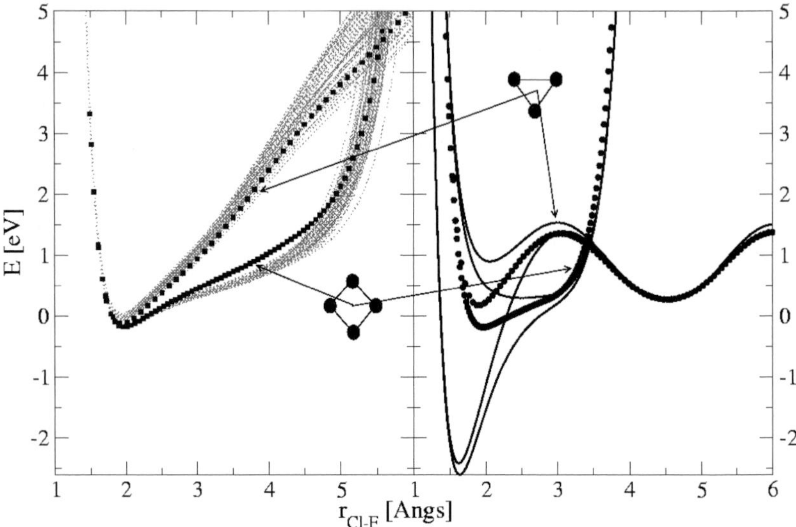

Fig. 4.12. DIM results for potential energy cuts demonstrates the dependence on orientation with respect to the cage. Left: B $^3\Pi_0$ state potential energy barriers for ClF bond stretch into the two Ar window directions. The curves with square symbols represent a fixed lattice and the swarms of dotted lines are for relaxed lattices at 5 K. Right: The two sets of lines plot the potential energy with solid lines for singlet states X $^1\Sigma_0$ and $^1\Pi_1$ and with circles for the B $^3\Pi_0$ when only the F atoms moves. Adapted from Fig. 5 in [24].

Figure 4.12 presents the difference in potential energy barriers between orientations toward tetra-atomic and triatomic windows for the stretching ClF molecule in Ar. The molecular center of mass is fixed at the substitutional site in the left panel of Fig. 4.12, whereas the Cl fragment is fixed and only F moves through the windows in the right-hand panel. The curves demonstrate the significantly different response of the solvation shells to the dissociative motion of the molecule with different orientation. Here, the interaction energies are obtained by a reduced DIM scheme in a molecular dynamics simulation core [24] (see Sect. 4.6). Most distinctly, the tetra-atomic window direction $\langle 100 \rangle$ is seen to allow for a longer range vibrational oscillation r_{ClF} in the B $^3\Pi_0$ state. Geminate recombination in $\langle 100 \rangle$ is the main outcome in the simulations for the dissociative excitations to B $^3\Pi_0$ and $^1\Pi_1$ states. On the other hand, the triatomic window direction $\langle 111 \rangle$ is seen to support a trapping site for the F fragment behind the cage wall at $r_{ClF} \approx 4.5$ Å; however, the penetration is inhibited by the height of the barrier.

Experimental investigations on photobleaching manifest the permanent dissociation of the ClF molecule in Ar and Kr solids [32]. This is illustrated in Fig. 4.13, where the A′ →X emission signal from remaining ClF molecules decays nearly exponentially to a static background for constant radiation flux

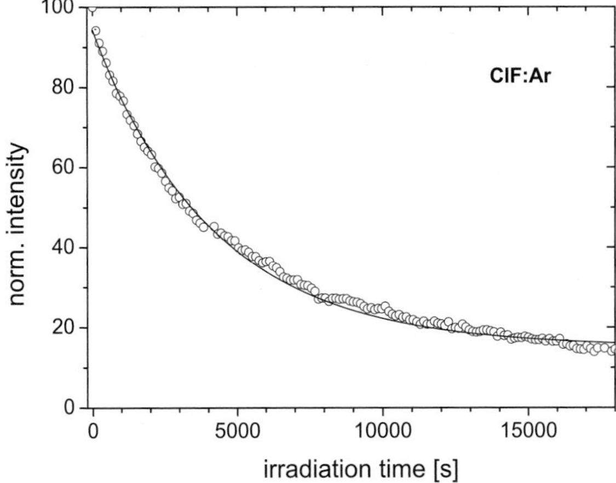

Fig. 4.13. Decay of the A$'$ →X emission intensity demonstrates the permanent dissociation of ClF in Ar. The signal is obtained by monitoring the laser-induced fluorescence bands during irradiation at 308 nm that excites into the repulsive $^1\Pi_1$ state. In a Kr matrix, the dissociation cross section is an order of magnitude larger. The quantum efficiencies are 5% in Ar and 50% in Kr [32].

at 308 nm. The slope in the exponential decay delivers a larger cross section for dissociation in Kr which exceeds that in Ar by a factor of ten [32]. These static measurements serve to prove in addition the mobility of the F fragment, since the background under the exponential decay in Fig. 4.13 increases with concentration and originates from a secondary formation of ClF molecules. These are photoproducts of species formed by migrating F atoms. In Kr, the fragments have a higher mobility and can form KrF complexes.

The dissociation of ClF in Kr is accompanied by an accumulation of $Kr_2^+F^-$ emission, when the laser beam is spatially overlapping with an additional 270 nm beam. The dissociation product KrF is excited to Kr^+F^- by the 270 nm light, which leads to formation of a $Kr_2^+F^-$ complex and local rearrangements in the matrix. These processes prove convenient for the ultrafast spectroscopy on this system as shall be presented below. In addition, the photochemical equilibrium of the system can be manipulated by controlling the dissociation versus the geminate recombination processes [17]. As the radiating $Kr_2^+F^-$ terminates on a repulsive part of the KrF potential, it dissociates and ClF can form again. Thereby, the F atom can be made to shuttle between the ClF and KrF configurations by a proper combination of two laser pulses. Using a 387 nm pulse operative for ClF dissociation via the $^3\Pi_0$ state and the 270 nm pulse for both the dissociation (via $^1\Pi_1$) and the probing of the KrF complex, the concentration of F fragments created was

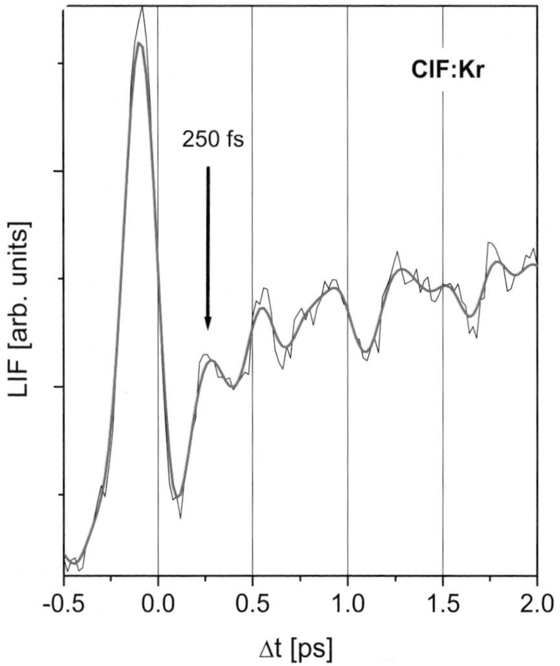

Fig. 4.14. Pump-probe spectrum demonstrates the ultrafast cage exit of F in Kr. The laser-induced fluorescence signal is emitted by the $Kr_2^+F^-$ exciplex formed upon irradiation of a $ClF:Kr$ sample at subsequent 387 nm (pump) and 270 nm (probe) pulses. After $\Delta t = 250$ fs, the first neutral F atoms have arrived at the nearest interstitial site and can be probed by $KrF \rightarrow Kr^+F^-$ excitation at 270 nm. Further modulation has been assigned to oscillatory lattice dynamics and delayed cage exits as the parent molecule continues to vibrate highly excited in the $^3\Pi$ states [17, 56]. Migration of F atoms also contributes to the rising signal. Upon inverting the pulse sequence, the barrier crossing proceeds via ionic Cl^+F^- pathway by absorbing at 270 nm to $^1\Pi$ prior to two-photonic resonance of the 387 nm pulse. The spectrum is recorded with lock-in technique at 4 K [56].

observed (emission from $Kr_2^+F^-$) to depend reversibly on time delays between the pulses. The probe pulse shifted the system toward ClF and dissociation toward KrF.

In the present pump-probe spectroscopy studies the ClF molecule was excited 0.6 eV above the $^3\Pi_0$ state dissociation limit by the pump pulse at 387 nm. Instead of probing the evolving wave packet on this potential, a 270 nm pulse was applied to excite the KrF complex formed when the F fragment has escaped the cage. The detected signal originates from the $Kr_2^+F^-$ emission at 460 nm. This scheme [56] enables a direct observation of the dissociation dynamics on the femtosecond timescale. The result is presented in Fig. 4.14, where the rising signal at $t > 0$ is assigned to the complexation mentioned above. The indicated peak at 250 fs corresponds to sudden cage exit of the F atom. Further modulation on top of the rising background indicates coherent motion in the system. Based on the observed timescales, the structures can be ascribed partially to dynamics of the cage atoms that are set to oscillation during the first $Cl - F$ bond elongation [17]. Delayed cage exits at subsequent elongations are more likely to contribute to the average signal rise, due to scatter-events that broaden the time resolution. An ionic pathway to $Kr_2^+F^-$ is observed for $t < 0$. In this case the 270 nm pulse arrives first and acts as the pump and excites ClF to its ${}^1\Pi_1$ state. A two-photon resonance with the second 387 nm pulse prepares a wave packet in the Cl^+F^- states. This wave packet promotes an F^- ion to escape the cage and $Kr_2^+F^-$ is formed subsequently by structural rearrangements.

A probable reason for the different ClF dissociation efficiencies in Ar and Kr can be the different ground-state alignment with respect to the cage. This was found in molecular dynamics simulations based on additive pair potentials within the DIM scheme [23, 24]. While in Ar, the preferred orientation appeared along $\langle 100 \rangle$ directions with a 100 cm^{-1} barrier to $\langle 110 \rangle$, the minima were along $\langle 110 \rangle$ in Kr. Although this qualitative difference was found between the two solids, the computation fails to reproduce the dissociation yields quantitatively. On the contrary, the alignment along nearest-neighbor directions in Kr prevented direct cage exits. The conflict reflects the sensitivity on the pair potentials utilized, as the resulting small, order of 10^2 cm^{-1}, barriers between the minima have drastic effects depending on the energetic order of the crystallographic directions.

4.3 Extraction and application of intramolecular coherences

M. Bargheer, M. Fushitani, M. Gühr, H. Ibrahim, and N. Schwentner

The coherent signature of the vibrational wave packet in pump-probe spectra can be used to determine vibrational relaxation, potentials and trajectories of the molecule (see Sect. 4.2). It was shown, that the vibrational coherence even survives strong energy relaxation and nonadiabatic transitions between different electronic states. The vibrational wave packet is formed of several *coherently* super imposed vibrational eigenstates. The intramolecular vibrational coherence decays due to the interaction with the environment and this process is called vibrational decoherence or irreversible vibrational dephasing. The phase in the coupling of two different electronic surfaces which is governed by the phase in the electric field of the coupling light pulse can be disturbed in an analogous way by the environment and one has to deal also with irreversible electronic dephasing. The vibrational wave packets in an excited electronic state are generated by an electronic transition. Therefore, in Sect. 4.3.1 electronic dephasing is investigated by applying phase-locked pulse sequences. In an extension to a three pulse experiment Sect. 4.3.2 demonstrates how this method can be exploited to control the chromophore - lattice coupling introduced in Sect. 4.2. Next methods are presented in Sect. 4.3.3 to determine the vibrational decoherence time, using pump-probe spectroscopy on anharmonic potentials. Further a new scheme is introduced based on chirped pump pulses [57]. Finally, it will be shown in Sect. 4.3.4 that in the dissociative regime vibrational coherence can survive the large energy losses and a transfer of coherence to lower lying eigenstates is discussed.

4.3.1 Phase-locked pulse sequences and electronic coherence

Experiments with femtosecond phase-locked pulse pairs (PLPP) provide a direct and flexible way to investigate electronic coherence. A PLPP can be generated by splitting a femtosecond pulse with a Michelson interferometer where the end mirror of one arm is controlled by a micrometer step motor to change the time delay $\Delta\tau$, between the PLPP while the other arm has a piezo steering mirror to adjust the relative phase ϕ, between the PLPP.

The first application of the PLPP experiment to a molecular system was carried out by Scherer et al. in 1990 [59] to study decoherence of rovibronic transitions of gaseous I_2 molecules [36, 60]. Later, other isolated systems [61–67] as well as systems in condensed phase [68–89] were studied. In molecular wave packet interferometry, the coherence in the electric field of the first pulse is imprinted to a molecular wave packet on the excited electronic state. A second phase-locked pulse creates another wave packet on the excited electronic state after a time delay $\Delta\tau$ (Fig. 4.15). The two wave packets can interfere on the excited state. If the first wave packet is located in the Franck-Condon range with the ground state, the interference leads to a change of the overall excited state population which one can record in a variation of the fluorescence intensity. Since the wave packet created by the first pulse leaves the original Franck-Condon region and evolves back and forth in the upper potential surface, the second pulse can modulate the population only when the wave packet returns to the original Franck-Condon region. Furthermore, if the electronic coherence of the ground and excited state is lost in the delay $\Delta\tau$, the interference phenomena are averaged out. Through the interference, population is transferred to the excited state in a constructive ($\phi = 0$) or destructive ($\phi = \pi$) way, according to the relative phase ϕ between the PLPP. The interference effect can be seen as vibrational recurrences in an interferogram if molecules preserve the electronic coherence. The decrease in modulation contrast of the recurrence provides the electronic decoherence time, in case the vibrational coherence lives much longer (which is often the case).

Fig. 4.16 shows a measured interferogram for Cl_2 in Ar by recording the intensity of the A'→X fluorescence for the accumulated population versus the

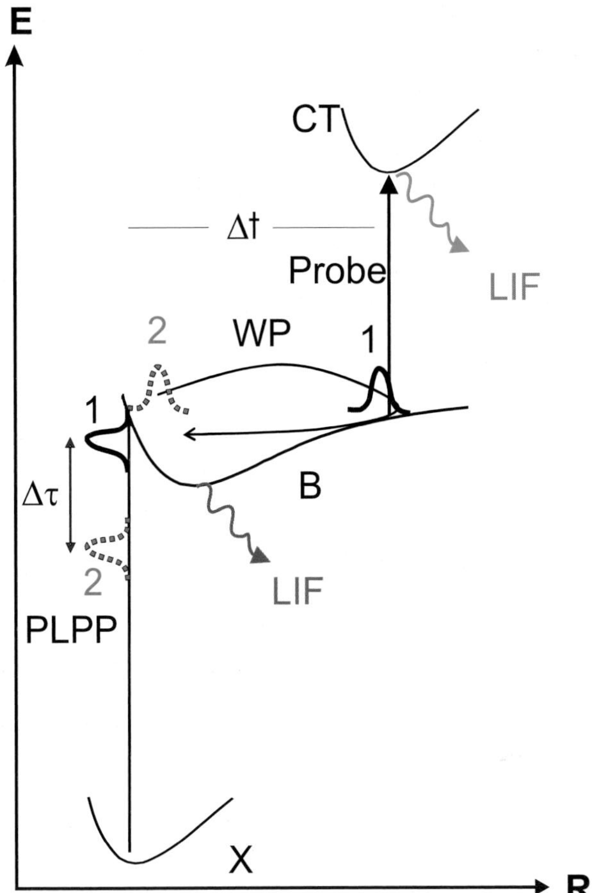

Fig. 4.15. Scheme for wave packet preparation (WP 1 and 2) on the B-state by phase-locked pulses 1 and 2 (PLPP), separated by the delay $\Delta\tau$. Laser induced fluorescence (LIF) is observed from the covalent B state and the lower lying A' state (not shown). Using a third pulse with delay Δt for probing one detects LIF from charge transfer (CT) states.

time difference $\Delta\tau$ of the two pulses. The open circles result from a tuning of the interferometer and thus of the relative phase ϕ between the electric field of the two pulses at every time step $\Delta\tau$. The envelope through the maximal values displays constructive interference i.e. ($\phi = 0$) and that through the minima a destructive one with ($\phi = \pi$). The pump pulse centered around 512 nm covers several vibrational levels around v'= 12 of the Cl_2 B state. The first recurrence at $\Delta\tau = 260 fs$ represents the vibrational round-trip time in this range [58].

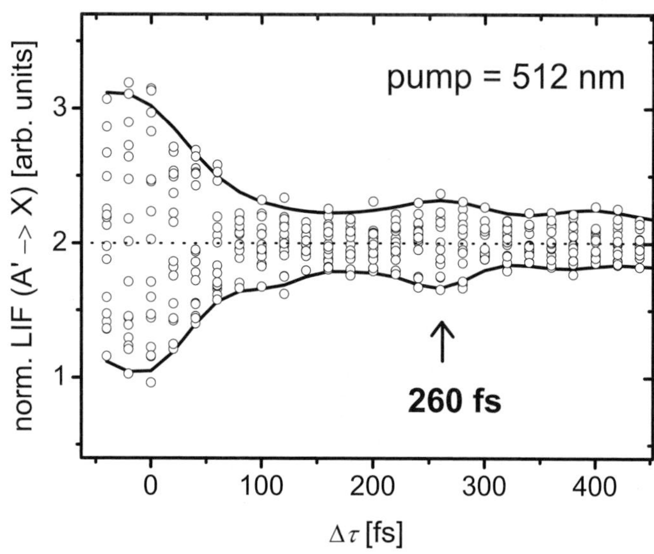

Fig. 4.16. LIF interferogram of Cl_2 in solid Ar. The A'→X fluorescence was measured for 16 different phases Φ at every delay $\Delta\tau$ between a PLPP at 512 nm(open circles). The solid curves show the maximum and minimum envelopes and the vibrational recurrence is seen at 260 fs. (Adapted from [58])

A similar experiment was carried out for $Br_2 : Ar$ with two laser pulses at 590 nm which couple coherently B state vibrational levels around v'= 8 according to the spectrum in Fig. 4.3. To investigate the dependence on phase ϕ in more detail the time delay $\Delta\tau$ between the two pulses has been fixed and the relative phase ϕ has been varied in small steps covering 8π. The expected sinusoidal modulation is maximal for $\Delta\tau = T_{vib}$ (Fig. 4.17), where the two wave packets meet each other at the inner turning point and can interfere optimally on the molecule, either constructively (maximum) or destructively (minimum). We observe a significant modulation contrast up to the sixth roundtrip time $\Delta\tau = 6T_{vib}$, however, with decreasing amplitude. For longer time delays the modulation lies within the noise level. Fig. 4.17 shows that this modulation is really caused by the interference. No modulation is observed for $\Delta\tau = 1.5T_{vib}$, where the two created wave packets are at the inner and outer turning points, respectively, which inhibits interference. Irregularities in the modulation period are caused by nonlinearities of the piezo drive, which controls the relative phase ϕ.

The modulation contrast in the B state fluorescence like in Fig. 4.17 for seven different time delays $\Delta\tau$ is collected in Fig. 4.18. The large drop from

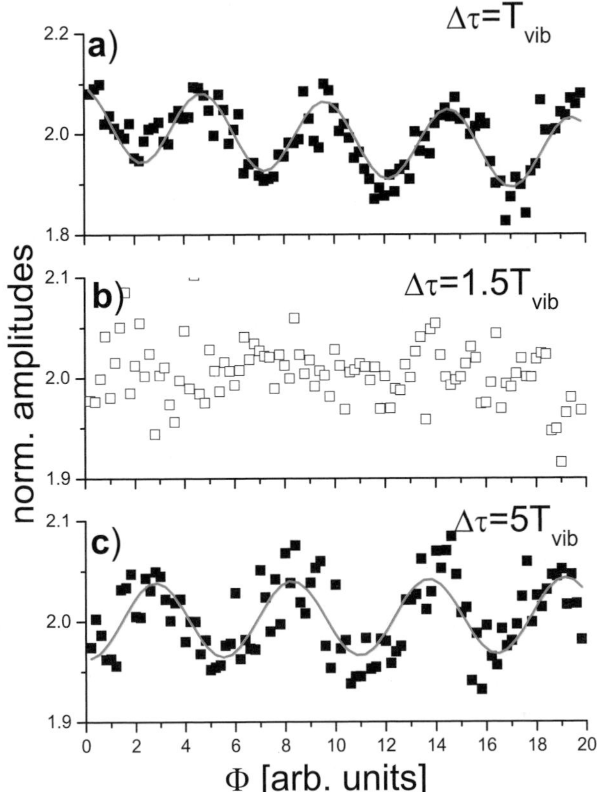

Fig. 4.17. Fluorescence intensity traces of Br_2 : Ar with fixed time delay $\Delta\tau$ versus phase Φ. We detect a modulation contrast in a) of 0.06 for $\Delta\tau = T_{vib}$. b): $\Delta\tau = 1.5T_{vib}$ the signal scatters statistically. The two wave packets are separated at the inner and outer turning point. c) When $\Delta\tau$ matches multiples of T_{vib} (here $5T_{vib}$) the modulation is visible again.

$\Delta\tau = 0$ to $\Delta\tau = T_{vib}$ is caused by the following fact: in case $\Delta\tau = 0$ one observes optical interference of light with a contrast of 0.9 representing the Michelson interferometer quality. This contrast is strongly reduced in the molecule (from $\Delta\tau = T_{vib}$ on and higher) for several reasons. Not all incoming light is absorbed by the B state. Those parts absorbed by the A state continuum do not contribute to the interference contrast. Furthermore, population that leaves the B state via predissociation to a repulsive state or suffers energy relaxation does not come back to the inner turning point to interfere with the second wave packet.

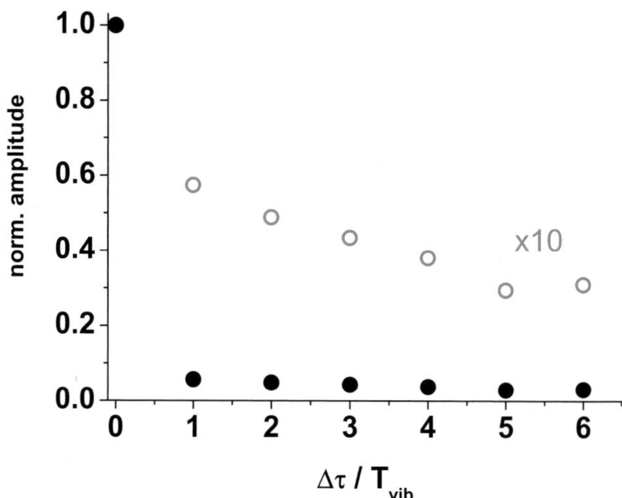

Fig. 4.18. Modulation contrasts of phase dependent signals ($Br_2 : Ar$ e.g. Fig. 4.17) taken at different T_{vib}. The amplitude is normalized to the one at $\Delta\tau$=0 (solid circles) and by a factor of ten magnified (open circles).

In nanosecond excitation spectra one observes strong features of population transfer from B state to the lower lying A' state due to predissociation via the repulsive $^1\Pi$ state. On the A' state the B state vibrational progression is seen at energies where the A' state contribution itself delivers a continuum. Since this B state fraction is in the range of 10-20% and the noise level in absorption around 8%, a PLPP dependence is not observed in the detection region of A' state's fluorescence.

The PLPP experiment strongly depends on the electronic coherence between the two electronic states since interference can only be observed if coherence is maintained. From vibrational wave packet revivals and focusing experiments in the following sections the vibrational coherence time of Br_2 in solid Ar is determined to be 3 ps [57]. The faster decaying modulation contrast in the PLPP spectrum of the B state indicates, therefore, a loss of electronic coherence in the range of 1.4 ps for Br_2 in Ar solid. For $Cl_2 : Ar$ [58] the electronic coherence was observed for more than 660 fs, while the vibrational coherence lasts for more than 3 ps.

4.3.2 Control of chromophore-lattice coupling by phase-locked pulse sequences

The superposition of the phase-locked pulses can be used to prepare vibrational wave packets which are coupled to the lattice with different strength. We explain this in the spectral domain which is more intuitive. Furthermore, by applying a third pulse, a probe pulse, which excites the wave packet resulting from the interference to charge transfer states (see Fig. 4.15), after a delay of Δt allows to record the difference in time evolution for the two types of wave packets. To illustrate the control of lattice coupling one should refer to experiments for $Cl_2 : Ar$ [58]. The center wavelength of the phase-locked

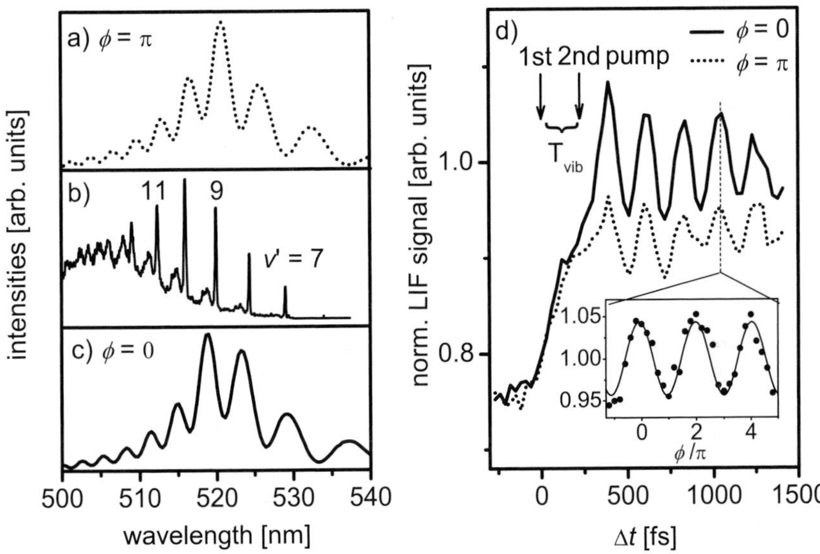

Fig. 4.19. a) Spectral fringes of a PLPP at λ =521 nm with $\Delta\tau = T_{vib} = 220$ fs and $\phi = \pi$ locked at 519nm. b) Excitation spectrum of Cl_2 in solid Ar. [3] c)Spectral fringe as in a) but with $\phi = 0$ d)Pump-probe spectra of Cl_2 in solid Ar for the PLPP with $\phi = 0$ (solid) and π (dotted). The inset shows the phase dependence of the LIF signal at the fifth oscillation maximum [15].

pulse pair (PLPP) is now tuned to 521 nm and the broad spectral distribution covers vibrational levels of v'= 7 – 13 in Fig. 4.19b. Spectral fringes of the phase locked pulses with a delay T_{vib} and the phase $\phi = 0$ or $\phi = \pi$ with respect to the wavelength at 519 nm are shown in Figs. 4.19a and c, respectively. The fringes can be recorded since the pulse length is elongated in a monochromator, even if the pulses are well separated in time at the sample position. Since the fringe spacing $\Delta\nu$ is given as $\Delta\nu = 1/\Delta\tau$, the Fourier transformation of the phase locked pulses with $\Delta\tau = T_{vib}$ has a fringe spacing equal to the vibrational splitting in the molecule. The relative phase of the PLPP moves the position of the fringes without changing the relative spacing, thus the phase can be used to excite predominantly ZPLs or PSBs (see Figs. 4.19a-c and Sect. 4.2). Besides the delay $\Delta\tau$ between the PLPP, a delay Δt between the first pulse of the PLPP and the third probe pulse has to be introduced. Figure 4.19d shows transient pump-probe spectra as a function of Δt for fixed $\Delta\tau = T_{vib}$ and the phase $\phi = 0$ (solid) or $\phi = \pi$ (dotted). When the second pulse interacts with the system, the interference is either constructive ($\phi = 0$) or destructive ($\phi = \pi$). Since position and spacing of the spectral fringes are adjustable via ϕ, one can manipulate the ratio of ZPL and PSB contributions. For instance, the PLPP in Fig. 4.19a and c excites predominantly ZPLs (PSBs) and prepares vibrational wave packets in a cold (hot) environment. As a result, such wave packets lead to a quantitative difference in vibrational energy relaxation. The different decay of the pump-probe spectra in Fig. 4.19d manifests such a wave packet dynamics coupled differently to the lattice; the averaged vibrational wave packet intensity including lattice oscillations (solid curve) decays after 0.5 ps due to the enhanced vibrational relaxation, while the wave packet with less phonon excitations (dotted curve) shows an almost constant mean value [58]. Such a control of phonon coupling by tuning the phase is general and can be applied to various phonon induced phenomena.

4.3.3 Coherence properties from wave packet focusing and revival control

The pump-probe method is predominantly sensitive to vibrational coherence, since population on an excited state is probed [2, 90] and here the focus will be on vibrational decoherence on the B state. A recent paper by Apkarian and coworkers [91] examines decoherence on the B state of I_2 in solid Ar alternatively using CARS. This method has been established on the ground state of I_2 in solid Ar before [2,48,92–94]. The ground state vibrational decoherence times can also be examined by Resonant Impulsive Raman Scattering (RISRS), as demonstrated for I_3^- in liquids [95–97]. Often, the loss of modulation in pump-probe spectra is dubbed dephasing, because the phases of the contributing wavefunctions do not add anymore constructively. Here it will be shown that two processes of different origin have to be distinguished which lead either to a reversible phase slip, i.e. dispersion or an irreversible phase scrambling, i.e. decoherence. Dispersion of wave packets is clearly distinct from processes where the phase information is lost irreversibly by decoherence. Dispersion of wave packets and optical pulses is analogous since it only reflects the different group-velocities of certain frequency components. Vibrational wave packets in harmonic potentials keep their shape after each vibrational round

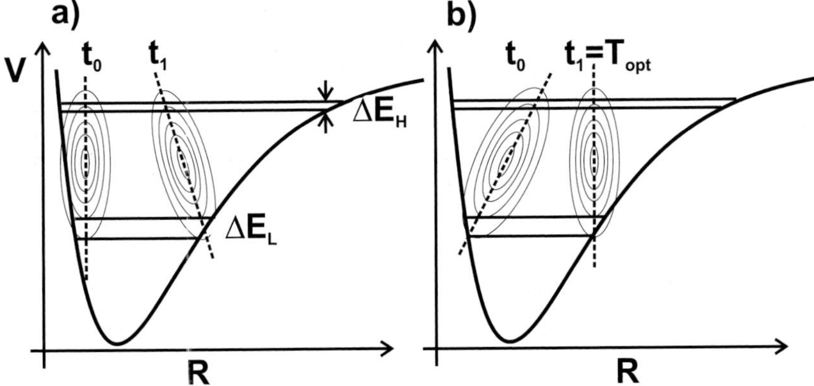

Fig. 4.20. Dispersion of a wave packet in an anharmonic oscillator. a) The wave packet is excited at t_0 by an unchirped laser pulse. The high vibrational energy (blue) and low vibrational energy (red) parts are therefore excited at the same time. The vibrational spacing in the high-energy range ΔE_H is smaller than the vibrational spacing in the low-energy range ΔE_L. The succeeding oscillation times T_H are therefore longer than T_L, and the low energy parts of the wave packet advance the high energy parts after some oscillations at a time $t_1 > t_0$. This is called wave packet dispersion. The wave packet is always plotted when moving from left to right. b) The dispersion can be suppressed by starting the "slow" blue components earlier than the "fast" red ones [98], as indicated by the dashed line at t_0. At a time T_{opt}, the red ones will have caught up with the blue wave packet components [15].

trip [99,100]. The pump-probe spectrum of such a wave packet shows the fundamental oscillation frequency (when probed at a turning point) [101]. If the wave packet suffers decoherence, the modulation will decay with the vibrational decoherence time $T_{\text{dec}}^{\text{vib}}$, and an unstructured transient will develop. As already stated in Sect. 4.2, the covalent states of the halogens like any other molecular potential are far from being harmonic. They can be approximated to a high accuracy by anharmonic Morse potentials and (4.1) gives the frequency of classical trajectories running on such potentials. Due to the change in oscillation frequency with excitation energy, the wave packets disperse, as demonstrated in Fig. 4.20a. Initially, all trajectories forming the "classical wave packet" are started at the same time t_0 by a Fourier transform limited laser pulse. The trajectories in the ensemble having a high vibrational energy E ("blue" ones) have a lower oscillation frequency than the low energy ("red") parts. Therefore, the trajectories excited in the red reach a specific internuclear separation at earlier times than the trajectories excited in the blue part of the oscillator. After a few oscillations the wave packet has "tilted" in the (R, V)-representation, as seen in Fig. 4.20a at t_1. What is the signature of this dispersion in pump-probe spectra? The modulation with the fundamental period decays after the dispersion time T_{disp}, and this decay is at first instance indistinguishable from vibrational decoherence. The full modulation contrast however revives after a time $T_{\mathrm{rev}} > T_{\mathrm{disp}}$, as the correct quantum calculations show. Two limiting cases can be distinguished:

- The decoherence time $T_{\mathrm{dec}}^{\mathrm{vib}}$ is on the same order as T_{disp}. No revivals can be observed in pump-probe spectra. In this case, one has to disentangle dispersion and decoherence by specially shaped wave packets.
- The decoherence time $T_{\mathrm{dec}}^{\mathrm{vib}}$ is longer than T_{disp} and in the range of T_{rev}. Thus, revivals show up and allow for determination of the decoherence time from the pump-probe modulation contrast.

The dispersion time is crucially depending on the anharmonicity $\omega_e x_e$ of the molecular potential energy surface:

$$T_{\mathrm{disp}} = \frac{\nu}{c\omega_e x_e \Delta E}. \qquad (4.5)$$

ΔE is the energetic width of the pump-pulse, and ν the central vibrational frequency of the wave packet. The light velocity c is introduced, because all numbers shall be given in units of wave numbers. $\omega_e x_e$ of the B states decreases from the lighter to the heavier halogens by one order of magnitude. Therefore, the dispersion times T_{disp} vary also by one order of magnitude from 400 fs (ClF) to 4 ps (I_2). Due to the variation in T_{disp}, we can test the two limiting decoherence cases given above.

In the first case ($T_{\mathrm{dec}}^{\mathrm{vib}} \approx T_{\mathrm{disp}}$), one needs to suppress dispersion, in order to find the vibrational decoherence time $T_{\mathrm{dec}}^{\mathrm{vib}}$. The mechanism for dispersion compensation is explained in Fig. 4.20b. The "slow" blue components are started earlier than the "fast" red components, as indicated by the tilted line in Fig. 4.20b. This can be accomplished by a negatively chirped laser pulse, in which the blue components of the pulse arrive earlier on the molecule than the red components [10, 57, 98, 102, 103]. The red components catch up with the blue ones after the focusing time:

$$T_{\mathrm{opt}} = -\frac{\beta' \nu^2}{4\pi\omega_e x_e}, \qquad (4.6)$$

where $\beta' = \frac{\beta(\nu)}{c}$ is the chirp parameter in units of fscm. The effect of focusing is demonstrated by a simulation for free Br_2 in Fig. 4.21b. If a wave packet is excited by an unchirped laser pulse, the modulation contrast in the pump-probe spectrum decays (solid line). When excited by a negatively chirped pulse of the same bandwidth, the modulation contrast in the transient spectrum (dashed line) is low in the beginning, since the pulse was artificially stretched. However, the vibrational levels composing the wave packet have the right phase relation to form a narrow wave packet at T_{opt}, according to (4.6). The experiment was performed with the same parameters as the simulation for Br_2 in solid Ar. The RKR potential of Br_2 in solid Ar was constructed and the anharmonicity is similar to the gas phase value in this

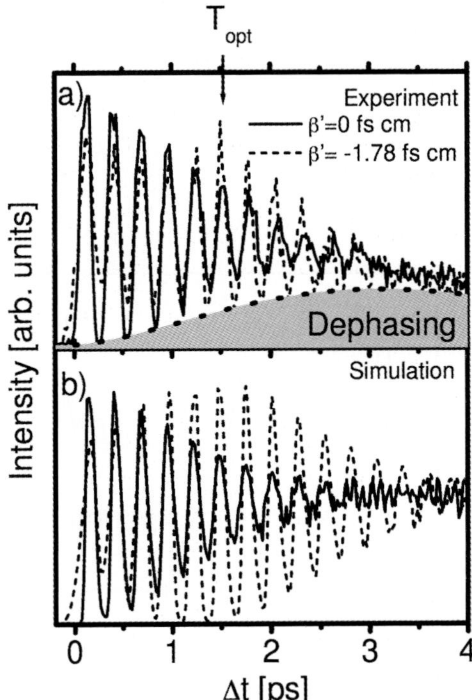

Fig. 4.21. a) Pump-probe spectra of Br_2 in Ar for chirps $\beta' = 0$ (solid) and -1.78 fs cm (dashed) excited at $\lambda_{\text{pump}} = 567$ nm and probed at 600 nm. The dashed gray shaded line gives the experimentally determined vibrational dephasing (decoherence) background. b) Simulations using the same laser parameters for a free Br_2 molecule [15].

range [14,104]. Besides the effect of dispersion and energy relaxation, also the decoherence by the matrix contributes to the shape of the transient spectrum. The unchirped excitation delivers a spectrum with decaying modulation contrast. With the negatively chirped excitation pulse, a maximum in contrast is observed at T_{opt}, in agreement with the simulation. However, the background at T_{opt} is not completely suppressed, contrary to the free Br_2 simulation. The background is due to population, that has undergone decoherence and thus cannot contribute to the focusing, for which phase memory is relevant [57]. The focusing time was systematically changed by varying chirp parameters. The shaded background could not be suppressed by focusing and thus its rise of 3 ps gives the vibrational decoherence time $T_{\text{dec}}^{\text{vib}}$. These are the first experiments on the $Br_2 : Ar$ system up to now, and one can compare the result with the electronic ground state vibrational decoherence of $I_2 : RGS$. In Kr and Ar, the decoherence time increases with quantum number v, however with different analytic behavior [94]. For $I_2 : Kr$, the decoherence time lies in the range from 100 to 33 ps [94], for $I_2 : Ar$ between 10 and 2.5 ps [93]. Taking the increased interaction with the environment in excited electronic states into account, the time constant for vibrational decoherence of 3 ps is consistent.

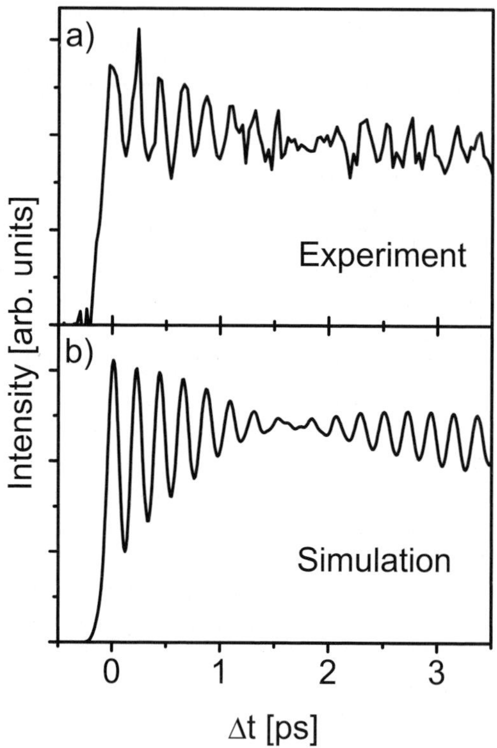

Fig. 4.22. a) Measured pump-probe spectrum for Cl_2 in solid Ar. The wave packet was excited with $\lambda_{\text{pump}} = 520$ nm and probed to the charge transfer states. After the decay of the fundamental period, it revives at 3.2 ps. This is the half revival showing the fundamental period. b) The spectrum is reproduced by a simulation. The modulation structure is damped with an exponential having a 3 ps time constant, reflecting the vibrational dephasing $T_{\text{rev}} = 2\pi/(\omega_e x_e)$. [15]

The influence of collisions on vibrational decoherence of I_2 in a high density rare gas was studied in a series of experiments by the Zewail group [33–35, 105–108] motivating theoretical studies by V. Engel, C. Meier and coworkers [19, 20, 109–111]. The authors determine vibrational energy relaxation times T_1 and vibrational dephasing times T_2 versus rare gas pressure. The decrease of T_2 (corresponds to $T_{\mathrm{dec}}^{\mathrm{vib}}$ in the nomenclature used here) with rising pressure is complex. Two processes induce a decoherence of the vibrational levels: collisions and vibration-rotation coupling. The free I_2 molecular rotation is not blocked in contrast to the model system described here. In the binary collision model, T_2 scales linearly with the collision time [112, 113], whereas the trend is reversed for the vibration-rotation dephasing [114]. The vibrational dephasing time scales from infinity at 0 bar rare gas pressure to 1 ps at 2 kbar pressure (examples for He as buffer gas).

The solid Ar environment used in the present experiments has a number density of 27 nm^{-3}. Extrapolating Zewail's results linearly to the solid Ar density, yields a time constant shorter than 250 fs [35]. This value is one order of magnitude shorter than the observed $T_{\mathrm{dec}}^{\mathrm{vib}} = 3$ ps for $Br_2 : Ar$. The stabilizing effect of a highly symmetric environment on coherence will also show up in the context of coherence transfer in the next section. Once more, the well defined symmetry of the RG host proves to preserve vibrational coherences for unexpected long times, when comparing to disordered environments.

Now, the second case ($T_{\mathrm{dec}}^{\mathrm{vib}} > T_{\mathrm{disp}}$) is discussed. If $T_{\mathrm{dec}}^{\mathrm{vib}}$ is on the order of the revival time $T_{\mathrm{rev}} = 2\pi/(\omega_e x_e)$, revivals can be recorded. The revivals can be described in a quantum treatment of vibrational wave packets [99, 101, 115–121]. An energy splitting of δE between two vibrational levels results in a wave packet oscillation time proportional to $1/\delta E$. In harmonic potentials, the splitting is constant, however in anharmonic ones, the splitting changes with the vibrational quantum number. For three subsequent level of a Morse oscillator two different oscillation periods T_A and T_B are active. This leads first to dispersion and after a number of n round trips to a revival according to $(n+1)T_A = nT_B = T_{\mathrm{rev}}$. Fig. 4.22 shows experimental and simulated spectra for Cl_2 with a revival structure. The revival time T_{rev} of Cl_2 should be 6.5 ps. Besides the full revival, so called fractional revivals exist, one of them the half revival shows up at $T_{\mathrm{rev}}/2$. The half revival appears with the fundamental frequency and full modulation contrast in the pump-probe spectrum. For Cl_2, the half revival is located at 3.2 ps, where we observe a recovery of the fundamental frequency in the experimental spectrum shown in Fig. 4.22a and the simulation shown in Fig. 4.22b. The modulation contrast in the simulation is in addition damped exponentially with a 3 ps time constant to reproduce the experiment. This decay reflects the vibrational decoherence of the fundamental with $T_{\mathrm{dec}}^{\mathrm{vib}} = 3$ ps as found for the case of Br_2 in solid Ar. At 1.6 ps, the so called 1/4 revival should appear with the doubled vibrational frequency. However, this short vibrational period is smeared out in the experiment due to the employed time resolution in agreement with the simulation. The focusing scheme and the scheme used here for determining the vibrational decoherence are sensitive on the fundamental period. Therefore, they characterize the coherence time of two subsequent vibrational levels, giving rise to the fundamental frequency.

The revival structure can be used to deduce even more detailed information on the vibrational coherence of a distinct group of vibrational levels. For example, the 1/4 revival at $T_{\rm rev}/4$ contains the doubled fundamental vibrational frequency. However, this revival can only appear, if three vibrational levels are still coupled coherently. The 1/6 revival at $T_{\rm rev}/6$ shows the tripled frequency and it requires four coherently coupled levels. The argument can be extended in an obvious way to even more levels [57]. Therefore, the observation of a distinct revival allows to deduce the multilevel coherence time. The scheme was demonstrated for the 1/6 revival of Br_2 in solid Ar. However, a serious problem had to be solved: The 1/6 revival appears at $T_{\rm rev}/6 = 3.5$ ps in Br_2. The two level vibrational coherence survives only for 3 ps, as stated above, thus a coherence of more levels cannot be observed afterwards. Therefore, a coherent control scheme using positively chirped pump-pulses was invented in order to shift the revival structure [57] to earlier times. With this scheme, the 1/6 revival was shifted to 1 ps. The observed features were analyzed with quantum simulations yielding a dephasing time of 1.2 ps [57].

In future experiments, several revivals should be recorded allowing for the determination of the n-level coherence. With a systematic approach, this would allow to distinguish between Poissonian and Gaussian dephasing as discussed in [122].

Focusing and fractional revivals of vibrational wave packets can be controlled by appropriate chirped pulses. Both phenomena are a consequence of dispersion and indicate maintenance of vibrational coherence during the evolution of the wave packet. However, in addition the electronic coherence is also required during preparation of the vibrational wave packets, since it is a prerequisite to control the initial phase distribution among the vibrational eigenfunctions [57]. Therefore, focusing and revivals of vibrational wave packets display electronic decoherence as well. The conditions that electronic coherence has to be preserved at least during the excitation pulse duration for a full imprint of the phase information allows to estimate a lower limit for the electronic coherence time. For example, the control of vibrational wave packets of Br_2 in Ar is achieved by chirped pulses whose duration corresponds to 300 fs. Therefore, the coherent wave packet evolution on the B state demonstrates that the electronic coherence between B and X states of Br_2 lasts at least 300 fs even in the presence of Ar atoms.

4.3.4 Coherence in the dissociative regime - the transition state analogue

We will now discuss in more detail, the evolution of molecular vibrational wave packets in the presence of weak and very strong dissipation. For the weak coupling case the well established concept of perturbations of the vibrational levels of the chromophore by fluctuations in the bath modes can be applied. The resulting fluctuations of the potential energy and equilibrium coordinate lead to dephasing of vibrational coherences with time constant T_2

and to vibrational energy relaxation with T_1, respectively. T_2 is expected to be faster than the time T_1 for depopulation of vibrational levels [123–127]. In the related energy gap picture [128] the population in vibrational level v relaxes stepwise to the next lower vibrational energy level $v-1$ (Sect.4.2). The concept of coherence transfer among vibrational levels in the master-equation description of anharmonic oscillators coupled to a harmonic bath is often applied to describe such a slow flowing down of population. The assumed functional form of the coupling determines, which vibrational levels are coupled. The dephasing and population relaxation of vibrational levels deep in a potential well are suitably treated in this way. The relaxation rates are slow in this region and are well in accordance with a perturbative treatment which leads to the energy gap law [128] and predicts an increase of the relaxation rate with the vibrational quantum number [4].

The trajectory constructed in Fig. 4.5 as well as the energy relaxation rates plotted in Fig. 4.6 show that one is able to observe also processes in the very strong coupling limit. Near the gas phase dissociation limit, more than 20 quanta of the molecular vibration can be lost within one vibrational period. This implies that the molecule loses much more energy than the spacing of vibrational levels, rendering this concept of stationary states invalid. The lifetime is much shorter than the round trip time. The opposite would be required to establish a "level" according to Heller's approach connecting high wavelength resolution spectroscopy with the time dependent wave packet dynamics [129–131]. The assumption of a stepwise relaxation process is not applicable anymore.

Instead, one can consider the hard collision of the molecule with the matrix cage in a short interaction time as an impact which creates coherences and populations in a manifold of previously unexcited molecular levels, as it could be achieved by an ultrashort infrared light pulse. For the preparation of a coherent superposition the scattering process must be shorter than the round-trip time of the populated levels. For the hard impact near the dissociation limit this prerequisite is indeed fulfilled. The reason why the creation of coherences by scattering has not been recognized more frequently is, that the timing and directionality of the scattering events is often not sufficiently well defined in the investigated ensembles. For a similar situation for example of diatomics in liquids, both the timing and the scattering angle can be distributed over a broad range due to the structural disorder and energetic fluctuations. For diatomics in solid rare gases the low temperature is of great advantage, because the low initial velocities of all cage atoms imply that the collision time is determined by the well triggered photoinduced chromophore motion. The weak modulation contrast of pump-probe spectra in the case of $ClF:Ar$, demonstrates the importance of restricted scattering angles. ClF takes a single substitutional site in the Ar matrix, and thus different angles are possible. In contrast, those molecules that occupy double substitutional sites ($I_2:Kr$, $Br_2:Ar$, $Cl_2:Ar$), and therefore have to move in a fixed direction like a piston, show pronounced wave packet oscillations. Again, the classical picture of scattering hard spheres is sufficient to point out the different amount of potential energy remaining in the molecular bond for different scattering angles. We can conclude that the conservation of vibrational coherence in the presence of dissipation depends crucially on order in the molecules environment. [13, 21].

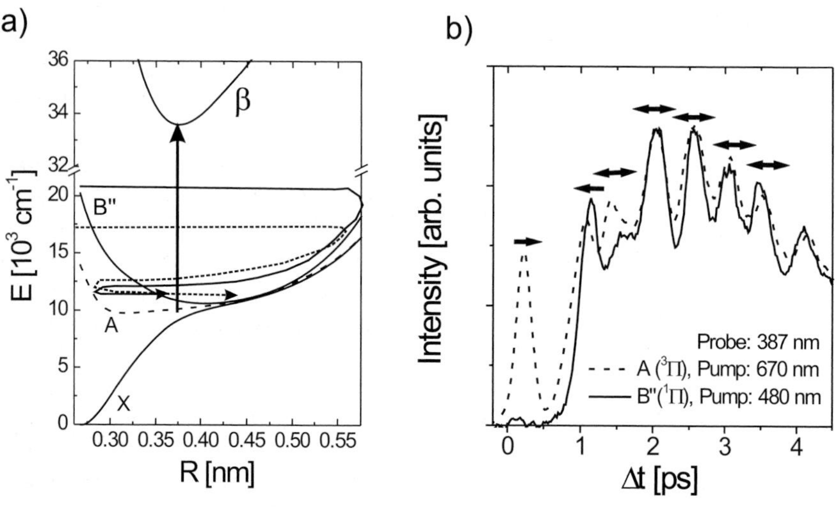

Fig. 4.23. a) Electronic potential energy scheme of $I_2 : Kr$. b) Pump-probe spectra of the A state upon A (dashed) and B" (solid) excitation. The B" wave packet undergoes an electronic transition and is probed in the A state (see text) [15].

Finally, the question is addressed, if vibrational coherence even survives a nonadiabatic transition between electronic states in combination with energy loss. A sketch of the process is displayed in the potential diagram of Fig. 4.23a. We compare pump-probe spectra for excitation to the repulsive B$''$ state (solid line) and the weakly bound A state (dashed line), respectively, while probing the population in the electronic A state with a pulse at $\lambda_{probe} = 387$ nm inducing a transition to β. The dashed line in Fig. 4.23b shows the spectrum for direct excitation to the A state at $\lambda_{pump} = 670$ nm and the solid line is for excitation to the repulsive B$''$ state ($\lambda_{pump} = 480$ nm). The spectra coincide after 2 ps, despite the fact that the excitation energies differ by 1.5 eV. The prominent first peak at 100 fs is missing in the solid line, because the wave packet is still in the B$''$ state and therefore not probed by the window in the A state. Here one can use the spectroscopic properties of the probe to distinguish wave packets on different PES. The positions of the two peaks near 1 ps still differ in the two spectra, displaying the different excitation conditions. After 2 ps both excitation pathways result in a similar coherent vibrational wave packet in the electronic A state, approx. 1000 cm^{-1} above the minimum of the potential, according to the probe wavelength of 387 nm.

The experimental results prove that the vibrational coherence is conserved in nonadiabatic transitions, even in the "unfavorable case" when the electronic potentials coincide and run parallel for a long time, i.e. do not have a well defined "timing" for the transition. Here the vibrational coherence is conserved because the outer limbs of the potential surfaces of the A and B" states are defined by the solvent cage and differ only slightly in this outer half of the vibrational motion (cf. Fig. 4.23a). We emphasize that the period of 550 fs, observed in the spectra of Fig. 4.23b, is clearly distinct from the 650 fs period that is attributed to the creation of zone-boundary phonons by impulsive excitation in an electronic transition treated in Sect. 4.4 [132]. The assignment of the 550 fs period to the molecular I_2 vibration in the A state is corroborated by the classical MD simulations on this system [6]. The concerted electronic and vibrational relaxation seems to occur unequivocally in all halogens in rare gas solids The same transition from the $^1\Pi$ to the $^3\Pi$ state has been experimentally observed and compared to results from DIM-trajectory simulations in the system ClF in solid Ar [18].

There are no obvious restrictions to the generation of secondary coherences via collisions, and is to be expected that the described scheme is universally applicable. However the question remains whether or not the coherences can be discerned. It is easy if the scattering parameters are similar for all molecules in the ensemble. In more complicated cases new tools involving for example higher order correlations may be applicable. These secondary coherences may play an important role in vibrationally coherent photochemistry. Photochemical reactions have to pass a transition state where due the intended rearrangement of the bonds large excursions are a prerequisite. This will inevitably lead to dissipation of energy, when the large-amplitude motion is abruptly stopped by the solvent. Nevertheless the process can continue coherently according to the presented scheme.

At present the theoretical approaches which describe most realistically such strong dynamical interactions of small molecules with a bath seem to be semi-classical molecular dynamics simulations [6,7] and path integral methods to include coherence [133,134] as well as extensions to several electronic states with the DIM method

4.4 Coherent matrix response

M. Bargheer, M. Fushitani, M. Gühr, H. Ibrahim, and N. Schwentner

4.4.1 Long-lasting zone boundary phonon (ZBP) oscillations

Up to now, only the intramolecular vibration has been treated and the coherent motion of rare gas atoms in the vicinity of the molecular guest will be discussed next. The motion will be attributed to coherent zone boundary phonons (ZBP), excited during the molecular electronic transition from the ground to another covalent electronic state. An excellent review of research activities in the field of such impulsively excited coherent phonons is given by Dekorsy [135].

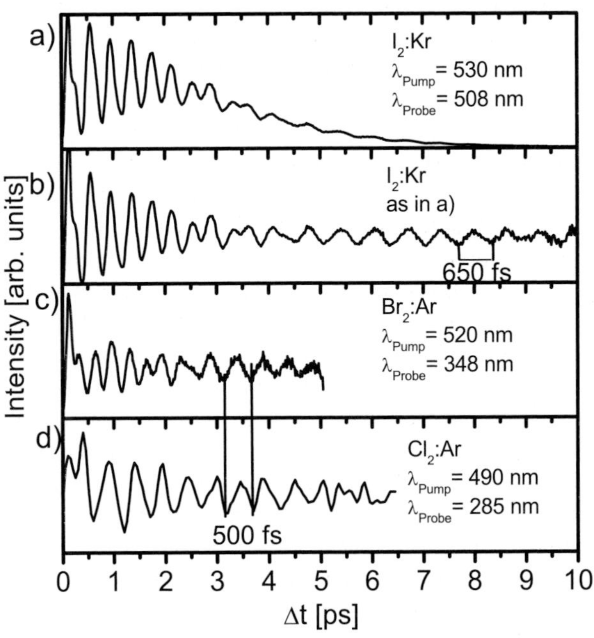

Fig. 4.24. Pump probe spectra of I_2:Kr (a) and b)). The upper panel a) shows an original spectrum, that decays due to energy relaxation. The effect has been removed in panel b) by a normalization to the average signal. c) Spectra of Br_2:Ar and d) Cl_2:Ar both corrected for energy relaxation. In the first few picoseconds, all spectra show features of the intramolecular vibrational wave packet. The effects of dispersion and energy relaxation manifest themselves in a broadening of the vibrational peaks and a shortening in the oscillation period. After some picoseconds, a new vibrational feature sets in, having a period of 650 fs for I_2:Kr and 500 fs for Br_2 and Cl_2 in solid Ar. [15]

Fig. 4.24a shows a fs pump-probe spectrum for I_2 in solid Kr. A vibrational wave packet in the B state was probed to the charge transfer manifold. From 0 to 4 ps, a modulation with T = 420 fs is visible, corresponding to the vibrational period of the B anharmonic oscillator at this particular excitation energy. The modulations loose contrast with propagation time, since the wave packet disperses and suffers vibrational decoherence (see Sect. 4.3). In addition the wave packet period shortens due to energy relaxation (see Sect. 4.2). Energy relaxation is also visible in the decay of the average signal in Fig. 4.24a. Since the wave packet is slipping to lower energies in the B state PES, only a decreasing fraction of the wave packet remains at the probe energy E_{win} and can be detected. After 4 ps, a new oscillatory pattern with a period of $T = 650$ fs is visible. It decays as the average signal. In order to amplify this new pattern, the pump-probe spectrum is normalized to its average decaying intensity. The result of the normalization is shown in Fig. 4.24b; the oscillations at late delay times Δt are now better visible. They are no longer discernible after about 10 ps, since also the overall noise is scaled up in the normalization procedure. The 650 fs oscillations are remarkably stable in frequency and an extrapolation of its phase results in a maximum at $\Delta t = 0$. The very same oscillation with $T = 650$ fs and the same phase has been observed when exciting to the A state of I_2 in solid Kr (see [132]).

Fig. 4.24c and d show fs pump-probe spectra for Br_2 and Cl_2 in solid argon. In both cases, the electronic B state has been excited and the wave packet is probed to the charge transfer states. The spectra are already normalized, as described for the spectrum in Fig. 4.24b. As in the case of I_2 in solid Kr, the initial dynamics is dominated by the intramolecular vibrational dynamics. Once more after some ps, a new oscillation period of now $T = 500$ fs is visible in both spectra. It has the same remarkable frequency stability and phase at $t = 0$ as the 650 fs feature in the I_2:Kr spectra. The 500 fs period also appears for the A state excitation of Br_2 in solid Ar [104].

Concerning the origin of the 650 fs (I_2:Kr) and 500 fs (Br_2:Ar, Cl_2:Ar) periods, one can exclude an intramolecular wave packet motion and attribute it to host dynamics for the following reasons:

(a) The oscillation does not show any sign of dispersion as expected from anharmonicity. This is in conflict with the well known intramolecular anharmonicity for I_2:Kr, Br_2:Ar and Cl_2:Ar and the dynamics cannot be assigned to a molecular state.

(b) The two frequencies for Br_2:Ar, Cl_2:Ar ($f_{\mathrm{P}} = 2$ THz) and I_2:Kr ($f_{\mathrm{P}} = 1.5$ THz) are quasi monochromatic and the line width is only limited by the observation time window, as a further analysis shows [132,136]. Energy relaxation shortens the periods in the anharmonic molecular potentials and would broaden the line in contrast to the observation.

(c) The 650 fs (I_2:Kr) or 500 fs (Br_2:Ar, Cl_2:Ar) oscillation is observed when exciting at many different energies in the B state of the respective molecule (see Figs. 2 and 7 in [136]) and also when exciting the electronic A state (see [132, 136]). Thus, the frequency is independent of the initial molecular electronic or vibrational state.

(d) The same 500 fs modulation is observed in the case of Br_2:Ar and Cl_2:Ar. The molecular masses and force constants are very different, the matrix, however, is the same. This once more favors an assignment of the late oscillation period to host dynamics.

Three questions are crucial: What kind of host motion is visible in the molecular pump-probe spectrum? How can this host motion be excited (see Sect. 4.4.2)? How does the host motion leave its fingerprint on the molecular pump-probe spectrum (see Sect. 4.4.3)?

To clarify the host motion assignment, in Fig. 4.25a the dispersion rela-

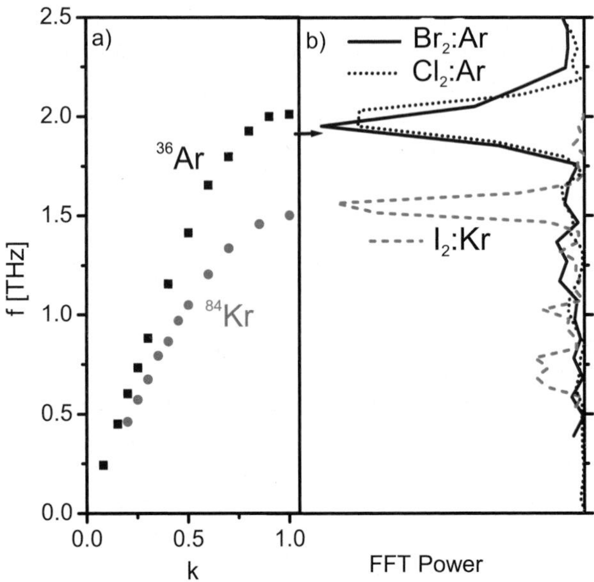

Fig. 4.25. Left side: <100> branch of the phonon dispersion relation of argon (blue squares) and of krypton (red circles). Only the frequencies of the longitudinal phonons are given. The arrow indicates the frequency of the zone boundary phonon of a ^{40}Ar host. Right side: The coherent host dynamics maxima of Br_2:^{40} and Cl_2:^{40}Ar (blue solid and dashed lines), and I_2:Kr (red line). The lines match exactly the frequency of the zone boundary phonon in the dispersion relation given on the right side. [15]

tion is shown for single crystals of solid ^{84}Kr (solid circles) [137] and solid ^{36}Ar (solid squares) [138] gained in neutron scattering experiments. Only the longitudinal parts of the <100> branch are shown. The Ar isotope used in those experiments was ^{36}Ar, however, in the present experiments Ar in natural abundance, consisting mostly of ^{40}Ar, was used. Accordingly the highest phonon frequency has to be shifted by $\sqrt{36/40}$ leading to the arrow on the right side of Fig. 4.25a. Comparing the phonon dispersion relations with the Fourier transformations of the pump-probe spectra in Fig. 4.25b reveals that the coherent host peaks match the frequency of the phonons with reduced wave vector $k = 1$, which are called *Zone Boundary Phonons (ZBP)*. Those have the shortest possible wavelength of all phonons. The ZBP are located at a frequency of 1.5 THz in the case of solid krypton and at about 2 THz for solid argon. Thus, the coherent dynamics observed in the pump-probe spectra can be attributed to *coherent zone boundary phonons* of the host crystal as was documented for $I_2{:}Kr$ in [132], in [136] for $Br_2{:}Ar$ and in [139] for $Cl_2{:}Ar$. In addition, coherent ZBP oscillations were also found on some pump-probe spectra for I_2 in solid Xe with a frequency of 1.25 THz corresponding to a period of 800 fs [48]. The ZBP have a vanishing group velocity $v_g = d\omega/dk$, since the slope in the phonon dispersion relation is zero near the edge of the first Brillouin Zone of the crystal. This fact will be quite crucial in clarifying the excitation and detection scheme.

4.4.2 Excitation scheme for coherent zone boundary phonons

One observes for the cases of iodine in solid krypton, bromine and chlorine in solid argon only one host induced frequency, the ZBP. The experimental observations also exclude phonon dynamics that is forced by the vibrational motion of the molecule. In such a *driven oscillation* scenario, the environment should show a mode being dependent on the exciting intramolecular motion. At the specific pump wavelengths used in [8], a 2:1 resonance of the I_2 intramolecular vibration period and the 650 fs coherent host vibration period was found. An impulsive creation (either electronic or vibronic in the first excursion of the intramolecular wave packet) was excluded in this reference. However, exploiting the tunability of the NOPA sources, the intramolecular vibrational oscillation period T is changed over a large range from $T = 1$ ps for excitations near the dissociation limits of I_2, and Br_2 down to 250 fs for Br_2 and about 350 fs for I_2. For all excitation conditions, only the coherent ZBP motion of only one frequency could be observed and no beating phenomena occurred. Thus, a driven oscillation cannot account for the observed phonon.

The relative phase of the host induced oscillation is stable when changing the excitation energy of the B state and when exciting the A state and comparing it to a B state excitation (see [132, 136]). Extrapolating the host modulation to $\Delta t = 0$ delivers the same phase under all excitation conditions. The fixed phase at $t = 0$ calls for an *impulsive* excitation of the environment correlated to the optical molecular excitation.

The variant of the general impulsive excitation scheme which is proposed here, bases on a model called Displacive Excitation of Coherent Phonons (DECP) [135], originally introduced for the excitation of zone center phonons in the case of semimetals and semiconductors [140–143]. If those materials are irradiated by an ultrafast laser pulse, an interband transition from bonding to antibonding orbitals occurs. The electronic system switches on timescales much faster than the nuclear response times and changes the forces between the nuclei. The system begins to oscillate around the new equilibrium geometry. Recent ultrafast X-ray diffraction experiments find this particular excitation mechanism in other systems [144,145]. The DECP assignment is detailed confirmed in [132, 136] and by specific calculations in Sect. 4.7.2 and [139]. Such an impulsive excitation drives the lattice to a new equilibrium position, while initially at $\Delta t = 0$, it is still arranged in the electronic ground state equilibrium position. The oscillation starts with the extreme amplitude accordingly and the atoms will oscillate around their new equilibrium position finally. Therefore, the DECP excitation results in a $\pm\cos(2\pi f_{\mathrm{P}}\mathrm{t})$ phonon motion. When extrapolating the phonon oscillation to the excitation time $t = 0$, the phonon amplitude has a maximum or minimum. In the present experiment a maximum occurs at $\Delta t = 0$.

A DECP mechanism was proposed by Chergui and coworkers for generation of coherent modes in rare gas and hydrogen hosts by excitation of NO molecules to Rydberg states [146–153]. In the present case, the molecules I_2, Br_2 or Cl_2 are initially in the $v = 0$ level of the electronic ground state X $^1\Sigma_g$. The fs pump pulse excites the molecule to a $^3\Pi_{\Omega u}$ state, *e.g.* A ($\Omega = 1$) or B ($\Omega = 0$). The internuclear distance of the molecule does not change during the short transition in which the electronic orbitals are switched. The environment remains also in its equilibrium around the electronic ground state. The excitation by a 20-50 fs light pulse is indeed *impulsive*, because it is much shorter than any phonon and intramolecular period.

To estimate the trend for the structural changes the potential is calculated for a single rare gas atom around the halogen molecule. The potential minima indicate the equilibrium distance of the rare gas atom from the molecule. The calculation uses the DIM formalism [154, 155], extensively applied in doped rare gases [21,22,39,50,156–163] (see Sect. 4.6).

The calculation shown in Fig. 4.26 is performed for the I_2:Kr case (calculations for Br_2:Ar and Cl_2:Ar are given in [14, 104, 139]. The interaction potential between halogen and rare gas atoms are known from scattering experiments [164]. The detailed description of the calculation can be found in [14,104]. Figure 4.26c shows the lines of vanishing force for the Kr atom in the vicinity of an I_2 molecule. The line of vanishing gradient for the B state lies outside of that for the ground state. Thus, when exciting the molecule from its electronic ground state to the B state, the molecule expands in the "eyes" of a rare gas atom.

One has now to search for those atoms decoupled from the molecular vibrational motion, since the coherent ZBP is not influenced by the molecular vibration. Fig. 4.26a and b show the (100) and (111) plane of the fcc Kr crystal respectively. The I_2 molecule on a double-substitutional site replaces two nearest neighbor atoms. The atoms are separated into four groups. Three groups of atoms are strongly coupled to the molecular vibrational motion. They are called head-on (3) , belt (5) and window atoms (4) . Due to the strong coupling to the intramolecular vibration, they are responsible for the energy

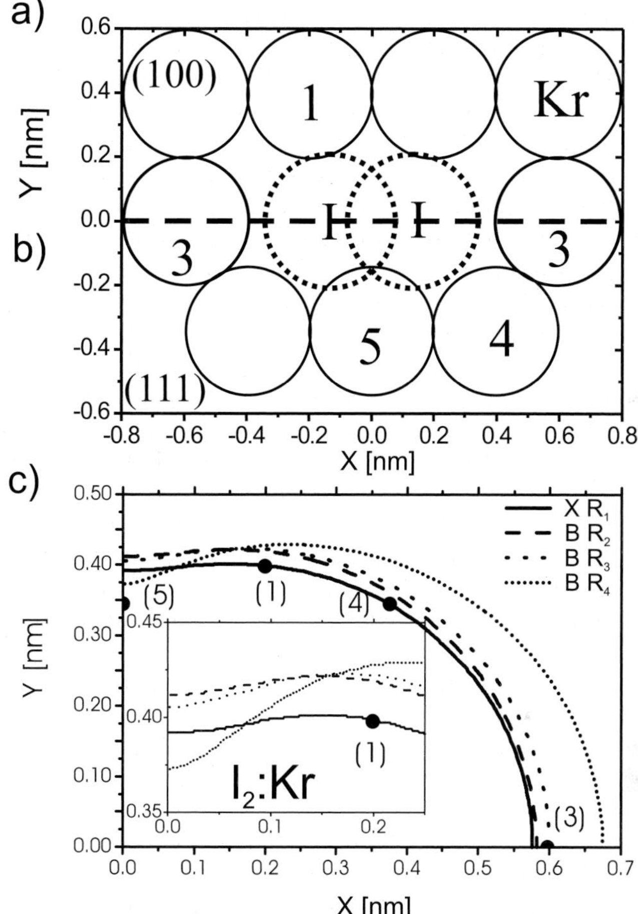

Fig. 4.26. a) (100) plane of the I_2:Kr system. b) (111) plane of the I_2:Kr system . Rare gas atoms are separated into groups: 1 coherent phonon atoms, 3 head-on atoms, 4 window atoms, 5 belt atoms. c) Equilibrium lines (local potential minima) of a Kr atom around the I_2 molecule for different electronic states and internuclear distances. Only one quadrant is shown for symmetry reasons. Solid line: electronic ground state X with $R_1 = 0.2666$ nm, dashed line: excited state B with $R_2 = 0.2666$ nm, dotted line: excited state B with $R_3 = 0.3024$ nm, short dotted line: B state with large internuclear distance $R_4 = 0.45$ nm. [15]

relaxation and wave packet retarding phenomena visible in the trajectory of Sect. 4.2.

Group (1) is decoupled from the molecular motion. As the molecule undergoes the B $\longleftarrow$ X transition, this group of atoms experiences a force in the

Y direction. The *Ar* atoms (1) are repelled to a position further away from the molecular axis in the Y direction (inset Fig. 4.26c). The new position is very close to a point where all B state equilibrium lines for different internuclear distances of the chromophore intersect. This indicates that no coupling to the molecular oscillation occurs. Thus, the coherent phonon oscillation has to be attributed essentially to the group (1) of atoms. A full calculation (see Sect. 4.7) on how all the cage atoms move in the course of time is available for Cl_2:*Ar* [165]. The results have been published recently together with experimental data for this system [139]. They confirm the single rare gas atom calculation presented here.

The phonons are excited symmetrically in one direction with wave vector **k** and in the other direction with a wave vector of $-\mathbf{k}$. The total wave vector of the phonons cancels. Therefore, *photons* in the visible light range with a very small wave vector $k = 2\pi/\lambda$ of typical values around $1 \cdot 10^7$ m^{-1} can excite zone boundary *phonons* with k around 10^{10} m^{-1}. The known examples for DECP processes usually lead to excitation of zone center phonons with small wave vector (see, for example, [144, 166–169]).

The phonons excited in this impulsive process travel away from the excitation source (molecule) according to their group velocity. They form a kind of wave packet that propagates and disperses since its constituents propagate with different velocities. That part of the acoustic wave packet belonging to zone boundary type phonons stays in the vicinity of the excited molecule since, as stated above, $v_g = 0$ holds at the boundary of the Brillouin zone. Thus, the dispersion relation of the crystal provides a sort of a filter mechanism. Zone center phonons propagate with the velocity of sound which is 1.64 nm/ps for longitudinal phonons in *Ar* and 1.375 nm/ps for longitudinal phonons in *Kr* (see, for example, [170] or [171]). Thus, after about 500 fs the zone center phonons have crossed the second solvation shell and leave the vicinity of the chromophore. If the phonons are probed later on, only the zone boundary phonons remain and contribute to the coherent signal.

4.4.3 Interrogation of coherent phonons

The RG host is completely transparent for the visible and UV laser pulses used in all experiments. The first absorption sets in above 10 eV for *Kr* and at 11 eV for *Ar*. Thus, the crystal dynamic cannot be directly probed by the ultrashort pulses. The coherent ZBP changes, however, the laser induced fluorescence intensity of the molecular charge transfer states, since this is the signal in the pump-probe measurement.

The signal is proportional to the overlap of the probe sensitivity curve with the vibrational wave packet, as sketched in Fig. 4.27. If the CT state shifts in energy, E_{win} and the whole sensitivity curve shift also. For a small shift the overlap does not change much for a wave packet located on the maximum of the sensitivity curve, since the probe sensitivity curve versus energy is flat. In contrast, when the wave packet is located partially below E_{win} and is thus

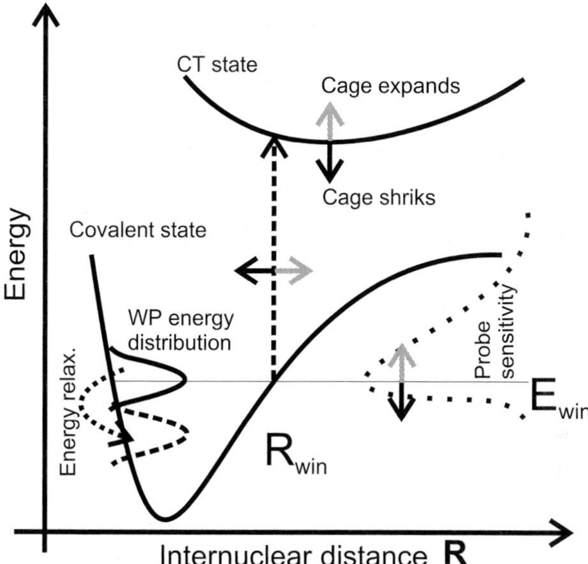

Fig. 4.27. Phonon detection scheme. The phonon modulates the charge-transfer (CT) state solvation energy. Thus, the probe resonance (R_{win}, E_{win}) and the schematic sensitivity curve (dashed, right hand side) shift accordingly, leading to a changing overlap of the probe sensitivity with the relaxing vibrational wave packet (WP) in the covalent state. [15]

near the edge of the sensitivity curve, the overlap changes quite dramatically with a variation of energy, due to the steep decay in this range. The wave packet can be either prepared in this range or it can relax to this region with time. If the wave packet is prepared significantly above E_{win}, a periodic shift of the sensitivity curve causes a weak modulation of the pump-probe spectrum, since the $1/\sqrt{\Delta E}$ decay is quite smooth. Only after energy relaxation has transferred it below E_{win}, the shift can be observed with high sensitivity on the pump-probe spectrum.

The coherent ZBP, that stay at the chromophore for several ps are a source for a periodic shift of the probe window. The solvation energy of the charge-transfer state of the halogens with several thousand cm^{-1} amounts to more than 10 % of the transition energy and is huge compared to that of the covalent B state. A local compression of the lattice around the halogen molecule increases this solvation energy of the charge-transfer state due to the shrinking cavity diameter d in the Onsager model [47, 172]. The model allows to estimate the energy shift ΔE of the dipole μ sitting in a spherical cavity with diameter d in a polarizable host with the dielectric constant ϵ. Stronger solvation pushes the ionic E state downwards in energy, the probe position R_{win} shifts inwards and E_{win} down in energy to accommodate for the fixed probe photon energy. Coherent oscillation of matrix atoms in the halogen vicinity generate such density modulations that can be decoded from the B state pump-probe signal. The process modulates the pump-probe spectrum in the decaying part, in accordance with the experimental evidence. Furthermore, the modulation depth is proportional to the average signal intensity, as all our spectra indicate.

The delay in the phonon contribution arising from the wave packet relaxing below the probe window energy E_{win} can be avoided. We prepare the wave packet below the probe window and observe a convolution of phonon dynamics and intramolecular wave packet dynamics from the first picosecond of the propagation on (see [136]).

As an alternative to the interpretation given here predissociation to lower lying and matrix stabilized states was proposed [48]. Those lower lying states should then be probed by two photons to the CT states, giving the typical modulation interpreted as ZBP motion here. In that case the results should strongly depend on the molecule. However, always a modulation with the ZBP frequency of the host has been found, regardless of the molecule used.

4.5 IR-driven photochemistry in rare gas matrices: The cis-trans isomerization of nitrous acid (HONO)

P. Hamm

An example of a larger chromophore is presented in this section, in order to illustrate the increase in complexity with the number of internal degrees of freedom. In addition, the feasibility of time resolved photochemistry in the electronic ground state is demonstrated by exploiting the advances in femtosecond IR technology. Chemistry in the electronic ground state plays a central role in practice and becomes accessible by these new tools.

From the experimental point of view, very little is known about the dynamics of ground state chemical reactions in the condensed phase on a microscopic, atomic level. Time resolved studies with high time resolution require a sharp trigger, which, in the case of electronically driven photochemistry, is a short pump laser pulse in the visible or UV spectral range. The same has become possible only recently in the IR spectral region. There are many examples known of IR-driven photochemical reactions, ranging from isomerization reactions such as in nitrous acid ($HONO$) [173] or formic acid [174], up to hydrogen bond dissociation reactions of, for example, $H_2O \cdots HI$ complexes [175] or bimolecular reactions like $CH_2 = C = CH_2 + F_2 \rightarrow CH_2 = C = CHF + HF$ [176] (see [177, 178] for reviews). By far the most prominent example is the cis $\rightarrow$ trans isomerization of $HONO$ in rare gas matrices (Kr, Ar) upon excitation of the OH stretch vibration, first described by Pimentel in the early 60's [173]. The cis-trans isomerization of $HONO$ is particularly interesting because a) it is small enough to describe it theoretically on the highest level [179–183], because (b) the cis $\rightarrow$ trans quantum yield is exceptionally high (close to 100%) [173, 184], and because (c) a single quantum excitation of one vibrational mode (the OH stretch vibration) is sufficient to trigger the reaction, reaching energy regions where the density of states is still small enough to be manageable. The energy lies well within the range of what might occur in 'real world' chemistry - i.e. without IR excitation, but at room temperature - due to thermal excitations with a Boltzmann factor of $e^{E_a/k_BT} = 5 \cdot 10^{-8}$. Hence, $HONO$ appears to be an ideal prototype molecule to study the mechanism of thermally driven chemical reactions, such as proton transfer reactions, on *electronic ground state surfaces.* The energy of the cis-configuration of $HONO$ lies 100-200 cm^{-1} above that of the trans configuration [179]. According to the highest level quantum chemistry calculation available (on the CCSD(T)/cc-pVQZ level), the transition state is located along the torsional coordinate at $\Phi = 86$ with an energy of 4105 cm^{-1} [179]. This value reduces to 3635 cm^{-1} when considering zero-point corrections [179, 182]. However, the effect of the matrix environment on the barrier is not known. Hence, the energy of one quantum of the OH-stretch vibration 3500 cm^{-1} is of the same order as that of the transition state, but it is not known exactly whether it is above or below, and whether tunneling effects need to be considered. Furthermore, the high cis $\rightarrow$ trans quantum yield of 100% is indeed remarkable, given the fact that it is not the reaction coordinate which is initially excited. This suggests, that the OH-stretching coordinate is efficiently coupled to the torsional coordinate. Experimental information about the photoisomerization of $HONO$ stems either from stationary spectroscopy, revealing for example the relative energetics of the cis and the trans state through thermodynamic measurements, or from kinetic measurements, revealing the quantum efficiency of the reaction [173, 184]. However, the mechanism of the photoisomerization, or merely the order of magnitude of its timescale (whether it is femtoseconds or milliseconds) was completely unknown until very recently. It is not known whether coherent wave packet motion or tunneling is involved, or whether it is a quasi-classical type random walk which gives rise to the cis-trans isomerization. The reaction has not been observed in the gas phase so far, which is why one may assume that the reaction is in some way environment assisted.

This remarkable model system was investigated recently with femto-to-nanosecond IR pump-probe spectroscopy [185, 186]. To address this questions, Fig. 4.28 shows a steady state absorption spectrum of $HONO$ in solid Krypton at 30K with two main bands, the OH-stretching band of the cis isomer at 3402 cm^{-1} and that of the trans isomer at 3552 cm^{-1}, respectively.

Fig. 4.29a shows typical transient difference spectra after selective excitation of the OH stretching band of the trans isomer. The dominating bands are the trans bleach/stimulated emission band at 3552 cm^{-1} and the trans excited state absorption band at 3390 cm^{-1}. The latter decays bi-exponentially on a 8 ps and 260 ps timescale reporting on energy relaxation out of the OH

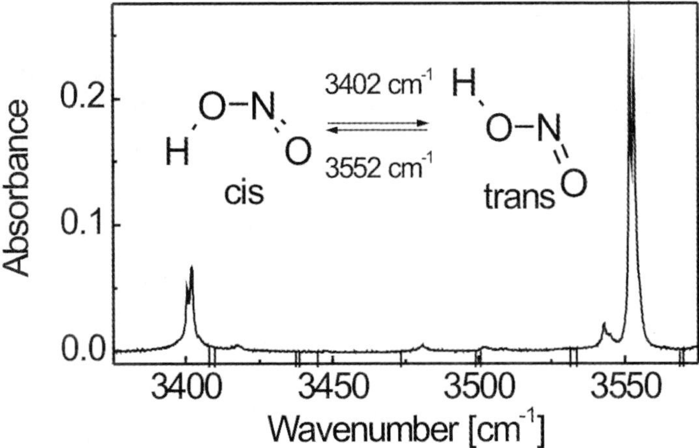

Fig. 4.28. Steady state absorption spectrum of $HONO$ in Kr Matrix at 30K, with OH-stretching bonds of the cis (3402 cm^{-1}) and the trans species (3552 cm^{-1}). Adapted from [185]

stretch vibration. Simultaneously with this decay, a third signal is growing in between 3515-3550 cm^{-1}, which is referred to as 'dark states'. It stems from trans molecules that have relaxed from the originally excited OH-stretching state, the bright state, to lower lying vibrational modes which are not observed in the steady state absorption spectrum and thus called dark states. These states can be overtones and/or combination modes, of which many exist in the molecule. As a consequence of anharmonic couplings between these dark states and the bright OH-stretching mode, the latter absorbs slightly red-shifted with respect to its original frequency, once one of the dark states is populated.

Fig. 4.29b shows transient difference spectra after selective excitation of the OH stretching band of the cis isomer. This time, only the strong cis bleach-stimulated emission signal at 3402 cm^{-1} is observed, while the cis excited state absorption is outside the spectral window of this measurement. The cis excited state is depopulated on a faster 20 ps timescale. Most important is a small, broad signal in the region between 3515 cm^{-1} and 3550 cm^{-1} (see arrow) which is attributed to dark states in the trans configuration. This signal directly reflects the cis-trans isomerization. The inset of Fig. 4.29b compares the dark state signal after pumping trans $HONO$ directly (gray line), with that after pumping cis $HONO$ (black line). As both are essentially the same, it is concluded that the same trans overtones and combination modes are populated after isomerization. Fig. 4.30a compares the build-up of the trans-$HONO$ signal after pumping cis-$HONO$ with the cis bleach/stimulated emission signal (Fig. 4.30b). The latter decays on two distinct timescales: On

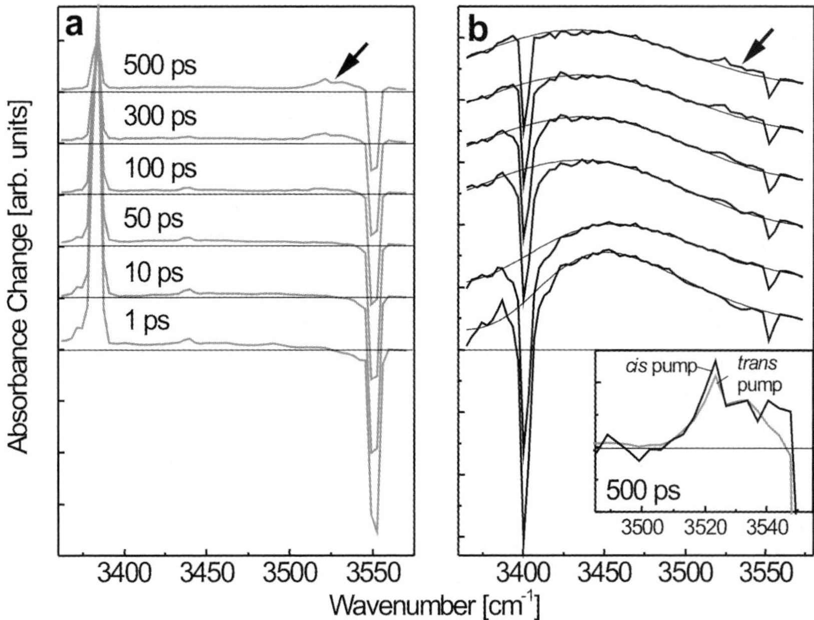

Fig. 4.29. (a) Transient response after selective excitation of trans *HONO* revealing a trans bleach/stimulated emission signal at 3552 cm^{-1}, a trans excited state absorption signal at 3390 cm^{-1}, and a trans dark state signal between 3515-3550 cm^{-1} (see arrow). (b) Transient response after selective excitation of cis *HONO* revealing a cis bleach/stimulated emission signal at 3402 cm^{-1} and a trans (!) dark state signal between 3515-3550 cm^{-1} (see arrow). The cis excited state absorption is outside the spectral window of this measurement. The signal is superimposed on a broad background which is due to a water film that is growing on the sample. Adapted from [185].

a fast, 20 ps timescale, one observes vibrational relaxation of the initially pumped *OH* stretch vibration into the other internal degrees of freedom of the *HONO* molecule (intramolecular vibrational redistribution IVR). On a significantly slower 20 ns timescale, on the other hand, cooling into the matrix environment is observed. Interestingly, isomerization follows the same two step process: As long as there is energy directly in the *OH* stretch vibration, isomerization is efficient, leading to 50% of the total quantum yield within the first 20 ps. However, as long as vibrational energy is still within the molecule (but no longer in the *OFH* stretch vibration) isomerization might still occur, albeit with much smaller rate. This cis-trans isomerization of *HONO* is an exceptional example of an IR driven photochemical reaction which has

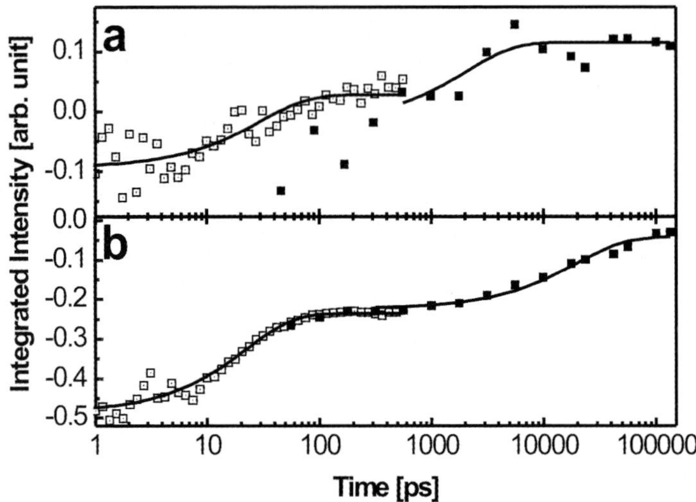

Fig. 4.30. (a) Build-up of the trans-*HONO* signal after pumping cis-*HONO* compared with (b) the cis bleach/stimulated emission signal. Open squares from [185] are completed with closed ones from [186].

been investigated in a time-resolved fashion [185, 186]. In that sense, it is presently not clear whether these results have universal character, and to what extent it can be transferred to more complex reactions like hydrogen bond dissociation or even bimolecular reactions. The picture emerging from combination of the available information is the following: from the theoretical point of view, the cis-trans isomerization has either been treated on a fully classical MD level, including the dissipative effect of the matrix [187], or on an fully quantum level, albeit in the gas phase [180–183]. However, in a classical picture, an energy equivalent of three quanta had to be deposited into the *OH* stretch vibration in order to overcome the reaction barrier [187]. In this case isomerization occurs on a few picosecond timescale. Since a classical proton cannot tunnel, it would never isomerize when putting only one quantum into the *OH* stretch vibration, like it is done in the experiment. A fully quantum-mechanical description, in contrast, is limited to very few degrees of freedom, which prohibits treating the interaction with the matrix environment on the same footing. The six internal degrees of freedom of a four-atomic molecule *HONO*, however, can be treated on the highest theoretical level: A full 6D potential energy surface has been calculated on the CCSD(T)/cc-pVQZ level [180] and the fully coupled 6D vibrational problem has been calculated on a numerically exact level [180, 181]. Yet, a coherent wave packet propagated on that fully coupled 6D potential surface would not isomerize on a 40 ns timescale, even when putting 2 quanta in the *OH* stretch vibration, high above the reaction barrier [182]. The eigenstates that carry oscillator strength, and hence might compose such a wave packet, perfectly localize on either the cis or trans side. Only eigenstates with high quantum numbers in the torsional mode tend to delocalize across the cis-trans barrier, however, these eigenstates have virtually no oscillator strength and hence cannot be excited optically. From the experiment one now knows that the final cooling time of the molecule is 20 ns [186], setting an absolute upper limit for any coherent wave packet picture. Hence, it has been found that either picture - fully classical and fully quantum-mechanical - is bound to fail. On the one hand it is clear that tunneling effects for the proton must be relevant, ruling out any classical treatment of, at least, the proton. On the other hand, the matrix environment, that perturbs the system so strongly that it cannot be neglected, can still not be treated quantum-mechanically. This situation calls for mixed quantum-classical simulations, which are currently pursued in our group. The density of states is too small in the gas phase to efficiently couple both reaction partners. In the gas phase, the proton states are discrete and are stationary eigenstates. This changes dramatically when the molecule is brought into the condensed phase where it is coupled to a quasi-continuum of states. Tunneling of the proton from the cis to the trans side is possible only when proton donor and acceptor states are in close resonance. In the gas phase, such a resonance would be accidental, and since the couplings between the various states are weak, such a coincidence is extremely unlikely. Matrix degrees of freedom, however, may fine-tune these resonances, as the system as a whole undergoes thermal motion. In the condensed phase, a resonance between two quantum states is no longer a matter of an unlikely coincidence, but will almost necessarily occur at a particular configuration of the molecule in the matrix cage. It is instructive to compare these conclusions with electronic photochemistry, which, of course, does occur in the gas phase, often very efficiently and on ultrafast timescales [188]. In the electronic case, the molecules own vibrational degrees of freedom serve as the quasi-continuum of states which is necessary to efficiently couple the electronic wave functions. The difference lies in the energy scales: In the electronic case, the density of states is gigantic. When taking $HONO$ as an example, the pure vibrational density of states at 30.000 cm^{-1} (the energy of the $S_0 - S_1$ electronic transition) is estimated to be $\approx 400/\text{cm}^{-1}$, compared to a density of states of $0.25/\text{cm}^{-1}$ at 3500 cm^{-1} (the energy of the OH stretch vibrational transition). Hence, in the case of IR driven photochemistry, it needs the coupling to a 'solvent' environment to provide a quasi-continuum of states to make jumps between proto-quantum states happen.

4.6 Interaction potentials: The diatomics-in-molecules approach

A. Borowski, A. Cohen, R. B. Gerber, and O. Kühn

The description of the interaction potentials for dihalogens in matrices requires an approach that can realistically deal with the many-body nature of these systems, that can address the electronically excited states as well as the ground state, which treats the calculation of nonadiabatic couplings between different electronic states as pertinent to the processes, and that is effective enough to be applied for multidimensional dynamics calculations. The model that was found to meet these multiple demands and served as a basis for the theoretical studies presented here is the Diatomics-in-Molecules (DIM) method [154]. The approach has had a diverse range of applications [189, 190] and in particular there have been numerous studies using DIM for impurities in a host noble-gas system [191–193]. A comprehensive review of DIM applications for guest species in matrices is given by Apkarian and Schwentner [43]. DIM potentials were used to describe photochemical processes of species such as HX and X_2 in noble-gas solids and several of these simulations included also the role of nonadiabatic transitions [22, 50, 157, 194, 195]. Specifically, during the last years DIM potentials have been developed for several systems of dihalogens in noble-gas clusters and matrices, including F_2 [21, 55], ClF [31], Cl_2 [139], and Br_2 [196] in Ar. In essence, the DIM approach combines a simplified electronic structure model with empirical input on the interactions involved. Consider, for example, the construction of the DIM potentials for $F_2 : Ar$ [21, 55]. To describe the chemical interactions between the two F atoms, a valence-bond model is employed, explicitly treating one effective, unpaired electron in the $2p$-orbital manifold (hole) of each of the two atoms. This yields, when considering the two spin states of each electron, a total of 36 states of the F_2 molecule, correlating asymptotically with two $F(^2P)$ atoms. For example, the singlet electronic ground state of the system is given by

$$^1\Sigma_g^+ = C[a_0(1)b_0(2) + b_0(1)a_0(2)] \times [\alpha(1)\beta(2) - \beta(1)\alpha(2)] , \qquad (4.7)$$

where C is a normalization factor and the notation $a_0(i)b_0(j)$ indicates that the ith electron is on the first (a) F atom, the second electron j is on the second (b) F atom, and the index 0 stands for the zero projection of the angular momentum onto the molecular axis. The DIM electronic Hamiltonian for $F_2 : Ar$ is given by

$$H^{el}_{F_2:Ar} = H_{F_2} + H_{F-Ar} + H_{Ar-Ar} + H_{SO} , \qquad (4.8)$$

where H_{F_2} is the electronic Hamiltonian of the isolated F_2 molecule, not including the spin-orbit interaction. H_{F_2} is determined using ab initio data and includes potential energy curves for the singlet ($X^1\Sigma_g^+$, $2^1\Sigma_g$, $^1\Sigma_u^-$, $^1\Pi_g$, $^1\Pi_u$, $^1\Delta_g$) and triplet ($^3\Sigma_g^-$, $^3\Sigma_u^+$, $2^3\Sigma_u^+$, $^3\Pi_u$, $^3\Pi_g$, $^3\Delta_u$) states of the free molecule [197]. The halogen-argon interaction is taken as the sum of pairwise potentials of the form

$$V_{F_i-Ar_j} = V_0(R_{F_i-Ar_j}) + V_2(R_{F_i-Ar_j})P_2(\cos\gamma_{ij}), \tag{4.9}$$

where $R_{F_i-Ar_j}$ denotes the distance between the ith F atom ($i = 1, 2$) and the jth Ar atom ($j = 1 \dots N$). γ_{ij} is the angle between the interatomic distance vector and the orientation of the p-orbital on the atom. The anisotropy of the $F - Ar$ interaction, contained in the dependence on the p-orbital orientation, is very important, especially for open shell (excited) states. The potentials $V_0(R_{F_i-Ar_j})$ and $V_2(R_{F_i-Ar_j})$ can be obtained from ab initio calculations of $F(^2P)/Ar$ interactions, as described in [21]. H_{SO} in (4.8) is the spin-orbit interaction, given by

$$H_{SO} = \Delta \boldsymbol{l} \cdot \boldsymbol{s} \tag{4.10}$$

where Δ is the F atom spin-orbit coupling parameter, and $\boldsymbol{l}$ and $\boldsymbol{s}$ are the orbital and spin angular momentum operators, respectively. A method for determining an effective Δ is described in Sect. 4.10. H_{SO} causes rotation of the p-orbital, compared with the situation in the free $F(^2P)$ atom in the absence of spin-orbit coupling. The electronic Hamiltonian (4.8) is diagonalized in the basis of atomic p-orbital (unperturbed by spin-orbit coupling). This results in a set of 36 adiabatic PES, $W_j(F_1, F_2, Ar_1, \dots, Ar_N)$ ($j = 1 \dots 36$) which carry the full dependence on the coordinates of the two F atoms and all the argon atoms of the system. The electronic states so generated include the effects of the electrostatic interaction between the F_2 subsystem and the argon atoms as well as the spin-orbit couplings. Notice, that the model described here does not include charge-transfer states. In principle, they can be treated by the Diatomics-in-Ionic-Systems (DIIS) extension of DIM [198,199], but the additional complexity associated with it is considerable.

Furthermore, it is important to note that for heavier halogens, e.g., the case of $Br_2 : Ar$ [196] where the spin-orbit coupling constant is 3685 cm^{-1}, a different spin-orbit coupling scheme according to Hund's case c is applicable. Here, spin-orbit coupling is incorporated in the coupled total electronic angular momentum basis used for the DIM Hamiltonian representation.

The PES provide in applications at least some insights also into the dynamics. Consider first Fig. 4.31, which shows the geometry of the F_2 molecule at the chemical equilibrium position in the argon lattice. The geometry has the F-F axis pointing toward a triangle of argon atoms, that form part of the cage that encloses the molecule. Upon promotion to a repulsive excited state, each F atom is accelerated toward the corresponding exit window shown in the figure. Consider now Fig. 4.32 [200] where panel (a) shows the potential energy curves of the bare F_2. All excited states shown are repulsive except $^3\Pi$, that has a weak attractive well. Fig. 4.32b shows the one-dimensional (1D) potential curves for F_2 in the Ar matrix, as a function of the $F - F$ distance. In this figure, the argon atoms and the center of mass of F_2 are kept fixed, and the $F - F$ points along the $\langle 111 \rangle$ direction in the crystal, as

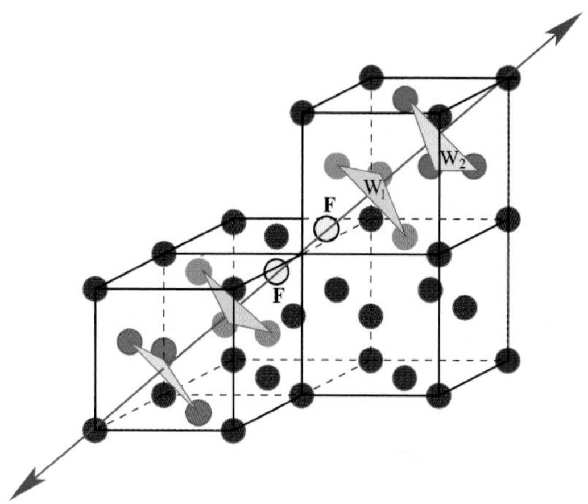

Fig. 4.31. F_2 in Argon oriented along the $\langle 111 \rangle$ direction. The argon windows which are important for the caging dynamics are marked W_1 and W_2. (adapted from [201])

in Fig. 4.31. The figure gives indications of the effect of the medium on the PES. In particular, the potential energy curves become more repulsive as the $F - F$ distance is increased, say beyond 5 a_0. This, of course, represents a "cage effect", that involves barriers, depending on the electronic state for exit of the $F(^2P)$ atoms from the cage after photoexcitation. The barriers depend strongly on the orientation of the molecule in the cage. This is demonstrated in Fig. 4.32c where the potential energy curves are shown for the F_2 molecule pointing along the $\langle 110 \rangle$ axis.

These DIM PES for the molecule in the (frozen) solid are very useful for understanding and analyzing cage exit of the photofragments upon photodissociation, and they are also very helpful for designing strategies for coherent control of cage exit, as shown in the study of Gerber et al. [54]. It goes without saying that rotation of the molecule, recoil of the cage atoms, and the multidimensionality of the PES are important in the actual dynamics, and these are not obvious from the potential curves. Nevertheless, much can already be learned from such figures. Furthermore, consideration of two-dimensional contours of the DIM potential makes it possible to incorporate at least a particular aspect of the cage atom dynamics and its role in cage exit. This was pursued in [54] and [55], in the context of reduced dimensionality quantum models, that will be discussed later in Sect. 4.10.

Another important advantage of the DIM methodology is that nonadiabatic couplings, leading to transitions between different adiabatic electronic states in dynamical processes, can be conveniently generated. The total Hamiltonian of the systems considered here can be written as

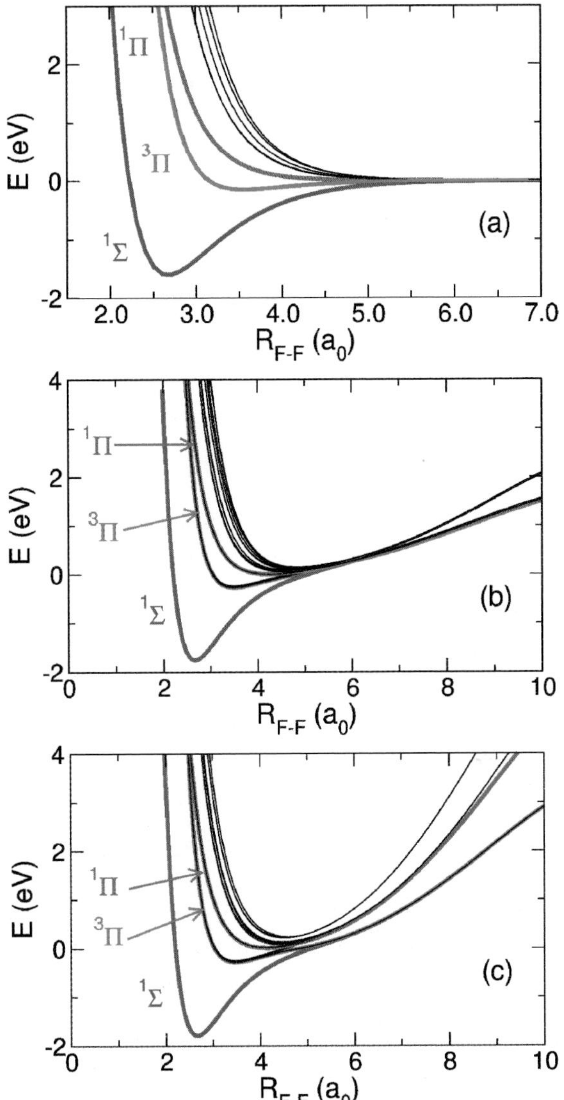

Fig. 4.32. (a) The ab initio F_2 free molecule potentials without spin-orbit coupling interactions. The graph highlights three states: the ground $X^1\Sigma_g^+$ state (blue), the attractive $^3\Pi_u$ state (red), and the excited $^1\Pi_u$ state (violet). Among the 36 electronic states of the F_2 molecule only two are bound: the ground and the $^3\Pi_u$ state. Panels (b) and (c) show the F_2 molecule potentials in a slab of 255 Ar atoms along the $\langle 111\rangle$ and $\langle 110\rangle$ directions, respectively, with spin-orbit interaction which lifts some of the degeneracies. In addition, the once repulsive excited states of the bare molecule, become attractive due to the matrix. The $F - Ar$ interaction which depends on the orientation of the $F(P)$ orbital causes a splitting in the $^3\Pi_u$ states potentials along the $\langle 110\rangle$ direction.

$$H = H^{el}(r, R) - \sum_{\alpha} \frac{\hbar^2}{2M_\alpha} \nabla_\alpha^2 , \tag{4.11}$$

where $H^{el}(r, R)$ is the electronic Hamiltonian (It is convenient to include all potential energy terms in (4.11). Specifically for DIM this means that the interactions between the Ar atoms are incorporated.). In (4.11), r denotes collectively the electronic degrees of freedom and R denotes the nuclear coordinates, $\mathbf{R}_\alpha$ is the position of nucleus α, having the mass M_α. To solve the time-dependent Schrödinger equation,

$$i\hbar \frac{\partial \Psi(r, R, t)}{\partial t} = H\Psi(r, R, t) , \tag{4.12}$$

the wavefunction is expanded in the electronic states, defined by

$$H^{el}(r, R)\varphi_n(r, R) = W_n(R)\varphi_n(r, R) . \tag{4.13}$$

The adiabatic electronic states and the potential energy surfaces $W_n(R)$ depend, of course, parametrically on R, the nuclear coordinates. The Born-Oppenheimer separation of electronic and nuclear degrees of freedom suggests the expression for $\Psi(r, R; t)$

$$\Psi(r, R, t) = \sum_n \varphi_n(r, R)\chi_n(R, t) . \tag{4.14}$$

Where the $\chi_n(R, t)$ are the wave packets that describe nuclear motion on the adiabatic potential W_n. Substitution of (4.14) into (4.8), yields after some algebra and approximations

$$i\hbar \frac{\partial \chi_n(R, t)}{\partial t} = \left[-\sum_\alpha \frac{\hbar^2}{2M_\alpha} \nabla_\alpha^2 + W_n(R) \right] \chi_n(R, t) - \sum_\alpha \sum_{m \neq n} C_{nm}^\alpha \nabla_\alpha \chi_m(R, t) , \tag{4.15}$$

where

$$C_{nm}^\alpha = \frac{\hbar^2}{M_\alpha} \langle \varphi_n(r, R) | \nabla_\alpha \varphi_m(r, R) \rangle_r . \tag{4.16}$$

The integral in (4.16) is over the electronic coordinates. The approximation involved neglects terms like

$$d_{mn}^\alpha = -\frac{\hbar^2}{M_\alpha} \langle \varphi_n(r, R) | \nabla_\alpha^2 \varphi_m(r, R) \rangle_r \tag{4.17}$$

which in nearly all test cases proved to be rather small. The kinetic coupling coefficients C_{nm}^α govern the dynamical transitions between different adiabatic states. They can be expanded according to the Hellmann-Feynman theorem by

$$C_{nm}^\alpha(R) = \frac{\left\langle \varphi_n(r, R) | \frac{\partial H^{el}(r,R)}{\partial R} | \varphi_m(r, R) \right\rangle}{W_m(R) - W_n(R)} . \tag{4.18}$$

These coefficients can be generated very effectively from DIM. In semiclassical treatments of nonadiabatic transitions, related coefficients appear, and are likewise conveniently generated from DIM [22,50,157,194,195]. Indeed, it will be pointed out that in the semiclassical approaches, all adiabatic PES, $W_n(R)$, and the nonadiabatic couplings, C^{α}_{nm}, can be generated on-the-fly, in the case of the trajectories that describe the dynamics.

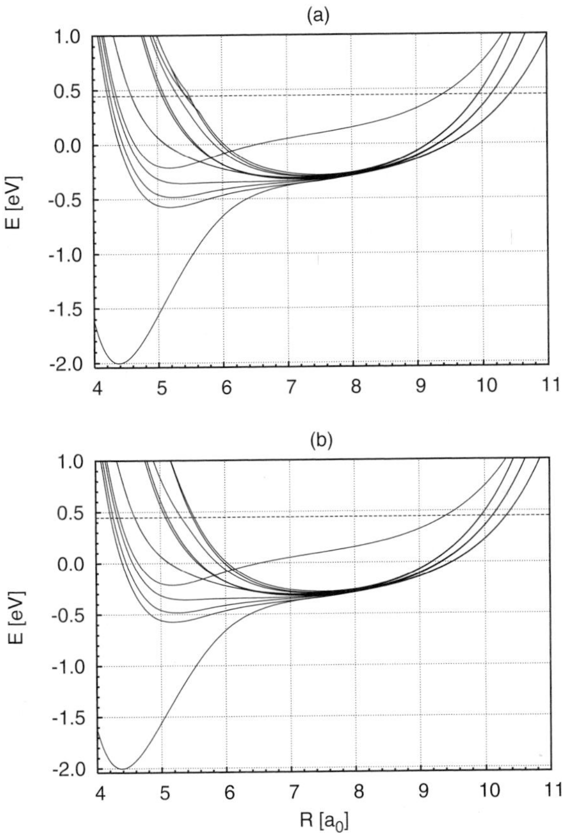

Fig. 4.33. Comparison of the lowest 17 adiabatic states (along the Br_2 bond distance and for a frozen Ar lattice) of a full (a) and a reduced (b) DIM description of $Br_2 : Ar$. The dashed line marks the energy of the Franck-Condon vertical $B \leftarrow X$ transition [196].

The DIM Hamiltonian discussed so far has the disadvantage that it requires the treatment of the complete set of valence states. On the other hand, in dynamics simulations often one is interested in a particular subset of electronic states only. Their relevance could be dictated by experimental conditions such as the selectivity provided by the laser spectrum (see Sect. 4.2). In order to reduce the numerical overhead in related dynamics simulations, a reduced DIM description has been proposed in [196]. Using $Br_2 : Ar$ as an example, it has been shown that the disturbance of the block-diagonal structure of the Hund's case c zero-order Hamiltonian due to the external potential provided by the matrix environment can be classified with respect to intra- and inter-symmetric couplings. Their interplay can be disentangled according to the different gas phase dissociation limits of Br_2.

In Fig. 4.33 the adiabatic potential energy curves along the Br_2 bond distance are compared for frozen Ar positions using the full 36 valence state DIM (panel (a)) and a reduced 17 valence state DIM (panel (b)) representation. The reduction has been targeted to a description of the B $\leftarrow$ X transition below the Franck-Condon vertical excitation such as to describe predissociation dynamics. The 17 states of the reduced model include all states within the ground state gas phase dissociation limit plus the B state which belongs to the second dissociation limit. Careful inspection of the energies reveals that the deviation from the full DIM description is below 10 % within the Franck-Condon vertical excitation range, even for the highest adiabatic states [196].

4.7 Classical simulations of adiabatic molecular dynamics

O. Kühn and M. Schröder

4.7.1 Energy redistribution in the lattice

The DIM method is ideally suited for the propagation of classical trajectories subject to forces generated "on-the-fly". For the time being transitions between different electronic states are neglected and the focus will be on the dynamics on adiabatic PES, W_n. Specifically, the system $Cl_2 : Ar$ [139] is considered which is shown in Fig. 4.34. In this figure different types of lattice atoms have been labeled: These are the so-called phonon atoms (1,2), the head-on atoms (3), the window atoms (4), and the belt atoms (5) which had already been introduced in Sect. 4.4.2. The naming is rather obvious except for the phonon atoms which will be discussed below. Overall these atoms play an important role for the initial dynamics after photoexcitation due to their proximity to the Cl_2.

First, let us consider excitation to the C state, i.e., the trajectory starts with momentum zero after being vertically displaced from the ground to the excited C state ($R_{Cl-Cl} = 3.98a_0$). This will cause rapid bond breaking and subsequent collision with the collision head-on atoms (3). Due to the similar masses the kinetic energy of the chlorine atoms is mostly transferred to the head-on atoms on impact. After the collision, the affected argon atoms start to move and hit the next argon atoms in $\langle 110 \rangle$ direction as shown in Fig. 4.35. In this way the kinetic energy is transported through the lattice, that is, a

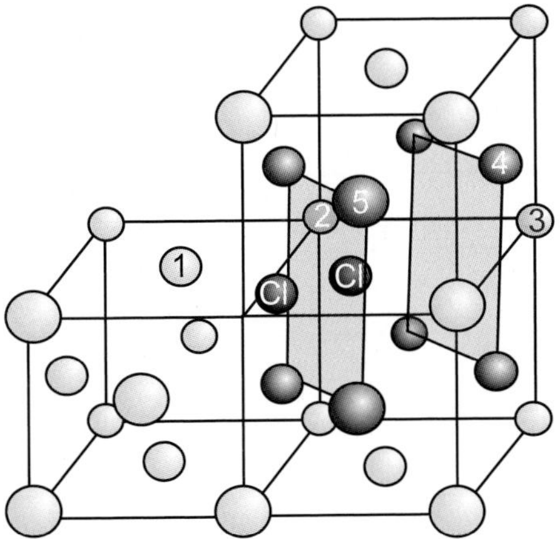

Fig. 4.34. Argon fcc lattice with a Cl_2 chromophore in a double substitutional site in a relaxed geometry. The following types of argon atoms have been marked: 1,2 phonon atoms, 3 head-on atoms, 4 window atoms, and 5 belt atoms (cf. Fig. 4.26). The belt and window atoms are spanning shaded rectangles for guidance. For the classical simulations a box containing besides the Cl_2 molecule 1272 argon atoms has been used with periodic boundary conditions.

shock wave is generated (see also simulations for $I_2 : Kr$ in [202]). It turns out that this imposes a practical constraint on the simulation time depending on the size of the box. In the present case the boundaries are reached after about 560 fs which roughly corresponds to the roundtrip time for the $Cl - Cl$ bond.

However, the energy redistributes not only in $\langle 110 \rangle$ direction. In Fig. 4.36 the displacement of lattice atoms in different layers is shown with respect to the Cl_2. The left column of Fig. 4.36 shows contour plots of the displacements of lattice atoms in the layer above the chlorine molecule. This layer contains belt and window atoms, see Fig. 4.34. Closest in distance to the center of mass of the chlorine molecule are the belt atoms. They are displaced because of the lack of repulsive forces, due to dissociation of the chlorine molecule and their movement into the empty space left by the Cl-fragments. When the chlorine returns to the inner turning point of the potential, according to the smallest Cl-Cl bond length, the now slightly closed belt will be opened by the chlorine again. The movement of the belt atoms is primarily perpendicular to the molecular axis. The largest displacement can be assigned to the window atoms, which will be pushed outwards by the chlorine atoms, when passing through the center of these windows. Upon collision of the Cl with the first head-on atom (3) the energy is transferred to the next Ar atom along this direction which starts to move thus passing through the same kind of window. This process continuous as seen from Fig. 4.35 and 4.36. The second layer will also be influenced by the dynamics of the chlorine molecule as seen in the

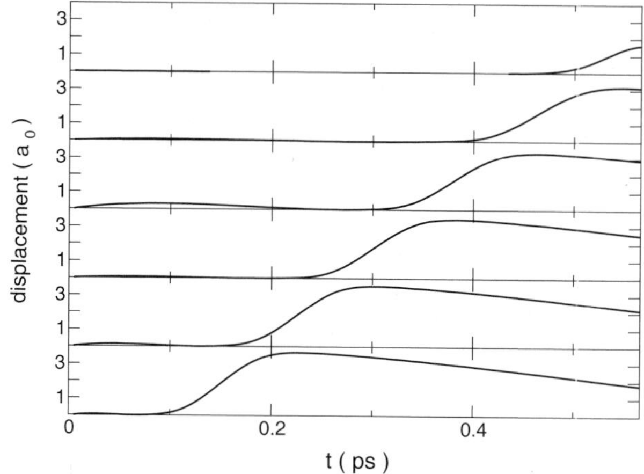

Fig. 4.35. Shock wave initialized by photoexcitation of Cl_2 to the C-state. The different panels show the displacements of the Argon atoms along ⟨110⟩ direction with respect to their equilibrium distance starting with the head-on atoms in the lowest panel (cf. Fig. 4.34).

right panel of Fig. 4.36. When belt, window or the other atoms in the first layer are displaced, they will affect the atoms in the next layer. Lattice atoms in the further layers are only marginally affected on this time scale.

Next the dynamics upon B state excitation is considered. Classical trajectories are launched on the $^3\Pi$-state PES assuming a vertical excitation from the minimum of the ground state PES ($R_{Cl-Cl} = 3.98a_0$). The time-dependence of the Cl-Cl bond length and of the displacement from equilibrium for some Ar atom groups is shown in Fig. 4.37 (dashed-dotted lines). The dynamics of the Cl-Cl bond appears to be free up to about $11a_0$ at ca. 150 fs. Afterwards one notices an appreciable motion of the lattice atoms and here primarily of the head-on atoms, similar to the C-state excitation in Fig. 4.35. The interaction with the other atoms is much weaker in this initial time interval such that the belt atoms tend to move slightly into the site emptied by the Cl_2 while the windows open, cf. C-state case in Fig. 4.36.

The dynamics upon excitation from a displaced geometry in the ground state ($R_{Cl-Cl} = 4.2a_0$) mimicking a vertical excitation from a vibrationally excited state to the $^1\Pi$-state is also shown in Fig. 4.37 (solid lines). The dynamics of the collision between the Cl atoms and the Ar atoms resembles that of the $^3\Pi$-state case. Inspecting Fig. 4.37 one notices that prior to the first collision with the solvent cage the stretching of the Cl-Cl bond is almost decoupled from that of the Ar lattice. In other words, during the first ca. 150 fs after Franck-Condon like photoexcitation a one-dimensional description of the dynamics is justified, where only the Cl-Cl bond changes in the frozen Ar lattice. This will be important for the model to be derived in Sect. 4.10.

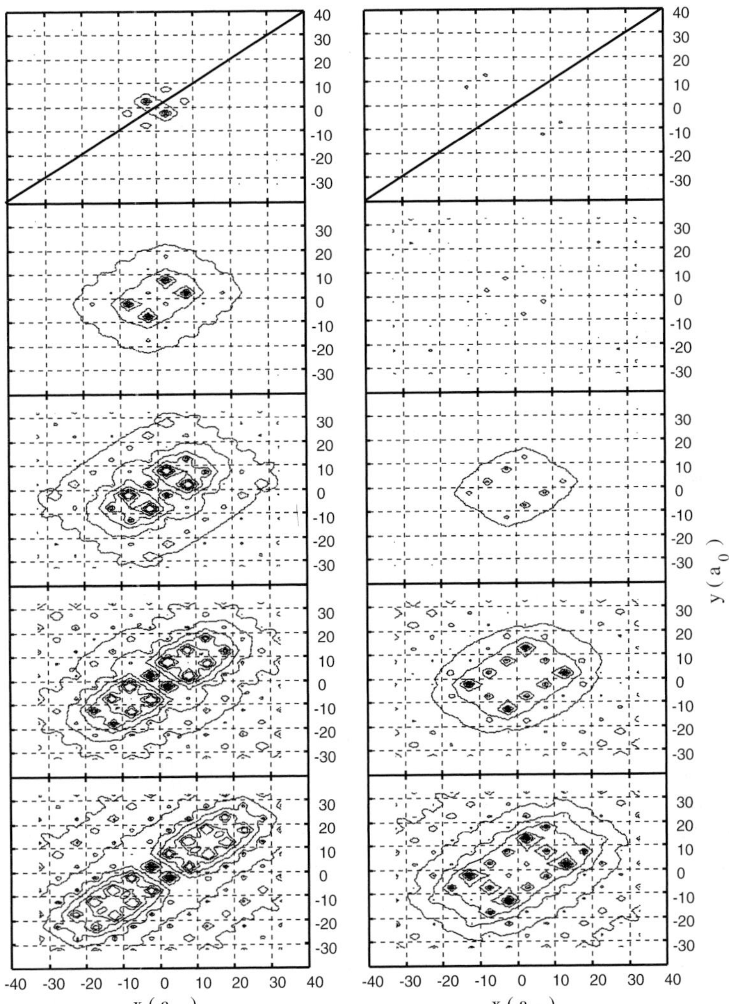

Fig. 4.36. Displacement of Ar atoms from their initial positions after $C \leftarrow X$ vertical transition in layers above/below the one containing the chlorine. Left column: one layer above chlorine layer, right column: two layers above chlorine layer. The rows are showing snapshots from top to bottom at 100, 200, 300, 400, and 500 fs. The solid line in the upper panels indicates the $\langle 110 \rangle$ direction. The contour lines show the displacements perpendicular to this axis, x and y are the coordinates in the respective layer [165].

4.7.2 Impulsive phonon mode excitation

In Sect. 4.4 experimental results on the impulsive excitation of coherent phonons have been presented. Since these zone boundary phonon vibrations

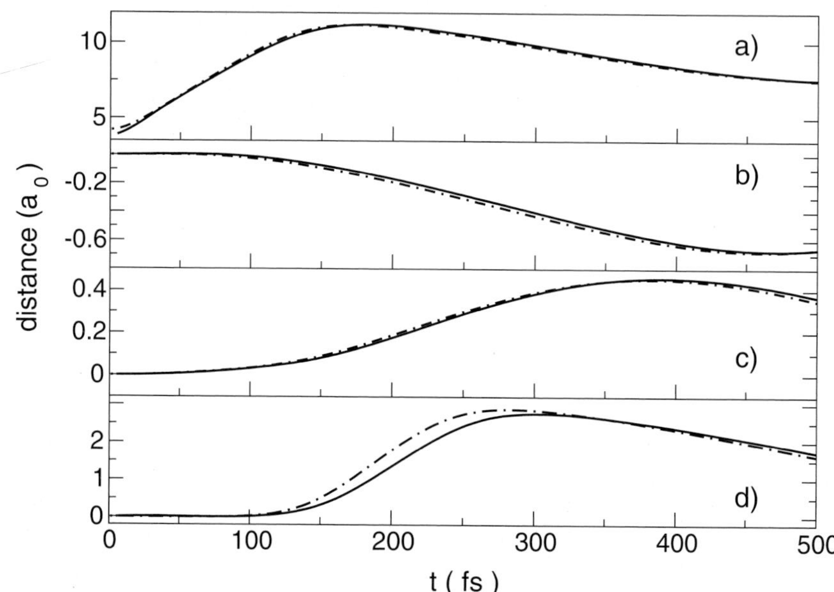

Fig. 4.37. Classical dynamics of interatomic distances after vertical Franck-Condon excitation from the potential minimum of Cl_2 in the ground state to the $^3\Pi$ state ($R_{Cl-Cl} = 3.98a_0$, dash- dotted lines) and to the $^1\Pi$-state from a displaced ground state geometry ($R_{Cl-Cl} = 4.2a_0$, solid lines): (a) *Cl-Cl*, (b) belt *Ar* atoms, (c) window *Ar* atom, (d) head-on *Ar* atoms. The matrix atom distances are defined by the deviation from the equilibrium position perpendicular to (b,c) and along (d) the Cl_2 axis (see, Fig. 4.34) [203].

are initiated in the moment of electronic excitation and not affected by the subsequent dynamics of the chromophore, they can be reasonably studied by adiabatic molecular dynamics [139]. At time zero, the phonon atoms (1,2) experience a sudden kick due to the change of the DIM state from X to B which comes along with a change of the anisotropic Cl_2-*Ar* interaction. In Fig. 4.38 typical trajectories are shown which are launched slightly above the gas phase dissociation limit of the B state. In Fig. 4.38a the *Cl-Cl* bond distance is shown for reference. After hitting the head-on atoms, the departing *Cl* atoms recombine and perform quasiperiodic motion in the bound part of the B state potential. It should be noted that in contrast to Fig. 4.35 there is no pronounced shock wave generation under the present excitation conditions. The energy redistributes over the whole simulation box thus minimizing artifacts due to the periodic boundary conditions. The belt atoms (Fig. 4.38b)

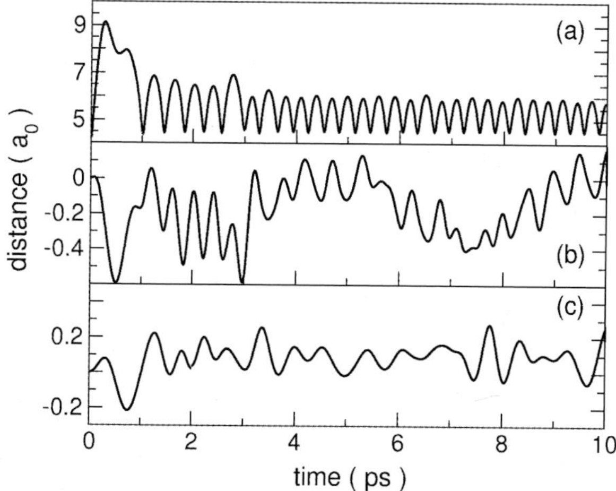

Fig. 4.38. Exemplary trajectories for a vertical excitation from the X to the B state where the *Cl-Cl* bond distance has been assigned the value of 4.2 a_0 which corresponds to an energy slightly above the gas phase dissociation limit of Cl_2. (a) *Cl-Cl* bond distance, (b) belt atoms, and (c) phonon atoms. The distance in panels (b) and (c) is defined with respect to the radial motion of the respective atoms away from the *Cl-Cl* bond axis [139].

essentially follow the dynamics of the *Cl-Cl* bond at least during the first 3-4 ps, that is, they perform driven oscillations as expected from the discussion of Fig. 4.36. The appearance of other frequencies such as the low-frequency modulation originates from the coupling to other lattice modes. Most notable, however, is the motion of the phonon atoms in Fig. 4.38c which appears to be essentially decoupled from the *Cl-Cl* trajectory. Initially the phonon atoms are pushed outwards due to the expansion of the electronic wavefunction. A detailed comparison between the trajectory in Fig. 4.38c and the oscillatory part of the pump-probe spectra in Fig. 4.24d reveals a good correlation between cage compression (minima in Fig. 4.38c) and maximum sensitivity in the pump-probe signal. Furthermore, a Fourier transformation of the adiabatic trajectory yields a pronounced amplitude at 75 cm^{-1} in fair agreement with the experimental value of 66.7 cm^{-1}, cf. Fig. 4.25. This supports the excitation mechanism discussed in Sect. 4.4.2.

4.8 Semiclassical simulations of nonadiabatic dynamics

A. Cohen, R. B. Gerber

4.8.1 Surface hopping

Extensive simulations were carried out for the ultrafast dynamics of the photochemistry of F_2 in an Ar matrix, including the effects of nonadiabatic transitions, by the semiclassical *surface hopping* method [21, 55, 204]. Physically, $F_2 : Ar$ is more attractive, at least from the perspective of the semiclassical simulations, as a model system providing useful insights also for related systems, such as $ClF : Ar$ and $Cl_2 : Ar$. The advantages of the surface hopping simulations are: (a) All 36 relevant electronic states and the nonadiabatic transitions between them can be treated. (b) The approach permits simulations employing a relatively large number of medium atoms, which throws light on the role of different degrees of freedom in the case of the solid. Specifically, calculations were carried out for F_2:Ar_{54} [21] and for F_2 in a slab of 255 argon atoms with periodic boundary conditions [204]. (c) The surface hopping algorithm was effectively combined with the DIM potentials and nonadiabatic couplings, in on-the-fly [205] calculations of the interactions. In particular, the nonadiabatic couplings can be computed without recourse to diabatic approximations. With all these ingredients, the surface hopping simulations have provided a flexible, versatile laboratory for the nonadiabatic photochemical dynamics of dihalogens in rare-gas solids. The semiclassical simulations have, however, also important disadvantages. Since the nuclear motions are treated classically in the surface hopping approach, it cannot be used directly to explore nuclear coherence effects, and possibilities of coherent control. It was found, however, that much can be learned by combining semiclassical simulations with reduced dimensionality quantum dynamics calculations. The semiclassical simulations showed that for a limited time window of the process, a feasible, reduced-dimensionality treatment should be justified, and the latter was pursued in quantum calculations that address coherence effects, as will be discussed in Sect. 4.10 [204].

Pioneering calculations on nonadiabatic dynamics of HX and X_2 (X-Halogen) in noble-gas solids, using a semiclassical treatment and DIM potentials, were presented by Gersonde and Gabriel [50]. In these simulations, spin-orbit coupling effects were not included. An important study of nonadiabatic dynamics of I_2 in solid and liquid xenon, using the surface hopping approach, is by Batista and Coker [22]. Relatively few trajectories were computed for this system, which limits the information obtained in the process involved. Algorithmically, the study on F_2:Ar benefited greatly from the surface hopping/DIM simulations by Krylov and Gerber on HCl in solid Ar [157], and by Niv, Krylov, and Gerber [158, 206], on HCl inside and on the surface of argon clusters.

The semiclassical surface hopping algorithm employed in the present simulations follows closely the approach of Tully [205]. Initially, the nuclei evolve on one of the adiabatic potential surfaces $W_j(\mathbf{R})$, where $\mathbf{R}$ denotes the nuclear position vector. As long as nonadiabatic transitions do not occur, the system is propagated by classical molecular dynamics on that surface. At each configuration $\mathbf{R}$, all adiabatic potential surfaces $W_j(\mathbf{R})$, $j = 1 \ldots 36$ and the corresponding electronic states $\varphi_j(\gamma_1, \gamma_2, \mathbf{R})$ are generated from DIM. Here, γ_1 and γ_2 denote the orientation angles of the relevant p-orbitals of the F atoms discussed in Sect. 4.6. The adiabatic states $\varphi_j(\gamma_1, \gamma_2, \mathbf{R})$ depend parametrically upon the nuclear configuration $\mathbf{R}$. Since these states are generated on the fly along a trajectory $\mathbf{R}(t)$, one can write $\varphi_j(\gamma_1, \gamma_2, t)$ for that trajectory. The electronic degrees of freedom are described in this approach by a wave function of the coordinates γ_i only, which evolves in time according to the electronic Schrödinger equation

$$i\hbar \frac{\partial}{\partial t} \Psi(\gamma_1, \gamma_2, t) = H^{el}(\gamma_1, \gamma_2, \mathbf{R}(t)) \Psi(\gamma_1, \gamma_2, t) \,. \tag{4.19}$$

The adiabatic states $\varphi_j(\gamma_1, \gamma_2, t)$ diagonalize the electronic Hamiltonian on the right-hand side of (4.19), yielding the adiabatic PES, $W_j(\mathbf{R}(t))$. We expand $\Psi(\gamma_1, \gamma_2, t)$ in the adiabatic basis $\varphi_j(\gamma_1, \gamma_2, t)$ as follows

$$\Psi(\gamma_1, \gamma_2, t) = \sum_{j=1}^{36} C_j(t) \varphi_j(\gamma_1, \gamma_2, t) \,. \tag{4.20}$$

This results in the following equations for the coefficients $C_j(t)$ given by Tully [205]

$$i\hbar \dot{C}_k(t) = -i\hbar \sum_j C_j(t) \langle \varphi_k | \dot{\varphi}_j \rangle + C_k(t) W_k(\mathbf{R}(t)) \,. \tag{4.21}$$

Using the "chain rule", one gets

$$\langle \varphi_k | \dot{\varphi}_j \rangle = \dot{\mathbf{R}} \langle \varphi_k(\gamma_1, \gamma_2, \mathbf{R}) \mid \nabla_{\mathbf{R}} \varphi_j(\gamma_1, \gamma_2, \mathbf{R}) \rangle = \dot{\mathbf{R}} \mathbf{D}_{kj} \,. \tag{4.22}$$

Applying the Hellmann-Feynman theorem [207], the adiabatic coupling can be calculated analytically (for $k \neq j$)

$$\mathbf{D}_{kj} = \frac{\langle \varphi_k(\gamma_1, \gamma_2, \mathbf{R}) \mid \nabla_{\mathbf{R}} H^{el}(\gamma_1, \gamma_2, \mathbf{R}) | \varphi_j(\gamma_1, \gamma_2, \mathbf{R}) \rangle_{\gamma_1, \gamma_2}}{W_j(\mathbf{R}) - W_k(\mathbf{R})} \,. \tag{4.23}$$

$H^{el}(\gamma_1, \gamma_2, \mathbf{R})$ in (4.23) is the full electronic Hamiltonian and $\dot{\mathbf{R}}$ is the velocity vector of the nuclei. Equations (4.22) and (4.23) provide a convenient way for an analytical on-the-fly evaluation of the time-dependent quantities $\langle \varphi_k | \dot{\varphi}_j \rangle$ which represent the nonadiabatic couplings. The diagonalization procedure, by which the adiabatic states defined here are constructed, leaves these states undefined to within a time-dependent phase factor. The treatment of the adiabatic eigenfunction phases used here was developed by Krylov et al. in [195] and previously used in [158, 194, 195, 205, 206]

The classical propagation of the nuclei together with equations (4.19-4.21) for the electronic state, are the basis for the surface hopping method. Suppose that initially $C_k(t_0) = 1$ and $C_i(t_0) = 0$ for $i \neq k$. The nuclei are then propagated classically on the adiabatic surface $W_k(\mathbf{R})$ using, for instance, the Gear algorithm [208], with a 0.01 fs time step. At each point along the trajectory, all 36 potential surfaces are constructed, and the nonadiabatic couplings $\langle \varphi_k | \dot{\varphi}_j \rangle$ are obtained. Equation (4.21) for the coefficients C_i ($i = 1, \ldots, 36$) is solved numerically using the Runge-Kutta algorithm, and their variation in time is followed. A "hop" from surface k to j takes place following Tullys "fewest switches" criterion [205]: If b_{kj} is the flux of the electronic population from state k to state j, so that the change in the population of the state k, $\dot{a}_{kk}$, is given by

$$\dot{a}_{kk} = \sum_{i \neq k} b_{ki} \,, \tag{4.24}$$

where b_{ki}, in the adiabatic representation used here, is given by

$$b_{ki} = -2\mathrm{Re}(a^*_{ki} \mathbf{R} \mathbf{D}_{ki}) \,, \tag{4.25}$$

and a_{kj} is defined as $C_k C^*_j$, and C_k are the time-dependent expansion coefficients defined in (4.20). A uniform random number ρ between 0 and 1 is generated at each integration time interval Δt. If

$$\frac{\Delta t b_{ki}}{a_{kk}} < \rho < \frac{\Delta t (b_{ki} + b_{kj})}{a_{kk}} \tag{4.26}$$

a hop from state k to j is performed, given that the energy conservation can be achieved via scaling of velocities. The adjustment of the velocities is performed as outlined by Coker and Xiao [209]. First, the velocities $\mathbf{V}^{pred}$ at time t are predicted based on the $\mathbf{V}$ at $t = t - \delta t$:

$$\mathbf{V}^{pred} = \mathbf{V} - \frac{(W_j - W_k)\mathbf{D}_{kj}}{M \mathbf{D}_{kj} \mathbf{V}} \,, \tag{4.27}$$

where M are the masses of the nuclei; the velocity changes parallel to the nonadiabatic coupling vector $\mathbf{D}_{kj}$ that was defined in (4.23). Second the scaling factor s is computed

$$s = \left(\frac{2(E^{tot} - W_j(t))}{M \mathbf{V}^{pred} \mathbf{V}^{pred}} \right)^{1/2} \,, \tag{4.28}$$

where E^{tot} is the total energy; W_j is the potential energy of the state into which the hopping is to be performed. Third, all the velocities are rescaled to give the exact energy conservation, $\mathbf{V} = s\mathbf{V}^{pred}$. Once a hopping event has occurred, the procedure continues with the nuclei being propagated on the surface $W_j(\mathbf{R})$. In earlier papers on HCl [158, 194, 195, 206] a slightly different version of surface hopping was used, where the probability for a hop was tested for a small time interval δt, chosen such that the results were stable with respect to changes in δt. Here, the "fewest switches" algorithm is chosen to avoid the computational cost associated with the time interval calibration procedure.

Note that one of the problems of the surface hopping approach is that it incorrectly overestimates the role of electronic dephasing: Electronic dephasing in these surface hopping simulations can be said to take place on the first surface hopping from the initial electronic state. There is a very different hybrid quantum (electronic)/classical (nuclear) approach, that is referred to as the mean field approach, the Quantum/Classical Time-Dependent Self-Consistent Field (Q/C-TDSCF) approach, and other titles [210–215]. In this approach, the nuclei move on the average field due to the electronic state, which evolves as a coherent superposition of the adiabatic states. Obviously, this underestimates (neglects) the electronic decoherence effect. This has been studied in depth by Truhlar and coworkers [216, 217] who provided alternative, in cases advantageous, surface hopping methods. This is still an open topic, and a very active field of research. A comprehensive, elegant review is by Stock and Thoss [218]. By the available indications, the Tully-style surface hopping approach seems, however, to work well both for $HCl{:}Ar$ and $F_2{:}Ar$. Partly, this could be due to the fact that these systems are different from the weak-coupling, rare-hopping type of systems for such the Tully approach may be problematic. It is also reasonable to expect that in condensed phase applications, with the extensive averaging that takes place, things are more favorable for the approach than for the single-collision gas-phase situations.

4.8.2 Electronic cage effects

As noted in Sect. 4.8.1, the semiclassical surface hopping simulations for F_2 in an argon medium provide a very useful "laboratory" for the dynamics of electronic states of dihalogens in a noble-gas environment. One must keep in mind, of course, that coherence effects of the nuclear motions are outside the scope of this approach since the nuclei are treated classically in the surface hopping method. However, the results provide a wealth of information on topics that include the dynamics of the electronic state populations, the nature and rates of the nonadiabatic transitions that occur, the propensities of different electronic states, and other important properties of the processes. Previous calculations were carried out for $F_2{:}Ar_{54}$ [21, 55] and very recent results were obtained for F_2 embedded in a slab of 255 argon atoms with periodic boundary conditions applied at the ends [204]. Results from both systems are included in the following discussion, but the focus is on the more recent findings.

4.8.2.1 The ultrafast spin-flip effect

A result of considerable interest is shown in Fig. 4.39 which presents the build up of population in the $^3\Pi$ state, following excitation of the F_2 into the $^1\Pi_1$ state. An instantaneous Franck-Condon promotion into the excited state was used in the model. Evidently, the population of the $^3\Pi$ state begins to rise very steeply a short time ($t \approx$ 25-30 fs) after the sudden excitation to the singlet state. The $^3\Pi$ population reaches a first maximum around 50 fs, decreases somewhat around 100 fs, then rises again and reaches a second and higher maximum near 200 fs. The result is surprising: F is a light atom of modest spin-orbit coupling strength of about 404 cm^{-1} only. For such species, transitions from singlet to triplet are typically much slower, with time scales in the nanoseconds or longer. The ultrafast spin-flip in this system is thus of much interest. The result of Fig. 4.39 is from calculations on F_2:Ar_{255} (with periodic boundary conditions at the ends of the slab), i.e. from a model of F_2 in a crystalline matrix [204]. The effect of ultrafast spin-flip was first discovered in surface hopping/DIM simulations of F_2:Ar_{54} by Niv, Bargheer and Gerber [21]. There are significant differences of details between the results of the simulations for the cluster and for the matrix which have interesting physical interpretations in their own right, but semi-quantitatively at least the findings are similar and consistent in both cases. The ultrafast spin-flip was experimentally observed by Bargheer and Schwentner for ClF in solid Ar [18] (see Sect. 4.2.4), and interpreted in that study on the basis of the simulation results for F_2:Ar_{54}. Reduced dimensionality quantum simulation, in which the argon atoms are held fixed, also show the ultrafast spin-flip effect, see Sect. 4.10 [31].

The physical interpretation of the ultrafast spin-flip is straightforward [204]: within a time-scale of 20fs or so after the excitation into the $^1\Pi_1$ state, the mutual distance between the F atoms reaches approximately 6 a_0, (more than twice the equilibrium distance in the ground state). At this configuration, the chemical valence interactions that lock together the spin of the electron pair in the $^1\Pi_1$ singlet are weakened to the point that the modest spin-orbit coupling interaction suffices for spin-flip. Another equivalent explanation is that at this configuration the singlet and triplet states are already near- degenerate and the spin-orbit coupling suffices to induce the transition.

The experiments on ClF demonstrate the occurrence of an ultrafast subpicosecond ($t \approx$ 250 fs) spin-flip in this system, but the time scale of this process is considerably longer than for the F_2 simulations. Thus the differences between the systems also affect considerably the details. In particular, the time for which the $Cl - F$ distance reaches the singlet-triplet near degeneracy range is much longer. This is partly due to the different potentials, and partly due to the fact that the rise time of the pulse in the experiments adds to the transition time, compared with the prompt Franck-Condon excitation used in the simulations. This development suggests to search for such ultrafast spin-flip also in other systems.

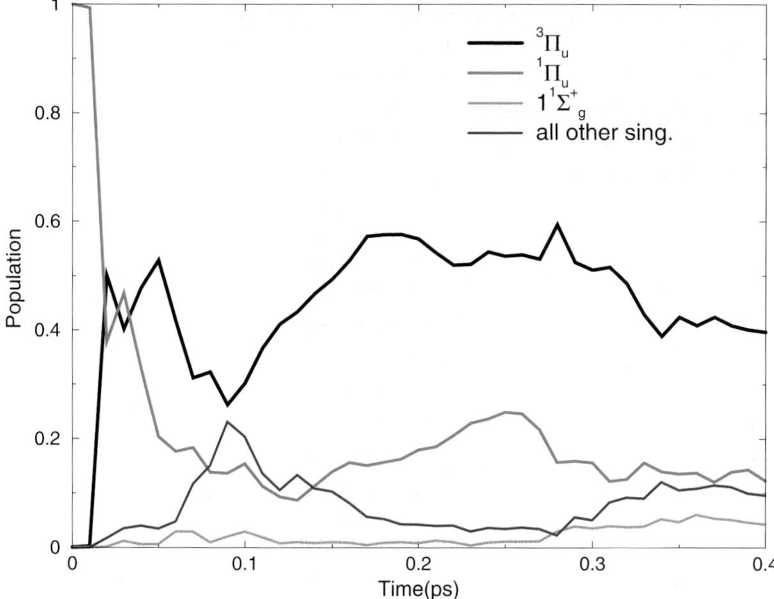

Fig. 4.39. F_2:Ar_{255} (at $T = 8K$): The $^3\Pi_u$, $^1\Pi_u$, $X^1\Sigma_g^+$ as well as the sum over all other singlet state populations vs. time, averaged over 35 caged trajectories.

Having established the ultrafast formation of $^3\Pi$ population, it is of interest to compare the population of this triplet with that of other states in the system. Fig. 4.39 also shows the population in time of the initially excited $^1\Pi_1$ state, the ground state $X^1\Sigma_g^+$ and the sum of all other singlets. It is evident from the results that shortly after the excitation the population of the $^3\Pi$ exceeds that of any of the singlets. In fact, for most of the interval $t \leq 0.4$ps, the population of $^3\Pi$ is larger than the populations of all singlets combined. The $^1\Pi_1$ initially excited state drops steeply in population after the first 100 fs, which demonstrates the efficiency of nonadiabatic transitions in this system. It has been (and still is) quite common in simulations of photochemistry in matrices to deal only with the initially excited state, assuming that this only ignores recombination events (see references in the review article [43]). For the present system such an approach is qualitatively wrong.

We note the small, but conceptually interesting, population in the ground state $X^1\Sigma_g^+$ on a subpicosecond timescale. This is ultrafast recombination onto the ground electronic state. The dominance of $^3\Pi$ over all other singlets, except the ground state is at least partly due to the attractive potential well of $^3\Pi$, as the excited singlet states are all repulsive. These results still await experimental confirmation. There is already evidence from the work of Schwentner and coworkers, that indeed for ClF the $^3\Pi$ population is considerably greater than that of all singlets in the ultrafast time domain studied.

Direct experimental studies of populations of the repulsive electronic states in systems of this type are not yet at hand.

4.8.2.2 Populations of states of different electronic angular momentum

Next the role of Ω is considered, that is, the projection of the electronic angular momentum on the molecular axis, upon the populations. Fig. 4.40 shows the populations in time of $^3\Pi_{0+}$, $^3\Pi_{0-}$, $^3\Pi_1$, and $^3\Pi_2$ as obtained from the surface hopping simulations of F_2 in a solid argon matrix. Since the populations are not equal during the time window shown, Ω clearly has an influence on the electronic state populations. Within the limited accuracy due to the modest number of only 35 trajectories calculated, $^3\Pi_{0-}$ and $^3\Pi_{0+}$ are equally populated throughout the process. This is expected to be the case, since the initial state is prepared without any bias with regard to the $\pm$ quantum number, and the interactions should preserve the symmetry. Both the spin-orbit and the electrostatic interactions between the argon atoms and the $F(^2P)$ atoms can influence the Ω populations. For changes in the triplet state populations for $t \geq 200$fs it seems possible that the electrostatic $Ar/F(^2P)$ interactions are the main factor influencing the Ω populations. Note that the build up of the $^3\Pi_1$ state is due to dynamical consequences of these interactions, since the $^3\Pi_2$ state has a somewhat deeper attractive well. Only at the very end of the simulation interval of 0.4 ps the populations suggest that the role of the corresponding attractive well-depth is decisive, that is, Pop $(^3\Pi_2) \approx$ Pop $(^3\Pi_{0+} + {}^3\Pi_{0-}) \geq$ Pop $(^3\Pi_1)$. Full understanding of the Ω state populations remains a challenge for the future.

4.8.2.3 Role of energy transfer to the matrix and the validity of a reduced-dimensionality treatment

Consider now energy transfer from the photoexcited F_2 molecule to the surrounding Ar environment [21, 204]. It is found that, subject to some dependence on the specific initial conditions of each trajectory, an F atom can strike an argon atom of the cage as early as 30 fs after the sudden (Franck-Condon) excitation. The kinetic energy acquisition of the argon atom after the collision is maximal around 70 fs. The energy of the Ar atom struck directly by the F can reach for this system values of the order of 1eV, and this indeed is the initiation of a shock wave propagating into the crystal. This topic has been discussed more extensively for another system in Sect. 4.7.

The above results have, however, important implications for the simulations of reduced dimensionality treatment of such systems: For the present system, for $t \leq 30$ fs, it should be justified to neglect energy transfer to the solid, and treat the lattice as static. With this, quantum mechanical modeling of the system, including only, say, the $F-F$ motion in the subspace of

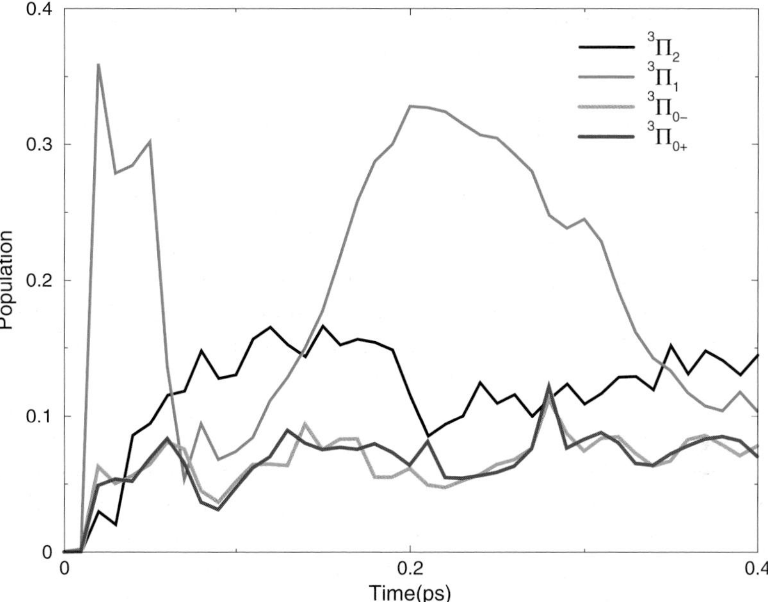

Fig. 4.40. F_2:Ar_{255} (at $T = 8K$): The $^3\Pi_\Omega$ ($\Omega = 2, 1, 0+, 0-$) states population vs. time as obtained from an average over 35 caged trajectories.

relevant electronic states becomes feasible. This has been turned into a basis for treating quantum-mechanical coherence effects on ultrafast timescales, including spin-flip, in such systems [31, 219], see Sect. 4.10. In fact the usefulness of the semiclassical simulation for the quantum modeling is relevant also to other manifolds of electronic states involved: the results of the surface hopping simulations show, that only a small part of the 36 electronic states play an important role. The computational effort of the quantum wave packet treatment scales as N^2, where N is the number of electronic states, hence considerable simplification can be gained. The combination of semiclassical simulations and reduced dimensionality treatments creates a powerful tool that addresses the many-body nature of the systems, yet makes possible the description of coherence effects of nuclear dynamics.

4.9 Alignment and orientation of molecules in matrices

T. Kiljunen and B. Schmidt

In the previous sections of this chapter the molecular photodynamics of matrix-isolated dihalogen molecules upon electronic excitation has been investigated. In all those studies the initial alignment or orientation of the guest molecule with respect to the crystallographic axes of the host lattice is essential and will now be treated in detail. Assuming alignment is justified for larger chromophores, occupying two (or more) adjacent substitutional sites (see Sect. 4.2.1). In that case, the guest rotational degrees of freedom are essentially frozen which allows, e.g., for the construction of low-dimensional models of (coupled) host–guest systems [54,55]. In contrast, smaller molecules occupying mono–substitutional sites are — to a certain extent — exhibiting rotational motion (see Sect. 4.2.3). The degree of hindering depends strongly on the nature of the guest–host interaction which shall be loosely termed "crystal field" throughout the following text. With increasing field strength, the rotational states undergo a gradual transition from free rotor states via hindered rotor states to librational states. In the latter case, the rotational motion is strongly frustrated and the molecular axes undergo angular vibrations about certain crystallographic directions. In general, the crystal field is determined by a subtle interplay of attractive and repulsive interaction of the guest molecule with the surrounding rare gas atoms. For example, the energetically preferred axes of the $HF : Ar$ and $HCl : Ar$ systems are along the $\langle 111 \rangle$ direction with the guest molecules pointing toward the triatomic windows formed by the nearest neighbors in the face–centered cubic rare gas crystal. [220,221]. In contrast, the $ClF : Ar$ system favors the $\langle 100 \rangle$ direction, i.e. tetra-atomic windows, while in a Kr matrix the ClF molecules are preferentially oriented directly toward the nearest neighbors located on the $\langle 110 \rangle$ axes [23, 24].

Obviously, the different directionality of rotational states of matrix-isolated molecules can have enormous implications on their photochemical properties: A preferred orientation of the guest molecules toward nearest neighbors typically results in strong reflections from the solvent cage eventually leading to recombination of the photofragments. In contrast, a molecular orientation toward windows of the first solvation shell is more likely to lead to cage exit and permanent separation of the photofragments (see Sect. 4.2.6). Hence, a promising approach to control the yield of matrix photodissociation proceeds via controlling the alignment and/or orientation of matrix-isolated molecules [222]. Similar approaches for librational control of photochemistry have already been suggested for diatomic molecules embedded in rare gas clusters [223–226]. It is the purpose of this section to review the state of the art in the manipulation of rotational degrees of freedom of matrix-embedded molecules, utilizing both time-independent and time-dependent methods, as documented in a recent series of articles [23, 24, 27–29]

For gas phase molecules the manipulation of rotational degrees of freedom has already reached a relatively mature state, for a review see [227]. Because the interaction of external fields with molecular dipole moments is limited to molecules with strong dipoles and requires very strong static fields [228, 229], efficient approaches to molecular alignment and orientation are nowadays based on the nonresonant interaction of intense laser fields with induced dipoles due to anisotropic molecular polarizabilities [230–234]. The use of linear, circular, and elliptic polarization allows to fix one, two, or all three axes of molecular rotation with respect to the laboratory frame. The corresponding molecular quantum states are termed pendular states which are defined as superpositions of field-free rotor states where the molecular axis librates about the polarization direction of the field [230, 235, 236]. In contrast to most other control mechanisms discussed in this book, light–induced alignment is usually achieved using nonresonant fields. This makes the control very robust with respect to variations of the laser pulse parameters, and in most cases excellent quantitative agreement between theory and experiment is found. Due to the cylindrical symmetry of the molecule–field interaction, pendular states occur in pairs of different parity which are connected by tunneling through the barrier provided by the $\cos^2$-shaped light-induced potential. This gives rise to a straight–forward approach of achieving molecular orientation: By applying a static field, oriented molecular states can be formed as superpositions of tunneling pairs of pendular states [237, 238]. The necessary electrostatic field strengths may be several orders of magnitude lower than in the pendular orientation approach using static fields alone [228, 229]. The concept of orientation by combined fields has been experimentally verified for molecules both in gas phase [239,240] and in small rare gas clusters [241,242].

In time-dependent molecular alignment, the rotational dynamics induced by pulsed light sources can be characterized by the following two limits: The adiabatic limit can be approached by using light pulses which are long compared to the rotational period of the molecule. In that case, however, alignment can only be achieved while the field is turned on [227]. Field-free alignment can be found in the nonadiabatic case, where the molecules interact with pulses much shorter than the rotational time scale. This leads to the formation of rotational wave packet states that can be considered as coherent superpositions of field-free states and the post-pulse evolution is dominated by corresponding quantum beats and rotational revivals [243]. Further "squeezing" of the rotational densities can be achieved by tailored "pulse trains", i.e., by suitable sequences of pulses [244–248]. In the present work the question shall be addressed in how far the results on the control of gas phase alignment and orientation can be transferred to the situation of matrix-isolated molecules. Based on the above concept of matrix-induced librational states, the following control objectives can be identified: (1) In a *cooperative* case, the targeted direction of alignment already coincides with the energetically preferred axes, i.e., the minima of the (internal) crystal field. In that case, the interaction with (external) laser fields can be utilized to further narrow the rotational distributions, possibly beyond the gas phase case. (2) In more challenging, *competitive* cases the control objective is to enforce the alignment of rotational states toward directions which correspond to saddles, or even maxima, of the crystal field [27]. To this end, the rotational densities will have to be turned by the external alignment field against the effect of the internal crystal field. Furthermore, the question is whether there exists a matrix analog to the gas phase mechanism of achieving additional "head vs. tail" orientation by means of combined fields.

4.9.1 Model and interactions

The following considerations shall be restricted to the case of a linear molecule in an electronically nondegenerate $^1\Sigma$ state occupying a mono-substitutional site in a fcc lattice, e.g., hydrogen halides or other small diatomic molecules in solid rare gases [43]. Furthermore, any coupling of the rotational to translational [249], vibrational [250, 251], or electronic [252] degrees of freedom of the guest molecule shall be neglected. While the latter two assumptions can be justified by the rather low temperatures considered in the present work, a noncoinciding center of mass and center of interaction could lead to rotation-translation coupling. However, the resulting eccentric rotational motion can be accounted for by appropriate down-scaling of the rotational constant B [253, 254]. Moreover, interaction between the external fields and the surrounding matrix can be neglected for rare gases as long as the field intensities are below damage thresholds of the respective rare gas crystal. Then the resulting Hamiltonian $\hat{H}$ for the rotational motion of the diatomic guest molecule can be written as

$$\hat{H}/B = (\hat{\mathbf{J}}/\hbar)^2 + \hat{V}_\kappa + \hat{V}_\alpha + \hat{V}_\mu \, , \tag{4.29}$$

where $\hat{\mathbf{J}}$ is the angular momentum operator of the embedded molecule and the remaining three terms stand for the interaction with the matrix ($\hat{V}_\kappa$) as well as with (nonresonant) laser fields ($\hat{V}_\alpha$) and electrostatic fields ($\hat{V}_\mu$).

Assuming a perfectly ordered fcc lattice environment, the guest–host interaction ("crystal field") is expanded in symmetry adapted spherical harmonics (SASHs) of the totally symmetric representation (A_{1g}) of the octahedral point group (O_h) [255]. Usually only the two lowest SASHs (V_4, V_6) are considered

$$V_\kappa(\theta, \phi; \kappa) = \kappa \left[K_4 V_4(\theta, \phi) + K_6 V_6(\theta, \phi) \right] \, , \tag{4.30}$$

where κ is a (dimensionless) scaling parameter. Building on the seminal studies of librational states in V_4 potentials [256], the effect of different shapes of hindering potentials was discussed in the literature [257, 258]. In the present work the parameters $K_4 = -52\sqrt{\pi}/(11\sqrt{21})$ and $K_6 = 16\sqrt{\pi}/(11\sqrt{26})$ are chosen such that the minima are along the six $\langle 100 \rangle$ directions with $V_{\text{min}}/\kappa = -1$, the saddles are along the twelve $\langle 110 \rangle$ directions with $V_{\text{sad}}/\kappa = 0$, and the maxima are along the eight $\langle 111 \rangle$ directions with $V_{\text{max}}/\kappa = 10/9$ [28, 29], see top panel of Fig. 4.41. This potential is very similar in shape to that obtained for a simplified DIM model of $ClF : Ar$ [23, 24]. According to the target directions relative to the extrema of the rotational potential, the case of alignment/orientation along $\langle 100 \rangle$ (minima) shall be termed cooperative, while the case of $\langle 110 \rangle$ (saddles) or $\langle 111 \rangle$ (maxima) shall be referred to as competitive in the following [27].

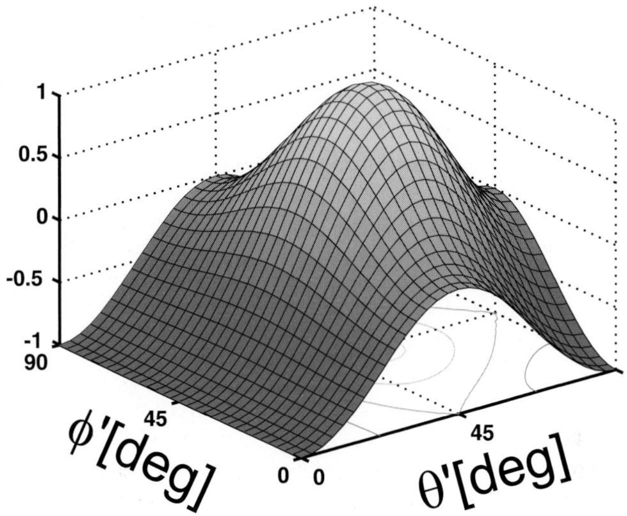

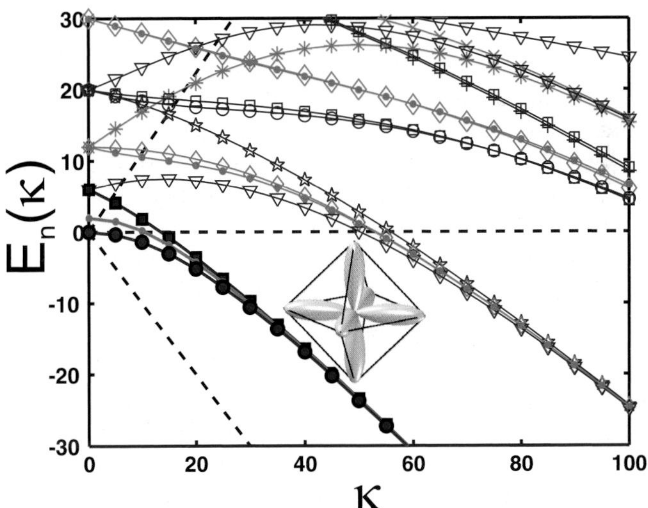

Fig. 4.41. Top: Octahedral "crystal field" model potential $V_\kappa(\theta,\phi)/\kappa$ of (4.30) with minima at the $\langle 100 \rangle$ crystallographic axes ($\theta = 0$), saddle points at the $\langle 110 \rangle$ axes ($\theta = \pi/2$, $\phi = \pi/4$), and maxima at the $\langle 111 \rangle$ axes ($\theta = \cos^{-1} 1/\sqrt{3}$, $\phi = \pi/4$). Bottom: Corresponding energy spectrum versus strength parameter of the potential. The dashed lines show the maxima, saddles, and minima of the potential energy surface. The various symbols denote different irreducible representations of the O_h point group. Ground state rotational density for $\kappa = 25$ is displayed in the octahedron. Reproduced with permission from [28].

The nonresonant interaction of a laser field $\mathcal{E}_\mathrm{L}$ with the induced dipole moment of the molecule leads to an effective, light–induced potential of the form

$$V_\alpha(\theta;\Delta\omega) = -(\Delta\omega\cos^2\theta + \omega_\perp)\,, \tag{4.31}$$

where the (dimensionless) interaction strength is related to the laser field and (parallel and perpendicular) components of the molecular polarizability through $\omega_{\|,\perp} = \mathcal{E}_\mathrm{L}^2\alpha_{\|,\perp}/(4B)$ with $\Delta\omega = \omega_\| - \omega_\perp$ and where θ is defined as the angle between the field direction and the molecular axis [237]. Note that polarization along fourfold $\langle 100\rangle$, threefold $\langle 111\rangle$, and twofold $\langle 110\rangle$ rotational axes reduces the octahedral (O_h) point group of the librational states to the D_{4h}, D_{3d}, and D_{2h} subgroups, respectively. Finally, the interaction of an electrostatic field $\mathcal{E}_\mathrm{S}$ with the dipole moment μ of the molecule is given by

$$V_\mu(\theta;\omega) = -\omega\cos\theta \tag{4.32}$$

with the dimensionless parameter $\omega = \mu\mathcal{E}_\mathrm{S}/B$. Due to the absence of inversion symmetry, the respective point groups are further reduced to C_{4v}, C_{3v}, and C_{2v}. The conversions of dimensionless units into practical ones are given by

$$\begin{aligned}
\mathcal{E}_S[\mathrm{kV\,cm^{-1}}] &= \frac{B[\mathrm{cm^{-1}}]}{0.0168\,\mu[\mathrm{D}]}\omega \\
\mathcal{I}_0[\mathrm{W\,cm^{-2}}] &= \frac{B[\mathrm{cm^{-1}}]}{2.11\times10^{-11}\Delta\alpha[10^{-30}\mathrm{m^3}]}\Delta\omega \\
T[\mathrm{K}] &= 1.44B[\mathrm{cm^{-1}}]\gamma \\
t[\mathrm{ps}] &= \frac{16.68}{B[\mathrm{cm^{-1}}]}\tau
\end{aligned} \tag{4.33}$$

with static field $\mathcal{E}_S$, laser intensity $\mathcal{I}_0$, temperature T, $\gamma = k_BT/B$ and time t, $\tau = 2Bt/h$.

4.9.2 From free rotors to librational states

The main distinction between the manipulation of external degrees of freedom of gas phase and matrix-isolated molecules is that in the latter case the rotational motion of molecules is governed by the crystal field, V_κ, describing the interaction with the surroundings. Upon increasing its strength, the rotational states are gradually changing from free rotor states to hindered states, until, finally, the limit of librational states is approached, i.e., angular oscillations about preferred crystallographic axes [259–261]. In Fig. 4.41 these three regimes can be approximately identified from their energies with respect to minima, saddles and maxima of the model potential given in (4.30). Starting from the free rotor states $|JM\rangle$ for $\kappa = 0$ with energies $E/B = J(J+1)$ and degeneracies $g = 2J+1$, essentially free rotation is still observed for $E > V_\mathrm{max}$.

The regime of $V_{\max} > E > V_{\text{sad}}$ is characterized by hindered rotor states with degeneracies $g = 1, 2, 3$ in case of A, E, T cubic symmetry, respectively. For $V_{\text{sad}} > E > V_{\min}$, the energy level scheme is dominated by librational manifolds of states which can be labeled by quantum numbers $n = 0, 1, \ldots$ with corresponding degeneracies $g = 6(n + 1)$ for rotational densities along $\langle 100 \rangle$ directionality.

Another aspect is the dependence of the energy levels on the crystal field parameter: Beyond certain thresholds for κ, all energy levels tend to decrease with increasing κ ("low-field seeking" states) and form manifolds of librational states. As can be seen in the bottom panel of Fig. 4.41, for the model potential (4.30) chosen here, the formation of the first and second manifold is found at $\kappa \approx 20$ and $\kappa \approx 60$, respectively. In the following, a fixed field strength, $\kappa = 25$, shall be considered where the potential wells are deep enough to support directional states as shown in the inserts in the bottom panel of Fig. 4.41. Here the first multiplet has already formed and the typical multiscale nature of librational states of matrix-isolated molecules already becomes apparent: While the crystal field splitting — as well as the zero point energy — is of the order of $10B$, the nearly degenerate ground state is comprised of a set of states (here A_{1g}, T_{1u}, and E_g) exhibiting tunnel splittings below $1B$. Accordingly, the fields used for alignment/orientation are termed as strong if they can achieve notable mixing of states belonging to different multiplets, while weak fields can only mix states inside a single set. Note that in the first case rotational densities are driven above potential energy saddles or barriers, while they can tunnel in the second case.

4.9.3 From librational to pendular states

In the following optically-dressed, pendular states for the combined action of the matrix- and light-induced potentials are discussed, $V_\kappa + V_\alpha$ for constant crystal field, $\kappa = 25$. The resulting energy levels, $E_n(\Delta\omega)$, for cooperative $\langle 100 \rangle$ alignment are shown in the upper part of Fig. 4.42.

Note that the angle–independent part of V_α in (4.31) is neglected by setting $\omega_\perp = 0$. First of all, the degeneracies of the E and T symmetry librational states are lifted as the field is switched on ($\Delta\omega > 0$) due to the reduction of the symmetry point group from O_h to D_{4h}, see the group theoretical considerations in [28]. Secondly, there are two classes of states: While one set of energy levels is essentially field–insensitive, the remaining states are high-field seeking, i.e., they gain energy for increasing laser intensity $\Delta\omega$. This different behavior is due to the directionality of the wave functions with their lobes pointing either perpendicular or parallel to the field. While the first ones essentially remain unchanged, the rotational densities of the latter ones become gradually more squeezed along the direction of the external field. Due to the noncrossing rule for states of equal symmetry avoided crossings appear.

The degree of alignment of pendular states is usually measured in terms of the alignment cosine, $\langle \cos^2 \theta \rangle$. This expectation value vanishes for a per-

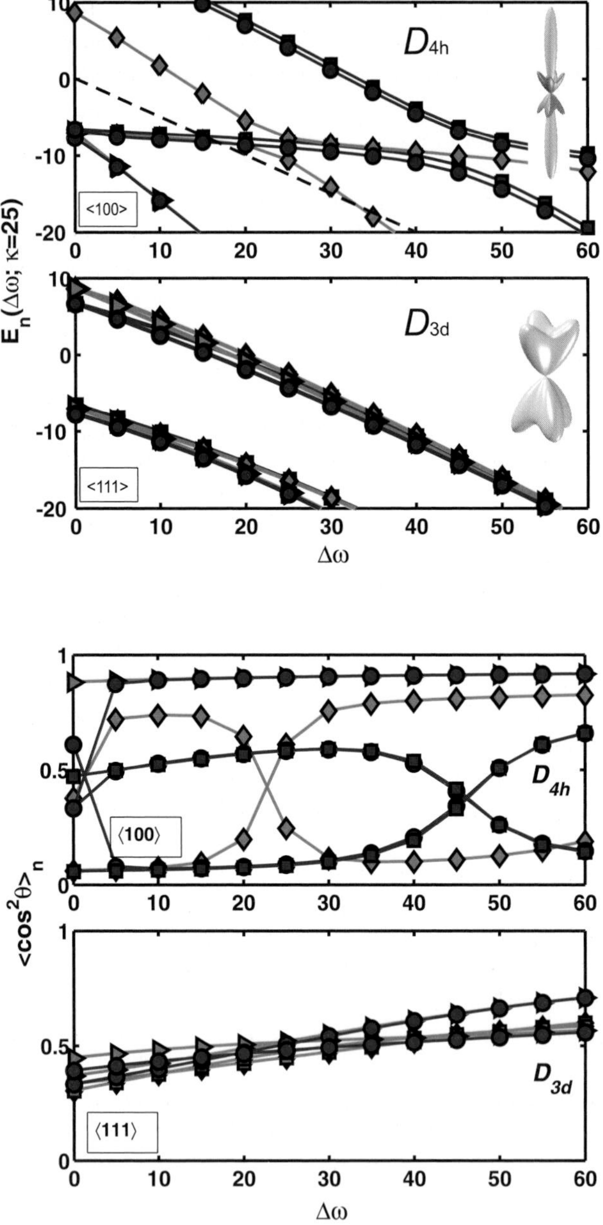

Fig. 4.42. Energy spectrum (top two panels) and alignment cosine (bottom two panels) versus strength of interaction $\Delta\omega$ of guest molecule with external laser field (4.31) for fixed internal field, $\kappa = 25$. Upper and lower figures correspond to cooperative $\langle 100 \rangle$ and competitive $\langle 111 \rangle$ polarization of the laser field, respectively. [15]

pendicular state, it is one third for an isotropic state, and it approaches unity for the limit of infinitely narrow wave functions. It is related to (the slopes of) the energy levels by virtue of the Hellmann–Feynman theorem

$$\langle \cos^2 \theta \rangle_n = -\frac{\partial E_n}{\partial(\Delta\omega)} . \tag{4.34}$$

This is visible in the third panel of Fig. 4.42, where the perpendicular states with $\langle \cos^2 \theta \rangle < 0.1$ can be readily distinguished from the parallel states with $\langle \cos^2 \theta \rangle > 0.9$, where the alignment cosine of the (parallel) ground state exceeds that obtained for the gas phase case [237]. Moreover, the pattern of avoided crossings of the energy levels is clearly reproduced for the alignment cosine. The sudden increase of alignment hints to an abrupt change of the character of the states rather than the smooth narrowing observed for the parallel states not involved in crossings. A qualitatively similar picture emerges for the competitive case of $\langle 110 \rangle$ alignment (not shown in the figure), but with reduced maximal alignment, $\langle \cos^2 \theta \rangle < 0.75$. The case of $\langle 111 \rangle$ alignment, however, is distinctly different, see lower part of Fig. 4.42: All states are high-field seeking, and there are no clear crossings visible. For vanishing crystal field, $\Delta\omega = 0$, the alignment is near the isotropic value of 1/3 because the librational states under investigation exhibit notable rotational density neither parallel nor perpendicular to the threefold $\langle 111 \rangle$ target. For the range of interactions considered here ($\Delta\omega \leq 60$), the alignment cosine $\langle \cos^2 \theta \rangle$ does not exceed a value of 0.7.

In [28] the degree of alignment for two thermal ensembles $\gamma = 1$ and $\gamma = 10$ is discussed in detail. For these temperatures the thermal energies are insufficient or sufficient to overcome rotational barriers, respectively. While equal weighting of pendular states would always result in an isotropic distribution, significant thermally averaged alignment is observed due to the higher Boltzmann weights of the energetically lower states usually bearing higher alignment. Moreover, the averaged results are insensitive to the dramatic changes of the components at avoided crossings, because parallel and perpendicular states merely change their role.

In the cooperative case, the parallel directions of crystal field minima and laser polarization lead to more effective alignment than in the gas phase: For the lower temperature, $\gamma = 1$, high degrees of alignment (> 0.9) can be achieved for fields one order of magnitude weaker than in the gas phase case. The enhancement is still significant for the higher temperature, $\gamma = 10$. In the competitive cases, where alignment along saddles or maxima of the crystal field is strived for, the maximal, thermally averaged alignment remains below the corresponding gas phase values. For the range of (dimensionless) laser intensities studied here ($\Delta\omega \leq 60$), only moderate values around 0.6–0.7 can be achieved for $\kappa = 25$, whereas the averaged alignment does not exceed the isotropic value notably for stronger matrix field, $\kappa = 60$.

In the case of a homo-nuclear diatomic molecule, the above-described techniques for alignment of the molecular axis relative to the crystallographic axes completely suffice to fix the rotational degrees of freedom. For a hetero-nuclear molecule, however, additional "head vs. tail" orientation may be desired in the control of photochemical events. In principle, this can be achieved by interaction V_μ of an electrostatic field with the molecular dipole moment, see (4.32). In practice, however, this "brute force" orientation is restricted to rather high values of the (dimensionless) interaction parameter ω [228, 229]. A more elegant approach to orientation of gas phase molecules circumvents this problem by utilizing combined laser and static field, $V_\alpha + V_\mu$. As mentioned before, the laser field serves to create (degenerate) g/u pairs of states, which can be split by rather weak static fields. In close analogy, a similar approach to orientation of molecules trapped in octahedral fields has been devised that builds on a combination of crystal and electrostatic fields, $V_\kappa + V_\mu$ [28]. As shown in the bottom panel of Fig. 4.41, the crystal field gives rise to the formation of a quasi-degenerate set of librational states which are connected to each other by two-dimensional tunneling. If the matrix-induced field is sufficiently strong, then the tunneling splittings of energy levels become sufficiently small, such that they can be efficiently manipulated by rather weak electrostatic fields.

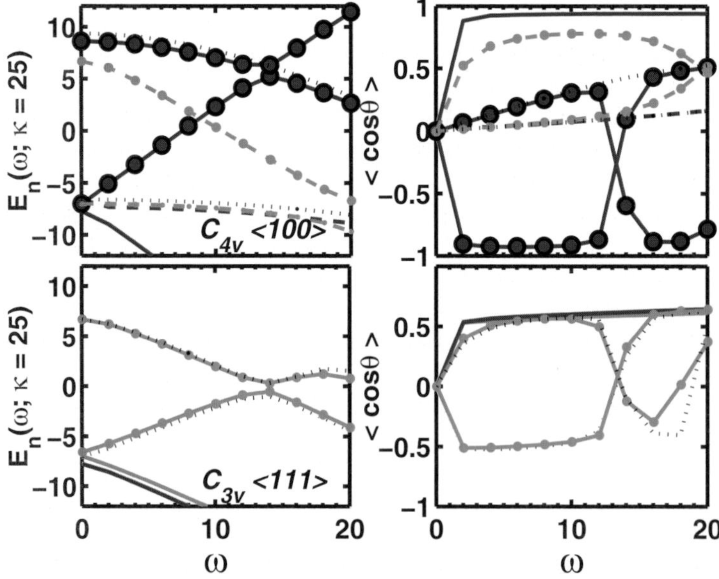

Fig. 4.43. Energy spectrum (left panel) and orientation cosine (right panel) versus strength of interaction ω of guest molecule with external static field (4.32) for fixed internal field, $\kappa = 25$. Upper and lower figures correspond to cooperative $\langle 100 \rangle$ and competitive $\langle 111 \rangle$ direction of the static field, respectively. [15]

Pendular, oriented states resulting from the combined action of the matrix-induced (crystal) field and electrostatic (orientation) fields, $V_\kappa + V_\mu$ are shown in Fig. 4.43, again for constant crystal field, $\kappa = 25$. The resulting energy levels, $E_n(\omega)$ are shown in the upper and lower left part of the figure for cooperative $\langle 100 \rangle$ and competitive $\langle 111 \rangle$ orientation. As in the case of laser-induced alignment, most of the degeneracies are lifted due to the lowering of the O_h to the C_{nv} point groups due to the $\cos\theta$ type interaction V_μ. In addition to field–insensitive (perpendicular) and high-field seeking (parallel) states, also low-field seeking states with orientation anti-parallel to the external static field are observed. Again, the noncrossing rule describes the behavior at the intersections: States transforming according to equal or unequal irreducible representations of the respective C_{nv} point groups are undergoing avoided or true intersections, respectively. Of particular interest are the avoided crossings of parallel and anti-parallel states observed both for the cooperative and competitive case. These events correspond to sudden 180 degree flips of oriented rotational wave functions. The slope of the energy plots is connected through a Hellmann–Feynman relation [similar to (4.34)] for the orientation cosine $\langle \cos\theta \rangle$ used to quantify orientation effects. The right hand side of Fig. 4.43 shows our results: While the qualitative picture is similar, it is obvious that the achieved maximal degree of orientation is higher in the cooperative case ($\langle \cos\theta \rangle \approx 0.95$) than in the competitive case ($\langle \cos\theta \rangle \approx 0.65$). When including the effect of thermal averaging, higher orientation than for gas phase molecules is found in the former case while the opposite is true for the latter case.

Another interesting aspect is the dependence of alignment and orientation on the strength of the crystal field: As discussed in more detail in [28], very *weak* laser or static fields are necessary to achieve very high degrees of alignment or orientation, respectively, for the cooperative case. For instance, for $\kappa \approx 100$, (dimensionless) interaction parameters $(\Delta\omega, \omega)$ of the order of 0.1 are sufficient to achieve near unity alignment or orientation if single pendular state are considered. However, as discussed above, *strong* fields with interaction parameters $(\Delta\omega, \omega)$ of the order of 10-100 are needed to achieve high alignment/orientation for a thermal ensemble. This is particularly true for the competitive cases, where strong fields are required even at rather low temperatures.

4.9.4 From analysis to control of rotational dynamics in matrices

So far, the manipulation of rotational degrees of freedom of matrix-isolated molecules has been discussed in terms of time-independent wave functions and their expectation values, i.e., alignment and orientation cosines. In realistic experiments, however, the respective control targets are achieved by means of pulsed light sources. As well as most other light-induced processes discussed in this book, the time-dependent picture of alignment/orientation of molecules in matrices can be categorized with respect to the adiabatic and nonadiabatic limit, depending whether the employed pulses are long or short with respect to the inherent time scales of the systems under investigation.

First, the nonadiabatic case is considered: The use of ultra-short laser pulses generally renders the system in a wave packet state, i.e., a coherent superposition of eigenstates of the quantum system under investigation. Upon time-evolution of a wave packet, each of the comprising states exhibits phase oscillations with its own frequency $E_n B/\hbar$. Hence, any time-dependent observable (expectation value) of a superposition is dominated by beating patterns $(E_n - E_m)B/\hbar$. In particular, recurrences occur, where two or more of the underlying oscillations are in phase. This principle is first illustrated for the rotational motion of gas phase molecules. The thermally averaged alignment signal, $\langle\langle\cos^2\theta\rangle\rangle(\tau)$, obtained for interaction of a linear molecule with an intense laser pulse ($\Delta\omega = 100$) and for low temperature ($\gamma = 1$) is shown in the top left panel of Fig. 4.44. The very short, Gaussian-shaped pulse (FWHM: $\sigma = 0.032$) creates a rotationally broad wave packet, and strongly nonadiabatic dynamics is observed with complex temporal oscillation patterns of the post-pulse alignment [262–264]. Note that on average the alignment cosine stays above the isotropic value 1/3. Signatures of recurrences are clearly visible: After $\tau = 1$ the wave packet is again spatially aligned due to complete phase matching of the rotational components and the initial alignment peak is regained. In addition, a free molecule exhibits fractional revivals, e.g. at $\tau = 1/2$, where the rotational density is delocalized in a plane perpendicular to the polarization of the field. This dip in the alignment signal (anti-alignment) [265] is surrounded by two peaks with high parallel alignment. In general, the simple energy level scheme of a free rotor leads to the observation of the (fractional) revival intervals. Obviously, the corresponding energy differences can be extracted from the complex beating pattern by virtue of a Fourier transform: The power spectrum in the upper right of Fig. 4.44 shows progression within gerade parity states ($J = 0, 2\cdots 8$) by distinct peaks ($4J + 6$) at 6, 14, 22, and 30, and the peak at $\Delta E = 26$ is due to the ungerade $J = 5$ and $J = 7$ states present in the wave packet. The intensities reflect both the excitation probabilities and populations of the states comprising the initial thermal ensemble [29].

Our results for the rotational motion molecules confined in octahedral crystal fields are presented in an analogous way in the middle and lower panel of Fig. 4.44 for the case of cooperative $\langle 100\rangle$ and competitive $\langle 110\rangle$ alignment, respectively. The transient signals $\langle\langle\cos^2\theta\rangle\rangle(\tau)$ exhibit global maxima near the center of the pulse. Unexpectedly, the value for the competitive case (≈ 0.7) exceeds that for the cooperative case (≈ 0.5). The further evolution is dominated by post-pulse oscillations for both cases considered here. However, the characteristic revival times found for gas phase molecules are absent and a highly irregular pattern is found instead. Local maxima of 0.5 are found for the cooperative case, while the averaged signal for the competitive case hardly exceeds the isotropic value of 1/3. For a more detailed study of effects

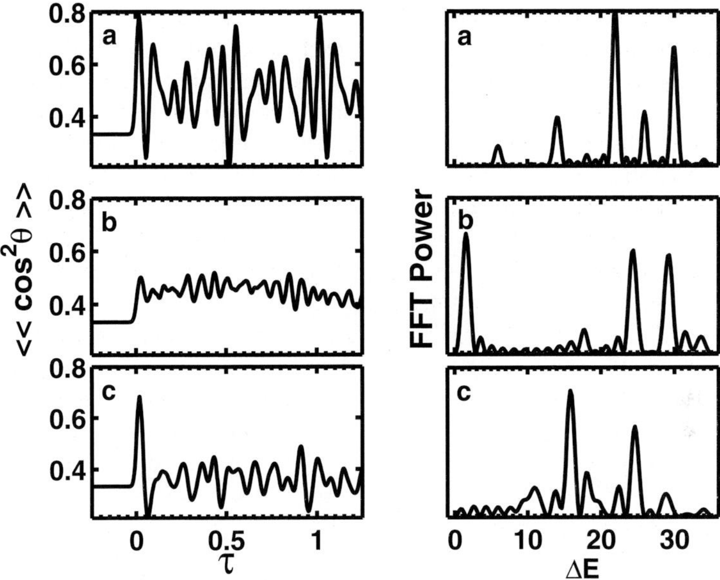

Fig. 4.44. Nonadiabatic alignment by short ($\sigma = 0.032$), intense ($\Delta\omega = 100$) pulses for low temperature ($\gamma = 1$). Time-dependent alignment signals (left panel) and corresponding Fourier transforms (right panel) for a) gas phase, b) cooperative $\langle 100 \rangle$, and c) competitive $\langle 110 \rangle$ polarization of the external field with respect to the internal field ($\kappa = 25$) [15].

of the pulse duration on the transient alignment signals, the interested reader is referred to [29].

A deeper understanding of the field-free evolution has to build on a closer spectral analysis of the post-pulse beating patterns. While wave packet states of gas phase molecules are composed of free rotor states, the wave packet states of matrix-isolated molecules can be interpreted as coherent superpositions of librational states. This can be seen from the power spectra of the transient alignment signals in the right panels of Fig. 4.44. The observed frequency patterns correspond to symmetry-allowed transitions between librational, i.e., matrix-induced rotational states. Consequently, all the peaks can be assigned to differences of the energy-levels of the trapped molecule, see Fig. 4.41. While all peaks with $\Delta E > 10$ correspond to transitions between different librational manifolds, the low energy structure found at $\Delta E \approx 1$ is caused by tunneling transitions within the ground librational manifold. The intensities are governed by the initial populations in the thermal ensemble as well as by rotational excitation probabilities. Since the transition rules and Hönl–London factors for excitations are different within the D_{4h} and D_{2h}

point groups, different intensities are observed for the equal frequencies in the cooperative and the competitive cases.

In summary, the analysis of power spectra of experimental nonadiabatic alignment signals provides a novel approach to the analysis of tunneling and librational energy levels of matrix-isolated molecules. In principle, comparison of experimental data with quantum simulations should, e.g., allow for validation of model potentials such as given in (4.30) as well as for a determination of the free potential parameters.

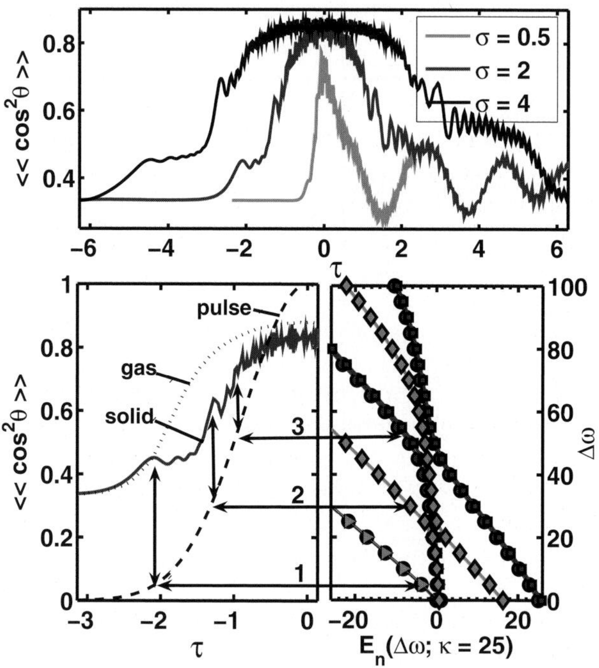

Fig. 4.45. Adiabatic alignment by long, intense ($\Delta\omega = 100$) pulses for low temperature ($\gamma = 1$) and for cooperative $\langle 100 \rangle$ polarization of the external field with respect to the internal field ($\kappa = 25$). Upper: Alignment signal in a time window spanning the whole pulse duration, $-6 \leq \tau \leq 6$. Lower: Stepwise enhancement (enumerated) of alignment signal during onset of the laser pulse (left) together with selected energy levels vs. alignment parameter $\Delta\omega$ (right), see also Fig. 4.42. Reproduced with permission from [29]

As an alternative to the post-pulse alignment created by nonadiabatic alignment, in the adiabatic case of longer pulses alignment is only achieved while the field is on. For gas phase molecules, the adiabatic limit is met if the pulse duration reaches (or exceeds) the rotational period of the molecules under investigation. For matrix-isolated molecules this condition has to be reformulated: Building on the above discussion of tunneling vs. librational energy scales, it is expected that, for the crystal fields considered here, the pulse lengths have to be one order of magnitude longer in order to reach the adiabatic limit with respect to tunneling motion. An example can be seen in Fig. 4.45 for various pulse widths σ (FWHM): In all curves ($0.5 \leq \sigma \leq 4$), the maximal alignment, here achieved very close to the center of the pulse, is near the value $\langle\cos^2\theta\rangle_{\max} \rightarrow 1 - \sqrt{1/\Delta\omega}$ found for light molecules at low temperatures [264]. However, all curves show also a certain degree of oscillatory post-pulse oscillations. For shorter pulses this is similar to the gas phase case, see our previous discussion on nonadiabatic alignment. For the longer pulses the adiabatic limit with respect to librational dynamics has already been reached; however, the oscillations indicate that the pulses are still nonadiabatic with respect to tunneling dynamics. Indeed, the periods of the field-free oscillations can be exactly assigned to certain tunneling transitions within the ground librational manifold. In principle these oscillations could be overcome by using even longer pulses. In practice, however, dissipation and de-phasing might become nonnegligible on these time scales. Nevertheless it is noted that adiabatic alignment presents a very promising approach for a highly effective control of rotational degrees of freedom, and can be used to steer the subsequent dynamics, e.g., in subsequent photoinduced processes.

The investigated systems serve as simplified models for molecules in rare gas matrices if couplings to translational [249], vibrational [250, 251], or electronic [252] degrees of freedom of the guest molecule are neglected. Building on the well-developed literature for gas phase molecules, alignment and orientation of matrix-isolated molecules induced by laser and static fields has been explored [23, 24, 27–29]. In summary, the resulting rotational states can be viewed as hybrids of librational (matrix-induced) or pendular (field-induced) states. If the external control fields are along energetically favored directions of the internal matrix-induced field, co-operative effects are found. Both the degrees of alignment and orientation exceed those obtained in gas phase studies. In general, weak fields are sufficient to align/orient single states, while higher fields are required to achieve the same for thermal ensembles. It is also shown that rotational densities can be controlled at will even though there is competition between internal and external fields, but then higher fields are mandatory. Nevertheless, the examined ranges of interaction strengths can be realized in future applications to demonstrate these effects in rare gas matrix experiments thus opening new approaches to study direction-dependent reaction dynamics of molecules in solids. Typically, such measurements would be performed in the time domain utilizing pulsed light sources where the distinction between nonadiabatic and adiabatic dynamics is based on the ratio of pulse duration and the time scales for tunneling and librational dynamics of the matrix-isolated molecules. In the nonadiabatic case, ultrashort pulses serve to create nonstationary wave packet states. The highly oscillating alignment signals in the post-pulse regime can be understood in terms of quantum beats of the comprising librational states. Hence, Fourier transforms of observed, transient alignment signals can be used to analyze the energy level scheme of trapped molecules in a solid which are hard to obtain by continuous infrared spectroscopy [254]. Long pulses close to the adiabatic limit can be used to achieve high degree of alignment during the pulse. Although strictly adiabatic dynamics would have to account for interfering tunneling states, for the purpose of control it is sufficient to choose pulse widths equal to the rotational period of the free molecule. For both the cooperative and competitive cases studied here, high alignment degrees close to the gas-phase results can be achieved by adiabatic techniques. This allows for preparing suitable precursor states to efficiently control subsequent direction–dependent reaction mechanisms.

4.10 Quantum simulations of wave packet dynamics in reduced dimensionality: From analysis to coherent spin control

M. V. Korolkov and J. Manz

4.10.1 Model and techniques

The classical and quasi-classical simulations, which have been presented in Sects. 4.7 and 4.8, respectively, show that the initial molecular dynamics of dihalogens in rare gas matrices following photoexcitation by short laser pulses may be described in terms of models with reduced dimensionality (cf. Fig. 4.37). In the very beginning, a one-dimensional (1D) model is sufficient, describing just the bond stretch of the molecule embedded in the frozen matrix. Subsequently, a two-dimensional (2D) model is required in order to describe the onset of motions of the system along another degree of freedom, e.g.,coherent motions of neighboring rare gas atoms or, in the case of a heteronuclear dihalogen, the motion of its center of mass in the matrix cage. Again, at this stage, all other degrees of freedom are considered as frozen. Later on, the dimensionality of adequate models increases with time. For the subsequent applications the focus, however, is on initial time evolutions when 1D or 2D models are adequate. These models may also serve as a reference, even during later times, in order to discover subsequent specific effects of additional degrees of freedom. For example, Fig. 4.46 illustrates 1D (q_1) and 2D (q_1, q_2) models for the early laser induced dynamics of F_2 oriented along the $\langle 111 \rangle$ direction in an Ar matrix (adapted from [201]). Here, $q_1 = X$ and $q_2 = Y$ describe the stretch of the F_2 molecule, and the concerted motions of six argon atoms which form two equivalent opposite "windows" (labeled W_1 in Fig. 4.46) for cage exit of dissociative F atoms, respectively. All comple-

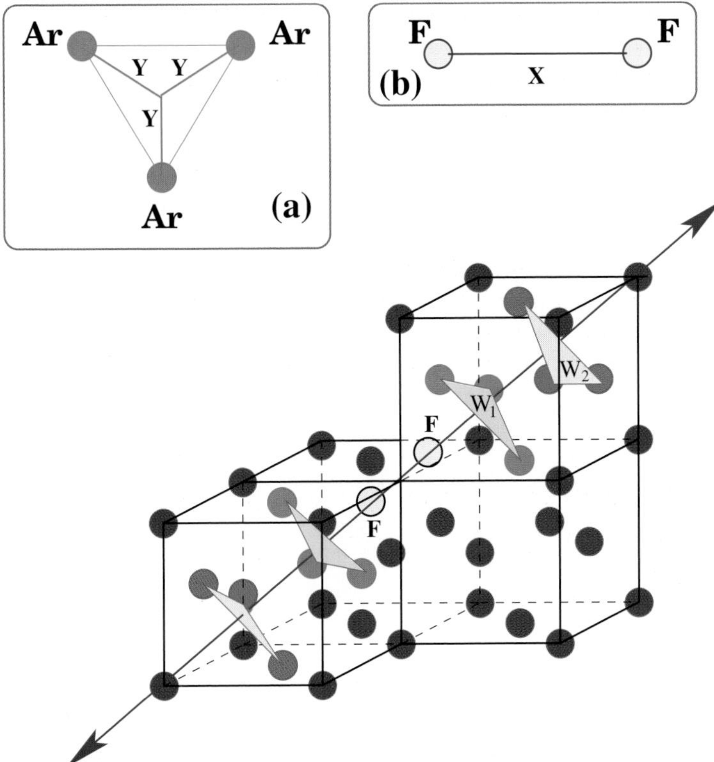

Fig. 4.46. One-dimensional (1D) and two-dimensional (2D) models for the early dynamics of laser-induced F_2 oriented along the $\langle 111 \rangle$ direction in an Ar matrix, with coordinates q_1 and q_1,q_2, respectively. The coordinates $q_1 = X$ and $q_2 = Y$ describe the F_2 bond stretch and the coherent motions of six Ar atoms which form two equivalent opposite triangular "windows" (labeled W_1, at $X \approx 11.5a_0$) for cage exit. All complementary degrees of freedom are frozen at their equilibrium values. Also indicated are six other Ar atoms which constitute the next equivalent opposite "windows" (labeled W_2, at $X \approx 23.0a_0$) for "second" cage exit (adapted from [201]).

mentary atoms are frozen in their equilibrium configurations, in these 1D or 2D models.

The present models are not only restricted in dimensionality, but also in the numbers of electronic states. Again, this is suggested by the quasi-classical simulations of Sect. 4.8 which show that populations are redistributed sequentially from the initially laser-excited state to others. Accordingly, a simple three-state model is employed in order to describe the time evolution at very early times. Subsequently, diabatic transitions to other electronic excited states would require that those should be included in adequate extended models. Usually, the electronic ground state $X = ^1\Sigma$ (labeled $e = 1$) is included, plus two excited triplet and singlet states ($^3\Pi$ and $^1\Pi$, labeled $e = 2$ and 3, respectively) which are populated immediately during or after the laser pulse. The corresponding potential energy surfaces $V_1(q)$, $V_2(q)$, $V_3(q)$ along the coordinates $q = q_1$ or $q = q_1, q_2$ of the 1D or 2D models are evaluated by means of the DIM method, as explained in Sect. 4.6. Examples of 1D and 2D potential energy surfaces will be shown in Figs. 4.52 and 4.53 (for ClF in Ar) or 4.51 (for F_2 in Ar) below, respectively. We employ the "adiabatic pure spin (singlet or triplet) presentation", i.e. the Hamiltonian of the model system $XY : RG$ (dihalogen XY in rare gas matrix) is written as

$$H_{XY:RG} = \begin{pmatrix} T + V_1 & 0 & 0 \\ 0 & T + V_2 & V_{SOC} \\ 0 & V_{SOC} & T + V_3 \end{pmatrix} \tag{4.35}$$

Here, T denotes the kinetic energy operator of the 1D or 2D models, i.e.

$$T = -\frac{\hbar^2}{2m_1}\frac{d^2}{dq_1^2} \qquad \text{(1D)} \tag{4.36}$$

with reduced mass $m = m_1 = m_X m_Y/(m_X + m_Y)$ for the stretch q_1 of the dihalogen XY, or

$$T = -\frac{\hbar^2}{2m_1}\frac{d^2}{dq_1^2} - \frac{\hbar^2}{2m_2}\frac{d^2}{dq_2^2} \qquad \text{(2D)} \tag{4.37}$$

with effective mass m_2 for the second degree of freedom q_2, e.g. $m_2 = m_X + m_Y$ if q_2 describes the motion of the center of mass of XY in its matrix cage, or $m_2 = 6m_{Ar}$ if q_2 describes the coherent motions of the six Argon atoms which form the opposite windows W_1, cf. Fig. 4.46. Moreover, V_{SOC} denotes the effective spin-orbit coupling between the excited singlet and triplet states. We employ empirical values of V_{SOC}, (see Fig. 4.48 below), neglecting any variations along the coordinate(s) q. Note that the Hamiltonians employed in Sects. 4.7 and 4.8 are based on the alternative adiabatic representation, i.e spin-orbit coupling is already included in those adiabatic potentials - this corresponds to diagonalization of the Hamiltonian (4.35). Both presentations are, of course, equivalent. We prefer the "adiabatic pure spin representation" with off-diagonal spin-orbit coupling because it lends itself for illuminating interpretations of singlet-to-singlet and singlet-to-triplet transitions induced by the laser fields and by spin orbit coupling, respectively. Kinetic couplings are neglected in expression (4.35).

The eigenstates of the model system are evaluated as eigenstates of the model Hamiltonian (4.35), yielding, e.g.,

$$\Psi_{1,v}(q) = \begin{pmatrix} \Phi_{1,v}(q) \\ 0 \\ 0 \end{pmatrix} \tag{4.38}$$

for the electronic ground state, where

$$(T + V_1)\Phi_{1,v} = E_{1,v}\Phi_{1,v} \tag{4.39}$$

with eigenenergies $E_{1,v}$ labeled by vibrational quantum number(s) $v = v_1$ (1D) or $v = v_1, v_2$ (2D). At very low temperatures, one can assume that the system is originally in its ground state $\Phi_{1,0}$. (At higher temperatures, one should employ a density matrix description, or equivalent thermal averages of the vibrational ground and excited states; exemplarily, the consequences of selective vibrational preexcitations will also be considered below). In special cases of small diatomic molecules which occupy a mono-substitutional site, such as F_2 or ClF in Ar (cf. Fig. 4.46), we know from the investigations presented in Sect. 4.9 that this vibrational initial state is coupled to rotational or librational motions of the diatomic molecule in its matrix cage. In the applications below, we assume, however, that the diatomic molecule has been preoriented along specific directions, e.g. $\langle 111 \rangle$ for the scenario of Fig. 4.46, using e.g. the methods described in Sect. 4.9, see also the review [227]. In the other cases of heavier dihalogens which occupy double substitutional sites, e.g. Cl_2 in Ar, preorientation is imposed already automatically by the matrix environment, e.g. $\langle 110 \rangle$ for Cl_2 in Ar, see Sect. 4.7.

In semiclassical dipole approximation, the time evolution of the system is described in terms of nuclear wave packets

$$\Psi(q,t) = \begin{pmatrix} \Psi_1(q,t) \\ \Psi_2(q,t) \\ \Psi_3(q,t) \end{pmatrix} \tag{4.40}$$

with components $\Psi_k(q,t)$ evolving on the adiabatic potentials $V_k(q)$. The $\Psi(q,t)$ are evaluated as solutions of the time-dependent Schrödinger equation. Using the semiclassical dipole approximation and compact matrix notation,

$$i\hbar \frac{d}{dt}\Psi(q,t) = (H_{XY:RG} - E(t)D(q))\Psi(q,t) \tag{4.41}$$

Here, $E(t)$ denotes the nonvanishing component of a laser field which is assumed to be linearly polarized. (In cases of two or more laser pulses with different polarizations, the term $-E(t)D(q)$ has to be replaced by corresponding sums of laser-dipole interactions). For example, for the scenario of Fig. 4.46, one can consider laser fields in the infrared (IR) and visible/ultraviolet (vis/UV) spectral domains propagating perpendicular to or along the $\langle 111 \rangle$ direction, with corresponding polarizations parallel or perpendicular to $\langle 111 \rangle$, respectively. In the applications below, the laser pulses are modeled as

$$E(t) = E_0 \cos(\omega t + \eta)s(t) \tag{4.42}$$

with amplitude E_0, carrier frequency ω, phase η, and shape function

$$s(t) = \sin^2(t\pi/t_p) \qquad \text{for} \quad 0 \le t \le t_p \tag{4.43}$$

and total pulse duration t_p. For convenience, all the phases are set equal to zero. The corresponding laser intensities (averaged over one cycle of the carrier frequency)

$$I(t) = 0.5\epsilon_0 c E_0^2 s^2(t) \tag{4.44}$$

have maximum values

$$I_{max} = 0.5\epsilon_0 c E_0^2, \tag{4.45}$$

full width at half height (FWHM) τ

$$\tau \approx 0.364\, t_p \tag{4.46}$$

and spectral widths Γ of the Fourier transform of the intensity, related to τ by the uncertainty type relation

$$\tau\Gamma \approx 3.295\hbar. \tag{4.47}$$

In the previous cases of laser control of reactions in the gas phase, I_{max} should be below the Keldysh limit in order to avoid competing ionization - as a rule of thumb this means $I_{max} \lesssim 10^{13} W/cm^2$, see Chapter 6 [266, 267]. For the present applications to molecules embedded in rare gas matrices, one has to observe the stronger restriction $I_{max} \lesssim 10^{12} W/cm^2$, because otherwise the laser pulse could destroy the matrix.

In some of the applications below, a sequence of laser pulses is applied with parameters labeled by $l = 1, 2$ etc, e.g. an initial pulse in the IR frequency domain ($l = 1$) is followed by another one in the vis/UV domain ($l = 2$). For simplicity, this sequence will be called IR + UV laser pulses. The expressions (4.41)-(4.45) are then extended to corresponding sums of laser dipole interactions with corresponding time delays t_d, e.g. the first and second pulses start at initial times $t_{i,l=1}$ and $t_{i,l=2} = t_{i,l=1} + t_d$, respectively. Zero phases of the IR and UV pulses implies phase-locking- this is important for the subsequent applications of coherent spin control.

The matrix $D(q)$ in (4.41) contains the transition dipole elements parallel to the laser field. All elements for transitions between pure singlet and triplet states vanish, i.e.

$$D(q) = \begin{pmatrix} d_1(q) & 0 & d_{1,3}(q) \\ 0 & d_2(q) & 0 \\ d_{3,1}(q) & 0 & d_3(q) \end{pmatrix} \tag{4.48}$$

Here, $d_{1,3}(q) = d_{3,1}(q)$ denotes the relevant component of the transition dipole elements between the electronic ground and excited singlet states. Usually, $d_{1,3}(q)$ for the isolated dihalogen is a good approximation in the Franck-Condon domain; else the Condon approximation is employed, i.e. $E_0 d_{1,3}(q)$ may be used as a parameter, independent of q. The $d_k(q)$ are the dipole functions in electronic states k. For homonuclear diatomic molecules, all $d_k(q)$ vanish, for symmetry reasons. For the subsequent applications, laser interactions with dipoles in excited states are irrelevant, i.e. one sets $d_2(q) = d_3(q) = 0$.

For the initial states, it is assumed that the system is in the electronic ground state, thus

$$\Psi(q, t = 0) = \begin{pmatrix} \Psi_1(q, t = 0) \\ 0 \\ 0 \end{pmatrix} \tag{4.49}$$

More specifically, the subsequent applications consider two scenarios, either starting from a vibrational eigenstate,

$$\Psi_1(q, t = 0) = \Phi_{1,v}(q), \tag{4.50}$$

typically the ground state ($v = 0$, i.e. $v_1 = 0$ or $v = v_1, v_2 = (0, 0)$ for the 1D or 2D models, respectively), or from a vibrational superposition state, also called a vibrational hybrid state

$$\Psi_1(q, t = 0) = c_0\Phi_{1,0}(q) + c_j\Phi_{1,j}(q) \tag{4.51}$$

with populations of the vibrationally ground ($v = 0$) and excited ($v = j \neq 0$) components,

$$P_{1,0} = |c_0|^2, \qquad P_{1,j} = |c_j|^2 \tag{4.52}$$

The corresponding ratio of the populations

$$r = P_{1,j}/(P_{1,0} + P_{1,j}) \tag{4.53}$$

varies from $r = 0$ (pure ground state $\Phi_{1,0}$) to $r = 1$ (pure vibrationally excited state $\Phi_{1,j}$).

These states (4.51) serve as important intermediate states for applications of coherent spin control, see below. For this purpose, one can also discuss various possibilities to prepare states (4.51) from the vibrational ground state (4.50).

The numerical methods for calculating the systems' eigenstates and time evolutions are similar to those which have been explained in Chapter 3, for applications to model systems with low dimensionality in the gas phase. For the details, see [31, 54, 55, 201, 203, 219, 268].

The resulting wave functions $\Psi_k(q, t)$ yield densities $|\Psi_k(q, t)|^2$ and populations of the electronic states k,

$$P_k(t) = \int dq |\Psi_k(q, t)|^2 . \tag{4.54}$$

4.10.2 Applications

This section for the applications starts with several tests of the validity of quantum simulations of dihalogens in rare gas matrices by models in reduced dimensionality. The tests for the validity of the models focus on convergence of the 1D versus 2D quantum results. Additional tests are provided by comparisons of the 1D or 2D quantum results against multi-D classical or

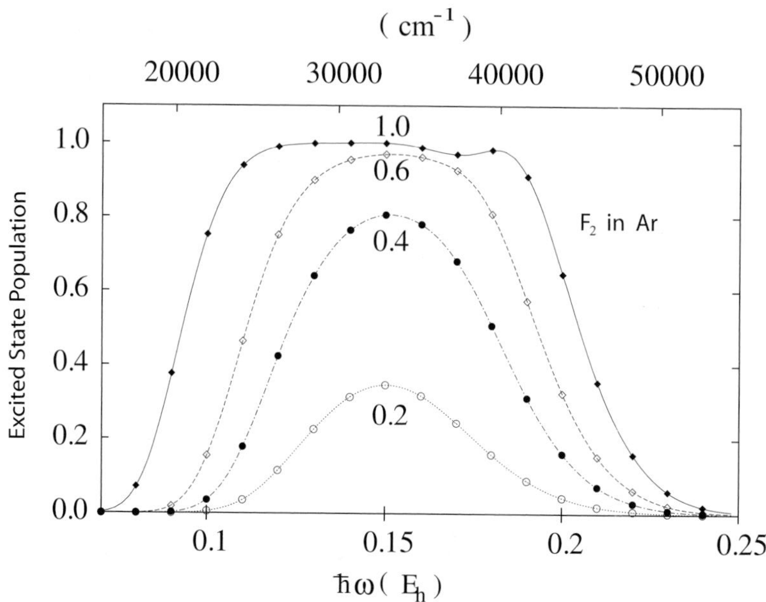

Fig. 4.47. Populations in the excited state of F_2 in an Ar matrix, achieved by laser pulses with durations $\tau \approx 36$ fs, depending on the laser frequency ω and field strength $E_0 = 0.2, ..., 1.0 E_h/(ea_0)$ marked on the lines. The results for the 1D and 2D models, as illustrated in Fig. 4.46, are indicated by lines and by points, respectively (adapted from [201]).

quasi-classical ones, during restricted initial time domains, cf. [55, 203]. Subsequently, the 1D or 2D models will be used for the empirical determination of effective spin-orbit couplings and for the investigation of coherence effects on cage exit versus recombination, relief reflections, cage exit supported by spin orbit coupling, and ultrafast spin flip.

Consider exemplarily the results for the 1D versus 2D quantum models of F_2 in Ar and ClF in Ar, with the dihalogen oriented along the $\langle 111 \rangle$ direction, cf. Fig. 4.46. Fig. 4.47 shows the populations in the excited states induced by laser pulses with durations $\tau \approx 36$ fs (t_p=100 fs, cf. eq. (4.46)), depending on the laser frequency ω and field strength E_0. Obviously, these absorption spectra for the 1D and 2D models agree within graphical resolution. This confirms the validity of a 1D model for describing absorption induced by these short laser pulses- the second degree of freedom, or others, are still frozen, and their effects are, therefore, negligible during these early times.

As an application, 1D model simulations of absorption spectra of dihalogens in rare gas matrices may be employed for comparison with experimental ones, and even for fitting the underlying empirical model parameters. As an example, Fig. 4.48 shows the absorption spectra of a 1D model of ClF in

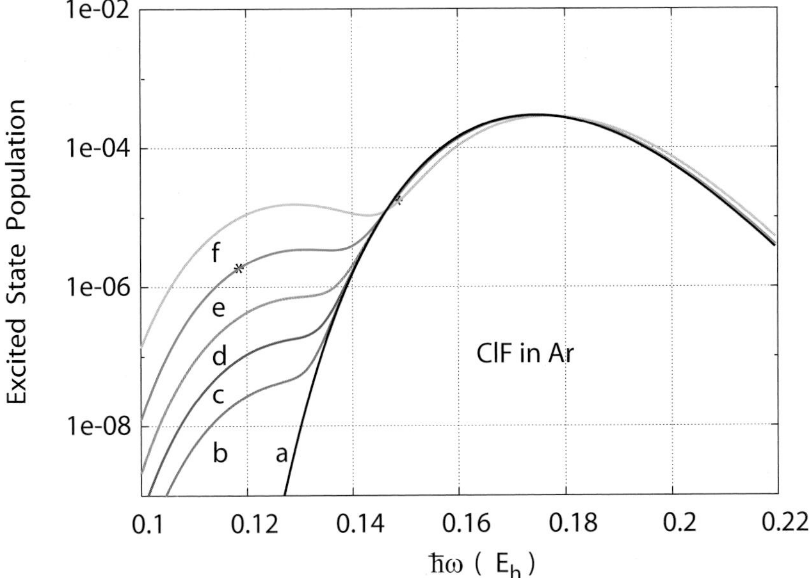

Fig. 4.48. Populations in the excited state of the model system, ClF in an Ar matrix, achieved by a laser pulse with duration $\tau \approx 18$ fs and field strength $E_0 = 0.0014 E_h/(ea_0)$ depending on the laser frequency ω. The results have been evaluated for a 1D model of ClF oriented along the $\langle 111 \rangle$ direction, analogous to the scenario of Fig. 4.46. The lines labeled a-f correspond to different values of the effective spin-orbit coupling $V_{SOC}/E_h = 0.0, 0.0005, 0.001, 0.002, 0.0045$, and 0.01. The theoretical results for $V_{SOC} = 0.0045 E_h$ agree with the experimental ones at $\omega \approx 0.117$ and $0.1475 E_h/\hbar$ [269, 270] (adapted from [219]).

Ar, for the scenario as depicted in Fig. 4.46, similar to the absorption spectra of Fig. 4.47 for F_2 in Ar for the same scenario, in the weak field limit (for the details, see the figure legend; note the logarithmic scale of Fig. 4.48). Close analysis of the spectra shows that they consist of a dominant peak for resonant or near resonant singlet-to-singlet transitions, plus a weak peak or shoulder for singlet-to-triplet transitions mediated by spin orbit coupling [219]. The ratio of the peaks depends, of course, on the value of V_{SOC}. As a consequence, $V_{SOC} = 0.0045 E_h$ is determined as empirical value of the spin-orbit coupling, because it yields the best agreement with the available experimental spectra [269,270]. (Similar determinations of V_{SOC} for Cl_2 in Ar are presented in [268]). This value has been used in [219]. It is larger than the value $V_{SOC} \approx 0.0013 E_h$ which has been adapted in [271] from the quantum chemical ab initio results for ClF in the gas phase [30].

Another convergence test for the 1D versus 2D models of F_2 in Ar is demonstrated in Fig. 4.49. Apparently, the 1D and 2D dynamics agree well

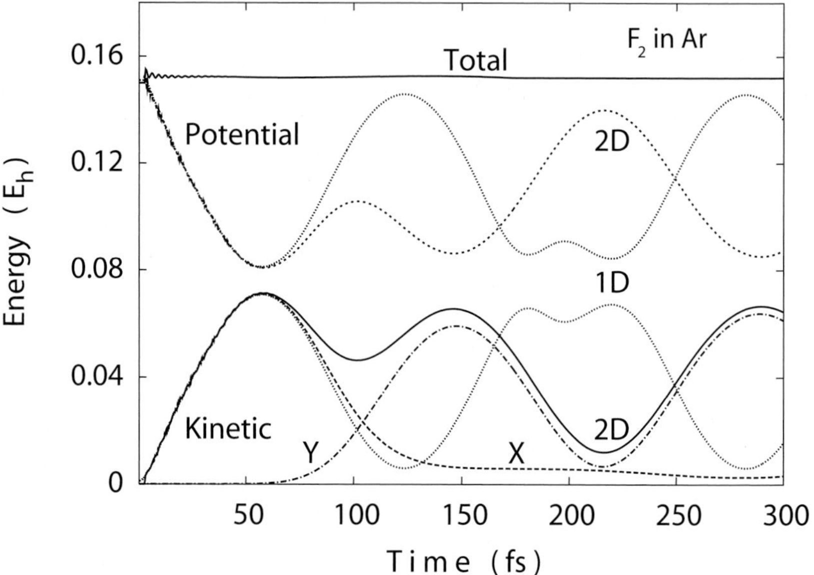

Fig. 4.49. Mean total, potential, and kinetic energies of the electronic excited states of F_2 in Ar, for the 1D and 2D scenarios of Fig. 4.46. The contributions from the motions of the F_2 molecule and of six Ar atoms constituting the opposite "windows" (cf. Fig. 4.46) are labeled by coordinates $q_1 = X$ and $q_2 = Y$, respectively. The laser parameters are $E_0 = 0.4E_h/(ea_0)$, $\tau \approx 36$ fs, and $\omega = 0.15\ E_h/\hbar$ (adapted from [201]).

during the first ca. 70 fs. During this period, the laser pulse ($\tau \approx 36$ fs) prepares the system in the electronic excited singlet state, inducing a sequence of various processes. Apparently, the first step is associated with mode selective energy transfer from potential energy to kinetic energy, exclusively for the bond stretch, implying frozen Ar atoms. After ca. 70 fs, one observes the onset of the second energy transfer, now preferably from kinetic energy of the F_2 stretch to the Ar atoms: the F atoms approach the opposite "windows" of the matrix cage, pushing aside the Ar atoms which form those windows. Accordingly, a 1D model for the F_2 bond stretch with frozen Ar atoms is adequate during the first 70 fs. Subsequently, energy transfer to the matrix atoms calls for a 2D model. The dimensionality of adequate models should then increase with time, depending on a cascade of energy transfers to more and more Ar atoms. In the following, it is assumed that the 2D model should be valid till slightly longer times, say ca. 100 fs. Models with larger dimensionality would be required for convergence tests of this working hypothesis.

The preceding convergence tests allow to use the 1D and 2D models for quantum simulations of the laser induced processes during at least 70 fs or slightly longer times, respectively. As first applications, let us use the 2D model for F_2 in Ar to discover laser induced coherence effects on cage exit versus recombination, cf. [201]. Successful cage exit requires that the F_2 bond stretch should exceed the value $X = 11.5a_0$ for the positions of the "windows" labeled W_1 in Fig. 4.46, in order to penetrate into the neighboring cage. Close analysis reveals a simple condition of coherent motions along the X and Y coordinates which is, however, not easy to satisfy: The bond stretch (X) must "open the window for cage exit" (Y), and moreover the window should be open at the moment when the halogen atoms approach in order to leave their initial cage. Specifically, the coordinate Y of the Ar atoms which form the windows for cage exit must increase from its equilibrium value $\approx 4.2a_0$ to at least to the limit ca. $5.5a_0$. As an example, the wave packet dynamics in the electronic excited singlet state induced by a laser pulse with parameters $E_0 = 0.4E_h$, $\tau \approx 36$ fs and $\hbar\omega = 0.1E_h$ is shown in Fig. 4.50. Apparently, the transfer of kinetic energy from the F_2 bond stretch to the Ar atoms (cf. Fig. 4.49) induces just a tiny increase of Y from ca. $4.2a_0$ to approximately $4.4a_0$ when the F atoms approach these windows, i.e. they are still closed thus imposing recombination, not cage exit. Nevertheless, efficient energy transfer from motions along X to Y causes continuous opening of the Ar windows till $Y \approx 5.8\ a_0$, i.e. even beyond the limit for cage exit- but that doesn't help any more for cage exit because at the same time, the two F atoms have already recombined to F_2. It turns out that, within the 2D model, both the windows and the bond stretch would have to oscillate several times till (at least part of) the wave packet arrives at the window at the time when it is open. This scenario of the 2D model is, of course, rather hypothetical beyond ca. 100 fs, but it is nevertheless useful because it points to the important condition for cage exit, i.e. good timing of the coherent coupled motions, i.e. $Y > 5.5a_0$ when $X \approx 12\ a_0$.

Based on the preceding analysis, systematic investigations have been carried out of the effects of the laser parameters, and also of vibrational preexcitations, in order to predict efficient scenarios for cage exits [54, 55, 201, 271]. One example is shown in Fig. 4.51, see also [54]: Assuming the initial state $\Phi_{1,v_1=0,v_2=3}$ (cf. (4.50)) which corresponds to vibrational preexcitation of the "window mode" of the six Ar atoms which form the opposite "first windows" for cage exit, the laser pulse with parameters $d_{1,3}E_0 \approx 0.03E_h$ (Condon approximation), $\omega = 0.26E_h/\hbar$, $\tau \approx 36$ fs, induces perfect cage exit through the preexcited "first window" (labeled W_1 in Fig. 4.46), at $t = 100$ fs. Subsequently, the two outer lobes of the initial wave function, which correspond to the largest preopening of the "first window" by matrix preexcitation, would even make it through the "second window" toward the next nearest cage of the matrix, within the frame of the 2D model (labeled W_2 in Fig. 4.46). The different time evolutions of alternative lobes of the initial wave function is determined by so-called "relief reflections" [54], i.e. by reflections from different parts of the walls of the potential energy surface, similar to rebounds of various billiard-balls from different parts of (hypothetically bent) cushions.

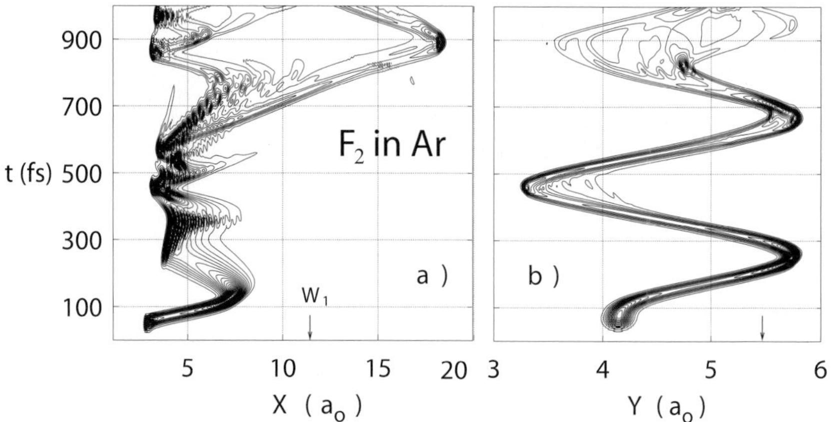

Fig. 4.50. Cage exit versus recombination of F_2 in Ar induced by a laser pulse, and depending on coherent opening or closing the windows toward cage exit. Panels a) and b) show the time evolutions of the densities of the laser driven wave packets in the excited singlet $^1\Pi$ state, for the 2D model, projected on coordinates $q_1 = X$ for the F_2 bond stretch , and $q_2 = Y$ for coherent motions of six Ar atoms which form the opposite windows labeled W_1 in Fig. 4.46, respectively. For cage exit, the bond length has to exceed $X > 12a_0$. It is blocked or supported if the window is closed ($Y \lesssim 5.5a_0$) or open ($Y \gtrsim 5.5a_0$), respectively, see the arrows. The laser parameters are $E_0 = 0.4E_h/ea_0$, $\tau \approx 36$ fs, and $\hbar\omega = 0.1E_h$. (adapted from [201].)

Another mechanism which supports cage exit is intersystem crossing mediated by spin-orbit coupling, as discovered in [271]. For illustration, Fig. 4.52 shows corresponding snapshots of the densities of the wavefunctions $\Psi_3(q,t)$ and $\Psi_2(q,t)$ representing the excited singlet (panels 4.52a,b) and triplet (panel c) states of ClF in Ar (2D model), superimposed on the potential energy surfaces $V_3(q)$ and $V_2(q)$, for two scenarios without (panel 4.52a, $V_{SOC} = 0$) and with (panels 4.52b,c, $V_{SOC} = 0.0013E_h$) spin-orbit coupling, at t=700 fs; for the details, see the legend of Fig. 4.52. Apparently, the laser excites the system with sufficiently high energy in order to overcome the barrier to cage exit, but nevertheless, without spin-orbit coupling, the wavefunction would remain trapped in its initial cage for very long times, due to the marginal coupling of the X and Y degrees of freedom in the domain of the initial cage. In contrast, spin-orbit coupling induces diabatic transitions from the initially excited singlet state to the triplet state and back, with two consequences: first, $V_2(q)$ has a lower barrier to cage exit than $V_3(q)$, cf. Fig. 4.52. Second, the loss of potential energy is compensated by higher kinetic energy in the triplet state, compared to the excited singlet state. Both effects support cage exit in the triplet state, from the domain of the initial cage to the neighboring one, cf.

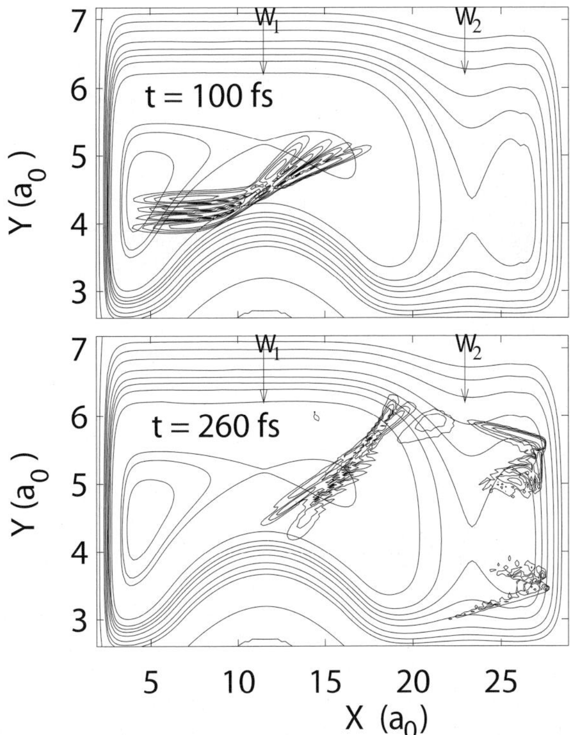

Fig. 4.51. Relief reflections of the wave packet representing F_2 in Ar in the excited singlet state leading to cage exits through the "windows" labeled W_1 and W_2 in Fig. 4.46, see the arrows (2D model). Snapshots of the wave packet are shown by its density superimposed on the potential energy surface $V_3(q)$. The wave packet dynamics is driven by a laser pulse with parameters $d_{1,3}E_0 = 0.03E_h$ (Condon approximation), $\omega = 0.26E_h/\hbar$, and $\tau \approx 36$ fs, starting form the initial state $\Phi_{1,v_1=0,v_2=3}(q)$ which corresponds to vibrational preexcitation ($v_2 = 3$) of the Ar atoms which form the first set of opposite "windows" (W_1).

Fig. 4.52. The snapshots at t =700 fs are of course beyond the validity of the 2D model, but this is just for illustration of the effect [271] using the rather small value $V_{SOC} = 0.0013E_h$ for ClF in the gas phase [30]. The larger empirical value $V_{SOC} = 0.0045E_h$ which has been determined for ClF in Ar later on in [219] (see Fig. 4.48) implies an increase of the singlet-to-triplet transition rate by a factor ca. 12, i.e. for the same laser parameters, one expects significant cage exit of ClF in Ar already at 100 fs, i.e. within the validity of the 2D model. By analogy, cage exit may also be supported by internal conversion mediated by kinetic couplings.

Efficient intersystem crossing, as discussed above [271], implies ultrafast spin-flip, on the time scale of ca. 100 fs. This has been discovered first by

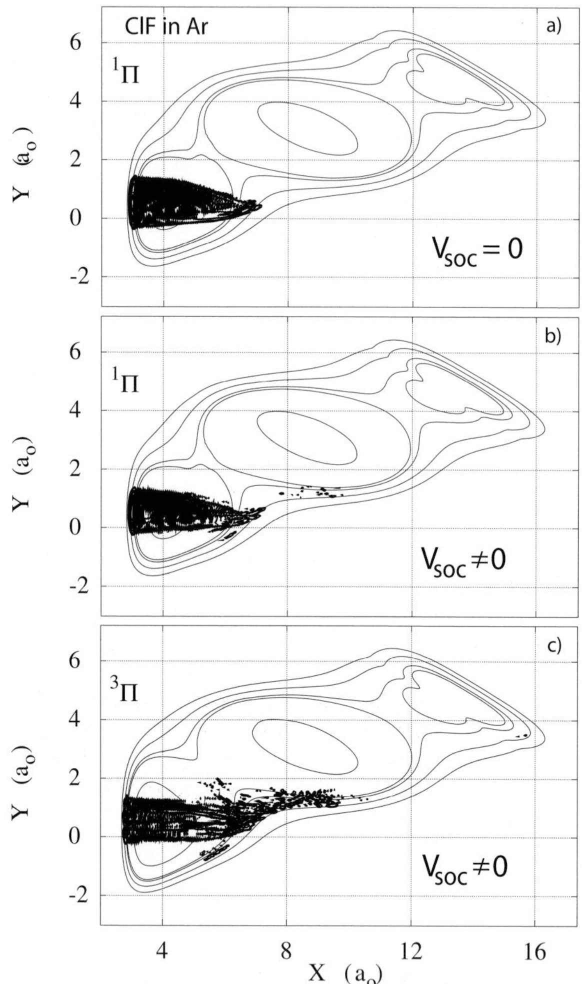

Fig. 4.52. Cage exit induced by a laser pulse and supported by spin-orbit coupling, for the system ClF in Ar. The snapshots show the densities of the wavefunctions superimposed on the potential energy surfaces of the excited singlet (a,b) and triplet (c) states, for two scenarios without (a, $V_{SOC} = 0$) and with (b,c, $V_{SOC} = 0.0013E_h$) spin orbit coupling, for the 2D model where X corresponds to the ClF bond stretch and Y to the motion of the center of mass of ClF along the preoriented $\langle 111 \rangle$ direction, at t =700 fs. The parameters of the laser pulse are $E_0 = 0.005E_h/(ea_0)$, $\omega = 0.19E_h/\hbar$, and $\tau \approx 36$ fs, starting from the initial ground state $\Phi_{1,v_1=0,v_2=0}(q)$. (adapted from [271]).

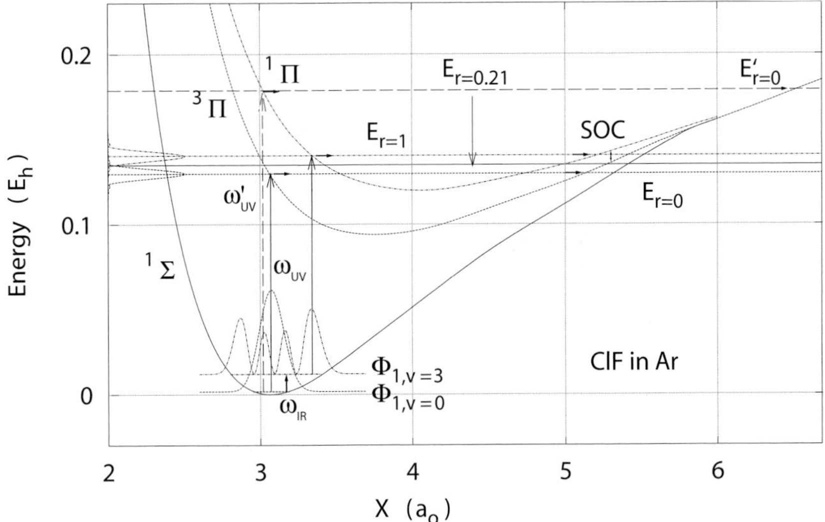

Fig. 4.53. 1D model for ultrafast spin flip and for coherent spin control of ClF in Ar. Shown are the potential energy surfaces $V_1(q)$, $V_2(q)$, $V_3(q)$ versus the ClF bond stretch $q = X$, for the electronic ground state ($^1\Sigma$) and the excited triplet ($^3\Pi$) and singlet ($^1\Pi$) states, respectively, together with the densities for two vibrational states $\Phi_{1,v=0}(q)$ and $\Phi_{1,v=3}(q)$ embedded in $V_1(q)$. The strength of the effective spin-orbit coupling $V_{SOC} = 0.0045E_h$ (cf. Fig. 4.47) is indicated by the label SOC at $X = q = q_{SOC}$. The mechanisms are indicated by vertical and horizontal arrows for the laser excitations and for the wave packet dynamics, respectively, see the text for the details. The energies $E'_{r=0}$ and $E_{r=0}$, $E_{r=1}$ or $E_{r=0.21}$ are the excitation energies for the UV laser pulses with frequencies ω'_{UV} and ω_{UV}, starting from $\Phi_{1,v=0}$ and from $\Phi_{1,v=0}$, $\Phi_{1,v=3}$ or the vibrational hybrid state (4.51) characterized by r=0.21, respectively, cf. (4.53). Also shown are the UV spectra for the cases $E_{r=0}$ and $E_{r=1}$. (adapted from [31] and [219]).

Schwentner et al. [18], cf. Sect. 4.2.4, 4.6, and 4.8. The observations have been explained in [31] using the simple 1D model, cf. Fig. 4.46, as illustrated in Fig. 4.53 (adapted from [31] and [219]; see also the discussion in Sect. 4.2.4). Essentially, the potential energy curves $V_2(q)$ and $V_3(q)$ of the excited triplet and singlet states form potential wells with large energy gaps between repulsive walls in the Franck-Condon domains of the UV laser pulse, and near degenerate attractive walls at larger values $q = X$ of the ClF bond stretches- this feature is an effect of the matrix cage, cf. Sects. 4.2.4 and 4.6. Indeed, for values of q larger than the limit $q_{SOC} \approx 5.2a_0$, the gap between the near-degenerate potential curves of the excited states becomes smaller than the effective spin-orbit coupling V_{SOC},

$$V_3(q) - V_2(q) < V_{SOC} \quad \text{for} \quad q \geq q_{SOC} \approx 5.2a_0. \tag{4.55}$$

As a consequence, spin orbit coupling induces efficient transitions between singlet and triplet states in the domain $q > q_{SOC}$ [31].

Now consider a pump UV laser pulse with frequency ω'_{UV} (illustrated by the dashed vertical arrow in Fig. 4.53) which induces a near-resonant Franck-Condon type transition at the FC window close to q_{FC}, from the initial wave packet $\Psi_1(q, t = 0) = \Phi_{1,v=0}(q)$ (i.e. $r = 0$ in (4.54)) e.g. to the excited ${}^1\Pi$ state,

$$\hbar\omega'_{UV} = V_3(q_{FC}) - V_1(q_{FC}) \approx E'_{r=0} \,. \tag{4.56}$$

Then the pump laser pulse generates the excited wave packet $\Psi_3(q, t)$ such that it evolves toward increasing values of the ClF bond stretch q, corresponding to vibration between classical inner and outer turning points, $q_{it} = q_{FC}$ and q_{ot}, respectively,

$$V_3(q_{FC} = q_{it}) = V_3(q_{ot}). \tag{4.57}$$

Now let us consider the scenario depicted in Fig. 4.53 where the mean energy $E'_{r=0}$, is sufficiently high such that the outer classical turning point q_{ot} exceeds q_{SOC},

$$q_{ot} > q_{SOC} \quad \text{for} \quad V_3(q_{ot}) - V_1(q_{FC}) \approx \hbar\omega. \tag{4.58}$$

Then the UV pump laser pulse creates the wave packet $\Psi_3(q, t)$ on $V_3(q)$ such that it enters the domain $q_{SOC} < q < q_{ot}$ on its way from $q_{FC} = q_{it}$ to q_{ot}. Here, efficient spin–orbit coupling causes separation of the wave function into two components $\Psi_3(q, t)$ and $\Psi_2(q, t)$ for the excited singlet and triplet states, respectively [31]. The generation of the triplet "partial wave" $\Psi_2(q, t)$ in the domain $q_{SOC} \lesssim q < q_{ot}$ from the original singlet wavefunction $\Psi_3(q, t)$ near $q \approx q_{FC}$ is a manifestation of spin-flip. The corresponding time evolution of the spin flip was measured by means of the pump-probe experiment, cf. Sect. 4.2.4. The quantitative results depend on the efficiency of V_{SOC}, on the parameters of the pump and probe pulses, and other details of the system. As a rule, the experimental time t_{sf} of spin flip should exceed the time $t(q_{SOC} - q_{FC})$ it takes $\Psi_3(q, t)$ to evolve from q_{FC} to the domain $q_{SOC} < q < q_{ot}$. For the present scenario of preorientation along the crystallographic $\langle 111 \rangle$ direction, typical times $t(q_{SOC} - q_{FC}) \approx$ 50-100 fs are predicted [31] . The experimental times t_{sf} are found to be systematically longer, i.e. ca. 150-250 fs, cf. Sect. 4.2.4 [18,31]. An effect which may contribute to the observation $t_{sf} > t(q_{SOC} - q_{FC})$ is different preferential preorientation along the crystallographic $\langle 110 \rangle$ direction, allowing larger domains $q_{ot} - q_{it}$ and longer times $t(q_{SOC} - q_{FC})$than for the present preorientation $\langle 111 \rangle$, see the discussion in Sect. 4.9 [24].

4.10.3 Coherent spin control

The observation of ultrafast spin flip (cf. Sects. 4.2.4, 4.10.2, and [18,31]) has served as motivation to predict coherent spin control, with spatial and time resolutions of ca. 1 a_0 and ca. 10 fs, respectively, cf. [203, 219, 268]. Coherent control was introduced by Brumer and Shapiro in 1986 in order to control two competing processes with degenerate energies by means of two interfering excitations pathways, e.g., single photon versus three photon excitations, or different two photon excitations. The original scheme of coherent control was designed for continuous wave lasers [272], but subsequently the method has also been extended to laser pulses, see Chapter 2, Sect. 2.4 and [273,274]. The present extension to coherent spin control adds a new dimension to coherent control, i.e. control of wave packets not only in time and coordinate or momentum space, but also concerning the spin. For this purpose, laser pulses are designed which induce two interfering pathways to near-degenerate wavefunctions with different spin multiplicities at specific times and locations. The effects of the matrix environment and the use of ultrashort laser pulses will turn out to be important for the present approach.

The proposed mechanism of coherent spin control consists of three steps, which are illustrated schematically by arrows in Fig. 4.53, for the 1D model of ClF in Ar. We assume that the system is originally in the ground state $\Phi_{1,v=0}$. The first step is the preparation of a vibrational superposition state, (4.51), with small population $r = P_{1,j}$ of the excited vibrational state (e.g. $v = j = 3$), by means of an IR laser pulse. Note that for the 1D model, the time for vibrational transitions is not restricted to ca. 100 fs- this limit applies just for processes induced by UV laser pulses, i.e. one may employ a rather long IR pulse with parameters $E_{0,IR} = 0.0026 E_h/a_0$, $\omega_{IR} = 0.0103 E_h/\hbar$, τ_{IR} =1 ps ($t_{p,IR} = 2.75$ ps), $t_{i,IR} = -2.75$ ps, and $\eta_{IR} = 0$.

The second step is electronic excitation from the $\Phi_{1,v=0}$ component of the vibrational hybrid state (4.51) to the triplet state $V_2(q)$, mediated by spin-orbit coupling, and simultaneously from the $\Phi_{1,j=3}$ component of state (4.51) to the singlet state $V_3(q)$, by means of a short ($\tau_{UV} < 50$ fs) special UV laser pulse. The UV laser parameters $E_{0,UV} = 0.0014 E_h/(ea_0)$, $\omega_{UV} = 0.128 E_h/\hbar$, $\tau_{UV} \approx 18$ fs, and $\eta_{UV} = 0$ are chosen such that the populations $P_2(t)$ and $P_3(t)$ of the wave functions $\Psi_2(t)$ and $\Psi_3(q,t)$ in the excited triplet and singlet state are about equal, albeit rather small, in order to allow optimal effects of interferences. For this purpose, the IR laser pulse excites only a small fraction (r) of $\Phi_{1,3}(q)$, followed by the UV pulse which transfers $\Phi_{1,3}(q)$ into $\Psi_3(q,t)$, with rather large transition probability. In contrast, the same UV pulse induces also the other transition from the complementary large fraction $(1-r)$ of $\Phi_{1,0}$ to $\Psi_2(q,t)$, but with rather low transition probability such that $P_2(\tau_{UV}) \approx P_3(\tau_{UV})$.

The third step consists of the laser induced, relatively slow and rapid evolutions of the wave packets $\Psi_2(q,t)$ and $\Psi_3(q,t)$ in the potentials $V_2(q)$ and $V_3(q)$, respectively, from their inner turning points to the outer ones, such that they overlap when they reach the domain of strong spin-orbit coupling, $q > q_{SOC}$. Here, as explained above for the mechanism of ultrafast spin flip, V_{SOC} transforms the original singlet wave function into two components with singlet and triplet spin multiplicity. By analogy, V_{SOC} also transforms the original triplet wave function into two triplet and singlet components. As a

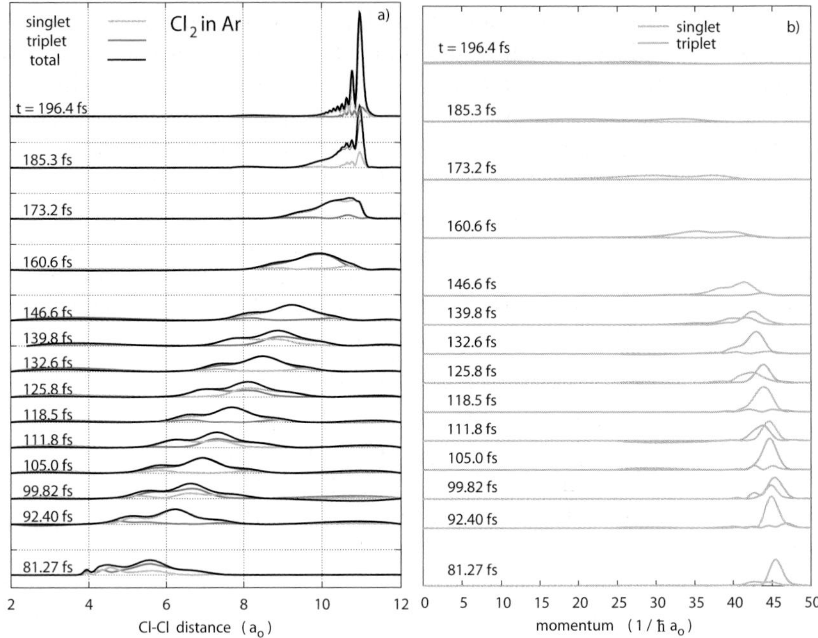

Fig. 4.54. Coherent spin control of Cl_2 in Ar. Panel a) shows the time evolutions of the densities of the wave packet $\Psi_2(q,t)$ and $\Psi_3(q,t)$ which evolve on potential energy surfaces of the excited triplet and singlet states, together with the total density of the excited states. Panel b) shows the densities of the corresponding wavefunctions for the triplet and singlet states in momentum space. This wave packet dynamics corresponds to the third step of the coherent spin control. The preceding first step has been excitation of the vibrational hybrid state, cf. (4.51) ($r = 0.01$), see [203,268] and Fig. 4.57. The second step is electronic excitation of the vibrational hybrid state to $\Psi_2(q,t)$ and $\Psi_3(q,t)$ by means of a UV pulse, with parameters $E_0 = 50MV/cm$, $\tau = 36$ fs, $\omega_{UV} = 0.103E_h$. (adapted from [203].)

consequence, steps 1-3 provide two different paths to the wave functions which represent the triplet and singlet components, in the domain $q > q_{SOC}$. Their interferences may be constructive and destructive. As a consequence, the total wave function may appear with dominant triplet or singlet character. Control of the spin multiplicity at a given time and geometry q in the domain $q > q_{SOC}$ may be controlled by specific choices of the parameters of the coherent IR+UV laser pulses, cf. Fig. 4.56 below.

The domain $q > q_{SOC}$ of coherent spin control by means of the coherent IR plus UV laser pulses and mediated by efficient spin orbit coupling is rather small for the 1D model of ClF oriented along $\langle 111 \rangle$ in Ar, cf. Fig. 4.53. Other preorientations, e.g. along $\langle 110 \rangle$ (cf. Sect. 4.9) or other systems offer larger domains $q_{SOC} < q < q_{ot}$ of coherent spin control. A prominent example is Cl_2 in Ar, with $\langle 110 \rangle$ orientation in its double substitutional site, cf. [203]. Fig. 4.54a illustrates the third step of coherent spin control, i.e. the wave packet dynamics of the Cl_2 bond in Ar, induced by an ultrashort UV excitation (step 2) from the superposition state (4.51), with $r = 0.01$ (step 1, see below); for the laser parameters, see the legend of Fig. 4.54. Apparently, the resulting wavefunctions $\Psi_2(q,t)$ and $\Psi_3(q,t)$ overlap in the entire domain $q_{SOC} \approx 6a_0 \leq q < q_{ot} \approx 11a_0$ of dominant spin-orbit coupling such that the total wavefunction appears alternatingly as dominant singlet or triplet wave function. The corresponding time resolution τ_{csc} for coherent spin control is imposed by the energy gap between the vibrational levels of the hybrid state (4.51) [203, 219, 268],

$$\tau_{csc}(E_{1,3} - E_{1,0}) \approx h \tag{4.59}$$

For Cl_2 in Ar, this implies coherent laser driven spin flips within $\tau_{csc} \approx 10$ fs- this is even 10 times faster than the "ultrafast" spin flip induced by a single UV pulse, as described in Sect. 4.10.2. The spatial resolution depends on the momenta and corresponding wavenumbers of the wave packets; for Cl_2 in Ar, it is ca. $1a_0$, cf. Fig. 4.54a.

A necessary condition for coherent spin control is that the coherent laser pulses generate two coherent paths to $\Psi_2(q,t)$ and $\Psi_3(q,t)$ which should overlap not only in coordinate space $q > q_{SOC}$, but also in momentum space. This is demonstrated exemplarily for Cl_2 in Ar, by means of snapshots of the wavefunctions in coordinate and momentum representations illustrated in Figs. 4.54a and b, respectively. In order to achieve the overlap in momentum space, the two UV spectra for the underlying transitions from the two components of the vibrational hybrid state, $\Phi_{1,0}(q)$ and $\Phi_{1,3}(q)$ to the excited triplet and singlet states, respectively, should overlap. These spectra are shown in Fig. 4.53, exemplarily for ClF in Ar. Apparently, the overlap is small, but nevertheless significant. Better overlaps and, therefore, even more efficient coherent spin control may be achieved by shorter UV laser pulses [203].

The alternating appearance of the total wavefunctions as dominant singlet or triplet state (cf. Fig. 4.54 for Cl_2 in Ar) corresponds to alternating values of the probabilities, $P_3(t) >$ or $< P_2(t)$, respectively. This phenomenon is illustrated in Fig. 4.55b, for the case of ClF in Ar driven by coherent IR + UV laser pulses, cf. Fig. 4.53. For comparison, Fig. 4.55a also shows the effect of the UV pulse, without preexcitation of the vibrational hybrid state (4.51) by the IR pulse ($r = 0$ in (4.51)-(4.53)). Of course, the pure UV pulse excites just a single path from $\Phi_{1,0}$ to $\Psi_2(q,t)$. As a consequence, it causes nothing but ultrafast spin flip similar to, but less efficient (ca. 10% within the time ca. 75 fs needed to enter the domain $q > q_{SOC}$) compared to the UV pulse with frequency ω', as discussed above. In contrast, the combined coherent IR+UV pulses excite two interfering paths causing much faster oscillations between dominant singlet and triplet states, on the time scale of $\tau_{SOC} \approx 12$ fs, in accord with (4.59) [219]. These oscillations are most prominent at approx. 75 fs, i.e. when the two wavefunctions $\Phi_2(q,t)$ and $\Phi_3(q,t)$ overlap in the domain

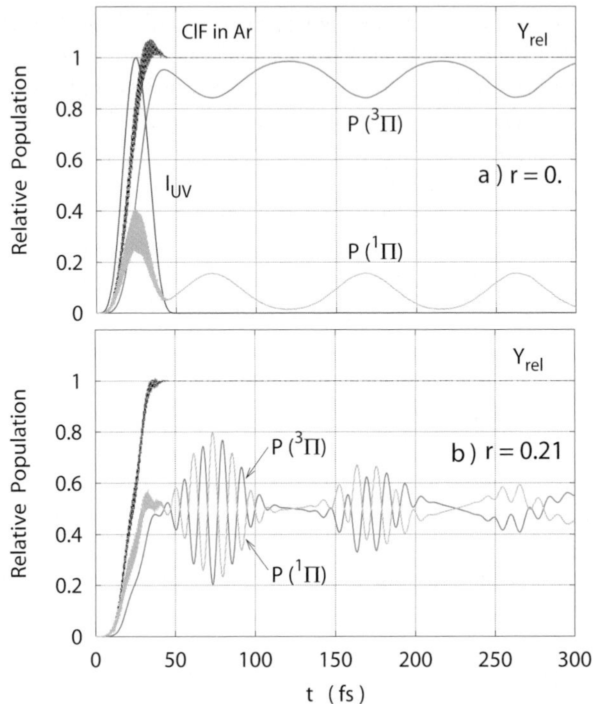

Fig. 4.55. From ultrafast spin flip to coherent spin control of ClF in Ar (cf. Fig. 4.53). Panels a and b compare the time evolutions of the relative populations $P_{rel,2}(t) = P_2(t)/Y_0$ and $P_{rel,3}(t) = P_3(t)/Y_0$ and the relative yields $Y_{rel}(t) = (P_2(t) + P_3(t))/Y_0$ of the excited triplet and singlet states, driven by the pure UV laser pulse ($r = 0$, cf. (4.51)-(4.53), $Y_0 = 3.35 \times 10^{-6}$), and by two coherent IR+UV laser pulses ($r = 0.21, Y_0 = 15.4 \times 10^{-6}$), respectively. The laser parameters are the same as in Fig. 4.53 (see text), for the case $E_{0,IR} \approx 0.022 E_h/(ea_0)$, $t_{i,UV} = 0$ fs and $\eta_{UV} = 0$.

of dominant effective spin-orbit coupling $q > q_{SOC}$ close to the outer turning point $q = q_{ot}$. The onset of these oscillations even before 75 fs allows the prediction of coherent spin control for ClF in Ar, within the validity of the 1D model.

For reference, Fig. 4.55 also shows the time evolutions of the probabilities $P_2(t)$ and $P_3(t)$ during longer times. Apparently, the phenomenon of oscillatory probabilities $P_2(t)$ and $P_3(t)$ appears not only close to 75 fs but also near 165 fs, 255 fs, etc. This is in accord with the ca. 90 fs period of oscillation of the wave packets for the 1D model of ClF in Ar, i.e. $P_2(t)$ and $P_3(t)$ oscillate rapidly with period τ_{csc} whenever the wavefunctions re-visit the domain $q > q_{SOC}$ of efficient spin-orbit coupling. The apparent coherent spin flips (τ_{csc} = ca. 12 fs) are most prominent, however, already during the first approach close to 75 fs. Subsequently, at 165 fs, 255 fs, etc. they tend to be washed-out. This is due to at least two effects: wave packet dispersion, and also the fact that $\Psi_2(q,t)$ and $\Psi_3(q,t)$ evolve on different PES, with different periods of oscillation. i.e. they run out of phase after few cycles. Additional effects of more degrees of freedom e.g. IVR will also contribute at times beyond the validity of the 1D model. In any case, the 1D model serves two purposes: prediction of coherent spin control, and prediction of optimal conditions for realization at rather early times, i.e. ca. 70 fs in the case of ClF in Ar.

An applications of coherent spin control to the 1D model of ClF in Ar is illustrated in Fig. 4.56. For this purpose, one employs the coherent, i.e. phase-

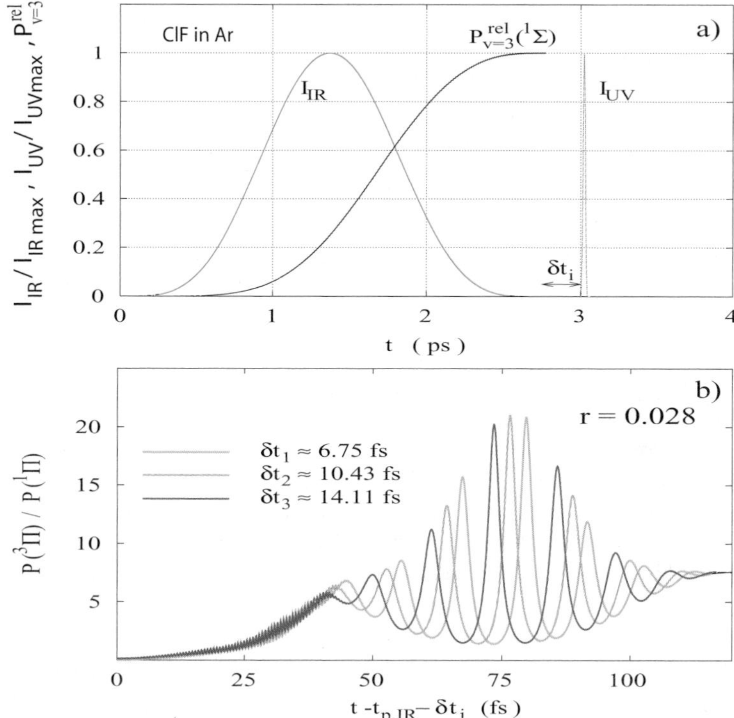

Fig. 4.56. Coherent spin control of ClF in Ar by coherent IR+UV laser pulses. The time evolutions of their relative intensities $I(t)/I_{max}$ (cf. (4.44) -(4.46)) are illustrated in panel a. Also shown in panel a is the time evolution of the relative population $P_{rel,v=3} = P_{1,v=3}(t)/r$ of the excited vibrational state $\Phi_{1,3}(q)$ in the electronic ground state, for the case $r = 0.0028$. The laser parameters are the same as in Fig. 4.53, for the case $E_{0,IR} = 0.0026E_h/(ea_0)$, except for the new definition $t_{i,IR} = 0$ and $t_{i,UV} = t_{p,IR} + \delta t_i$, with variable time delays. Panel b shows the resulting ratio $P(^3\Pi)/P(^1\Pi)$ of the populations $P_2(t)$ and $P_3(t)$ of the excited triplet and singlet states, for three different values of the time delay parameter δt_i. (adapted from [219].)

locked IR + UV laser pulses with the parameters as designed for the scenario of Fig. 4.53; for further details, see legend of Fig. 4.56. The time evolutions of the laser intensities, (4.44), are illustrated in panel 4.56a, depending on the time delay $t_d = t_{i,UV} - t_{i,IR} = t_{p,IR} + \delta t_i - t_{i,IR}$. Also shown in panel a is the effect of the IR pulse, i.e. preparation of the vibrational hybrid state (4.51) with population $r = P_{1,3} = 0.0028$ in the vibrationally excited state $\Phi_{1,3}(q)$. The resulting ratios of the populations of the excited triplet versus singlet states are shown in panel b, depending on the parameter δt_i for the time delay of the laser pulses. Apparently, one can employ the time delay as control parameter in order to achieve say dominant triplet character at a given time, e.g. at $t - \delta t_i = 60$ fs, with time resolution better than $\tau_{csc} \approx 12$ fs, cf. (4.59). The example illustrated in Fig. 4.54 shows that the corresponding spatial resolution is of the order of ca. 1 a_0. The effect may be rationalized by the fact that the control parameter δt_i implies the change of the phase shift, $(E_{1,3} - E_{1,0})\delta t_i/\hbar$ of the two components of the vibrational hybrid state (4.51), cf. [219]. It allows time-, coordinate-, and spin selective control of sequel processes, e.g. one may design another, third laser pulse which is fired at the time when the system is controlled to be in its triplet state, for selective excitations of ionic triplet states, and these in turn might induce selective preparation of new molecules, as discussed in the conclusion section.

Coherent spin control may employ not only coherent IR+UV laser pulses, as in the preceding example of ClF in Ar, cf. Figs. 4.53, 4.54, 4.56, but also alternative excitation schemes, e.g. two UV+UV laser pulses, as demonstrated in Fig. 4.57, cf. [268]. Here, two pulses labeled UV_1 replace the previous IR pulse for step 1, i.e., preparation of the vibrational hybrid state (4.51), whereas the last pulse labeled UV_2 fulfills the same task as the previous UV pulse, i.e. step 2 followed by step 3. Note that an IR pulse cannot be employed for step 1 because Cl_2 in Ar does not possess any dipole, for symmetry reasons. In order to generate the vibrational superposition state (4.51) the first UV pulse is designed such that it induces intrapulse Raman-type pump-dump processes from the original ground state $\Phi_{1,0}(q)$ via a transient electronic excited state to the intermediate vibrationally excited hybrid state (4.51). For this purpose, the laser pulse UV_1 has two overlapping components labeled $\mathrm{UV}_{1,a}$ and $\mathrm{UV}_{1,b}$, with the same durations $\tau_{UV,1a} = \tau_{UV,1b}$ but different frequencies and field strengths; for the details, see legend of Fig. 4.57. Specifically, the frequency difference matches the energy gap of the two components of the hybrid state (4.51),

$$\omega_{1a} - \omega_{1b} \approx (E_{1,j=3} - E_{1,v=0})/\hbar\,, \qquad (4.60)$$

and the laser field strength are chosen such that the two components UV_{1a} and UV_{1b} transfer the designated population from the initial state $\Phi_{1,0}(q)$ via an electronic excited state to the excited state component $P_{1,3}$ of the vibrational hybrid state (4.51), without leaving any population in the excited state; for further details, see [268]. The latter criterion is important because otherwise, any population which might be left behind in the electronic excited

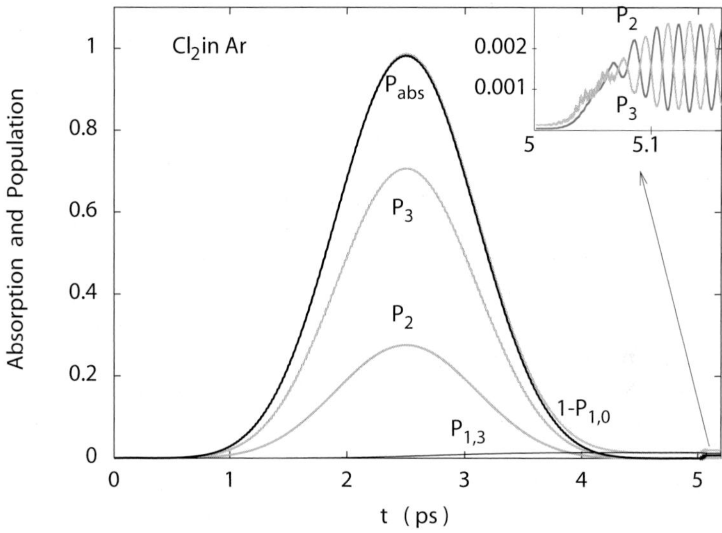

Fig. 4.57. Coherent spin control of Cl_2 in Ar by two coherent laser pulses in the UV domain. The first one consists of two overlapping sub-pulses with parameters $E_{0,1a} = 0.0035E_h/(ea_0)$, $\omega_{1a} = 0.142825E_h/\hbar$, τ_{1a}=1.8 ps ($t_{p,1a}$=5 ps), and $E_{0,1b} = 0.0020E_h/(ea_0)$, $\omega_{1b} = 0.136240E_h/\hbar$, τ_{1b}=1.8 ps. The second pulse has the same parameters as the UV pulse used in Fig. 4.54. The figure shows the preparation of the vibrational hybrid state (4.51) with the corresponding population dynamics $P_{1,3}(t)$ of the vibrationally excited state $\Phi_{1,3}(q)$, together with the population $1 - P_{1,0}(t)$ complementary to the ground state. Also shown are the transient populations $P_2(t)$ and $P_3(t)$ of the excited singlet and triplet states, together with the total absorption probability $P_{abs}(t) = P_2(t) + P_3(t)$. The insert shows the effect of the second UV laser pulse yielding oscillatory $P_2(t)$ and $P_3(t)$, in accord with the results shown in Fig. 4.54, and similar to Figs. 4.55 and 4.56 for ClF in Ar.

states, at the end of step 1, would disturb the interferences of the singlet and triplet state components during the subsequent step 3 of coherent spin control. Note that this criterion has not been applied previously, because it has not been considered important for complementary Raman type excitations of vibrations of matrix isolated molecules [275]. Fig. 4.57 shows the population dynamics during coherent spin control of the 1D model Cl_2 in Ar, including the populations $P_{1,3}$ and $1-P_{1,0}$ for the components of the vibrational hybrid state (4.51), as well as the components $P_2(t)$ and $P_3(t)$ of the excited triplet and singlet states. The resulting rapid oscillations of $P_2(t)$ and $P_3(t)$ correspond to the alternating appearance of the total wave function as dominant singlet or triplet state, cf. Fig. 4.55, similar to the results for coherent spin control shown in Figs. 4.54 and 4.56 for ClF in Ar.

The example of coherent spin control of Cl_2 in Ar shown in Fig. 4.57 employs the laser pulse UV_1 with duration $\tau = 1$ ps, well beyond the validity of the 1D model. Different schemes of coherent spin control of Cl_2 in Ar, using even longer UV pulses, are discussed in [203, 268]. It is a challenge to design shorter UV laser pulses for the first step of coherent spin control of Cl_2 in Ar, within the validity of the 1D model. The results shown in Fig. 4.57 for the 1D model of Cl_2 in Ar are, nevertheless, useful i.e. they should stimulate analogous applications in other systems with longer validities of adequate models in restricted dimensionality. A promising candidate is Br_2 in Ar, see Sect. 4.6. Indeed, the rather narrow widths Γ of the zero phonon lines of the high resolution spectra of Br_2 in Ar imply rather long (> 1 ps) life times of the corresponding vibronic excited states (cf. Fig. 4.3). Accordingly, we suggest that those vibronic states may be employed as transient electronic excited states for the preparations of the vibrational hybrid states (4.51) by means of the Raman type pump-dump mechanism which has been explained above for coherent spin control Cl_2 in Ar by means of coherent UV+UV laser pulses.

Concluding this section the present quantum simulations of dihalogens in rare gas matrices show that the initial dynamics induced by ultrashort laser pulses may be described in terms of models with reduced dimensionality. The results allow to analyze or predict various effects such as cage exits supported by spin-orbit coupling, ultrafast spin flip within ca. 100 fs, or coherent spin control, with spatial and time resolutions of ca. 1 a_0 and 10 fs, respectively. So far, these effects have been discovered for dihalogens in rare gas matrices, and the effects of the matrix environment as well as the use of ultrashort laser pulses have been seen to be important, both qualitatively and quantitatively. For example, the first analyses or predictions of effects of spin orbit coupling on the wave packet dynamics of molecules in the gas phase have shown that these proceed on much longer times, from several 100 fs to ps and beyond [276], see also [277]. It is easy to predict that extensions of the present 1D and 2D quantum models to higher dimensionality should allow to discover new effects of analysis and control of molecules in matrices.

4.11 Outlook

The overview presented in the previous sections illustrates that a rather complete and consistent characterization of energetics and nonradiative transitions of small molecules in rare gas matrices has been achieved in experiment and theory. This situation lays a very important ground for a future exploration of the, compared to other condensed phase systems, extraordinarily pronounced and lasting coherence effects in chromophore and cage coordinates. The results in Sect. 4.3 indeed call for a systematic development of vibrational wave packet interferometry in dissipative surroundings based on these prototypical systems to pave the ground for applications in other media. Both directions, the dispersion-revival method and the coherent pulse sequence technique should be pursued experimentally and be accompanied by Wigner type wave packet simulations which include degrees of freedom of the cage. An exceptional sophisticated theoretical treatment of nonradiative transitions induced by interaction with the environment is displayed in Sects. 4.6, 4.8, and 4.10. Merging these achievements with the experimental developments (see Sect. 4.3) will trigger new types of coherent control schemes. Coherent control of environment induced predissociation by wave packet interferometry can be expected. Implementation of control by interference of wave packets on different electronic surfaces as derived in Sect. 4.10 presents a challenge for the experimental community.

Dynamics in the angular coordinate θ in Fig. 4.1 delivers surprises according to Sect. 4.2.3. The experimental characterization of librational dynamics is, however, not as mature as for the internal vibrational coordinate R. Here a successful conversion of the theoretical concepts outlined in Sect. 4.9 into experiments can be ground braking. Preparation and analysis of librational wave packets can provide a new spectroscopic method. In addition it bears a route to coherent control by positioning molecules relative to the cage using post pulse alignment. For this direction of research parahydrogen matrices may become important. The spherical shape of the parahydrogen molecule in the $J = 0$ state and the lattice parameter in the crystal resemble a rare gas matrix. The softness with respect to librational motion due to the large zero point amplitude, however, favors its use in experiments attempting alignment parallel to the electric field of a nonresonant light pulse.

Coherences in the environment induced by excitation of the chromophore are a fascinating subject which merits future extensive investigation. The results from these very well defined systems may have a fundamental impact in other areas. The coherent matrix response presented in Sect. 4.4 which manifests displacive excitation of zone boundary phonons opens up a connection to several hot research topics in other systems. The controversy of displacive versus Raman type excitations is invoked. A further clarification of crystal phonons versus local modes is on the agenda. The dominance of zone boundary versus zone center phonons and the energy flow dynamics are topics which are intimately related to a fundamental issue of lasting coherence in chromophore and cage coordinates despite exchange of large amounts of energy (see Sect. 4.3.4). The mutual flow of coherence and energy is especially pronounced in CARS-spectroscopy of impulsively excited vibrational wave packets near the dissociation limit [91]. It leads us into the very heart of quantum mechanics and decoherence as it is caught by the phrase of "Schrödingers kitten" coined by A. Apkarian in [91]. The systems of this chapter are especially prone to drive progress in theory as evident from Sects. 4.6 to 4.10. This important topic of decoherence will challenge theory further in order to include more and more quantum mechanically treated degrees of freedom within a framework of classically handled coordinates. The necessity and also difficulty is illustrated by the bunch of questions arising from a mild increase in number of internal coordinates (see Sect. 4.5).

Another route deals with reactivity of the rare gas cage atoms induced by photo excitation, keeping in mind that ionic rare gas halogen compounds are the excimers per se. Experiments in this direction are discussed in Sect. 4.2.6. Recently, there have been interesting developments in noble gas chemistry, in which photodissociation of small molecules in a noble-gas host was found to result in formation of new, noble-gas containing molecules [278]. It seems that the conditions are now ripe for ultrafast analysis and control of the new noble-gas molecules. In the future this should become useful and may have a major impact on noble-gas chemistry. Some important aspects in this context are: (i) It was found that new noble-gas molecules, such as $HXeCl$, can be formed in ultrafast photochemical processes (e.g. photolysis of HCl in Xe), in a picosecond timescale [278]. Thus, ultrafast techniques should be very suitable for studying these processes. (ii) The formation of the noble-gas molecules can be successfully described by extension of DIM potentials [279]. (iii) The dynamics of $F(^2P)$ atoms in an argon medium should be directly relevant to the formation of several interesting new noble-gas molecules, including $HArF$, $HKrF$ [280], the theoretically predicted $HHeF$ in pressurized solid helium [281], and the predicted argon compounds: $FArCCH$ and $FArSiF_3$ [282]. (iv) The group -$C \equiv CH$ was found to behave as a "pseudo-halogen" in matrix photochemistry in this study. Photolysis of $HCCH$ in Xe and Kr was found to produce $HXeCCH$, $HKrCCH$, which were predicted by theoretical work [283]. Some of the findings for halogens may thus prove relevant to acetylenic groups in organic photochemistry.

Acknowledgement

PH wishes to thank Roland Schanz and Virgiliu Botan for essential contributions to the work presented in Sect. 4.5 and the Swiss Science Foundation (SNF) for financial support. JM thanks the Fonds der Chemischen Industrie for continuous support. The authors of Sects. 3, 4, and 5 thank Mónika Héjjas for the extensive editing support and acknowledge the intense discussions with A. Apkarian, M. Chergui, D. Coker, V. Engel, and D. Tannor.

References

1. M. Bargheer, J. Pietzner, P. Dietrich, N. Schwentner, J. Chem. Phys. **115**, 9827 (2001)
2. Z. Bihary, M. Karavitis, V.A. Apkarian, J. Chem. Phys. **120**, 8144 (2004)
3. V.E. Bondybey, S.S. Bearder, C. Fletcher, J. Chem. Phys. **64**, 5243 (1976)
4. R. Englman, *Non-radiative decay of ions and molecules in solids* (North-Holland Publishing Company, Amsterdam, 1979)
5. N. Schwentner, E.E. Koch, J. Jortner, *Electronic excitations in condensed rare gases* (Springer Verlag, Berlin, Heidelberg, 1985)
6. R. Zadoyan, P. Ashjian, C.C. Martens, V.A. Apkarian, Chem. Phys. Lett. **218**, 504 (1994)
7. R. Zadoyan, M. Sterling, V.A. Apkarian, J. Chem. Soc., Faraday. Trans. **92**, 1821 (1996)
8. R. Zadoyan, J. Almy, V.A. Apkarian, Faraday Discuss. **108**, 255 (1997)
9. R. Zadoyan, M. Sterling, M. Ovchinnikov, V.A. Apkarian, J. Chem. Phys. **107**, 8446 (1997)
10. M. Bargheer, P. Dietrich, K. Donovang, N. Schwentner, J. Chem. Phys. **111**, 8556 (1999)
11. M. Bargheer, M. Gühr, P. Dietrich, N. Schwentner, Phys. Chem. Chem. Phys. **4**, 75 (2002)
12. L.D. Landau, E.M. Lifschitz, *Lehrbuch der theoretischen Physik*, vol. I: Mechanik (Akademie-Verlag, Berlin, 1979)
13. N. Yu, C.J. Margulis, D.F. Coker, J. Phys. Chem. B **105**, 6728 (2001)
14. M. Gühr, N. Schwentner, J. Chem. Phys. **123**, 244506 (2005)
15. M. Gühr, M. Bargheer, M. Fushitani, T. Kiljunen, N. Schwentner, Phys. Chem. Chem. Phys. **123**, accepted. Several figures from this review are reproduced with kind permission. (2006)
16. M. Ovchinnikov, V.A. Apkarian, J. Chem. Phys. **108**, 2277 (1998)
17. M. Bargheer, *Ultrafast photodynamics in condensed phase: ClF, Cl_2, and I_2 in solid rare gases* (Shaker Verlag, Aachen, 2002)
18. M. Bargheer, M.Y. Niv, R.B. Gerber, N. Schwentner, Phys. Rev. Lett. **89**, 108301 (2002)
19. C. Meier, J.A. Beswick, J. Chem. Phys. **121**, 4550 (2004)
20. V.A. Ermoshin, V. Engel, A.K. Kazanky, J. Phys. Chem. A **105**, 7501 (2001)
21. M.Y. Niv, M. Bargheer, R.B. Gerber, J. Chem. Phys. **113**, 6660 (2000)
22. V.S. Batista, D.F. Coker, J. Chem. Phys. **106**, 6923 (1997)
23. T. Kiljunen, M. Bargheer, M. Gühr, N. Schwentner, B. Schmidt, Phys. Chem. Chem. Phys. **6**, 2932 (2004)
24. T. Kiljunen, M. Bargheer, M. Gühr, N. Schwentner, Phys. Chem. Chem. Phys. **6**, 2185 (2004)
25. A.C. Albrecht, J. Mol. Spectrosc. **6**, 84 (1961)
26. M. Bargheer, M. Gühr, N. Schwentner, J. Chem. Phys. **117**, 5 (2002)
27. T. Kiljunen, B. Schmidt, N. Schwentner, Phys. Rev. Lett. **94**, 123003 (2005)
28. T. Kiljunen, B. Schmidt, N. Schwentner, Phys. Rev. A **72**, 053415 (2005)
29. T. Kiljunen, B. Schmidt, N. Schwentner, J. Chem. Phys. **124**, 164502 (2006)
30. A.B. Alekseyev, H. Liebermann, R.J. Buenker, D.B. Kokh, J. Chem. Phys. **112**, 2274 (2000)
31. M. Bargheer, R.B. Gerber, M.V. Korolkov, O. Kühn, J. Manz, M. Schröder, N. Schwentner, Phys. Chem. Chem. Phys. **4**, 5554 (2002)

32. M. Bargheer, P. Dietrich, N. Schwentner, J. Chem. Phys. **115**, 149 (2001)
33. C. Lienau, A.H. Zewail, J. Phys. Chem. **100**, 1829 (1996)
34. A. Materny, C. Lienau, A.H. Zewail, J. Phys. Chem. **100**, 18650 (1996)
35. Q. Liu, C. Wan, A.H. Zewail, J. Phys. Chem. **100**, 18666 (1996)
36. N.F. Scherer, L.D. Ziegler, G.R. Fleming, J. Chem. Phys. **96**, 5544 (1992)
37. O. Roncero, N. Halberstadt, J.A. Beswick, Chem. Phys. Lett. **226**, 82 (1994)
38. O. Roncero, B. Lepetit, J.A. Beswick, N. Halberstadt, A.A. Buchachenko, J. Chem. Phys. **115**, 6961 (2001)
39. V.S. Batista, D.F. Coker, J. Chem. Phys. **105**, 4033 (1996)
40. A.A. Buchachenko, N.F. Stepanov, J. Chem. Phys. **104**, 9913 (1996)
41. M. Gühr, M. Bargheer, P. Dietrich, N. Schwentner, J. Phys. Chem. A **106**, 12002 (2002)
42. I. Horenko, C. Salzmann, B. Schmidt, C. Schütte, J. Chem. Phys. **117**, 11075 (2002)
43. V.A. Apkarian, N. Schwentner, Chem. Rev. **99**, 1481 (1999)
44. H. Kunz, J.G. McCaffrey, N. Schwentner, J. Chem. Phys. **94**, 1039 (1991)
45. J.G. McCaffrey, H. Kunz, N. Schwentner, J. Chem. Phys. **96**, 155 (1992)
46. R. Alimi, R.B. Gerber, J.G. McCaffrey, H. Kunz, N. Schwentner, Phys. Rev. Lett. **69**, 856 (1992)
47. J. Helbing, M. Chergui, J. Chem. Phys. **115**, 6158 (2001)
48. Z. Bihary, R. Zadoyan, M. Karavitis, V.A. Apkarian, J. Chem. Phys. **120**, 7576 (2004)
49. J.G. McCaffrey, H. Kunz, N. Schwentner, J. Chem. Phys. **96**, 2825 (1992)
50. I.H. Gersonde, H. Gabriel, J. Chem. Phys. **98**, 2094 (1993)
51. N. Schwentner, V.A. Apkarian, Chem. Phys. Lett. **154**, 413 (1989)
52. H. Kunttu, V.A. Apkarian, Chem. Phys. Lett. **171**, 423 (1990)
53. H. Kunttu, J. Feld, R. Alimi, A. Becker, V.A. Apkarian, J. Chem. Phys. **92**, 4856 (1990)
54. R.B. Gerber, M.V. Korolkov, J. Manz, M.Y. Niv, B. Schmidt, Chem. Phys. Lett. **327**, 76 (2000)
55. G. Chaban, R.B. Gerber, M.V. Korolkov, J. Manz, M.Y. Niv, B. Schmidt, J. Phys. Chem. A **105**, 2770 (2001)
56. M. Bargheer, N. Schwentner, J. Low Temp. Phys. **29**, 165 (2003)
57. M. Gühr, H. Ibrahim, N. Schwentner, Phys. Chem. Chem. Phys. **6**, 5353 (2004)
58. M. Fushitani, M. Bargheer, M. Gühr, N. Schwentner, Phys. Chem. Chem. Phys. **7**, 3143 (2005)
59. N.F. Scherer, A.J. Ruggiero, M. Du, G.R. Fleming, J. Chem. Phys. **93**, 856 (1990)
60. N.F. Scherer, R.J. Carlson, A. Matro, M. Du, A.J. Ruggiero, V. Romero-Rochin, J.A. Cina, G.R. Fleming, J. Chem. Phys. **95**, 1487 (1991)
61. J.F. Christian, B. Broers, J.H. Hoogenraad, W.J. van der Zande, L.D. Noordam, Optics Communications **103**, 79 (1993)
62. R.R. Jones, D.W. Schumacher, T.F. Gallagher, P.H. Bucksbaum, J. Phys. B: At. Mol. Opt. Phys. **28**, L405 (1995)
63. M.A. Bouchene, C. Nicole, B. Girard, J. Phys. B: At. Mol. Opt. Phys. **32**, 5167 (1999)
64. Y. Liau, A.N. Unterreiner, Q. Chang, N.F. Scherer, J. Phys. Chem. B **105**, 2135 (2001)

65. M. Wollenhaupt, A. Assion, D. Liese, C. Sarpe-Tudoran, T. Baumert, S. Zamith, M.A. Bouchene, B. Girard, A. Flettner, U. Weichmann, G. Gerber, Phys. Rev. Lett. **89**, 173001 (2002)
66. S. Sato, Y. Nishimura, Y. Sakata, I. Yamazaki, J. Phys. Chem. A **107**, 10019 (2003)
67. C. Petersen, E. Péronne, J. Thogersen, H. Stapelfeldt, M. Machholm, Phys. Rev. A **70**, 033404 (2004)
68. P.C.M. Planken, I. Brener, M.C. Nuss, M.S.C. Luo, S.L. Chuang, Phys. Rev. B **48**, 4903 (1993)
69. Q. Hong, J. Durrant, G. Hastings, G. Porter, D.R. Klug, Chem. Phys. Lett. **202**, 183 (1993)
70. A.P. Heberle, J.J. Baumberg, K. Köhler, Phys. Rev. Lett. **75**, 2598 (1995)
71. H. Petek, A.P. Heberle, W. Nessler, H. Nagano, S. Kubota, S. Matsunami, N. Moriya, S. Ogawa, Phys. Rev. Lett. **79**, 4649 (1997)
72. M. Woerner, J. Shah, Phys. Rev. Lett. **81**, 4208 (1998)
73. N.H. Bonadeo, J. Erland, D. Gammon, D. Park, D.S. Katzer, D.G. Steel, Science **282**, 1473 (1998)
74. P.I. Tamborenea, H. Metiu, Phys. Lett. A **240**, 265 (1998)
75. Y. Mitsumori, M. Mizuno, S. Tanji, T. Kuroda, F. Minami, J. Luminescence **76-77**, 113 (1998)
76. C. Leichtle, W.P. Schleich, I.S. Averbukh, M. Shapiro, J. Chem. Phys. **108**, 6057 (1998)
77. V. Szöcs, H.F. Kauffmann, J. Chem. Phys. **109**, 7431 (1998)
78. A. Tortschanoff, K. Brunner, C. Warmuth, H.F. Kauffmann, J. Phys. Chem. A **103**, 2907 (1999)
79. A. Tortschanoff, K. Brunner, C. Warmuth, H.F. Kauffmann, J. Chem. Phys. **110**, 4493 (1999)
80. J. Bok, A. Tortschanoff, F. Sanda, V. Capek, H.F. Kauffmann, Chem. Phys. **244**, 89 (1999)
81. T. Yoda, T. Fuji, T. Hattori, H. Nakatsuka, J. Opt. Soc. Am. B **16**, 1768 (1999)
82. C. Warmuth, A. Tortschanoff, F. Milota, M. Shapiro, Y. Prior, I.S. Averbukh, W. Schleich, W. Jakubetz, H.F. Kauffmann, J. Chem. Phys. **112**, 5060 (2000)
83. Y. Mitsumori, T. Kuroda, F. Minami, J. Luminescence **87-89**, 914 (2000)
84. H. Petek, H. Nagano, M.J. Weida, S. Ogawa, J. Phys. Chem. B **105**, 6767 (2001)
85. Q. Luo, D.C. Dai, G.Q. Wang, V. Ninulescu, X.Y. Yu, L. Luo, J.Y. Zhou, Y.J. Yan, J. Chem. Phys. **114**, 1870 (2001)
86. C. Warmuth, A. Tortschanoff, F. Milota, M. Leibscher, M. Shapiro, Y. Prior, I.S. Averbukh, W. Schleich, W. Jakubetz, H.F. Kauffmann, J. Chem. Phys. **114**, 9901 (2001)
87. J. Sperling, F. Milota, A. Tortschanoff, C. Warmuth, B. Mollay, H. Bäsler, H.F. Kauffmann, J. Chem. Phys. **117**, 10877 (2002)
88. F. Milota, J. Sperling, A. Tortschanoff, V. Szöcs, L. Kuna, H.F. Kauffmann, J. Luminescence **108**, 205 (2004)
89. F. Milota, A. Tortschanoff, J. Sperling, L. Kuna, V. Szöcs, H. Kauffmann, Appl. Phys. A **78**, 497 (2004)
90. S. Mukamel, *Principles of nonlinear optical spectroscopy* (Oxford University Press, New York, 1995)
91. D. Segale, M. Karavitis, E. Fredj, V.A. Apkarian, J. Chem. Phys. **122**, 111104 (2005)

92. M. Karavitis, R. Zadoyan, V.A. Apkarian, J. Chem. Phys. **114**, 4131 (2001)
93. M. Karavitis, D. Segale, Z. Bihary, M. Pettersson, V.A. Apkarian, Low Temp. Phys. **29**, 814 (2003)
94. M. Karavitis, V.A. Apkarian, J. Chem. Phys. **120**, 292 (2004)
95. U. Banin, A. Bartana, S. Ruhman, R. Kosloff, J. Chem. Phys. **101**, 8461 (1994)
96. Z.H. Wang, T. Wasserman, F. Gershgoren, S. Ruhman, J. Mol. Liqu. **86**, 229 (2000)
97. E. Gershgoren, J. Vala, R. Kosloff, S. Ruhman, J. Phys. Chem. A **105**, 5081 (2001)
98. B. Kohler, V.V. Yakovlev, J. Che, J.L. Krause, M. Messina, K. Wilson, N. Schwentner, R.M. Whitnell, Y. Yan, Phys. Rev. Lett. **74**, 3360 (1995)
99. D.J. Tannor, *Introduction to Quantum Mechanics, A time-dependent perspective* (University Science Books, Sausalito, 2003)
100. B.M. Garraway, K.A. Suominen, Rep. Prog. Phys. **58**, 365 (1995)
101. T. Lohmuller, V. Engel, J.A. Beswick, C. Meier, J. Chem. Phys. **120**, 10442 (2004)
102. M. Sterling, R. Zadoyan, V.A. Apkarian, J. Chem. Phys. **104**, 6497 (1996)
103. J. Cao, K.R. Wilson, J. Chem. Phys. **107**, 1441 (1997)
104. M. Gühr, *Coherent dynamics of small molecules in rare gas crystals* (Cuvillier Verlag, Göttingen, 2005)
105. C. Wan, M. Gupta, J.S. Baskin, Z.H. Kim, A.H. Zewail, J. Chem. Phys. **106**, 4353 (1997)
106. J.S. Baskin, M. Gupta, M. Chachisvilis, A.H. Zewail, Chem. Phys. Lett. **275**, 437 (1997)
107. J.S. Baskin, M. Chachisvilis, M. Gupta, A.H. Zewail, J. Phys. Chem. A **102**, 4158 (1998)
108. M. Chachisvilis, I. Garcia-Ochoa, A. Douhal, A.H. Zewail, Chem. Phys. Lett. **293**, 153 (1998)
109. V.A. Ermoshin, A.K. Kazansky, V. Engel, J. Chem. Phys. **111**, 7807 (1999)
110. A.K. Kazansky, V.A. Ermoshin, V. Engel, J.Chem. Phys. **113**, 8865 (2000)
111. V. Ermoshin, V. Engel, C. Meier, J. Chem. Phys. **113**, 6585 (2000)
112. D.W. Oxtoby, D. Levesque, J.J. Weis, J. Chem. Phys. **68**, 5528 (1978)
113. D.W. Oxtoby, J. Chem. Phys. **70**, 2605 (1979)
114. K.S. Schweizer, D. Chandler, J. Chem. Phys. **76**, 2296 (1982)
115. I.S. Averbukh, N. Perel'man, Phys. Lett. A **139**, 449 (1989)
116. I.S. Averbukh, N.F. Perel'man, Sov. Phys. JETP **69**, 464 (1989)
117. I.S. Averbukh, N.F. Perel'man, Sov. Phys. Usp. **34**, 572 (1991)
118. S.I. Vetchinkin, A.S. Vetchinkin, V.V. Eryomin, Chem. Phys. Lett. **215**, 11 (1993)
119. S.I. Vetchinkin, V.V. Eryomin, Chem. Phys. Lett. **222**, 394 (1994)
120. C. Leichtle, I.S. Averbukh, W.P. Schleich, Phys. Rev. A **54**, 5299 (1996)
121. R.W. Robinett, Phys. Lett. **392**, 1 (2004)
122. E. Gershgoren, Z. Wang, S. Ruhman, J. Vala, R. Kosloff, J. Chem. Phys. **118**, 3660 (2003)
123. A.G. Redfield, IBM J. Res. Dev. **1**, 19 (1957)
124. R. Kubo, in *Fluctuations relaxation and resonance in magnetic systems*, ed. by O.T. Haar (Plenum Press, New York, 1962)
125. D.W. Oxtoby, Adv. Chem. Phys. **40**, 1 (1979)
126. F. Figueirido, R.J. Levy, Chem. Phys. **97**, 703 (1992)

127. J.S. Bader, B.J. Berne, E. Pollak, P. Hänggi, J. Chem. Phys. **104**, 1111 (1996)
128. A. Nitzan, S. Mukamel, J. Jortner, J. Chem. Phys. **63**, 200 (1975)
129. M.J. Davis, E.J. Heller, J. Chem. Phys. **75**, 794 (1981)
130. E.J. Heller, Acc. Chem. Res. **14**, 368 (1981)
131. E.J. Heller, S. Tomsovic, Physics Today **46**, 38 (1993)
132. M. Gühr, M. Bargheer, N. Schwentner, Phys. Rev. Lett. **91**, 085504 (2003)
133. M. Ovchinnikov, V.A. Apkarian, J. Chem. Phys. **105**, 10312 (1996)
134. M. Ovchinnikov, V.A. Apkarian, J. Chem. Phys. **106**, 5775 (1997)
135. T. Dekorsy, G.C. Cho, H. Kurz, in *Light scattering in solids VIII*, ed. by M. Cardona, G. Güntherodt (Springer Verlag, Berlin, Heidelberg, 2000), pp. 169–209
136. M. Gühr, N. Schwentner, Phys. Chem. Chem. Phys. **6**, 760 (2005)
137. J. Skalyo, Y. Edoh, G. Shirane, Phys. Rev. B **9**, 1797 (1974)
138. Y. Fujii, N.A. Lurie, R. Pynn, G. Shirane, Phys. Rev. B **10**, 3647 (1974)
139. M. Fushitani, N. Schwentner, M. Schröder, O. Kühn, J. Chem. Phys. **124**, 024505 (2006)
140. H.J. Zeiger, J. Vidal, T.K. Cheng, E.P. Ippen, G. Dresselhaus, M.S. Dresselhaus, Phys. Rev. B **45**, 768 (1992)
141. T.K. Cheng, S.D. Brorson, A.S. Kazretoonian, J.S. Moodera, G. Dresselhaus, M.S. Dresselhaus, E.P. Ippen, Appl. Phys. Lett. **57**, 1004 (1990)
142. T.K. Cheng, J. Vidal, H.J. Zeiger, G. Dresselhaus, M.S. Dresselhaus, E.P. Ippen, Appl. Phys. Lett. **59**, 1923 (1991)
143. A.V. Kutsnetsov, C.J. Stanton, Phys. Rev. Lett. **73**, 3243 (1994)
144. K. Sokolowski-Tinten, C. Blome, J. Blums, A. Cavalleri, C. Dietrich, A. Tarasevitch, I. Uschmann, E. Förster, M. Kammler, M. Horn-von-Hoegen, D. von der Linde, Nature **422**, 287 (2003)
145. M. Bargheer, N. Zhavoronkov, Y. Gritsai, J.C. Woo, D.S. Kim, M. Woerner, T. Elsaesser, Science **306**, 1771 (2004)
146. S. Jimenez, A. Paquarello, R. Car, M. Chergui, Chem. Phys. **233**, 343 (1997)
147. C. Jeannin, M.T. Porella-Oberli, S. Jimenez, F. Vigliotti, B. Lang, M. Chergui, Chem. Phys. Lett. **316**, 51 (2000)
148. S. Jimenez, M. Chergui, G. Rojas-Lorenzo, J. Rubayo-Soneira, J. Chem. Phys. **114**, 5264 (2001)
149. M. Chergui, *Structural dynamics in quantum solids* (Academie des sciences Paris, 2001), chap. Trends in femtosecond lasers and spectroscopy, pp. 1453–1467
150. F. Vigliotti, L. Bonacina, M. Chergui, G. Rojas-Lorenzo, J. Rubajo-Soneira, Chem. Phys. Lett. **362**, 31 (2002)
151. F. Vigliotti, L. Bonacina, M. Chergui, J. Chem. Phys. **116**, 4553 (2002)
152. F. Vigliotti, L. Bonacina, M. Chergui, Phys. Rev. B **67**, 115118 (2003)
153. G. Rojas-Lorenzo, J. Rubayo-Soneira, F. Vigliotti, M. Chergui, Phys. Rev. B **67**, 115119 (2003)
154. F.O. Ellison, J. Am. Chem. Soc. **85**, 3540 (1963)
155. J.C. Tully, J. Chem. Phys. **58**, 1396 (1973)
156. M.Y. Niv, A.I. Krylov, R.B. Gerber, Farad. Diss. **108**, 243 (1997)
157. A.I. Krylov, R.B. Gerber, J. Chem. Phys. **106**, 6574 (1997)
158. M.Y. Niv, A.I. Krylov, R.B. Gerber, U. Buck, J. Chem. Phys. **110**, 11047 (1999)
159. T. Kiljunen, J. Eloranta, J. Ahokas, H. Kunttu, J. Chem. Phys. **114**, 7144 (2001)

160. T. Kiljunen, J. Eloranta, J. Ahokas, H. Kunttu, J. Chem. Phys. **114**, 7157 (2001)
161. F.Y. Naumkin, D.J. Wales, Comp. Phys. Comm. **145**, 141 (2002)
162. C.R. Gonzales, S. Fernandez-Alberti, J. Echave, M. Chergui, Chem. Phys. Lett. **367**, 651 (2003)
163. N. Yu, D.F. Coker, Mol. Phys. **102**, 1031 (2004)
164. P. Casavecchia, G. He, R.K. Sparks, Y.T. Lee, J. Chem. Phys. **77**, 1878 (1982)
165. M. Schröder, Ph.D. thesis, Freie Universität Berlin (2004)
166. K.M. M. Hase, H. Harima, S. Nakashima, M. Tanbi, K. Sakai, M. Hangyo, Appl. Phys. Lett. **69**, 2474 (1996)
167. M. Hase, I. Ishioka, M. Kitajima, S. Hishita, K. Ushida, Appl. Surf. Science **197-198**, 710 (2002)
168. M. Hase, M. Kitajima, S. Nakashima, K. Mizoguchi, Phys. Rev. Lett. **88**, 067401 (2002)
169. G.A. Garrett, T.F. Albrecht, J.F. Whitaker, R. Merlin, Phys. Rev. Lett. **77**, 3661 (1996)
170. M.L. Klein, J.A. Venables, *Rare gas solids Vol. I + II* (Academic Press, London, 1976)
171. I.Y. Fugol, Adv. Phys. **27**, 1 (1978)
172. L. Onsager, J. Am. Chem. Soc. **58**, 1486 (1936)
173. R. Hall, G. Pimentel, J. Chem. Phys. **38**, 1889 (1963)
174. E.M.S. Macoasa, L. Khriachtchev, M. Pettersson, J. Juselius, R. Fausto, M. Räsänennen, J. Chem. Phys. **119**, 11765 (2003)
175. Y. Hannachi, L. Schriver, A. Schriver, J.P. Perchard, Chem. Phys. **135**, 285 (1989)
176. A.K. Knudsen, G.C. Pimentel, J. Chem. Phys. **78**, 6780 (1983)
177. H. Frei, G.C. Pimentel, Ann. Rev. Phys. Chem. **36**, 491 (1985)
178. A.J. Barnes, Faraday Discuss. Chem. Soc. **86**, 45 (1988)
179. G. DeMare, Y. Moussaoui, Int. Rev. Phys. Chem. **18**, 91 (1999)
180. F. Richter, M. Hochlaf, P. Rosmus, F. Gatti, H.D. Meyer, J. Chem. Phys. p. 1306 (2004)
181. D. Luckhaus, J. Chem. Phys. **118**, 8797 (2003)
182. D. Luckhaus, J. Chem. Phys. **304**, 79 (2004)
183. F. Richter, P. Rosmus, F. Gatti, H.D. Meyer, J. Chem. Phys. **120**, 6072 (2004)
184. L. Khriachtchev, J. Lundell, E. Isoniemi, M. Räsänennen, J. Chem. Phys. **113**, 4265 (2000)
185. R. Schanz, V. Bolan, P. Hamm, J. Chem. Phys. **122**, 044509 (2005)
186. V. Botan, R. Schanz, P. Hamm, J. Chem. Phys. **124**, 234511 (2006)
187. P.M. Agrawal, D.L. Thompson, L.M. Raff, J. Chem. Phys. **102**, 7000 (1995)
188. A.H. Zewail, J. Chem. Phys. A. **104**, 5660 (2000)
189. J.C. Tully, *Semiempirical mehods of electronic structure calculations* (Plenum Press, New York, 1977)
190. P.J. Kuntz, in *Atom-molecule collision theory - A guide for experimentalists*, ed. by R.B. Bernstein (Plenum Press, New York, 1979), p. 79
191. H.H. van Grunberg, H. Gabriel, J. Chem. Phys. **105**, 4173 (1996)
192. A.A. Buchachenko, N.F. Stepanov, J. Chem. Phys. **106**, 4358 (1997)
193. B.V. Grigorenko, A.V. Nemukhin, V.A. Apkarian, J. Chem. Phys. **108**, 4413 (1998)
194. A.I. Krylov, R.B. Gerber, Chem. Phys Lett. **231**, 395 (1994)

195. A.I. Krylov, R.B. Gerber, R.D. Coalson, J. Chem. Phys. **105**, 5626 (1996)
196. A. Borowski, O. Kühn, Theor. Chem. Acc. (2007)
197. D.C. Cartwright, P.J. Hay, J. Chem. Phys. **70**, 3191 (1979)
198. I. Last, T.F. George, J. Chem. Phys. **89**, 3071 (1988)
199. B.L. Grigorenko, A.V. Nemukhin, V.A. Apkarian, J. Chem. Phys. **104**, 5510 (1996)
200. A. Cohen, R.B. Gerber, unpublished
201. M.V. Korolkov, J. Manz, Z. Phys. Chem. **217**, 115 (2003)
202. A. Borrmann, C.C. Martens, J. Chem. Phys. **102**, 1905 (1995)
203. A.B. Alekseyev, M.V. Korolkov, O. Kühn, J. Manz, M. Schröder, J. Photochem. Photobiol. A **180**, 262 (2006)
204. M. Bargheer, A. Cohen, R.B. Gerber, M. Gühr, M.V. Korolkov, J. Manz, M. Niv, M. Schröder N. Schwentner, in preparation
205. J.C. Tully, J. Chem. Phys **93**, 1061 (1990)
206. R. Baumfalk, N.H. Nahler, U. Buck, M.Y. Niv, R.B. Gerber, J. Chem. Phys. **113**, 329 (2000)
207. R.P. Feynman, Phys. Rev. **56**, 340 (1938)
208. C.W. Gear, *Numerical initial value problems in ordinary differential equations* (Prentice Hall, Englewood Cliffs, 1971)
209. D.F. Coker, L. Xiao, J. Chem. Phys. **102**, 496 (1995)
210. R.B. Gerber, V. Buch, M.A. Ratner, J. Chem. Phys. **77**, 3022 (1982)
211. D.A. Micha, J. Chem. Phys. **78**, 7138 (1983)
212. R. Alimi, A. Garcia-Vela, R.B. Gerber, J. Chem. Phys. **96**, 2034 (1992)
213. G. Stock, J. Chem. Phys. **103**, 2888 (1995)
214. C. Zhu, A.W. Jasper, D.G. Truhlar, J. Chem. Phys. **120**, 5542 (2004)
215. J. Fany, C.C. Martens, J. Chem. Phys. **104**, 3684 (1996)
216. M. Hack, D.G. Truhlar, J. Chem. Phys. **104**, 7917 (2000)
217. A.W. Jasper, D.G. Truhlar, J. Chem. Phys. **116**, 5424 (2002)
218. G. Stock, M. Thoss, Adv. Chem. Phys. **131**, 243 (2005)
219. M.V. Korolkov, J. Manz, J. Chem. Phys. **120**, 11522 (2004)
220. B. Schmidt, P. Jungwirth, R.B. Gerber, in *Ultrafast chemical and physical processes in molecular systems*, ed. by M. Chergui (World Scientific, Singapore, 1996), pp. 637–640
221. B. Schmidt, P. Jungwirth, Chem. Phys. Lett. **259**, 62 (1996)
222. J. Manz, P. Saalfrank, B. Schmidt, J. Chem. Soc. Faraday Trans. **93**, 957 (1997)
223. P. Jungwirth, Chem. Phys. Lett. **289**, 324 (1998)
224. P. Jungwirth, P. Žďánská, B. Schmidt, J. Phys. Chem. **102**, 7241 (1998)
225. B. Schmidt, Chem. Phys. Lett. **301**, 207 (1999)
226. P. Žďánská, B. Schmidt, P. Jungwirth, J. Chem. Phys **110**, 6246 (1999)
227. H. Stapelfeldt, T. Seideman, Rev. Mod. Phys. **75**, 543 (2003)
228. B. Friedrich, D. Herschbach, Nature **353**, 412 (1991)
229. H.J. Loesch, Ann. Rev. Phys. Chem. **46**, 555 (1995)
230. B. Friedrich, D. Herschbach, Phys. Rev. Lett. **74**, 4623 (1995)
231. B. Friedrich, D. Herschbach, J. Phys. Chem. **99**, 15686 (1995)
232. T. Seideman, J. Chem. Phys. **103**, 7887 (1995)
233. T. Seideman, J. Chem. Phys. **106**, 2881 (1997)
234. J.J. Larsen, K. Hald, N. Bjerre, H. Stapelfeldt, T. Seideman, Phys. Rev. Lett. **85**, 2470 (2000)

235. W. Kim, P.M. Felker, J. Chem. Phys. **104**, 1147 (1996)
236. W. Kim, P.M. Felker, J. Chem. Phys. **108**, 6763 (1998)
237. B. Friedrich, D. Herschbach, J. Chem. Phys. **111**, 6157 (1999)
238. B. Friedrich, D. Herschbach, J. Phys. Chem. A **103**, 10280 (1999)
239. S. Minemoto, H. Nanjo, H. Tanji, T. Suzuki, H. Sakai, J. Chem. Phys. **118**, 4052 (2003)
240. H. Sakai, S. Minemoto, H. Nanjo, H. Tanji, T. Suzuki, Phys. Rev. Lett. **90**, 083001 (2003)
241. N.H. Nahler, R. Baumfalk, U. Buck, Z. Bihary, R.B. Gerber, B. Friedrich, J. Chem. Phys. **119**, 224 (2003)
242. B. Friedrich, N.H. Nahler, U. Buck, J. Mod. Opt. **50**, 2677 (2003)
243. T. Seideman, E. Hamilton, Adv. At. Mol. Opt. Phys. **52**, 289 (2005)
244. I.S. Averbukh, R. Arvieu, Phys. Rev. Lett. **87**, 163601 (2001)
245. M. Leibscher, I.S. Averbukh, H. Rabitz, Phys. Rev. Lett. **90**, 213001 (2003)
246. C.Z. Bisgaard, M.D. Poulsen, E. Péronne, S.S. Viftrup, H. Stapelfeldt, Phys. Rev. Lett. **92**, 173004 (2004)
247. P. Marquetand, A. Materny, N.E. Henriksen, V. Engel, J. Chem. Phys. **120**, 5871 (2004)
248. J. Ortigoso, Phys. Rev. Lett. **93**, 073001 (2004)
249. H. Friedmann, S. Kimel, J. Chem. Phys. **47**, 3589 (1967)
250. S. Hennig, A. Cenian, H. Gabriel, Chem. Phys. Lett. **205**, 354 (1993)
251. D.M. Villeneuve, S.A. Aseyev, P. Dietrich, M. Spanner, M.Y. Ivanov, P.B. Corkum, Phys. Rev. Lett. **85**, 542 (2000)
252. S. Estreicher, T.L. Estle, Phys. Rev. B **30**, 7 (1984)
253. J. Manz, J. Am. Chem. Soc. **102**, 1801 (1980)
254. V. Berghof, M. Martins, B. Schmidt, N. Schwentner, J. Chem. Phys. **116**, 9364 (2002)
255. B. Schmidt, P. Žďánská, Comp. Phys. Comm. **127**, 290 (2000)
256. A.F. Devonshire, Proc. Roy. Soc. (London) A **153**, 601 (1936)
257. H.U. Beyeler, J. Chem. Phys. **60**, 4123 (1974)
258. G.K. Pandey, K.L. Pandey, M. Massey, R. Kumar, Phys. Rev. B **34**, 1277 (1986)
259. H. Bethe, Ann. Phys. **3**, 132 (1929)
260. L. Pauling, Phys. Rev. **36**, 430 (1930)
261. R.E. Miller, J.C. Decius, J. Chem. Phys. **59**, 4871 (1973)
262. J. Ortigoso, M. Rodriguez, M. Gupta, B. Friedrich, J. Chem. Phys. **110**, 3870 (1999)
263. F. Rosca-Pruna, M.J.J. Vrakking, J. Chem. Phys. **116**, 6567 (2002)
264. T. Seideman, J. Chem. Phys. **115**, 5965 (2001)
265. F. Rosca-Pruna, M.J.J. Vrakking, J. Chem. Phys. **116**, 6579 (2002)
266. L.V. Keldysh, Sov. Phys. JETP **20**, 1307 (1965)
267. K. Mishima, K. Nagaya, M. Hayashi, S.H. Lin, J. Chem. Phys. **122**, 104312 (2005)
268. M.V. Korolkov, J. Manz, J. Chem. Phys. (submitted)
269. G.P. Zhineva, Kinet. Katal. p. 690 (1986)
270. D. Philipovich, H.H. Rogers, R.D. Wilson, Inorg. Chem. **11**, 2192 (1972)
271. M.V. Korolkov, J. Manz, Chem. Phys. Lett. **393**, 44 (2004)
272. P. Brumer, M. Shapiro, Chem. Phys. Lett. **126**, 541 (1986)
273. S.A. Rice, M. Zhao, *Optimal control of molecular dynamics* (John Wiley and Sons, New York, 2001)

274. M. Shapiro, P. Brumer, *Principles of the quantum control of molecular processes* (Wiley, Hoboken, 2003)
275. R. Pausch, M. Heid, T. Chen, H. Schwoerer, W. Kiefer, J. Raman Spectroscopy **31**, 7 (2000)
276. C. Daniel, M.C. Heitz, J. Manz, C. Ribbing, J. Chem. Phys. **102**, 905 (1995)
277. S. Rutz, R. de Vivie-Riedle, E. Schreiber, Phys. Rev. A **54**, 306 (1994)
278. R.B. Gerber, Ann. Rev. Phys. Chem. **55**, 55 (2004)
279. A. Cohen, M.Y. Niv, R.B. Gerber, Faraday Discuss. Chem. Soc. **118**, 269 (2001)
280. M. Petterson, L. Khriachtchev, A. Lingell, M. Räsänen, Z. Bihary, R. Gerber, J. Chem. Phys. **116**, 5521 (2002)
281. Z. Bihary, G.M. Chaban, R.B. Gerber, J. Chem. Phys. **117**, 5105 (2002)
282. A. Cohen, J. Lundell, R.B. Gerber, J. Chem. Phys. **119**, 6415 (2003)
283. J. Lundell, A. Cohen, R.B. Gerber, J. Phys. Chem. A **106**, 11950 (2002)

5

Ultrafast dynamics of photoinduced processes at surfaces and interfaces

Christian Frischkorn[1], Martin Wolf[1], Ulrich Höfer[2], Jens Güdde[2], Peter Saalfrank[3], Mathias Nest[3], Tillmann Klamroth[3], Frank Willig[4], Ralph Ernstorfer[4], Lars Gundlach[4], Volkhard May[5], Luxia Wang[5], Walter R. Duncan[6], and Oleg V. Prezhdo[6]

[1] Fachbereich Physik, Freie Universität Berlin, Germany
[2] Fachbereich Physik, Philipps-Universität, Marburg, Germany
[3] Institut für Chemie, Universität Potsdam, Germany
[4] Hahn-Meitner-Institut, Berlin, Germany
[5] Institut für Physik, Humboldt Universität zu Berlin, Germany
[6] University of Washington, Department of Chemistry, Seattle, USA

Coordinated by: Christian Frischkorn

5.1 Introduction

C. Frischkorn

While femtochemistry in the gas and solution phase has led to enormous progress in the understanding and even control of chemical reactions over the past decades, a comparable level of sophistication in the analysis of surface chemical reactions has not been achieved. In part, this originates from the additional complexity of energy dissipation channels introduced by the presence of a solid interface interacting with the reactants. Reactive processes at surfaces are of fundamental importance for technological applications such as heterogeneous catalysis. Here, metals are often investigated as model substrates since the interaction of the adsorbed reaction partners with the substrate may cause a favorable energy landscape, e.g. a reduced reaction barrier compared to the gas phase. Moreover, the ability to bring the reactants on a surface like on a template into close proximity provides additional control of the reaction dynamics. On the other hand, coherent control schemes which exploit the phase of the exciting laser light field mostly fail in photoinduced surface processes due to ultrafast dephasing caused by coupling to the underlying substrate. In addition, femtochemistry at surfaces and interfaces is closely related to the break-down of the Born-Oppenheimer approximation, a key concept in chemical reaction dynamics whereby electrons are assumed to follow the nuclear motion instantaneously. Thus being of general importance, nonadiabatic coupling between nuclear motion and electronic degrees of freedom is one of the fundamental ingredients to the dynamics of ultrafast processes at surfaces and interfaces. Although nonadiabatic phenomena are crucial also in various other surface science fields [1], e.g. in the detection of chemicurrents, chemiluminescence or exoelectron emission [2] as well as in chemical reactions induced by scanning tunneling microscopy (STM) [3], ultrafast-laser initiated surface chemistry has the main advantage of direct access to nonadiabatic coupling effects in the time domain.

The present Chapter comprises two fields; (i) ultrafast chemical reactions on *metal surfaces* and (ii) charge transfer processes across molecule-*semiconductor interfaces*. For both types of interfaces (with a metal surface being a gas-solid interface), electronic coupling and charge transfer between the adsorbate and substrate play key roles. On metal surfaces, absorption of a femtosecond laser pulse results in a continuum of electron-hole pair excitations in the substrate, which enables charge transfer to energetically low-lying resonance levels of adsorbed molecules. On the other hand, the existence of a band gap of several eV in semiconductor substrates allows for intramolecular excitations of the adsorbate followed by charge injection into the conduction band of the substrate.

In the following, this Chapter starts with a brief overview of surface femtochemistry with the description of general concepts common to all femtochemical surface reactions. Subsequently in separate sections, two experimental examples are given; the recombination of two adsorbed atoms which form a molecule and desorb into the gas phase and the diffusion of atomic adsorbates which represents one of the most fundamental motions involved in any chemical process where interatomic bonds are formed and cleaved. A profound theoretical account of these kinds of ultrafast nonadiabatic processes will complete the femtosecond laser induced chemistry on metal surfaces. The second part of this Chapter concentrates on the ultrafast heterogeneous electron transfer in dye-sensitized metal oxide semiconductors. Starting with a presentation of relevant concepts, a qualitative description and discussion of basic phenomena follow. Particular systems of organic molecule-sensitized semiconductors, which undergo ultrafast charge transfer after femtosecond-laser excitation, will be discussed both experimentally and theoretically. Finally, ab initio time-domain simulations of the photoinduced electron transfer will complement this field of ultrafast dynamics at interfaces.

5.2 Surface femtochemistry: Basic concepts

C. Frischkorn

Surface femtochemistry is initiated by ultrashort-laser pulse excitation of an absorbate-covered metal surface [4]. The subsequent elementary processes can be grouped into two parts. The first field covers processes within the substrate where the laser pulse is absorbed and the energy dissipated, since substrate-mediated reactions typically dominate (i.e. direct optical absorption is mostly negligible in atomically thin adsorbate layers). The second field comprises the subsequent energy transfer to the adsorbate system which involves the coupling of electronic and nuclear degrees of freedom. Figure 5.1a illustrates the primary excitation process together with the subsequent energy flow between the different subsystems. Characteristic time constants for the respective energy transfer are given. In the following subsection, the processes of intrasubstrate and substrate-adsorbate energy transfer are expanded in more detail.

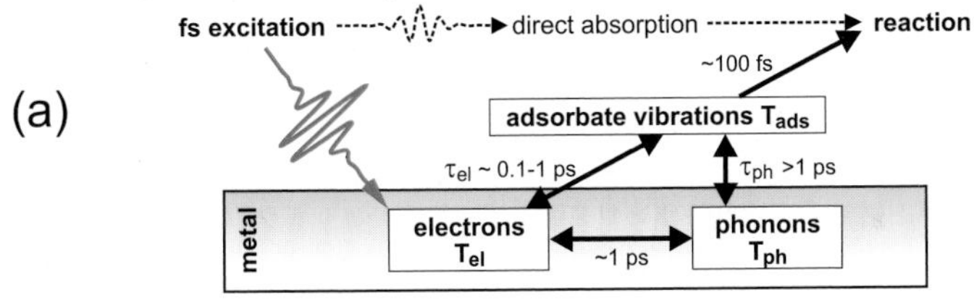

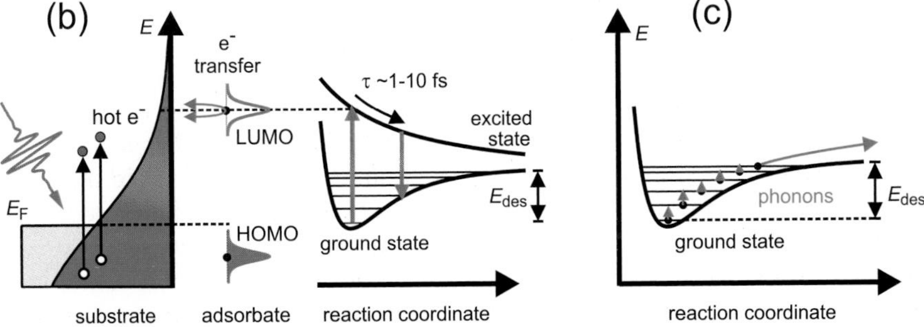

Fig. 5.1. (a) Schematic diagram of energy flow upon femtosecond-laser excitation of a metal surface. (b) Hot-substrate electron-mediated excitation of the adsorbed reactants involving (multiple) electronic transitions. (c) Phonon-mediation of a surface reaction, which proceeds adiabatically on the electronic ground state via vibrational ladder climbing [5].

5.2.1 Excitation and subsequent energy flow within the substrate

From a simplified point of view, a metal substrate consists of two heat baths; the ion cores (lattice) and the surrounding electron gas. Excitations of either of these subsystems, i.e. collective lattice vibrations (phonons) and electron gas excitations, respectively, constitute the energy content of the substrate. Consequently, the degree of excitation can be described by two population distributions each characterized (in the limit of thermalized distributions) by a temperature T_{el} or T_{ph} for the electron and phonon subsystem, respectively (see Fig. 5.1a). In thermal equilibrium or when the substrate is heated conventionally (i.e. not by ultrafast-laser excitation), equal temperatures for both heat baths prevail; $T_{\mathrm{el}} = T_{\mathrm{ph}}$ due to electron-phonon coupling with a typical equilibration time in the few picosecond (ps) range. However, excitation with a femtosecond (fs) laser pulse will drive the system out of equilibrium. The pulse energy is deposited into the electron system and due to the small heat capacity of the electrons compared to the lattice T_{el} rises within the pulse width to levels far above the melting point of the lattice. This electronic excitation energy is then dissipated either by electron diffusion into the bulk or by energy transfer into the phonon subsystem via electron-phonon coupling. This gives rise to an increase of the phonon temperature T_{ph}, however, on a much slower time scale than the electronic response. Within a time span of approximately the electron-phonon coupling time of the substrate, both the electron and phonon heat bath equilibrate. The so-called two-temperature model (2TM) has been established to quantitatively describe such a system of two coupled heat baths using the following coupled differential equations [6,7]

$$C_{\mathrm{el}} \frac{\partial}{\partial t} T_{\mathrm{el}} = \overbrace{\frac{\partial}{\partial z} \kappa_{\mathrm{el}} \frac{\partial}{\partial z} T_{\mathrm{el}}}^{\text{therm. diffusion}} \overbrace{- g(T_{\mathrm{el}} - T_{\mathrm{ph}})}^{\text{el-ph.coupling}} + \overbrace{S(z,t)}^{\text{opt. excitation}} \tag{5.1}$$

$$C_{\mathrm{ph}} \frac{\partial}{\partial t} T_{\mathrm{ph}} = +g(T_{\mathrm{el}} - T_{\mathrm{ph}}) \tag{5.2}$$

Here, $C_{\mathrm{el}} = \gamma\, T_{\mathrm{el}}$ and C_{ph} are the electron and ion heat capacities, respectively. $\kappa_{\mathrm{el}} = \kappa_0\, T_{\mathrm{el}}/T_{\mathrm{ph}}$ is the electronic thermal conductivity, and g is the electron-phonon coupling constant. Heat conduction by phonons can be neglected in metals due to the fact that the respective mean velocities of electrons and phonons enter the heat conductivity quadratically, (note that in metals the Fermi velocity exceeds the speed of sound by far). The term $S(z,t)$ finally describes the optical excitation of the electrons and can be expressed by [8]

$$S(z,t) = \frac{\delta^{-1}(1-R)I(t)\exp(-z/\delta)}{1-\exp(-d/\delta)}, \tag{5.3}$$

where δ, R and $I(t)$ stand for the optical penetration depth, the substrate reflectivity and the time profile of the laser intensity, respectively. Details on the electron and lattice dynamics following ultrafast optical excitation of metals can be found, for instance, in [8–10].

5.2.2 Coupling between substrate and adsorbate

The two energy reservoirs of the substrate, i.e. the electron and phonon subsystem, respectively, may couple energy independently into the adsorbate system. After accumulation of sufficient energy in the coordinate relevant to a reaction, the adsorbate may undergo desorption, dissociation or reactions between coadsorbed species (see Fig. 5.1a). The energy transfer from the initially excited electronic degrees of freedom of the substrate to the nuclear motion of the reactants occurs either directly through electronically nonadiabatic substrate-adsorbate coupling or indirectly via equilibration with the lattice and subsequent coupling to the adsorbate. In surface femtochemistry, two conceptually different frameworks have been developed to describe these transfer mechanisms. One approach treats the energy transfer in terms of frictional forces [11, 12] with frictional coefficients[1] $\eta_{\mathrm{el}} = 1/\tau_{\mathrm{el}}$ and $\eta_{\mathrm{ph}} = 1/\tau_{\mathrm{ph}}$ for both electrons and phonons, respectively, which determine how fast energy flows into the adsorbate system. In the reverse process of vibrational energy relaxation, these coupling times can be interpreted as the vibrational lifetime (T_1) and are therefore connected to the IR linewidth of the respective vibration. The origin of electronic friction can be explained within the Anderson-Newns model [14], whereby an adsorbate-derived affinity level shifts downwards and broadens for decreasing adsorbate-substrate distances. If the level is transiently populated by substrate electrons the charge-induced adsorbate motion results in a level shift as the adsorbate starts moving along the corresponding reaction coordinate. Electron flow back and forth between the metal substrate and the adsorbate is intrinsically affected by damping, i.e. friction. To which extent adsorbate levels might be populated depends on the electronic temperature of the substrate, usually low-lying energy levels are involved in excitations via electronic friction.

The second substrate-adsorbate coupling scenario accounts only for purely electron-mediated excitation processes and invokes "desorption (or more generally: dynamics) induced by multiple electronic transitions" (DIMET) [15]. In the DIMET process as illustrated in Fig. 5.1b, hot substrate electrons transiently populate a normally unoccupied affinity level transferring the absorbate-substrate complex to an electronically excited potential energy surface (PES), which can be either anti-bonding, i.e. repulsive, as in the Menzel-Gomer-Redhead [16, 17] (MGR) picture or bonding as proposed by Antoniewicz [18]. Similar to the Anderson-Newns model, in the DIMET picture, the new charge distribution resulting from the transient electron transfer initiates nuclear motion converting potential energy into kinetic energy. After relaxation back to the electronic ground state, the system has acquired vibrational energy. At high excitation densities, additional excitation/de-excitation cycles might occur before vibrational energy relaxation takes place on the ground state PES, thus enabling the adsorbate to accumulate sufficient energy

[1] η_{el} is related to the friction coefficient f of a classical velocity-proportional friction force $F = -fv$ by $f = \eta_{\mathrm{el}} M$ with M being the adsorbate mass [13].

in the relevant coordinate to overcome the reaction barrier. Experimentally, a nonlinear dependence of the reaction yield on the absorbed laser fluence is a characteristic consequence of multiple repetition of such excitation/de-excitation cycles [15, 19, 20]. An adsorbate-mass dependent reaction yield, i.e. isotope effect, can also be rationalized in the DIMET picture, in which the lighter reactant will have gained more vibrational energy after relaxation back to the ground state than its heavier counterpart due to the mass-dependent acceleration on the excited PES. The two concepts (friction model and DIMET) incorporate similar physical processes in a different mechanistic description. The DIMET process would correspond to a strongly temperature dependent friction coefficient in the electronic friction picture. At high excitation densities, i.e. for multiple electronic transitions between the ground and excited state PES, both scenarios are physically equivalent; a unifying formalism is given in Ref. [12].

For a quantitative theoretical description of the energy transfer from the laser-excited substrate to the reactants in the adsorbate layer, frictional coupling between the electron and phonon heat bath to an harmonic oscillator of the adsorbate motion is typically used. Based on a master equation formalism [13], the time evolution of the energy content of the adsorbate is represented by [21]

$$\frac{d}{dt}U_{\mathrm{ads}} = \eta_{\mathrm{el}}(U_{\mathrm{el}} - U_{\mathrm{ads}}) + \eta_{\mathrm{ph}}(U_{\mathrm{ph}} - U_{\mathrm{ads}}), \tag{5.4}$$

with the Bose-Einstein distributed mean vibrational energy

$$U_{\mathrm{x}} = \frac{h\nu_{\mathrm{ads}}}{\mathrm{e}^{h\nu_{\mathrm{ads}}/k_{\mathrm{B}}T_{\mathrm{x}}} - 1} \tag{5.5}$$

of an oscillator at temperature T_{x}. ν_{ads} refers to the frequency of the vibration along the reaction coordinate. In this so-called empirical friction model accounting for both electronic and phononic contributions, the adsorbate temperature T_{ads} is obtained by solving (5.4) with T_{el} and T_{ph} computed with the two-temperature model (5.1),(5.2). The reaction rate R and, finally, to compare with the experiment, the reaction yield Y as the time integral of R are calculated with an Arrhenius-type expression

$$R(t) = -\frac{d}{dt}\theta(t) = \theta^{n}(t)\, k_0\, \mathrm{e}^{-E_{\mathrm{a}}/k_{\mathrm{B}}T_{\mathrm{ads}}(t)}, \tag{5.6}$$

where θ and n denote the coverage and the order of the reactions kinetics, respectively. As an alternative, a modified friction model has been proposed by Brandbyge and co-workers [12], in which a purely electronic frictional coupling is incorporated. Here, the frictional force originates from coupling of Langevin noise of the electron heat bath (T_{el}) into the adsorbate center of mass coordinate (T_{ads}). Based on the same master equation as in the former case, one obtains the adsorbate temperature T_{ads} by solving

$$\frac{d}{dt}T_{\mathrm{ads}}(t) = \eta_{\mathrm{el}}(t)[T_{\mathrm{el}}(t) - T_{\mathrm{ads}}(t)]. \tag{5.7}$$

Equation (5.7) can be formally derived from the high-temperature limit of (5.4) (with electronic contributions only) if $h\nu_{\mathrm{ads}}/k_{\mathrm{B}}T_{\mathrm{x}} \ll 1$ and therefore $U_{\mathrm{x}} \approx k_{\mathrm{B}}T_{\mathrm{x}}$. In principle, η_{el} depends on time and space and has to be calculated by microscopic theories [22–24], see Sect. 5.3.5. The rate in this modified electronic friction model scales proportionally with the reaction probability which in turn depends on a Boltzmann factor in a similar way as in the empirical model

$$R(t) \propto P_{\mathrm{rxn}} = E_{\mathrm{a}} \int_0^{\infty} (t)\, dt \frac{\eta_{\mathrm{el}}}{T_{\mathrm{ads}}(t)}\, e^{-E_{\mathrm{a}}/k_{\mathrm{B}}T_{\mathrm{a}}(t)}. \tag{5.8}$$

However, in contrast to (5.6) the friction coefficient η_{el}, T_{ads} and the energy E_{a} enter the preexponential factor. Taking into account the mass dependence of the friction coefficient $\eta_{\mathrm{el}} \propto 1/m$, the Brandbyge model [12] directly leads to an isotope effect in the reaction yield for isotopically substituted reactants in contrast to the empirical model discussed above. It should be noted that within both frictional models the energy E_{a} is the well depth of a truncated one-dimensional harmonic oscillator. However, it turns out that values for E_{a} extracted from experimental data typically exceed the measured activation energies for desorption. It has been speculated that either the dynamics on a multidimensional potential energy surface is the origin of this discrepancy [4] or that E_{a} should be regarded as a modified activation energy which is larger than the depth of the adsorption well indicating the population of electronically excited states [25].

5.3 Ultrafast dynamics of associative H_2 desorption from Ru(001)

C. Frischkorn and M. Wolf

For a microscopic understanding of chemical reactions at surfaces, it is essential to obtain detailed knowledge on the underlying elementary processes. The reaction mechanism, the pathways and time scales of energy flow and the energy partitioning between different degrees of freedom of the reaction products are of key interest. The recombination of two hydrogen atoms forming an H_2 molecule, which leaves the surface, represents one of the most basic surface reactions one could think of and thus may serve as a prototype system for fs-laser induced surface chemistry. Hydrogen and its interaction, in particular, with transition-metal surfaces has attracted significant attention in both experimental and theoretical work due to its importance to technological applications such as catalytic reactions [26] and hydrogen storage in metals [27]. Fundamental research on hydrogen at surfaces has been carried out to address a variety of different aspects like H-induced surface reconstruction [28, 29], substrate-mediated interaction between coadsorbed H

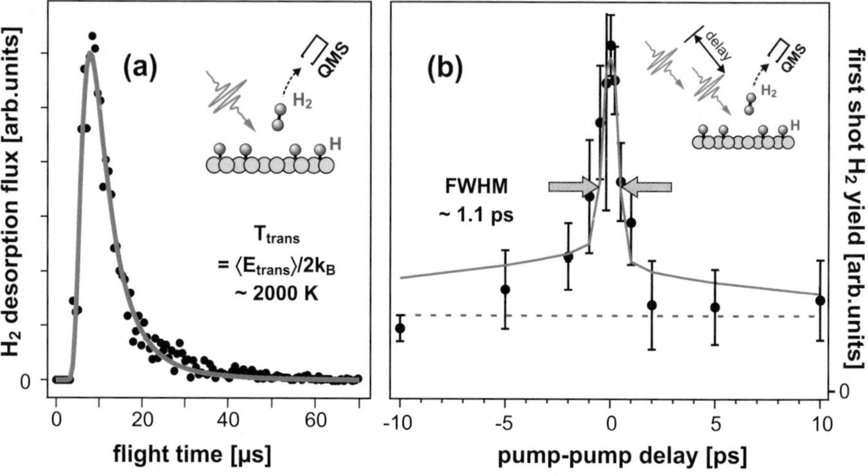

Fig. 5.2. (a) TOF spectrum of H_2 desorbing from a Ru(001) surface after fs-laser excitation with a fluence $\langle F \rangle = 60\,\mathrm{J/m^2}$ together with a modified Maxwell-Boltzmann distribution (solid line). (b) Two-pulse correlation of the H_2 recombination yield exhibiting a narrow FWHM of ~1 ps, i.e. a hot-substrate electron-mediated excitation mechanism. The solid line marks the outcome of the theoretical modeling (two-temperature model with subsequent electronic friction) [25, 31].

atoms [30, 31] and quantum delocalization [32, 33], which might impose additional complexity of the adsorbate-substrate system despite the structural simplicity of the adsorbate.

5.3.1 Experimental considerations

In contrast to gas phase experiments, femtochemistry studies on surfaces face the challenge that the sample under investigation cannot not refreshed between subsequent laser shots as in molecular beam arrangements. If applicable, redosing of the adsorbed reactants via the background pressure is one option to provide experimental conditions to repeat otherwise time-consuming preparation/laser excitation cycles. In addition, accumulation of unwanted side products on the sample surface has to be avoided. Finally, sufficiently high excitation densities to initiate an ultrafast surface process might not be applicable without exceeding the damage threshold of the underlying substrate. For experimental details to successfully perform typical surface femtochemistry experiments, in particular those for the fs-laser induced associative desorption of H_2 from Ru, the reader is referred to [4, 25, 34].

5.3.2 Excitation mechanism

Hydrogen recombination on ruthenium, $H_{ad} + H_{ad} \rightarrow H_{2,gas}$ on Ru(001), may be initiated thermally (i.e. under conditions of thermal equilibrium between all degrees of freedom), but if induced by fs-laser excitation several characteristic differences are observed which will be discussed in detail in the following subsections. A first indication that a different reaction mechanism is operative in the fs-laser driven H_2 recombination than in the thermally initiated process is obtained from the time-of-flight (TOF) measurements. As seen in Fig. 5.2a, a mean translational energy of about 2000 K is derived for the desorbing H_2 molecules, which is a much higher value than one would expect from thermal desorption spectroscopy (TDS) with H_2 desorbing in a temperature range from 250 - 450 K. Consequently, the high T_{trans} value indicates that after fs-laser excitation the hydrogen molecules leave the surface excited without complete equilibration with the heat bath of the solid. The appropriate experimental approach to unambiguously distinguish a phonon- from an electron-mediated reaction mechanism is to measure the two-pulse correlation (2PC) of the reaction yield [19]. In such an experiment, two equally intense fs-laser pulses excite the sample with a variable time delay Δt between them. The reaction yield is then detected as a function of Δt. In the case of H_2 desorption from Ru(001) (see Fig. 5.2b), a narrow yield correlation is observed with a full width at half maximum (FWHM) of $\sim$1 ps indicating an energy pathway predominantly driven by hot substrate *electrons*. If the time separation of the two pulses does not exceed the electron-phonon coupling time ($\sim$1 ps for Ru), the combined effect of both pulses causes exceedingly high electronic temperatures and hence a narrow correlation function. In contrast, if the H_2 recombination reaction were dominated by *phonon*-mediated energy flow, the slower cooling of phonons ($\sim$10 to 100 ps) would cause a much wider 2PC [35].

Figure 5.3a displays the fluence dependence of the H_2 desorption yield. The clear nonlinear relationship reflects that in accordance with a DIMET process multiple excitation/de-excitation cycles occur until desorption is completed. In addition, this nonlinearity underlines how essential it is to account for an energetically nonuniform spatial beam profile of the exciting laser. Beam sections of higher intensity contribute to the overall yield to a nonlinearly higher extent than portions of lower intensity do. Therefore a yield-weighting procedure [21] based on a power-law parameterization is usually applied to obtain the absorbed yield-weighted fluence $\langle F \rangle$, which characterizes the experimental laser conditions more reliably.

Further corroboration for a reaction mechanism driven by hot substrate electrons is found when one compares absolute desorption yields per laser shot from H/Ru(001) and D/Ru(001), respectively. It is significantly easier to form the lighter isotope H_2 by fs-laser excitation than the heavier D_2. In Fig. 5.3b, this isotope yield ratio $Y(H_2){:}Y(D_2)$ is plotted versus $\langle F \rangle$ exhibiting a remarkable fluence dependence. The exceedingly high values for $Y(H_2){:}Y(D_2)$, at low fluences even >20, reflect the extraordinary mass ratio of the reactants of 2:1. As already qualitatively rationalized in the DIMET picture, a clear isotope effect is a general phenomenon for an electron-driven reaction mechanism. In contrast, phonon-mediated processes typically exhibit -if at all- only

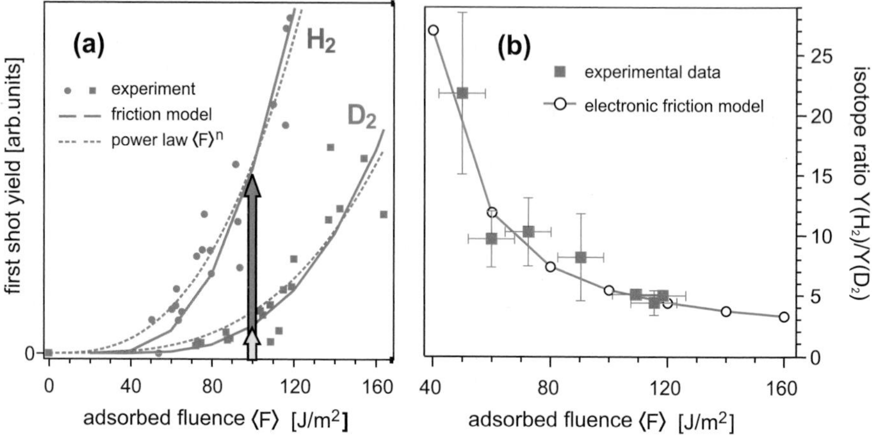

Fig. 5.3. Femtosecond-laser induced hydrogen desorption from a saturated Ru(001) surface. (a) Fluence dependence of the H_2 and D_2 yield together with friction model calculations and power-law parameterizations. Arrows indicate the observed isotope effect. (b) Fluence dependence of the isotope yield ratio between H_2 and D_2 (experimental data points and friction calculations) [25].

very small isotope effects [36], as also seen in TDS experiments with H and D on Ru(001) [31]. The solid lines in Figs. 5.2b, 5.3a and b represent the outcome of theoretical modeling the data using the two-temperature model(5.1),(5.2) together with the modified electronic friction model (5.8). All experimental findings (2PC data, yield fluence dependencies and isotope effect) can be well reproduced within the framework of these models with a single parameter set consisting of an effective activation energy E_a of 1.35 eV and energy coupling times of 180 and 360 fs for H_2 and D_2, respectively. Note that τ_{el} for D_2 was scaled in accordance to Brandbyge et al.'s mass proportionality $\tau_{el} \propto m$ [12], which provides a very good fit to the D_2 data.

Figure 5.4 illustrates the origin of the isotope effect between the fs-laser induced H_2 and D_2 desorption yield from Ru(001) from the perspective of the frictional description of a femtochemical reaction. Due to the electronic nature of the excitation mechanism (i.e. the different coupling times for H and D, respectively), the adsorbate temperature T_{ads} of both layers (H and D) follow the electronic temperature T_{el} with a certain delay Δt_X, X=H;D. However, the faster coupling time for an H layer causes the transient adsorbate temperature $T_{ads}(H)$ of an H layer to rise earlier and reach higher values as compared to $T_{ads}(D)$ for a D layer. The slower coupling time for the heavier D atoms is responsible for the fact that the electronic temperature T_{el} has already passed its maximum when the D layer starts being excited. As a consequence of the nonlinear dependence of the reaction rate on the adsorbate temperature, the difference in the desorption yields between both isotope is even enhanced

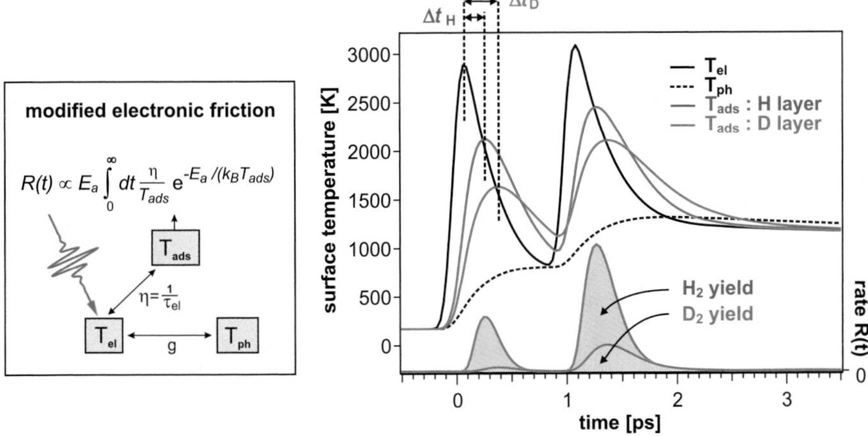

Fig. 5.4. Temperature and yield transients for the associative H_2 and D_2 desorption upon fs-laser excitation (here: sequence of two pulses separated in time by 1 ps). The diagram at the left illustrates the used theoretical modeling with two heat baths within the metal substrate and a third one for the mode relevant to reaction which is coupled to the substrate via electronic friction [31].

resulting in pronounced isotope effects. This applies in particular to the H_2 vs D_2 recombination reaction with the largest possible mass ratio.

5.3.3 Adsorbate-adsorbate interactions

Reactants adsorbed on a solid surface interact with their surroundings, both with other neighboring reactants, which eventually leads to the chemical reaction, and with those adsorbates, which do not directly participate in the chemical reaction but modify the electronic structure of the adsorbate-substrate complex. Experiments with isotopically mixed hydrogen adlayers reveal an intriguing consequence of adsorbate-adsorbate interactions. Starting with a saturation coverage, but with varying proportions of both isotopes H and D, the total yield of all three product molecules H_2, D_2 and HD is measured for both excitation methods. Under thermal equilibrium condition, typical TD spectra are obtained whose integrated yield is plotted in Fig. 5.5a as a function of the D concentration in these mixed layers, i.e. the D fraction of the total coverage. According to second-order kinetics, the rate equations

$$\frac{d}{dt}[H_2] = k[H]^2, \quad \frac{d}{dt}[D_2] = k[D]^2, \text{ and } \frac{d}{dt}[HD] = 2k[HD] \qquad (5.9)$$

describe the change of each hydrogen isotope with time. While observing the mass continuities for H and D [31], numerical integration of (5.9) yields excellent agreement with the experimental data of the thermally induced hydrogen formation. In strong contrast, analogous experiments performed under

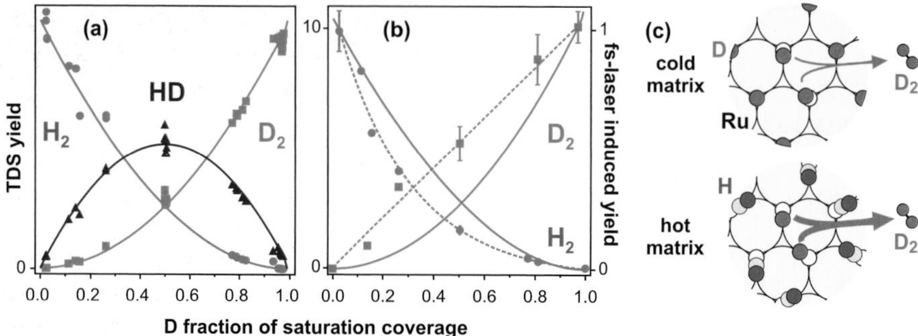

Fig. 5.5. Hydrogen desorption from a saturation coverage H/D[1×1] on Ru(001) with varying proportions of H and D. Mixtures are characterized by the D fraction whereby 0 corresponds to a pure H layer while 1 marks the pure D layer. (a) Thermally induced yields for all three isotopomers H_2, D_2 and HD, respectively, as a function of the D fraction. Data points follow second order reaction kinetics (solid lines) (b) Femtosecond-laser induced recombination yields. Solid lines again describe the outcome of second order rate equations, here, however, based on the time-dependent rate constants obtained with the friction model. Dashed lines are guide to the eye. Note that more D_2 and less H_2 are formed in the experiment than theoretically expected, which is attributed to dynamic promotion effects. (c) Illustration of the dynamic promotion; the more rapid excitation of H atoms creates a hot matrix, in which the recombination of D_2 molecules is enhanced as compared to surroundings where the D reactants are embedded in a relatively cold matrix of D atoms [31].

fs-laser excitation indicate a much more complicated recombination process than a simple bimolecular reaction. A concerted action of more partners than only both reactants forming the respective hydrogen molecule seems to be involved.

Figure 5.5b shows the respective results of the fs-laser induced desorption yields as a function of the D fraction. The yield ratio from pure adlayers, i.e. D fraction equals 0 and 1, respectively, amounts to 10:1 in consistence with the isotope effect for a fluence of $\langle F \rangle = 60\,\mathrm{J/m^2}$ (see Fig. 5.3b). If one now compares the experimental data with results of a rate equation modeling similar to (5.9), however, with time-dependent rate constants $k[T_{\mathrm{ads}}(t)]$ [see (5.8)], clear differences are observed. Significantly more D_2 is formed than predicted by the rate equation modeling. The obvious promotion effect is attributed to the faster energy transfer from the Ru substrate to an H adsorbate than to the heavier D [see above $\tau_{\mathrm{el}}(H_2) = 1/2\,\tau_{\mathrm{el}}(D_2)$]. Hence, a surrounding consisting of H starts earlier exploring locations on the surface which favor the recombination of neighboring D reactants than a D matrix, which is still relatively cold due to slower excitation as illustrated in Fig.5.5c. A physical picture of this effect would involve attractive and/or repulsive interactions between the faster excited H atoms and the two D reactants. Also electronic changes in the adsorbate-substrate complex, e.g. changes in the reaction barrier and/or energetic shifts in the excited PES, might contribute. In such a dynamic promotion process, the transient influence of the reactants' surrounding may reduce the activation energy for the desorption reaction and hence increase the rate constant. In an even more general way, a bimolecular surface reaction/desorption can be reformulated as $A_{ads} + B_{ads} \xrightarrow{k[U(t)]} AB$ with $U(t)$ describing the time-dependent surroundings. In moments of certain adsorbate-substrate conditions causing a reduced barrier height, altered excited state conditions, and/or attractive/repulsive adsorbate-adsorbate interactions, a favorable energy landscape is created which leads to reaction. This microscopic picture also complies with thermally initiated surface reactions, in which statistical fluctuations cause the respective surroundings conditions.

5.3.4 Energy partitioning

Investigations on the energy partitioning between different (translational, vibrational, rotational) degrees of freedom of the reaction product in a surface reaction offer additional insights into the underlying excitation mechanism and the pathway of energy flow. Under reaction conditions close to thermal equilibrium, e.g. in thermal desorption or ns-laser pulse excitation, nonactivated reaction systems typically show an equally balanced energy partitioning, while the reaction proceeds adiabatically on the electronic ground state. In contrast, activated systems usually exhibit an energy content of the reaction product which is unequally distributed between the different degrees of freedom. Depending on the location of a reaction barrier in the entry or exit (with respect to *ad*sorption) channel of the electronic ground state, translational or vibrational excitation may facilitate the reactants to overcome the transition state [37], hence the terms "translational" and "vibrational" barrier, respectively. The topology of the PES also determines to which extent in a recombinative desorption reaction the initial excitation normal to the surface at an early stage of the reaction might be converted to lateral and ultimately to interatomic motion, i.e. vibration. Nonadiabatic effects, however, can also result in an unequal energy transfer into different degrees of freedom of the reaction product as seen, for instance, in experiments on the associative desorption of N_2 from Ru(001) by Diekhöner et al. [38]. In these studies, contrary to expectations for a vibrational barrier, the nascent N_2 molecules carry only little vibrational energy. Apparently, they lose most of their energy on their way beyond the reaction barrier which was explained by strong nonadiabatic coupling of the vibrational coordinate to electron-hole pairs.

As already shown in Fig. 5.2a, rather high translational energies are obtained for the desorbing product molecules in the fs-laser induced associative desorption of H_2 from Ru(001). Additional TOF measurements on both isotopes H_2 and in particular D_2 revealed a pronounced fluence dependence (see Fig. 5.6a). For D_2, translational temperatures $T_{trans} = E_{trans}/2k_B$ extracted from the second moment of the experimental data points range from 2000 K

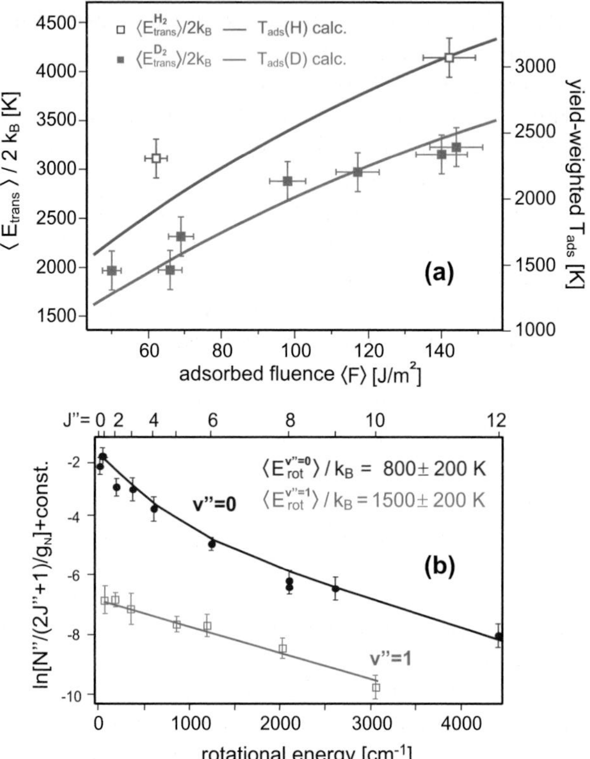

Fig. 5.6. Energy partitioning in the fs-laser induced associative hydrogen desorption from Ru(001). (a) Mean translational energies $\langle E_{\mathrm{trans}} \rangle / 2k_{\mathrm{B}}$ (left axis) for D_2 and H_2 as a function of the adsorbed laser fluence $\langle F \rangle$. Solid lines represent an averaged adsorbate temperature $T_{\mathrm{ads}}^{\mathrm{YW}}$ (right axis), see text. Note the qualitative agreement despite the scaling factor between both axes. (b) Rotational population for the vibrational ground and first excited state obtained by state-resolved detection based on the REMPI technique [39].

at low absorbed fluence $\langle F \rangle$ of 50 J/m^2 to over 3200 K at $\langle F \rangle = 140$ J/m^2. The lighter H_2 even reaches temperatures T_{trans} of more than 4000 K at the highest fluence applied (140 J/m^2,) underlining a clear kinetic isotope effect [39]. Both the trend of increasing $\langle E_{\mathrm{trans}} \rangle / 2k_{\mathrm{B}}$ with increasing $\langle F \rangle$ and the isotope effect in the translational energies between D_2 and H_2 are qualitatively reproduced by friction calculations whereby an averaged adsorbate temperature $T_{\mathrm{ads}}^{\mathrm{YW}}$ is used. This $T_{\mathrm{ads}}^{\mathrm{YW}} = \int_0^\infty T_{\mathrm{ads}} R(t) dt / (\int_0^\infty R(t) dt)$ with T_{ads} and $R(t)$ as the adsorbate temperature and rate obtained from the friction model (5.8), respectively, may be rationalized in a similar way as the weighting procedure for the laser fluence as described in Sect. 5.3.2, for details see [39].

Information on the energy content transferred to internal degrees of freedom during the formation reaction of D_2 is gained through state-resolved detection of the desorbing molecules in a resonance-enhanced multiphoton ionization (REMPI) process using $B^1\Sigma_u^+(v', J') \leftarrow X^1\Sigma_g^+(v'', J'')$ Lyman bands as resonant transitions [39]. Figure 5.6b displays Boltzmann plots for the vibrational ground and first excited state revealing a nonthermal rotational population distribution. In particular, low J states in the $v = 0$ vibrational state seem to be overpopulated. Weighting with the population of the corresponding quantum states, a common rotational temperature $T_{\text{rot}} = E_{\text{rot}}/k_{\text{B}}$ is extracted which amounts to 910 K. In addition, if a Boltzmann-like vibrational distribution is assumed for both vibrational states ($v = 0$ and 1) detected in the experiment, a vibrational temperature of T_{vib} of 1200 K is derived. Together with these values and the respective translational temperature $T_{\text{trans}} = E_{\text{trans}}/2k_{\text{B}} = 2500\,\text{K}$ and $T_{\text{vib}} = E_{\text{vib}}/k_{\text{B}} = 1200\,\text{K}$ (for $\langle F \rangle = 85\ \text{J/m}^2$), one may then establish a balance of the total energy according to $E_{\text{D}_2}^{\text{flux}} = 2k_{\text{B}}T_{\text{trans}} + k_{\text{B}}T_{\text{rot}} + k_{\text{B}}T_{\text{vib}}$ [40]. The factor of 2 in the term for the translational contribution originates from the density-to-flux conversion the measured TOF spectra to account for particle velocity dependent ionization probability of the mass detector [41]. For equilibrium conditions with all temperatures equal, comparison of $E_{\text{D}_2}^{\text{flux}}$ with the energy content of the adsorbate layer $E_{\text{ads}} = 4k_{\text{B}}T_{\text{ads}}$ yields an experimental value for T_{ads} of 1780 K. This is in excellent agreement with the calculated yield-weighted adsorbate temperature $T_{\text{ads}}^{\text{YW}} = 1810\,\text{K}$. Given that in the electronic friction model one assumes a constant and one-dimensional friction coefficient together with a single adsorbate temperature which uniformly characterizes the adlayer, these results of the energy balancing appear rather astonishing. Experimental facts are, however, that energy partitioning in the hydrogen recombination is significantly different for the various degrees of freedom. The energy contents in translational, vibrational and rotational degrees approximately scale as $6:2:1$. This obvious preference for translational excitation of the reaction product could, in principle, originate from two contributions. (i) In the case the ground state PES governs the reaction dynamics, the location of a possible reaction barrier crucially influence the energy partitioning as mentioned before. (ii) A multidimensional frictional coupling to the hot electron distribution might also account for the unequal energy transfer between different degrees of freedom of the reaction product. In other words, a faster and preferential energy flow into the height coordinate z between the adsorbate and substrate versus the interatomic-distance coordinate d could also explain the enhanced translational energy of the desorbing D_2 molecules as seen in the experiment. In the following final subsection on the femtochemistry of H/Ru(001), calculations on the multidimensional dynamics of this system will be discussed in more detail.

5.3.5 Multidimensional dynamics

The quantitative description of nonadiabatic coupling between the substrate and the adsorbate via (electronic) friction (as outlined in Sect. 5.2.2) was originally developed to describe fs-laser induced desorption of diatomic molecules along the center-of-mass coordinate, which can be reduced to a one-dimensional (1D) problem [11, 12]. Thus, it was not clear at all, why such a 1D model should be appropriate for a associative desorption reaction, since this process has to be viewed at least as a two-dimensional (2D) problem comprising the interatomic distance, i.e. the bond length d and the distance of the center of the diatomic product molecule from the surface z. Nonetheless, the 1D model has been applied with (almost surprisingly) great success also to association reactions like the CO + O oxidation [35] and the H + H recombination on Ru(001) [25, 31, 34, 39], see Figs.5.2b,5.3,5.4 and 5.6a. After a first successful application of frictional coupling along *two* dimensions to the H_2/Cu(111) and N_2/Ru(001) by Luntz and Persson [24], the same concept was also applied to the H/Ru(001) system very recently [42]. Here, a three-dimensional (3D) model was introduced with two coordinates d and

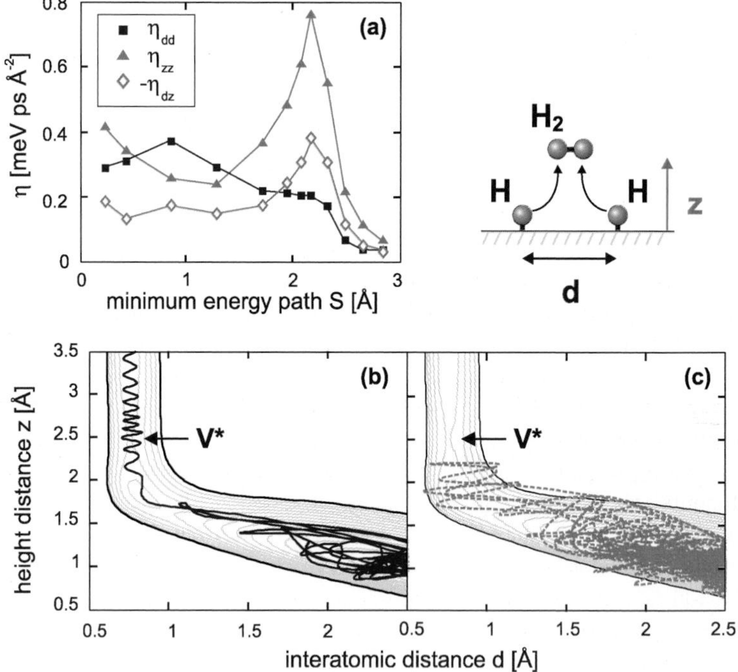

Fig. 5.7. Multidimensional friction calculations based on density functional theory for the fs-laser induced H_2 desorption from Ru(001). (a) Frictional coefficients η_{ij} along the minimum energy pathway S, where S = 0 Å corresponds to the adsorbed state before excitation, [1×1]H/Ru(001), and S = 3 Å to the H_2 + H/Ru(001) asymptote. (b) Typical trajectory for a nascent H_2 molecule leaving the Ru surface after ultrafast laser excitation with $\langle F \rangle = 140$ J/m^2 overlaid on a two-dimensional contour plot of the potential energy surface $V(z, d)$ with transition state V^*. (c) Same as (b), however, with a unsuccessful trajectory, i.e. not leading to desorption [42].

z representing the nascent hydrogen molecule, whereas the third dimension with a single phonon coordinate q described the coupling to the Ru lattice by dynamic recoil. Both the potential energy surface and the electronic friction tensor were calculated by DFT so that there are no adjustable parameters in the comparison of this model with the wide range of experimental data available for the H/Ru(001) system. Based on the molecular dynamics with electronic friction by Head-Gordon and Tully [23] the 3D classical equation of motion on the PES $V(z, d, q)$ are given by

$$\mu\ddot{d} = -\frac{\partial V}{\partial d} - \eta_{dd}\dot{d} - \eta_{dz}\dot{z} + F_d(t) \tag{5.10a}$$

$$m\ddot{z} = -\frac{\partial V}{\partial z} - \eta_{zz}\dot{z} - \eta_{dz}\dot{d} + F_z(t) \tag{5.10b}$$

$$M_s\ddot{q} = -\frac{\partial V}{\partial q} - \eta_q\dot{q} + F_q(t) \tag{5.10c}$$

where μ, m and M_s are the reduced mass of the vibration, the molecular mass and the surface mass of a Ru atom, respectively. The molecular modes z and d are coupled to a thermalized electron distribution at temperature T_{el} via the frictional tensor $\eta_i j$ which causes damping and induces fluctuating forces $F_i(t)$, $(i = z, d)$ according to the second fluctuation-dissipation theorem [43]

$$\langle F_i(t)F_i(t')\rangle = 2k_B T_{el}\eta_{ii}\delta(t - t') \tag{5.11}$$

Equation (5.11) and the analog version for the phonon coordinate q relate the transient electronic and phononic temperatures $T_{el}(t)$ and $T_{ph}(t)$ obtained from the 2TM to the forces driving the molecular dynamics of the photodesorption process. Detailed background of this and other theoretical modeling concepts are discussed further below in Sect. 5.5.

Figure 5.7a shows the calculated frictional tensor elements η_{ij}, which are plotted along the minimum energy path S toward desorption. The frictional coefficients for the different coordinates are largely similar at $S = 0$, the position on the PES which corresponds to the initially adsorbed state where both hydrogen atoms reside on the Ru surface in equilibrium before the laser excitation occurs. On the contrary, near the transition state V^* at $S \approx 2$Å, $3\eta_{dd} \approx \eta_{zz}$. Also by DFT calculations the 2D PES $V(z, d, q = 0)$ of the H[1×1]/Ru(001) ground state was obtained using a 2×2 (and partially 4×4) Ru unit cells for desorption of a single H_2 molecule. Depending on the excitation density, i.e. laser fluence, and hence the desorption probability, classical trajectories were run for this system in molecular dynamics calculations. In Figs. 5.7b and c, two exemplary trajectories of two H atoms forming a H–H complex, one successful and one unsuccessful with respect to desorption, are overlaid onto the 2D contour plot of the calculated PES. By evaluating an appropriate number of trajectories successfully leading to desorption, most of the experimental results could be reproduced with remarkably good agreement;

the two-pulse correlation, the nonlinear fluence dependence of the desorption yield and the isotope effect [42].

Analyzing individual trajectories (like those of Figs. 5.7b and c) reveals the following intriguing conclusion: Most of the primary electron mediated excitation of nuclear coordinates occurs during or shortly after the laser pulse when the system is still deep in the H–H adsorption well and much below the reaction barrier. In this region, although $\eta_{dd} \approx \eta_{zz}$, most of the nuclear excitation occurs through the vibrational coordinate because of the four times smaller reduced mass along the coordinate d versus that along z [42]. By the time the H–H approaches the barrier where $\eta_{zz} \gg \eta_{dd}$, the electron temperature T_{el} has already cooled down so that the frictional force F_i is small. However, even though initially most of the excitation occurs through the vibrational coordinate, the rapid energy exchange, i.e. "thermalization", between d and z coordinates along the trajectory conserves little memory of the mode of excitation. Hence the observed differences in the energy partition between translational and vibrational degrees of freedom originate predominantly from the topology of the ground state PES. In particular, the small but distinct barrier in the translational channel plays a crucial role for the excess energy in the translational degree of freedom.

The H + H $\rightarrow$ H_2/Ru system provides an example for multidimensional dynamics, where the rapid energy exchange between modes allows a reduction to a single reaction coordinate. A further scenario for a complex multidimensional reaction, which involves anharmonic coupling between the initially excited motion and a second mode relevant to the actual reaction, will be described in detail in Sect. 5.4.

In summary, Sect. 5.3 has detailed the reaction dynamics of a prototypical surface reaction induced by ultrafast laser excitation, $H_{\mathrm{ads}} + H_{\mathrm{ads}} \rightarrow H_{2,\mathrm{gas}}$ on Ru(001). The findings of an ultrafast, hot substrate electron-mediated reaction pathway, significant adsorbate-adsorbate interactions as manifested in dynamic promotion effects and energy partitioning between different degrees of freedom with preferentially translational excitation of the reaction product provide a comprehensive understanding of many microscopic aspects of this reaction. Multidimensional friction calculations reveal that initially energy in transferred to nuclear motion by nonadiabatic coupling, but that subsequently the ground state potential energy surface governs the ultrafast reaction dynamics of the H_2 recombination. The rapid energy exchange between different modes of the nascent H_2 molecule is the origin of the remarkable success of the usually applied one-dimensional friction model in describing this inherently multidimensional reaction.

5.4 Diffusion of O on Pt(111) induced by femtosecond laser pulses

U. Höfer and J. Güdde

Diffusion is an important elementary step of many surface processes such as epitaxial growth or catalytic reactions. Usually, surface diffusion is a thermally activated process that is initiated by heating the substrate. At sufficiently high temperature, the thermal population of frustrated translations or rotations enables a small fraction of the adsorbates to overcome the barrier E_{diff} for lateral motion and to hop to the next adsorption site. This process requires a minimum temperature in the order of $k_{\mathrm{B}}T \approx E_{\mathrm{diff}}/20$, and it generally strongly favors the diffusion pathway with the lowest barrier height over any other one. In some cases, it would be desirable to have more control over migration pathways or to enable diffusion of a particular species at a lower temperature where competing surfaces reaction have not yet set in. For these and other purposes, one would like to induce diffusion by electronic instead of thermal excitation of the adsorbate-substrate system. From an energetic point of view, lateral motion is easier to excite than desorption since diffusion barriers are generally much lower than chemisorption energies. Similarly, one can expect that the amount of electronic excitation required to initiate diffusion should be less than the amount required for desorption. In fact, the concept of electronic friction [44] to describe the electronic coupling of an adsorbate with a metal surface was first discussed in the context of diffusion [45, 46].

Experimentally, however, the observation of lateral motion following electronic excitation is more difficult to detect than desorbing atoms or molecules into the gas phase. One option is to use scanning tunneling microscopy as was demonstrated by Bartels et al. for the CO/Cu(110) system [47]. They were able to show that electronic excitation of the substrate induced by absorption of ultrashort laser pulses gives rise to diffusion of CO parallel and perpendicular to the close-packed rows, while thermal excitation leads to diffusion only along the rows. In the study presented here on the O diffusion on Pt, the sensitivity of second harmonic generation (SHG) on surface symmetry is employed to monitor in situ the fs-laser induced diffusion of the atomic adsorbate from step sites onto the terraces of a vicinal (111) surface [48]. While this technique cannot detect individual hopping-events, it has advantages in terms of extracting time-domain information which usually requires averaging over individual events [49]. Systematic studies of the hopping-rate for step-terrace diffusion as a function of fluence and delay time of two pump laser pulses clearly showed that the diffusion process is driven by the laser-excited electrons and not meditated by substrate phonons [48]. The idea of employing a vicinal Pt surface to distinguish between different adsorption sites was also used by Backus et al. [50]. In their time-resolved sum-frequency generation experiment, it is the site-dependent stretch frequency of adsorbed CO that allowed them to monitor lateral motion from step sites onto the terraces.

5.4.1 Experimental findings

The choice of atomic oxygen on a vicinal Pt(111) surface for the study of laser induced diffusion is motivated by its model character for diffusion of a strongly chemisorbed atomic adsorbate and by experimental reasons. Due to the importance of Pt as a catalyst in oxidation reactions, the dissociative adsorption of O_2 on Pt has been well characterized by a variety of methods (see, for example, [51,52] and references therein). Steps increase the reactivity of the substrate dramatically [52–54] such that dissociative adsorption takes place directly at the step edges as shown by STM investigations at temperatures lower than 160 K, where atomic oxygen is immobile [55]. Hence, it is possible to easily generate a well-defined initial distribution of oxygen atoms which occupy all available step sites. In the present femtosecond laser experiments on O/Pt(111), hopping from step sites onto the initially empty terraces was induced by absorption of fs laser pulses as sketched in the inset of Fig. 5.8. The relatively large diffusion barrier ensured that diffusion was only induced by the laser pulses and not initiated by thermal activation at a temperature of 80 K. The depopulation of the step sites was monitored via SHG that owes part of its sensitivity to the break of the centro-symmetry at surfaces [56,57]. Since the presence of regular steps on a vicinal surface breaks the symmetry parallel to the surface, step sites can be a very efficient source of SHG [58–60]. The capability of SHG to monitor step coverage in situ allows for the determination of diffusion rates as a function of laser fluence and delay between two pump pulses.

Experiments were performed in ultrahigh vacuum with a Pt crystal miscut by 4° from the (111) plane in the $[1\bar{1}0]$ direction. The resulting (10 12 11) surface consists of 12-unit-cell wide terraces and step edges along the $[11\bar{2}]$ direction [61]. The sample was attached to a liquid-nitrogen cryostat at a base temperature of 80 K. A gas-dosing system allowed for the exposure of controlled amounts of oxygen through a micro-channel plate. Sample preparation and verification of sufficient surface cleanness have been performed by standard procedures [49]. In particular, the reproducibility of the thermal desorption spectra for recombinatively desorbed oxygen is a very sensitive probe of surface cleanness since even small amounts of contaminations strongly suppress dissociated adsorption [49, 62]. Diffusion was induced by 50-fs pulses from a kHz Ti:sapphire amplifier system operating at 800 nm. The output was split into two orthogonally polarized beams, combined collinearly with variable time delay, incident on the sample at 40° from the surface normal and focused onto the sample. As the input radiation for SHG, fs-laser pulses also at 800 nm were used which were p-polarized and incident at 45° in a plane parallel to the step edges. The spot size was 10 times smaller than that of the pump beam, and the absorbed fluence of the probe pulses was kept below 0.5 mJ/cm^2 in order to exclude any influence on the diffusion process. The p-polarized component of the SH radiation originates predominantly from steps, thus providing a sensitive probe of the step coverage.

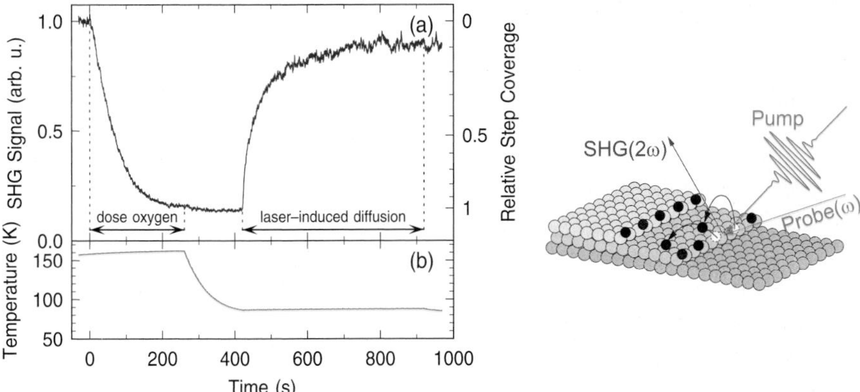

Fig. 5.8. Second-harmonic signal (a) and sample temperature (b) as a function of time. First, oxygen is dosed at a temperature of 160 K. After cooling to 80 K, diffusion is induced by applying the pump laser with an absorbed fluence of 5.6 mJ/cm^2 at a repetition rate of 1 kHz leading to a recovery of the SH signal due to the depopulation of the step edges. The scheme on the right sketches the optical excitation by an intense femtosecond pump pulse and the monitoring of the step coverage by SHG [62].

Figure 5.8 shows typical raw SHG data as a function of time during dosage and subsequent laser induced diffusion of oxygen. First, the sample was kept at 160 K and exposed to constant flux of molecular oxygen. At this temperature, chemisorbed O_2 is not stable on the terraces. It desorbs or it diffuses to the step edges where it preferentially dissociates and forms strongly-bound atomic oxygen on top of the step edges [55]. Filling of the step sites with atomic oxygen leads to a strong reduction of the SHG signal until the steps are saturated (see "dosing" period in Fig. 5.8). This strong signal reduction with oxygen coverage demonstrates the high sensitivity of the SHG detection even for a low step density of 1/12 as is the case of the crystal used in the experiments. The monotonous decrease of the SHG signal makes it possible to relate the SHG signal to the relative coverage of the step sites θ_s (right axis in Fig. 5.8) using a simple model for the coverage dependence of the nonlinear susceptibility [49]. Subsequent to the step decoration with atomic oxygen, the sample was cooled down under UHV to 80 K where oxygen is immobile even on the terraces. Partial depletion of the steps by thermal diffusion has been observed for sample temperatures exceeding 260 K. After reaching 80 K the sample was irradiated with femtosecond laser pulses at a repetition rate of 1 kHz (see "laser-induced diffusion" period in Fig. 5.8). A continuous recovery of the SHG signal is observed for pulses that exceed an absorbed fluence of 3.5 mJ/cm^2. The recovery of the SHG signal is due to the depletion of the step sites by oxygen diffusion onto the terraces and *not* due to desorption. This was verified by slowly scanning the spot of the pump beam with a fluence of 6 mJ/cm^2 over the whole sample surface. The subsequently recorded

TD spectra show no indication of a laser-induced decrease of the total oxygen coverage. Since thermal desorption of atomic oxygen takes place recombinatively around 800 K, i.e. at a much higher temperature than diffusion, it was expected that only laser pulses exceeding the damage threshold of the sample would induce desorption in the used experimental setup. The fact that almost complete step depletion at higher laser fluences was observed indicates that the laser-induced diffusion process was not defect-mediated but affects all step sites.

The saturation of the SH signal for large times below the initial value reflects the equilibrium distribution of oxygen at steps and on terraces under conditions of intense laser excitation. In thermal equilibrium, the ratio between step- and terrace-bound O atoms depends on the temperature and on the difference between the binding energy for step and terrace sites. At first glance, one would expect a rather large energy difference since oxygen binds preferentially at the step edges even at room temperature [55, 63]. However, the decoration of the step edges is primarily due to the enhanced reactivity for dissociative adsorption at the step edges and does not require any difference in the binding energies as long as the establishment of a thermal equilibrium distribution is hindered by a large enough diffusion barrier. First-principle calculations [63] for 2×1 superlattices of O on Pt(211) and Pt(322) yielded a binding energy difference of 0.4 - 0.6 eV between step- and terrace-bound O depending on step type. This difference is expected to be considerably lower when isolated O atoms on the terraces are considered. At saturation coverage, the repulsive interaction between O atoms leads to a reduction of the adsorption energy by about 20% [53, 64]. The presence of a relatively small binding energy difference between step and terrace adsorption under experimental conditions reported here is supported by the fact that terraces can well be populated by thermal induced diffusion for temperatures around 300 K.

The step-selective detection of the SHG signal makes it possible to extract diffusion rates on an atomic scale, even though this optical technique averages over a large surface area. The quantity discussed in the following is the hopping probability p_{dif} per laser shot for migration from the step sites onto the terraces. In principle, this quantity can be determined directly from the initial slope $d\theta_{\mathrm{s}}/dt$ of the SHG data recorded after switching on the pump pulses. The data shown in Fig. 5.8 have roughly an initial slope of $d\theta_{\mathrm{s}}/dt \approx -0.5/10$ s which gives $p_{\mathrm{dif}} \approx 5 \times 10^{-5}$ for a repetition rate of 1 kHz. In order to exploit the excellent statistics of all data sets, p_{dif} was determined by describing the diffusion kinetics using a simple 1D rate equation model [49]. It should be noted that this method of continuously monitoring the step coverage with SHG yields accuracies for the laser-induced diffusion rates that are comparable with or even higher than the best measurements of desorption rates using mass spectrometry detection [21, 35, 65].

As already pointed out in Sect. 5.2.1 ("Coupling between Substrate and Adsorbate"), one characteristic feature of surface femtochemistry in the DIMETregime is a nonlinear dependence of the reaction yield on laser flu-

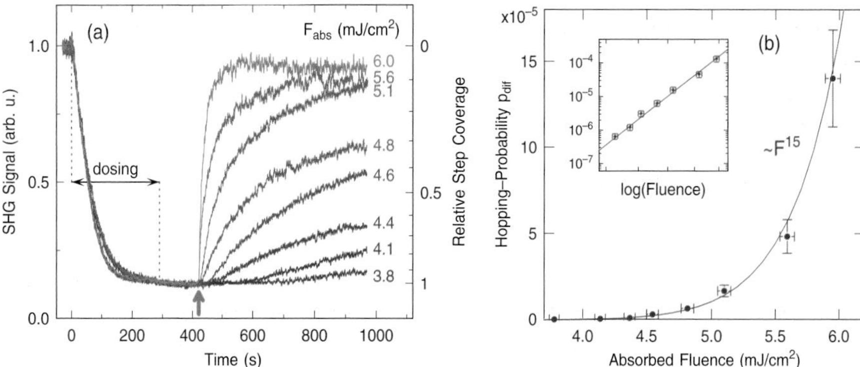

Fig. 5.9. (a) Second-harmonic response of the Pt sample during dissociative adsorption of O_2 at the steps (dosing period marked) and diffusion of atomic oxygen induced by fs laser pulses of various absorbed fluences F_{abs}. Arrow indicates begin of laser irradiation. The right y axis is converted into the relative step coverage. (b) Hopping probability per laser shot p_{dif} for the diffusion of oxygen from the filled step sites onto the empty terraces as a function of absorbed laser fluence F. Inset shows the same data in a logarithmic scale together with a fit to a power law $\propto F^n$ with $n = 15$ (solid line) [62].

ence due to repetitive electronic excitation [15]. In contrast, a reaction governed by a single nonthermal excitation is characterized by a linear fluence dependence [66], hence the term DIET without the "M" of DIMET is used. Figure 5.9a displays several sets of raw SHG data during oxygen dosage and induced diffusion for various absorbed laser fluences. This plot demonstrates the high reproducibility of the experiment and the strong dependence of the diffusion rate on laser fluence. Variation of the laser fluence by only 50 % covers the whole accessible dynamic range of the diffusion rate. Plotting the initial hopping probability per laser shot p_{dif} as a function of absorbed fluence reveals an extremely strongly nonlinear fluence dependence (Fig. 5.9b). In the fluence range investigated in the experiment, it can be described by a power law $p_{\text{dif}} \propto F^n$ similar to desorption probabilities in fs-laser experiments like the H_2 recombination described in the previous Sect. 5.3. However, the nonlinearity shown here of $n = 15$ is much larger than in all laser-induced desorption experiments reported so far, where an exponent in the range of 3 to 8 has typically been observed [21, 25, 65, 67, 68]. Even for such a large nonlinearity, the factor-of-ten smaller spot size of the probe beam ensures that the yield is spatially uniform over the diameter of the probe beam within 10%. The nonlinear detection further narrows the effective probe diameter by a factor of $\sqrt{2}$, which finally reduces the nonuniformity to 5%. Thus, yield averaging of the laser fluence like the yield-weighting procedure applied in the H_2 experiments (see Sect. 5.3.2) is not necessary here.

As in other femtochemical surface reactions, one would like to obtain detailed information on the underlying mechanism for the fs-laser induced diffu-

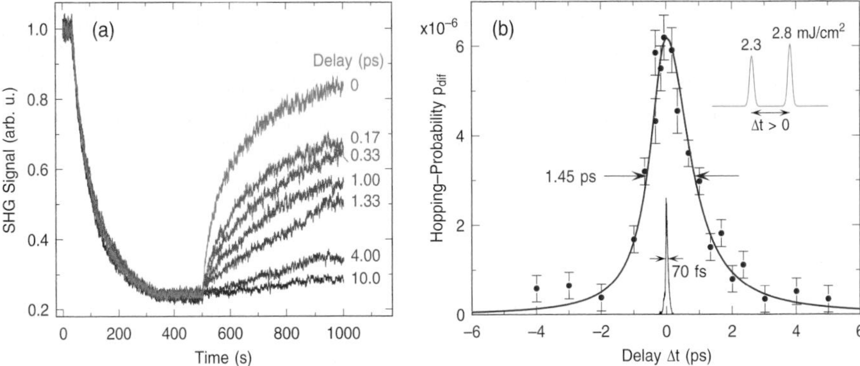

Fig. 5.10. (a) Second-harmonic response during dosing and induced diffusion for different delays between two cross-polarized pump-laser pulses. (b) Hopping probability per laser shot $p_{\rm dif}$ as a function of delay between the p- and s-polarized pump beams with absorbed fluences of 2.3 and 2.8 mJ/cm^2, respectively (symbols). At positive delays, the weaker excitation precedes the stronger one (inset). The thick solid line is a guide to the eye. The thin line shows the SHG cross correlation of the two pump-pulses generated at the sample surface [62].

sion process, i.e. the pathway of the energy transfer from the optically excited substrate electrons to the diffusive motion of the O atoms. Consequently, a two-pulse correlation scheme as described for the H_2 recombination reaction in Sect. 5.3.2 had to be applied, in which the step-site decorated Pt sample has to be excited by a pair of two pump pulses with a variable time delay. The hopping probability as the experimental observable was then extracted again by the recovery rate of the SHG signal from the probe beam. Figure 5.10 shows such a two-pulse correlation of the hopping probability of the O atoms (panel b) together with a set of raw SHG data (panel a) for various delays between the two pump pulses. $p_{\rm dif}$ shows a large contrast between small and large delays, which is related to the high nonlinearity of the fluence dependence. (Note the significant difference in the nonlinearity of the fluence dependent yields, $n \sim 3$ for the H_2 desorption versus $n = 15$ for the O diffusion.) For the O/Pt system, the 2PC FWHM amounts to 1.45 ps and is much larger than the cross-correlation of the two laser pulses and has the value of a typical electron-phonon coupling time. This unambiguously shows that the diffusive motion of atomic oxygen on a stepped Pt surface is driven by the laser-excited electrons of the metallic substrate and not by the phonon bath. The two pump beams were cross-polarized in order to suppress a coherent interference around zero delay. However, even small deviations of the polarization alignment result in an enhanced hopping-rate if the pump pulses are overlapping in time. This can be seen for the zero-delay curve in Fig. 5.10a (not shown in panel b). This might point to a contribution of a direct excitation of the adsorbate, which would result in a very narrow two-pulse correlation due to the short lifetime of adsorbate resonances. *Direct* excitation has been observed in photochemistry of molecular adsorbates using UV light [69, 70]. Polarization-dependent experiments, or better two-photon photoemission, could unambiguously identify this mechanism [71–73]. Here, however, the following sections will focus on the *indirect* excitation via the laser-excited hot substrate electrons, which cause an ultrafast energy transfer.

5.4.2 Model calculations

The two-pulse correlation as well as the fluence dependence of the oxygen hopping probability show the typical characteristics of a process which is induced by multiple electronic transitions. Thus, this fs-laser induced diffusion process should be modeled within the same formalisms used in the description of desorption reactions like the associative H_2 desorption of Sect. 5.3. Both the first elementary steps which involve the primary excitation of the metal substrate, the subsequent energy flow within the substrate as well as the energy transfer from the substrate to the adsorbate via nonadiabatic coupling were outlined in their basic concepts in Sect. 5.2.1 and 5.2.2.

In the following, model calculations will be presented which were performed within the empirical [see (5.6)] and the modified electronic friction model [see (5.8)] in order to reproduce simultaneously the observed fluence dependence and the two-pulse correlation data of the hopping probability (Figs. 5.9b and 5.10b, respectively). It should be noted beforehand that a temporally constant frictional coefficient η_{el}, which is usually assumed in various studies modeling experimental data, e.g. [31] and many other references, is not necessarily to be expected. It has been shown that the friction coefficient has a weak dependence on temperature and thereby on time if the adsorbate resonance is broad and close to the Fermi level. This case corresponds to the theory of adsorbate vibrational damping. The DIMET limit of a high-lying and well-defined affinity level, on the other hand, is characterized by a friction coefficient which is small at low temperatures and increases strongly if the electron temperature is large enough to populate the adsorbate resonance significantly [12]. Thus, the dynamics of the reaction rate $R(t)$ due to the different analytical structure of (5.6) and (5.8) might be very different for either excitation limit, the low-temperature friction and a DIMET-type excitation with high-lying, narrow adsorbate resonances. However, a dependence of η_{el} on electron temperature leads in any case to faster dynamics, as has been discussed for the desorption of NO from Pd(111) [12].

For the two-temperature modeli(5.1),(5.2), material parameters for Pt as reported in [74] ($g = 6.76 \times 10^{17}$ $\mathrm{WK^{-1}m^{-3}}$, $\gamma = 748$ $\mathrm{JK^{-2}m^{-3}}$, $\kappa_0(77\ \mathrm{K}) = 71.6$ $\mathrm{WK^{-1}m^{-1}}$, Debye temperature $T_{\mathrm{D}} = 240$ K) were used. In this study by Lei et al., the calculated electron temperature has been verified by comparing it with the dynamics of the transient electron distribution which has been observed by time-resolved photoelectron spectroscopy using a 1.5-eV pump pulse and a 42-eV probe pulse. These observations showed that the electron distribution is completely thermalized within 250 fs after the pump pulse. Thermally activated hopping has been observed at 275 K with a rate of 5×10^{-4} s^{-1}, which increases to 5×10^{-3} s^{-1} at 305 K [75]. From theses experiments, the diffusion barrier from step to terrace sites could be estimated as $E \approx 0.8$ eV which is by ≈ 0.3 eV larger than the barrier for hopping between terrace sites [76, 77]. The frustrated translation mode of O on Pt(111) has been observed by IR spectroscopy at a frequency of 1.2×10^{13} s^{-1} [78]. X-ray absorption spectroscopy of O/Pt(111) by Puglia et al. [79] revealed a broad unoccupied resonance very close to the Fermi level, which has been assigned to an antibonding state due to hybridization of the O 2p$_z$ level and the Pt 5d band. A second weaker and broader resonance centered at 8 eV above the latter has been assigned to the hybridization of the Pt 6sp and the O 2p$_{xy}$ states. Thus, energy transfer to the oxygen atoms, which weakens the O-Pt bond, can be expected even at low electron temperatures and without a strong temperature dependence. This has been confirmed by recent ab initio calculations of the friction coefficient for atomic oxygen in the fcc hollow site on a flat Pt(111) surface [80]. The lateral friction for a fixed distance of the oxygen atom at low temperatures has been calculated to $f = 2.5$ meV ps/Å^2 ($\eta_{\mathrm{el}} = 1.5$ ps^{-1}) which changes by less than 10% for temperatures of up to 3000 K.

Figure 5.11 shows the results of the model calculations for a single-pulse excitation with 50 fs pulse duration. For an absorbed laser fluence of 5 mJ/cm^2, the electron temperature rises up to 2200 K while the phonon temperature remains below 500 K. The adsorbate temperature T_{ads} and the diffusion rate R have been calculated for the empirical as well as for the modified friction model using a constant electronic friction of $\eta_{\mathrm{el}} = 1.5$ ps^{-1}. This results in a transient of the adsorbate temperature, which lies between the dynamics of the electron and the phonon temperature. The maximum adsorbate temperature of 1200 K is reached at about 800 fs after the pump pulse. The empirical and the modified electronic friction model show nearly identical results not only for T_{ads}, but also for R. At first glance, the agreement for the latter is surprising, since the prefactor of the reaction rate show different dependencies for both cases [as mentioned above, see (5.6) and (5.8)]. This is due to the fact that the Boltzmann factor contributes mainly for temperatures for which $E_{\mathrm{ads}}/k_{\mathrm{B}}T$ is of the order of 10. For these temperatures the prefactor is comparable in both cases. However, qualitative differences appear in the two-pulse correlations shown in Fig. 5.12b. The contrast between p_{dif} at zero and large delays is slightly smaller for the modified friction model. This is mainly caused by the fact that the adsorbate temperatures enters in the denominator of (5.8), which leads to a reduced rise of p_{dif} at small delays where the adsorbate temperature is large. Compared to the experimental data, however, the calculated hopping probabilities are too large by 2 to 3 orders of magnitude, and the strong nonlinearity of the fluence dependence cannot be reproduced for a diffusion barrier of $E_{\mathrm{dif}} = 0.8$ eV (Fig. 5.12a). On the other hand, the normalized two-pulse correlations are only slightly broader than observed in

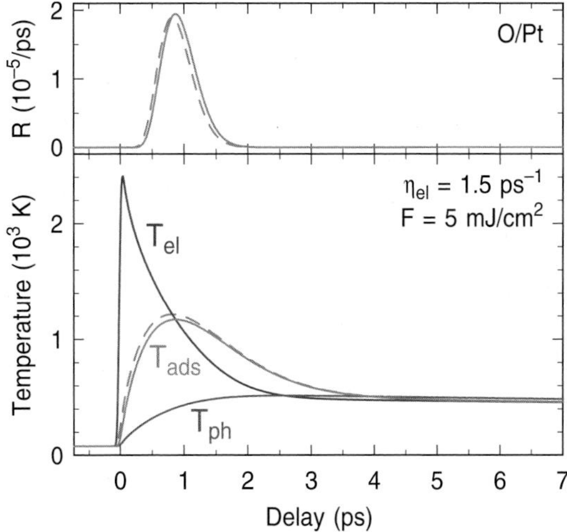

Fig. 5.11. Time dependence of electron temperature T_{el}, phonon temperature T_{ph}, adsorbate temperature T_{ads}, and diffusion rate R as derived from the two-temperature model in conjunction with the empirical as well as with the modified electronic friction model for a constant electron friction $\eta_{\mathrm{el}} = 1.5\ \mathrm{ps}^{-1}$, a laser pulse length of 50 fs, an absorbed laser fluence of 5 mJ/cm^2, and a diffusion barrier of $E_{\mathrm{dif}} = 1.4$ eV. In both cases, the same T_{el} and T_{ph} were used. The solid magenta line show the results for T_{ads} and R derived from the modified friction model, while the dashed magenta line show the results calculated within the empirical friction model [62].

the experiment. This suggests that the energy transfer time, which is associated with the friction coefficient, is in the correct range. However, the huge discrepancy between the absolute values cannot be reduced by variation of the friction coefficient. A larger friction narrows the 2PC traces, but even further increases p_{dif}, while a smaller friction results in an even broader 2PC.

A rather good description of the experimental fluence dependence and two-pulse correlation can be achieved if the diffusion barrier is arbitrarily increased. This is shown in Figs. 5.12c,d which display the results of the modified friction model with constant friction and varying diffusion barrier E_{dif}. With increasing E_{dif} the width of the two-pulse correlation narrows, and the fluence dependence becomes steeper since the variation of the Boltzmann factor in (5.8) dominates the more the larger E_{dif} is compared to $k_{\mathrm{B}}T_{\mathrm{ads}}$. The best agreement is achieved for $E_{\mathrm{dif}} = 1.4$ eV which is, however, much larger than the barrier observed for thermally activated diffusion.

As discussed in Sect. 5.2.2, an activation energy extracted from experimental data fitting which significantly exceeds the barrier value obtained for thermally induced reactions might have different reasons. Besides the already mentioned multidimensionality of the potential energy landscape and the influence of electronically excited states, a further origin might be related to

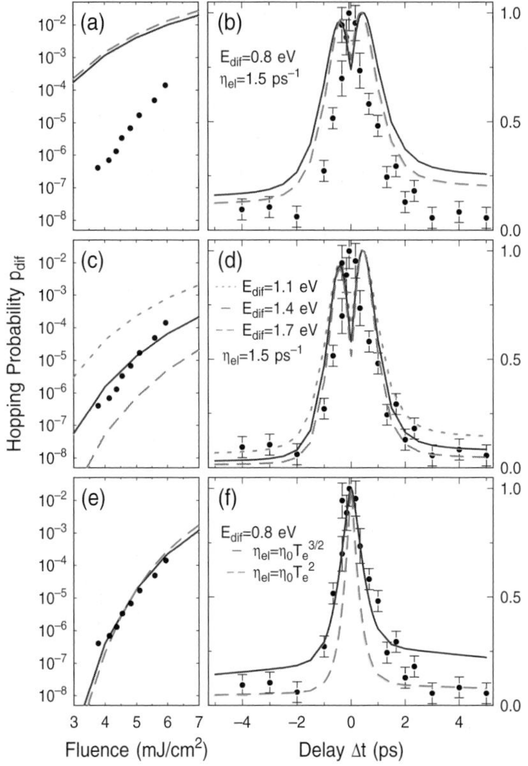

Fig. 5.12. Calculated hopping probability $p_{\rm dif}$ (left) as a function of absorbed fluence and (right) as a function of time delay between two pump pulses. Experimental data are indicated by symbols. The 2PC traces are normalized to unity. (a) and (b) show the results of the empirical (dashed lines) and the modified (solid line) friction model for $E_{\rm dif} = 0.8$ eV and $\eta_{\rm el} = 1.5$ ps^{-1}. (c) and (d) illustrate the influence of the diffusion barrier height on the results of the modified friction model using a constant friction of $\eta_{\rm el} = 1.5$ ps^{-1}. The solid and dashed lines in (e) and (f) are the results for two different dependencies of the friction coefficient on electron temperature [62].

an electron-temperature dependent friction. This is only accounted for in the modified (and not in the empirical friction model) in order to describe an (activated) population of an electronic state which cannot be significantly occupied at low temperatures. Thus, an empirical dependence of the electronic friction on the electron temperature has been introduced in order to reproduce the experimental data presented here with a reasonable value for the diffusion barrier.

Figures 5.12e and f show the results of the modified friction model using a diffusion barrier of $E_{\rm dif} = 0.8$ eV and a temperature dependence of the friction according to a power law $\eta_{\rm el} = \eta_0 T^x$ with $x = 3/2$ and $x = 2$. Indeed, this parameterization reproduces particularly well the high nonlinearity of the fluence dependence and results in a narrow two-pulse correlation even for a low barrier. The exact shape of the two-pulse correlation apparently depends on the analytic form of the assumed temperature dependence. For the chosen cases, the two-pulse correlation is either somewhat smaller than the experimental data ($x = 2$) or does not reach the experimentally observed contrast ratio of $p_{\rm dif}$ between zero and large delays ($x = 3/2$). Nevertheless, these calculations demonstrate that a temperature-dependent friction can reproduce the main features of the experimental observations without choosing an unphysical barrier height. As discussed in reference [49], the temperature-dependent friction results in different dynamics of the adsorbate temperature T_{ads}. It shows only a fast rise time and cools down slowly since the coupling strength rapidly decreases with electron temperature. Nevertheless, the hopping rate R has dynamics on the timescale of one picosecond since $\eta_{\mathrm{el}}[T_{\mathrm{el}}(t)]$ enters into the prefactor of R [see (5.8)]. The dependence of R on electron temperature via η_{el} is responsible for the narrow two-pulse correlation shown in Fig. 5.12f, which has its maximum at zero delay.

The dip of the calculated hopping probability around zero delay in Figs. 5.12b and d results from the competition between electron-phonon coupling and diffusive hot-electron transport [81]. The former tends to localize the heat at the surface while the latter leads to heat dissipation into the bulk. For delays which are small compared to the electron-phonon coupling time, the enhanced electron temperature leads to a larger temperature gradient between surface and bulk. This results in a faster diffusion of the hot electrons into the bulk. Thus, the maximum phonon temperature at the surface is reduced for small delays between the pump pulses. The more the dynamics of adsorbate and phonon temperature become similar, i.e. for small friction, the more this reduction affects the adsorbate temperature. The hopping probability should be even more sensitive on this effect since it depends exponentially on the adsorbate temperature. This consequence of the two-temperature model has been experimentally verified for the surface temperature of the substrate in time-resolved reflectivity measurements on various metal surfaces [81], but has not yet been observed in femtosecond surface photochemistry experiments. Even the laser-induced desorption of CO from Ru(001) [35], which is believed to be mediated almost solely by substrate phonons, does not show this proposed reduction. This, however, might be related to the data scattering in typical two-pulse correlation measurement based on mass spectrometer detection and in particular to the fact that the expected width of the dip is rather small for a metal like ruthenium due to the strong electron-phonon coupling in this material [81].

5.4.3 Discussion

The model calculations of the previous subsection show that the experimental data can only be reproduced with a reasonable value for the diffusion barrier if a dependence of the friction on the electron temperature is assumed. However, the electronic structure of O on Pt(111) and ab initio calculations do not support a temperature dependence of the electronic excitation of low-energy vibrational modes. Thus, the question arises if the pathway of the energy transfer to the diffusive motion is in fact more complicated than a direct coupling of the electronic excitation to the frustrated translation mode. To resolve this, an indirect excitation mechanism is suggested which requires an effective dependence of η_{el} on electron temperature within in the description of the friction model. A possible mechanism is sketched in Fig. 5.13. The O-Pt

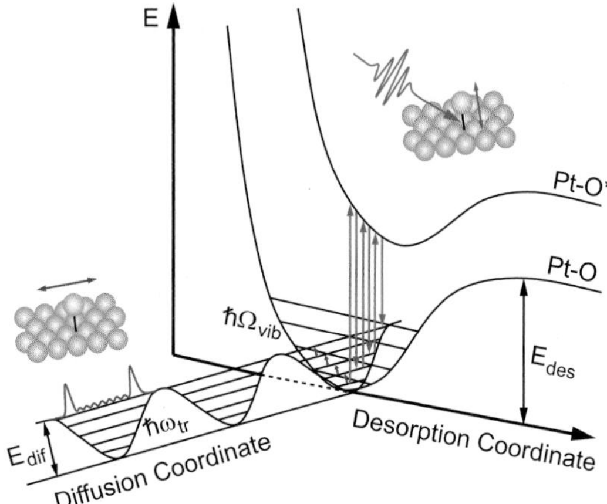

Fig. 5.13. Schematic potential energy scheme for an indirect excitation of diffusion via anharmonic coupling of the frustrated translational mode (diffusion coordinate) to the Pt-O vibration (desorption coordinate) [62].

stretch vibration is excited first via hot substrate electrons, and subsequently the frustrated translation is populated via anharmonic coupling to that primary mode. Since the anharmonicity of vibrations generally increases with amplitude, the corresponding coupling strength increases with the vibrational temperature of the O-Pt stretch. Its quantum energy (60 meV) is only slightly higher than the respective energy of frustrated O-Pt translation (50 meV) [78]. Therefore, many quanta of the stretch vibration mode need to be excited by repetitive electronic excitation cycles before the vibrational motion can couple efficiently to the lateral mode and the diffusion barrier of $\sim 0.8\,\mathrm{eV}$ can be overcome. If such a scenario is described by a single coupling strength, it will effectively depend on the electron temperature, while the primary electronic excitation of the O-Pt stretch vibrations could still be mediated by a constant electronic friction η_{el}, as illustrated in Fig. 5.14.

This proposed model of indirect excitation of the lateral motion is motivated by recent STM experiments of Komeda et al. [82] and Pascual et al. [83] who found a threshold energy for inducing diffusion of CO/Pd(110) and NH_3/Cu(100) by inelastic electron tunneling, which coincides with the internal CO and NH stretch vibration, respectively. The lateral motion is shown to be initiated by anharmonic coupling between the high-frequency internal and the low-frequency frustrated translation mode. For these systems, even one quantum of internal vibrational energy exceeds the diffusion barrier. Thus, an indirect excitation of lateral motion is possible, even if the low current in the STM experiments of < 0.05 e^-/ps inhibits ladder climbing due to the short lifetimes of the vibrational modes on metal surfaces. Komeda et al. [82] demonstrated the importance of anharmonic coupling in experiments with CO/Pd(110) and CO/Cu(110). Even if the diffusion barrier of the lat-

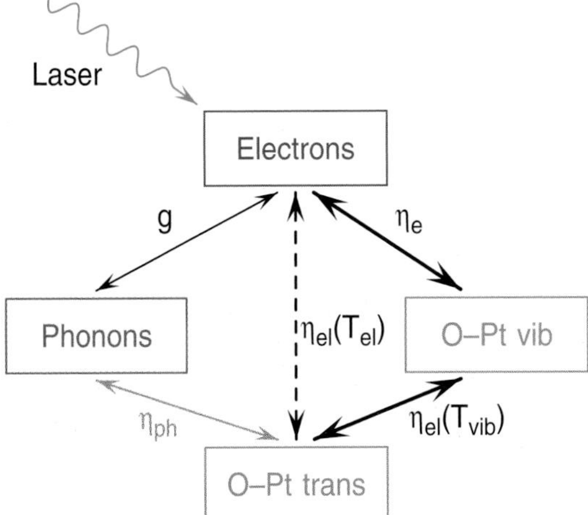

Fig. 5.14. Scheme of the energy transfer from the initial excitation of the electrons by an ultrashort laser pulse to the frustrated translational mode which initiates the diffusive motion of the oxygen atoms. An effective dependence of the electronic friction η_{el} on the electron temperature T_{el} can result from an indirect excitation via anharmonic coupling between the O-Pt stretch and the O-Pt translational mode [62].

ter system is smaller by a factor of two, no lateral hopping by excitation of the internal stretch was observed. This has been explained by the different strength of the anharmonic coupling [84,85] which is larger by a factor of >20 in the case of CO/Pd(110) as compared to CO/Cu(110). The anharmonic coupling $\hbar\delta\omega$ of vibrational modes can be estimated from the temperature dependence of their frequency and linewidth [86]. If this is applied to the temperature dependent IR data of Engström *et al.* [78], an anharmonic coupling $\hbar\delta\omega \approx 2$ meV is obtained for the O-Pt stretch vibration. Based on these IR data, it is believed that a relevant decay mechanism for the O-Pt stretch vibration is the excitation of a parallel adsorbate mode by anharmonic coupling which increases with temperature [87].

The proposed mechanism suggests a stronger electronic coupling to the O-Pt stretch vibration than to the frustrated translation. Anisotropic electronic friction has been found, for example, for H_2 on Cu(111) and N_2 on Ru(001) [24], which implies an initially strong adsorbate-substrate vibration perpendicular to the surface.[2] For oxygen on a flat Pt(111) surface, however, the electronic friction calculated for the direction perpendicular to the surface is only 20% larger than for the lateral direction [80]. On the other hand, this

[2] Note that in association reactions also the different masses (reduced mass along the vibrational coordinate versus the molecular mass for desorption) need to be considered in the overall frictional coupling. Moreover, the time the reactants take to overcome the barrier determines to which extent anisotropic electronic coupling might play a role (see Sect. 5.3.5 and [42]).

difference does not need to be very large in order to result in a temperature-dependent coupling strength. In any case, the anharmonic coupling of the stretch vibration provides an additional pathway for the energy transfer, which becomes more efficient with increasing excitation density. Thus, such a mechanism should be included in an accurate description of the energy transfer dynamics. The anisotropy of the friction may also be different for oxygen atoms at step sites due to their different coordination, which might even enhance the importance of the anharmonic coupling in the case of the fs-laser induced diffusion reported here.

In summary, Sect. 5.4 has discussed femtosecond laser-induced diffusion of atomic oxygen on a vicinal Pt(111) surface. This system is used to study the energy transfer dynamics from the optical excitation to the frustrated translation for a strongly chemisorbed atomic adsorbate. The experimental results (two-pulse correlation measurements) showed that the diffusive motion is driven by the laser-excited electrons of the metallic substrate. A consistent description of the experimental data within the modified electronic friction model cannot be achieved with a constant electronic friction and a value for the diffusion barrier consistent with other experiments. A temperature-dependent electronic friction coefficient, however, is suggested which allows to reproduce the experimental data. As the origin of this temperature-dependence an indirect excitation mechanism is proposed. Anharmonic coupling of the initially excited O-Pt stretch vibration to the frustrated translational mode via hot substrate electrons depends on the excitation density and indeed would explain the observed effective dependence of the electronic friction on the electron temperature.

5.5 Theory of ultrafast nonadiabatic processes at surfaces and their control

P. Saalfrank, M. Nest, and T. Klamroth

In this section, the models used in Sects. 5.3 and 5.4 to describe femtosecond-laser induced processes at surfaces will be put in a broader theoretical perspective. The section also serves to address possibilities and limitations to control the reactivity of adsorbates at metal surfaces by tailored light pulses.

5.5.1 From "weakly adiabatic" to "strongly nonadiabatic" dynamics: Methods

Weakly nonadiabatic dynamics

The experimental examples mentioned so far are "weakly nonadiabatic" processes in the sense that the dynamics is dominated by nuclear motion on a single (ground state) potential energy surface, with electronic transitions serving "merely" to drive the dynamics on that surface. As outlined above (Sect. 5.2.1), hot-electron mediated femtochemistry at metal surfaces is frequently modeled with coupled heat baths of the electron, phonon, and adsorbate system, respectively, in conjunction with (multidimensional) classical Langevin dynamics, master equation approaches, or with Arrhenius-type rate equations. All of these are based on the concepts of electronic friction, and well-defined temperatures for the various subsystems (substrate electrons, phonons, and adsorbate vibrations, respectively) of the adsorbate-substrate complex, see (5.1) through (5.8).

However, several problems and limitations of these models arise: The Arrhenius models have the disadvantages of being one-dimensional, classical, and they assume electronic and vibrational temperatures for a nonequilibrium situation. The electron temperature is questionable at least within the first few 100 fs or so after the pulse according to experimental [88] and theoretical investigation [89, 90]. The concept of an adsorbate temperature can be inapplicable for even much longer timescales [91, 92].

Some of the restrictions of above can be overcome by Langevin molecular dynamics with electronic friction [93], which was used above for H_2/Ru(001). Here nuclear motion is still classical, and the electronic degrees of freedom are expressed in the form of friction and fluctuating forces. If q is the only degree of freedom considered, e.g., the molecule-surface distance Z, mass m_q), the equation of motion is [see (5.10) and (5.11) in Sect. 5.3]

$$m_q \frac{d^2q}{dt^2} = -\frac{dV}{dq} - \eta_{qq} \frac{dq}{dt} + F_q(t) \quad . \tag{5.12}$$

Here, V is the ground state potential, and η_{qq} is related to the electronic friction coefficient of above through $\eta_{qq} = m_q \eta_{el}$. $F_q(t)$ is a fluctuating force that obeys a fluctuation-dissipation theorem, and depends on the electronic temperature, $T_{el}(t)$, obtained from the 2TM.i Equation (5.12) can be easily generalized to more than one degree of freedom, by interpreting the friction coefficient as a tensor with elements $\eta_{qq'}$ [24, 93].

Various methods based on, e.g., Hartree-Fock cluster calculations [93] or on periodic density functional theory [24] have been suggested of how to calculate the friction tensor from first principles. All of these methods are based on a perturbative, Golden Rule type treatment of the vibration-electron coupling. At zero temperature, the electronic friction coefficient is directly related to the vibrational relaxation rate $\Gamma^{\text{vib}}_{1\to 0}$ of the first excited level, and thus to the lifetime τ_{vib} of the mode of interest:

$$\eta_{\text{el}} = \Gamma^{\text{vib}}_{1\to 0} = \tau^{-1}_{\text{vib}} \quad . \tag{5.13}$$

τ_{vib} ranges typically from several hundred fs to several ps for metal surfaces [94].

Finally, the classical approximation can be overcome with the help of master equations,

$$\frac{dP_\alpha}{dt} = \sum_\beta W_{\beta\to\alpha} P_\beta(t) - \sum_\beta W_{\alpha\to\beta} P_\alpha(t) \quad , \tag{5.14}$$

giving the population P_α of state $|\alpha\rangle$ of the ground state potential. The latter can be derived from (multidimensional) system Hamiltonians. The populations are governed by interlevel transition rates,

$$W_{\alpha\to\beta} = \Gamma^{\mathrm{vib}}_{\alpha\to\beta} + W^{\mathrm{ex}}_{\alpha\to\beta} \quad , \tag{5.15}$$

where $\Gamma^{\mathrm{vib}}_{\alpha\to\beta}$ denotes a vibrational relaxation rate (or reexcitation rate if the temperature is finite), and $W^{\mathrm{ex}}_{\alpha\to\beta}$ stands for a rate caused by an external energy source which drives the "ladder climbing". The latter proceeds typically by transient population of electronically excited states, for example a "negative-ion resonance", i.e. by temporary attachment of an electron to the adsorbate. For example, for femtosecond-laser induced, hot-electron mediated reactions the transition rates, calculated from perturbation theory, are a function of the electron temperature $T_{\mathrm{el}}(t)$, and of the energetic position ε_a and width Δ_a of the adsorbate resonance [92, 95]. By solving (5.14), a reaction probability can be defined by analyzing those populations which reach the desorption continuum, with an energy $E_\alpha > D$ where D is again the binding energy of the adsorbate if "desorption" is the reaction of interest. Such a damped ladder climbing process is schematically illustrated in Fig.5.1c (Sect. 5.2.1). In certain limits, the master equation (5.14) simplifies to reaction rates R of the Arrhenius-type [see (5.8) [95]].

Strongly nonadiabatic dynamics

In contrast to the weakly nonadiabatic dynamics which predominantly proceed on a single potential energy surface, "strongly nonadiabatic" dynamics cannot be idealized in this way. Examples of this type of reactions are (i) resonant charge transfer during atom-surface scattering, (ii) DIET(single-excitation limit), and (iii) reactions proceeding through long-lived excited states, such as adsorbate excitations at insulators. The explicit inclusion of more than one electronic state can also be useful for modeling the "weakly nonadiabatic" processes listed above.

The most general approach to treating strongly nonadiabatic situations is through a coupled, multi-state time-dependent Schrödinger equation (TDSE). Assuming that the system is initially in its electronic ground state $|g\rangle$ associated with a nuclear wavefunction $\psi_g(R,t)$ (where R denotes the nuclear coordinates), and assuming that only a single adsorbate resonance $|a\rangle$ is of importance, the TDSE reads

$$i\hbar\frac{\partial}{\partial t}\begin{pmatrix}\psi_a\\ \psi_g\\ \psi_{k_1}\\ \psi_{k_2}\\ \vdots\end{pmatrix} = \begin{pmatrix}\hat{H}_a & \tilde{V}_{ag} & \tilde{V}_{ak_1} & \tilde{V}_{ak_2} & & \cdots\\ \tilde{V}_{ga} & \hat{H}_g & \tilde{V}_{gk_1} & \tilde{V}_{gk_2} & & \cdots\\ \tilde{V}_{k_1a} & \tilde{V}_{k_1g} & \hat{H}_{k_1} & \tilde{V}_{k_1k_2} & \tilde{V}_{k_1k_3} & \cdots\\ \tilde{V}_{k_2a} & \tilde{V}_{k_2g} & \tilde{V}_{k_2k_1} & \hat{H}_{k_2} & \tilde{V}_{k_2k_1} & \cdots\\ \vdots & \vdots & \vdots & \vdots & \vdots & \ddots\end{pmatrix}\begin{pmatrix}\psi_a\\ \psi_g\\ \psi_{k_1}\\ \psi_{k_2}\\ \vdots\end{pmatrix} \quad . \tag{5.16}$$

In (5.16), a quasi-continuum of substrate-excited states $|k_i\rangle$ was also included, which applies to metal surfaces. The diagonal elements of the Hamiltonian matrix are $\hat{H}_n = \hat{T}_{nuc} + V_n(R)$, where $\hat{T}_{nuc}$ is the nuclear kinetic energy operator, and $V_n(R)$ is the potential energy curve for state $|n\rangle$ (in diabatic representation). The off-diagonal elements $\tilde{V}_{nm}(R,t)$ stand for possible (direct) dipole couplings, and non-Born-Oppenheimer couplings $V_{nm}(R)$, i.e.

$$\tilde{V}_{nm}(R,t) = V_{nm}(R) - \left\langle \Psi_n(r,R) \middle| \hat{\mu} E(t) \middle| \Psi_m(r,R) \right\rangle_r \quad . \tag{5.17}$$

Here $\Psi_m(r,R)$ denotes an electronic wavefunction which depends on electron coordinates r and parametrically on R; $\hat{\mu}$ is the molecular dipole operator, and E the electric field. In (5.17), the semiclassical dipole approximation is used for matter-field coupling; the vector character of dipole moment and field are neglected here.

Equation (5.16) can hardly ever be solved, in particular for metals. An alternative is to map the closed-system, N-state model (5.16) to an open-system density matrix model with two states, where the excited state (e.g., the negative ion state) is treated as a (nonstationary) *resonance*. The resonance width, Δ_a, can in principle be calculated from the non-Born Oppenheimer coupling matrix elements V_{ak}, for instance within the Anderson-Newns model [14, 96].

Within an open-system two-state model of DIET, for example, a Liouville-von Neumann (LvN) equation [97, 98]

$$\frac{\partial}{\partial t}\begin{pmatrix} \hat{\rho}_a & \hat{\rho}_{ag} \\ \hat{\rho}_{ga} & \hat{\rho}_g \end{pmatrix} = -\frac{i}{\hbar}\left[\begin{pmatrix} \hat{H}_a & \tilde{V}_{ag} \\ \tilde{V}_{ga} & \hat{H}_g \end{pmatrix}, \begin{pmatrix} \hat{\rho}_a & \hat{\rho}_{ag} \\ \hat{\rho}_{ga} & \hat{\rho}_g \end{pmatrix}\right] + \sum_s \frac{\partial}{\partial t}\begin{pmatrix} \hat{\rho}_a & \hat{\rho}_{ag} \\ \hat{\rho}_{ga} & \hat{\rho}_g \end{pmatrix}_{D,s} \tag{5.18}$$

has to be solved. In (5.18), the $\hat{\rho}_i$ and $\hat{\rho}_{ij}$ are operators in the vibrational space of the ground and excited state vibrational functions $\{\phi_n^g\}$ and $\{\phi_n^e\}$, respectively. The last term in (5.18) accounts for energy and phase relaxation in general, but also for substrate-mediated excitation, with s denoting various "dissipative" channels, i.e. ways to exchange energy and phase with an environment.

As an example, energy relaxation of an excited state $|a\rangle$ with width Δ_a and electronic relaxation rate $\Gamma_{a\to g}^{\mathrm{el}} = \frac{\Delta_a}{\hbar}$ can be modeled by the choice

$$\frac{\partial}{\partial t}\begin{pmatrix} \hat{\rho}_a & \hat{\rho}_{ag} \\ \hat{\rho}_{ga} & \hat{\rho}_g \end{pmatrix}_{D,1} = -\Gamma_{a\to g}^{el}\begin{pmatrix} \hat{\rho}_a & \frac{\hat{\rho}_{ag}}{2} \\ \frac{\hat{\rho}_{ga}}{2} & -\hat{\rho}_a \end{pmatrix} . \tag{5.19}$$

which transfers population from $|a\rangle$ to $|g\rangle$ (and destroys possible coherences at the same time). DIET, i.e. the single-excitation limit, is treated by a single, initial Franck-Condon excitation of the ground state wave function ϕ_0^g to the excited state, i.e. $\hat{\rho}_0 = |a\rangle\langle a| \otimes |\phi_0^g\rangle\langle\phi_0^g|$. Vibrationally excited states (see below) or thermal initial states may be used instead, depending on which

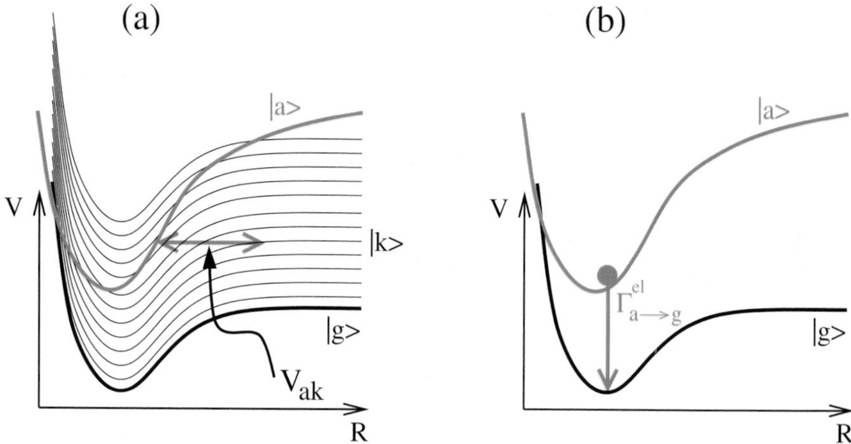

Fig. 5.15. (a) The closed-system, coupled N-state model of a nonadiabatic surface process, with ground state $|g\rangle$, a photoactive excited state $|a\rangle$, and a (quasi-) continuum of substrate-excited states $|k\rangle$. A specific non-Born Oppenheimer coupling element V_{ak} is indicated. (b) Illustration of the corresponding open-system two-state model with transition rate $\Gamma^{\mathrm{el}}_{a\to g} = \Delta_a/\hbar$.

experimental situation is to be modeled. In general, the resonance width Δ_a depends on the nuclear coordinates R, and only if this dependence is neglected, the resonance decays strictly exponentially with an electronic lifetime $\tau_{\mathrm{el}} = \frac{\hbar}{\Delta_a}$. τ_{el} can be as short as a few fs, as outlined earlier. The closed-system N-state model is contrasted with the open-system 2-state model in Fig.5.15.

When intense femtosecond lasers are used for photodesorption, DIMET becomes possible. In the two-state density matrix model, the hot-electron substrate-mediated excitation can be treated by an extra term in (5.18) [97,98] which depopulates the ground state

$$\frac{\partial}{\partial t}\begin{pmatrix} \hat{\rho}_a & \hat{\rho}_{ag} \\ \hat{\rho}_{ga} & \hat{\rho}_g \end{pmatrix}_{D,2} = \Gamma^{el}_{g\to a}(t)\begin{pmatrix} \hat{\rho}_g & -\frac{\hat{\rho}_{ag}}{2} \\ -\frac{\hat{\rho}_{ga}}{2} & -\hat{\rho}_g \end{pmatrix}. \tag{5.20}$$

For DIMET, the initial condition is $\hat{\rho}_0 = |g\rangle\langle g| \otimes |\phi_0^g\rangle\langle\phi_0^g|$ if $T = 0$ initially. Further,

$$\Gamma^{\mathrm{el}}_{g\to a}(t) = \Gamma^{\mathrm{el}}_{a\to g}\exp\left\{-\frac{V_a - V_g}{k_B\,T_{\mathrm{el}}(t)}\right\} \tag{5.21}$$

is a time-dependent upward rate that obeys detailed balance. Other relaxation channels (e.g., vibrational relaxation) can easily be incorporated in the model [91].

Note that in (5.21) the assumption of an electronic temperature has been made, similar to the weakly nonadiabatic friction models of above. However, no adsorbate vibrational temperature $T_{ads}(t)$ needs to be assumed in contrast to Arrhenius-type treatments, a feature which the present method shares with molecular dynamics with friction, and the master equation approach. Nevertheless, there are also differences between the "weakly nonadiabatic" friction, and the "strongly nonadiabatic" excitation-deexcitation models of femtosecond laser chemistry, leading sometimes to different conclusions/interpretations [91]. For example, the two- and multi-state models account for dynamical details which follow from the topology of the excited state potential. An example is the observed, pronounced vibrational excitation of NO molecules desorbing from Pt(111) [99], which is thought to proceed through a negative-ion resonance state NO^-Pt^+ with an elongated NO bond.

5.5.2 Controlling nonadiabatic surface processes: Strategies

The open-system density matrix theory just scratched will now be applied to the question as to whether *control* of laser-induced desorption of admolecules from metal surfaces is possible.

Since in substrate-mediated photochemistry the adsorbate is excited only indirectly, *coherent control* is evidently feasible. Here, an *incoherent* control scheme may be more promising, for instance for DIMET through the control of $T_{el}(t)$. This type of control can be achieved in various ways, for example through the laser fluence (see above), or by changing the width of the laser pulse envelope at constant fluence. The latter possibility was suggested [100] and experimentally supported [101] for DIMET of CO from Pt(111).

An alternative is to nanostructure a surface. It has been found that NO molecules adsorbed on amorphous or ordered aggregates of Pd atoms on alumina surfaces can be efficiently desorbed by nanosecond UV laser pulses in contrast to single crystal surfaces [102]. The enhanced reactivity can be due to electronic effects, e.g., additional binding sites with lower adsorption energies, field enhancement at rough surfaces, or new reactive intermediates such as plasmons. One further possible enhancement factor for nanostructured materials, when driven by femtosecond laser pulses, is that the photon energy is now localized in a smaller region of space thus leading to larger electronic temperatures and excitation rates $\Gamma^{el}_{g\to a}$. This latter mechanism is again incoherent and will be further exploited in Sect. 5.5.4.

Another control mechanism for substrate-mediated photochemistry is a "hybrid scheme", in which the adsorbate is vibrationally excited by an IR pulse prior to electronic excitation. The first (IR) step is direct, while the latter is incoherent. It was suggested to use an IR+UV/vis strategy to control the photodesorption of molecules from surfaces [98, 103–105], similar to "vibrationally mediated chemistry" in gas phase dynamics [106, 107]. The IR preexcitation will not only lead to larger UV/vis desorption yields, but may also be useful for isomerization reactions [108, 109] and for isotope-selective chemistry [110]. Such a hybrid scheme will now be applied to the case of associative desorption of H_2 from Ru(001) under DIET conditions.

5.5.3 "Hybrid control": Vibrationally enhanced DIET of H_2 from Ru(001)

In the DIET limit, the reaction is enforced by single excitations, possibly with a low-intensity nanosecond laser. In this regime the friction models used in Sect. 5.3 for DIMET are not applicable, mostly because the concept of an electron temperature becomes inadequate, and we use a two-state density matrix model instead. The proposed "hybrid" control scheme is illustrated in Fig. 5.17a below.

Ground state potential and IR excitation

Similar to the theoretical modeling of the DIMET experiments of Sect. 5.3 by electronic friction approaches, we employ a two-dimensional model with r (H-H distance) and Z (the H_2-surface distance) modes also for the density matrix description. We assume that the two H atoms, initially in adjacent face centered cubic (fcc) hollow sites, remain in the plane spanned by these two sites an the direction to the surface. For the ground state potential energy function, $V_g(r, Z)$, a functional form similar to the one adopted for the electronic friction model in Sect. 5.3 has been used. This surface is based on the six-dimensional potential generated by Baerends [111] for a (2× 2) unit cell, i.e. coverage 1/2, obtained from periodic DFT calculations within a generalized gradient approximation (GGA) for the exchange-correlation functional. The modifications of the original potential as introduced here pertain to the extension of the potential to regions which were not calculated in [111] and to adding smooth repulsive walls behind the transition states toward diffusion and subsurface adsorption. This allows us to model a high-coverage situation H-(1×1) $(\theta = 1)$ as used in [31, 39], and to exclude subsurface adsorption after excitation. The resultant two-dimensional PES predicts an adsorption minimum at $Z_0 = 1.06$ Å and $r_0 = 2.75$ Å (the shortest distance between two fcc sites on the surface), with a binding energy of 0.85 eV (i.e., 0.42 eV per H atom). Further, in the 2D model, an incoming H_2 molecule would have to overcome a classical barrier of 0.18 eV. The barrier is located "early" on the reaction path for dissociative adsorption and hence "late" on the reaction path for associative desorption, i.e. $Z^{\ddagger} = 2.24$Å and $r^{\ddagger} = 0.77$Å [112] – the bond length of free H_2 is 0.74 Å, i.e. only slightly shorter.

The potential $V_g(r, Z)$ supports bound vibrational states, the lowest of which can be easily classified according to their quantum numbers in the r and Z modes, i.e. $\psi_{n_r,n_Z}(r, Z)$. By solving the time-independent vibrational Schrödinger equation

$$\left[-\frac{\hbar^2}{2M}\frac{\partial^2}{\partial Z^2} - \frac{\hbar^2}{2\mu}\frac{\partial^2}{\partial r^2} + V_g(r, Z)\right] \psi_{n_r,n_Z}(r, Z) = \varepsilon_{n_r,n_Z}\psi_{n_r,n_Z}(r, Z), \quad (5.22)$$

one finds about 30 bound vibrational states for H_2/Ru(001). The vibrational energies of the fundamentals are $\hbar\omega_r = \varepsilon_{1,0} - \varepsilon_{0,0}$ and $\hbar\omega_Z = \varepsilon_{0,1} - \varepsilon_{0,0}$ of

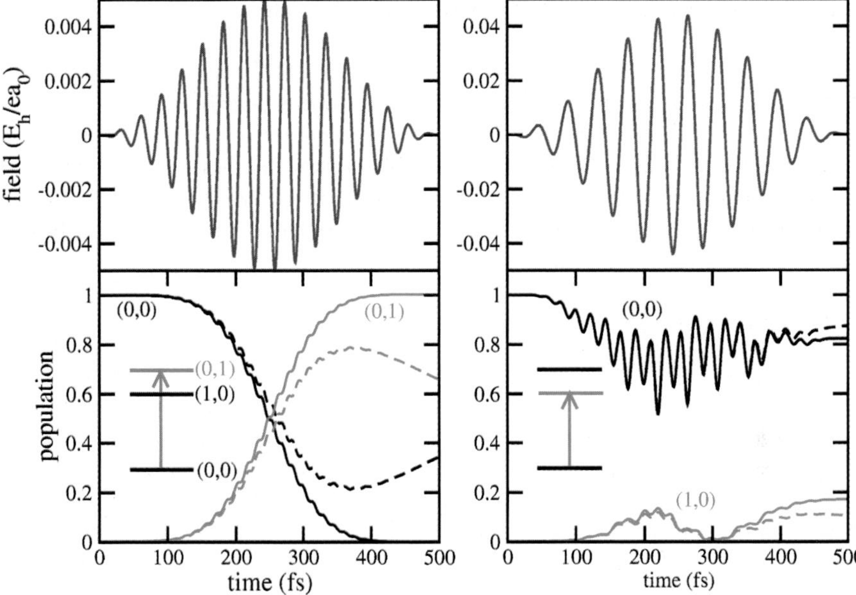

Fig. 5.16. Excitation of the (0,1) vibrational level (Z mode, left), and the (1,0) vibrational level (r mode, right), from the vibrational ground state (0,0), using infrared π pulses (upper panels). Initial and target state populations are shown in the lower panels together with a schematic representation of the process. Solid curves are for the dissipation-free case, dashed ones for assuming a vibrational lifetime of 500 fs for each target. A 3-level LvN equation was solved to obtain populations – see [114] for an analogous model.

$\hbar\omega_r = 94$ meV and $\hbar\omega_Z = 136$ meV. This is in reasonable agreement with experimental data of Jacobi et al. [113], who find for the H-(1×1) covered Ru(001) surface $\hbar\omega_r = 85$ meV and $\hbar\omega_Z = 140$ meV. It should be noted, however, that in the experiment the generation of such laser frequencies in the IR spectral range remains a challenge.

In order to IR-excite selected vibrational levels, we assume that the dipole moment of adsorbed H_2 couples to the z component of an external field. From a cluster model (see below), the transition dipole moments $\langle\psi_{n_r,nZ}|\mu_z(r,Z)|\psi_{n_r',n_Z'}\rangle$ can be computed by quantum chemical methods, where $\mu_z(r,Z)$ is the z component of the ground state dipole moment. For the fundamental excitations of the r and Z modes, we find $|\mu_{(00,10)}| = 5.84 \times 10^{-32}$ Cm and $|\mu_{(00,01)}| = 5.15 \times 10^{-31}$ Cm, i.e. the Z mode will be easier to excite. The two states $\psi_{10} := (1,0)$ and $\psi_{01} := (0,1)$ are also directly coupled through a small dipole moment $|\mu_{(10,01)}| = 2.22 \times 10^{-32}$ Cm.

That in contrast to the r mode, the Z mode can be easily excited is demonstrated in Fig. 5.16 where infrared π pulses of $\sin^2$ form, i.e.

$$E_{IR} = E_0 \sin^2(\pi t/t_f) \cos(\omega_0 t) \tag{5.23}$$

are used to populate the (1,0) state or the (0,1) state from the ground state (0,0). Here, E_0 is the field amplitude, t_f the pulse length, and ω_0 the carrier frequency of the laser. The carrier frequencies were chosen to match the transition frequencies ω_Z and ω_r in the left and right columns of Fig. 5.16, respectively. The pulse length was $t_f = 500$ fs in both cases. The field amplitudes were chosen as $E_0 = \frac{2\pi\hbar}{t_f|\mu_{if}|}$, which is the π pulse condition for a $\sin^2$ pulse [114]. π pulses lead to a population inversion in an ideal two-level system, connected by a transition dipole moment $|\mu_{if}|$.

Using a three-level open-system density matrix model comprising of the (0,0), (1,0) and (0,1) states, and the π pulses for excitation of (1,0) and (0,1) as shown in the left and right upper panels of Fig. 5.16, we obtain the populations of vibrational levels as a function of time. For an analogous recent work and model, see [114]. The solid lines in the lower panels of Fig. 5.16 lines refer to initial and target state populations for the dissipation-free case, i.e. no vibrational relaxation accounted for. The dashed lines stand for the dissipative case, where a finite vibrational lifetime of $\tau_{\text{vib}} = 500$ fs was assumed for each of the two vibrations. From the left panels of Fig. 5.16, we note that an ideal π pulse works in fact well for the Z mode, i.e., excitation of (0,1), in particular, in the dissipation-free case. In contrast, a simple π pulse is not very effective for the r excitation simply because the corresponding dipole moment is small such that the nearby (0,1) state is also excited with a pulse which is only 500 fs long and, therefore, energetically broad. Using longer pulses is found to enhance selectivity in the dissipation-free case. On the other hand, from the electronic friction models, we know that vibrational relaxation is important for H_2/Ru(001), thus longer pulses are impractical. Other strategies (e.g. multipulse excitations including higher-lying states), and/or pulse optimization techniques (e.g. optimal control theory [115]) are needed for selectively exciting vibrational levels – see [112].

Desorption induced by electronic transition

To treat DIET, it is assumed that a single, "representative" excited state, $|a\rangle$, is sufficient to promote the desorption. One can then solve (5.18) with (5.19), using different initial vibrational states ψ_{n_r,n_Z}, under tacit assumption that those can be selectively populated. A particular efficient method to solve (5.18) and (5.19) (with direct couplings neglected) was suggested by Gadzuk some time ago [116]. In his "jumping wave packet" model, DIET is treated in two steps. In step one, the initial wave function, a pure state ϕ_0, is projected on the excited state $|a\rangle$, propagated there for some residence time τ_R, transferred back to the ground state $|g\rangle$, and propagated to a final time t

$$|\psi(t;\tau_R)\rangle = \exp\left\{-\frac{i\hat{H}_g(t-\tau_R)}{\hbar}\right\} \Big|g\Big\rangle\Big\langle a\Big| \exp\left\{-\frac{i\hat{H}_a\tau_R}{\hbar}\right\} |\phi_0\rangle \quad . \tag{5.24}$$

From this, operator expectation values for single trajectories are computed as $A(t;\tau_R)= \left\langle \psi(t;\tau_R)|\hat{A}|\psi(t;\tau_R)\right\rangle$. In the second step, an incoherent averaging scheme is adopted which, if the decay of the resonance is strictly exponential with an excited state lifetime τ_{el} reads

$$\left\langle \hat{A} \right\rangle (t) = \frac{1}{\tau_{el}} \int_0^\infty e^{-\tau_R/\tau_{el}} \; A(t;\tau_R)\; d\tau_R \quad . \tag{5.25}$$

It has been shown that this approach, in which only repeated wave packet propagations have to be carried out, is in fact rigorously equivalent to an open-system density matrix treatment [117]. It was also shown how to extend the approach to, e.g., a nonexponential decay of the resonance state [118,119].

A critical point is the choice of the excited state potential appearing in $\hat{H}_a = \hat{T}_{nuc} + V_a(r,Z)$, and also of the excited state lifetime, τ_{el}. Reliable ab initio calculations of (adsorbate) excited states at metal surfaces and their lifetime are not yet available. Recently, attempts have been made to calculate excited states of metal/adsorbate systems using a cluster approach in combination with time-dependent DFT (TD-DFT [120]). Here, we employ clusters H_2Ru_n (with $n = 3$ and $n = 12$), for which TD-DFT gives a multitude of excited states in the DIET-relevant energy regime up to about 2 eV [112]. (Already for H_2Ru_3, there are about 40 states with excitation energy ≤ 2 eV.) Closer inspection shows that most of these states are excitations with a topology very similar to the ground state surface, several of them, however, with their minimum simultaneously shifted along the r and Z modes. In particular, it is found that (i) $\Delta r \approx -\Delta Z$, (ii) both positive and negative Δr (ΔZ) occur, and (iii) shifts $\Delta := |\Delta_r| \approx |\Delta_Z|$ up to $\Delta \approx 0.2 - 0.3$ Å are observed. Based on this information, the excited state potential was taken as

$$V_a(r,Z) = V_g(r-\Delta, Z+\Delta) + E_{\mathrm{ex}} \quad , \tag{5.26}$$

where Δ was chosen from reasonable ranges. E_{ex} is the electronic excitation energy, whose actual choice, however, plays no role in the Gadzuk DIET scheme. The excited state lifetime τ_{el} was varied between 1 and about 10 fs, and thus treated as an empirical parameter. As an important "side product", the cluster calculations gave the dipole function $\mu_z(r,Z)$ which was used for the transition dipole moments for IR excitation of above. (The dipole function was computed for H_2Ru_{12} and fitted to an analytic form.) Vibrational relaxation on the ground state potential due to vibration-electron coupling was neglected in the DIET calculation. Note that neither the exact nature of the excited states, nor their precise couplings are of concern in the spirit of the effective, dissipative two-state model. The Gadzuk algorithm was realized below by using standard wave packet propagation techniques on a grid [121], and analyzing the flux going through a dividing line Z_{des} in front of the surface, from where the density can be classified as desorbed.

When using the ground vibrational state as initial state, $\Delta = 0.2$ Å, and a lifetime of $\tau_{\mathrm{el}} = 2$ fs, one finds that associative desorption also takes place in

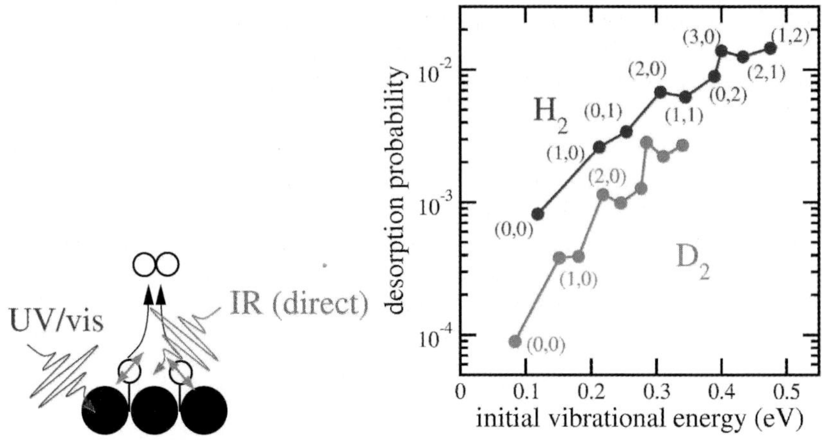

Fig. 5.17. Desorption probability for H_2 and D_2 per excitation event. Vibrational eigenstates have been used as initial states, with quantum numbers indicated for H_2. Further, $\Delta = 0.2$ Å and $\tau_{\rm el} = 2$ fs were used. The IR+UV hybrid control strategy is indicated schematically (left). Accordingly, an IR pulse excites the r and/or Z mode of adsorbed hydrogen (open, small balls), leading to vibrational excitation (green arrows) and an enhanced desorption probability after the Ru surface (black, filled balls) is hit by UV/vis radiation.

the DIET case, with computed observables similar to those found in DIMET. In particular, an isotope effect of $I_{des} = Y(H_2)/Y(D_2) = 9.2$ is found. In addition, the products are vibrationally and translationally "hot" with a clear propensity for more translational energy, due to the "late" barrier along the desorption path. For D_2/Ru(001), for example, computed DIET translational energy is about seven times higher than the vibrational energy. All quantities depend on the shift Δ and the lifetime $\tau_{\rm el}$ quantitatively, however, both positive and negative Δ lead to desorption and the qualitative features are independent of the particular choice of those parameters, when taken within reasonable ranges. For example, with $\Delta = -0.2$ Å (excited state shifted toward the surface) one obtains $I_{\rm des} = 12.6$, and $E_{\rm trans}(D_2)/E_{\rm vib}(D_2) \approx 4.3$.

The dependence of the desorption probability for H_2 and D_2 on the initial state is shown in Fig. 5.17b, where the desorption yield is plotted as a function of the initial vibrational energy. In all cases, vibrational eigenstates have been used as initial states.

First of all, one notes that with (0,0) initial states, the desorption probability per excitation event is 8.1×10^{-4} for H_2 and 8.8×10^{-5} for D_2, resulting in the isotope effect of $I_{\rm des} \approx 9$ as mentioned above. For both H_2 and D_2, it is found that the desorption yield is considerably enhanced with vibrationally excited initial states. For example, using the r-excited (1,0) initial state increases Y for D_2 by a factor of 4.3, and using (0,2) as an initial state gives a factor of about 15 relative to the ground state. Excitation of the Z mode is even slightly more efficient. It is also observed that both the translational and vibrational energies of the desorbing particles increase with the level of initial vibrational excitation. Note that "isotope-selective" desorption, i.e. the control of the isotope effect should be possible, because one or the other isotopomer can be selectively IR-excited due to their distinct vibrational energies. Of course, the highest vibrational levels shown in Fig. 5.17b will not be easy to excite in practice, but it is already clear that even the low-lying excited levels have a strong impact on the reaction yield. It is also expected that, using vibrational wave packets rather than eigenstates, offers a route to further increase the photodesorption cross section.

5.5.4 Incoherent control: DIMET of NO from Pt(111) metal films

As outlined above, nanostructuring a metal surface can lead to enhanced photoreactivity by various mechanisms. Here, we will concentrate on thin metal films with thicknesses on the order of 10 to several hundred nm. We also focus on the effect that in "hot-electron"-mediated femtosecond-laser chemistry the confinement has an influence on the $T_{\rm el}(t)$ curve, neglecting other enhancement factors, i.e. electronic or lifetime effects [122].

In Fig. 5.18a, we show the calculated hot-electron temperature $T_{\rm el}(t)$ at the surface of Pt(111) films of various thicknesses after excitation by Gaussian 80 fs laser pulses with a fluence of 6 mJ/cm^2 and $\lambda = 619$ nm [100]. Similar pulses have been used in DIMET experiments of NO/Pt(111) [123]. For the calculation, the two-temperature model was used with a thickness-dependent source term $S(t)$ according to (5.3).

First, it is observed that with decreasing film thickness the maximum electronic temperature $T_{\rm el}^{\rm max}$ increases. This is due to the fact that the absorbed photon energy is confined in a smaller region of space, thus leading to a higher electron temperature. More quantitatively, one finds an approximate relation $T_{\rm el}^{\rm max} \propto \sqrt{1/d}$ [8]. It is also observed that at around one hundred nm the "bulk limit" is reached, with $T_{\rm el}^{\rm max} \approx 2700$ K. The dependence of $T_{\rm el}^{\rm max}$ on d is shown in Fig. 5.18b.

In Fig. 5.18c, we show the maximum electronic temperature $T_{\rm el}^{\rm max}$ for the bulk material (i.e. $d \to \infty$) as a function of laser fluence, again with 80 fs Gaussian pulses. Here it is found, in excellent approximation, that $T_{\rm el}^{\rm max}$ is proportional to $\sqrt{F}$. This scaling is well-known from experiment and earlier theory [8, 124]. As a consequence, one has for metal films

$$T_{\rm el}^{\rm max} \propto \sqrt{\frac{F}{d}} \quad , \tag{5.27}$$

which shows that both lowering d and increasing F enhance $T_{\rm el}^{\rm max}$. Similar to other systems, the DIMET yield for NO/bulk Pt(111) is known to increase superlinearly with laser fluence, $Y \propto F^n$ where $n = 6 \pm 1$ is found experimentally [123]. Accordingly, one would expect an approximate scaling law

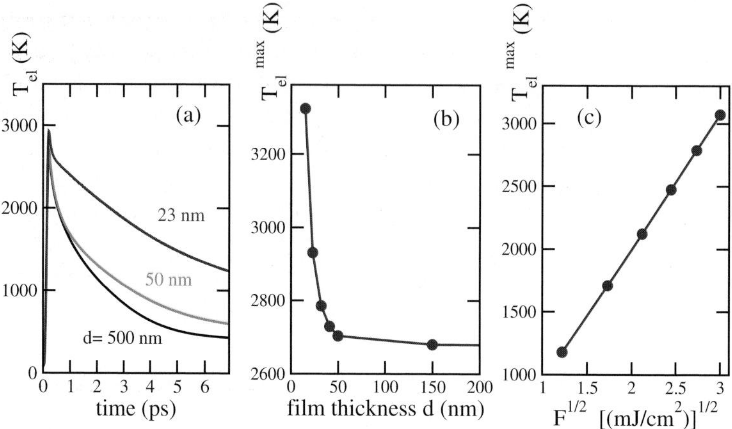

Fig. 5.18. (a) $T_{\rm el}(t)$ curves for a Pt film, 500 nm thick (indistinguishable from the bulk), and two films with $d = 50$ and $d = 23$ nm, respectively. An 80 fs Gaussian laser pulse polarized perpendicular to the substrate with fluence $F = 6$ mJ/cm^2 and $\lambda = 619$ nm was used. (b) Maximum electronic temperature $T_{\rm el}^{\rm max}$ as a function of film thickness d for the same pulse [100]. (c) Maximum electronic temperature $T_{\rm el}^{\rm max}$ for a semiinfinite Pt(111) surface as a function of the square root of the laser fluence $F^{1/2}$, [91].

$$Y \propto \left(\frac{F}{d}\right)^n \tag{5.28}$$

with F and d being independent control parameters.

For femtosecond-laser induced desorption of NO from Pt(111), the non-linear scaling law $Y \propto F^n$ could be reproduced with a two-state open-system density matrix model [91]. The two-state model consists of a Morse-type ground state potential $V_g(Z)$ (with Z = molecule surface distance), and a negative-ion resonance state $V_a(Z)$ with a lifetime $\tau_{\rm el}$ of a few fs [116]. The excited state potential was of the simple model form

$$V_a(Z) = V_g(Z) + \Phi - \mathrm{EA} - \frac{e^2}{4Z} \quad . \tag{5.29}$$

For $Z \to \infty$, $V_a(Z)$ gives an energy difference between ionic and neutral (ground) state of Φ – EA, the difference between the metal work function and the electron affinity of the molecule. Close to the metal surface, the ionic state is stabilized by image charge attraction; see the last term in (5.29). As a consequence, a photoexcited adsorbate moves initially *inward* – this is the so-called Antoniewicz model of photodesorption [18]. The two potential curves are similar to those shown in Fig. 5.15b.

The hot-electron excitation was modeled by (5.20) and (5.21) with $T_{\rm el}(t)$ calculated from the 2TM. The desorption probability and other properties

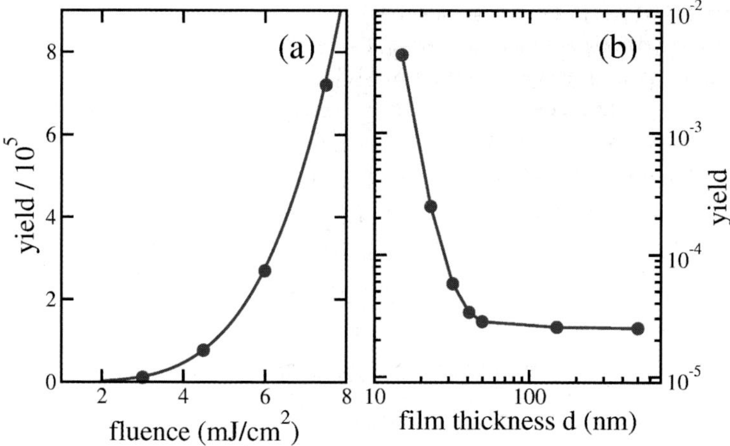

Fig. 5.19. (a) DIMET yield for NO/Pt(111) as a function of laser fluence F, when 80 fs Gaussian pulses with $\lambda = 619$ nm were assumed. Symbols: Calculated from the 2-state density matrix model, solid line: best fit $Y \propto F^{4.4}$ (after [100]). (b) DIMET yield for NO/Pt(111) as a function of film thickness d (80 fs Gaussian pulse with $F = 6$ mJ/cm^2 and $\lambda = 619$ nm), after [91]).

were computed by analyzing the flux going through an asymptotic point Z_{des} on the ground state potential, at which the density can be considered desorbed. For the lifetime, $\tau_{\text{el}} = 2$ fs was assumed.

As demonstrated in Fig. 5.19a, the desorption yield is very small for low fluences, but increases nonlinearly with a scaling exponent of $n \approx 4.4$, in reasonable agreement with the experimental $n = 6 \pm 1$ [123]. For a hot-electron mediated process, n has no simple physical interpretation since the excitation rate $\Gamma^{\text{el}}_{g \to a}$ depends exponentially on $T_{\text{el}}(t)$ according to (5.21), which itself is a nonlinear function of the laser field. For multidimensional extensions of the model, see [118, 119, 125].

The computed scaling of Y with d, on the other hand, is not $Y \propto d^{-4.4}$ as one might naively expect from (5.28). From Fig. 5.19b, we find $Y \approx$ const. for $d > 50$ nm, and $Y \propto d^{-6.6}$ for thinner films. In the thin-film regime, d is obviously a more sensitive control parameter than the fluence F. This can be understood from Fig. 5.18a, where it is evident that thin films not only lead to increasing peak temperatures, but also cause $T_{\text{el}}(t)$ to remain high for longer times. As a consequence, the probability for multiple excitations increases.

To summarize, several concepts to control photochemical reactions at metal surfaces have been outlined. However, due to the indirect, i.e. substrate-mediated excitation mechanism and dissipation, the same selectivity and efficiency as in gas phase examples cannot be expected, and other, at least partially "incoherent" control schemes are necessary.

The last sections have discussed recent experimental and theoretical advances in surface femtochemistry, whereby nonadiabatic coupling between electronic excitations of the substrate and adsorbate nuclear degrees of freedom leads to reactions like desorption or diffusion processes. The adsorbate-substrate coupling is often assumed to be mediated by electron transfer of excited substrate electrons into unoccupied adsorbate resonances. In the following sections, experimental and theoretical studies of the reverse process, the electron injection into the substrate conduction band following intra-adsorbate excitation will be discussed.

5.6 Ultrafast photoinduced heterogeneous electron transfer: Overview

F. Willig, O. V. Prezhdo, and V. May

Electron transfer (ET) is a ubiquitous phenomenon in physics, chemistry, and biology which has attracted continuous interest over the last six decades [126–130]. Heterogeneous electron transfer (HET) plays a key role in catalytic reactions, in electrochemistry, and in photoelectrochemistry. Since more than two decades, there have been considerable efforts toward developing the field of molecular electronics [131,132], where HET will also be of fundamental importance. In this context, current flow through single molecules placed between the tip of an STM and a planar metal electrode has been investigated to approach the field of molecular electronics. In addition, HET has been studied in nano-hybrid systems, mostly with the aim of practical applications [133–138]. Since HET reveals unique properties of the electron transfer process it can be considered also as a research topic in its own rights [139–141].

ET from a molecule to a solid, either a metal or a semiconductor electrode, is conventionally defined as HET. It is distinguished from homogeneous ET, in which the electron hops between two molecules in solution or in a solid matrix. HET becomes ultrafast if the ground state donor orbital of the molecule is lifted sufficiently high above the Fermi level. This can be achieved by setting up a suitable overvoltage at the interface. In the latter case, the electron injected by the molecular donor can be accommodated in a wide continuum of electronic acceptor states of the electrode that might span an energy range of up to 1 eV (wide band limit). It is virtually impossible to elucidate the corresponding dynamics of hot electron injection on the basis of steady state current measurements.

Photoinduced HET (PHET) can occur if a molecule that is adsorbed or in close proximity to an electrode is promoted via photon absorption from its electronic ground state to an electronically excited state (see Fig. 5.20). With suitable energy level matching at the interface, the electron injected from the excited donor orbital of the molecule can again be accommodated in a wide continuum of electronic acceptor states of the electrode. The PHET process can be utilized best if the electrode is a semiconductor with a wide band gap, larger than the photon energy for exciting the molecular donor state. In contrast to HET in the dark, PHET can realize the wide band limit already in the absence of an applied voltage. PHET is in general more complicated at the surface of a metal electrode than at a semiconductor electrode with a wide band gap, since hot electrons are produced at the metal electrode also via photon absorption in the bulk. PHET at the surface of a wide band gap semiconductor is the primary event in the well-known dye-sensitized AgBr photographic process [142] and also in the more recently developed dye-sensitized electrochemical solar cell, where conversion of solar light has been achieved with efficiencies around 10 percent [133].

The study of PHET involves many different fields and concepts: molecular science, surface science, the dynamics of ultrafast reactions, the dynamics of electron transfer from surface to bulk states, and topics from solid state physics like electronic band structure, charge carrier transport, and electron-phonon scattering. PHET can occur as an instantaneous process or with a finite injection time, the first case is direct optical interfacial charge transfer and the second one is charge injection from a local, excited molecular state. In the latter situation, the electron transfer time can range from a few femtoseconds to several picoseconds, or even longer depending, among other parameters, on the distance between the donor orbital of the molecule and the surface atoms of the solid.

PHET has been investigated in the last years with femtosecond spectroscopy methods like transient absorption and two-photon photoemission (2PPE). Other techniques, e.g. fluorescence up-conversion or frequency mixing appear feasible but hitherto have not been used very often for probing PHET. Transient absorption spectroscopy has the virtue that it can address both, the molecular reactant state, i.e. the excited chromophore, and the molecular product state, i.e. the ionized chromophore, for a suitable chromophore like perylene at two separate wavelengths. Transient absorption, however, is not yet sufficiently sensitive to probe a sub-monolayer coverage of dye molecules on a planar single crystal surface. This difficulty has been partially overcome by adsorbing dye molecules in nm cavities that are formed in a several μm-thick layer of TiO_2 colloids [139]. This nm-structured TiO_2 film was developed for the dye–sensitized electrochemical solar cell by Grätzel and coworkers [143] because it has a several hundred times larger surface area than a planar electrode. Since the nm cavities cannot be cleaned with surface science preparation techniques, like sputtering and annealing cycles, and they expose several different crystal planes, it remained unclear for a long time how to compare data for dye molecules adsorbed in nm cavities with data for dye molecules with well-defined adsorption geometry on the surface of single crystals. Recently, injection times were measured with the same dyes on these two different surfaces [144, 145] in ultrahigh vacuum (UHV). UHV conditions prevent the deterioration of the system via slow side reactions that are known to occur in

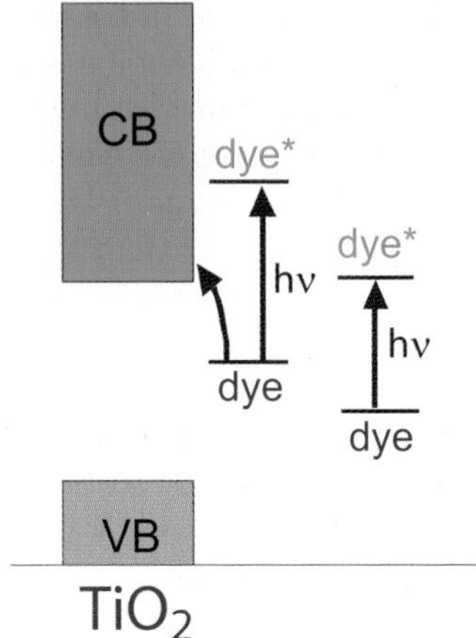

Fig. 5.20. Dye-sensitized TiO_2. Energy levels of the chromophore-semiconductor system. The ground state of the chromophore is placed inside the gap between the valence band (VB) and the conduction band (CB) and below the Fermi level. The excited state is in resonance with the CB. Upon photoexcitation, the chromophore transfers an electron into the semiconductor. The electron then relaxes to the edge of the CB. Photoexcitation directly into the CB becomes possible with strong chromophore-semiconductor coupling.

the excited state of most organic chromophores in the presence of H_2O and O_2. The injection times measured with the two different surfaces showed good agreement, except for the case of long rod-shaped rigid molecules. Differences for the latter molecules were attributed to adsorption at edge and corner sites that are only available in the nm cavities.

The easiest preparation of chromophores on an electrode is simply evaporating the molecules onto the surface. This can lead to direct bond formation between the chromophore and the electrode, e.g. via -O- (or -S-) bonds to metal atoms. Correspondingly, these adsorption systems show the most complicated and strongest electronic interaction. Such systems have been studied recently with DFT-slab calculations [146]. They can give rise to direct photoinduced interfacial charge transfer transitions, where the electron is lifted from the molecular ground state directly to electronic acceptor states of the semiconductor, e.g. in the case of catechol on TiO_2 [147].

The strength of the electronic interaction can be varied systematically in a PHET system by inserting different suitable bridge–anchor groups between the organic chromophore and the surface of the semiconductor. This strategy requires demanding synthetic chemistry because the different bridge–anchor groups have to be bonded covalently to the chromophore. Such complicated organic molecules cannot be evaporated without fragmentation and thus have to be adsorbed from solution. To this end, a specific UHV chamber was developed that allows for an easy switch from a solvent or gas phase environment to UHV conditions [139]. In this way, complex molecules can be adsorbed from solution and subsequently the sample brought back to UHV so that all the tools of surface science are available to be utilized for sample characterization and carrying out fs 2PPE measurements. Depending on whether the bridge group consists of saturated or conjugated bonds, the bridge will function as an electronic tunneling barrier or an conducting wire, respectively. Corresponding systematic studies with the chromophore perylene [144, 145] will be described in the next section.

Two different theoretical approaches have been adopted in the last years to explain the trend in the experimental results obtained with different bridge–anchor groups. One is based on the solution of the time-dependent Schrödinger equation for a reduced dimensionality model of the whole dye-semiconductor system [148]. The phenomenological parameters were obtained from a fit to linear absorption curves of the systems that display ultrafast PHET [148, 149]. The other approach uses a fully dimensional atomistic description of the systems and applies time-dependent DFT in order to model the injection dynamics [150].

It is worth noting that in the wide band limit PHET shows a mono-exponential decay of the donor state irrespective of the strength of the electronic interaction (as in the Fermi's Golden Rule perturbation treatment for weak interaction). This is the case as long as the donor state can deliver the electron to a continuum of electronic acceptor states that spans the energy range of all the Franck-Condon factors (wide band limit). The initial energy distribution of the injected electron is controlled by these Franck-Condon factors irrespective of whether the electron transfer time is much shorter or much longer than the oscillation period of the vibrational modes.

Electron transfer times for PHET in the experimentally studied systems have also been predicted with DFT cluster calculations [151]. After determining the equilibrium adsorption geometry, the combined density of states was calculated for the adsorbate system. Assuming a Lorentzian lineshape, the width of the chromophore LUMO in the adsorbed state can be interpreted with the Newns-Muscat model [152] as the lifetime of the corresponding LUMO state. Even though the LUMO is not identical to the excited state of the chromophore, the corresponding lifetimes agree very well with the trend in the experimental data for different bridge–anchor groups and show also reasonable agreement with the absolute values of the measured injection times. The electron transfer time for PHET from the same chromophore perylene has been increased from less than 10 fs to more than 1 ps by changing the bridge–anchor group. It should be kept in mind that for a very long saturated bridge group, providing for a large tunneling barrier, the PHET dynamics should reach the well-known thermally activated case that is described by the Marcus theory and relevant semi-classical treatments (see, for example, [130]).

The above mentioned experimental data were all collected in UHV. Therefore, a solvent environment is not expected to change the dynamics of ultrafast

PHET, but it can shift the energy levels at the interface and can increase the reorganization energy, thereby changing the injection time. Unfortunately, it can enhance the probability of side reactions. For the Ru dyes with bipyridyl ligands that are employed in the dye-sensitized electrochemical solar cell, the fastest injection time has been reported as $\approx$50 fs [153]. This is still 5 times slower than the 10 fs electron transfer for the perylene chromophore with the identical anchor group formic acid [144]. So far, it is not clear whether the longer injection time of the Ru dye is due to a closer position of the donor orbital to the lower edge of the conduction band compared to the perylene chromophore, or to another, yet unknown effect.

Isolated chromophores adsorbed to semiconductors can be treated with DFT-cluster calculations but not so easily with DFT-slab calculations. Several groups have carried out the former calculations in recent years, in particular for dye molecules adsorbed on various surfaces of TiO_2 [150, 151, 154, 155]. After calculating the adsorption geometries of the ground state molecules, they have simulated the dynamics of PHET by applying very demanding time-dependent DFT calculations to the electronic degrees of freedom, combined with classical calculations for the phonon part of the system [150, 155]. These time-dependent DFT calculations have also given insight into the time scale for the release of the electrons from the TiO_2 surface to electronic bulk states. The predicted escape scenario starting with a few fs time constant and turning into a decay with a 100 fs time constant is in good qualitative agreement with recent experimental results obtained for the same systems with fs 2PPE [145]. Further theoretical work can help in understanding the experimental femtosecond 2PPE signals for PHET.

Time-dependent DFT model calculations of ultrafast PHET by Prezhdo et al. have shown the influence of vibrations and phonons on the injection time, even when the latter is much shorter than the oscillation period of the vibrational modes [156]. This effect can be attributed to advantageous level crossing brought about by the nuclear motion when the injection occurs close to the conduction band edge. Some fingerprints of nuclear motion have been observed in PHET signals [157], but the latter data do not provide sufficient experimental evidence for the theoretical prediction of a decisive influence of nuclear motion on the injection time. More experimental data have to be collected to clarify the influence of nuclear motions on PHET, in particular, in situations where the molecular donor orbital comes close to the lower edge of the conduction band. Another pending question concerns the relevance of direct optical charge transfer, as compared to PHET, from the excited state of the chromophore. To answer this question, one has to investigate suitably synthesized molecules where both the different injection mechanisms can be directly compared in the same molecule.

The following section will explain how a general theory for PHET can be applied to elucidate systematic trends in measured injection times and absorption spectra. In the final section of the chapter, the real-time ab initio

DFT approach to PHET is applied providing atomistic details of the injection dynamics.

5.7 Ultrafast photoinduced electron transfer from anchored molecules into semiconductors

L. Wang, V. May, R. Ernstorfer, L. Gundlach, and F. Willig

Perylene attached to TiO_2 will be of particular interest in the following. Since the first excited singlet state of perylene is energetically positioned about 1 eV above the conduction band edge of TiO_2 thus realizing a mid-band charge injection situation (see Fig. 5.20), this system is well suited for a systematic study of ultrafast PHET. Introducing different bridge–anchor groups the transfer coupling initiating HET can be tuned from a strong coupling situation (with charge injection times of 10 fs) down to weaker coupling strengths (with charge injection times of up to 1 ps [158]).

There have been earlier attempts of time-resolving HET already with picosecond laser pulses, e.g. in order to elucidate the primary step in the well-known process of AgBr photography [142]. With this time-resolution, however, reliable experimental data could only be obtained for those HET reactions where a thermal activation step slowed down HET by several orders of magnitude [159]. PHET without thermal activation could be time-resolved after laser pulses became available with a few femtoseconds duration and tunable wavelengths. Electron transfer times can be measured in an unambiguous way with transient absorption signals probing the decay of the reactant as well as the rise of the product state. Until now, transient absorption is not sensitive enough to probe HET with femtosecond resolution for dye molecules adsorbed at sub-monolayer coverage on the surface of single crystals.

This sensitivity problem of HET measurements has been overcome initially by probing dye molecules that were adsorbed on the inner surface of nm-structured colloidal layers of anatase TiO_2. Light passing through such a nm-structured layer of a few μm thickness probes a hundred times more dye molecules than on the planar surface of a single crystal. This surface enhancement effect has been utilized by Grätzel for the dye-sensitized solar cell [133]. Employing transient absorption and also luminescence up-conversion, HET was measured in such a nm-structured layer for the first time with femtosecond resolution under UHV conditions between room temperature and 20 K [139].

For basic studies of HET, such layers have the complication that the individual colloidal particles contribute with different crystal surfaces to the nm cavities which make up the inner surface and thus give rise to different adsorption sites [143]. In addition, the nm cavities offer a high percentage of adsorption sites at edges and corners. Moreover, a considerable degree of uncertainty has remained concerning the chemical environment and the geometrical orientation of dye molecules adsorbed in these nm cavities. The preparation procedure involves a heating period in the presence of oxygen to burn away organic contaminants in the cavities prior to dye adsorption. Since these cavities are hidden inside the layer they cannot be cleaned with the preparation techniques of surface science, e.g. sputtering and annealing cycles. In view of these complications, it was not too surprising that HET data obtained with nm-structured layers have shown a considerable spread in the values for electron injection times, when the same dye molecules were measured by different laboratories and even when the same nm-structured system was prepared by different people in the same laboratory [144]. Hence, the criterion seemed reasonable to trust the specific preparation procedure that reproducibly resulted in the shortest injection time for the same dye molecules. To prevent photoinduced side reactions and further contamination, the measurements were always carried out with the samples mounted in UHV [139, 140, 144].

The long standing question concerning the reliability of injection times measured for dye molecules adsorbed in nm cavities of anatase TiO_2 layers has been clarified only very recently by measuring the injection times for the same molecules on the surface of rutile $TiO_2(110)$ single crystals employing fs 2PPE [145]. The single crystal surfaces were prepared with the established techniques of surface science and characterized with LEED, UPS, XPS, before adsorbing the chromophores equipped with specific anchor groups from solution in a specially designed UHV chamber. Such a chamber allowed for an easy switch from UHV conditions to gaseous or liquid environments for the samples [144]. Recent experiments have shown that HET is not influenced by the specific choice of the solvent from which the organic molecules are adsorbed onto the single crystal surface, at least if HET is measured in UHV [160]. The fs 2PPE measurements offer superior sensitivity compared to transient absorption measurements and can probe adsorbates with femtosecond resolution at sub-monolayer coverage on the planar surface of a single crystal. Angle and polarization dependence of 2PPE signals have revealed the orientation of chromophores like perylene with respect to the crystal surface [161].

Interpretation of the time-resolved 2PPE signals, however, is more involved than that of transient absorption signals, provided the latter can address different stages of HET at different wavelengths, as is the case for a chromophore like perylene [140]. In the time-resolved 2PPE signals, one has to separate contributions originating from the excited state of the adsorbed dye molecules and those from the injected electrons. It is important to note that virtually identical injection times have been found with 2PPE for the same organic molecules on the surface of rutile single crystals [145] as had been measured before with the same molecules adsorbed in nm cavities of the anatase TiO_2 layers [144]. This agreement between the two sets of data collected in the two different experimental systems holds true as long as the molecules are anchored on the respective surface with short bridge–anchor groups. Edge and corner sites in the nm cavities lead to severe complications when HET is to be studied with long bridge–anchor groups that can function as tunneling barriers. Instead of HET along the direction of the bridge–anchor group, in nm cavities electron transfer might occur to the nearest TiO_2 surface at an edge or corner adsorption site. Such complications can be avoided on the surface of single crystals where 2PPE measurements can be employed. Considering all possible complications, a set of reliable injection times has now been collected under UHV conditions for HET of the perylene chromophore with different bridge–anchor groups. These experimental data form a solid basis for theoretical model calculations and mechanistic scenarios.

The following subsections will summarize theoretical [148, 149, 162–168] and experimental [139, 140, 144, 145, 158, 160] studies on HET with particular emphasis on the perylene–TiO_2 system. To account for optical excitation, subsequent electron-vibrational quantum dynamics, the formation of an electron distribution in the semiconductor band continuum, its detection by 2PPE spectroscopy, and for fs laser pulse control a model of reduced dimensionality has to be applied. This reduction concerns the electronic levels involved as well as the number of intramolecular vibrational coordinates. Similar computations have been carried out in [169, 170] but with a more involved account of the vibrational dynamics, however, at the same time without any detailed consideration of the state of the electron injected into the conduction band continuum. Such considerations have been in the focus of DFT-based electronic structure calculations of dye-sensitized semiconductor surfaces as, for example, in [156, 171–174]. In particular, a combination with molecular dynamics simulations has been presented in [156,172,173]. So far, these sophisticated computations could not be developed such that data from femtosecond optical experiments might be fitted.

In the following, the experimental systems as well as the HET model are introduced in Sect. 5.7.1 with specifications for the simple charge transfer model used for the perylene–TiO_2 system. The way to simulate femtosecond charge injection dynamics is explained in Sect. 5.7.2 and is illustrated with some experimental results valid for perylene compounds anchored on the surface of TiO_2. Preliminary results on 2PPE spectra are presented in Sect. 5.7.2.2. In Sect. 5.7.2.3, the computation of the linear absorption coefficient is demonstrated and applied to experimental data for parameter adjustment. Finally, the discussion includes some speculations on the role of vibrations in special injection scenarios and on laser pulse control of HET.

5.7.1 Experimental systems and theoretical model

For complicated systems as considered here the establishment of systematic changes and trends in experimental data can probably give better insight into the reaction mechanism than the exact fit to a few signals that are collected for only one or two specific experimental systems. In the case of large molecules, an exact fit requires the adjustment of a large number of parameter values.

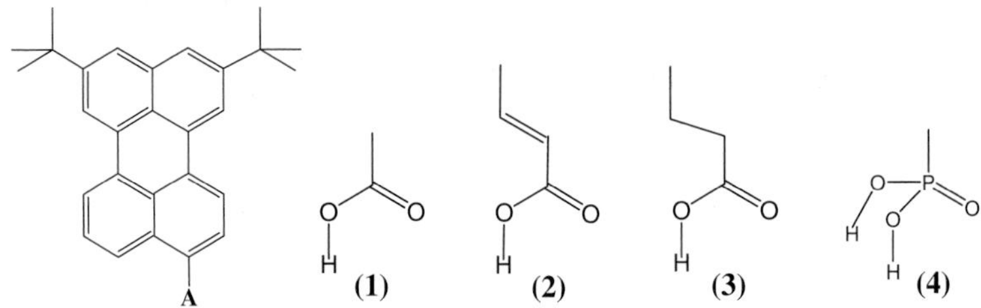

Fig. 5.21. Core chromophore DTB–Pe–A (di–tertiary–butyl perylene, left) with bridge–anchor groups (right) in position A as studied in [148]. **1**: –COOH (carboxylic acid), **2**: $-(CH)_2$–COOH (acrylic acid), **3** $-(CH_2)_2$–COOH (propionic acid), and **4**: $-P(O)(OH)_2$ (phosphonic acid).

To establish the desired systematic trend, several different molecules were synthesized using the same chromophore perylene but attaching different bridge–anchor groups. Anchor groups were chosen such that they can form stable chemical bonds on the crystal surface that persist well above room temperature. The bridge–anchor groups determine the distance and orientation of the chromophore with respect to the surface of the semiconductor. Depending on whether the bridge–anchor groups contain saturated or conjugated bonds they can function either as electronic tunneling barriers or as conducting wires. To a first approximation, the main effect of different bridge–anchor groups is the change in the electronic coupling strength for interfacial electron transfer. In Fig. 5.21, a selection of the used bridge–anchor groups is shown together with the perylene chromophore. In these synthesized molecules, the perylene core carries two additional bulky side groups that are not part of the actual chromophore. The side groups prevent close contact of neighboring perylene chromophores when adsorbed on the surface and thus prevent complications that can arise from the formation of perylene dimers.

The properties that are relevant for interfacial electron transfer were determined by applying different measuring techniques to the adsorbed molecule-semiconductor system. The HOMO energy level of the chromophore perylene attached to the surface of TiO_2 has been determined from UPS data with respect to the band edges of TiO_2 [145]. The position of the excited singlet state, which functions as the electron donor in the system, has been deduced from 2PPE data [145]. A typical line-up of the energy levels at the interface as derived from such measurements is illustrated in Fig. 5.22 for perylene with the tripod bridge–anchor group [158].

Polarization and angle resolved 2PPE data have revealed the orientation and alignment of the chromophore perylene when the molecules of Fig. 5.21 were anchored on the (110) surface of a TiO_2 rutile single crystal electrode [145]. In the case of a rigid rod-like bridge–anchor group, distance and orientation of the chromophore with respect to the crystal surface is determined by the bonds formed between the anchor group and the surface atoms of TiO_2.

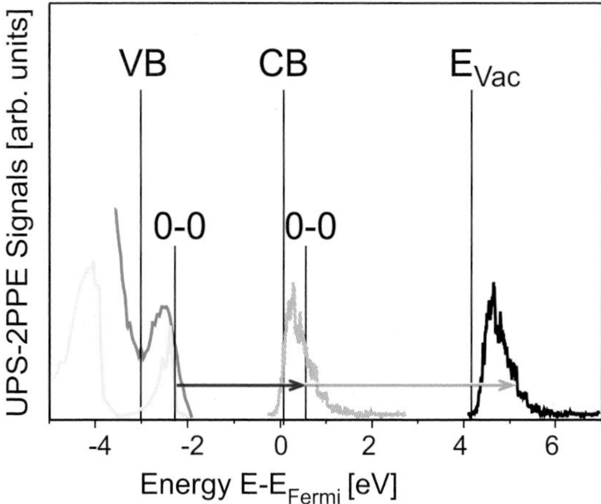

Fig. 5.22. Alignment of the ground state and excited state of perylene with the tripod bridge–anchor group on the (110) TiO_2 surface [158]. The lower curve at the left shows the HOMO–1 and HOMO peaks obtained from an UPS difference signal. The signal above the vacuum level E_{vac} is measured by 2PPE. The curve above the conduction band edge CB represents the projected density of injected electrons. The two 0–0 markers indicate the energy of the electronic ground and first excited singlet state of the perylene chromophore, respectively.

Adsorption geometries deduced from 2PPE measurements [145] agreed well with those predicted by DFT calculations [175].

Standard ET is described as the transition from a single donor state into a single acceptor state, with the donor and acceptor state potential energy surface (PES) written as $E_D + U_D(Q)$ and $E_A + U_A(Q)$, respectively.[3] As it is well known, the electronic coupling strength V_{DA} and the relative position of the donor with respect to the acceptor PES (leading to a fixation of the driving force and the reorganization energy) are crucial for the exact type of ET (adiabatic or nonadiabatic ET, as well as ET of the normal, activationless or inverted region, see, e.g. [129, 130]).

In contrast to this standard picture of ET, the peculiarity of HET from a surface attached molecule into a semiconductor lies in the presence of a continuum of acceptor states formed by the conduction band. This is shown in Fig. 5.23, where the description is reduced to the presence of the electronic ground state and a single excited state of the molecule with PES $E_g + U_g(Q)$ and $E_e + U_e(Q)$, respectively. The semiconductor has been described by a single conduction band. The electron in the excited state of the molecule may be transferred into the semiconductor since its empty conduction band is

[3] Note that the energy of the PES at the equilibrium position of the vibrational coordinate has been separated into E_D and E_A. Q denotes the set of vibrational coordinates.

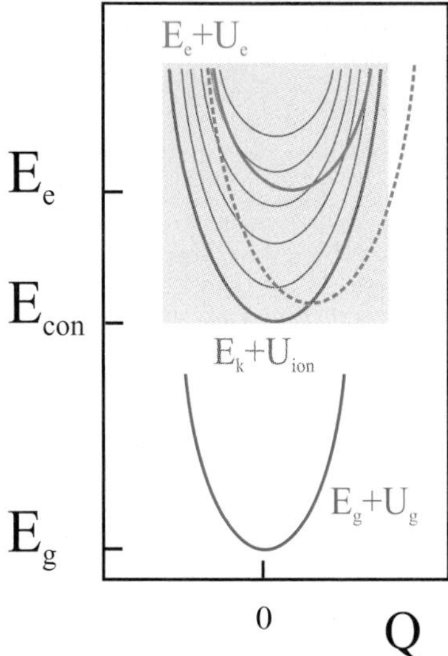

Fig. 5.23. PES of the molecule-semiconductor system $E_a + U_a$ versus a single vibrational coordinate Q (the vertical position of the PES is determined by the electronic energy E_a, here without the inclusion of the vibrational zero-point energy). $a = g$ and $a = e$ correspond to the ground and excited state of the molecule, respectively, whereas $a = \mathbf{k}$ characterizes the semiconductor conduction band states. The gray box indicates the continuum of band states starting at the lower band edge with energy E_{con}. The position of the excited state PES, i.e. $E_e + U_e$, drawn by a full line is typical for perylene on TiO_2, while the PES drawn by a dashed curve resembles a near band edge position.

degenerate with the excited molecular level. Thus, the accepting levels of the ET are formed by the continuous band structure $E_{\mathbf{k}}$ of the semiconductor, where $\mathbf{k}$ denotes the quasi-momentum (if a nano-cluster is considered, $\mathbf{k}$ has to be replaced by other quantum numbers). At the same time the vibrational motion of the ionized molecule is determined by its cationic PES denoted as U_{ion} (see Fig. 5.23). Therefore, also the complete PES $E_{\mathbf{k}} + U_{\text{ion}}$ forms a continuum, as indicated by the shaded area in Fig. 5.23.

As already indicated for standard ET, the position of the donor with respect to the acceptor PES fixes the type of ET proceeding either as normal ET, activationless ET, or ET of the inverted region. In the present case of HET, however, all types may appear simultaneously. This is particularly the case if the injecting level is in a mid-band position (far away from the edges of the semiconductor conduction band, full line in Fig. 5.23). Now, the donor PES has arbitrary crossing points with the multitude of acceptor PES, i.e. ET occurs at every part of the donor PES.

If the ET is ultrafast as it is the case for perylene with a short bridge–anchor attached to TiO_2 one has to consider ET in terms of vibrational wave packets. The fs photoexcitation results in a vibrational wave packet in the PES U_e moving forth and back. Simultaneously, at every step of this motion a transfer to the acceptor PES is possible. Therefore, the decay of the overall donor population denoted here as P_e will be smooth and structureless. But P_e should show superimposed oscillations, if the injection level is in a near-band edge position (dashed curve in Fig. 5.23). This results from the fact that the donor PES crosses only a part of the multitude of acceptor PES.

Let us now return to HET with the molecular injection level in a rather mid-band position. For this case, Fig. 5.24 shows the excited molecule PES and some PES of the continuum of product states. For every PES, the energetic positions of the vibrational eigenstates are also shown (here levels corresponding to a single coordinate). The vibrational levels of the PES $E_{\mathbf{k}} + U_{\text{ion}}$ of the band continuum (see Fig. 5.23) are degenerate with the levels of the molecular PES $E_e + U_e$. Thus, the scheme indicates that completely resonant transitions are possible from a certain excited molecular electron-vibrational level into different vibrational levels of the molecular cation, with the electron in the respective band states.

These transitions are the only ones which remain if the transfer coupling is weak and the Golden Rule of quantum mechanics suffices to describe the transition. Consequently, electronic band states of the semiconductor are populated around the injection energy E_e shifted by multiples of the vibrational quanta. For stronger transfer coupling, however, also transitions into states which are not completely degenerate with the excited molecular level are possible. Nevertheless, one again may expect structures around E_e reflecting the vibrational progression of the molecule but, this time, with a broadening which also reflects the strength of transfer coupling.

While so far the discussion has concentrated on the temporal evolution of HET, the effect of HET on steady state properties like the linear absorbance will be considered next. Neglecting vibrational contributions, one arrives first at the scheme of a single excited molecular level (with energy E_e) coupled to the continuum of conduction band levels (with energy $E_{\mathbf{k}}$). Accordingly, a shift and a broadening of the molecular absorbance are expected. Taking the vibrational degrees of freedom into account, a respective vibrational progression seen at the isolated molecule may survive the attachment to the semiconductor surface. In particular, this depends on the coupling strength to the band continuum but also on the relative positions of the two PES (the reorganization energy of the charge injection). The qualitative discussions will be substantiated next by respective quantitative considerations based on a uniform theory of ultrafast PHET.

The molecule-semiconductor system discussed in the following is based on a diabatic-state like description distinguishing between molecular and semi-

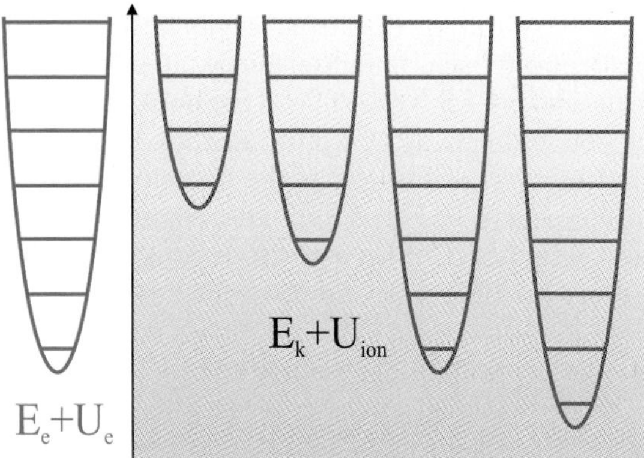

Fig. 5.24. PES $E_e + U_e$ of the excited molecular state (left) and of some selected PES $E_{\mathbf{k}} + U_{\text{ion}}$ corresponding to the molecular cation and the injected electron at a particular band state (right part, the relative horizontal shifts have been introduced for clarity). All PES have been drawn together with the respective level positions of vibrational eigenstates. The PES at the right act as acceptor levels and are simultaneously addressed in the transfer process.

conductor states.[4] Moreover, a description is introduced which concentrates on the state of a single electron to be transferred. In addition, a common description of the semiconductor states is applied. This results in an identification of the states before charge injection by φ_a $(a = g, e)$ and afterwards by $\varphi_{\mathbf{k}}$. The respective electronic energies are $\hbar\varepsilon_a$, here with a also including the quasi-wave vector $\mathbf{k}$ as an electronic quantum number. It is not required to distinguish whether the $\varphi_{\mathbf{k}}$ are bulk or surface states since we use for actual computations the density of states (DOS)

$$\mathcal{N}(\Omega) = \sum_{\mathbf{k}} \delta(\Omega - \omega_{\mathbf{k}}) \, . \tag{5.30}$$

Here and in the following, $\hbar\Omega$ labels the semiconductor band energy. Choosing a particular form of $\mathcal{N}(\Omega)$ it should cover all semiconductor band states in the vicinity of the molecular injection level. The use of an averaged DOS $\bar{\mathcal{N}} = N_{\Delta\Omega}/\Delta\Omega$ with the level number $N_{\Delta\Omega}$ in the energy interval $\Delta\Omega$ should be sufficient (wide-band approximation).

According to what has been discussed before the Hamiltonian reads

$$H_{\text{mol-sem}} = \sum_{a=g,e,\mathbf{k}} (\hbar\varepsilon_a + H_a)|\varphi_a\rangle\langle\varphi_a| + \sum_{\mathbf{k}} \big(V_{\mathbf{k}e}\,|\varphi_{\mathbf{k}}\rangle\langle\varphi_e| + \text{h.c.}\big) \, . \tag{5.31}$$

[4] When ab initio calculations are carried out, diabatic states are not directly obtained and a diabatization procedure becomes necessary. It results in separate molecular and semiconductor states as used in [148, 149, 163–167].

For the following, it is useful to write the band energies $\hbar\varepsilon_{\mathbf{k}}$ as $\hbar\varepsilon_{\mathrm{con}} + \hbar\omega_{\mathbf{k}}$, with the lower band edge $\hbar\varepsilon_{\mathrm{con}}$ and $\hbar\omega_{\mathbf{k}}$ running over the conduction band with width $\hbar\Delta\omega_{\mathrm{con}}$. Again, all levels are understood as given by the minima of the related PES plus the zero-point vibrational energy. H_g and H_e denote the vibrational Hamiltonian (with a spectrum starting at zero energy) for the electronic ground state and the first excited state of the molecule, respectively, and H_{ion} is the respective Hamiltonian of the ionized molecule if charge injection into the conduction band took place. The related eigenvalues and vibrational wave functions (with vibrational quantum numbers M) are denoted as $\hbar\omega_{aM}$ and χ_{aM}, respectively. Charge injection from φ_e into the manifold of states $\varphi_{\mathbf{k}}$ is realized by the transfer coupling $V_{\mathbf{k}e}$ whose $\mathbf{k}$ dependence will be later replaced by a frequency dependence leading to $V_e(\Omega)$; (within the wide-band approximation it can be substituted by the frequency independent, averaged quantity $\bar{V}_e$).

The coupling to the radiation field is considered by the standard expression

$$H_{\mathrm{field}} = -\mathbf{E}(t)\hat{\mu} \,. \tag{5.32}$$

The electric field strength is denoted by $\mathbf{E}$, and $\hat{\mu}$ is the transition dipole operator which may account for an exclusive excitation of the molecule. But also direct transitions from the molecular ground state into the semiconductor band states as well as photoionization transitions into the semiconductor vacuum states can be included. If it is assumed that optical excitation exclusively takes place in the molecule between the ground and the excited state, the dipole operator introduced in (5.32) reads

$$\hat{\mu} = \mathbf{d}_{eg}|\varphi_e\rangle\langle\varphi_g| + \mathrm{h.c.} \,, \tag{5.33}$$

with the transition-dipole matrix element denoted as $\mathbf{d}_{eg}$.

5.7.2 Experimental data on electron transfer dynamics analyzed with the theoretical model

5.7.2.1 Electron injection times

Measured electron injection times for the perylene/TiO_2 system were analyzed making use of the electron injection model described in Sect. 5.7.1. Since the photoinduced dynamics are considered in a 100 fs time window, it is reasonable to neglect any relaxation effect. Therefore, it is sufficient to propagate the time-dependent Schrödinger equation related to the Hamiltonian introduced in (5.31) together with the field part (5.32). Its solution is carried out by using an expansion with respect to the diabatic electron-vibrational states $\varphi_a\,\chi_{aM}$, which reads ($a = g, e, \mathbf{k}$)

$$|\Psi(t)\rangle = \sum_{a,M} C_{aM}(t)|\varphi_a\rangle|\chi_{aM}\rangle \,. \tag{5.34}$$

The given state expansion is fairly standard except for the presence of the band continuum leading to a continuous set $C_{\mathbf{k}M}(t)$ of expansion coefficients. This problem will be tackled as described in [148, 149, 165–167]. The **k** dependence of the $C_{\mathbf{k}M}(t)$ is replaced by a frequency dependence leading to the quantities $C_M(\Omega;t)$. These $C_M(\Omega;t)$ in turn will be expanded by the functions $u_r(\Omega)$ forming an orthogonal set. The latter is complete over the energy range of the conduction band, in the present case characterized by the frequency interval $[0, \Delta\omega_{\text{con}}]$ (from the lower to the upper conduction band edge).

Once the $C_{aM}(t)$ are determined, different observables can be computed. The populations of the molecular states follow as $(a = g, e)$

$$P_a = \sum_M |C_{aM}(t)|^2 . \tag{5.35}$$

That of the ionized state may be obtained from

$$P_{\text{ion}}(t) = \sum_{\mathbf{k},M} |C_{\mathbf{k}M}(t)|^2 \equiv \int d\Omega\, \mathcal{N}(\Omega) P_{\text{el}}(\Omega;t) . \tag{5.36}$$

Here, the electron distribution $P_{\text{el}}(\Omega;t) = \sum_M |C_M(\Omega;t)|^2$ versus the band energies was introduced.

In order to characterize the ultrafast charge injection process, we present the solution of the time-dependent Schrödinger equations starting at the vibrational ground state χ_{g0} of the electronic ground state and including a laser field of 10 fs duration (FWHM). Moreover, the wide-band approximation has been used. Resulting charge injection dynamics related to two of the four bridge–anchor groups shown in Fig. 5.21 (see also Table 5.2) are displayed in Fig. 5.25. In the strong coupling case, the excited state population P_e follows the laser pulse envelope accompanied by a direct charge transfer into the conduction band continuum. The respective overall band population is identical with the population of the ionized molecular state P_{ion}. One may consider the laser pulse excitation as a direct population of the semiconductor states. In the other case of a weaker coupling, the excited state population starts to decay into the band continuum when the laser pulse excitation is over, indicating the separation of excited state preparation and charge injection.

Decay into the band continuum

The results of the foregoing paragraphs are complemented by calculations referring to the decay of the population $P_{eM}(t)$ of an excited molecular electron-vibrational state upon charge injection (starting at $t = 0$)

$$P_{eM}(t) = |\langle \varphi_e \chi_{eM} | e^{-iH_{\text{mol-sem}}t/\hbar} | \chi_{eM} \varphi_e \rangle|^2 \equiv \left| \int \frac{d\omega}{2\pi} e^{-i\omega t} G_{eM,eM}(\omega) \right|^2 . \tag{5.37}$$

Detailed considerations of the Green's function $G_{eM,eM}(\omega)$ can be found in [168]. As a main ingredient of these computations the self-energy due to the coupling of the excited molecular level to the band continuum appears

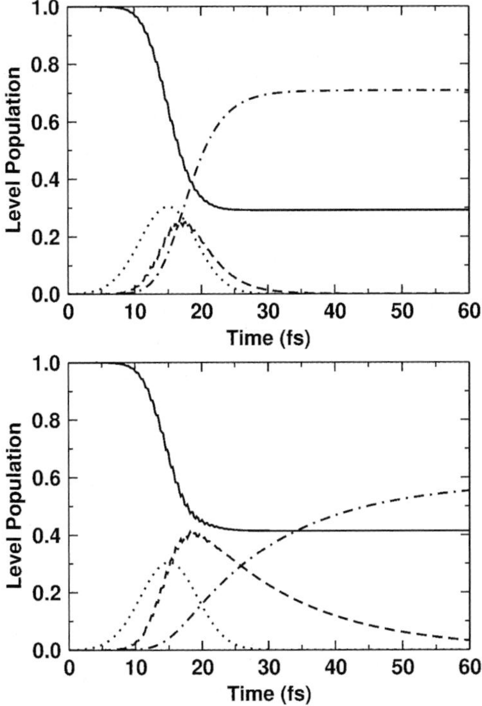

Fig. 5.25. Electronic level population after a 10 fs (FWHM) laser pulse excitation. (top) DTB–Pe–COOH system (for parameters see Table 5.2), (bottom) DTB–Pe–$(CH_2)_2$–COOH system (for parameters see Table 5.2). Solid lines: ground state population P_g of the molecule, dashed lines: excited state population P_e of the molecule, dashed-dotted line: population P_{ion} of the ionized molecular state (what equals the total conduction band population), dotted lines: shape of the laser pulse envelope (in arbitrary units).

$$\Sigma(\omega) = \frac{1}{\hbar^2} \sum_{\mathbf{k}} \frac{|V_{\mathbf{k}e}|^2}{\omega - \varepsilon_{\mathbf{k}} + i\epsilon} \,. \tag{5.38}$$

Using the DOS and (5.30), and changing from $V_{\mathbf{k}e}$ to $V_e(\Omega)$, the imaginary part of $\Sigma(\omega)$ is denoted as

$$-\mathrm{Im}\Sigma(\omega) = \Gamma(\omega) = \frac{\pi}{\hbar^2}\mathcal{N}(\omega)|V_e(\omega)|^2 \,. \tag{5.39}$$

In the general case, an analytical expression for $P_{eM}(t)$, see (5.37), is hardly obtainable. Applying, however, the wide-band approximation (where the frequency dependence of the self-energy is neglected) the standard expression

$$P_{eM}(t) = e^{-k_{\mathrm{HET}}t} \tag{5.40}$$

follows with the rate of HET obtained as

$$k_{\mathrm{HET}} = 2\bar{\Gamma} = \frac{2\pi}{\hbar^2}\bar{\mathcal{N}}|\bar{V}_e|^2 , \tag{5.41}$$

$\bar{\mathcal{N}}$ and $\bar{V}_e$ denote the mean (frequency averaged) DOS and the mean transfer coupling, respectively. Note that this wide-band approximation suppresses any vibrational contributions.

Two-photon photoemission spectra

Transient spectra may be related to optical transitions from the excited molecular state or from the ground state of the cation. Moreover, the transition is of interest addressing the other product state, driven by a fixed, higher photon energy that lifts the injected electrons from the different states at the surface of the semiconductor. Here, quasi free electron states above the vacuum level are excited and the corresponding kinetic energy distribution of the emitted electrons is measured. This type of pump-probe measurement is known as the 2PPE process.

In the following, a preliminary description based on the injection dynamics as studied above will be given. This includes a complete account of the laser pulse with field strength $\mathbf{E}_1$ initiating charge injection in the time-dependent Schrödinger equation. In contrast, the photoemission caused by the second laser pulse with field strength $\mathbf{E}_2$ will be described in perturbation theory. Accordingly, the state of the system corresponding to the action of $\mathbf{E}_2$ can be written as (see, for example [130]):

$$|\Psi^{(1)}(t)\rangle = \frac{i}{\hbar}\int_{t_0}^{t} d\bar{t}U(t,\bar{t};\mathbf{E}_1)\hat{\mu}\mathbf{E}_2(\bar{t})|\Psi(\bar{t};\mathbf{E}_1)\rangle . \tag{5.42}$$

For the following, nonoverlapping pulses are assumed, i.e. U can be replaced by the field-independent expression $\exp(-iH_{\mathrm{mol-sem}}(t-\bar{t})/\hbar)$. Moreover, direct molecular contributions are neglected and $\Psi^{(1)}$ is expanded with respect to the states $\phi_\kappa^{(-)}$ characterizing the freely moving electron (see Sect. 5.7.1). It follows

$$\chi_\kappa(t) = \frac{i}{\hbar}\int_{t_0}^{t} d\bar{t}e^{i\varepsilon_\kappa(t-\bar{t})}\mathbf{E}_2(\bar{t})e^{iH_{\mathrm{ion}}(t-\bar{t})/\hbar}\langle\phi_\kappa^{(-)}|\hat{\mu}|\Psi(\bar{t};\mathbf{E}_1)\rangle , \tag{5.43}$$

what describes the distribution of the emitted electron versus the quasi-free states $\phi_\kappa^{(-)}$ (with energies $\hbar\varepsilon_\kappa$) and the vibrational state of the molecular cation.

A detailed description of the charge injected state Ψ can be achieved with the tight-binding model, i.e. using the atomic orbital basis for the semiconductor part. As a first estimate, the states $\varphi_{\mathbf{k}}$ are used and an impulse excitation is assumed ($\mathbf{E}_2(\bar{t}) = \mathbf{E}_2(t_2)\tau_2\delta(\bar{t}-t_2)$ with the pulse duration τ_2)

$$\chi_\kappa(t) = \frac{i}{\hbar}\Theta(t-t_2)e^{i\varepsilon_\kappa(t-t_2)}\mathbf{E}_2(t_2)\tau_2\sum_{\mathbf{k}M}\langle\phi_\kappa^{(-)}|\hat{\mu}e^{i\omega_{\text{ion}M}(t-t_2)}|\varphi_\mathbf{k}\rangle C_{\mathbf{k}M}(t_2)\chi_{\text{ion}M}\ . \tag{5.44}$$

If furthermore simple plane waves for the description of the emitted electron are used and $\langle\phi_\kappa^{(-)}|\hat{\mu}|\varphi_\mathbf{k}\rangle \sim \delta_{\kappa,\mathbf{k}}$ is assumed, the overall free electron distribution becomes

$$P_\mathbf{k}^{(\text{vac})} = \langle\chi_\kappa(t)|\chi_\kappa(t)\rangle \sim \sum_M |C_{\mathbf{k}M}(t_2)|^2\ . \tag{5.45}$$

Use of (5.36) yields

$$P^{(\text{vac})} = \sum_\mathbf{k} P_\mathbf{k}^{(\text{vac})} \sim P_{\text{ion}}(t)\ . \tag{5.46}$$

If dispersed with respect to energy $\hbar\Omega$, one may conclude that $P^{(\text{vac})}(\Omega)$ is proportional to the energetic distribution $P_{\text{el}}(\Omega;t_2)$, (5.36) of the injected electron versus the band states times the DOS $\mathcal{N}(\Omega)$ shown in Fig. 5.27. Note here that vibrational signatures may become observable even though the electron transfer proceeds on a time scale faster than 10 fs.

Experimental data

Electron transfer times have been measured with a few fs resolution by applying two different experimental techniques. Transient absorption was applied when the molecules were adsorbed on the inner surface of a nano-structured anatase TiO_2 layer of typically a few μm thickness [140, 176, 177]. 2PPE was employed when the molecules were adsorbed on the (110) surface of a rutile TiO_2 single crystal [145, 160]. Both sets of data for the electron transfer times measured in these closely related but distinctly different experimental systems showed rather good agreement.

The change in the electronic coupling strength arising with different bridge–anchor groups can be measured as the corresponding change in the electron transfer time [148]. As long as the role of the bridge–anchor group can be approximated either by that of an electronic tunneling barrier or as extension of the molecular donor orbital the change in the measured injection time can be identified with the change in just one parameter, i.e. the electronic coupling strength. Analysis of the experimental data has essentially substantiated this expectation of a change in the electronic coupling strength due to the use of different bridge–anchor groups. In addition, however, different bridge–anchor groups can bring about changes also in other parameters, i.e. the energy of the electronic transition to the excited singlet state of the chromophore perylene or the reorganization energy which is defined as the characteristic energy for the change in nuclear equilibrium coordinates upon ionization of the excited state of the chromophore due to electron injection into TiO_2 (see, e.g. [130]). Thus, identifying the changes seen in experimental

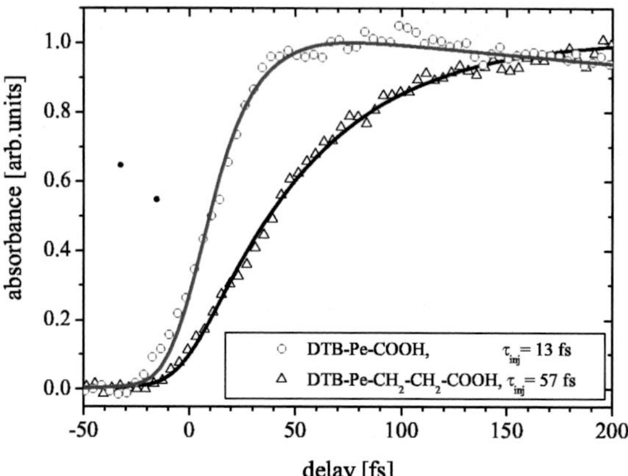

Fig. 5.26. Rise of the cation absorption which probes the population of ionized perylene with two different bridge–anchor groups, i.e. carboxylic and propionic acid.

data for different bridge–anchor groups with the change in just one experimental parameter would be an oversimplification. It is not only the electronic coupling strength that is changed but in addition there are smaller changes of several properties of the perylene chromophore when it is attached with different bridge–anchor groups to the surface of the semiconductor. The most meaningful criterion when comparing theory with experiment appears to be a qualitative trend that can be seen in the experimental data. The trend can be quantified through an analysis of the experimental data by applying the above theoretical model to the molecules with the different bridge–anchor groups.

Figure 5.26 shows the rise of transient absorption signals that probe the population of the ionized perylene chromophore [139,140] when anchored with either the formic acid group or the propionic acid group, respectively, on the surface of anatase TiO_2 colloidal particles. The latter surfaces are exposed in the nm-scale cavities that are formed when the nm-scale colloids are glued together to form a TiO_2 layer of several μm thickness. From the experimental curves, it is clear that the electron transfer time is considerably longer in the system with the propionic acid compared to that with the formic acid as the anchor group.

Corresponding experimental data have been collected also for several other bridge–anchor groups. As first approximation, assuming a simple, exponential rise, the corresponding fits to the curves in Fig. 5.26 yield electron transfer times of 13 fs and of 57 fs for the formic acid and propionic acid as anchor group, respectively (compare Figs. 5.25 and 5.26). The different injection times reflect the different electronic tunneling barriers realized by the two different bridge–anchor groups [144,174]. The inner surface of the nano-structured anatase TiO_2 films offers different adsorption sites. There is a considerable concentration of edge and corner sites and there are adsorption sites on

Table 5.1. Comparison between electron injection times measured via 2PPE [145] and transient cation absorption spectroscopy [144].

compound	2PPE injection time [fs]	transient absorption injection time [fs]
Pe'-COOH	8.4	13
Pe'-CH=CH-COOH	13.5	10
Pe'-CH_2-CH_2-COOH	47.2	57
Pe'-$PO(OH)_2$	23.5	28
Pe'-CH_2-$PO(OH)_2$	35.9	63

different crystal surfaces [143]. It is therefore quite possible that the assumption of a single exponential rise ignores finer details, e.g. a likely distribution of slightly different electron transfer times. Nevertheless, the exponential fits to the rise of the curves in Fig. 5.26 appear to be an acceptable approximation. Electron transfer times determined from 2PPE signals for the same molecules (Fig. 5.26) anchored on the flat (110) surface of rutile TiO_2 gave very similar absolute values (Table 5.1), and a very similar trend is seen in the electron transfer times for the different bridge–anchor groups [145]. At first glance, this close similarity of the injection times obtained with these two related but also clearly different surfaces onto which the molecules were anchored (i.e. the inner surface of the nano-structured anatase film versus the flat (110) surface of the rutile single crystal) appeared somewhat surprising since there are already clear differences in the electronic density of bulk states for the two different TiO_2 modifications. This point will be resumed in the next subsection.

5.7.2.2 Energy distribution of injected electrons

To characterize the time evolution of the injected electron in more detail, we now consider the probability distribution $P_{\rm el}(\Omega;t)$ of the electron in the band continuum, see (5.36). Results are shown in Fig. 5.27 for the two bridge–anchor groups also used in Figs. 5.25 and 5.26. It has already been indicated that $P_{\rm el}(\Omega;t)$ displays the vibrational progression of the involved coordinate [163]. Figure 5.27 demonstrates that after a certain time interval (reflecting the energy-time uncertainty) the broad distribution decays into different peaks. They correspond to transitions from the excited molecular state with energy $\hbar\varepsilon_e + \hbar\omega_{eM}$ into the conduction band continuum with energy $\hbar\varepsilon_{\rm con} + \hbar\Omega + \hbar\omega_{{\rm ion}N}$. The possible energy values $\hbar\Omega$ in the band follow as $\hbar\varepsilon_e - \hbar\varepsilon_{\rm con} + \hbar(\omega_{eM} - \omega_{{\rm ion}N})$ reflecting an inelastic charge injection accompanied by the creation or annihilation of quanta of the vibrational coordinate (see also Fig. 5.24).

If the vibrational ground state of the excited molecular state were to be populated only, $P_{\rm el}(\Omega;t)$ should extend to an energy range below $\hbar\varepsilon_e - \hbar\varepsilon_{\rm con}$.

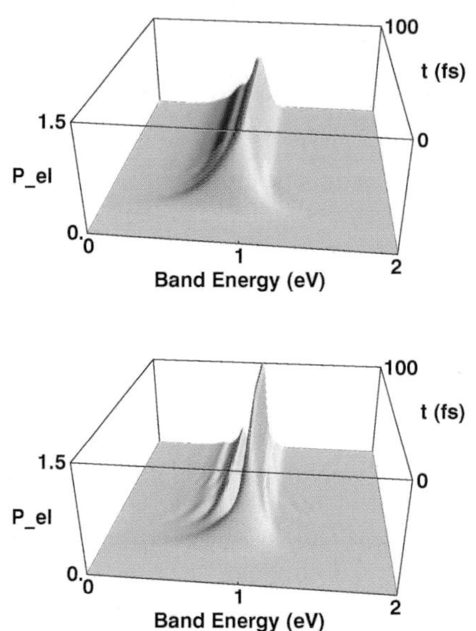

Fig. 5.27. Probability distribution of the injected electron $P_{\rm el}(\Omega;t)$, (5.36) versus the energy $\hbar\Omega$ within the conduction band and versus time (the origin of the energy axis is given by $\hbar\varepsilon_{\rm con}$, the excitation conditions and parameters are identical with those of Fig. 5.25). (top) DTB–Pe–COOH system, (bottom) DTB–Pe–$(CH_2)_2$–COOH system (for parameters see Table 5.2).

The simultaneous population of excited vibrational states may cause also structures in $P_{\rm el}(\Omega;t)$ above $\hbar\varepsilon_e$. This would be the case after an ultrashort optical excitation. In Fig. 5.27, it is less obvious since the vibrational energy is larger than 0.1 eV (see Table 5.2). Note the similarity between the probability distribution for the injected electron and the spectrum of the linear absorbance as will be discussed further below in Sect. 5.7.2.3 where also a strong transfer coupling leads to a strong broadening of the vibrational progression (compared with the case of the molecule in solution) and a weak coupling to a less pronounced broadening. In any case, the whole energetic extension of $P_{\rm el}(\Omega;t)$ reflects the distribution of Franck-Condon overlap integrals $\langle\chi_{{\rm ion}N}|\chi_{e0}\rangle$. It should be mentioned that the energetic dispersion of $P_{\rm ion}$, (5.36), by introducing $P_{\rm el}(\Omega;t)$ resolves vibrational state contributions although the HET proceeds on a time scale of a few femtoseconds. Analyzing exclusively $P_{\rm ion}$, such detailed information would be not available.

Experimental results

The energy distribution of the injected electrons has been probed with 2PPE for the molecules adsorbed on the (110) surface of a rutile TiO_2 single crystal

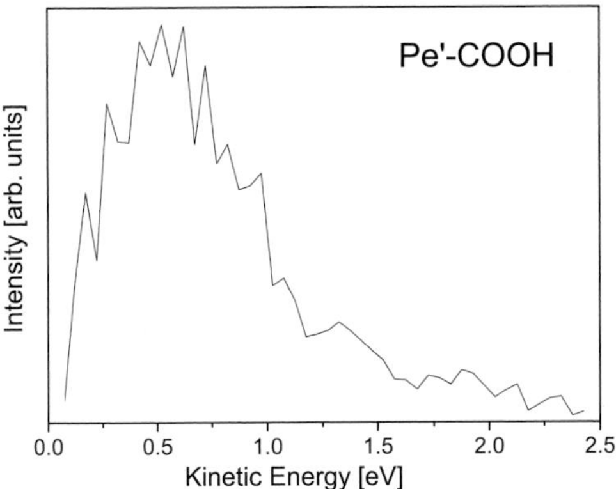

Fig. 5.28. Experimental energy distribution curve for the electrons injected from anchored Pe'-COOH into rutile TiO_2.

[145, 160]. The nature and density of the electron acceptor states have not yet been identified in any detail. DFT-cluster calculations suggest the involvement of surface resonances in the electron transfer reaction that arise in the vicinity of the bonds formed by the anchor groups on the surface of TiO_2 [156,172,173]. The above mentioned anchor groups bind by either forming bonds between the O atoms of the acid group and the Ti atoms on the crystal surface or bonds with O atoms on the crystal surface [171,174,178]. In addition to photoemission from the excited singlet state of the chromophore perylene the 2PPE signals contain a second contribution due to photoemission of the electrons already injected into the TiO_2. From the latter contribution measured in the case of a direct optical charge transfer transition (see below), the time scale of escape of the injected electrons from the surface into bulk states of TiO_2 was derived [160]. The measured energy distribution of the injected electrons covers a wide spectral range of typically 0.5 eV width (FWHM), that is controlled by the Franck-Condon envelope for the transition from the excited state to the ionized state of the chromophore. This was also predicted by the theoretical model in the case of the so-called wide-band limit [163, 164].

Figure 5.28 gives the experimental energy distribution curve for the electrons injected from anchored Pe'-COOH into rutile TiO_2. The wide band limit should be fulfilled by all the perylene derivatives of Fig. 5.21 since the donor state of the perylene chromophore is located far above the conduction band edge of both anatase and rutile TiO_2 (Fig. 5.22). The exact physical nature of the initial electronic acceptor states, however, has not yet been clarified. DFT-cluster calculations suggest an important role of localized states that are formed around the surface bonds formed by the anchor groups [173,174]. Surface resonances have been shown by DFT calculations to be connected with the formation of surface bonds due to alien molecules in the case of InP(100) [179], and this prediction has been confirmed experimentally [180, 181]. Surface reconstructions can change the situation significantly compared to the vacuum interface when alien molecules form chemical bonds on the crystal surface. The atomic and consequently also the electronic structure of the surface is altered in such a case compared to the surface terminated by the vacuum. In the present systems, the alien molecules are the anchor groups that form strong chemical surface bonds.

The effect of long, rigid bridge–anchor groups on the electron injection times can be investigated in an unambiguous way only when such molecules are anchored on the flat surface of a single crystal. Optical measurements which address an ensemble of such very long and rigid molecules in the cavities of the nano-structured inner surface of the anatase TiO_2 layer have always shown several very different injection times [144]. They are ascribed to short-cuts for HET to the nearest TiO_2 surface at edge and corner sites. In contrast, only one, i.e. the slower injection time, has been seen in the 2PPE signals when molecules were attached with the same long, rigid bridge–anchor group that contains saturated C-C bonds to the flat surface of a rutile single crystal [158]. Observation of only one injection time in this situation is in agreement with the well-defined adsorption geometry deduced for such molecules from angle and polarization dependent 2PPE signals on such a single-crystalline surface [145].

5.7.2.3 Steady-state absorption spectra

Linear absorption spectra are of particular importance since their detailed analysis offers a rather unique way to specify all parameters of the model presented here (see also [148]). The computation of linear absorption spectra of molecular systems represents a standard task (see, e.g. [130]) and is based on the following expression

$$\alpha(\omega) = \frac{4\pi\omega n_{\mathrm{mol}}|\mathbf{d}_{eg}|^2}{3\hbar c}\mathrm{Re}\int_0^\infty dt\, e^{i(\omega+\varepsilon_g)t}\langle\chi_{g0}\varphi_e|e^{-iH_{\mathrm{mol-sem}}t/\hbar}|\varphi_e\chi_{g0}\rangle\,. \tag{5.47}$$

Besides n_{mol} that denotes the volume density of the absorbing molecules all other expressions have already been introduced in Sect. 5.7.1.

The absorption formula indicates that the state vector $\varphi_e\chi_{g0}$ has to be propagated under the action of the complete Hamiltonian $H_{\mathrm{mol-sem}}$ (5.31).[5] We introduce the expansion (5.34) with respect to the electron-vibrational states $\varphi_a\chi_{aM}$ (with a restricted here to e and $\mathbf{k}$) and carry out the time propagation. As it is well known, this procedure avoids any calculation of system eigenstates and eigenfunctions. In particular, a full account for the

[5] The absence of any ground state-excited state coupling in $H_{\mathrm{mol-sem}}$ ensures that the propagation of $\varphi_e\chi_{g0}$ does not include any contribution proportional to the electronic ground state φ_g.

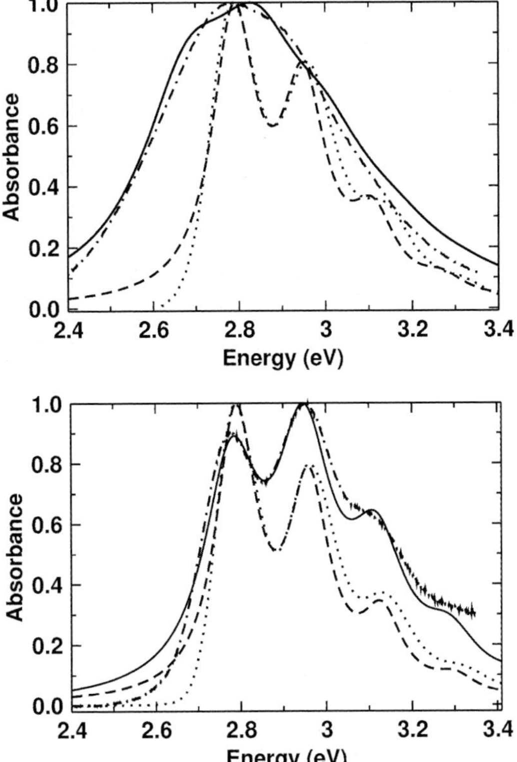

Fig. 5.29. Rescaled linear absorption spectrum of the DTB–Pe–COOH system (top) and the DTB–Pe–$(CH_2)_2$–COOH system (bottom). Dotted lines: experimental data for the system in the solvent, dashed-dotted lines: experimental data for the system adsorbed at a TiO_2 surface, dashed lines: calculated absorbance for the system in the solvent, full lines: calculated absorbance for the system adsorbed at a TiO_2 surface (for the used parameters see Table 5.2).

frequency dependence of the DOS and the transfer integral is equivalent to a complete consideration of the self-energy, (5.38) (which, of course, is not calculated explicitly here). To account for line broadening (dephasing) due to IVR, the absorbance, (5.47), is calculated by additionally introducing the factor $\exp(-\gamma t)$ with the overall dephasing rate γ.[6]

[6] A more involved description via density matrix theory would offer a microscopic description of IVR resulting in relaxation processes among different vibrational states. If these processes are fast enough, they may influence the HET rate. In the present description, however, the latter quantity, (5.41), has to be distinguished from γ.

Table 5.2. Parameters of the DTB–Pe–COOH and DTB–Pe–$(CH_2)_2$–COOH system, respectively, at the TiO_2 surface (for explanations see text).

	DTB–Pe–COOH	DTB–Pe–$(CH_2)_2$–COOH
$\hbar\varepsilon_e$	2.79 eV	2.79 eV
$\hbar\omega_{\text{vib}}$	0.16 eV	0.17 eV
λ_{eg}, $(Q_e - Q_g)$	0.116 eV, (1.7)	0.187 eV, (2.1)
d_{eg}	3 D	3 D
$\hbar\gamma$	0.062 eV	0.058 eV
$\hbar\varepsilon_{\text{con}}$	1.79 eV	1.79 eV
$\hbar\Delta\omega_{\text{con}}$	6.0 eV	6.0 eV
$\hbar\Gamma$	0.094 eV	0.0213 eV
$\bar{V}_e$ $(\mathcal{N}/\hbar)$	0.1 eV, (2/eV)	0.058 eV, (2/eV)
$\lambda_{\text{ion}\,e}$, $(Q_{\text{ion}} - Q_e)$	0.014 eV, (-0.6)	0.014 eV, (-0.6)

Perylene anchored on TiO_2

Equation (5.47) has been used to compute the absorbance related to perylene which is attached to the surface of TiO_2 nanocrystals via different bridge–anchor groups. Respective experimental spectra are displayed in Fig. 5.29 for the DTB-Pe-COOH-TiO_2 and the DTB-Pe-$(CH)_2$-COOH-TiO_2 system as well as for the dye in a solvent with the respective bridge–anchor groups. The measured spectra for the molecules in a solvent show a vibrational progression which was related to a perylene in-plane C–C stretch vibration with quantum energy of 1370 cm^{-1} [148]. The 0–0 transition as well as the 0–1, 0–2, and 0–3 transitions are clearly resolved. The solvent spectra were used to achieve a first fixation of some internal perylene parameters (energetic position of the excited state $\hbar\varepsilon_e$, vibrational energy $\hbar\omega_{\text{vib}}$, reorganization energy accompanying the excitation λ_{eg}, and dephasing rate γ, see Table 5.2).

We now turn to the spectra of the molecules attached to the TiO_2 surface. For the DTB-Pe-COOH-TiO_2 system the vibrational progression found in the solvent is lost in the adsorbed state and an almost structureless absorption band appears instead. In contrast, the system of DTB-Pe-$(CH_2)_2$-COOH-TiO_2 retains the vibrational progression in the adsorbed states but with the 0–1 transition stronger than the 0–0 transition. The trend observed in the absorption spectra, i.e. the different degrees of broadening, in the surface-attached case follows the intuitive expectation based on the molecular structure of the different bridge–anchor groups (see Fig. 5.21).

As indicated in Fig. 5.29, the solvent spectra as well as those for the case of perylene attached to TiO_2 could be rather well reproduced what gave the basis to fix all parameters as presented in Table 5.2.[7] Respective parameters deduced from a fit of the absorption spectra are shown for the first system (with the formic acid bridge–anchor group) and the last mentioned system (with the propionic acid bridge–anchor group) in Table 5.2. Besides the electronic parameters (excited molecular level $\hbar\varepsilon_e$, lower conduction band edge $\hbar\varepsilon_{\text{con}}$, band width $\hbar\Delta\omega_{\text{con}}$, averaged DOS $\bar{\mathcal{N}}$, and transfer coupling $\bar{V}_e$), Table 5.2 contains a single vibrational frequency ω_{vib} and two reorganization energies λ (see below) as well as the energy broadening $\hbar\bar{\Gamma}$ (Sect. 5.7.2.1) and the dephasing rate. For the transition-dipole matrix element d_{eg}, there does not exist a univocal value. In our case, a value of 3 Debye was used (see [148]).

In general, several vibrational modes will contribute to the absorption spectrum of aromatic chromophores, which holds true also in the case of perylene [182,183]. The spectra at room temperature, however, could be simulated rather well by a single-mode description.[8] They display the dominance of a single vibrational mode (also when the molecule is in a solvent) with a quantum energy $\hbar\omega_{\text{vib}}$ of about 0.17 eV (1370 cm^{-1}), which corresponds to an in-plane C–C stretching vibration. Such a conclusion as drawn in [148] had also been taken as a justification of the single-mode description already used in [163,165–167]. This allows to write the involved PES as ($a = g, e$, ion)

$$U_a(Q) = \hbar\omega_{\text{vib}}\left(\frac{1}{4}(Q - Q_a)^2 - \frac{1}{2}\right), \tag{5.48}$$

with the vibrational frequency ω_{vib} common to all considered electronic states. The notation removes the zero-point energy and is based on the use of a dimensionless coordinate Q (Q_a denotes the respective equilibrium position). Reorganization energies for transitions among the states simply follow as

$$\lambda_{ab} = \frac{\hbar\omega_{\text{vib}}}{4}(Q_a - Q_b)^2 \,. \tag{5.49}$$

They coincide with half of the respective Stokes shift.

We next consider the transfer integral. Once the line broadening $\bar{\Gamma}$ [see (5.39) and (5.41)] is determined and an estimate for the mean DOS $\bar{\mathcal{N}}$ is taken, one may deduce the mean transfer integral $\bar{V}_e$. An actual value for $\bar{\mathcal{N}}$ can be obtained from the calculations of [171,174], which have been restricted to rather small TiO_2 clusters, i.e. to the localized states around the binding site. While a determination of the transfer coupling from DFT data would require a so-called diabatization (the used DFT calculations only offer common molecule TiO_2 nanoparticle levels), $\bar{\mathcal{N}}$ is directly obtained from calculations

---

[7] Note that the replacement of $\bar{\mathcal{N}}$ by a frequency dependent DOS does not change the spectra (see [149]), indicating the validity of the wide-band approximation at the present mid-band position of the injection level.

[8] Including more vibrational modes which couple to the electronic transition is possible and may improve the fit to the measured data. But at the same time, this would require the introduction of many more fit parameters which makes the procedure rather ambiguous.

Table 5.3. Comparison of charge injection times for all perylene bridge–anchor group TiO_2 systems shown in Fig. 5.21. The time constants $\tau_{\rm inj}^{\rm (exp)}$ follow from a rate equation fit of measured transient absorption data (see [184]) and $1/k_{\rm HET}$ are the inverse of the HET rates, see (5.41).

	$\tau_{\rm inj}^{\rm (exp)}$ [fs]	$1/k_{\rm HET}$ [fs]
DTB-Pe-COOH–TiO_2	13	5
DTB-Pe-$(CH)_2$-COOH–TiO_2	10	6
DTB-Pe-$(CH_2)_2$-COOH–TiO_2	57	16
DTB-Pe-P(O)$(OH_2)_2$–TiO_2	28	9

neglecting the presence of the molecule on the nanoparticle. As a rough estimate, these calculations suggest a Gaussian-like DOS extending across the conduction band (with width of about 6 eV). An averaged value for the DOS can be obtained by counting the number of TiO_2 levels per eV. This leads to rather reasonable values of $\bar{V}_e$, which are summarized in Table 5.2. Since the transfer coupling connects the excited molecular state to some atoms of TiO_2 around the binding site of perylene only, the used small DOS seems to be adequate.

Table 5.3 relates measured injection time constants $\tau_{\rm inj}^{\rm (exp)}$ obtained from data of the cation transient absorption to those derived from the simulation of steady-state absorption spectra. The latter are given as the inverse of the HET rates $k_{\rm HET}$, see (5.41). The time constants derived from the calculations reproduce the qualitative trend of the measured injection time constants. However, the $k_{\rm HET}$ are always too small which might be explained by the neglect of structural and energetic disorder or of the broadening introduced by other normal modes not included so far.

Frequency domain description

In the following paragraph, the direct computation of the absorption coefficient in the frequency domain will be discussed. This will offer a simple picture for the influence of conduction band coupling of the excited molecular state, which is rather hidden in the time-dependent description. The Green's operator technique explained in detail in [168] is used, which then yields in the wide-band approximation

$$\alpha(\omega) = \frac{4\pi\omega n_{\rm mol} \mid \mathbf{d}_{eg} \mid^2}{3\hbar c} \sum_{N,K} \frac{f(\hbar\omega_{gN}) |\langle\chi_{gN}|\chi_{eK}\rangle|^2 \mid {\rm Im}\bar{\Sigma} \mid}{(\omega - \varepsilon_{eg} - \omega_{eK,gN} - {\rm Re}\bar{\Sigma})^2 + ({\rm Im}\bar{\Sigma})^2} \ . \quad (5.50)$$

Note the introduction of transition frequencies $\varepsilon_{eg} = \varepsilon_e - \varepsilon_g$ and $\omega_{eK,gN} = \omega_{eK} - \omega_{gN}$. As a result, the absorbance follows as an expression with Lorentzian line shape for every ground-to-excited state transition. The broadening originates from the imaginary part of the self-energy [see (5.38) and (5.39)], whereas the real part of the self-energy induces a shift of the transition frequencies.

The frequency-domain formulation for the absorbance according to (5.50) is compared with the time-dependent formulation displayed in Fig. 5.29. Therefore, we concentrate on the strong-coupling case (top panel of Fig. 5.29). The used values of $\bar{\Gamma} \equiv | \mathrm{Im}\bar{\Sigma} |$ are taken from Table 5.2. To achieve complete agreement, however, a transition frequency shift due to $\hbar\mathrm{Re}\bar{\Sigma}$ of about -0.05 eV has to be introduced. Interestingly, the combined effect of strong line broadening and a red shift of the transition frequencies gives the impression that the absorbance peak for the case of perylene attached to TiO_2 stays at the same position which respect to the case of perylene in a solvent.

Finally, the effect of structural and energetic disorder presumably present in all measured spectra will be estimated. Here, the easiest way to do this is using (5.50) and restricting to fluctuations of the excited molecular level only. Fluctuations of the molecular orientation at the surface might be possible (leading to fluctuations of the transfer coupling) and the surface structure of the semiconductor around the molecular binding site might also vary. However, concentrating on the simplest case of molecular *on-site* disorder, the respective disorder-averaged absorbance follows by integrating (5.50) with respect to the disorder distribution of ε_e. If the absorption spectra of Fig. 5.29 are affected by inhomogeneous broadening, the presented values of $\bar{\Gamma}$ [see (5.41)] and the HET rate k_{HET} are somewhat too large for the perylene-TiO_2 systems (Table 5.3). The differences between $1/k_{\mathrm{HET}}$ and the injection times derived from transient absorption data, which are less affected by disorder, may be diminished.

Contributions of charge transfer (CT) states

Direct optical charge transfer from the ground state of an adsorbed molecule to electronic acceptor states of a semiconductor has been established already many years ago for the experimental system catechol/TiO_2 [185]. A possible adsorption geometry is illustrated in the top panel of Fig. 5.30. The absorption spectrum of this system is shown in the bottom panel of this Figure. It appears most plausible that the 0,0 transition occurs at the red edge of the absorption spectrum around 600 nm [160, 178] and that the CT transition is controlled by a large reorganization energy. Since there is a continuum of electronic acceptor states in the semiconductor, e.g. in the conduction band, the absorption spectrum shown in Fig. 5.30 results from a superposition of many CT transitions.

When the CT transition occurs spectrally far away from the strong transition to a local, excited state in the organic chromophore, the former will give rise to a much weaker absorption compared to the local molecular transition. This fact, for example, has prevented the use of CT transitions in the dye-sensitized solar cell. Electron injection in the catechol/TiO_2 system had been

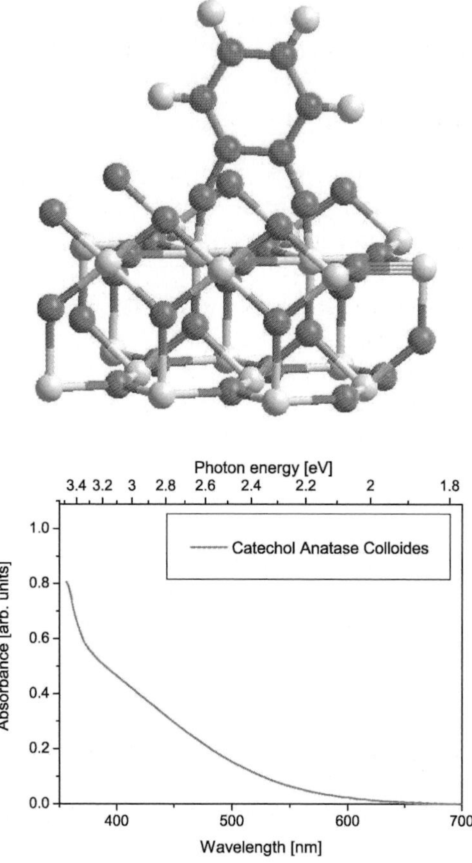

Fig. 5.30. Proposed binding geometry of catechol adsorbed on the rutile TiO_2 (110) surface (top), and linear absorption spectrum of catechol adsorbed on colloidal anatase TiO_2 (bottom).

tested by Grätzel and coworkers [147] in their early investigations directed at the dye-sensitized solar cell. With a low absorption coefficient, to reach the saturation range which is necessary for absorbing maximal light from the solar spectrum would require the use of a rather thick catechol/TiO_2 layer. In such a layer, the transport time of the injected electrons to reach the charge collecting electrode would become too long compared to the recombination time for the injected electrons. The efficiency of the corresponding photovoltaic device will certainly be rather low. It remains, however, an interesting question with respect to the mechanism of ultrafast charge injection from molecules whether direct optical charge transfer will occur in the spectral vicinity of a strong transition to a local, excited state like in the perylene-COOH/TiO_2 system. This would give rise to interferences with the latter transition.

5.7.3 Role of vibrations and the possibility of coherent control

The present scenario for ultrafast electron transfer in the wide-band limit assumes that electron transfer occurs with the same time constant from every excited vibrational level of the donor state which is photoinduced by the chromophore perylene. In the wide-band limit, i.e. in the presence of a broad continuum of electronic acceptor states below the donor state, the whole width of the envelope over all the possible Franck-Condon factors can be accommodated. This possibility of realizing all the Franck-Condon factors in parallel leads to the electron transfer time independent of any individual Franck-Condon factor (see Sect. 5.7.2.1). Consequently, a temperature independent electron transfer time results, which indeed was observed down to ~20 K [139]. The obtained mono-exponential decay of the donor state is valid as long as the wide-band limit is fulfilled, but does not originate from a perturbation treatment as in the Fermi's Golden Rule approximation.[9] The hitherto measured electron transfer times and the measured linewidth broadenings for the perylene derivatives appear to agree with the above mentioned injection scenario in the wide-band limit.

The time evolution of the injection process obeys the energy-time uncertainty, i.e. the energy distribution of the injected electrons is smeared out for sufficiently short injection times right after absorption of the laser pump pulse. The time evolution of the injection process can show the influence of vibrational wave packet motion. Some indication of this has been seen in the experiments for moderately fast electron injection [163]. It is, however, not clear whether the observed wave packet motion had any influence on the actual injection time or took place independently in the molecular donor and final states primarily due to excitation by an ultrashort laser pulse. At early times, interference effects between neighboring electronic acceptor states into which the electron is injected should be observable in the time-dependence of the injected electron distribution and thus in the 2PPE signal. Yet, it appears a rather demanding task to unambiguously identify such details in the complicated experimental system.

There is another very interesting scenario for ultrafast electron injection from a donor state close to the conduction band edge of the semiconductor, in particular, for ultrafast adiabatic electron transfer promoted by nuclear

[9] The model assumes nuclear tunneling from the initially excited vibrational state of the excited neutral chromophore which functions as molecular electron donor to all vibrational excited states of the ionized chromophore, i.e. the molecular product state. Energy is conserved in the electron injection reaction by depositing the electron in an electronic acceptor level of the semiconductor that is lower than the donor state by the same amount of energy as is spent for the vibrational excitation of the molecular final state, i.e. the ionized chromophore.

motion. The corresponding scenario is based on time-dependent DFT calculations where nuclear motion is treated with a classical model. This scenario has been advanced by Prezhdo and coworkers [156, 173] (see also Sect. 5.8). It is not yet clear whether injection at the band edge can be utilized for irreversible electron injection or whether escape from the surface into bulk states requires the injection of hot electrons. These would quickly reach a spatially sufficient separation from the geminate recombination center while losing some of the excess energy via phonon emission. The geminate recombination center is the ionized chromophore generated in the interfacial electron transfer reaction.

Finally, some speculation on femtosecond laser pulse control of HET shall be presented. Laser pulse guided molecular dynamics within closed-loop control experiments represents one frontier in ultrafast optical spectroscopy (for recent overviews see, e.g., [186, 187]). The whole research field is based on the vision to tailor femtosecond laser pulses in the optical and infrared region in order to drive the molecular wave function in a desired way. In the case of HET, the control of the vibrational motion in the excited molecular state was discussed in [167]. However, the strong coupling of this level to the semiconductor band continuum suppressed any efficient manipulation of the vibrational motion. The control of the vibrational states in the electronic ground state has been more efficient.

For the control scenario to be discussed here, it suffices to use a scheme of laser pulse control of molecular dynamics where one asks to realize a certain state $|\Psi_{\text{tar}}\rangle$ (the target state) at time t_f (final time of laser pulse action) [188]. Figure 5.31 displays the results of such a control task where the target state Ψ_{tar} is given by the vibrational ground state χ_{g0} in the electronic ground state but shifted away from its equilibrium position ($Q_{\text{shift}} = -1$, for more details see [167]). Therefore, the control pulse introduces an excitation to the excited electronic level and back to the ground state (pump-dump scheme of laser pulse control, top right panel of Fig. 5.31). Since excitation and de-excitation covers a 20 fs interval, the coupling to the continuum does not restrict the overall control yield. The dependence on the injection position, however, indicates that the control is more efficient for a near band-edge injection position (top left panel of Fig. 5.31).

The present section has given a survey of recent theoretical studies and experimental results on ultrafast heterogeneous electron transfer with emphasis on the perylene/TiO_2 system. The approach accounts for molecular degrees of freedom as well as the band continuum of the semiconductor and allows to describe different spectroscopic excitation and detection processes. Such a uniform description of photon absorption, electron transfer, and detection of HET required the use of a reduced dimensionality model. Since the studied HET proceeds on a femtosecond time scale it suffices to solve the respective time-dependent Schrödinger equation governing the electron-vibrational wave function. All parameters of the used model could be specified by a comparison with measured steady state absorption spectra. Qualitative trends in the

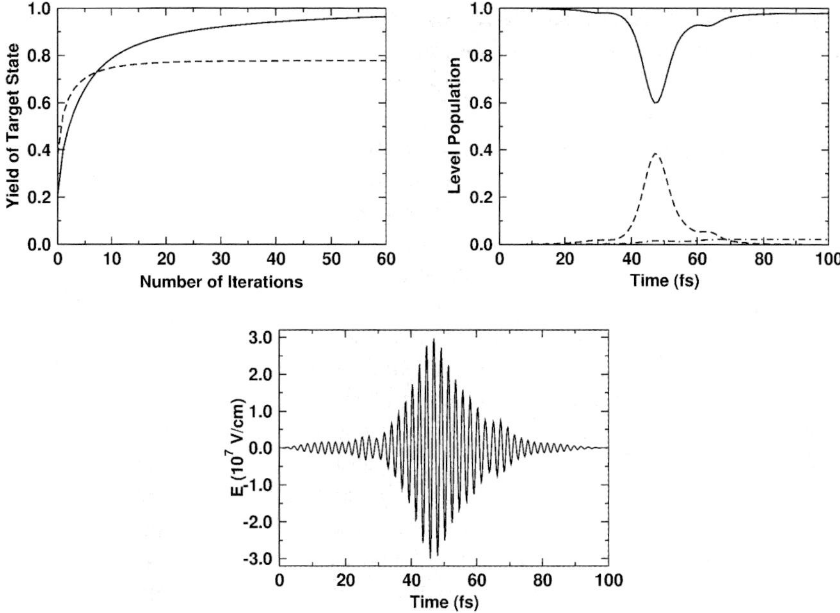

Fig. 5.31. Laser pulse control of the vibrational motion in the electronic ground state. Variation of the injection position $\hbar\varepsilon_e$ into the band continuum. (top left) Control efficiency versus the number of iteration steps, solid line: $\hbar\varepsilon_e = \hbar\varepsilon_{\text{con}}$ (band-edge injection), dashed line: $\hbar(\varepsilon_e - \varepsilon_{\text{con}}) = 1$ eV. (top right) Level population P_g (solid line), P_e (dashed line) and P_{ion} (dashed-dotted line) versus time for the band-edge injection position. (bottom) Temporal behavior of the optimal pulse for the band-edge injection position.

measured electron transfer times and corresponding line broadening in the absorption spectra are in fair agreement.

First results have been presented concerning the energetic distribution of the injected electrons measured with two-photon photoemission signals. In this context, the possible observation of vibrational signatures in the two-photon photoemission signal should be noted although the electron transfer proceeds on a 10 fs time scale. Some speculations on femtosecond laser pulse control of the injection process have been given at the end.

However, despite the rather complete description of ultrafast HET, there do exist different routes for further investigations. The used model might be extended in different respects, for example, by including additional intramolecular vibrational modes. Yet, it appears even more important that the present studies will be extended to a picosecond time scale. This requires the inclusion of different relaxation channels, in particular, the consideration of electron relaxation with respect to the conduction band states caused by electron-phonon scattering. As an additional challenge, further theoretical elaboration of two-photon photoemission processes at surfaces might be considered.

5.8 Ab initio time-domain simulations of photoinduced electron transfer in dye-sensitized TiO_2

W. R. Duncan and O. V. Prezhdo

The fully quantum-mechanical treatment of the HET dynamics described in the previous section was performed with simplified models of the interface [149, 162, 164, 166, 167, 169, 170]. The current section focuses on the explicit atomistic treatment of the interface, combined with quantum-classical electron-vibrational dynamics [156, 172, 173, 189–195]. The quantum-classical approximation is justified by the separation of electronic and nuclear masses and time-scales. The two treatments complement each other. Quantum-mechanical models of the interface provide exact dynamics, transparent interpretations of the parameters entering the model, and the ability to vary these parameters in order to probe and characterize various injection regimes. Fully atomistic descriptions, on the other hand, model the actual interface as closely as possible. By taking into account the chemical structure of the chromophores, the chromophore-semiconductor binding details and the conditions of the surface, they give rise to a spatial-temporal picture of ET with explicit electron and nuclear motions, conformational and chemical changes, disorder, and so on.

5.8.1 Electronic structure of the interface

The chromophores that are used in the chromophore-sensitized TiO_2 photovoltaic cell harvest visible light and are selected both to match the semiconductor energy levels, and also to be photochemically and thermally stable [196]. They fall into two broad categories: purely organic, conjugated molecules and transition-metal/ligand complexes. The photoactive ligands in the transition-metal-based chromophores are also organic, conjugated molecules. The ground state of the purely organic chromophores is a π-state, while the ground state of the transition metal based chromophores is localized on the n-orbitals occupied by the metal's undivided electron pairs. The photoexcited states are the donor states in the interfacial ET and are formed by the π^*-orbitals of the conjugated systems in both chromophore types. Therefore, the study focuses on conjugated molecules.

The photoexcitations encountered in the dye-sensitized TiO_2 systems can be classified into several types, see Fig. 5.20, all of which have the chromophore ground state located within the band gap of TiO_2 and the photoexcited state in resonance with the TiO_2 conduction band (CB). In the first and most

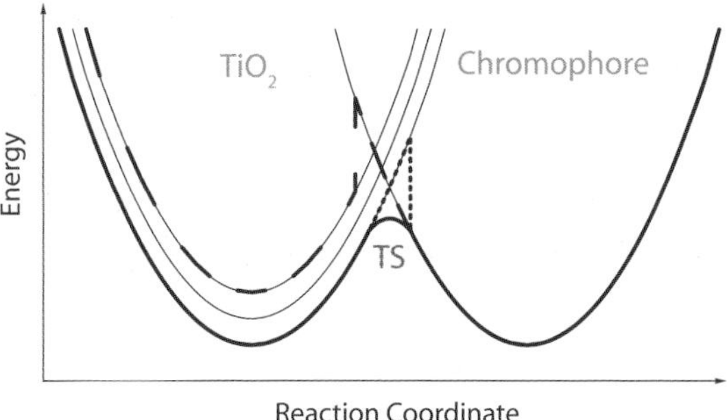

Fig. 5.32. Electron injection mechanisms. The chromophore and TiO_2 electronic energies are functions of atomic coordinates, and are represented by the parabolas, as in the Marcus model. In the case of a strong chromophore-semiconductor coupling, an atomic fluctuation across the transition state (TS) gives adiabatic electron transfer. If the coupling is weak, the electron remains in the chromophore state even after the TS has been crossed (dashes). In the weak-coupling limit, the injection occurs nonadiabatically by "hops" between the states (dashed and dotted lines).

common type of photoexcitation, the photoexcited states are well localized on the chromophore and do not appreciably mix with the CB and surface states. Under these circumstances, the injection dynamics proceed in two distinct steps: photoexcitation followed by ET. The second type of photoexcitation involves systems with exceptionally strong chromophore-semiconductor coupling, such as catechol/TiO_2, see Fig. 5.33. In such systems the lowest energy photoexcited state is already a surface state [170, 172, 173, 189, 197]. The third photoexcitation type occurs in some chromophores such as alizarin. In this case, the photoexcited states are located close to the bottom of the TiO_2 CB [141, 156, 173, 193–195]. Combined with strong chromophore-semiconductor coupling, chromophore photoexcited states located at the CB edge present the most complicated injection dynamics, involving states that are both localized on the chromophore and delocalized into the surface.

Transition metal chromophores exhibit strong spin-orbit coupling, promoting intersystem crossing into triplet states. Apart from the fact that triplet excited states have lower energy than singlets, similar ET mechanisms can be expected from states of either spin.

Two competing ET mechanisms have been proposed to explain the observed ultrafast injection events [198, 199]. These mechanisms have drastically different implications for the variation of the interface conductance and solar cell voltage with system properties. In the adiabatic mechanism, the coupling between the dye and the semiconductor is large, and ET occurs through a transition state (TS) along the reaction coordinate that depends on a concerted motion of nuclei, Fig. 5.32. During adiabatic transfer the electron remains in the same Born-Oppenheimer (adiabatic) state, which continuously changes its localization from the dye to the semiconductor along the reaction coordinate. The adiabatic mechanism is a typical chemical phenomenon and forms the basis of TS theories, in which the reaction rate is determined by the probability of finding and crossing the TS. [200] A small TS barrier relative to the nuclear kinetic energy gives fast adiabatic ET.

Nonadiabatic effects (NA) decrease the amount of ET that happens at the TS, but also open up a new channel involving direct transitions from the dye into the semiconductor that can occur at any nuclear configuration, as represented by the dashed and dotted lines in Fig. 5.32. NA transfer becomes important when the dye-semiconductor coupling is weak and is often described by perturbation theory, as in the case of Fermi's Golden Rule [200], wherein the rate of transfer is proportional to the acceptor density of states (DOS). NA ET is a quantum effect and, similar to tunneling, shows exponential dependence on the donor-acceptor separation. The question of which mechanism is at work is of practical concern because of the design implications. NA transfer does not require a strong donor-acceptor interaction, but relies rather on a high density of states in the CB. Since the DOS increases with energy [201], an increase of the chromophore excited state energy relative to the edge of the CB will accelerate the transfer. At the same time, the photoexcitation energy and solar cell voltage will be lost, due to the relaxation of the injected electron to the bottom of the CB. In the event of NA ET it is also important to minimize chromophore intramolecular vibrational relaxation, which lowers the chromophore energy and thereby the accessible DOS. The rate of NA ET will decrease exponentially with increasing distance between the donor and acceptor species. The adiabatic ET requires strong donor-acceptor coupling, but depends much less on the density of acceptor states. Since adiabatic transfer requires an energy fluctuation that can bring the system to the TS, a fast exchange of energy between vibrational modes of the chromophore will increase the likelihood of adiabatic ET.

5.8.2 Density functional theory of electron-vibrational dynamics

Real-time atomistic dynamics simulations provide unprecedented amounts of detailed temporal and spatial information. The simulations are especially valuable for the description of the interfacial electron injection between the molecular chromophores and the TiO_2 surface, since the conventional reaction rate theories have only limited application in this case and cannot possibly capture all aspects of the ultrafast, nonequilibrium process. The key components of the simulation include the quantum-mechanical many-body theory of the electronic structure that responds to vibrational motions and the prescription for the vibrational trajectory that, generally, is sensitive to the changes in the electronic degrees of freedom. The simulations are performed using the ab initio time-dependent Kohn-Sham (TDKS) theory developed in our group [195].

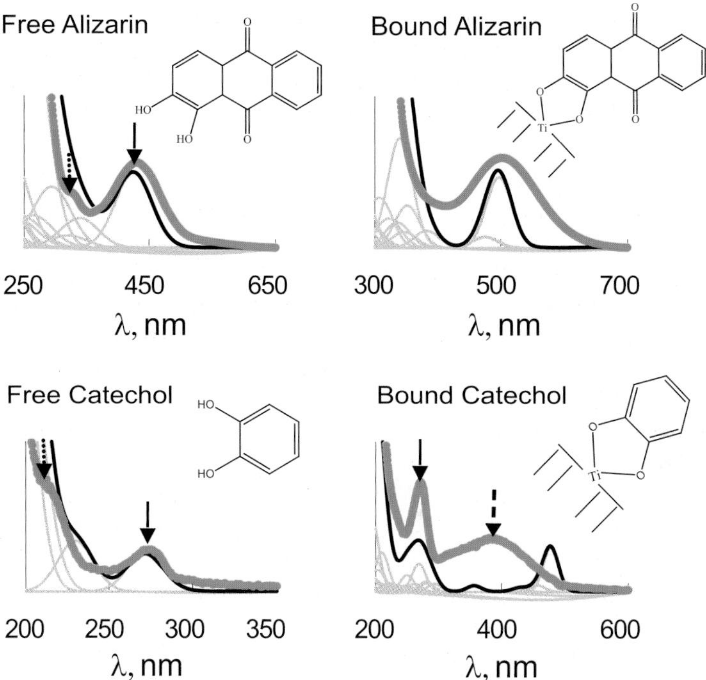

Fig. 5.33. Spectra of alizarin and catechol chromophores in the free state and bound to titanium. The thick red line represents the experimental data; the gray lines give the calculated spectra. Upon absorption, alizarin's spectrum slightly red-shifts as a result of the interaction of the TiO_2 states with the chromophore excited state, which is located at the CB edge, see Fig. 5.20. In contrast, the spectrum of catechol attached to TiO_2 exhibits an entirely new band in the long wavelength region. This transition involving direct ET from catechol to TiO_2 is optically active, due to strong chromophore-semiconductor coupling. The bound catechol spectrum also contains the optical transitions seen with free catechol. These intramolecular photoexcitations are followed by distinct injection events.

Time-dependent Kohn-Sham theory for electron-nuclear dynamics

The TDKS approach is a variant of density functional theory (DFT), in which the electron density is represented by the sum over single-electron orbitals [202] $\varphi_p(x,t)$

$$\rho(x,t) = \sum_{p=1}^{N_e} |\varphi_p(x,t)|^2 , \tag{5.51}$$

where N_e is the number of electrons. The time-evolution of $\varphi_p(x,t)$ is determined by applying the Dirac TD variational principle to the KS energy

$$E\{\varphi_p\} = \sum_{p=1}^{N_e} \langle \varphi_p | K | \varphi_p \rangle + \sum_{p=1}^{N_e} \langle \varphi_p | V | \varphi_p \rangle + \frac{e^2}{2} \int \int \frac{\rho(x',t)\rho(x,t)}{|x-x'|} d^3x d^3x' + E_{xc}. \tag{5.52}$$

The right-hand side of (5.52) gives the kinetic energy of noninteracting electrons, the electron-nuclear attraction, the Coulomb repulsion of density $\rho(x,t)$, and the exchange-correlation energy that accounts for the residual many-body interactions. Application of the variational principle leads to a system of single-particle equations [203–207]

$$i\hbar \frac{\partial \varphi_p(x,t)}{\partial t} = H(\varphi(x,t))\varphi_p(x,t), \; p = 1, \ldots, N_e, \tag{5.53}$$

where the Hamiltonian H depends on the KS orbitals. In the generalized gradient approximation [208] that is used in the current simulations, E_{xc} depends on both density and its gradient, and the Hamiltonian is written as

$$H = -\frac{\hbar^2}{2m_e}\nabla^2 + V_N(x;\mathbf{R}) + e^2 \int \frac{\rho(x')}{|x-x'|} d^3x' + V_{xc}\{\rho, \nabla\rho\}. \tag{5.54}$$

The KS energy (5.52) may be related to the expectation value of the Hamiltonian with respect to the Slater determinant (SD) formed with the KS orbitals [202]. The TD KS orbitals $\varphi_p(x,t)$ are expanded in the basis of adiabatic KS orbitals $\widetilde{\varphi}_k\,(x;\mathbf{R})$ that are the single-electron eigenstates of the KS Hamiltonian (5.54) for fixed vibrational coordinates $\mathbf{R}$

$$\varphi_p(x,t) = \sum_k^{N_e} c_{pk}(t) \big| \widetilde{\varphi}_k\,(x;\mathbf{R}) \big\rangle. \tag{5.55}$$

The adiabatic KS orbital basis is readily available from a time-independent DFT calculation [209, 210]. In the adiabatic KS basis, the TDKS equation (5.53) transforms into the equation for the expansion coefficients

$$i\hbar \frac{\partial}{\partial t} c_{pk}(t) = \sum_m^{N_e} c_{pm}(t) \Big(\epsilon_m \delta_{km} + \mathbf{d}_{km} \cdot \dot{\mathbf{R}} \Big). \tag{5.56}$$

The nonadiabatic (NA) coupling

$$\mathbf{d}_{km} \cdot \dot{\mathbf{R}} = -i\hbar \langle \widetilde{\varphi}_k\,(x;\mathbf{R}) | \nabla_{\mathbf{R}} |\, \widetilde{\varphi}_m\,(x;\mathbf{R}) \rangle \cdot \dot{\mathbf{R}} = -i\hbar \langle \widetilde{\varphi}_k \,|\frac{\partial}{\partial t}|\, \widetilde{\varphi}_m \rangle \tag{5.57}$$

arises from the dependence of the adiabatic KS orbitals on the nuclear trajectory, and is computed numerically from the right-hand-side of (5.57) [211].

Classical trajectory

The prescription for the classical trajectory $\mathbf{R}$ constitutes the fundamental problem in coupling quantum and classical mechanics, which is also known as the quantum back-reaction problem. The most common models include the classical path approximation [212, 213] (CPA), the Ehrenfest [214] or mean-field (MF) dynamics, and surface hopping [215–217] (SH), all of which have been used in studies of the chromophore-semiconductor interface [156, 173, 190–195].

The CPA provides the simplest solution by ignoring the back-reaction and assuming that the classical path is predetermined and independent of the electronic evolution. [212,213] It is the most computationally efficient approximation, and is a valid approach if the nuclear dynamics are not sensitive to changes in the electronic subsystem. The classical nuclear trajectory associated with the electronic ground state can be used in cases where excited state PES are similar to the ground state PES, and where the nuclear kinetic energy and thermal fluctuations of the nuclei are large in comparison to the differences in the PES.

The MF [214] approximation is the simplest form of the back-reaction of electrons on nuclei. The classical variables couple to the expectation value of the quantum force operator [212, 213]

$$M\ddot{\mathbf{R}} = -Tr_x \rho(x,t) \nabla_{\mathbf{R}} H(x;\mathbf{R}). \tag{5.58}$$

The gradient $\nabla_{\mathbf{R}}$ is applied directly to the Hamiltonian according to the TD Hellmann-Feynman theorem [213]. Many authors have thoroughly investigated the Ehrenfest method and found it to be valid under conditions similar to those needed for the CPA, but it requires modification when electron-nuclear correlations [216] and detailed balance must be taken into account [217–219]. Advanced versions of the Ehrenfest approach include "quantum fluctuation variables" [218–228].

SH generates electron-vibrational correlations and detailed balance by a stochastic prescription for the nuclear trajectory to "hop" between electronic states [215–217]. One of many flavors of SH, the fewest-switches (FS) SH is designed to minimize the number of hops and to satisfy a number of other key physical criteria. [216] SH can be viewed as a quantum master equation with the transition probabilities computed nonperturbatively and on-the-fly for the current nuclear configuration. In contrast to the traditional quantum master equations, SH is capable of describing the short-time Gaussian component of quantum dynamics that is responsible for the quantum Zeno effect and related phenomena [229–233].

5.8.3 Atomistic time-resolved electron injection

The details of the photoinduced interfacial ET provided by the ab initio time-domain simulation are exemplified below with the alizarin-TiO_2 system, Fig. 5.34, which generates rich dynamics with a variety of injection processes and mechanisms. Particular attention is devoted to the photoexcited state, the acceptor state density, the vibrational motions, the ET mechanism, and the distinct ET steps.

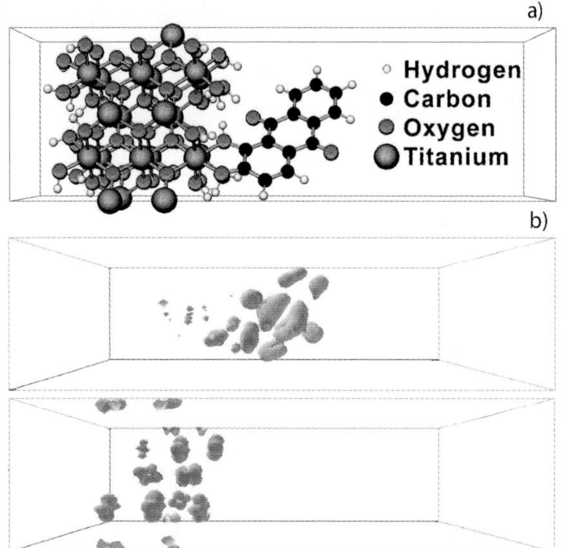

Fig. 5.34. (a) Alizarin-TiO_2 simulation cell. The cell includes five layers of TiO_2. Alizarin is bound to TiO_2 chemically via the Ti-O bonds. The Ti_2 surfaces are terminated with H and OH groups, eliminating dangling bonds. (b) Electron densities of the donor and acceptor states. π electrons from the conjugated electron system of alizarin form the donor state. The d orbitals of Ti atoms create the acceptor state, which is contained within the first three layers of TiO_2.

Roles of vibrational dynamics

Nuclear dynamics have a two-fold influence on the ultrafast electron injection process. On the one hand, thermal fluctuations of the nuclei create an ensemble of initial conditions with slightly different geometries and photoexcitation energies. On the other hand, upon photoexcitation, nuclei drive ET by moving along the reaction coordinate or, alternatively, by inducing direct quantum transitions between the donor and acceptor states. Fig. 5.35 plots the evolution of the excited state energies of the combined alizarin-TiO_2 system. The majority of the states are bulk and surface states that represent the CB of the semiconductor. Only the first excited state of the dye falls within the energy range shown in Fig. 5.35.

The energies of the photoexcited states are sensitive to the positions of the ions of the chromophore-TiO_2 system and oscillate with the amplitude of several tenths of an eV. At room temperature the energy fluctuations are sufficient to move the photoexcited states into different regions of the TiO_2 CB. The amplitude of the photoexcited state energy oscillation is small in relation to the several eV excitation energy of the dye, but it has a substantial impact on positioning of the dye state in the CB of TiO_2. The DOS of the

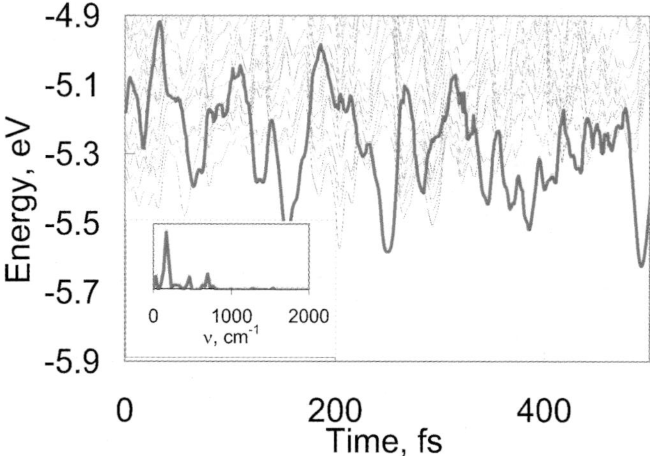

Fig. 5.35. Time-evolution of the photoexcited state of alizarin (bold red line) shown in relationship to the TiO_2 CB states (gray lines). Atomic motions modulate the state energies, such that the alizarin state crosses in and out of the CB. Inset: The Fourier transform of the photoexcited state energy shows that the ET occurs primarily by coupling to low frequency vibrational modes.

CB increases with energy [201], such that an excited state of the dye near the oscillation minimum can interact with substantially fewer semiconductor states than a state that is near the oscillation maximum. The effect of the energy fluctuation on the ET is particularly pronounced in the alizarin system, where the molecular photoexcited state constantly crosses the CB edge, thereby generating two ET regimes. Outside of the band, the coupling of the chromophore excited state to the semiconductor states is small. Inside the band, the chromophore excited state interacts with a large number of TiO_2 states.

The Fourier transform (FT) of the evolution of the photoexcited state energy shown in the inset of Fig. 5.35 has several peaks in the region of 700 cm^{-1} and below. This frequency range is associated with the bending and torsional motions of the dye as well as with the TiO_2 modes. Small peaks are seen up to 1600 cm^{-1}, characteristic of the C-C and C=O stretches. Vibrations above 1600 cm^{-1} do not contribute to the oscillation of the photoexcited state energy, although they do contribute to the fluctuation of the photoexcited state localization [156]. Due to the delocalized nature of the chromophore's excited state, multiple vibrations modulate its energy.

Examples of single-molecule injection events

Thermal fluctuations of atomic coordinates produce a distribution of the photoexcited state energies and localizations that creates an inhomogeneous ensemble of initial conditions for the electron injection. The great variation in the individual ET events is illustrated in Fig. 5.36. Part a of this Figure shows the most typical example; the photoexcited state is located within the TiO_2 CB, and the injection occurs ultrafast and primarily by the adiabatic mechanism due to the strong chromophore-semiconductor coupling. Over time, the injected electron delocalizes from the original surface acceptor state into the bulk by NA hopping between the states, thereby increasing the contribution of the NA mechanism. Even though the photoexcited state exits the CB after 40 fs of this simulation, the electron remains delocalized inside the semiconductor. This first type of injection dominates the averages and is responsible for the sub-10 fs injection time, Fig. 5.37.

In the second example, shown in Fig. 5.36b, the initial injection is also fast and occurs adiabatically. The difference between it and the first example arises from the fact that the photoexcited state is close to the edge of the CB and is able to leave the band before the injected electron has had a chance to delocalize into the bulk by the NA mechanism. Adiabatic crossing of the same TS that has lead to the electron injection now results in the transfer of the electron back onto the chromophore. This pattern of adiabatic injection followed by adiabatic back-transfer occurs several times over the course of the simulation and is superimposed on a much slower NA ET that is practically irreversible.

In the third example, Fig. 5.36c, the photoexcited state is well below the TiO_2 CB edge, and essentially nothing happens with the electron until the photoexcited state crosses into the band. Only a small and slow NA injection component is seen during the first 20 fs of the simulation. Crossing the CB edge results in a rapid adiabatic transfer. It is quite remarkable that photoexcitation below the CB can lead to fast and efficient electron injection. [156,193]

Mechanism of electron injection

The time dependence of the averaged ET coordinate is presented in Fig. 5.37a by a thin line, together with the averaged adiabatic and NA contributions to the overall ET, Fig. 5.32. The overall ET rate is fit with the exponential

$$\mathrm{ET} = 1 - \exp\left[(t + t_0)/\tau\right]. \tag{5.59}$$

The fit, shown by a solid line, takes into account the fact that photoexcitation has already caused a partial ET prior to the simulated excited state dynamics, with t_0 representing the time the system is advanced along the ET reaction coordinate by the photoexcitation. On average the photoinduced ET dynamics start with 30% of the electron pretransferred by the photoexcitation. The value for an individual run varies substantially, see examples in Fig. 5.36, depending on the strength of the chromophore-semiconductor coupling, the relative energies of the chromophore, the semiconductor states, and other factors detailed in [156].

The simulation results reproduce the experimental data; the sub-10 fs time scale of ET obtained in the simulation of the alizarin-TiO_2 system agrees with

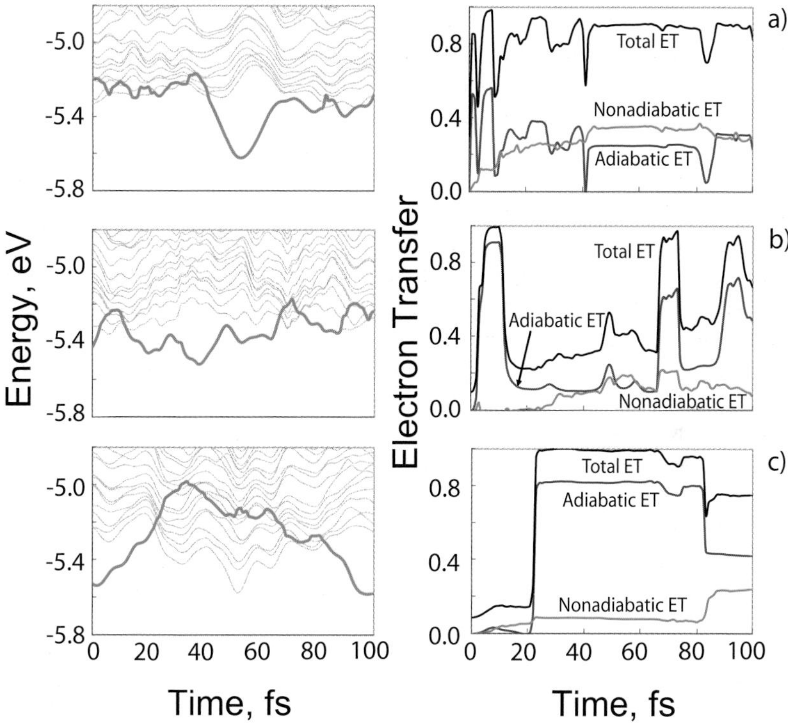

Fig. 5.36. Examples of individual ET events in the alizarin system. The left frames show evolutions of the photoexcited state energies, as in Fig. 5.35. The right panels present the ET progress, as in Fig. 5.37a below. For details see text.

the experimentally observed 6 fs injection time [141] and the small oscillations in the data that result from coherent nuclear vibrations are similar to those observed by Willig and co-workers with perylene [166]. The decomposition of the total ET into the NA and adiabatic contributions shows that the adiabatic ET mechanism, which requires strong chromophore-semiconductor coupling and relies on thermal fluctuations that drive the system over the TS, see Fig. 5.32, dominates the NA mechanism in the alizarin-TiO_2 system. Adiabatic transfer both is faster and has a larger contribution to the overall transfer. This is in contrast to the low temperature simulation of the isonicotinic acid chromophore [190], where the chromophore-semiconductor coupling is weaker and the NA ET mechanism is more prominent than the adiabatic mechanism.

The following general picture of the ET mechanism has emerged: The NA pathway is always present and contributes to the overall ET immediately following the photoexcitation. The efficiency of the NA transfer depends on the number of semiconductor states that are localized close to the molecular state, both in space and energy, and is only slightly dependent on the temperature.

NA ET electron acceptor states are weakly coupled to the donor. Adiabatic transfer, on the other hand, occurs between the donor state and a small number of acceptor states to which the donor state is strongly coupled. Adiabatic ET is temperature dependent and is more sensitive to the ET conditions. At low temperatures adiabatic ET is typically slow and is observed only when the NA transfer efficiency is reduced, but at high temperatures adiabatic ET dominates. Since both weakly and strongly coupled acceptor states are found in TiO_2, the two pathways are largely independent, and it is possible to observe a significant variation in the ET times between individual ET events. The alizarin-TiO_2 system presents a case of strong chromophore-semiconductor coupling. In other systems the dye is bound to the semiconductor through a longer bridge, resulting in a weaker coupling and a stronger NA component.

Electron evolution after injection

Following the photoexcitation and the chromophore-semiconductor ET, the electron delocalizes into the semiconductor bulk and, eventually, finds its way back to the chromophore ground state. The electron injection occurs between the dye and a TiO_2 state that is primarily restricted to the first 3 surface layers, Fig. 5.34b. In some cases, e.g. in the presence of isonicotinic acid [190,191], a single Ti atom can constitute 20% of the acceptor surface state. A rigorous description of electron delocalization into the bulk requires large simulations cells. However, even the small cell used in our simulations, Fig. 5.34a, can sufficiently detect this process. Fig. 5.37b shows the time evolution of the five largest adiabatic state occupations averaged over all initial conditions. The highest occupation decreases over the length of the run as the electron spreads to other adiabatic states. This effect cannot be described by a single exponential decay and is instead fit by a double exponential. The faster 7.9 fs component represents 38% of the total fit and corresponds to the electron injection across the interface. The longer 104 fs component of the fit describes the NA dynamics that follow the ultrafast injection and reflects the spreading of the electron population from the surface into the bulk. The simulated time scale provides a lower bound on the delocalization process and is in agreement with the available experimental data [234, 235].

Due to the high-surface area of dye-sensitized TiO_2, an electron delocalized inside bulk TiO_2 has a high probability of finding a surface. Trapping at the surface can result in ET back onto the chromophore or the electrolyte mediator, which normally brings the electron from the counter electrode to the dye. Figure 5.37c addresses the back-transfer process, assuming that the electron is trapped inside the first five surface layers, as represented by the simulation cell, Fig. 5.34. The results reported in Fig. 5.37c are obtained using the SH approach [215–217] that properly describes the energy relaxation, as discussed briefly in the theory section. The simulation shows that the back-transfer to the ground state of the dye (HOMO) occurs two orders of magnitude slower than the injection and is preceded by a 100 fs relaxation to the bottom of the

TiO_2 CB (LUMO). The back-transfer of the trapped electron is much faster than the time needed by the electron to travel through nano-porous TiO_2 to the primary electrode.

The data of Fig. 5.37 allows us to quantify the complete sequence of events: the ultrafast photoinduced electron injection from a chromophore to TiO_2 is followed by a rapid delocalization of the electron into the bulk that competes with electron relaxation to the bottom of the TiO_2 CB. Both delocalization and relaxation occur on a 100 fs time scale. If the electron is trapped at the surface, either immediately after the injection or following some evolution in the bulk, it returns to the dye ground state on a picosecond time scale.

In summary, Sect. 5.8 has shown that the chromophore-TiO_2 interface provides an excellent case study for elucidating the issues that arise when localized molecular species are combined with extended bulk materials. Such configurations have become increasingly common in recent years, as molecular and solid-state domains have converged, with molecules assembled into ever more complicated mesoscopic structures, and periodic systems miniaturized on the nanoscale. Understanding the molecule-bulk interfaces is one of the most challenging problems in a variety of fields and applications, including photovoltaics, photo- and electrochemistry, molecular electronics, photography, detection tools, bio-analytical chemistry, and biomechanics. Molecules and bulk materials are opposites in nearly every respect: the former have discrete electronic states, while the latter form energy bands; the fraction of high frequency vibrational modes is much higher in molecules than in inorganic semiconductors such as TiO_2; and the electron-phonon coupling and excitonic effects are much stronger in finite systems.

The analysis of the structure, electronic properties, and electron-vibrational dynamics in the chromophore-TiO_2 system indicates that the chromophore creates a local perturbation within the extended TiO_2 system. Most of the experimental data characterizing the effect that bulk TiO_2 has on chromophores can be modeled and understood with relatively small-scale calculations that include moderate-sized portions of the semiconductor. A cluster representation of the semiconductor can often be sufficient, and sometimes even the crudest few-atom representation of TiO_2 captures the essential phenomena.

The photoexcited states leading to the interfacial ET can be classified into three types. *First*, most commonly, the chromophore excited state is well inside the TiO_2 CB. This situation results in efficient electron injection even if the chromophore-semiconductor coupling is weak. The injection is facilitated by a high density of semiconductor states and proceeds by the NA mechanism. *Second*, if the chromophore-semiconductor coupling is strong, a new low energy photoexcitation involving direct ET from the molecule into the surface becomes possible, as exemplified with catechol. *Third*, the chromophore excited state can be near the TiO_2 CB edge, as in the alizarin system. Efficient injection at the low DOS region of the CB also requires a strong coupling and proceeds by the adiabatic mechanism.

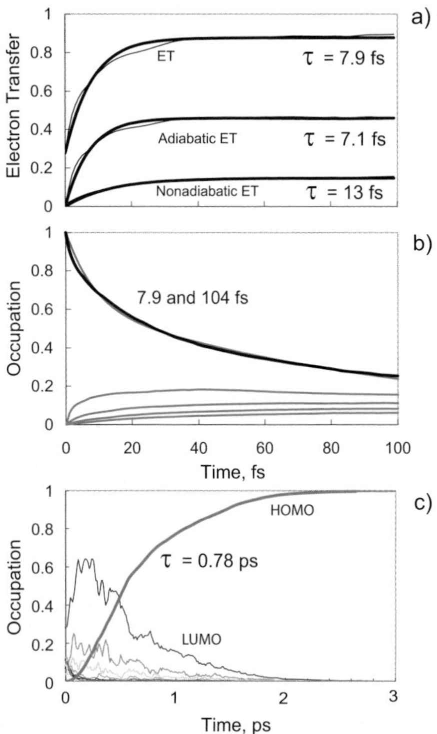

Fig. 5.37. (a) Average ET in the alizarin-TiO_2 system, separated into the adiabatic and nonadiabatic contributions. (b) Evolution of the average five largest occupations of the adiabatic states. The occupation of the initial state is fit with the sum of two exponents. The 7.9 fs contribution to the fit corresponds to the interfacial ET detailed in part (a). After the initial injection the electron spreads from surface to bulk, as reflected in the 104 fs fit component. (c) Back ET from TiO_2 to alizarin. The electron that is trapped within the first five surface layers returns to the chromophore ground state (HOMO) within 1 ps. Electron relaxation to the bottom edge of the TiO_2 CB (LUMO) occurs on a 100 fs time scale.

Real-time modeling of electron-vibrational dynamics is particularly valuable for understanding the interfacial electron injection, since it occurs on an ultrafast time scale and shows a variety of individual injection events with well-defined dynamical features that an average rate description cannot make apparent. Such simulations are still rare, but computationally demanding, state-of-the-art techniques are currently being developed in several groups throughout the world. As this research progresses, larger systems and longer time scales will become accessible, allowing one to probe more examples and finer details of the interfacial ET and to study different surfaces, surface defects, bridges, temperature and solvent dependence.

5.9 Outlook

C. Frischkorn and M. Wolf

The work described in the present Chapter on femtosecond-laser induced processes at surfaces and interfaces predominantly has focused on the *analysis* of ultrafast charge transfer and reaction dynamics. It is now well established that nonadiabatic coupling between the transient photoexcited electron distribution in a metal substrate and adsorbate nuclear degrees freedom can initiate ultrafast surface chemistry like desorption, diffusion, dissociation or association reactions. Traditionally, these processes have been described within a one-dimensional friction model, which reduces the multidimensional dynamics to an effective reaction coordinate coupled to the heat bath of substrate electrons. Recently, there is growing experimental evidence for more complex dynamical processes including coupling between different vibrational modes, as exemplified Sects. 5.3 and 5.4. Recent theoretical developments to describe such multidimensional nonadiabatic dynamics as outlined in Sects. 5.3 and 5.5 pave the way for an in depth microscopic understanding of surface reaction dynamics. The process of interfacial electron transfer (see Sects. 5.6 through 5.8) provides a basis to study the electronic-nuclear coupling which is underlying most nonadiabatic reaction dynamics at surfaces. However, this coupling leads also to ultrafast loss of electronic coherence and vibrational damping, which hinders the experimental realization of coherent and optimum *control* schemes in the field of surface femtochemistry [4]. Despite first attempts to successfully apply shaping of laser pulses to influence photoinduced surface chemistry [236], control of surface reaction by light still remains a challenge. One attractive route predicted by theoretical modeling may be optimum control in the electronic ground state by tailored infrared laser pulses. With the advances in ultrashort pulse generation, the required light fields should become available on a routine basis.

Acknowledgments

CF and MW want to thank Daniel Denzler, Steffen Wagner and Marco Rutkowski for their valuable contributions to the hydrogen recombination experiments and Helmut Zacharias, Alan C. Luntz, and Mats Persson for successful collaborations. JG and UH thank Krisztina Stépán and Michael Dürr for their contributions to the oxygen diffusion project and acknowledge funding by the Deutsche Forschungsgemeinschaft through SPP 1093 and GK

790, the German Israel Science Foundation and the Marburg Center for Optodynamics. PS, TK and MN acknowledge the computational work of Tijo Vazhappilly, and fruitful discussions with Stephanie Beyvers. LG and LW are grateful to the Deutsche Forschungsgemeinschaft for financial support through SPP 1093. Finally, OVP and WRD thank William Stier and Colleen Craig for implementing the classical path and surface hopping approaches within DFT and gratefully acknowledge financial support through awards from the USA National Science Foundation, Petroleum Research Fund and Department of Energy.

References

1. A.M. Wodtke, J.C. Tully, D.J. Auerbach, Int. Rev. Phys. Chem. **23**, 513 (2004)
2. H. Nienhaus, Surf. Sci. Rep. **45**, 1 (2002)
3. W. Ho, Acc. Chem. Res. **31**, 567 (1998)
4. C. Frischkorn, M. Wolf, Chem. Rev. **106**, 4207 (2006)
5. D.N. Denzler, C. Hess, S. Funk, G. Ertl, M. Bonn, C. Frischkorn, M. Wolf, in *Femtochemistry and Femtobiology - Ultrafast dynamics in molecular science*, ed. by A. Douhal, J. Santamaria (World Scientific, Singapore, 2002), p. 653
6. M.I. Kaganov, I.M. Lifshitz, L.V. Tanatarov, Sov. Phys. JETP **4**, 173 (1957)
7. S.I. Anisimov, B.L. Kapeliovich, T.L. Perel'man, Sov. Phys. JETP **39**, 375 (1974)
8. J. Hohlfeld, S.S. Wellershoff, J. Güdde, U. Conrad, V. Jähnke, E. Matthias, Chem. Phys. **251**, 237 (2000)
9. J. Hohlfeld, PhD thesis, Freie Universität Berlin (1998)
10. M. Lisowski, P.A. Loukakos, U. Bovensiepen, J. Stähler, C. Gahl, M. Wolf, Appl. Phys. A **78**, 165 (2004)
11. F. Budde, T.F. Heinz, A. Kalamarides, M.M.T. Loy, J.A. Misewich, Surf. Sci. **283**, 143 (1993)
12. M. Brandbyge, P. Hedegard, T.F. Heinz, J.A. Misewich, D.M. Newns, Phys. Rev. B **52**, 6042 (1995)
13. D.M. Newns, T.F. Heinz, J.A. Misewich, Prog. Theor. Phys. **106**, 411 (1991)
14. D.M. Newns, Phys. Rev. **178**, 1123 (1969)
15. J.A. Misewich, T.F. Heinz, D.M. Newns, Phys. Rev. Lett. **68**, 3737 (1992)
16. D. Menzel, R. Gomer, J. Chem. Phys. **41**, 3311 (1964)
17. P.A. Redhead, Can. J. Phys. **42**, 886 (1964)
18. P.R. Antoniewicz, Phys. Rev. B **21**, 3811 (1980)
19. F. Budde, T.F. Heinz, M.M.T. Loy, J.A. Misewich, F. de Rougemont, H. Zacharias, Phys. Rev. Lett. **66**, 3024 (1991)
20. J.A. Misewich, T.F. Heinz, A. Kalamarides, U. Höfer, M.M.T. Loy, J. Chem. Phys. **100**, 736 (1994)
21. L.M. Struck, L.J. Richter, S.A. Buntin, R.R. Cavanagh, J.C. Stephenson, Phys. Rev. Lett. **77**, 4576 (1996)
22. J.C. Tully, M. Gomez, M. Head-Gordon, J. Vac. Sci. Technol. A **11**, 1914 (1993)
23. M. Head-Gordon, J.C. Tully, J. Chem. Phys. **103**, 10137 (1995)
24. A.C. Luntz, M. Persson, J. Chem. Phys. **123**, 074704 (2005)

25. D.N. Denzler, C. Frischkorn, M. Wolf, G. Ertl, J. Phys. Chem. B **108**, 14503 (2004)
26. G. Ertl, H. Knözinger (eds.), *Handbook of heterogeneous catalysis* (VCH, Weinheim, 1997)
27. C.G.V. de Walle, J. Neugebauer, Nature **423**, 626 (2003)
28. K. Christmann, Prog. Surf. Sci. **48**, 15 (1995)
29. G.A. Somorjai, J. Phys. Chem. B **106**, 9201 (2002)
30. K. Christmann, Surf. Sci. Rep. **9**, 1 (1988)
31. D.N. Denzler, C. Frischkorn, C. Hess, M. Wolf, G. Ertl, Phys. Rev. Lett. **91**, 226102 (2003)
32. C.M. Mate, G.A. Somorjai, Phys. Rev. B **34**, 7417 (1986)
33. N. Takagi, Y. Yasui, T. Takaoka, M. Sawada, H. Yanagita, T. Aruga, M. Nishijima, Phys. Rev. B **53**, 13767 (1996)
34. C. Frischkorn, Surf. Sci. **593**, 67 (2005)
35. M. Bonn, S. Funk, C. Hess, D.N. Denzler, C. Stampfl, M. Scheffler, M. Wolf, G. Ertl, Science **285**, 1042 (1999)
36. M. Yada, R.J. Madix, Surf. Sci. **328**, 171 (1995)
37. H. Eyring, M. Polanyi, Z. Phys. Chem. B **12**, 279 (1931)
38. L. Diekhöner, L. Hornekaer, H. Mortensen, E. Jensen, A. Baurichter, V.V. Petrunin, A.C. Luntz, J. Chem. Phys. **117**, 5018 (2002)
39. S. Wagner, C. Frischkorn, M. Wolf, M. Rutkowski, H. Zacharias, A. Luntz, Phys. Rev. B **72**, 205404 (2005)
40. K.W. Kolasinski, W. Nessler, A. de Meijere, E. Hasselbrink, Phys. Rev. Lett. **72**, 1356 (1994)
41. E. Hasselbrink, in *Laser spectroscopy and photochemistry on metal surfaces*, ed. by H.L. Dai, W. Ho (World Scientific, Singapore, 1995), p. 685
42. A.C. Luntz, M. Persson, S. Wagner, C. Frischkorn, M. Wolf, J. Chem. Phys. **124**, 244702 (2006)
43. J.C. Tully, J. Chem. Phys. **73**, 1975 (1995)
44. P.N.J. Persson, *Sliding friction: Physical principles and applications* (Springer, Berlin, 1998)
45. K.P. Bohnen, M. Kiwi, H. Suhl, Phys. Rev. Lett. **34**, 1512 (1975)
46. A. Nourtier, Journal De Physique **38**, 479 (1977)
47. L. Bartels, F. Wang, D. Moller, E. Knoesel, T.F. Heinz, Science **305**, 648 (2004)
48. K. Stépán, J. Güdde, U. Höfer, Phys. Rev. Lett. **94**, 236103 (2005)
49. K. Stépán, M. Dürr, J. Güdde, U. Höfer, Surf. Sci. **593**, 54 (2005)
50. E.H.G. Backus, A. Eichler, A.W. Kleyn, M. Bonn, Science **310**, 1790 (2005)
51. H. Wang, R.G. Tobin, D.K. Lambert, C.L. DiMaggio, G.B. Fisher, Surf. Sci. **372**, 267 (1997)
52. A.T. Gee, B.E. Hayden, J. Chem. Phys. **113**, 10333 (2000)
53. J.L. Gland, B.A. Sexton, G.B. Fisher, Surf. Sci. **95**, 587 (1980)
54. A. Winkler, X. Guo, H.R. Siddiqui, P.L. Hagans, J.T. Yates, Surf. Sci. **201**, 419 (1988)
55. P. Gambardella, Z. Sljivancanin, B. Hammer, M. Blanc, K. Kuhnke, K. Kern, Phys. Rev. Lett. **87**, 056103 (2001)
56. Y.R. Shen, Annu. Rev. Phys. Chem. **40**, 327 (1989)
57. G.A. Reider, T.F. Heinz, in *Photonic probes of surfaces*, ed. by P. Halevi (North-Holland, Amsterdam, 1995), pp. 3–66
58. G. Lüpke, D.J. Bottomley, H.M. van Driel, J. Opt. Soc. Am. B **11**, 33 (1994)

59. P. Kratzer, E. Pehlke, M. Scheffler, M.B. Raschke, U. Höfer, Phys. Rev. Lett. **81**, 5596 (1998)
60. M.B. Raschke, U. Höfer, Phys. Rev. B **59**, 2783 (1999)
61. J.E. Reutt-Robey, D.J. Doren, Y.J. Chabal, S.B. Christman, J. Chem. Phys. **93**, 9113 (1990)
62. J. Güdde, U. Höfer, J. Phys.: Condens. Matter **18**, S1409 (2006)
63. P.J. Feibelman, S. Esch, T. Michely, Phys. Rev. Lett. **77**, 2257 (1996)
64. C.T. Campbell, G. Ertl, H. Kuipers, J. Segner, Surf. Sci. **107**, 220 (1981)
65. S. Deliwala, R.J. Finlay, J.R. Goldman, T.H. Her, W.D. Mieher, E. Mazur, Chem. Phys. Lett. **242**, 617 (1995)
66. D.G. Busch, W. Ho, Phys. Rev. Lett. **77**, 1338 (1996)
67. J.A. Prybyla, T.F. Heinz, J.A. Misewich, M.M.T. Loy, J.H. Glownia, Phys. Rev. Lett. **64**, 1537 (1990)
68. F.J. Kao, D.G. Busch, D. Cohen, D.G. Dacosta, W. Ho, Phys. Rev. Lett. **71**, 2094 (1993)
69. X.Y. Zhu, S.R. Hatch, A. Campion, J.M. White, J. Chem. Phys. **91**, 5011 (1989)
70. K. Watanabe, K. Sawabe, Y. Matsumoto, Phys. Rev. Lett. **76**, 1751 (1996)
71. M. Bauer, S. Pawlik, R. Burgermeister, M. Aeschlimann, Surf. Sci. **404**, 62 (1998)
72. S. Ogawa, H. Nagano, H. Petek, Phys. Rev. Lett. **82**, 1931 (1999)
73. M. Wolf, A. Hotzel, E. Knoesel, D. Velic, Phys. Rev. B **59**, 5926 (1999)
74. C. Lei, M. Bauer, K. Read, R. Tobey, Y. Liu, T. Popmintchev, M.M. Murnane, H.C. Kapteyn, Phys. Rev. B **66**, 245420 (2002)
75. K. Stépán, PhD thesis, Philipps-Universität Marburg (2006)
76. J. Wintterlin, R. Schuster, G. Ertl, Phys. Rev. Lett. **77**, 123 (1996)
77. A. Bogicevic, J. Stromquist, B.I. Lundqvist, Phys. Rev. B **57**, R4289 (1998)
78. U. Engström, R. Ryberg, Phys. Rev. Lett. **82**, 2741 (1999)
79. C. Puglia, A. Nilsson, B. Hernnas, O. Karis, P. Bennich, N. Martensson, Surf. Sci. **342**, 119 (1995)
80. A.C. Luntz, private communications
81. M. Bonn, D.N. Denzler, S. Funk, M. Wolf, S.S. Wellershoff, J. Hohlfeld, Phys. Rev. B **61**, 1101 (2000)
82. T. Komeda, Y. Kim, M. Kawai, B.N.J. Persson, H. Ueba, Science **295**, 2055 (2002)
83. J.I. Pascual, N. Lorente, Z. Song, H. Conrad, H.P. Rust, Nature **423**, 525 (2003)
84. B.N.J. Persson, H. Ueba, Surf. Sci. **502**, 18 (2002)
85. H. Ueba, T. Mii, N. Lorente, B.N.J. Persson, J. Chem. Phys. **123**, 084707 (2005)
86. B.N.J. Persson, R. Ryberg, Phys. Rev. Lett. **54**, 2119 (1985)
87. U. Engström, R. Ryberg, J. Chem. Phys. **112**, 1959 (2000)
88. W.S. Fann, R. Storz, H.W.K. Tom, J. Bokor, Phys. Rev. Lett. **68**, 2834 (1992)
89. F. Weik, A. de Meijere, E. Hasselbrink, J. Chem. Phys. **99**, 682 (1993)
90. R. Knorren, G. Bouzerar, K.H. Bennemann, Phys. Rev. B **63**, 094306 (2001)
91. M. Nest, P. Saalfrank, J. Chem. Phys. **116**, 7189 (2002)
92. S. Gao, B.I. Lundquist, W. Ho, Surf. Sci. **341**, L1031 (1995)
93. J.C. Tully, M. Gomez, M. Head-Gordon, J. Vac. Sci. Technol. A **11**, 1914 (1993)
94. M. Morin, N.J. Levinos, A.L. Harris, J. Chem. Phys. **96**, 3950 (1992)

95. S. Gao, Phys. Rev. B **55**, 1876 (1997)
96. P.W. Anderson, Phys. Rev. **124**, 41 (1961)
97. P. Saalfrank, R. Baer, R. Kosloff, Chem. Phys. Lett. **230**, 463 (1994)
98. P. Saalfrank, R. Kosloff, J. Chem. Phys. **105**, 2441 (1996)
99. S.A. Buntin, L.J. Richter, R.R. Cavanagh, D.S. King, Phys. Rev. Lett. **61**, 1321 (1988)
100. M. Nest, P. Saalfrank, Phys. Rev. B **69**, 235405 (2004)
101. L. Cai, X. Xiao, M.M.T. Loy, Surf. Sci. **464**, L727 (2000)
102. M. Kampling, K. Al-Shamery, H.J. Freund, M. Wilde, K. Fukutani, Y. Murata, Phys. Chem. Chem. Phys. **4**, 2629 (2002)
103. G. Boendgen, P. Saalfrank, J. Phys. Chem. B **102**, 8029 (1998)
104. P. Saalfrank, G. Boendgen, C. Corriol, T. Nakajima, Faraday Discuss. **117**, 65 (2000)
105. K. Nakagami, Y. Ohtsuki, Y. Fujimura, Chem. Phys. Lett. **360**, 91 (2002)
106. V.S. Letokhov, Science **180**, 451 (1973)
107. F.F. Crim, Science **249**, 1387 (1990)
108. A. Abe, K. Yamashita, P. Saalfrank, Phys. Rev. B **67**, 235411 (2003)
109. G.K. Paramonov, P. Saalfrank, Chem. Phys. Lett. **301**, 509 (1999)
110. G.K. Paramonov, P. Saalfrank, J. Chem. Phys. **110**, 6500 (1999)
111. J.K. Vincent, R.A. Olsen, G.J. Kroes, M. Luppi, E.J. Baerends, J. Chem. Phys. **122**, 044701 (2005)
112. T. Vazhappilly, T. Klamroth, P. Saalfrank, (to be published)
113. H. Shi, K. Jacobi, Surf. Sci. **313**, 289 (1994)
114. S. Beyvers, Y. Ohtsuki, P. Saalfrank, J. Chem. Phys. **124**, 234706 (2006)
115. Y. Ohtsuki, W. Zhu, H. Rabitz, J. Chem. Phys. **110**, 9825 (1999)
116. J.W. Gadzuk, L.J. Richter, S.A. Buntin, D.S. King, R.R. Cavanagh, Surf. Sci. **235**, 317 (1990)
117. P. Saalfrank, Chem. Phys. **211**, 265 (1996)
118. K.F. und P. Saalfrank, Chem. Phys. Lett. **268**, 291 (1997)
119. P. Saalfrank, G. Boendgen, K. Finger, L. Pesce, Chem. Phys. **251**, 51 (2000)
120. N.A. Besley, Chem. Phys. Lett. **390**, 124 (2004)
121. V. Mohan, N. Sathyamurthy, Comput. Phys. Rep. **48**, 213 (1988)
122. U. Thumm, P. Kürpick, U. Wille, Phys. Rev. B **61**, 3067 (2000)
123. W. Ho, Surf. Sci. **363**, 166 (1996)
124. P.B. Corkum, F. Brunel, N.K. Sherman, T. Srinivasan-Rao, Phys. Rev. Lett. **61**, 2886 (1988)
125. A. Abe, K. Yamashita, J. Chem. Phys. **119**, 9710 (2003)
126. D. DeVault, *Quantum–mechanical tunneling in biological systems*, (Cambridge Univ. Press: London, 1984)
127. E.G. Petrov, *Physics of charge transfer in biosystems* (Naukowa Dumka: Kiev, 1984) (in Russian)
128. R.A. Marcus, N. Sutin, Biochim. Biophys. Acta **811**, 265 (1985)
129. J. Jortner, M. Bixon, Adv. Chem. Phys. **106**, 35 (1999)
130. V. May, O. Kühn, *Charge and energy transfer dynamics in molecular systems* (Wiley–VCH, Berlin, 2004)
131. F.L. Carter (ed.) *Molecular electronic devices* (Marcel Dekker, New York, 1982).
132. A. Nitzan, Mark A. Ratner. Science, **300**, 1384 (2003)
133. B. O'Regan, M. Grätzel, Nature **353**, 737 (1991).

134. R.J.D. Miller, G. McLendon, A. Nozik, W. Schmickler, F. Willig, *Surface electron transfer processes* (VCH Publishers, New York, 1995)
135. J.B. Asbury, E. Hao, T. Wang, H. N. Ghosh, T. Lian, J. Phys. Chem. B **105**, 4545 (2001)
136. J. Kallioinen, G. Benkö, V. Sundström, J.E.I. Korppi-Tommola, A.P. Yartsev, J. Phys. Chem. B **106**, 4396 (2002)
137. A. Hagfeldt, M. Grätzel, Chem. Rev. **95**, 49 (1995)
138. K. Schwarzburg, R. Ernstorfer, S. Felber, F. Willig, Coord. Chem. Rev. **248**, 1259 (2004)
139. B. Burfeindt, T. Hannappel, W. Storck, F. Willig, J. Phys. Chem. **100**, 16463, (1996)
140. C. Zimmermann, F. Willig, S. Ramakrishna, B. Burfeindt, B. Pettinger, R. Eichberger, W. Storck, J. Phys. Chem. B **105**, 9245, (2001)
141. R. Huber, J.–E. Moser, M. Grätzel, J. Wachtveitl, J. Phys. Chem. B **106**, 6494 (2002)
142. T.H. James (ed.) *The theory of the photographic process* (McMillan Publishing Co., New York, 1977)
143. V. Shklover, M.-K. Nazeeruddin, S. M. Zakeeruddin, C. Barbe, A. Kay, T. Haibach, W. Steurer, R. Hermann, H.–U. Nissen, M. Grätzel, Chem. Mater. **9**, 430 (1997)
144. R. Ernstorfer, PhD Thesis, Free University Berlin, (2004)
145. L. Gundlach, PhD Thesis, Free University Berlin, (2005)
146. A. Hermann, W.G. Schmidt, F. Bechstedt, J. Phys. Chem. B **109**, 7928 (2005)
147. J. Moser, S. Punchihewa, P.P. Infelta, M. Grätzel, Langmuir **7**, 3012 (1991)
148. L. Wang, R. Ernstorfer, F. Willig, V. May, J. Phys. Chem. B **109**, 9589 (2005)
149. L. Wang, F. Willig, V. May, J. Chem. Phys. **124**, 014712 (2006)
150. W.R. Duncan, O.V. Prezhdo, J. Phys. Chem. B **109**, 17998 (2005), and references given in this paper.
151. P. Persson, M.J. Lundqvist, R. Ernstorfer, W.A. Goddard III, F. Willig, J. Chem. Theory. Comput. **2**, 441 (2006), and references given in this paper.
152. J.P. Muscat, D.M. Newns, Prog. Surf. Sci. **9**, 1 (1978)
153. G. Benkö, J. Kallioinen, J.E.I. Korppi–Tommola, A.P. Yartsev, V. Sundström, J. Am. Chem. Soc. **124**, 489 (2002)
154. P. C. Redfern, P. Zapol, L. A. Curtiss, T. Rajh, M. C. Thurnauer, J. Phys. Chem. B **107**, 11419 (2003)
155. S. G. Abuabara, L.G.C. Rego, V.S. Batista, J. Am. Chem. Soc. **127**, 18234 (2005), and references given in this paper.
156. W.R. Duncan, W.M. Stier, O.V. Prezhdo, J. Am. Chem. Soc. **127**, 7941 (2005)
157. C. Zimmermann, F. Willig, S. Ramakrishna, B. Burfeindt, B. Pettinger, R. Eichberger, W. Storck, J. Phys. Chem. B **105**, 9245 (2001)
158. L. Gundlach, S. Felber, W. Storck, E. Galoppini, Q. Wei, F. Willig, Res. Chem. Intermed. **31**, 39 (2005)
159. B. Trösken, F. Willig, K. Schwarzburg, A. Ehert, M. Spitler, J. Phys. Chem. **99**, 5152 (1995)
160. L. Gundlach, R. Ernstorfer, F. Willig, Phys. Rev. B **74**, 035324 (2006)
161. L. Gundlach, J. Szarko, L. D. Socaciu-Siebert, A. Neubauer, R. Ernstorfer, F. Willig. Phys. Rev. B (submitted)
162. S. Ramakrishna, F. Willig, J. Phys. Chem. B **104**, 68, (2000)
163. S. Ramakrishna, F. Willig, V. May, Phys. Rev. B **62**, R16330, (2000)

164. S. Ramakrishna, F. Willig, V. May, J. Chem. Phys. **115**, 2743, (2001)
165. S. Ramakrishna, F. Willig, V. May, Chem. Phys. Lett. **351**, 242 (2002)
166. S. Ramakrishna, F. Willig, V. May, A. Knorr, J. Phys. Chem. B **107**, 607 (2003)
167. L. Wang, V. May, J. Chem. Phys. **121**, 8039 (2004)
168. L. Wang, F. Willig, V. May, in *Molecular simulation, special issue on Electron Transfer*, ed. by D.N. Beratan, S.S. Skourtis (in press)
169. M. Thoss, I. Kondov, H. Wang, Chem. Phys. **304**, 169 (2004)
170. I. Kondov, M. Thoss, H. Wang, J. Phys. Chem. A **110**, 1364 (2006)
171. P. Persson, R. Bergström, S. Lunell, J. Phys. Chem. B **104**, 10348, (2000)
172. L.G.C. Rego, V.S. Batista, J. Am. Chem. Soc. **125**, 7989, (2003)
173. W. R. Duncan, O. V. Prezhdo, J. Phys. Chem. B **109**, 365 (2005)
174. P. Persson, M.J. Lundqvist, R. Ernstorfer, W.A. Goddard III, F. Willig, J. Chem. Theory Comp. **2**, 441, (2006)
175. M. Nilsing, S. Lunell, P. Persson, L. Ojamäe, Surf. Sci. **582**, 49 (2005)
176. F. Willig, C. Zimmermann, S. Ramakrishna, W. Storck, Electrochim. Acta **45**, 4565 (2000)
177. B. Burfeindt, C. Zimmermann, S. Ramakrishna, T. Hannappel, B. Meißner, W. Storck, F. Willig, Z. Phys. Chem. **212**, 67 (1999)
178. P.C. Redfern, P. Zapol, L.A. Curtiss, T. Rajh, M.C. Thurnauer, J. Phys. Chem. B **107**, 11419 (2003)
179. P.H. Hahn, W.G. Schmidt, Surf. Rev. Lett. **10**, 163 (2003)
180. W.G. Schmidt, P.H. Hahn, F. Bechstedt, N. Esser, P. Vogt, A. Wange, W. Richter. Phys. Rev. Lett. **90**, 126101 (2003)
181. T. Letzig, H.-J. Schimper, T. Hannappel, F. Willig, Phys. Rev. B, **71**, 033308 (2005)
182. T.M. Halasinski, J.L. Weisman, R. Ruiterkamp, T.J. Lee, F. Salama, M. Head-Gordon, J. Phys. Chem. A **107**, 3660 (2003)
183. K. K. Ong, J.O. Jensen, H.F. Hameka, J. Mol. Str. (Theochem). **459**, 131 (1999)
184. R. Ernstorfer, L. Gundlach, S. Felber, W. Storck, R. Eichberger, F. Willig, J. Phys. Chem. B (accepted)
185. R. Rodriguez, M.A. Blesa, A.E. Regazzoni, J. Coll. Int. Sci. **177**, 122, (1996)
186. J.L. Herek (ed.), *Coherent control of photochemical and photobiological processes*, special issue, J. Photochem. Photobiol. (2006)
187. T. Halfmann (ed.), *Quantum control of light and matter*, special issue, Opt. Comm. (2006).
188. A. P. Pierce, M.A. Dahleh, H. Rabitz, Phys. Rev. A **37**, 4950 (1988)
189. S.G. Abuabara, L.G.C. Rego, V.S. Batista, J. Am. Chem. Soc. **127**, 18234 (2005)
190. W. Stier, O.V. Prezhdo, J. Phys. Chem. B **106**, 8047 (2002)
191. W. Stier, O.V. Prezhdo, Isr. J. Chem. **42**, 213 (2003)
192. W. Stier, W.R. Duncan, O.V. Prezhdo, SPIE Proc. **5223**, 132 (2003)
193. W. Stier, W.R. Duncan, O.V. Prezhdo, Adv. Mat. **16**, 240 (2004)
194. W.R. Duncan, O.V. Prezhdo, J. Phys. Chem. B **109**, 17998 (2005)
195. C.F. Craig, W.R. Duncan, O.V. Prezhdo, Phys. Rev. Lett. **95**, 163001 (2005)
196. R. Argazzi, N.Y.M. Iha, H. Zabri, F. Odobel, C.A. Bignozzi, Coord. Chem. Rev. **248**, 1299 (2004)
197. Y. Wang, K. Hang, N.A. Anderson, T. Lian, J. Phys. Chem. B **107**, 9434 (2003)

198. J.B. Asbury, R.J. Ellingson, H.N. Ghosh, S. Ferrere, A.J. Nozik, T.Q. Lian, J. Phys. Chem. B **103**, 3110 (1999)
199. J.B. Asbury, E.C. Hao, Y.Q. Wang, H.N. Ghosh, T.Q. Lian, J. Phys. Chem. B **105**, 4545 (2001)
200. R. Memming, *Semiconductor Electrochemistry* (Wiley-VCH, Weinheim, 2001)
201. C. Kittel, *Introduction to solid state Physics* (Wiley, New York, 1996)
202. W. Kohn, L.J. Sham, Phys. Rev. A **4**, 1133 (1965)
203. T. Frauenheim, G. Seifert, M. Elstner, T. Niehaus, C. Köhler, M. Amkreutz, M. Sternberg, Z. Hajnal, A. DiCarlo, S. Suhai, J. Phys.: Condens. Matter **14**, 3015 (2002)
204. M.A.L. Marques, E.K.U. Gross, Annu. Rev. Phys. Chem. **55**, 427 (2004)
205. R. Baer, T. Seideman, S. Ilani, D. Neuhauser, J. Chem. Phys. **120**, 3387 (2004)
206. I. Franco, S. Tretiak, J. Am. Chem. Soc. **126**, 12130 (2004)
207. M.A.L. Marques, X. López, D. Varsano, A. Castro, A. Rubio, Phys. Rev. Lett. **90**, 258101 (2003)
208. J.P. Perdew, K. Burke, M. Ernzerhof, Phys. Rev. Lett. **77**, 3685 (1996)
209. G. Kresse, J. Furthmüller, Comput. Mater. Sci. **6**, 15 (1996)
210. G. Kresse, J. Furthmüller, Phys. Rev. B **54**, 11169 (1996)
211. S. Hammes-Schiffer, J.C. Tully, J. Chem. Phys. **101**, 4657 (1994)
212. G.D. Billing, Int. Rev. Phys. Chem. **13**, 309 (1994)
213. J.C. Tully, in *Classical and quantum dynamics in condensed phase simulations*, ed. by B.J. Berne, G. Ciccotti, D.F. Coker (World Scientific, Singapore, 1998), pp. 489–514
214. P. Ehrenfest, Z. Physik **45**, 455 (1927)
215. J.C. Tully, R.K. Preston, J. Chem. Phys. **55**, 562 (1971)
216. J.C. Tully, J. Chem. Phys. **93**, 1061 (1990)
217. P.V. Parahdekar, J. C. Tully, J. Chem. Phys. **122**, 094102 (2005)
218. T.N. Todorov, J. Hoekstra, A.P. Sutton, Phys. Rev. Lett. **86**, 3606 (2001)
219. A.P. Horsfield, D.R. Bowler, A.J. Fisher, T.N. Todorov, C.G. Sanchez, J. Phys. Cond. Mat. **16**, 8251 (2004)
220. O.V. Prezhdo, Y.V. Pereverzev, J. Chem. Phys. **113**, 6557 (2000)
221. C. Brooksby, O.V. Prezhdo, Chem. Phys. Lett. **346**, 463 (2001)
222. O.V. Prezhdo, Y.V. Pereverzev, J. Chem. Phys. **116**, 4450 (2002)
223. E. Pahl, O.V. Prezhdo, J. Chem. Phys. **116**, 8704 (2002)
224. O.V. Prezhdo, J. Chem. Phys. **117**, 2995 (2002)
225. C. Brooksby, O.V. Prezhdo, Chem. Phys. Lett. **378**, 533 (2003)
226. D. Kilin, Y.V. Pereverzev, O.V. Prezhdo, J. Chem. Phys. **120**, 11209 (2004).
227. E. Heatwole, O.V. Prezhdo, J. Chem. Phys. **121**, 10967 (2004).
228. E. Heatwole, O.V. Prezhdo, J. Chem. Phys. **122**, 234109 (2005)
229. O.V. Prezhdo, P.J. Rossky, Phys. Rev. Lett. **81**, 5294 (1998)
230. O.V. Prezhdo, J. Chem. Phys. **111**, 8366 (1999)
231. O.V. Prezhdo, Phys. Rev. Lett. **85**, 4413 (2000)
232. A. Luis, Phys. Rev. A **67**, 062113 (2003)
233. P. Exner, J. Phys. A: Math. Gen. **38**, L449 (2005)
234. S. Ramakrishna, F. Willig, A. Knorr, Surf. Sci **558**, 159 (2004)
235. L. Toben, L. Gundlach, R. Ernstorfer, R. Eichberger, T. Hannappel, F. Willig, A. Zeiser, J. Forstner, A. Knorr, P. H. Hahn, M.G. Schmidt, Phys. Rev. Lett. **94**, 067601 (2005)
236. P. Nuernberger, D. Wolpert, H. Weiss, G. Gerber, in *Ultrafast Phenomena XV*, ed. by P.B. Corkum, D.M. Jonas, R.J.D. Miller, and A.M. Weiner (Springer Series of Chemical Physics, Springer, Berlin,) Vol. 88

6

Molecules and clusters in strong laser fields

Claus Peter Schulz[1], Tobias Burnus[2], Alberto Castro[3], E. K. U. Gross[3], Andreas Heidenreich[4], Ingolf V. Hertel[1,5], Joshua Jortner[4], Tim Laarmann[1], Isidore Last[4], Robert J. Levis[6], Miguel A. L. Marques[7], Dmitri A. Romanov[6], and Alejandro Saenz[8]

[1] Max Born Institute, Berlin, Germany
[2] II.Physikalisches Institut, Universität zu Köln, Germany
[3] Institut für Theoretische Physik, Freie Universität Berlin, Germany
[4] School of Chemistry, Tel Aviv University, Israel
[5] Institut für Experimentalphysik, Freie Universität Berlin, Germany
[6] Temple University, Philadelphia, USA
[7] Departamento de Física, Universidade de Coimbra, Portugal
[8] Institut für Physik, Humboldt-Universität zu Berlin, Germany

Coordinated by: Claus Peter Schulz

6.1 Light matter interaction in the high field regime

In light matter interaction a 'strong' laser field will be defined differently depending on the process under investigation. In the context of this chapter a laser field is considered to be strong if the potential energy surfaces of the irradiated molecule or cluster are altered considerably. On an absolute scale one can compare the electromagnetic field of a laser pulse with inner-atomic fields. The electric field amplitude F of a laser pulse with the intensity I is given by

$$F = \sqrt{\frac{2I}{\varepsilon_0 c}} \tag{6.1}$$

with ε_0 being the vacuum dielectric constant, and c the velocity of light. For a ground state H-atom (with its electron in a distance of 1 Bohr radius a_0 from the nucleus) the field is strong if the intensity is $I \geq 3.5 \times 10^{16}\,\mathrm{Wcm}^{-2}$, while above some $10^{18}\,\mathrm{Wcm}^{-2}$ relativistic effects as well as the magnetic field component of the field become important. However, it turns out that molecules and clusters already completely disintegrate at intensities above $10^{16}\,\mathrm{Wcm}^{-2}$, ejecting fast highly charged ions. This interesting new research field of cluster dynamics initiated by ultraintense laser pulses will be presented in Sect. 6.6.

To study the chemical aspects one has to consider the typical energy level spacing of polyatomic molecules and clusters for the definition of the 'strong field regime'. Laser intensities in excess of $10^{12}\,\mathrm{Wcm}^{-2}$ already induce substantial Stark shifting, polarization, and disturbance of the field free electronic states occurs to produce a quasi-continuum of new states in the molecule. Thus, intense short pulse lasers have led to the observation of many interesting strong field phenomena in atoms, molecules and clusters including: X-ray generation from high harmonics [1], above threshold ionization [2], above threshold dissociation [3], multiple electron emission from molecules [4], intact ionization of large polyatomic molecules [5–7], forced molecular rotation in an optical centrifuge [8], production of extremely high charge states from molecular clusters [9], production of highly energetic ions [10], and neutrons from clusters [11].

6.1.1 Experimental aspects

It is useful to mention some experimental aspects of strong field studies here, although the various laser systems and experimental methods of ultrafast physics will not be discussed. Suffice it to say that typically the basis of all table top short pulse laser systems used in these studies is the Ti:Sapphire laser at 800 nm which may be converted conveniently to the second harmonics and by using other nonlinear conversion schemes visible or near infrared laser wavelength can be employed. Present state of the art high intensity laser facilities can provide intensities up to $10^{20}\,\mathrm{Wcm}^{-2}$, using 800 nm Ti:Sapphire lasers with sub-50 fs pulse duration at a comfortable 10 Hz repetition rate (see e.g. [12]). Techniques to manipulate, stretch, shape and characterize the laser pulses have already been described in Chapter 2. However, a few subtleties concerning the definition of intensities, and calibrating them have to be discussed here.

Intensity calibration with short pulse lasers is a non trivial problem. What can be measured directly is the pulse energy E_P. But since one never has a constant spatial and temporal distribution of intensity it is important to define clearly to which quantity a given intensity refers. In the most favorable case the spatial and temporal distribution is Gaussian and the intensity I (mostly given in Wcm^{-2}) is

$$I(r,t) = I_m \exp(-(r/w)^2)\exp(-(t/\tau)^2) \tag{6.2}$$

with I_m giving the maximum intensity at $t = r = 0$ with τ and w characterizing the temporal and spatial beam profile – quantities which can in principle be measured experimentally by determining the autocorrelation function (for τ) and, e.g., the total energy passing a knife edge which is moved into the beam (for w). For a Gaussian one easily works out that

$$I_m = \frac{E_p}{\sqrt{\pi}\tau\pi w^2} \simeq \frac{E_p}{t_h\,(\pi w^2)} = 0.83\frac{E_p}{t_h d_h^2} \tag{6.3}$$

with t_h and d_h being FWHM of the temporal and spatial pulse width, respectively. We note that the maximum intensity at the pulse center corresponds to a hypothetical cylindrical beam of constant intensity of radius w. Note, however, that w is the $1/e$ beam waist radius, not the $1/e^2$ radius often given for laser beams! Since the beam parameters are difficult to determine quantitatively, the intensity in typical ion yield experiments is usually not measured absolutely. Rather one calibrates the intensity with a known intensity dependence for a standard ion (typically Xe^+) or compares photoelectron spectra with known data from the literature [13, 14].

6.1.2 Processes in strong laser fields: An overview

The response of an atomic or molecular system to a laser field can approximately be divided into two different regimes, the multi photon and the quasi-static regime. One relevant quantity to define the boundary is the average oscillation energy which a free electron acquires in the radiation field of the laser pulse. This *ponderomotive potential* is given by

$$U_p = \frac{e_0^2}{2m_e \varepsilon_0 \, c} \frac{I}{\omega_L^2} \propto \lambda^2 I \tag{6.4}$$

with m_e the mass of the electron, ω_L the angular frequency of the laser radiation. Note that the ponderomotive potential depends quadratically on the laser wavelength λ, i.e., is most significant in the infrared wavelength region. To determine whether a field is strong one has thus to compare U_p with the atomic or molecular energy in question. When discussing ionization processes one would consider the *quasi-static regime* to begin at

$$U_p > E_I \tag{6.5}$$

where E_I is the ionization potential of the system. If one takes for example C_{60} with $E_I = 7.58\,\text{eV}$ an 800 nm Ti:Sapphire pulse would in that sense be 'strong' when $I > 1.3 \times 10^{14}\,\text{Wcm}^{-2} = I_{crit}$.

For low intensities and high frequencies ($I \ll I_{crit} \propto \omega_L^2 E_I$) the interaction can be well described by lowest-order perturbation theory (LOPT). The order N of the process is then given by the least number of photons required to reach the ionic ground state from the neutral initial state, i.e., N is the smallest integer fulfilling $N\hbar\omega_L > E_I$. This multi photon regime is illustrated for an atom in Fig. 6.1 (left panel). Note, that in LOPT approximation only a single photo electron peak with the kinetic energy $E_0 = N\hbar\omega_L - E_I$ exists.

Within the LOPT approximation the ionization rate Γ_{LOPT} is proportional to the Nth power of the laser intensity I, $\Gamma_{\text{LOPT}} \propto \sigma^{(N)} I^N$ where $\sigma^{(N)}$ is the generalized N-photon cross-section that depends on the atomic or molecular system under consideration. The ab initio evaluation of $\sigma^{(N)}$ for systems only slightly more complicated than hydrogen-like atoms is, however, a very demanding task, since it implies a summation over all field-free eigenstates, including all continua. Converged 2-, 3-, and 4-photon cross-sections for atomic helium were presented in [15].

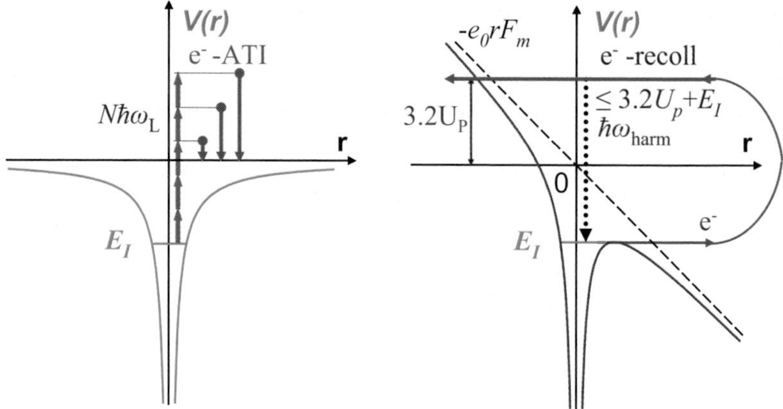

Fig. 6.1. Schematic illustration of an atom in a strong field with electron energetics and high harmonic generation. Left: multi photon ionization (upward arrows indicate photon energies), electron energies (full downward arrows) in above threshold ionization (ATI). Right: above barrier ionization in a strong field, indication the recollision electron with a maximum energy of $3.2\,U_p$, and high harmonics generation (dotted downward arrow).

The absence of spherical symmetry makes molecular calculations even more demanding. Within the fixed-nuclei approximation 2-, 3-, and 4-photon cross-sections for H_2^+ at the equilibrium distance $R = 2.0\,a_0$ were presented in [16]. These results were confirmed and extended to different R values and alignment of the molecular axis with respect to a linear polarized laser field in [17]. The molecular results for fixed R are qualitatively very similar to the ones obtained for atoms: a rather smooth variation of $\sigma^{(N)}$ as a function of the photon frequency underlying very sharp resonant peaks (that within LOPT actually diverge). These resonances indicate the presence of resonantly enhanced multiphoton ionization (REMPI) and are a consequence of intermediate states coming into resonance with M $(< N)$ photons. If this occurs, the intermediate state becomes populated. A proper treatment (within LOPT) requires the solution of the corresponding rate equations considering population and depopulation of the intermediate state by an M- and an $N-M$-photon process, respectively. The presence of REMPI peaks are also the reason that inclusion of vibrational motion in LOPT calculations of molecules is a nontrivial task [17]. While for processes like single-photon ionization it is possible to obtain approximate results by simply averaging over R weighted vibrational wave function of the initial state (usually the ground state), such a procedure works only in a very limited frequency regime for N photon processes. The reason is that for most photon frequencies the intermediate states are embedded in dissociative continua and the LOPT amplitudes are thus divergent. A proper treatment of this problem was then adopted in [18]. A further technical problem encountered for molecules is the dependence of $\sigma^{(N)}$ on the molecular alignment [19].

If the laser intensity is increased, the LOPT approximation becomes increasingly inaccurate. As a consequence, the ionization rate shows an intensity dependence differing from I^N. Usually, the increase with intensity is smaller than predicted by LOPT. Often this is called the nonperturbative multi photon regime. One characteristic of this regime is visible, if photoelectron spectra are recorded. By construction, the LOPT approximation assumes that only a single photoelectron peak exists, its energetic position being given by $E_0 = N\hbar\omega - E_I$. However, with increasing intensity additional peaks occur in the photoelectron spectrum at $E_n = (N+n)\,\hbar\omega - E_I$, $n = 1, 2, \dots$ These so called above-threshold ionization (ATI) peaks are also indicated in Fig. 6.1 (left panel). In the language of perturbation theory they result from additional absorption of photons within the electronic continuum of states. With increasing laser intensity the relative importance of the higher-order ATI peaks increases and they can even provide the dominant path to ionization. As a consequence, LOPT fails to predict even the integrated ionization spectrum.

At intensities $I \gtrsim I_{crit}$ the atomic or molecular potentials are significantly deformed by the field as depicted in Fig. 6.1 (right panel). The electron e^- can escape from the atom in an *above barrier* process and is accelerated in the oscillating laser field. While a completely free electron would not be able to gain energy from the laser pulse (for energy and momentum conservation reasons) with the atom present this is possible. As first pointed out by [20], the electron can accumulate energy and – depending on its starting phase with respect to the laser field – may return to the atom with kinetic energies up to about $3.2\,U_p$. This is schematically illustrated by the trajectory in Fig. 6.1 (right panel) showing an electron starting at the barrier. This so-called *recollision process* is instrumental in high harmonic generation (HHG) and explains the experimentally observed plateau and cutoff of HHG toward high energies [20, 21].

Traditionally, one defines a dimensionless quantity, the so-called *Keldysh parameter*, introduced in [22],

$$\gamma = \sqrt{\frac{E_I}{2U_p}} = \sqrt{\frac{\varepsilon_0\, m_e\, c}{e_0^2}\,\frac{E_I\,\omega_L^2}{I}} \propto \frac{t_T}{t_L} \tag{6.6}$$

to differentiate between the high and low intensity regime. In the high-frequency low-intensity multi photon regime one has $\gamma \gg 1$, while $\gamma \ll 1$ indicates the validity range of the low-frequency high-intensity quasi-static regime. The Keldysh parameter γ may be understood as an adiabaticity parameter. If the laser frequency is small and thus the periodic change of the electric field $t_L = 2\pi/\omega_L$ is slow, the atomic or molecular systems responds to the field similar as to a slowly varying electric field. For sufficiently low laser frequencies the system can follow adiabatically and thus it is at every instant of time in an eigenstate of the total field dressed Hamiltonian. For sufficiently high intensity, ionization in an electric field occurs via tunneling through the field deformed Coulomb barrier with the tunneling time t_T or even by escape over the completely suppressed barrier (see Fig. 6.1). On the opposite side, in the high-frequency limit, the field changes too quickly for the system to follow and thus it remains in its field-free eigenstate. In this case, the system does not react on the oscillating field amplitude in terms of the quasi-static picture but on the presence of a perturbation given by the pulse envelope. Transitions are then (formally) induced by photon absorption, since the fast changing direction of the electric field vector does not allow the electron to escape through or over the field deformed barrier before the barrier closes.

This clear picture of the boundary between the multi photon and the quasi static regime at $\gamma = 1$ becomes blurred when the size of the molecular system is increased. Especially for molecules with extended π electrons (e.g., benzene, anthracene, unsaturated hydrocarbons, and fullerenes) [23–26] but also for metal clusters [27] it has been found that the ionization rate deviates from the power law (I^N) at much lower intensities than estimated from the Keldysh parameter. In contrast to atoms and small molecules, where the excitation and ionization processes are typically described by a single active electron, in larger molecular system multi electron effects cannot be neglected mainly for two reasons: The motion of the active electrons is no longer fast compared to the period of the laser field and doubly excited states exists below the ionization limit. More details and examples on this issue are presented in Sects. 6.3 and 6.4.

6.1.3 Control in strong fields

In recent years, studies in the strong field regime have been extended by using tailored laser pulses to interact with molecules and clusters. In contrast to the control in the linear or nonlinear perturbative regime discussed in Chapter 2 "strong- or intense-field control" relates to intensities and frequencies where the potential energy surfaces are disturbed by the laser field and a quasi-continuum of new states is produced. This allows the laser field to "reprogram" the molecular Hamiltonian in a time-dependent manner, creating a new quantum system while the laser pulse exists. For small molecules the intensities for strong field control lead to Keldysh parameters $\gamma < 1$ and the quasi-static picture is appropriate in those cases. However, not only the laser intensity itself determines the strength of the interaction with respect to the underlying physics, but also the ratio of the photon energy to the ionization potential of the atomic or molecular system considered. More details and an example for the control of wave packets in the ground state of H_2 is given in Sect. 6.2.

For larger molecular systems one might suspect the degree of chemical control using pulses with an intensity of 10^{12} to $10^{14}\,\mathrm{Wcm^{-2}}$ to be rather limited as a consequence of the highly nonlinear processes induced in the molecule. However, by using short pulse durations ($\approx 50\,\mathrm{fs}$), the exciting laser couples primarily into the electronic system modes of the molecule because it has they have similar characteristic response times. This limits the intuitively expected catastrophic decomposition to atomic fragments and ions. For example, in the strong field excitation of benzene [5], up to intensities of 10^{14} Wcm^{-2} exclusively ionization of the parent species was observed, with little induced dissociation. The observation of a single dominant channel (intact parent ionization) suggested that most of the possible final state channels (i.e., the large manifold of dissociative ionization states) may be suppressed in the short pulse strong field regime. In this high intensity regime there is opportunity to substantially manipulate the molecular wave function with suitably shaped laser pulses to induce and manage photochemical reactivity and products. Section 6.3 will give more details and examples of successful experiments in "strong field chemistry". The control of photo fragmentation in fullerenes is presented in Sect. 6.4.

One big challenge in this context is a detailed theoretical description of strong field induced processes in complex molecular system. A new theoretical method, which allows the *time-resolved* observation of the formation, the modulation, and the breaking of chemical bonds is presented in Sect. 6.5. This method provides a visualization of complex reactions involving the dynamics of excited states.

6.2 Strong-field control in small molecules

Alejandro Saenz

In view of the goal to control nuclear motion and since the laser field interacts primarily with the electrons, it is of interest to concentrate on the interplay between vibrational and electronic motion and thus it is important to understand the influence of vibrational motion on the strong-field behavior of molecules. It will be demonstrated below that strong fields allow for control of vibrational motion in a different way than low-intensity laser pulses. While control schemes based on multi photon processes can be understood as being based on resonant transitions between field-free states, the schemes in the quasi-static regime are often nonresonant and involve field-modified states. However, the strong-field control schemes will have to compete with ionization that is increasingly important as the laser intensity increases. In fact, most of the strong-field molecular experiments indicating field-induced formation of vibrational wave packets concentrated on the nuclear motion induced in molecular ions created by strong-field ionization. Similarly to the case of forming coherent vibrational wave packets in excited electronic states by a sufficiently short laser pulse that transfers the initial-state wave packet onto the excited state potential curve, a sufficiently short intense laser field can transfer the initial vibrational wave packet onto the ground-state potential curve of the molecular ion. A lot of experimental work has been devoted to the simplest possible molecule that is easily experimentally accessible: H_2. Due to its high ionization potential, it is a perfect candidate for studying molecules in the quasi-static regime.

Very early and long time before intense-field two-pulse pump-probe experiments became available, the behavior of H_2 (and other mostly diatomic molecules) in strong short laser pulses were analyzed assuming the following sequential process. During the very first phase of the laser pulse the molecules are aligned. In the case of a linear polarized laser pulse and a diatomic molecule like H_2 it should align with its molecular axis parallel to the electric field vector. It should be noted, however, that with the occurrence of short laser pulses (below 50 fs) the rotational period of the molecules is too long for alignment to occur. More accurately, the pulse creates a rotational wave packet, but due to the shortness of the pulse with respect to the rotational period, the induced rotational wave packet motion takes place on a time scale much larger than the pulse length. As a consequence, the subsequent field-induced processes like ionization occur while the molecules are still in their initial rotational wave packet. In a second step, but still during the rise of the laser-pulse intensity, H_2 is ionized due to tunneling ionization. Since the tunneling process is very fast (it should occur within half an optical cycle), the nuclei are practically frozen during the ionization process and thus the H_2^+ ion was predicted to be generated in a vibrational wave packet that is simply given by the projection of the vibrational ground state of the neutral onto the vibrational states of the ion, the Franck-Condon factors. While this wave packet starts to move, the laser intensity increases further and also the ion may be ionized. If the ionization stage has reached a certain level (in the case of H_2 evidently 2), the Coulomb repulsion of the nuclei in the resulting ion is not any longer compensated by the remaining electrons. Coulomb explosion occurs. From the analysis of the kinetic energy distribution of the fragments (protons in the case of H_2) three molecular strong-field effects were discovered: bond softening, bond hardening, and enhanced ionization. In fact, bond softening had already been predicted and experimentally observed long time before lasers were available that allowed the investigation of Coulomb explosion of H_2.

6.2.1 Potential-curve distortion: bond softening and hardening

Hiskes [28] investigated theoretically the behavior of H_2 and H_2^+ in a strong electric field. For H_2^+ the electronic ground and first excited states $1\,\sigma_g$ and $1\,\sigma_u$, respectively, are degenerate in the limit $R \rightarrow \infty$. Since an electric field parallel to the molecular axis breaks the *gerade/ungerade* symmetry, the degeneracy must be lifted (Wigner-von Neumann noncrossing rule for states of the same symmetry). In fact, the dipole moment between the two states (and thus the electric-field induced coupling) diverges for $R \rightarrow \infty$. The two potential curves repel each other and go asymptotically to $\pm\infty$. As a result of the distortion of the ground-state potential curve the high-lying vibrational bound states become unbound (dissociative) or can tunnel through the distorted potential curve (predissociation). This effect is called bond softening and is a general feature for diatomic molecular ions with an odd number of electrons (a condition that is fulfilled for most singly ionized molecular ions). In a static electric field this effect was experimentally observed almost immediately after its theoretical prediction [29]. Since the adiabatic field dressed curve of the upper electronic state bends upwards, it can support vibrational bound states, while it is purely repulsive in the field-free case. The appearance of field-induced bound states is called bond hardening or occurrence of light-induced states.[1]

The occurrence of bond softening also in laser fields in the quasi-static regime explained the experimental result that the fragments of Coulomb explosion of H_2 had a kinetic energy that corresponded to an ionization of H_2^+ (and thus to a transition onto the purely repulsive curve of H^++H^+) at relatively large values of R. These R values were on the other hand not reachable, if the wave packet created in the first ionization step would be propagating on the field-free potential curves. This indicates an important ingredient for strong-field control schemes: the in comparison to multi photon control additional ability of distorting potential curves allows to reach with a given wave packet R regions that would otherwise be very difficult to access.

Bond softening alone was, however, not sufficient to explain the experimental data, since according to the kinetic energy distribution there was not only a relatively large fraction of H_2^+ molecular ions produced at quite large R values, but the energy distribution of this (overall dominant) part of the spectrum was in addition very narrow. This indicated that the ionization of H_2^+ occurred in a very small R interval. Enhanced ionization, i.e., a pronounced increase of the ionization rate at a certain (typically quite large) value of R, turned out to be the reason for the observed narrow kinetic energy distribution. The first semiclassical models [30, 31] were later on also confirmed by full ab initio calculations [32,33]. In the framework of the quasi-static approximation one should analyze the ionization rates on the adiabatic field dressed potential curves. As is discussed in [33] and in agreement with simple pictures explaining enhanced ionization, the lowest-lying adiabatic curve does not show enhanced ionization. However, the first excited (in the field upwards bending) state shows pronounced peaks in the ionization rate at specific large R values. The reason for this behavior are (field-induced) avoided crossings of this state with other higher lying but not (or not as strongly) upwards bending states. Since those higher lying states usually possess higher ionization rates due to their lower ionization potential and since this ionization rate is shared between two states showing an avoided crossing occurs, the upward bending state shows a pronounced peak in its ionization rate close to the avoided crossing. It is important to again emphasize that enhanced ionization occurs only

[1] Although the name bond hardening appears to be more popular as it implies the opposite effect to bond softening, it is not really appropriate, since the repulsive field-free state does not support any bond that can be "hardened".

in the excited adiabatic field dressed state. Since this state is supposed to be not noticeably populated in the first ionization step ($H_2 \rightarrow H_2^+$), the occurrence of enhanced ionization requires some nonadiabatic coupling between the two lowest states of H_2^+.

In view of the fact that the energy difference between the ground and first excited field dressed adiabatic states of H_2^+ is much smaller than the gap between H_2 and H_2^+ it is in the case of many experiments more appropriate to interpret the ionization of H_2 in the quasi-static picture, but the transition from the ground to the first-excited state of H_2^+ (though field distorted!) by means of (multi-)photon absorption. A recent work [34] has nicely demonstrated how the transition between the quasi-static and the multi photon picture, especially the multi photon transitions between field-distorted potential curves arises in the Floquet picture. With the aid of the very recently developed pump-probe experiments [35,36] with laser pulses of a pulse length of 12 fs and less it became now possible to monitor the wave packet motion of H_2^+ in real time. These experiments show clearly the effect of bond softening, i.e., the motion of the wave packet on field-distorted potential curves. The recent experiments [36] also indicate the occurrence of more than a single enhanced ionization peak in agreement with the older theoretical predictions.

In view of these strong-field effects specific to molecules (and absent for atoms) it is of course of interest whether they occur also for neutral molecules like H_2. Hiskes considered in fact this question with respect to bond softening and concluded that it should be absent for a molecule like H_2 (and all other typical covalent diatomic neutral molecules with an even number of electrons). The reason is quite straightforward. Although the electronic ground state of H_2 is also degenerate with an excited state for $R \rightarrow \infty$, the two degenerate states are not coupled by an electric field, since the excited state has triplet symmetry. This conclusion was seemingly not questioned for many years. In [37] it was, however, found from a numerical calculation for a 1D model of H_2 that enhanced ionization may occur for H_2.

A full ab initio treatment of molecules in laser fields is nontrivial, even if the quasi-static approximation is adopted. The reason is the requirement to evaluate the eigensolutions of the Hamiltonian describing the molecule and the laser field (usually only its dominant electric component) at a given instant of time. This is equivalent to the calculation of the eigenstates of a molecule in a strong static electric field. While standard quantum chemistry codes nowadays allow to calculate the molecular states in weak electric fields, the adopted procedures do not work for strong electric fields. The reason is that in such fields the states are unstable with respect to ionization. Thus one has to treat the metastability of the molecular states in a proper way. A correlated, fully three-dimensional calculation of the eigenstates of H_2 exposed to strong electric fields has been presented in [38]. Using a complex-scaled geminal approach both potential curves and ionization rates were given as a function of R. In a later, more detailed work also rovibrational states (within the Born-Oppenheimer approximation) were calculated and the problem of

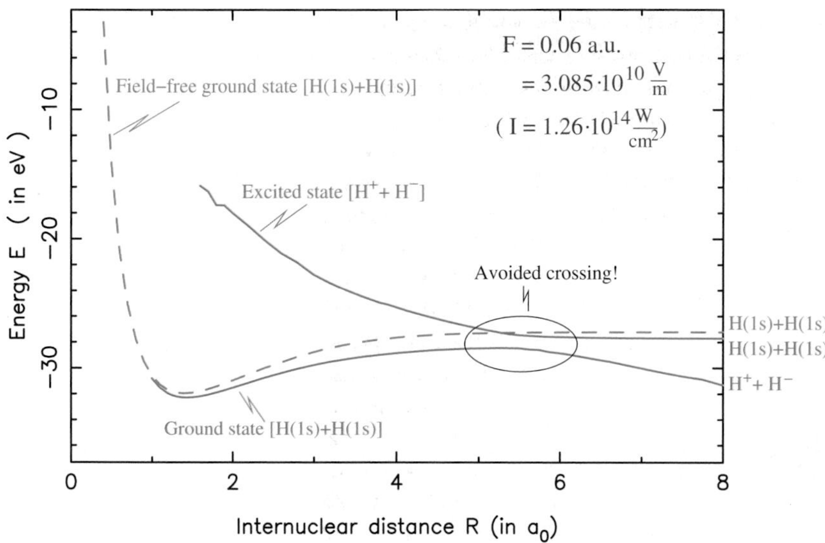

Fig. 6.2. The two energetically lowest lying adiabatic potential curves of H_2 (green) in a static field with strength $F = 0.06\,a_0$ (corresponding to a laser intensity of $I = 1.26 \times 10^{14}\,\mathrm{Wcm}^{-2}$) (cf. [38, 39]). For comparison, the adiabatic field-free potential curve is also shown (dashed red). A dominant covalent or ionic character is indicated by the corresponding dissociation limits [H(1s)+H(1s)] and [H^++H^-], respectively.

standard quantum chemistry approaches to treat the metastable states was discussed [39, 40].

In Fig. 6.2 the lowest-lying adiabatic potential curves of H_2 in an electric field with a field strength $F = 0.06\,a_0$ are shown. Clearly, in contrast to the prediction of Hiskes, also neutral H_2 shows the effect of bond softening. In view of possible control applications this is very important, since it implies that a substantial deformation of potential curves is possible, even if there is no degeneracy of two field-free potential curves that is lifted by the field, as was the case for H_2^+. The origin of bond softening for H_2 is also evident from Fig. 6.2. While the covalent ground state of H_2 dissociating into H+H is in fact almost not influenced by the electric field (only a small Stark shift is visible), the state that dissociates into H^++H^- is strongly affected by the field, as it represents a (diverging) dipole for $R \to \infty$. It has to be reminded that this ionic state does not exist in the adiabatic picture. In this picture the ionic character is contained for small R mainly in the $\mathrm{B}\,^1\Sigma_u$ state, but for increasing R it is partially transferred to other adiabatic states. The double-well character of some of the excited electronic states is a consequence of the avoided crossings that occur due to the passing of the diabatic ionic state through the sequence of adiabatic states. Since the in the field-free case quite energetic ionic state represents an electric dipole, its energy decreases (to a good approximation) linearly for $R \to \infty$. As a consequence, in a diabatic picture the ionic state will for a sufficiently large value of R cross the covalent ground state. If the covalent ground state and the ionic state possess (at least in the field that breaks parity) the same symmetry, an avoided crossing occurs. This avoided crossing is the one shown in Fig. 6.2. It is responsible for bond softening. As a further consequence, the adiabatic ground-state curve changes its character from covalent (dissociating into H+H) to ionic ($H^+ + H^-$).

6.2.2 *R* dependence of the ionization yield

In Fig. 6.3 the ionization rate of H_2 in an electric field ($F = 0.08\,a_0$) is shown as a function of R. The avoided crossing with the ionic state leads to a pronounced peak in the ionization rate. This confirms the occurrence of enhanced ionization also for neutral H_2 for which an earlier indication was already given by a 1D model calculation [37]. On the basis of the accompanying analysis of the potential curves the origin of enhanced ionization for H_2 is of course evident: at the avoided crossing the adiabatic eigenstate is a mixture of the covalent and the ionic state. Since the latter has a much smaller ionization potential and thus larger ionization rate, the covalent state acquires ionization

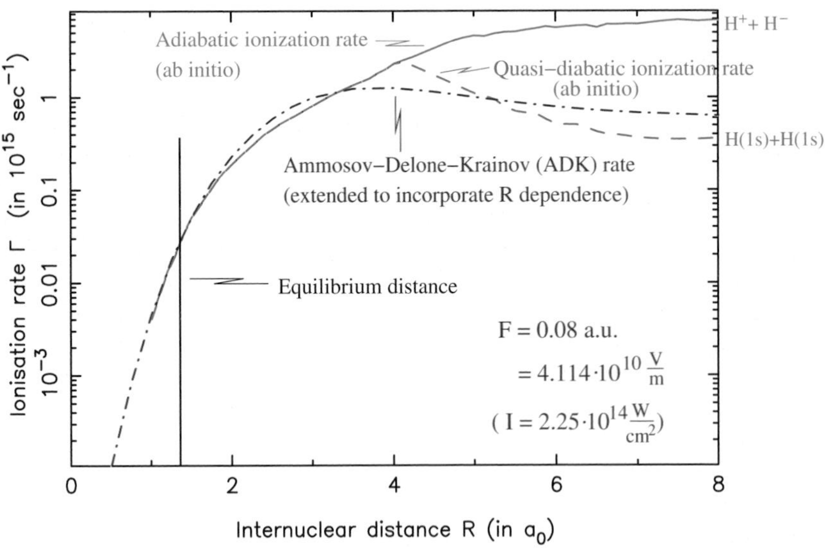

Fig. 6.3. The ionization rate of H_2 in a static field with strength $F = 0.08\,a_0$ (corresponding to a laser intensity of $I = 2.25 \times 10^{14}\,\mathrm{Wcm}^{-2}$) is shown as a function of the internuclear separation R (cf. [38,40]). The ab initio results within the adiabatic (red) and a quasi-diabatic (green) approximation are compared to a simple atomic tunneling model (ADK approximation) that is extended to the molecular case by using an R-dependent ionization potential (blue).

rate from the ionic state [38, 39]. If a wave packet moves very slowly over the avoided crossing, it would follow the adiabatic potential curve and thus change character from the covalent to the ionic state. In such a case a step-like feature in the ionization rate occurs at the crossing. If the wave packet moves very fast over the avoided crossing, it will follow the diabatic curves and thus regain its covalent character (and ionization rate) after the crossing. In this case a peak appears in the ionization rate. A time-dependent calculation (for short laser pulses with 800 nm wavelength) has shortly thereafter confirmed the possible occurrence of enhanced ionization in H_2 and its origin due to a change from the covalent to the ionic character [41].

A lesson that can be learned from the examples of H_2^+ and H_2 for strong-field control is the occurrence of potential-curve deformation, if either the field-free degeneracy of states is lifted by the field or if field-induced avoided crossings occur. Clearly, molecular states dissociating into cation-anion pairs will be most strongly affected by the external field. As these states are usually not existent as well-defined states in the adiabatic formulation, but are admixtures of different states, it is clear that already weaker fields than the ones discussed here for the H_2 ground state should have a strong effect on the excited states potential curves. Furthermore, the average energy difference between excited states is usually much smaller than the gap between the electronic ground and the first excited state. As is discussed in [42] a large number of field-induced avoided crossings occur in the electronically excited state spectrum of H_2.

Returning to Fig. 6.3 it may be observed that even in the R range where no avoided crossing occurs, the ionization rate is strongly R dependent (note the logarithmic scale). Already about thirty years ago it was pointed out by Hanson that in the case of H_2 the ionization in a strong electric field should dominantly occur at a larger internuclear distance than the equilibrium one [43]. A more general discussion of the influence of vibrational motion on the laser-field ionization in the quasi-static regime was then given in [44]. In the quasi-static picture (appropriate for sufficiently long pulses for which the ionization during the turn-on and -off process is a negligibly small fraction) and assuming validity of the Born-Oppenheimer approximation the ionization process may be described as a transition from the rovibrational initial state of the neutral to one of the rovibrational states of the ion. This transition is mediated by some electronic transition moment. In the quasi-static approximation this transition moment is the R-dependent ionization amplitude describing the ionization probability (at fixed R) in an electric dc field. As was discussed above, for molecules such ionization amplitudes are difficult to obtain from ab initio treatments. For H_2 ionization rates of the type shown in Fig. 6.3 can be used. In order to obtain approximate R-dependent ionization rates for larger molecules, simpler approaches are required. The probably simplest model one can think of is the very popular Ammosov-Delone-Krainov (ADK) [45] model.

In order to adopt this atomic model to molecules it was proposed to use in the ADK formula an R-dependent ionization potential $E_I(R)$ instead of the physical one [44]. While $E_I(R)$ is the energy difference between the ground-state potential curves of the ion and the neutral, the physical ionization potential is of course the R-independent difference between the rovibrational ground states of ion and neutral. The concept of a substitution of the physical by an R-dependent ionization potential can also be used in other originally atomic laser-field ionization models like the strong-field approximation (SFA), as was mentioned already in [44]. The later on performed full ab initio calculation of the static ionization rate for H_2 [38, 46] confirmed, however, the validity of the ADK model to predict the R dependence of the ionization rate [40] at the static limit. This is also apparent from Fig. 6.3 where both ab initio and ADK (with R-dependent ionization potential) rates are compared. In fact, even the quantitative agreement is very good, as long as the parameters belong to the validity regime of the ADK model. Since the ADK approximation is based on a tunneling model, this approximation is only applicable, if the field distortion of the Coulomb barrier is not so strong that the electron can escape over the distorted barrier, but has to tunnel through it. From classical arguments the field strength where over-the-barrier ionization sets in is given by $F_{\mathrm{BSI}} = E_I^2/4$. Using the ADK tunneling formula outside its validity regime, i.e., for $F > F_{\mathrm{BSI}}$ yields a clear overestimation of the ionization rate. Using the R dependent ionization potential, one obtains $F_{\mathrm{BSI}}(R) = E_I^2(R)/4$ and thus deviations from the ADK behavior occur not only as a function of F but also of R [40]. Of course, ADK is also only valid in the quasi-static limit and not in the multi photon regime. This sets an intensity and wavelength limit from below for the applicability of the ADK model.

Based on the ADK approximation the influence of vibrational motion on strong-field ionization can be visualized with the aid of the example of three prototype diatomic molecules [44]. Assuming for simplicity the same shape of the potential curves of the electronic ground states of the neutral and the ion, one may still distinguish three generic cases. The equilibrium distance of the neutral and the ion ($R_{\mathrm{eq}}^{\mathrm{ion}}$ and $R_{\mathrm{eq}}^{\mathrm{neut}}$, respectively) are either almost identical, or one of the curves is shifted horizontally with respect to the other. Example molecules would be N_2 where the removal of a nonbonding electron results in $R_{\mathrm{eq}}^{\mathrm{ion}} \approx R_{\mathrm{eq}}^{\mathrm{neut}}$, H_2 where removal of a bonding electron gives $R_{\mathrm{eq}}^{\mathrm{ion}} > R_{\mathrm{eq}}^{\mathrm{neut}}$, and O_2 where an antibonding electron is removed and thus $R_{\mathrm{eq}}^{\mathrm{ion}} < R_{\mathrm{eq}}^{\mathrm{neut}}$. As a consequence of the exponential dependence of the tunneling ionization rate on the ionization potential even small shifts of R_{eq} have relatively drastic effects, see Fig. 6.4. Compared to an atom (with R-independent transition rate) the case $R_{\mathrm{eq}}^{\mathrm{ion}} \neq R_{\mathrm{eq}}^{\mathrm{neut}}$ yields a smaller ionization probability (suppressed ionization), but there is a substantial cancelation effect and thus the overall ionization rate (to all possible rovibrational final states) is only slightly suppressed for the given examples (when using realistic molecular data for the mentioned diatomic example systems H_2, N_2, and O_2). More pronounced effects can be expected for nonsequential multiple ionization processes. It has to be emphasized that these three examples serve just as an illustration of

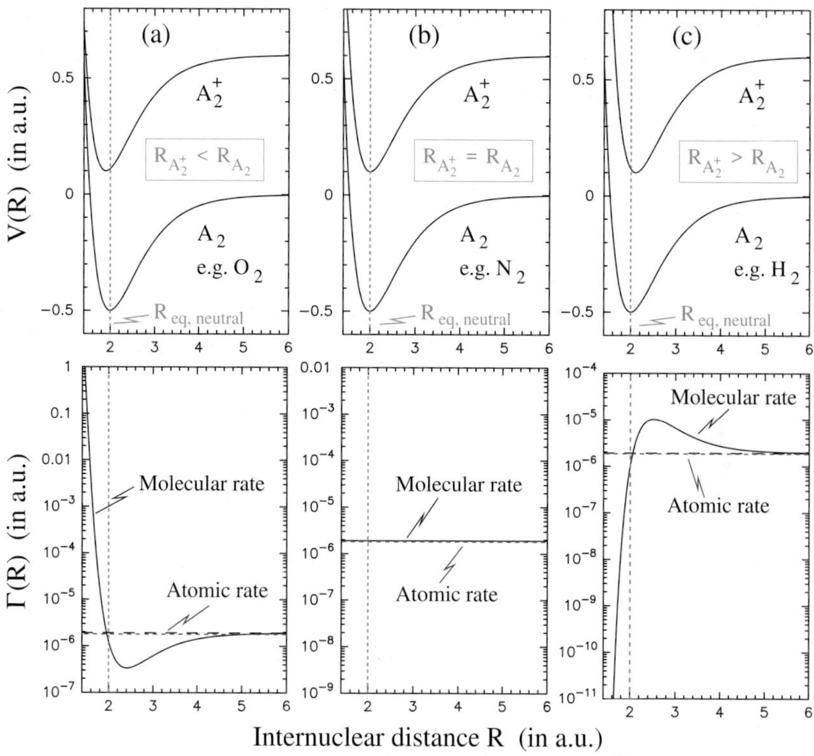

Fig. 6.4. The potential curves (upper row) and ionization widths (lower row) of three prototype diatomic molecules A_2 are shown [44]. The three cases differ in the equilibrium distance of the cation relative to the one of the neutral. If these distances agree (2 nd column) the ionization rate predicted by an R-dependent ADK model agrees with the one of an atom with the same ionization potential. Otherwise, a strong R dependence of the ionization rate is predicted.

how the ionization process can be influenced by structural properties closely related to the vibrational dynamics. As is discussed below, this can serve as a tool for control of vibrational motion. The model proposed in [44] is on the other hand very simple and thus can easily be applied to many molecular systems (using realistic potential curves for neutral and ion), including (especially nonsequential) multiple ionization.

Before discussing consequences of the within the quasi-static approximation predicted pronounced R dependence for molecules like H_2 and O_2, it is of course an important question whether the quasi-static approximation itself is at all applicable for realistic laser pulses. In fact, it was pointed out in [48] that already for the simplest atomic system, the hydrogen atom, the quasi-static approximation is only applicable for extremely high intensities in which for femtosecond lasers saturation occurs. In a very recent work the

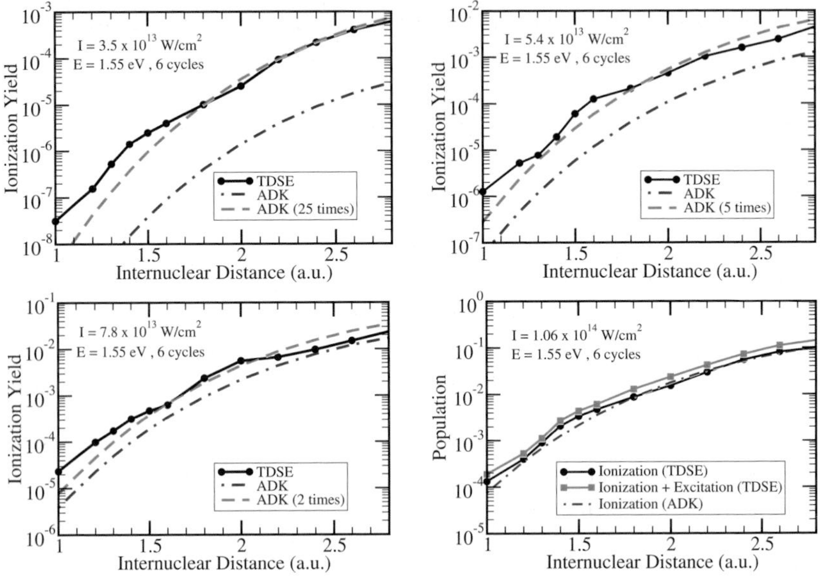

Fig. 6.5. Comparison of the R dependent ionization yield of H_2 in 800 nm laser pulses of 6 cycle duration ($\cos^2$ envelope) and four different peak intensities as specified in the plots [47]. The results of the full TDSE solution (black) is compared to the prediction of the (R-dependent) unscaled (blue) and scaled(red) ADK model that is equivalent to the quasi-static approximation. For 1.06×10^{14} Wcm^{-2} only the unscaled ADK result is shown together with the TDSE result (red) showing population of all but the initial H_2 ground state.

R dependence of the ionization rate of H_2 in a 800 nm laser pulse of 15 fs duration was investigated using a full solution of the TDSE (describing the two electrons in full dimensionality and including correlation, the technical details being described in [49]). It was found that the quasi-static approximation fails in most cases to predict the ionization yield *quantitatively*, but especially the R dependence is very well *qualitatively* described for intensities above 10^{13} Wcm^{-2} [47]. This is seen in Fig. 6.5 that shows the R dependence of the ionization yield in 800 nm laser pulses of about 16 fs duration for different intensities around 10^{13} to 10^{14} Wcm^{-2}. The results are compared to the ones obtained with the ADK model (that for the considered intensities agrees well with the complete quasi-static approximation, cf. Fig. 6.3). If the ADK results are multiplied by an overall multiplicative factor, the agreement to the full TDSE calculations is very good. In fact, for a peak intensity of 1.06×10^{14} Wcm^{-2} even the quantitative agreement is very good, but this is to some extent accidental. For this intensity not only the ion yield (= population of ionic states) but also the population of all but the initial ground state is shown. In agreement with simple strong-field models the excitation of neutral excited states is small at these intensities, and shows almost the same R dependence as the ionization yield.

6.2.3 Deviations from the Franck-Condon approximation

An immediate consequence of a pronounced R dependence of the strong-field ionization rate is the fact that the wave packet created by means of a corresponding laser in the molecular ion will not be correctly described by Franck-Condon factors. The reason is simple. In the derivation of the Franck-Condon factors it is assumed that the electronic transition moment is sufficiently R independent (within the R range covered by the initial vibrational wave packet) to be taken out of the integral over R. This point was already remarked in [44]. Based on the model proposed in that work, the vibrational wave packet created in intense laser fields was predicted and compared to experiment in [50]. The results are shown in Fig. 6.6. The (simple) theoretical model predicts a strong deviation from the Franck-Condon distribution, since the latter is rather broad and has its maximum for $v = 2$. The distributions predicted for strong laser fields (in between 10^{13} and $10^{14}\,\mathrm{Wcm}^{-2}$) are on the other hand much narrower and have their maxima at $v = 0$ (or 1). Within the model it is also predicted that the deviation from the Franck-Condon distribution is more pronounced for lower intensities. In fact, Fig. 6.6 shows for the theoretical data two distributions for every intensity. One simulates a parallel orientation of the field and the molecular axis, the other one a perpendicular orientation. Within the simple model the difference is given by the fact that only for parallel orientation there is a noticeable potential-curve distortion at the considered field intensities. As a consequence of the distortion, the higher lying vibrational states of H_2^+ become unbound and thus yield dissociation. In Fig. 6.6 only the nondissociated H_2^+ ions are shown, since the experiment was only able to measure those.

In Fig. 6.6 also experimental results are given for three peak intensities (covering the range of the theoretical data) for a laser pulse of about 45 fs pulse length. The experimental results are in very good qualitative agreement with the theoretical predictions. One clearly sees a very similar deviation from the Franck-Condon distribution, and the same trend as a function of intensity (stronger deviation for lower intensities) is also visible. Finally, Fig. 6.6 shows also the result obtained with a ns laser pulse. According to the discussion given above, one expects that in the ns pulse the molecules will be quite well aligned, while in the 45 fs laser pulse they will be more or less randomly oriented. The sharper cut-off (due to dissociation) predicted by the theory for a parallel aligned sample is also seen in the experiment. In conclusion, the experiment confirms the predicted pronounced R dependence of the strong-field ionization rate of molecules like H_2. This is in contrast to the multi photon regime (LOPT or weakly perturbative) that shows only very little R dependence, if no resonances are involved. As a consequence of the R dependence in the

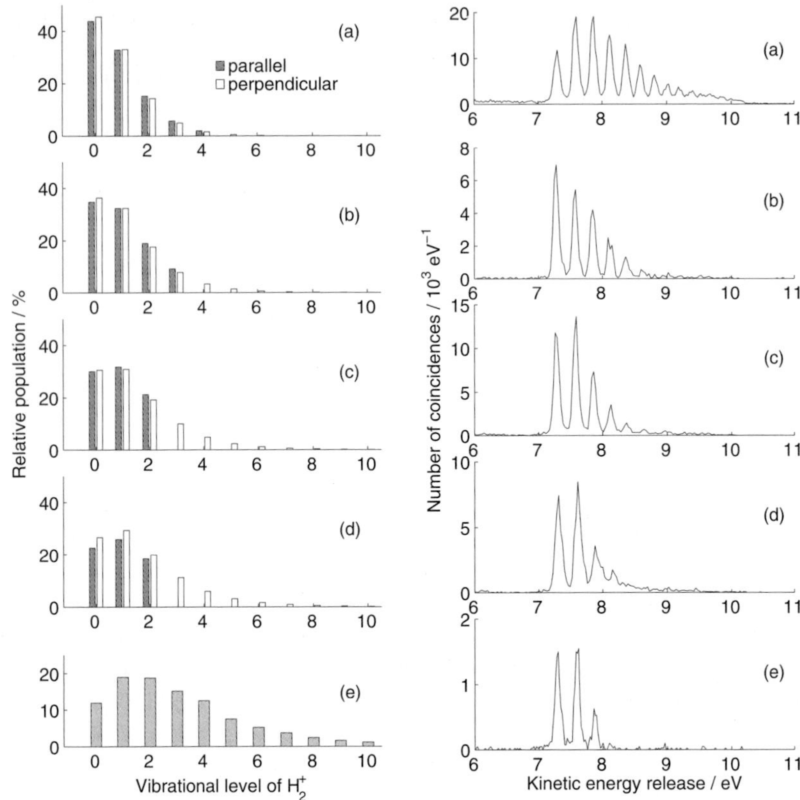

Fig. 6.6. Distribution over bound vibrational states of H_2^+ formed in strong laser-field ionization of H_2. Left panel: theoretical prediction on the basis of the quasi-static approximation for laser peak intensities (a) 3.5×10^{13}, (b) 5.4×10^{13}, (c) 7.8×10^{13}, and (d) 1.1×10^{14} Wcm^{-2}. The dark-gray bars correspond to a parallel, the white bars to a perpendicular orientation of the molecular axis with respect to the linear-polarized laser field. Right panel: Experimental results for laser peak intensities (a) 3×10^{13}, (b) 4.8×10^{13}, and (c) 1.5×10^{14} Wcm^{-2} and pulses of about 50 fs FWHM pulse duration. In (d) the distribution measured with a laser peak intensity of 1×10^{14} Wcm^{-2} but a pulse length in the ns range is shown.

strong-field regime it is possible to control the shape of the formed vibrational wave packet by varying the intensity of the laser.

6.2.4 Formation of vibrational wave packets in the nonionized neutral initial state

The pronounced R dependence of the strong-field ionization rate of molecules like H_2 may also be used in a different way to influence and control vibrational motion. In [51] it is proposed that with the aid of a sufficiently short pulse it

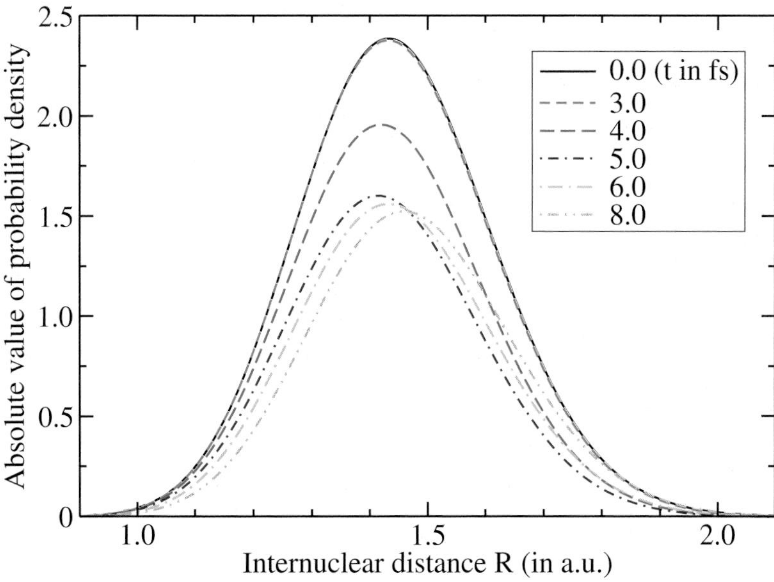

Fig. 6.7. The vibrational wave packet in the electronic ground state of H_2 is shown at different times for a laser pulse with 800 nm, 8 fs duration ($\cos^2$-envelope), and peak intensity $6 \times 10^{14}\,\mathrm{Wcm}^{-2}$ [51].

is possible to create a vibrational wave packet in the nonionized neutral H_2 molecules. For the theoretical modeling of this process the model proposed in [44] was extended in such a way that the vibrational motion was treated explicitly. For this purpose, the time-dependent Schrödinger equation describing vibrational motion of H_2 was solved. The ionization process was described with the aid of an imaginary (optical) potential. For the considered intensities the influence of potential-curve distortion was assumed to be negligible and thus omitted for clarity. The resulting wave packet motion, i.e., snapshots at specific time intervals, is shown in Fig. 6.7. Besides the overall decrease of the amplitude of the wave packet due to ionization one notices that in fact, as expected from the R dependence of the ionization rate, the wave packet is preferentially ionized and thus decreases more pronounced at larger values of R. If this R-dependent depletion process occurs fast enough to not be washed out by vibrational motion (for H_2 the ionization should occur within at least 5 to 10 fs), the laser pulse leaves behind a vibrational wave packet in the neutral H_2 molecules. It should be emphasized that this is a purely quantum-mechanical effect that cannot be explained with any ensemble statistics, since in a (semi-)classical world the molecules that happen to have a larger R value when being hit by the laser pulse would only be ionized with a higher probability. However, the molecule would be either ionized or not. In both cases no wave packet in the nonionized molecules can be created.

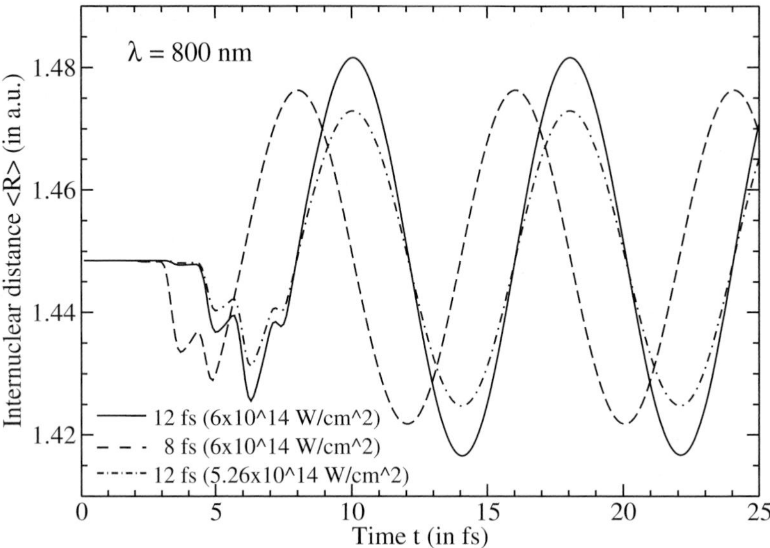

Fig. 6.8. The variation of the mean value of the internuclear separation $\langle R \rangle$ as a function of time in laser pulses with $\cos^2$ envelope, 800 nm, peak intensity $I = 6 \times 10^{14}\,\mathrm{Wcm}^{-2}$, and a duration of 12 fs (solid) or 8 fs (dashed) [51]. Also shown (chain) is the result for a 12 fs pulse with the same ionization rate as the 8 fs pulse which is achieved by lowering the intensity to $I = 5.26 \times 10^{14}\,\mathrm{Wcm}^{-2}$.

The creation of the wave packet is better monitored by plotting the time-dependent mean value of R as is shown in Fig. 6.8. In the beginning of the pulse (once the intensity is sufficient to ionize) R decreases due to the R-dependent depletion. However, while the linearly polarized laser changes field direction, its field component passes through zero. During this period of low intensity no further ionization occurs and the already created wave packet starts to swing back (R increases). In the following half cycle further ionization occurs, and R decreases even more. This continues until the pulse envelope decreases so much that no further ionization occurs. Now the created wave packet swings back and forth. Since the wave packet is formed from vibrational states of the same electronic state of the homonuclear H_2 molecule, the decoherence time is expected to be extremely long. The reason is that the wave packet can only relax via spontaneous two-photon emission, because one-photon transitions are dipole forbidden. The wave packet shows also a surprisingly clean oscillation without evident collapses and revivals etc. This is a consequence of the fact that in the considered intensity regime the wave packet consists almost exclusively of $v = 0$ and $v = 1$ components. The contributions of higher v components are to a good approximation negligible. Therefore the oscillation period is very accurately given by the one determined by the energy difference between $v = 0$ and $v = 1$ of H_2 which is about 8 fs.

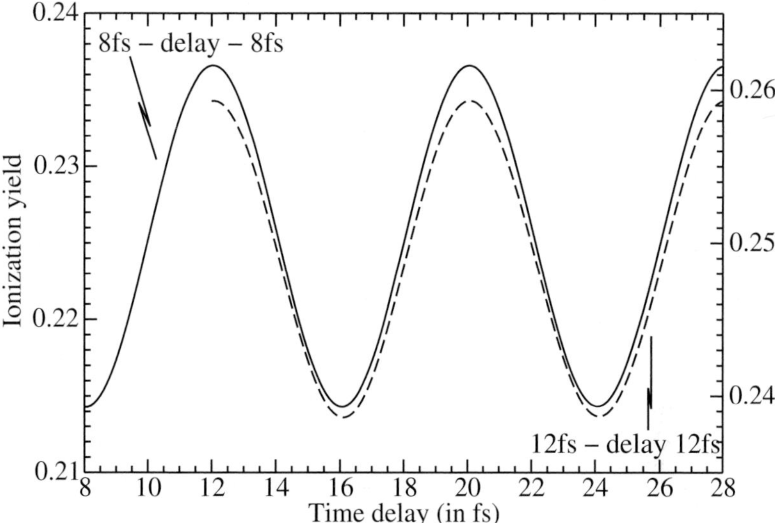

Fig. 6.9. Theoretically predicted total ionization yield measured in a pump-probe experiment on H_2 using either two identical laser pulses with 8 fs (solid) or 12 fs (dashed) duration. The pulses possess peak intensity 6×10^{14} Wcm^{-2}, a $\cos^2$ envelope, and 800 nm wavelength [51].

Most surprisingly, the wave packet formation process happens to be very robust with respect to the field parameters. The amplitude of the created wave packet depends on the exact form of the pulse, but its shape is almost independent of the exact pulse parameters. For example, changing the wavelength (as long as the quasi-static description is valid) from 800 to 1064 nm or the absolute carrier-envelope phase by $\pi/2$ does practically not influence the final form of the wave packet. Therefore, no stabilization of the carrier-envelope phase is, e.g., required for the wave packet formation and also the finite bandwidth (quite substantial for ultrashort laser pulses) does not pose any problem. The remaining question is, of course, whether the predicted wave packet can be experimentally verified. As is clear from Fig. 6.8, the oscillation takes place in a very small interval of R values. However, the pronounced R dependence of the ionization rate that was the origin of the wave packet may also be used for its detection. As is discussed in [51], performing a pump-probe experiment with a second very short pulse (possibly a replica of the first one, since this is most easily experimentally accessible) uses also in the probe step the strongly nonlinear R dependence. In Fig. 6.9 the result of such a simulated pump-probe experiments is shown.

Already very shortly after its prediction, the wave packet formation was experimentally confirmed by a pump-probe experiment as the one proposed in [51]. Performing the experiment on D_2 and using laser pulses of a FWHM

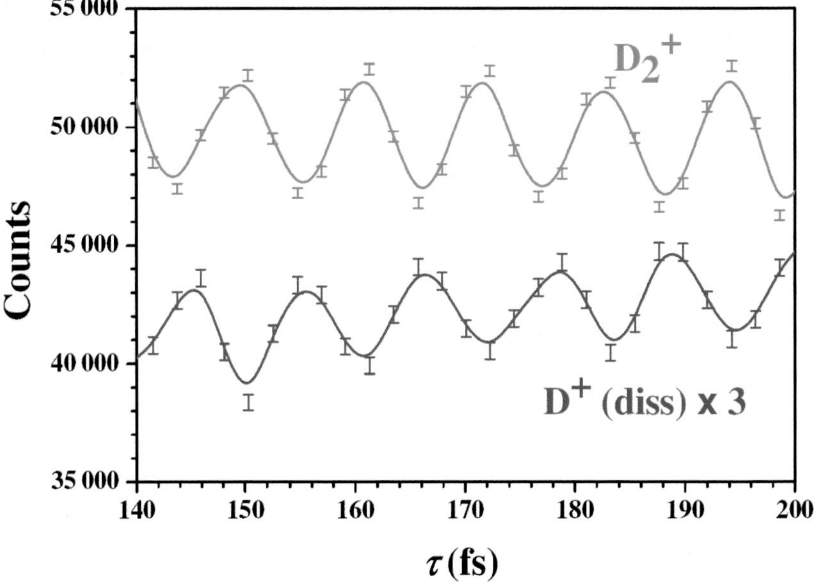

Fig. 6.10. Experimentally measured ion counts measured in a pump-probe experiment on D_2 using two identical 780 nm laser pulses with 7 fs FWHM duration. The peak intensity is $4\,(\pm 1)\times 10^{14}\,\mathrm{Wcm}^{-2}$. Shown are the counts for nondissociated D_2^+ (red) and dissociated D^+ (blue) ions [52].

of about 8 fs a clear oscillation signal of both ionization and dissociative ionization was found as can be seen in Fig. 6.10 [52]. The oscillation period is in very good agreement with the predicted one (when taking into account the mass scaling between H and D) and even the oscillation depth is in reasonable agreement with the theoretical prediction. Since the experiment was capable of following the oscillation over a very long time interval of up to 1200 fs, the expected long decoherence time is also confirmed.

The intensities used in the experiment were, however, slightly lower than the ones used in the theoretical work. Therefore, it was less clear, whether the wave packet formation was really due to the selective R-dependent depletion or due to bond softening. The latter, already discussed effect theoretically predicted in [38, 39] has so far not been experimentally verified for neutral molecules like H_2, and thus even this would certainly be an interesting outcome. However, since the general effect of bond softening is already known (though only for ions like H_2^+ but not for covalent neutral molecules like H_2), the selective depletion (termed *Lochfrass*) would be even more spectacular. In any case, it is of course very interesting to determine the exact mechanism of the wave packet formation. It turns out that the already discussed robustness of the formation process is the key, together with the impressive stability of the experimental set-up. As is shown in Fig. 6.9 for different pulse lengths and was also demonstrated with respect to the absolute carrier-envelope phase or wavelength of the laser pulse, the absolute phase of the oscillation signal of the pump-probe experiment is independent on the mentioned laser parameters. The same is valid for wave packets formed by bond softening. However, there is an almost maximum phase lag in between the wave packets created by the two different formation processes. As a consequence, it is possible to distinguish the two mechanisms by the absolute phase. In a very simple picture the origin of this phase lag may be understood in the following way, illustrated schematically with Fig. 6.11.

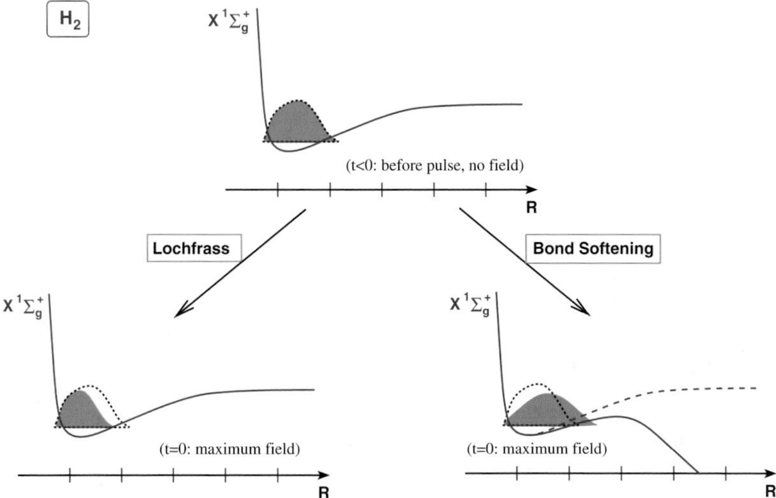

Fig. 6.11. Sketch of the two possible mechanisms responsible for the formation of a vibrational wave packet in the H_2 electronic ground state and their possible distinction by the absolute phase. In the case of *Lochfrass* the initial wave function is depleted at large R values and thus at $t = 0$ (maximum field strength) it is created at its inner turning point, moving initially to larger R. In the case of bond softening the wave packet flows over the field-distorted (lowered) potential well. At $t = 0$ (maximum field strength) it is at its outer turning point, moving initially to smaller R.

In the case of *Lochfrass*, the laser pulse depletes first the wave packet at the outer turning point. In the case of bond softening the wave packet escapes during the pulse over the field-distorted barrier. At the end of the laser pulse the wave packet created by bond softening is thus located with its maximum at the outer turning point and moves toward the inner one. In the case of *Lochfrass* the wave packet was depleted at the outer turning point and thus its maximum is located at the inner side of the potential well. The wave packet then starts moving toward the outer turning point. Therefore, the two creation mechanisms are distinguishable by the phase of the oscillation at zero delay time. This phase is, however, not directly experimentally accessible, since delay time zero means pump and probe pulses overlap completely in time. However, scaling the oscillation signal by the oscillation period and thus plotting the oscillation as a function of the number of oscillations, it is possible to extrapolate to time zero. Of course, this is only unambiguously possible, if the oscillation signal is very stable over many oscillation periods. Since this is the case, the experiment revealed that *Lochfrass* appears to be the dominant creation mechanism, see Fig. 6.12. In this figure the results of a theoretical simulation based on the bond-softening proposed in [38] and *Lochfrass* discussed in [51] are compared (on a relative scale) to the experimental results. In order to obtain very good agreement between experiment and theory, a small contribution of bond softening (in agreement with the theoretical prediction) has to be assumed.

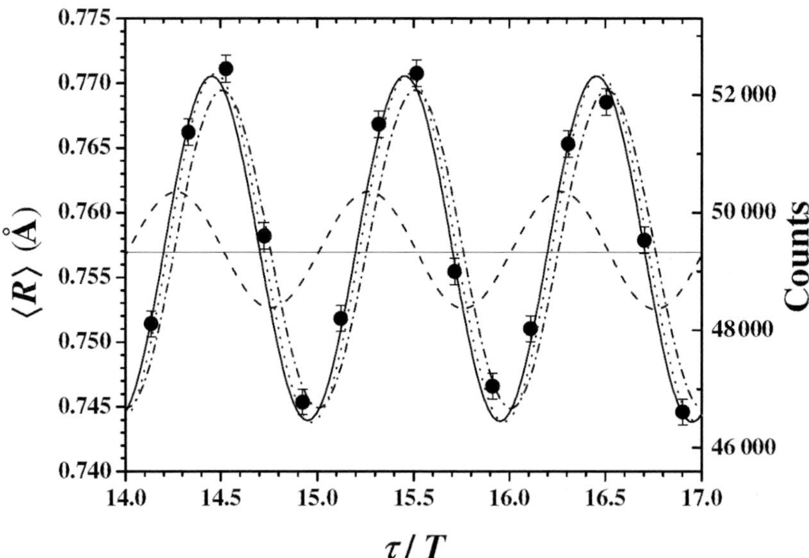

Fig. 6.12. Comparison of a numerical simulation of the R variation expected for *Lochfrass* (chain) or bond softening (dashed) only, and both effects (dotted) together. Also shown is the experimentally measured oscillation signal (dots) of the pump-probe experiment [52]. To guide the eye, a best-fit of the experimental data to a sinusoidal curve is also shown (solid).

Returning to Fig. 6.10, one notices a clear out-of-phase oscillation of the ionization (D_2^+) and the dissociative ionization (D^+) signals. Clearly, it is possible to control the two product channels by varying the delay time be-

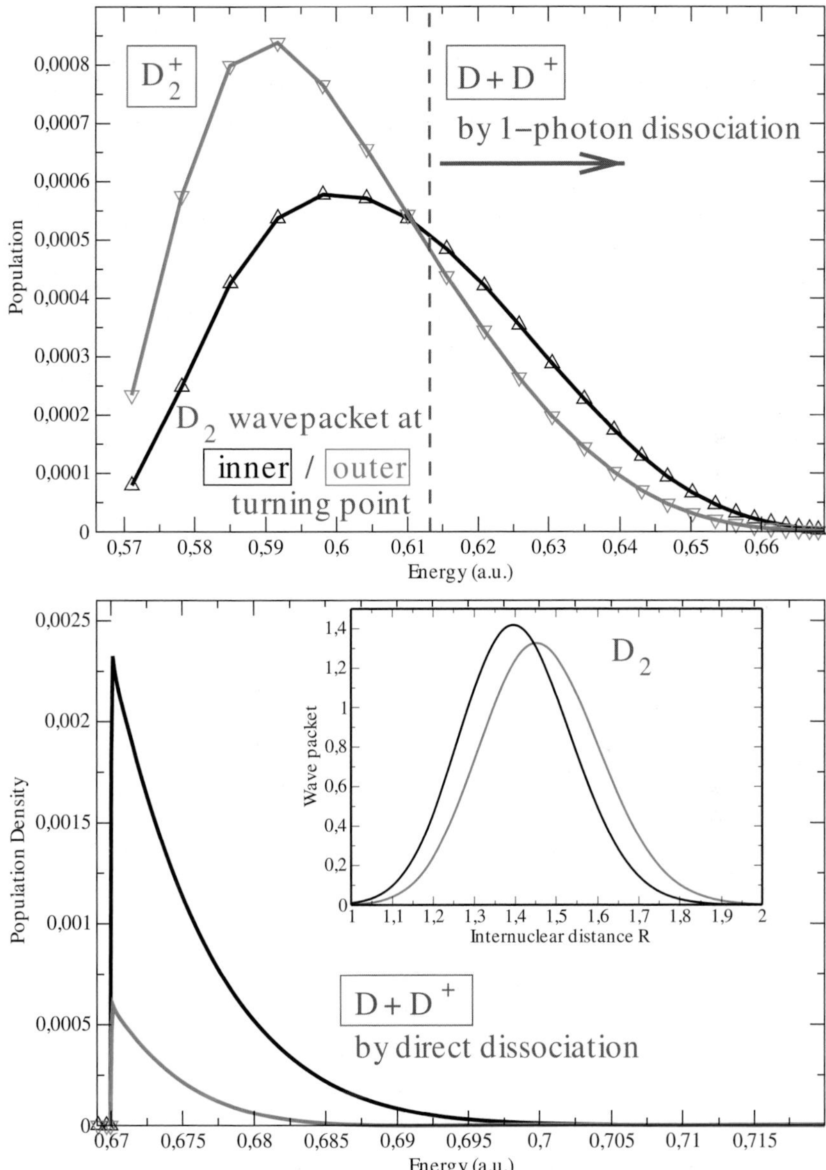

Fig. 6.13. Distribution over the vibrational states of D_2^+ following the laser-field ionization of the wave packet in the electronic ground state of D_2 (shown in the insert) at its inner (black) or outer turning (red) point. The upper spectrum shows the population of the field-free bound (nondissociative) states, the lower spectrum the continuous distribution in the dissociative continuum. The vertical dashed blue line indicates those states that can dissociate by one-photon absorption. (To guide the eye, also in the upper spectrum the discrete points are connected by lines.)

tween the pump and the probe pulses. Notably, depending on when the H_2 wave packet is probed, different products are preferentially produced. Even this phenomenon can be explained with the aid of the simple model proposed in [44] that describes the interplay between electronic and vibrational motion in strong fields. Analyzing not only the total ion rate of the pump-probe experiment as was done in [51], but predicting the vibrational distribution of the H_2^+ ion for the H_2 wave packet created by *Lochfrass* at its inner and outer turning points, it is possible to show that at the outer turning point not only the total ionization rate is larger than at the inner turning point, but also the vibrational distribution shifts to lower vibrational states, see Fig. 6.13. Therefore, at the outer turning point H_2^+ is preferentially formed in the low lying vibrational states that are harder to dissociate. At the inner turning point the total ionization rate is smaller, but the formed H_2^+ ions are preferentially formed in dissociative states or those states that can more easily dissociate. Since the H_2^+ wave packet formed in the pump pulse has practically no time to dissociate by passing over the field-suppressed barrier, it is most likely that (besides direct dissociation of those wave packet components that lie above the field-free potential barrier) dissociation occurs via one-photon absorption in the pedestal of the probe pulse. This is again in agreement with the simple theoretical model that predicts that the H_2^+ wave packet created at the inner turning point of the H_2 wave packet will populate preferentially those vibrational states that can directly or by one-photon absorption dissociate. Therefore, the dissociation signal has its maximum at the minimum of the ionization signal. In this context it may be noted that in another very recent theoretical work the enhancement and control of dissociative vs. nondissociative ionization of H_2 in intensive ultrashort laser pulses has been reported [53]. In contrast to the example discussed above, the process described in [53] takes place in the multi photon regime and involves VUV photons. In any case, these examples demonstrate the richness of the control scenarios in the strong-field regime.

6.3 Strong field chemistry and control

Dmitri A. Romanov and Robert J. Levis

6.3.1 Molecules in intense laser fields

When a 3-atomic or larger molecule interacts with an intense laser pulse a number of product channels may be accessed. Some of the potential outcomes are listed in Fig. 6.14 where coupling into the nuclear, electronic and nonlinear optical channels are delineated. Initial intuition suggested, incorrectly, that intense, short duration laser pulses interacting with polyatomic molecules would result primarily in multiphoton dissociation as shown in the first channel. Early experiments using intense nanosecond, picosecond and femtosecond pulses provided ample evidence for the second and third coupling channels in Fig. 6.14, which may be described as dissociative ionization and Coulomb explosion [54], respectively. Pulses of femtosecond duration have been shown to couple into electronic channels resulting in ionization without nuclear fragmentation for molecules like benzene and naphthalene [5]. In such experiments the energy in excess of the ionization potential (up to 50 eV!) [55, 56] couples mainly into the kinetic energy of the photoelectron. In terms of control experiments, the ability to produce intact ions at such elevated laser intensities suggested the possibility that intense lasers could be used to guide the dynamics of a molecule into a desired channel.

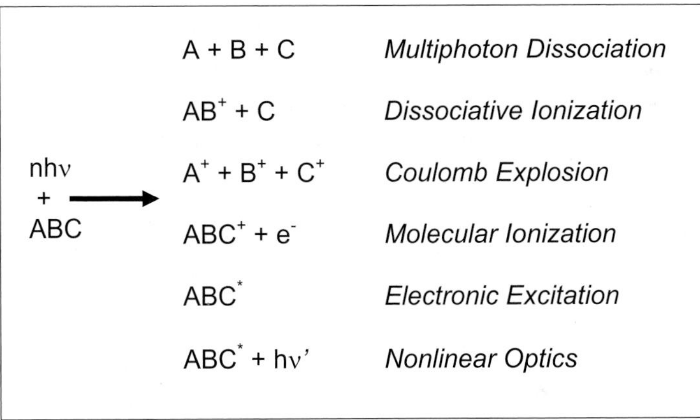

Fig. 6.14. Potential outcomes of the interaction of intense laser radiation with a molecule. At the present time the wavelengths used for the interaction range between $10\,\mu$m and 200 nm. The wavelengths employed in the studies reported here range between 750 and 850 nm with intensities of 10^{13}-10^{15} Wcm^{-2}

The relative importance of each product channel shown in Fig. 6.14 is dictated by the Hamiltonian for the molecule-radiation system. The understanding of the Hamiltonian for polyatomic molecules in general, and the more complex Hamiltonian for the interaction between strong fields and molecules in particular, is rather limited at the present time [7]. One would like to have high quality time-dependent calculations to model the strong field interaction, but these are simply intractable with current computational technology. Calculations for simple systems containing up to three protons and one or two electrons have been performed and these systems are reasonably well-understood [57–59]. For polyatomic systems, the number of degrees of freedom is too large for first-principles calculations. Thus, simple models have been employed to gain some insight into the mechanisms of interaction between intense laser pulses and atoms.

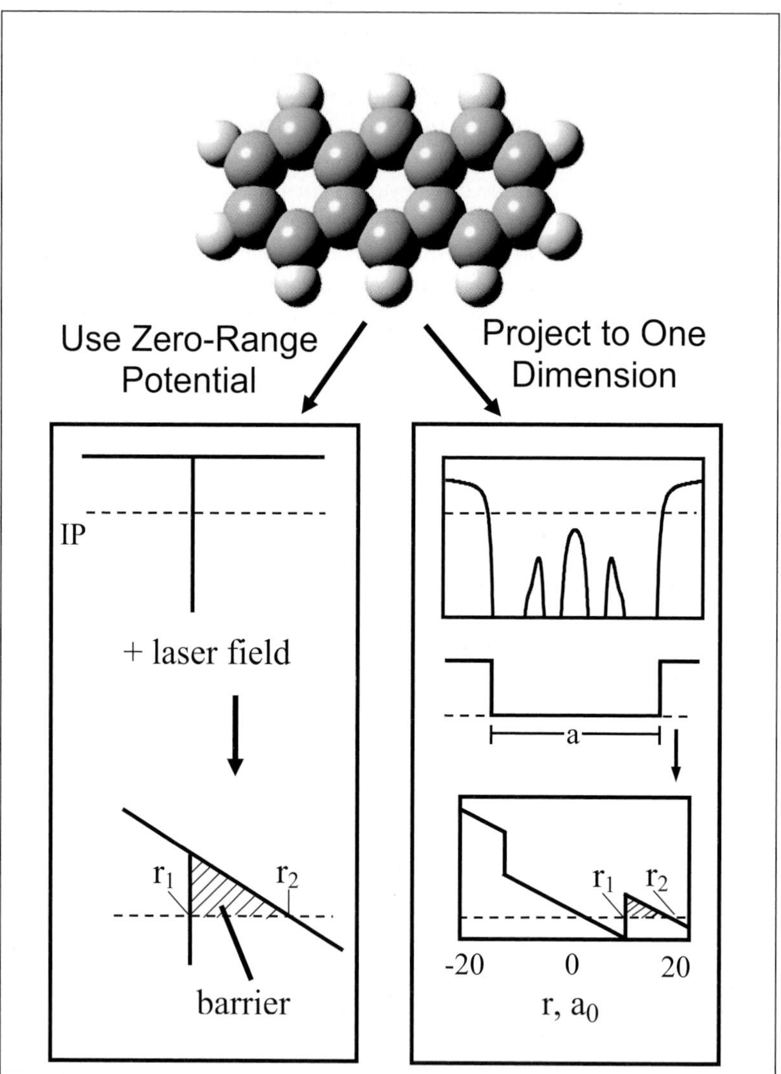

Fig. 6.15. A schematic of the structure-based model for representing molecules in intense fields. The presentation in the right hand panel is the zero-range model where only the ionization potential of the system is employed in calculations. The presentation in the left hand panel represents the use of the electrostatic potential of the molecule in determining an appropriate one-dimensional rectangular well to represent the spatial extent of the system. To compare the models an electric field of 1 V/Å is superimposed on each potential to reveal the barrier for tunnel ionization.

There is a hierarchy of models for representing molecules interacting with intense laser fields. The earliest models viewed the potential energy of interaction between the electron and the core as a delta function having a single state at the ionization potential of the system (called a zero-range potential) [22]. Subsequently, a Coulomb potential was employed for calculations in atoms [45, 60]. This was followed by a rectangular potential for molecules defined within the context of the structure-based model as shown in Fig. 6.15 [7, 23, 55, 61, 62]. The rectangular potential approximates the delocalization of electrons over the length scale of the molecular dimension by defining the width of the well to be equal to the characteristic length of the molecule. The characteristic length is defined as the largest distance between classical turning points in the three-dimensional electrostatic potential energy surface at the ionization potential of the molecule. The height of the rectangular well is the ionization potential of the molecule. A further advance incorporated time-dependence into the radiation-molecule interaction to go beyond the quasi-static regime [24].

6.3.1.1 Electronic dynamics of molecules in intense laser fields

To describe the mechanisms of strong field control of chemical processes it is important to consider the influence of the intense laser field on electrons in the molecule. For instance, we will see that bound electrons can gain significant ponderomotive energy ($\approx$ 1-5 eV) during the pulse and eigenstates can shift by similar energies [63]. In the case of the interaction of a laser pulse with a molecule, the appropriate starting point is the Hamiltonian for a multielectron system interacting with an electromagnetic field:

$$H = \frac{\boldsymbol{P}_c^2}{2M} + \frac{1}{2m_e}\sum_{I=1}^{Z} \boldsymbol{P}_I^2 + \frac{1}{m_n}\sum_{i>j=1}^{Z} \boldsymbol{P}_i \cdot \boldsymbol{P}_j + V(x_1, \ldots, x_i, \ldots, x_c) +$$
$$\frac{e}{m_e c}\boldsymbol{A}(\boldsymbol{x_c}, t) \times \sum_{i=1}^{Z} \boldsymbol{P}_i + \frac{Ze^2}{2m_e c^2}\boldsymbol{A}^2(\boldsymbol{x_c}, t) \tag{6.7}$$

where $\boldsymbol{P}$ is momentum, V is the potential energy as a function of position, Z is the nuclear charge, and $\boldsymbol{A}(\boldsymbol{x_c}, t)$ is the vector potential of the laser radiation. The first four terms describe the field free motion of the system. The last two terms describe the effect of the laser radiation on the population of eigenstates and corresponding shifts in the eigenstates of the system. In the electric field gauge the last term becomes:

$$\frac{Ze^2}{2m_e\omega_L^2}\boldsymbol{F}^2(\boldsymbol{x_c}, t) \tag{6.8}$$

where $\boldsymbol{F}$ is the electric field of the laser, and ω_L is the frequency of the laser. The average of this term over the period of oscillation for linearly polarized light is known as the ponderomotive potential U_p (see (6.4) in Sect. 6.1). In strong fields this term shifts all eigenstates upward in energy equally by U_p. A differential shifting of eigenstates results from the $\boldsymbol{A} \cdot \boldsymbol{P}$ term. To first and higher order, the $\boldsymbol{A} \cdot \boldsymbol{P}$ term may used to describe allowed transitions of amplitude between eigenstates. To second and higher order this term will describe differential shifting of the eigenstates. The magnitude and sign of the shift of a given state is dependent on the wavelength and the electronic structure of the system. Pan et al. [64] have derived expressions for the shifting of the ground state and Rydberg/continuum states of a model system. A lowest nonvanishing order perturbation theory treatment yields the ground (ΔE_g) and Rydberg level (ΔE_R) energy shifts as [64]:

$$\begin{aligned} \Delta E_g &= -\frac{Ze^2F_0^2}{4m_e\omega_L^2} - \frac{1}{2}\alpha F_0^2 \\ \Delta E_R &\approx 0 \end{aligned} \tag{6.9}$$

where α is the ground state polarizability. The first term in (6.9) is the negative of the ponderomotive potential U_p. The second term is equivalent to the dc Stark shift. This treatment is valid when the ground state is deeply bound and separated from adjacent eigenstates by many times the photon energy, $h\nu$ (the low frequency approximation). This is valid for most atoms and molecules investigated with near infrared or longer wavelength light. High lying bound states and all continuum states experience no $\boldsymbol{A} \cdot \boldsymbol{P}$ shift whereas deeply bound states of the atom experience a much greater, negative shift [65]. The pertinent shift in the states as a function of the terms in the Hamiltonian in the long wavelength limit is summarized in Fig. 6.16.

The laser intensities employed in recent high field experimental manipulation of chemical reactivity range up to $5 \times 10^{14}\,\mathrm{Wcm^{-2}}$. This corresponds to ponderomotive shifts up to 10 eV with similar shifts in the separation of the ground and excited state potential energy levels. The laser employed in these investigations has a period of 2.5 fs and an envelope with FWHM of 60-170 fs corresponding to at least a several hundred significant oscillations in the electric field vector interacting with the molecule. The states of the molecule undergo an associated oscillation in the splitting between energy levels that may result in periodic excitation on a time scale of the period of the laser. This dynamic shifting of energy levels implies that there will be transient field-induced resonances (or Freeman resonances) [66]. Evidence for these resonances in the case of molecules has been obtained by measuring the strong field photoelectron spectroscopy of a number of molecules including acetone, acetylene [63], water, benzene and naphthalene [67]. The oscillatory nature of the intense laser excitation also leads to above threshold ionization (ATI) peaks in the photoelectron spectrum [2]. These are denoted by peaks spaced by the photon energy extending to many photons above the minimum number required for ionization.

An important consideration for the control of chemical reactivity in the strong field regime is the order of the multiphoton process during excitation.

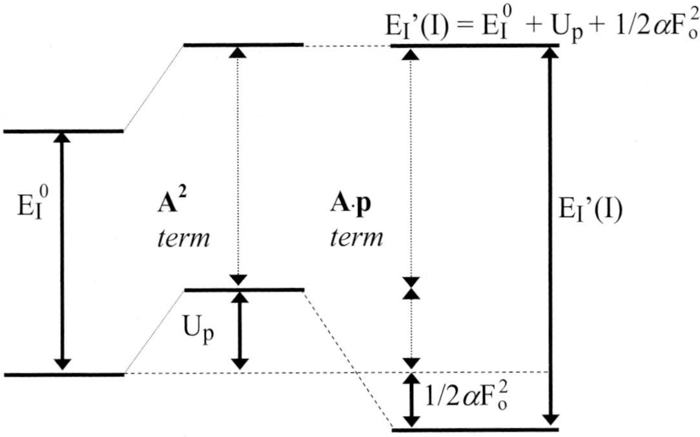

Fig. 6.16. The effect of various terms in the Hamiltonian for a charged particle in an oscillating electromagnetic field is shown. The ionization potential of the system remains unchanged by the $\boldsymbol{A}^2$ term as all states are raised equally. The $\boldsymbol{A} \cdot \boldsymbol{P}$ term lowers the ground state of the system by an amount equal to the $\boldsymbol{A}^2$ term plus an additional amount due to the induced polarization of the system. The net result is an increase in the ionization potential by an amount approximately equal to the ponderomotive potential of the laser pulse.

This order indicates the maximum number of photons that are available to drive a chemical reaction. Some indication of the number of photons involved in the strong field excitation process can be gleaned from measurements of strong field photoelectron spectra. Figure 6.17 displays the photoelectron kinetic energy distribution for benzene with the energy axis rotated by 90 degrees. The energy scale has been offset to include the energy of the ground and ionization potential of the molecule in the absence of the strong electric field. The arrows on the Figure represent the photons involved in both exceeding the ionization potential and in creating the ATI photoelectron distribution. At least six photons are required to surmount the ionization potential of benzene. Recall that in the presence of the strong electric field, the ionization potential will increase by an amount greater than the ponderomotive potential, further increasing the actual number of photons involved in the excitation process. At the intensity of 10^{14} Wcm^{-2} in this measurement, on the order of 10 photons may be absorbed to induce the photoelectron spectra shown. Including the photons required to reach the ionization potential, this means that approximately 20 photons may be involved in the excitation process. With the shaped pulses used in the experiments described in Sect. 6.3.2, the intensities are lower and on the order of 10 or fewer photons are likely involved in the excitation process.

Several other methods have been developed to predict the ionization probability of molecules. One is based on discretizing a molecule into a collection

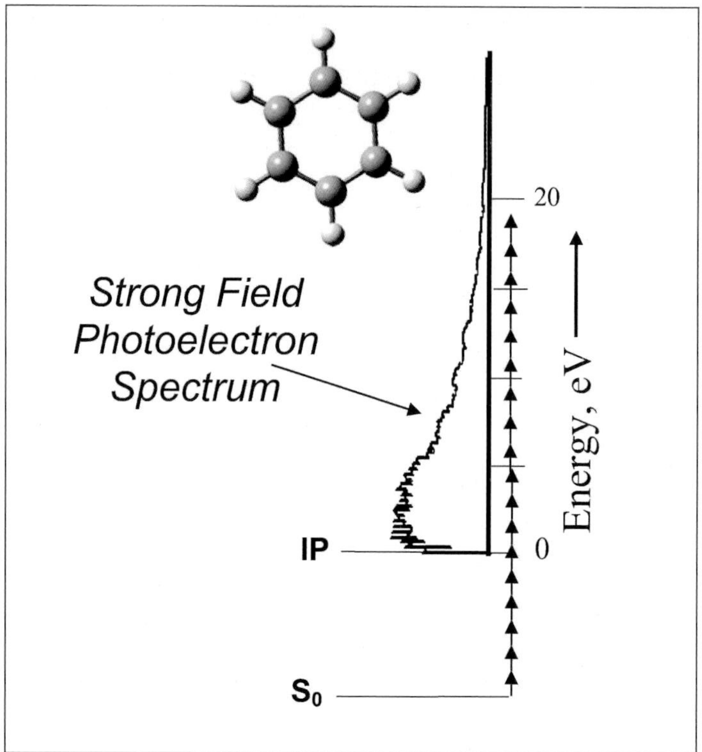

Fig. 6.17. The strong field photoelectron spectrum for benzene shown on an energy axis that includes the photons necessary to induce ionization. The photoelectron spectrum was obtained using $2 \times 10^{14}\,\mathrm{Wcm^{-2}}$, 800 nm radiation of duration 80 fs. The quantum energy of the photons are shown to scale and indicate that 10-20 photons are available to drive excitation processes in the strong field excitation regime. In addition, uncertainty broadening of the pulse will also produce a distribution of allowed photon energies that approaches the photon energy when multiphoton processes of order ten are approached.

of atomic cores that individually interact with the strong laser field and emit electrons [68, 69]. In this model, a carbon atom, for instance, is represented by an atom with an effective potential. The ionization probability is then a function of the individual ionization probabilities from atoms with opportunity for quantum interference during the ionization event. Unfortunately, the method must be parameterized for each molecule at the present time. The second method under development employs S-matrix theory [70] to calculate the ionization probability for atoms and molecules. This method focuses on the interference of the outgoing electron wave. Predictions about relative ionization probabilities are based on the symmetries of the highest occupied molecular orbital.

6.3.1.2 Nuclear dynamics of molecules in intense laser fields

The response of a molecule to a time-dependent electric field is the means by which chemical reactivity is controlled in these experiments. In the case of weak laser fields the response can be calculated with reasonable accuracy [71–73]. In the case of strong fields, the situation is much more complex, but the dynamical possibilities are much richer. In principle, the nuclear dynamics in strong laser fields could be determined using exact numerical solutions of the time-dependent Schrödinger equation. Such solutions are possible only for the simplest of molecules at the present time [57–59]. In fact, the bulk of such simulations have been performed using a one-dimensional model for the H_2^+ system [31, 74, 75]. These calculations show the presence of non-Born-Oppenheimer electron-nuclear dynamics. Since the nuclei move considerably on the time scale of the laser pulse, electronic wave functions are necessarily coupled with nuclear motion. Three distinct final states have been observed in strong field (no pulse shaping) mass spectra of polyatomic molecules: production of intact molecular ion, ionization with molecular dissociation, and removal of multiple electrons to produce Coulomb explosion [7]. The hallmark of the latter process is production of ions substantial ($> 5\,\mathrm{eV}$) kinetic energy. The presence of Coulomb explosion has been shown to depend on charge resonance-enhanced ionization (CREI) [76] which becomes the dominant mechanism at large critical internuclear distances. Interestingly, the production of high charge states in molecular clusters can be controlled using pump-probe excitation schemes [77].

At intensities that are lower than the threshold for multielectron ionization, the majority of molecules display some fraction of intact ionization. This phenomenon is not expected intuitively because the ionization processes are not resonant with low order multiples of the fundamental frequency implying that intense pulses must be employed for excitation. None the less, many molecules have been investigated to date and all appear to provide some degree of intact molecular ionization when 800 nm excitation is employed. The mechanism behind this ionization appears to involve suppression of ladder switching coupled with coherent excitation of electronic wave functions. The state of this subject has been reviewed in [7, 78, 79].

To measure the amount of energy that may couple into the nuclear degrees of freedom during the intense laser excitation event, we have investigated [56] the kinetic energy release in H^+ ions using both time-of-flight and retarding field measurements. A typical time-of-flight mass spectroscopy apparatus employed to make such measurements is shown in Fig. 6.18. In the series benzene, naphthalene, anthracene, and tetracene the most probable kinetic energy in the measured distributions was observed to increase as the characteristic length of the molecules increased as shown in Fig. 6.19. The corresponding retarding field measurements are shown in Fig. 6.20. Again the coupling into nuclear degrees of freedom was observed to increase in the larger molecules. The most probable kinetic energies increased from 30 eV for benzene to 60 eV

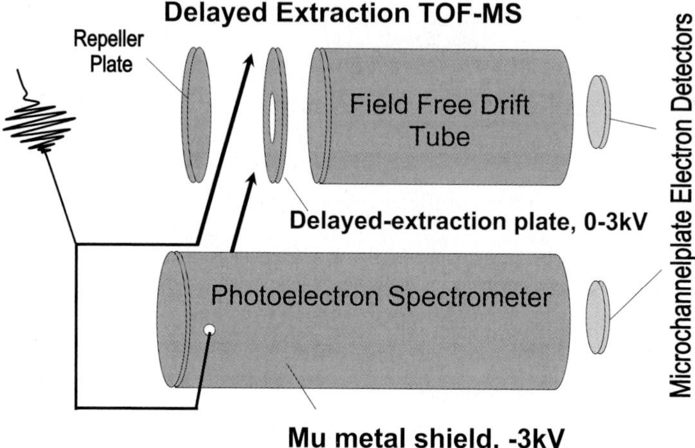

Fig. 6.18. Schematics of the photoelectron spectrometer and the time-of-flight ion detector used for measuring the kinetic energy distribution and molecular weight of the product ions.

for tetracene when a $1.2 \times 10^{14}\,\mathrm{Wcm}^{-2}$ laser excited the molecules. In terms of providing an enabling capability for strong field control, these results suggest that up to 80 photons may be involved in the excitation process when a molecule such as tetracene is excited under strong field conditions.

A general observation after ionization of large polyatomic molecules is the measurement of an enhanced degree of dissociation as the length of the molecule increases. This was first attributed to field-induced effects [5] without a quantitative model. Recently, a strong field nonadiabatic coupling model has been introduced to account for the enhanced coupling into nuclear modes in molecules with increasing characteristic length [24]. This excitation is akin to plasmon excitation where the precise energy of the resonance depends on the coherence length and binding energy of the electrons and the strength and frequency of the driving field. The model considers the amplitude of electron oscillation in comparison with the length of the molecule. If the amplitude of oscillation is small, the molecule may first absorb energy nonresonantly and then ionize from the excited states. The amplitude of the electron oscillation in an laser field is given by $a_{osc} = F/\omega_L^2$. In the event that the $a_{osc} < \ell$, where ℓ is the characteristic length of the molecule, the electron gains ponderomotive energy from the laser. Given an energy level spacing of Δ_0, the probability of nonadiabatic excitation within the Landau-Zener model becomes $\exp(-\pi\Delta_0^2/4\omega_L F\ell)$. As described in [22], the threshold for nonadiabatic excitation (when $\Delta_0^2 = \omega_L F\ell$) of a 4 eV transition for a system having $\ell = 13.5$ Å with 700 nm radiation occurs at $5.6 \times 10^{12}\,\mathrm{Wcm}^{-2}$. This theory implies that the probability for exciting nuclear modes in large molecules with delocalized electronic orbitals increases monotonically with characteristic length as

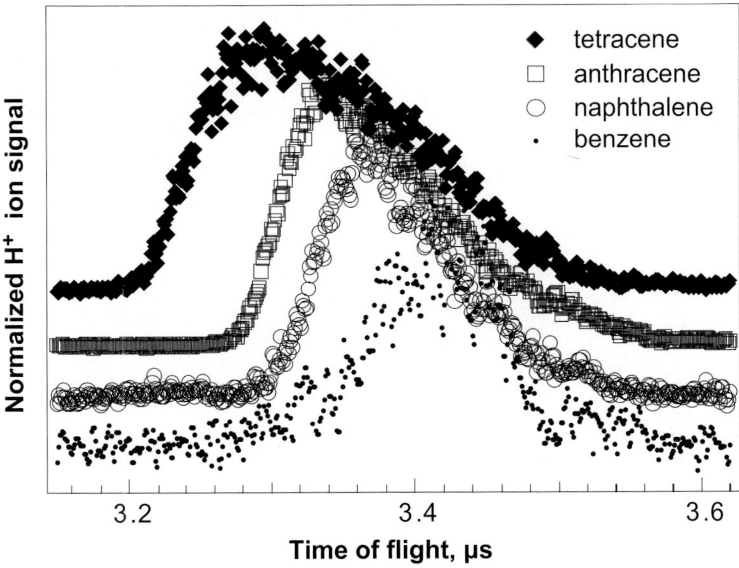

Fig. 6.19. Time-of-flight distributions for the H^+ ions for benzene, naphthalene, anthracene and tetracene after excitation using 2×10^{14} Wcm^{-2}, 800 nm radiation of duration 80 fs. The time of arrival distributions were measured by allowing the ions to drift in a field free zone of length 1 cm prior to extraction into the drift tube. In this experiment, earlier arrival times denote higher kinetic energies.

observed experimentally [5, 24]. The theory also suggests that intact molecular ionization will increase with increasing excitation wavelength for large molecules and this has been confirmed [24].

In this simple model, the extent of the excitation and ionization processes is singularly determined by the spatial size of the molecule. Experiments, however, reveal a much more complex picture of the nonadiabatic excitation mechanism and the corresponding products in real polyatomic molecules. One reason for this complexity is that a large molecule actually has two energy scales: (i), the gap between the ground state and the excited state manifold; and (ii), the gap between energy levels in the excited state manifold. The former is usually much greater than the inter-level distances in the manifold. As a result, the first step in the excitation process requires a special treatment. Accordingly, a consistent model of dissociative ionization caused by nonadiabatic excitation has been developed [80, 81] that is based on three major elements: (i), the doorway state for the nonadiabatic transition into the excited state manifold (the state that has the maximum transition dipole matrix element with the ground state); (ii), multielectron polarization of the ground state and the doorway state (that is, the dynamic Stark shift that strongly modifies the transition rate); (iii), sequential energy deposition in the neutral

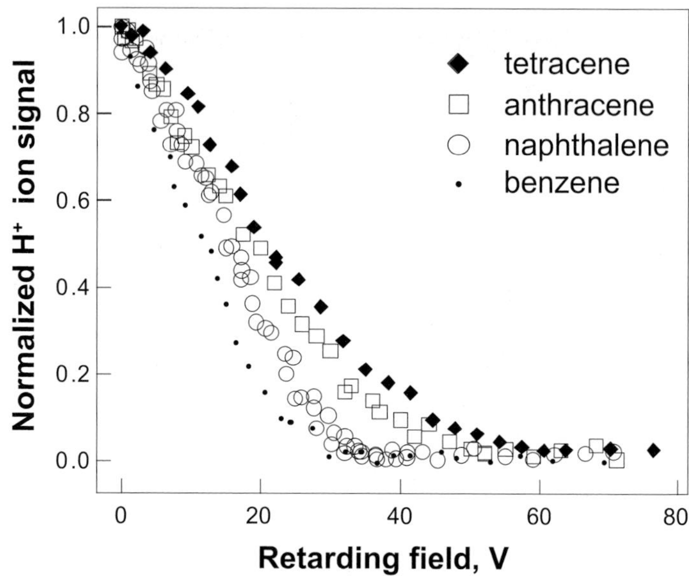

Fig. 6.20. Retarding field measurement of the H^+ ion kinetic energy distributions arising from benzene, naphthalene, anthracene and tetracene after excitation using 2×10^{14} Wcm^{-2}, 800 nm radiation of duration 80 fs. The measurements reveal that as the characteristic length of the molecule increases, the cutoff energy increases monotonically.

molecules and corresponding molecular ions (for large molecules, the ionic transition dipole and the dynamic Stark shift are usually greater than those for the neutral molecule, while the bottleneck energy gap between the ground state and the doorway state is typically smaller) [82]. In this model, the first excitation stage leads to ionization; the second (and subsequent) stages result in the molecular ion fragmentation.

The predictions of the model have been compared with experimental data on dissociative ionization for two series of related molecules as a function of laser intensity. In Series a, benzene, naphthalene, anthracene, and tetracene, the characteristic length of the aromatic molecules increases from benzene to tetracene; along with the extent of π-electron delocalization that should directly affect the dipole transition matrix element and the energy distance for the electronic excitation from the ground state to the doorway state. In Series b, 1,2,3,4,5,6,7,8-octahydroanthracene (OHA), 9,10-dihydroantracene (DHA), and anthracene, the characteristic lengths are similar but the extent of π-delocalization nevertheless increases from OHA to anthracene, with increasing number of unsaturated aromatic rings. The mass spectra were obtained at laser intensities between 0.1×10^{13} Wcm^{-2} and 25.0×10^{13} Wcm^{-2}. The extent of nonadiabatic energy transfer and the subsequent molecular frag-

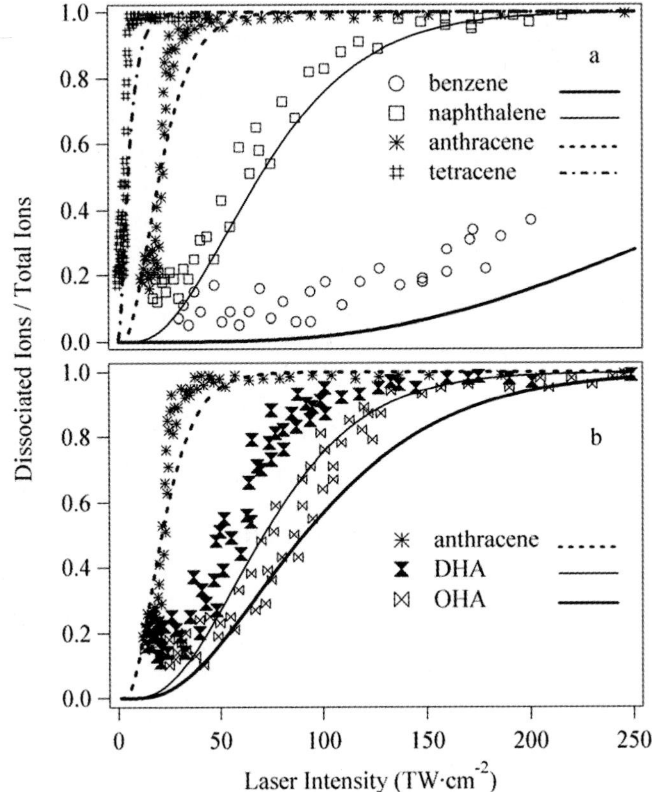

Fig. 6.21. Fragmentation fraction and nonadiabatic multielectron dynamics calculation: a) benzene-naphthalene-anthracene-tetracene series; b) anthracene-DHA-OHA series. The curves show the calculated fraction of the molecular ions excited nonadiabatically by the end of a laser pulse (integrated conditional probabilities of two-stage nonadiabatic excitation).

mentation was quantified by plotting the ratio of fragment ion signal to the total ion signal vs. the laser intensity. The plots in Fig. 6.21 reveal that the onset of extensive dissociation occurs at lower laser intensities with increasing molecular size for Series a and increasing degree of unsaturation in Series b. Note that this observation apparently runs contrary to the conventional multiphoton perturbative picture. Indeed, for larger molecules (tetracene, anthracene) the number of photons required for electronic excitation decreases and thus the intensity dependence should be of lower order than for smaller molecules (naphthalene, benzene). However, the molecular excitation/double ionization curves calculated as functions of the laser intensity according to the above-delineated model, agree quantitatively well with the experimental data.

(Fig. 6.21) This agreement is especially remarkable because the all the model operates with the transition dipoles, energy gaps, and dynamic polarizabilities taken from ab initio calculations and thus does not contain any fitting parameters.

Additional information on the process of the nuclear subsystem excitation and possibilities to control this process can be gleaned from the kinetic energy distributions of the ionized fragments released after excitation [83, 84]. From this standpoint, the most informative fragments are the positive hydrogen ions (protons). Because of their small mass, they (i) move substantially on the timescale of the pulse duration; and (ii) acquire more kinetic energy than their massive counterparts. In large molecules protons usually occupy peripheral positions thus having an unobstructed outgoing trajectory. As an example of such proton-related information, we present in Fig. 6.22 the energy distributions of protons resulting from the dissociative ionization (Coulomb explosion) of anthracene subjected to intense laser pulses ($\sim 10^{14}\,\mathrm{Wcm}^{-2}$ intensity, 800 nm wavelength, and 60 fs duration) and demonstrate counterintuitive details of the pulse-driven fragmentation process. Two distinct regimes of proton ejection dynamics were observed: at lower laser intensities the proton kinetic energy release increases rapidly with the laser intensity, only to saturate at higher laser intensities. Most surprisingly, the proton kinetic energies occur to exceed 30 eV; actually, the cutoff of the energy distribution reaches 52 eV. To account for this excessive energy, a strong-field charge localization model was suggested. It assumes that nonadiabatic dynamics of charge distribution in a large (multiply) ionized molecule leads to charge localization on one side of the molecule, sustained through successive ionizations of the molecular ion. The model explains quantitatively the dependence of the proton kinetic energy on the laser intensity (Fig. 6.22). Dissociative ionization of a polyatomic molecule enabled by long-lived charge localization is a specific strong-field phenomenon that can well serve as a useful physical mechanism of electron-nuclear dynamics control.

Yet another type of strong-field electron-nuclear dynamics emerges from comparison of proton kinetic energy distributions of two similar molecules: anthracene and 9,10-anthraquinone as illustrated in Fig. 6.23 [84]. These distributions are similar at lower laser intensities but differ significantly at higher intensities: starting at $\sim 9.0 \times 10^{13}\,\mathrm{Wcm}^{-2}$, a high-energy mode with a cutoff value extending to approximately $83 \pm 3\,\mathrm{eV}$ forms in the anthraquinone spectra. These higher kinetic energies are not due to higher degree of ionization, because the rate of nonadiabatic excitation and ensuing ionization of anthraquinone is actually even smaller than that for anthracene. Instead, the high-energy mode is explained by restructuring of the anthraquinone molecule prior to its Coulomb explosion. Model dynamical calculations based on Gaussian 03 geometry optimization and local charge distributions show that anthraquinone can form a field-dressed enol zwitterion where one of the "inner" protons (1,8,4, or 5 in Fig. 6.23) migrates to oxygen creating an O-H bond. The strong-field polarization of the zwitterion in the O-O direction provides

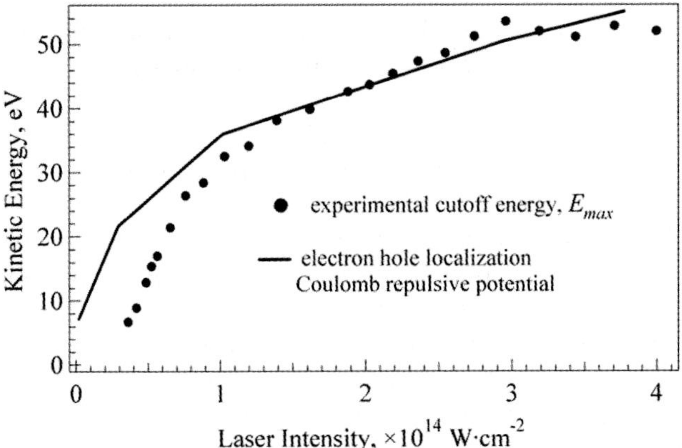

Fig. 6.22. The cutoff values of the proton kinetic energy distributions and model maximum Coulomb potential expelling protons, as functions of the laser intensity.

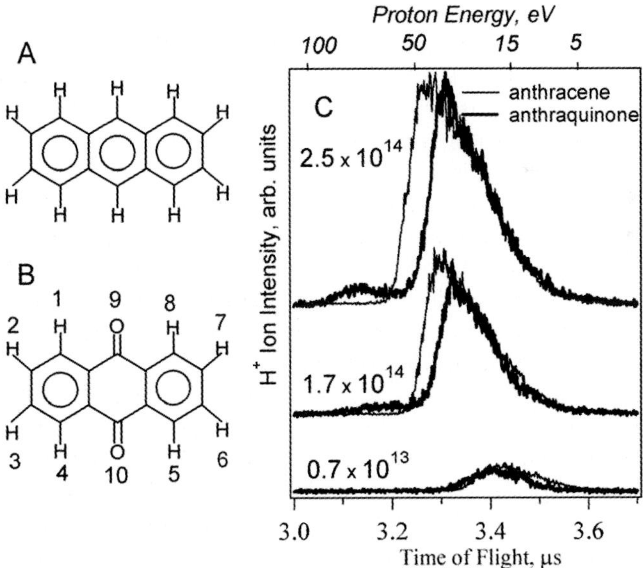

Fig. 6.23. Structures of anthracene (A) and 9,10-anthraquinone (B) with labeled proton positions. C: time-of-flight distributions of protons ejected from anthracene (thin line) and anthraquinone (thick line) at three different laser intensities (Wcm^{-2}).

the necessary degree of charge separation and ensuing nonadiabatic charge localization to eject the observed high-energy protons. These results demonstrate that modification of nuclear potentials of a polyatomic molecule by a strong oscillating electric field can force dynamic nuclear rearrangement into metastable positions that are quasi-bound in the presence of the field. (Note that this effect differs essentially from weak-field rearrangements, where one-photon electronic transition is followed by slow internal conversion on excited potential energy surfaces; it is rather analogous to bond softening in diatomic molecules during an intense laser pulse.) Thus, direct manipulation of intramolecular nuclei motion can be achieved in polyatomic molecules by strong laser fields.

Whether the nonadiabatic excitation can be controlled remains an open question at the present time. The present successes [85–87] in controlling chemical reactivity suggest that nonadiabatic processes either are not significant or that the closed loop control method is able to effectively deal with this excitation pathway.

6.3.2 Strong field control using tailored laser pulses

The use of strong fields to control chemistry is quite new, while the area of coherent control research has broad foundations [88–90] (see also Chapter 2, Sect. 2.4). The essence of the control concept in terms of optical fields and molecules is captured by the following transformation goal:

$$|\psi_i\rangle \xrightarrow{F(t)} |\psi_f\rangle \tag{6.10}$$

where an initial quantum state $|\psi_i\rangle$ is steered to a desired final state $|\psi_f\rangle$ via interaction with some external field $F(t)$. As a problem in quantum control, the goal is typically expressed in terms of seeking a tailored laser electric field $F(t)$ that couples into the Schrödinger equation:

$$\mathrm{i}\hbar\frac{\partial}{\partial t}|\psi\rangle = [H_0 - \mu F(t)]\,|\psi\rangle \tag{6.11}$$

through the dipole μ. The goal is to create maximum constructive interference in the state $|\psi_f\rangle$ according to (6.10), while simultaneously achieving maximal destructive interference in all other states $|\psi_{f'}\rangle$, $f' \neq f$ at the desired target time T. A simple analogy to this process is the traditional double slit experiment [91]. However, a wave interference experiment with two slits will lead to only minimal resolution. Thus, in the context of quantum control, two pathways can produce limited selectivity when there are many accessible final states for discrimination. Rather, a multitude of effective slits should be created at the molecular scale in order to realize high quality control into a single state [92], while eliminating the flux into all other states.

The requirement of optimizing quantum interferences to maximize a desired product leads to the need for introducing an adjustable control field $F(t)$

having sufficiently rich structure to simultaneously manipulate the phases and amplitudes of all of the pathways connecting the initial and final states. Construction of such a pulse is currently possible in the laboratory using the technique of spatial light modulation [93, 94]. However, calculation of the time-dependent electric fields to produce the desired reaction remains a problematic issue for chemically relevant reactions. Unfortunately, solution of the Hamiltonian at the Born-Oppenheimer level remains largely unknown for polyatomic molecules, and this severely limits the ability to perform a priori calculations at the present time. Even if the field free molecular Hamiltonian were known, the highly nonlinear nature of the strong field excitation process effectively removes all possibility of calculating an appropriate pulse shape in this regime. Thus we are left with the following conundrum: If the design can be carried out reliably, then the physical system will likely not be of much interest, while for interesting physical systems, reliable designs can not be performed. The method of closed-loop control for laser-induced processes [95] offers a way to surmount our lack of knowledge of the Hamiltonian to find appropriate pulse shapes, $F(t)$.

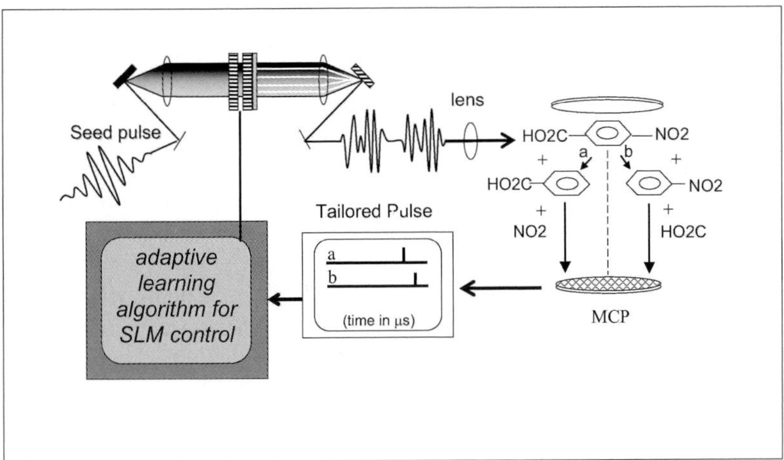

Fig. 6.24. A schematic of the closed loop apparatus for tailoring the time-dependent laser fields to produce the desired reaction product. In this scheme an algorithm controls the spatial light modulator that produces a well-defined waveform. The tailored light pulse interacts with the molecular sample to produce a particular product distribution. The product distribution is rapidly measured using time-of-flight mass spectrometry and the results are fed back into the control algorithm. The same closed-loop concept with other sources or detectors can be applied to control a broad variety of quantum phenomena.

To implement the optical control experiment (OCE) closed-loop control paradigm in the strong field regime three technologies are combined: (1) regenerative amplification of ultrashort pulses; (2) pulse shaping using spatial light modulation; and (3) some feedback detection system, (i.e. time-of-flight mass spectral detection in the experiments presented here). An overview of this implementation of the closed-loop control experiment is shown in Fig. 6.24. Briefly, the experiment begins with a computer generating a series of random, time-dependent laser fields (forty such control pulses are employed in the experiments presented here). In some cases prior estimates for fields might be available by design or from related systems to introduce specific trial field forms. Each of the control pulses is amplified into the strong field regime and subsequently interacts with the gas phase sample under investigation. Products are measured using time-of-flight mass spectrometry and this requires approximately 10 μs to detect all of the ion fragments. The mass spectra are signal averaged with a number of repeats for the same pulse shape and analyzed by the computer to determine the quality of the match to the desired goal. The remainder of the control fields sequentially interact with the sample and the fitness of the products are also stored on a computer. After each of the forty control fields have been analyzed in terms of the product distribution, the results of the fitness are employed to determine which fields will be used to create the next set of laser pulses for interaction with the sample. The system iterates until an acceptable product distribution has been achieved.

6.3.2.1 Trivial control of photochemical ion distributions

We first consider whether manipulation of the dissociation distribution can be achieved by simple alteration of either pulse energy or pulse duration. These are termed trivial control methods and in either case, there is no need to systematically manipulate the relative phases of the constituent frequency components. Pulse energy modulation is achieved here using a combination of a polarization rotator and beam splitter or by the use of thin glass cover slips to reflect away several percent of the beam. Pulse duration control can be implemented by either restricting the bandwidth of the seed laser or by placing a chirp onto the amplified pulse in the compressor optics.

Investigations of trivial control suggest that the ionization/fragmentation distribution can often be manipulated by altering either pulse energy or pulse duration. As an example, Fig. 6.25 shows the mass spectral distributions measured for p-nitroaniline as a function of either pulse duration (Fig. 6.25a) or pulse energy (Fig. 6.25b). In the case of the transform limited mass spectrum at 10^{14} Wcm^{-2}, there are many features in the mass spectrum corresponding to production of the $C_{1-5}H_x^+$ fragments. There is a minor peak at $m/e = 138$ amu corresponding to formation of the parent molecular ion. We observe that when the pulse duration is increased the fragmentation distribution shifts toward lower mass fragments. This indicates an enhanced opportunity for ladder switching during the excitation process. Ladder switching allows facile excitation of the internal modes of the molecule [7]. Increasing the pulse duration also leads to lowering the pulse intensity. Alternatively to

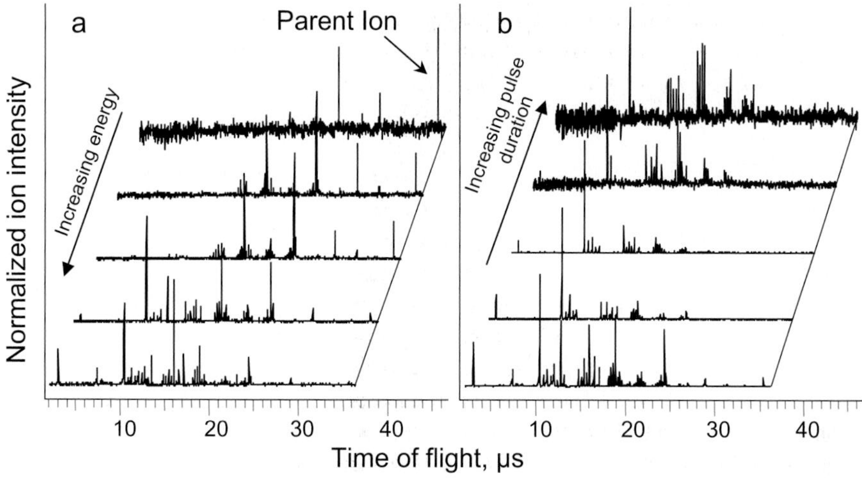

Fig. 6.25. Time-of-flight ion spectra of p-nitroaniline after excitation using pulses centered at 790 nm, of duration 80 fs. In panel a the pulse energy was varied from 0.60 to 0.10 mJ/pulse, the pulse duration was 80 fs. In panel b the pulse duration was varied from 100 fs to 5 ps, the pulse energy was 0.60 mJ/pulse.

lower the pulse intensity, the pulse energy can be reduced. When this form of trivial control is implemented, a completely different mass spectral distribution is obtained, as shown in Fig. 6.25b. When the intensity is reduced by a factor of 5 the parent molecular ion becomes one of the largest features in the mass spectrum. These results suggest that in any control experiment a series of reference experiments probing the products as a function of pulse energy and duration are necessary to rule out the possibility of trivial effects.

6.3.2.2 Closed-loop control of selective bond cleavage processes

Closed-loop control in the strong field regime has now been demonstrated on a series of ketone molecules [85]. We begin with acetone as a simple polyatomic system. Fig. 6.26 displays the transform limited mass spectrum resulting from the interaction of acetone vapor with a pulse of duration 60 fs and intensity 10^{13} Wcm^{-2}. There are a number of mass spectral peaks corresponding to various photoreaction channels as summarized in scheme I. Channel (a) corresponds to simple removal of an electron from the molecule to produce the intact acetone radical cation at $m/e = 58$. As noted in Sect. 6.1, the ability to observe the intact molecule in the mass spectrum reveals that not all of the excitation energy necessarily couples into nuclear modes. The second pathway, (b), observed is cleavage of one methyl group to produce the CH_3CO and methyl ions. The third pathway corresponds to the removal of two methyl species to produce the CO and methyl ions. Only one of the product species

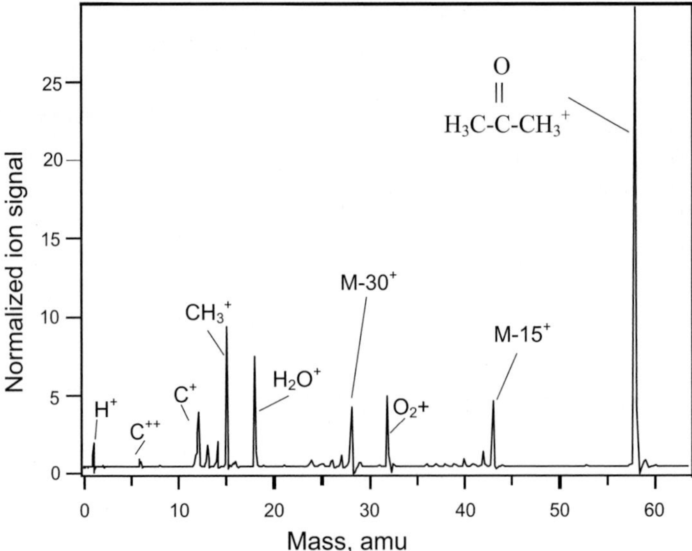

Fig. 6.26. The time-of-flight mass spectrum for acetone after excitation using $5 \times 10^{13}\,\mathrm{Wcm}^{-2}$, 800 nm radiation of duration 60 fs. The prominent peaks in the mass spectrum are marked.

in each channel is shown with a positive charge. Clearly there will be a probability for each of the product species to be ionized that depends on the details of the laser pulse, the fragment's electronic and nuclear structure, and the dissociation pathway.

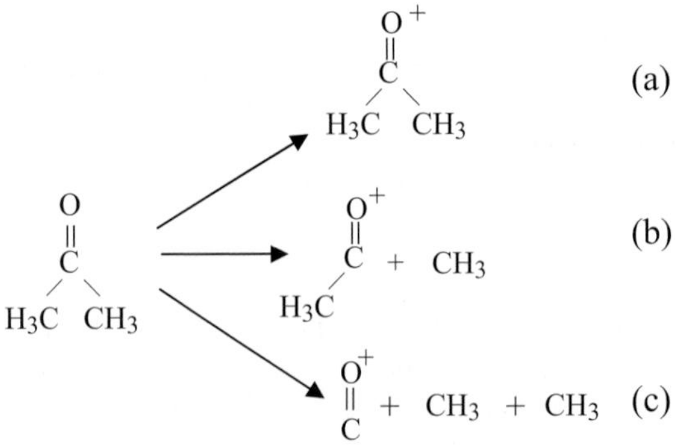

$(CH_3)_2C{=}O \rightarrow (CH_3)_2C{=}O^+$ (a)

$(CH_3)_2C{=}O \rightarrow H_3C{-}C{=}O^+ + CH_3$ (b)

$(CH_3)_2C{=}O \rightarrow C{=}O^+ + CH_3 + CH_3$ (c)

Scheme I

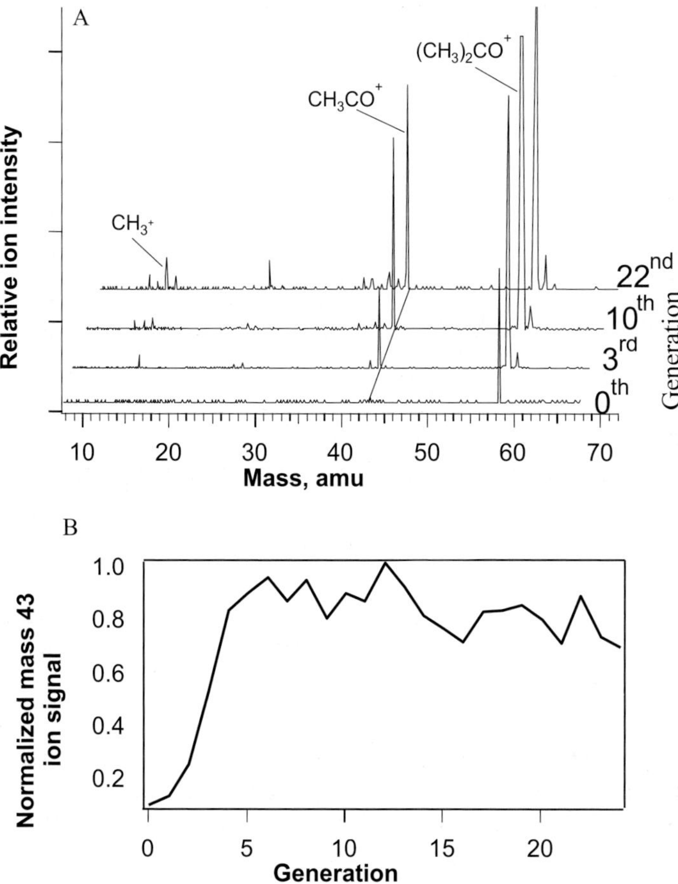

Fig. 6.27. (A) A representative mass spectra of acetone (CH_3-CO-CH_3) for the initial 0th, 3rd, 10th and 22nd generations of the laboratory learning process when maximization of the CH_3CO^+ ion from acetone is specified; (B), The CH_3CO^+ signal as a function of generation of the genetic algorithm. In (B) and the following plots of this type, the average signal for the members of the population at each generation is shown.

One of the simplest illustrations of the OCE closed-loop control algorithm is the case of enhancing the CH_3CO ion signal from acetone. This corresponds to specifying optimization of the second pathway (b) shown in scheme I. Using this criteria, representative mass spectra are shown as a function of generation in Fig. 6.27 when the algorithm has been directed to increase the intensity of the methyl carbonyl ion at $m/e = 43\,\text{amu}$. The intensity of this ion increases by an order of magnitude by the 5th generation in comparison with the initial randomly generated pulses and is seen to saturate shortly thereafter. The modulation in the signal in subsequent generations is largely due to the algorithm searching new regions of amplitude and phase control field space through the operations of mutation and crossover. The experiment demonstrated two important features of the closed-loop control. The first was that the algorithm was capable of finding suitable solutions in a reasonable amount of laboratory time (10 minutes in this case). The second was that the shaped strong field pulses were able to dramatically alter the relative ion yields and thus the information content in a mass spectrum. We anticipate that the method will have important uses as an analytical tool based on this capability. Finally, the control exerted in this case is of the trivial form, and is due to intensity control as indicated by the masks showing that the optimal pulse was near transform limited and of full intensity. The reference experiments also demonstrated that intense transform limited pulses resulted in a similar fragmentation distribution.

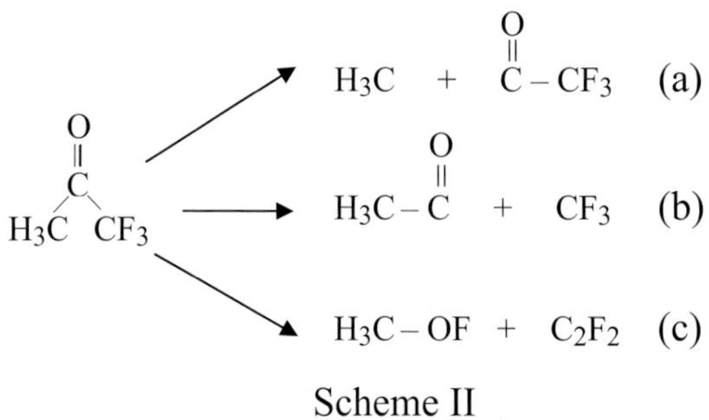

Scheme II

The control over the selective cleavage of various functional groups has been investigated using the molecules trifluoroacetone and acetophenone. Trifluoroacetone was investigated because there are two distinct unimolecular decomposition routes as shown in scheme II (a) and (b). Fig. 6.28 displays the mass spectrum associated with the transform limited, intense laser excitation of trifluoroacetone. The ions of importance in the spectrum include peaks at $m/e = 15, 28, 43, 69$, and 87 corresponding to CH_3, CO, CH_3CO, CF_3 and CF_3CO. These peaks are associated with cleavage of the methyl, fluoryl or both species from the carbonyl group as indicated in scheme II. Interestingly, there is also a feature at $m/e = 50$ amu which can only be assigned to CH_3OF shown in pathway (c). This species must be formed by an intense field rearrangement process and has not been observed in the weak field regime of photochemical reactivity.

The ability of the closed-loop control to cleave a specific bond is demonstrated in Fig. 6.29 where we have specified that the algorithm search for solutions enhancing the signal at $m/e = 69$. This ion corresponds to the CF_3 species. Fig. 6.29 demonstrates that the closed-loop OCE method may be

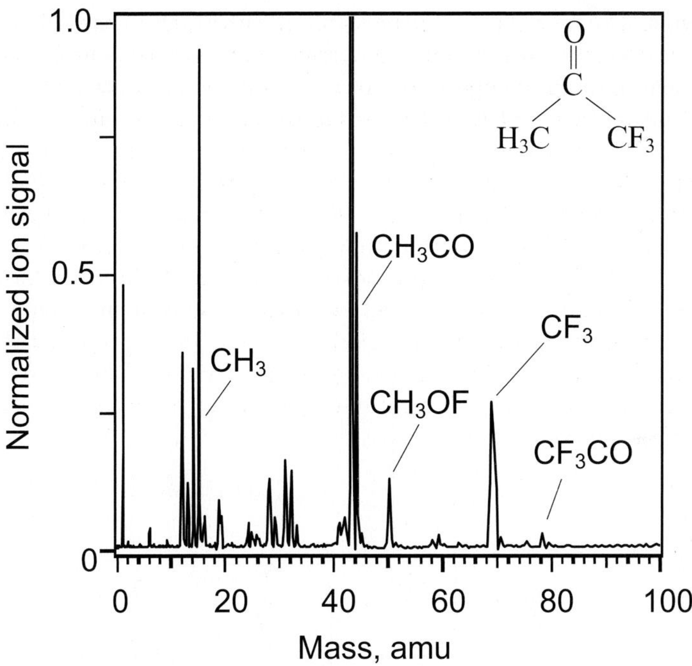

Fig. 6.28. The time-of-flight mass spectrum for trifluoroacetone (CF_3-CO-CH_3) after excitation using 5×10^{13} Wcm^{-2}, 800 nm radiation of duration 60 fs. The prominent peaks in the mass spectrum are marked.

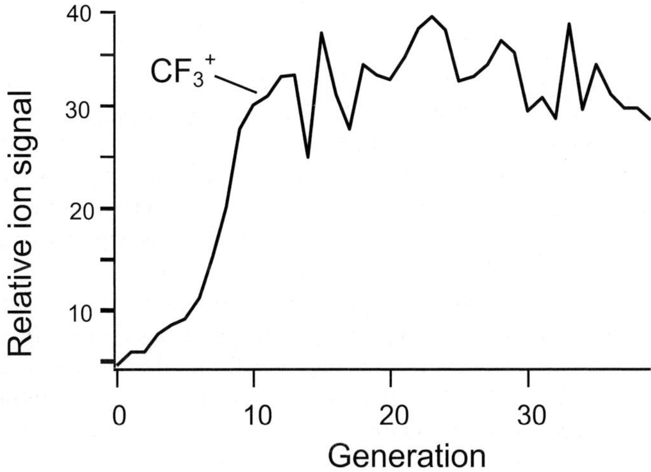

Fig. 6.29. The CF_3^+ signal as a function of generation of the genetic algorithm. In this experiment the cost functional was designed to simply optimize this signal.

used to enhance the desired ion signal by a factor of approximately thirty in comparison with the initial random pulses. While this experiment was successful in enhancing the desired ion yield, it does not necessarily demonstrate control. Control is achieved when one channel is enhanced at the expense of another.

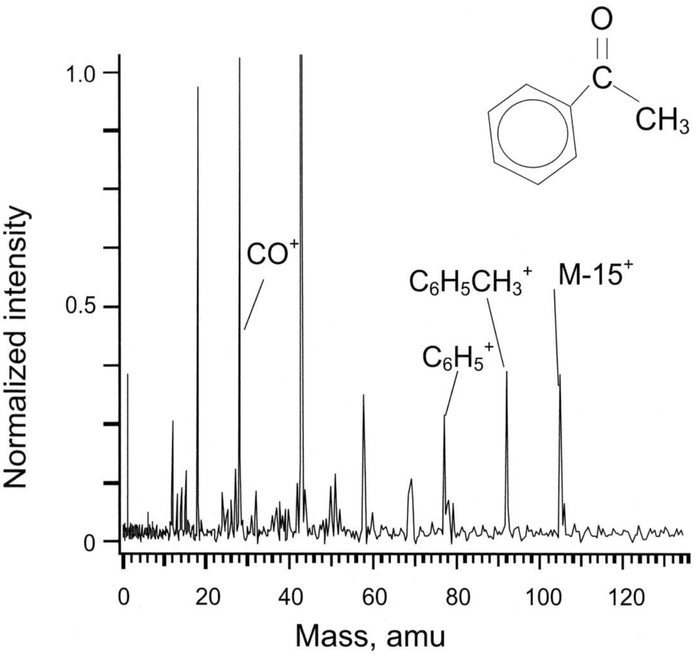

Fig. 6.30. The time-of-flight mass spectrum for acetophenone (C_6H_5-CO-CH_3) after excitation using 5×10^{13} Wcm^{-2}, 800 nm radiation of duration 60 fs. The prominent peaks in the mass spectrum are marked.

To demonstrate control over selective cleavage of specified bonds in a molecule we consider acetophenone, a system that has a carbonyl species bound to methyl and phenyl functional groups. The transform limited mass spectrum for acetophenone is shown in Fig. 6.30. There are numerous peaks detected in the spectrum revealing that there are a multitude of decomposition paths available after excitation. The ions observed at 15 and 105 amu correspond to the species obtained after cleavage of the methyl group. The pair of ions at 77 and 43 amu correspond to cleavage of the phenyl group. The dissociation and rearrangement reactions investigated for this molecule are shown in scheme III. Scheme III(c) implies the rearrangement of acetophenone to produce toluene and CO and this is signified in the mass spectrum by peaks at 92 and 28 amu respectively. To determine whether a path can be

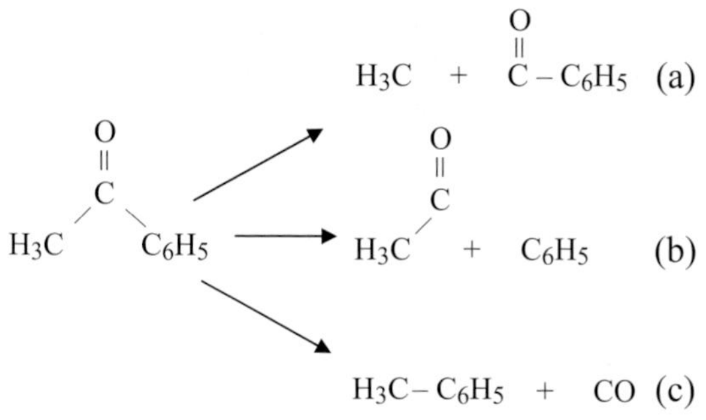

$H_3C - C(=O) - C_6H_5 \rightarrow H_3C + C(=O)-C_6H_5$ (a)

$H_3C - C(=O) - C_6H_5 \rightarrow H_3C-C(=O) + C_6H_5$ (b)

$H_3C - C(=O) - C_6H_5 \rightarrow H_3C-C_6H_5 + CO$ (c)

Scheme III

selectively enhanced we specified enhancement of the ion ratio for the species C_6H_5CO/C_6H_5. This denotes selective cleavage of the methyl group at the expense of the phenyl group. Note that we do not stipulate how the ratio should be increased, i.e. increase C_6H_5CO or decrease C_6H_5. Picking a particular path could be done with another cost functional. The ratio as a function of generation is shown in Fig. 6.31. The ratio increases by approximately a factor of 2 after 20 generations. Other ions could have been chosen to control the cleavage reaction, the two chosen happen to be experimentally convenient. Thermodynamically, the goal of enhancing methyl dissociation is the favored cleavage reaction because the bond strength of the methyl group is 15 kcal less than that of the phenyl group [96]. The ratio of phenyl ion to phenyl carbonyl can also be enhanced as shown in Fig. 6.32. The learning curve for this experiment reveals that the phenyl carbonyl ion remains relatively constant while the phenyl ion intensity increases. This is interesting because the energy required to cleave the phenyl-CO bond is 100 kcal while the methyl-CO bond requires 85 kcal. Thus the ratio of these ions can be controlled over a dynamic range of approximately five in the previously reported experiment [85] and a dynamic range of up to 8 has been recently observed.

The goal of laser control of chemical reactivity transcends the simple unimolecular dissociation reactions observed to date [85–87, 97, 98] Observation of the toluene ion in the strong-field acetophenone mass spectrum suggests that control of molecular dissociative rearrangement may be possible. To test this hypothesis we specified the goal of maximizing the toluene yield from acetophenone, as shown in scheme IV. For toluene to be produced from acetophenone, the loss of CO from the parent molecule must be accompanied by formation of a bond between the phenyl and methyl substituents. The closed-loop control procedure produced an increase in the ion yield at 92 amu of a factor of 4 as a function of generation as shown in Fig. 6.33. As a further test, we specified maximization of the ratio of toluene to phenyl ion and observed a similar learning curve to that in Fig. 6.32; with an enhancement in the toluene to phenyl ratio of a factor of 3. Again, the final tailored

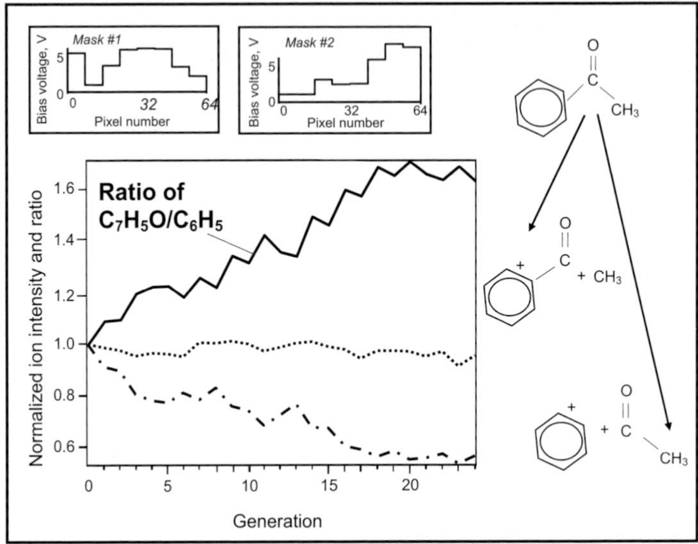

Fig. 6.31. The relative ion yield for phenylcarbonyl (dotted) and phenyl (dashed) and the $C_6H_5CO^+/C_6H_5^+$ ratio (solid) as a function of generation when maximization of this ratio is the specified goal in the closed-loop experiment. The optimal masks resulting from the closed loop process are shown in the inset.

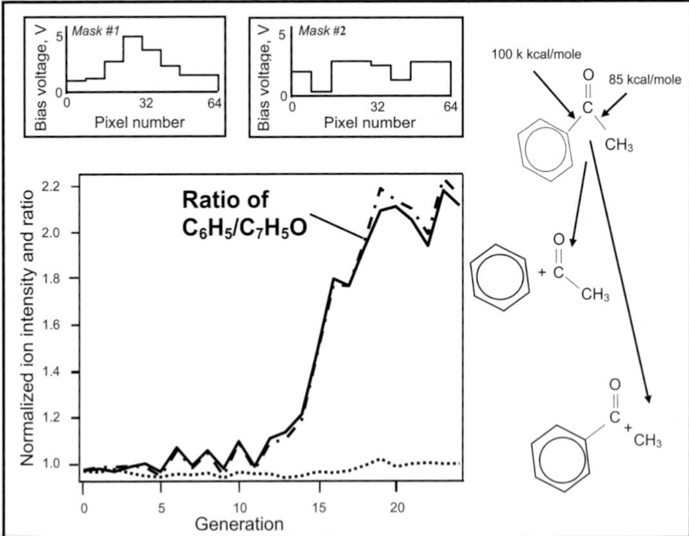

Fig. 6.32. The relative ion yield for phenylcarbonyl (dotted) and phenyl (dashed) and the $C_6H_5^+/C_6H_5CO^+$ ratio (solid) as a function of generation when maximization of this ratio is specified. . The optimal masks resulting from the closed loop process are shown in the inset.

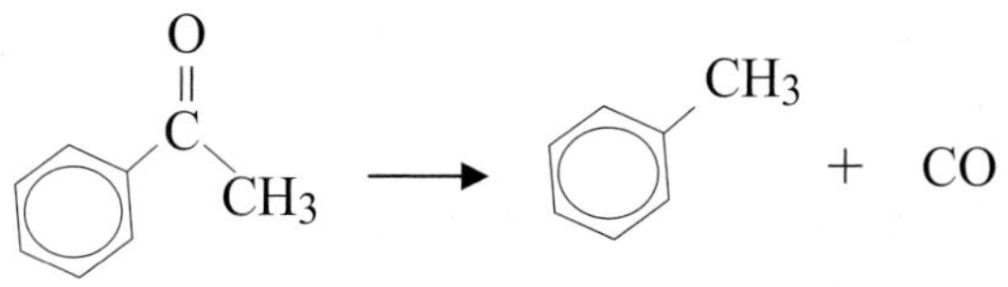

Scheme IV

pulse does not resemble the transform-limited pulse. In order to confirm the identity of the toluene product, measurements on the deuterated acetophenone molecule $C_6H_5COCD_3$ were carried out and the $C_6H_5CD_3^+$ ion was the observed product in an experiment analogous to Fig. 6.33. The observation of optically-driven dissociative rearrangement represents a new capability for strong field chemistry. In fact conventional electron-impact mass spectrometric analysis of acetophenone is incapable of creating toluene in the cracking pattern. In strong-field excitation, the molecular electronic dynamics during the pulse is known to be extreme, and substantial disturbance of the molecular eigenstates can produce photochemical products, such as novel organic radicals, that are not evident in the weak-field excitation regime. Operating in the strong field domain opens up the possibility of selectively attaining many new classes of photochemical reaction products.

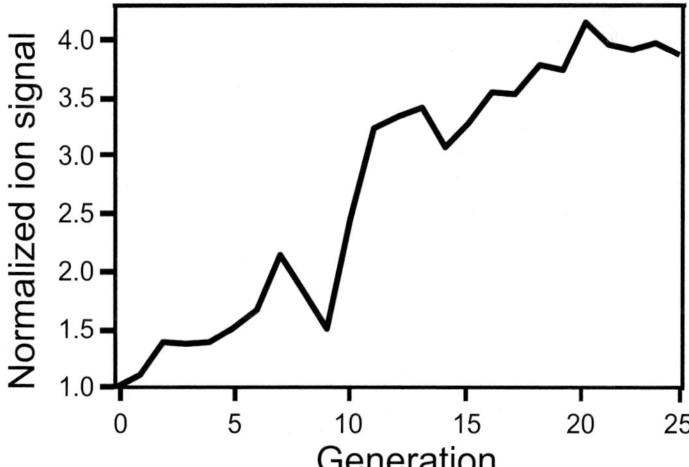

Fig. 6.33. The average signal for toluene, 92 amu, as a function of generation when maximization of the ion signal for this reaction product was specified for optimization. Corresponding electron-impact-ionization mass spectrometry revealed no evidence for toluene in the sample.

Extensive manipulation of mass spectra is possible when shaped, strong field laser pulses interact with molecules under closed-loop control. The control pulses occur with intensity of $\approx 10^{13}\,\mathrm{Wcm}^{-2}$ where the radiation significantly disturbs the field-free eigenstates of the molecule. Even in this highly nonlinear regime, the learning algorithm can identify pulse shapes that selectively cleave and rearrange organic functionality in polyatomic molecules. These collective results suggest that closed-loop strong field laser control may have broad applicability in manipulating molecular reactivity. The relative ease in proceeding from one parent molecule to another should facilitate the rapid exploration of this capability [85].

The limit on the range in control in the examples shown here may be due to a number of factors. The first is that we have employed a limited search space by ganging series of 8 collective pixels in each of the two masks to produce a total of 16 variable elements. We have observed that relaxing this restriction leads to a much longer convergence time, and while a better result is expected, we have not observed such to date. However, other researchers have employed schemes using all pixels, as well as schemes to constrain the amplitude and phase search space [99, 100]. Furthermore, the mass spectrometer was limited to eight averages for these experiments so that convergence can occur on a reasonable time scale. Obviously longer averaging will require longer experiment times. This parameter is under investigation at the present time. Another reason for limited dynamic range is the requirement that the same pulse used to alter the nuclear dynamics also must produce ionization. Each of these processes requires a different pulse timescale. In the case of ionization, the shortest pulse possible, $\approx$ tens of fs, is best for high ionization rates with little dissociation. For the control of the nuclear wave packet it is expected that a pulse with duration on the time scale of nuclear motion, $\sim$ps, should be optimal. Thus separation of these two processes should lead to a higher dynamic range.

In summary, recent progress in the understanding of fundamental quantum control concepts and in closed-loop laboratory techniques opens the way for coherent laser control of a variety of physical and chemical phenomena. Ultrafast laser pulses, with shapes designed by learning algorithms, already have been used for laboratory control of many quantum processes, including unimolecular reactions in the gas and liquid phases, formation of atomic wave packets, second harmonic generation in nonlinear crystals, and high harmonic generation in atomic gases. One may expect a further increase in the breadth of controlled quantum phenomena, as success in one area should motivate developments in others. The various applications of coherent laser control, no matter how diverse, all rely on the same principal mechanism: the quantum dynamics of a system is directed by the tailored interference of wave amplitudes, induced by means of ultrafast laser pulses of appropriate shape. An important question is whether applications exist for which coherent laser control of molecular reactions offers special advantages (e.g., new products or better performance) over working in the traditional fully incoherent kinetic regime. Finding these applications will be of vital importance for the future progress of coherent control in chemistry and physics.

In addition to the practical utilization of laser control, the ultimate implications for controlling quantum processes may reside in the fundamental information extracted from the observations about the interactions of atoms. The following is intuitively clear, the more complete our knowledge of a quantum system, the better our ability to design and understand successful controls. But, is it possible to exchange the tools and the goals in this logical relationship, and use control as a means for revealing more information on properties of microscopic systems? A challenging objective is to use observations of the controlled molecular dynamics to extract information on the underlying inter-atomic forces. Attaining precise knowledge of inter-atomic forces [101] has been a long-standing objective in the chemical sciences, and the extraction of this information from observed coherent dynamics requires finding the appropriate data inversion algorithms.

Traditionally, the data from various forms of continuous wave spectroscopy have been used in attempts to extract intramolecular potential information. Although such spectroscopic data are relatively easy to obtain, serious algorithmic problems have limited their inversion to primarily diatomic molecules or certain special cases of polyatomics. Analyses based on traditional spectroscopic techniques suffer from a number of serious difficulties, including the need to assign the spectral lines and to deal with inversion instabilities. An alternative approach to the inversion problem is to use an excited molecular wave packet that scouts out portions of the molecular potential surfaces. The sensitive information about the intramolecular potentials and dipoles may be read out in the time domain, either by probing the wave packet dynamics with ultrashort laser pulses or via measurements of the emitted fluorescence. A difficulty common to virtually all inverse problems is their ill-posedness (i.e., the instability of the solution against small changes of the data) which arises because the data used for the inversion are inevitably incomplete. Recent studies suggest that experiments in the time domain may provide the proper data to stabilize the inversion process [102, 103]. In this process, the excitation of the molecular wave packet and its motion on a potential energy surface may be guided by ultrafast control laser fields. Control over the wave packet dynamics in this context can be used to maximize the information on the molecular interactions obtained from the measurements. The original suggestion [104] for using closed-loop techniques in quantum systems was for the purposes of gaining physical information about the system's Hamiltonian. Now that closed-loop OCE is proving to be a practical laboratory procedure, the time seems right to consider refocusing the algorithms and laboratory tools to reveal information on fundamental physical interactions.

6.4 Ionization and fragmentation dynamics in fullerenes

T. Laarmann, C. P. Schulz, and I. V. Hertel

Fullerenes are a special form of carbon clusters, which have been discovered by Curl, Kroto, and Smalley in the mid 80th of the last century [105]. Their discovery has opened a new rapidly growing interdisciplinary research field (see e.g. [106] and references therein). Many of the interesting properties of C_{60} have their origin in its special geometric structure, a truncated icosahedron belonging to the I_h symmetry point group. This unique, football like structure with 12 pentagons and 20 hexagons makes C_{60} the most stable one of the fullerene family. Experimental studies got a strong boost after a method to produce C_{60} in macroscopic quantities was at hand [107]. Ever since, C_{60} became a model for a large finite molecular system with many electronic and nuclear degrees of freedom. Especially, structural and dynamical studies in the gas phase offer a direct way to focus on the properties of isolated C_{60} molecules free from environmental effects. A wide range of processes has been studied leading to a detailed understanding of the mechanism involved in the energy deposition, redistribution, ionization, fragmentation and finally cooling of C_{60}. Just a few early and some recent examples are mentioned out of a wealth of experimental and theoretical studies ranging from thermal heating [108], single-photon [109–111] or multiphoton absorption [112], electron impact [113], collisions with neutral particles [114], atomic ions, including highly charged ions [115–120] as well as molecular ions [121, 122], cluster ions [123] to surface collisions [124–126]. All of these studies have shown that C_{60} is very resilient and can accommodate a substantial amount of energy before it disintegrates. This is mainly due to its highly symmetric structure with 174 nuclear degrees of freedom and 240 valence electrons comprising 60 essentially equivalent delocalized π- and 180 structure defining, localized σ-electrons. The investigation of photon-induced energetics and dynamics have revealed that C_{60} shows atomic properties such as ATI as well as bulk properties such as thermionic electron emission (delayed ionization) [127]. In this sense, photo physical studies of fullerenes cover the whole range from atomic over molecular to solid state physics. The broad band width of responses of C_{60} to strong laser fields and their dependence on the intensity and pulse duration will be discussed in this section.

As has been shown in Sect. 6.3 the photophysics of large finite systems is already at laser intensities below 10^{15} Wcm^{-2} dominated by the nonadiabatic multielectron dynamics (NMED), which leads to size and intensity dependent nuclear dynamics and also opens the possibility to control molecular reactions in strong tailored laser fields. These studies have been extended to the C_{60} fullerene and some of the results will be presented in this section. Section 6.4.1 will focus on the ionization process, charge states and fragmentation as observed by mass spectroscopy. Also in this section ATI will be discussed, which has been observed experimentally in photoelectron spectra at different laser intensities. These results will be compared to recent theoretical calculations leading to a critical discussion of the primary excitation mechanism in an intense laser pulse. The single active electron (SAE) picture which is generally used to describe atoms interacting with intense laser light is no longer adequate when describing a system with many almost equivalent electrons. It turns out that many electrons may be excited during the laser pulse. The description of this process has similarities to photo induced processes in the band structure of semiconductors. This will be illustrated in the Sect. 6.4.2 by three characteristic examples: the nonresonant excitation of Rydberg states in C_{60}, the fast fragmentation processes of C_{60} beyond the well established statistical fragmentation processes known from experiments with ns lasers, and the excitation of C_{60} on a time scale below electron-electron and electron-phonon coupling. At the end of this section, experiments to control the energy redistribution in C_{60} using self-learning algorithms with temporally shaped laser pulses will be presented.

6.4.1 Ionization and fragmentation of C_{60} revisited

One of the surprising "early" observations was the delayed ionization of neutral C_{60} on a μs time scale upon irradiation with ns laser pulses [128]. This has been explained by statistical, thermionic electron emission from vibrationally excited molecules. The strong electron-phonon coupling leads to energy exchange between the nuclear and electronic system. Due to the low ionization potential of C_{60} (7.58 eV) compared to the barrier for C_2 loss (> 10 eV), electron emission is the main channel for cooling [129, 130]. Recently, it was found that the ionization behavior sensitively depends on the excitation time scale [127]. The spectacular difference observed in the mass spectra when changing the pulse duration $\Delta\tau$ from 25 fs to 5 ps is illustrated in Fig. 6.34. These mass spectra were obtained for nearly equal laser pulse energies (fluences) of about 20 $\mathrm{Jcm^{-2}}$, the corresponding intensities being 1×10^{15} $\mathrm{Wcm^{-2}}$ and 3.2×10^{12} $\mathrm{Wcm^{-2}}$, respectively. A strong contribution of multiply-charged C_{60}^{q+} ions together with their large fragments (C_2 evaporative cooling) is very clearly seen in the 25 fs spectrum. However, extremely little fragmentation is detected for singly charged C_{60}^{+} – as illustrated by the insert – and only a few small fragments if any. In contrast, only singly charged ions and massive fragmentation are observed with 5 ps pulses. The large fragment ions in both case are highly vibrationally excited up to an effective temperature of 4000 K and undergo metastable fragmentation μs-ms after the initial energy deposition has occurred [131]. The corresponding mass peaks are marked with asterisks. Fig. 6.34 also shows the typical delayed ionization tail in the 5 ps mass spectrum on the C_{60}^{+} mass peak which is not present for 25 fs.

On first sight, the large finite molecular system behaves as one might intuitively expect: For short pulses of 25 fs length, one (active) electron is ionized by the absorption of many photons and carries most of the energy. In contrast, energy can be transferred efficiently into vibrational modes during a

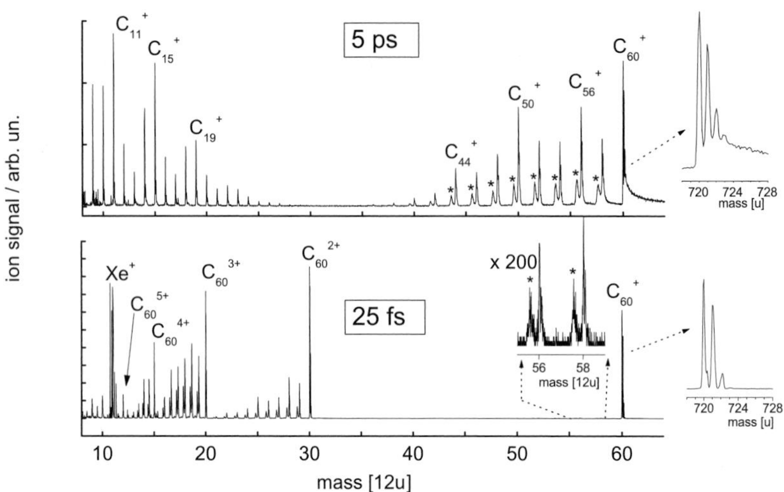

Fig. 6.34. Typical mass spectra taken from [26] obtained from C_{60} by ionizing with Ti:Sa laser pulses of 5 ps (top) and 25 fs duration (bottom) at equal laser fluence. For details, see the text.

laser pulse of 5 ps, since electron-phonon coupling is on the order of 200-300 fs [129]. However, important details remain unexplained in this intuitive picture: Can the different magnitudes of the ion signals be explained quantitatively? Why are multiply charged fragments so dramatically more abundant than singly charged ones – a prominent phenomenon observed for all pulse durations below a few 100 fs and a wide range of intensities? Several mechanisms might be held responsible but one may be related to another important question: How many electrons are actually excited when the electronic ground state is coupled to the continuum by means of the intense, ultrashort laser pulse? This will determine whether the electronic system of the remaining C_{60}^{+} ion core is hot or cold after the first electron has been ejected in a strong fs laser field. If the molecular ion is mainly in its electronic ground state then a theoretical description of the ultrafast perturbation using a single active electron model for the ionization process might be a valid approximation.

Photoelectron spectra can give a complementary and more detailed view of laser induced electron and nuclear dynamics in strong fields compared to mass spectroscopy. Fig. 6.35 shows photoelectron spectra recorded with laser pulses of different duration. Below ∼500 fs the excitation energy remains mainly in the electronic system and ionization is due to statistical electron emission after equilibrium among the electronic degrees of freedom [129]. Thermalization within the electron bath due to electron-electron scattering occurs on a time-scale below ca. 70 fs. The photoelectron spectra recorded with very short pulses of $\Delta\tau < 70$ fs at a few 10^{13} Wcm^{-2} clearly show an atom-like behavior

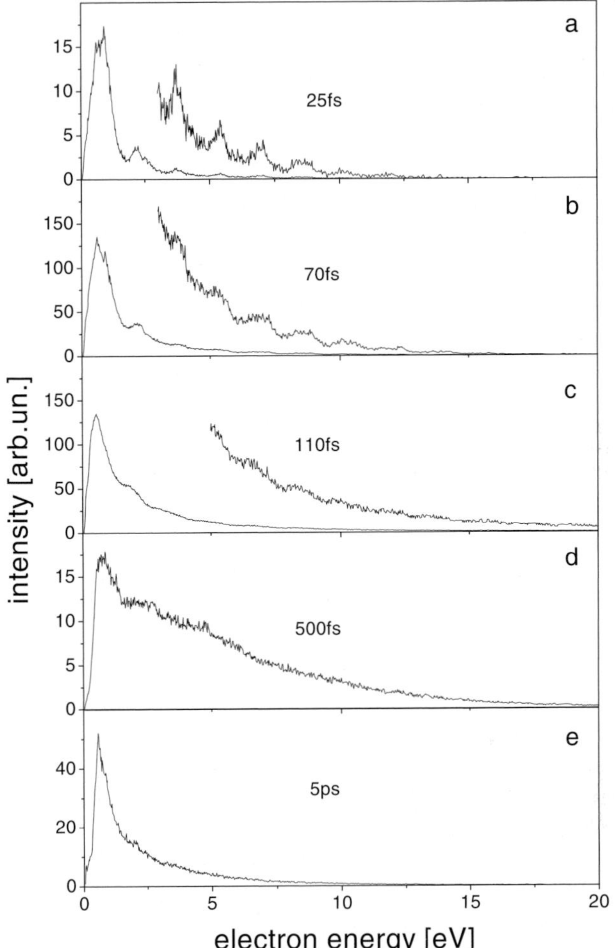

Fig. 6.35. Photoelectron spectra from C_{60} as a function of pulse duration (a)-(d) $8 \times 10^{13}\,\mathrm{Wcm}^{-2}$, and (e) $5 \times 10^{12}\,\mathrm{Wcm}^{-2}$, taken from [127].

of C_{60} with the characteristic ATI structure [132]. This is a fingerprint for direct multi photon ionization in which one active electron absorbs more laser photons than necessary to overcome the ionization potential. Consequently, a kinetic energy distribution of photoelectrons is observed, which exhibits a series of equally spaced maxima separated by the photon energy $h\nu$, as well known from atomic systems [2]. From this point of view, the SAE description of photoinduced processes in the limit of ultrashort laser pulses is appropriate. However, it should be recalled that even ATI in C_{60}, a genuine SAE effect, exhibits multielectron signatures according to recent time-dependent density functional theory (TDDFT) calculations by Bauer et al. [26, 133]. The starting point of these calculations is a jellium-like potential, which is then used to derive Kohn-Sham orbitals for all relevant π and σ electrons.

With this approach it is possible to distinguish between ionization, single particle transition, and plasmon excitation, and also to account for higher order processes beyond single particle-hole excitations. One nice advantage of theoretical simulations is that one can easily switch certain interactions on and off. Doing so, one can either propagate all Kohn-Sham orbitals in time (many active electron, MAE picture), or "freeze" all orbitals except the outermost one, suppressing all MAE effects and following the SAE dynamics exclusively. It turns out that in the SAE model the degree of C_{60} ionization is higher because energy cannot be transferred to the other electrons and the ATI lines are much narrower due to the lack of electron-electron interaction. It seems that for a complete description of the photoinduced dynamics in C_{60} the full MAE picture is needed.

6.4.2 Multielectron excitation, energy dissipation and coupling to the nuclear backbone

As already discussed in the previous Sect. 6.4.1, one of the interesting but also difficult to analyze facets of intense laser field interaction with C_{60} fullerenes is the large variety of potential responses ranging form atom-like to solid-like behavior such as ATI on one side and thermionic electron emission on the other side depending on the laser pulse duration. This raises the question when the SAE dynamics dominating the strong field response of atoms [22, 45, 60] passes over to the multielectron response in large finite systems [134–137]?

While such information cannot be extracted from presently available experimental data, one can try to identify specific aspects of the response of C_{60} to strong fields as being attributable to the one or the other of these "two faces". One example is the observation of Rydberg states [138]: while the population mechanism of these states is clearly driven by multielectron excitation, the binding energies of the Rydberg states themselves can be derived in a very simple SAE approach describing the almost atom-like single Rydberg electron in its orbital far away from the C_{60} ion core [139, 140]. Consequently, this is an ideal observable to address these questions, which will be discussed in the following Sect. 6.4.2.1.

It has been shown in Sect. 6.4.1, that the efficient excitation of the electronic system and the subsequent heating of the nuclear backbone lead to extensive fragmentation depending on the laser parameters. Many aspects of this process such as the high excitation threshold for fragmentation (kinetic shift) and the bimodal fragment distribution at high excitation energies can be explained very well in terms of statistical theory, essentially on the basis of knowing the energetics of the system as described, e.g., in [131]. However, recent experiments give also evidence to direct, nonstatistical processes driven by bond-softening and/or repulsive state crossings induced by the strong laser field [141]. This leads back to a more molecular description of dissociation, where the system "surfs" on potential energy surfaces rather than being exclusively controlled by statistics. This coexistence will be discussed in the Sect. 6.4.2.2 underlying the complex energetic and dynamics of fullerenes.

Interesting parallels can be found when comparing collision studies on C_{60} with fs laser excitations when looking at the ultrafast electronic and nuclear response (see e.g., [142, 143]). Ultrashort pulses as well as fast collisions deposit energy predominantly into the electronic system. Naively, one could imagine that the shorter the ultrafast perturbation of the C_{60} molecule the easier the absorption process can be understood. Of course, this is partly true since energy redistribution processes such as (i) electron-electron scattering and (ii) electron-phonon coupling increase the complexity of the energy absorption process if the laser pulse is still "on". The characteristic coupling time constants estimated experimentally are for process (i) < 70 fs and for process (ii) 200-300 fs [127, 129]. On the other hand, rather complex MAE effects might come into play in the limit of ultrashort (sub-10 fs) pulses. This issue is addressed in detail in the last Sect. 6.4.2.3, where time-of-flight mass spectroscopic data will be discussed, which were recorded upon irradiation of C_{60} with intense laser pulses down to 9 fs pulse duration.

6.4.2.1 Population of C_{60} Rydberg states beyond the single active electron picture

Sharp peaks were discovered in photoemission studies of C_{60} on top of the ATI series and the thermal electron contribution after Ti:Sa laser excitation at a few 10^{12} Wcm^{-2} as shown in Fig. 6.36. By solving the Schrödinger equation for a single active electron in a jellium-like potential [144], this structure could be clearly assigned to the population of several Rydberg series with binding energies E_b between 0.5 and 1.5 eV [138]. By studying the effect of different laser parameters such as excitation wavelength, intensity, polarization, and positive, respectively negative chirp on the excitation dynamics of Rydberg states further insight into the underlying processes was obtained [140]. The results from single pulse spectroscopy can be summarized as follows. The *excitation* of Rydberg states occurs mainly during the first part of the laser pulse while the *ionization* takes place toward the end of the pulse. The spectra recorded for different Fourier-limited pulse durations $\Delta\tau$ and corresponding bandwidths ΔE – albeit broadened in accord with the bandwidth – indicate that the excitation mechanism must be very fast: traces of a Rydberg population can be observed even for pulses as short as 30 fs [140]. The final single photon ionization step in the cascade is supported by studying details of the photoelectron spectra depending on the laser photon energy. The kinetic energy of photoelectrons converges toward the respective photon energy, i.e., the accessible excited state for ionization is limited by the photon energy [140].

The observation of Rydberg peaks seems to be a clear fingerprint of the SAE picture. However, some important aspects warrant further discussion. Most critical is the energy mismatch between the observed excitation energy

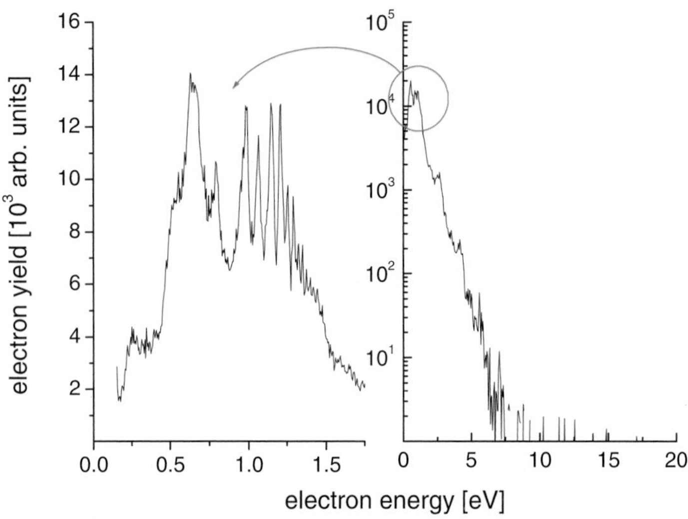

Fig. 6.36. Photoelectron spectra of C_{60} for 800 nm, 1.5 ps laser excitation at 1.1×10^{12} Wcm^{-2}, in log-lin scale to show the signal over a wide range of electron energies (right) and with linear scales to show the pronounced Rydberg structure on top of the first ATI peak (left). Reproduced from [138]

of the Rydberg states ($E_I - E_b$) and a multiple of the photon energy: It is simply not possible to be in resonance with all observed Rydberg states simultaneously through the absorption of n photons with a given energy. The Fourier-limited energy bandwidth of the up to 2 ps long laser pulse is much too narrow to allow for the excitation of Rydberg states covering 1-2 eV in energy. Moreover, in this intensity regime the field-induced ponderomotive shift of the energy levels is also too small (< 100 meV) to account for the observed energy mismatch. Thus, key mechanisms such as line broadening and energy sweeping, known from atomic systems in strong laser fields [145], cannot explain the Rydberg excitation process under the present conditions in C_{60} fullerenes. In contrast, a plausible explanation may be to invoke excitation of intermediate (doorway) states during the laser pulse by single or multi photon processes. The concept of doorway electronic states originates from the fact that the initial step in the excitation cascade is rate limiting and can be considered as a bottleneck for energy coupling into the electronic system [80,81]. Such processes have recently received great attention in the literature and a number of theoretical models have been discussed. Two of them are mentioned explicitly, (i) the nonadiabatic multielectron dynamic (NMED) model introduced by Stolow and collaborators [24,25] and (ii) time-dependent adiabatic potential energy crossings suggested by Kono et al. [146,147]. NMED has been used successfully to describe the dissociative ionization dynamics of different aromatic molecules as a function of their characteristic length and the excitation of the π-electron delocalization. The latter has been applied to lighter molecules in comparison with the NMED studies.

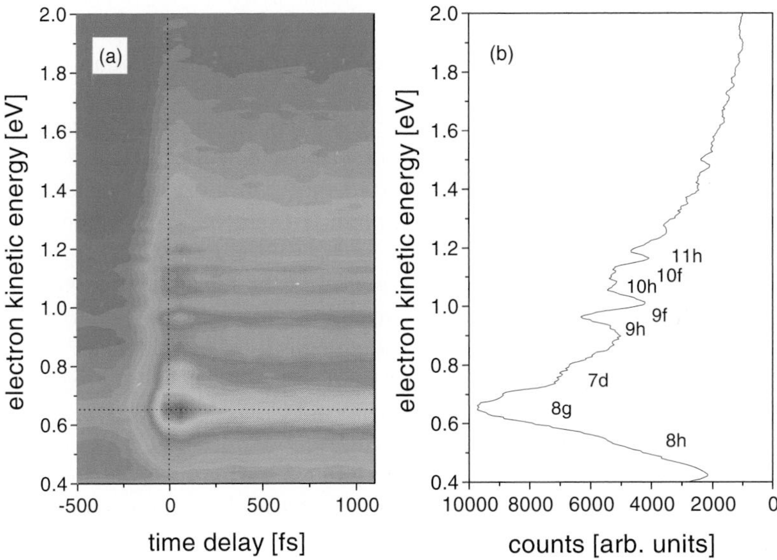

Fig. 6.37. (a) Contour plot of the photoelectron signal as a function of the time-delay between 400 nm pump $1 \times 10^{11}\,\mathrm{Wcm}^{-2}$ and 800 nm probe pulse $2 \times 10^{12}\,\mathrm{Wcm}^{-2}$. (b) Kinetic energy distribution of photoelectrons for zero delay time, which corresponds to a vertical cut in (a) along the dotted line, taken from [140].

It is suggested that the key to understand the population mechanism of the Rydberg series is indeed the MAE/NMED picture which is supported by recent, more detailed studies applying two-color pump-probe spectroscopy [139, 140]. The photoelectron spectra recorded as a function of the time-delay between 400 nm pump ($1 \times 10^{11}\,\mathrm{Wcm}^{-2}$) and 800 nm probe-pulse ($2 \times 10^{12}\,\mathrm{Wcm}^{-2}$) are shown by the contour plot in Fig. 6.37a. A blue 100 fs pump pulse of relatively low intensity, which is resonant to the dipole-allowed HOMO (h_u) → LUMO+1 (t_{1g}) transition, was used to deposit energy efficiently into the electronic system. The dynamics of the energy redistribution within the electronic system and the accompanied coupling to the nuclear motion is then probed by a time-delayed 100 fs red probe pulse. Thus, several steps of the excitation and detection process are separated. A cut through this contour plot for zero time-delay along the vertical dotted line is given in Fig. 6.37b. It corresponds to a photoelectron spectrum which essentially reproduces the Rydberg series obtained in the one color (800 nm) experiments (Fig. 6.36a) – except for a poorer spectral resolution due to the shorter pulses.

At negative time delay when the red pulse leads, almost no photoemission signal from excited Rydberg states is observed. Once pump and probe pulse overlap the photoelectron yield increases dramatically and a maximum population of the Rydberg series is found at a time delay of 50-100 fs. It can be inferred from this observation that the resonant preexcitation of the LUMO+1 (t_{1g}) state by the weak blue laser pulse is essential to populate Rydberg states. At time delays longer than 400 fs the photoelectron spectra remain nearly the same for several picoseconds.

In a classical molecular picture one would typically invoke doubly excited states and internal conversion (IC) to describe such processes. Indeed, similar Rydberg structures have been reported for several organic molecules and the excitation mechanism has been explained there by such "superexcited" states [148, 149]. In the context of the large finite system C_{60} exposed to fs laser radiation the MAE/NMED processes may be considered to be the adequate equivalent to Rydberg state excitation via such superexcited states. This interpretation is confirmed in the calculations by Zhang et al. [135], predicting multielectron excitation of the LUMO+1 level of C_{60} that is accompanied by strong vibrational excitation and massive energy exchange of ~ 1 eV per electron with the $a_g(1)$ breathing mode.

The experimental results point toward an excitation mechanism including four main steps [140]: (i) At the beginning of the laser pulse nonadiabatic multielectron excitation from the HOMO (h_u) leads to a very efficient population of the LUMO+1 (t_{1g}), which is considered to be the doorway state for all subsequent processes. (ii) The rapid thermalization within the electronic system on a time scale below 100 fs and the coupling of the electronic excitation to nuclear motion of the molecule results in the population of a broad energy band of 1-2 eV depending on the photon energy. The energy is stored for at least several ps in the doorway state without discernable relaxation. (iii) The "level broadening" allows the population of Rydberg states via multi photon absorption. (iv) This is followed by single photon ionization from the excited states resulting in a characteristic sequence of photoelectron peaks.

Investigation of cold C_{60} molecular beams with reduced vibrational energy content and hence, reduced phonon density highlight the importance of electron-phonon coupling in the excitation process of Rydberg states. Due to the reduced vibrational coupling, the characteristic signature of populated Rydberg levels in the photoelectron spectra is absent [139]. Time-resolved photoion spectroscopy shows that these mechanisms are also active in multiple ionization and fragmentation of the molecule [140], as will be discussed in the following Sects. 6.4.2.2, 6.4.2.3, and 6.4.3.

6.4.2.2 Ultrafast fragmentation of C_{60} beyond purely statistical, unimolecular decay

The dynamics of the fragmentation in C_{60} following the strong field excitation is far from being fully understood. While it is clear that large fragments arise essentially from evaporative cooling of hot C_{60}^{q+} ions it is not obvious how the substantial amount of internal energy needed for fragmentation is deposited into the system [150]. Absorption bands in the cations have been held responsible [151, 152], excitation of the plasmon resonance [153], or even recollision of the emitted electrons [154]. While neither of these processes explains the general trend to more extensive fragmentation at higher charge states, also observed in fast collisions, shake processes in the ionic system might eventually lead to a more consistent picture [111]. Even less obvious are the pathways to form small carbon cluster ions C_n^+ with odd and even numbers of carbon atoms during longer ps pulses shown in Fig. 6.34. Many different processes and their combination have to be considered, such as asymmetric fission of multiply charged ions [155], complete breakup of highly excited C_n^+, dissociative ionization, postionization of neutral fragments during the laser pulse and photofragmentation of small neutral and ionic clusters C_n, C_n^+ ($n \leq 20$).

In this section first results are reported from an effort to shed light onto this dynamics, focussing mainly on the formation of small C_n^+ [141]. An earlier, pioneering study of Lykke and Wurz [156, 157] may be seen as a precursor of this work: they used ns-laser pulses to preexcite and/or ionize C_{60} and probed the interaction products with a second, postionizing laser, detecting C^+, C_2^+, C_3^+, and C_4^+. Since the fragmentation pattern of larger fullerenes shows only even masses C_{60-2n}^+ one may safely assume that these small ions arise as final products from a series of fragmentation processes, concurrent with the above mentioned studies [158, 159]. No temporal information on the underlying fast dynamics could be derived on the ns time scale. Hence, the basic idea is to use a one color pump-probe scheme with 800 nm laser pulses of 50 fs pulse duration to simulate in a controlled way the effect of broadening the laser pulse which, as shown in Fig. 6.34, generates small fragments. Since the majority of fragmentation channels results in at least one neutral fragment (typically the smaller fragment), the pump-probe postionization method is a useful technique to study directly their formation dynamics. In these experiments a pump pulse at an intensity of $5 \times 10^{13}\,\mathrm{Wcm^{-2}}$ deposits energy into the electronic system by exciting one or more electrons into higher lying states (see Sect. 6.4.2.1). The multielectron dynamics initiated is probed by a weaker probe pulse of $1.8 \times 10^{13}\,\mathrm{Wcm^{-2}}$, which further excites and ionizes by a multi photon process. Fig. 6.38 shows the formation of C^+, C_2^+, C_3^+, and C_4^+.

For negative time delays, the weak pulse leads the strong pulse and a constant signal for each fragment is observed. For positive time delays, when the strong pulse initiates the multielectron dynamics, a dramatic increase in the C^+–C_3^+ ion yields is observed as the separation of the pulses increases, whereas C_4^+ exhibit nearly no dynamic behavior. At time delays > 50 ps the signal remains almost constant up to the longest time scales studied in these experiments (~ 100 ps). It is possible to fit the dynamics using single exponential curves with time constants of 11 ps (C^+), 12 ps (C_2^+) and 18 ps (C_3^+). The

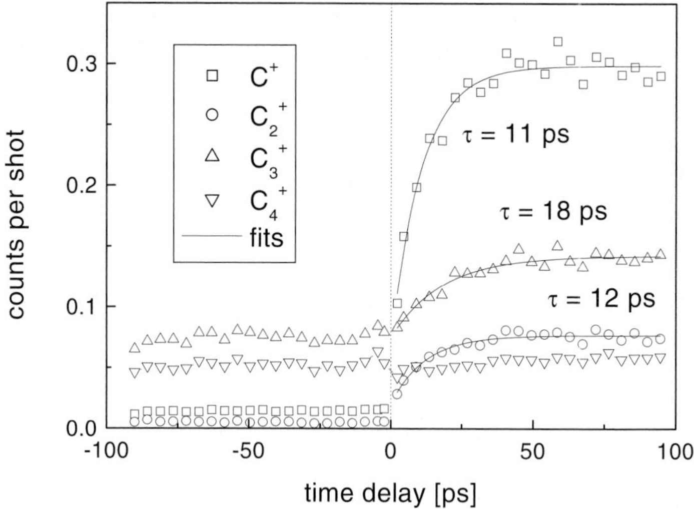

Fig. 6.38. Time-dependent C^+, C_2^+, C_3^+, and C_4^+ ion signals formed by Ti:Sa laser pulses (800 nm, 100 fs) interaction of C_{60} at 5×10^{13} Wcm^{-2} (pump) and ionization at 1.8×10^{12} Wcm^{-2} (probe). For positive delay times, the stronger pulse leads the weaker pulse. From [141]

time constants are found to be almost independent of the weak probe pulse energy, which indicates that the weak probe pulse is not active in the formation process of small neutral fragments [141]. Furthermore, the absence of small fragments in single pulse experiments indicates that the small fragments are initially uncharged.

Clearly, the observed fragmentation times on the order of some 10 ps indicate a non statistical decay: for comparison one estimates from simple RRK considerations [160] that, e.g., a unimolecular C_2 evaporation from C_{60} would require internal energies as high as about 200 eV - while the very low abundance of C_{60-2m}^+ fragments detected shows that only a small part of all parent molecules contains energies above 100 eV.

6.4.2.3 Excitation of C_{60} on a time scale below electron-electron and electron-phonon coupling

The ultrafast response of C_{60} fullerenes to intense, short laser pulses with a duration down to 9 fs has been investigated with pump-probe photoion spectroscopy [153]. The irradiation of a beam of C_{60} with such ultrashort pulses allows one to separate the energy deposition into the electronic system in time clearly from the energy redistribution among the manifold of electronic and nuclear degrees of freedom, because the excitation time lies well below the characteristic time scales for electron-electron and electron-phonon coupling. The goal is to *directly* observe fingerprints of multielectron effects in the initial excitation steps of C_{60} irradiated with ultrashort 9 fs pulses. More specifically, the aim is to find indications for a remaining excited electron cloud after the first electron has been "kicked-out". In general, the coupling of excited electrons to atomic motion leads to nuclear rearrangement in the ionic or in the neutral molecular system. According to recent theoretical work on C_{60} [135, 161], this results in characteristic oscillations, discussed already in the context of the population mechanism of Rydberg states. Both, multielectron excitation and the characteristic oscillation may be observed with time-resolved mass spectroscopy, since the density of excited electrons and the nuclear geometry are expected to affect the photoionization yield of C_{60} in a time-dependent study. The ultrashort pump pulse with an intensity of $7.9 \times 10^{13}\,\mathrm{Wcm}^{-2}$ solely deposits the energy in the electronic systems during the interaction. The energy redistribution within the electronic and nuclear degrees of freedom is then probed by a delayed, slightly less intense probe pulse ($6.8 \times 10^{13}\,\mathrm{Wcm}^{-2}$).

Fig. 6.39a shows the measured time dependence of the normalized C_{60}^+ ion signal. Particularly, the comparison with the simultaneously measured Xe^+ signal included in the figure is instructive. Xe^+ formation constitutes a genuine direct MPI process with probably only one active electron determining the systems response and, thus, can be taken as an auto-correlation measurement. The C_{60}^+ ion signal is clearly broadened at the bottom of the spectrum. As shown in Fig. 6.39b, the deconvolution of the total ion yield results into two main contributions: direct MPI of C_{60} from the neutral ground state to the continuums state (dark gray-shaded), which essentially follows the Xe auto-correlation plus a significant contribution exhibiting dynamics on a sub-100-fs time scale (light gray-shaded) which is slightly shifted toward positive time delays, when the stronger pump pulse leads the weaker probe pulse. This deviation of the C_{60} ion pump-probe signal from the auto-correlation function can be interpreted as a clear indication of multielectron excitation in a sub-ensemble of C_{60} during the laser interaction. Supported by recent theoretical work, [134, 135, 161] one believes that in addition to the direct MPI process there is a probability to initially excite two or more electrons via the t_{1g} resonant state, which in turn acts as a doorway (bottleneck) to ionization. The observed dynamics is comparable to the characteristic time for thermalization within the electronic system due to inelastic electron-electron scattering ($<$ 70 fs), as previously concluded from single pulse experiments [127, 129]. As intuitively expected the density of the hot electron cloud depends on the laser intensity, and its time evolution on the electron-electron scattering time constant [135]. The excited electron density in the doorway state determines the transition probability into the ionic continuum. Since pump and probe pulse have slightly different intensities (7.9:6.8) the ion distribution due to doorway state excitation is slightly shifted to positive time-delays, as shown in Fig. 6.39b. Based on a rough fit with two response functions for the undelayed, direct SAE/MPI process (proportional to the acf signal) and the MAE/NMED with its memory effect (taken as exponential decay), respectively, an estimate

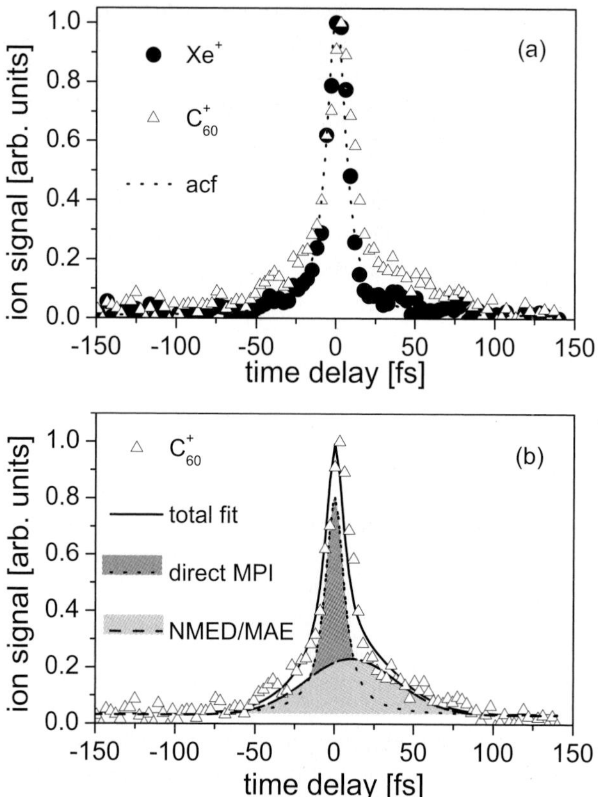

Fig. 6.39. a) C_{60}^+ ion yield (open triangles) as a function of the time-delay between pump ($7.9 \times 10^{13}\,\mathrm{Wcm^{-2}}$) and probe pulse ($6.8 \times 10^{13}\,\mathrm{Wcm^{-2}}$), normalized to the maximum signal. $t = 0$ is defined by the auto-correlation function (acf, dotted line) derived from a fit to the simultaneously measured Xe^+ signal (closed circles). (b) Contributions from direct SAE/MPI (dark gray-shaded) and MAE/NMED (light gray-shaded) refer to our tentative deconvolution of the C_{60}^+ photoion yield, for details see the text. (from [153])

of 65% to 35% for the contribution of SAE and NMED processes to the signal have been obtained.

6.4.3 Control of energy dissipation processes using temporally shaped laser pulses

The previous sections were focused on the *analysis* of photophysical processes in C_{60} by comparing photoelectron or mass spectra taken with different characteristic laser parameters, such as a intensity, pulse duration, photon energy or pump-probe delay. In the following results are presented with the goal to

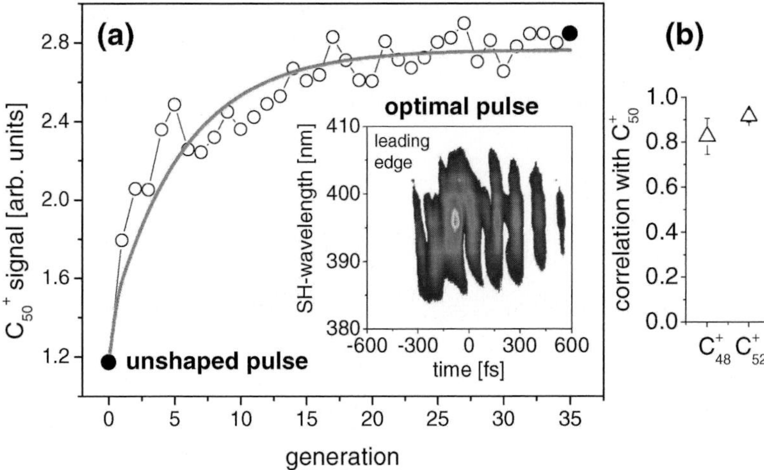

Fig. 6.40. (a) C_{50}^+ signal as a function of generation of the evolutionary algorithm. The inset shows the SH-XFROG trace of the optimal solution. (b) Correlation with neighboring fragment masses C_{48}^+ and C_{52}^+ taking into account 7 independent optimization processes [162].

control the molecular response with suitably tailored fs laser pulses [162]. Details of the self-learning closed-loop adaptive feedback technique can be found in Chapter 2, Sect. 2.4 and also in Sect. 6.3 of the present Chapter. Here, the selective enhancement of C_2 evaporation is reported, a typical energy loss channel upon laser excitation of fullerenes as discussed in previous sections. The learning curve for maximization of the C_{50}^+ fragment ion yield is plotted in Fig. 6.40a. The thus determined optimal pulse shape for this specific target is characterized by means of second-harmonic, cross-correlation frequency-resolved optical gating (SH-XFROG) shown in the inset. As a result, the mass peak increased by a factor of $S \sim 2.0$ compared to the signal recorded with unshaped pulses given as 0th generation. The height of the C_{50}^+ peak was chosen as fitness criterion because its abundance is a measure for the temperature of the nuclear backbone, i.e., indicates efficient energy coupling into nuclear motion. It is well-known that cooling of highly excited C_{60} proceeds mainly via sequential evaporation of C_2 units in a statistical process. This explains why a strong correlation of the C_{50}^+ enhancement with neighboring fragment masses C_{48}^+ and C_{52}^+ is observed when comparing 7 independent optimization runs in Fig. 6.40b. It has to be pointed out that the optimal control scheme applied here is selective for depositing energy into the C_{60} system, and not for selective bond-breaking. The key result is that a sequence of pulses is best suited for most efficient energy coupling into vibrational motion of C_{60}. It gives a direct fingerprint of the laser induced electron and nuclear dynamics with high mode-selectivity as seen in the SH-XFROG trace in Fig. 6.40.

This microscopic view goes beyond the common wisdom where the response of fullerenes to intense laser fields was assumed to be mainly determined by the interaction time scale, i.e., electron-electron and electron-phonon coupling. One may call the observed process "coherent heating".

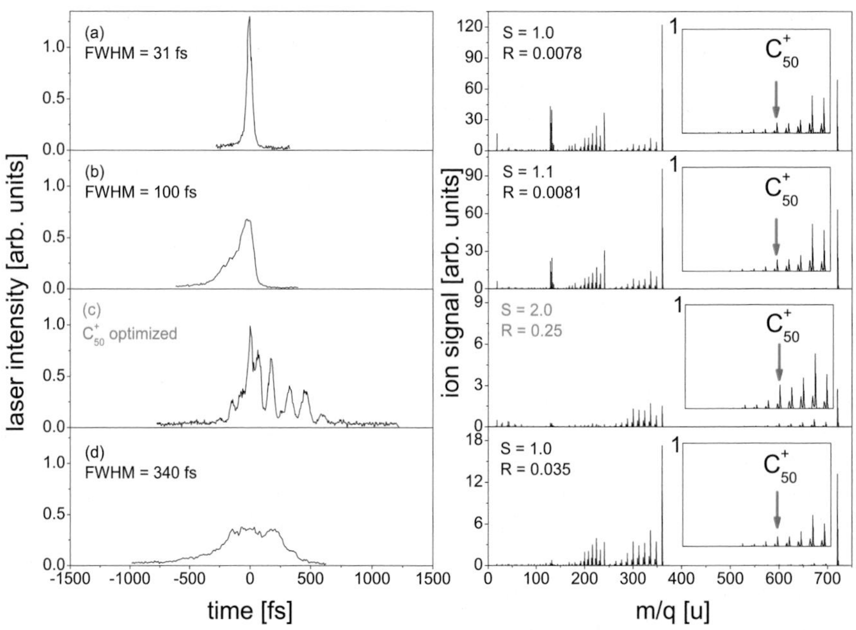

Fig. 6.41. Mass spectra (right panel) recorded with constant pulse energy (260 μJ but different pulse shapes given as projections of the corresponding SH-XFROG-traces (left panel): (a) original pulse (31 fs), (b) stretched pulse to 100 fs, (c) optimal pulse shape, and (d) 340 fs. The insets show the mass range of singly charged, large fragments plotted on the same scale. S gives the enhancement of the C^+_{50} signal, and R denotes the ratio C^+_{50}/C^+_{60}. (From [162])

Fig. 6.41a-d compares mass spectra recorded for stretched pulses with the optimal control result regarding the formation of C^+_{50} fragments in order to proof the relevance of the pulse sequence for most efficient coherent heating. Pulse broadening was achieved by applying parabolic spectral phase functions that keep the energy constant at 260 μJ. The temporal shapes are given as projections of the corresponding SH-XFROG traces on the left. From the mass spectra shown on the right, where the singly charged ion signals are all plotted on the same scale (insets), it is obvious that the pulse sequence of the optimal control field (c) is the key for enhanced energy coupling followed by statistical evaporation of C_2 units and not simply the increased overall pulse width. Both, stretched pulses of (b) 100 fs and (d) 340 fs duration result in significantly less singly charged fragments.

The combination of optimal control with comprehensive studies using 2-color pump-probe spectroscopy (not shown here) allows us to pinpoint the mechanism of optimal heating nuclear motion in C_{60} fullerenes, namely (multi)electron excitation via the t_{1g} doorway state followed by efficient coupling to the $a_g(1)$ breathing mode of the nuclear backbone.

This section has touched some aspects of the present state-of-the-art of research on the ultrafast laser interaction with C_{60} as a model for large finite systems with many active electrons and vibrational degrees of freedom. The comparison of experimental results using time-resolved photoelectron and mass spectroscopy with recent theoretical work gives a strong indication that nonadiabatic multielectron dynamics (NMED) plays a key role for the understanding of the molecular response to short-pulse laser radiation. Nevertheless, one is still far from fully understanding the intricacies of intense field interaction with such a complex system. Rigorous theoretical efforts are needed to quantitatively explain the key aspects of the experimental observations presented here and those to emerge in the near future: the nature of the ionization and fragmentation mechanisms which produce predominantly multiply charged fragments, the excitation dynamics for the population of Rydberg states, the long lifetimes observed in the doorway state and the ultrafast fragmentation mechanism. On the other hand, further experimental work is needed – preferentially with even shorter pulses (and better tunability of the fs light sources) – to perform sophisticated and direct multicolor pump-probe experiments. Experimental detection schemes need to become more sophisticated, e.g., the ion imaging technique promises a new view into the dynamics discussed here. This will, in connection with coincidence techniques, allow to follow fragmentation cascades directly and to separate prompt ionization from postionization processes. Furthermore, new laser schemes for intense radiation at shorter wavelength such as high-harmonic generation, table-top plasma sources, and Free-Electron Lasers, are expected to open completely new horizons for strong field laser-matter interaction.

6.5 Time-dependent electron localization function: A tool to visualize and analyze ultrafast processes

A. Castro, T. Burnus, M. A. L. Marques, and E. K. U. Gross

The classical picture of chemical bonding in terms of electron pairs that are shared by atoms in order to form molecules was nicely systematized by G. N. Lewis, in his seminal work entitled "The Atom and the Molecule" [163], dated 1916. Lewis noticed the overwhelming evidence pointing to the "pairing" of the electrons, as well as the preference to close "shells" of eight electrons. Soon afterwards, the pairing of electrons was explained in terms of the Pauli exclusion principle together with the electronic intrinsic one-half spin, whereas the number eight in fact emanates from both Pauli's principle and the spherical symmetry of atoms in a three dimensional world. Lewis, however, was some years too early, and designed "the theory of the cubical atom", with the electrons occupying the vertex of a cube (although he acknowledged the picture to be more methodological than fact-founded), and pointed to a breakdown of Coulomb law at short distances in order to explain the electron pairs. Despite these exotic suggestions, the usefulness of Lewis model has persisted even until today's textbooks.

The reason is that electrons do indeed "localize" in pairs when forming molecules, and a big amount of the basic machinery of Chemistry is rather well explained with Lewis arguments. In fact, more generally, Chemistry is intuitively understood in terms of "localized" groups of electrons, either pairs of electrons shared between atoms ("bonds"), nonbonding pairs of electrons ("lone pairs"), and also larger groups – double, triple bonds –, atomic inner shells, π electronic systems, etc.

With the advent, in the past years, of sources of coherent light featuring high intensity and ultrafast pulses (in the femtosecond [164], or already below the femtosecond limit [165]), it has become possible to *time resolve* the intermediate steps of chemical reactions – paving the way to the possibility of analyzing and controlling chemical reactions. These technical advances stress the need of understanding how the electrons rearrange, forming and destroying bonds, in the midst of a laser pulse, and during the possible ionic recombination. The chemical concepts of bonds, lone pairs, etc. have to be fathomed also for time-dependent phenomena.

Unfortunately, the transformation of these concepts into a mathematically rigorous scheme for classifying the elements of the chemical bonding turns out to be astonishingly difficult. The canonical single-particle orbitals that stem from Hartree-Fock (HF) calculations are not very helpful, since they, typically, have sizable contributions from many regions in space. Moreover, they are only one possible choice, since unitary transformations within the subspace of solutions yield equally legitimate orbitals. There are several ways in which one can perform these unitary transformations in order to obtain localized functions [166], but these methods are also not unique, and may result in qualitatively different information.

In any case, HF is but one of the possible schemes to obtain an approximate solution to the many-body problem. A definition based on the HF solution would always be affected by the HF error – absence of correlation effects. It is desirable to have a scheme that does not rely on a particular method. Kohn-Sham (KS) [167] density functional theory (DFT) [168–170] also provides single-particle orbitals (in this case unique, except for degenerate ground-states), but they are usually also very delocalized in real-space. The electronic density is an observable, and thus independent of the method. Moreover, it contains all the information of the system by virtue of Hohenberg-Kohn theorem [171]. Unfortunately, the density itself is not suitable to visualize chemical bonding: It does not peak in the position of the bonds, it does not show the shell structure of atoms, and lone pairs, also, are poorly represented.

The key to comprehending electron localization is, in fact, Pauli's exclusion principle, and, relatedly, the Fermi hole: Bader and collaborators [172] demonstrated how all manifestations of the spatial localization of an electron of a given spin are the result of corresponding localizations of its Fermi hole. An appropriate localization function should be closely related to this Fermi hole or to an analysis of Pauli's principle. This is indeed the case for the function to which we devote this section: Becke and Edgecombe's electron localization function [173] (ELF), as generalized by Burnus, Marques, and Gross for time-dependent cases [174]. Section 6.5.1 will show how the Fermi hole appears naturally in the derivation of the ELF.

An alternative way to rationalize the ELF definition is to think in terms of how Pauli's exclusion principle affects the kinetic energy. This principle applies to fermionic systems; the kinetic energy of a bosonic system is a lower bound to the local kinetic energy of a fermionic one [175]. Thus we can define an *excess kinetic energy*, which would be the difference between the two of them. Intuitively, in a region of electron localization (electrons forming pairs, isolated electrons), their behavior is more bosonic-like. So we will require, to define localization, that the excess kinetic energy is minimized. This is indeed the case for the ELF, as it will be demonstrated later.

The ELF, as introduced by Becke and Edgecombe, involved two approximations: (i) First, it assumed that the many-electron wave function is a single Slater determinant. The natural choice is the Hartree-Fock solution. (ii) Secondly, it assumed that the single-particle orbitals that form the single Slater determinant are *real functions*. This prevents its validity in a time-dependent formalism, or for static but current-carrying states. A generalized derivation that lifted this restriction was presented by Dobson [176], and later by Burnus, Marques, and Gross [174] who demonstrated how this general form could be applied for time-dependent processes. The observation of this function is useful for the study of chemical reactions and for processes that involve the interaction of molecular systems with high-intensity ultra-short laser pulses (femtosecond or even attosecond regime), or collision processes between molecules and/or ions. In this time scale, and for these probably violent deformations of the molecular fields, the electrons are bound to exhibit a complex behavior: bonds are destroyed or created, bond types change as the molecules isomerize, dissociate, or recombine in chemical reactions. These events are especially patent in the evolution of the ELF.

Next subsection is dedicated to the definition of the (possibly time-dependent) ELF. In Sect. 6.5.2, some examples of the ELF for systems in the ground state are shown, in order to illustrate the association between ELF topological features and Chemistry bonding elements. Sect. 6.5.3 provides examples of time-dependent calculations in which the TDELF is monitored: collision processes leading to chemical reactions, and interaction of molecules with laser pulses. The chapter closes, in Sect. 6.5.4, with an example in which the coupled evolution of electrons and nuclei, both treated quantum-mechanical, is computed for a model system. The ELF is then used to learn about the strength of nonadiabatic effects.

6.5.1 The time-dependent electron localization function

6.5.1.1 General definition

We depart from the definitions of the one and two-body density matrices for a system of N electrons [177, 178], whose evolution is described by the wave function $\Psi(\boldsymbol{r}_1\sigma_1, ..., \boldsymbol{r}_N\sigma_N; t)$:

$$\Gamma^{(1)}_{\sigma_1|\sigma_1'}(\boldsymbol{r}_1|\boldsymbol{r}_1'; t) = N \sum_{\sigma_2,\dots,\sigma_N} \int \mathrm{d}^3 r_2 \dots \int \mathrm{d}^3 r_N \, \Psi^\star(\boldsymbol{r}_1\sigma_1, \boldsymbol{r}_2\sigma_2, ..., \boldsymbol{r}_N\sigma_N; t) \times \Psi(\boldsymbol{r}_1'\sigma_1', \boldsymbol{r}_2\sigma_2, ..., \boldsymbol{r}_N\sigma_N; t) , \tag{6.12}$$

$$\Gamma^{(2)}_{\sigma_1,\sigma_2|\sigma_1'\sigma_2'}(\boldsymbol{r}_1, \boldsymbol{r}_2|\boldsymbol{r}_1'\boldsymbol{r}_2'; t) = N(N-1) \sum_{\sigma_3,\dots,\sigma_N} \int \mathrm{d}^3 r_3 \dots \int \mathrm{d}^3 r_N \, \Psi^\star(\boldsymbol{r}_1\sigma_1, \boldsymbol{r}_2\sigma_2, ..., \boldsymbol{r}_N\sigma_N; t)\Psi(\boldsymbol{r}_1'\sigma_1', \boldsymbol{r}_2'\sigma_2', ..., \boldsymbol{r}_N\sigma_N; t) . \tag{6.13}$$

The spin-densities are defined in terms of the diagonal one-body density matrix:

$$n_\sigma(\boldsymbol{r}, t) = \Gamma^{(1)}_{\sigma|\sigma}(\boldsymbol{r}|\boldsymbol{r}; t) . \tag{6.14}$$

For equal spin ($\sigma_1 = \sigma_2 = \sigma$), the diagonal of the two-body density matrix, that is, $\Gamma^{(2)}_{\sigma\sigma|\sigma\sigma}(\boldsymbol{r}_1, \boldsymbol{r}_2|\boldsymbol{r}_1\boldsymbol{r}_2; t)$, is the same-spin pair probability function, $D_\sigma(\boldsymbol{r}_1, \boldsymbol{r}_2; t)$. Its value is the probability of finding one electron at $\boldsymbol{r}_1$ and another electron at $\boldsymbol{r}_2$, both with the same spin σ:

$$D_\sigma(\boldsymbol{r}_1, \boldsymbol{r}_2; t) = \Gamma^{(2)}_{\sigma\sigma|\sigma\sigma}(\boldsymbol{r}_1, \boldsymbol{r}_2|\boldsymbol{r}_1, \boldsymbol{r}_2; t) . \tag{6.15}$$

If the electrons were *uncorrelated*, the probability of finding the pair of electrons at $\boldsymbol{r}_1$ and $\boldsymbol{r}_2$ would be the product of the individual probabilities: $D_\sigma(\boldsymbol{r}_1, \boldsymbol{r}_2; t) = n_\sigma(\boldsymbol{r}_1; t)n_\sigma(\boldsymbol{r}_2; t)$. Electrons are, however, correlated, and the same-spin pair density is less than that value by a factor that is defined as the *pair correlation function*:

$$D_\sigma(\boldsymbol{r}_1, \boldsymbol{r}_2; t) = n_\sigma(\boldsymbol{r}_1; t)n_\sigma(\boldsymbol{r}_2; t)g_{\sigma\sigma}(\boldsymbol{r}_1, \boldsymbol{r}_2; t) . \tag{6.16}$$

The difference between the correlated and the uncorrelated case is also contained in the *Fermi hole* function $h_\sigma(\boldsymbol{r}_1, \boldsymbol{r}_2; t)$:

$$D_\sigma(\boldsymbol{r}_1, \boldsymbol{r}_2; t) = n_\sigma(\boldsymbol{r}_1; t)\left(n_\sigma(\boldsymbol{r}_2; t) + h_{\sigma\sigma}(\boldsymbol{r}_1, \boldsymbol{r}_2, t)\right) . \tag{6.17}$$

The same-spin *conditional* probability function, $P_\sigma(\boldsymbol{r}_1, \boldsymbol{r}_2; t)$ is then defined as the probability of finding a σ-spin electron at $\boldsymbol{r}_2$, knowing that there is one σ-spin electron at $\boldsymbol{r}_1$. It can be expressed in terms of the previous definitions:

$$P_\sigma(\boldsymbol{r}_1, \boldsymbol{r}_2; t) = \frac{D_\sigma(\boldsymbol{r}_1, \boldsymbol{r}_2; t)}{n_\sigma(\boldsymbol{r}_1; t)} = n_\sigma(\boldsymbol{r}_2; t) g_{\sigma\sigma'}(\boldsymbol{r}_1, \boldsymbol{r}_2; t)$$
$$= n_\sigma(\boldsymbol{r}_2; t) + h_{\sigma\sigma}(\boldsymbol{r}_1, \boldsymbol{r}_2; t) \,. \qquad (6.18)$$

From this equation, the meaning of the Fermi hole (a negative function at all points) is more transparent: it is a measure of how probability at $\boldsymbol{r}_2$ is reduced due to the spreading out of the same spin density originated at $\boldsymbol{r}_1$.

However, it will be more useful to define an alternative same-spin conditional pair probability function: given a reference electron of σ-spin at $\boldsymbol{r}$, we are interested in the probability of finding a same-spin electron at a distance s. This involves taking a spherical average on a sphere of radius s around point $\boldsymbol{r}$, $S(s, \boldsymbol{r})$:

$$p_\sigma(\boldsymbol{r}, s; t) = \frac{1}{4\pi} \int_{S(s,\boldsymbol{r})} \mathrm{d}S P_\sigma(\boldsymbol{r}, \boldsymbol{r}'; t) \,. \qquad (6.19)$$

The integration is done for the $\boldsymbol{r}'$ variable. For small values of s one can obtain the following Taylor expansion:

$$p_\sigma(\boldsymbol{r}, s; t) = \frac{1}{3} \left[\frac{1}{2} \frac{\left[\nabla^2_{\boldsymbol{r}'} D_\sigma(\boldsymbol{r}, \boldsymbol{r}'; t)\right]_{\boldsymbol{r}'=\boldsymbol{r}}}{n_\sigma(\boldsymbol{r}, t)} \right] s^2 + \mathcal{O}(s^3) \,. \qquad (6.20)$$

In this expansion, the term in s^0 is absent due to the Pauli exclusion principle. The linear term in s is also null [179]. The coefficient of s^2 (except for the one-third factor) thus tells us about the same-spin pair probability in the vicinity of $\boldsymbol{r}$:

$$C_\sigma(\boldsymbol{r}) = \frac{1}{2} \frac{\left[\nabla^2_{\boldsymbol{r}'} D_\sigma(\boldsymbol{r}, \boldsymbol{r}'; t)\right]_{\boldsymbol{r}'=\boldsymbol{r}}}{n_\sigma(\boldsymbol{r}, t)} \,. \qquad (6.21)$$

This function is an *inverse* measure of localization: it tells us how large the same-spin conditional probability function is at each point in space. The smaller this magnitude is, the more likely than an electron *avoids* electrons of equal spin.

In addition to having an inverse relationship to localization – for example, it is null for perfect localization –, C_σ is not bounded by above. Visually, it does not mark the chemical structure with great contrast. These reasons led Becke and Edgecombe to suggest a re-scaling, noticing that, for the homogeneous electron gas, C_σ is nothing else than the kinetic energy density (atomic units will be used in all equations of this section):

$$C_\sigma^{\mathrm{HEG}} = \tau_\sigma^{\mathrm{HEG}} = \frac{3}{5}(6\pi^2)^{(2/3)} n_\sigma^{(5/3)} \,. \qquad (6.22)$$

One may then refer the value of C_σ at each point to the value that the homogeneous electron gas would have for the density of that point at that time t,

$C_\sigma^{\mathrm{HEG}}(\boldsymbol{r};t)$. Moreover, since there is an inverse relationship between C_σ and localization, it is useful to invert it. The final expression for the "electron localization function", $\eta_\sigma(\boldsymbol{r})$, is

$$\eta_\sigma(\boldsymbol{r};t) = \frac{1}{1 + (C_\sigma(\boldsymbol{r};t)/C_\sigma^{\mathrm{HEG}}(\boldsymbol{r};t))^2} \,. \tag{6.23}$$

6.5.1.2 Expression for one-determinantal wave functions

Up to this point, the equations allow for complete generality. Equation (6.23) in particular, together with (6.21), defines the ELF for any system, either in the ground state or in a time-dependent situation, and regardless of which scheme is chosen to approximate a solution to the many electron problem. However, the ELF was originally introduced assuming a Hartree-Fock formulation (one determinantal character of the many-body wave function). The formulation may thus be translated to the Kohn-Sham (KS) formulation of density-functional theory (DFT).

For one-determinantal wave functions, the function C_σ (6.21) may be explicitly calculated. Let us assume the Slater determinant to be formed of the orbitals $\{\varphi_{i\uparrow}\}_{i=1}^{N_\uparrow}$ and $\{\varphi_{i\downarrow}\}_{i=1}^{N_\downarrow}$, for spin up and down, respectively $(N = N_\uparrow + N_\downarrow)$. In this case, one can use the two following identities:

$$\Gamma^{(1)}(\boldsymbol{r}_1\sigma|\boldsymbol{r}_2\sigma;t) = \sum_{i=1}^{N_\sigma} \varphi_{i\sigma}^*(\boldsymbol{r}_2;t)\varphi_{i\sigma}(\boldsymbol{r}_1;t)\,. \tag{6.24}$$

$$\text{(This implies immediately:} \quad n_\sigma(\boldsymbol{r},t) = \sum_{i=1}^{N_\sigma} |\varphi_{i\sigma}(\boldsymbol{r},t)|^2\,.)$$

$$D_\sigma(\boldsymbol{r}_1.\boldsymbol{r}_2;t) = n_\sigma(\boldsymbol{r}_1;t)n_\sigma(\boldsymbol{r}_2;t) - |\Gamma^{(1)}(\boldsymbol{r}_1\sigma|\boldsymbol{r}_2\sigma;t)|^2\,. \tag{6.25}$$

Equations (6.24) and (6.25) are then introduced in the expression for C_σ, (6.21):

$$C_\sigma(\boldsymbol{r};t) = \frac{1}{2}\left[\nabla_{\boldsymbol{r}'}^2 n_\sigma(\boldsymbol{r}';t)\right]_{\boldsymbol{r}'=\boldsymbol{r}} - \frac{1}{2}\left[\nabla_{\boldsymbol{r}'}^2 \frac{|\Gamma^{(1)}(\boldsymbol{r}'|\boldsymbol{r};t)|^2}{n_\sigma(\boldsymbol{r};t)}\right]_{\boldsymbol{r}'=\boldsymbol{r}} \,. \tag{6.26}$$

And after some algebra [180]:

$$C_\sigma(\boldsymbol{r};t) = \tau_\sigma(\boldsymbol{r};t) - \frac{1}{4}\frac{(\nabla n_\sigma(\boldsymbol{r};t))^2}{n_\sigma(\boldsymbol{r};t)} - \frac{j_\sigma^2(\boldsymbol{r};t)}{n_\sigma(\boldsymbol{r};t)}\,. \tag{6.27}$$

where $\tau_\sigma(\boldsymbol{r};t)$ is the kinetic energy density,

$$\tau_\sigma(\boldsymbol{r};t) = \sum_{i=1}^{N_\sigma} |\nabla\varphi_{i\sigma}(\boldsymbol{r};t)|^2\,, \tag{6.28}$$

and $j^2_\sigma(\boldsymbol{r};t)$ is the squared modulus of the current density:

$$\boldsymbol{j}_\sigma(\boldsymbol{r};t) = \langle\Psi(t)|\ \frac{1}{2m}\sum_{i=1}^{N}[\delta(\boldsymbol{r}-\hat{\boldsymbol{r}}_i)\delta_{\sigma\sigma_i}\hat{\boldsymbol{p}}_i + \hat{\boldsymbol{p}}_i\delta(\boldsymbol{r}-\hat{\boldsymbol{r}}_i)\delta_{\sigma\sigma_i}]\ |\Psi(t)\rangle =$$

$$\frac{1}{2\mathrm{i}}\sum_{i=1}^{N_\sigma}[\varphi^*_{i\sigma}(\boldsymbol{r};t)\boldsymbol{\nabla}\varphi_{i\sigma}(\boldsymbol{r};t) - \varphi_{i\sigma}(\boldsymbol{r};t)\boldsymbol{\nabla}\varphi^*_{i\sigma}(\boldsymbol{r};t)]\ . \tag{6.29}$$

Expression (6.27), upon substitution in (6.23), leads to the general form for the ELF, if one assumes one-determinantal wave functions. In the original derivation, however, a further restriction was introduced from the beginning: the system is assumed to be in the a stationary state, and the single-particle orbitals are real, which implies zero current. The derivation presented above [174,180], however, allows for time-dependent Slater determinants (and complex ground-states with non-null current).

The original, "static" ELF, is simply obtained by eliminating the current term from the expression for C_σ (6.27):

$$C^{\mathrm{static}}_\sigma(\boldsymbol{r}) = \tau_\sigma(\boldsymbol{r}) - \frac{1}{4}\frac{(\nabla n_\sigma(\boldsymbol{r}))^2}{n_\sigma(\boldsymbol{r})}\,, \tag{6.30}$$

and plugging this formula in the ELF definition, (6.23).

At this point, it is worth noting that this expression is nothing else than the "excess kinetic energy" mentioned in the introduction of this Section. The first term, $\tau(\boldsymbol{r})$ (summing over the two spins) is the local kinetic energy of the electronic system. A bosonic system of equal density n, at its ground state, will concentrate all particles at the ground state orbital, $\sqrt{n}/N$. From this fact it follows that the second term of the previous equation is the kinetic energy density of the bosonic system. It is thus clear how the high localization corresponds to a minimization of the excess kinetic energy.

6.5.1.3 Density-functional theory approximation to the ELF

It is useful to briefly recall here the essential equations of DFT [168–170] and of TDDFT [181–186], since these are the theories that are employed to obtain the orbitals from which the ELF is calculated in the examples presented in the following subsection.

There exists a one-to-one correspondence between the ground-state density of a many electron system, n, and its external potential v. This permits to write every observable as a functional of the density. For each interacting system, there also exists an auxiliary noninteracting system of fermions, subject to an external potential different to the one in the original system, such that the densities of the two systems are identical. One can then solve this noninteracting system, and obtain any observable of the interacting system by using the appropriate functional of the density.

The one-particle equations that provide the single-particle orbitals that conform the one-determinantal solution to the noninteracting problem are the so-called Kohn-Sham equations:

$$\{-\frac{1}{2}\nabla^2 + v_{\mathrm{KS}}(\boldsymbol{r})\}\ \varphi_i(\boldsymbol{r}) = \varepsilon_i \varphi_i(\boldsymbol{r})\,, \quad i = 1, ..., N. \tag{6.31}$$

The density of both the interacting and noninteracting system is then simply:

$$n(\boldsymbol{r}) = \sum_{i=1}^{N} |\varphi_i(\boldsymbol{r})|^2\,. \tag{6.32}$$

The problem lies in the calculation of the Kohn-Sham potential, $v_{\mathrm{KS}}(\boldsymbol{r})$, itself a functional of the density. For this purpose, it is usually split into a known and an unknown part – the latter being the so-called exchange and correlation potential $v_{\mathrm{xc}}(\boldsymbol{r})$:

$$v_{\mathrm{KS}}(\boldsymbol{r}) = v(\boldsymbol{r}) + \int \mathrm{d}^3 r'\, \frac{n(\boldsymbol{r})}{|\boldsymbol{r} - \boldsymbol{r}'|} + v_{\mathrm{xc}}(\boldsymbol{r})\,. \tag{6.33}$$

TDDFT extends the parallelism between the interacting and the noninteracting system to time-dependent systems [181]. One then has to deal with time-dependent Kohn-Sham equations:

$$\mathrm{i}\frac{\partial \varphi_i}{\partial t}(\boldsymbol{r};t) = \{-\frac{1}{2}\nabla^2 + v_{\mathrm{KS}}(\boldsymbol{r};t)\}\ \varphi_i(\boldsymbol{r};t)\,, \quad i = 1, ..., N. \tag{6.34}$$

Once again, an approximation to a time-dependent exchange and correlation potential is needed.

The ELF is calculated in terms of spin-orbitals, and is not an explicit functional of the density. One may then approximate the ELF of the interacting system by considering the ELF of its corresponding Kohn-Sham system – whose state is a Slater determinant, and can be calculated using the previous equations. Note that this is a completely different approximation to the one taken by considering the Hartree-Fock ELF – even if it leads to an analogous expression. However, it has been shown that the main features of the ELF are rather insensitive to the method utilized in its calculation [187, 188], even for more approximate schemes such as the extended Hückel model.

6.5.2 Examples in the ground-state

This subsection will present some applications of the ELF for systems in the ground state. All calculations have been done within the KS/DFT formalism. For the exchange-correlation potential, the local-density approximation (LDA) has been employed in all cases, except for the water molecule and the hydroxide ion, for which – both in the ground state calculations and in the collision processes presented in the next subsection – the self-interaction correction was added. The resulting functional is orbital dependent, and in order to calculate it, one has to make use of the optimized effective potential theory – together, in this case, with the approximation of Krieger, Li, and Iafrate [189]. The functions are represented on a real-space regular rectangular grid (base-less approach). The pseudopotential approach is taken for the ion-electron interaction in order to avoid the explicit treatment of the chemically inert core electrons.[2] The motion of the cores is treated classically. The computations have been carried out with the `octopus` code [190, 191].

In order to appreciate the usefulness of the ELF to monitor fast, time-dependent molecular processes, it is important to learn the characteristics of the ELF in the ground state. Silvi and Savin [192] outlined a proposal for the classification of chemical bonds based on the topological analysis of the ELF. Let us recall here some basic ideas, illustrated below with some examples. The ELF is a scalar real function, bounded between zero and one – the value one corresponding to maximum localization. The *attractors* are the points where it has maxima; to each attractor corresponds a *basin*, the set of points whose gradient field drives to the attractor. The shape of the isosurfaces of the ELF is also informative: as we change the isosurface value, it may or may not change – when it does, we have a *bifurcation*, which occurs at ELF *critical values*. The attractors may have zero, one, or two dimensions: In general, only zero dimensional attractor are allowed; however system with spherical symmetry (atoms) will have spherical (2D) attractor manifolds, whereas $C_{\infty v}$ (or higher) systems (linear molecules) may have one-dimensional sets of attractors, forming a ring around the molecular axis.

To each attractor one may associate an *irreducible f-localization domain.* An f-localization domain is the set of connected points for which the ELF is larger than f. It is irreducible if it only contains one attractor. The spatial arrangement of these domains is the key to classify chemical bonds: there are three types of attractors: core (its domain contains a nuclei), bonding (located between the core attractors of different atoms) and nonbonding (the rest, that contain the so-called lone pairs). All atoms will have an associated core attractor, except hydrogen.

In each domain, one may integrate the electronic density, and obtain a number of electrons. In the absence of symmetry, at most two electrons with opposite spins should be found in a basin. An attractor for which the number of electrons in its associated domain is less than two is an *unsaturated attractor.* A multiple bond is created when there is more than one bonding attractor between two core attractors. A ring attractor containing six electrons is also a multiple bond.

---

[2] It may be argued that the ELF that we depict, is, in fact, a pseudo-ELF. The effect of removing the core electrons in the ELF is the removal of localized electrons in the vicinity of the nuclei. This is irrelevant if one is interested in learning about the chemical properties of the systems.

Fig. 6.42. ELF isosurfaces ($\eta = 0.85$) of ethane (left), ethene (center) and ethyne (right).

A first illustrative example is the clear distinction between the single, double and triple bonds of the ethane, ethene and ethyne molecules, as presented in Fig. 6.42. The ethyne (acetylene) molecule is an example of linear molecule ($D_{\infty h}$ symmetry), which allows for continuous ring attractors. These may occur specially for cases in which one expects a triple bond, such as is the case in acetylene. However, other textbook "Lewis" triple bonds do not show a ring attractor: the nitrogen molecule presents only one point attractor between the nuclei, and two other point attractor at their sides. The double bond of ethene (center in Fig. 6.42) is clearly manifested by the presence of two attractors between the carbons. This leads to isosurfaces with a characteristic "eight" shape. The ethane molecule (left), presents only one attractor between the carbons (single bond), and the six domains corresponding to the CH bonds.

It is known that the ring isomer of C_{20} (see Fig. 6.43, left side) does not have a 20th order axis of symmetry, due to the presence of alternating bonds, which reduces the molecule symmetry group to C_{10h}. The different nature of the bonds ("single-triple alternation", in the Lewis picture), is clearly patent in the ELF: the continuous ring of attractors for the triple bonds, whereas one single point attractor for the single bonds. In the case of the C_{60} fullerene (see Fig. 6.43, right side) due to its high symmetry, there are also in principle two possibly different kinds of bonds: the ones for which the bond line is separating two hexagons, and the ones for which the bond line is separating one pentagon and one hexagon. A look at the ELF tells us that the character of these bonds is, however, very similar.

The usefulness of the ELF is specially patent for the analysis of nonbonding electron groups [193]. In Fig. 6.44 two examples are shown: the hydroxide (OH^-) ion, and the water molecule. In the first case (right), there is once again a continuous ring attractor, that contains six electrons. This reflects in the torus-like shape of the isosurfaces defined in its domain. The water molecule, on the contrary, breaks the linear symmetry, and thus does not permit for continuous attractors. In this case one can see, in addition to two isosurfaces in the CH bond basins, one "bean"-shaped isosurface, that contains two point attractors on each side of the oxygen atom. Each irreducible domain, corresponding to each of these two attractors, contains two electrons.

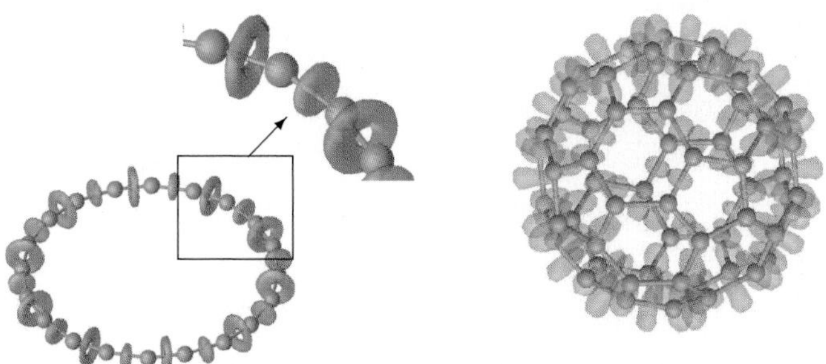

Fig. 6.43. ELF isosurfaces ($\eta = 0.85$) for the ring isomer of C_{20} (left), and for the C_{60} fullerene.

Fig. 6.44. ELF isosurfaces ($\eta = 0.85$) of the water molecule (left), and of the hydroxide ion (right), showing the very different shape of the lone pair basin with four electrons (two point attractors, as it is the case for water), and with six electrons (ring-shaped attractor, as it is the case for the hydroxide ion).

Figs. 6.45 and 6.46 present another case: the formaldimine molecule (also referred to as the smallest imine, or as the smallest unprotonated Schiff base). This molecules presents a double bond between carbon and nitrogen, and a lone pair attached to the nitrogen atom. The upper figures of Fig. 6.45 depict the electronic density: an isosurface on the left, and a logarithmic color map on the plane of the molecule on the right. Below, the figures depict the ELF in the same way – although the scale of the colormap in this case is not logarithmic. Both the bond (and its type) and the lone pair are clearly visible in the ELF, whereas the density presents much less structure.

Fig. 6.46 displays the same formaldimine molecule; however, it shows the gradient lines of the ELF, which converge in the attractors. This alternative pictorial representation is also helpful to identify the positions of the attractors.

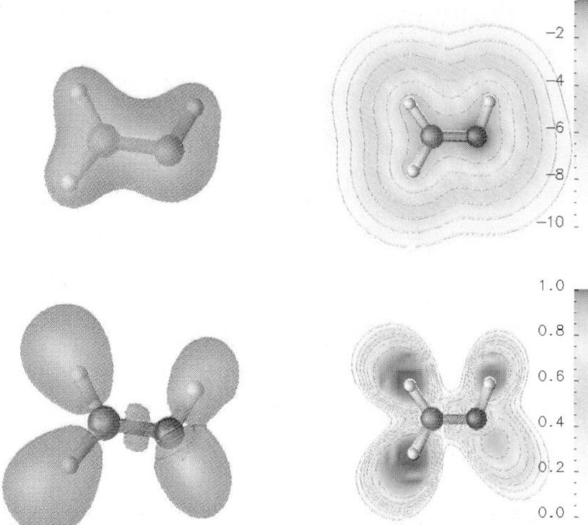

Fig. 6.45. Electronic density (top) and ELF (bottom, see text for its definition) of the ground state of the formaldimine molecule. Left figures show one three dimensional isosurface, whereas the right figures show a color-mapped two dimensional plane. Note that the scale in the case of the density is logarithmic; the values in the legend reflect the exponent.

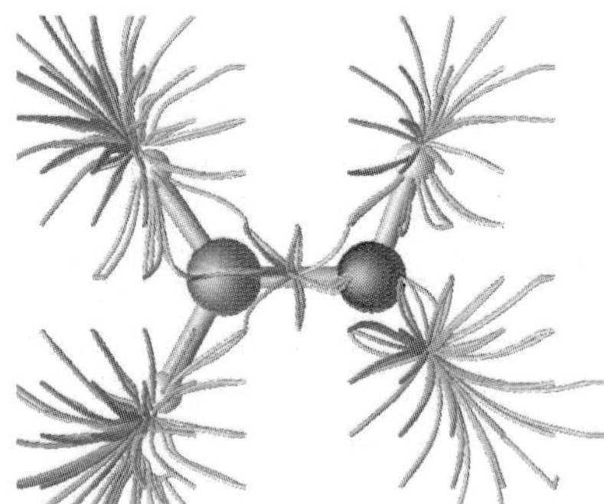

Fig. 6.46. Streamlines running through the gradient field of the ELF of formaldimine, and meeting at the basin attractors – the ELF local maxima.

6.5.3 Fast processes

The following time-dependent calculations of the ELF have been done by making use of TDDFT to describe the many-electron system. On top of this, the ions are also allowed to move. These are treated classically as point particles (the next subsection describes a model in which this restriction is lifted). The forces that define the ionic movement are calculated through Ehrenfest's theorem. It amounts to the simultaneous and coupled evolution of both a classical and a quantum system. The resulting Molecular Dynamics is nonadiabatic, since the electrons may occupy any excited state, and change these occupations.

6.5.3.1 The $H^+ + OH^- \rightarrow H_2O$ reaction.

In the following, the TDELF is used to monitor, "in real time", the chemical behavior of the electrons involved in a chemical reaction. In this case, a specially simple one: the formation of a water molecule after the collision of a proton and a OH^- group.

One should recall, first of all, the topological differences between the lone-pair basin in the water molecule, containing two pairs, and the ring-shaped basin of the hydroxide ion (see Fig. 6.44). The chemical reaction that produces water should involve the transformation of this lone-pair basin. The collision of the two reactants produces different results depending on the original velocities and orientations; two typical outcomes are presented here: a successful event (meaning formation of water), and an unsuccessful collision, leading to three isolated nuclei.

Fig. 6.47 shows the first of these two cases. At time zero, one can identify the characteristic ELF of the ground-state hydroxide ion. Note that this figure depicts isosurfaces of the ELF at a value of $\eta = 0.8$, and these isosurfaces are color-coded: an intense red means a region of high electronic density, whereas the whitish areas of the isosurfaces correspond to regions of almost negligible density. This is done in order to make apparent one of the less intuitive features of the ELF: it may have large values in regions of low electronic density.

The proton and the hydroxide group initially approach each other with a velocity of 10^{-2} a.u., or 0.21 Å/fs. The proton is directed to the middle point of the ion. As the proton approaches the hydroxide group in the first snapshots, an accumulation of ELF becomes apparent near it. This corresponds to a small transfer of electronic density – even if this density will be strongly localized and very large in size (see that snapshot taken at 9.7 fs), the amount of charge transfer is minute. This fact may be learnt from the lack of red color in this isosurface.

In the snapshots of the second row, the proton collides with the hydroxide group, and as a result the two protons jump away off the oxygen atom. Each proton has now its associated ELF basin, whereas the lone pairs basin associated to oxygen is already distorted. The last snapshots in the third row show the return of the protons to the influence of the oxygen core, which demonstrates that water has been formed. The very last snapshot, some 30 fs after the process was initiated, clearly depicts the lone-pairs basin with the typical "bean" shape corresponding to two electron pairs. Note, however, that both nuclear and electronic degrees of freedom are in highly excited state, and thus the final picture is not a steady structure.

Fig. 6.48 shows another possibility, which occurs for higher proton velocities. In this case, the simulation is illustrated with a different representation

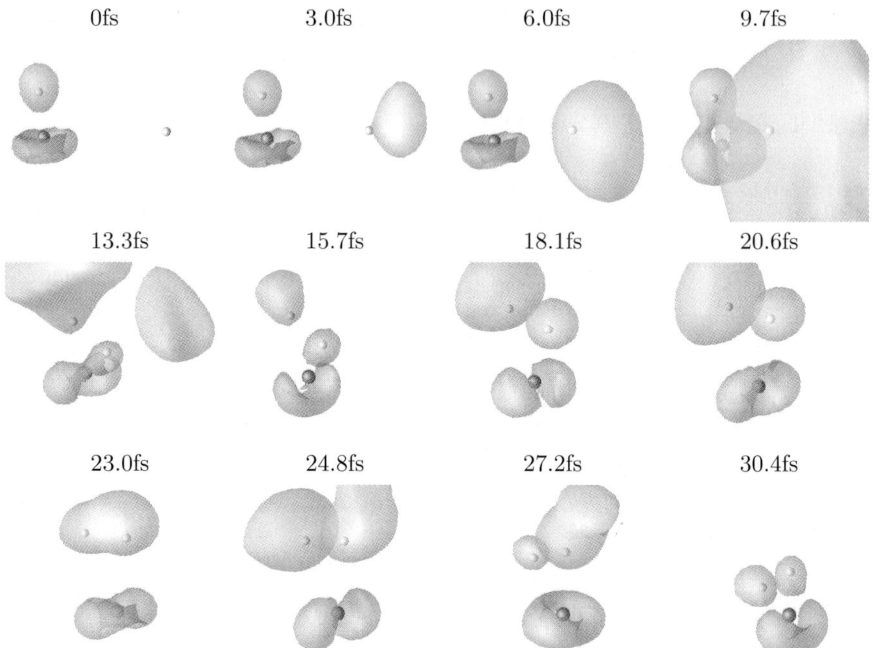

Fig. 6.47. Snapshots taken during the formation of a water molecule due to the collision of a proton and a OH^- group. Isosurfaces for the ELF at a value of $\eta = 0.8$ are shown in red. This red color, however, is graduated depending on the local value of the electronic density: more intense red means higher density. The white areas, thus, correspond to regions of high electronic localization but low density. The oxygen core is colored in red, whereas the protons are colored in white.

procedure: a color map on the plane in which the three atoms move. The initial geometry is similar, but in this case the relative velocity is 5×10^{-2} a.u., or 1.1 Å/fs. Once again, the second snapshot shows how a cloud of localized electrons develops around the proton as it approaches the anion. It becomes specially large after 2.9 fs; note however that it does not mean a large electronic transfer; to learn about that one needs to look at the density. In the fourth snapshot, the incoming proton cleanly passes through the bond. The original shape of the ELF is completely distorted; however the speed of the process did not allow yet for fast movements of the nuclei – except the straight line movement due to their original velocities.

In the second row one may see the proton scatter away from the anion; it does so at an angle from its initial trajectory. The bond of the anion is broken; as a consequence the two nuclei separate from each other. Each of the three nuclei carries away an electronic cloud: a spherical crown in the case of the oxygen atom (corresponding to the typical two dimensional spherical attractor of an isolated many electron atom), and spatially large accumulations of localized electrons for the protons (note, once again, that this does not imply a large number of electrons. In order to learn about the electronic charge carried away by each of the ions, it is necessary to integrate the density in each of the localization domains).

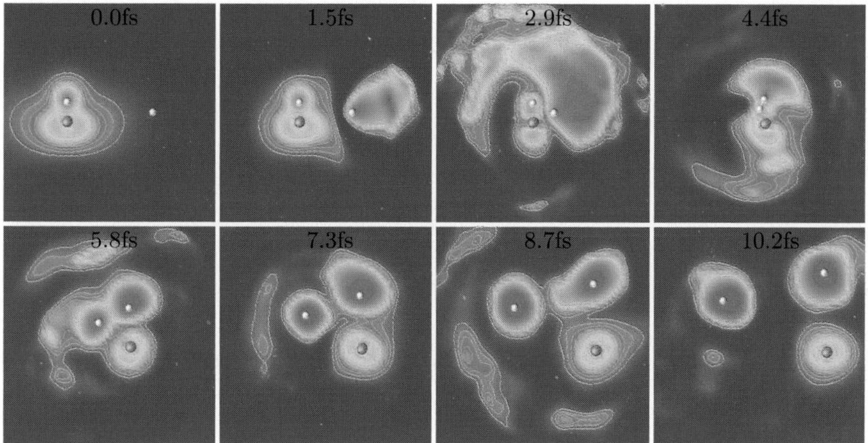

Fig. 6.48. Snapshots taken during the collision of a hydroxide ion with a proton, leading to the dissociation of the hydroxide group.

6.5.3.2 Proton capture by a lone pair

The next case focuses in the formaldimine molecule, Fig. 6.46. It presents one lone pair, which chemically may behave as a possible anchorage for a radical. For example, it may attract a "traveling" proton in an acid environment. This is demonstrated in the simulation depicted in Fig. 6.49.

In the first snapshot, the formaldimine molecule is in its ground state, both its electronic and nuclear degrees of freedom. The topology of the ELF for this particular case was discussed in the previous subsection. A proton travels with a velocity of 5.2×10^{-3} atomic units (corresponding to an energy of $0.673\,\mathrm{eV}$), in the plane of the molecule, and initially aiming to the center of the CN double bond. The lone pair, however, attracts the proton to its basin. As a result, the proton drifts to the right, in the direction of the nitrogen atom, accelerating its movement. The molecule itself also rotates as the nitrogen atom attempts to approach the incoming proton. This enters the nonbonding basin, and transforms it into a bonding NH loge. The ensuing collision results in the proton quickly accelerating out of the molecule; however, the bond has been established, and soon it is driven back. The result is a highly excited molecule: the nuclei will vibrate, whereas the electronic state will also be a mixture of the ground state and higher lying states. Of course, eventually it could relax upon photon emission; this is however not included in the model.

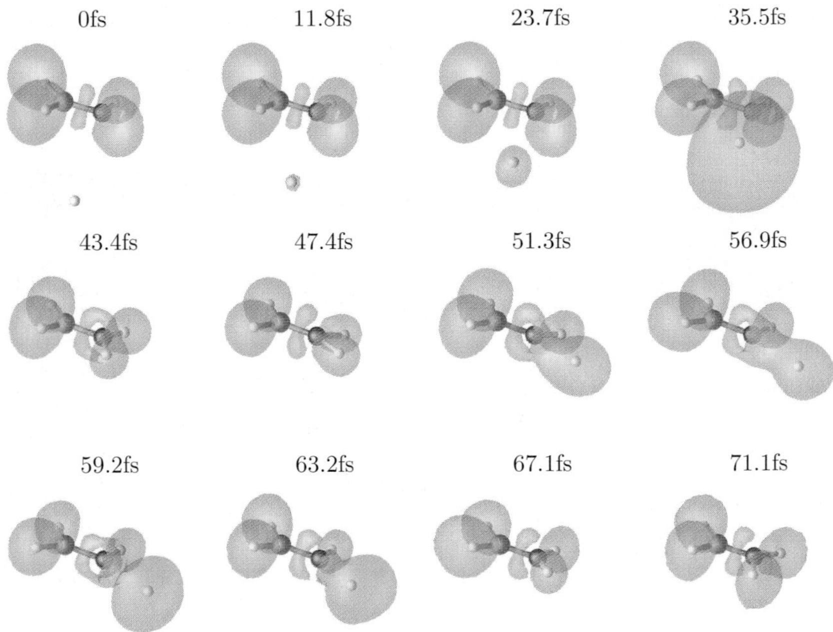

Fig. 6.49. Snapshots taken during the capture of a proton by a formaldimine molecule. Isosurfaces for the ELF at a value of $\eta = 0.8$ are shown in red. The carbon and nitrogen cores are colored in green and blue respectively, whereas the protons are colored in white.

6.5.3.3 Bond-breaking by an intense, ultrafast laser pulse

The next example shows the excitation of the ethyne molecule by means of a strong laser. The aim is especially the triple bond. The laser is polarized along the molecular axis; it has a frequency of 17.15 eV ($\lambda = 72.3\,\text{nm}$) and a maximal intensity of $I_m = 1.19 \times 10^{14}\,\text{Wcm}^{-2}$. Fig. 6.50 depicts snapshots of the ELF of acetylene in form of slabs through a plane of the molecule. At the beginning (a) the system is in the ground state and the ELF visualizes these features: The torus between the carbon atoms, which is typical for triple bonds, and the blobs around the hydrogen atoms. As the intensity of the laser increases, the system starts to oscillate and then ionizes (Fig. 6.50b,c). Note that the ionized charge leaves the system in fairly localized packets (the blob on the left in b, and on the right in c). The central torus then starts to widen (Fig. 6.50d) until it breaks into two tori centered around the two carbon atoms (Fig. 6.50e,f). This can be interpreted as a transition from the π bonding to the $\pi^\star$ nonbonding state. The system then remains in this excited state, and eventually dissociates, after the laser has been switched off. In the process,

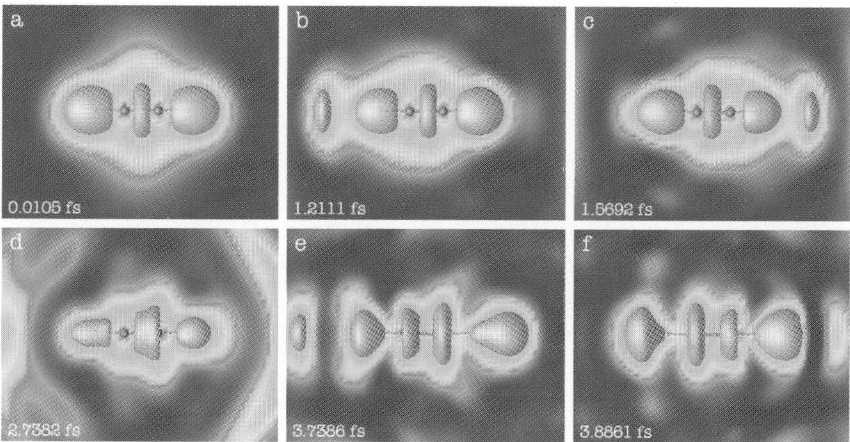

Fig. 6.50. Snapshots of the time-dependent ELF for the excitation of ethyne (acetylene) by a 17.15 eV ($\lambda = 72.3$ nm) laser pulse. The pulse had a total length of 7 fs, a maximal intensity of 1.2×10^{14} Wcm^{-2}, and was polarized along the molecular axis. Ionization and the transition from the bonding π to the anti-bonding $\pi^\star$ are clearly visible.

the molecule absorbs about 60 eV of energy, and looses 1.8 electrons through ionization.

6.5.4 TDELF for coupled nuclear-electronic motion

The examples presented in the previous subsection neglected the quantum nature of the atomic nuclei. Erdmann, Gross, and Engel [194] have presented one application of the TDELF for a model system in which one nucleus is treated quantum mechanically, and the full Schrödinger equation is computed exactly. This model is specially suited to study, from a fundamental point of view, the effects of nonadiabaticity. It is instructive to see how the ELF may help for this purpose.

The model is depicted in Fig. 6.51: two electrons and a nucleus that move in a single dimension between two fixed ions. Its Hamiltonian is:

$$H(x, y, R) = T(x) + T(y) + T(R) + V(x, y, R)\,, \tag{6.35}$$

where $T(x)$, $T(y)$ and $T(R)$ are the kinetic energy operator of the two electrons and of the moving ion, respectively. The potential is:

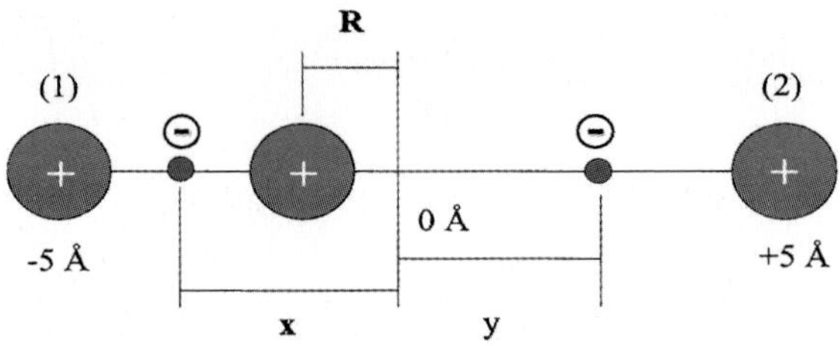

Fig. 6.51. Configuration of the model system: An ion (coordinate R) and two electrons (at x and y) are allowed to move between two fixed ions (1) and (2), fixed at a distance of 10Å.

$$\begin{aligned} V(x,y,R) = {} & \frac{Z_1 Z}{|R_1 - R|} + \frac{Z_2 Z}{|R_2 - R|} + \frac{\mathrm{erf}(|x-y|)}{R_e|x-y|} \\ & - \frac{Z_1 \mathrm{erf}(|R_1 - x|)}{R_f|R_1 - x|} - \frac{Z_2 \mathrm{erf}(|R_2 - x|)}{R_f|R_2 - x|} - \frac{Z \mathrm{erf}(|R - x|)}{R_c|R - x|} \\ & - \frac{Z_1 \mathrm{erf}(|R_1 - y|)}{R_f|R_1 - y|} - \frac{Z_2 \mathrm{erf}(|R_2 - y|)}{R_f|R_2 - y|} - \frac{Z \mathrm{erf}(|R - y|)}{R_c|R - y|} \, . \end{aligned} \tag{6.36}$$

Note that the interactions are screened; The values of the screening are modulated by the parameters R_f (for the interaction electron – fixed ions), R_c (for the interaction electron – moving ion), and R_e (for the electron – electron interaction). By tuning these parameters, the nonadiabatic couplings may be reduced or enhanced [195–199].

The degree of diabaticity is qualitatively pictured in the adiabatic potential energy surfaces (PES) – which show the eigenvalues, parameterized with the nuclear coordinate R, of the electronic equation:

$$\{T(x) + T(y) + V(x,y,R)\}\phi_n^{\sigma\tau}(x,y;R) = V_n^{\sigma\tau}(R)\phi_n^{\sigma\tau}(x,y;R) \, , \tag{6.37}$$

so that $\phi_n^{\sigma\tau}(x,y;R)$ are the electronic eigenfunctions in state n. Two different initial configurations are possible: the two electrons are in the same spin state – corresponding to spatial functions of gerade symmetry –, or in opposite spins – corresponding to ungerade spatial functions. (Note that since the full Hamiltonian does not contain the spin, the system will remain in the same spin configuration during any evolutions). The adiabatic PES are depicted in Fig. 6.52 for the anti-parallel spin (top) and parallel spin (bottom) cases, and for the ground state, and the first three excited states.

In the anti-parallel case, the ground state and the first excited state show an avoided crossing, so we should expect clear nonadiabatic behavior in that region. In the parallel spin case, however, the ground state and the first excited state are well separated from each other and from the higher states, whereas the second and third excited states again show avoided crossings.

The localization functions for this particular model have to be defined. The full time-dependent density matrix is given by:

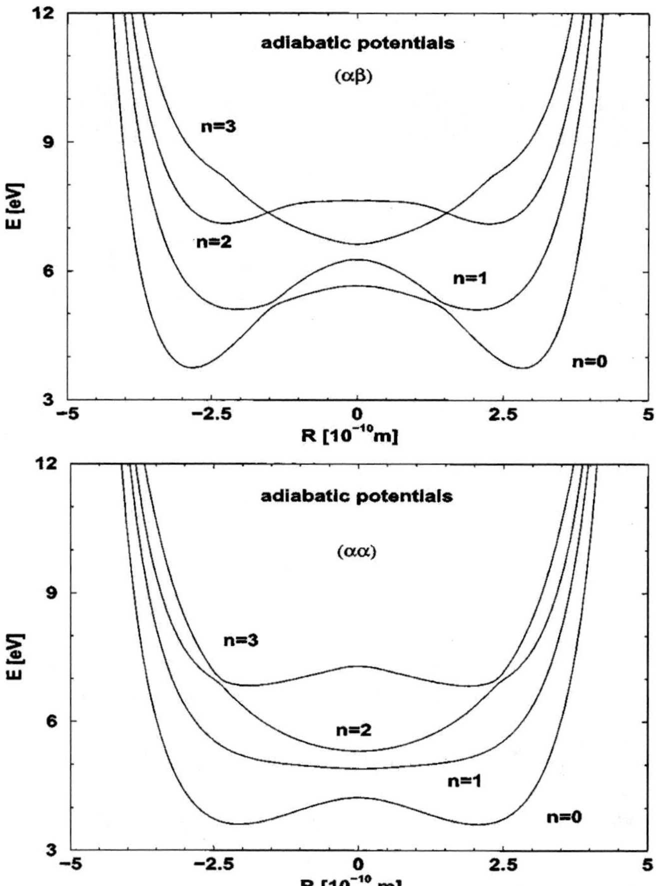

Fig. 6.52. Adiabatic potentials for the anti-parallel (top panel) and parallel spin case (bottom panel). Different parameters were used in the parameterization of the interaction energy: $R_c = R_f = 1.5$ Å; $R_e = 2.5$ Å(left panel), and $R_c = R_f = R_e = 1.5$ Å(right panel).

$$D_{\sigma\tau}(x, y, R; t) = |\Psi(x\sigma, y\tau, R; t)|^2 , \tag{6.38}$$

where Ψ is the full wave function. Integrating out the nuclear degree of freedom, one obtains the density matrix for the two electrons:

$$D_{\sigma\tau}(x, y; t) = \int d^3R \, D_{\sigma\tau}(x, y, R; t) , \tag{6.39}$$

and one may then define the conditional pair probability function:

$$P_{\sigma\tau}(x, y; t) = \frac{D_{\sigma\tau}(x, y; t)}{\rho_\sigma(x; t)} , \tag{6.40}$$

where ρ_σ is the electronic one-particle spin-density. Two cases have to be distinguished:

1. Anti-parallel spins: $P_{\alpha\beta}(x, x; t)$ is the conditional probability to find one electron at time t at point x, if we know with certainty that other electron with opposite spin is in the same place. This is an indirect measure of localization. One may define, in analogy to the usual ELF, a time-dependent *anti-parallel spin* electron localization function (TDALF), η^{ap}, as:

$$\eta^{\mathrm{ap}}(x;t) = \frac{1}{1 + |P_{\alpha\beta}(x, x; t)/F_\alpha(x;t)|^2} \,. \tag{6.41}$$

   $F_\alpha(x;t) = (4/3)\pi^2\rho_\alpha^3(x;t)$ is the Thomas-Fermi kinetic energy density for anti-parallel spins and 1D systems.
2. Parallel spins: This would correspond to the usual ELF, presented previously. However, the one-dimensionality of the model changes the derivation since the spherical average is not necessary. Defining $s = x - y$, one may expand $P_{\alpha\alpha}(x, s; t)$ in a Taylor series up to second order around $s = 0$:

$$P_{\alpha\alpha}(x, s; t) = \frac{1}{2}\frac{\partial^2 P_{\alpha\alpha}}{\partial s^2}(x, 0; t)s^2 + \mathcal{O}(s^3) \,. \tag{6.42}$$

   The constant term is null due to Pauli's principle, whereas the linear term also vanishes since, according to Kato's cusp theorem [200], the wave function is proportional to s. The s^2 coefficient, $a_{\alpha\alpha}(x;t)$, is now used to define the TDELF with the usual re-normalization precautions:

$$\eta(x;t) = \frac{1}{1 + |a_{\alpha\alpha}(x;t)/F_\alpha(x,t)|^2} \,. \tag{6.43}$$

   In this case, $F_\alpha(x) = (16/3)\pi^2\rho_\alpha^3(x)$.

The nuclear movement is investigated through the time-dependent nuclear density:

$$\Gamma_{\sigma\tau}(R;t) = \int \mathrm{d}^3x \int \mathrm{d}^3y \, D_{\sigma\tau}(x, y, R; t) \,. \tag{6.44}$$

The time-evolution of the system is then initiated from an initial state with the form:

$$\Psi(x\sigma, y\tau, R; t = 0) = e^{-\gamma(R-R_0)^2}\phi_n^{\sigma\tau}(x, y; R) \,, \tag{6.45}$$

that is, from the first electronic excited state, and from a Gaussian nuclear distribution around some initial point – in this case, $R_0 = -3.5$ Å.

Once again, two possible spin configurations for the initial state have to be distinguished:

1. Anti-parallel spins.
   This case is shown in Fig. 6.53, left side. The top graph represents the *nuclear* time-dependent density. This density, initially localized around -3.5 Å, travels toward its turning point, while it strongly disperses. Soon,

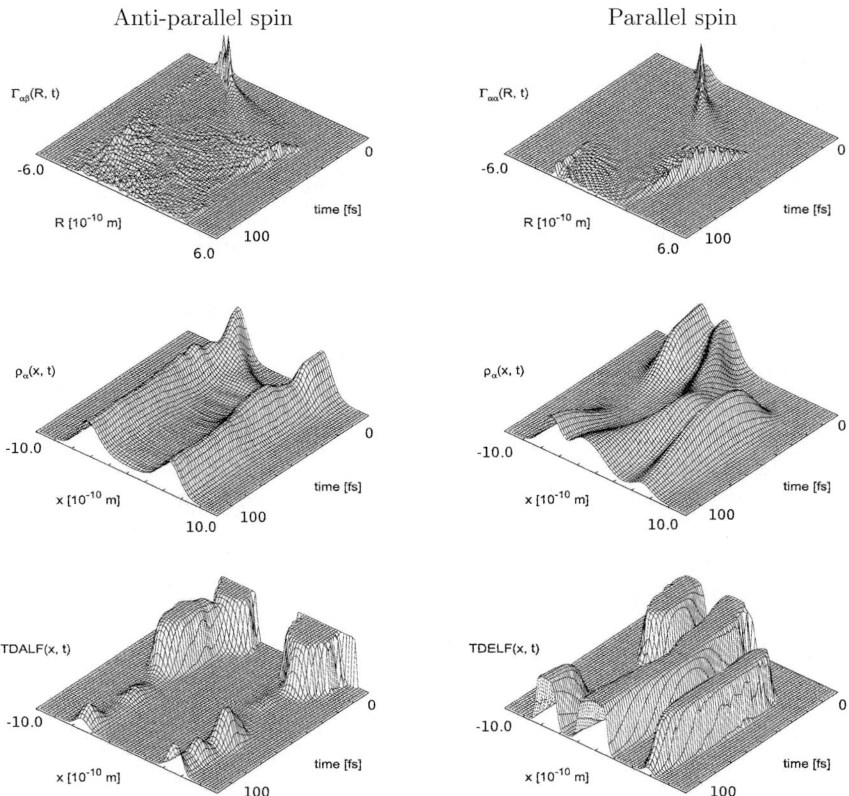

Fig. 6.53. Quantum dynamics of the model system presented in Sect. 6.5.4, for the anti-parallel spin (left) and the parallel spin cases (right). The upper panel shows the nuclear density. The time-dependent electron density and TDELF are shown in the middle and lower panels, respectively.

as a consequence of the strong nonadiabatic coupling, the nuclear wave packet becomes extremely broad and a defined structure can no longer be seen.

The electron density (middle panel) seems to be unaware of the nuclear motion. This does not mean that electrons are static; its behavior may be best analyzed by looking at the TDALF (lower panel). We have two localization domains, which correspond also with the initial areas of high density. It may be seen how, as the nucleus transverses this area, the localization amplitude diminishes, and almost vanishes for those two areas. This illustrates how the strong nonadiabatic coupling is effective in decreasing localization.

2. Parallel spins.
   This case is shown in Fig. 6.53, right side. Also, the nuclear time-dependent density is on the top and the time-dependent electron density is in the middle, although in this case it is the usual TDELF (parallel-spins) which is shown in the bottom panel.
   This case has been tailored to avoid the presence of nonadiabatic effects (the first excited state is well separated from the others). As a result, with the chosen initial conditions, the motion takes place exclusively in a single electronic state. The nuclear wave packet is initially localized in the left half of the potential well, and starts moving to the right side where it is repelled by the right side fixed ion at about 40 fs. The wave packet then shows an oscillatory structure, and broadens due to the anharmonicity of the potential.
   The electronic density reflects a charge transfer from the left fixed ion to the right one, with the moving ion acting as an "electron carrier". Initially, there are two maxima in the vicinity of the left fixed ion and on the moving one. After the nucleus crosses the origin, the initial density drops to zero and the new two maxima are on top of the moving ion and on the right side. If the nucleus were not affected by dispersion, the process would reverse with each half-cycle of the nuclear vibration.
   The behavior of the TDELF is now very different with respect to the TDALF in the anti-parallel spin case. The localization remains high at all times, and the transfer of electrons from left to right is clear: Initially there are two localization domains; one around the fixed ion, and another near the origin. As the nuclear movement starts, the first domain vanishes, and a third domain appears near the right fixed ion. After the vibrational period of the nucleus is finished, this third domain disappears, and the initial ELF is however restored. The vanishing of the first domain and the appearance of a third domain indicates that one electron must have been removed from the left fixed nucleus and dragged to the right.

In conclusion, the handful of examples presented in this section illustrate the amount of information that can be gained from the time-dependent ELF in theoretical studies of ultrafast phenomena. One can learn about the time scales of the processes, and/or about how the various sub-events that make up a complex reaction are ordered in time: which bonds break first, which second, how the new links are created, etc. One can observe and interpret intermediate electronic structure that may be short lived but relevant for the overall outcome. This information starts to become available to experimentalists, as the time resolution of the sub-femtosecond laser sources increases.

6.6 Cluster dynamics in ultraintense laser fields

A. Heidenreich, I. Last, and J. Jortner

6.6.1 How intense is ultraintense?

Table-top lasers are currently characterized by a maximal intensity of $\sim 10^{22}$ $\mathrm{Wcm^{-2}}$ [201,202], which constitutes the highest light intensity on earth. Such ultrahigh intensity corresponds to an electric field of $\sim 10^{12}\,\mathrm{Vcm^{-1}}$, a magnetic field of $\sim 10^9$ Gauss and an effective temperature of $\sim 10^8\,\mathrm{K}$, which exceeds that in the interior of the sun and is comparable to that prevailing in the interior of hot stars. The interaction of ultraintense (peak intensity $I_m = 10^{15}$–$10^{20}\,\mathrm{Wcm^{-2}}$), ultrafast (temporal length 10–100 fs) laser pulses with matter drives novel ionization phenomena [25,203–251], attosecond electron dynamics [20,154,252,253], the production of high energy particles (i.e., electrons, x-rays, and ions) [9,241,242,245–251,254–268] in atoms, molecules, clusters, plasmas, solids, and liquids. The coupling of macroscopic dense matter with ultraintense laser fields is blurred by the effects of inhomogeneous dense plasma formations, isochoric heating, beam self-focusing and radiative continuum production [269,270]. To circumvent the debris problem from macroscopic solid targets, it is imperative to explore efficient laser energy acquisition and disposal in clusters, which constitute large, finite systems, with a density comparable to that of the solid or liquid condensed phase and with a size that is considerably smaller than the laser wavelength. This section addresses electron and nuclear dynamics driven by ultraintense laser–cluster interaction [9,25,203–251,254–267].

The Rabi frequency for the interaction of an ultraintense laser ($I_m = 10^{20}\,\mathrm{Wcm^{-2}}$), with an atom or molecule with a transition moment of 1–5 Debye, falls in the range of 2–10 keV. Such high values of the Rabi frequency signal the breakdown of the perturbative quantum electrodynamics approach for the ultraintense laser–atom/molecule interaction. The perturbative quantum dynamic approaches are applicable only for 'ordinary' fields (i.e., $I_m < 10^{12}\,\mathrm{Wcm^{-2}}$, where the Rabi frequency is lower than $\sim 0.1\,\mathrm{eV}$), while for strong laser fields, whose frequency is considerably lower than the atomic/molecular ionization potential, the ionization process can be described as electron removal through an electrostatic barrier in a static electric field. The potential for a q-fold ionized atom (or neutral atom), formed by an electric field of charge $(q+1)$ and an external electric field F, is characterized by a high U_b of the potential barrier [232,243,271]

$$U_b = -2[eF\bar{B}(q+1)]^{1/2} \tag{6.46}$$

where $\bar{B} = 14.4\,\mathrm{eV}$ and eF is given in units of $\mathrm{eV\AA^{-1}}$. The barrier is located at the distance

$$r_b = [\bar{B}(q+1)/eF]^{1/2} \tag{6.47}$$

from the ion center along the electric field direction. When the tunneling through such a barrier is of minor importance [243], as realized for ultraintense laser fields (whose intensity domain will be specified below), a classical barrier suppression ionization (BSI) mechanism can be applied. The BSI of a single ion of charge q is realized when the barrier height, (6.46), is equal, with an opposite sign, to the ionization potential E_I^{q+1} of this ion. The threshold field for inducing ionization is

$$eF = (E_I^{q+1})^2/4\bar{B}(q+1) \tag{6.48}$$

where $F = F_\ell$ is the laser field, with $|eF_\ell| = 2.745 \times 10^{-7} I_m^{1/2}$ eVÅ^{-1}, where I_m is given in Wcm^{-2}. The threshold laser intensity for BSI is then given by

$$I_m = 8.295 \times 10^{11} (E_I^{q+1})^4/\bar{B}^2(q+1)^2 \tag{6.49}$$

The barrier distance for the threshold field, (6.47), assumed the form

$$r_b = 2\bar{B}(q+1)/E_I^{q+1} \tag{6.50}$$

The BSI, (6.49), describes multielectron ionization of Xe atoms in the intensity range of $I_m = 10^{15}$ Wcm^{-2} (where Xe^{3+} is produced) up to $I_m = 10^{20}$ Wcm^{-2} (where Xe^{36+} is produced), with the calculated results in the range $I_m = 10^{16}$–10^{18} Wcm^{-2} being in good agreement with the available experimental data [272, 273].

To account for tunneling effects in multielectron atomic ionization through the barrier U_b, (6.46), the Amosov–Delone–Krainov (ADK) model [45] gives the ionization probability $W(I(t))$ at the laser intensity $I(t) = I_m \cos^2(2\pi\nu t)$ (where ν is the laser frequency). The peak intensity I_m was determined from the single-cycle averaging $\int_0^{1/\nu} dt W(I(t)) = 1$. In the intensity range $I_m = 10^{15}$–10^{19} Wcm^{-2}, the BSI and the ADK results for a single Xe atom agree within 10% and are close to the available experimental data [272, 273]. For lower intensities ($I_m < 10^{15}$ Wcm^{-2}), the BSI model is no longer applicable, with the intensity $I_m \geq 10^{15}$ Wcm^{-2} marking the lower limit of the ultraintense laser domain.

The laser driven one-step BSI mechanism for the ionization of a single atom requires a significant extension when applied to cluster ionization. Of considerable interest is the situation when the cluster size characterized by n atomic/molecular constituents and cluster radius $R_0 = r_0 n^{1/3}$ (where r_0 is the constituent radius) significantly exceeds the size of the single constituent barrier distance r_b, (6.50). Under these circumstances a compound cluster ionization mechanism is manifested, which occurs via a sequential-simultaneous inner-outer ionization process [205, 206, 211, 212, 217–219, 231, 243–245]. Electron dynamics triggers nuclear dynamics, with outer ionization being accompanied with and followed by cluster Coulomb explosion (CE), which results in the production of high-energy (keV–MeV) multicharged ions on the (10–500 fs) time scale of nuclear motion.

The realm of ultrafast phenomena in molecular science currently moves from femtosecond dynamics on the time scale of nuclear motion [274–277] toward attosecond electron dynamics [20, 154, 252, 253]. This new attosecond temporal regime for dynamics constitutes a "spin off" of ultraintense laser–matter interactions. In the attosecond domain nonperturbative effects are fundamental and new mechanisms of ionization and of multielectron dynamics in atoms, molecules, clusters, plasmas and condensed matter are unveiled. In this context this section will address the response of clusters to ultraintense laser fields that induces novel ionization processes and manifests new features of electron dynamics [211, 222, 238–244, 267, 278–281], which drives nuclear dynamics of CE [224, 228, 245–251, 254–256, 259].

6.6.2 Extreme cluster multielectron ionization

The cluster response to ultraintense laser fields triggers well-characterized ultrafast electron dynamics (on the time scale of $< 1\,\mathrm{fs}$–$100\,\mathrm{fs}$). The compound, extreme multielectron ionization mechanism of clusters involves three sequential-parallel processes of inner ionization, nanoplasma formation and outer ionization. Inner ionization results in the formation of a charged, energetic nanoplasma within the cluster or in its vicinity, which is followed by the partial or complete outer ionization of the nanoplasma. Extreme multielectron ionization of elemental and molecular clusters, e.g., Ar_n [203, 207, 234–236], Xe_n [25,216,223,231,233], $(\mathrm{H}_2)_n$ [237], $(\mathrm{D}_2)_n$ [221,254,257], $(\mathrm{H_2O})_n$ [230,234], $(\mathrm{D_2O})_n$ [224,230,234], $(\mathrm{CH}_4)_n$ [226], $(\mathrm{CD}_4)_n$ [213,226,259], and $(\mathrm{HI})_n$ [9,258, 260], in ultraintense laser fields leads to the production of highly charged ions. These involve the stripping of the valence electrons, or even all the electrons from light first-row atoms, e.g., H^+ and D^+ [208,211,221,226,244,261], O^{q+} ($q = 6$–8) [211,224,241], C^{q+} ($q = 4$–6) [213,221,226,230,242,246,262], as well as the production of highly charged heavy ions, e.g., Xe^{q+} ($q = 3$–26) [216,217,223,239–244,246,257,261,265,278,279,282,283]. These unique inner/outer ionization processes and nanoplasma dynamics and response driven by ultraintense laser–cluster interactions were explored by theoretical models and by computer simulations.

The laser electric field acting on the elemental or molecular cluster is taken as $F_\ell(t) = F_{\ell 0}(t)\cos(2\pi\nu t)$, where ν is the laser frequency and $F_{\ell 0}(t)$ is the pulse envelope function. Molecular dynamics simulations (including magnetic field and relativistic effects) and analyses of high-energy electron dynamics and nuclear dynamics in a cluster interacting with a Gaussian laser field $F_{\ell 0}(t) = F_m \exp[-2.773(t/\tau)^2]$, F_m being the electric field at the pulse peak. The infrared laser parameters used for the simulations reported in this section are $\nu = 0.35\,\mathrm{fs}^{-1}$ (photon energy $1.44\,\mathrm{eV}$ and pulse temporal width (FWHM) $\tau = 10$–$100\,\mathrm{fs}$) [238, 243, 279]. The laser pulse is defined in the time domain $t \geq -\infty$ and the peak of the laser pulse is attained at $t = 0$. An initially truncated laser pulse was used for the simulations, with the initial laser field (corresponding to the threshold of single electron/molecule ionization in the cluster) being located at the (negative) time $t = t_s$, which is laser intensity and pulse width dependent [243, 279]. The end of the pulse was taken at $t = -t_s$ [279]. The simulations of electron dynamics [238, 243–246, 279] elucidated the time dependence of inner ionization, the formation, persistence and decay of the nanoplasma, and of outer ionization.

The cluster inner ionization is driven by two processes:

(A) The BSI mechanism, which is induced by a composite field $\boldsymbol{F} = \boldsymbol{F}_\ell + \boldsymbol{F}_i$, where $\boldsymbol{F}_i$ is the inner field generated by electrostatic interactions with the ions (ignition effects) [210, 243] and with the nanoplasma electrons (screening effects) [243, 282]. The BSI level and time-resolved dynamics were evaluated from (6.48) for the threshold composite field [243, 279].
(B) Electron impact ionization (EII), which involves inelastic, reactive impact ionization of ions by the nanoplasma electrons [223, 243, 278, 283]. EII in Xe_n clusters was explored using experimental data [284–288] for the energy dependence of ionization cross sections of Xe^{q+} ions ($q = 1$–10), which were fit by a three-parameter Lotz-type equation [289]. The proper parameterization [278, 279] of the EII cross sections led to reliable information on the EII ionization levels and their relative contribution to inner ionization and to the nanoplasma populations. At the lower intensity domain of $I_m = 10^{15}$–$10^{16}\,\mathrm{Wcm^{-2}}$, the EII contribution to the inner ionization yield is substantial ($\simeq 40\%$ for Xe_{2171} at $\tau = 25\,\mathrm{fs}$), increases with increasing the cluster size and manifests a marked increase with increasing the pulse length [279]. The EII yield and the EII level enhancement markedly decrease with increasing the laser intensity. The EII involves reactive dynamics of nanoplasma electrons driven by the laser field, and will be further considered in Sect. 6.6.3.

Elemental Xe_n clusters provide benchmark systems for the theoretical and experimental studies of electron and nuclear dynamics. A Xe_{2171} cluster coupled to Gaussian laser pulses ($\tau = 25\,\mathrm{fs}$) of intensities $I_m = 10^{15}\,\mathrm{Wcm^{-1}}$ (Fig. 6.54) and $I_m = 10^{18}\,\mathrm{Wcm^{-2}}$ (Fig. 6.55) reveals the following dynamic processes:

A. Electron dynamics involving sequential–parallel multielectron inner ionization (represented by the color coding of atom charges in Figs. 6.54 and 6.55), nanoplasma formation (represented by the electron cloud and by the positive ions in Figs. 6.54 and 6.55), and outer ionization (which corresponds to complete depletion of the nanoplasma for $I_m = 10^{18}\,\mathrm{Wcm^{-2}}$, Fig. 6.55).
B. Nuclear dynamics of CE, which is spatially nonuniform and relatively slow ($\sim 120\,\mathrm{fs}$) for $I_m = 10^{15}\,\mathrm{Wcm^{-2}}$ (where complete outer ionization does not prevail).

The extreme ionization level of Xe_n clusters results in the production of $(Xe^{q+})_n$ multicharged ions (characterized by an average charge q_{av} on each ion). The cluster size and laser intensity dependence of q_{av} produced from

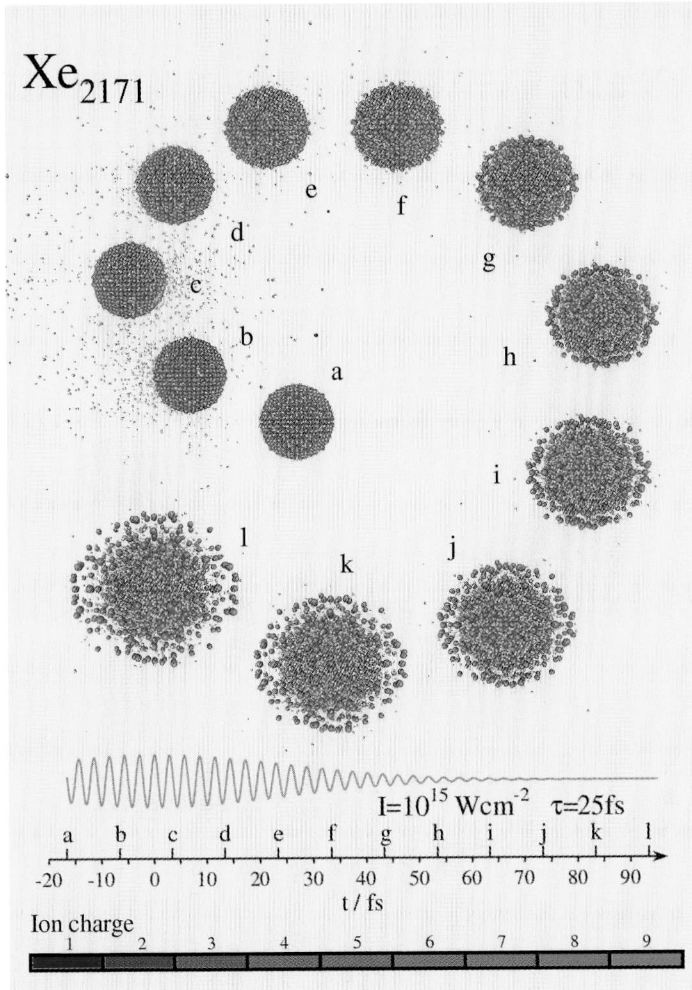

Fig. 6.54. Snapshots of the time-resolved inner ionization, nanoplasma charge distribution, outer ionization and structures of Xe_{2171} clusters induced by Gaussian laser pulses with a peak intensity of $I_M = 10^{15}$ Wcm^{-2} and a pulse width of $\tau = 25$ fs. The lower part of the panel portrays the electric field of the laser and the time axis $t - t_s$ (where t_s is the onset time for the laser field [85]). The snapshot instants are marked a–l on the time axes. The Xe ions are color coded according to their charge: blue corresponds to the initial charge +1 and red to the maximum charge +9, which can be obtained at this intensity (and τ). A map of the color coding of the ionic charges is given in the panel at the bottom of the figure. The electrons are represented by light gray spheres.

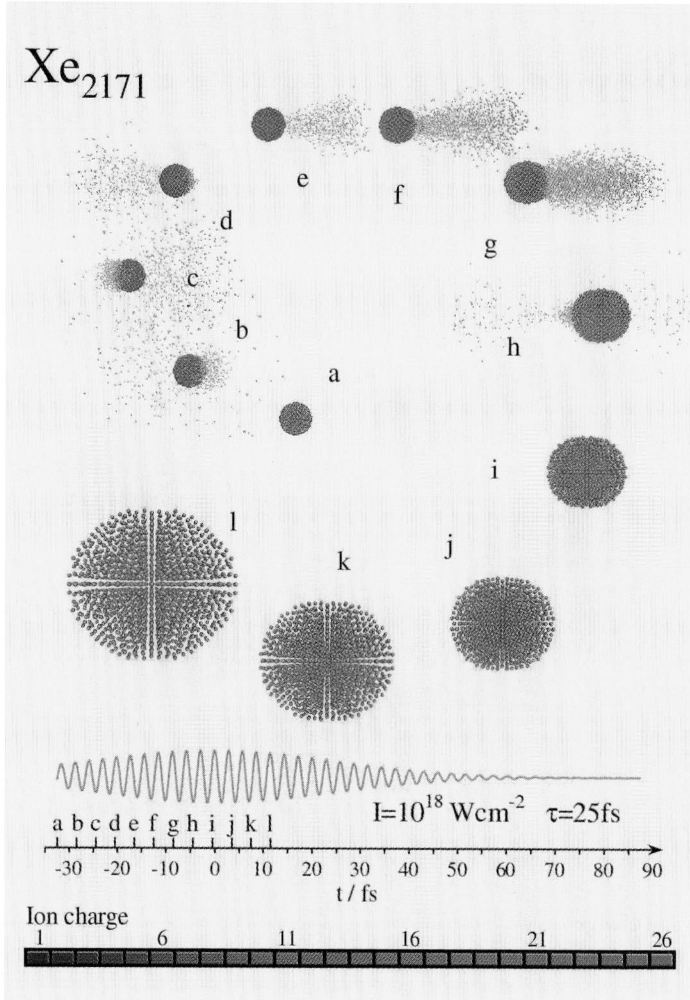

Fig. 6.55. Snapshots of the time-resolved inner ionization, nanoplasma charge distribution, outer ionization and structures induced by a Gaussian laser pulse with a peak intensity of $I_m = 10^{18}\,\mathrm{Wcm}^{-2}$ and a pulse width of $\tau = 25\,\mathrm{fs}$. Presentation and notation as in Fig. 6.54, with the color coding of the Xe ions from blue for the initial charge +1 to deep red for the maximal charge +26.

$n = 55$–2171 clusters (Fig. 6.56) exhibits the formation of highly charged ($q_{av} = 6$–36) clusters. For any cluster size, q_{av} increases with increasing I_m, essentially manifesting the contribution of the laser field to the BSI. The cluster size dependence at fixed I_m exhibits an intensity specific behavior. The highest intensity domains, $q_{av} = 26$ at $I_m = 10^{19}\,\mathrm{Wcm^{-2}}$ (corresponding to ionization of the $4s^24p^64d^{10}5s^25p^6$ shells) and $q_{av} = 36$ at $I_m = 10^{20}\,\mathrm{Wcm^{-2}}$ (corresponding to the ionization of the $3d^{10}4s^24p^64d^{10}5s^25p^6$ shells), are independent of the cluster size at fixed I_m. For $I_m = 10^{18}\,\mathrm{Wcm^{-2}}$, q_{av} increases with increasing R_0, achieving the value of $q_{av} = 24$ at $n = 2171$, which presumably manifests ignition effects. For $I_m = 10^{17}\,\mathrm{Wcm^{-2}}$, q_{av} first increases (in the range $n = 55$–459), and subsequently decreases (in the range $n = 459$–2171) with increasing R_0 (and n), apparently manifesting the interplay between ignition effects (lower n) and screening effects (higher n). In the lowest intensity range of $I_m = 10^{15}\,\mathrm{Wcm^{-2}}$ $q_{av} = 5.8$–6.8, while at $I_m = 10^{16}\,\mathrm{Wcm^{-2}}$ $q_{av} = 8$, with q_{av} being nearly cluster size independent. In the intensity range of $I_m = 10^{18}$–$10^{20}\,\mathrm{Wcm^{-2}}$, q_{av} converges for clusters to the single atom–value with decreasing R_0, indicating that ignition effects are minor for moderately small $n = 55$ clusters in this highest I_m range. At $I_m = 10^{16}$–$10^{17}\,\mathrm{Wcm^{-2}}$, the values of q_{av} for Xe_{55} are somewhat larger (by 0.5–1.0 electrons per atom) than the single-atom values. A dramatic enhancement of q_{av} for Xe_{55} at $I_m = 10^{15}\,\mathrm{Wcm^{-2}}$ is exhibited, where q_{av} increases by a numerical factor of 2 relative to the single–atom value.

The inner ionization level $n_{ii}^L = q_{av}$ (per constituent with the indexsubject L representing an asymptotic long-time value) is $n_{ii}^L = n_{BSI}^L + n_{imp}^L$ being given by the sum of the BSI contribution n_{BSI}^L and the EII contribution n_{imp}^L. From the simulation data for n_{ii}^L, n_{BSI}^L and n_{imp}^L (Fig. 6.57), complete information emerges concerning the interplay between the contributions of BSI, and ignition and screening, together with the contribution of EII to inner ionization, which reveals the following features: (1) Ignition effects on n_{BSI}^L for very small Xe_n clusters ($n = 2$–13) are manifested at the lowest intensity of $I_m = 10^{15}\,\mathrm{Wcm^{-2}}$ (Fig. 6.57a), where the inner field is comparable to the laser field. This ignition effect for $n = 2$–13 is not operative at $I_m \geq 10^{16}\,\mathrm{Wcm^{-2}}$ (Figs. 6.57b–d), where the laser field overwhelms the inner field. (2) Ignition effects for large clusters ($n > 55$), manifested by the increase of q_{av} with increasing n, are exhibited in the higher intensity domain $I_m = 10^{17}$–$10^{18}\,\mathrm{Wcm^{-2}}$ (Figs. 6.57c and 6.57d). (3) Screening effects, manifested by the decrease of n_{BSI}^L with increasing n, are exhibited at $I_m = 10^{15}$–$10^{16}\,\mathrm{Wcm^{-2}}$ for $n \geq 13$ (Figs. 6.57a and 6.57b) and for $n = 459$–2171 at $I_m = 10^{17}\,\mathrm{Wcm^{-2}}$ (Fig. 6.57c). At $I_m = 10^{18}\,\mathrm{Wcm^{-2}}$ (Fig. 6.57d) screening effects are not operative. (4) The EII contribution of n_{imp}^L to $n_{ii}^L = q_{av}$ at fixed I_m increases with increasing n for large clusters in the intensity range $I_m = 10^{15}$–$10^{18}\,\mathrm{Wcm^{-2}}$ (Figs. 6.57a–d). This significant issue will be further discussed in Sect. 6.6.3. (5) The maximal ionic charge q_{max} (Figs. 6.57a–d) exceeds the average charge, i.e., $q_{max} > n_{ii}^L = q_{av}$. For $I_m = 10^{15}\,\mathrm{Wcm^{-2}}$, a flattening at $q_{max} = 8$ is ex-

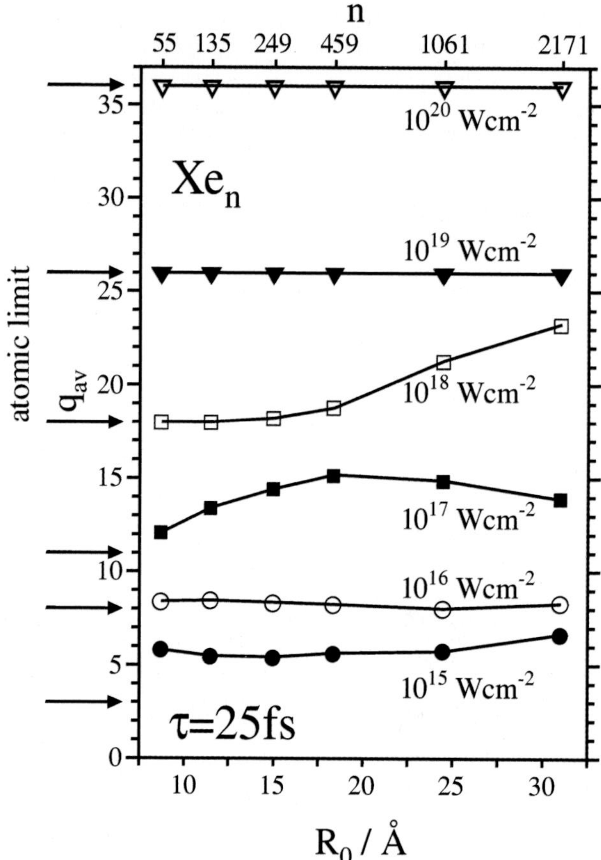

Fig. 6.56. Cluster size and laser intensity dependence of inner ionization levels (expressed by the average final charge $q_{av} = n_{ii}^L$ of the $(\mathrm{Xe}^{q+})_n$ ions) of Xe_n clusters (n = 55–2171) over the intensity range $I_m = 10^{15}$–$10^{20}\,\mathrm{Wcm}^{-2}$ (marked on the curves) with a laser pulse width of $\tau = 25$ fs. The horizontal arrows (marked atomic limit) represents the single atom ionization level calculated by the BSI model [228].

hibited over a broad size domain, followed by EII at $n = 2171$ (Fig. 6.57a), while for $I_m = 10^{17}\,\mathrm{Wcm}^{-2}$ a flattening at $q_{max} = 18$ is observed (Fig. 6.57c). (6) 'Magic numbers' in cluster multielectron ionization are observed in the cluster size domain where laser induced BSI dominates over ignition, screening and EII. The 'magic numbers' are $q_{av} = 8$ at $I_m = 10^{16}\,\mathrm{Wcm}^{-2}$ for $n = 2$–55 (Fig. 6.57b) corresponding to the ionization of the $5\mathrm{s}^2 5\mathrm{p}^6$ shells, $q_{av} = 18$ at $I_m = 10^{18}\,\mathrm{Wcm}^{-2}$ for $n = 2$–135 (Fig. 6.57d) corresponding to the ionization of the $4\mathrm{d}^{10}5\mathrm{s}^2 5\mathrm{p}^6$ shells, $q_{av} = 26$ at $I_m = 10^{19}\,\mathrm{Wcm}^{-2}$ for $n = 55$–2171 (Fig. 6.56) corresponding to the ionization of the $4\mathrm{s}^2 4\mathrm{p}^6 4\mathrm{d}^{10} 5\mathrm{s}^2 5\mathrm{p}^6$ shells, and

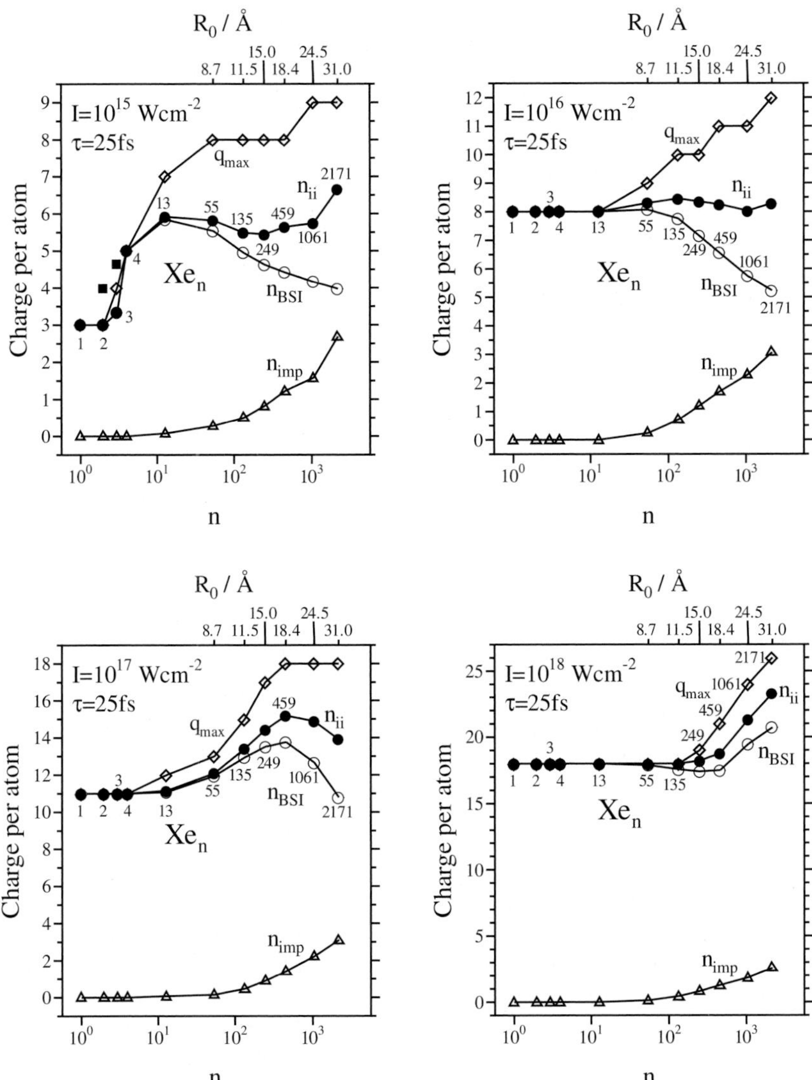

Fig. 6.57. The cluster size dependence of the long-time BSI level n^L_{BSI} (O), the EII level n^L_{imp} (Δ), and the total inner ionization level n^L_{ii} (● and ■) for the maximal ionic charge $q_{\max}$ (◇) from Xe$_n$ (n = 2–2171) clusters. (a) $I_m = 10^{15}$ Wcm^{-2} (τ = 25 fs). The $n^L_{BSI} = n^L_{ii}$ data for small (n = 2 and 3) clusters at a fixed nuclear configuration are dependent on the direction of the laser field, with (■) for the laser field being parallel to the molecular axis and (●) for the laser field being perpendicular to the molecular axis. (b) $I_m = 10^{16}$ Wcm^{-2} (τ = 25 fs). (c) $I_m = 10^{17}$ Wcm^{-2} (τ = 25 fs). (d) $I_m = 10^{18}$ Wcm^{-2} (τ = 25 fs).

$q_{av} = 36$ at $I_m = 10^{20}\,\mathrm{Wcm}^{-2}$ for $n = 55$–2171 (Fig. 6.56) corresponding to the ionization of the $3d^{10}4s^24p^64d^{10}5s^25p^6$ shells.

The cluster size, laser intensity, and laser pulse shape dependence of inner ionization levels of Xe_n clusters are induced by a complex superposition of laser-induced BSI, inner field ignition effects and nanoplasma screening effects, as well as by the contribution of EII. The inner field ignition and screening effects, in conjunction with EII, constitute collective effects, which preclude the description of cluster inner ionization in terms of an additive contribution of the constituents. The laser intensity dependent ionization levels of Xe_n clusters, with the dramatic enhancement of q_{av} and q_{max} with increasing I_m, originate from the laser field contribution to the BSI. This enhancement is characteristic for partial multielectron inner ionization of heavy atoms, with no production of nuclei in the currently available laser intensity domain ($I_m \leq 10^{21}\,\mathrm{Wcm}^{-2}$). On the other hand, for $(H_2)_n$ and $(D_2)_n$ clusters, complete ionization can be achieved at $I_m \simeq 2 \times 10^{14}\,\mathrm{Wcm}^{-2}$ [221, 243], which is below the lowest limits of the intensity range used herein. For first-row molecular heteroclusters consisting of light atoms, e.g., $(CA_4)_n$, $(A_2O)_n$ (A = H, D), complete multielectron ionization, with the production of H^+, D^+, C^{6+} and O^{8+} nuclei, can be realized in the currently available laser intensity domain.

Table 6.1. Multielectron ionization of Xe_n clusters

EXPERIMENT						THEORY				
	n (R_0)	I_m Wcm^{-2}	τ fs	q_{av}	$q_{\max}$	n (R_0)	I_m Wcm^{-2}	τ fs	q_{av}	$q_{\max}$
(a) [280]	2×10^6 (270 Å)	5×10^{17}			25	2171 (31 Å)	10^{18}	25	23	25
(b) [281]	5×10^4 (79 Å)	2×10^{17}	20	12	18	2171 (31 Å)	10^{17}	25	14	18
(c) [267]	10^5–10^6 (100 Å–210 Å)	10^{18}			26–30	2171 (31 Å)	10^{18}	25	23	26
(d) [222]	1.2×10^4 (49 Å)	10^{15}	100		11	2171 (31 Å)	10^{15}	100		10

In Table 6.1 the ionization levels are compared for the largest Xe_n ($n = 2171$) cluster sizes calculated by us with the available experimental results [222, 267, 280, 281]. The experimental values of q_{av} and q_{max} at $I_m = 2 \times 10^{17}\,\mathrm{Wcm}^{-2}$ [data set (b)] [281] and at $I_m = 10^{15}\,\mathrm{Wcm}^{-2}$ [data set (d)] [222] are in good agreement with experiment. For the experimental results of $I_m = 5 \times 10^{17}\,\mathrm{Wcm}^{-2}$ [data set (a)] [280] and $I_m = 10^{18}\,\mathrm{Wcm}^{-2}$ [data set (c)] [267], the experimental cluster radius $R_0 = r_0 n^{1/3}$ ($r_0 = 2.16$ Å being the constituent radius) is considerably larger than the corresponding

R_0 values used for the simulations. However, the relevant cluster size domain at this high intensity of $I_m = 10^{18}\,\mathrm{Wcm}^{-2}$ is determined by the border radius [245, 246, 279] $R_0^{(I)}$ for the complete sweeping of the nanoplasma from the cluster and for cluster vertical ionization, which prevails for $R_0 \leq R_0^{(I)}$ (see Sect. 6.6.4). For $R_0 > R_0^{(I)}$ the inner ionization levels are weakly cluster size dependent. For Xe_n, in the intensity range of $I_m = 10^{18}\,\mathrm{Wcm}^{-2}$, $R_0^{(I)} = 35 \pm 5\,\text{Å}$ [279]. Thus the experimental data (a) and (c) correspond to $R_0 \gg R_0^{(I)}$, while the simulation data correspond to $R_0 \leq R_0^{(I)}$, allowing for an approximate comparison between theory and experiment. The computational results reported in this section account for the gross features of the cluster ionization levels.

6.6.3 The nanoplasma

The nanoplasma is formed from the unbound electrons, which are confined to the cluster and to its vicinity, and from the ion cluster. The life story of the nanoplasma is portrayed in Figs. 6.54 and 6.55 for Xe_{2171} clusters and in Fig. 6.58 for $(\mathrm{D}_2)_{2171}$ and $(\mathrm{HT})_{2171}$ clusters. These snapshots portray the formation of the nanoplasma, its response to the laser field, followed by its complete depletion at $I_m = 10^{18}\,\mathrm{Wcm}^{-2}$ (Figs. 6.55 and 6.58) or its partial depletion at $I_m = 10^{15}\,\mathrm{Wcm}^{-2}$ (Fig. 6.54). The time-dependent nanoplasma population is characterized by the number $n_p(t)$ of nanoplasma electrons (per atomic constituent), which is given by $n_p(t) = n_{ii}(t) - n_{oi}(t)$. The number $n_{ii}(t)$ ($n_{oi}(t)$) of depleted electrons (per constituent) for inner (outer) ionization exhibits a gradual increase and long-time saturation [244, 279]. At long times, after the termination of the laser pulse, the nanoplasma population is finite at lower intensities of $I_m = 10^{15}$–$10^{16}\,\mathrm{Wcm}^{-2}$, exhibiting only partial depletion (Fig. 6.59), with the long-time population n_p^L manifesting a marked increase with increasing the pulse length. The cluster size and laser parameters in Fig. 6.59 provide the conditions for effective reactive dynamics of the nanoplasma [244, 279].

The electron dynamics of the nanoplasma (Figs. 6.54, 6.55 and 6.58–6.60) reveals the following features:

1. Composition of the nanoplasma. It consists of the electron cloud and of the positive Xe^{q+} ions produced by inner ionization. The nanoplasma responds to the laser field, which strips electrons from the cluster by outer ionization. Accordingly, the nanoplasma is positively charged.
2. Formation time. The time scales for the near completion of inner ionization decrease with increasing I_m, assuming the approximate values of $t - t_s \sim 35\,\mathrm{fs}$ at $I_m = 10^{15}\,\mathrm{Wcm}^{-2}$, down to $t - t_s \sim 15\,\mathrm{fs}$ for $I_m = 10^{19}\,\mathrm{Wcm}^{-2}$ for Xe_{2171}.
3. Electron energies. The electron cloud is characterized by high (average) energies $\mathcal{E}$. For Xe_{2171} we found $\mathcal{E} = 53\,\mathrm{eV}$, $150\,\mathrm{eV}$, $930\,\mathrm{eV}$, $72\,\mathrm{keV}$ and $100\,\mathrm{keV}$ at $I_m = 10^{15}, 10^{16}, 10^{17}, 10^{18}$, and $10^{19}\,\mathrm{Wcm}^{-2}$, respectively. The

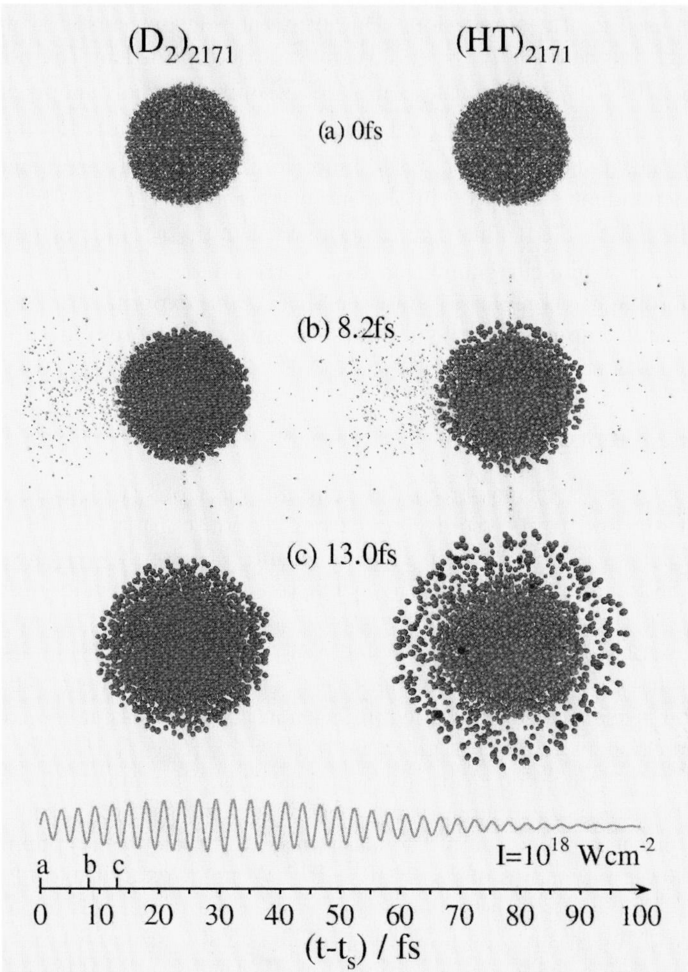

Fig. 6.58. Snapshots of the time-resolved structures of $(D_2)_{2171}$ and $(HT)_{2171}$ clusters in a Gaussian laser field ($I_m = 10^{18}$ Wcm^{-2} at $\tau = 25$ fs marked on the images), at three different times $t - t_s$. The lowest part of the panel portrays the time axis and the electric field of the laser. The instants of the snapshots are marked on the time axis by a, b and c. H atoms are represented in blue, T atoms in red, and electrons in light gray. (a) The initial nanoplasma at $t - t_s = 0$. (b) At $t - t_s = 8.2$ fs, the beginning of spatial expansion of the clusters is manifested. In case of the $(HT)_n$ cluster, a shell of H^+ ions is displayed. At this time, a large number of electrons is stripped by outer ionization, which occurs repeatedly when the electric field of the laser is close to a maximum. (c) At $t - t_s = 13.2$ fs, the spatial expansion and shell formation of the HT cluster is pronounced. Also, all nanoplasma electrons have been removed at this time.

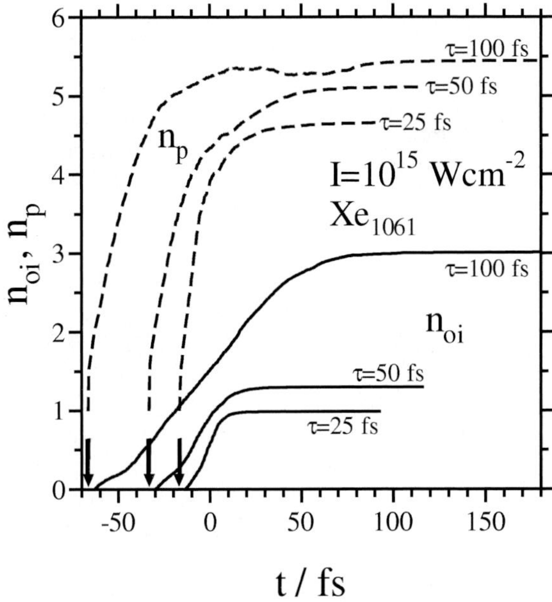

Fig. 6.59. The time dependence of the nanoplasma population $n_p(t) = n_{ii}(t) - n_o(t)$ (dashed lines) and the outer ionization levels n_{oi} (solid lines) for Xe_n clusters ($n = 1061$) at $I_m = 10^{15}\,\mathrm{Wcm^{-2}}$. The laser pulse lengths are $\tau = 25\,\mathrm{fs}$, $50\,\mathrm{fs}$, and $100\,\mathrm{fs}$, as marked on the curves. The vertical arrows represent the times for the onset of the pulse ($t = t_s$).

electron energies increase by 4 orders of magnitude with increasing I_m in this intensity domain.

4. Spatial inhomogeneity and angular anisotropy of the nanoplasma. For $I_m = 10^{15}\,\mathrm{Wcm^{-2}}$ the electron cloud is nearly spatially isotropic, with the majority of the electrons being located within the cluster. For higher intensities of $I_m = 10^{18}\,\mathrm{Wcm^{-2}}$ and $I_m = 10^{19}\,\mathrm{Wcm^{-2}}$, the electron angular distribution is spatially anisotropic, assuming a "sausage type" shape along the laser electric field direction.
5. Attosecond response of the nanoplasma. At intensities of $I_m = 10^{18}$–$10^{19}\,\mathrm{Wcm^{-2}}$ the "sausage type" shape of the electron cloud oscillates along the electric field direction on the time scale ν^{-1} of the laser period, manifesting ultrafast electron dynamics.
6. The outer ionization of the nanoplasma can be either partial (at intensities of 10^{15}–$10^{16}\,\mathrm{Wcm^{-2}}$) or complete (at highest intensities of 10^{18}–$10^{20}\,\mathrm{Wcm^{-2}}$).
7. Persistent and transient nanoplasmas. At intensities of $I_m = 10^{15}$–$10^{16}\,\mathrm{Wcm^{-2}}$, where outer ionization is partial, a persistent nanoplasma on the time scale of $t - t_s > 100\,\mathrm{fs}$ exists, while for higher intensities of

$I_m = 10^{18}$–$10^{19}\,\mathrm{Wcm}^{-2}$, a transient nanoplasma is formed, being completely depleted on the time scale of $t - t_s \simeq 15$–$25\,\mathrm{fs}$. The persistent nanoplasma exists over the cluster size domain and intensity range for which $R_0 > R_0^{(I)}$, while the transient nanoplasma prevails for $R_0 < R_0^{(I)}$.

8. The 'metallic' nanoplasma. The average time-dependent electron density in the nanoplasma is

$$\rho_e(t) = n\, n_p(t)/(4\pi/3)R(t)^3 \tag{6.51}$$

where $R(t)$ is the cluster radius, with $R(0) = R_0$ at the onset of the pulse and $R(t) > R_0$ at longer times, due to CE. For the entire I_m range $\rho_e(t)$ first increases with increasing t, reaching a maximum of $\rho_e^{MAX} = 0.08$–$0.09\,\text{Å}^{-3}$ (Fig. 6.60). The weak dependence of ρ_e^{MAX} on I_m (Fig. 6.60) can be traced to the weak intensity dependence of the maximal value of $n_p(t)$, both for the persistent and for the transient nanoplasma. At $I_m = 10^{18}\,\mathrm{Wcm}^{-2}$ $\rho_e(t)$ vanishes for $t \geq 0$, as appropriate for a transient nanoplasma. For the lowest intensity of $I_m = 10^{15}\,\mathrm{Wcm}^{-2}$, $\rho_e(t)$ decreases gradually with increasing t on the time scale of 10–90 fs, due to Coulomb explosion, retaining a long-time ($t_L = 90\,\mathrm{fs}$) electron density of $\rho_e^L = 0.02\,\text{Å}^{-3}$ for Xe_{2171}. The nanoplasma electron densities at the maximum, i.e., $\rho_e^{MAX} = 8 \times 10^{22}$–$9 \times 10^{22}\,\mathrm{cm}^{-3}$ for the entire I_m domain, and the long–time electron density of $\rho_e^L = 2 \times 10^{22}\,\mathrm{cm}^{-3}$ at $I_m = 10^{15}\,\mathrm{Wcm}^{-2}$, are comparable to electron densities in metals.
It is instructive to establish contact between the microscopic nanoplasma model used herein and a macroscopic 'plasma model' for the nanoplasma response and outer ionization, considering the enhancement of light absorption by resonance effects. The frequency of the linear oscillations for a thermally equilibrated and uniform nanoplasma is [203, 290]

$$\omega_p = (4\pi e^2 \rho_e / 3m_e)^{1/2} \tag{6.52}$$

The maximal (nearly intensity independent) electron density in the nanoplasma is $\rho_e^{MAX} = 0.08$–$0.09\,\text{Å}^{-3}$ (at $I_m = 10^{15}$–$10^{19}\,\mathrm{Wcm}^{-2}$, $\tau = 25\,\mathrm{fs}$, Fig. 6.60). The nanoplasma energy is $\hbar\omega_p = 6.1$–$6.4\,\mathrm{eV}$. This value of $\hbar\omega_p$ is considerably larger than the photon energy of 1.44 eV. The simulation results for the persistent nanoplasma at $I_m = 10^{15}$–$10^{16}\,\mathrm{Wcm}^{-2}$ over the time scale of 100 fs (Fig. 6.59) do not reveal any steep temporal decrease of $n_p(t)$ or an increase of the electron energy, which could be interpreted as resonance generation of nanoplasma oscillations, precluding the possibility of such excitations. The role of the macroscopic 'plasma model' [212, 233] is not borne out by the simulations.

9. Attosecond oscillations of the nanoplasma population. In the intensity range $I_m = 10^{17}$–$10^{18}\,\mathrm{Wcm}^{-2}$ $\rho_e(t)$ exhibits an oscillatory time dependence during the temporal rise of the inner/outer ionization levels (Fig. 6.60). The period of these temporal oscillations is close to the laser period ν^{-1}, manifesting the attosecond response and driving of outer ionization by the ultraintense laser field.

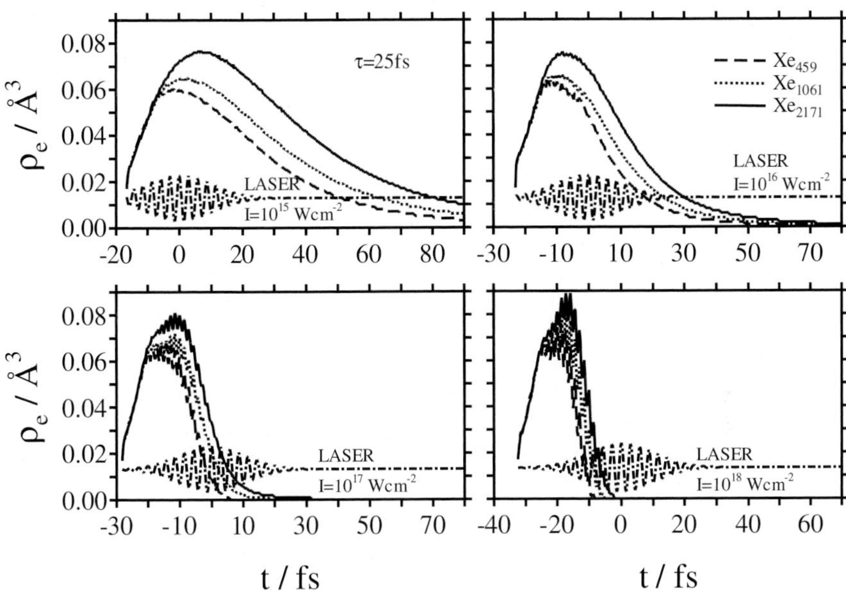

Fig. 6.60. The time dependent electron density in the nanoplasma $\rho_e^{(t)} = n_p^{(t)}/(4\pi/3)R^3$ in Xe_n clusters ($n = 459$, 1061 and 2171, as marked on the curves). R is the cluster radius obtained from molecular dynamics simulations of CE. Data are presented for $I_m = 10^{15}$, 10^{16}, 10^{17}, and 10^{18} Wcm^{-2}, as marked on the panels. The Gaussian laser fields $(- \cdot - \cdot -)$, expressed in arbitrary units for $t \geq t_s$, are presented on the panels.

The reactive dynamics of the nanoplasma is manifested by EII, which is important for clusters of heavy multielectron atoms or molecules, e.g., Xe_n, where the corresponding cross sections are large [278,279,284–288]. The cluster size dependence of n_{imp}^L (at fixed I_m) reveals an increase with increasing n, with the largest EII yields being exhibited at $I_m = 10^{15}$–10^{16} Wcm^{-2}, where the persistent nanoplasma prevails (Fig. 6.61). Significantly, in the persistent nanoplasma domain (at $I_m = 10^{15}$ –10^{16} Wcm^{-2}), the EII yields and the total ionization levels manifest a marked increase with increasing the laser pulse width. As is evident from the data for Xe_{2171} at $I_m = 10^{15}$ Wcm^{-2} and $\tau =$ 10–100 fs (Fig. 6.62), the BSI yield exhibits a weak pulse length dependence, while the EII yield increases from $n_{imp}^L = 1.7$ at $\tau = 10$ fs to $n_{imp}^L = 5.2$ at $\tau = 100$ fs. This results in a marked increase in n_{ii}^L, with the EII becoming the dominant ionization mechanism at $\tau = 100$ fs (Fig. 6.62). Another interesting effect pertains to 'laser free' EII by the persistent nanoplasma (with a modest yield of 10%), which was documented. On the other hand, at higher intensities ($I_m > 10^{17}$ Wcm^{-2}), where the nanoplasma is transient and the cross sections

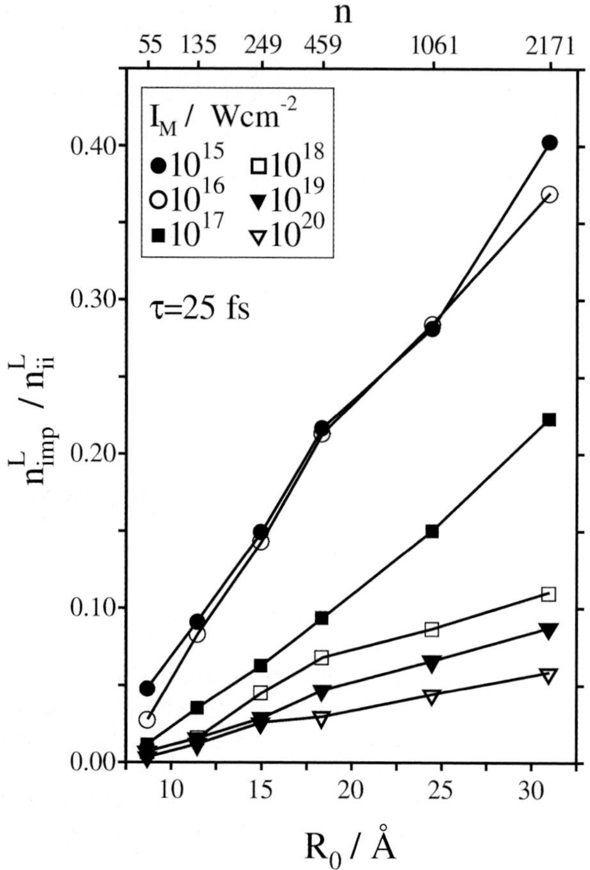

Fig. 6.61. The cluster size dependence of the relative EII yield n^L_{imp}/n^L_{ii} for Xe_n ($n = 55$–2171) clusters in the intensity range $I_m = 10^{15}$–10^{20} Wcm^{-2} ($\tau = 25$ fs).

for EII are reduced (due to the increase in the electron energy), EII competes with BSI, but does not lead to a net effect on the inner ionization levels.

6.6.4 Outer ionization

Cluster outer ionization manifests the nanoplasma response to the laser field, due to barrier suppression of the entire cluster [243] and due to quasiresonance effects [214, 239, 248]. The outer ionization removes all, or part, of the nanoplasma electrons by the laser field. In the simulations outer ionization is described in terms of a cluster barrier suppression ionization (CBSI) model, which involves the balancing between the cluster exterior Coulomb potential and the laser field potential at the cluster boundary. The long time outer

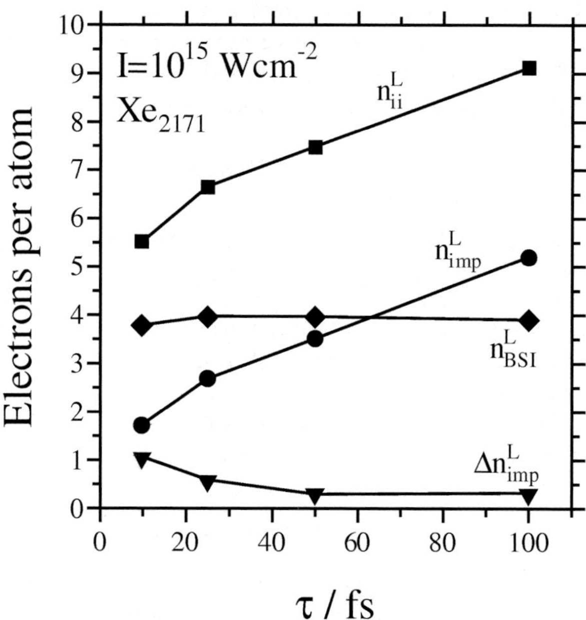

Fig. 6.62. The laser pulse length dependence of the long-time ionization levels of Xe_{2171} coupled to a laser field at $I_m = 10^{15}\,\mathrm{Wcm}^{-2}$ ($\tau = 10\,\mathrm{fs}$–$100\,\mathrm{fs}$). Ionization levels are presented for BSI (n^L_{BSI}, ♦), for EII (n^L_{imp}, ●), for inner ionization (n^L_{ii}, ■), and for "laser free" EII (Δn_{imp}, ▼) in the time domain after the termination of the laser pulse. The marked increase of the inner ionization yield with increasing τ marks control by laser pulse shaping.

ionization level was expressed in the form [279]

$$n^L_{oi} = F_m \gamma \xi^2 / \left(\frac{4\pi\sqrt{2}}{3} \right) \overline{B} \rho_{mol} R_0 \tag{6.53}$$

where F_m is the laser electric field for peak intensity I_m, $\rho_{mol} = 3n/4\pi R_0^3$ is the initial atomic / molecular constituent density expressed in terms of the initial cluster radius, $\xi = R(t)/R_0$ is the cluster expansion parameter due to CE and $\gamma \simeq 4$ is a numerical correction factor. This result is applicable for the intensity range and cluster size domain where a persistent nanoplasma prevails, i.e., $n^L_p = n^L_{ii} - n^L_{oi} > 0$ (or $n^L_{ii} > n^L_{oi}$). The use of (6.53) gives [279]

$$n^L_{oi} = A\sqrt{I_m}/R_0 \tag{6.54}$$

where

$$A = 2.745 \times 10^{-7} \gamma \xi^2 / \left(\frac{4\pi\sqrt{2}}{3} \right) \overline{B} \rho_{mol} \tag{6.55}$$

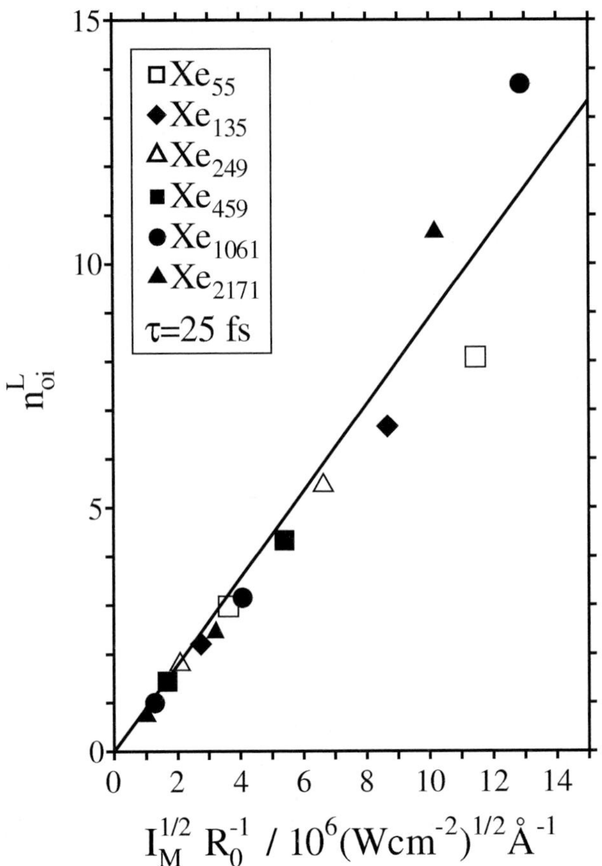

Fig. 6.63. A test of the electrostatic model for outer ionization, which predicts a linear dependence of n^L_{oi} vs $\sqrt{I_m}/R_0$ over broad cluster size ranges and laser intensities at constant τ, when the persistent nanoplasma prevails in Xe_n clusters. $\tau = 25$ fs, $n = 55$–2171 at $I_m = 10^{15}$–10^{16} Wcm^{-2} and $n = 1061$, 2171 at $I_m = 10^{17}$ Wcm^{-2}.

In (6.54) and (6.55) I_m is presented in Wcm^{-2}, R_0 in Å, ρ_A in Å^{-3} and $\overline{B} = 14.4$ eV. The most striking prediction of the CBSI model, which is confirmed by the simulation data, is the linear dependence of n^L_{oi} on $\sqrt{I_m}/R_0^{(I)}$ over a broad laser intensity and cluster size range (at fixed τ), where the persistent nanoplasma exists. A typical example is portrayed in Fig. 6.63 for $\tau = 25$ fs with $n = 55$–2171, and $I_m = 10^{15}$–10^{17} Wcm^{-2}. From the dependence of the compound parameter $\gamma^{1/2}\xi$ on τ it was inferred that the outer ionization level for Xe_n clusters is [279]

$$n^L_{oi} = 1.06 \times 10^{-7} \tau^{0.64} \sqrt{I_m}/R_0 \tag{6.56}$$

where R_0 is given in Å, I_m in Wcm^{-2} and τ in fs. An identical function of the form $n_{oi}^L \propto \tau^{0.62}\sqrt{I_m}/R_0$ was obtained for outer ionization of $(D_2)_n$ clusters pointing toward the generality of the electrostatic CBSI model. It is also gratifying that the relation between the electron outer ionization levels and the expansion parameter, $\xi = R(t)/R_0$, provides information on nuclear CE dynamics.

Complete outer ionization of a cluster can be specified by the cluster border radius $R_0^{(I)}$ at the intensity I_m, and prevails for $R_0 < R_0^{(I)}$, where cluster vertical ionization (CVI) is attained. The border radius for molecular clusters can be obtained from three independent sources.

1. The electrostatic model. Using (6.53) it was inferred that the maximal cluster radius $R_0^{(I)}$ for the attainment of the conditions $n_p^L = 0$ and $n_{oi}^L = n_{ii}^L = q_{av}$ is given by [279]

$$R_0^{(I)} = |F_m|\,\gamma\xi^2 \Big/ \left(\frac{4\pi\sqrt{2}}{3}\right) \overline{B}\,\rho_{mol}\,q_{av} \tag{6.57}$$

   Using (6.55) one gets

$$R_0^{(I)} = \frac{A\sqrt{I_m}}{q_{av}}\,. \tag{6.58}$$

2. Electron dynamics. $R_0^{(I)}$ can be inferred for the cluster size that exhibits complete (taken as 95%) outer ionization at the peak of the laser pulse, as obtained from molecular dynamics simulations [245, 246, 279].
3. The border radius $R_0^{(I)}$ is central in the characterization of the nuclear dynamics and energetics of CE. In the cluster size domain and in the laser intensity range where $R_0 \le R_0^{(I)}$, the CVI is applicable, with the energetics (e.g., the average ion energy E_{av}) of CE being characterized by the cluster size scaling equation $E_{av} \propto q_{av}^2 R_0^2$, being explicitly independent of I_m and of other laser parameters. $R_0^{(I)}$ can be estimated from the deviation of the energetics of CE from the scaling equation [245, 246, 279].

The border radii $R_0^{(I)}$ for electron and nuclear dynamics in Xe_n, $(CD_4)_n$ and $(D_2)_n$ molecular clusters (Table 6.2) exhibit good agreement (within $\sim 20\%$) between the electrostatic model and the simulations of electron dynamics. From this agreement it can be inferred, on the basis of (6.54), (6.57) and (6.58), that for electron and nuclear explosion dynamics in molecular clusters [279]

$$R_0^{(I)} \propto \frac{\sqrt{I_m}\xi^2}{q_{mol}\,\rho_{mol}} \tag{6.59}$$

where q_{mol} is the average charge per molecule (or atomic constituent) and ρ_{mol} is the initial molecular density. The dependence $R_0^{(I)} \propto \sqrt{I_m}/q_{av}$ implies that the border radii for $(D_2)_n$ clusters are considerably larger than for $(CD_4)_n$ and Xe_n clusters at the same intensity. For $(D_2)_n$ clusters, over the entire

Table 6.2. Border radii $R_0^{(I)}$(Å) for electron dynamics and nuclear CE dynamics of multicharged molecular clusters.

I_M / Wcm^{-2}	Xe_n [279]				$(\mathrm{CD}_4)_n$ [246]			$(\mathrm{D}_2)_n$ [245]			
	q_{av}	$R_0^{(I)}$ / Å			q_{av}	$R_0^{(I)}$ / Å		q_{av}	$R_0^{(I)}$ / Å		
$\tau = 25$fs		(1)	(2)	(3)		(1)	(2)		(1)	(2)	(3)
10^{15}	6	5.6						1	6.5	5.8	<8
10^{16}	8	11.0	8.0		8	10.0	10.1	1	20.4	20.4	≈ 25
10^{17}	15	18.5	15.2	17.3	8	31.4	31.4	1	64.5	76.5	
10^{18}	23	38.0	32.4		8	99.5		1	204		
10^{19}	26	107			10	251		1	640		
10^{20}	36	244						1			

(1) Electrostatic model.
(2) Simulations of electron dynamics.
(3) Simulations of CE nuclear dynamics.

intensity range, and for $(\mathrm{CD}_4)_n$ clusters at $I_m = 10^{15}$–$10^{18}\,\mathrm{Wcm}^{-2}$, q_{mol} is constant (but distinct) for each cluster resulting in a $R_0^{(I)} \propto \sqrt{I_m}$ intensity dependence. For Xe_n the increase of $R_0^{(I)}$ with increasing $\sqrt{I_m}$ is sublinear, reflecting on the increase of $q_{mol} = q_{av}$ with increasing I_m. The scarce data for CE nuclear dynamics are in good agreement with the results for electron dynamics as well. The border radius $R_0^{(I)}$ builds a bridge between electron (outer ionization) dynamics and nuclear (CE) dynamics, which will now be considered.

6.6.5 Uniformity, energetics, kinematics, and dynamics of Coulomb explosion

Cluster electron dynamics triggers nuclear dynamics, with the outer ionization being accompanied and followed by CE [241, 242, 245, 246, 254, 271]. The multicharged (totally or partially ionized) metastable clusters undergo CE whose notable applications pertain to ion imaging [291], accelerator technology [292] of high-energy (keV–MeV) ions or nuclei, and extreme ultraviolet lithography [293].

The traditional view of CE under CVI conditions involves uniform ion expansion. This prevails for homonuclear clusters with an initially uniform, constant charge and spherically symmetric ion distributions, which retain the succession of the ion distances from the cluster center throughout the expansion [245, 246]. Time-dependent structures of the $(\mathrm{D}_2)_{2171}$ cluster at $I_m = 10^{18}\,\mathrm{Wcm}^{-2}$ (Fig. 6.58) manifest uniform CE, which corresponds to complete stripping of all the electrons from $(\mathrm{D}_2)_{2171}$. The CE exhibits a unimodal spatial expansion of the D^+ ions, as evident from the time-resolved

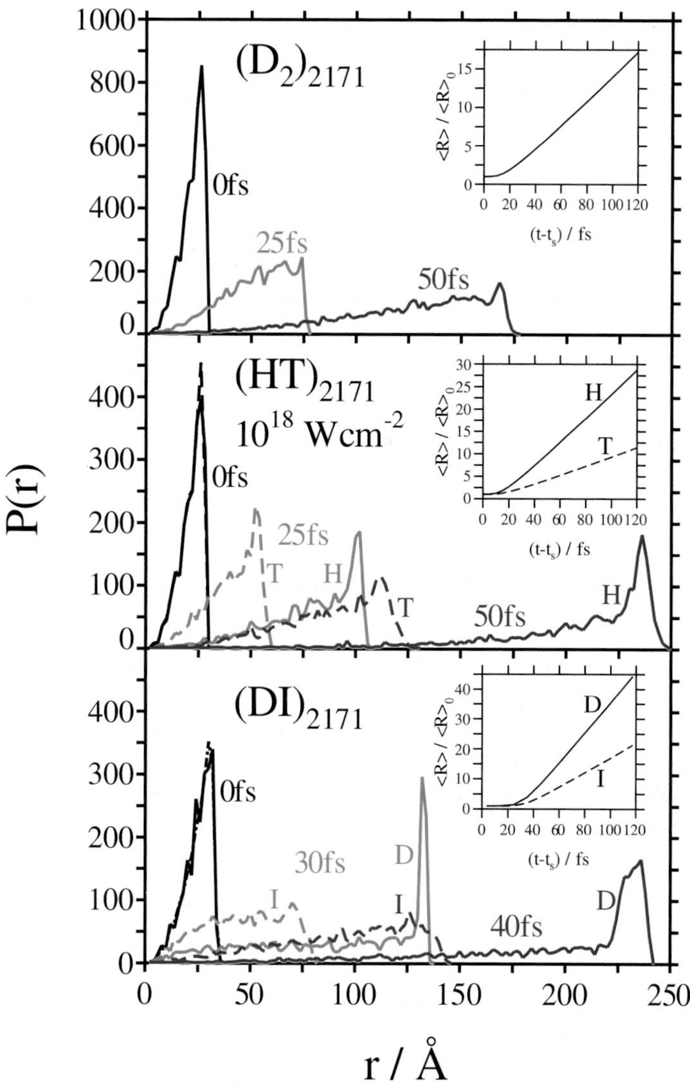

Fig. 6.64. The radial distribution functions $P(r)$ for the $(D_2)_{2171}$, $(HT)_{2171}$ and $(DI)_{2171}$ clusters in a Gaussian laser field ($I_m = 10^{18}\,\mathrm{Wcm}^{-2}$ and $\tau = 25\,\mathrm{fs}$) at various times $t - t_s$. t_s is the starting time of the simulation with respect to the maximum of the Gaussian laser field envelope located at $t = 0$. For the heteronuclear $(HT)_{2171}$ and $(DI)_{2171}$ clusters, $P(r)$ is drawn separately for each ion (in blue and red,as marked on the panels), exhibiting shell formations of the non overlapping distributions of different isotope/element ions at times $t - t_s > 0$. These shells expand with different velocities. The insets portray the time-dependent increase of the first moments $\langle R \rangle$ of $P(r)$ relative to the first moment $\langle R \rangle_0$ at $t - t_s = 0$. Data are presented for different ions, as marked on the insets.

structures (Fig. 6.58) and from the single, broad time-dependent spatial distributions $P(r)$ at each t (Fig. 6.64). Even for the CE of ions from highly multicharged elemental clusters, e.g., $(Xe^{+q})_n$ (Figs. 6.54 and 6.55), the explosion is nonuniform. At lower intensities of $I_m = 10^{15}$ Wcm^{-2} outer ionization is incomplete, and the screening of CE by the persistent nanoplasma in the center of the cluster results in nonuniformity in the exterior explosion (Fig. 6.54). At $I_m = 10^{18}$ Wcm^{-2}, where outer ionization is complete, a spatial anisotropy in the angular distribution of the Xe^{q+} ($q = 22$–26) ions is exhibited (Fig. 6.55), with a velocity increase along the polarization axes of the laser field.

The dynamics of uniform CE (inset to $(D_2)_{2171}$ in Fig. 6.64) will be characterized by the time dependence of $\langle R\rangle/\langle R\rangle_0$, where $\langle R\rangle$ is the first moment of the spatial distribution of the A^{q_A+} light ions at time t, while $\langle R\rangle_0$ is the initial value of $\langle R\rangle$. The CE dynamics is described by the near-linear time dependence [245, 246, 294]

$$\frac{\langle R\rangle}{\langle R\rangle_0} = a\,(t - t_{onset}) \tag{6.60}$$

at $t > t_{onset}$ (inset for $(D_2)_{2171}$ in Fig. 6.64). The onset time t_{onset} in the linear dependence of $\langle R\rangle$ vs t (e.g., $t_{onset} = 10$ fs for $(D_2)_{2171}$) is due to the switching-off of acceleration effects in CE. From the electrostatic model of uniform CE under CVI conditions [245, 246, 294] $(a/\mathrm{fs})^{-1} = 1.074(\rho_{mol}\, q_A/m_A)^{1/2}$ where q_A is the ion charge, $q_{mol} = kq_A$, and m_A is ion mass. For the uniform CE of $(D_2)_n$ clusters, the results of the electrostatic model (6.60), are in good agreement with the fit of the simulation data by (6.59).

Nonuniform CE under CVI conditions occurs in $A_k^{q_A+}B_l^{q_B+}$ light-heavy heteroclusters, which consist of k light A^{q_A+} ions of mass m_A and charge q_A, and l heavy B^{q_B+} ions of mass m_B and charge q_B (with $m_A < m_B$). The CE dynamics is governed by the kinematic parameter $\eta_{AB} = q_A m_B/q_B m_A$ [241, 242, 245, 246]. The nonuniform CE of $(H^+T^+)_{2171}$ heteroclusters (Fig. 6.58), for which $\eta_{HT} = 3$, manifests kinematic run-over effects of the H^+ ions relative to the T^+ ion. These kinematic effects are characterized by a spatial segregation of the exterior distribution of H^+ ions relative to an interior distribution of the T^+ ions (Fig. 6.58). The distinct spatial distributions of the H^+ and T^+ ions overlap at short times and separate at longer times (Fig. 6.64). The case of CE of extremely charged light-heavy $(A_k^{q_A+}B_l^{q_B+})_n$ heteroclusters (ECLHH) (corresponding to $m_A \ll m_B$ and $kq_A \ll lq_B$) exhibits the formation of exterior spherical nanoshells of the light ions, which manifest the attainment of transient self-organization driven by repulsive Coulomb interactions [250]. The time-dependent structures (Fig. 6.65) of CE of $(DI)_{2171}$ clusters at $I_m = 10^{18}$ Wcm^{-2} correspond to $\eta_{DI} = 2.5$ and $q_I = 21$–23, with $q_D = 1 \ll q_I$, where q_I increases with increasing n due to ignition effects induced by the inner field, as is the case for extreme multielectron ionization. The fs CE dynamics reveals an extreme case of spatial segregation between the light D^+ ions and the heavy I^{q_I+} ions, with the formation of a transient halo of the expanding light D^+ ions, which surrounds the inner subcluster of

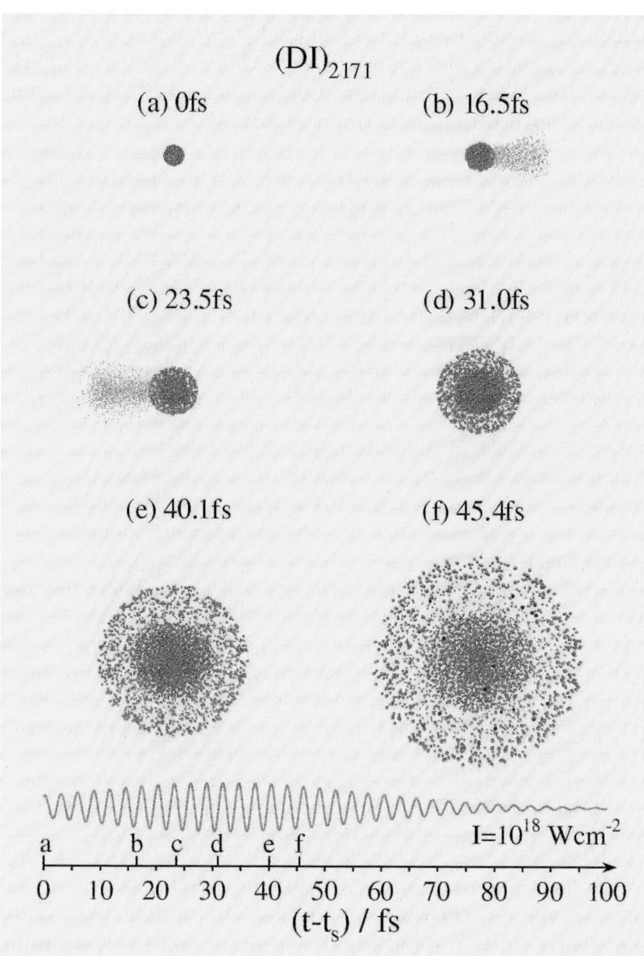

Fig. 6.65. Snapshots of the time-resolved structures in the Coulomb explosion of the $(DI)_{2171}$ cluster induced by a Gaussian laser pulse ($I_m = 10^{18}$ Wcm^{-2}, $\tau = 25$ fs). The lowest part of the panel portrays the time axis $t - t_s$ and the electric field of the laser. The instants of the snapshots are marked by a–f on the time axis. The deuterons are represented by green and the electrons by light gray spheres. The iodine atoms are color coded according to their charge; blue corresponds to the initial charge +1, deep red to the maximum charge +25, which can be obtained at this laser intensity. (a) The initial nanoplasma at $t - t_s = 0$, with the cluster radius $R_0 = 34.8$ Å. (b) At $t - t_s = 16.9$ fs the onset of the deuteron shell expansion becomes apparent. The radius of the deuteron shell, defined by the outermost D atoms, is 39.1 Å. The average charge per iodine atom is 16.0. (c) At 21.5 fs the average charge per iodine atom reaches 24.3 and the radius of the deuteron shell is 51 Å. (d) At 26.4 fs the outer ionization is complete. The radius of the deuteron shell grows to 76 Å and the expansion of the iodine shell sets in. At even longer times, (e) 31.8 fs and (f) 37.2 fs, the deuteron shell radius is 111 Å and 126 Å, respectively.

Table 6.3. Energetics and dynamics of Coulomb explosion of deuterium containing homo-nuclear $(D_2)_n$ clusters and heteronuclear $(A_k^{q_A+}B_l^{q_B+})_n$ or $(A_k^{q_A+}B_l^{q_B+}C_p^{q_C+})_n$ clusters at $I_m = 10^{18}\,\mathrm{Wcm}^{-2}$. Simulation data (marked SIM) are compared with the results of the electrostatic model (marked EML).

Cluster	ρ_{mol} (Å^{-3}) [a]	q_{mol} or q_B [c]	Z (eV) [d] SIM [e]	Z (eV) [d] EML [f]	κ [d] SIM [e]	κ [d] EML [f]	a (fs^{-1}) SIM [g]	a (fs^{-1}) EML [f]
$(D_2)_n$	0.025	2	12.5	13.6	0.61	0.60	0.16	0.17
$(CD_4)_n$	0.016	8	42.5	46.7	0.70	0.60		0.27
$(DI)_n$	0.013	22	115	165	0.80	0.83	0.50	0.45
$(CD_3I)_n$	0.010	26	130	181	0.80	0.83		

[a] Initial molecular density of molecular ions in the cluster.
[b] The cluster initial radius is related to n by $R_0 = (3n/4\pi\rho_{mol})^{1/3}$.
[c] Ion charge $q_{mol} = kq_A + lq_B + pq_C$ for cases (A) and (B), and $q_{mol} = lq_B + pq_C$ for ECLHH, where $q_I = 22$ is an average charge in the size domain $n = 1061$–2171 and $q_C = 4$ for $(CD_3I)_{2171}$.
[d] $E_M(n) = Zn^{2/3}$ and $\kappa = E_M(n)/E_{av}(n)$, with Z and κ being independent of n.
[e] Fig. 6.66.
[f] See text. For the ECLHHs we neglect a weak cluster size dependence of Z, due to the dependence of q_I on n, which arises from ignition and screening effects on inner ionization (6.57,6.62).
[g] From the time dependence of the first moment of the distribution of the light ions, $\langle R\rangle = \langle R\rangle_0 = a(t - t_{onset})$.

the I^{q_I+} ions (Fig. 6.64). The dynamics of nonuniform CE is given by (6.59) and (6.60) with $q_{mol} = kq_A + lq_B$, while for ECLHH $q_{mol} = lq_B$. Note that a is independent of the cluster size at fixed I_m. As evident from Table 6.3, the results of the electrostatic model, (6.60), account well for the simulation data (insets to Fig. 6.64). i.e., $a = 0.16\,\mathrm{fs}^{-1}$ $(0.17\,\mathrm{fs}^{-1})$ for H^+ ions from $(HT)_{2171}$, and $a = 0.56\,\mathrm{fs}^{-1}$ $(0.45\,\mathrm{fs}^{-1})$ for D^+ ions from $(DI)_{2171}$ at $I_m = 10^{18}\,\mathrm{Wcm}^{-2}$. The agreement between theory and simulation provides benchmark reference data for CE in the CVI domain. The maximization of the energies of the light ions in the CE of ECLHHs requires the applicability of the CVI ($I_m \gtrsim 10^{17}$–$10^{18}\,\mathrm{Wcm}^{-2}$), and the use of the highest attainable laser intensities for the maximization of the heavy atom charge q_B for effective energetic driving.

The maximum energy E_M and the average energy E_{av} of the light ions in the uniform CE of homonuclear clusters and in the nonuniform CE of heteronuclear clusters can be obtained from electrostatic models, which in the CVI limit result in the general expressions for the cluster size dependence [241, 242, 294]

$$E_M(n) = XR_0^2 \tag{6.61}$$

where $R_0 = (3n/4\pi\rho_{mol})^{1/3}$ is the initial cluster radius, whereupon [241, 242, 250, 251, 294]

$$E_M(n) = Zn^{2/3} \tag{6.62}$$

The ratio between E_M and E_{av} is [241, 242, 250, 251, 294]

$$E_{av}(n) = \kappa E_M(n) \tag{6.63}$$

The parameters X, Z and κ assume the following forms:

(A) For homonuclear $(A_2)_n$ clusters, $X = (4\pi/3)\overline{B}\rho_{mol}q_A q_{A_2}$ with $\overline{B} = 14.40\,\text{eV\AA}$, while $Z = (4\pi/3)^{1/3}\overline{B}q_{A_2}\rho_{mol}^{1/3}$ and $\kappa = 3/5$.
(B) For $(A_k^{q_A+}B_l^{q_B+})_n$ heteronuclear clusters with $m_A < m_B$ and $\eta_{AB} = 1$, $X = (4\pi/3)\overline{B}\rho_{mol}q_A q_{mol}$ where $q_{mol} = kq_A + lq_B$, $Z = (4\pi/3)^{1/3}\overline{B}q_A q_{mol}\rho_{mol}^{1/3}$, and $\kappa = 3/5$.
(C) For nonuniform CE of an ECLHH $(A_k^{q_A+}B_l^{q_B+}C_p^{q_C+})_n$ with $m_A < m_C \ll m_B$, $kq_A \ll lq_B$ and $q_{mol} \simeq lq_B + pq_C$, $X = 2\pi\overline{B}\rho_{mol}q_{mol}q_A$, $Z = (9\pi/2)^{1/3}\overline{B}q_A q_{mol}\rho_{mol}^{1/3}$ and $\kappa = 4/5$.

The simulation results for cluster CE at $I_m = 10^{18}\,\text{Wcm}^{-2}$ for $(D_2)_n$ (case (A)), for $(CD_4)_n$ (case (B) with $q_C = 4$) and for $(DI)_n$ and $(CD_3I)_n$ (case (C)

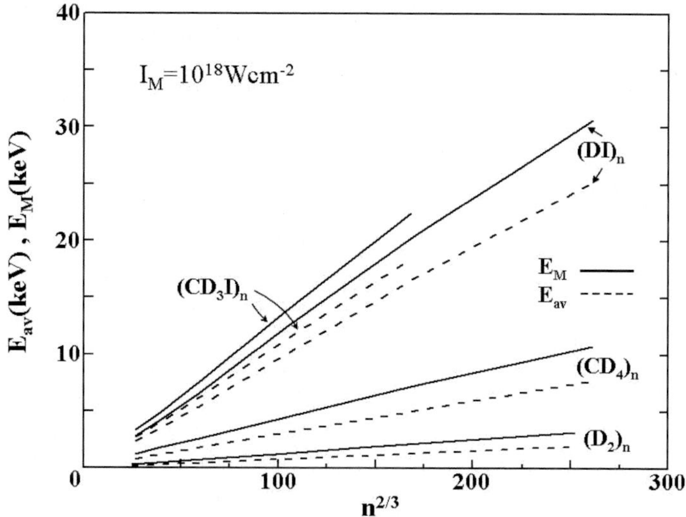

Fig. 6.66. The cluster size dependence of the maximal energies E_M (solid lines) and average energies E_{av} (dashed lines) of D^+ ions in the uniform CE of $(D_2)_n$ clusters and in the nonuniform CE of $(CD_4)_n$, $(DI)_n$ and $(CD_3I)_n$ clusters at $I_m = 10^{18}\,\text{Wcm}^{-2}$ and $\tau = 25\,\text{fs}$. The simulation data manifest the (divergent) power law $E_M, E_{av} \propto n^{2/3}$. Slight deviations from this scaling dependence for small $(DI)_n$ and $(CD_3I)_n$ ECLHHs originate from ignition and screening effects for inner ionization. The inset shows the dependence of E_M and I_m for clusters marked on the curves.

with $q_I = 21$–23), obey the size dependence E_M, $E_{av} \propto n^{2/3}$ (Fig. 6.66). The agreement between the Z parameters obtained from the simulations and the predictions of the electrostatic model (Table 6.3) is better than 30%, while the κ parameters are accounted for within 10% by the electrostatic model. The marked increase of E_M and E_{av} of D^+ in the series $(D_2)_n \ll (CD_4)_n \ll (DI)_n \ll (CD_3I)_n$ (at fixed n) exhibited in Fig. 6.66 manifests energy driving (i.e., energy boosting) by the multicharged heavy ions, which is determined by the ionic charge q_{mol} (e.g., at $I_m = 10^{18}\,\mathrm{Wcm}^{-2}$, $q_{mol} = 8$ for $(CD_4)_{2171}$, while $q_{mol} = 22$ for $(DI)_{2171}$ and $q_{mol} = 26$ for $(CD_3I)_{2171}$). Of considerable interest are the energy distributions of the high-energy D^+ ions and how they are affected by kinematic effects [241, 242, 245, 246, 250, 251]. All the kinetic energy distributions $P(E)$ of the product D^+ ions (Fig. 6.67) from CE of $(D_2)_n$ and from several heteroclusters, exhibit a maximal cut-off energy E_M analyzed below. For $(D_2)_n$ and $(CD_4)_n$ clusters the onset of $P(E)$ occurs at $E = 0$, while for the $(DI)_n$ and $(CD_3I)_n$ ECLHHs a narrow distribution of $P(E)$ is exhibited with a relative energy spread $\Delta E/E_{av} \simeq 0.2$. The EML for uniform CE results in $P(E) = (3/2E_M)(E/E_M)^{1/2}$ $(E \leq E_M)$, in agreement with the simulated energy distribution for CE of $(D_2)_n$ (inset for $(D_2)_{16786}$ to Fig. 6.67). For $(CD_4)_n$ clusters a marked deviation of $P(E)$ from the $E^{1/2}$ relation is exhibited with about 75% of the D^+ ions lying in a narrow energy interval $\Delta E/E_{av} \simeq 0.4$, below E_M, manifesting kinematic run-over effects (inset for $(CD_4)_{4213}$ to Fig. 6.57). The EML for CE of the extremely charged light–heavy heteroclusters for a frozen subcluster of the I^{q_I+} ions predicts a low-energy onset of the energy distribution at $E_{min} = (4\pi/3)^{1/3}\overline{B}q_A q_B \rho_{mol} n^{2/3}$ with $P(E) = (3/E_{min})[3 - (2E/E_{min})]^{1/2}$ for $E_{min} \leq E \leq 3E_{min}/2$, where $E_M = 3E_{min}/2$ and $E_{av} = 6E_{min}/5$. The narrow distribution of $P(E)$ for CE of $(DI)_{2171}$ and of $(CD_3I)_{2171}$ (Fig. 6.67) is in accord with these predictions. This CE of $(DI)_n$ and $(CD_3)_n$ light–heavy heteroclusters constitutes an extreme manifestation of kinematic run-over effects, resulting in a narrow, high-energy distribution of the light ions.

6.6.6 Nuclear fusion driven by cluster Coulomb explosion

Eighty years of search for table-top nuclear fusion driven by bulk or surface chemical reactions, which involved the production of deuterons by catalytic processes [295] or by electrochemical methods [296], reflect on a multitude of experimental and conceptual failures [297]. This is not surprising, as the typical deuteron kinetic energies of 10 keV–100 keV are required for dd nuclear fusion, i.e., $D^+ + D^+ \rightarrow {}^3He + n + 3.3\,\mathrm{MeV}$ and $T^+ + H^+ + 4.0\,\mathrm{MeV}$. This characterizes the lower limit of the D^+ energy domain for the accomplishment of nuclear fusion, which cannot be attained in ordinary chemical reactions in macroscopic bulk or surface systems. CE of multicharged clusters produces high-energy (1 keV–1 MeV) ions in the energy domain of nuclear physics. Nuclear fusion can be driven by energetic deuterons produced by CE of multicharged deuterium containing homonuclear $(D_2)_n$ clus-

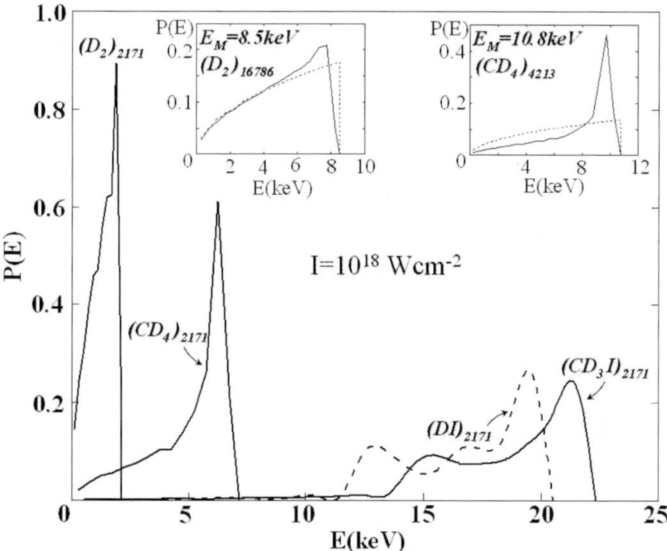

Fig. 6.67. The kinetic energy distributions of D^+ ions from several clusters (marked on the curves) at $I_m = 10^{18}$ Wcm^{-2} and $\tau = 25$ fs. The two insets show the simulated data (solid curves) and the results of the ELM (dashed curves) for CE of $(D_2)_{16786}$ and $(CD_4)_{4213}$.

ters [221,241,254,255,257] and heteronuclear (e.g., $(D_2O)_n$, [224,230,233,241] $(CD_4)_n$ [213, 226, 242, 259], $(DI)_n$) [250, 251]) clusters. CE in an assembly of deuterium containing homonuclear or heteronuclear clusters produces a plasma filament within the laser focal region, which constitutes a source of high-energy deuterons for nuclear fusion. Compelling experimental [254, 255] and theoretical [211, 245] evidence was advanced for nuclear fusion driven by CE (NFDCE) in an assembly of $(D_2)_n$ clusters. It was proposed and demonstrated [211, 241, 242, 246, 250, 251, 294] a marked enhancement of yields for NFDCE of deuterium containing heteroclusters due to energetic driving and kinematic effects.

While the quest for table-top nuclear fusion was realized for cluster CE, as well as for neutron production driven by a macroscopic piezoelectric crystal in a deuterium gas [298], these constitute low-yield processes. The neutron yield, Y, experimentally observed by Ditmire et al. [254, 255] for NFDCE of $(D_2)_n$ clusters ($n = 10^3$–2×10^4), is $Y \simeq 10^3$–10^4 per laser pulse (at $I_m = 10^{17}$ Wcm^{-2}). We demonstrated a seven orders of magnitude enhancement of Y in the NFDCE of ECLHHs, e.g., $(DI)_n$ and $(CD_3I)_n$ as compared to Y from $(D_2)_n$ clusters of the same size [294].

CE of multicharged deuterium containing heteroclusters and, in particular, of ECLHHs, manifests a marked increase of the average and maximal ener-

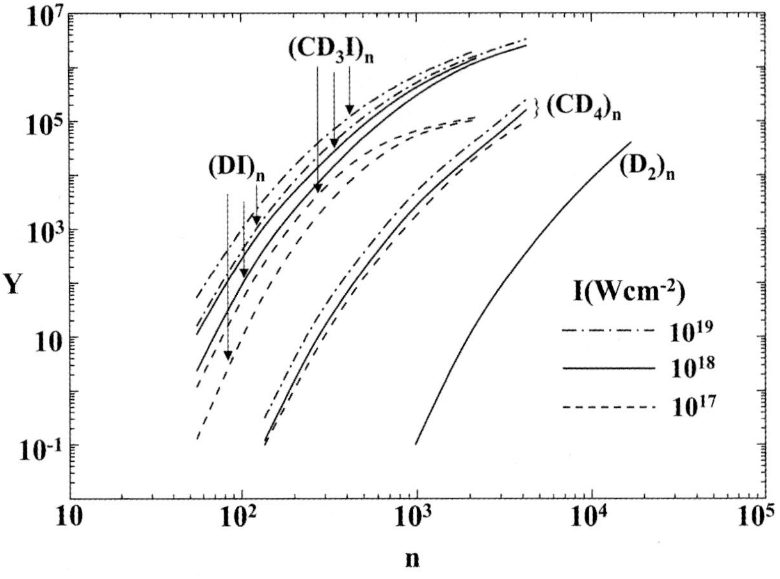

Fig. 6.68. Neutron yields per laser pulse (see text) for NFDCE of $(D_2)_n$, $(CD_4)_n$, $(DI)_n$ and $(CD_3I)_n$ clusters in the intensity range 10^{17}–$10^{19}\,\mathrm{Wcm^{-2}}$.

gies of the D^+ ions. This is due to energetic driving effects by multicharged heavy ions and to the narrowing of the energy redistribution at high energies (just below E_M) due to kinematic run-over effects, (6.61). These energetic and kinematic driving effects will result in a marked enhancement of the neutron yields Y from NFDCE of these heteroclusters, in comparison to NFDCE of $(D_2)_n$ clusters of the same size. The fusion yield per laser pulse in a plasma filament (produced by a laser of intensity $I_m = 10^{16}$–$10^{18}\,\mathrm{Wcm^{-2}}$) is given by $Y = (1/2)\rho_d^2 V_f(\overline{\ell}/\overline{v})\langle\sigma v\rangle$ where ρ_d is the deuteron density within the (cylindrical) reaction volume V_f, v is the relative velocity of the colliding nuclei, $\overline{v}$ is their average velocity, σ the fusion cross section, $\overline{\ell}$ is the deuterons mean free path, while $\langle\ \rangle$ denotes an average over the energy distribution. Using the conditions of the Lawrence-Livermore experiment, $\rho_d = 2 \times 10^{19}\,\mathrm{cm^{-3}}$ and $V_f = 6 \times 10^{-5}\,\mathrm{cm^3}$, while $\overline{\ell} = 0.016\,\mathrm{cm}$. The neutron yields Y (per laser pulse) calculated under the conditions of the Lawrence-Livermore experiment [254] for $I_m > 10^{17}\,\mathrm{Wcm^{-2}}$ are higher by 2–3 orders of magnitude for CE of $(CD_4)_n$ clusters than for $(D_2)_n$ clusters of the same size (Fig. 6.68). The theoretical predictions [211, 241, 242, 246, 250, 251, 294] were experimentally confirmed in Sacley [259], the Lawrence-Livermore laboratory [256], and in the Max Born Institute [224]. Moving to NFDCE of extremely charged light-heavy hetero-rocluster, e.g., $(D^+I^{25+})_n$ and $(CH_3I)_n$ (Fig. 6.68), Y can be enhanced by another 2–3 orders of magnitude over $(CD_4)_n$ clusters of the same size. For $(DI)_n$ and $(CD_3I)_n$ extremely charged light–heavy heteroclusters in the size domain $n = 1000$–2000 at $I_m = 10^{18}$–$10^{19}\,\mathrm{Wcm^{-2}}$, a dramatic increase of the neutron yields in the NFDCE of ECLHHs manifests energetic driving and kinematic effects. The realization of dd "hot–cold" nuclear fusion driven by CE made an 80 years old quest [297] for table–top nuclear fusion come true. An interesting application of NFDCE (with neutron yields of up to $\sim 10^9$ per laser pulse, see Fig. 6.68) pertains to the production of 100 ps–1 ns neutron pulses, which will be of interest for the exploration of time–resolved structures. Another "spin off" of nuclear reactions in cluster beams driven by ultraintense lasers pertains to nuclear astrophysics.

6.6.7 Table-top nucleosynthesis

Cluster dynamics transcends molecular dynamics toward nuclear reactions in ultraintense laser fields (Sect. 6.6.6). On the basis of theoretical and computational studies Last and Jortner proposed and demonstrated [299] that CE of molecular clusters will drive astrophysical nucleosynthesis [299,300] of protons with heavier nuclei

$$^{12}\mathrm{C}^{6+} + \mathrm{H}^{+} \rightarrow {}^{13}\mathrm{N}^{7+} + \gamma\,; \quad {}^{12}\mathrm{C}(\mathrm{p},\gamma){}^{13}\mathrm{N} \tag{6.64}$$

$$^{14}\mathrm{N}^{7+} + \mathrm{H}^{+} \rightarrow {}^{15}\mathrm{O}^{8+} + \gamma\,; \quad {}^{14}\mathrm{N}(\mathrm{p},\gamma){}^{15}\mathrm{N} \tag{6.65}$$

$$^{16}\mathrm{O}^{8+} + \mathrm{H}^{+} \rightarrow {}^{17}\mathrm{F}^{9+} + \gamma\,; \quad {}^{16}\mathrm{O}(\mathrm{p},\gamma){}^{17}\mathrm{N} \tag{6.66}$$

These reactions are part of the CNO cycle that constitutes the energy source of hot stars, which results in the fusion of four protons into $^4\mathrm{He}^{2+}$, with $^{12}\mathrm{C}^{6+}$ serving as a regenerable catalyst in this set of reactions [300,301]. For the realization of the astrophysical nucleosynthesis reactions (6.64), (6.65) and (6.66), the reagent nuclei will be produced by CE of the completely ionized $(\mathrm{CH_4})_n$, $(\mathrm{H_2O})_n$ and $(\mathrm{NH_3})_n$ molecular clusters in ultraintense laser fields. The realization of nucleosynthesis of protons with $^{12}\mathrm{C}^{6+}$, $^{14}\mathrm{N}^{7+}$, and $^{16}\mathrm{O}^{8+}$ nuclei in exploding cluster beams requires the fulfillment of the following conditions:

1. Cluster sizes. One has to utilize the largest cluster size at the given (very high) laser intensity that allows for the formation of bare nuclei and for the attainment of the highest energies of the nuclei. This requires extreme inner ionization, together with complete cluster outer ionization. Complete outer ionization involves CVI for subsequent–parallel CE, being achieved for the cluster border radius (section 6.6.6), i.e., $R_0 \simeq R_0^{(I)}$. On the basis of the electrostatic CBSI model, $R_0^{(I)}$ (at $I_m = 10^{20}\,\mathrm{Wcm^{-2}}$) assumes the values:

$$R_0^{(I)} = 750\,\text{Å}\ (n^{(I)} = 2.7 \times 10^7)\ \text{for}\ (\mathrm{CH_4})_{n^{(I)}}\,,$$
$$R_0^{(I)} = 500\,\text{Å}\ (n^{(I)} = 1.3 \times 10^7)\ \text{for}\ (\mathrm{NH_3})_{n^{(I)}}\ \text{and,}$$
$$R_0^{(I)} = 360\,\text{Å}\ (n^{(I)} = 6.3 \times 10^6)\ \text{for}\ (\mathrm{H_2O})_{n^{(I)}}\,.$$

   Such large cluster (droplet) sizes are amenable for experimental preparation [224].

2. Complete inner ionization. The BSI and the ADK models (Sects. 6.6.1 and 6.6.2) were utilized for the estimates of the laser intensity thresholds required for the C, N and O single atoms, which are accomplished at $I_m \geq 4 \times 10^{19}\,\text{Wcm}^{-2}$. The laser threshold intensity for the complete ionization of the corresponding molecular clusters is lower than that of a single atom, due to ignition effects (see Sect. 6.6.2). For the very large clusters of radius $R_0^{(I)}$, the intensity threshold values are $I_m = 10^{18}\,\text{Wcm}^{-2}$ for $(\text{NH}_3)_{n^{(I)}}$ and $I_m = 3 \times 10^{18}\,\text{Wcm}^{-2}$ for $(\text{H}_2\text{O})_{n^{(I)}}$. The $(\text{CH}_4)_n$ cluster requires a more detailed treatment of the optimal largest cluster size, in view of the resonance structure [300] in the energy-dependent cross sections for reaction (6.64).
3. Highest possible energies of the nuclei. The scaling law $E_M \propto R_0^2$ (Sect. 6.6.5) for the energetics of cluster CE under CVI conditions requires the use of the largest cluster size in the $R_0 \leq R_0^{(I)}$ domain. The energetics of the bare nuclei from exploding clusters of size $R_0^{(I)}$ at $I_m = 10^{20}\,\text{Wcm}^{-2}$ are $E_M \simeq 3\,\text{MeV}$ for protons and $E_M \simeq 30\,\text{MeV}$ for the C^{6+}, N^{7+} and O^{8+} heavy nuclei. These energies are high enough to drive nucleosynthesis. The optimal conditions for the attainment of table-top nucleosynthesis reactions (6.64)–(6.66) driven by CE of molecular clusters, involve laser intensities of $I_m = 10^{19}$–$10^{20}\,\text{Wcm}^{-2}$ and cluster sizes of $\text{R}_0 = R^{(I)} = 400$–800Å in this intensity range. The high-energy Coulomb exploding nuclei produce a macroscopic plasma filament [254]. The nucleosynthesis reactions take place both inside the plasma filament (IF) where high-energy nuclei collide [245,246,254,255,294] and outside the plasma filament (OF), where the energetic nuclei produced inside the plasma filament collide with the nuclei of clusters in the cluster beam outside the filament [224] (Fig. 6.69). In the intensity range $I_m = 10^{16}$–$10^{18}\,\text{Wcm}^{-2}$, where the volume of the plasma filament is large, i.e., $\simeq 10^{-3}$–10^{-4} cm^3, the IF mechanism dominates [254]. On the other hand, in the highest intensity range $I_m = 10^{19}$–$10^{20}\,\text{Wcm}^{-2}$, which is of interest to us, the small values of $V_f \simeq 10^{-9}$–$10^{-7}\,\text{cm}^3$, and the low values of the path of the energetic nuclei inside the plasma filaments, result in a dominating OF mechanism. The cluster size dependence of the nucleosynthesis yields (per laser pulse) were calculated using the experimental cross sections [301] and the theory of CE energetics under CVI conditions (Sect. 6.6.4). Figure 6.69 portrays the dependence of the nucleosynthesis yields vs the number n_A of atoms in the cluster. For $^{14}\text{N}(\text{p},\gamma)^{15}\text{O}$ and $^{16}\text{O}(\text{p},\gamma)^{17}\text{F}$ reactions (6.65) and (6.66), where no resonances are exhibited in the cross sections, a smooth dependent Y increases smoothly with increasing n_A, while for the $^{18}\text{C}(\text{p},\gamma)^{13}\text{N}$ reaction (6.64) Y shows two peaks due to the resonance structure of the cross sections. At $I_m = 10^{20}\,\text{Wcm}^{-2}$ ($\tau = 25\,\text{fs}$), the maximal values of $Y = 50$–100 for γ production (per laser pulse) are predicted. Table-top astrophysical nucleosynthesis reactions are amenable to experimental observation.

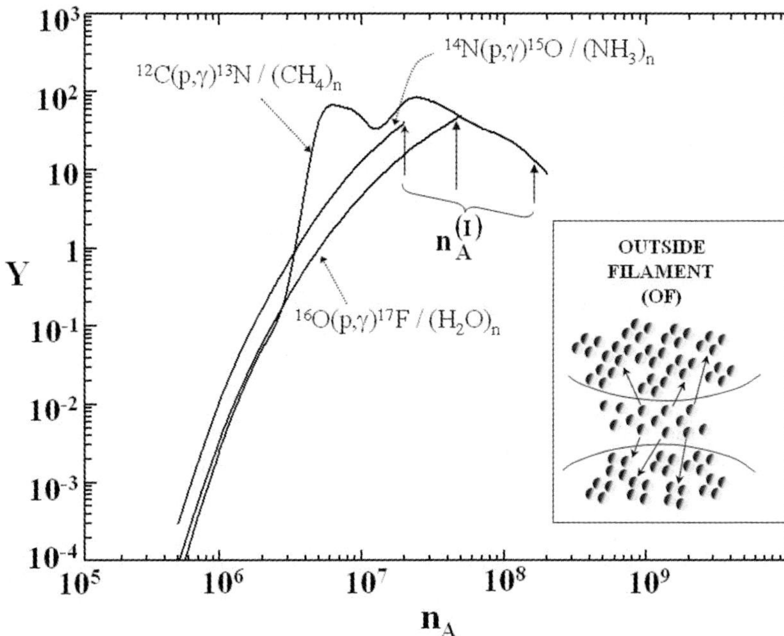

Fig. 6.69. Cluster size dependence of γ-ray yields (per laser pulse) for the nucleosynthesis reactions $^{12}C(p,\gamma)^{13}N$, $^{14}N(p,\gamma)^{15}O$, and $^{16}O(p,\gamma)^{17}F$, with the H^+, C^{6+}, N^{7+} and O^{8+} nuclei being produced by SE of $(CH_4)_n$, $(NH_3)_n$ and $(H_2O)_n$ clusters. The Y values are presented vs n_A, the total number of atoms in the cluster. The values of $n_A^{(I)}$, corresponding to the border radius $R_0^{(I)}$ for this intensity, are marked by vertical arrows. The dominating reaction mode involves the outside filament (OF) nucleosynthesis mechanism (schematically portrayed in the inset at the RHS of the figure).

6.6.8 Control in ultraintense laser fields

The control of reaction products in ultraintense laser fields ($I_m \geq 10^{15}$ Wcm^{-2}) is technically and conceptually different from the exploration of control in ordinary fields. At low laser intensities ($I_m \leq 10^{12}$ Wcm^{-2}), where perturbative treatment of radiative interactions is applicable, remarkable progress was made by pursuing control via pump-dump, phase control and optimal laser pulse shaping mechanisms. Control by laser pulse shaping is applicable up to $I_m = 10^{14}$ Wcm^{-2}. In this context, Vrakking et al. [222, 223] advanced and explored optimal control of the ionization level of Xe$_n$ clusters by shaping the laser pulse train at $I_m = 10^{14}$ Wcm^{-2}, below the lowest limits of the ultraintense intensity domain. In ultraintense laser fields, where nonperturbative effects are fundamental, new mechanisms of control based on novel ionization mechanisms, electron dynamics and high–energy CE nuclear dynamics will be of considerable interest. In the new field of control by ultraintense lasers ($I_m \geq 10^{15}$ Wcm^{-2}), the concept of pump-dump low–field control seems to be inapplicable and optimal laser pulse shaping via learning algorithms is fraught with considerable technical difficulties. Ultraintense field control can be achieved by changing the laser parameters, i.e., intensity, pulse length, shape and phase. In what follows we shall discuss two scenarios for control of cluster extreme ionization and nucleosynthesis driven by CE in ultraintense laser fields:

1. Control of extreme multielectron ionization of clusters [279]. From the analysis of the complex cluster size and laser parameter dependence of the ionization level of Xe_n clusters (Sect. 6.6.2) we can infer that EII reactive dynamics opens avenues for the control of the ionization products (Sect. 6.6.3). The dependence of the long-time inner ionization yield on the pulse length, with $q_{av}(= n_{ii}^L)$ for Xe_{2171} at $I_m = 10^{15}$ Wcm^{-2} (Fig. 6.62), constitutes control of extreme ionization in ultraintense laser fields. Of course, the ionization levels can markedly increase by raising the laser intensity, but this constitutes a trivial mode of control. More significant control involves product changes with changing the laser parameters (e.g., the pulse length). The variation of the laser pulse length leads to a marked change in the ionization products at a fixed laser intensity. This control mechanism for Xe_n clusters is induced by EII and is expected to be effective in the intensity range and cluster size domain where EII by the persistent nanoplasma does prevail. This will be realized in the intensity range $I_m = 10^{15}$–10^{16} Wcm^{-2} for the large cluster size domain of Xe_n ($n = 459$–2171) presented in Sects. 6.6.2 and 6.6.3. The results presented in this section established an ultraintense laser pulse length control mechanism driven by EII in the persistent nanoplasma within clusters of heavy atoms. It will be instructive to provide a further analysis of single pulse and multiple pulse nanoplasma electron dynamics and ionization levels in Xe_n clusters, by assessing the contribution of EII to optimal control in ultraintense laser fields.
2. Branching ratios in nucleosynthesis. CE of extremely ionized molecular heteroclusters containing carbon, deuterium and hydrogen, will result in branching between the $^{12}C(p,\gamma)^{13}N$ nucleosynthesis and the $^{2}D(d,n)^{3}He$ dd fusion. To be more explicit, consider the CE of a completely ionized $(CH_3D)_n$ cluster
$$(CH_3D)_n \rightarrow C^{6+} + D^+ + 3H^+ \tag{6.67}$$
Two parallel nuclear reactions are expected to occur inside and/or outside the nanoplasma filament
$$^{12}C^{6+} + {}^{1}H^{+} \rightarrow {}^{13}N^{7+} + \gamma\ ; \quad {}^{12}C(p,\gamma)^{13}N \tag{6.68}$$
$$^{2}D^{+} + {}^{2}D^{+} \rightarrow {}^{3}He^{2+} + n\ ; \quad {}^{2}D(d,n)^{3}He \tag{6.69}$$
Control of the branching ratio between reactions (6.68) and (6.69) (interrogated by monitoring the ratio of the numbers of (γ rays)/(neutrons))

can be induced by changing the cluster size and the laser parameters. Lowering the laser intensity in the threshold region for complete C atom ionization can reduce the contribution of the $^{12}C(p,\gamma)^{13}N$ reaction, while changing the pulse shape may modify the energetics of CE. Perspectives will also be explored for laser control by a further increase in the energies of the H^+, D^+ and C^{6+} nuclei by CE of the light–heavy $(CDH_2I)_n$ heteroclusters, where the highly charged I^{25+} ion (at $I_m = 10^{19}\,Wcm^{-2}$) or I^{35+} ion (at $I_m = 10^{20}\,Wcm^{-2}$) will act as an energetic booster for those nuclei participating in the parallel $^{12}C(p,\gamma)^{13}N$ and $^{2}D(d,n)^{3}He$ nuclear reactions.

We addressed some new research directions in the realm of laser–cluster interaction in ultraintense laser fields ($I_m \geq 10^{15}\,Wcm^{-2}$). In this new physical/chemical world, all the conventional 'rules of the game' regarding laser energy acquisition, storage and disposal in large finite systems have to be modified, differing quantitatively from the situation in ordinary, 'weak' laser fields. Two major novel research directions pertaining to control in ultraintense laser fields are emerging. First, a conceptual basis is being developed for control of exotic products (e.g., extreme ionization levels) in ultraintense laser fields. Second, cluster dynamics transcends molecular dynamics in large finite systems toward nuclear reactions (e.g., dd fusion and nucleosynthesis) driven by extreme cluster multielectron ionization in ultraintense laser fields, with the formation of unique clusters that consist of bare nuclei undergoing high-energy CE.

References

1. A. L'Huillier, P. Balcou, Phys. Rev. Lett. **70**, 774 (1993)
2. P. Agostini, F. Fabre, G. Mainfray, G. Petite, N.K. Rahman, Phys. Rev. Lett. **42**, 1127 (1979)
3. A. Zavriyev, P.H. Bucksbaum, J. Squier, F. Saline, Phys. Rev. Lett. **70**, 1077 (1993)
4. C. Kosmidis, P. Tzallas, K.W.D. Ledingham, T. McCanny, R.P. Singhal, P.F. Taday, A.J. Langley, J. Phys. Chem. A **103**, 6950 (1999)
5. M.J. DeWitt, R.J. Levis, J. Chem. Phys. **102**, 8670 (1995)
6. M.J. DeWitt, D.W. Peters, R.J. Levis, Chem. Phys. **218**, 211 (1997)
7. R.J. Levis, M.J. DeWitt, J. Phys. Chem. A **103**, 6493 (1999)
8. D.M. Villeneuve, S.A. Aseyev, P. Dietrich, M. Spanner, M.Y. Ivanov, P.B. Corkum, Phys. Rev. Lett. **85**, 542 (2000)
9. J. Purnell, E.M. Snyder, S. Wei, A.W. Castleman, Chem. Phys. Lett. **229**, 333 (1994)
10. M. Schmidt, D. Normand, C. Cornaggia, Phys. Rev. A **50**, 5037 (1994)
11. T. Ditmire, J. Zweiback, V.P. Yanovsky, T.E. Cowan, G. Hays, K.B. Wharton, Nature **398**, 489 (1999)
12. M.P. Kalachnikov, P.V. Nickles, H. Schönnagel, W. Sandner, Nucl. Instrum. Meth. A **472**, 254 (2001)

13. S. Larochelle, A. Talebpour, S.L. Chin, J. Phys. B **31**, 1201 (1998)
14. V. Schyja, T. Lang, H. Helm, Phys. Rev. A **57**, 3692 (1998)
15. A. Saenz, P. Lambropoulos, J. Phys. B **32**, 5629 (1999)
16. M.G. Baik, M. Pont, R. Shakeshaft, Phys. Rev. A **54**, 1570 (1996)
17. A. Apalategui, A. Saenz, P. Lambropoulos, J. Phys. B **33**, 2791 (2000)
18. A. Palacios, S. Barmaki, H. Bachau, F. Martin, Phys. Rev. A **71**, 063405 (2005)
19. A. Apalategui, A. Saenz, J. Phys. B **35**, 1909 (2002)
20. P.B. Corkum, Phys. Rev. Lett. **71**, 1994 (1993)
21. K.J. Schafer, B. Yang, L.F. Dimauro, K.C. Kulander, Phys. Rev. Lett. **70**, 1599 (1993)
22. L.V. Keldysh, Sov. Phys. JETP **20**, 1307 (1965)
23. M.J. DeWitt, R.J. Levis, J. Chem. Phys. **110**, 11368 (1999)
24. M. Lezius, V. Blanchet, D.M. Rayner, D.M. Villeneuve, A. Stolow, M.Y. Ivanov, Phys. Rev. Lett. **86**, 51 (2001)
25. M. Lezius, V. Blanchet, M.Y. Ivanov, A. Stolow, J. Chem. Phys. **117**, 1575 (2002)
26. I.V. Hertel, T. Laarmann, C.P. Schulz, in *Advances in atomic, molecular, and optical Physics, Vol. 50*, ed. by B. Bederson, H. Walter (Elsevier Academic Press, San Diego, 2005), pp. 219–286
27. M. Smits, C.A. de Lange, A. Stolow, D.M. Rayner, Phys. Rev. Lett. **93**, 203402 (2004)
28. J.R. Hiskes, Phys. Rev. **122**, 1207 (1961)
29. A.C. Riviere, D.R. Sweetman, Phys. Rev. Lett. **5**, 560 (1960)
30. T. Seideman, M.Y. Ivanov, P.B. Corkum, Phys. Rev. Lett. **75**, 2819 (1995)
31. T. Zuo, A.D. Bandrauk, Phys. Rev. A **52**, R2511 (1995)
32. Z. Mulyukov, M. Pont, R. Shakeshaft, Phys. Rev. A **54**, 4299 (1996)
33. M. Plummer, J.F. McCann, J. Phys. B **29**, 4625 (1996)
34. T.T. Nguyen-Dang, C. Lefebvre, H. Abou-Rachid, O. Atabek, Phys. Rev. A **71**, 023403 (2005)
35. A.S. Alnaser, M. Zamkov, X.M. Tong, C.M. Maharjan, P. Ranitovic, C.L. Cocke, I.V. Litvinyuk, Phys. Rev. A **72**, 041402(R) (2005)
36. T. Ergler, A. Rudenko, B. Feuerstein, K. Zrost, C.D. Schröter, R. Moshammer, J. Ullrich, Phys. Rev. Lett. **95**, 093001 (2005)
37. H.T. Yu, T. Zuo, A.D. Bandrauk, Phys. Rev. A **54**, 3290 (1996)
38. A. Saenz, Phys. Rev. A **61**, 051402(R) (2000)
39. A. Saenz, Phys. Rev. A **66**, 063407 (2002)
40. A. Saenz, Phys. Rev. A **66**, 063408 (2002)
41. K. Harumiya, I. Kawata, H. Kono, Y. Fujimura, J. Chem. Phys. **113**, 8953 (2000)
42. A. Saenz, J. Phys. B **35**, 4829 (2002)
43. G.R. Hanson, J. Chem. Phys. **62**, 1161 (1975)
44. A. Saenz, J. Phys. B **33**, 4365 (2000)
45. M.V. Ammosov, N.B. Delone, V.P. Krainov, Sov. Phys. JETP **64**, 1191 (1986)
46. A. Saenz, J. Phys. B **33**, 3519 (2000)
47. M. Awasthi, A. Saenz, J. Phys. B **39**, S389 (2006)
48. A. Scrinzi, M. Geissler, T. Brabec, Phys. Rev. Lett. **83**, 706 (1999)
49. M. Awasthi, Y.V. Vanne, A. Saenz, J. Phys. B **38**, 3973 (2005)
50. X. Urbain, B. Fabre, E.M. Staicu-Casagrande, N. de Ruette, V.M. Andrianarijaona, J. Jureta, J.H. Posthumus, A. Saenz, E. Baldit, C. Cornaggia, Phys. Rev. Lett. **92**, 163004 (2004)

51. E. Goll, G. Wunner, A. Saenz, Phys. Rev. Lett. **97**, 103003 (2006)
52. T. Ergler, B. Feuerstein, A. Rudenko, K. Zrost, C.D. Schröter, R. Moshammer, J. Ullrich, Phys. Rev. Lett. **97**, 103004 (2006)
53. A. Palacios, H. Bachau, F. Martin, Phys. Rev. Lett. **96**, 143001 (2006)
54. C. Cornaggia, J. Lavancier, D. Normand, J. Morellec, P. Agostini, J.P. Chambaret, A. Antonetti, Phys. Rev. A **44**, 4499 (1991)
55. M.J. DeWitt, R.J. Levis, Phys. Rev. Lett. **81**, 5101 (1998)
56. A.N. Markevitch, N.P. Moore, R.J. Levis, Chem. Phys. **267**, 131 (2001)
57. A.D. Bandrauk, J. Ruel, Phys. Rev. A **59**, 2153 (1999)
58. A.D. Bandrauk, S. Chelkowski, Chem. Phys. Lett. **336**, 518 (2001)
59. A.D. Bandrauk, S. Chelkowski, Phys. Rev. Lett. **84**, 3562 (2000)
60. A.M. Perelomov, V.S. Popov, M.V. Terent'ev, Sov. Phys. JETP **23**, 924 (1966)
61. M.J. DeWitt, R.J. Levis, J. Chem. Phys. **108**, 7045 (1998)
62. M.J. DeWitt, B.S. Prall, R.J. Levis, J. Chem. Phys. **113**, 1553 (2000)
63. N.P. Moore, R.J. Levis, J. Chem. Phys. **112**, 1316 (2000)
64. L. Pan, L. Armstrong, J.H. Eberly, J. Opt. Soc. Am. B **3**, 1319 (1986)
65. P.H. Bucksbaum, R.R. Freeman, M. Bashkansky, T.J. McIlrath, J. Opt. Soc. Am. B **4**, 760 (1987)
66. R.R. Freeman, P.H. Bucksbaum, J. Phys. B **24**, 325 (1991)
67. N.P. Moore, A.N. Markevitch, R.J. Levis, J. Phys. Chem. A **106**, 1107 (2002)
68. T.D.G. Walsh, F.A. Ilkov, J.E. Decker, S.L. Chin, J. Phys. B **27**, 3767 (1994)
69. A. Talebpour, S. Larochelle, S.L. Chin, J. Phys. B **31**, 2769 (1998)
70. J. Muth-Bohm, A. Becker, S.L. Chin, F.H.M. Faisal, Chem. Phys. Lett. **337**, 313 (2001)
71. S. Mukamel, S. Tretiak, T. Wagersreiter, V. Chernyak, Science **277**, 781 (1997)
72. S. Tretiak, V. Chernyak, S. Mukamel, Phys. Rev. Lett. **77**, 4656 (1996)
73. S. Tretiak, V. Chernyak, S. Mukamel, Chem. Phys. Lett. **259**, 55 (1996)
74. H.T. Yu, T. Zuo, A.D. Bandrauk, J. Phys. B **31**, 1533 (1998)
75. H.T. Yu, A.D. Bandrauk, Phys. Rev. A **56**, 685 (1997)
76. S. Chelkowski, A. Conjusteau, T. Zuo, A.D. Bandrauk, Phys. Rev. A **54**, 3235 (1996)
77. E.M. Snyder, S.A. Buzza, A.W. Castleman, Phys. Rev. Lett. **77**, 3347 (1996)
78. K.W.D. Ledingham, R.P. Singhal, Int. J. Mass Spectrom. Ion Process. **163**, 149 (1997)
79. T. Baumert, G. Gerber, Phys. Scr. **T72**, 53 (1997)
80. A.N. Markevitch, S.M. Smith, D.A. Romanov, H.B. Schlegel, M.Y. Ivanov, R.J. Levis, Phys. Rev. A **68**, 011402 (2003)
81. A.N. Markevitch, D.A. Romanov, S.M. Smith, H.B. Schlegel, M.Y. Ivanov, R.J. Levis, Phys. Rev. A **69**, 013401 (2004)
82. S.M. Smith, A.N. Markevitch, D.A. Romanov, X.S. Li, R.J. Levis, H.B. Schlegel, J. Phys. Chem. A **108**, 11063 (2004)
83. A.N. Markevitch, D.A. Romanov, S.M. Smith, R.J. Levis, Phys. Rev. Lett. **92**, 063001 (2004)
84. A.N. Markevitch, D.A. Romanov, S.M. Smith, R.J. Levis, Phys. Rev. Lett. **96**, 163002 (2006)
85. R.J. Levis, G.M. Menkir, H. Rabitz, Science **292**, 709 (2001)
86. A. Assion, T. Baumert, M. Bergt, T. Brixner, B. Kiefer, V. Seyfried, M. Strehle, G. Gerber, Science **282**, 919 (1998)

87. N.P. Moore, G.M. Menkir, A.N. Markevitch, P. Graham, R.J. Levis, in *Laser control and manipulation of molecules, ACS symposium series*, vol. 821, ed. by A.D. Bandrauk, Y. Fujimura, R.J. Gordon (Amer. Chem. Soc., Washington, 2002), pp. 207–220
88. D.J. Tannor, S.A. Rice, Adv. Chem. Phys. **70**, 441 (1988)
89. P. Brumer, M. Shapiro, Laser Part. Beams **16**, 599 (1998)
90. W.S. Warren, H. Rabitz, M. Dahleh, Science **259**, 1581 (1993)
91. M. Shapiro, P. Brumer, in *Advances in atomic, molecular, and optical Physics*, vol. 42 (Academic Press, San Diego, 2000), pp. 287–345
92. H. Rabitz, R. de Vivie-Riedle, M. Motzkus, K. Kompa, Science **288**, 824 (2000)
93. A.M. Weiner, Opt. Quantum Electron. **32**, 473 (2000)
94. J.X. Tull, M.A. Dugan, W.S. Warren, in *Advances in magnetic and optical resonance*, vol. 20 (Academic Press, New York, 1997), p. 1
95. R.S. Judson, H. Rabitz, Phys. Rev. Lett. **68**, 1500 (1992)
96. J. Berkowitz, G.B. Ellison, D. Gutman, J. Phys. Chem. **98**, 2744 (1994)
97. S. Vajda, A. Bartelt, E.C. Kaposta, T. Leisner, C. Lupulescu, S. Minemoto, P. Rosendo-Francisco, L. Wöste, Chem. Phys. **267**, 231 (2001)
98. C. Daniel, J. Full, L. Gonzalez, C. Kaposta, M. Krenz, C. Lupulescu, J. Manz, S. Minemoto, M. Oppel, P. Rosendo-Francisco, S. Vajda, L. Wöste, Chem. Phys. **267**, 247 (2001)
99. T. Hornung, R. Meier, M. Motzkus, Chem. Phys. Lett. **326**, 445 (2000)
100. B.J. Pearson, J.L. White, T.C. Weinacht, P.H. Bucksbaum, Phys. Rev. A **63**, 063412 (2001)
101. H. Rabitz, W.S. Zhu, Acc. Chem. Res. **33**, 572 (2000)
102. W.S. Zhu, H. Rabitz, J. Chem. Phys. **111**, 472 (1999)
103. Z.M. Lu, H. Rabitz, Phys. Rev. A **52**, 1961 (1995)
104. H. Rabitz, S. Shi, in *Advances in molecular vibrations and collision dynamics*, vol. 1A, ed. by J. Bowman (JAI Press, Greenwich, 1991), pp. 187–214
105. H.W. Kroto, J.R. Heath, S.C. O'Brien, R.F. Curl, R.E. Smalley, Nature **318**, 162 (1985)
106. M.S. Dresselhaus, G. Dresselhaus, P.C. Eklund, *Science of fullerenes and carbon nanotubes* (Academic Press, San Diego, 1996)
107. W. Krätschmer, L.D. Lamb, K. Fostiropoulos, D.R. Huffman, Nature **347**, 354 (1990)
108. E. Kolodney, A. Budrevich, B. Tsipinyuk, Phys. Rev. Lett. **74**, 510 (1995)
109. I.V. Hertel, H. Steger, J. De Vries, B. Weisser, C. Menzel, B. Kamke, W. Kamke, Phys. Rev. Lett. **68**, 784 (1992)
110. R.K. Yoo, B. Ruscic, J. Berkowitz, J. Chem. Phys. **96**, 911 (1992)
111. A. Reinköster, S. Korica, G. Prümper, J. Viefhaus, K. Godehusen, O. Schwarzkopf, M. Mast, U. Becker, J. Phys. B **37**, 2135 (2004)
112. S.C. O'Brien, J.R. Heath, R.F. Curl, R.E. Smalley, J. Chem. Phys. **88**, 220 (1988)
113. V. Foltin, M. Foltin, S. Matt, P. Scheier, K. Becker, H. Deutsch, T.D. Märk, Chem. Phys. Lett. **289**, 181 (1998)
114. M. Takayama, Int. J. Mass Spectrom. Ion Process. **121**, R19 (1992)
115. B. Walch, C.L. Cocke, R. Völpel, E. Salzborn, Phys. Rev. Lett. **72**, 1439 (1994)
116. S. Martin, J. Bernard, L. Chen, A. Denis, J. Desesquelles, Eur. Phys. J. D **4**, 1 (1998)
117. T. Schlathölter, O. Hadjar, J. Manske, R. Hoekstra, R. Morgenstern, Int. J. Mass Spectrom. **192**, 245 (1999)

118. A. Reinköster, B. Siegmann, U. Werner, H.O. Lutz, Radiat. Phys. Chem. **68**, 263 (2003)
119. J. Jensen, H. Zettergren, H.T. Schmidt, H. Cederquist, S. Tomita, S.B. Nielsen, J. Rangama, P. Hvelplund, B. Manil, B.A. Huber, Phys. Rev. A **69**, 053203 (2004)
120. H. Brauning, A. Diehl, R. Trassl, A. Theiss, E. Salzborn, A.A. Narits, L.P. Presnyakov, Fullerenes Nanotubes and Carbon Nanostructures **12**, 477 (2004)
121. E.E.B. Campbell, V. Schyja, R. Ehlich, I.V. Hertel, Phys. Rev. Lett. **70**, 263 (1993)
122. R. Vandenbosch, B.P. Henry, C. Cooper, M.L. Gardel, J.F. Liang, D.I. Will, Phys. Rev. Lett. **81**, 1821 (1998)
123. B. Farizon, M. Farizon, M.J. Gaillard, R. Genre, S. Louc, J. Martin, J.P. Buchet, M. Carré, G. Senn, P. Scheier, T.D. Märk, Int. J. Mass Spectrom. Ion Process. **164**, 225 (1997)
124. H.G. Busmann, T. Lill, B. Reif, I.V. Hertel, H.G. Maguire, J. Chem. Phys. **98**, 7574 (1993)
125. C. Yeretzian, K. Hansen, R.D. Beck, R.L. Whetten, J. Chem. Phys. **98**, 7480 (1993)
126. A. Bekkerman, A. Kaplan, E. Gordon, B. Tsipinyuk, E. Kolodney, J. Chem. Phys. **120**, 11026 (2004)
127. E.E.B. Campbell, K. Hansen, K. Hoffmann, G. Korn, M. Tchaplyguine, M. Wittmann, I.V. Hertel, Phys. Rev. Lett. **84**, 2128 (2000)
128. E.E.B. Campbell, G. Ulmer, I.V. Hertel, Phys. Rev. Lett. **67**, 1986 (1991)
129. K. Hansen, K. Hoffmann, E.E.B. Campbell, J. Chem. Phys. **119**, 2513 (2003)
130. F. Rohmund, M. Héden, A.V. Bulgakov, E.E.B. Campbell, J. Chem. Phys. **115**, 3068 (2001)
131. E.E.B. Campbell, R.D. Levine, Annu. Rev. Phys. Chem. **51**, 65 (2000)
132. E.E.B. Campbell, K. Hoffmann, H. Rottke, I.V. Hertel, J. Chem. Phys. **114**, 1716 (2001)
133. D. Bauer, F. Ceccherini, A. Macchi, F. Cornolti, Phys. Rev. A **64**, 063203 (2001)
134. B. Torralva, T.A. Niehaus, M. Elstner, S. Suhai, T. Frauenheim, R.E. Allen, Phys. Rev. B **64**, 153105 (2001)
135. G.P. Zhang, X. Sun, T.F. George, Phys. Rev. B **68**, 165410 (2003)
136. M. Kitzler, J. Zanghellini, C. Jungreuthmayer, M. Smits, A. Scrinzi, T. Brabec, Phys. Rev. A **70**, 041401 (2004)
137. T. Brabec, M. Côté, P. Boulanger, L. Ramunno, Phys. Rev. Lett. **95**, 073001 (2005)
138. M. Boyle, K. Hoffmann, C.P. Schulz, I.V. Hertel, R.D. Levine, E.E.B. Campbell, Phys. Rev. Lett. **87**, 273401 (2001)
139. M. Boyle, M. Heden, C.P. Schulz, E.E.B. Campbell, I.V. Hertel, Phys. Rev. A **70**, 051201 (2004)
140. M. Boyle, T. Laarmann, K. Hoffmann, M. Heden, E.E.B. Campbell, C.P. Schulz, I.V. Hertel, Eur. Phys. J. D **36**, 339 (2005)
141. M. Boyle, T. Laarmann, I. Shchatsinin, C.P. Schulz, I.V. Hertel, J. Chem. Phys. **122**, 181103 (2005)
142. T. Kunert, R. Schmidt, Phys. Rev. Lett. **86**, 5258 (2001)
143. F. Alvarado, R. Hoekstra, R. Morgenstern, T. Schlathölter, J. Phys. B **38**, L55 (2005)

144. M.J. Puska, R.M. Nieminen, Phys. Rev. A **47**, 1181 (1993)
145. R.R. Freeman, P.H. Bucksbaum, H. Milchberg, S. Darack, D. Schumacher, M.E. Geusic, Phys. Rev. Lett. **59**, 1092 (1987)
146. H. Kono, Y. Sato, Y. Fujimura, I. Kawata, Laser Physics **13**, 883 (2003)
147. H. Kono, Y. Sato, N. Tanaka, T. Kato, K. Nakai, S. Koseki, Y. Fujimura, Chem. Phys. **304**, 203 (2004)
148. C.P. Schick, P.M. Weber, J. Phys. Chem. A **105**, 3725 (2001)
149. N. Kuthirummal, P.M. Weber, Chem. Phys. Lett. **378**, 647 (2003)
150. K. Gluch, S. Matt-Leubner, O. Echt, B. Concina, P. Scheier, T.D. Mark, J. Chem. Phys. **121**, 2137 (2004)
151. S.A. Trushin, W. Fuß, W.E. Schmid, J. Phys. B **37**, 3987 (2004)
152. M. Murakami, R. Mizoguchi, Y. Shimada, T. Yatsuhashi, N. Nakashima, Chem. Phys. Lett. **403**, 238 (2005)
153. I. Shchatsinin, T. Laarmann, G. Stibenz, G. Steinmeyer, A. Stalmashonak, N. Zhavoronkov, C.P. Schulz, I.V. Hertel, J. Chem. Phys. **125**, 194320 (2006)
154. V.R. Bhardwaj, P.B. Corkum, D.M. Rayner, Phys. Rev. Lett. **93**, 043001 (2004)
155. A. Rentenier, A. Bordenave-Montesquieu, P. Moretto-Capelle, D. Bordenave-Montesquieu, J. Phys. B **37**, 2429 (2004)
156. K.R. Lykke, P. Wurz, J. Phys. Chem. **96**, 3191 (1992)
157. K.R. Lykke, Phys. Rev. A **52**, 1354 (1995)
158. M.E. Geusic, M.F. Jarrold, T.J. McIllrath, R.R. Freeman, W.L. Brown, J. Chem. Phys. **86**, 3862 (1987)
159. P.P. Radi, T.L. Bunn, P.R. Kemper, M.E. Molchan, M.T. Bowers, J. Chem. Phys. **88**, 2809 (1988)
160. H. Hohmann, R. Ehlich, S. Furrer, O. Kittelmann, J. Ringling, E.E.B. Campbell, Z. Phys. D **33**, 143 (1995)
161. G.P. Zhang, T.F. George, Phys. Rev. Lett. **93**, 147401 (2004)
162. T. Laarmann, I. Shchatsinin, A. Stalmashonak, M. Boyle, N. Zhavoronkov, J. Handt, R. Schmidt, C.P. Schulz, I.V. Hertel, Phys. Rev. Lett. (submitted)
163. G.N. Lewis, J. Am. Chem. Soc. **38**, 762 (1916)
164. T. Brabec, F. Krausz, Rev. Mod. Phys. **72**, 545 (2000)
165. P.M. Paul, E.S. Toma, P. Breger, G. Mullot, F. Auge, P. Balcou, H.G. Muller, P. Agostini, Science **292**, 1689 (2001)
166. C. Edmiston, K. Ruedenberg, Rev. Mod. Phys. **35**, 457 (1963)
167. W. Kohn, L.J. Sham, Phys. Rev. **140**, A1133 (1965)
168. C. Fiolhais, F. Nogueira, M.A.L. Marques (eds.), *A primer in Density Functional Theory, Lecture notes in Physics*, vol. 620 (Springer, Berlin, 2003)
169. R.M. Dreizler, E.K.U. Gross, *Density Functional Theory* (Springer, Berlin, 1990)
170. R.G. Parr, W. Yang, *Density Functional Theory of atoms and molecules* (Oxford University Press, New York, 1989)
171. P. Hohenberg, W. Kohn, Phys. Rev. B **136**, B864 (1964)
172. R.F.W. Bader, S. Johnson, T.H. Tang, P.L.A. Popelier, J. Phys. Chem. **100**, 15398 (1996)
173. A.D. Becke, K.E. Edgecombe, J. Chem. Phys. **92**, 5397 (1990)
174. T. Burnus, M.A.L. Marques, E.K.U. Gross, Phys. Rev. A **71**, 010501 (2005)
175. Y. Tal, R.F.W. Bader, Int. J. Quantum Chem. **12**, 153 (1978)
176. J.F. Dobson, J. Chem. Phys. **98**, 8870 (1993)

177. P.O. Löwdin, Phys. Rev. **97**, 1474 (1955)
178. R. McWeeny, Rev. Mod. Phys. **32**, 335 (1960)
179. A.D. Becke, Int. J. Quantum Chem. **23**, 1915 (1983)
180. T. Burnus, Time-dependent electron localization function. Diploma thesis, Freie Universität Berlin (2004)
181. E. Runge, E.K.U. Gross, Phys. Rev. Lett. **52**, 997 (1984)
182. M.A.L. Marques, C. Ullrich, F. Nogueira, A. Rubio, K. Burke, E.K.U. Gross, *Time-dependent Density Functional Theory, Lecture notes in Physics*, vol. 706 (Springer, Berlin, 2006)
183. M.A.L. Marques, E.K.U. Gross, Annu. Rev. Phys. Chem. **55**, 427 (2004)
184. R. van Leeuwen, Int. J. Mod. Phys. B **15**, 1969 (2001)
185. E. Gross, J. Dobson, M. Petersilka, in *Topics in current Chemistry*, vol. 181, ed. by R.F. Nalewajski (Springer, Berlin, 1996), pp. 81–172
186. E.K.U. Gross, W. Kohn, Adv. Quantum Chem. **21**, 255 (1990)
187. A. Savin, R. Nesper, S. Wengert, T.F. Fässler, Angew. Chem. Int. Ed. Engl. **36**, 1809 (1997)
188. M. Kohout, A. Savin, J. Comp. Chem. **18**, 1431 (1997)
189. J.B. Krieger, Y. Li, G.J. Iafrate, Phys. Rev. A **46**, 5453 (1992)
190. A. Castro, H. Appel, M. Oliveira, C.A. Rozzi, X. Andrade, F. Lorenzen, M.A.L. Marques, E.K.U. Gross, A. Rubio, phys. stat. sol. (b) **243**, 2465 (2006)
191. M.A.L. Marques, A. Castro, G.F. Bertsch, A. Rubio, Comput. Phys. Commun. **151**, 60 (2003)
192. B. Silvi, A. Savin, Nature **371**, 683 (1994)
193. D.B. Chesnut, J. Phys. Chem. A **104**, 11644 (2000)
194. M. Erdmann, E.K.U. Gross, V. Engel, J. Chem. Phys. **121**, 9666 (2004)
195. S. Shin, H. Metiu, J. Chem. Phys. **102**, 9285 (1995)
196. S. Shin, H. Metiu, J. Phys. Chem. **100**, 7867 (1996)
197. M. Erdmann, P. Marquetand, V. Engel, J. Chem. Phys. **119**, 672 (2003)
198. M. Erdmann, V. Engel, J. Chem. Phys. **120**, 158 (2004)
199. M. Erdmann, S. Baumann, S. Gräfe, V. Engel, Eur. Phys. J. D **30**, 327 (2004)
200. T. Kato, Commun. Pure Appl. Math. **10**, 151 (1957)
201. G.A. Mourou, C.P.J. Barty, M.D. Perry, Phys. Today **51**, 22 (1998)
202. G.A. Mourou, T. Tajima, S.V. Bulanov. Rev. Mod. Phys. **78**, 309 (2006)
203. T. Ditmire, T. Donnelly, A.M. Rubenchik, R.W. Falcone, M.D. Perry, Phys. Rev. A **53**, 3379 (1996)
204. T. Ditmire, J.W.G. Tisch, E. Springate, M.B. Mason, N. Hay, R.A. Smith, J. Marangos, M.H.R. Hutchinson, Nature **386**, 54 (1997)
205. T. Ditmire, Phys. Rev. A **57**, R4094 (1998)
206. K. Ishikawa, T. Blenski, Phys. Rev. A **62**, 063204 (2000)
207. J.S. Liu, R.X. Li, P.P. Zhu, Z.Z. Xu, J.R. Liu, Phys. Rev. A **64**, 033426 (2001)
208. P.B. Parks, T.E. Cowan, R.B. Stephens, E.M. Campbell, Phys. Rev. A **63**, 063203 (2001)
209. V.P. Krainov, A.S. Roshchupkin, J. Phys. B **34**, L297 (2001)
210. C. Rose-Petruck, K.J. Schafer, K.R. Wilson, C.P.J. Barty, Phys. Rev. A **55**, 1182 (1997)
211. I. Last, J. Jortner, Phys. Rev. A **64**, 063201 (2001)
212. V.P. Krainov, M.B. Smirnov, Phys. Rep.-Rev. Sec. Phys. Lett. **370**, 237 (2002)
213. D.A. Card, E.S. Wisniewski, D.E. Folmer, A.W. Castleman Jr., J. Chem. Phys. **116**, 3554 (2002)

214. U. Saalmann, J.M. Rost, Phys. Rev. Lett. **89**, 143401 (2002)
215. D. Bauer, A. Macchi, Phys. Rev. A **68**, 033201 (2003)
216. J. Schulz, H. Wabnitz, T. Laarman, P. Gürtler, W. Laasch, A. Swiderski, T. Möller, A.R.B. de Castro, Nucl. Instrum. Meth. A **507**, 572 (2003)
217. C. Siedschlag, J.M. Rost, Phys. Rev. A **67**, 013404 (2003)
218. C. Siedschlag, J.M. Rost, Phys. Rev. Lett. **93**, 043402 (2004)
219. C. Siedschlag, J.M. Rost, Phys. Rev. A **71**, 031401 (2005)
220. T. Laarmann, A.R.B. de Castro, P. Gürtler, W. Laasch, J. Schulz, H. Wabnitz, T. Möller, Phys. Rev. Lett. **92**, 143401 (2004)
221. K.W. Madison, P.K. Patel, D. Price, A. Edens, M. Allen, T.E. Cowan, J. Zweiback, T. Ditmire, Phys. Plasmas **11**, 270 (2004)
222. S. Zamith, T. Martchenko, Y. Ni, S.A. Aseyev, H.G. Muller, M.J.J. Vrakking, Phys. Rev. A **70**, 011201(R) (2004)
223. T. Martchenko, C. Siedschlag, S. Zamith, H.G. Muller, M.J.J. Vrakking, Phys. Rev. A **72**, 053202 (2005)
224. S. Ter-Avetisyan, M. Schnürer, D. Hilscher, U. Jahnke, S. Busch, P.V. Nickles, W. Sandner, Phys. Plasmas **12**, 012702 (2005)
225. D.M. Niu, H.Y. Li, F. Liang, L.H. Wen, X.L. Luo, B. Wang, H.B. Qu, J. Chem. Phys. **122**, 151103 (2005)
226. M. Hohenberger, D.R. Symes, K.W. Madison, A. Sumeruk, G. Dyer, A. Edens, W. Grigsby, G. Hays, M. Teichmann, T. Ditmire, Phys. Rev. Lett. **95**, 195003 (2005)
227. S. Ter-Avetisyan, M. Schnürer, P.V. Nickles, M. Kalashnikov, E. Risse, T. Sokollik, W. Sandner, A. Andreev, V. Tikhonchuk, Phys. Rev. Lett. **96**, 145006 (2006)
228. I. Last, J. Jortner, Phys. Rev. A **73**, 013202 (2006)
229. C. Siedschlag, J.M. Rost, Phys. Rev. Lett. **89**, 173401 (2002)
230. V. Kumarappan, M. Krishnamurthy, D. Mathur, Phys. Rev. A **67**, 063207 (2003)
231. E. Springate, S.A. Aseyev, S. Zamith, M.J.J. Vrakking, Phys. Rev. A **68**, 053201 (2003)
232. M. Rusek, A. Orlowski, Phys. Rev. A **71**, 043202 (2005)
233. G.M. Petrov, J. Davis, Phys. Plasmas **13**, 033106 (2006)
234. M. Krishnamurthy, D. Mathur, V. Kumarappan, Phys. Rev. A **69**, 033202 (2004)
235. J.S. Liu, C. Wang, B.C. Liu, B. Shuai, W.T. Wang, Y. Cai, H.Y. Li, G.Q. Ni, R.X. Li, Z.Z. Xu, Phys. Rev. A **73**, 033201 (2006)
236. M. Hirokane, S. Shimizu, M. Hashida, S. Okada, S. Okihara, F. Sato, T. Iida, S. Sakabe, Phys. Rev. A **69**, 063201 (2004)
237. S. Sakabe, S. Shimizu, M. Hashida, F. Sato, T. Tsuyukushi, K. Nishihara, S. Okihara, T. Kagawa, Y. Izawa, K. Imasaki, T. Iida, Phys. Rev. A **69**, 023203 (2004)
238. I. Last, J. Jortner, Phys. Rev. A **62**, 013201 (2000)
239. I. Last, J. Jortner, Phys. Rev. A **60**, 2215 (1999)
240. I. Last, J. Jortner, Z. Phys. Chemie-Int. J. Res. Phys. Chem. Chem. Phys. **217**, 975 (2003)
241. I. Last, J. Jortner, Phys. Rev. Lett. **87**, 033401 (2001)
242. I. Last, J. Jortner, J. Phys. Chem. A **106**, 10877 (2002)
243. I. Last, J. Jortner, J. Chem. Phys. **120**, 1336 (2004)

244. I. Last, J. Jortner, J. Chem. Phys. **120**, 1348 (2004)
245. I. Last, J. Jortner, J. Chem. Phys. **121**, 3030 (2004)
246. I. Last, J. Jortner, J. Chem. Phys. **121**, 8329 (2004)
247. I. Last, Y. Levy, J. Jortner, Proc. Natl. Acad. Sci. USA **99**, 9107 (2002)
248. I. Last, Y. Levy, J. Jortner, J. Chem. Phys. **123**, 154301 (2005)
249. Y. Levy, I. Last, J. Jortner, Mol. Phys. **104**, 1227 (2006)
250. I. Last, J. Jortner, Proc. Natl. Acad. Sci. USA **102**, 1291 (2005)
251. I. Last, J. Jortner, Phys. Rev. A **71**, 063204 (2005)
252. J. Jortner, Proc. R. Soc. London, Ser. A **356**, 477 (1998)
253. F. Krausz, Optics and Photonic News **13**, 62 (2002)
254. J. Zweiback, T.E. Cowan, R.A. Smith, J.H. Hartley, R. Howell, C.A. Steinke, G. Hays, K.B. Wharton, J.K. Crane, T. Ditmire, Phys. Rev. Lett. **85**, 3640 (2000)
255. J. Zweiback, T.E. Cowan, J.H. Hartley, R. Howell, K.B. Wharton, J.K. Crane, V.P. Yanovsky, G. Hays, R.A. Smith, T. Ditmire, Phys. Plasmas **9**, 3108 (2002)
256. K.W. Madison, P.K. Patel, M. Allen, D. Price, R. Fitzpatrick, T. Ditmire, Phys. Rev. A **70**, 053201 (2004)
257. M. Isla, J.A. Alonso, Phys. Rev. A **72**, 023201 (2005)
258. T.E. Dermota, D.P. Hydutsky, N.J. Bianco, A.W. Castleman, J. Chem. Phys. **123**, 214308 (2005)
259. G. Grillon, P. Balcou, J.P. Chambaret, D. Hulin, J. Martino, S. Moustaizis, L. Notebaert, M. Pittman, T. Pussieux, A. Rousse, J.P. Rousseau, S. Sebban, O. Sublemontier, M. Schmidt, Phys. Rev. Lett. **89**, 065005 (2002)
260. J.W.G. Tisch, N. Hay, E. Springate, E.T. Gumbrell, M.H.R. Hutchinson, J.P. Marangos, Phys. Rev. A **60**, 3076 (1999)
261. M. Eloy, R. Azambuja, J.T. Mendonca, R. Bingham, Phys. Plasmas **8**, 1084 (2001)
262. E.S. Toma, H.G. Muller, Phys. Rev. A **66**, 013204 (2002)
263. Q. Zhong, A.W. Castleman, Chem. Rev. **100**, 4039 (2000)
264. V. Mijoule, L.J. Lewis, M. Meunier, Phys. Rev. A **73**, 033203 (2006)
265. C. Jungreuthmayer, M. Geissler, J. Zanghellini, T. Brabec, Phys. Rev. Lett. **92**, 133401 (2004)
266. S. Ter-Avetisyan, M. Schnürer, H. Stiel, U. Vogt, W. Radloff, W. Karpov, W. Sandner, P.V. Nickles, Phys. Rev. E **6403**, 036404 (2001)
267. P.V. Nickles, S. Ter-Avetsyan, H. Stiehl, W. Wandner, M. Schnürer, in *Superstrong fields in plasmas, AIP conference proceedings*, vol. 611, ed. by M. Lontano, G. Mourou, O. Svelto, T. Tajima (Am. Inst. Phys., 2002), p. 288
268. A. Youssef, R. Kodama, M. Tampo, Phys. Plasmas **13**, 030701 (2006)
269. R.A. Snavely, M.H. Key, S.P. Hatchett, T.E. Cowan, M. Roth, T.W. Phillips, M.A. Stoyer, E.A. Henry, T.C. Sangster, M.S. Singh, S.C. Wilks, A. MacKinnon, A. Offenberger, D.M. Pennington, K. Yasuike, A.B. Langdon, B.F. Lasinski, J. Johnson, M.D. Perry, E.M. Campbell, Phys. Rev. Lett. **85**, 2945 (2000)
270. U. Andiel, K. Eidmann, K. Witte, I. Uschmann, E. Förster, Appl. Phys. Lett. **80**, 198 (2002)
271. S. Augst, D. Strickland, D.D. Meyerhofer, S.L. Chin, J.H. Eberly, Phys. Rev. Lett. **63**, 2212 (1989)
272. G. Gibson, T.S. Luk, C.K. Rhodes, Phys. Rev. A **41**, 5049 (1990)
273. M. Dammasch, M. Dörr, U. Eichmann, E. Lenz, W. Sandner, Phys. Rev. A **64**, 061402 (2001)

274. A.H. Zewail, *Femtochemistry: Ultrafast dynamics of the chemical bond* (World Scientific, Singapore, 1994)
275. J. Manz, L. Wöste (eds.), *Femtosecond Chemistry* (VCH, Weinheim, 1995)
276. M. Chergui (ed.), *Femtochemistry: Ultrafast chemical and physical processes in molecular systems* (World Scientific, Singapore, 1996)
277. V. Sundström (ed.), *Femtochemistry and Femtobiology: Ultrafast reaction dynamics at atomic-scale resolution.* Nobel Symposium 101 (Imperial College Press, London, 1997)
278. A. Heidenreich, I. Last, J. Jortner, Eur. Phys. J. D **35**, 567 (2005)
279. A. Heidenreich, I. Last, J. Jortner, (to be published)
280. M. Lezius, S. Dobosz, D. Normand, M. Schmidt, Phys. Rev. Lett. **80**, 261 (1998)
281. Y. Fukuda, K. Yamakawa, Y. Akahane, M. Aoyama, N. Inoue, H. Ueda, Y. Kishimoto, Phys. Rev. A **67**, 061201(R) (2003)
282. R. Santra, C.H. Greene, Phys. Rev. Lett. **91**, 233401 (2003)
283. F. Megi, M. Belkacem, M.A. Bouchene, E. Suraud, G. Zwicknagel, J. Phys. B **36**, 273 (2003)
284. C. Achenbach, A. Müller, E. Salzborn, R. Becker, J. Phys. B **17**, 1405 (1984)
285. D.C. Griffin, C. Bottcher, M.S. Pindzola, S.M. Younger, D.C. Gregory, D.H. Crandall, Phys. Rev. A **29**, 1729 (1984)
286. D.C. Gregory, D.H. Crandall, Phys. Rev. A **27**, 2338 (1983)
287. M.E. Bannister, D.W. Mueller, L.J. Wang, M.S. Pindzola, D.C. Griffin, D.C. Gregory, Phys. Rev. A **38**, 38 (1988)
288. G. Hofmann, J. Neumann, U. Pracht, K. Tinschert, M. Stenke, R. Völpel, A. Müller, E. Salzborn, in *The physics of highly charged ions, AIP conference proceedings*, vol. 274, ed. by M. Stockli, P. Richard (AIP Press, New York, 1993), p. 485
289. W. Lotz, Z. Phys. **216**, 241 (1968)
290. K.J. Mendham, N. Hay, M.B. Mason, J.W.G. Tisch, J.P. Marangos, Phys. Rev. A **64**, 055201 (2001)
291. S. Chelkowski, P.B. Corkum, A.D. Bandrauk, Phys. Rev. Lett. **82**, 3416 (1999)
292. J.M. de Conto, J. Phys. IV **9**, 115 (1999)
293. D. Atwood, *Soft X-rays and extreme ultraviolet radiation: Principles and applications* (Cambridge University Press, Cambridge, 1999)
294. A. Heidenreich, I. Last, J. Jortner, Proc. Natl. Acad. Sci. USA **103**, 10589 (2006)
295. F. Paneth, Nature **119**, 706 (1927)
296. J.R. Huizenga, *Cold fusion: The scientific fiasco of the century* (Oxford University Press, Oxford, 1992)
297. J. Jortner, I. Last, ChemPhysChem **3**, 845 (2002)
298. B. Naranjo, J.K. Gimzewski, S. Putterman, Nature **434**, 1115 (2005)
299. I. Last, J. Jortner, Phys. Rev. Lett. (in press)
300. D.A. Ostlie, B.W. Carroll, *An introduction to modern stellar Astrophysics* (Addision-Wesley Publ. Comp. Inc., New York, 1996)
301. E.G. Adelberger, S.M. Austin, J.N. Bahcall, A.B. Balantekin, G. Bogaert, L.S. Brown, L. Buchmann, F.E. Cecil, A.E. Champagne, L. de Braeckeleer, C.A. Duba, S.R. Elliott, S.J. Freedman, M. Gai, G. Goldring, C.R. Gould, A. Gruzinov, W.C. Haxton, K.M. Heeger, E. Henley, C.W. Johnson, M. Kamionkowski, R.W. Kavanagh, S.E. Koonin, K. Kubodera, K. Langanke, T. Motobayashi,

V. Pandharipande, P. Parker, R.G.H. Robertson, C. Rolfs, R.F. Sawyer, N. Shaviv, T.D. Shoppa, K.A. Snover, E. Swanson, R.E. Tribble, S. Turck-Chieze, J.F. Wilkerson, Rev. Mod. Phys. **70**, 1265 (1998)

7

Vibrational dynamics of hydrogen bonds

Erik T. J. Nibbering[1], Jens Dreyer[1], Oliver Kühn[2], Jens Bredenbeck[3], Peter Hamm[3], Thomas Elsaesser[1]

[1] Max-Born-Institut für Nichtlineare Optik und Kurzzeitspektroskopie, Berlin, Germany
[2] Institut für Chemie und Biochemie, Freie Universität Berlin, Germany
[3] Physikalisch Chemisches Institut, Universität Zürich, Switzerland

Coordinated by: Erik T. J. Nibbering

7.1 Significance of hydrogen bonding

Hydrogen bonds are ubiquitous and of fundamental relevance in nature. Representing a local attractive interaction between a hydrogen donor and an adjacent acceptor group, they result in the formation of single or multiple local bonds with binding energies in the range from 4 to 50 kJ mol^{-1}, much weaker than a covalent bond, but stronger than most other intermolecular forces and thus decisive for structural and dynamical properties of a variety of molecular systems [1–4]. Disordered extended networks of intermolecular hydrogen bonds exist in liquids such as water and alcohols, determining to a large extent their unique physical and chemical properties. At elevated temperatures, such liquids undergo pronounced structural fluctuations on a multitude of time scales, due to the limited interaction strength. In contrast, well-defined molecular structures based on both intra- and intermolecular hydrogen bonds exist in polyatomic molecules, molecular dimers and pairs and – in particular – in macromolecules such as DNA and other biomolecular systems.

Vibrational spectra of hydrogen-bonded systems reflect the local interaction strength and geometries as well as the dynamics and couplings of nuclear motions. Steady-state infrared and Raman spectra have been measured and modeled theoretically for numerous systems, making vibrational spectroscopy one of the major tools of hydrogen bond research [5]. In many cases, however, conclusive information on structure and – in particular – dynamics is difficult to derive from stationary spectra which average over a multitude of molecular geometries and time scales. In recent years, vibrational spectroscopy in the ultrafast time domain [6] is playing an increasingly important role for observing hydrogen bond dynamics in real-time, for separating different types of molecular couplings, and for determining them in a quantitative way [7]. Using such sophisticated probes, it has been established that the basic dynamic properties of hydrogen bonds are determined by processes in the femto- to picosecond time domain. The acquired knowledge about both potential energy surfaces of hydrogen-bonded systems and their dynamical properties may prove to be useful to design optical control schemes to steer reactive processes – that is hydrogen or proton transfer – by interaction with tailored coherent infrared pulses or pulse sequences.

In this Chapter, an overview of this exciting new field is presented with emphasis on both experimental and theoretical studies of the coherent response and of incoherent relaxation dynamics of hydrogen bonds in the electronic ground state [8–32]. In Sect. 7.2, the characteristic features of the vibrational spectra of hydrogen-bonded systems, the underlying coupling mechanisms, and the resulting molecular processes are introduced. Section 7.3 describes the experimental and theoretical methods applied in ultrafast vibrational spectroscopy of hydrogen bonds. Vibrational dephasing and incoherent processes of population relaxation and energy dissipation in disordered hydrogen-bonded systems such as water are discussed in Sect. 7.4. Section 7.5 is devoted to coherent nuclear motions and relaxation processes in intra- as well as intermolecular hydrogen-bonded structures with well-defined geometries in a liquid environment. An outlook is presented in Sect. 7.6.

7.2 Molecular vibrations as probe of structure and dynamics of hydrogen bonds

Vibrational spectroscopy provides direct insight into the couplings between normal modes as governed by the potential energy surfaces of hydrogen bonds. Here the characteristics of vibrational motions in hydrogen-bonded molecular systems are introduced and aspects of the theoretical modeling are discussed.

7.2.1 Vibrational modes of hydrogen-bonded systems

The formation of hydrogen bonds results in pronounced changes of the vibrational spectra of the molecules involved [5]. In a X–H$\cdots$Y hydrogen bond, with X and Y usually being electronegative atoms such as O or N, the absorption band of the stretching mode of the X–H donor group (ν_{XH}) displays the most prominent modifications, a red-shift and – in most cases – a substantial spectral broadening and reshaping as well as a considerable increase in intensity (Fig. 7.1).

The red-shift reflects the reduced force constant of the oscillator and/or the enhanced anharmonicity of the vibrational potential along the X–H stretching coordinate, i.e. an enhanced diagonal anharmonicity. The red-shift has been used to characterize the strength of hydrogen bonds [33–35]. Spectral broadening can originate from different types of interactions, among them anharmonic coupling of the high-frequency X–H stretching mode to low-frequency

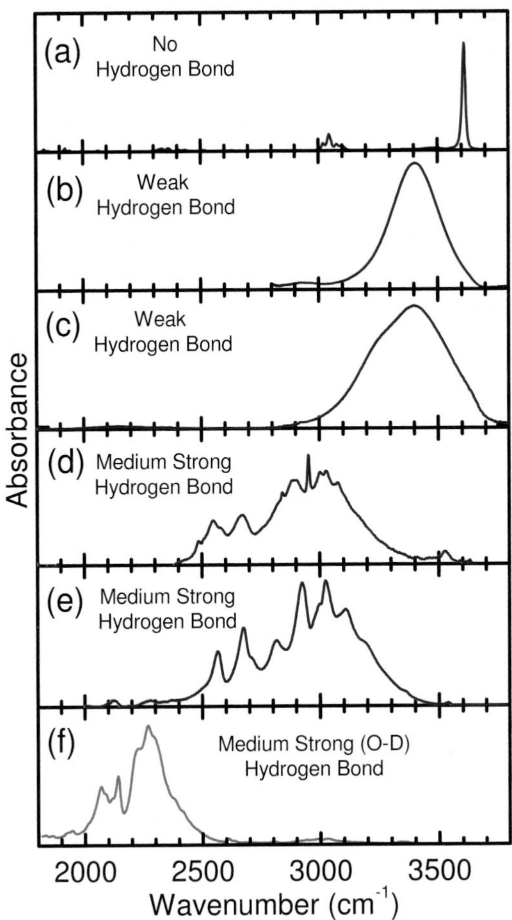

Fig. 7.1. Infrared absorption spectra showing O–H stretching absorption bands of (a) uncomplexed phenol in C_2Cl_4, (b) the weak hydrogen bond of HOD in D_2O, (c) neat H_2O, (d) the medium strong hydrogen bond of PMME-H, (e) acetic acid dimer $(CD_3{-}COOH)_2$, and (f) the O–D stretching band of acetic acid dimer $(CD_3{-}COOD)_2$.

modes, Fermi resonances with overtone and combination tone levels of fingerprint modes, vibrational dephasing, and inhomogeneous broadening due to different hydrogen bonding geometries in the molecular ensemble [36–39]. The substantial enhancement of the spectrally integrated absorption intensity accompanying spectral broadening is related to changes of the electronic structure. In contrast to the stretching mode, X–H in-plane bending modes (δ_{XH}), with transitions located in the fingerprint region of the vibrational spectrum, undergo small blue-shifts upon hydrogen bonding, but usually without appreciable effects on the line shape [5].

In the weak attractive potential between hydrogen donor and acceptor groups, modes which are connected with motions of the heavy atoms thereby affecting the hydrogen bond distance, can be identified. Formation of intermolecular hydrogen-bonded dimers gives rise to six new vibrational low-frequency modes out of translational and rotational degrees of freedom, among them the dimer hydrogen bond stretching (ν_{dimer}) and bending (δ_{dimer}) modes. In hydrogen bond networks, e.g., in water, intermolecular modes often extend over several molecules and dominate the low-frequency vibrational spectrum. The small force constants and the large reduced mass of such hydrogen bond modes result in low-frequency modes typically located below 200 cm^{-1}. Corresponding intramolecular hydrogen bond modes, on the other hand, are often found up to 400 cm^{-1}. Thus, these hydrogen bond motions with vibrational periods of about 660 to 80 fs are clearly separated in time from high-frequency X–H stretching motions with vibrational periods on the order of 10 to 15 fs.

7.2.2 Vibrational coupling mechanisms

Hydrogen bonding enhances the anharmonicity of the potential energy surface resulting in strengthened mechanical coupling of different vibrational modes. Anharmonic mode coupling is crucial for the line shape of vibrational absorption bands, hydrogen bond dynamics, and for vibrational energy transfer. In the vibrational spectra, anharmonic mode couplings are manifested by the appearance of over- and combination tones, band splittings and/or frequency shifts of harmonic transitions. In the following different mechanisms of anharmonic coupling are discussed.

7.2.2.1 Anharmonic coupling with low-frequency hydrogen bond modes

Anharmonic coupling between the high-frequency X–H stretching mode and low-frequency hydrogen bond modes has been considered a potential broadening mechanism of the X–H stretching band [36, 37, 39]. The separation of time scales between low- and high-frequency modes allows for a theoretical description in which the different states of the X–H stretching oscillator define adiabatic potential energy surfaces for the low-frequency modes (Fig. 7.2a), similar to the separation of electronic and nuclear degrees of freedom in the Born-Oppenheimer picture of vibronic transitions. Vibrational transitions from different levels of the low-frequency oscillator in the $v(\nu_{XH}) = 0$ state to different low-frequency levels in the $v(\nu_{XH}) = 1$ state with a shifted origin of the potential result in a progression of lines which, for moderate displacements, is centered at the pure X–H stretching transition and displays a mutual line separation by one quantum of the low-frequency mode (as depicted in Fig. 7.2b). The absorption strength is determined by the dipole moment of the $v(\nu_{XH}) = 0 \rightarrow 1$ transition of the X–H stretching mode and

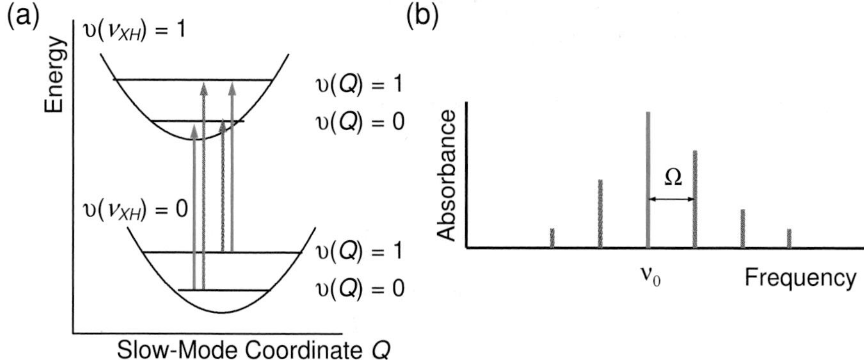

Fig. 7.2. (a) Potential energy surfaces for the hydrogen bond low-frequency mode in a single hydrogen bond, showing a displacement along the low-frequency (slow-mode) coordinate Q as function of the quantum state of the X–H/X–D high-frequency (fast-mode) stretching coordinate q (ν_{HX}). The X–H/X–D stretching transition is accompanied by changes in the low-frequency mode quantum state governed by Franck-Condon factors; (b) The corresponding stick spectrum of the X–H/X–D stretching transition shows the Franck-Condon progression centered at ν_0 ($v(\nu_{XH}) = 0 \rightarrow 1$) with frequency separations Ω of the low-frequency mode.

the Franck-Condon factors between the optically coupled levels of the low-frequency mode. With increasing difference in quantum number of the low-frequency mode in the $v(\nu_{XH}) = 0$ and 1 states, the Franck-Condon factors decrease and the progression lines become weaker for larger frequency separation from the progression center. For each low-frequency mode coupling to an X–H stretching oscillator, an independent progression of lines occurs.

Excitation of anharmonically coupled oscillators with a broadband ultrashort laser pulse resonant to the X–H stretching band can create a phase-coherent superposition of several levels of the low-frequency mode making up a vibrational wave packet. The wave packet can be generated in the $v(\nu_{XH}) = 1$ excited state by direct excitation, in the $v(\nu_{XH}) = 0$ ground state through a Raman-like process, or by coherence transfer from the $v(\nu_{XH}) = 1$ to the $v(\nu_{XH}) = 0$ state. These vibrational wave packets can be observed by ultrafast nonlinear vibrational spectroscopy as coherent dynamics [8,9].

7.2.2.2 Fermi resonances of the X–H stretching $v(\nu_{XH}) = 1$ state with overtone or combination levels of fingerprint vibrations

Coupling of the $v(\nu_{XH}) = 1$ state with higher lying states of fingerprint vibrations, such as the X–H bending mode, through Fermi resonances can lead to level splittings [38]. In this way the over- and combination tones, which are harmonically forbidden, gain in cross section. For Fermi resonances in the weak coupling regime, a vibrational energy redistribution channel is facilitated (Fig. 7.3a). For the strong coupling limit a level splitting is observed

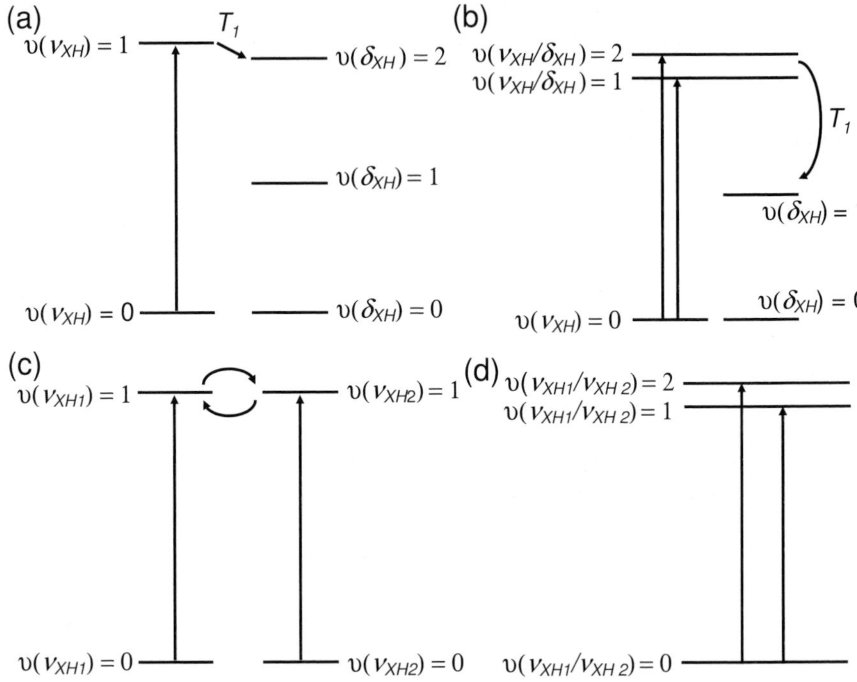

Fig. 7.3. Fermi resonance scheme in the weak (a) and strong (b) coupling limits between the X–H/X–D stretching $v(\nu_{XH}) = 1$ and the $v(\delta_{XH}) = 2$ bending levels. In (a) the weak Fermi resonance facilitates an efficient relaxation channel for energy redistribution, whereas case (b) explains the observation of additional transitions within the X–H/X–D stretching bands. The Davydov coupling scheme in the weak (c) and strong (d) limits between $v(\nu_{XH}) = 1$ levels of different X–H/X–D oscillators explain the phenomenon of vibrational excitation energy transfer (c) and excitation delocalization (d).

in the X–H stretching spectral region, indicative of states with mixed quantum character and concomitant similar absorption cross sections (Fig. 7.3b). In Sect. 7.5 experimental and numerical evidence of this mechanism will be provided for a few selected cases.

7.2.2.3 Davydov coupling between local X–H stretching oscillators

If several local oscillators can be found in one molecular assembly, for instance X–H stretching modes in acetic acid dimers or nucleic acid base pairs, or amide I vibrations in peptides, an excitonic type of interaction denoted as Davydov coupling may occur between the respective $v(\nu_{XH}) = 1$ states of the local oscillators. Small couplings enable vibrational excitation energy transfer between nearby X-H oscillators (Fig. 7.3c), whereas strong coupling leads to new combinations of quantum states (Fig. 7.3d), in a similar fashion as for amide I vibrations in proteins [40]. Line splittings with altered absorption cross sections of the individual transitions are dictated by the relative orientations of the transition dipole moments of the single X–H oscillators. In the case of the cyclic acetic acid dimer (see Sect. 7.5), selection rules based on the C_{2h} symmetry govern the IR- or Raman-activity of the transitions [37].

7.2.2.4 Coupling with fluctuating solvent modes

In the condensed phase the hydrogen bond is subject to fluctuating forces exerted by the solvent bath. These solvent bath modes, typically being overdamped, lead to spectral diffusion and broadening of the vibrational transition lines. Depending on the modulation strength and the fluctuation time scales, the spectra may vary between a distribution of transition frequencies corresponding to different hydrogen bond configurations (inhomogeneous broadening) or an averaged motionally narrowed transition (homogeneous broadening). Different theoretical approaches to describe the frequency modulation have been used: either the dipole moment of the X–H stretching oscillator is directly coupled to the local electric field induced by the solvent [41,42], or the low-frequency modes coupled to the X–H oscillator are stochastically modulated [43–45]. For an extensive discussion on the different underlying line broadening mechanisms, both using classical and quantum approaches, the reader is referred to a review by Henri-Rousseau, Blaise, and Chamma [39].

Using the density operator approach, the vibrational absorption line shape $\alpha(\omega)$ for a $v = 0 \rightarrow 1$ transition is given by:

$$\alpha(\omega) \propto \int_{-\infty}^{\infty} dt\, e^{-i(\omega-\omega_0)t} \langle \boldsymbol{\mu}_{10}(t)\boldsymbol{\mu}_{10}(0)\rangle \,, \tag{7.1}$$

with the transition dipole moment correlation function:

$$\begin{aligned} \langle \boldsymbol{\mu}_{10}(t)\boldsymbol{\mu}_{10}(0)\rangle &= |\mu_{10}|^2 \left\langle \exp\left[-i\int_0^t d\tau\, \delta\omega_{XH}(\tau)\right]\right\rangle \\ &\cong |\mu_{10}|^2 \exp\left[-i\int_0^t d\tau_2 \int_0^t d\tau_1\, \langle \delta\omega_{XH}(\tau_1)\delta\omega_{XH}(\tau_2)\rangle\right] , \end{aligned} \tag{7.2}$$

where $\boldsymbol{\mu}_{10}$ is the transition dipole moment of the vibrational transition centered at ω_0. In the interpretation of linear spectra, but even more so for results obtained with nonlinear experiments, the quantity governing the line broadening is the X–H transition frequency fluctuation correlation function $C(t)$:

$$C(t) = \langle \delta\omega_{XH}(t)\delta\omega_{XH}(0)\rangle \,, \tag{7.3}$$

where $\omega_{XH}(t)$ is the time-dependent X–H transition frequency. Solvent-induced line broadening for hydrogen-bonded systems has been modeled using

$C(t)$ subject to the assumption of Gauss-Markov modulation of the transition frequency:

$$C(t) = \Delta^2 \exp\left(-\frac{t}{t_c}\right) , \tag{7.4}$$

with Δ being the modulation strength (a measure for the distribution of possible transition frequencies for hydrogen-bonded systems) and t_c the frequency fluctuation correlation time (indicating the average time a particular hydrogen-bonded system resides at a particular frequency), with which one can interpolate between the inhomogeneous static and homogeneous fast modulation limits [46]. In the case of an infinitely short correlation time t_c, $C(t)$ can be approximated as a δ-function, i.e. $C(t) = \Delta^2 t_c \delta(t) = \delta(t)/T_2$, and a Lorentzian line shape dictated by the transverse dephasing time T_2 results. A homogeneous broadening described by a single T_2 time has been used for hydrogen-bonded systems with well-defined geometries in weakly interacting nonpolar solvents (see Sect. 7.5). However, for disordered highly fluctuating hydrogen-bonded systems often a sum of exponentially decaying functions is being used to mimic the more complex temporal behavior suggesting a multitude of characteristic fluctuation time scales underlying the microscopic dynamics (see the discussion in Sects. 7.4.1 and 7.4.2). More extensive expressions for $C(t)$ using Brownian oscillators have also been used [47]. A special case is the stochastic exchange between a limited number of hydrogen bond configurations, a case that has also been treated in the review by Kubo [46]. Here, the frequency separation between vibrational transitions indicative of the different hydrogen bond configurations and the exchange rate between these configurations dictates the absorption line shapes. In Sect. 7.4.4 an example will be provided of how chemical exchange influences the outcome of nonlinear infrared experiments.

Summarizing, the discussed mechanisms result in spectral substructures and/or a strong broadening of the overall O–H stretching band, even for a small number of absorption lines with large Franck-Condon factors. The different anharmonic couplings transform the hydrogen stretching oscillator into a vibrational multilevel system with a manifold of transition lines, each broadened by the solvent modulations.

7.2.3 Calculation of vibrational couplings in hydrogen bonds

Hydrogen bond potential energy surfaces are often dominated by electrostatic interactions and exchange repulsions, but depending on the type of hydrogen bond, polarization, charge-transfer, covalent or van der Waals forces contribute as well [48]. Thus, only a well-balanced theoretical description accurately accounts for the subtle interplay between these physical interactions. Numerical treatments to account for the anharmonicity of potential energy surfaces are introduced next, followed by a discussion of the suitability of different quantum chemical methods to describe hydrogen bonding.

7.2.3.1 Anharmonic coupling of vibrational modes

Most commonly anharmonically coupled vibrational force fields are calculated perturbatively by expanding the potential energy surface in a Taylor series with respect to a suitable set of coordinates $\{q_i\}$ around a reference geometry usually taken as the equilibrium geometry [49]:

$$\begin{aligned} V = V_0 &+ \sum_{i}^{3N-6} \Phi_i \, q_i + \sum_{i,j}^{3N-6} \Phi_{ij} \, q_i q_j \\ &+ \sum_{i,j,k}^{3N-6} \Phi_{ijk} \, q_i q_j q_k + \sum_{i,j,k,l}^{3N-6} \Phi_{ijkl} \, q_i q_j q_k q_l \cdots , \end{aligned} \tag{7.5}$$

with the nth order force constants given by

$$\Phi_{ijk\ldots} = \frac{1}{n!} \left(\frac{\partial^n V}{\partial q_i \partial q_j \partial q_k \ldots} \right)_0 . \tag{7.6}$$

Cubic anharmonic force constants Φ_{ijk} describe the coupling between fundamental modes and overtone and combination bands (Fermi resonance), whereas the coupling between overtone or combination states themselves (Darling-Dennison coupling) is characterized by quartic force constants Φ_{ijkl}. The nuclear displacement coordinates q_i are usually chosen to be Cartesian, internal, or normal coordinates. Cartesian coordinates are well-defined and easy to use but molecular motion may be better visualized using internal coordinates. Internal coordinates such as bond lengths, bond angles, and dihedral angles are inherently localized as they are confined to 2, 3 and 4 atoms, respectively. The resulting force constants are usually diagonally dominant and facilitate transfer and comparison between related molecules. Isotope effects can be readily studied as effective masses for internal coordinates enter the vibrational Hamiltonian directly. However, curvilinear internal coordinates introduce strong couplings in the kinetic energy operator that are difficult to calculate for larger systems. When the force constants are related to spectroscopic observables, (dimensionless) normal modes are the most common choice. As normal modes are often delocalized, it is, however, difficult to relate them to microscopic (local) properties of the molecule. Delocalized force constants are not easily transferred from one molecule to another and – in contrast to Cartesian and internal coordinates – the mass-weighted normal modes only provide isotope dependent force fields.

While first and second order analytical derivatives of the potential energy surface are available for most popular quantum chemical methods [50], higher order force constants mostly have to be determined numerically, either by least-square fitting of pointwise calculated multidimensional potential energy surfaces [51] or by finite difference procedures [52–54]. Nonperturbative approaches to calculate anharmonic coupling such as the vibrational self-consistent field approach (VSCF) [55, 56], diffusive Monte Carlo methods [57, 58] or grid methods [59–61] have also been put forward.

The computational treatment of anharmonic coupling in hydrogen-bonded systems depends on the type of hydrogen bond. Weak hydrogen bonds with X–H stretching frequencies above $3200\,\mathrm{cm}^{-1}$ can often be sufficiently described in the harmonic approximation applying an empirical scaling factor, or, if not, by an one-dimensional anharmonic correction in the proton coordinate. Hydrogen bonds of intermediate strengths exhibit X–H stretching frequencies in the range of $2800\text{-}3100\,\mathrm{cm}^{-1}$. Here, the potential energy surface along the X–H$\cdots$Y coordinate develops a shoulder or even a shallow well for a tautomeric X$\cdots$H–Y configuration. The potential energy surface along the proton coordinate becomes broader, the barrier for proton transfer decreases, and the vibrational energy levels are spaced more closely leading to a larger frequency shift. Strong hydrogen bonds with X–H frequencies below $2700\,\mathrm{cm}^{-1}$ may even develop a single minimum potential energy surface with the zero vibrational energy level being above the barrier. In both medium-strong and strong cases, an explicit anharmonic treatment of at least the X–H and the X$\cdots$Y coordinates is advisable. The latter coordinate reflects hydrogen bond stretching and/or bending modes.

Potential energy surfaces for nonreactive hydrogen bonds have been discussed until now. The description of reactive dynamics involving large-amplitude motion is conveniently done within the Reaction Surface Hamiltonian approach (for a recent review, see [32]). It combines the exact treatment of a few large-amplitude coordinates with the harmonic approximation for the majority of the degrees of freedom. In terms of the anharmonic expansion of the potential in the vicinity of a minimum this represents an approximation, that is, only certain terms in (7.5) are preserved [19]. The advantage, however, is that a full-dimensional simulation becomes possible as long as respective ab initio computations of the second derivative matrix can be carried out.

Derivatives of the dipole moment $\boldsymbol{\mu}$ in the coordinates q_i determine infrared intensities [62]:

$$\boldsymbol{\mu} = \boldsymbol{\mu}_0 + \sum_i^N \left(\frac{\partial \boldsymbol{\mu}}{\partial q_i}\right)_0 q_i + \sum_{i,j}^N \left(\frac{\partial \boldsymbol{\mu}}{\partial q_i q_j}\right)_0 q_i q_j + \ldots \quad . \tag{7.7}$$

Linear spectroscopy in the dipole approximation truncates the expansion after the linear term, an approximation that is also frequently adopted in nonlinear vibrational spectroscopy [6], although the influence of higher order terms on infrared intensities has been considered [63–65].

7.2.3.2 Quantum chemistry of hydrogen bonds

A large variety of theoretical methods including mechanical force fields, semi-empirical and ab initio methods have been applied to describe hydrogen bonding. The conceptually simplest approach is provided within the framework of classical mechanical force fields [66–68]. Some empirical mechanical force fields simply rely upon electrostatic and van der Waals interactions without explicit hydrogen bonding terms. Others modify the parameters and functional form of the van der Waals interaction, commonly a Lennard-Jones potential. In some cases directional terms such as $(\cos\theta^{XHY})^4$ are included as factors on the distance dependent parts in the Lennard-Jones potential to account for the hydrogen bond angle θ^{XHY}.

Among semiempirical electronic structure methods, older methods such as CNDO, INDO, MNDO and AM1 are considered inappropriate due to an overestimation of the exchange repulsion, whereas more recently developed semiempirical methods such as SAM1 and PM3 are regarded as reasonable methods for the description of hydrogen bond interactions [48, 69].

For ab initio methods the inclusion of electron correlation has been shown to be essential for an accurate treatment of hydrogen bonds [3, 70]. Hartree-Fock methods underestimate hydrogen bond energies and are thus inappropriate for a reliable description. In contrast, methods covering correlation effects explicitly such as second-order Møller-Plesset perturbation theory (MP2), configuration interaction (CI) or coupled-cluster (CC) methods are well suited to describe hydrogen bonding. In addition, sufficiently large basis sets, preferably of polarized triple-zeta quality with diffuse functions, have to be applied. The finite size of the basis set leads to a source of error in supermolecular calculations of intermolecular hydrogen bonding, the basis set superposition error, an artificial overestimation of the complexation energy. For very accurate treatments, this error may be approximately accounted for by the counterpoise correction [71, 72].

MP2, CI, and in particular CC methods such as CCSD(T) or CCSDT are capable of achieving very high accuracy, they are, however, computationally quite demanding and thus only affordably for smaller systems [73]. Therefore, approaches are being pursued to study hydrogen bonding with similar accuracy to MP2 or higher levels of theory, but with less computational effort. In particular, density functional theory (DFT) methods became very popular. Unfortunately, the accuracy of DFT to describe hydrogen bonding depends on the functional used to approximate the electronic exchange and correlation. DFT methods applying the generalized gradient approximations (GGA) or hybrid functionals have been shown to perform rather well in describing thermochemical, structural, and vibrational properties of hydrogen-bonded systems provided that sufficiently flexible basis sets are used [74]. DFT methods suffer from the inability to describe energy contributions stemming from dispersion forces [74]. This applies mainly to weak hydrogen bonds, hydrogen bond geometries deviating from linearity and extended hydrogen-bonded complexes [73, 75, 76]. However, attempts to explicitly include dispersion interaction into DFT methods are underway [77]. Some recently developed DFT functionals have also been shown to provide reasonably accurate energetic and geometric predictions for hydrogen bonds to π acceptors [78].

Approximate resolution of the identity (RI) MP2 methods, which are about one order of magnitude faster than exact MP2 methods, are capable of an accurate description of hydrogen bonds as well as systems, for which the proper treatment of dispersion interaction is essential, e.g. stacked base pairs [79].

An important and likewise challenging field is the simulation of extended hydrogen-bonded networks, e.g. liquids such as water or alcohols [80]. Many studies have used classical molecular dynamics with empirical potentials, often optimized to reproduce experimentally observable bulk properties [66,81,82]. In recent years, ab initio molecular dynamics have developed into a very powerful tool, which combines molecular dynamics with forces calculated on-the-fly from electronic structure calculations, in particular with DFT methods [83–86]. By this means ab initio molecular dynamics of liquid water [87–89], methanol [90] or hydrated protons in water [91–93] have been investigated. Because of considerable computational costs, ab initio molecular dynamics methods are restricted to small system sizes and short trajectories. Therefore, hybrid high-level/low-level approaches have been designed [94–96]. Hybrid methods include quantum effects for the electronically active region, whereas the environment is treated at a lower level of theory, either by lower level quantum mechanical methods (QM/QM) or by classical molecular mechanics (QM/MM).

7.3 Ultrafast nonlinear infrared spectroscopy

Ultrafast nonlinear infrared spectroscopy requires femtosecond pulses for resonant excitation and probing of vibrational transitions. Pulses of microjoule energy are necessary to induce a nonlinear vibrational response because of the comparably small cross sections of vibrational absorption on the order of $\sigma = 10^{-18} - 10^{-19}\,\text{cm}^2$. So far, ultrafast dynamics of hydrogen bonds have been studied in the frequency range from 500 to 4000 cm^{-1} (wavelength range 20 to 2.5 μm) with the main emphasis on O–H and N–H stretching vibrations. Microjoule pulses with a duration of 40 - 250 fs have been generated throughout this range by nonlinear optical frequency conversion. A brief review of the relevant techniques has been given in [7].

Ultrafast vibrational spectroscopy does not only enable the determination of real time dynamics of vibrational states. With the bandwidth of the laser pulses the potential of exciting a collection of vibrational states into a coherent superposition can be explored as well [8,9,11,13,20–22,24,27]. The multilevel character and the coherent nature of X–H stretching excitations in hydrogen bonds have to be taken into account to describe the observed nonlinear signals.

Theoretical descriptions of nonlinear experiments often rely on perturbation theory [40,47]. In addition to the material response, the applied light fields $E_i(t)$ interacting at three different times govern the third-order nonlinear polarization $P^{(3)}(t)$:

$$P^{(3)}(t) = \int_0^\infty dt_3 \int_0^\infty dt_2 \int_0^\infty dt_1 \; S^{(3)}(t_3,t_2,t_1) \times E_3(t-t_3)E_2(t-t_3-t_2)E_1(t-t_3-t_2-t_1)\,. \qquad (7.8)$$

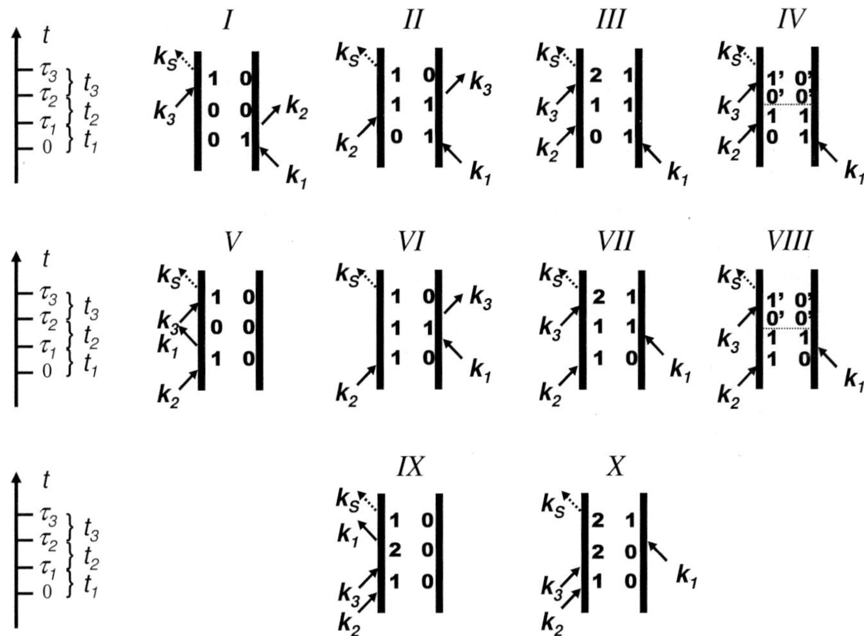

Fig. 7.4. Double sided Feynman diagrams used to describe nonlinear infrared experiments. Diagrams *I-IV* describe echo-generating (rephasing) Liouville space pathways, while diagrams *V-X* symbolize nonrephasing pathways. For two-level systems only diagrams *I, II, V,* and *VI* contribute, whereas diagrams *III, VII, IX* and *X* have to be added for three-level (vibrational ladder) systems. Diagrams *IV* and *VIII* take into account the population relaxation to a hot ground state, as indicated by the horizontal dashed line.

Here, $S^{(3)}(t_3, t_2, t_1)$ is the third-order response function:

$$S^{(3)}(t_3, t_2, t_1) = \sum_{\alpha=I}^{X} R_\alpha(t_3, t_2, t_1) , \tag{7.9}$$

which is a system property containing all information that potentially can be obtained from a system by means of third-order spectroscopy (for an application in electronic spectroscopy see Chapter 8) . To determine the different light-matter interactions that are possible in nonlinear vibrational experiments, double-sided Feynman diagrams are a standard means of depicting the possible Liouville space pathways. Due to the multilevel nature of hydrogen-bonded X−H oscillators 10 different diagrams often have to be considered (Fig. 7.4) [12,40,97].

Diagrams *I-IV* are denoted as rephasing pathways due to the time reversal of the system evolution in period t_3 after a coherence period t_1 and a population period t_2. Diagrams *V-X*, on the other hand, do not cause a rephasing.

Diagrams *III, VI, IX,* and *X* have to be implemented for vibrational ladder systems. Diagrams *II, III, VI,* and *VII* diminish in importance upon T_1 (population) relaxation. Diagrams *IV* and *VIII* have been implemented to describe an effective five level system of hydrogen-bonded X–H stretching vibrations (see Sect. 7.3.1). For each Liouville space pathway $\alpha = I, .., X$ a specific material response function $R_\alpha(t_3, t_2, t_1)$ can be defined, depending on the particular diagonal and off-diagonal elements of the density operator through which the system evolves in the three different time periods. Explicit equations can be found in Sect. 3.2 in the overview [7]. In the following different forms of nonlinear vibrational spectroscopy that have been pursued in the investigation of hydrogen bond dynamics are described.

7.3.1 Pump-probe spectroscopy

Pump-probe spectroscopy is conventionally used to follow the population kinetics upon an excitation of the $v(\nu_{XH}) = 0 \rightarrow 1$ transition. This implies that measurements are made in the well-separated pulse regime, i.e. when pump and probe pulses do not overlap in time and additional signal contributions caused by alternate order in the field interactions can be discarded [47]. In this well-separated pulse regime the time-dependent populations can be followed using a probe pulse in spectrally integrated fashion (with a single detector) or, by taking advantage of the large pulse bandwidth, after spectral dispersion with a monochromator (Fig. 7.5a-c). In the case of Fig. 7.5a the spectrally integrated absorption change $\Delta\alpha(\omega_{pr})$ measured by a probe pulse with carrier frequency ω_{pr} is proportional to [47,98]:

$$\Delta\alpha(\omega_{pr}) \propto \frac{\operatorname{Im} \int_{-\infty}^{\infty} dt\, E_{pr}^*(t) \times P^{(3)}(t)}{\int_{-\infty}^{\infty} dt\, |E_{pr}(t)|^2} . \tag{7.10}$$

Here $E_{pr}(t)$ is the time-dependent electrical field of the probe pulse and $P^{(3)}(t)$ is given by (7.8). For spectrally resolved detection of the probe pulse (Fig. 7.5b,c), the measured signal is given by:

$$\Delta\alpha(\omega_{pr}) \propto \frac{|E_{pr}(\omega)|^2 \operatorname{Im}\left(P^{(3)}(\omega)/E_{pr}^*(\omega)\right)}{\int_0^{\infty} d\omega\, |E_{pr}(\omega)|^2} . \tag{7.11}$$

As a consequence of a significantly large diagonal anharmonicity of X–H stretching vibrations, the $v(\nu_{XH}) = 0 \rightarrow 1$ and $v(\nu_{XH}) = 1 \rightarrow 2$ transitions are centered at different spectral positions [40], separated by an anharmonic shift of up to several hundreds of wavenumbers (see Fig. 7.6). Thus, after excitation of the X–H stretching vibration, one observes the ground state $v(\nu_{XH}) = 0 \rightarrow 1$ bleaching and excited state $v(\nu_{XH}) = 1 \rightarrow 0$ stimulated emission, both leading to a decrease of absorbance at the frequency position of the X–H stretching band observed in the linear IR spectrum. In addition,

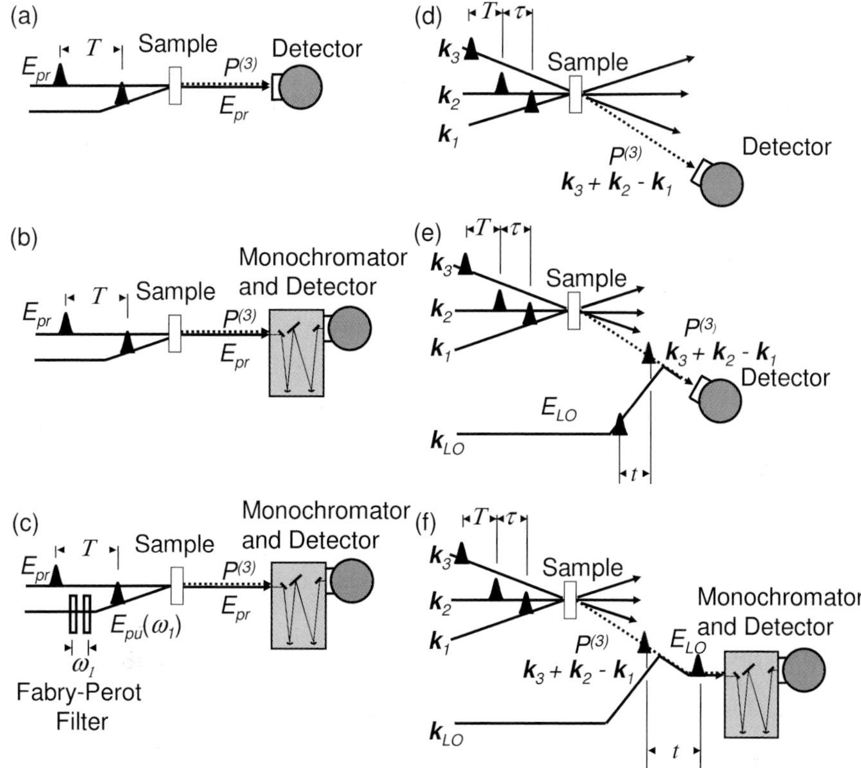

Fig. 7.5. Schematics of nonlinear spectroscopy techniques: (a) spectrally integrated pump-probe; (b) spectrally resolved pump-probe; (c) 2D-IR pump-probe; (d) homodyne (intensity) detected photon echo; (e) heterodyne (amplitude) detected photon echo with local oscillator LO; (f) heterodyne (amplitude) detected photon echo using spectral interferometry with local oscillator (2D-FT-IR).

a transient red-shifted absorption occurs on the $v(\nu_{XH}) = 1 \rightarrow 2$ transition, decaying upon $v(\nu_{XH}) = 1$ population relaxation.

From a large collection of experimental work the conclusion has been drawn that the $v(\nu_{XH}) = 1$ state does not relax directly back to the $v(\nu_{XH}) = 0$ state, leading to a disappearance of the bleach signals [8,9,12,13,20,24,99–101]. Instead, the short-lived $v(\nu_{XH}) = 1$ state, often with sub-picosecond lifetimes, redistributes vibrational excess energy into other modes actively involved in the hydrogen bond motions, such as the X–H bending and the hydrogen bond low-frequency modes (Fig. 7.6). Subsequently, energy dissipation to the solvent (vibrational cooling) occurs on longer time scales on the order of picoseconds to hundreds of picoseconds. In this hot ground state configuration the X–H stretching mode is not excited, but its transition $v(\nu_{XH}) = 0' \rightarrow 1'$ is frequency up-shifted due to anharmonic coupling to the transiently highly

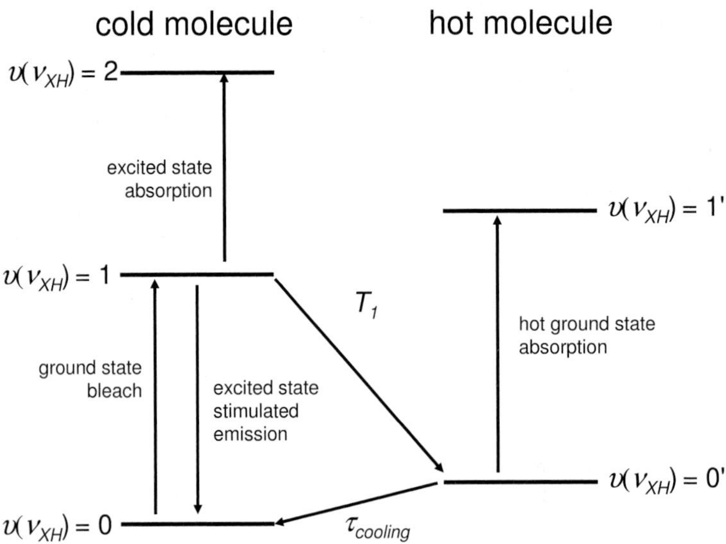

Fig. 7.6. Level structure implementing the $v(\nu_{XH}) = 0$, $v(\nu_{XH}) = 1$, and $v(\nu_{XH}) = 2$ states of the X–H/X–D stretching mode, as well as the two levels of the hot molecule generated upon population relaxation of the $v(\nu_{XH}) = 1$ state.

populated hydrogen bond modes. In contrast, when X–H bending vibrations relax into the hot ground state upon intramolecular vibrational energy redistribution (IVR), red-shifting is observed [25, 28, 31], in accordance with more typically found negatively valued anharmonic couplings between intramolecular vibrational modes [62, 102, 103].

Besides the investigation of population dynamics, pump-probe spectroscopy can be used to examine coherent vibrational dynamics. The anharmonic coupling between the X–H stretching and low-frequency hydrogen bond modes allows for the preparation of a coherent superposition of vibrational eigenstates of these low-frequency modes by exciting the stretching vibration with a broadband pump pulse (see Fig. 7.2a). This effect has been explored in medium strong hydrogen-bonded molecular systems (see Sect. 7.5).

Finally, the probe pulse can also be tuned to the fingerprint region [23, 25]. In this two-color pump-probe experiment one can follow the dynamics of fingerprint vibrations upon excitation of the X–H stretching mode. Here the idea is to explore the population kinetics after the decay of the $v(\nu_{XH}) = 1$ state, where the IVR pathways and vibrational energy dissipation into the solvent are of main interest. Population relaxation of X–H stretching vibrations in hydrogen-bonded systems is of ultrafast nature, because the $v(\nu_{XH}) = 1$ state is close in energy to the bending $v(\delta_{XH}) = 2$ overtone state (Fig. 7.3a,b), but also other overtone and combination levels may play a role. These overtone and combination states may facilitate vibrational energy relaxation pathways. It should be noted that changes of vibrational population induced by an IR pump pulse have also been probed by spontaneous anti-Stokes Raman scattering of a visible or near-infrared pulse [104–106]. The detection of the weak Raman signals remains a challenge, even when resonance enhancement is used by tuning the Raman probe pulse to an electronic transition [106].

7.3.2 Photon-echo spectroscopy

Femtosecond four-wave mixing spectroscopy allows for directly monitoring the macroscopic nonlinear polarization dynamics [107, 108], with which the coherence dynamics of X–H stretching vibrations can be followed in real-time [10,12,21,22,26,27,109–121]. In general, vibrational photon echo spectroscopy involves the resonant interaction of two or three laser pulses with carrier frequencies ω_i and propagation directions k_i (see Fig. 7.5d-f). Here the first pulse generates a coherent superposition of the $v(\nu_{XH}) = 0$ and $v(\nu_{XH}) = 1$ states, which then evolves during the first pulse delay period (coherence time τ). A second pulse converts the phase information into a frequency population grating in the $v(\nu_{XH}) = 0$ and $v(\nu_{XH}) = 1$ states. After a second pulse delay period (population time T) a third pulse converts the phase information again into a coherence. When the phase information has not been lost due to dephasing and spectral diffusion a macroscopic photon echo signal appears in the phase-matched directions $k_3 + k_2 - k_1$ and $k_3 - k_2 + k_1$. The photon echo signal can either be homodyne-detected in a time integrated way (Fig. 7.5d):

$$I_{hom}(\tau, T) \propto \int_{-\infty}^{\infty} dt \left| P^{(3)}(\tau, T, t) \right|^2 , \tag{7.12}$$

or heterodyne-detected by convoluting it with a fourth phase-locked light pulse serving as local oscillator (LO). The heterodyne-detected signal can either be recorded by scanning the pulse delay between echo and local oscillator (Fig. 7.5e) [122]:

$$I_{het}(\tau, T, t) \propto \mathrm{Im} \int_{-\infty}^{\infty} dt' \left\{ E_{LO}^{*}(t' - t) P^{(3)}(\tau, T, t') \right\} , \tag{7.13}$$

or by Fourier transforming the signal with a monochromator and recording it as a spectral interferogram (Fig. 7.5f) [123]:

$$S_{het}(\tau, T, \omega_3) \propto \mathrm{Re} \left\{ E_{LO}^{*}(\omega_3) P^{(3)}(\tau, T, \omega_3) \right\} , \tag{7.14}$$

that can be converted into a two-dimensional spectrum:

$$S_{het}(\omega_1, T, \omega_3) \propto i \, \mathrm{sign}(\omega_3) \int_{-\infty}^{\infty} d\tau P^{(3)}(\tau, T, \omega_3) \exp(i\omega_1 \tau) , \tag{7.15}$$

provided the relative phase between the LO and signal fields has been determined.

Different aspects of the vibrational dynamics can be probed by a specific choice of pulse delays τ and T. These different situations are:

1) *Transient grating (TG) scattering,* where the first pulse delay $\tau = 0$. A population grating is generated by pulses 1 and 2, and the signal diffracted from this grating as function of pulse delay T, is a direct measure of processes affecting this grating, such as population relaxation or rotational diffusion [124–126].

2) *Two pulse photon echo (2PE),* where the second pulse delay $T = 0$. A macroscopic polarization due to a coherent superposition generated by the first pulse is allowed to dephase during the coherence time τ, after which the interactions with pulses 2 and 3 invert the coherent superposition, and a rephasing of the macroscopic polarization can occur [10,127–129]. Only when inhomogeneity exists during the pulse sequence, a macroscopic photon echo is generated.

3) *Three pulse photon echo (3PE),* where both pulse delays τ and T are scanned. With a 3PE photon echo signal one can fully explore the dephasing and spectral diffusion dynamics affecting the spectral line shapes [130].

4) *Three pulse echo peak shift (3PEPS),* where the maximum of the echo signal along the coherence time τ is monitored as function of the population time T [12,131,132]. The decrease of this signal directly reflects the diminishing of the spectral inhomogeneity in time. For two-level systems it has been shown that the 3PEPS-signal mimics the frequency fluctuation correlation function $C(t)$ for pulse delays T at which temporal overlap between the three pulses is negligible.

5) *Two-color three pulse photon echo* (2ω*3PE),* where phase information is written into the inhomogeneously broadened transition of one particular transition, and afterwards detected by generating an echo from a second molecular transition by the third pulse [133]. This experimental configuration may provide insight into correlated frequency fluctuations for two molecular vibrations [120, 134, 135], e.g., the same microscopic solvent motions may be responsible for phase memory loss for the X–H stretching and X–H bending vibrations in a particular hydrogen bond.

7.3.3 Multidimensional spectroscopy

Until now well-established nonlinear spectroscopic techniques have been described providing insight into the vibrational dynamics of hydrogen-bonded systems, such as IVR, population relaxation, coherent wave packet motions, dephasing and spectral diffusion. Typically the information obtained is depicted in the time domain. However, the multilevel nature of the molecular vibrations that characterize hydrogen bonds has still many features that necessitate novel methodology, e.g., to elucidate questions such as the connectivity of energy levels, the cross relaxation between levels and the discrimination between different mechanisms of dephasing. Recent developments in multidimensional infrared spectroscopy have led to significant advances in deciphering the dynamics of complex multilevel systems. In these multidimensional nonlinear techniques the information on the molecular systems is typically mapped out in the frequency domain. Most papers in the field of multidimensional infrared spectroscopy treat the response of amide I vibrations in peptides and proteins [40, 122, 136], as well as other systems [65, 137], and recent extensions use the structure resolving method on transient states [138, 139]. The purpose of this section is to indicate different methodological approaches for multidimensional spectroscopy and to provide the underlying similarities in terms of information obtained, and differences with respect to experimental limitations.

In NMR spectroscopy multidimensional nonlinear techniques have been extremely successful in the study of such properties [140]. The transferability of the concepts of multidimensional NMR spectroscopy to IR spectroscopy has been postulated already in the earliest publications on two-dimensional NMR [141], however, the first two-dimensional IR spectrum was measured only recently [142]. The essential idea of 2D-IR spectroscopy is the following: A relatively narrow band pump-pulse, or a sequence of two subsequent broad band pulses, excite a vibrational state, and a delayed probe pulse measures the response of the directly pumped transition, as well as that of any other transition. The first response gives rise to a so-called diagonal contribution, whereas the second response gives rise to a cross-peak (off-diagonal peak), which reports on the connectivity of pumped and probed transitions. It has been shown that the connectivity between two states can be related to the geometry of a molecule, as well as to exchange processes between various conformations of a molecular system. In particular, the connectivity can be related to local contacts, which is the basic principle of structure determination both in 2D-NMR and 2D-IR spectroscopy. All basic NMR experiments (EXSY, NOESY) have been demonstrated also in the IR range, with the important difference that the IR experiment is intrinsically the – by many orders of magnitudes – faster method. This allows one to follow or to freeze in even the fastest motions of molecular systems on a picosecond or even faster timescale.

While this first 2D-IR experiment was a quasi-frequency domain double resonance experiment, a technique that has been developed further [135, 136, 138, 139, 143–147], time domain pulsed Fourier transform IR techniques have been devised meanwhile, mostly applied to amide I and carbonyl stretching vibrations [65, 122, 137, 148–153], but also on X–H/X–D stretching vibrations [26, 27, 97, 109–121, 154–158]. In the following the frequency and time domain approaches of 2D-IR spectroscopy, concerning the potential information content as well as regarding technical issues, are compared.

The principles of the two experimental implementations of 2D-IR spectroscopy are shown in Fig. 7.5:

1) *Double-resonance spectroscopy (also called dynamic hole burning):* The double-resonance experiment is essentially a conventional pump-probe experiment (Fig. 7.5c). An intense ultrashort (typically 100 fs) IR laser pulse

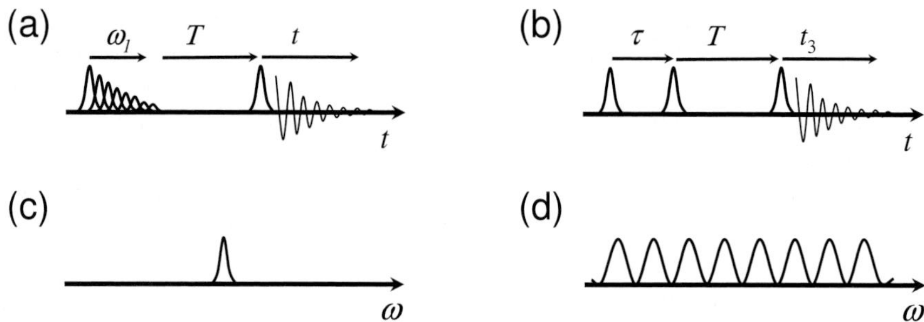

Fig. 7.7. Frequency vs. time domain 2D-IR method. Pulse sequence of the double resonance experiment of Fig. 7.5c in the time domain (a) and in the frequency domain (c). The corresponding representation of the pulsed Fourier transform experiment of Fig. 7.5e is shown in (b) in the time domain and in (d) in the frequency domain.

[159, 160], the bandwidth of which ($\approx$200 cm^{-1}) covers the whole spectral range of interest, is split into a pump and a probe beam. The difference between the double-resonance experiment and the conventional pump-probe experiment of Fig. 7.5b is the adjustable Fabry-Perot filter which the pump beam passes before reaching the sample. It consists of two partial reflectors separated by a distance which is regulated by a feedback-controlled piezoelectric mount. It slices out a narrow-band pump pulse (typical bandwidth 5-15 cm^{-1}), the center frequency of which is controlled by the computer. In this way two frequencies are defined, the center frequency of the pump pulse and the probe frequency. These are the frequency axes used in the 2D-IR spectrum. Hence, in a 2D-IR spectrum, each (horizontal) cut in the probe frequency direction represents a transient absorption spectrum obtained by pumping at the frequency on the pump (vertical) axis.

2) *Pulsed Fourier transform 2D spectroscopy (also called heterodyne-detected photon echo spectroscopy):* The pulsed Fourier transform experiment is based on a three-pulse photon echo experiment, as described in the previous section (Fig. 7.5e,f). The generated third-order field is 2D-Fourier transformed with respect to times τ (the time between the first and the second pulse) and t (the time after the third pulse), generating a 2D-IR spectrum as a function of two frequencies ω_1 and ω_3. In order to perform the Fourier transform, one needs to know the electric field irradiated by the third-order polarization (see (7.8)), rather than the time-integrated intensity, which is what 'normal' square law detectors measure. This is the reason for using the heterodyne-detected version of photon echo spectroscopy (Fig. 7.5e,f). The interferometric superposition of echo signal and local oscillator can either be performed in the time domain (by scanning the time and phase of the local oscillator) or in the frequency domain (by spectrally dispersing both beams in a spectrograph). The outcome of the interferometric superposition depends on the optical phase

between first and second pulse as well as of the third and local oscillator pulse, $(\phi_1 - \phi_2) + (\phi_3 - \phi_{LO})$, which is why the requirements on the mechanical stability of the setup and an accurate measurement of the phase are high. Schemes utilizing intrinsic [123] and active [161] phase stabilization have been proposed recently.

Principally speaking, the two types of experiments can be looked at both in the time (Fig. 7.7a,b) and in the frequency (Fig. 7.7c,d) domains and both are connected through a simple Fourier transformation. Yet, the more intuitive picture is the frequency domain picture for double-resonance spectroscopy and the time domain picture for pulsed Fourier transform spectroscopy. In the double-resonance experiment, a (relatively) narrow band pump pulse frequency selectively excites a transition (i.e. burns a hole), and a broad band probe beam probes the response of other (but the excited) transitions. If such a response exists, a cross peak shows up in the 2D-IR spectrum, which reports on the connectivity of pumped and probed transition.

For pulsed Fourier transform spectroscopy, where the natural language works in the time domain, a frequency domain picture can also be adapted: The pulse sequence of first two pulses produces a frequency grating with a spacing that depends on the coherence time τ. That is, the pulse sequence produces a 'hole-burning' pulse with sinusoidal shape (Fig. 7.7d), and one needs a Fourier transform to regain the spectral information. In the same way, one could also look at the double resonance experiment in the time domain, where the Fabry-Perot filter produces a sequence of exponentially decaying phase-locked pulses (Fig. 7.7a).

Almost all optical 2D experiments published so far are third-order experiments in the weak field regime, in which a power expansion of the nonlinear response in terms of the electric field of the incident pulses is well justified [47] (exception to the rule is the reported optical 2D spectral study of atomic Rb vapor that is clearly not in the weak field regime [162]). It is important to note that the emitted polarization $P^{(3)}(t)$ of (7.8) is *linear* with respect to the field of each interaction E_i. Therefore, at least principally speaking, the information content of frequency and time domain approaches are absolutely identical and are connected through a simple Fourier transformation. It is mostly practical issues by which both methods differ.

In the case of NMR, pulsed Fourier transform methods became much more widespread than frequency domain methods soon after their introduction, partly due to greater sensitivity because of their multiplex advantage, but also due to their greater experimental versatility. However, it is important to keep in mind that modern NMR spectroscopy works in the limit of strong fields ($\pi/2$ and π pulses in most cases). Hence, in contrast to 2D-IR spectroscopy, a simple Fourier relation between time-domain and frequency domain experiments does in general *not* exist.

7.4 Disordered hydrogen-bonded liquids

Most ultrafast infrared spectroscopic studies on ultrafast structural dynamics of hydrogen-bonded liquids have until now dealt with isotopically diluted samples, e.g. either X–H stretching vibrations in a solvent bath consisting of X–D oscillators, or vice versa. Isotopically diluted samples avoid the experimental limitation of handling extremely thin samples due to the large IR cross sections in neat liquids. However, isotopic dilution implies different vibrational systems, accompanied by different energy levels and couplings that may have different vibrational dynamics. In this Section the discussion on disordered hydrogen-bonded liquids first focuses on results obtained on the coherent response of isotopically diluted water and neat liquid water, before discussing vibrational energy redistribution and relaxation of water. The Section concludes with an example of chemical exchange in hydrogen-bonded systems. An extensive discussion on other hydrogen-bonded liquids is given in the recent overview [7].

7.4.1 Coherent response of isotopically diluted water

Protic solvents, such as water and alcohols, form extended hydrogen-bonded networks connecting the molecules that are, however, continuously changing in configuration and hydrogen bond strengths. As a result, the correlation between hydrogen bond distance (strength) and IR transition frequency of the X–H stretching oscillator [33–35, 163, 164] becomes somewhat blurred [165–167], and additional orientational parameters, such as the hydrogen bond angle, play a role [165, 166]. Nevertheless, when a particular molecule has its X–H stretching frequency at a particular value, it corresponds to a particular hydrogen bond structure. Dynamical properties of the X–H stretching oscillator, such as dephasing, spectral diffusion, anisotropy decay, and population relaxation, are thus intimately correlated to fluctuations in the hydrogen bond network.

Water has attracted the most attention in the family of protic solvents, because of its utmost important role in nature. The structure and dynamics of water as a liquid remains an intense subject of debate as indicated by recent numerous studies ranging from experiment [168–172] to theory [81, 173–176]. Until now most studies on the vibrational dynamics of hydrogen bonds in water have focused on the dephasing and spectral diffusion of isotopically diluted water, i.e. on the O–H stretching transition of HOD in D_2O, and on the O–D stretching band of HOD in H_2O. The first photon echo on HOD in D_2O was reported by Stenger et al. (Fig. 7.8) [10], where the fast dephasing time of 90 fs was ascribed to the combined effect of large diagonal anharmonicity of the O–H stretching oscillator and the fast solvent fluctuations of D_2O [177, 178]. For the HOD/D_2O system a full determination of the frequency fluctuation correlation function $C(t)$ has been achieved with three pulse photon echo peak shift (3PEPS) measurements using homodyne detection. In the early work by

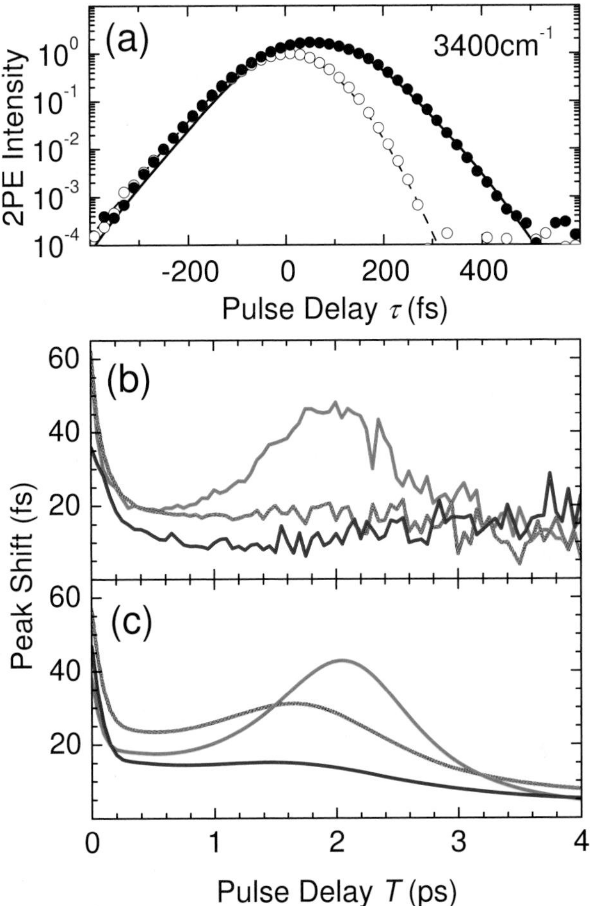

Fig. 7.8. (a) Homodyne detected 2PE signal of HOD in D_2O (*solid dots*), as recorded by tuning the excitation pulses near the maximum of the O–H stretching band. The *open circles* indicate the time resolution as measured in a CaF_2 sample. The *solid line* is a fit of the echo data assuming homogeneous broadening ($T_2 = 90$ fs) and frequency fluctuations on (sub)picosecond time scales; (b) 3PEPS signals of HOD in D_2O recorded by excitation at 3300 (*red line*), 3400 (*green line*) and 3500 cm^{-1} (*blue line*); (c) Calculated 3PEPS signals, where the frequency correlation function $C(t)$ was described using an instantaneous δ-component (leading to an effective $T_2 = 90$ fs), and two components with 700 fs and 15 ps decay constants.

Stenger et al. the bandwidth of the laser pulses was smaller than the O–H stretching line width of HOD in D_2O (pulse duration ~ 150 fs) [12]. A time dependence of the 3PEPS-signal was reported with a delayed rise on a time scale of 1-2 ps for excitation on the low-frequency side of the O–H stretching band that clearly does not mimic the temporal behavior of $C(t)$ (see Fig. 7.8).

Using the five level system of Fig. 7.6, this effect has been explained to be the result of a cancellation effect, decreasing its magnitude as function of the population lifetime T_1, of signal contributions described by Liouville space pathways that follow the ground state (diagram *I* of Fig. 7.4) and those that follow the excited state (diagrams *II* and *III* of Fig. 7.4) during the population time period T. In contrast, when much larger pulse bandwidths are used (pulse duration $\sim$ 50 fs), covering approximately the full fundamental $v(\nu_{OH}) = 0 \rightarrow 1$ O–H stretching band as well as most of the $v(\nu_{OH}) = 1 \rightarrow 2$ excited state absorption, a delayed rise has not been detected in 3PEPS measurements [111, 112, 179]. It appears that the cancellation effect will be less pronounced when the laser bandwidth covers the resonances of the fundamental $v(\nu_{OH}) = 0 \rightarrow 1$ and excited state $v(\nu_{OH}) = 1 \rightarrow 2$ transitions. The same approach of Stenger et al. has recently been used to interpret the results of a photon echo spectroscopic study of the dephasing and spectral diffusion dynamics of the N–H stretching mode of DCONHD in deuterated formamide $DCOND_2$ [97].

From the 3PEPS experiment by Stenger et al. a temporal behavior $C(t)$ has been derived that has been approximated with three time components ($\delta(t)$ (i.e. an infinitely small time constant), 700 fs and 5-15 ps). In the same work it has been noted that an equally satisfying simulation result had been obtained by using two time components ($\delta(t)$ and 2 ps) [12]. Recent echo measurements on the O–H stretching transition of HOD/D_2O with improved time resolution have led to the determination of the multicomponent temporal behavior of $C(t)$, with the shortest characteristic time constant ranging from 50-130 fs, whereas the long tail approximates an exponential decay behavior with a 1-1.5 ps time constant [109, 111, 112, 179]. Similar behavior has been found for the O–D stretching band for HOD in H_2O [113–115]. Since the O–D stretching lifetime T_1 of HOD in H_2O is $\sim$ 1.8 ps as opposed to $\sim$ 0.7 ps for the O–H stretching excited state for HOD in D_2O, the echo decay can be probed for a larger time range before the effects of the energy redistribution and thermalization set in.

Only in the case of HOD/D_2O 3PEPS measurements have been reported to show a recurrence with a period of 180 fs, suggesting that the O–H stretching mode of HOD is anharmonically coupled to the low-frequency hydrogen bond stretching mode that is underdamped with a frequency of 160 cm^{-1} [111, 112, 179]. A low-frequency recurrence has not been reported for the case of HOD/H_2O even though a similar time resolution was effective [113–115], which may hint at a smaller magnitude of anharmonic coupling between the O–D stretching vibration and the hydrogen bond stretching mode with the surrounding solvent.

More insight can be obtained by comparison of experiment and theory. For the latter mixed quantum/classical studies have been pursued. Until now only numerical studies of vibrational dynamics in isotopically diluted water have been reported. Here the intramolecular modes of the solute HOD (O–H and O–D stretching and the H–O–D bending modes) have been treated quantum mechanically, whereas the remaining degrees of freedom, in particular the hydrogen bond modes, have been treated classically. Whereas it seems plausible to describe the low-frequency translational hydrogen bond stretching modes between solute and solvent classically, this is not warranted for the librational (hindered rotational) modes, that typically have frequencies between 500-800 cm^{-1}, with a tail extending up to 1600 cm^{-1}. Several numerical approaches have been reported with which one can describe the vibrational dynamics. In one class of simulations the same molecular dynamics (MD) force field has been applied to describe solvent, solute and the solute-solvent interaction. This has been used to describe the water monomer and static water clusters [180–183]. In another class of calculations a classical MD field describes the solvent and solute-solvent interaction, whereas the solute vibrations are subject to a different field adapted to fit experimental findings [165–167, 184, 185]. In this approach polarizable force fields can be implemented [115, 186]. In a third class of modeling an MD force field is applied on the solvent only. The solute-solvent interaction and the solute have been estimated by a map accounting for the electrostatic field generated by the solvent field [111, 112, 187, 188]. The importance of electrostatic fields on the vibrational line shapes and nonlinear response has been recognized [111, 112, 189, 190]. Quantum corrections need to be considered in the calculation of vibrational line shapes and the nonlinear response [191]. Comparison between experiment and numerical simulations have led to the conclusion that the temporal behavior of $C(t)$, in particular the extended tail on picosecond time scales, is better described with polarizable water models [115, 186, 188].

Comparison between experiment and theory provides insight into the underlying mechanisms that dictate the temporal behavior of $C(t)$. It should be emphasized that dissection of $C(t)$ into a number of exponentially decaying functions does not necessarily imply an equal amount of underlying microscopic processes. The long time tail of $C(t)$ has been associated with more drastic rearrangements of hydrogen bonds, such as hydrogen bond breaking and formation [113, 115, 166] and rotational diffusion [166, 192] of the solute HOD molecule, but also more collective hydrogen bond network rearrangements [112, 188]. The recurrences at early times have been ascribed to underdamped motion of the translational stretching mode [111, 112, 165, 166]. Faster dynamics on a time scale of ~50 fs also appears in the calculated correlation functions. However, as librational modes have not been explicitly implemented in the numerical routines, one cannot directly assign these early time dynamics in the calculated $C(t)$ to be due to librational motions. Thus, to validate the proposed explanation for the fastest dynamics observed in 3PEPS experiments on HOD/D_2O [112, 179], and in 2D photon echo experiments on neat H_2O [26] (see below), it is important to include librational modes explicitly in the numerical studies.

As a final note it should be mentioned that work has been reported on two-dimensional infrared photon echo spectroscopy of the hydrogen bond dynamics of methanol-OD oligomers [118, 119, 155] and of N,N-dimethylacetamide and pyrrole in benzene solution [120].

7.4.2 Coherent response of neat liquid water

The vibrational dephasing and spectral diffusion of neat water has only recently been addressed in a heterodyne detected photon echo experiment [26]. The high optical density of the O–H stretching mode in neat H_2O and parasitic window signals in conventional samples have so far hindered attempts to access the fastest relaxation processes of liquid water. In the photon echo study of neat liquid water it has been shown that thin (800 nm) Si_3N_4 windows negligibly contribute to the nonlinear response. As a result the fastest dynamics of water (500 nm thickness) at room temperature has now been determined with 70 fs time resolution. From these experiments it has been possible to temporally resolve for the first time the anisotropy decay (occurring with a 75 fs time constant), providing insight into the coupling strengths between neighboring water molecules (Fig. 7.9). Earlier reported work has only indicated an upper limit for the anisotropy decay in neat water [193]. The spectrally resolved grating scattering and the 2D correlation spectra (Fig. 7.10) show that spectral diffusion fully occurs within 50 fs. This is much shorter than the time scale on which the hydrogen bond stretching mode can respond to excitation of the hydrogen O–H stretching oscillator (~200 fs period of a low-frequency of 170 cm^{-1}), whether underdamped or overdamped. Only librational modes, strongly coupled to the O–H stretching oscillator, and highly susceptible to

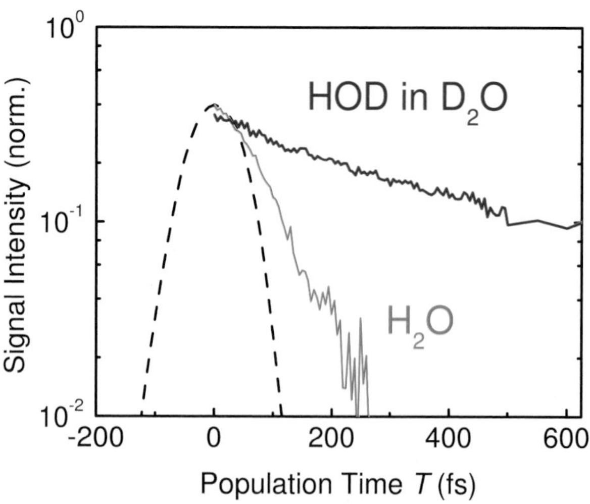

Fig. 7.9. Anisotropy decay of the transient grating scattering signal of the O–H stretching transition of neat H_2O (*red line*) and of 13 M HOD in D_2O (*blue line*). The temporal resolution is indicated by the *black dashed curve*.

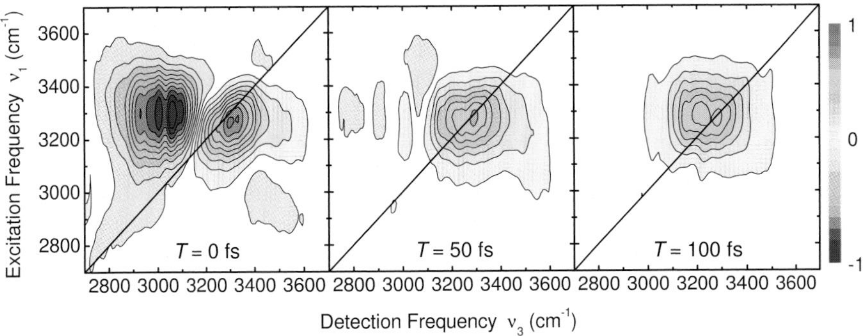

Fig. 7.10. Absorptive components of the three-pulse photon echo signal of pure H_2O for population times $T = 0, 50$ and 100 fs. The inhomogeneity, as indicated by the skewness of the positive peak along the diagonal in the $T = 0$ fs spectrum, has almost completely disappeared in the $T = 50$ fs plot, clearly indicative of the rapid dephasing and ultrafast memory loss of the O–H stretching polarization in H_2O.

the fluctuating hydrogen bond network, can be assigned as the reason for this ultrafast spectral diffusion. These recent experimental results on neat water emphasize the necessity of theoretical studies on the determination of the couplings of librational modes with the O–H stretching oscillator [194], as well as intermolecular resonant vibrational energy transfer [195, 196], fluctuations of the hydrogen bond network of neat water [176, 197], and reorientational dynamics [198].

7.4.3 Vibrational energy redistribution and relaxation of O–H/O–D stretching and bending vibrations in water

When a molecular vibration is excited, couplings with the other molecular normal modes cause IVR, and intermolecular vibrational excitation exchange (vibrational energy relaxation, VER) [199–201]. IVR may lead to recurrences in vibrational excitation for isolated small molecules in the gas phase. The sizeable number of degrees of freedom in large molecules provide an efficient intramolecular heat bath. For VER to surrounding molecules, e.g., solvent shells, the vibrational energy transfer occurs in one direction and a true dissipation takes place. The anharmonicities of molecular vibrations involved in the hydrogen bond are significant. As a result, these vibrational modes (e.g., in-plane and out-of-plane X–H bending, librational and hydrogen bond stretching modes) are the first candidates to accept the vibrational energy corresponding to one quantum of X–H/X–D stretching vibration. This vibrational redistribution process occurs on ultrafast time scales: the X–H/X–D stretching oscillator in a hydrogen bond has a vibrational population time T_1 of ~ 1 ps or less [7], typically much faster than VER to the solvent, and certainly faster than the time scale of a full thermalization of the vibrational excess energy, i.e., when a Boltzmann distribution is reached over all vibrational modes of solute and solvent with an elevated temperature. Despite the fact that only in the case when a full equilibration is reached, and a vibrational temperature can be defined, the situations of elevated population numbers of a fraction of intramolecular vibrational modes – while the X–H/X–D stretching vibration has returned to the $v(\nu_{OH}) = 0$ state – have been described as hot ground states (cf. Sect. 7.3.1 and Fig. 7.6).

The magnitude of anharmonic couplings of the X–H/X–D stretching oscillator with other modes dramatically increase with increasing hydrogen bond strength. The increased frequency down-shift of the X–H/X–D stretching band with increasing hydrogen bond strength may lead to a more efficient Fermi coupling with the X–H/X–D first overtone band, and often this has been assumed to be the relaxation pathway for the X–H/X–D stretching excitation, although direct evidence that IVR occurs along this pathway has until now been only scarcely reported (see Sect. 7.5.2) [25].

Water has three intramolecular degrees of freedom [62,104]. For HOD these are the O–H and O–D stretching and the H–O–D bending modes, each with distinct transition frequencies. In H_2O, on the other hand, the antisymmetric and symmetric O–H stretching modes have similar transition frequencies, whereas the O–H bending fundamental frequency is about half of the stretching frequencies. In the case of HOD in H_2O or D_2O a significant frequency mismatch of 500 cm^{-1} exists between the $v(\nu_{OH}) = 1$ and $v(\delta_{HOD}) = 2$ levels [184, 201, 202]. Here the O–H stretching vibration may dissipate the excitation energy through the bending pathway, but the involvement of the hydrogen bond modes (librations, hydrogen bond stretching modes with the H_2O or D_2O solvent bath) should play an active acceptor role in the primary step in energy redistribution as well. In contrast, for neat liquid H_2O the energy mismatch between the $v(\nu_{OH}) = 1$ and $v(\delta_{HOD}) = 2$ levels is minor, and within the spectral width of the O–H stretching vibration a clear overlap exists. Energy redistribution through the bending pathway is - thus - much more facilitated by the vibrational mode structure of the hydrogen-bonded H_2O molecule, although here also the excess energy has eventually to be channeled into the hydrogen bond network vibrational modes.

The bending vibration of any isotopomer of water (HOD, H_2O, D_2O) is the intramolecular degree of freedom with the lowest fundamental frequency. As a result, population relaxation of the bending mode can only proceed by dissipation into the intermolecular modes of the hydrogen bond network, making the bending vibration an ideal means for studies of energy transport through the hydrogen bond network triggered by initial local vibrational excitation.

Measurements of the O–H stretching population relaxation for isolated non-hydrogen-bonded H_2O molecules dissolved in $CDCl_3$ have resulted in a T_1 time of 36 ps for the asymmetric O–H stretching vibration [203, 204]. Isolated HOD molecules dissolved in acetonitrile have an O–H stretching lifetime value of $T_1 = 12$ ps, whereas an increased HOD fraction leads to a gradual decrease of this value approaching that of HOD in D_2O ($T_1 = 0.7$ ps) [205].

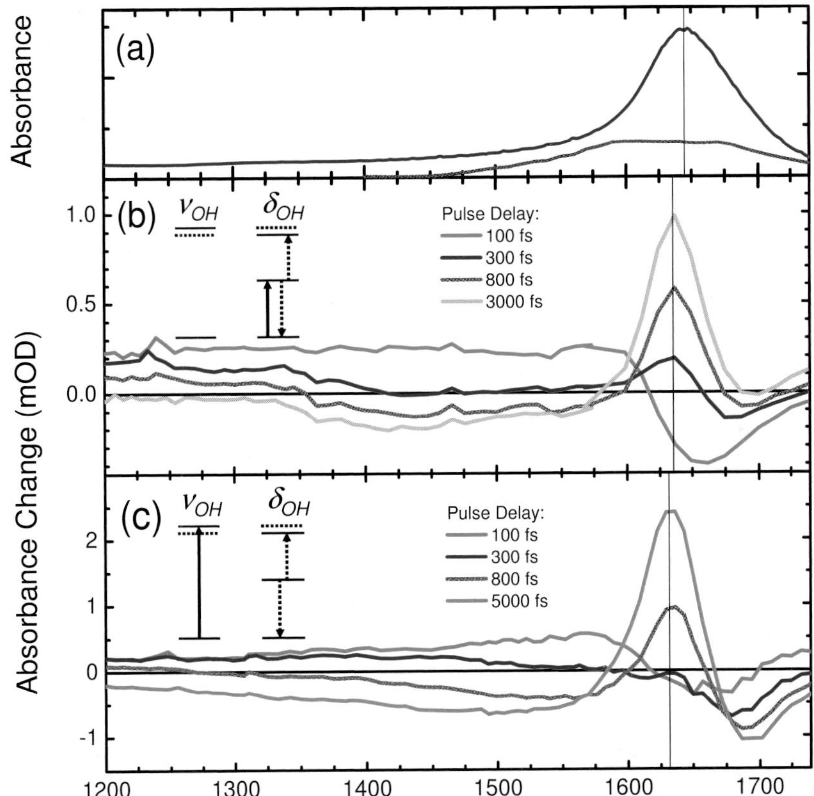

Fig. 7.11. (a) Linear vibrational absorption spectrum of pure H_2O, showing the O–H bending band and a broad librational absorption (*blue line*) as well as the spectral bandwidth of the IR pump pulse (*gray curve*); (b) Transient infrared spectra of H_2O measured with pump pulses centered at $1650\,\mathrm{cm}^{-1}$, showing the $v(\delta_{OH}) = 0 \rightarrow 1$ bleaching and an extremely broad $v(\delta_{OH}) = 1 \rightarrow 2$ excited state absorption. After population relaxation of the bending vibration one finds a frequency red-shifted $v(\delta_{OH}) = 0' \rightarrow 1'$ absorption of the hot bending ground state, as well as a red-shifted librational band; (c) Transient infrared spectra of H_2O in the bending and librational frequency region measured with pump pulses centered at $3150\,\mathrm{cm}^{-1}$, pumping the O–H stretching band.

This decrease with increasing water fraction is accompanied by a frequency down-shift of the O–H stretching band. Here it has been argued that the frequency red-shift of the O–H band upon hydrogen bonding results into a more efficient Fermi coupling with the bending overtone, with a shorter O–H stretching T_1 population lifetime as a result. This argument has also been used to explain the decrease of the O–H stretching lifetime for liquid H_2O, measured to be 200 fs for neat water [26, 100, 206]. The bending overtone for the H_2O is located around $3250\,\mathrm{cm}^{-1}$, whereas for HOD the mismatch of around $500\,cm^{-1}$ between bending overtone and O–H stretching fundamental occurs. An increased red-shift of the O–H stretching band with more overlap with the O–H bending overtone is, however, not the only prerequisite for a faster O–H stretching population decay. Solvent fluctuations are necessary for enabling all members of the ensemble to sample with its O–H stretching oscillator the right resonance frequency space facilitating efficient energy redistribution [207], for instance, through Fermi coupling with the bending overtone.

Another example providing evidence that solvent fluctuations are playing a key role in the relaxation of the O–H stretching vibration are water molecules bound in solvation shells of ions. These exhibit longer O–H stretching population lifetimes (for an overview see Sect. 6.2 of [7]); for HOD up to 9 times longer than in bulk water [208], which cannot be explained by the frequency shifts of the O–H stretching bands only. Here it is known that water near ions has a lower mobility, as indicated by the decreased anisotropy decay of the O–H stretching oscillator of water in ion shells [209] and by recent molecular dynamics calculations [210]. A lifetime increase of O–H stretching vibrations of up to 0.8 ps has also been observed for H_2O molecules confined into micelle surroundings [211, 212]. Here the large number of ionic heads and counter ions lead to a lower water mobility (as exemplified by the inhomogeneity and anisotropy decay of the hydrogen stretching vibrations persisting well into the picosecond domain [116, 213]).

To test the validity of the O–H/O–D stretching to bending relaxation pathway mechanism, one has to consider population kinetics using the experimentally determined O–H/O–D bending lifetime as input. Single H_2O molecules dissolved in $CDCl_3$ have an O–H bending lifetime of 28.5 ps [214]. When water is hydrogen bonded to nearby solvent molecules, hydrogen bond stretching and librational modes enhance the bending relaxation rates significantly. When one quantum of O–H/O–D stretching modes is redistributed over such vibrational modes as O–H/O–D bending, librational and hydrogen bond stretching modes, making these highly excited, anharmonic couplings of these modes with the O–H/O–D stretching oscillator lead to transient frequency shifts [215]. The O–H/O–D stretching band appears frequency up-shifted when a molecule is in a "hot" ground state (see Sect. 7.3.1), clearly different from the $v(\nu_{OH}) = 1 \rightarrow 2$ excited state absorption that appears frequency down-shifted due to the diagonal anharmonicity. The O–H/O–D bending vibration exhibits the more typical transient frequency down-shifting for "hot" ground states (see Fig. 7.11) that are more difficult to distinguish from the vibrationally excited $v(\delta_{OH}) = 1$ states that also have red-shifted $v(\delta_{OH}) = 1 \rightarrow 2$ excited state absorption contributions [28, 31, 216, 217].

Only recently the population lifetime of the bending vibration of HOD in D_2O [217], of H_2O in D_2O [216], as well as of H_2O in neat water [28] have been measured with IR-pump/IR-probe spectroscopy. For HOD in D_2O a value of 390 fs was found [217], very close to recent theoretical estimations [218, 219]. For the determination of the bending vibration of H_2O it is important to measure with a time resolution as short as 100 fs. To eliminate unwanted cross phase modulation by cell window material that prevents analysis of the response at early pulse delays [216], a recent study on neat H_2O using thin Si_3N_4 windows with negligible nonresonant response, has led to a value of 170 fs for the population lifetime of the water bending vibration [28]. In this work also the important signal contributions of the nonlinear response of librational excitations in the frequency range between 1200 and 1700 cm^{-1} have been indicated (Fig. 7.11). A H_2O bending lifetime of 170 fs seems to be in contradiction with the previously reported value of 1.4 ps, found with IR-pump/anti-Stokes Raman probe measurements [220]. Measurements on the intramolecular (stretching, bending) modes of neat H_2O, however, are always intertwined with frequency shifts caused by anharmonic couplings to the intermolecular hydrogen bond network librational and hydrogen bond stretching modes, since the vibrational energy accepting molecules are of the same species as the initially excited H_2O molecule carrying one quantum of the O–H stretching or bending vibration. This may be the reason for the apparent discrepancy between the recent IR-pump/IR-probe and the IR-pump/anti-Stokes Raman probe measurements, where a larger effect on the measured signals is expected in the latter case due to a substantial fraction of excited molecules [221–223]. For H_2O a fully completed cooling cannot be observed until heat transport out of the laser beam interaction zone has finished that occurs well into the microsecond time regime. Only in the case of water confined in micelles vibrational excitation transfer to the micelle molecular units, and eventually to the organic phase lead to a full cooling of water back to room temperature with micelle size-dependent picosecond time scales [224–226].

In light of the 170 fs population relaxation time of the bending vibration of H_2O, and the 200 fs lifetime of the O–H stretching vibration in H_2O, one can estimate that the transient population of the bending $v(\delta_{HOH}) = 1$ state will not reach more than a maximum value of 0.35 compared to the hypothetical case where the bending vibration has an infinite lifetime, provided all excited molecules relax through the bending vibrational pathway. As a result, one has to be careful in concluding that a red-shifting of the bending vibration upon excitation of the O–H stretching is a direct indication of the transient population of the bending mode [205,212,223,227–229], as excitation of librational and hydrogen bond stretching modes [28], and even the O–H stretching vibration [31], also will play a role in the observed features. Figure 7.11c shows how upon O–H stretching excitation of H_2O, the O–H bending shows an instantaneous response, pointing to a Fermi resonance between the $v(\nu_{OH}) = 1$ and $v(\delta_{OH}) = 2$ states. As a result the $v(\delta_{OH}) = 2 \rightarrow 3$ and $v(\delta_{OH}) = 1 \rightarrow 2$ transitions of the bending oscillator are strongly broadened, extending over more than 200 cm^{-1}. A delayed increase in stimulated emission of the $v(\delta_{OH}) = 1 \rightarrow 0$ transition observed at 1650 cm^{-1}, before contributions of a "hot" bending ground state begin to dominate, is in perfect agreement with a sequential population transfer from the $v(\nu_{OH}) = 1/v(\delta_{OH}) = 2$ manifold to the $v(\delta_{OH}) = 1$ state (with partial vibrational energy release into librational degrees of freedom with a 200 fs time constant), followed by bending relaxation to the $v(\delta_{OH}) = 0$ ground state, again with energy dissipation to librational modes.

7.4.4 Chemical exchange

The basic idea of exchange spectroscopy, both in NMR and in the IR spectral range, is the following: A particular group in the molecule is tagged, (i.e. a nuclear spin in the NMR case and a local normal mode in the IR case) by excitation. As long as that excitation lives, and does not decay due to T_1 relaxation, the molecular group will carry it around. When the molecular group changes its spectroscopic properties due to chemical exchange, i.e. changes its chemical environment due to some diffusive process and as a result of that changes its absorption frequency, it will just take its tag (i.e. its excitation) with it.

The basic assumption of exchange spectroscopy is that the excitation of the molecular group does *not* influence the way how the molecules move in time; the excitation is just used as a label to follow how molecules move in time. They move not because of the excitation, but due to room temperature diffusive motion. This assumption certainly is valid in the case of NMR, where the excitation energy is far below $k_B T$ and hence, very unlikely to influence the course of a reaction just because of an energy argument. However, in the case of an IR excitation, this assumption might sometimes be questionable. Energy, in principle, would be sufficient to change the outcome of a diffusive process, but what helps is the fact that vibrational transitions, which are harmonic oscillators to a very good approximation, tend to decouple from each other. However, in the case of strongly coupled states, such as in hydrogen-bonded systems, this assumption might break down [230].

2D spectroscopy is ideally suited to study ultrafast chemical exchange processes. This was first demonstrated by Woutersen and Hamm, studying the exchange between hydrogen bonded and non-hydrogen bonded N-methylacetamide (NMA) dissolved in methanol [231]. More recently and almost in parallel, chemical exchange was also studied by Hochstrasser and co-workers for hydrogen-bond exchange between CH_3OH linked to the cyano-site of CH_3CN [121], as well as by Fayer and co-workers [157,158] for exchange between hydrogen-bonded phenol-benzene complexes as well as for rotational isomerization in ethane derivatives [232]. When phenol-*d* and benzene are mixed together in CCl_4, both exist in an equilibrium between a complexed and a dissociated form (Fig. 7.12a), forming a weak hydrogen bond between the OD group of phenol and the benzene ring. Upon hydrogen bonding, the O–D stretch mode changes its frequency from about 2670 cm^{-1} to 2530 cm^{-1}. Hence, in the linear absorption spectrum, two peaks appear which represent the equilibrium ratio between both configurations (which can be tuned by adjusting the benzene concentration). However, the linear spectrum gives no

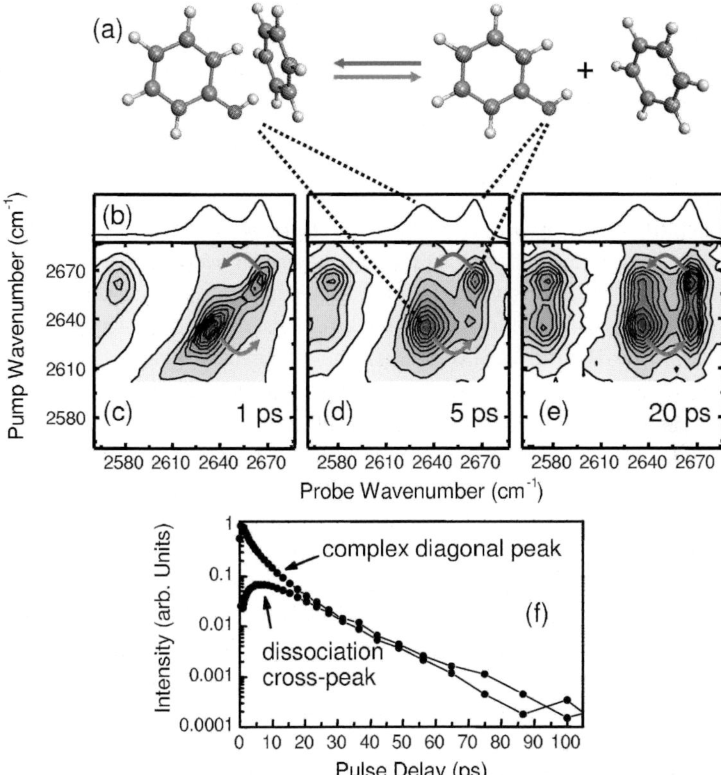

Fig. 7.12. (a) Chemical exchange between uncomplexed phenol and phenol hydrogen-bonded to benzene; (b) linear IR spectrum; (c-e) 2D-IR measurements using the double resonance approach measured at three different pulse delays show the onset of exchange between the two conformations; (f) the magnitude of the diagonal and off-diagonal peaks as function of pulse delay show that full equilibration occurs within 20 ps.

clue about the exchange rate between both configurations, except that it is slower than some threshold value, in which case both bands would start to merge due to motional narrowing effects.

The 2D spectrum, in contrast, provides exactly this information. The essential parameter varied in an exchange experiment is the so-called waiting or mixing time τ_m, i.e. the time between labeling and probing the O–D vibrator. Initially, after 1 ps, essentially no exchange had occurred. As a result, the 2D spectrum is still diagonal, saying that each labeled molecule, complexed or dissociated, is still in the configuration in which it was when exciting it. As times go on, however, labeled O–D vibrators change from complexed to dissociated form and vice versa, carrying their tag with them. As a result, cross peaks appear in the 2D spectrum which report on molecules which were in one configuration when pumping it, but jumped to the other configuration during the waiting time before probing it.

The intensities of diagonal and cross peak directly report on the time scale of the exchange process (Fig. 7.12e). Initially, the cross peak intensity is zero, but as both configurations equilibrate, the intensities of diagonal and cross peak approach each other. In fact, the ratio between both intensities would directly reflect the equilibration time. Figure 7.12f also demonstrates a limitation of exchange spectroscopy: The exchange process can be observed only as long as the vibrational excitation, that is the tag, lives. As T_1 relaxation times are relatively short for vibrational transitions in the solution phase (typically 1-20 ps), the time window accessible to 2D-IR spectroscopy is relatively short. In that sense, 2D-IR and 2D-NMR exchange spectroscopy are complimentary, as they address completely different time regimes. Improving the signal-to-noise ratio of the 2D apparatus may increase the accessible time window. For example, in Fig. 7.12 the exchange process has been followed up to 100 ps although the T_1 relaxation time is $\approx$ 12 ps, but T_1 relaxation will eventually stop the possibility to observe the process. For hydrogen-bonded systems the vibrational cooling rates will delimit the observation time window, as these systems relax into spectrally shifted hot ground states (Fig. 7.6). If one were to remove the vibrationally excited molecules by, for example, a photochemical process in the vibrationally excited state so that they could not fall back into the initial state, one could follow chemical exchange for unlimited time for the hole burned into the ensemble.

It is worth noting that Fig. 7.12 and the data shown in [157] have been measured with the two complimentary techniques of 2D-IR spectroscopy discussed in Sect. 7.3.3, one in the frequency domain (Fig. 7.12) using the double resonance scheme, and one in the time domain [157] using the heterodyne detected photon echo scheme. Besides this, the conditions were exactly the same. The comparison shows that in principle the information content of both techniques is indeed exactly the same.

7.5 Hydrogen bonds with ordered molecular topologies

Inter- as well as intramolecular hydrogen bonds with well-defined geometries appear in a number of different classes of compounds including numerous ordinary organic molecules, carbohydrates, carboxylic acids, amino acids, nucleosides and nucleotides as well as biomolecular macromolecules such as proteins or nucleic acid structures [48]. Hydrogen bonding crucially determines the molecular structure and thus, the functionality in these systems. In this Section two examples of medium-strong hydrogen bonds with well-defined geometries are presented. Several ultrafast nonlinear spectroscopic techniques are employed to discern the different hydrogen bond vibrational coupling mechanisms.

7.5.1 Coherent nuclear motions and relaxation processes in intramolecular hydrogen bonds

The anharmonic coupling of the X–H stretching oscillator to low-frequency hydrogen bond modes results in a vibrational multilevel system and progressions of vibrational transitions within the X–H stretching absorption band. Broadband excitation of such a system by a femtosecond pulse generates a nonstationary superposition of vibrational levels of the low-frequency mode, i.e. a vibrational wave packet that propagates in the potential defined by the high frequency X–H stretching vibrations (see Fig. 7.2). Oscillatory wave packet motions have been observed for different molecular systems forming a medium-strong intramolecular hydrogen bond of well-defined geometry. As a prototype example, results are presented here of a detailed pump-probe study of *o*-phthalic acid monomethylester (PMME-H) and its deuterated analog (PMME-D) (Fig. 7.13).

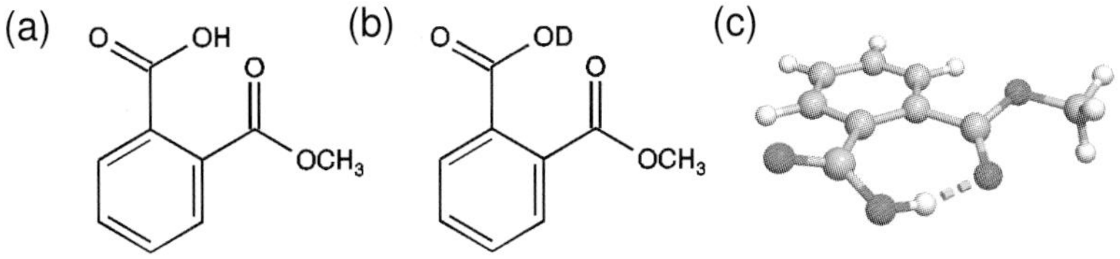

Fig. 7.13. *o*-Phthalic acid monomethylester (PMME); (a) PMME-H; (b) PMME-D; (c) lowest energy conformer.

The structure and the vibrational spectra of PMME were analyzed by quantum chemical calculations which are discussed in [13]. The lowest energy conformer of PMME consists in a quasi-planar structure in which the hydrogen bond is part of a seven-membered ring. This 7-ring does not entirely fit into the plane of the benzene ring so that the substituents are slightly tilted out-of-plane (Fig. 7.13c). The heavy atom hydrogen bond O$\cdots$O distance of 256 pm and the H$\cdots$O distance of 160 pm indicate an intramolecular hydrogen bond of intermediate strength and a well-defined geometry.

The steady-state O–H and O–D stretching bands of PMME-H and PMME-D are shown in Fig. 7.14a. In the femtosecond pump-probe experiments with PMME-H, the infrared pulses were tuned throughout the O–H stretching band, as indicated by the pulse spectra in Fig. 7.14b, and the spectrally integrated probe pulses transmitted through the sample were detected. The time evolution of the nonlinear change of absorbance at the different spectral positions is plotted in Fig. 7.15. Apart from small solvent and sample cell window contributions around delay zero, the signal originates from changes of the O–H stretching absorption of PMME-H. For probe frequencies between 2500 and 3200 cm^{-1}, a pronounced decrease of infrared absorption is observed that is strongest around the maximum of the steady-state infrared band. The

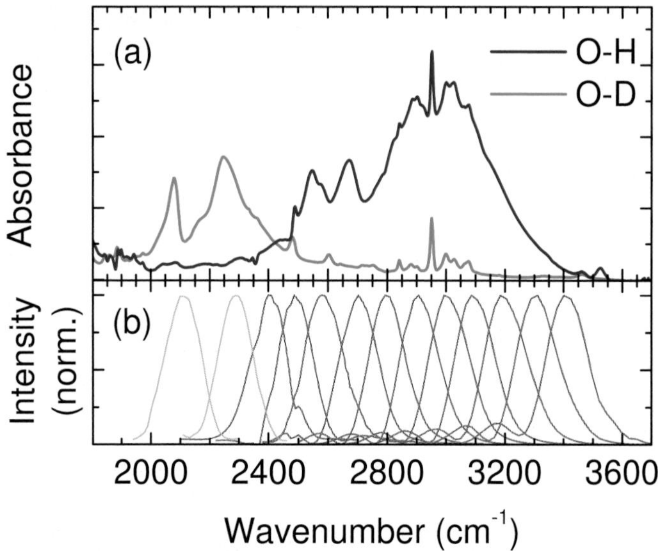

Fig. 7.14. (a) Steady-state O–H stretching band of PMME-H (*blue line*) and O–D stretching band of PMME-D (*red line*). The sharp line at 2950 cm^{-1} is due to C–H stretching absorption; (b) Spectra of the femtosecond pulses used in the pump-probe experiments with spectrally integrated detection (*green lines*) and with spectrally resolved detection (*orange lines*).

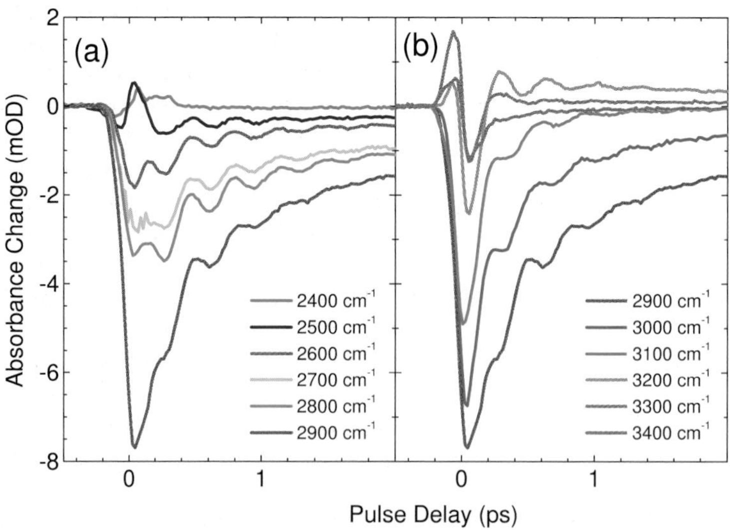

Fig. 7.15. PMME-H: Pump-probe transients measured using spectrally integrated detection. The absorbance change is plotted as function of delay between pump and probe pulses at the frequencies indicated.

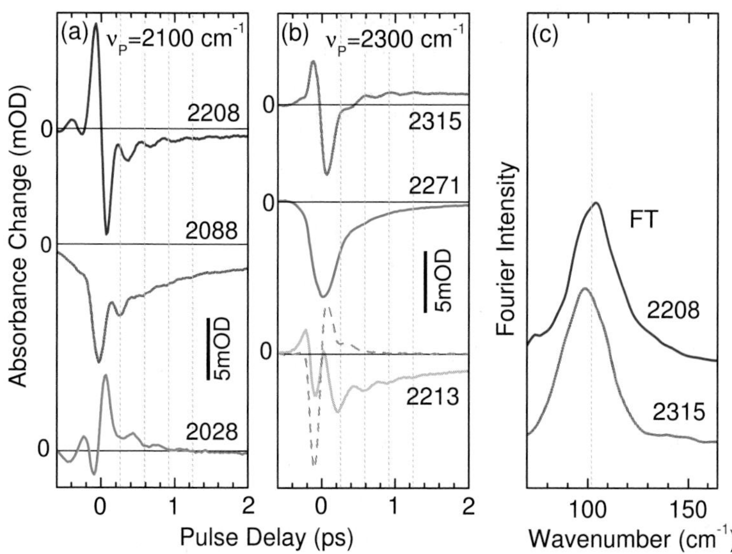

Fig. 7.16. PMME-D: (a,b) Time-resolved absorbance changes measured using spectrally resolved detection (ν_p: center frequency of the pump pulses, probe frequencies are given in cm^{-1}). The *dashed line* in (b) represents the response of the solvent; (c) Fourier transforms (FT) of the oscillatory signals.

absorption changes show a fast decay within the first 1 ps, followed by a slower relaxation on a time scale of 30 ps (not shown). At 3300 and 3400 cm^{-1}, an absorption increase is found which rises within the first 200-300 fs and shows a slow decay within 30 ps. All transients are superimposed by strong oscillatory signals with a frequency of 100 cm^{-1} and a damping time of approximately 500 fs.

The transient O–D stretching absorption of PMME-D reveals a similar behavior. In Fig. 7.16, results are presented for pump pulses centered at 2100 cm^{-1} (a) and 2300 cm^{-1} (b) and spectrally dispersed probe pulses. Depending on the probe frequency, one observes either bleaching or enhanced absorption. At 2028 cm^{-1} (Fig. 7.16a), the enhanced absorption decays with a time constant of about 400 fs. The bleaching transients (except for a probe frequency of 2271 cm^{-1}) show a fast decay with 400 fs, followed by a slower decay on a 20 to 30 ps time scale (not shown). Again, pronounced oscillations with a frequency of 100 cm^{-1}(Fig. 7.16c), i.e. identical to that in PMME-H, are superimposed on the incoherent pump-probe signals.

For an analysis of the pump-probe data, the level scheme in Fig. 7.6 has been applied (A refined level scheme including the potential curves for the low-frequency hydrogen bond motion will be presented in the following Subsection). In addition to the vibrational transition from the $v(\nu_{OH}) = 0$ to the $v(\nu_{OH}) = 1$ state, the red-shifted $v(\nu_{OH}) = 1$ to $v(\nu_{OH}) = 2$ transition and the blue-shifted $v(\nu_{OH}) = 0'$ to $v(\nu_{OH}) = 1'$ transition of molecules with a vibrationally hot ground state are taken into account. The pump pulse promotes part of the molecules from the $v(\nu_{OH}) = 0$ to the $v(\nu_{OH}) = 1$ state, resulting in a bleaching of the $v(\nu_{OH}) = 0 \rightarrow 1$ transition due to the depletion of the $v(\nu_{OH}) = 0$ state and stimulated emission from the $v(\nu_{OH}) = 1$ state. Such two contributions dominate the bleaching signals observed at delay times < 1 ps in Figs. 7.14 and 7.15. The transient population of the $v(\nu_{OH}) = 1$ state gives also rise to a red-shifted $v(\nu_{OH}) = 1 \rightarrow 2$ absorption which is evident from the PMME-D transient at 2028 cm^{-1} (Fig. 7.16a) and – much less pronounced – from the PMME-H data taken at 2400 cm^{-1} (Fig. 7.15). The decay of excited state absorption reflects the depopulation of the $v(\nu_{OH}) = 1$ level with time constants of $\sim$400 fs for PMME-D and < 250 fs for PMME-H. Population relaxation is due to a coupling of the $v(\nu_{OH}) = 1$ state to other vibrational modes at lower frequency, transferring the energy of the O–H (O–D) excitation into the vibrational manifold (the details of which will be discussed in the following Subsection). As a result, molecules with a hot vibrational system are created in which the O–H (O–D) stretching mode is in the $v(\nu_{XH}) = 0'$ ground state. In the hot molecules, the H-bond is weakened and, thus, the O–H (O–D) stretching absorption is blue-shifted compared to the initial absorption. This mechanism gives rise to (i) the enhanced absorption of PMME-H at 3300 and 3400 cm^{-1} and of PMME-D at 2315 cm^{-1}, and (ii) to the bleaching signals extending to delay times longer than 1 ps. Dissipation of the vibrational excess energy into the surrounding solvent – a process occurring on a multitude of time scales extending up to several tens of picoseconds – leads to the slow nonexponential decay of the absorption changes. It should be noted that the fast rise of the blue-shifted absorption of PMME-H at 3400 cm^{-1} gives independent support for a decay of the $v(\nu_{OH}) = 1$ state with a sub-250 fs time constant.

The oscillatory signals present in the pump-probe transients are discussed next. In both PMME-H and PMME-D, the oscillations display a frequency of 100 cm^{-1}. The oscillations are due to wave packet motions along a low-frequency mode which couples anharmonically to the fast stretching modes. Wave packet generation is visualized with the help of the potential energy scheme in Fig. 7.2a which is based on an adiabatic separation of the time scales of the fast stretching and the slow low-frequency motion. In the experiments, the spectral field envelope of the pump pulse covers several levels of the low-frequency mode, allowing for the excitation of a coherent superposition of low-frequency states, i.e. of wave packets. Wave packets in the $v(\nu_{OH}) = 1$ state of the stretching oscillator are created via vibrational absorption whereas wave packet motion in the $v(\nu_{OH}) = 0$ state is induced by a Raman-like process within the bandwidth of the pump pulse. This Raman process is resonantly enhanced by the vibrational transition.

The enhanced absorption of PMME-D at 2028 cm^{-1} (Fig. 7.16a) is due to the $v(\nu_{OH}) = 1 \rightarrow 2$ transition and – thus – the oscillations on this transient reflect wave packet motion in the $v(\nu_{OH}) = 1$ state. The damping time of the oscillations of 500 fs is close to the $v(\nu_{OH}) = 1$ lifetime, suggesting that vibrational population relaxation represents a major damping mechanism. In PMME-H, the $v(\nu_{OH}) = 1$ lifetime of the O–H stretching mode has a substantially smaller value of 200 to 250 fs, i.e. close to the oscillation period of the $100\,\mathrm{cm}^{-1}$ mode. As a result, oscillations in the $v(\nu_{OH}) = 1$ state are not observed.

Wave packet motions in both the $v(\nu_{OH}) = 0$ and $v(\nu_{OH}) = 1$ states are relevant for the analysis of the oscillations found on top of the bleaching signals. As shown schematically in Fig. 7.2a, the $v(\nu_{OH}) = 0 \rightarrow 1$ transition frequency of the O–D/O–H stretching mode depends on the elongation along the low-frequency mode. Wave packet motion along this coordinate results in a periodic modulation of the O–D/O–H transition frequency. In PMME-D (Fig. 7.16), this mechanism leads to the oscillatory blue-shift of the O–D stretching band which is evident from the oscillatory enhancement of absorption at $2315\,\mathrm{cm}^{-1}$. Consistent with this picture, there are no oscillations around the maximum of the O–D band at $2271\,\mathrm{cm}^{-1}$ and oscillations of opposite phase at $2213\,\mathrm{cm}^{-1}$. PMME-H (Fig. 7.15) shows the same behavior. Here, however, only wave packet motion in the $v(\nu_{OH}) = 0$ state is relevant [13]. As frequency shifts of the O–D/O–H stretching bands are a measure of the length of the hydrogen bond, the oscillatory blue-shift is equivalent to a modulation of the O$\cdots$O distance in PMME, determining the strength of the hydrogen bond.

A normal mode analysis of PMME provides insight into the microscopic nature of the low-frequency mode displaying a frequency of $100\,\mathrm{cm}^{-1}$. As discussed in detail in [8, 13], the periodic modulation of the O$\cdots$O distance is attributed to an out-of plane torsional motion by which the two ring substituents in PMME move relative to each other, thus changing the O$\cdots$O distance. This picture has been confirmed by anharmonic potential energy surface calculations and quantum dynamics simulations [14–17]. The frequency of this mode remains unchanged upon H/D exchange and – thus – identical oscillation frequencies occur in the PMME-H and PMME-D data. This intramolecular mode displays a comparably long damping time on the order of 0.5 ps, resulting in an underdamped character of this mode. Note, that the population relaxation time of the low-frequency mode in the $v(\nu_{OH}) = 0$ ground state is about 1.7 ps as obtained from classical molecular dynamics simulations of the force-force correlation function [18]. This hints at a more complex damping mechanism including, e.g., intramolecular anharmonic couplings. This is in contrast to many theoretical studies of H-bond dynamics in which overdamped low-frequency motions have been assumed [39, 44, 233]. It should be noted that there are modes at higher frequencies also modulating the length of the hydrogen bond [14]. The spectral bandwidth of the generated pump pulses and the time resolution of the experiment, however, were not sufficient to excite wave packets along such modes and to follow their time evolution.

7.5.2 Intramolecular vibrational redistribution of the O–H stretching excitation

In the case of PMME-H the fate of the O–H stretching excitation after T_1 relaxation has also been investigated. Two-color pump-probe spectroscopy on the O–H stretching, C=O stretching and O–H in-plane bending vibrations of PMME-H, have demonstrated that excitation energy of the O–H stretching oscillator is redistributed on a subpicosecond time scale along the O–H bending vibration (Fig. 7.17) [25].

Here the dynamics of the O–H bending vibration is measured, by following the decay of the O–H bending bleach at 1415 cm^{-1}, and the excited state absorption at 1390 cm^{-1}, as observed for three different pumping frequencies: excitation of the O–H stretching vibration at 2900 cm^{-1} (Fig. 7.17a,b), excitation of the C=O stretching vibration at 1740 cm^{-1} (Fig. 7.17e,f), and excitation of the bend vibration itself at 1400 cm^{-1} (Fig. 7.17c,d). In all cases a red-shifted feature appears within temporal resolution, albeit with significant differences in shape. Red-shifted vibrational resonances are expected both for excitation of the O–H bending mode (when $v(\delta_{OH}) = 1$), and for cases when other - in particular low-frequency - modes are excited. Assignment of the spectral contributions due to these clearly distinct situations is not trivial in itself, but it is the dynamics of these different states that enable a clear interpretation. Whereas excitation of the O–H stretching and O–H bending leads to similar behavior of the transient absorbance in the bend region, with both 800 fs and 7 ps temporal components, excitation of the C=O stretching vibration only results in the 7 ps component in the response of the O–H bending vibration. From these results it can be concluded that with an O–H stretching lifetime of 220 fs and an O–H bending lifetime of 800 fs it is experimentally feasible to detect a transient population build-up of the bending vibration after O–H stretching excitation.

In order to get a better idea of the multidimensional aspects of IVR in PMME quantum dynamical model calculations of the energy flow have been performed. A full-dimensional description of the intramolecular dynamics driven by an IR laser field has been given in [14] by combining a reaction surface Hamiltonian with a time-dependent self-consistent field propagation of nuclear wave packets. In agreement with experiment (cf. Figs. 7.15 and 7.16) this parameter-free simulation revealed the periodic modulation of the transient absorption signal due to the anharmonic coupling between the excited O–H stretching vibration and a low-frequency hydrogen bond mode. Despite its high-dimensional nature the model did not show any signatures of intramolecular energy randomization during the first 3 ps. Subsequently, in [16, 17] reduced 9- to 19-dimensional reaction surface models have been investigated on the basis of a multiconfiguration time-dependent Hartree approach [234]. Here, it could be demonstrated that the additional flexibility of the multiconfiguration wave function can capture the onset of IVR with an estimated time constant of about 20 ps [17]. However, this is still in strong

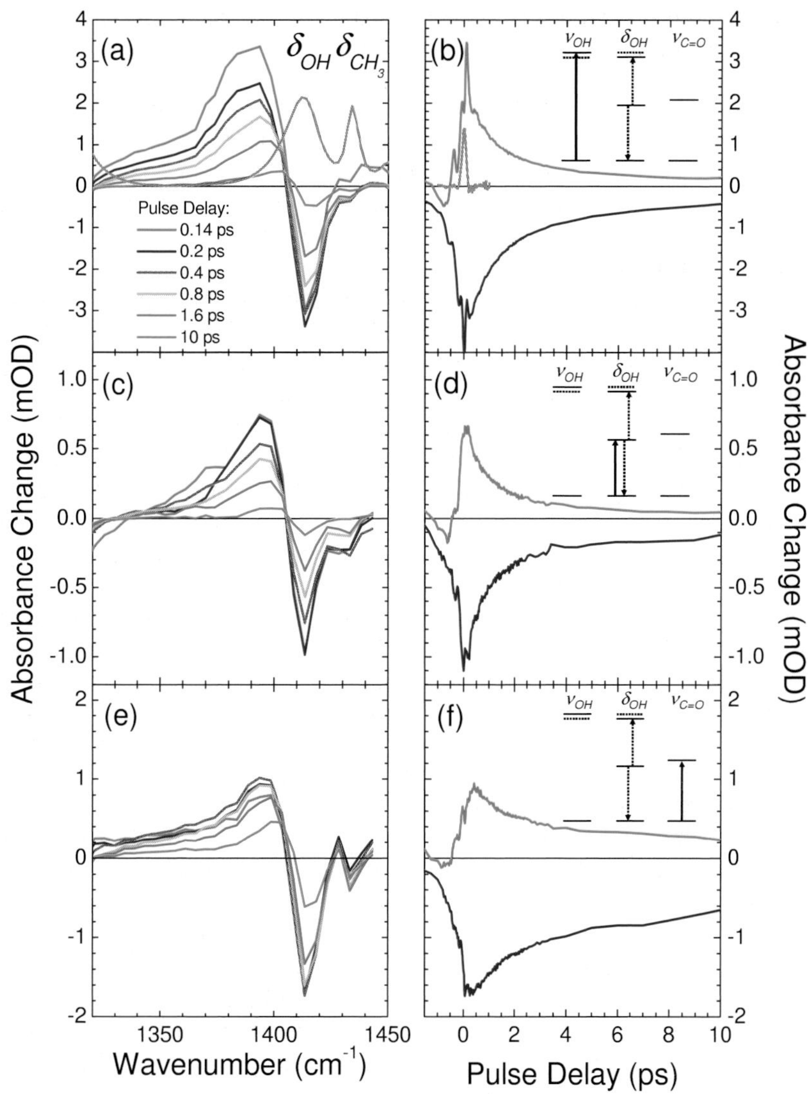

Fig. 7.17. PMME-H: Two color pump-probe results. The *gray curve* in (a) represents the steady-state IR absorption spectrum showing the O–H bending (at $1415\,\text{cm}^{-1}$) and the C–H bending of the CH_3-group ($1435\,\text{cm}^{-1}$). The *gray curve* in (b) indicates the temporal resolution in the experiments. Here the bending vibration is probed for three excitation conditions: excitation of the O–H stretching vibration at $2900\,\text{cm}^{-1}$ (a,b), the bending vibration at $1400\,\text{cm}^{-1}$ (c,d) and the C=O stretching vibration at $1700\,\text{cm}^{-1}$ (e,f). The frequency-resolved transient spectra are shown in panels (a,c,e) whereas the transient absorbance changes of the $v(\delta_{OH}) = 0 \rightarrow 1$ ground state bleach at $1415\,\text{cm}^{-1}$ (negative signals) and the $v(\delta_{OH}) = 1 \rightarrow 2$ excited state absorption at $1390\,\text{cm}^{-1}$ (positive signals) are shown in panels (b,d,f).

contrast to the observed time scale of about 200 and 400 fs for the O–H and O–D stretching vibrational relaxation, respectively, what led to the conclusion that the solvent interaction contributes considerably to these rapid relaxation processes.

The mechanistic details have been simulated by a dissipative system-bath model [235] specified to the case where the bath consists of intramolecular as well as solvent degrees of freedom. Because of its simpler absorption spectrum in the O–D stretching region, PMME-D was the first target for a system-bath description. In [15] a 3-dimensional model system was proposed which included the O–D stretching (ν_{OD}) and bending (δ_{OD}) coordinates as well as a low-frequency hydrogen bond mode (ν_{HB}). Due to its single minimum character the anharmonic potential energy surface could be spanned by the respective normal mode coordinates. Instead of the Taylor expansion in (7.5), the potential was calculated on a grid using DFT/B3LYP. The double peak structure of the IR spectrum in Fig. 7.14a could be explained as being due to a Fermi resonance between the $v(\nu_{OD}) = 0 \rightarrow 1$ fundamental and the $v(\delta_{OD}) = 0 \rightarrow 2$ overtone transition. The combination transitions with the low-frequency mode, e.g. $v(\nu_{OD}) + nv(\nu_{HB})$, are masked by the broadening due to the system-bath coupling. In the model of [15] the latter comprised linear and nonlinear terms responsible for energy and phase relaxation. A key feature of the model which proved to be vital for the explanation of the rapid $v(\nu_{OD}) = 1$ decay has been a solvent-assisted intramolecular relaxation pathway. Using classical molecular dynamics simulation of force-force correlation functions the relaxation time for the low-frequency hydrogen bond mode was determined to be about 1.7 ps in CCl_4 at room temperature [18]. In [18] it was also shown that the dissipative model supports coherence transfer with respect to the hydrogen bond mode between the ν_{OD} fundamental excitation and its ground state. This contributes to hydrogen bond wave packet dynamics in the ν_{OD} ground state which is additionally triggered by the initial IR-laser excitation (as has been discussed in Sect. 7.5.1).

The two-color IR pump-probe experiments of Fig. 7.17 described above provided a more detailed picture of the rapid IVR process in PMME-H [25]. Building on [18] a five-dimensional dissipative model has been derived whose relevant coordinates are shown in Fig. 7.18 in terms of their normal mode displacement vectors. Besides the O–H stretching (ν_{OH}) and bending (δ_{OH}) modes calculation of the anharmonic force constants suggests to include two modes of approximately out-of-plane O–H-bending character (γ_{OH1} and γ_{OH2}) into the model. These four fast coordinates are dressed by the slow coordinate of the hydrogen bond motion (ν_{HB}). It is instructive to inspect the diabatic potential energy curves for the motion of the hydrogen bond coordinate in the different states of the fast coordinates (see Fig. 7.18). Starting from the ground state there are certain "bands" of transitions such as the γ_{OH1} and γ_{OH2} fundamentals, the δ_{OH} fundamental as well as the γ_{OH1} and γ_{OH2} combination and first overtones, their second overtones as well as combinations with the δ_{OH} fundamental, and finally the ν_{OH} fundamental which

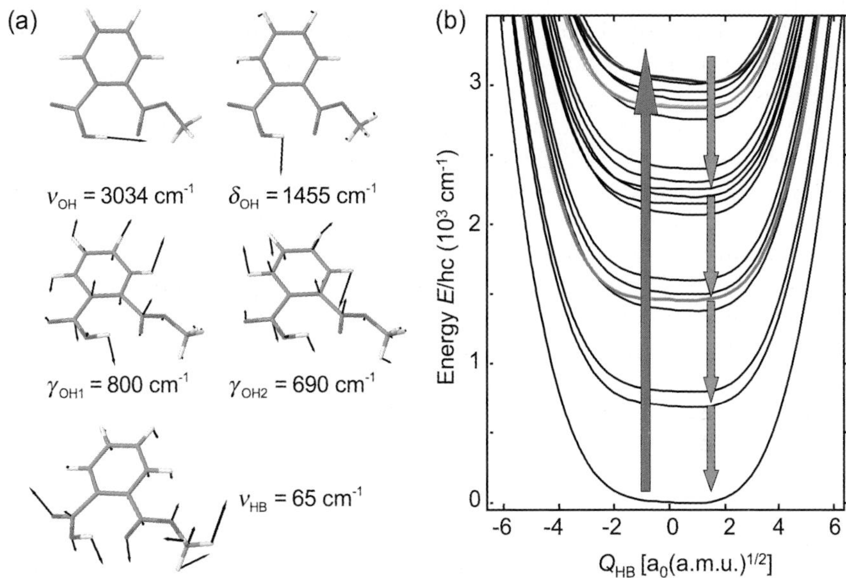

Fig. 7.18. (a) Normal mode displacement vectors of the 5 modes of the dissipative PMME-H model; (b) Uncoupled diabatic potential energy curves governing the motion of the low-frequency hydrogen bond mode for the different quantum states of the fast modes. These states as well as their coupling have been obtained by expressing the 5-dimensional Hamiltonian in the basis of the fast-coordinate states defined at the equilibrium geometry of the hydrogen bond mode (red - δ_{OH}, blue - ν_{OH}). The red arrow indicates the O−H stretching excitation by the IR laser field, whereas the gray arrows represent the energy cascading process.

is immersed into the manifold of states coming from the various combination transitions of the bending modes as well as the first δ_{OH} overtone. Notice that each potential curve in Fig. 7.18 contains the vibrational ladder of the low-frequency hydrogen bond mode and that there are diabatic state coupling between the different curves as well.

The quantum dynamics of this 5D dissipative model has been modeled using the Quantum Master equation approach [235]. From the density of states in the ν_{OH} region one would expect a rapid energy redistribution among several zero-order states upon excitation in this spectral range. However, simulations show that this is not sufficient to account for the 200 fs energy relaxation time scale. Modeling this relaxation requires to include other intramolecular as well a solvent degrees of freedom. Since the hydrogen bond mode has the lowest intramolecular frequency it can only relax into the solvent; the respective spectral density obtained from classical molecular dynamics has been discussed above [18]. As far as the modes γ_{OH1} and γ_{OH2} are concerned,

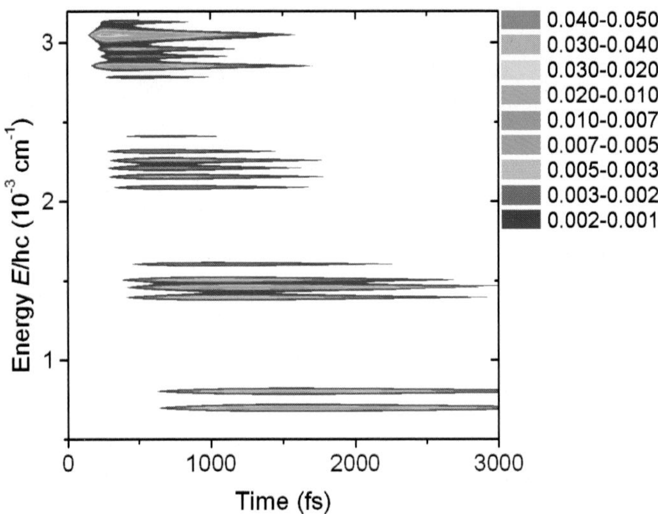

Fig. 7.19. Population dynamics of the diabatic states of PMME-H as indicated in Fig. 7.18 summed over all respective states of the hydrogen bond mode for the 5-dimensional dissipative model of PMME-H. Populations are shown at the energy of the corresponding diabatic states (ground state not shown).

there are different possibilities, such as the decay via modes of comparable frequency or via excitation of two quanta in a lower-frequency mode. According to the calculated anharmonic force constants the former model is to be favored. However, due to a slight energy mismatch, assistance of the transition by low-frequency solvent modes is required similar to what had been proposed for PMME-D in [15]. Notice that the strength of the solvent-assisted coupling to the intramolecular bath is the only free parameter of the model.

The resulting dynamics upon O–H stretching excitation with a 100 fs IR pulse is shown in Fig. 7.19 in terms of the diabatic state populations of the fast modes. The cascading of the energy down the vibrational ladder formed by the manifolds of different bending mode states is clearly discernible (see arrows in Fig. 7.18). The time scale of the ν_{OH} and δ_{OH} decay is about 200 fs and 800 fs, respectively, in good agreement with the experiment of Fig. 7.17, thus supporting the proposed cascading mechanism via the δ_{OH} bending vibration for the ultrafast ν_{OH} relaxation in intramolecular hydrogen bonds of this medium-strong type.

7.5.3 Coherent and relaxation dynamics of hydrogen-bonded dimers

Cyclic dimers of carboxylate acids represent important model systems forming two coupled intermolecular hydrogen bonds (Fig. 7.20). The linear vibrational

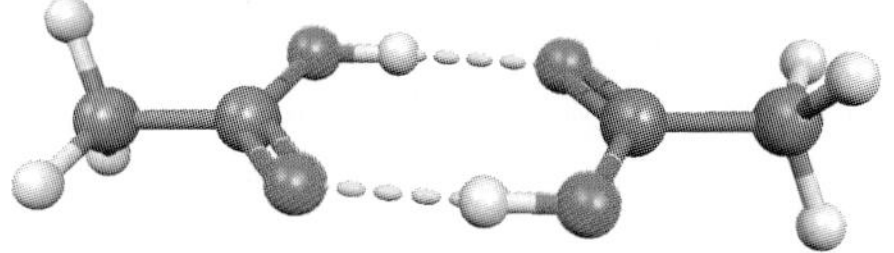

Fig. 7.20. Acetic acid dimer $(CX_3-COOX)_2$, X = H, D.

spectra of carboxylate acid dimers have been studied in detail, both in the gas and liquid solution phase, and a substantial theoretical effort has been undertaken to understand the line shape of their O–H and/or O–D stretching bands. In contrast, only few experiments have been reported on the nonlinear vibrational response. The coupling of the two carbonyl oscillators in acetic acid dimers has been investigated by femtosecond pump-probe and photon echo measurements [236], and vibrational relaxation following O–H stretching excitation has been addressed in picosecond pump-probe studies [237].

In the following, recent extensive pump-probe studies of cyclic acetic acid dimers in the femtosecond time domain are presented [20, 21, 24], followed by a discussion of results from photon echo measurements [22, 27], and a presentation of quantum chemical calculations of the anharmonic couplings and the resulting linear spectra and nonlinear signals [24, 27, 29, 30]. From the interplay of experiment and theory a quantitative estimation of the couplings as well as the linear and nonlinear infrared response is achieved. In order to disentangle the different coupling mechanisms (as described in Sect. 7.2.2) of the O–H/O–D stretching oscillator in acetic acid dimer, dimer structures containing two O–H$\cdots$O (OH/OH dimer) or two O–D$\cdots$O (OD/OD dimer) hydrogen bonds as well as dimers with one O–H$\cdots$O and one O–D$\cdots$O hydrogen bond (mixed dimers) have been explored. The combination of different nonlinear spectroscopic techniques enable the extraction of information on particular coupling mechanisms.

The steady state and the transient O–H stretching absorption spectra of OH/OH dimers are displayed in Fig. 7.21a,b. The transient spectra show a strong bleaching in the central part of the steady-state band and enhanced absorption on the red and blue wing, in accordance with the level scheme of Fig. 7.6. The bleach, consisting of a series of comparably narrow spectral dips, originates from the depopulation of the $v(\nu_{OH}) = 0$ state and stimulated emission from the $v(\nu_{OH}) = 1$ state. The enhanced absorption at small frequencies is due to the $v(\nu_{OH}) = 1 \rightarrow 2$ transition and decays by depopulation of the $v(\nu_{OH}) = 1$ state with a lifetime of $T_1 = 200\,\text{fs}$. The latter is evident from time-resolved measurements at a probe frequency of $2250\,\text{cm}^{-1}$ (not shown). The enhanced absorption on the blue side is caused by the vibrationally hot ground state formed by relaxation of the $v(\nu_{OH}) = 1$ state, followed by vibrational cooling on a 10 to 15 ps time scale. Transient spectra measured for the OD/OD and the mixed dimers – both on the O–H and O–D stretching bands – display a very similar behavior.

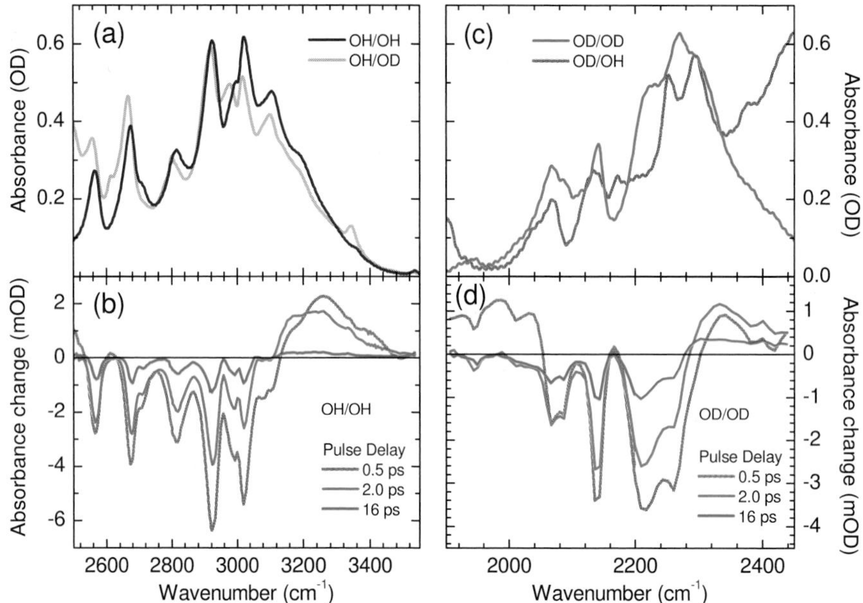

Fig. 7.21. (a) O–H stretching band of acetic acid dimer $(CD_3-COOH)_2$ (*blue line*: OH/OH-dimer) and of (CD_3-COOH)-(CD_3-COOD) (*orange line*: OH/OD-mixed dimer); (b) Transient spectra of the OH/OH dimer measured at selected pulse delay between pump and probe; (c) O–D stretching band of acetic acid dimer $(CH_3-COOD)_2$ (*red line*: OD/OD-dimer) and of (CH_3-COOD)-(CH_3-COOH) (*green line*: OD/OH-mixed dimer); (d) Transient spectra of the OD/OD dimer measured at selected pulse delay between pump and probe.

The time evolution of the nonlinear O–H stretching absorption (Fig. 7.22) shows pronounced oscillatory signals for all types of dimers studied. In contrast to the intramolecular hydrogen bond of PMME, and of other similar systems [8,9,13] the time-dependent amplitude of the oscillations displays features of a beatnote, demonstrating the presence of more than one oscillation frequency. In Fig. 7.23 the Fourier transforms of the oscillatory signals are plotted as obtained for the uniform isotopomers of acetic acid dimer. There are three prominent frequency components, a strong doublet with maxima at 145 and $170\,\text{cm}^{-1}$ and a much weaker component around $50\,\text{cm}^{-1}$.

In principle the observed oscillatory components could originate from Davydov, Fermi or from coupling to low-frequency modes. The two local stretching oscillators in the OH/OH and OD/OD dimers are subject to Davydov coupling resulting in a splitting of the $v(\nu_{OH}) = 1$ states, on top of the anharmonic coupling to low-frequency modes. In the linear absorption spectrum of the ensemble of dimers, this results in two separate low-frequency progressions originating from the $v(Q_u) = 0$ and $v(Q_u) = 1$ levels in the $v(\nu_{OH}) = 0$ state of the O–H stretching vibrations (Fig. 7.24). In thermal

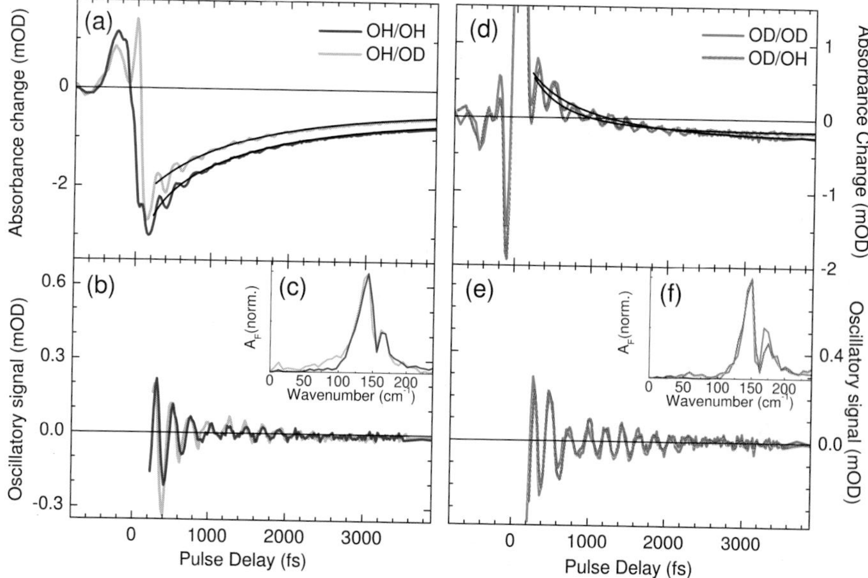

Fig. 7.22. (a) Transient absorbance changes in the bleach region of the O–H stretching band of acetic acid dimer $(CD_3{-}COOH)_2$ (*blue line*: OH/OH-dimer) and of $(CD_3{-}COOH)$-(CD_3COOD) (*orange line*: OH/OD-mixed dimer) as recorded with pump pulses centered at $2950\,\mathrm{cm}^{-1}$ and a probe detection at $2880\,\mathrm{cm}^{-1}$; (b) Oscillatory contribution of the pump-probe signals of the OH/OH and OH/OD dimers of panel (a); (c) Fourier transform of the oscillatory signals of (b) of the OH/OH and OH/OD dimers showing two low-frequency components; (d) Transient absorbance changes in the bleach region of the O–D stretching band of acetic acid dimer $(CH_3{-}COOD)_2$ (*red line*: OD/OD-dimer) and of $(CH_3{-}COOD)$-$(CH_3{-}COOH)$ (*green line*: OD/OH-mixed dimer) as recorded with pump pulses centered at $2100\,\mathrm{cm}^{-1}$ and a probe detection at $2020\,\mathrm{cm}^{-1}$; (e) Oscillatory contribution of the pump-probe signals of the OD/OD and OD/OH dimers of panel (d); (f) Fourier transform of the oscillatory signals of (e) of the OD/OD and OD/OH dimers showing two the same low-frequency components.

equilibrium, a particular dimer populates only one of the $v(Q_u)$ levels at a certain instant in time and – thus – only one of the progressions can be excited. Consequently, a quantum coherent nonstationary superposition of the split $v^{\pm}(\nu_{OH}) = 1$ states of the stretching mode cannot be excited in an individual dimer and quantum beats due to Davydov coupling are absent in the pump-probe signal.

This behavior is evident from a comparison of transients recorded with OH/OD and OH/OD dimers, the latter displaying negligible Davydov coupling because of the large frequency mismatch between the O–H and O–D stretching oscillators. The time-resolved change of O–D stretching absorption

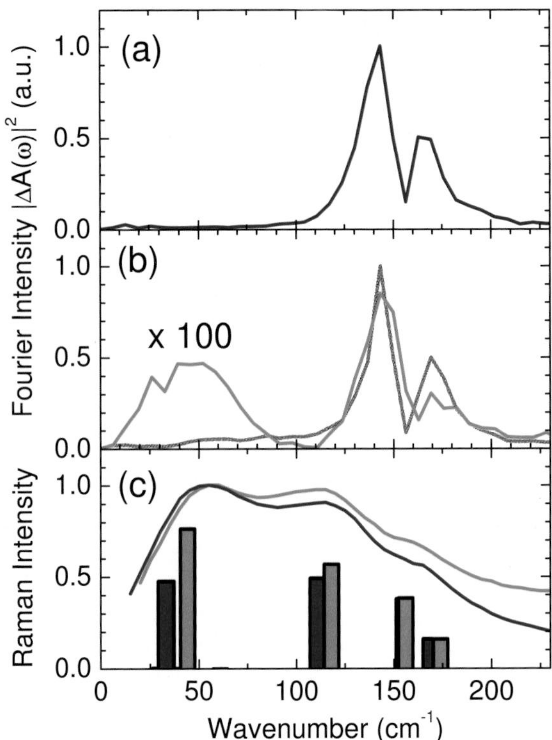

Fig. 7.23. Comparison of the Fourier intensities of the oscillatory components in the frequency-resolved pump-probe data of the acetic acid dimer isotopomers. (a) $(CD_3{-}COOH)_2$, measured at $3040\,cm^{-1}$; (b) $(CH_3{-}COOD)_2$, measured at $2300\,cm^{-1}$ (*green line*) and $2250\,cm^{-1}$ (*red line*); (c) experimental (*lines*) and calculated (*bars*) Raman spectra of $(CD_3{-}COOH)_2$ (*blue line/bars*) and of $(CH_3{-}COOD)_2$ (*red line/bars*).

of the two types of dimers shows a very similar time evolution, and the Fourier spectra agree within experimental accuracy.

Group theoretical analysis (Fig. 7.24c) and calculations of IR-active progressions originating from a ground state with the Q_u low-frequency mode being excited suggest small IR cross sections for this series [29].

A contribution of quantum beats between states split by Fermi resonances can be ruled out as well. There are different Fermi resonances within the O–H and O–D stretching bands. Depending on the spectral positions of pump and probe, this should lead to a variation of the oscillation frequencies, in particular when comparing O–H and O–D stretching excitations. Such a behavior is absent in the experiment, which shows identical oscillation frequencies for O–H and O–D stretching excitation that remain unchanged for pumping throughout the respective absorption band.

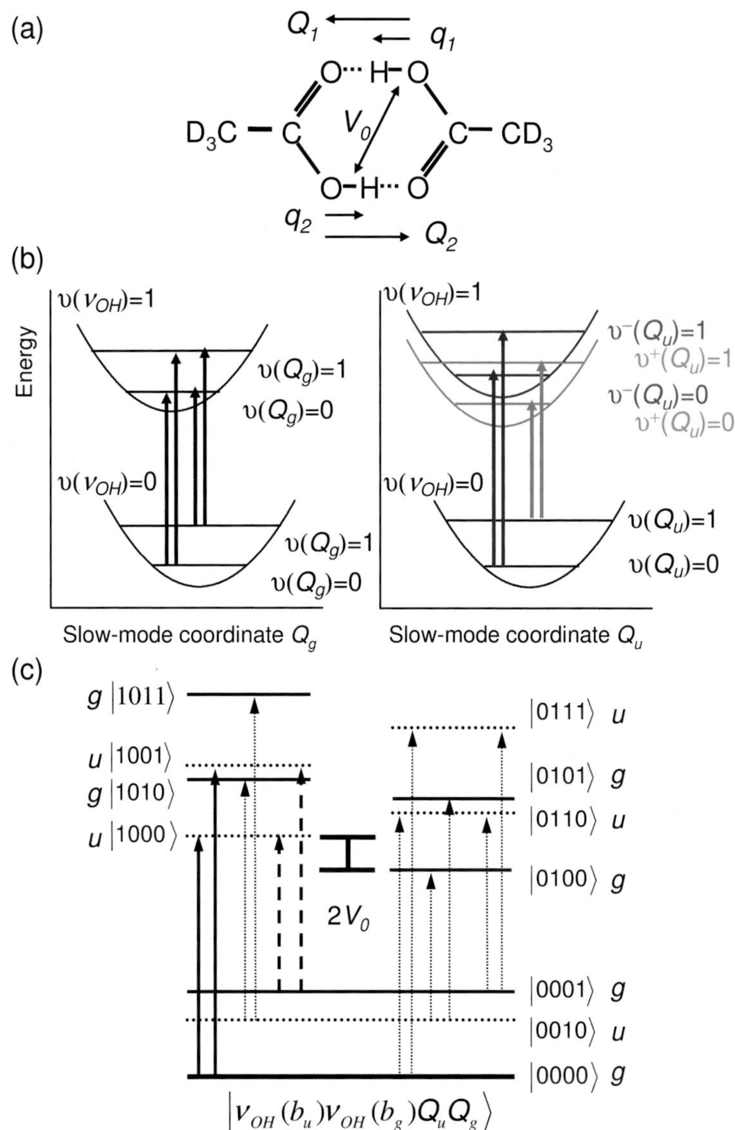

Fig. 7.24. (a) Schematic showing acetic acid dimer with local high-frequency (q_1, q_2) O–H stretching and low-frequency (Q_1, Q_2) O$\cdots$O hydrogen bond modes; (b) Potential energy surfaces of the low-frequency modes anharmonically coupled to the O–H stretching vibrations, shown using symmetrized Q_g and Q_u low-frequency modes. Davydov coupling (V_0) leads to a splitting of the energy levels of the asymmetric Q_u mode, for which a priori selection rules apply. (c) A refined analysis of the IR cross sections for the different transitions within the O–H stretching manifold (*solid lines*: strong ground state absorption; *dashed lines*: hot transitions; *dotted lines*: weak transitions; for details see [29]).

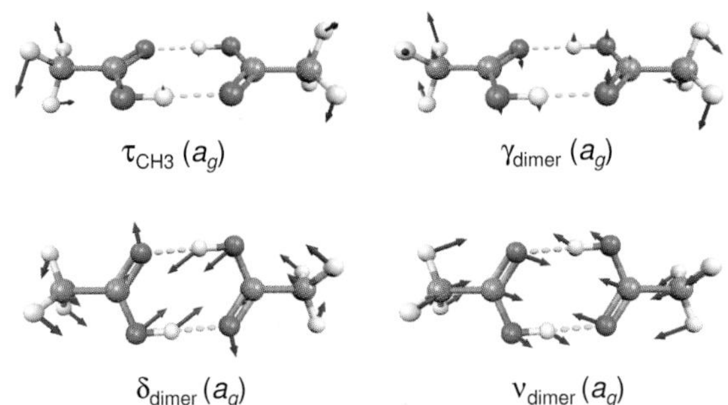

Fig. 7.25. Raman-active low-frequency vibrational modes of acetic acid dimer.

Instead, the oscillatory absorption changes are due to coherent wave packet motions along several low-frequency modes which anharmonically couple to the stretching modes. Wave packets in the $v(\nu_{OH/OD}) = 0$ states that are generated through an impulsive resonantly enhanced Raman process, govern the oscillatory response, whereas wave packets in the $v(\nu_{OH/OD}) = 1$ states are strongly damped by the fast depopulation process. Low-frequency modes of acetic acid have been studied in a number of Raman experiments. The spectrum in Fig. 7.23c was taken from [238] and displays three maxima around 50, 120 and 160 cm^{-1}. The number of subbands in such strongly broadened spectra and their assignment have remained controversial [239]. The character of the different low-frequency modes and their anharmonic coupling to the O–H stretching mode have now been analyzed in recent normal mode calculations based on density functional theory [24]. In Fig. 7.23c the calculated Raman transitions are shown for the OH/OH and OD/OD dimers.

There are four Raman-active vibrations (Fig. 7.25): the methyl torsion at 44 cm^{-1} (τ_{CH3}), the out-of-plane wagging mode at 118 cm^{-1} (γ_{dimer}), the in-plane bending mode around 155 cm^{-1} (δ_{dimer}) and the dimer stretching mode at 174 cm^{-1} (ν_{dimer}). In this set of vibrations, the in-plane bending and the dimer stretching modes couple strongly to the O–H/O-D stretching mode via a third-order term in the vibrational potential that dominates compared to higher order terms. The coupling of the methyl torsion is much weaker, and that of the out-of-plane wagging mode even negligible. Such theoretical results are in good agreement with the experimental findings: the strong doublet in the Fourier spectra (Fig. 7.23a,b) is assigned to the in-plane bending and the dimer stretching mode, the weak band around 50 cm^{-1} to the methyl torsion. The out-of-plane wagging is not observed at all. It should be noted that the spectra derived from the oscillatory pump-probe signals, i.e. from time domain data, allow for a much better separation of the low-frequency mode coupling than the steady-state spontaneous Raman spectra.

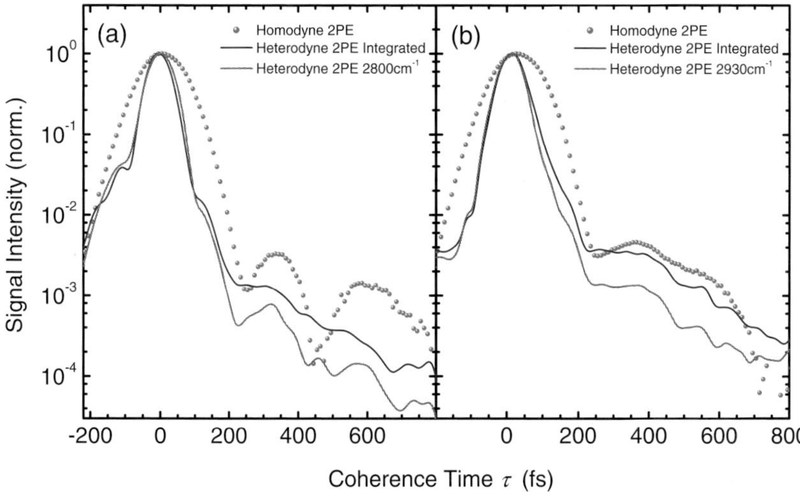

Fig. 7.26. Homodyne and heterodyne detected 2PE signals of acetic acid dimer $(CH_3{-}COOH)_2$ (a) and mixed dimers $(CD_3{-}COOH)$-$(CD_3{-}COOD)$ (b). The *red dotted trace* has been recorded using homodyne detection, whereas the *solid curves* have been derived from the heterodyne detected 2D-IR spectrum, for a particular detection frequency ν_3 and for spectral integration over ν_3.

In the case of photon echo spectroscopy on the acetic acid dimer, the number of coherences contributing to the nonlinear signals is large as a consequence of the multilevel structure of the O–H/O–D stretching band. Here in principle all coupling mechanisms described in Sect. 7.2.2 should affect the outcome of the echo experiments. Homodyne detected 2PE measurements (Fig. 7.26) have indicated the importance of the quantum beat contributions of the low-frequency mode couplings [22, 27], as indicated by the recurrences with an oscillation period given by the low-frequency in-plane stretching and in-plane deformation hydrogen bond modes. In addition Davydov coupling appears to be of minor importance. The low-frequency wave packet motions also lead to beating contributions in the integrated 3PEPS signals [22]. The homodyne detected photon echo measurements, in conjunction with a refined analysis of the fine structure of the bleach signals in the pump-probe measurements (Fig. 7.21) have led to the conclusion that the individual transitions in the O–H stretching manifold have a homogeneous broadening corresponding to a dephasing time $T_2 = 200\,\text{fs}$.

Two-color pump-probe experiments have indicated the importance of anharmonic couplings between the IR-active O–H stretching and O–H bending and C–O stretching modes [23]. The effects of these couplings between the O–H stretching and the fingerprint vibrations are even more pronounced in heterodyne-detected 2D infrared photon echo signals (Fig. 7.27). For the 2D spectrum with a population waiting time $T = 0\,\text{fs}$, transitions between

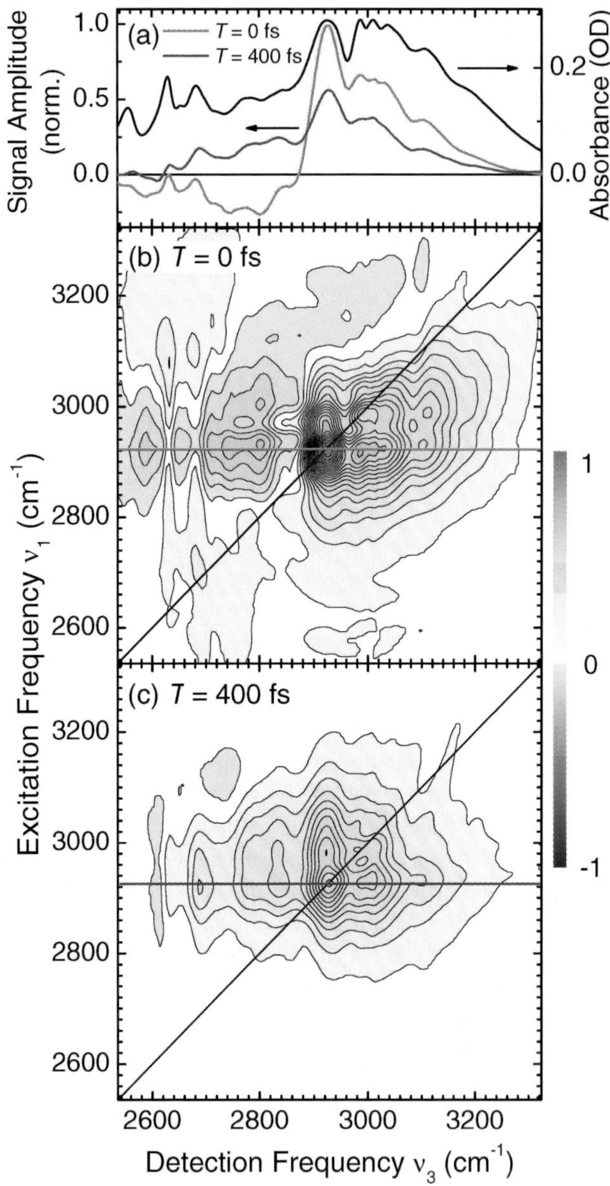

Fig. 7.27. (a) Linear spectrum of the O–H stretching band of cyclic dimers of acetic acid in CCl_4 (*black line*); Cross sections through the 2D-IR spectra for population times $T = 0\,\mathrm{fs}$ (*red line*) and $T = 400\,\mathrm{fs}$ (*blue line*); (b,c) 2D-IR spectra of acetic acid dimer for population times $T = 0$ and $400\,\mathrm{fs}$. The amplitude of the photon echo signal is plotted as a function of the excitation frequency ν_1 and the detection frequency ν_3.

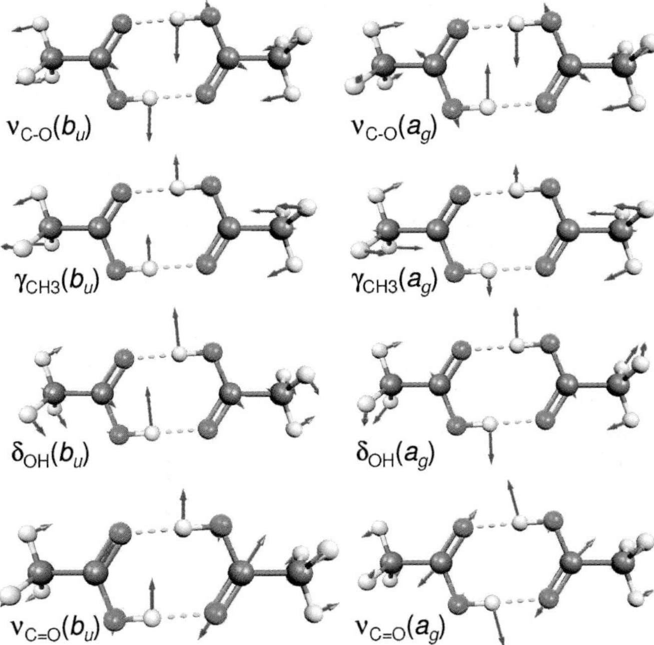

Fig. 7.28. Vibrational modes of acetic acid dimer $(CH_3{-}COOH)_2$, strongly coupled to the IR-active O–H stretching mode. IR- (b_u) and Raman-active (a_g) fingerprint modes.

$v(\nu_{OH}) = 0 \to 1$ and $v(\nu_{OH}) = 1 \to 2$ contribute with diagonal and off-diagonal peaks. Since the excited state of the O–H stretching vibration (lifetime $T_1 = 200\,\text{fs}$ [24]) decays into a hot ground state (Fig. 7.6) with complete loss of phase memory [22], the 2D spectrum for a population waiting time $T = 400\,\text{fs}$ is determined by the nonlinear signal contributions of the fundamental $v(\nu_{OH}) = 0 \to 1$ transition only, as described by the Liouville space pathway I of Fig. 7.4. In particular overtone and combination levels of the IR- and Raman-active O–H bending and C–O stretching modes contribute, but in addition the roles of the C=O stretching and C–H bending vibrations cannot be discarded (see Fig. 7.28) [27,30]. More details on this are described in the following section.

7.5.4 Anharmonic coupling in acetic acid dimers: Simulation of linear and two-dimensional infrared spectra

Several attempts have been made to analyze the lineshape of the O–H and/or O–D stretching band of acetic acid dimers in the gas and liquid phase as well as in matrices [37,38,240–246]. Besides the vibrational coupling mechanisms discussed in Sect. 7.2.2 inhomogeneous broadening resulting from different dimer structures, monomers or oligomers [247, 248] may contribute to the spectral width.

In this Section results are summarized on theoretical calculations based on DFT to investigate three mechanisms for vibrational coupling within the two intermolecular hydrogen bonds: (i) anharmonic coupling of the high-frequency O–H stretching mode to low-frequency modes, (ii) Fermi resonance coupling the of O–H stretching $v(\nu_{OH}) = 1$ state to fingerprint mode combination tones, and (iii) a combination of both [29]. Linear IR absorption spectra of the O–H stretching mode are simulated and compared to high-resolution IR spectra measured by supersonic jet Fourier transform spectroscopy [246, 249]. These cold gas phase spectra uncover much of the vibrational fine structure and can thus serve as a reference for zero temperature gas phase calculations (Fig. 7.29a). With subsequent extension to simulations of two-dimensional IR spectra [47, 250] the investigation of multidimensional signatures of the different coupling mechanisms has become possible [27, 30].

All quantum chemical calculations were carried out using DFT applying the B3LYP/6-311+G(d,p) method as implemented in Gaussian03 [251]. This study is restricted to the most commonly appearing C_{2h}-symmetric cyclic dimer structure (Fig. 7.20) the optimized structure of which is found to be in good agreement with earlier calculations [246, 249, 252].

Anharmonic multidimensional potential energy surfaces were calculated in the basis of dimensionless normal modes by expanding the potential energy V up to sixth-order in a selected set of modes. First-order and harmonic second-order off-diagonal force constants vanish at the equilibrium geometry. For higher-order force constants up to three-body interactions were included. Dipole moment derivatives were expanded to first-order only, thereby neglecting electrical anharmonicities, which are usually much smaller than mechanical ones [253,254]. Cubic and higher-order force constants were calculated numerically by finite differences from analytical second-order force constants [52]. For benzoic acid dimers the validity of second-order perturbation theory with respect to vibrational modes in the fingerprint region has been critically examined [61]. It was found that standard perturbation theory accounts for anharmonic shifts of the vibrational modes in almost all cases. A notable exception, however, is the out-of-plane O–H bending mode, the accurate treatment of which requires multidimensional nonperturbative calculations at high quantum-chemical level.

The cyclic dimer exhibits C_{2h} symmetry and is thus subject to the IR-Raman exclusion rule. Upon dimerization all monomeric vibrations, which would be pairwise degenerate without coupling, experience excitonic or Davydov coupling and split into a set of *ungerade* (u) IR active and a set of *gerade* (g) Raman-active dimer modes [38, 243]. It should be noted that the one-dimensional potential energy surfaces for u modes are symmetric, whereas g modes have asymmetric potentials. For symmetric potential energy surfaces odd-order force constants vanish. Thus, the first anharmonic correction to IR active fundamental vibrations is in quartic order. The number of nonzero force constants is further reduced by symmetry. Only force constants, for which the product of irreducible representations of the respective coordinates involved in the force constant is totally symmetric, can be different from zero [62].

Vibrational eigenstates were calculated from an effective vibrational exciton Hamiltonian with up to 4 excitation quanta [29, 253, 255]. Calculated harmonic vibrational frequencies ω_i are usually too high with respect to experimental values. The deviation is mainly caused by missing anharmonic corrections as well as insufficient treatment of electron correlation of the quantum chemical method. Here, anharmonic corrections were included, but the imperfect treatment of electron correlation by the DFT B3LYP/6-311+G(d,p) method had to be accounted for. Therefore, harmonic force constants were scaled such that the frequencies of the fundamental vibrational modes resulting from the anharmonic vibrational Hamiltonian match experimental gas phase values [246, 249], which leads to scaling factors of about 0.98. The harmonic force constant for the $\nu_{OH}(b_u)$ mode was set to $2915\,\mathrm{cm}^{-1}$. This corresponds to a scaling factor of 0.90, which shows that the B3LYP method underestimates the strength of hydrogen bonding in acetic acid dimers. The assignments of the major subbands of the O–H stretching mode vibrational spectrum are relatively insensitive to a variation of scaling factors in a reasonably range, but for an accurate determination of the fine structure, scaling is important.

Femtosecond three-pulse photon echo studies have shown that the O–H stretching absorption spectrum is predominantly homogeneously broadened, whereas inhomogenous broadening can be neglected [22]. From the observed dephasing time $T_2 \approx 200\,\mathrm{fs}$ a homogeneous linewidth $\Gamma \approx 50\,\mathrm{cm}^{-1}$ is derived. In the following, calculated IR spectra with homogeneous linewidths (full width at half maximum) of $\Gamma = 1$ and $40\,\mathrm{cm}^{-1}$ will be presented. The latter value was found to fit the experimental spectrum measured in CCl_4 more appropriately than the value derived from the photon echo study.

Early treatments suggested anharmonic coupling to a single low-frequency mode, an intermolecular hydrogen bond stretching mode, together with Davydov coupling to be exclusively responsible for the vibrational line shape of the O–H stretching bands [37]. Dimerization gives rise to 6 new intermolecular vibrational modes. Because of the C_{2h} inversion symmetry of the acetic acid dimer, 3 of them are IR- (u) and 3 are Raman-active (g) (Fig. 7.25). They are characterized by distance as well as in- and out-of-plane angle modulations between the two hydrogen bonded monomers. These hydrogen bond modes are thus associated with relatively weak hydrogen bond forces and large reduced masses. Therefore, they all appear at low frequencies, that is below $200\,\mathrm{cm}^{-1}$ for acetic acid dimers [25, 238, 239, 256].

Upon coupling of the IR active ν_{OH} mode with Raman-active low-frequency modes, an IR active Franck-Condon series with one or several quanta of low-frequency modes arises [36]. Combining coherent femtosecond IR-pump–IR-probe spectroscopy and quantum chemical calculations 2 modes

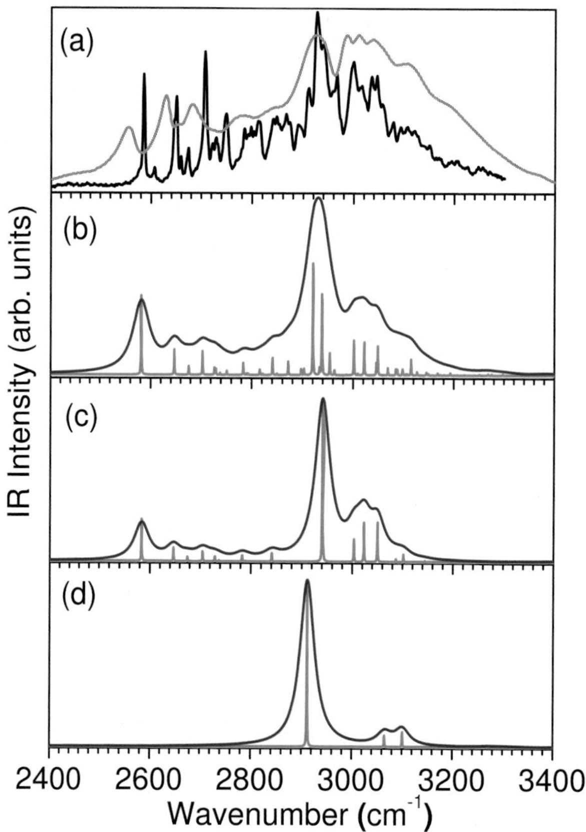

Fig. 7.29. $(CH_3{-}COOH)_2$: (a) Experimental IR absorption spectra: Cold gas phase spectrum (*black*) taken from [246] with permission and spectrum in CCl_4 (*gray*) measured at 298 K; (b-d) Calculated spectra (1 K) using homogeneous line widths of $\Gamma = 1$ (*red*) and $40\,\mathrm{cm}^{-1}$ (*blue*) including the following modes: $\nu_{OH}(b_u)$ + (b) fingerprint modes and $\delta_{dimer}(a_g) + \nu_{dimer}(a_g)$; (c) fingerprint modes only; (d) $\delta_{dimer}(a_g)$ and $\nu_{dimer}(a_g)$ only.

were identified, which couple strongly to ν_{OH} [25]. These are the hydrogen bond bending ($\delta_{dimer}(a_g)$) and stretching ($\nu_{dimer}(a_g)$) modes absorbing at 145 and $170\,\mathrm{cm}^{-1}$, respectively, which both modulate the hydrogen bond distance strongly (Fig. 7.25). Cubic coupling constants, which contribute most to anharmonic coupling, are about 150 and $-145\,\mathrm{cm}^{-1}$ for $\delta_{dimer}(a_g)$ and $\nu_{dimer}(a_g)$, respectively. The IR spectrum based on 3 modes, $\nu_{OH}(b_u)$, $\delta_{dimer}(a_g)$ and $\nu_{dimer}(a_g)$, is shown in Fig. 7.29d. On a linear intensity scale a progression with only a single quantum of each low-frequency mode becomes visible. It can be concluded that a spectrum solely based on anharmonic coupling of $\nu_{OH}(b_u)$ to low-frequency modes can not account for the substructure and total width of the experimentally observed spectra shown in Fig. 7.29a.

Instead of coupling to low-frequency modes, multiple Fermi resonance coupling with combination tones of fingerprint vibrations has been considered to account for the width and substructure of the O–H stretching absorption spectrum [38, 242–245]. The relevance of Fermi resonances contributing to the substructure of the IR spectrum is supported by recent model calculations [246, 257]. With the exception of methyl group rocking and scissoring modes all IR active (b_u) fingerprint modes are included as well as their Raman-active (a_g) counterparts as only ungerade combination tones can directly interact with the $\nu_{OH}(b_u)$ high-frequency mode. These are the C–O (ν_{C-O}) and C=O ($\nu_{C=O}$) stretching modes, the O–H in-plane bending (δ_{OH}) and the methyl (γ_{CH3}) wagging modes (Fig. 7.28). It should be noted that none of the normal modes is local. Instead, in addition to a dominating contribution, O–H in-plane bending and methyl deformation motions mix into the normal coordinates. The O–H out-of-plane bending mode (γ_{OH}) was initially taken into account, but later on rejected from the Hamiltonian, because combination tones of this mode were shown not to contribute to the spectrum to any appreciable extent [29]. A linear IR spectrum for this model is shown in Fig. 7.29c. This spectrum accounts for major parts of the substructure of the experimental gas as well as solution phase spectra [22,29,246]. The vibrational fine structure, however, which is evident from the gas phase spectrum, is still lacking.

As neither anharmonic coupling to low-frequency modes nor Fermi resonance coupling to fingerprint combination bands sufficiently account for the observed lineshape of the O–H stretching band absorption, both mechanisms were merged into an 11-mode model. The resulting linear spectrum (Fig. 7.29b) shows a rich fine structure in good agreement with the experimental high-resolution gas phase and solution phase spectra (Fig. 7.29a) [22, 29, 246]. The more intense peaks in the linear spectrum are mostly dominated by Fermi resonances similar to the pure Fermi resonance spectrum without coupling to low-frequency modes (Fig. 7.29c). The low-frequency mode progressions build up on all Fermi resonance peaks forming the vibrational fine structure. Some low-frequency modes even gain intensities comparable to Fermi resonance bands [29]. The experimental spectrum measured in CCl_4 (Fig. 7.29a) shows broadening compared to the gas phase spectrum, whereby the overall width of the solution phase spectrum is reasonably well described by the calculated spectrum using a homogeneous linewidth of $\Gamma = 40\,\mathrm{cm}^{-1}$. Individual peaks observed in the gas phase spectrum merge into broader subbands. These solution phase subbands partly coincide with spectral features in the gas phase spectrum, but, as a consequence of shifted fundamental fingerprint vibrations, some of combination bands are also shifted. This is particularly obvious in the low-energy side of the spectrum below $2700\,\mathrm{cm}^{-1}$, where the solution phase bands are red-shifted with respect to their counterparts in the gas phase spectrum. All of the bands gain intensity from resonance coupling with $\nu_{OH}(b_u)$. However, as the $\nu_{OH}(b_u)$ intensity is spread over so many bands, its own contribution to individual states is relatively small. It is largest for the peak at 2920 cm^{-1}, which accordingly is the most intense peak in the spectrum with a contribution of about 15% for $\nu_{OH}(b_u)$. This corroborates the relevance of Fermi resonances in the O–H stretching IR absorption spectrum that was pointed out before [38, 246, 257]. However, a prominent role of a single Fermi resonance, namely the O–H stretching/bending coupling [257], which contributes to the medium intense band around 2840 cm^{-1} cannot be confirmed. The strongest anharmonic couplings are calculated for the combination bands on the red side of spectrum. This explains their relatively high peak intensities despite less fortunate resonance conditions.

The important role of low-frequency hydrogen bond bending and stretching modes in characterizing the O–H stretching IR absorption band as well as in providing efficient energy relaxation channels should be emphasized. This argument is supported by the recent experimental observation of coherent low-frequency motions after femtosecond excitation of the O–H stretching mode [25]. The strength of anharmonic coupling to low-frequency modes is on the same order of magnitude as Fermi resonance couplings. Low-frequency mode progressions in the IR spectra start at 2728 cm^{-1} and then insert over the remaining blue part of the spectrum. Some of the peaks gain intensities comparable to the weaker Fermi resonances.

To investigate the multidimensional signatures of the different coupling mechanisms, 2D-IR spectra were simulated based on the vibrational Hamiltonians described above. Thus, the estimated anharmonic coupling constants between the IR-active ν_{OH} mode and the fingerprint and low-frequency modes depicted in Figs. 7.25 and 7.28 have been used as input in the calculation of the nonlinear signals. Two-dimensional photon echo signals in the phase matching direction $k_{echo} = -k_1 + k_2 + k_3$ were calculated with parallel linear polarization applying the sum-over-states formalism [47, 250]. The signal displayed in the frequency domain is given by:

$$S(\omega_3, T, \omega_1) = \int_{-\infty}^{\infty} dt_3 \int_{-\infty}^{\infty} dt_1\, R_a(t_3, T, t_1) \exp(-i\omega_3 t_3 - i\omega_1 t_1)\,. \quad (7.16)$$

In the rotating wave approximation, the response function is given by a sum of Liouville space pathways (diagrams I, II, and III in Fig. 7.4), that is $R_a(t_3, T = 0, t_1) = R_I(t_3, T = 0, t_1) + R_{II}(t_3, T = 0, t_1) - R^*_{III}(t_3, T = 0, t_1)$. Liouville space pathways are calculated as a sum over all vibrational eigenstates [47, 258]. They are proportional to a population factor of the initially occupied vibrational state, a polarization factor, a product of 4 transition dipole moments related to 4 infrared transitions, and a lineshape function, chosen here to be in the homogeneous limit. R_I and R_{II} describe transitions involving the ground and the first excited vibrational states giving rise to diagonal and cross peaks, while R_{III} describes transitions from one- to two-quantum states thereby adding excited state absorption to the spectra. For the calculation of 2D spectra, the three incident laser pulses were tuned to 2900 cm^{-1} assuming a rectangular electric field spectrum of $\pm$400 cm^{-1} width

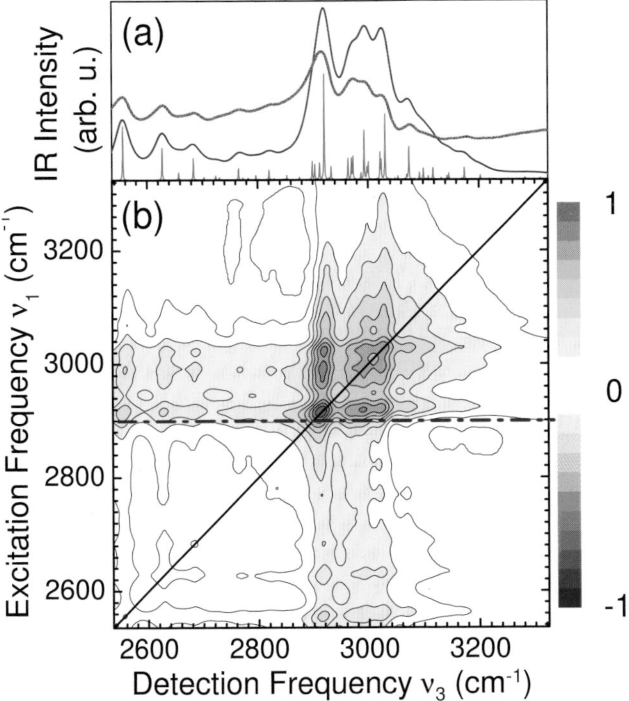

Fig. 7.30. (a) Calculated one-dimensional linear spectra with homogeneous linewidth of $\Delta\nu(fwhm) = 1$ (*red*) and $36\,\mathrm{cm}^{-1}$ (*blue*); cross section for an excitation frequency of $\nu_1 = 2921\,\mathrm{cm}^{-1}$ for ground state absorption (*green*); (b) Two-dimensional spectrum for a population time with zero- to one-quantum pathways only, with $\Delta\nu(fwhm) = 36\,\mathrm{cm}^{-1}$ (temperature 0 K), mimicking the population time $T = 400$ fs.

for selecting resonant transitions. A homogeneous linewidth of $\Delta\nu = 36\,\mathrm{cm}^{-1}$ was used.

A spectrum including ground state absorption only ($R_a = R_I$, $R_{II} = R_{III} = 0$, Fig. 7.30b) resembles conditions, where the excited state has decayed already into the hot ground state without conservation of phase memory [12]. This spectrum, which can be compared to the experimental 2D spectrum measured at $T = 400$ fs (Fig. 7.27c), contains only diagonal and cross peaks. The most striking diagonal and cross peaks are observed for the most intense bands in the linear spectrum, that is at $2921\,\mathrm{cm}^{-1}$ which is mainly due to the $\nu_{C-O}(b_u)/\nu_{C=O}(a_g)$ combination band with a cubic coupling constant with $\nu_{OH}(b_u)$ of $\Phi = -86\,\mathrm{cm}^{-1}$, at $2993\,\mathrm{cm}^{-1}$ ($\nu_{C-O}(a_g)/\nu_{C=O}(b_u)$, $\Phi = 48\,\mathrm{cm}^{-1}$) and at $3022/3031\,\mathrm{cm}^{-1}$ ($\gamma_{CH3}(b_u)/\nu_{C=O}(a_g)$, $\Phi = 62\,\mathrm{cm}^{-1}$; $\nu_{C-O}(a_g)/\delta_{OH}(b_u)/\delta_{dimer}(a_g)/\nu_{dimer}(a_g)$) [29]. In addition, prominent cross peaks exist for bands in the low-energy part of the spectrum, that is for the peak at $2555\,\mathrm{cm}^{-1}$ ($\nu_{C-O}(a_g)/\nu_{C-O}(b_u)$, $\Phi = 150\,\mathrm{cm}^{-1}$), at $2627\,\mathrm{cm}^{-1}$

($\nu_{C-O}(a_g)/\gamma_{CH3}(b_u)$, $\Phi = -118\,\mathrm{cm}^{-1}$) and at $2684\,\mathrm{cm}^{-1}$ ($\nu_{C-O}(a_g)/\delta_{OH}(b_u)$, $\Phi = -126\,\mathrm{cm}^{-1}$). This coupling pattern within the O–H stretching band is also evident from a comparison of the cross section at $\nu_1 = 2921\,\mathrm{cm}^{-1}$ to the linear spectrum (Fig. 7.29a), where the off-diagonal bands, the peak intensities of which revealing the coupling strengths, match the resonance frequencies. From the good agreement with experimental results (Fig. 7.27) it can be concluded that Fermi resonances also dominate the $T = 400$ fs experimental 2D spectrum. Low-frequency progressions make a minor contribution to the 2D spectra because of their substantially smaller transition dipoles. In going from anharmonic coupling of $\nu_{OH}(b_u)$ to low-frequency modes (3-mode model) to Fermi resonance coupling with fingerprint mode combination tones (9-mode model) to a combined mechanism (11-mode model) the number of individual 2D peaks grows considerably and the complexity of the vibrational signatures increases substantially. Introduction of homogeneous broadening causes many individual peaks to overlap to a single spectral feature. A comparison of the calculated linear spectra with high-resolution gas and solution phase spectra shows that only the combined mechanism accounts for the experimentally observed vibrational fine structure of the O–H stretching band demonstrating that both anharmonic coupling to low-frequency modes and Fermi resonance coupling are important vibrational coupling mechanisms for hydrogen bonds in acetic acid dimers. The corresponding 2D IR spectra, however, are dominated by vibrational signatures originating from cross peaks due to Fermi resonance coupling.

7.6 Outlook

The results presented in this chapter demonstrate how ultrafast nonlinear infrared spectroscopy allows for a separation of different microscopic couplings present in hydrogen-bonded systems. The X–H stretching oscillator is a sensitive probe for hydrogen bonding and the elucidation of its dynamics provides insight into the anharmonic couplings to other vibrational modes, such as X–H bending and hydrogen bond stretching modes, as well as to the (fluctuating) surroundings. Femtosecond infrared nonlinear spectroscopy and ab initio quantum chemical molecular dynamics have revealed for neat protic liquids, in particular water, the time scales for local fluctuations and associated loss of structural memory in the hydrogen-bonded networks. Energy redistribution and energy transport into the hydrogen bond stretching and librational modes deserve more attention of both experiment and theory as these modes are directly connected to the hydrogen bond networks. Future activities may involve the study of these hydrogen-bonded liquids in different situations, e.g. water in direct contact with biomolecular systems such as proteins and DNA.

In the case of the investigated medium-strong hydrogen-bonded molecular systems with well-defined geometries, the anharmonic couplings of high-frequency X–H stretching modes with low-frequency vibrations have been found to underly oscillatory wave packet motions contributing to the pump-probe signals. Coherent multidimensional spectroscopy, which was applied in the investigation of acetic acid dimers, has revealed the key role of Fermi and Davydov resonance couplings in the observed nonlinear signals.

The results obtained for PMME and cyclic acetic acid dimers demonstrate that coherent intermolecular motions persist for several picoseconds. This should allow for the development of control schemes using tailored vibrational wave packets prepared by excitation with phase-shaped infrared pulses. Moreover, as fingerprint vibrations such as the X–H bending mode couple strongly to X–H stretching motions, multicolor spectroscopy has to be explored further. For instance, both X–H stretching and X–H bending modes could be manipulated simultaneously to achieve control. With this the potential of optimal steering of hydrogen bond motions, and associated with that the control of hydrogen or proton transfer in double well potential systems should be explored.

Until now, the majority of studies have focused on the vibrational dynamics of hydrogen donating hydroxyl groups. Equally important are however amino groups, the hydrogen bonds of which can be found in nucleic acid base pairs in DNA and RNA, and in the formation of secondary structures in proteins, such as α-helices and β-sheets. Future application of ultrafast vibrational spectroscopy will provide a wealth of new information on hydrogen bonds in these biomolecular systems.

Cordial acknowledgment is being due to Jens Stenger, Dorte Madsen, Karsten Heyne, Nils Huse, R. J. Dwayne Miller, Barry. D. Bruner, Michael A. Cowan, Jason R. Dwyer, Satoshi Ashihara, Agathe Espagne, Holger Naundorf, Milena Petković, and Gennady K. Paramonov for their important contributions to the work describe here.

References

1. G.C. Pimentel, A.L. McClellan, *The hydrogen bond.* A Series of Chemistry Books (W.H. Freeman and Company, San Francisco and London, 1960)
2. P. Schuster, G. Zundel, C. Sandorfy (eds.), *The hydrogen bond: Recent developments in theory and experiments*, vol. I-III (North Holland, Amsterdam, 1976)
3. D. Hadži (ed.), *Theoretical treatments of hydrogen bonding* (John Wiley & Sons, Chichester, 1997)
4. S. Scheiner, *Hydrogen bonding: A theoretical perspective* (Oxford University Press, New York, 1997)
5. D. Hadži, S. Bratos, in *The hydrogen bond: Recent developments in theory and experiments*, vol. II. Structure and spectroscopy, ed. by P. Schuster, G. Zundel, C. Sandorfy (North Holland, Amsterdam, The Netherlands, 1976), p. 565
6. M.D. Fayer (ed.), *Ultrafast infrared and Raman spectroscopy.* Practical Spectroscopy Series Vol. 26 (Marcel Dekker, Inc., New York, 2001)

7. E.T.J. Nibbering, T. Elsaesser, Chem. Rev. **104**, 1887 (2004)
8. J. Stenger, D. Madsen, J. Dreyer, E.T.J. Nibbering, P. Hamm, T. Elsaesser, J. Phys. Chem. A **105**, 2929 (2001)
9. D. Madsen, J. Stenger, J. Dreyer, E.T.J. Nibbering, P. Hamm, T. Elsaesser, Chem. Phys. Lett. **341**, 56 (2001)
10. J. Stenger, D. Madsen, P. Hamm, E.T.J. Nibbering, T. Elsaesser, Phys. Rev. Lett. **87**, 027401 (2001)
11. J. Stenger, D. Madsen, J. Dreyer, P. Hamm, E.T.J. Nibbering, T. Elsaesser, Chem. Phys. Lett. **354**, 256 (2002)
12. J. Stenger, D. Madsen, P. Hamm, E.T.J. Nibbering, T. Elsaesser, J. Phys. Chem. A **106**, 2341 (2002)
13. D. Madsen, J. Stenger, J. Dreyer, P. Hamm, E.T.J. Nibbering, T. Elsaesser, Bull. Chem. Soc. Jpn. **75**, 909 (2002)
14. G.K. Paramonov, H. Naundorf, O. Kühn, Eur. J. Phys. D **14**, 205 (2001)
15. O. Kühn, J. Phys. Chem. A **106**, 7671 (2002)
16. H. Naundorf, G.A. Worth, H.D. Meyer, O. Kühn, J. Phys. Chem. A **106**, 719 (2002)
17. H. Naundorf, O. Kühn, in *Femtochemistry and Femtobiology*, ed. by A. Douhal, J. Santamaria (World Scientific, Singapore, 2002), p. 438
18. O. Kühn, H. Naundorf, Phys. Chem. Chem. Phys. **5**, 79 (2003)
19. M. Petković, O. Kühn, Chem. Phys. **304**, 91 (2004)
20. K. Heyne, N. Huse, E.T.J. Nibbering, T. Elsaesser, Chem. Phys. Lett. **369**, 591 (2003)
21. K. Heyne, N. Huse, E.T.J. Nibbering, T. Elsaesser, J. Phys.: Condens. Matter **15**, S129 (2003)
22. N. Huse, K. Heyne, J. Dreyer, E.T.J. Nibbering, T. Elsaesser, Phys. Rev. Lett. **91**, 197401 (2003)
23. K. Heyne, N. Huse, E.T.J. Nibbering, T. Elsaesser, Chem. Phys. Lett. **382**, 19 (2003)
24. K. Heyne, N. Huse, J. Dreyer, E.T.J. Nibbering, T. Elsaesser, S. Mukamel, J. Chem. Phys. **121**, 902 (2004)
25. K. Heyne, E.T.J. Nibbering, T. Elsaesser, M. Petković, O. Kühn, J. Phys. Chem. A **108**, 6083 (2004)
26. M.L. Cowan, B.D. Bruner, N. Huse, J.R. Dwyer, B. Chugh, E.T.J. Nibbering, T. Elsaesser, R.J.D. Miller, Nature **434**, 199 (2005)
27. N. Huse, B.D. Bruner, M.L. Cowan, J. Dreyer, E.T.J. Nibbering, R.J.D. Miller, T. Elsaesser, Phys. Rev. Lett. **95**, 147402 (2005)
28. N. Huse, S. Ashihara, E.T.J. Nibbering, T. Elsaesser, Chem. Phys. Lett. **404**, 389 (2005)
29. J. Dreyer, J. Chem. Phys. **122**, 184306 (2005)
30. J. Dreyer, Int. J. Quant. Chem. **104**, 782 (2005)
31. S. Ashihara, N. Huse, A. Espagne, E.T.J. Nibbering, T. Elsaesser, Chem. Phys. Lett. **424**, 66 (2006)
32. K. Giese, M. Petković, H. Naundorf, O. Kühn, Phys. Rep. **430**, 211 (2006)
33. A. Novak, Structure and Bonding (Berlin) **18**, 177 (1974)
34. W. Mikenda, J. Mol. Struct. **147**, 1 (1986)
35. W. Mikenda, S. Steinböck, J. Mol. Struct. **384**, 159 (1996)
36. B.I. Stepanov, Nature **157**, 808 (1946)
37. Y. Marechal, A. Witkowski, J. Chem. Phys. **48**, 3697 (1968)

38. Y. Maréchal, J. Chem. Phys. **87**, 6344 (1987)
39. O. Henri-Rousseau, P. Blaise, D. Chamma, Adv. Chem. Phys. **121**, 241 (2002)
40. P. Hamm, R.M. Hochstrasser, in *Ultrafast infrared and Raman spectroscopy*, ed. by M.D. Fayer (Marcel Dekker, Inc., New York, 2001), p. 273
41. R. Janoschek, E.G. Weidemann, G. Zundel, B. Johnson, J. Chem. Soc. Faraday Trans. II **69**, 505 (1973)
42. N. Rösch, M.A. Ratner, J. Chem. Phys. **61**, 3344 (1974)
43. S. Bratos, J. Chem. Phys. **63**, 3499 (1975)
44. G.N. Robertson, J. Yarwood, Chem. Phys. **32**, 267 (1978)
45. B. Boulil, O. Henri-Rousseau, P. Blaise, Chem. Phys. **126**, 263 (1988)
46. R. Kubo, Adv. Chem. Phys. **15**, 101 (1969)
47. S. Mukamel, *Principles of nonlinear optical spectroscopy, Oxford Series in Optical and Imaging Sciences*, vol. 6 (Oxford University Press, Oxford, 1995)
48. G.A. Jeffrey, *Introduction to hydrogen bonding* (Oxford University Press, New York, 1997)
49. A.G. Császár, in *The encyclopedia of Computational Chemistry*, ed. by P. Schleyer, N. Allinger, T. Clark, J. Gasteiger, P.A. Kollman, H.A. Schaefer III, P.R. Schreiner (John Wiley and Sons, Chichester, 1998), p. 13
50. J. Gauss, in *Modern methods and algorithms in Quantum Chemistry, NIC*, vol. 3, ed. by J. Grotendorst (John von Neumann Institute for Computing, Jülich, 2000), p. 541
51. G.C. Schatz, in *Reaction and molecular dynamics, Lecture notes in Chemistry*, vol. 14, ed. by A. Lagana, A. Riganelli (Springer, Berlin, 2000), p. 15
52. W. Schneider, W. Thiel, Chem. Phys. Lett. **157**, 367 (1989)
53. S. Dressler, W. Thiel, Chem. Phys. Lett. **273**, 71 (1997)
54. A.D. Boese, W. Klopper, J.M.L. Martin, Mol. Phys. **103**, 863 (2005)
55. J.M. Bowman, S. Carter, N.C. Handy, in *Theory and applications of Computational Chemistry: The first forty years*, ed. by C. Dykstra, G. Frenking, K.S. Kim (Elsevier Science Publishing Company, 2005), p. 251
56. R.B. Gerber, G.M. Chaban, B. Brauer, Y. Miller, in *Theory and applications of Computational Chemistry: The first forty years*, ed. by C. Dykstra, G. Frenking, K.S. Kim (Elsevier Science Publishing Company, 2005), p. 165
57. W.A. Lester, Jr., B.L. Hammond, Annu. Rev. Phys. Chem. **41**, 283 (1990)
58. J.B. Anderson, Rev. Comp. Chem. **13**, 132 (1999)
59. Z. Bačić, J.C. Light, Annu. Rev. Phys. Chem. **40**, 469 (1989)
60. M.V. Vener, O. Kühn, J.M. Bowman, Chem. Phys. Lett. **349**, 562 (2001)
61. J. Antony, G. von Helden, G. Meijer, B. Schmidt, J. Chem. Phys. **123**, 014305 (2005)
62. G. Herzberg, *Molecular spectra and molecular structure. II. Infrared and Raman spectra of polyatomic molecules* (Van Nostrand, Princeton, 1945)
63. J.R. Fair, O. Votava, D.J. Nesbitt, J. Chem. Phys. **108**, 72 (1998)
64. M. Khalil, A. Tokmakoff, Chem. Phys. **266**, 213 (2001)
65. M. Khalil, N. Demirdöven, A. Tokmakoff, J. Phys. Chem. A **107**, 5258 (2003)
66. M.P. Allen, D.J. Tildesley, *Computer simulation of liquids* (Oxford University Press, New York, 1987)
67. W. Wang, O. Donini, C.M. Reyes, P.A. Kollman, Annu. Rev. Biophys. Biomol. Struct. **30**, 211 (2001)
68. A.D. Mackerell, Jr., J. Comput. Chem. **25**, 1584 (2004)

69. J.J.P. Stewart, in *The encyclopedia of Computational Chemistry*, ed. by P. Schleyer, N. Allinger, T. Clark, J. Gasteiger, P.A. Kollman, H.A. Schaefer III, P.R. Schreiner (John Wiley and Sons, Chichester, 1998), p. 1263
70. J.E. Del Bene, M.J.T. Jordan, Int. Rev. Phys. Chem. **18**, 119 (1999)
71. S.F. Boys, F. Bernardi, Mol. Phys. **19**, 553 (1970)
72. F.B. van Duijneveldt, J.G.C.M. van Duijneveldt-van de Rijdt, J.H. van Lenthe, Chem. Rev. **94**, 1873 (1994)
73. P. Hobza, Annu. Rep. Prog. Chem. Sect. C **100**, 3 (2004)
74. W. Koch, M.C. Holthausen, *A chemist's guide to density functional theory*, 2nd edn. (Wiley-VCH, Weinheim, 2001)
75. J. Ireta, J. Neugebauer, M. Scheffler, J. Phys. Chem. A **108**, 5692 (2004)
76. J. Černý, P. Hobza, Phys. Chem. Chem. Phys. **7**, 1624 (2005)
77. E.R. Johnson, A.D. Becke, J. Chem. Phys. **124**, 174104 (2006)
78. Y. Zhao, O. Tishchenko, D.G. Truhlar, J. Phys. Chem. B **109**, 19046 (2005)
79. P. Jurečka, P. Nachtigall, P. Hobza, Phys. Chem. Chem. Phys. **3**, 4578 (2001)
80. B.M. Ladanyi, M.S. Skaf, Annu. Rev. Phys. Chem. **44**, 335 (1993)
81. I. Ohmine, S. Saito, Acc. Chem. Res. **32**, 741 (1999)
82. G. Sutmann, in *Quantum simulations of complex many-body systems: From theory to algorithms*, *NIC series*, vol. 10, ed. by J. Grotendorst, D. Marx, A. Muramatsu (John von Neumann Institute for Computing, Jülich, 2002), p. 211
83. D. Marx, J. Hutter, in *Modern methods and algorithms of Quantum Chemistry*, *NIC series*, vol. 1, ed. by J. Grotendorst (John von Neumann Institute for Computing, Jülich, 2000), p. 1
84. R. Iftimie, P. Minary, M.E. Tuckerman, Proc. Natl. Acad. Sci. USA **102**, 6654 (2005)
85. D. Marx, in *Computational nanoscience: Do It yourself!*, *NIC series*, vol. 31, ed. by J. Grotendorst, S. Blügel, D. Marx (John von Neumann Institute for Computing, Jülich, 2006), p. 195
86. D. Marx, ChemPhysChem **7**, 1848 (2006)
87. K. Laasonen, M. Sprik, M. Parrinello, R. Car, J. Chem. Phys. **99**, 9080 (1993)
88. P.L. Silvestrelli, M. Parrinello, J. Chem. Phys. **111**, 3572 (1999)
89. P.L. Geissler, C. Dellago, D. Chandler, J. Hutter, M. Parrinello, Science **291**, 2121 (2001)
90. E. Tsuchida, Y. Kanada, M. Tsukada, Chem. Phys. Lett. **311**, 236 (1999)
91. M.E. Tuckerman, D. Marx, M.L. Klein, M. Parinello, Science **275**, 817 (1997)
92. D. Marx, M.E. Tuckerman, J. Hutter, M. Parinello, Nature **397**, 901 (1999)
93. G.A. Voth, Acc. Chem. Res. **39**, 143 (2006)
94. M. Eichinger, P. Tavan, J. Hutter, M. Parrinello, J. Chem. Phys. **110**, 10452 (1999)
95. J. Gao, D.G. Truhlar, Annu. Rev. Phys. Chem. **53**, 467 (2002)
96. R.A. Friesner, V. Guallar, Annu. Rev. Phys. Chem. **56**, 389 (2005)
97. J. Park, J.H. Ha, R.M. Hochstrasser, J. Chem. Phys. **121**, 7281 (2004)
98. W.T. Pollard, R.A. Mathies, Annu. Rev. Phys. Chem. **43**, 497 (1992)
99. H.K. Nienhuys, S. Woutersen, R.A. van Santen, H.J. Bakker, J. Chem. Phys. **111**, 1494 (1999)
100. A.J. Lock, H.J. Bakker, J. Chem. Phys. **117**, 1708 (2002)
101. M.A.F.H. van den Broek, M.F. Kropman, H.J. Bakker, Chem. Phys. Lett. **357**, 8 (2002)

102. P. Hamm, S.M. Ohline, W. Zinth, J. Chem. Phys. **106**, 519 (1997)
103. E.T.J. Nibbering, H. Fidder, E. Pines, Annu. Rev. Phys. Chem. **56**, 337 (2005)
104. J.C. Deàk, S.T. Rhea, L.K. Iwaki, D.D. Dlott, J. Phys. Chem. A **104**, 4866 (2000)
105. D. Dlott, Chem. Phys. **266**, 149 (2001)
106. V. Kozich, J. Dreyer, S. Ashihara, W. Werncke, T. Elsaesser, J. Chem. Phys. **125**, 074504 (2006)
107. G.R. Fleming, M. Cho, Annu. Rev. Phys. Chem. **47**, 109 (1996)
108. W.P. de Boeij, M.S. Pshenichnikov, D.A. Wiersma, Annu. Rev. Phys. Chem. **49**, 99 (1998)
109. S. Yeremenko, M.S. Pshenichnikov, D.A. Wiersma, Chem. Phys. Lett. **369**, 107 (2003)
110. S. Yeremenko, M.S. Pshenichnikov, D.A. Wiersma, Phys. Rev. A **73**, 021804 (2006)
111. C.J. Fecko, J.D. Eaves, J.J. Loparo, A. Tokmakoff, P.L. Geissler, Science **301**, 1698 (2003)
112. C.J. Fecko, J.J. Loparo, S.T. Roberts, A. Tokmakoff, J. Chem. Phys. **122**, 054506 (2005)
113. J.B. Asbury, T. Steinel, C. Stromberg, S.A. Corcelli, C.P. Lawrence, J.L. Skinner, M.D. Fayer, J. Phys. Chem. A **108**, 1107 (2004)
114. T. Steinel, J.B. Asbury, S.A. Corcelli, C.P. Lawrence, J.L. Skinner, M.D. Fayer, Chem. Phys. Lett. **386**, 295 (2004)
115. J.B. Asbury, T. Steinel, K. Kwak, S.A. Corcelli, C.P. Lawrence, J.L. Skinner, M.D. Fayer, J. Chem. Phys. **121**, 12431 (2004)
116. H.S. Tan, I.R. Piletic, R.E. Riter, N.E. Levinger, M.D. Fayer, Phys. Rev. Lett. **94**, 057405 (2005)
117. J.B. Asbury, T. Steinel, C. Stromberg, K.J. Gaffney, I.R. Piletic, A. Goun, M.D. Fayer, Chem. Phys. Lett. **374**, 362 (2003)
118. J.B. Asbury, T. Steinel, C. Stromberg, K.J. Gaffney, I.R. Piletic, A. Goun, M.D. Fayer, Phys. Rev. Lett. **91**, 237402 (2003)
119. J.B. Asbury, T. Steinel, M.D. Fayer, J. Phys. Chem. B **108**, 6544 (2004)
120. I.V. Rubtsov, K. Kumar, R.M. Hochstrasser, Chem. Phys. Lett. **402**, 439 (2005)
121. Y.S. Kim, R. Hochstrasser, Proc. Natl. Acad. Sci. USA **102**, 11185 (2005)
122. M.C. Asplund, M.T. Zanni, R.M. Hochstrasser, Proc. Natl. Acad. Sci. USA **97**, 8219 (2000)
123. M.L. Cowan, J.P. Ogilvie, R.J.D. Miller, Chem. Phys. Lett. **386**, 184 (2004)
124. H. Eichler, H. Stahl, J. Appl. Phys. **44**, 3429 (1973)
125. H.J. Eichler, Optica Acta **24**, 631 (1977)
126. M.D. Fayer, Annu. Rev. Phys. Chem. **33**, 63 (1982)
127. N.A. Kurnit, S.R. Hartmann, I.D. Abella, Phys. Rev. Lett. **13**, 567 (1964)
128. T.J. Aartsma, D.A. Wiersma, Phys. Rev. Lett. **36**, 1360 (1976)
129. D. Zimdars, A. Tokmakoff, S. Chen, S.R. Greenfield, M.D. Fayer, T.I. Smith, H.A. Schwettman, Phys. Rev. Lett. **70**, 2718 (1993)
130. K. Duppen, D.A. Wiersma, J. Opt. Soc. Am. B **3**, 614 (1986)
131. T. Joo, Y. Jia, J.Y. Yu, M.J. Lang, G.R. Fleming, J. Chem. Phys. **104**, 6089 (1996)
132. W.P. de Boeij, M.S. Pshenichnikov, D.A. Wiersma, Chem. Phys. Lett. **253**, 53 (1996)

133. K. Duppen, D.P. Weitekamp, D.A. Wiersma, Chem. Phys. Lett. **108**, 551 (1984)
134. R.M. Hochstrasser, N.H. Ge, S. Gnanakaran, M.T. Zanni, Bull. Chem. Soc. Jpn. **75**, 1103 (2002)
135. I.V. Rubtsov, J. Wang, R.M. Hochstrasser, J. Chem. Phys. **118**, 7733 (2003)
136. P. Hamm, M. Lim, W.F. DeGrado, R.M. Hochstrasser, Proc. Natl. Acad. Sci. USA **96**, 2036 (1999)
137. M. Khalil, N. Demirdöven, A. Tokmakoff, Phys. Rev. Lett. **90**, 047401 (2003)
138. J. Bredenbeck, J. Helbing, R. Behrendt, C. Renner, L. Moroder, J. Wachtveitl, P. Hamm, J. Phys. Chem. B **107**, 8654 (2003)
139. J. Bredenbeck, J. Helbing, P. Hamm, J. Am. Chem. Soc. **126**, 990 (2004)
140. R.R. Ernst, G. Bodenhausen, A. Wokaun, *Principles of nuclear magnetic resonance in one and two dimensions* (Clarendon, Oxford, 1987)
141. W.P. Aue, E. Bartholdi, R.R. Ernst, J. Chem. Phys. **64**, 2229 (1976)
142. P. Hamm, M. Lim, R.M. Hochstrasser, J. Phys. Chem. B **102**, 6123 (1998)
143. S. Woutersen, P. Hamm, J. Phys. Chem. B **104**, 11316 (2000)
144. S. Woutersen, Y. Mu, G. Stock, P. Hamm, Proc. Natl. Acad. Sci. USA **98**, 11254 (2001)
145. S. Woutersen, R. Pfister, P. Hamm, Y. Mu, D.S. Kosov, G. Stock, J. Chem. Phys. **117**, 6833 (2002)
146. S. Woutersen, P. Hamm, J. Phys.: Condens. Matter **14**, R1035 (2002)
147. K. Kwac, M. Cho, J. Chem. Phys. **119**, 2256 (2003)
148. O. Golonzka, M. Khalil, N. Demirdöven, A. Tokmakoff, Phys. Rev. Lett. **86**, 2154 (2001)
149. M.T. Zanni, N.H. Ge, Y.S. Kim, R.M. Hochstrasser, Proc. Natl. Acad. Sci. USA **98**, 11265 (2001)
150. M.T. Zanni, S. Gnanakaran, J. Stenger, R.M. Hochstrasser, J. Phys. Chem. B **105**, 6520 (2001)
151. M. Zanni, R.M. Hochstrasser, Curr. Opin. Struct. Biol. **11**, 516 (2001)
152. A.T. Krummel, P. Mukherjee, M.T. Zanni, J. Phys. Chem. B. **107**, 9165 (2003)
153. M. Khalil, N. Demirdöven, A. Tokmakoff, J. Chem. Phys. **121**, 362 (2004)
154. I.R. Piletic, K.J. Gaffney, M.D. Fayer, J. Chem. Phys. **119**, 423 (2003)
155. J.B. Asbury, T. Steinel, C. Stromberg, K.J. Gaffney, I.R. Piletic, M.D. Fayer, J. Chem. Phys. **119**, 12981 (2003)
156. I.R. Piletic, H.S. Tan, M.D. Fayer, J. Phys. Chem. B **109**, 21273 (2005)
157. J. Zheng, K. Kwak, J. Asbury, X. Chen, I.R. Piletic, M.D. Fayer, Science **309**, 1338 (2005)
158. J. Zheng, K. Kwak, X. Chen, J.B. Asbury, M.D. Fayer, J. Am. Chem. Soc. **128**, 2977 (2006)
159. P. Hamm, R.A. Kaindl, J. Stenger, Opt. Lett. **25**, 1798 (2000)
160. R.A. Kaindl, M. Wurm, K. Reimann, P. Hamm, A.M. Weiner, M. Woerner, J. Opt. Soc. Am. B **17**, 2086 (2000)
161. V. Volkov, R. Schanz, P. Hamm, Opt. Lett. **30**, 2010 (2005)
162. P. Tian, D. Keusters, Y. Suzaki, W.S. Warren, Science **300**, 1553 (2003)
163. G.M. Gale, G. Gallot, F. Hache, N. Lascoux, S. Bratos, J.C. Leicknam, Phys. Rev. Lett. **82**, 1068 (1999)
164. S. Bratos, J.C. Leicknam, G. Gallot, H. Ratajczak, in *Ultrafast hydrogen bonding dynamics and proton transfer processes in the condensed phase, Understanding chemical reactivity*, vol. 23, ed. by T. Elsaesser, H.J. Bakker (Kluwer Academic Publishers, Dordrecht, 2002), p. 5

165. R. Rey, K.B. Møller, J.T. Hynes, J. Phys. Chem. A **106**, 11993 (2002)
166. K.B. Møller, R. Rey, J.T. Hynes, J. Phys. Chem. A **108**, 1275 (2004)
167. C.P. Lawrence, J.L. Skinner, J. Chem. Phys. **118**, 264 (2003)
168. R. Torre, P. Bartolini, R. Righini, Nature **428**, 296 (2004)
169. P. Wernet, D. Nordlund, U. Bergmann, M. Cavalleri, M. Odelius, H. Ogasawara, L.A. Naslund, T.K. Hirsch, L. Ojamae, P. Glatzel, L.G.M. Pettersson, A. Nilsson, Science **304**, 995 (2004)
170. J.D. Smith, C.D. Cappa, K.R. Wilson, B.M. Messer, R.C. Cohen, R.J. Saykally, Science **306**, 851 (2004)
171. A. Nilsson, P. Wernet, D. Nordlund, U. Bergmann, M. Cavalleri, M. Odelius, H. Ogasawara, L.A. Naslund, T.K. Hirsch, L. Ojamae, P. Glatzel, L.G.M. Pettersson, Science **308**, 793 (2005)
172. J.D. Smith, C.D. Cappa, B.M. Messer, R.C. Cohen, R.J. Saykally, Science **308**, 793 (2005)
173. A. Luzar, D. Chandler, Nature **379**, 55 (1996)
174. A. Luzar, D. Chandler, Phys. Rev. Lett. **76**, 928 (1996)
175. G. Reddy, C.P. Lawrence, J.L. Skinner, A. Yethiraj, J. Chem. Phys. **119**, 13012 (2003)
176. D. Xenides, B.R. Randolf, B.M. Rode, J. Chem. Phys. **122**, 174506 (2005)
177. D.W. Oxtoby, D. Levesque, J.J. Weis, J. Chem. Phys. **68**, 5528 (1978)
178. D.W. Oxtoby, Adv. Chem. Phys. **40**, 1 (1979)
179. J.J. Loparo, C.J. Fecko, J.D. Eaves, S.T. Roberts, A. Tokmakoff, Phys. Rev. B **70**, 180201 (2004)
180. C.J. Burnham, S.S. Xantheas, J. Chem. Phys. **116**, 5115 (2002)
181. C.J. Burnham, S.S. Xantheas, M.A. Miller, B.E. Applegate, R.E. Miller, J. Chem. Phys. **117**, 1109 (2002)
182. G.S. Fanourgakis, E. Apra, W.A. de Jong, S.S. Xantheas, J. Chem. Phys. **122**, 134304 (2005)
183. A. Lagutschenkov, G.S. Fanourgakis, G. Niedner-Schatteburg, S.S. Xantheas, J. Chem. Phys. **122**, 194310 (2005)
184. C.P. Lawrence, J.L. Skinner, J. Chem. Phys. **117**, 5827 (2002)
185. C.P. Lawrence, J.L. Skinner, J. Chem. Phys. **117**, 8847 (2002)
186. S.A. Corcelli, C.P. Lawrence, J.B. Asbury, T. Steinel, M.D. Fayer, J.L. Skinner, J. Chem. Phys. **121**, 8897 (2004)
187. S.A. Corcelli, C.P. Lawrence, J.L. Skinner, J. Chem. Phys. **120**, 8107 (2004)
188. T. Hayashi, T. la Cour Jansen, W. Zhuang, S. Mukamel, J. Phys. Chem. A **109**, 64 (2005)
189. S.A. Corcelli, J.L. Skinner, J. Phys. Chem. A **109**, 6154 (2005)
190. J.R. Schmidt, S.A. Corcelli, J.L. Skinner, J. Chem. Phys. **123**, 044513 (2005)
191. C.P. Lawrence, J.L. Skinner, Proc. Natl. Acad. Sci. USA **102**, 6720 (2005)
192. S. Bratos, J.C. Leicknam, S. Pommeret, G. Gallot, J. Mol. Struct. **708**, 197 (2004)
193. S. Woutersen, H.J. Bakker, Nature **402**, 507 (1999)
194. W. Amir, G. Gallot, F. Hache, J. Chem. Phys. **121**, 7908 (2004)
195. J.A. Poulsen, G. Nyman, S. Nordholm, J. Phys. Chem. A **107**, 8420 (2003)
196. D. Laage, H. Demirdjian, J.T. Hynes, Chem. Phys. Lett. **405**, 453 (2005)
197. R. DeVane, B. Space, A. Perry, C. Neipert, C. Ridley, T. Keyes, J. Chem. Phys. **121**, 3688 (2004)
198. D. Laage, J.T. Hynes, Science **311**, 832 (2006)

199. T. Elsaesser, W. Kaiser, Annu. Rev. Phys. Chem. **42**, 83 (1991)
200. J.C. Owrutsky, D. Raftery, R.M. Hochstrasser, Annu. Rev. Phys. Chem. **45**, 519 (1994)
201. R. Rey, K.B. Møller, J.T. Hynes, Chem. Rev. **104**, 1915 (2004)
202. R. Rey, J.T. Hynes, J. Chem. Phys. **104**, 2356 (1996)
203. H. Graener, G. Seifert, A. Laubereau, Chem. Phys. **175**, 193 (1993)
204. H. Graener, G. Seifert, J. Chem. Phys. **98**, 36 (1993)
205. D. Cringus, S. Yeremenko, M.S. Pshenichnikov, D.A. Wiersma, J. Phys. Chem. B **108**, 10376 (2004)
206. A.J. Lock, S. Woutersen, H.J. Bakker, J. Phys. Chem. A **105**, 1238 (2001)
207. D.W. Oxtoby, Annu. Rev. Phys. Chem. **32**, 77 (1981)
208. M.F. Kropman, H.J. Bakker, J. Am. Chem. Soc. **126**, 9135 (2004)
209. A.W. Omta, M.F. Kropman, S. Woutersen, H.J. Bakker, J. Chem. Phys. **119**, 12457 (2003)
210. K.B. Møller, R. Rey, M. Masia, J.T. Hynes, J. Chem. Phys. **122**, 114508 (2005)
211. A.M. Dokter, S. Woutersen, H.J. Bakker, Phys. Rev. Lett. **94**, 178301 (2005)
212. D. Cringus, J. Lindner, M.T.W. Milder, M.S. Pshenichnikov, P. Vöhringer, D.A. Wiersma, Chem. Phys. Lett. **408**, 162 (2005)
213. H.S. Tan, I.R. Piletic, M.D. Fayer, J. Chem. Phys. **122**, 174501 (2005)
214. G. Seifert, T. Patzlaff, H. Graener, J. Chem. Phys. **120**, 8866 (2004)
215. Z.H. Wang, Y.S. Pang, D.D. Dlott, J. Chem. Phys. **120**, 8345 (2004)
216. O.F.A. Larsen, S. Woutersen, J. Chem. Phys. **121**, 12143 (2004)
217. P. Bodis, O.F.A. Larsen, S. Woutersen, J. Phys. Chem. A **109**, 5303 (2005)
218. C.P. Lawrence, J.L. Skinner, J. Chem. Phys. **119**, 1623 (2003)
219. C.P. Lawrence, J.L. Skinner, J. Chem. Phys. **119**, 3840 (2003)
220. A. Pakoulev, Z. Wang, Y. Pang, D.D. Dlott, Chem. Phys. Lett. **380**, 404 (2003)
221. H.J. Bakker, A.J. Lock, D. Madsen, Chem. Phys. Lett. **385**, 329 (2004)
222. A. Pakoulev, Z.H. Wang, Y.S. Pang, D.D. Dlott, Chem. Phys. Lett. **385**, 332 (2004)
223. H.J. Bakker, A.J. Lock, D. Madsen, Chem. Phys. Lett. **384**, 236 (2004)
224. T. Patzlaff, M. Janich, G. Seifert, H. Graener, Chem. Phys. **261**, 381 (2000)
225. G. Seifert, T. Patzlaff, H. Graener, Phys. Rev. Lett. **88**, 147402 (2002)
226. J.C. Deàk, Y. Pang, T.D. Sechler, Z. Wang, D.D. Dlott, Science **306**, 473 (2004)
227. Z.H. Wang, Y. Pang, D.D. Dlott, Chem. Phys. Lett. **397**, 40 (2004)
228. Z.H. Wang, A. Pakoulev, Y. Pang, D.D. Dlott, J. Phys. Chem. A **108**, 9054 (2004)
229. J. Lindner, P. Vöhringer, M.S. Pshenichnikov, D. Cringus, D.A. Wiersma, M. Mostovoy, Chem. Phys. Lett. **421**, 329 (2006)
230. T. Steinel, J.B. Asbury, J. Zheng, M.D. Fayer, J. Phys. Chem. A **108**, 10957 (2004)
231. S. Woutersen, Y. Mu, G. Stock, P. Hamm, Chem. Phys. **266**, 137 (2001)
232. J. Zheng, K. Kwak, J. Xie, M.D. Fayer, Science **313**, 1951 (2006)
233. P. Blaise, O. Henri-Rousseau, A. Grandjean, Chem. Phys. **244**, 405 (1999)
234. M.H. Beck, A. Jäckle, G.A. Worth, H.D. Meyer, Phys. Rep. **324**, 1 (2000)
235. V. May, O. Kühn, *Charge and energy transfer dynamics in molecular systems, 2nd revised and enlarged edition* (Wiley–VCH, Weinheim, 2004)
236. M. Lim, R.M. Hochstrasser, J. Chem. Phys. **115**, 7629 (2001)
237. G. Seifert, T. Patzlaff, H. Graener, Chem. Phys. Lett. **333**, 248 (2001)

238. O. Faurskov Nielsen, P.A. Lund, J. Chem. Phys. **78**, 652 (1983)
239. T. Nakabayashi, K. Kosugi, N. Nishi, J. Phys. Chem. A **103**, 8595 (1999)
240. M. Haurie, A. Novak, J. Chim. Phys. **62**, 146 (1965)
241. Y. Greenie, J.C. Cornut, J.C. Lassegues, J. Chem. Phys. **55**, 5844 (1971)
242. R.L. Redington, K.C. Lin, J. Chem. Phys. **54**, 4111 (1971)
243. D. Chamma, O. Henri-Rousseau, Chem. Phys. **248**, 53 (1999)
244. D. Chamma, O. Henri-Rousseau, Chem. Phys. **248**, 71 (1999)
245. D. Chamma, O. Henri-Rousseau, Chem. Phys. **248**, 91 (1999)
246. C. Emmeluth, M.A. Suhm, D. Luckhaus, J. Chem. Phys. **118**, 2242 (2003)
247. C. Emmeluth, M.A. Suhm, Phys. Chem. Chem. Phys. **5**, 3094 (2003)
248. J. Chocholoušová, J. Vacek, P. Hobza, J. Phys. Chem. A **107**, 3086 (2003)
249. T. Häber, U. Schmitt, C. Emmeluth, M.A. Suhm, Faraday Discuss. **118**, 331 (2001)
250. S. Mukamel, Annu. Rev. Phys. Chem. **51**, 691 (2000)
251. M.J. Frisch, et al. Gaussian 03, Revision C.02. Gaussian, Inc., Wallingford, CT, 2004
252. A. Burneau, F. Génin, F. Quilès, Phys. Chem. Chem. Phys. **2**, 5020 (2000)
253. A.M. Moran, J. Dreyer, S. Mukamel., J. Chem. Phys. **118**, 1347 (2003)
254. K. Park, M. Cho, S. Hahn, D. Kim, J. Chem. Phys. **111**, 4131 (1999)
255. T. Hayashi, S. Mukamel, J. Phys. Chem. A **107**, 9113 (2003)
256. H.R. Zelsmann, Z. Mielke, Y. Marechal, J. Mol. Struct. **237**, 273 (1990)
257. G.M. Florio, T.S. Zwier, E.M. Myshakin, K.D. Jordan, E.L. Sibert III, J. Chem. Phys. **118**, 1735 (2003)
258. J. Dreyer, A.M. Moran, S. Mukamel, J. Phys. Chem. B **107**, 5967 (2003)

8

Observing molecular structure changes and dynamics in polar solution

Alexander L. Dobryakov[1], Nikolaus P. Ernsting[1], Wojciech Gawelda[2], Christian Bressler[2], Majed Chergui[2]

[1] Department of Chemistry, Humboldt-Universität zu Berlin, Germany
[2] Institut des Sciences et Ingénierie Chimiques, École Polytechnique Fédérale de Lausanne, Switzerland

Coordinated by: Nikolaus P. Ernsting

Most chemical reactions and all biological processes take place in solution. The solvent influences the energetics of chemical reactivity because the reactants, transition state, and products may be stabilized differently by the solvent, and it creates fluctuating external forces for the dynamics. Here charge-transfer reactions of large molecules in polar media are considered. The principal questions in such cases are: what is the molecular reaction coordinate, which intramolecular spectator modes are involved, and how - in terms of energy and dynamics - is the solvent involved? Because of strong solvent coupling the distinction between molecular and external coordinates may no longer be possible and effective coordinates have to be found for an adequate description. Femtosecond optical pulses are used to initiate the reaction photochemically. It is often assumed that the polarity of the environment is in equilibrium with the charge distribution during the reaction in the excited state of the optical chromophore. But ultrafast reactions may even "leave the solvent behind", the polarity can no longer stay adjusted, and the energy barrier is crossed in a non-equilibrated environment. In such cases a retarding force builds up between solute and solvent along the reaction coordinate. That force appears as friction which does not only act instantaneously but also depends on time. The polar solvent thus also enters the dynamics of charge-transfer reactivity.

Methods are clearly needed to observe structural changes in the molecule and in its environment as the reaction unfolds. Abundant water is present in biological work. Because bulk water absorbs strongly throughout the infrared region, it becomes difficult to follow the reaction of dissolved molecules by time-resolved infrared spectroscopy. Two methodical directions from the realm of pump-probe spectroscopy are instead explored in this chapter. Common to both is the exploitation of spectral information carried by the probe pulse. The first contribution summarizes the state of the art of optical probing with a UV/vis supercontinuum - a classic which needs theory for quantitative analysis. In a second contribution, exciting first steps are presented to observe structural changes *directly* by time-resolved X-ray absorption spectroscopy.

8.1 Analysis of electronic and vibrational coherence in transient absorption spectroscopy

A. L. Dobryakov and N. P. Ernsting

8.1.1 Scope for transient absorption spectroscopy in solution

Many interesting reactions can be achieved only in solution, for example when polar product states must be stabilized by the solvent to be accessible. New limitations and concepts come into play in solution at room temperature. The main limitation is that optical transitions of a solute chromophore become blurred into bands in principle. This is because electronic energy levels are no longer well defined but fluctuate on all time scales, due to interaction with the solvent, just as the solvent configuration does. Even if it were possible to excite - with short pulses - only those chromophores that have a specific solvent configuration, one never observes that the original configuration is reached again some time later, simultaneously, for all members of the ensemble. In other words, the complete recurrences which are characteristic (and informative) for small molecules in the gas phase are absent in solution. From the chromophore's point of view the solvent motion is not determined but has a statistical, dissipative nature. The evolution of the chromophore is governed by the molecular Hamiltonian as before, but in addition the solvent motion, through the resultant perturbations, induces energy flow, dephasing, and coherence transfer in the molecule. A stochastic perturbation by the solvent can be characterized by its time correlation function. The task of electronic and vibrational spectroscopy in the condensed phase is to find the relevant Hamiltonian and characteristic correlation times [1].

A simple example is shown in Fig. 8.1a. The molecule has only one optically active mode with internal coordinate q'. High-frequency motion along q' is quantized in vibrational states which are separated by more than the spectral width of the laser pulse. Solvent coordinate q represents the configurational state of the environment and is treated classically. Before femtosecond laser excitation the solute is in the equilibrated ground-state S_0 and the absorption band expands over a Franck-Condon progression for upward optical transitions. In the first excited state S_1 the interaction with the solvent is assumed to be different, resulting in a displacement of the classical parabola along q for the free energy of the solute/solvent system. First consider the vibronic $0-0$ transition as seen by stationary absorption spectroscopy from S_0. For a single molecule, the solvent coordinate $q(t)$ will fluctuate in a stochastic manner around $q = 0$. This is shown as a noisy green line below. (If the fluctuations are completely uncorrelated in time, the solvent acts like a Brownian harmonic oscillator. It is more likely that the fluctuations have some time structure since they are caused by the solvent molecular mechanics. In such cases $q(t)$ can always be described as a multimode Brownian oscillator.) Because the S_1 solvation parabola is displaced, the $S_1 - S_0$ energy gap is linear in q. This means that the frequency ω_{00} of the vibronic origin transition also fluctuates, by $\delta\omega_{00}(t) \propto q(t)$ (noisy line at left which samples the absorption lineshape). $q(t)$ can also be explored by starting a chromophore/solvent trajectory at $t = 0$ with $q = 0$ in the excited state. But usually an ensemble of chromophores is examined. In such an experiment one would excite in the $0-0$ band with femtosecond pulses and then monitor the red-shift and broadening of an emission band. A typical spectroscopic task could be to find the underlying autocorrelation function $\langle q(t)q(0)\rangle$.

An example for an intramolecular reaction is shown in Fig. 8.1b. In this case there are two excited electronic states. When S_1 is excited near the origin, the reaction to S_2 can not occur because the latter is not solvated enough and therefore higher in energy. But as solvation proceeds in S_1, the S_2 state is lowered. While the states are isoenergetic, the reaction takes place efficiently depending on the coupling strength V. One may even observe a population oscillation between them with frequency $2V/\hbar$. It is readily seen that time-resolved observables should depend on intramolecular and solvent dynamics in a complicated manner.

Several nonlinear optical techniques explore the spectral changes which were mentioned above on femtosecond to picosecond time scales. They differ by the composition of the optical field from component pulses: their frequencies, pulse-to-pulse delays, wave and electric field vectors, as well as by the mode of detection. In this way one may obtain information on vibronic energy levels and achieve complementary sensitivity to different relaxation pathways [1]. Here only effects are considered which scale with the third order of the applied electric field $\mathbf{E}$. The most general architecture is realized by multicolor four-wave mixing, shown in Fig. 8.2a. In this case three interactions (white bars) with the applied field are separately controlled, i.e. they can be assigned to the component pulses by detection at a combination frequency in a phase-matched direction. The electric polarization which the field generates in the material (wavy line) may be detected by the light it generates (red arrow) or by heterodyning with another field. A much simpler architecture is used for transient absorption, Fig. 8.2b. Now two interactions are derived from a pump pulse having wave vector $\mathbf{k}_1$. The third interaction is contributed by the delayed probe pulse with wavevector $\mathbf{k}_2$. By detecting at the probe frequency and direction, phase-matching is fulfilled. The induced polarization is heterodyned by the probe field itself, resulting in probe absorption.

This contribution explores possibilities for transient absorption in condensed-phase spectroscopy. The fact that phase matching is always fulfilled allows to scan the probe frequency without having to change the detection

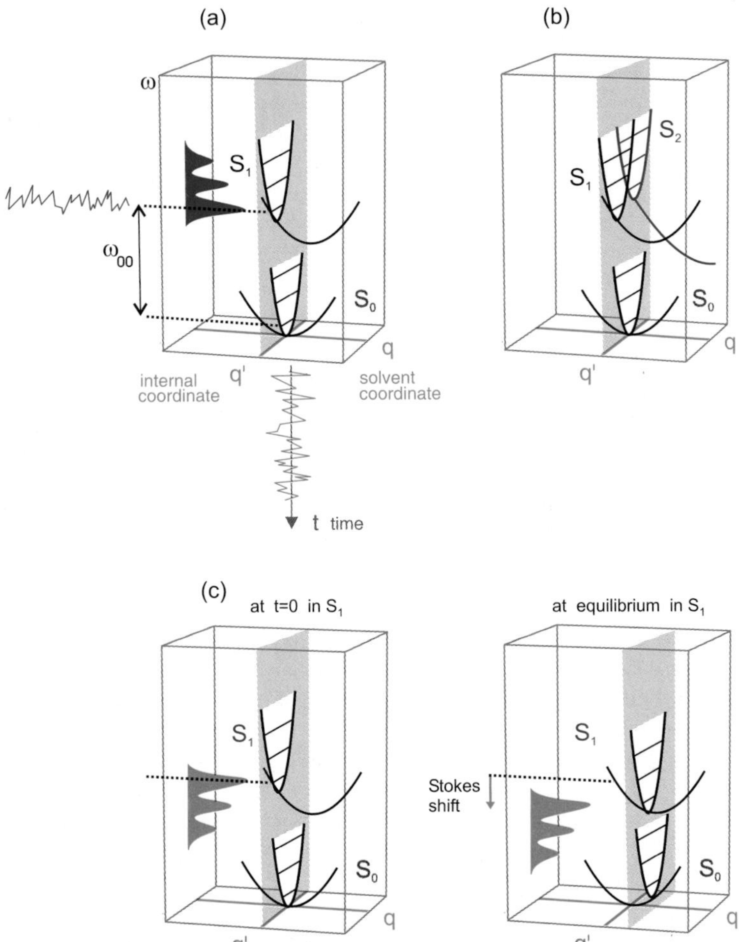

Fig. 8.1. (a) The bandshape of a vibronic transition (here the origin 0–0) in solution reflects the stochastic perturbation by solvent motion. (b) The chemical reactivity of the S_1 state, for example through vibronic coupling with S_2, may depend on the solvent. Complicated changes of the emission and excited-state absorption bandshapes are thus caused by solvent dynamics. Broad spectral coverage is required to characterize the evolution. (c) After ultrafast excitation, the solvent dynamics may be monitored by the time-resolved Stokes shift of fluorescence.

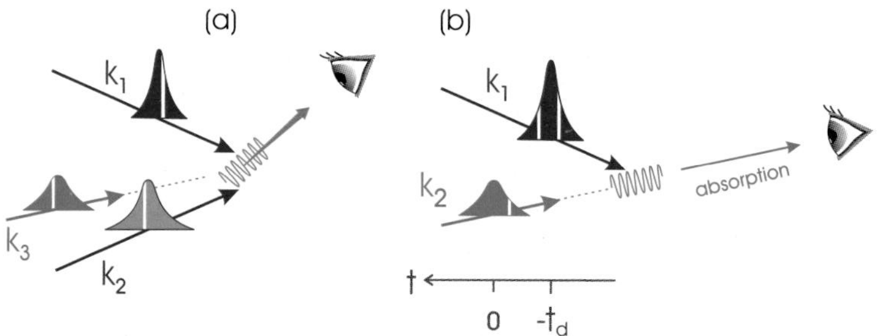

Fig. 8.2. (a) Four-wave mixing allows complete control over three interactions. The polarization at a phase-matched direction (red wave) generates signal light (red arrow) for background-free detection. The interactions (white bars) can be assigned to the component pulses which make up the full field. (b) Transient absorption uses a degenerate scheme. The two light-matter interactions with the pump light at wavevector $\mathbf{k}_1$ can no longer be controlled separately. The probe pulse ($\mathbf{k}_2$) causes a polarization of the same frequency and direction (green wave). It is heterodyned with the probe light itself, registering absorption.

geometry during measurement. In this way the Franck-Condon envelopes for electronic transitions which become visible after optical pumping, for example excited-state absorption $S_n \leftarrow S_1$ or stimulated emission $S_1 \rightarrow S_0$, may be monitored in time. The precision of such measurements is much improved if the induced absorption is recorded at all relevant frequencies simultaneously. This broadband approach was introduced by Alfano and coworkers ([2] and references therein) who probed with a supercontinuum generated by picosecond pulses in about 10 cm of water. Supercontinuum probing has been used extensively since then ([3–8] represent a variety of recent applications). The present work tries to improve on previous achievements in the sense that the near-UV to NIR spectral range extending over 20000 cm^{-1} is covered continuously with photometric accuracy. Only in this case can band shapes be observed completely and their integrals be obtained. Time-dependent band integrals may then be used for comparison with quantum-chemical calculations. The corresponding experimental setup will be described in detail.

A serious limitation of transient absorption spectroscopy, on the other hand, is due to incomplete control over the three field-matter interactions which was already mentioned. As a consequence a "femtosecond artifact" or "coherent spike" appears when pump and probe pulses overlap in time. It arises from coherent terms when the order of interactions is mixed (*pump-probe-pump* or *probe-pump-pump*) while for well-separated pulses only sequential terms apply (*pump-pump-probe*). The solvent contributes a nonresonant part which can be measured separately and subtracted. But the resonant contribution from solute chromophores is often seen to have rich spectral structure; it distorts weak sequential transient spectra at earliest time and renders data analysis ambiguous there.

The main concern of this contribution is therefore to "look under the femtosecond artifact", to observe molecular evolution in the excited state at earliest time. The conceptual frame for this work was outlined above: an essential molecular Hamiltonian in conjunction with relaxation and dephasing processes. This model is combined with the composite electric field which includes the chirp of the supercontinuum, and all terms inherent to transient absorption spectra are calculated. In this way the "coherence spectrum" is extracted while electronic and vibrational dephasing times are estimated at the same time. Indeed, coherence analysis offers the key for a quantitative understanding of the entire transient signal. The aim is to demonstrate that observed spectra can thus be described quantitatively for all delay times and wavelengths. The presentation combines the developments in [9–11] and contains new equations which speed up data fitting considerably. The ultimate aim is to enable practical use in other laboratories.

8.1.2 Experimental setup for supercontinuum probing

The experimental setup is sketched in Fig. 8.3. The probe pulses generate white light in a 1 mm CaF_2 plate. This supercontinuum [2] is multifilamented and extends approximately $\pm 4^\circ$ from the optical axis. Rays near the center of the axis are blocked because they are too dominated by generating light. The supercontinuum is then passed through a color filter (200 μm flowing dye solution between 200 μm quartz windows) and focused 3-5 mm before the thin beam splitter (BS). The sample beam is made from the front reflection and the transmitted light constitutes the reference beam. All imaging (1:1) is performed with off-axis Schwarzschild objectives (mirrors with radii R_1=-406.4 mm and R_2=+9400 mm). The diameter of the light spot on the sample is 50 μm and the probe pulse energy less than 100 nJ. Light from either beam is dispersed in special spectrographs, by an imaging grating as shown (Zeiss wide-range MCS, 248 mm^{-1}) or by a quartz prism in an Ebert mount if the range 350-1100 nm must be covered simultaneously. Spectra are registered on a photodiode array (PDA, Hamamatsu S3904-512Q). The spectra shown later were recorded with a probe repetition rate of 140 Hz, in which case the CaF_2 plate can be kept fixed during measurement. For every second shot, a pump pulse excites the liquid sample in a flow cell. Microscopes can be swiveled into place in order to view the supercontinuum spot in the input plane of the spectrographs. After centering the spot on a pinhole, the latter is removed for measurement to avoid spatial - and hence spectral - filtering. It turned out to be vital that the sample and reference spectra are monitored in real time on a dual-beam oscilloscope, allowing to fine-tune the photometric balance across the entire spectral range simultaneously. The chirp and duration of the supercontinuum is measured with the nonresonant signal from pure solvents (see below).

While the spectra are being registered, two events are distinguished: either the pump beam is "off" or "on". Figure 8.4 shows how the pump-induced

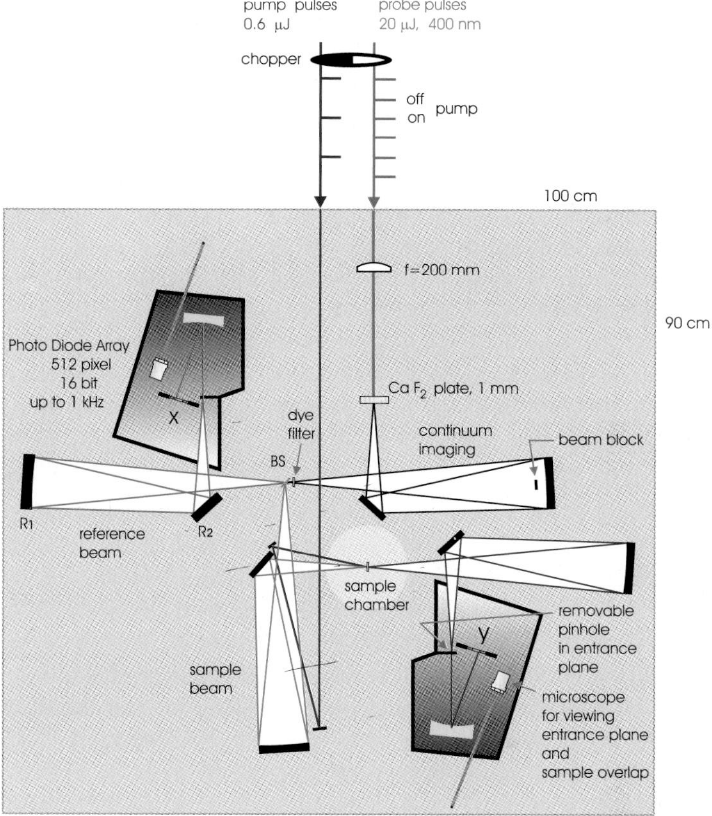

Fig. 8.3. Experimental setup for transient absorption with a supercontinuum. It contains 4 off-axis Schwarzschild objectives (distinguished by color for convenience; see text).

optical density $\Delta D(\lambda)$ is calculated from the readings at pixels corresponding to λ on the two photodiode arrays. After n (typically 100) events "off" and "on" have been recorded in alternate succession, the 2n data points for every λ are grouped as shown in the figure. For perfect division of the probe light by the beam splitter and with perfect imaging, the points in the left panel (for example, for which the pump was off) fall on a straight line which passes through the origin. The readings from one spectrograph alone may fluctuate (e.g. with standard deviation s_x around mean $\bar{x}$) due to small changes of the laser pulse energy, but the quotient y_i/x_i $(i = 1, ...n)$ should be constant. In practice the measurement points deviate from the straight line and an optimal quotient or slope Q is formed by minimizing their distances (red lines). By comparing optimal quotients with and without pump one obtains for the considered wavelength

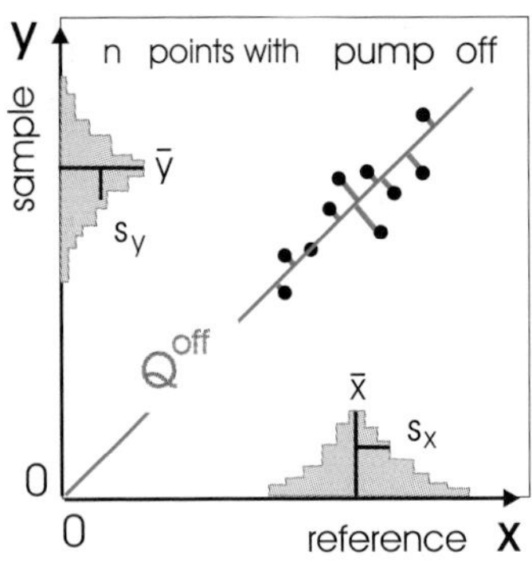

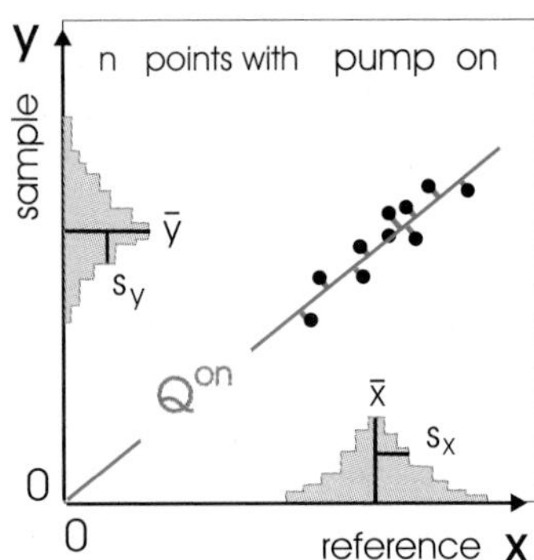

Fig. 8.4. $\Delta D(\lambda)$ is calculated from readings at the corresponding pixel in the reference (x-values) and sample spectrograph (y-values). For every wavelength two plots are made, for events with pump pulse off and on. Lines through the origin with optimal slope Q^{off} and Q^{on} minimize the deviations from the data points (red bars).

$$\Delta D = -\log\left(Q^{on}/Q^{off}\right). \tag{8.1}$$

Dual-beam performance depends, obviously, on how well readings y_i versus x_i fall on a straight line. A measure for this is given by the correlation

$$g = \frac{\sum (x_i - \bar{x})(y_i - \bar{y}) \ / \ (\text{n-1})}{s_x \ s_y} . \tag{8.2}$$

The parameters g, s_x and s_y are all obtained from the data set in the left panel of Fig. 8.4. With the setup shown in Fig. 8.3 the correlation is typically $g \approx 0.993$ even if $s_x/\bar{x} \approx s_y/\bar{y} \approx 0.20$. The standard error or confidence interval for a quotient $Q \approx 1$ can be estimated as

$$CI_q \approx \frac{s_x}{\bar{x}} \ \sqrt{2\,(1-g)\,/n} \tag{8.3}$$

The confidence interval for ΔD is readily found from (8.1) if the two beams are balanced, $Q^{off} \approx Q^{on}$, i.e. when the transient absorption is small:

$$CI_{\Delta D} \approx \frac{s_x^A}{\bar{x}\sqrt{n}} \ \sqrt{1-g} \ \frac{2}{\ln(10)} \tag{8.4}$$

This formula allows to compare same-shot referencing (as done here) with the more common setup which uses only one spectrograph. Then alternate supercontinuum pulses must be used for reference. This situation may be simulated in Fig. 8.4, left panel, if y is now the reading on the reference spectrograph, but for the next pulse without pump. In this "alternate" measurement mode, but otherwise under the same conditions as before, one finds

$g_{alt} \approx 0.75$ with a multifilament continuum. From a comparison of $CI_{\Delta D}$ using either g or g_{alt} it follows that the alternate mode needs 37 times more data points to reach the same signal-to-noise as the same-shot mode here, or continuum fluctuations must be reduced by a factor 0.16.

8.1.3 Theoretical description of pump-probe transient absorption spectroscopy

8.1.3.1 The molecular model

The model for the molecular energy levels is shown in Fig. 8.5. The pump excites from the electronic ground state $|g\rangle$ to the first excited state $|e\rangle$. A higher electronic state $|f\rangle$ is needed to treat excited-state absorption. The potential energy for the electronic states is assumed to be separable into normal harmonic modes with frequencies $\nu', \nu'', ...$ along intramolecular coordinates q', $q''....$ Modes in e are displaced relative to g by dimensionless Δ', $\Delta'', ...$ The geometry of the final state is taken to be identical to that of the ground state for simplicity.

The notation $|g\rangle$ etc corresponds to the electronic states in their zero-point vibrational state. The vibrational term values become [15] $G(\mathrm{v}', \mathrm{v}'', ...) = \mathrm{v}'\nu' + \mathrm{v}''\nu'' + ...$ where $\mathrm{v}', \mathrm{v}'', ...$ are the vibrational quantum numbers. All vibronic transitions are organized in Franck-Condon progressions. A specific transition from the vibrationless ground state, for example, to an excited vibronic state may be written as $|g(0,0)\rangle \rightarrow |e(\mathrm{v}', \mathrm{v}'')\rangle$. Within the Born-Oppenheimer approximation the square of transition dipole moment is given by Franck-Condon factors $|\langle g(0,0)|e(\mathrm{v}', \mathrm{v}'')\rangle|^2 = F_{0\,\mathrm{v}'}(\Delta')\ F_{0\,\mathrm{v}''}(\Delta'')$. When a molecular model $\Sigma_{egf}(\nu', \nu'', ...)$ with vibrational manifolds $|g(\mathrm{v}', \mathrm{v}'', ...)\rangle$, $|e(\mathrm{v}', \mathrm{v}'', ...)\rangle$ and $|f(\mathrm{v}', \mathrm{v}'', ...)\rangle$ is discussed, the treatment of all vibronic transitions is implicitly understood.

Dephasing and energy relaxation within a mode and between electronic states are induced by the fluctuating external forces which were already mentioned (red noisy line in Fig. 8.1a. With reasonable assumptions, all relevant relaxation rates could be calculated from Redfield theory ([16] and references therein). Here the empirical Bloch model [17] is used instead which neglects coherence transfer and assigns lifetime parameters to the remaining processes. In the present application an additional simplification is made for intramolecular vibrational relaxation (IVR), by letting population of excited vibrational states decay directly to the vibrationless state (red downward arrows in Fig. 8.5). The corresponding lifetimes of vibrational states $|a\rangle$ and $|b\rangle$ are denoted T_{aa}, T_{bb}. Dephasing times T_{ab} for the transition dipole between both states are given by $1/T_{ab} = (1/T_{aa} + 1/T_{bb})/2 + 1/\hat{T}_{ab}$. Pure dephasing times $\hat{T}_{ab}$ are indicated by dashed red lines in the figure. They are organized in several groups. The first group describes vibrational dephasing in ground and excited electronic states, $\hat{T}_{vib} \equiv \hat{T}_{g(\mathrm{v}',\mathrm{v}''...)g(0,0...)}, \hat{T}_{e(\mathrm{v}',\mathrm{v}''...)e(0,0...)}, \hat{T}_{f(\mathrm{v}',\mathrm{v}''...)f(0,0...)}$. The groups $\hat{T}_{BL} \equiv \left\{ \hat{T}_{g(0,0...)e(\mathrm{v}',\mathrm{v}''...)} \right\}$ and $\hat{T}_{SE} \equiv \left\{ \hat{T}_{e(\mathrm{v}',\mathrm{v}''...)g(\mathrm{w}',\mathrm{w}''...)} \right\}$

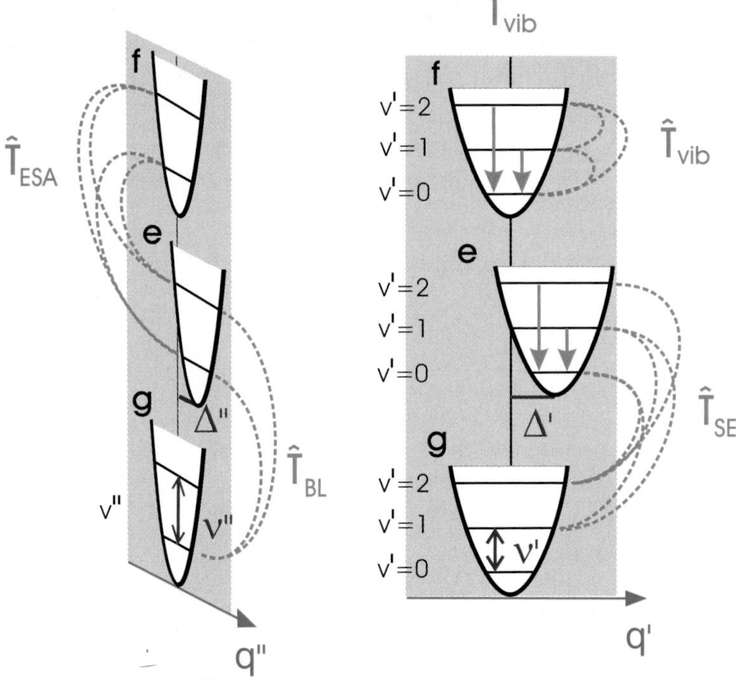

Fig. 8.5. Vibrational population relaxation (solid red arrows) and dephasing between vibronic states (dashed red lines) are described by empirical lifetimes (Bloch approximation [1, 17]). The latter are collected in groups as indicated and all lifetimes in a group are assigned an identical value. Here two orthogonal modes are shown.

of electronic dephasing times are connected with the homogeneous width of vibronic bands in the absorption and stimulated emission spectra; for simplicity it is assumed that $\hat{T}_{BL} = \hat{T}_{SE}$. The last "dephasing" group $\hat{T}_{ESA} \equiv \left\{ \hat{T}_{e(\mathrm{v}',\mathrm{v}''\ldots)f(\mathrm{w}',\mathrm{w}''\ldots)} \right\}$ contributes to vibronic bandwidths in excited-state absorption. The population lifetime of a vibrationless ground state is taken to be infinite,

$$\left\{ T_{g(0,0..)g(0,0..)}, T_{e(0,0..)e(0,0..)}, T_{f(0,0..)f(0,0..)} \right\} \to \infty \, .$$

The vibrational population lifetimes are grouped in

$$T_{vib} \equiv \left\{ T_{g(\mathrm{v}',\mathrm{v}'',\ldots)g(\mathrm{v}',\mathrm{v}'',\ldots)}, T_{e(\mathrm{v}',\mathrm{v}'',\ldots)e(\mathrm{v}',\mathrm{v}'',\ldots)}, T_{f(\mathrm{v}',\mathrm{v}'',\ldots)f(\mathrm{v}',\mathrm{v}'',\ldots)} \right\} \, .$$

In the remainder of this section the molecular model is driven by an optical field which is composed from the pump and chirped supercontinuum probe pulses. The induced transient absorption spectra as function of delay time are calculated analytically as far as possible. In the final section the results will be explored by fitting transient absorption spectra of a dye in solution (Rhodamine 110 in methanol, Fig. 8.6a). In this way the molecular modes together with effective population and pure dephasing times can be extracted.

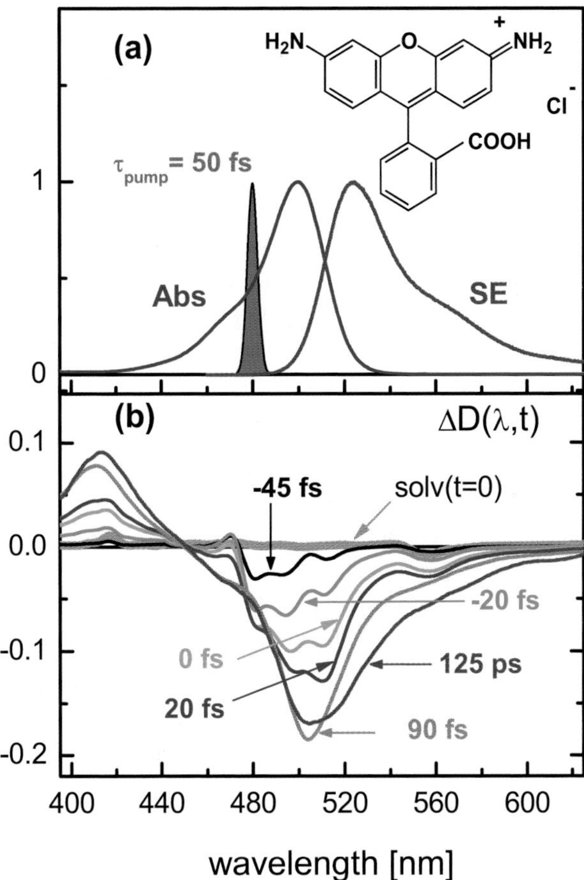

Fig. 8.6. Experimental results for Rhodamine 110 in methanol. (a) Linear absorption and stimulated emission. Spectra are normalized at the peak. (b) Transient absorption spectra induced by 50 fs pump pulses at 480 nm. Vibronic structure is observed during pump-probe overlap. Absorption changes $\Delta D < 0$ correspond to bleach (BL) or stimulated emission (SE) and $\Delta D > 0$ to excited-state absorption (ESA). Solvent signal at $t = 0$ is presented by a thick gray line, and all spectra have been time corrected.

8.1.3.2 The electric field, and basic calculation of the transient spectrum [9–11]

The perturbation formalism developed by Mukamel and co-workers [1] will be used throughout. In a pump-probe experiment the system is subjected to two light pulses and the electric field strength is written as

$$E(\mathbf{r},t) = E_1(t)\exp(i\mathbf{k}_1\mathbf{r}) + E_2(t)\exp(i\mathbf{k}_2\mathbf{r}) + c.c. \tag{8.5}$$

Here $E_1(t)$ and $E_2(t)$ represent the pump and probe pulse with wave vectors $\mathbf{k}_1$ and $\mathbf{k}_2$, respectively, and c.c is the complex conjugate. It is assumed that the pump pulse is a nonchirped Gaussian with duration τ_1 whose frequency is centered at Ω_1:

$$E_1(t) = \exp\left(-(t+t_d)^2/2\tau_1^2 - i\Omega_1(t+t_d)\right) \tag{8.6}$$

In the experiment the probe path is usually fixed. The time delay t_d between pump and probe pulses is increased by shortening the optical path for the pump beam, and this is reflected in (8.6). For optical probing, white-light or supercontinuum pulses are used here which cover the entire near-UV - visible range [12]. From studies of such pulses by nonresonant transient absorption spectroscopy of pure liquids with 10-20 fs time resolution it was concluded that, for this application, the probe may still be modeled by a single chirped pulse [11], i.e.

$$E_2(t) = \exp\left(-t^2/2\tau_2^2 - i(\Omega_2 t + \beta t^2)\right) = \exp\left(-\alpha t^2/2\tau_2^2 - i\Omega_2 t\right). \tag{8.7}$$

Here Ω_2, τ_2 are the central frequency and pulse duration, β is the chirp parameter or rate, and $\alpha = 1 + i2\beta\tau_2^2$. (Later the probe pulse duration will be referred to that of the pump through the parameter $\xi = \tau_2/\tau_1$). The sample is treated as a thin optical medium and therefore propagation effects are neglected.

The quantity of interest is the transient absorption spectrum $\Delta D(\omega_2, t_d)$ induced by pumping and measured at delay t_d as function of probe frequency ω_2; it can be written as [1]

$$\Delta D(\omega_2, t_d) = 2\omega_2\, Im\left(E_2^*(\omega_2)P^{(3)}(\omega_2, t_d)\right) / \left|E_2(\omega_2)\right|^2. \tag{8.8}$$

The Fourier transform ($FT(...)$) giving the third-order polarization spectrum is considered to be taken in the probe direction $\mathbf{k_2} = -\mathbf{k_1} + \mathbf{k_1} + \mathbf{k_2}$,

$$P^{(3)}(\omega_2, t_d) = FT\left(P^{(3)}(\mathbf{k_2}, t, t_d)\right) \equiv \int_{-\infty}^{\infty} dt P^{(3)}(\mathbf{k_2}, t, t_d)\exp(i\omega_2 t) \tag{8.9}$$

so that the spatial component with wavevector $\mathbf{k}_2$ of the probe beam must be extracted from the third-order polarization $P^{(3)}(\mathbf{r}, t, t_d)$. This is achieved by considering the spatial properties of the terms generated by expansion of $E(t - t_3)E(t - t_3 - t_2)E(t - t_3 - t_2 - t_1)$. Collecting all terms with $\exp(i\mathbf{k_2 r})$ one finds that the corresponding factor, i.e. the effective field product $E^{(3)}(\mathbf{k_2},\ t, t_3, t_2, t_1)$, is

$$E^{(3)}(\mathbf{k_2}, t, t_3, t_2, t_1) = E_S^{(3)}(t, t_3, t_2, t_1) + E_C^{(3)}(t, t_3, t_2, t_1) + E_D^{(3)}(t, t_3, t_2, t_1), \tag{8.10}$$

$$E_S^{(3)}(t, t_3, t_2, t_1) = E_S(t, t_3, t_2, t_1, \Omega_1) + E_S(t, t_3, t_2, t_1, -\Omega_1), \tag{8.11}$$

$$E_C^{(3)}(t, t_3, t_2, t_1) = E_C(t, t_3, t_2, t_1, \Omega_1) + E_C(t, t_3, t_2 + t_1, -t_1, \Omega_1), \tag{8.12}$$

$$E_D^{(3)}(t, t_3, t_2, t_1) = E_D(t, t_3, t_2, t_1, \Omega_1) + E_D(t, t_3, t_2 + t_1, -t_1, \Omega_1), \tag{8.13}$$

where

$$E_S(t, t_3, t_2, t_1, \Omega_1) = E_2(t-t_3)E_1^*(t-t_3-t_2+t_d)E_1(t-t_3-t_2-t_1+t_d), \tag{8.14}$$

$$E_C(t, t_3, t_2, t_1, \Omega_1) = E_1(t-t_3+t_d)E_1^*(t-t_3-t_2+t_d)E_2(t-t_3-t_2-t_1), \tag{8.15}$$

$$E_D(t, t_3, t_2, t_1, \Omega_1) = E_1^*(t-t_3+t_d)E_1(t-t_3-t_2+t_d)E_2(t-t_3-t_2-t_1). \tag{8.16}$$

The pump center frequency Ω_1 is noted explicitly here because this will ease manipulations later. The first term in (8.10) corresponds to sequential contributions (with subscript S) when the sample first interacts twice with the pump and then with the probe field. The last two terms represent coherent contributions (subscript C and D) when the sample interacts with the interference between pump and probe fields. Accordingly a transient absorption spectrum may be separated into sequential and coherent spectra:

$$\Delta D(\omega_2, t_d) = \Delta D_S(\omega_2, t_d) + \Delta D_C(\omega_2, t_d) + \Delta D_D(\omega_2, t_d). \tag{8.17}$$

Here the contributions $\Delta D_{S,C,D}(\omega_2, t_d) = 2\omega_2\, Im\left(P^{(3)}_{S,C,D}(\omega_2, t_d)\big/E_2(\omega_2)\right)$ are determined by the corresponding third-order polarization spectra

$$P^{(3)}_{S,C,D}(\omega_2, t_d) = \int_0^\infty dt_3 \int_0^\infty dt_2 \int_0^\infty dt_1 R(t_3, t_2, t_1) FT\left(E^{(3)}_{S,C,D}(t, t_3, t_2, t_1)\right). \tag{8.18}$$

The nonlinear response $R(t_3, t_2, t_1)$ is the sum of eight dipole correlation functions [1]

$$R_1\,(t_3, t_2, t_1) = \sum_{a,b,c,d} p(a)\mu_{ab}\mu_{bc}\mu_{cd}\mu_{da} I_{dc}(t_3) I_{db}(t_2) I_{da}(t_1), \tag{8.19}$$

$$R_2\,(t_3, t_2, t_1) = \sum_{a,b,c,d} p(a)\mu_{ab}\mu_{bc}\mu_{cd}\mu_{da} I_{dc}(t_3) I_{db}(t_2) I_{ab}(t_1), \tag{8.20}$$

$$R_3\,(t_3, t_2, t_1) = \sum_{a,b,c,d} p(a)\mu_{ab}\mu_{bc}\mu_{cd}\mu_{da} I_{dc}(t_3) I_{ac}(t_2) I_{ab}(t_1), \tag{8.21}$$

$$R_4(t_3,t_2,t_1) = \sum_{a,b,c,d} p(a)\mu_{ab}\mu_{bc}\mu_{cd}\mu_{da} I_{ba}(t_3) I_{ca}(t_2) I_{da}(t_1), \tag{8.22}$$

$$R_5(t_3,t_2,t_1) = \sum_{a,b,c,d} p(a)\mu_{ab}\mu_{bc}\mu_{cd}\mu_{da} I_{cb}(t_3) I_{db}(t_2) I_{ab}(t_1), \tag{8.23}$$

$$R_6(t_3,t_2,t_1) = \sum_{a,b,c,d} p(a)\mu_{ab}\mu_{bc}\mu_{cd}\mu_{da} I_{cb}(t_3) I_{db}(t_2) I_{da}(t_1), \tag{8.24}$$

$$R_7(t_3,t_2,t_1) = \sum_{a,b,c,d} p(a)\mu_{ab}\mu_{bc}\mu_{cd}\mu_{da} I_{cb}(t_3) I_{ca}(t_2) I_{da}(t_1), \tag{8.25}$$

$$R_8(t_3,t_2,t_1) = \sum_{a,b,c,d} p(a)\mu_{ab}\mu_{bc}\mu_{cd}\mu_{da} I_{ad}(t_3) I_{ac}(t_2) I_{ab}(t_1). \tag{8.26}$$

The time evolution in the absence of external fields is assumed to follow the reduced (Bloch) equations of motion [1] which treat relaxation through population relaxation rates Γ_{aa} and pure dephasing rates Γ_{ab} for density matrix elements, so that

$$I_{ab}(t) = \exp\left[-i\omega_{ab}t - \Gamma_{ab}t\right] \equiv \exp\left[-i\tilde{\omega}_{ab}t\right] \tag{8.27}$$

with complex frequency $\tilde{\omega}_{ab} = \omega_{ab} - i\Gamma_{ab}$. Two subscripts indicate the frequency difference, $\omega_{ab} \equiv \varepsilon_a - \varepsilon_b$, and dephasing rate Γ_{ab} between states $|a>$ and $|b>$. Finally, $p(a) = \exp\left(-\varepsilon_a/k_BT\right) \Big/ \sum_a \exp\left(-\varepsilon_a/k_BT\right)$ is the thermal population of the state $|a>$.

Each material response function $R_i(t_3,t_2,t_1)$ in (8.19-8.26) is a sum over possible vibronic pathways. For the moment assume that just one vibrational mode is active and that only vibrationless states $|g(0)>, |e(0)>, |f(0)>$ and vibrationally excited states $|g(1)>, |e(1)>, |f(1)>$ are involved. Furthermore let all vibronic transitions $|g(0)>, |g(1)> \leftrightarrow |e(0)>, |e(1)>$ and $|e(0)>, |e(1)> \leftrightarrow |f(0)>, |f(1)>$ be allowed and set the vibrational energy $\hbar\omega_{10} >> k_BT$. The pathways inherent in (8.19-8.26) for this typical case are presented in Fig. 8.7 as energy ladder diagrams [9,13]. Time moves from left to right in a diagram, straight vertical arrows mark a dipole transition moment from the molecular ket and bra state (solid or dashed lines, respectively) and a wavy arrow corresponds to the dipole transition needed to evaluate the third-order polarization. The pathways can be classified in several groups according to the way in which the system is prepared when some arbitrary field is applied. The first group (left column, Fig. 8.7) represents processes when two field interactions first alter the population of the ground/excited state, after which the third interaction induces electronic coherence and thereby leads to an emissive/absorptive polarization wave. This group collects *bleach* (BL, from response functions $R_{3,4}$), *stimulated emission* (SE, from $R_{1,2}$) and *excited-state absorption* (ESA, from $R_{5,6}$) processes, respectively. By contrast the following group [9,10] (middle column, Fig. 8.7) involves the creation of vibrational coherence by the first two interactions, off which the field is scattered

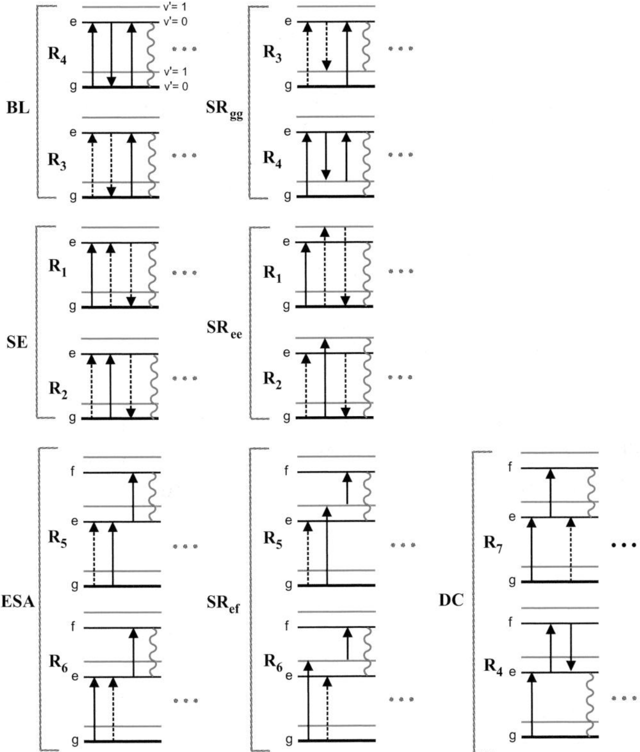

Fig. 8.7. Energy ladder diagrams for molecular model $\{g(\mathrm{v}', \mathrm{v}'', ...), e(\mathrm{v}', \mathrm{v}'', ...), f(\mathrm{v}', \mathrm{v}'', ...)\}$ where optically active vibrational modes are distinguished by primes. Vibronic pathways are classified by their characteristic nonlinear-optical process: *bleach* (*BL*), *stimulated emission* (*SE*), *excited-state absorption* (*ESA*), impulsive *stimulated Raman* (SR_{gg}, SR_{ee}, SR_{ef}) and *double coherence* (*DC*). The high-frequency limit $\hbar\omega_{10} >> k_B T$ is assumed for the figure so that $g(v' = 1)$ is not populated initially. Each diagram in fact represents a family of diagrams which differ only in the way the vibrational states $\{g(\mathrm{v}', \mathrm{v}'', ...), e(\mathrm{v}', \mathrm{v}'', ...), f(\mathrm{v}', \mathrm{v}'', ...)\}$ are involved. Wavy lines indicate electric polarization at the corresponding frequencies. R_8 does not appear because it requires a further excited state above f for resonance.

to give *resonant impulsive stimulated Raman scattering* - here labeled as SR_{gg}, SR_{ee} and SR_{ef}. The last group (right column, Fig. 8.7) collects processes when two fields create *double coherence* (DC) between g, e, and f states. Note that BL, SE, ESA, and SR pathways appear not only with sequential field $E_S^{(3)}(t, t_3, t_2, t_1)$ but also with coherent $E_C^{(3)}(t, t_3, t_2, t_1)$, while DC pathways appear with coherent field $E_D^{(3)}(t, t_3, t_2, t_1)$ only. The overall transient signal is the sum of all resonant pathways. Compact forms will be given for the sequential and coherent transient spectra as measured with linearly-chirped supercontinuum probe pulses. As regards the influence of the environment, remember that this was already taken into account by the empirical population and pure dephasing times. The corresponding decay rates are the result of fast perturbations by the solvent. The remaining influence on the chromophore is considered here in the limit of an infinitely long bath correlation time, i.e. for a frozen solvent (solvent relaxation will be treated empirically later). This static inhomogeneous broadening may be incorporated [1] by convoluting the homogeneous response functions with a static (Gaussian) distribution of transition frequencies:

$$< \Delta D(\omega_2, t_d) >= (1/\sqrt{2\pi}\delta\omega_{eg}) \int_{-\infty}^{\infty} d\omega' \Delta D(\omega_2 + \omega') \; \exp\left(-\tfrac{1}{2}(\omega'/\delta\omega_{eg})^2\right). \tag{8.28}$$

8.1.3.3 Sequential spectral contributions for linearly chirped probe pulses

To calculate the sequential contribution the reduced response functions contained in the sums (8.19-8.26) are discussed in their general form

$$R_{[k]}(t_3, t_2, t_1) = -i\; p(a) \exp\left[-i\tilde{\omega}_{ab} t_3 - i\tilde{\omega}_{cd} t_2 - i\tilde{\omega}_{ef} t_1\right], \tag{8.29}$$

The pathway $[k]$ is given by subscripts $\{ab, cd, ef\}$. For example, the pathways corresponding to the BL-process in Fig. 6 can be written in (8.29) with subscripts $\{ab = eg,\ cd = gg,\ ef = eg\}$ and $\{ab = eg,\ cd = gg,\ ef = ge\}$ for response functions R_4 and R_3, respectively. The third-order polarization spectrum $P_S^{(3)}(\omega_2, t_d)$ may be written as the sum over all possible (i.e. near-resonant) pathways :

$$P_S^{(3)}(\omega_2, t_d) = \sum_{[k]} \int_0^{\infty} dt_3 \int_0^{\infty} dt_2 \int_0^{\infty} dt_1 \; R_{[k]}(t_3, t_2, t_1) FT\left(E_S^{(3)}(t, t_3, t_2, t_1)\right). \tag{8.30}$$

The delay time t_d at left comes in because it is contained in the structure of the electric field. The spectrum $P_S^{(3)}(\omega_2, t_d)$ may be separated into $P_{S_1}^{(3)}(\omega_2, t_d)$ and $P_{S_2}^{(3)}(\omega_2, t_d)$ terms which are generated by the field components $E_S(t, t_3, t_2, t_1, \Omega_1)$ and $E_S(t, t_3, t_2, t_1, -\Omega_1)$, respectively (see (8.11)). Here only the $P_{S_1}^{(3)}(\omega_2, t_d)$-contribution is considered because the result for $P_{S_2}^{(3)}(\omega_2, t_d)$ is obtained by substituting $\Omega_1 \rightarrow -\Omega_1$. After substitution of $FT\left(E_S(t, t_3, t_2, t_1, \Omega_1)\right)$ (whose detailed expression is given in [9]) into (8.30), the integration over t_3 and t_2 can be carried out analytically. Then by some rearrangements, the contribution of polarization $P_{S_1}^{(3)}(\omega_2, t_d)$ to the sequential absorption term is captured through

$$P_{S_1}^{(3)}(\omega_2,t_d)/E_2(\omega_2) = D_0 \sum_{[k]} F_{S_1}^{[k]}(\omega_2,t_d) \tag{8.31}$$

with

$$F_{S_1}^{[k]}(\omega_2,t_d) = \frac{\sqrt{\pi}}{2} p(a) L(\omega_2 - \tilde{\omega}_{ab}) \exp\left(-\eta\left(t_d + t_0(\omega_2)\right)^2/\tau_1^2\right) \times J_{S_1}\left(z_{S_1}^{(1)}, z_{S_1}^{(2)}, k_{S_1}^{(1)}, k_{S_1}^{(2)}\right). \tag{8.32}$$

Here the time-zero function

$$t_0(\omega_2) = i\tau_1^2(\omega_2 - \Omega_2)/(2\delta\eta) \tag{8.33}$$

was introduced and the other parameters are $\delta = 1 + \alpha/2\xi^2$, $\varepsilon = 1 - 1/2\delta$, $\eta = 1 - 1/\delta$, $D_0 = -i\pi\tau_1^3\sqrt{\alpha/\delta\varepsilon\xi^2}$, $\xi = \tau_2/\tau_1$. The function $L(\omega_2 - \tilde{\omega}_{ab}) = i/(\pi\tau_1(\omega_2 - \tilde{\omega}_{ab}))$ is the dimensionless complex Lorentzian whose real part has its maximum at the optical resonance frequency ω_{ab} with width Γ_{ab}. The remaining time-integrals were placed into the auxiliary function

$$J_{S_1}\left(z_{S_1}^{(1)}, z_{S_1}^{(2)}, k_{S_1}^{(1)}, k_{S_1}^{(2)}\right) = \int_0^\infty dx \exp\left(-k_{S_1}^{(1)}x^2 - 2z_{S_1}^{(1)}x\right) W\left(ik_{S_1}^{(2)}(x + z_{S_1}^{(2)})\right). \tag{8.34}$$

Compared to [9], (8.34) only needs one integration which speeds up spectral fitting substantially. It depends jointly on the pathway and on the parameters through

$$z_{S_1}^{(1)} = -\eta(t_d + t_0(\omega_2))/\tau_1 + i\tau_1\tilde{\omega}_{cd}/2\eta, \tag{8.35}$$

$$z_{S_1}^{(2)} = -(t_d + t_0(\omega_2))/\tau_1 + i\tau_1(\tilde{\omega}_{ef} - \Omega_1)/\eta, \tag{8.36}$$

$$k_{S_1}^{(1)} = \eta,\ k_{S_1}^{(2)} = \eta/\sqrt{2\varepsilon}. \tag{8.37}$$

Here $W(z)$ is expressed through the complementary error function [14] of the complex argument z:

$$W(z) = (i/\pi)\int_{-\infty}^{\infty} dt\left(\exp(-t^2)\right)/(z-t) = \exp(-z^2)erfc(-iz). \tag{8.38}$$

Below, for brevity, the arguments $z_{S_1}^{(1)}$, $z_{S_1}^{(2)}$, $k_{S_1}^{(1)}$, $k_{S_1}^{(2)}$ of this function are omitted and instead the short notation $J_{S_1} \equiv J\left(z_{S_1}^{(1)}, z_{S_1}^{(2)}, k_{S_1}^{(1)}, k_{S_1}^{(2)}\right)$ is used. The same notation applies to the other integrals in (8.18) but with subscripts $C_{1,2}$, accordingly. Remember that $P_{S_2}^{(3)}(\omega_2,t_d)$ itself is obtained from (8.31), (8.32), and (8.34) by the substitution $\Omega_1 \to -\Omega_1$ in (8.36), i.e.

$$P_{S_2}^{(3)}(\omega_2,t_d) = P_{S_1}^{(3)}(\omega_2,t_d)\,|_{\Omega_1\to-\Omega_1}\,. \tag{8.39}$$

The sequential contribution $\Delta D_S(\omega_2,t_d)$ is defined as the sum over all possible pathways.

8.1.3.4 Coherent spectral contributions for linearly chirped probe pulses

The coherent contribution $\Delta D_C(\omega_2, t_d)$ to the transient absorption signal is obtained in a manner similar to the sequential term. The polarization $P_C^{(3)}(\omega_2, t_d)$ can be cast into the form

$$P_{C_1;C_2}^{(3)}(\omega_2, t_d)/E_2(\omega_2) = D_0 \sum_{[k]} F_{C_1;C_2}^{[k]}(\omega_2, t_d), \tag{8.40}$$

$$F_{C_1;C_2}^{[k]}(\omega_2, t_d) = p(a) L(\omega_2 - \tilde{\omega}_{ab}) \exp\left(-\eta\left(t_d + t_0(\omega_2)\right)^2 / \tau_1^2\right) \mathrm{J}_{C_1;C_2}, \tag{8.41}$$

with the auxiliary functions J_{C_1,C_2} defined earlier in (8.34). Now $J_{C_1}$is a function of $z_{C_1}^{(1)}, z_{C_1}^{(2)}$ and $k_{C_1}^{(1)}, k_{C_1}^{(2)}$ which in turn depend jointly on the pathway and on the parameters for the electric field $E_C(t, t_3, t_2, t_1, \Omega_1)$ (see (8.12)):

$$z_{C_1}^{(1)} = \eta(t_d + t_0(\omega_2))/\tau_1 + i\tau_1(\tilde{\omega}_{ef} - \omega_2)/2, \tag{8.42}$$

$$z_{C_1}^{(2)} = (t_d + t_0(\omega_2))/\tau_1 + i\tau_1(\tilde{\omega}_{cd} + \Omega_1 - \omega_2)/\eta, \tag{8.43}$$

$$k_{C_1}^{(1)} = \eta, \; k_{C_1}^{(2)} = \eta/\sqrt{2\varepsilon}. \tag{8.44}$$

Auxiliary $J_{C_2}$is similarly a function of $z_{C_2}^{(1)}, z_{C_2}^{(2)}, k_{C_1}^{(1)}, k_{C_1}^{(2)}$ through the parameters for the field product $E_C(t, t_3, t_2 + t_1, -t_1, \Omega_1)$:

$$z_{C_2}^{(1)} = \eta(t_d + t_0(\omega_2))/2\tau_1 + i\tau_1(\tilde{\omega}_{cd} + \Omega_1 - \omega_2)/2, \tag{8.45}$$

$$z_{C_2}^{(2)} = -2\delta\eta\left((t_d + t_0(\omega_2))/\tau_1 - i\tau_1(\tilde{\omega}_{ef} + \Omega_1)/\eta\right), \tag{8.46}$$

$$k_{C_2}^{(1)} = \varepsilon/2, \; k_{C_2}^{(2)} = 1/(2\delta\sqrt{2\varepsilon}). \tag{8.47}$$

As follows from (8.15) and (8.16) the coherent contribution $\Delta D_D(\omega_2, t_d)$ may be calculated as

$$\Delta D_D(\omega_2, t_d) = \Delta D_C(\omega_2, t_d)\,|_{\Omega_1 \to -\Omega_1} . \tag{8.48}$$

The coherent contribution $\Delta D_{C,D}(\omega_2, t_d)$ is again obtained as the sum over all possible pathways.

8.1.3.5 Analytical results for well-separated pulses

Well-separated pump and probe pulses, $t_d >> \tau_1$, are considered next. In this case, as seen from (8.10), only the sequential contribution remains. The sequential term is calculated starting from (8.34) which is valid for all delays. After substitution $x \to x + z_{S_1}^{(2)}$ the appropriate auxiliary function may be written as

$$J_{S_1} = \exp\left[\left(z_{S_1}^{(1)}\right)^2 / k_{S_1}^{(1)}\right] \int_{z_{S_1}^{(2)}}^{\infty} dx \exp\left[-k_{S_1}^{(1)}\left(x + z_{S_1}^{(1)}/k_{S_1}^{(1)} - z_{S_1}^{(2)}\right)^2\right] W\left(ik_{S_1}^{(2)}\right). \tag{8.49}$$

For well-separated pulses the lower limit in (8.49) can be expanded to minus infinity since $z_{S_1}^{(2)} \to -\infty$ in (8.36) for $t_d >> \tau_1$. This integral can be calculated with the relation

$$\int_{-\infty}^{\infty} dx \exp\left[-k_{S_1}^{(1)}\left(x + z_{S_1}^{(1)}/k_{S_1}^{(1)} - z_{S_1}^{(2)}\right)^2\right] W\left(ik_{S_1}^{(2)}x\right) = \sqrt{2\pi\varepsilon/\eta}\, W\left(-i\left(z_{S_1}^{(1)} - k_{S_1}^{(1)} z_{S_1}^{(2)}\right)\right) \tag{8.50}$$

Then the transient signal, corresponding to the [k]-pathway, can be written exactly as

$$\begin{aligned} \varDelta D_{S_1}^{[k]}(\omega_2, t_d) = {} & D_0 Re\left\{L(\omega_2 - \tilde{\omega}_{ab}) W(iz_0)\right. \\ & \left.\times \exp\left(-\tau_1^2 \tilde{\omega}_{cd}^2/4\eta - i\tilde{\omega}_{cd}\left(t_d + t_0(\omega_2)\right)\right)\right\} \end{aligned} \tag{8.51}$$

with $D_0 = 2\omega_2\pi^2\tau_1^3 p(a)$, $z_0 = i\tau_1\left(\tilde{\omega}_{ef} - \tilde{\omega}_{cd}/2 - \varOmega_1\right)$ and $t_0(\omega_2)$ defined by (8.33).

8.1.3.6 Time correction of transient spectra

The time behavior of the sequential and coherent transient spectra is characterized by the factor $\exp\left(t_d + t_0(\omega_2)\right)$ in (8.32), (8.41). The function $t_0\left(\omega_2\right)$ (8.33) can also be written as

$$t_0(\omega_2) = i\tau_1^2\left(\omega_2 - \varOmega_2\right)\xi^2/\alpha = \tau_2^2\left(\omega_2 - \varOmega_2\right)2\beta\tau_2^2\left(1 + i/2\beta\tau_2^2\right)/\alpha\alpha^* \quad . \tag{8.52}$$

In PSCP experiments the chirp rate is usually large, $2\beta\tau_2^2 >> 1$, and in this case one may approximate

$$t_0(\omega_2)\Big|_{2\beta\tau_2^2>>1} \approx \left(\omega_2 - \varOmega_2\right)/2\beta + i\left(\omega_2 - \varOmega_2\right)/4\beta^2\tau_2^2 \quad . \tag{8.53}$$

The other extreme is presented by nonchirped probe pulses for which

$$t_0(\omega_2)\,|_{\beta=0} = i\tau_1^2\,(\omega_2 - \Omega_2)\,\xi^2 \quad . \tag{8.54}$$

Note that (8.53) coincides with the so-called time-zero function $t_0(\omega_2)$ which was introduced for the *nonresonant* case in [11]. There it was shown that by nonresonant transient absorption measurements, for example on pure solvents, the supercontinuum probe can be characterized through its $t_0(\omega_2)$-dependence and its spectral shape. Returning to resonant spectroscopy, the difference between chirped to nonchirped probing disappears if the substitution

$$t_d + Re\,\{t_0(\omega_2)\} \rightarrow t \tag{8.55}$$

is made. This is seen for the sequential contribution with well-separated pulses in (8.51): the temporal behavior for the pathway after time shift (8.55) is proportional to $\propto \exp\left(-(i\omega_{cd} + \Gamma_{cd})t\right)$ as expected for a Fourier transform-limited probe pulse.

An illustration of the time correction procedure for resonant PSCP spectroscopy was given before, by examining an electronic two-level system (see Fig. 3 in [9]). By a temporal shift (8.55) the recorded signal can be time-corrected. The same behavior holds for any resonantly excited system. Thus the time correction routine is identical for the resonant and for the nonresonant case and the methodology of [11] may be employed throughout. In the remainder the transient data are always treated in this way and PSCP transient signals shown should be considered as having been time-corrected.

8.1.4 Simulations of coherent and sequential spectra

Characteristic features of transient absorption spectra can be understood with the help of simulations. In this way a preview on the real example in the next section is obtained. Consider a basic system $\Sigma_{egf}(\nu', \nu'')$ with parameters $\lambda_{eg} = 510$ nm, $\lambda_{fe} = 420$ nm, $\nu' = 800$ cm^{-1}, v$' = 0, 1$, $\nu'' = 1400$ cm^{-1}, v$'' = 0, 1$, $\Delta_{\nu'} = \Delta_{\nu''} = 1$, $\hat{T}_{vib} = 1$ ps, $\hat{T}_{el} \equiv \hat{T}_{BL} = \hat{T}_{SE} = \hat{T}_{ESA} = 10$ fs, and $T_{vib} = 1$ ps. The spectra are inhomogeneously broadened with $\delta\omega = 500\,cm^{-1}$ (It is assumed that $\delta\omega = \delta\omega_{eg} = \delta\omega_{fe}$.). With these parameters the contribution $\Delta D_D(\omega_2, t_d)$ to (8.17) is small and not considered here. The system is excited at $\lambda_{pump} = 480$ nm with pulse duration τ_1=50 fs and probed with a supercontinuum. (The supercontinuum probe light is characterized by the central wavelength 480 nm (as the pump), probe pulse duration 250 fs, and chirp rate $\beta = 1.7 \times 10^{-3} fs^2$. These parameters are determined from measurements with the pure solvent as reported in [2-4] and are kept fixed.) The simulated signal is plotted against probe wavelength (because this is what experiments usually deliver) and normalized to -1 for the most negative signal. Fig. 8.8 shows the transient absorption at t=0 and at 1 ps. Because of the pulse duration, high-frequency vibrational coherence can not be seen as oscillations in the time domain, instead the vibrational structure shows up as coherent features in the frequency domain. The overall transient absorption signal (total), sequential (seq), and coherent contributions (coh) are shown for t=0.

The coherent part is calculated through (8.41, 8.40, 8.8). It has a dispersive shape near the pump frequency which is rooted in BL+SE+ESA pathways, and negative and positive vibrational satellites (on the Stokes and anti-Stokes side, respectively) are characteristic for stimulated Raman processes. The coherent spectrum reaches its maximum value at t=0 and disappears with the pump-probe overlap. The sequential spectrum is calculated through (8.32, 8.31, 8.8) or through (8.51) for separated pulses.

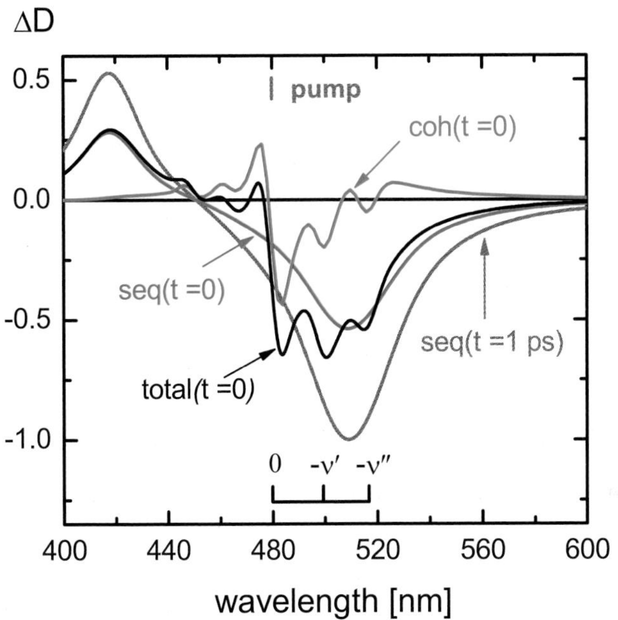

Fig. 8.8. Simulation of coherent and sequential spectra including excited-state absorption, for an example close to Rh110/methanol. The system is excited by 50 fs pulse at 480 nm and probed by a supercontinuum which is chirped (see text).

It was shown in [9,10] how electronic and vibrational dephasing, detuning, and inhomogeneous broadening affect the coherent component of femtosecond transient absorption spectra, and the reader is referred to theses papers for details. An additional tool to enhance or suppress a coherent component in early transient spectra is varying the pump-pulse duration. This is demonstrated in Fig. 8.9 where the spectral composition of the overall signal is plotted for different pump-pulse durations. One can see that with increasing duration, the broad coherent signal becomes narrow, observable in sharp features corresponding to the pump frequency and frequencies which are resonant with vibrational modes, i.e. $\omega = \omega_{pump} \pm \mathrm{v}'\nu' \pm \mathrm{v}''\nu''$. When the frequencies of the vibrations are comparable with the pump pulse width, vibrational activity is seen as oscillations in the time domain. But, although 5-10 fs pulses are readily available to molecular spectroscopist, still they are not sufficiently short to coherently excite a complete progression of high-frequency vibrational states. PSCP spectroscopy observes activity of high-frequency modes in the frequency domain instead (Fig. 8.9c).

8.1.5 Fit of transient absorption spectra from Rhodamine 110 in methanol

8.1.5.1 Measured transient spectra

Rhodamine 110 chloride dissolved in methanol was excited by 50 fs pulses centered at 480 nm. The structure of the molecule and its steady-state absorption and emission spectra were shown in Fig. 8.6a. The measured fluorescence quantum distribution has been converted to cross sections for stimulated emission so that comparison with transient absorption spectra may be made. For femtosecond measurements the optical path length was typically 0.2-0.3 mm and the optical density at the pump wavelength was adjusted to 0.3-0.6. The pump-induced absorption $\Delta D(\lambda, t)$ was monitored with a supercontinuum probe in the range 360-780 nm. All transient data which are presented and discussed here have been time-corrected using nonresonant signal from the pure solvent [11]. The pump-probe intensity cross-correlation time τ_{cc} was 65-70 fs and the temporal resolution after deconvolution from the pump pulse is estimated to be $\sim$ 10 fs over the entire probe range.

Transient absorption spectra of Rh110 were also already shown, in Fig. 8.6b, for overlapping and for well-separated pulses. Coherent features during pump-probe overlap around t=0 are a dispersive shape at the pump wavelength and local minima which indicate Franck-Condon vibronic states of the chromophores. The minima at ~500 nm and ~515 nm correspond to high frequency modes, $\nu' \sim$800 cm^{-1} and $\nu'' \sim$1400 cm^{-1}, in agreement with resonance Raman (RR) scattering studies [18] of Rhodamine dyes (see below). These frequencies will be used for a start when fitting with the molecular model. A negative and positive peak at 560 nm and 420 nm, respectively, at t=0 are due to nonresonant stimulated Raman scattering of the solvent methanol [11]. With increasing time delay all structure disappears and for separated pump and probe pulses the transient spectra consist entirely of the sequential component. Negative induced optical density ΔD corresponds to bleach (BL) or stimulated emission (SE) while positive ΔD indicates excited-state absorption (ESA). The SE band shifts to the red by $\sim$ 400 cm^{-1} without visible change of shape up to the maximal time delay (125 ps) due to solvation [12, 21–26] (see below).

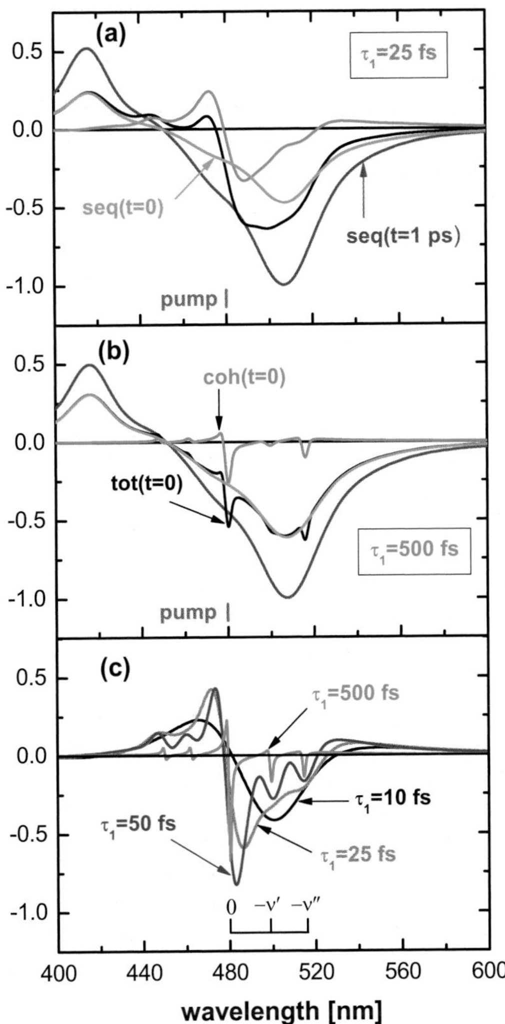

Fig. 8.9. Pump duration effect: Overall transient spectra and their decomposition for 25 fs (a) and 500 fs (b) pump pulse. (c) Dependence of the coherent contribution on the pump pulse duration.

8.1.5.2 Their global decomposition

The spectral features of the model and their dependence on parameters were studied qualitatively at the end of the previous section. Let us proceed to fit the measured transient absorption of $Rh110$ in methanol. The aim is to obtain a minimal set of parameters which describes consistently all available experimental data. Of interest are the pure homogeneous dephasing times, both electronic $\hat{T}_{el} \equiv \left\{\hat{T}_{BL} = \hat{T}_{SE}, \hat{T}_{ESA}\right\}$ and vibrational $\hat{T}_{vib}$, and the population decay times T_{vib} for excited vibrational states. For this purpose all measurements are jointly fitted with the nonlinear response function model of the previous section: the linear absorption/emission spectra (Fig. 8.10a), the transient absorption spectra of the whole time window (Fig. 8.10b), and the nonresonant solvent signal which was recorded separately (not shown). A global fit [19] provides the following optimal values: $\nu' = 780$ cm^{-1}, $\nu'' = 1520$ cm^{-1}, $v', v'' = 0 - 3$, $\Delta' = 0.8$, $\Delta'' = 0.6$, $\delta\omega = 400cm^{-1}$, $\hat{T}_{el} \equiv \left\{\hat{T}_{BL} = \hat{T}_{SE} = 40 \text{ fs}, \hat{T}_{ESA} = 5 \text{ fs}\right\}$, $\hat{T}_{vib} = 1.2$ ps, $T_{vib} = 20fs$, $\lambda_{eg} = 500nm$, $\lambda_{SE(\infty)} = 522$ nm. In Fig. 8.10 and 8.11 the fit results are drawn as black lines while thick gray lines represent measurement results. It allows to identify the transient spectra with coherent and sequential parts and to understand the latter in terms their transient components, i.e. stimulated emission, Raman, bleach, and excited-state absorption pathways. The accuracy of parameter values will be examined with the help of Fig. 8.12 and Table 1.

The linear absorption and emission spectra are reproduced in Fig. 8.10a. Their vibronic structure is exposed by the homogeneous spectra derived from the fit. Excitation at 480 nm is seen to access primarily the vibrational level $\mathrm{v}' = 1$ of the 780 cm^{-1} mode in $|e(\mathrm{v}', \mathrm{v}'')\rangle$. At this point one should mention vibrational and solvent relaxation. Fully relaxed emission is recorded by stationary measurements and its description requires (i) that the excited population is in the vibrational ground state and (ii) that the electronic $|e(0,0)\rangle - |g(0,0)\rangle$ energy gap is reduced by the Stokes shift [12, 21–26]. At intermediate times, for example at $t = 0.3$ ps as in Fig. 8.10c, vibrational relaxation is completed (since T_{vib}=20 fs) while solvent reorganization is still proceeding. The situation is modeled by a time-dependent, empirical position of the stimulated emission band as will be discussed below. Due to fast repopulation of $|e(0,0)\rangle$ a simulation in terms of a time-shifted stimulated emission band should be adequate. The pump and probe pulses are well separated and the recorded signal ΔD in the lower panel is decomposed into sequential BL, SE, and ESA components. But here interest is mainly focused on the initial time window when pump- and probe pulses act simultaneously on the material system. For example consider the transient absorption spectrum at $t = 0$ which is shown in Fig. 8.10b. From the global fit the coherent and sequential contributions to the time-zero spectrum, $\Delta D(t = 0)$, can be obtained separately. Most impor-

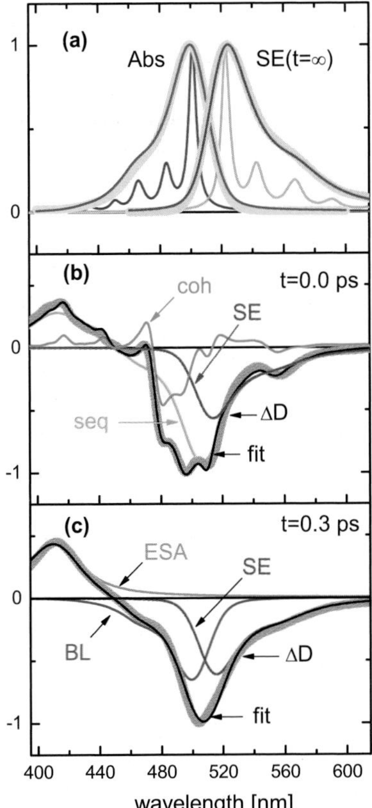

Fig. 8.10. Global fit (black) of measured spectra (thick gray lines) for $Rh110$/methanol. (a) Linear absorption and stimulated emission and their homogeneous kernels. (b) $\Delta D(t=0)$ is decomposed into coherent and sequential components, and the latter contains the stimulated emission band SE. (c) $\Delta D(t=0.3\,ps)$ consists of bleach BL, stimulated emission SE, and excited-state absorption ESA bands. The most negative transient signal during evolution was set to -1.

tantly, the latter contains the time-zero stimulated emission band $SE(t=0)$ (brown line) which defines the starting point for solvent relaxation studies.

The coherence at $t=0$ is examined in more detail with the help of Fig. 8.11. In the top panel the spectrum labeled *"coh"* of Fig. 8.10b is reproduced as a red line. It consists of a smooth envelope from $BL+SE+ESA$ pathways (black) and weak signal from the solvent (thick gray). The remainder is vibronic structure which stems from stimulated Raman pathways; it is shown separately in the bottom panel as a thick gray line and decomposed into

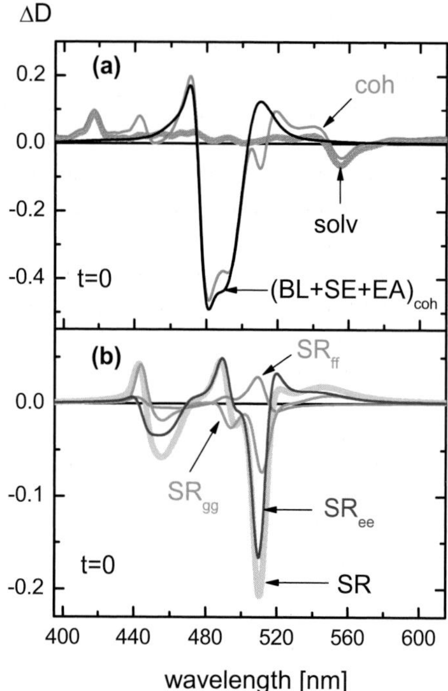

Fig. 8.11. Quantitative explanation of the observed coherent transient at $t = 0$ by model Σ_{gef} with optimized parameters. (a) The nonresonant solvent signal (thick gray line) is relatively weak. The coherent spectrum from the chromophore (red line) contains a contribution due to $BL + SE + ESA$ pathways (black line). (b) The remaining part (thick gray) is due to stimulated Raman processes between vibrational levels in the ground state SR_{gg} , between excited states SR_{ef}, and in the excited state SR_{ee}. The latter is responsible for most of the observed vibronic structure.

processes between vibrational levels in the ground state (SR_{gg}), in the excited state (SR_{ee}), and between excited states (SR_{ef}). By comparison, SR_{ee} is found to be responsible for most of the observed vibrational structure at $t = 0$.

8.1.5.3 How many modes?

Resonance Raman scattering (RR) studies of Rhodamine dyes [18] showed that the strongest bands are the aromatic stretching vibrations at 1365, 1509,

1575 and 1650 cm^{-1} (mean frequency 1525 cm^{-1}). Strong bands were also observed at 614 and 776 cm^{-1} and assigned to bending modes. A number of RR-bands which appear above 1700 cm^{-1} can be attributed to overtones and combination bands, and their intensity is about 10-15 times smaller. In the experiments the pump pulse is ~200 cm^{-1} wide and therefore the above-mentioned Raman bands can not be resolved . Instead Raman frequencies which are mentioned in this work should be considered as a mean over groups of modes, from which a specific mode may deviate by to 100 cm^{-1}. In this sense, how many vibrational modes are needed to describe the experimental data? It was shown [10] that two vibrational modes describe consistently all pertinent experimental data. To answer this question the electronic model Σ_{gef} is considered and either one, or two, or three vibrational modes are assumed to be active. Correspondingly all measurements were fitted by models $\Sigma_{gef}(\nu')$, $\Sigma_{gef}(\nu',\nu'')$, and $\Sigma_{gef}(\nu',\nu'',\nu''')$, and significant improvements were noted. The stationary absorption/stimulated emission spectra, as calculated from the appropriated fit parameters, and the time-resolved sequential contribution are practically indistinguishable between the mode schemes. Differences are seen only in coherent spectrum at t=0 which is shown in Fig. 8.12.

8.1.5.4 Accuracy of fit parameters

The accuracy by which model parameters can by extracted from the measurements is discussed next. For this purpose the fit by model $\Sigma_{gef}(\nu',\nu'')$, $v' = v'' = 0 - 3$ was repeated with a genetic algorithm [20]. Compared to methods that use gradients or higher derivatives, genetic algorithms are less sensitive to the starting point in parameter space. Results are summarized in Table 1. Averaging along 20 optimal sets provides mean values together with their standard deviations. It gives a reliable lower estimate for pure homogeneous dephasing times of $Rh110$.

8.1.5.5 Treatment of solvation relaxation in the excited state

Optical excitation alters the charge distribution of the solute chromophore and causes a polar solvent to reorganize. That process in turn stabilizes the excited solute molecule. Initially the emission is observed at shorter wavelengths than the steady state spectrum to which it shifts progressively as a function of time (Fig. 8.1c). What is being discussed here is *not* the fast fluctuation of the solvent coordinate around its initial value $q \approx 0$ in S_1 , on a time scale well below the pulse widths. Their effect on the solute is already covered by the empirical treatment of IVR and dephasing as outlined before. But in addition, and related to fluctuations on time scales that can be resolved, all trajectories experience a common drift toward lower solvation energy. The

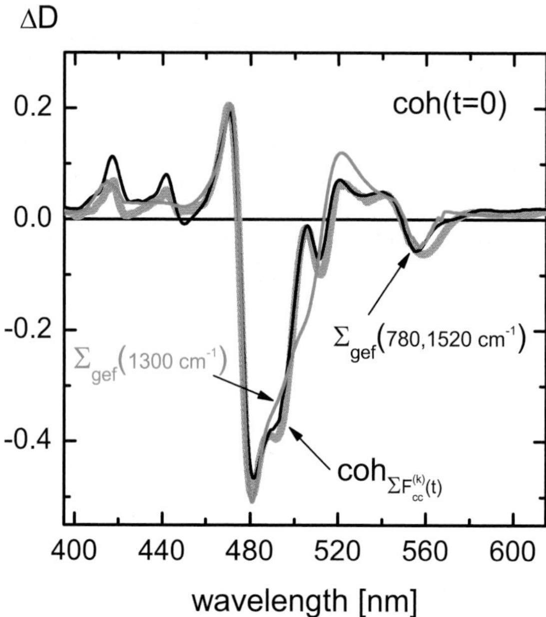

Fig. 8.12. Coherent spectrum at t=0 obtained by global fits with different number of vibrational modes (their frequencies were also optimized).

	ν_{abs} cm^{-1}	$\nu_{SE}(\infty)$ cm^{-1}	ν' cm^{-1}	ν'' cm^{-1}	Δ'	Δ''	$\delta\omega$ cm^{-1}	$\hat{T}_{BL} = \hat{T}_{SE}$, fs	$\hat{T}_{ESA}$ fs	$\hat{T}_{vib}$ ps	T_{vib} fs	$\nu_{SE}(0)$ cm^{-1}	$\Delta\nu$ cm^{-1}
$\bar{\mathbf{m}}$	19.96 $\times 10^3$	19.13 $\times 10^3$	740	1551	0.86	0.61	356	44	5.0	1.2	19	19.54 $\times 10^3$	410
std	21	20	68	43	0.04	0.03	22	7	0.2	$\geq$	6	26.5	27

Table 8.1. Averaged parameters $\bar{m}$ and their standard deviations std, from 20 optimal sets obtained with a genetic algorithm [20]. ν_{abs}, $\nu_{SE}(\infty)$, $\nu_{SE}(0)$ are the position of the absorption, relaxed fluorescence, and stimulated emission band at t=0. ν', ν'' are the vibrational frequencies of $Rh110$ with dimensionless displacements Δ', Δ''. $\delta\omega$ controls inhomogeneous broadening. Pure electronic $\hat{T}_{el}$=$\{\hat{T}_{BL} = \hat{T}_{SE}, \hat{T}_{ESA}\}$ and vibrational $\hat{T}_{vib}$ dephasing times are also given. $\Delta\nu = \nu_{SE}(\infty) - \nu_{SE}(0)$ is Stokes shift due to solvation dynamics (see also Fig. 8.10).

time-dependence of the resulting spectral shift is expressed in terms of a normalized relaxation function $C_\nu(t) = (\nu_{SE}(t) - \nu_{SE}(\infty))/(\nu_{SE}(0) - \nu_{SE}(\infty))$. Here $\nu_{SE}(0)$, $\nu_{SE}(\infty)$, and $\nu_{SE}(t)$ are the optical frequencies of the peak or first moments of the emission band at zero time (just after excitation), at infinite time (when the solvent polarization has relaxed to equilibrium, i.e., steady state stimulated emission), and at intermediate times t (during solvent relaxation). Femtosecond solvation dynamics in polar liquids has been investigated intensely (so that only representative references can be given here) by transient absorption [21], transient fluorescence [22], dynamical hole burning [23], and photon echo measurements [24–26].

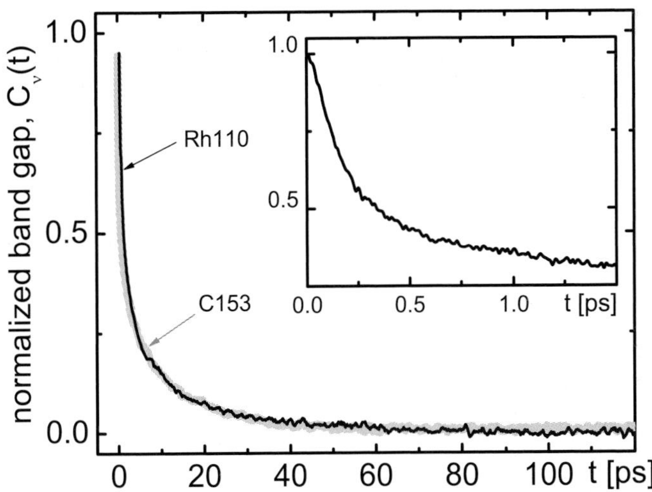

Fig. 8.13. Normalized solvation relaxation function for *Rh*110 in methanol (solid). Remarkably, it traces the function for the solvatochromic dye Coumarin 153 (thick gray line) (from an empirical analysis of unpublished data)

For *Rh*110 in methanol the full Stokes shift can be estimated from the stationary spectra in Fig. 8.10a. The induced change of charge distribution is multipolar in this case and therefore not much felt by solvent molecules beyond the first solvation shell; this is why the Stokes shift is small compared to that of dipolar "solvatochromic" dyes. But the shift is still large compared to the spectral resolution and therefore the effect must be included in the molecular model. This was done within model $\Sigma_{gef}(\nu', \nu'')$ by treating the frequencies of electronic-origin transitions as free parameters. The result is given, in essence, by the time-zero stimulated emission spectrum which was shown in Fig. 8.10b together with the relaxation function $C(t)$ in Fig. 8.13.

The Stokes shift is obtained as $410\,cm^{-1}$. The relaxation function is practically identical to that of solvatochromic Coumarin 153 and can be characterized by a multiexponential fit [22]. The initial portion of the solvation response (insert) is best described with an additional Gaussian time function so that $C_\nu(t) \approx b_G \exp[-(t/\tau_G)^2] + \sum_j b_j \exp[-t/\tau_j]$. The optimal parameters from the global fit are τ_G=0.12 ps, b_G=0.31, and τ_j/ps ={0.99, 11.9}, b_j={0.35, 0.34}.

It should be mentioned that the response-function approach which was reviewed here solves the "time-zero analysis problem" of solvation dynamics monitored by transient absorption. When pump and probe pulses are well separated, a transient spectrum at given delay time can be decomposed [21] into the BL, ESA, and SE components, each of which is described by an appropriate bandshape function. At long delays when solvation is complete, the SE spectrum can be obtained from independent stationary fluorescence measurements. The analysis usually begins with this kind of tail-fitting and is then carried forward toward $t = 0$. But when pump and probe pulses overlap in time the most interesting region is obscured by coherence. For this reason the initial SE spectrum at $t = 0$ is not directly available from transient absorption experiments and $C(t)$ is correspondingly uncertain. Modeling the entire evolution including the coherence spectrum and solvation dynamics solves this problem.

8.2 A novel experimental approach: Ultrafast X-ray absorption spectroscopy

W. Gawelda, C. Bressler, and M. Chergui

The previous sections demonstrated the power of nonlinear optical absorption spectroscopy for understanding molecules in solution. In particular, dephasing and vibrational relaxation times can be extracted and information about interaction with the solvent is also obtained, in the form of a solvation correlation function. In all these processes it is the rearrangement of charges that changes the force field within the solute and between the solute and solvent species, and which thereby drives the chemical dynamics leading to structural changes. It is therefore desirable to have a means to observe directly, in real time, the light-induced charge rearrangements and the ensuing structural changes. With optical probe pulses one may achieve complete spatial information only for a few rare cases, such as small diatomic or triatomic molecules in the gas phase, for which an a-priori knowledge of the potential energy surfaces is available [1, 27]. This is usually not the case for larger assemblies of atoms (e.g. polyatomic molecules) or when dealing with condensed phases or biological samples, in which case just a few dominant displacements can be determined.

In order to overcome these limitations, the pump-probe approach must be extended to the X-ray domain, since X-ray techniques (diffraction and absorption) are routinely used in structural analysis. Several groups are presently active in implementing ultrafast X-ray diffraction [47–51]. X-ray absorption spectroscopy (XAS), the alternative technique, lends itself ideally to the study of electronically excited solutions, and this is why an optical pump/X-ray absorption probe scheme was developed over the past few years [28]. Here results are presented on the photoinduced intramolecular electron transfer in aqueous $[Ru^{II}(bpy)_3]^{2+}$ [29] which demonstrate the capability of the method.

8.2.1 X-ray absorption spectroscopy

Many excellent books and reviews exist, which give a detailed description of XAS, see e.g. [30–33]. Here the basic features of X-ray absorption spectroscopy are briefly outlined.

Typical transitions from a core shell of an atom into the continuum give rise to saw-tooth like features, called absorption edges, which coincide with ionization thresholds and sit on a continuous absorption background that decreases with increasing energy. The nomenclature for these X-ray absorption features reflects the core orbital from which the absorption originates. For example K edges refer to transitions from the innermost $n = 1$ electron orbital, L edges refer to the $n = 2$ absorbing electrons (L_I to $2s$, L_{II} to $2p_{1/2}$ and L_{III} to $2p_{3/2}$ orbitals), and M, N, etc., to the corresponding higher-lying bound core shells. The transitions are always referred to partially or fully unoccupied states, just below and above the ionization limit, leaving behind a core hole. Above the ionization limit the excited electron is often referred to as a photoelectron, and depending on its kinetic energy E_{kin} it can propagate more or less freely through the molecule as a spherical wave with:

$$E_{kin} = h\nu - E_B, \tag{8.56}$$

where $h\nu$ is the incident X-ray energy and E_B the binding energy (or ionization potential). The photoelectron wave vector is then defined as

$$k = \frac{2\pi}{h} \cdot \sqrt{2m(h\nu - E_B)}. \tag{8.57}$$

Zooming into one of the absorption edges reveals fine structure which contains information about the electronic and geometric structure around the absorbing atom. Typical absorption spectra for the K-edge of Iron in $[Fe^{II}(bpy)_3]^{2+}$ complexes in aqueous solution are shown in Fig. 8.14. The region around the absorption edge is the so-called XANES (X-ray absorption near-edge structure) region, while that well above the edge is the EXAFS (Extended X-ray absorption fine structure) region.

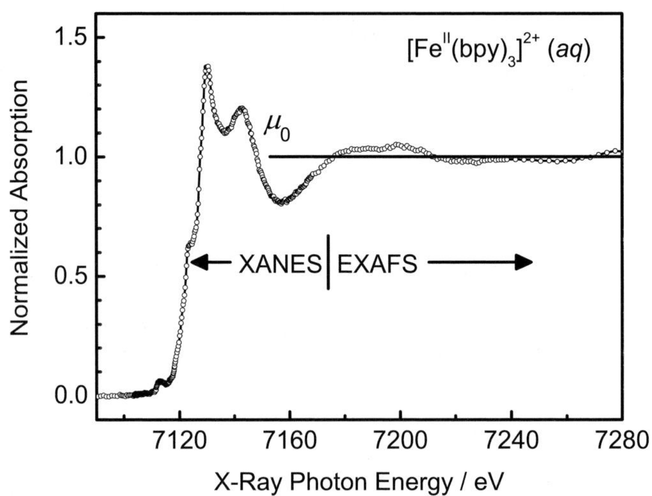

Fig. 8.14. K-edge x-ray absorption spectra of iron in aqueous $[\mathrm{Fe^{II}(bpy)_3}]^{2+}$. The relative absorption with respect to the high energy background is plotted.

The XANES is due to absorption from core shells to the occupied or fully unoccupied *valence orbitals* of the atom. Just above the edge, a photoelectron is created and resonant features may appear due to multiple scattering events (Fig. 8.15b), which arise from the high scattering cross-section of the low energy photoelectron with the surrounding atoms. These features contain information about bond distances, bond angles and coordination number around the absorbing atom [30, 32, 33].

Well above the edge, the photoelectron has large E_{kin}, and weak modulations appear in the EXAFS region, which arise from the interference between the outgoing photoelectron wave and the backscattered one on the surrounding neighbor atoms. Since the scattering cross section decreases with energy of the electron, EXAFS is dominated by single scattering events, and it contains information about bond distances and coordination number of the atomic neighbors around the absorbing atom.

The way the structural information is retrieved from the XANES and/or the EXAFS features is the subject of intense studies. Standard programs (FEFF [34, 35], TT-multiplet code [33, 36], MXAN [37, 38], etc.) are available which are constantly being improved in performance and precision. However, EXAFS is more commonly used as a structural analytical tool compared to XANES, because it is dominated by single scattering and therefore allows the analysis of the data by a simple Fourier transformation [39]. For this purpose

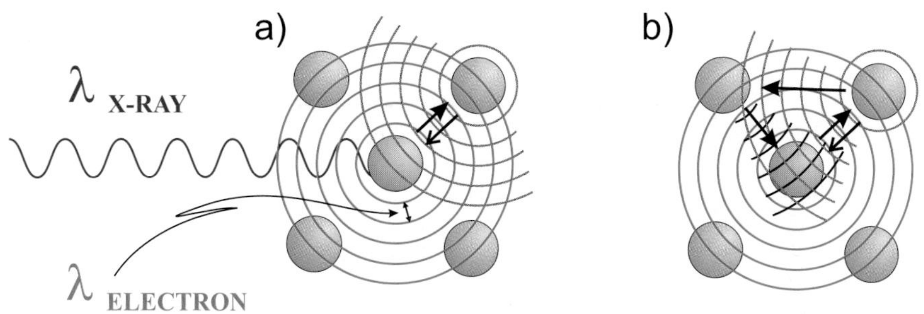

Fig. 8.15. Patterns of outgoing and backscattered photoelectron waves in the case of EXAFS (left - single scattering events) and of XANES (right - multiple scattering events) .

one has to generate a normalized X-ray absorption spectrum (normalized to the absorption edge jump under consideration), which is defined as the normalized oscillatory part of μ(E) (the X-ray absorption coefficient), i.e. the EXAFS, via

$$\chi(E) = \frac{[\mu(E) - \mu_0(E)]}{\Delta\mu_0(E)} \tag{8.58}$$

with $\mu_0(E)$ being the smoothly varying atomic-like background absorption. $\Delta\mu_0$ is a normalization factor that arises from the net increase in the total atomic background (or simply the absorption edge jump). Above the edge under consideration, using (8.57) to substitute E with the photoelectron wave vector, $\chi(k)$ can be rewritten as

$$\chi(k) = \sum_j S_0^2 N_j \frac{|f_j(k)|}{kR_j^2} \sin\left[2kR_j + 2\delta_e + \Phi\right] \cdot e^{-\frac{2R_j}{\lambda(k)}} \cdot e^{-2\sigma_j^2 k^2} \tag{8.59}$$

which is the standard EXAFS formula [52]. The structural parameters (the subscript j refers to the group of N_j atoms with identical properties, e.g. bond distance and chemical species) are: a) the interatomic distances R_j, b) the coordination number (or number of equivalent scatterers) N_j, c) the temperature-dependent *rms* fluctuation in bond length σ_j which should also include effects due to structural disorder. $f_j(k) =| f_j(k) | e^{i\phi(k)}$ is the backscattering amplitude, δ_e the central-atom partial-wave phase shift of the final state, $\lambda(k)$ the energy-dependent photoelectron mean free path (not to be confused with its de Broglie wavelength), and S_0^2 is the overall amplitude reduction factor. In a very transparent and simple form, (8.59) contains all of the key elements that provide a convenient parameterization for fitting the local atomic structure around the absorbing atom to the measured EXAFS data.

From the above, it appears that XAS is a particularly attractive technique for probing the electronic and geometric structure in solutions in a pump-probe configuration:

(i) It is highly selective since one can interrogate a specific type of atom, e.g. the physically, chemically or biochemically significant one, by simply tuning to its characteristic core absorption edges.

(ii) XANES interrogates the (partially or fully unoccupied) *valence orbitals* of the atom of interest, which are precisely those *causing bond formation, bond breaking and bond transformation.* Thus one can detect the *underlying electronic changes that drive the structural ones.* In particular, the degree of oxidation of an atom and the occupancy of its valence orbitals are reported in XANES by shifts of the edge. In addition, for atoms embedded in an ordered atomic environment, the selection rules of core transitions can be altered by the local symmetry so that structural information can be retrieved as well.

(iii) Since *short time scales correspond to short distance scales*, XANES and EXAFS are ideal tools for the ultrafast structural dynamics because they probe the local environment of the atom of interest.

(iv) The precision of structural determination by EXAFS is on the order of $10^{-2} - 10^{-3}$ Å [40], which is ideal for observing transient structures resulting from the photoinduced reactions.

(v) It can detect optically silent species which may result from a photoinduced process.

8.2.2 Experimental method

A number of prerequisites need to be fulfilled in order to implement XAS in the ultrashort time domain:

a) A source of tunable X-rays.

b) Ultrashort pulses of X-rays.

c) As described in [28,41], a high X-ray flux on the sample is also needed in order to capture changes in the sample transmission.

A compromise between these requirements is given by Third Generation synchrotron sources, which combine the advantages of a stable, tunable and high flux source of X-rays, with pulses in the 50-100 ps range. While this may not be enough to probe the femtosecond dynamics, it has to be stressed that there exist to date, *no sources of intense tunable femtosecond X-ray pulses.* Many schemes are being developed to produce such pulses, but pending their full implementation, it is necessary to establish methods and techniques with present-day technology.

The experimental strategy is based on the pump-probe scheme, using an ultrashort laser pulse to excite the system, and a hard X-ray pulse from a synchrotron to monitor the photoinduced changes in the system at variable time delay after the pump pulse. Details of the method, including the data acquisition scheme, are given elsewhere [42,43]. In particular, the transient XAS data are recorded at twice the repetition rate of the pump laser exciting the sample. Thus the transmitted X-ray pulse intensity is recorded alternatively for the unexcited and the laser-excited sample. These are then appropriately sorted and averaged for a given data (point) accumulation time and stored together with their in-situ measured standard deviation. This scheme permits to record the transient absorption spectrum by step-scanning the X-ray monochromator at a fixed pump-probe time delay and subtracting adjacent data points.

By sorting the data according to the laser-excited and the dark sample, the static and photoexcited transient transmission spectra are obtained for a given photoexcitation yield. Alternatively, we can record the temporal evolution of the X-ray transmission by scanning the laser/X-ray time delay at a fixed energy. The recorded transmission spectra of the unexcited sample are transformed into absorption spectra via normalization with a transmission spectrum through the neat solvent, and the excited state spectrum is obtained from the measured transient absorption spectrum at each fixed time delay.

8.2.3 Results

In order to demonstrate the capability of picosecond XAS for the study of electronically excited molecules in solution, recent results on Ruthenium(II)-tris-2,2'-bipyridine ($[Ru^{II}(bpy)_3]^{2+}$) are briefly presented. Its photochemical cycle is sketched as a simple level scheme in Fig. 8.16. Light excitation of a metal-centered valence electron from its singlet ground state (^{1}GS) (originating from the ligand-field split $4d$ level) into the lowest-energy absorption band (400-500 nm) leads to the formation of a Franck-Condon singlet Metal-to-Ligand Charge Transfer (^{1}MLCT) state, and localization of the electron on one of the bipyridine ligands, which undergoes intersystem crossing to a long-lived triplet state (^{3}MLCT) in < 100 fs. At room temperature and in aqueous solutions the emission of the ^{3}MLCT state exhibits a measured lifetime of about 600 ns [29].

The photoinduced electronic changes (change of oxidation state of the Ru atom, localization of the electron on a ligand) should have consequences on the molecular structure, and on the Ru-N bond distances. Here it is shown how these features appear in the spectra and how the structural parameters can be quantitatively extracted.

The static XAS spectrum of the ground state complex is shown in Fig. 8.17a. It exhibits bands labeled B, C and D at both L edges. To explain their origin, the electronic structure of the metal atom in the field of the bpy ligands must be considered : The Ru atom has a $4d^6$ configuration in the ground state of $[Ru^{II}(bpy)_3]^{2+}$, and in the presence of an octahedral crystal field the $4d$ orbitals transform into t_{2g} and e_g orbitals, separated in energy

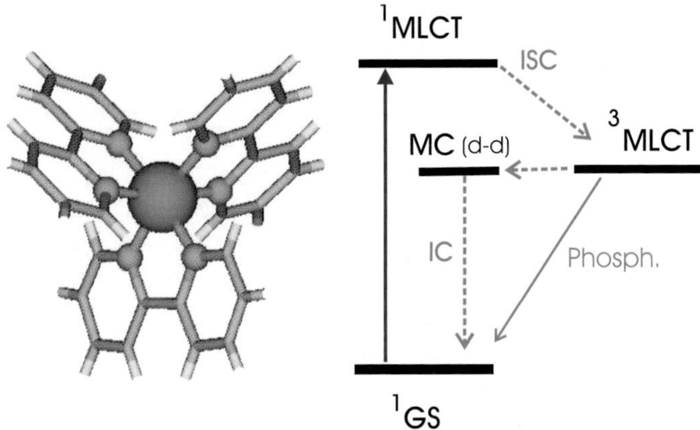

Fig. 8.16. Photochemical cycle of $[Ru^{II}(bpy)_3]^{2+}$ in a simplified energy level scheme. Absorption of visible light removes a metal-centered (MC) $4d$ electron in its singlet ground state (^{1}GS) into the singlet Metal-to-Ligand Charge Transfer (^{1}MLCT) state, where it undergoes ultrafast intersystem crossing into the triplet ^{3}MLCT state, localized on the bipyridine ligand system. The ^{3}MLCT state decays to the ground state with an emission lifetime of $\sim$600 ns at room temperature.

by the octahedral crystal field splitting. Since this is a low spin compound, all 6 electrons fill up the t_{2g} orbitals, while the e_g orbitals are empty. The L edges arise from atomic-like electric dipole transitions (change of angular momentum $\Delta l = \pm 1$) from the $2p_{1/2}$ (L_2) and $2p_{3/2}$ (L_3) core orbitals to unoccupied orbitals of both s and d symmetry. Excitation of the $2p$ electron is only possible to the empty e_g-states, giving rise to the B-band at both L_3 and L_2 edges. The C-band has recently been shown to be a multiple scattering above-ionization resonance [44], while the D-feature is an EXAFS feature reflecting the molecular structure. Fig. 8.17b shows the difference spectrum between the unexcited ($\Delta t < 0$) and the excited sample transmission spectra, at a time delay of 50 ps after laser excitation. Clear photoinduced changes appear at the L_3 and L_2 edges, while a weak change is also observed in the region of the D-band between the two edges. These changes are signatures of electronic and structural changes induced by the laser excitation. From Fig. 8.17a and Fig. 8.17b, and with the photolysis yield determined from laser-only pump-probe experiments [29], one can retrieve the X-ray absorption spectrum of the compound in its ^{3}MLCT state which is shown in Fig. 8.17c. Similar features (as in Fig. 8.17a) appear labeled with primes, but this time an additional band A' shows up below the B'-band at both edges. Fig. 8.18 zooms into details of the ground and excited state (time delay: 50 ps) L_3-edge

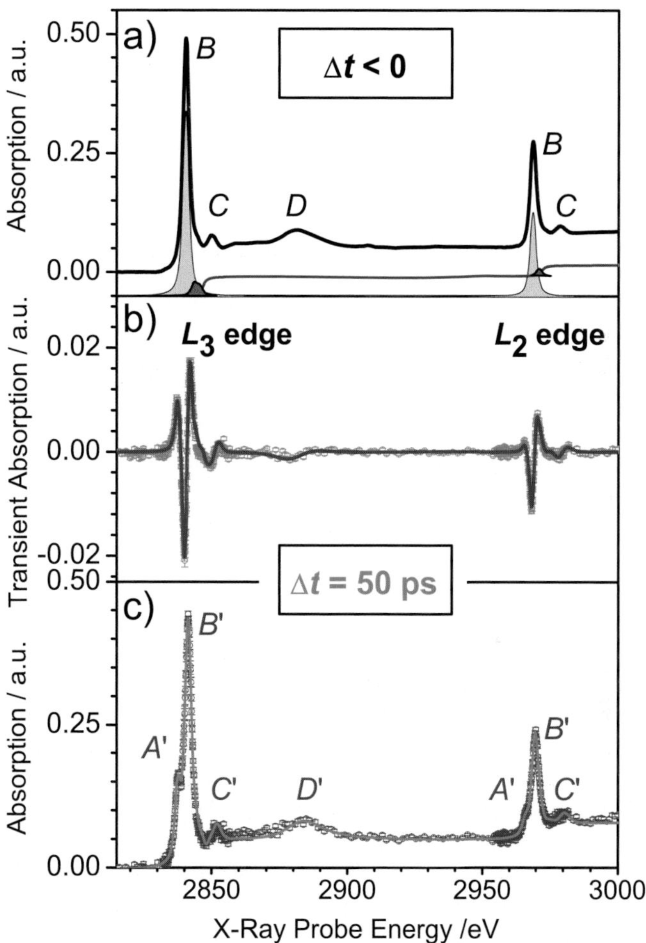

Fig. 8.17. a) Static absorption spectrum of aqueous $[Ru^{II}(bpy)_3]^{2+}$ in the region of the Ru L_3 and L_2 edges. b) Transient difference absorption spectrum measured 50 ps after photoexcitation (the blue trace results from a global fit of the data, see [29] for details). c) Excited state XAS spectrum extracted from spectra a) and b) and the fractional population of the excited state [29]. The red trace is the result of a fit. Note that compared to (a), an additional band (A') shows up.

absorption. The following information can be retrieved from the XANES and EXAFS domains:

XANES: It can be seen that the B-band shifts to the blue by 0.9 eV. This shift results from the change of the oxidation state of the Ru atom, in good agreement with previous experimental studies on ruthenium compounds of different valencies [45] and with quantum chemical calculations [46]. The new feature A', seen in Fig. 8.17c and 8.18, originates from the light-driven charge transfer process (changing the Ru atom occupancy from $4d^6$ to $4d^5$) which creates a vacancy in the previously fully occupied t_{2g} orbital. The A'-B' splitting amounts to 3.75 eV, which is close to the value found in purely octahedral compounds [45], suggesting that the trigonal distortion (D_3) and the axial distortion (leading to C_2 symmetry) of the photoexcited complex are minor perturbations to the dominant O_h ligand field. Since the ligand-field splitting depends on the Ru-N bond distances, the full analysis of the XANES line shapes delivers information about the multiplet electronic sublevels that cause the splitting and about the bond distances. The analysis is described in detail in [29], and is beyond the scope of the present contribution. Here the focus is on the more direct extraction of structural parameters from the EXAFS domain.

EXAFS: EXAFS is sensitive to the wave vector of the X-ray generated free photoelectron wave (thus above the IP). The D feature, which represents a clear EXAFS modulation, is blue shifted by 1 eV (after correcting for the IP shift, which can be read off from the shift of band C, see [29] for details), pointing to a bond contraction of the Ru-N nearest-neighbor distance. A full calculation of the ground and excited state EXAFS using the FEFF 8.20 code was performed (see [29] for details), based on (8.59) to extract the Ru-N bond distances, which were all treated equally (i.e. in D_3 symmetry). Fig. 8.19a and Fig. 8.19b (Figs. 8.19c and 8.19d), respectively, show the ground state (excited state) experimental and optimized EXAFS spectra in q-space and their Fourier transform power spectra.

The power spectra represent a Pseudo Radial Distribution Function (PRDF), which is characterized by a main peak near 1.8 Å and a weaker one near 3.3 Å. (not phase-corrected for the central atom phase shift, therefore the actual peaks show up at R values shorter than those corresponding to nearest neighbor distances). The first peak corresponds to single scattering contributions by the nearest shells of N and C atoms around the Ru atom (N/N', C2/C2' and C6/C6' shown in the inset of Fig. 8.19b), while the second peak is largely due to contributions of more distant shells of atoms (C3/C3' to C5/C5'). Only the first peak of the PRDF was considered in the EXAFS analysis. Once again, the pseudoradial distribution function for the excited complex reflects a small contraction with respect to the ground state. The simulated amplitudes of the three dominant single scattering (SS) shells in the 1-2.8 Å range of neighboring N and C atoms are shown, and their sum is compared with the experimental Fourier transform for the ground state (Fig. 8.19a) and excited complex (Fig. 8.19d). Bond lengths for the ground state

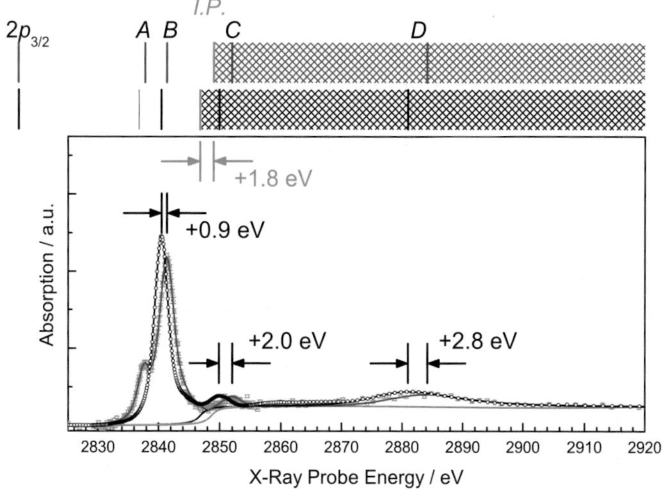

Fig. 8.18. Ground (black dots) and excited state (red dots with error bars) XAS spectra of $[Ru(bpy)_3]^{2+}$ at the Ru L_3 edge, together with a fit of the dominant features (solid curve through data points). The hatched areas at the top of the figure represent the ionization threshold (IP) and continuum of the ground (black) and excited complex (blue). [29]

are derived, and the resulting bond contraction for the excited state is found to be -0.037±0.014 Å. This relatively weak bond contraction, despite a dramatic change of electronic structure, results from a balance between strong attractive forces between metal and ligand on one hand, and steric effects on the other, due to the fact that the three byp ligands are already in a strained geometry in the ground state.

In conclusion, the electronic and geometric structure of an electronically excited complex is captured 50 ps after excitation. The experiment demonstrates the capacity of X-ray absorption to observe both the geometric changes and the underlying electronic structure changes that cause them. With the present state of the technology, this type of study is being extended to a large variety of molecular systems in solution. An area where significant progress is anticipated is the study of solvation dynamics around atomic ions, since one may visualize the structural rearrangement of the solvent shells after electronic excitation of the solute. In this sense, X-ray spectroscopy is powerful in that it can detect optically silent species, such as neutral atoms produced, and thus allow experiments which avoid the interference of intramolecular modes, present in molecular solutes, with intermolecular modes.

Extending time-resolved XAS to the femtosecond domain is only limited by the lack of adequate sources of ultrashort, tunable, femtosecond X-ray pulses.

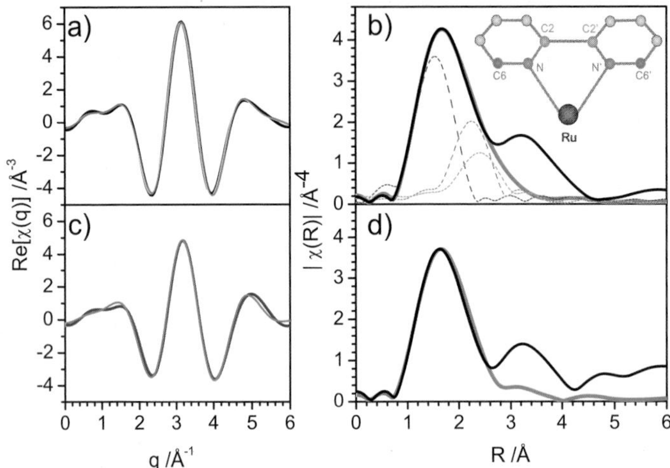

Fig. 8.19. Wave vector spectra and pseudoradial distribution functions (PRDF) of the ground state (a,b) and the excited state complex (c,d). The black lines represent the experimentally derived traces, while the red ones represent those simulated using the FEFF 8.20 code. The PRDF's are decomposed in terms of the various single scattering contributions due to the N, C2 and C6 atoms as seen in the inset (colors correspond between atom and trace), which shows the Ru atom and one bpy ligand. Note that the PRDFs were not phase-corrected for the central atom phase shift, therefore the actual peaks show up at R values shorter than those corresponding to nearest neighbor atoms (the central atom phase shift can be included in the Fourier transform and it results in ca. 0.3-0.5 Å displacement of the radial distribution function). [29]

However, recent developments at 3rd generation sources and the construction of free electron lasers are very promising steps that will extend the technique to the femtosecond time domain. This is shown in Fig. 8.20 where the actual pulse intensity (in photons per 0.1 % Bandwidth) is plotted as a function of temporal pulse width for the present and future tunable X-ray sources. It can be seen that lab-based plasma sources are already in the sub-picosecond time domain. However, these sources are not suitable for X-ray spectroscopy since they consist of spectral lines. More promising in this respect is the Femtosecond Slicing scheme [53], but at the cost of a decrease in flux. Still, this scheme is the only one that delivers a continuum X-radiation with femtosecond time structure. Coming up in the next few years are Energy Recovery Linacs, and the Free Electron Lasers that are being built in Stanford and Hamburg. These represent 6 orders of magnitude increase in flux compared to synchrotrons. Considering that our data presented here were recorded with an accumulated

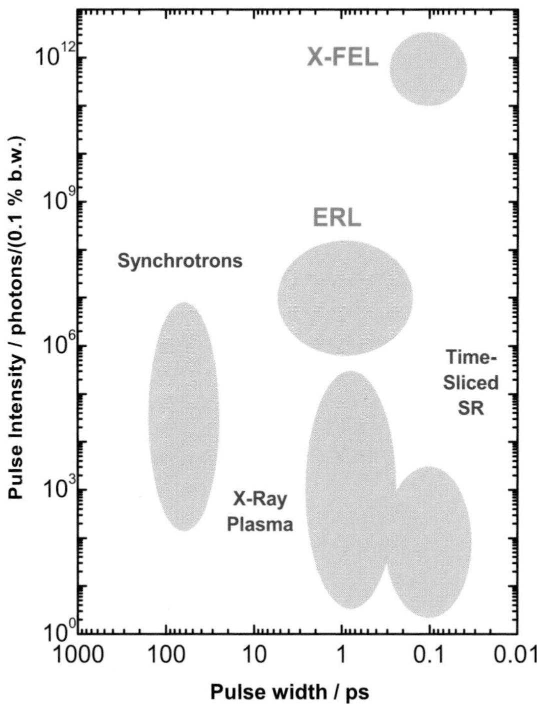

Fig. 8.20. Pulse intensity (in photons per 0.1 % Bandwidth) as function of temporal pulse width for the present and future tunable X-ray sources; see text.

(2-6 sec) photon flux of 10^6 to 10^8 X-ray photons in a 50-100 ps pulse, it is clear that the advent of the new sources open exciting possibilities in the field of ultrafast X-ray spectroscopy.

References

1. S. Mukamel, *Principles of nonlinear optical spectroscopy* (Oxford University, New York, 1995)
2. *The supercontinuum laser source*, ed. by R. R. Alfano (Springer, Berlin, Heidelberg 1989)
3. T. Cordes, D. Weinrich, S. Kempa, K. Riesselmann, S. Herre, C. Hoppmann, K. Rck-Braun, W. Zinth, Chem. Phys. Lett. **428**, 167 (2006)
4. G. Vogt, G. Krampert, P. Niklaus, P. Nuernberger, G. Gerber, Phys. Rev. Lett., **94**, 068305 (2005)
5. D. S. Larsen, E. Papagiannakis, I. H.M. van Stokkum, M. Vengris, J.T.M. Kennis, R. van Grondelle, Chem. Phys. Lett. **381**, 733 (2003)

6. S. Mitra, N. Tamaib, Phys. Chem. Chem. Phys., **5**, 4647 (2003)
7. D. W. McCamant, Ph. Kukura, S. Yoon, R. A. Mathies, Rev. Sci. Instr. **75**, 4971 (2004)
8. A. Espagne, P. Changenet-Barret, P. Plaza, M.M. Martin, J. Phys. Chem. A **110**, 3393 (2006)
9. A.L. Dobryakov, S.A. Kovalenko, N.P. Ernsting, J. Chem. Phys. **119**, 988 (2003)
10. A.L. Dobryakov, S.A. Kovalenko, N.P. Ernsting, J. Chem. Phys. **123**, 044502 (2005)
11. S.A. Kovalenko, A.L. Dobryakov, J. Ruthmann, N.P. Ernsting, Phys. Rev. A **59**, 2369 (1999)
12. J.L.P. Lustres, S.A. Kovalenko, M. Mosquera, T. Senyushkina, W. Flasche, N.P. Ernsting, Angew. Chem. Int. Ed. **44**, 5635 (2005)
13. D. Lee, A.C. Albrecht, in *Advances in Infrared and Raman spectroscopy, vol.12*, (Wiley-Heyden, Chichester, 1985)
14. Eds. M. Abramowitz, I.A. Stegun, *Handbook of mathematical functions* (Dover Publ., New York, 1972)
15. G. Herzberg, *Molecular spectra and molecular structure. II. Infrared and Raman spectra of polyatomic molecules* (Krieger publishing company: Malabar, Florida, 1991)
16. W.T. Pollard, A. K. Felts, R.A. Friesner, Adv. Chem. Phys., **93**, 77 (1996)
17. F. Bloch, Phys. Rev., **70**, 460 (1946)
18. G. Li, H. Li, Y. Mo, X. Huang, L. Chen, Chem. Phys. Lett. **330**, 249 (2000); P. Hildebrandt, M. Stockburger, J. Phys. Chem. **88**, 5953 (1984)
19. *Optimization toolfox user's guide, version 3* (MathWorh, Inc., 1990-2004)
20. *Genetic algorithm and direct search toolbox user's guide, version 1* (MathWorks, Inc., 2004)
21. S. A. Kovalenko, J. Ruthmann, N.P. Ernsting, Chem. Phys. Lett. **271**, 40, (1997)
22. (a) M. L. Horng, J. A. Gardecki, A. Papazyan, M. Maroncelli, J. Phys. Chem. **99**, 17311 (1995); (b) L. Zhao, J.L.P. Lustres, V. Farztdinov, N.P. Ernsting, PCCP, **7**, 1716 (2005)
23. D. Bingemann, N. P. Ernsting, J. Chem. Phys. **102**, 2691 (1995)
24. (a) E. T. J. Nibbering, D. A. Wiersma, K. Duppen, Chem. Phys. **183**, 167 (1994); (b) W. P. de Boeij, M. S. Pshenichnikov, D. A. Wiersma, Chem. Phys. Lett. **247**, 264 (1995)
25. T. Joo, Y. Jia, G. R. Fleming, J. Chem. Phys. **102**, 4063 (1995)
26. P. Vöhringer, D. C. Arnett, R. A. Westervelt, M. J. Feldstein, N. F. Scherer, J. Chem. Phys. **102**, 4027 (1995)
27. A. H. Zewail, J. Phys Chem A. **104**, 5660 (2000)
28. C. Bressler, M. Chergui, Chem. Rev. **104**, 1781 (2004)
29. W. Gawelda, M. Johnson, F.F.M. de Groot, R. Abela, C. Bressler, M. Chergui, J. Am. Chem. Soc. **128**, 5001 (2006)
30. A. Bianconi, In *X-ray absorption: Principles, applications, techniques of EXAFS, SEXAFS, and XANES*, ed. by D. C. Köningsberger, R. Prins (Wiley: New York etc., 1988) Chapter XI, pp. 573
31. E. A. Stern, In *X-ray absorption: Principles, applications, techniques of EXAFS, SEXAFS, and XANES*, ed. by D.C. Köningsberger, R. Prins (Wiley, New York etc., 1988) Chapter I, pp. 3

32. J.J. Rehr, R. C. Albers, Rev. Mod. Phys. **72**, 621 (2000)
33. F. de Groot, Coordin. Chem. Rev. **249**, 31 (2005)
34. A. L. Ankudinov, C. E. Bouldin, J. J. Rehr, J. Sims, H. Hung, Phys. Rev. B, **65**, 104107, (2002)
35. A. L. Ankudinov, B. Ravel, J. J. Rehr, S. D. Conradson, Phys. Rev. B, **58**, 7565 (1998)
36. F. M. F. de Groot, J. Vogel, (Oxford University Press, Oxford, 2004)
37. T. A. Tyson, K. O. Hodgson, C. R. Natoli, M. Benfatto, Phys. Rev. B, **46**, 5997 (1992)
38. M. Benfatto, A. Congiu-Castellano, A. Daniele, S. D. Longa, J. Synchrotron Radiat., **8**, 267 (2001)
39. D. E. Sayers, E. A. Stern, F. W. Lytle, Phys. Rev. Lett., **27**, 1204 (1971)
40. A. Filipponi, P. D. D' Angelo, J. Chem. Phys. **109**, 5356 (1998)
41. C. Bressler, M. Saes, M. Chergui, D. Grolimund, R. Abela, P. Pattison, J. Chem. Phys., **116**, 2955 (2002)
42. M. G. Saes, M. Kaiser, A. Tarnovsky, Ch. Bressler, M. Chergui, S. L. Johnson, D. Grolimund, R. Abela, Synchrotron Radiation News, **16**, 12 (2003)
43. M. Saes, C. Bressler, F. van Mourik, W. Gawelda, M. Kaiser, M. Chergui, C. Bressler, D. Grolimund, R. Abela, T. E. Glover, P. A. Heimann, R. W. Schoenlein, S. L. Johnson, A. M. Lindenberg, R. W. Falcone, Rev. Sci. Instrum. **75**, 24 (2004)
44. H. Benfatto, S. Della Longa, K. Hakada, K. Hayakawa, W. Gawelda, Ch. Bressler, M. Chergui, J. Phys. Chem., **B110**, 14035 (2006)
45. T. K. Sham, J. Am. Chem. Soc., **105**, 2269 (1983)
46. G. Calzaferri, R. Rytz, J. Phys. Chem. **99**, 12141 (1995)
47. M. Bargheer, N. Zhavoronkov, Y. Gritsai, J. C. Woo, D. S. Kim, M. Woerner, T. Elsaesser, Science **306**, 1771 (2004)
48. M. Bargheer, N. Zhavoronkov, M. Woerner, T. Elsaesser, ChemPhysChem **7**, 783 (2006)
49. A. Cavalleri, R. W. Schoenlein, Top. in Appl. Phys. **92**, 309 (2004).
50. B. Krenzer, A. Janzen, P. Zhou, D. von der Linde, M. Horn-von Hoegen, New J. Phys. **8**, 190 (2006)
51. A. M. Lindenberg, J. Larsson, K. Sokolowski-Tinten, K. J. Gaffney, C. Blome, O. Synnergren, J. Sheppard, C. Caleman, A. G. MacPhee, D. Weinstein, D. P. Lowney, T. K. Allison, T. Matthews, R. W. Falcone, A. L. Cavalieri, D. M. Fritz, S. H. Lee, P. H. Bucksbaum, D. A. Reis, J. Rudati, P. H. Fuoss, C. C. Kao, D. P. Siddons, R. Pahl, J. Als-Nielsen, S. Duesterer, R. Ischebeck, H. Schlarb, H. Schulte-Schrepping, Th. Tschentscher, J. Schneider, D. von der Linde, O. Hignette, F. Sette, H. N. Chapman, R. W. Lee, T. N. Hansen, S. Techert, J. S. Wark, M. Bergh, G. Huldt, D. van der Spoel, N. Timneanu, J. Hajdu, R. A. Akre, E. Bong, P. Krejcik, J. Arthur, S. Brennan, K. Luening, J. B. Hastings, Science **308**, 392 (2005)
52. E. A. Stern, D. E. Sayers, F. W. Lytle, Phys. Rev. B **11**, 4836 (1975)
53. R. W. Schoenlein, S. Chattopadhyay, H. H. W. Chong, T. E. Glover, P. A. Heimann, C. V. Shank, A. A. Zholents, and M. S. Zolotorev, Science **287**, 2237 (2000)

9

Biological systems: Applications and perspectives

Henk Fidder[1], Karsten Heyne[1], Selma Schenkl[5], Frank van Mourik[2], Gert van der Zwan[5], Stefan Haacke[2], Majed Chergui[2], Mikas Vengris[3], Delmar S. Larsen[4], Emmanouil Papagiannakis[5], John T.M. Kennis[5], Rienk van Grondelle[5], Ben Brüggemann[6], Volkhard May[7], Inés Corral[8], Leticia González[8], Alexandra Lauer[1], Eike Meerbach[9], Christof Schütte[9], Illia Horenko[9], Burkhard Schmidt[9], and Jean-Pierre Wolf[10,11]

[1] Institut für Experimentalphysik, Freie Universität Berlin, Germany
[2] École Polytechnique Fédérale de Lausanne, Switzerland
[3] Vilnius University Faculty of Physics Laser Research Center, Lithuania
[4] Department of Chemistry University of California, USA
[5] Faculty of Sciences Vrije Universiteit Amsterdam, The Netherlands
[6] Chemical Physics, Lund University, Sweden
[7] Institut für Physik, Humboldt-Universität zu Berlin, Germany
[8] Institut für Chemie und Biochemie, Freie Universität Berlin, Germany
[9] Institute of Mathematics II, Freie Universität Berlin, Germany
[10] GAP-Biophotonics, University of Geneva, Switzerland
[11] LASIM (UMR 5579), Université Claude Bernard Lyon 1, France

Coordinated by: Karsten Heyne

9.1 Investigation of diverse biological systems

H. Fidder and K. Heyne

Viable biological systems occur on a wide scale of dimensions ranging from sizes of 30 meters for mammals (blue whale), and 120 meters for plants (sequoia tree), down to 10^{-6} meters for single cellular organisms. These systems consist either of networks of eucaryotic cells, in which specialized cells take over unique functions, or of single procaryotic as well as eucaryotic cells. Procaryotic cells exhibit a simple build without a cell nucleus, whereas the build of eucaryotic cells is of a higher complexity, including a cell nucleus with a double membrane. Every cell consists of a cell membrane which encloses the cytoplasm, the DNA (desoxyribonucleic acid) and RNA (ribonucleic acid) which acts as the carrier of the genetic information, and proteins and enzymes which catalyze cellular reactions and build up substructures inside the cell. Ribosomes represent the protein production machine and are found in both procaryotes and eucaryotes. Ribosomes are responsible for processing the genetic instructions and for converting the genetic code into the exact sequence of amino acids that make up a protein. Structures of higher complexity are, for example, the endoplasmic reticulum and the Golgi apparatus. The endoplasmic reticulum is the transport network for molecules targeted for certain modifications and specific destinations. The Golgi apparatus is the central delivery system for the cell, and is a site for protein processing, packaging, and transport.

The cellular membrane is a lipid bilayer impermeable to ions and polar molecules, which acts as a barrier to define a biochemical environment inside the cytoplasm that differs from outside. Permeability is conferred by two classes of membrane proteins, channels and pumps. Channels enable ions to flow rapidly through membranes in a thermodynamically downhill direction. Pumps, by contrast, use a source of free energy such as ATP (adenosine triphosphate) or light to drive the uphill transport of ions or molecules. One example for such a pump is bacteriorhodopsin (bR), which pumps protons across the cellular membrane against a proton gradient upon light absorption. The proton gradient can be harnessed to generate ATP from ADP (Adenosine diphosphate) by another membrane protein called ATP synthase. This process allows the bacterium *Halobacterium salinarium*, which is sketched in Fig. 9.1, to survive under unfavorable conditions. In addition to proteins, peptidoglycan polymers consisting of sugars and amino acids, can form a homogeneous layer outside the plasma membrane giving the wall shape and structural strength, as well as counteracting the osmotic pressure of the cytoplasm. The peptidoglycan layer is substantially thicker in Gram-positive bacteria (20 to 80 nm) than in Gram-negative bacteria (7 to 8 nm). Gram-positive bacteria are classified as bacteria that retain a crystal violet dye during the gram stain process. Gram-positive bacteria will appear blue or violet under a microscope, whereas gram-negative bacteria will appear red or pink. The difference in classification is largely based on a difference in the bacteria's cell wall structure and therefore in its composition (see Sect. 9.7). This is highly relevant in medical treatments, because antibacterial drugs, such as penicillin, interfere with the formation of the peptidoglycan layer, and therefore with cell reproduction.

The key events of biological processes originate from protein functions and their interactions with cofactors and substrates. There is a variety of important cofactors such as: i) ATP (adenosine triphosphate), the universal "currency" of free energy in biological systems; ii) NADH (nicotinamide adenine dinucleotide) and $FADH_2$ (flavin adenine dinucleotide), the major electron carriers in the oxydation of "fuel" molecules, such as glucose and fatty acids; and iii) NADPH (nicotinamide adenine dinucleotide phosphate), the major electron donor in reductive biosynthesis. Proteins play crucial roles in virtually all biological processes. In enzymatic catalysis proteins usually increase reaction rates by at least a millionfold. They transport and store small molecules and ions, e.g., oxygen in haemoglobin. Proteins are responsible for coordinated motion, such as muscle contraction, by the sliding motion of two kinds of protein filaments, or the swimming of bacteria by rotation of flagellae (e.g., Fig.9.1). Immune protection is accomplished by antibodies, which

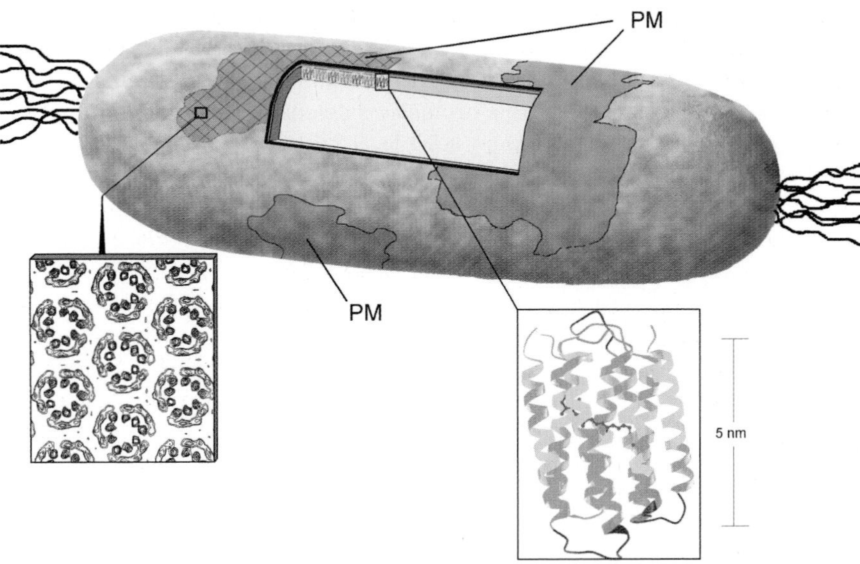

Fig. 9.1. Model of the *Halobacterium salinarium.* The bacterium spanning a length of some micrometers is sketched with its flagella on both sides; regions of the purple membrane (PM) containing the transmembrane protein bacteriorhodopsin (bR); lower panel; left side: top view of the bacteriorhodopsin proteins in the PM measured by X-ray diffraction [1,2]; right side: side view of bacteriorhodopsin with the retinal chromophore (purple) measured by electron-diffraction [3].

are highly specific proteins that recognize and combine with viruses, bacteria, and cells from other organisms. Moreover, mechanical support, such as the strength of the skin and bone, is due to the presence of the fibrous protein collagen. Generation and transmission of nerve impulses is mediated by receptor proteins. For example, rhodopsin, a protein related to bacteriorhodopsin, is the light-sensitive protein in retinal rod cells in the eye, which undergoes a cis to trans isomerization of its chromophore retinal, and triggers a reaction cascade leading to a nerve impulse from the eye. Growth and differentiation of cells is controlled by proteins, while the coordination of different activities of cells is regulated by hormones, which are often proteins themselves. Proteins serve in all cells as sensors that control the flow of energy and matter and thus they are the machinery of the cell.

All proteins are built from a repertoire of 20 amino acids. An amino acid consists of an amino group (NH_2), a carboxyl group ($COOH$), a hydrogen atom, and a R group, called residue or side chain. All groups are bonded to a carbon atom called α-carbon; and the tetrahedral array of the four different groups confers optical activity to amino acids, with two mirror-image forms called the L isomer and the D isomer. Almost only L isomers are constituents of natural proteins. The side chains R of the 20 amino acids differ in size, shape, charge, hydrogen-bonding capacity, and chemical reactivity. Two examples of amino acids are shown in Fig. 9.2. Glycine has the simplest side chain with just a hydrogen atom. Tryptophan, on the other hand, has a longer side chain belonging to the aromatic side chains, and consisting of an indole joined to a methylene group (see Fig. 9.2). In proteins, the α-carboxyl group of one amino acid is joined to the α-amino group of another amino acid

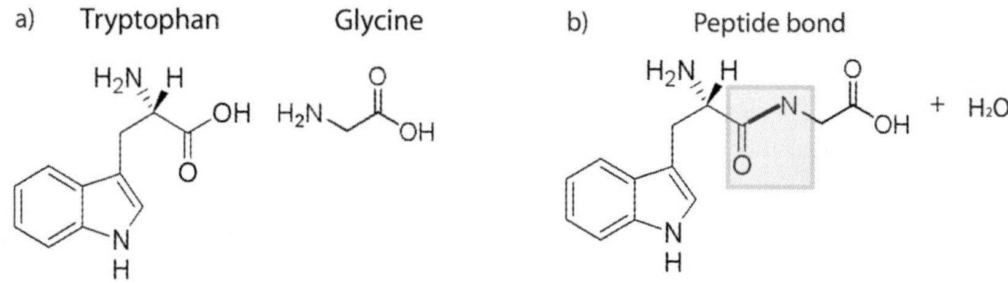

Fig. 9.2. Amino acids tryptophan and glycine; and the dipeptide with the planar peptide bond.

forming a peptide bond (see Fig. 9.2). The protein consists of a repetitive and a variable region. The main chain or protein backbone is a repetition of peptide bonds, whereas the side chains constitute the variable part. Most natural proteins contain anywhere from about ten to several thousands of amino acid residues and the sequence of each protein is unique. The sequence is called the primary structure, where small variations can lead to widely different three dimensional structures and biological functions. Three dimensional structural motifs are called secondary structures. They are determined by the interaction of peptide units $-C_\alpha - (C = O) - (N - H) - C_\alpha-$ of a peptide chain. The peptide units are rigid and planar, narrowing down the possibilities of peptide conformations. The planarity and the possibility of forming hydrogen bonds between the carboxy and amino groups of peptide units lead to regular protein structures, as α-helices and β-sheets. An example of an α-helical protein is the photoreceptor bacteriorhodopsin plotted in Fig. 9.1. It has seven α-helices reaching from inside the cell through the cellular membrane to the other side. Many photoreceptor proteins exhibit a chromophore covalently bound to a specific amino acid called the binding site. This specific location inside the protein predominantly has a well-defined three-dimensional structure with optimized interactions between the chromophore and the protein surroundings. These surroundings can give rise to Coulomb interactions and interactions by hydrogen bonds, π-stacking, steric hindrance, etc. In bacteriorhodopsin these interactions result in an optimized trans to cis isomerization process of the retinal chromophore, with an increased quantum yield and an isomerization dynamics of 500 fs, which is at least 3 times faster than the isomerization reaction in solution [4–7]. A mechanism for reaction optimization in bacteriorhodopsin is presented in Sect. 9.2.

Since living plants and organisms are continuously undergoing many structural changes and meanwhile perform tremendously complex chemical reactions, spectroscopic techniques have to be selected to answer specific questions in such systems. The many chemical and physical events at the basis of the functioning of any living organism occur on a hierarchy of timescales ranging from femtoseconds to seconds. A real-time study of the mechanisms and dynamics of rudimentary chemical reaction steps and energy flow in plants, bacteria, and living beings requires a well-defined trigger to set off a particular chain of events. Therefore, ultrafast time-resolved investigations are mainly concerned with photoinitiated reactions and processes. Despite this limitation, this subcategory includes some of the most important and interesting biological processes, particularly in light of nature's ability to harvest the sun's energy as fuel being the basis for virtually all life on our planet. Despite the plurality in photoinitiated processes taking place around us, it is remarkable how nature's design performs these tasks with a very restricted variety of molecules.

Chlorophylls, carotenoids, xantopsins, and flavins are the work horses in harvesting solar energy for biological usage. The absorption wavelengths of these "light-harvesting" molecules are tuned through variations in side-groups, interactions with protein residues, and interactions with other photoactive molecules. For instance, in oxygenic photosynthesis (splitting water in hydrogen and oxygen), vast amounts of chlorophylls are arranged in the so-called light-harvesting antennae, whereas the core reaction center PSI (Photosystem I) contains about 100 chlorophyll and 20 carotenoid molecules. Depending on intermolecular distances and orientations the absorbed excitation energy is funneled to the reaction centers by a multitude of energy transfer steps, at times residing on a single molecule, at times existing as delocalized excitonic states.

The origin of both energy transfer and exciton formation lies in intermolecular Coulomb interaction. Even neutral molecules represent an overlay of positive and negative charge density patterns, and the full Coulomb interaction between molecules therefore involves an expansion into a series of dipolar, quadrupolar, etc. interaction terms. Typically, only the dipolar interaction terms are considered, which fall off with the third power of the intermolecular separation; see Sect. 9.2 for the explicit form of this interaction. Whether these dipolar interaction terms give rise to Förster energy transfer or Frenkel exciton formation (in both intermolecular electron exchange is excluded), depends on the strength of these terms relative to that of other factors, such as IVR, intramolecular population decay (radiative or nonradiative), and optical (pure) dephasing (by coupling to the environment), which may disrupt the intermolecular coupling. If the dipolar coupling is large relative to these dynamic disturbances, electronic excited states of the ensemble of N molecules will be linear combinations of the N-molecule direct product states with one or more of the molecules singly excited. These linear combinations represent delocalized optical excitations known as Frenkel excitons, and form manifolds labeled the one-exciton manifold, two-exciton manifold, etc., depending on the number of singly excited molecules in the N-molecule basis states forming the exciton states. The combined oscillator strength of the N molecules is redistributed over the one-exciton states leading to enhanced radiative decay rates for some of these exciton states. Note that the delocalized exciton states are eigenstates of the N-molecule Hamiltonian, and therefore stationary solutions.

In Sect. 9.4 a theoretical study demonstrates the possibility of shaped pulses (optimal control theory) to manipulate excitonic wave packets, in order to localize an exciton on a specific molecule in the Fenna-Matthews-Olson (FMO) complex, which is part of the photosynthetic apparatus of green bacteria. A mechanism for reaction optimization by the protein environment is presented in Sect. 9.2. There, transient electric field effects on the absorption band of tryptophan units upon optical excitation of retinal inside bacteriorhodopsin is investigated and modeled with excitonic coupling. If a coherent intermolecular coupling is not established one obtains incoherent intermolecular Förster energy transfer [8], which shows a r^{-6} dependence, instead of r^{-3}, because one now considers population transfer (wave function squared), whereas in exciton formation wave function amplitudes are coupled. If the molecules are in close proximity energy transfer involving simultaneous intermolecular electron exchange is possible, which is referred to as Dexter transfer. The strength of this process drops off exponentially with the intermolecular separation (as all wave functions have this distance dependence for their amplitude in common).

Energy transfer is also widely used in medical and analytical applications. In cancer treatment (photodynamic light therapy) energy is used to generate locally high concentrations of singlet oxygen, ultimately leading to cell death. The mechanism of releasing oxygen from molecules with endoperoxide groups is studied in Sect. 9.5. Photoreceptor proteins absorb light by chromophore molecules and convert the energy into biologically useful properties, e.g. proton gradient. The wide use of the green fluorescent protein (GFP) as biological label (see Sect. 9.3), relies on using the extent of fluorescence quenching for conversion of fluorescence quantum yields into a ruler for determining distances on atomic scales. As mentioned above, various chemical and structural changes within the complete chain of a biological process can stretch over a wide range of timescales from femtoseconds to seconds. Many of these changes, which lead to biomolecular conformations or metastable structures are not accessible with optical spectroscopy, but could lead to characteristic changes in the vibrational spectra. The protein dynamics described by all possible conformations, time scales and couplings would contain a complete set of information to understand the function of proteins. However, an obstacle to calculate metastable conformations is the curse of dimensionality, as typical biomolecular systems contain hundreds or thousands of degrees of freedom. In Sect. 9.6 the Hidden Markov Model approach is applied to reduce the dimensionality of biological systems to a few essential degrees of freedom and to identify metastable conformational structures in time series. The recent developments in time-resolved vibrational spectroscopy make the prospect of combining experimental time-resolved vibrational data with molecular dynamical calculations of metastable conformation an exciting direction for future investigations.

Section 9.7 deals with real-time identification of airborne bacteria using multiphoton excitation, multiphoton ionization and laser induced breakdown with femtosecond pulses to optimize the fluorescence emission of aerosol microparticles. The higher order interaction of the electric field with the bacteria reveals specific spectroscopic data not available in linear spectroscopy and is a key feature for identifying bacteria in urban aerosols. These techniques may prove very useful in future attempts to contain disease outbreaks by bacteria or viruses, such as SARS or anthrax, or straightforward monitoring of levels of typical highly antibiotic resistant bacteria in hospitals.

9.2 Measuring transient electric fields within proteins

S. Schenkl, F. van Mourik, G. van der Zwan, S. Haacke, and M. Chergui

"Electrostatic interactions are the common denominator and probably the most important element in structure-function correlation in biological systems". This citation by A. Warshel [9] stresses the importance of electrostatic interactions for functional dynamics in Biology, previously highlighted by M. Perutz [10]. NMR and X-ray experiments give insight in the dynamical aspects acting on the molecular level by probing the flexibility and highlighting large amplitude motion of atoms. UV-VIS spectroscopy probes electronic transitions and is thus sensitive to light-induced charge redistributions. These determine the dynamic force fields at all time scales, which in return drive the structural changes needed for biological function. The description of electric fields in proteins is still a timely issue to this day [11], in part due to the difficulty of measuring them within proteins. A further degree of complexity is added, when it comes to measuring these electric fields in the ultrashort time domain, where fundamental phenomena occur, that govern the fate of photobiological functions. Indeed, there is no direct experimental evidence for charge displacements in the femtosecond time regime, let alone within a protein, because photovoltaic techniques cannot be used to detect them if they occur faster than picoseconds, and less so on a sub-nanometer distance scale.

Cohen et al. [12] designed a modified, fluorescent, aminoacid that can be incorporated into proteins. However, the use of naturally occurring molecular groups as local probes of electrostatic effects is preferable. In heme proteins, the CO-ligand has been used as a molecular probe of the electrostatic potential in the distal pocket of myoglobin [13], though not in the time domain. In photosynthesis, the electrochromic response of carotenoid molecules is often used, to measure locally electric fields from the millisecond [14] to the femtosecond [15] time regime.

Systems for which the relation between electrostatic interactions and function is most discussed are retinal proteins and excitonic complexes involved in light-harvesting for photosynthesis. The former play an essential role in a broad range of light-driven biological processes, such as vision, energy transduction and circadian control. All these functions involve both the conversion of light energy into charge separation and retinal isomerization, but the interplay of these processes is the subject of intense debate. The retinal protein of which the initial photochemistry is most extensively studied is the photosynthetic protein bacteriorhodopsin (bR). This protein acts as a light-driven proton pump by means of a multistep process, the so-called photocycle. In the traditional model, the light driven trans-cis isomerization of retinal (Fig. 9.3), is the key primary step in this cycle. However, it is also known that Franck-Condon excitation of all-trans retinal results in an immediate change of permanent isomerization dipole moment by more than 12 D [16,17], causing a sudden polarization of retinal. It has been suggested that the dielectric response of the protein, might be the primary light-induced event [18,19] that could drive structural changes on longer time scales [20]. In addition, it is well established that the specific protein environment enhances the yield and the stereo-selectivity of the isomerization [21], when compared to retinal in solvents. However, the actual charges and dipoles that create these favorable electrostatic properties in bR still need to be identified.

Ultrafast transient absorption studies of bR and its mutants with non-isomerizable retinal chromophores, show identical dynamics in the first 200 fs after optical excitation [22]. Isomerization, i.e. a C13-C14 dihedral angle > 90 , thus occurs on a time scale longer than 200 fs. This conclusion is supported by transient absorption studies on wild type bR using 5 fs pulses [23], suggesting that a skeletal change occurs in the first 200 fs, which corresponds to a large twisting about the C13-C14 bond, prior to isomerization. Similar conclusions were drawn from stimulated Raman studies showing a decay of the high frequency C=C, C-C and methyl rock modes in $\sim$250 fs [24]. Femtosecond vibrational spectroscopy shows that isomerization occurs indeed later ($\leq$500 fs) and is followed by a slower vibrational relaxation of the isomerized photoproduct (the so-called J to K transition) [7,24–26].

González-Luque et al. [27] computed the distribution of charges in the S_1 excited state, as a function of the isomerization coordinate of model chro-

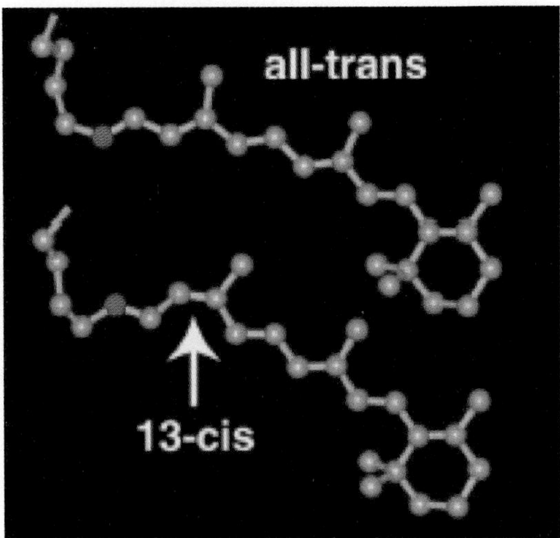

Fig. 9.3. Cis-Trans light-induced isomerization of the retinal chromophore in bacteriorhodopsin

mophores of rhodopsin and bacteriorhodopsin. The computed dipole moment of the S_1 state, increases along the isomerization coordinate of the chromophore, confirming an earlier suggestion by Salem and Bruckman [28]. It thus appears that while there is no isomerization in the first 200 fs or so, the system undergoes a large dipole moment change upon photoexcitation, followed by a translocation of charge. However, the time scale of the latter and its interplay with the initial twisting and subsequent isomerization are still unknown. Direct measurements of the charge separation process using photovoltaic techniques have been performed [29, 30]. However, the intrinsic temporal resolution of these techniques is limited to several picoseconds, i.e. slower than the ≈500-fs isomerization process. To visualize the electronic response of bR on a short time scale, Groma et al. [31] recently reported a different approach, based on a nonlinear technique: interferometric detection of optical rectification. The second order susceptibility $\chi^{(2)}$ of the retinal chromophore is unusually high, especially in its native protein environment, and this property has been shown to be a requirement for functioning of bR [32]. These strong nonlinear properties give rise to optical rectification [33], among others, in non-centrosymmetric crystals. Groma et al. [31] used it in the case of oriented bR crystal films and measured an impulsive macroscopic polarization, which they associated with the charge separation.

In order to measure more specifically the translocation of charge within a protein, we recently implemented an approach that uses the natural tryptophan (Trp) residues of the protein (bR in this case), as a sensor of the electric field changes associated with excitation of retinal and translocation of charge. Literally, all proteins contain Trp amino-acid residues, which are highly polarizable and can be used as sensors for intra-protein electric field changes. Therefore, the description of our results on bR should represent a test case that can be extended to other classes of proteins.

9.2.1 Origin of the transient electric fields

Four tryptophans are in the vicinity of retinal in the binding pocket of bR (Fig. 9.4A) [34, 35]: Trp86 and Trp182 sandwich retinal, while Trp138 and Trp189 are located in the vicinity of the β-ionone ring. The absorption spectrum of bR (Fig. 9.4B) exhibits a band in the visible, due to retinal, and a band near 280 nm, predominantly due to the eight Trps present in the protein. The Trp absorption is due to transition from the ground state to the lowest two close-lying excited states (labeled L_a and L_b) of the indole moiety. The transition into the L_a state implies a large difference dipole moment with respect to the ground state [36], making the Trp's particularly well-suited molecular-level sensors of electric field changes within the protein, as the latter are expected to cause large changes in their spectral features. Note that a $\sim$ 10 D dipole on retinal creates a field of $\sim$10 MV/cm on the nearby amino acids.

In retinal the positive charge from the protonated Schiff base linkage with Lys216 (Fig. 9.4) is partially moved toward the ionone ring upon excitation. To probe the electric field changes associated with the charge translocation in retinal, we have excited the latter at 560 nm, and interrogated the Trp absorption changes with 80-90 fs time resolution, using tuneable near-UV pulses, from a noncollinear optical parametric oscillator. The details of the set-up are given in [37, 38]. Our approach has two advantages: (a) in contrast to experiments in the VIS or IR, that probe retinal directly [7, 23–25], the charge displacements become detectable with high sensitivity through an electric dipole interaction in the UV, (b) the local electric field changes are followed in real-time from femtoseconds to picoseconds (ps) or more.

Results and discussion

Figure 9.5A shows the transient absorption changes, detected between 265 nm and 280 nm. The traces are normalized, highlighting their similar shape over the whole range of probe wavelengths. The presence of a rise time is directly deduced from the temporal derivative of the bleach transient (Fig. 9.4B). Its full width at half maximum (FWHM) of 150 fs is significantly larger than the 80-90 fs pump-probe cross-correlation.

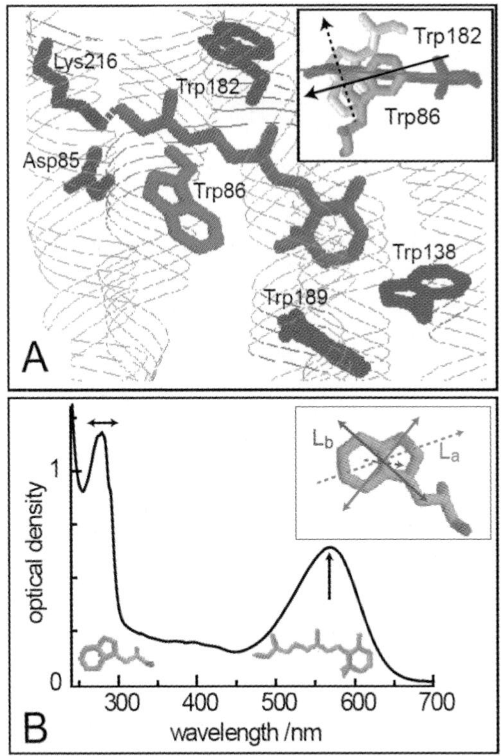

Fig. 9.4. Structure and spectral properties of bacteriorhodopsin. A) The retinal binding pocket of bacteriorhodopsin [35]. Retinal (purple) is covalently bound to Lys216 through a Schiff base linkage. The four nearest Tryptophan (Trp) residues are shown. Insert: Relative orientations of the difference dipole moments for retinal (red arrow), of Trp 86 (full black arrow) and Trp182 (dashed black arrow). B) Absorption spectrum of bR in the purple membrane. The band with maximum at 568 nm is due to the S_0-S_1 transition of the protonated Schiff base form of retinal, while the near-UV band at 280 nm is dominated by the S_0-L_a and S_0-L_b transitions of eight Trp's. The vertical arrow indicate the excitation wavelength and the horizontal one the range of probe wavelengths. Insert: Orientations of the transition (full arrows) and difference (dashed arrows) dipole moments of tryptophan for S_0 - L_a (red) and S_0 - L_b (blue) [36]. Arrows point to the positive end of the difference dipole moment.

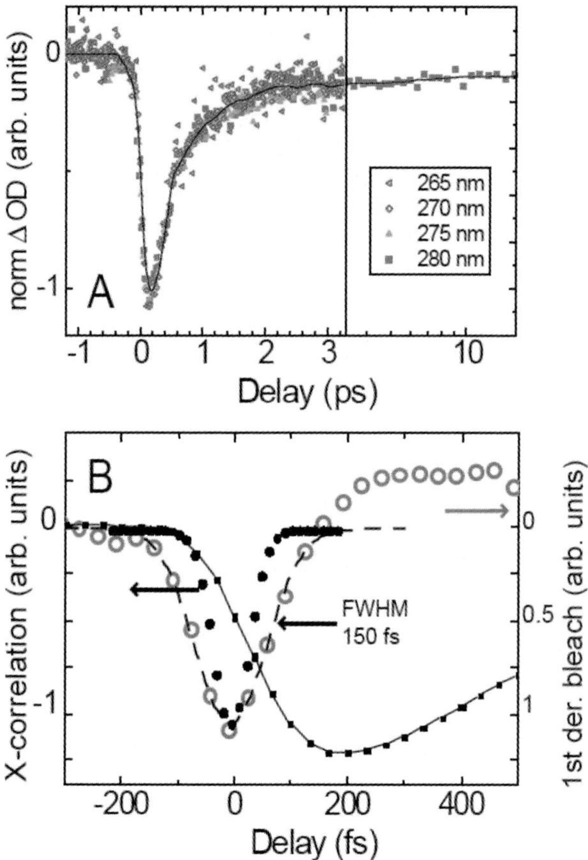

Fig. 9.5. Ultrafast response of bacteriorhodopsin in the near-UV. A) Transient bleach signals at 265-280 nm after excitation with 25 fs pulses at 560 nm at 1 kHz repetition rate. UV probe pulses (80-90 fs) were obtained by second harmonic generation of the output of a tunable noncollinear OPA [37, 38]. The transients are normalized in amplitude. The time axis of the left panel is linear, while the right panel has a logarithmic one. B) The early time portion of the bleach transient in fig. 9.4A (black squares), and its temporal derivative (red circles) showing a full width at half maximum (FWHM) of 150 fs. The latter is clearly larger than the experimental time resolution of 85 fs (black dots) determined by the pump-probe cross-correlation [38].

For longer times, we observe a biexponential recovery of the absorption with time scales: $\tau_1 = 420$ fs, $\tau_2 = 3.5$ ps, which have been reported in experi-

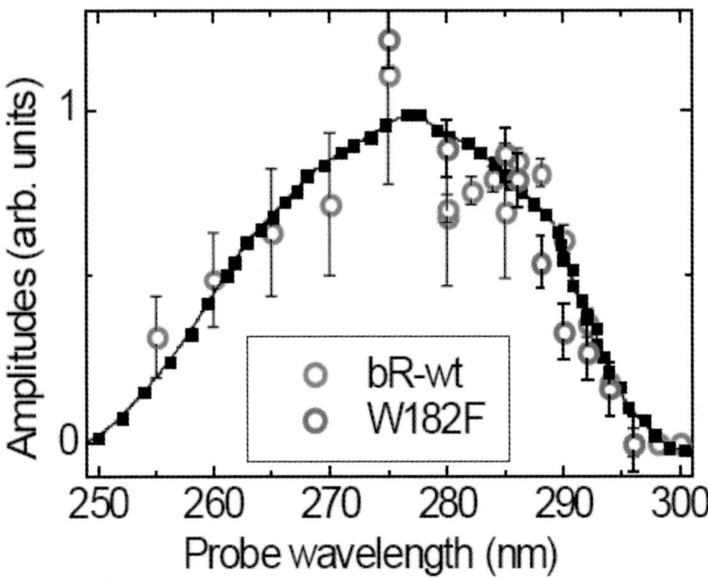

Fig. 9.6. Maximum amplitude of the bleach transients of wild type bR and of the W182F mutant, as a function of the probe wavelength. Experimental data (circles) are compared with the L_a absorption component of Trp in propylene glycol at - 50 C [39] (black squares). To account for the apolar character of the binding pocket, the L_a absorption was shifted by 5 nm to the blue. The data have been normalized to the Trp L_a spectrum at 285 nm.

ments that probe retinal directly [7, 24, 25]. A constant, weak bleach signal is also observed at the longest delay times (τ_∞).

In [37], several arguments are presented, which allowed an unambiguous assignment of the bleach signal to the response of Trp residues, not the least of them being the fact that the wavelength dependence of the bleach amplitude maps reasonably well the Trp L_a absorption spectrum (Fig.9.6). More specifically, we could show that the signal was due to Trp86, by repeating the experiment with a mutant that lacked Trp182, obtaining the same results as with the wild type (Fig. 9.6). This makes sense since the linear polarization anisotropy of the signal is high (0.30 ± 0.05), meaning that the responding Trp transition dipole is almost parallel to the transition dipole moment of retinal along the conjugate chain, as is the case for the L_a state of Trp86 (insert Fig. 9.4A).

A dielectric response of Trps through reorientation of their dipoles is unlikely to account for the rise time, as the polarization anisotropy is constant throughout the time scale of the transients. In addition, the fast inertial relaxation of indole-like molecules is slower, even in liquids [40]. Rather, we attribute the bleach to the above-mentioned large dipole moment change of retinal. The rise time of the bleach signal (Fig. 9.5B) then reflects the instantaneous dipole moment change and an ensuing gradual charge translocation along retinal. However, the Trp response does not resemble a difference spectrum that was to be expected in the case of a pure Stark shift of the L_a transition. As a matter of fact, the bleach component does not have a positive counterpart with comparable amplitude at nearby wavelengths. As shown in [37], and briefly described below, the difference spectrum is rather dominated by the effects of excitonic coupling resulting from the resonance interaction of retinal and Trp at ~280 nm, meaning that oscillator strength borrowing can occur between retinal and the Trp chromophores.

9.2.2 A model of three coupled chromophores

In our model, we consider a system consisting of three coupled chromophores, having three electronic levels each: S_0, S_1, S_n for retinal and, S_0, L_a and L_b for Trp86 and Trp182. They form an excitonic complex, whose proper electronic states are linear combinations of the singly-excited near UV states of $|L_{a,b},0,0>$, $|0,L_{a,b},0>$ and $|0,0,S_n>$ (product basis notation |Trp86, Trp182, retinal>), of which three are important (X_1 trough X_3, see Fig. 9.7), as a result of favorable relative orientations of the transition dipole moments. The doubly excited states (XX_1 and XX_2) with retinal in the S_1 state, arise from linear combinations of $|L_{a,b},0,S_1>$ and $|0,L_{a,b},S_1>$ (Fig. 9.7).

Since the L_a and L_b levels of Trp are nearly degenerate, we took their transition energies to be both at 280 nm (35 714 cm^{-1}). Retinal was modeled, with all dipole moments along the retinal backbone. S_1 corresponds to the first excited state and S_n to the excited state resonant with the Trp absorption (also fixed at 280 nm [41]). The interaction between the chromophores can be modeled by their dipole moment operators. The full dipole operator can be expressed as

$$\begin{aligned} \hat{\boldsymbol{\mu}}_{trp} = \boldsymbol{\mu}_0 \mid 0 < 0 \mid + \boldsymbol{\mu}_a \mid L_a >< L_a \mid + \boldsymbol{\mu}_b \mid L_b >< L_b \mid + \\ \boldsymbol{\mu}_{0a}[\mid 0 >< L_a \mid + \mid L_a >< 0 \mid] + \boldsymbol{\mu}_{0b}[\mid 0 >< L_b \mid + \mid L_b < 0 \mid] \end{aligned} \tag{9.1}$$

where the first three terms correspond to the permanent dipole moments, and the next two to transitions between the ground and the excited states. The dipolar interaction is accounted for by the usual interaction energy between two dipoles $\boldsymbol{\mu}_1$ and $\boldsymbol{\mu}_1$ at positions $\boldsymbol{r}_1$ and $\boldsymbol{r}_2$ respectively:

$$V_{12} = \frac{1}{4\pi\varepsilon_0\varepsilon_r r_{12}^3}\left(\boldsymbol{\mu}_1 \cdot \boldsymbol{\mu}_2 - 3\frac{(\boldsymbol{\mu}_1 \cdot \boldsymbol{r}_{12})(\boldsymbol{r}_{12} \cdot \boldsymbol{\mu}_2)}{r_{12}^2}\right) \tag{9.2}$$

where $\boldsymbol{r}_{12} = \boldsymbol{r}_1 - \boldsymbol{r}_2$ and $r_{12} = \mid \boldsymbol{r}_1 - \boldsymbol{r}_2 \mid$, and ε_r is the relative dielectric constant of the protein environment (in these calculations we set $\varepsilon_r = 1.5$,

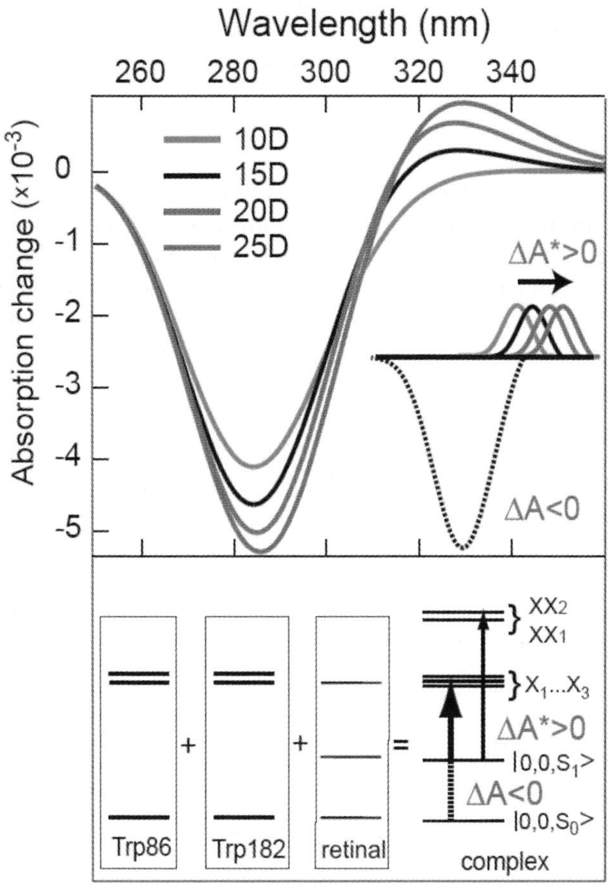

Fig. 9.7. Top panel: Calculated difference absorption spectra as a function of the retinal difference dipole moment $\Delta\mu$ and its decomposition (scheme on the right) in terms of a bleach contribution (ΔA$<$ 0) and a $\Delta\mu$-dependent absorption (ΔA*>0). Bottom panel: Dipole-dipole coupling of Trp86, Trp182 and retinal leads to the formation of an excitonic complex with proper electronic states [37]. The X_i's represent singly excited states of the excitonic complex, with one excitation per chromophore, while the XX_i's represent doubly excited states with retinal in the S_1 state. Photoexcitation of retinal attenuates the transition to the $X_1 \dots X_3$ states (ΔA<0), due to the bleach of the $\mid 0, 0, S_0$ (product basis notation $\mid$ Trp86, Trp182, retinal $>$) ground state of the excitonic complex. The photoinduced absorption from $\mid 0, 0, S_1 >$ reaches the XX_1 and XX_2 states (in which both Trp and retinal are excited), giving rise to an absorption (ΔA*>0). As the difference dipole moment $\Delta\mu$ increases, ΔA* red-shifts due to a lowering of the XX_1 and XX_2 levels (attractive interaction between retinal and Trp86). This leads to an increasing bleach around 280 nm, which is a measure of the retinal difference dipole moment $\Delta\mu$.

to account for the hydrophobic binding pocket). The full interaction operator corresponds to all possible combinations of dipoles between the three pigments, i.e. we evaluate the interaction energies [36] with the dipole operators $\hat{\mu}_{\mathbf{trp86}}$, $\hat{\mu}_{\mathbf{trp182}}$, $\hat{\mu}_{\mathbf{ret}}$ expressed in analogy to [42]. In the case of the three chromophores, the total dipolar interaction is taken as the sum of pairwise interactions. The first three terms in (1) will give rise to the Stark effect, while the last four give rise to the resonance interaction. Thus we can form the full interaction Hamiltonian H = H_0 + V of the three (27x27 matrix) or of two (9x9 matrix) interacting chromophores. H_0 is the Hamiltonian of each isolated chromophore and V represents the dipole-dipole interaction between the chromophores. Diagonalization of this Hamiltonian by standard methods gives a set of eigenvalues, from which new transition energies can be found, and eigenvectors, from which new transition dipole moments can be determined. As illustrated in Fig. 9.7 (bottom panel), this gives rise to a single ground state ($\mid 0, 0, S_0$) and a single excited state ($\mid 0, 0, S_1 >$) close to 570 nm (17 540 cm^{-1}). The exciton manifold consists of the five states $\mid L_{a,b}, 0, 0 >$, $\mid 0, L_{a,b}, 0 >$ and $\mid 0, 0, S_n >$ (in the product basis notation |Trp86,Trp182, retinal>), derived from the single excitations to the original levels absorbing at 280 nm (35 714 cm^{-1}). It is represented by the three dominant transitions X_1, X_2, and X_3, which are separated by a Davydov splitting of ~270 cm^{-1}. The double excitation of retinal and Trp's ($\mid L_{a,b}, 0, S_1 >$ and $\mid 0, L_{a,b}, S_1 >$) leads to an exciton manifold of four states around 52 000 cm^{-1} with two dominant ones, labeled XX_1 and XX_2.

Stick spectra are generated from the transition energies and intensities. The latter are determined by the respective oscillator strengths, which are proportional to the square of the transition dipole moments. Gaussian broadening of 3500 cm^{-1} (FWHM) is assumed, a spectral broadening commonly observed for Trp and retinal transitions in proteins. Thus, the UV ground state absorption spectrum is dominated by the transition from $\mid 0, 0, S_0 >$ to X_1, X_2 and X_3, while the excited state spectrum, i.e. with retinal in the S_1 state, is due to the transitions from $\mid 0, 0, S_1 >$ to XX_1 and XX_2 (Fig. 9.7, bottom). The difference absorption spectrum is obtained by subtracting the ground state spectrum from the excited state spectrum. For low $\Delta\mu$ ($\leq 10D$), the ground and excited state spectra almost compensate, leading to a small negative difference signal around 280 nm (Fig. 9.7, top).

The simulated absorption changes are plotted in Fig. 9.7 (top panel) for various values of the retinal difference dipole moment $\Delta\mu$. The photoinduced signal in the region of Trp absorption indeed appears as a bleach of the excitonic transitions into the X_i levels (the bleach character arises from the depletion of ground state retinal due to excitation with the pump pulse), while transitions from $\mid 0, 0, S_1 >$ to XX_i levels show up as a positive signal on the red side. As the retinal difference dipole moment $\Delta\mu$ acts only when the system is in the $\mid 0, 0, S_1 >$ state, the position of the induced absorption ΔA* red-shifts with increasing difference dipole moment, while the ground state bleach A is $\Delta\mu$-independent. The red shifting of ΔA* is due to an attractive interaction between the excited state dipole moments of Trp86 and retinal (insert Fig. 9.3), with the latter's varying in time. The signal is the sum of ΔA and ΔA*, and it results in an increasing bleach signal with increasing $\Delta\mu$ (Fig. 9.7), as observed. The bleach amplitude at 280 nm is therefore a measure for the retinal difference dipole moment $\Delta\mu$.

Based on the calculations that link the amplitude of the bleach signal to the retinal difference dipole moment, the additional rise time of the experimental bleach transients appears to reflect a progressive increase of $\Delta\mu$, on a time scale of 200 fs after excitation. The dipole moment change is due to the translocation of charge, for which we measure the time scale for the first time. Given that the events at $t < 200$ fs are also characterized by large amplitude torsional motion of the retinal skeleton, prior to isomerization [23,24], we conclude that the dipole moment increase is related to these structural changes, as also predicted by quantum chemistry calculations [27]. In an intuitive picture, the torsional motion breaks the conjugation at the C13-C14 bond and thus localizes the charges on the left and right of that bond, which leads to a dipole moment increase. A similar situation is encountered in donor-acceptor molecules like DMABN, where the build-up of a charge transfer state involves rotation of the dimethyl amine with respect to the phenol moiety [43,44]. In the protein environment, the gradual dipole moment change of retinal in the excited state then acts as driving force for the torsional motion, because the electrostatic interactions are attractive [45,46].

At later times (400-500 fs), the difference dipole moment decreases. This is the time scale of formation of the 13-cis isomer [7], pointing to a weaker dipolar interaction in the ground state 13-cis retinal than in excited all-trans retinal. Since the isomerization leads to an electronic ground state, with a mainly covalent character as in the all-trans adduct state [27], the dipole moment change is small, and so is the remaining bleach signal. The additional 3.5 ps decay component, which is similar to the vibrational relaxation in the 13-cis photoproduct found in experiments probing retinal [24–26], suggests a further decrease of the dipolar interaction. Vibrational relaxation leads to a weak displacement of the center of gravity of the charge, which may cause the decrease in dipole strength. Alternatively, energy dissipation may cause structural changes in the retinal binding pocket, which modify the field detected at the location of Trp86. Finally, the weak bleach component at longer times (τ_∞) points to long-lived changes in the structure of the pocket and/or the dielectric constant, which may relate to changes detected in the picosecond range in mid-IR experiments [7].

In summary, we have measured the light-induced electric field changes within a protein (bacteriorhodopsin), following them over the whole course of the dynamics from the Franck-Condon region to the vibrationally relaxed photoproduct. Our observations allow us to establish the connection between the translocation of charge [27] and the skeletal changes of the conjugate chain [23]. The present work also stresses the role of the dynamic electric force fields, which drive structural dynamics and govern enzymatic reactions.

9.3 Multipulse transient absorption spectroscopy: A tool to explore biological systems

M. Vengris, D. S. Larsen, E. Papagiannakis, J. T.M. Kennis, and R. van Grondelle

Light-driven events in biological systems are triggered by the absorption of a photon in pigment-protein complexes, which eventually produces a physiological response. Even though the light is absorbed by the pigments (chlorophylls, carotenoids, xantopsins, flavins, etc.), protein environment plays a key role in the absolute majority of these processes: the protein determines the course of the photoinduced reactions by 'guiding' the excited pigments along the reaction pathway, and, at some stage, taking over all the reaction functions, which then no longer rely on the physicochemical properties of the pigment. For the photoinduced reactions to be efficient, their rates must compete effectively with the lifetime of electronic excitation in the molecules (usually of the order of several ns for singlet excited states), which makes these reactions the subject of ultrafast laser spectroscopy. In photoreceptors the protein environment plays a key role in the absolute majority of light-related biological processes: the protein determines the course of the photoinduced reactions by 'guiding' the excited pigments along the reaction pathway, and, at some stage, takes over all the reaction functions, which then no longer rely on the physicochemical properties of the pigment.

Pump-probe (PP) spectroscopy (see Chaps. 2 and 8), is probably the most widely used ultrafast technique to investigate photoinduced reactions [47]. Pump-probe data are usually presented in the form of time-resolved difference spectra, i.e. $\Delta OD = \Delta OD(t, \lambda)$, where ΔOD is the difference between the sample absorbance with and without the pump pulse (Fig. 9.8). The absorption-difference (ΔOD) spectra measured in pump-probe experiments are shaped by a number of qualitatively different contributions that may originate from transitions between the different states of the system (Fig. 9.8), e.g. ground state bleach (GSB), stimulated emission (SE) and induced absorption (excited state absorption, ESA, or absorption by nonequilibrated ground state). Without additional experimental information, when no additional control is exerted on the sample, these contributions often become impossible to distinguish one from another and the data interpretation becomes ambiguous.

9.3.1 Multipulse transient absorption spectroscopy: The principles

The difficulties of traditional PP experiments can be addressed using multipulse transient absorption spectroscopies (MPTAS), such as the dispersed pump-dump-probe (PDP) and the pump-repump-probe (PrPP) techniques, which involve the introduction of a third pulse to the PP experiment (Fig. 9.9)

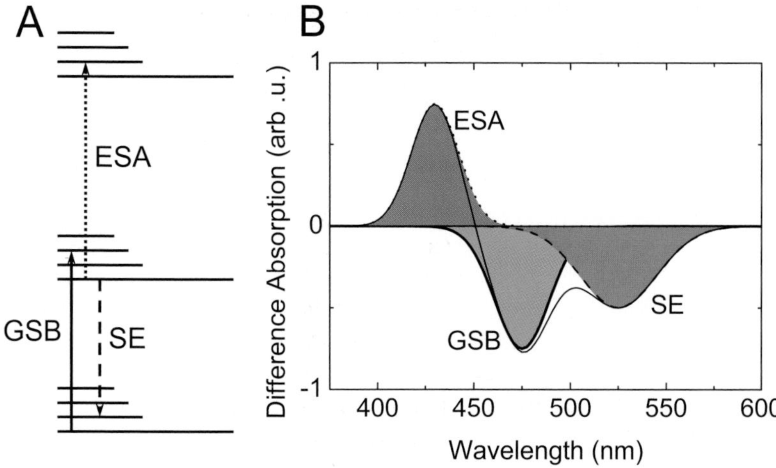

Fig. 9.8. The energy level scheme (A) of a hypothetical molecule and different transitions influencing the pump-probe spectrum (B): ground state bleach (GSB), stimulated emission (SE) and excited state absorption (ESA). By tuning the third pulse of a laser into the resonance with ESA and SE bands, a means of controlling the photoreaction can be established. In the case of the third pulse acting on ESA, pump-repump-probe technique (PrPP) is obtained, whereas if the third pulse is in resonance with SE, the technique is called pump-dump-probe (PDP).

[48,49]. This additional pulse can be appropriately wavelength-tuned and delayed to selectively interact with a 'targeted' electronic transition. In PDP experiments, this added pulse interacts with the SE of an excited state and results in the de-excitation of molecules and the transfer of population from the excited state to the ground state. In PrPP experiments, the additional pulse interacts with an ESA band and causes the redistribution of population within the excited state manifold whilst the total amount of excitations is preserved, i.e. the ground state population remains unaffected.

Ultrafast single-wavelength PDP experiments were previously used to study the properties of the excited states in bacteriorhodopsin [50, 51], and to explore ground state dynamics in calmodulin [52]; asymptotic-limited dispersed PDP has been used to study ground state liquid dynamics [53]. The higher excited-state dynamics of bacteriorhodopsin has been studied by single-wavelength PrPP [54] and the combined use of PDP and PrPP allowed the distinction of overlapping bands in the PP signals [50]. In the recent years, the MPTAS techniques have been further developed to include dispersed detection [48, 49], which allows the simultaneous probing of complete spectra with an unprecedented signal-to-noise ratio, that subsequently can be analyzed globally [55–57]. This development, which allowed to test explicit physical models for light-driven biological processes, moved ultrafast research on biological systems to a qualitatively new level.

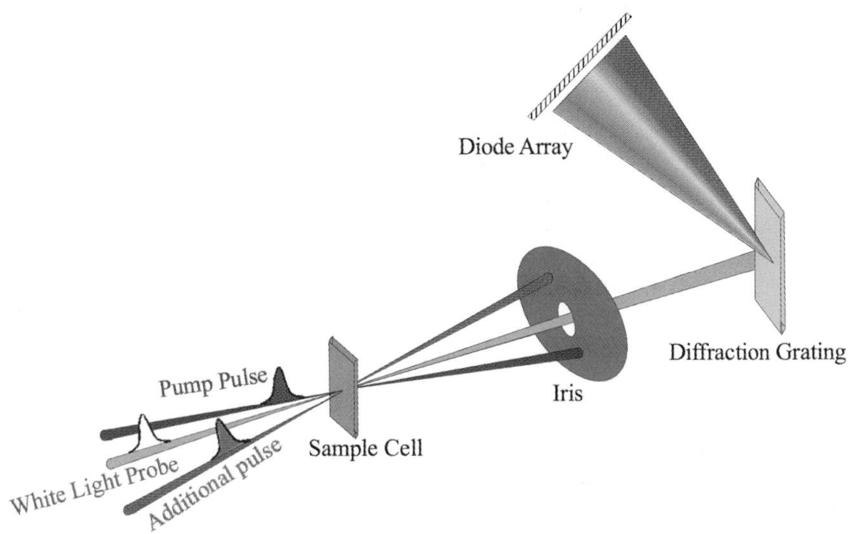

Fig. 9.9. Laser beam arrangement in the multipulse transient absorption experiment. The additional pulse can be tuned to be in resonance with excited state absorption or stimulated emission (see Fig. 9.8).

The utility of MPTAS spectroscopies lies in their ability to control reactions as they evolve, by manipulating the population of transient species with applied laser pulses. This is, to a certain extent, related to coherent control mechanisms where reactions are manipulated via complex processes such as vibrational wave packet motion, quantum interferences and electronic coherences [58]. A more appropriate term for describing MPTAS spectroscopies is incoherent control, because the control effects are achieved and the data is interpreted solely in terms of manipulated electronic state populations. This simplification, as will be shown in the following applications, allows for clear, concise and biologically relevant interpretations of the measured data.

The important aspects of biological pigment-protein systems, where MPTAS can resolve issues that remain obscure in conventional ultrafast spectroscopic experiments, can be summarized in the following list:

• MPTAS can help to attribute specific bands in the transient absorption spectrum to either ground or excited-state reaction intermediates. In a PDP experiment, excited state population is diminished via the dumping of SE by the second pulse. If a specific spectral band measured by the third pulse is diminished as well, one can safely conclude that this band originates from the excited state; if, however, it increases in magnitude, it must come from the ground state.

• MPTAS can be used to determine the pathway of the excited-state evolution of a light-driven biosystem by providing the ability to perturb it (both dumping and re-pumping the excited state). For example, if the excitation pulse initiates two parallel processes, only one of which leads to a specific photoproduct, the additional pulse can be used to selectively perturb one 'branch' of such a reaction and observe which spectroscopic intermediates originate from that particular pathway.

• If some intermediates of a photoreaction are very short-lived and do not accumulate significant population during the reaction course, the excited and ground state populations can be 'artificially tweaked' using pump and dump pulses and thereby used to selectively populate specific intermediates. This allows monitoring the desired photoreaction steps (and measuring the spectra of transient products) with great precision and selectivity.

• Finally, since MPTAS involves multiple interactions between the sample and the excitation pulses, it can be used as a tool to explore the interaction between excited states in multichromophore systems [59, 60].

In the following subsections we will discuss a variety of biological applications of MTAPS, where all of the mentioned advantages of these spectroscopies will be exploited, but first we will discuss several issues of experimental arrangement, data collection and analysis.

Experimental layout

The principal scheme of MPTAS experiment is shown in Fig. 9.9. Because two excitation pulses are used along with dispersed detection, MPTAS measurements generate multidimensional data (two time delays and one wavelength dimension), which can be collected and presented in two different ways (Fig. 9.10) [48,49]. In a kinetic trace measurement (Fig. 9.10B), the additional pulse is placed at the selected delay (called 'dump-delay' in the case of PDP measurement) after the pump pulse and remains fixed during the measurement. The dump-delay corresponds to a specific distribution of excited-state population and thus determines the effect that will be induced. The delay of the probe pulse is scanned as in PP measurements and records the effect of the dump pulse on the spectrum and dynamics of the PP signal. Alternatively, the dump-induced dynamics can be explored by measuring an action trace (Fig. 9.10C) [48,49], in which the delay of the additional pulse is scanned whilst the probe pulse is maintained at a selected delay. An action trace monitors how the PP spectrum at the chosen probe delay changes as a consequence of the interaction of the additional pulse with the system at variable instances. An additional advantage of an action trace is that it is not affected by time-zero artifacts (e.g. cross-phase modulation, stimulated Raman scattering, 2-photon absorption etc.) because the probed signal is at a much longer delay, allowing clear observation of the dump-induced dynamic changes at early times.

Data presentation and analysis

Here we will briefly discuss the data of PDP experiments only; the same discussion is valid for PrPP experiments, because the two experiments do not

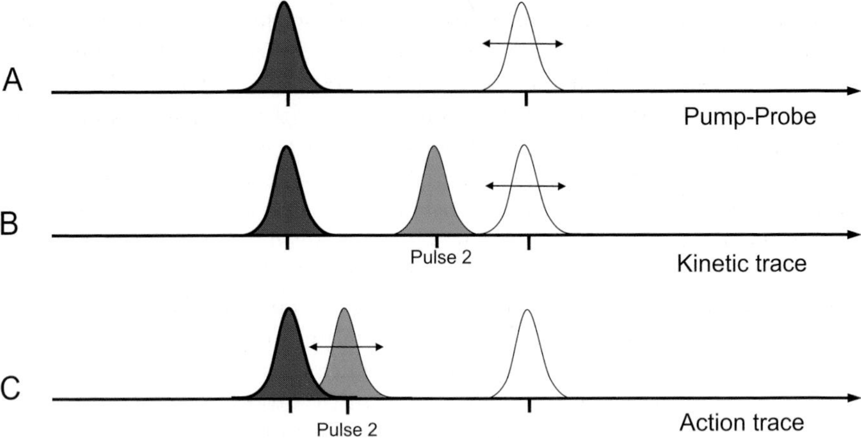

Fig. 9.10. The timing schemes in MPTAS experiments. The pump pulse is represented in blue, the probe pulse is shown in white, and the additional pulse in green. In the pump probe experiment (A) and MPTAS kinetic trace (B), the delay of the probe is varied whereas the actinic pulses are kept at fixed delays. In the action trace (C), pump and probe pulses are fixed at certain time delays and the delay of the additional pulse is varied.

differ in the way the data is collected, but in the nature of interaction between the additional pulse and the sample. Along with the ΔOD signals measured without (PP) and with (PDP) the dump pulse, the double-difference signals are constructed

$$\Delta\Delta OD(\lambda, t, \tau) = PDP(\lambda, t, \tau) - PP(\lambda, t)$$

where λ and t represent the probe wavelength, and delay and τ is the delay of the dump pulse. The $\Delta\Delta OD$ signal is often easier to interpret as it has non-zero amplitude only when there is a dump-induced effect on the PP measurement. However, the time dependence of $\Delta\Delta OD$ signals inherently contains the dynamics of the underlying PP signal; to examine the occurrence of additional, dump-induced dynamics, we can construct the relative double-difference signal, representing a fractional change of PP signal due to additional pulse:

$$\Delta\Delta OD_{rel}(\lambda, t, \tau) = \Delta\Delta OD(\lambda, t, \tau)/PP(\lambda, t)$$

In order to take full advantage of the time- and wavelength resolved experimental data and quantify the observed dynamics, global and target analysis methodologies are instrumental [55–57]. They allow the concise parameterization of results, as well as testing a specific physical model to describe the observed dynamics. For the preliminary parameterization description of the PP signals alone, a sequential scheme is used, where the components interconvert unidirectionally with increasing exponential lifetimes; it provides estimates of the characteristic timescales for spectral evolution and the corresponding evolution-associated difference spectra (EADS) [56,57]. The EADS estimated from such a scheme do not necessarily reflect pure transient states occurring in the experiment but rather provide a phenomenological description of the observed spectral evolution. To estimate the species-associated difference spectra (SADS) that portray the real states of the system, the PDP data is subsequently analyzed simultaneously with the PP data by applying a specific physical model in a so-called 'target analysis' [56,57]. In this description of the data, a connectivity scheme (physical model) is chosen with each of its compartments depicting a true physical state of the system. The model applied in a target analysis is designed to make physical sense, to produce plausible spectra and (of course) to represent the experimental data well.

In the following we will discuss three applications of MTAPS to biological systems, including proton transfer in the green fluorescent protein, the excited state decay in photosynthetic carotenoids and the initial events in photoactive yellow protein, leading to formation of the signaling state. These cases are selected to illustrate different aspects of how MPTAS can be used to get new insights in the photoinduced dynamics of the molecules.

9.3.2 Uncovering the hidden ground state of green fluorescent protein

In the last decade, green fluorescent protein (GFP) has become the fluorescent label of choice in cell biology and biophysics [61]. GFP absorbs primarily in the near-UV (the main absorption band attributed to a so-called state A of GFP peaks at around 400 nm), but fluoresces in the green (the fluorescence maximum lies near 510 nm). It is generally accepted that the large shift of the emission to longer wavelengths after near-UV absorption is caused by a deprotonation reaction of the neutral chromophore in the excited state [62]. Upon excitation of the state A band with near-UV light, the chromophore deprotonates on a picosecond timescale and subsequent proton transport occurs, most likely via a hydrogen-bonded network that involves an internal water molecule (W25) a serine residue (S205) and a terminal proton acceptor, an ionized glutamate (E222) [63]. The thus formed intermediate state, with a deprotonated, anionic chromophore and a protonated, neutral glutamate is generally referred to as I^* and is the emitting state of GFP, responsible for the green fluorescence around 510 nm. An anionic ground state species, which has been labeled I, can be stabilized and photoconverted back to A at cryogenic temperatures [64]. It has been postulated that the molecular ground state of the chromophore in this anionic form acts as an intermediate in the GFP photocycle [62, 64–67]. However, with time-resolved spectroscopy only the excited states A^* and I^* of GFP could be detected [68], and the molecular nature and dynamics of any ground state intermediate remained elusive.

MPTAS in combination with global analysis was successfully used as a tool to directly measure the ground-state proton transfer dynamics in GFP [69]. By manipulating the excited and ground state populations using an additional femtosecond laser pulse that dumped the excited state I^*, the so far hidden anionic ground state of GFP was shown to be an integral part of the photocycle.

Pump-probe spectra recorded on GFP with excitation at 400 nm are shown in Fig. 9.11. The first spectrum (black line), taken at a delay of 400 fs, exhibits a ground-state bleaching near 400 nm, an ESA band at 440 nm and a broad SE band ranging from 450 to 550 nm. This stimulated emission can primarily be assigned to the excited state of the neutral chromophore, A^*. At progressively increasing time delays, a pronounced SE band with maximum intensity around 509 nm and a shoulder at 545 nm develops. The intensity of the SE signal near 509 nm is a direct measure of the population of I^*, and these results demonstrate the formation of I^* on a picosecond timescale, which corresponds to the deprotonation reaction of the phenolic oxygen of the chromophore [62]. Concomitant with the rise of SE, the excited-state absorption around 440 nm increases, indicating that like A^*, I^* has a pronounced ESA in this spectral region. The transient absorption signal then slowly decays to zero amplitude on a nanosecond timescale (spectra not shown). These dynamics are also reflected in Fig. 9.12A, where the kinetic trace probed at 509 nm is shown (blue squares). The trace shows little signal at early times after excitation. On a timescale of several picoseconds the SE signal rises to a maximum (negative) amplitude. The SE signal subsequently decays to zero in

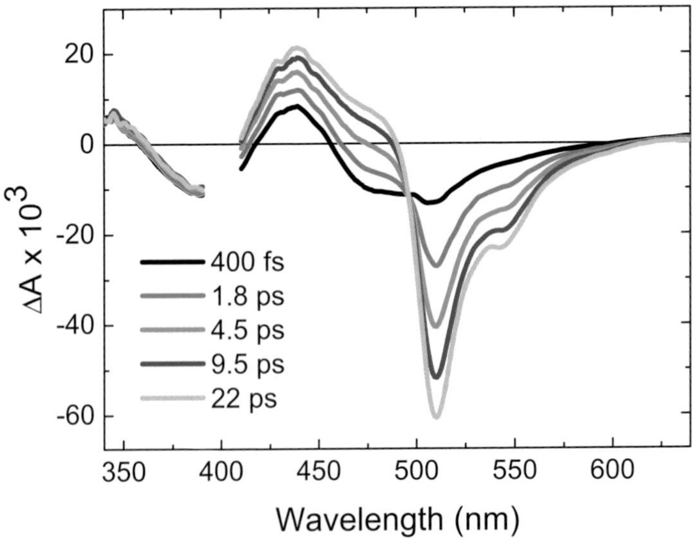

Fig. 9.11. Time-resolved absorbance difference spectra recorded in GFP upon excitation at 400 nm at the time delays indicated.

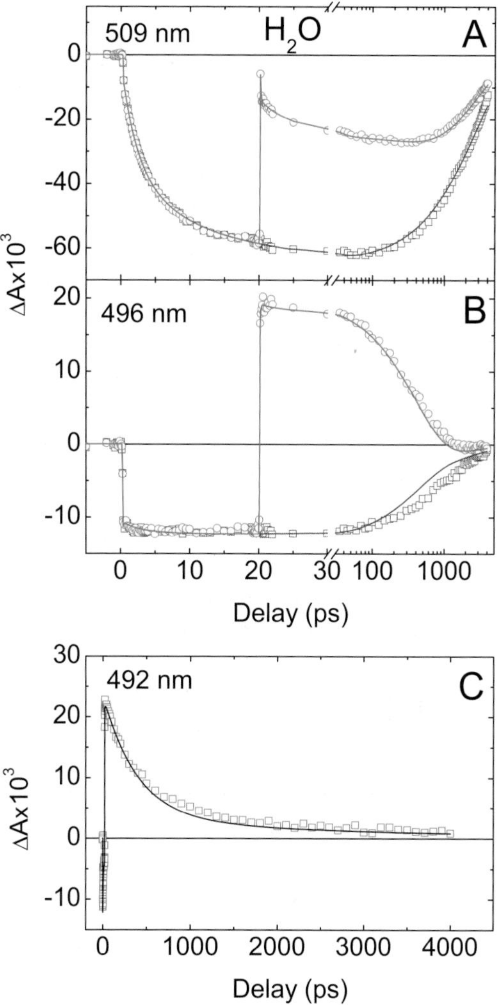

Fig. 9.12. Transient absorption trace at 509 nm (A), 496 nm (B) and 492 nm (C) with excitation of GFP at 400 nm in the absence (blue symbols) and presence (red symbols) of a dump pulse at 540 nm at 20 ps delay. The black thin lines show the result of the global fitting. The time axis is linear up to 30 ps, and logarithmic thereafter in (A) and (B), and linear throughout (C).

about 3 ns, which corresponds mainly to radiative decay of I^* to the ground state [67, 70].

Application of MPTAS to GFP

Pump-dump-probe experiments were performed with the aim to transfer population on a femtosecond timescale from the excited state to the ground state, thereby overruling the 'slow' 3 ns radiative decay [69]. The red circles in Fig. 9.12A show the kinetic trace at 509 nm in the presence of a short 540 nm dump pulse, which overlaps with a shoulder of GFP's fluorescence spectrum and is fired 20 ps after the 400 nm excitation flash. Up to 20 ps, the trace overlaps with an unmodified (i.e. without dump-pulse) transient absorption trace. At 20 ps, a sudden decrease of the SE signal indicates that the excited-state population has decreased as a result of the action of the dump pulse. Figure 9.12B contains the corresponding kinetic traces measured at 496 nm, showing the appearance of a large induced absorption at 20 ps. This indicates that as a result of the action of the dump pulse, a product species that absorbs at 496 nm is formed concomitantly with the depletion of the excited state. This species corresponds to the ground state of the deprotonated chromophore, I, and therefore this experiment clearly demonstrates, for the first time, a ground state intermediate in the photocycle of GFP. Figure 9.12C shows the kinetic trace at 492 nm, where I^* shows an isosbestic point. At this wavelength, the signals exclusively represent the dynamic behavior of the ground state species I (red squares). We observe that after its formation by the dump pulse, the I state disappears rapidly; the smooth curve represents a fit of the data with a single exponential decay of 400 ps. Without the dump pulse the I state remains invisible because it decays about ten times faster than it is formed. To illustrate the spectral evolution induced by the dump pulse, time-resolved spectra at selected delays are shown in Fig. 9.13, measured in the presence (dashed lines) and absence (solid lines) of the dump pulse at 20 ps. As can be seen from the difference spectrum taken at a delay of 22 ps (i.e., 2 ps after the dump pulse, Figure 9.12A), the stimulated emission band of I^* from 509 to 600 nm has been significantly depleted. Moreover, a pronounced new absorption band is observed near 500 nm in the PDP spectrum, along with a 15 nm red-shift of the isosbestic point (at 492 nm in the PP spectrum), indicating the formation of the ground state species upon dumping. Thus, the dumped spectrum is a superposition of the excited state species I^* and the newly formed ground state species I. To assess the spectral properties of the ground state species I, the difference spectrum of I^* was scaled by a factor of 0.57 (to reflect that an estimated fraction of population 0.43 was dumped to the ground state) and subtracted from the dumped spectrum. The resulting double-difference spectrum (dotted line) corresponds to the 'pure' signal of I. At subsequent delays of the probe pulse (450 ps and 900 ps, Fig. 9.13 B and C), the absorption feature near 500 nm decreases significantly and the 'dumped' spectrum starts to resemble the undumped spectrum, indicating that the newly formed ground state species indeed has a significantly shorter lifetime than I^*.

MPTAS data measured in both water and D_2O buffer solutions were globally analyzed. The analysis revealed a pronounced effect of deuteration on the observed dynamics, but virtually no influence on the spectra, as was expected for a proton (or deuterium) transfer reaction [69]. The model used to interpret and quantify the data is shown in Fig. 9.14 (upper panel) along with the tran-

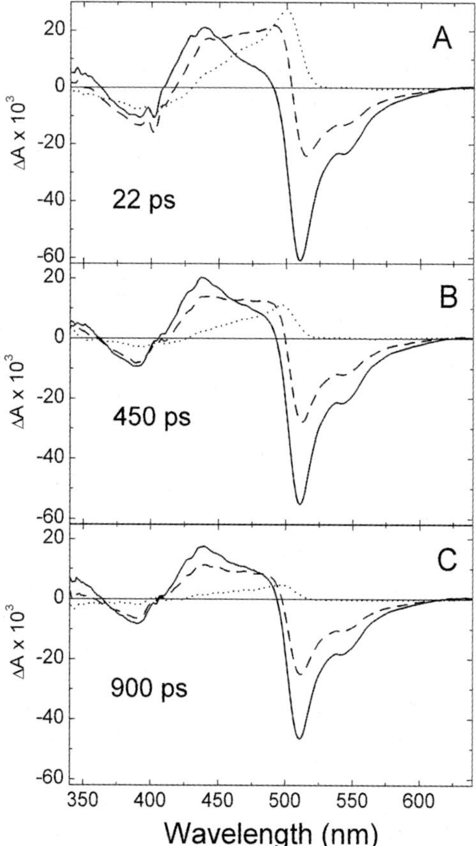

Fig. 9.13. Time-resolved absorbance difference spectra recorded for GFP at the delays indicated after the excitation at 400 nm in the absence (solid line) and presence (dashed line) of a 'dump' pulse at 540 nm, applied 20 ps after the excitation. The dotted lines denote a double-difference spectrum, in which a fraction of 0.57 of the undumped spectrum is subtracted from the dumped spectrum, and denote the 'pure' difference spectra corresponding to the anionic ground state *I*.

sient spectra of the GFP photocycle intermediates (Fig. 9.14, lower panel). Both proton transfer events in the excited and the ground state proceed in a double-exponential fashion, with the resulting species-associated difference spectra virtually identical for H_2O and D_2O solutions. The substantial difference is in the kinetics: the excited- and ground-state proton transfer is 2 and 12 times slower, respectively, when the proton is substituted to deuterium. The estimated proton transfer time in the ground state is 440 ps. The importance of this observation is in the fact that the proton back transfer in GFP occurs in the molecular ground state, as do the vast majority of proton transfers in biology. Thus using MPTAS the 'true', intrinsic rate constant of

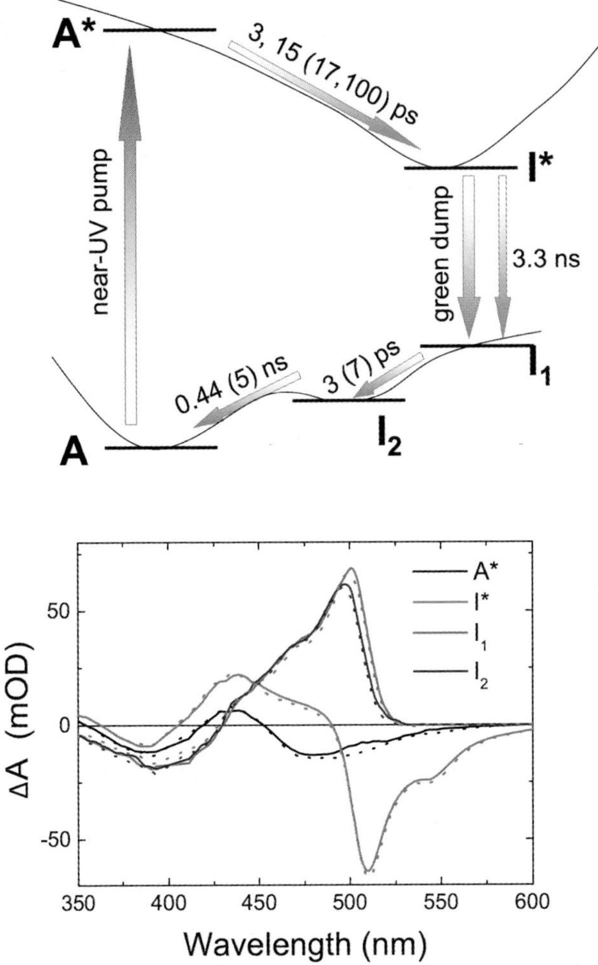

Fig. 9.14. Upper panel: kinetic scheme and potential energy level surfaces representing the photocycle of GFP, to fit the multipulse and traditional transient absorption data of GFP in H_2O and D_2O. The time constants by which the states evolve into one another as estimated from the target analysis are indicated at the arrows, with the values in parentheses representing those obtained for D_2O. Lower panel: Species-associated difference spectra (SADS, [19, 20]) of A^*, I^*, I_1, and I_2 species relative to the A state that result from a target analysis of the time-resolved spectra recorded for GFP in H_2O (solid lines) and D_2O (dotted lines).

the ground state proton transfer in a protein was determined by moving the proton with the first femtosecond pulse to a terminal acceptor, and then 'instantaneously' switching the pK of the donor back to its original value (by applying the dump pulse and returning the donor to the ground state) and watching the proton migrate back. It turns out that the characteristic time of such migration is less than a nanosecond. When using other (chemical) methods than MPTAS, this measured rate is prone to be estimated slower because some steps in the experimental sequence will probably be diffusion-limited and much slower than the intrinsic rate.

9.3.3 Untangling the excited state dynamics of carotenoids

Carotenoids are vital for photosynthetic organisms for light-harvesting and photoprotection, functions that are dictated by their conjugated π-electron backbone. Nevertheless, the molecular mechanisms behind these functions are not fully understood and in fact a concise description of their properties is lacking. Carotenoids typically have a π-conjugation length between 7 and 13 (Fig. 9.15) and their spectroscopic properties are interpreted on the basis of what is known for linear polyenes. Traditionally, their electronic structure has been described as a manifold with two low-lying electronic states: S_2 and S_1 (the $1B_u^+$ and $2A_g^-$ polyene states) [71]. The S_1 state has the same electronic symmetry as the ground electronic state S_0 ($1A_g^-$) and therefore is dark, i.e. it is not observed in the linear ground-state absorption spectrum. The strong absorption of blue-green light corresponds to the transition from S_0 to S_2. The S_2 state is easily observed in fluorescence-upconversion and pump-probe (PP) measurements as it emits in the visible region and absorbs in the near-infrared. S_2 decays by ultrafast ($\sim$200 fs) internal conversion (IC) to the S_1 state, which internally converts to the ground state, with a lifetime that depends on the conjugation length ($\sim$1 to $\sim$100 ps). The S_1 state is easily observed and characterized in PP experiments because of its distinctly strong ESA in the visible [71].

The improvement of ultrafast spectroscopy led to the identification of several features that can not be described in the frame of this simple model and the assignment of newly observed spectral and temporal features in a variety carotenoid-containing systems has been hotly debated. These features include the vibrational cooling of the S_1 and S_0 states, the interference of additional states, such as the $1B_u^-$ and the $3A_g^+$, with the S_2-S_1 relaxation, the branching of S_2 to populate a state S*, which most likely is the precursor to the triplet generation by singlet fission, etc. [71–75]. The electronic character and the spectroscopic properties of these new carotenoid excited states have still to be characterized and integrated into a comprehensive model linking carotenoid structure with function. Clearly, identifying the underlying connectivity of temporally overlapping states is crucial for interpreting their origin and understanding how and why Nature selected and shaped these molecules into

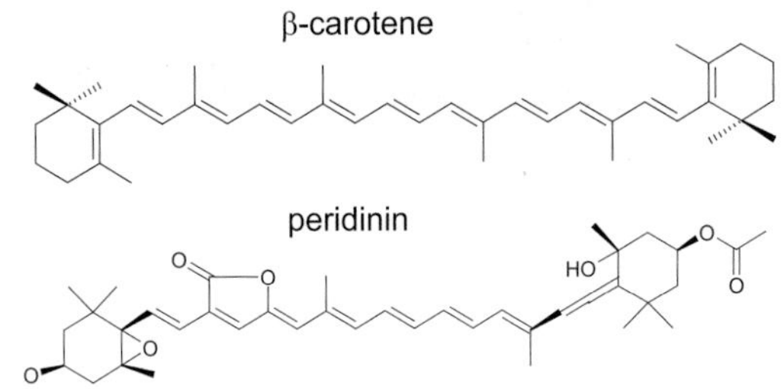

Fig. 9.15. The structural formulae of carotenoids discussed here: β-carotene (top) and peridinin (bottom).

the multifunctional biological pigment as we know them; MPTAS techniques provide the experimental tools to further this understanding.

9.3.3.1 An additional excited state in β-carotene

A comparison of the transient absorption spectra of β-carotene measured 3 ps after excitation at 400 and 500 nm, shows that the excitation at the higher energy side of the S_2 state (400 nm) generates a spectrum with a pronounced shoulder on the blue side (500-525 nm) of the strong ESA of S_1 which peaks at $\sim$550 nm (green line in Fig. 9.16). This shoulder is absent when 500 nm light is used for the excitation (black line in Fig. 9.16) To address this feature, a series of multipulse experiments were performed [49].

The first was a PrPP experiment; an 800-nm pulse was used to repump population from the S_1 state to a higher excited state, 1 ps after 500 and 400-nm excitation. The effect of the additional 800 nm pulse on the transient absorption spectra is shown in Fig. 9.16 (red and blue lines). When comparing $\Delta\Delta OD$ spectra (red and blue lines) to their respective PP spectra (green and black lines), the effect is markedly different for the two pump wavelengths that were applied. The $\Delta\Delta OD$ spectrum measured with 500-nm excitation (red line) is similar to the respective PP signal in the S_1 region (black line), whereas after 400-nm excitation the $\Delta\Delta OD$ (blue line) lacks the blue shoulder observed in the PP spectrum (green line). If the entire ESA band from 500 nm to 600 nm were due to S_1 state, the shapes of $\Delta\Delta OD$ and PP spectra should be identical. This different response to the 800-nm pulse indicates that the ESA shoulder originates from a species other than the S_1 state. This species is denoted $S^{\ddagger}$ [49]. The second MPTAS experiment performed was the addition of a 530-nm pulse to the PP measuring scheme [49]. This pulse is resonant with both the stimulated emission of S_2 and the absorption of S_1; its effect on the observed PP kinetics depends on the exact timing (i.e. on the state of the molecule that it finds it in). Figure 9.17 illustrates the effect of the

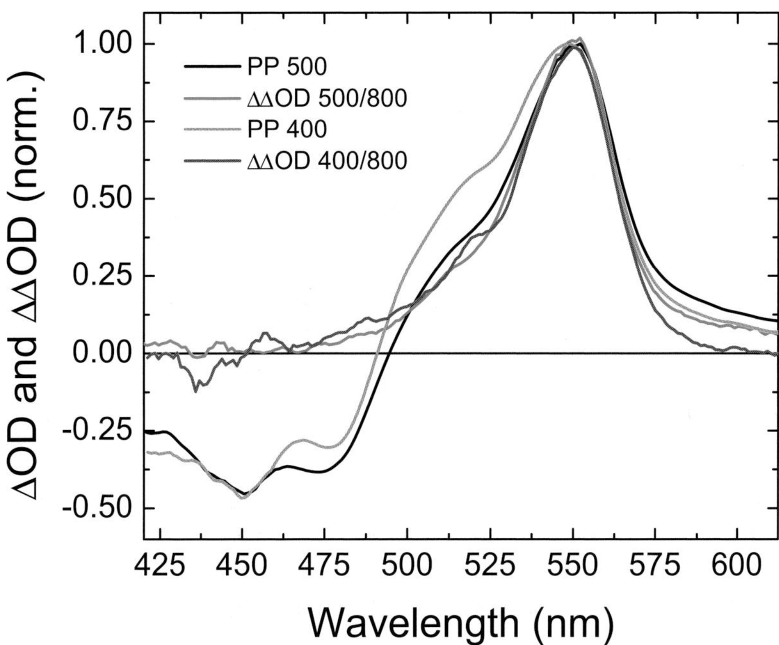

Fig. 9.16. Pump-probe spectra of β-carotene measured 3 ps after excitation at 400 nm (green line) and 500 nm (black thick line). The spectra have been normalized on the S_1 ESA maximum (550 nm). Blue and red lines show the double difference spectra (the difference between pump probe spectra in the presence and absence of the 800 nm repump pulse). The red line to a pump wavelength of 500 nm, the blue line - to a pump wavelength of 400 nm. The double difference spectra are also normalized at 550 nm.

530-nm pulse on the population of S_1: the S_1 signal is largely depleted upon interaction with the 530 nm pulse at 3 ps; the ensuing dynamics illustrates that population returns back to S_1 only partly. The complex dynamics is observed at 580 nm, where the ESA is due to vibrationally hot S_1 [72]. From the observed kinetics, it can be inferred that the back relaxation from the higher states involves the transient population of higher vibrational levels within S_1, before relaxing to the cool S_1 within 500 fs.

An action trace measurement was performed to investigate the changes induced on the 3-ps PP spectrum as the delay of the 530-nm pulse is varied (see Fig. 9.10C). Figure 9.18 shows the action traces measured at two wavelengths: 450 nm (GSB) and at 550 nm, (S_1 ESA peak). The effects differ significantly and two timescales can be separated. At 450 nm and early times (when the additional pulse is in resonance with the SE of S_2), the $\Delta\Delta OD$ signal exhibits a loss of GSB which results from the dumping of excited state population from the S_2 to the ground state. At 550 nm and early delays of the additional pulse, a similar effect is observed, because the S_1 population is also reduced

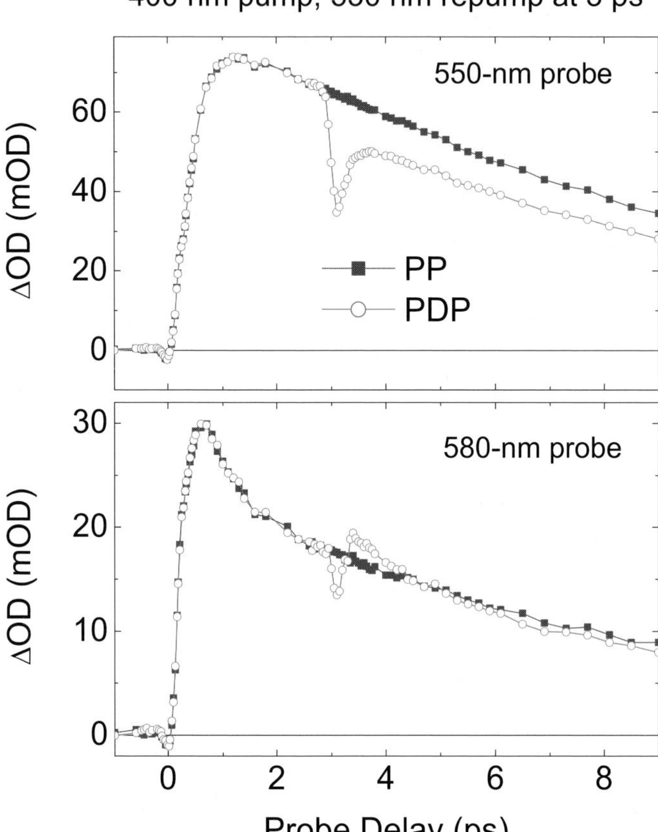

Fig. 9.17. The effect of a 530 nm pulse on the pump-probe kinetics of β-carotene. Excitation wavelength was 400 nm, the traces shown were probed at 550 nm (top) and 530 nm (bottom). Blue squares show unperturbed pump signals, red open circles depict the difference absorption signal in the presence of additional 530 nm pulse.

as a result of the S_2 dump. An additional, longer, timescale is also present in the 550-nm trace, describing the removal of population from the S_1 state by repumping its ESA at 530 nm.

Using global analysis, timescales and spectra that correspond to these two processes were estimated [49] (see Fig. 9.18 bottom panel). A fast, 300 fs, component which corresponds to the dumping of S_2 describes the full loss of the ESA band (also shown for comparison), i.e. due to both the S_1 and the $S^{\ddagger}$ bands, whereas the slow, 10 ps, component, describes the preferential loss of the S_1 ESA. The fact that $S^{\ddagger}$ is depleted after dumping S_2 shows it corresponds to an excited state which is formed via S_2, and which, as both 800 nm and 530 nm repump experiments illustrate, is separate from S_1.

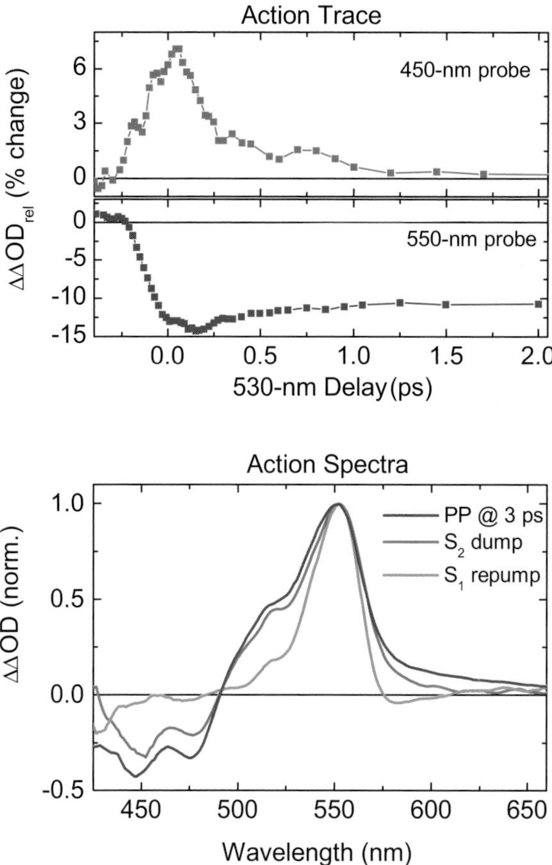

Fig. 9.18. Action traces (top) and spectra (bottom) measured for β-carotene with a 400 nm pump and 530 nm dump/repump pulse. The probe delay was 3 ps (see Figure 3C). In the top panel, the relative $\Delta\Delta OD$ signal is shown as a percentage of PP signal at two different probe wavelengths (indicated on the figure). The bottom panel contrasts the pump-probe spectrum measured at 3 ps probe delay (blue line), action spectrum of S_2 (solid line) and action spectrum of S_1 (green line).

The $S^{\ddagger}$ is not dissimilar from the S^{*} state, which was observed in several other carotenoids; however, unlike $S\ddagger$, high-energy excitation is not required to create S^{*} and moreover its ESA is not as blue-shifted and broad as that of $S^{\ddagger}$ [73, 76]. The nature and relationship of these states remains elusive. Calculations on polyenes [77] predict two electronic states, apart from S_1, under S_2, the $1B_u^-$ and the $3A_g^-$ states. However, in order to assign the observed $S\ddagger$ (or the S^{*}) to either of them, determination of their 0-0 energy is required. It is also unclear to what degree the spatial configuration of carotenoids influences their excited-state manifold; it has been proposed that the relaxation in the excited state manifold of β-carotene includes minor structural changes [72] and that in light-harvesting complexes, the protein-imposed configuration controls an additional deactivation channel [78]. Thus, it was concluded that these new states result from overcoming configurational barriers in the S_2 state when excited with additional excitation energy [49].

9.3.3.2 Excited state equilibrium between charge transfer state and S_1 in peridinin

Responding to the presence of only blue/green light underwater, photosynthetic marine organisms often rely on carotenoids rather than chlorophylls for light harvesting; the highest carotenoid/chlorophyll ratio (4:1) is found in the water-soluble peridinin chlorophyll protein (PCP) of the dinoflagellate *Amphidinium carterae* [79].

Peridinin is special among carotenoids because its carbonyl group induces a large dependence of its excited state manifold and dynamics on the solvent polarity and, in the case of protic solvents, also on the excitation wavelength [74, 79]. In polar solvents, a distinct SE band is observed in the near-IR, which is absent in nonpolar media [74]. This band has been attributed to an intramolecular charge transfer (ICT) state of peridinin. The exact relationship of the ICT state with the S_1 state, the only state observed in nonpolar solvents, has been debated, both in experimental [79,80] and theoretical [80,81] works, and is not elucidated yet: are they separate states or are they the same entity? This complex excited state structure makes peridinin a challenge that MPTAS techniques can help to address.

To investigate the complex excited state dynamics, a PDP experiment on peridinin in methanol was performed; the molecule was excited at 530 nm and an 800-nm pulse was then used to dump the near infrared SE of the ICT state [82]. Figure 9.19A shows the normalized PP spectra measured 1 ps and 10 ps after excitation. The region below 550 nm, which has been associated with the ESA of the S_1 state [79,80], is enhanced in the 10-ps spectrum. This implies that S_1 is populated on the timescale of a few picoseconds after the ICT state. Figure 9.19B contains the $\Delta\Delta OD$ traces measured in the PDP experiment with the dump pulse placed 3 ps after the pump pulse. Before the interaction of the dump pulse with the sample (probe delay less than 3 ps), the signal is zero - the dump pulse has not arrived yet and therefore the PP and PDP data are identical. The induced change appears upon the arrival of dump pulse and has a temporal profile that depends on the probe wavelength: the peak of the ICT ESA at 590 nm is lost instantaneously, whereas the loss at 535 nm (S_1 ESA) and 435 nm (GSB) shows slower, and more complex, dynamics. Figure 9.19C contrasts the $\Delta\Delta OD$ spectra to the PP spectra at two probe delays after the 3-ps dump. At a probe delay of 3.5 ps, the $\Delta\Delta OD$ spectrum does not have any amplitude in the S_1 region and indicates that the

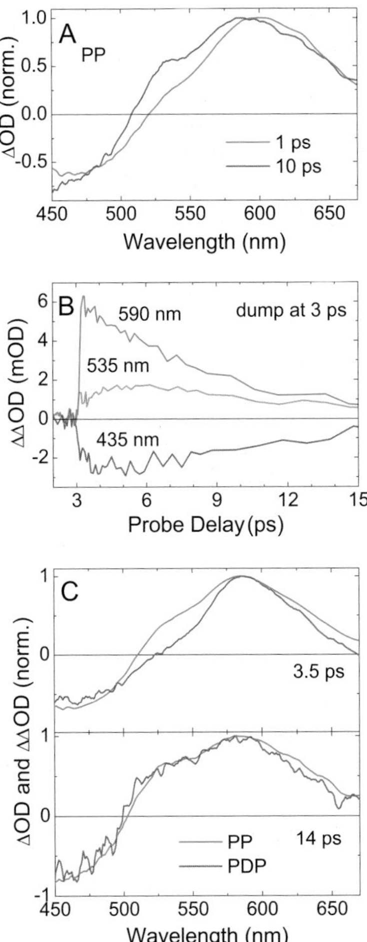

Fig. 9.19. Transient absorption measurements for peridinin in methanol after 530-nm excitation. *A*. pump-probe spectra measured at different probe delays; *B*. Pump-dump-probe $\Delta\Delta$OD kinetics with the 800-nm dump pulse placed at 3 ps; *C*. The $\Delta\Delta$OD spectra (blue lines) measured at probe delays of 3.5 ps and 14 ps contrasted to the respective pump-probe spectra (red lines).

dump pulse selectively depletes the ESA of the ICT-state; in contrast, the loss found at a probe delay of 14 ps is uniform across the complete ESA as the $\Delta\Delta OD$ spectrum overlaps with the PP spectrum. This leads to a conclusion that the S_1 state is separated from the ICT state: the dump does not affect it in the same instantaneous fashion. However, the fact, that after a certain time delay S_1 does exhibit a response, indicates that the S_1 and the ICT state are in equilibrium with each other.

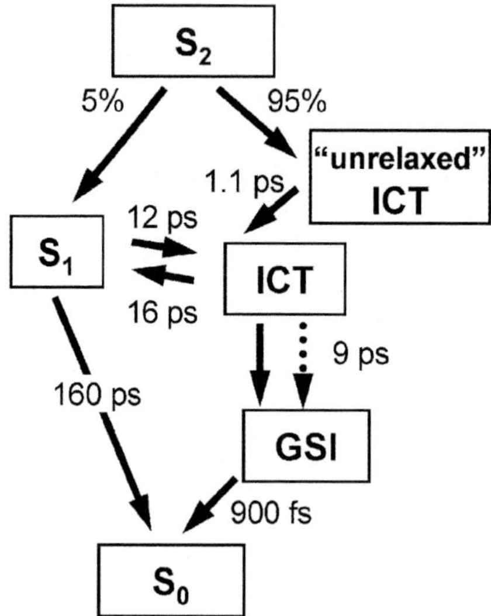

Fig. 9.20. Kinetic scheme used to model the excited states of peridinin [82]. GSI denotes the unequilibrated ground state that is produced by the relaxation (or dumping) of the ICT state.

The connectivity scheme shown in Fig. 9.20 was proposed, that allowed to explain the observed PP and PDP data on peridinin [82]. Upon excitation, population is transferred to the S_2 state, which then relaxes through two separate (branched) pathways, populating either the S_1 state or the ICT state. This relaxation involves an unrelaxed, 'hot' intermediate similar to other carotenoids [71, 72]. Both the natural and the dumped evolution of the ICT state populates an unequilibrated ground state, which in turn decays into the equilibrium ground state. The single trace that shows this most clearly is the 435 nm trace in Fig. 9.19B (blue): upon dumping, the bleach does not recover instantaneously, but ground state equilibration dynamics takes place. In addition to their own decay pathways, the populations of the S_1 and ICT states are in equilibrium with each other - another result that was deduced from MPTAS experiments.

9.3.4 Resolving the primary steps of photoactive yellow protein photocycle

The Photoactive Yellow Protein (PYP) is a small, 125 amino acid containing, water-soluble protein found in the bacterium *Halorhodospira halophila* and is responsible for triggering the negative phototaxis response of the organism to blue light [83–85]. PYP contains an intrinsic chromophore, para-coumaric acid, that is covalently bound to the protein backbone via a cysteine residue [86–89]. Upon photoexcitation, this chromophore undergoes a trans-cis isomerization around its double bond [89–91] and initiates a complicated series of reversible reactions that extends over 15 decades in time extending from femtoseconds to seconds [92–95]. These isomerization-induced processes include chromophore protonation [92, 96], protein unfolding [93], hydrogen-bond disruption [97, 98], and their corresponding recovery reactions. Hence, PYP is an excellent system for studying the complex relationship between protein dynamics and chemical reaction dynamics and more specifically how Nature has tuned both to produce biological function. The properties and the photocycle of PYP have been extensively studied (see [99–101] for review), here we will mostly concentrate on the application of MPTAS to investigate the primary events occurring in PYP after the excitation.

Ultrafast PP transient spectra of wt-PYP, excited with ultrafast 395-nm excitation pulses, are shown in Fig. 9.21, spanning probing times from 180 fs to 4 ns [48]. The collected spectra and dynamics of the PYP system are similar to the data measured in other dispersed PP measurements on PYP [94,102–104]. All transient spectra show GSB, peaking at the maximum of the absorption spectrum (446 nm). The 180 fs and 2 ps spectrum also exhibit a broad negative band at 500 nm and a pronounced positive band at 370 nm, which are ascribed to the SE and the ESA band, respectively. In contrast, the 35 ps, 500 ps and 4 ns spectra no longer exhibit clear SE bands, but instead show positive product state absorption bands peaking at 500 nm and 480 nm, respectively, which are ascribed to the product state absorptions of the initial red-shifted intermediates in the PYP photocycle: I_0 and pR [99, 100]. The 35 ps, 500 ps and 4 ns PP spectra also exhibit a noticeable positive sharply peaked band at 360 nm that has not been previously ascribed to either the I_0 or the pR product state band. Since the ultrafast fluorescence signals on PYP show that no appreciable excited state population remains at this time [105–107], this sharply featured band is not ascribed to the similar UV ESA band observed in the 180 fs and 2 ps spectra. Furthermore, a weak broad absorption is observed in the 4 ns spectrum, which extends beyond 550 nm. Excitation power-dependence of these bands along with the MPTAS (PDP) measurements allowed to attribute these bands to a photoionization of the PYP chromophore, resulting in hydrated electron responsible for the broad induced absorption band at wavelengths longer than 550 nm, and the radical of the chromophore, featuring the sharp induced absorption band around 360 nm [48].

From the spectra shown in Fig. 9.21, it is obvious that the pump-probe spectrum is composed out of largely overlapping bands (both spectrally and temporally) due to different initial intermediates in the PYP photocycle. Getting a picture of how precisely the initial steps of PYP photocycle are organized is impossible using pump-probe and fluorescence upconversion data alone. For example, the near complete overlap of the SE band with the absorption of the first photoproduct, I_0, precludes directly correlating the quenching timescales of the excited state with the timescale of photocycle initiation [94, 104, 108]. Because of this ambiguity, different kinetic schemes

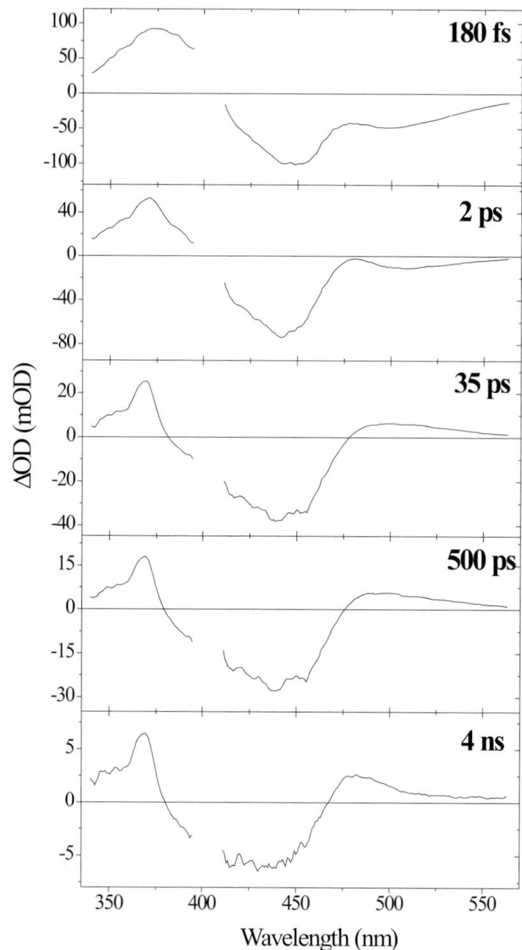

Fig. 9.21. Representative dispersion corrected transient PP spectra. The region around the pump excitation wavelength (395 nm) is corrupted due to scatter from the pump pulse.

have been proposed for modeling the initial dynamics of the PYP photocycle [102,104]. MPTAS was applied to get a better understanding of the initial dynamics of PYP [48].

Representative kinetic traces from two PDP data sets dumped at 500 fs (green circles) and 2 ps (solid blue circles) are shown in Fig. 9.22, overlapping the PP signals (red triangles). The dump pulse shifts part of the population from the excited state to the ground state, which is observed as a loss of SE (550 nm) and ESA (375 nm) with a concomitant recovery in the bleach region

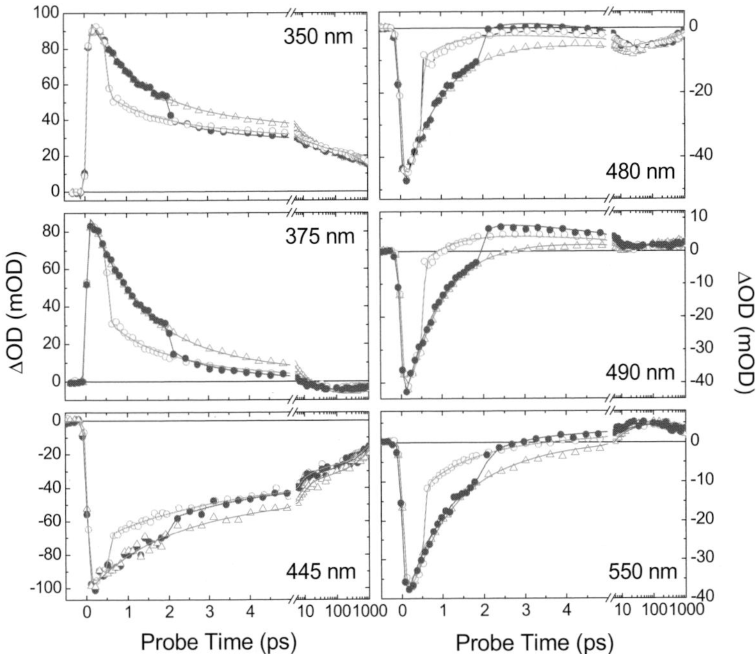

Fig. 9.22. Selected PP (red triangles) and PDP traces dumped at 500 fs (green circles) and 2 ps (blue circles). Symbols are the experimental data and the solid lines are the results of the global fits to these data. Note that the time axis is linear up to 5 ps, and then logarithmic to 1 ns.

(445 nm). However, the relative magnitude of the GSB recovery at 445 nm (∼25%) is not comparable with the depletion of the SE and ESA (∼50%), suggesting the involvement of a third intermediate state that temporarily stores ground-state population before refilling the bleach on a longer timescale. This ground state intermediate (GSI) is more clearly observed as an increase in the dump-induced absorption in the 480 nm and 490 nm traces, where a local maximum in both the PP and the PDP signals is observed. The structural basis of this intermediate state is not known; it may originate from a competing structural rearrangement motion (e.g. rotation about one of the sigma bonds, instead of the double bond of the chromophore) or from a local minimum in the ground state potential energy along the isomerization coordinate. The smaller dump effect in the 350-nm trace is indicative of the insensitivity of the overlapping radical contribution to the dump pulse; this observation is useful for characterizing photoionization properties. Such characterization was performed by measuring the action trace with the probe pulse placed at 500 ps after the excitation and scanning the dump pulse. This action trace is shown in Fig. 9.23, which shows the dump effect as a function of dump delay on the GSB (red triangles), I_0 induced absorption (green diamonds),

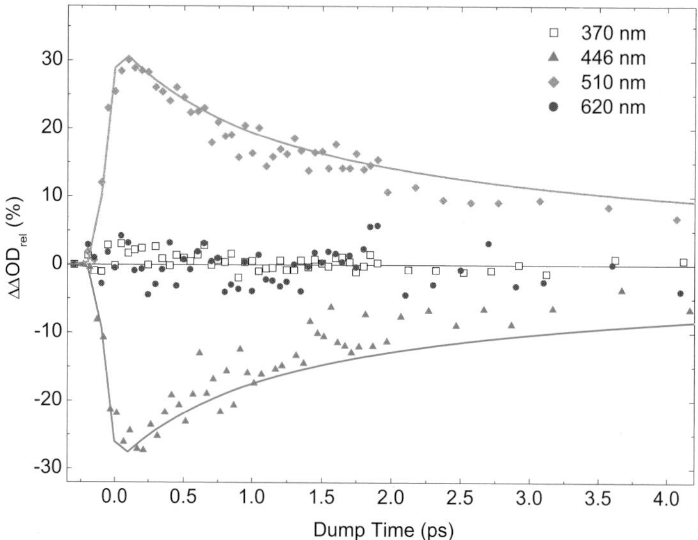

Fig. 9.23. Action trace of PYP, measured with probe time at 500 ps. Plotted is the $\Delta\Delta OD_{rel}$ signal (PDP-PP)/PP *vs.* the dump time. The red triangles are the dump-induced depletion of the bleach (446 nm) and the green diamonds are the induced-depletion of the I_0 photoproduct absorption (510 nm). The black squares and blue circles are the depletion of the radical (370 nm) and the ejected electron (620 nm) respectively. The amplitude of the PP signals were 13 mOD (radical), 30 mOD (bleach), 12 mOD (I_0) and 4 mOD (electron). Overlapping the bleach and I_0 signals is the fitted excited state population resulting from global analysis.

combined ESA and radical band (black squares) and hydrated electron band (blue circles). As expected for photoionization, which is a near-instantaneous process, the radical and electron bands in 500 ps pump-probe spectrum remain completely unaffected by the dump pulse, regardless of its timing. The action traces measured in the spectral regions of GSB and I_0 induced absorption allow to monitor the efficiency of entering PYP photocycle as a function of time [48]. By comparing them to excited state population (solid lines in Fig. 9.23), which can be estimated from the intensity of SE or ESA, one can see, that in the early times, the probability of producing an I_0 intermediate is surprisingly much higher.

On the basis of these PDP data and the PP data two self-consistent kinetic models were constructed to describe the observed dynamics (Fig. 9.24) [48]. Both the homogeneous (Fig. 9.24B) and the inhomogeneous (Fig. 9.24A) models describe the observed PDP and PP dynamics as an evolution between discrete interconnected transient states that are separated into four categories: i) excited-state dynamics, ii) ground-state dynamics, iii) photocycle dynamics and iv) an ionization channel. Whilst both models produce similar quality fits to the measured PDP data, their respective interpretations differ. The kinetic

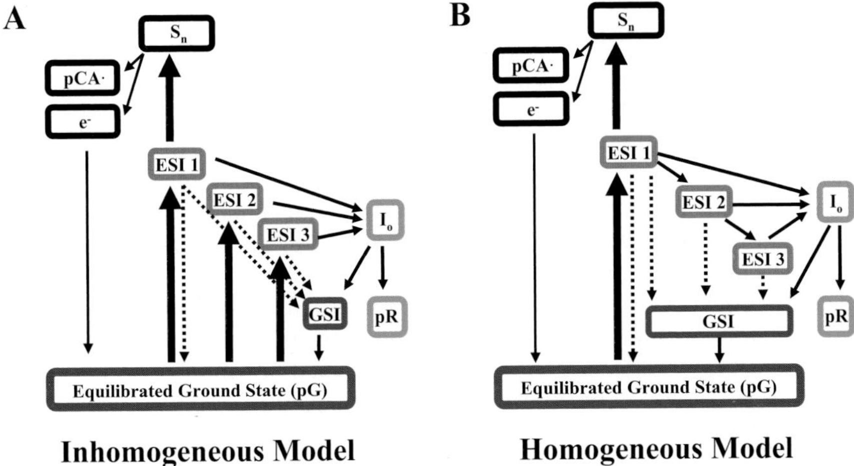

Fig. 9.24. Connectivity schemes compatible with the data and used in the global analysis for the PDP: A) inhomogeneous model and B) homogeneous model. Dynamical states are separated into four classes: excited state (red), ground state (blue), photocycle products (green) and two-photon ionization dynamics (black). ESI1, ESI2, and ESI3 refer to the excited state lifetimes τ_1, τ_2 and τ_3 respectively. pG is the equilibrated ground state species, and GSI is the ground state intermediate. Thick solid arrows represent the initial excitation process from the laser pulse and thin solid arrows dynamics represent the natural PP population dynamics. The dashed arrows represent the population transfer dynamics that may be enhanced with the dump pulse. S_n is a higher lying electronic state and $\text{pCA}^{\bullet}$ is *p*-coumaric acid PYP chromophore radical after ionization.

models differ primarily in the connectivity scheme for the excited-state evolution, and with respect to which state(s) is (are) initially excited by the applied laser pulse (thick black lines). Whilst the inhomogeneous model ascribes the multiexponential excited-state behavior to a superposition of multiple subpopulations with differing decay times, the homogeneous model ascribes the multiexponential decay to evolution along the excited-state potential energy surface.

The PDP technique probes the reaction yield of each ESI by dumping the ESI at different times and then probing the magnitude of the change of the I_0 photoproduct absorption. The PDP data favor the inhomogeneous over the homogeneous model since each ESI does not have the same yield in initiating the PYP photocycle. This observation, in combination with the global analysis of the kinetic trace PDP signals, supports the observation that ESI1 has the highest yield ($\sim$40%) followed by ESI2 (20%), whereas ESI3 has a near negligible yield ($\sim$1%) in initiating the photocycle. A homogenous model would require that the excited state population would have a time-dependent yield and evolve across a complex potential energy surface with quenching pathway(s) that compete with photocycle generation. The inhomogeneous model, in contrast, is simpler to interpret.

In summary, photoinduced dynamics of biological systems presents a significant challenge to laser spectroscopists. Besides the problems related to the nature of the samples (biological material is 'soft', difficult to obtain in large quantities, and often unstable under laser light illumination), the dynamic properties of biological molecules are inherently complex. Photoactive proteins and pigments use the light in a variety of ways, their principle reaction pathways occur in parallel with the cul-de-sac events and the spectral features of different photoreaction intermediates co-exist in time and spectrally overlap with reaction 'byproducts'. These properties often prevent unambiguous interpretations of ultrafast spectroscopic data obtained in 'traditional' ways, such as pump-probe and fluorescence up-conversion. Multi-pulse transient absorption spectroscopies, together with the state-of-the-art achievements in laser technology, open up the possibilities of bringing the understanding the light-induced biological reactions to a qualitatively new level.

9.4 Laser pulse control of excitation energy dynamics in biological chromophore complexes

B. Brüggemann and V. May

The investigation of electronic excitations in chromophore complexes known as Frenkel excitons represents one major application of femtosecond spectroscopy (for recent introductions into this field see [109, 110]). Of particular interest have been studies of light harvesting antennae belonging to the photosynthetic apparatus of bacteria or higher plants (cf. the overview in [111,112]). Although femtosecond laser pulse control techniques are widely used meanwhile (see [113, 114] and Chapter 2 for an up to date overview), there only exist a single example where these techniques have been applied to chromophore complexes. Reference [115] describes such an application to discriminate between internal conversion and excitation energy transfer taking place among a carotenoid and bacteriochlorophyll (BChl) molecule in the light harvesting antenna LH2 of purple bacteria.

This concept of guiding excitation energy into one of two particular transfer channels will be put here into a more general frame. In the following, a theoretical analysis is presented of laser pulse controlled excitation energy motion and localization in systems of strongly coupled chromophores like the FMO–complex or the PS1 (for both antenna systems see [111]). While general aspects have been already discussed in [116–118] emphasis is put here on polarization control [119] (see also [120]). This is of particular interest since recent pulse shaping technology allows for a simultaneous and independent manipulation of the two different polarization directions of the laser beam [121].

When studying Frenkel excitons one is faced with spatially delocalized excited states with the basic electronic excitations, however, completely localized at the individual chromophores of the complex. The respective state vector for such a localized excitation will be denoted as $|\phi_m\rangle$ with the index m indicating the excited chromophore. The strong coupling among different chromophores results in the formation of delocalized (single)–exciton states $|\alpha_1\rangle = \sum_m C(\alpha_1; m)\, |\phi_m\rangle$ with energy $\hbar\Omega_{\alpha_1}$. This energy is much larger than the thermal energy (at room temperature conditions) and if the coupling to vibrational coordinates remains weak, as it is often the case, excitation energy transfer may proceed coherently up to some 100 fs.

The particular control task which will be discussed in the following aims at an excitation energy localization at a single chromophore at a definite time. The localization has to be achieved against the tendency to form delocalized states. It requires the photo–induced formation of an excitonic wave packet, i.e. the time–dependent superposition $\sum_{\alpha_1} A_{\alpha_1}(t)|\alpha_1\rangle$ of the various exciton states. This has to be done in such a way that at the final time t_f of the control task the superposition corresponds to excitation energy localization at a particular chromophore m, i.e. $\sum_{\alpha_1} A_{\alpha_1}(t = t_f)|\alpha_1\rangle = |\phi_m\rangle$. Of course, it would be of interest to study a situation where the time t_f to reach the target state is replaced by a time–interval around t_f as it would be necessarily the case in an experiment which proofs the suggested excitation energy localization [122–124]. If the detection interval, however, amounts to clearly less than 100 fs it would not change the outcome so much, since the exciton dynamics in the considered systems is slow on a sub–100 fs time–scale.

To form the superposition state $\sum_{\alpha_1} A_{\alpha_1}(t)|\alpha_1\rangle$ all exciton states in a control task have to be addressed, therefore the oscillator strength should be distributed over all exciton states $|\alpha_1\rangle$ which excludes the use of highly symmetric complexes for such studies. For an appropriate non–regular structure, however, different transition dipole moments $\mathbf{d}_{\alpha_1}$ may also posses different spatial orientations. This would favor the use of polarization shaped control fields. The spatial orientation of $\mathbf{E}(t)$ (perpendicular to the propagation direction) would increase the flexibility for putting the various coupling expressions $\mathbf{d}_{\alpha_1}\mathbf{E}(t)$ in the right order of magnitude at the right time interval to achieve the proper wave packet formation. (Of course, a random spatial orientation of the complexes has to be included into these considerations.)

The description of excitation energy localization discussed so far has to be generalized to the inclusion of exciton relaxation and dephasing originated by the presence of exciton–vibrational coupling. This requires an approach based on open system dynamics techniques, i.e. by introducing the (reduced) exciton density matrix. Moreover, one has to account for structural and energetic disorder. Such a combination of density matrix propagation, polarization shaping, and disorder to solve the control task using the Optimal Control Theory (OCT) is new and will be explained in detail in the following sections.

First, in the subsequent section, the exciton model and the density matrix description are introduced. Afterwards, a specific implementation of the OCT is shortly explained. Some selected results are presented in the last section.

9.4.1 Excitons in biological chromophore complexes

When studying chromophore complexes by ultrafast and intense laser pulses higher excited states of single molecules as well as of the whole complex have to be taken into account. It results the multiexciton scheme which found various applications in different non–biological complexes as well as photosynthetic antenna systems [111,112]. The respective basic model applied in the following is explained in Fig. 9.25.

Based on this model one may introduce singly excited states $|\phi_m\rangle$ of the complex as well as doubly excited states $|\phi_{mn}\rangle$, corresponding to a double excitation of a single molecule ($m = n$) or the simultaneous single excitation of two different chromophores ($m \neq n$). Their superposition may result in single–exciton states $|\alpha_1\rangle = \sum_m C(\alpha_1; m)|\phi_m\rangle$, with quantum numbers α_1

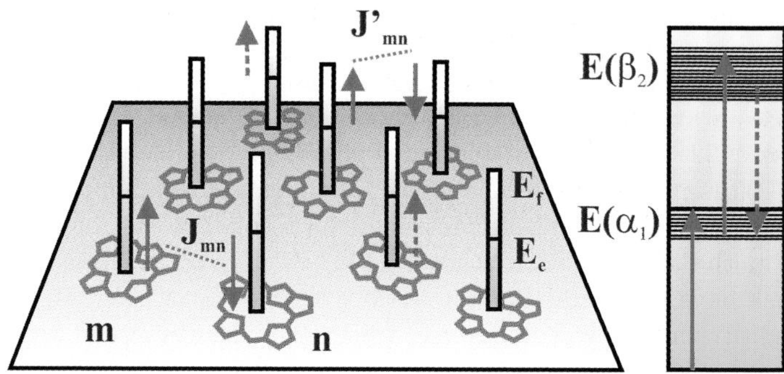

Fig. 9.25. Excitation energy transfer in a chromophore complex (left part) with the chromophores (tetrapyrrole type molecules) arranged to a planar complex and described by a three–level model. Beside the electronic ground–state energy E_g (not shown) every chromophore is characterized by a first excited state with energy E_e and a higher excited state with energy E_f. Excitation energy transfer is possible via de–excitation of chromophore n (from E_e to E_g, shown by vertical arrow) and excitation of chromophore m (from E_g to E_e). Both processes are caused by the Coulombic inter–chromophore coupling J_{mn} (broken line). Transfer of the double excitation of a single chromophore is realized by J'_{mn} (double excitation via a fusion of two single excitations is also possible). A single chromophore may be excited by photon absorption (vertical dotted arrows). The right part shows the manifold of single and two–exciton levels with energies $E(\alpha_1) \equiv \hbar\Omega_{\alpha_1}$ as well as $E(\alpha_2) \equiv \hbar\Omega_{\alpha_2}$ and resulting from a superposition of singly and doubly excited states of the complex, respectively (full arrows: optical excitation, broken arrow: radiationless decay of two–exciton state via exciton–exciton annihilation).

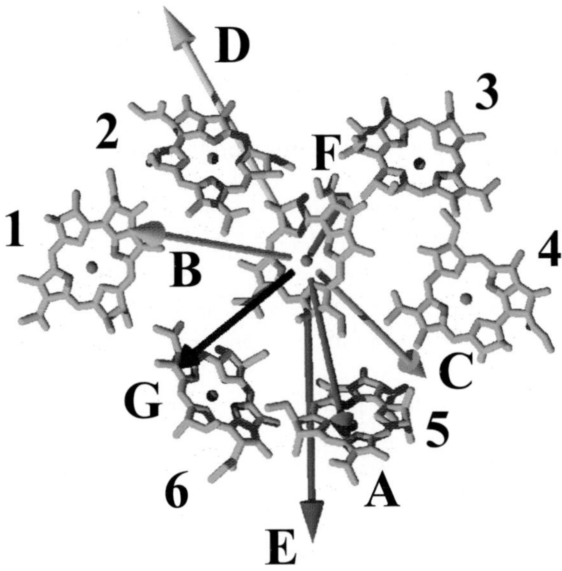

Fig. 9.26. Spatial arrangement of the seven BChls in the monomeric FMO complex of *Prosthecochloris aestuarii* (without protein matrix, the used counting scheme labels BChl 1 to 6, BChl 7 is positioned in the center) [125]. Atoms in line of the respective Q_y dipole moments are marked (all parameters used in the calculations can be found in [116, 126]). The different arrows display the magnitude and spatial orientation of the exciton transition dipole moments $\mathbf{d}_{\alpha_1}$ (to distinguish the labeling from those of the BChl, capital letters A to G have been used for $\alpha_1 = 1$ to 7). Note, that the exciton levels 3 and 5 mainly contribute when BChl 7 should be exclusively excited [116].

as well as in two–exciton states $|\alpha_2\rangle = \sum_{m,n} C(\alpha_2; m, n)|\phi_{mn}\rangle$, with quantum numbers α_2. Using these states and accounting for all inter–chromophore Coulombic interactions (often used in the form of a dipole–dipole coupling, cf., for example, [110]) the complete electronic part of the overall Hamiltonian can be diagonalized (with the vibrational coordinates fixed at their ground–state equilibrium configuration). Both types of states together with the ground–state $|\alpha_0\rangle \equiv |\phi_0\rangle$ define the multiexciton part $H_{\rm mx}$ of the overall Hamiltonian $H_{\rm mx} = \sum_N \sum_{\alpha_N} \hbar\Omega_{\alpha_N} \, |\alpha_N\rangle\langle\alpha_N|$. Here, $N = 0, 1, 2$ counts the so–called exciton manifolds and the $\hbar\Omega_{\alpha_N}$ define the respective spectrum.

Besides $H_{\rm mx}$ the chromophore complex Hamiltonian

$$H_{\rm CC}(t) = H_{\rm mx} + H_{\rm mx-vib} + H_{\rm vib} + H_{\rm field}(t)$$

includes the coupling to intra- and inter-molecular vibrations ($H_{\rm mx-vib}$) and the coupling to external laser fields ($H_{\rm field}(t)$). Neglecting inter–manifold coupling, the multiexciton–vibrational coupling is given by:

$$H_{\text{mx-vib}} = \sum_N \sum_{\alpha_N,\beta_N} H^{(\text{mx-vib})}_{\alpha_N\beta_N} |\alpha_N\rangle\langle\beta_N|$$

with the matrix elements expanded with respect to the vibrational coordinates q_ξ, i.e. $H^{(\text{mx-vib})}_{\alpha_N\beta_N} = \sum_\xi \hbar\omega_\xi g_{\alpha_N\beta_N}(\xi) q_\xi$. This represents the most basic expression. Its introduction can be understood as the result of a normal–mode description of all vibrations (with parabolic PES). Accordingly, the Hamiltonian H_{vib} is that of decoupled harmonic oscillators. Often, a vibrational modulation of the inter–chromophore coupling is neglected. More involved formulas as well as the derivation of the matrix elements $g_{\alpha_N\beta_N}(\xi)$ can be found in, e.g., [110,112,127,128] ([127] puts also emphasis on internal conversion processes being responsible for exciton–exciton annihilation, see below). The Hamiltonian $H_{\text{field}}(t)$ describing the coupling to the radiation field $\mathbf{E}(t)$ is written in the standard form $-\hat{\mu}\mathbf{E}(t)$ where the chromophore complex dipole operator $\hat{\mu}$ includes ground–state single–exciton transitions with dipole matrix elements $\mathbf{d}_{\alpha_1}$ and transitions from the single into the two–exciton states with dipole matrix elements $\mathbf{d}_{\alpha_2\beta_1}$.

In order to specify the multiexciton approach to light harvesting antenna complexes a three–level model (see Fig. 9.25) is introduced for every Chlorophyll (Chl) or Bacteriochlorophyll (BChl) molecule. The so–called Q_y–state of tetrapyrroles is taken as the first excited state. The higher excited state has to be considered as a representative of the multitude of higher excited electron–vibrational states (it has to be introduced to allow for intra–chromophore excitations in the same energetic range where the energy of two singly excited states at different chromophores is positioned). Besides their energy level structure every chromophore has to be characterized by transition dipole moments. If data on the electronic inter–chromophore couplings are available they can be used to define the respective expressions in the Hamiltonian. Otherwise one applies the dipole–dipole approximation (necessary in any case if the higher excited states are concerned). Finally, the spectral densities resulting from the coupling to the multitude of vibrational coordinates (see below) are fixed by fitting optical absorption or luminescence. Independent calculations are not available so far (cf. also [111, 112]).

To describe the light–driven exciton dynamics in the presence of relaxation and dephasing the multiexciton density matrix is introduced

$$\rho(\alpha_M, \beta_N; t) = \langle\alpha_M|\hat{\rho}(t)|\beta_N\rangle \; , \tag{9.3}$$

where $\hat{\rho}(t)$ denotes the related reduced density operator. The diagonal elements $\rho(\alpha_M, \alpha_M; t)$ define the exciton level populations and the off-diagonal ones the different coherences, which may refer to the same manifold or to different manifolds. There are different ways to compute $\rho(\alpha_M, \beta_N; t)$. Here, a description is chosen based on the following Markovian quantum master equation

$$\frac{\partial}{\partial t}\hat{\rho}(t) = -\frac{i}{\hbar}[H_{\text{mx}} + H_{\text{field}}(t), \hat{\rho}(t)] - \mathcal{R}_{\text{mx-vib}}\hat{\rho}(t) \; . \tag{9.4}$$

The multiexciton Hamiltonian $H_{\rm mx}$ together with the coupling to the external laser pulses $H_{\rm field}(t)$ determine the reversible part of the quantum master equation, whereas energy relaxation and dephasing are accounted for by the superoperator $\mathcal{R}_{\rm mx-vib}$. The latter is assumed to take the following form

$$\mathcal{R}_{\rm mx-vib}\,\hat{\rho}(t) = \sum_{N=1,2}\sum_{\alpha_N,\beta_N} k^{\rm (mx-vib)}_{\alpha_N\to\beta_N}$$

$$\left\{\frac{1}{2}\left[|\alpha_N\rangle\langle\alpha_N|, \hat{\rho}(t)\right]_+ - |\beta_N\rangle\langle\alpha_N|\hat{\rho}(t)|\alpha_N\rangle\langle\beta_N|\right\}. \tag{9.5}$$

The transition rates $k^{\rm (mx-vib)}_{\alpha_N\to\beta_N}$ are mainly determined by the spectral densities $\mathcal{J}_{\alpha_N\beta_N,\beta_N\alpha_N}$ taken at the transition frequencies $\Omega_{\beta_N} - \Omega_{\alpha_N}$. The single–exciton spectral density, for example, is given by the expression $\mathcal{J}_{\alpha_1\beta_1,\beta_1\alpha_1}(\omega) = \sum_m \mid C(\alpha_1;m)C(\beta_1;m)\mid^2 J_e(\omega)$, including chromophore local spectral functions $J_e(\omega)$ (a more involved description can be found in [112]). Notice that the spectral densities account for delocalized multiexciton states since they are defined in using the related expansion coefficients. Often it becomes also necessary to account for exciton–exciton annihilation as the dominant non–radiative two–exciton decay channel. For respective considerations see [111, 112, 127, 128].

The sequence of approximations necessary to arrive at (9.4) and (9.5) is well documented in the literature (for a recent overview see [129, 130]). One has to carry out a second order perturbation theory with respect to the exciton–vibrational coupling. The correlation time of the vibrational equilibrium correlation function should be short enough to allow for a neglect of non–Markovian contributions, the laser pulse field strength has to be small enough to exclude contributions in $\mathcal{R}_{\rm mx-vib}$ and, finally, the spectra of single and two–exciton states have to be anharmonic to neglect the coupling between diagonal and off–diagonal density matrix elements.

9.4.2 Theory of femtosecond laser pulse control

Theoretical simulations of laser pulse control experiments are mainly carried out in the framework of the Optimal Control Theory (OCT) [110,131,132] and Chapter 2. It allows one to compute the laser pulse (the control field) **E** which optimizes the observable $\mathcal{O}$ measured in the particular control experiment (taking place in the time interval from t_0 up to t_f and under the constraint of a finite laser pulse intensity). Therefore, one searches for an $\mathbf{E}(t)$ which leads to an extremum of the overall control functional:

$$J[\mathbf{E}] = \mathcal{O}[\mathbf{E}] - \lambda\Big(\frac{1}{2}\int_{t_0}^{t_f} dt\,\mathbf{E}^2(t) - I_0\Big). \tag{9.6}$$

The second term represents the constraint to ensure finite control field intensity fixed by the value I_0, and the quantity λ is a Lagrange multiplier. (Often one avoids to adapt the field to the predetermined intensity but fixes λ by a reasonable value and determines the related I_0 after the control task has been solved.) The field $\mathbf{E}$ resulting in an extremum of $\mathcal{O}[\mathbf{E}]$ will be called the *optimal* field. This scheme has been widely applied to the description of pure state molecular dynamics. Generalizations to open molecular system have been also worked out, but found less applications (cf. [133–136]).

A rather general expression for $\mathcal{O}$ defined via a trace expression with respect to the multiexciton levels and the multiexciton density operator is given by

$$\mathcal{O}[\mathbf{E}] = \int_{t_0}^{\infty} dt_f \int dp \; \mathrm{tr}_{\mathrm{mx}}\{\hat{O}(t_f;p)\hat{\rho}(t_f;p)\} \; . \tag{9.7}$$

It accounts for a distribution of the final time t_f where the observable (represented by the operator $\hat{O}(t_f;p)$) has to be maximized (for example, optimization of the probe–pulse signal in a pump–probe scheme [122–124]). p is a parameter or set of parameters which should refer to a particular property changing among the individual molecules. Therefore, $\mathcal{O}[\mathbf{E}]$ not only accounts for a distribution of the operator $\hat{O}$ in time but also in the space of the parameters p. Then, $\mathcal{O}$ is ready to describe, for example, inhomogeneous broadening present in the considered molecular system. In this case p counts all molecules in the sample, where the properties of the single molecules, for example, their excitation spectrum change from molecule to molecule.

The determination of the optimal control field is achieved by searching for the extremum of J, i.e. the solution of $\delta J/\delta \mathbf{E} = 0$. The solution of the resulting it gives the temporal behavior of the optimal pulse (self–consistency condition for the optimal field)

$$\mathbf{E}(t) = \frac{i}{\hbar\lambda} \int dp \int_{t}^{\infty} dt_f \; \mathrm{tr}_{\mathrm{S}}\{\hat{O}(t_f;p)\mathcal{U}(t_f,t;p;\mathbf{E})[\hat{\mu},\hat{\rho}(t;p)]\} \; . \tag{9.8}$$

The field at time t is determined by the commutator of the dipole operator with the density operator taken also at time t, but further propagated up to t_f. The latter procedure is indicated by the action of the time–propagation superoperator $\mathcal{U}(t_f,t;\mathbf{E})$ (depending on the parameter set p and at the presence of the field $\mathbf{E}$). Notice also the integration with respect to t_f which originates from the time distribution of $\hat{O}$. To solve (9.8) one rearranges the trace expression in order to replace $\int_t^{\infty} dt_f \, \hat{O}(t_f;p)\mathcal{U}(t_f,t;p;\mathbf{E})$ by the auxiliary density operators $\hat{\theta}(t;p)$ propagated backwards in time (see also [134, 135]). The respective equation of motion is different from that with $\mathcal{R}_{\mathrm{mx-vib}}$. A changed expression appears (cf. [116]) which enables a stable propagation backward in time (in the presence of dissipation and of the field). The control task is solved by a combined iterative forward and backward propagation [133] (see also [134, 135]).

If energetic disorder is considered a discrete set of different $\hat{\rho}(t;p)$ and $\hat{\theta}(t;p)$ appears with p counting the different complexes in the ensemble. Accounting also for polarization shaping ($\mathbf{E}$ consists of an x– and y–component perpendicular to the propagation direction), (9.8) reads ($j = x, y$):

$$E_j(t) = \frac{1}{N_{\text{CC}}} \sum_p \frac{i}{\hbar\lambda} \text{tr}_{\text{mx}}\{\hat{\theta}(t;p)[\hat{\mu}_{pj}, \hat{\rho}(t;p)]\} \,. \tag{9.9}$$

Here, $\hat{\mu}_{pj}$ denotes the x– or y–component of the dipole operator and N_{CC} is the number of different chromophore complexes considered in the disordered ensemble. Now, the optimal pulse $\mathbf{E}(t)$ represents a compromise with respect to the driven dynamics in the individual chromophore complexes of the disordered ensemble. In order to calculate $\mathbf{E}(t)$ one has to solve simultaneously the coupled equations of motion of the various $\hat{\rho}(t;p)$ and $\hat{\theta}(t;p)$.

9.4.3 Controlling excitonic wave packet motion in the FMO complex

The multiexciton density matrix theory in combination with OCT has been used to study fs laser pulse induced excitation energy localization in the PS1 [118] and the FMO complex [116, 119]. Here, emphasis is put on the latter example. Focusing on energy localization at a particular time t_f, the operator $\hat{O}$ introduced in (9.7) which characterizes the target of the control task is given by the projector $|\phi_{m_{\text{tar}}}\rangle\langle\phi_{m_{\text{tar}}}|$ times $\delta(t_f - \tau_f)$. Then, $\mathcal{O}[\mathbf{E}]$ of identified with the population of the target state P_{tar} at $t = \tau_f$ and can be calculated from the single–exciton density matrix according to

$$P_{\text{tar}}(\tau_f) = \frac{1}{N_{\text{CC}}} \sum_p \sum_{\alpha_1,\beta_1} C_p(\alpha_1; m_{\text{tar}}) C_p^*(\beta_1; m_{\text{tar}}) \rho_p(\alpha_1, \beta_1, \tau_f; p) \,. \tag{9.10}$$

Fig. 9.27 displays respective results for excitation energy localization at chromophore 7 of the monomeric FMO–complex (cf. Fig 9.26), but in the absence of two–exciton states and disorder. Most efficient localization is achieved for a control pulse length of around 600 fs. As has to be expected an increase in temperature acts counterproductive. The effect of transitions among the single and two–exciton manifold is demonstrated in Fig. 9.28. It results in a reduction of $P_{\text{tar}}(\tau_f)$ from a value above 0.6 to a value somewhat below 0.4. To correctly rate this value one has to notice that the probability to stay in the chromophore complex ground–state is larger than 0.4. Therefore, it is advisable to consider the renormalized target state population $P_{\text{tar}}(\tau_f)/P_0(\tau_f)$ which amounts to a value of about 0.6 ($P_0(\tau_f)$ is the ground–state population).

This justifies in part to neglect the two–exciton states when considering the influence of disorder (note also that an iteration of the OCT equations combined with an ensemble averaging now refers to less than 60 density matrix

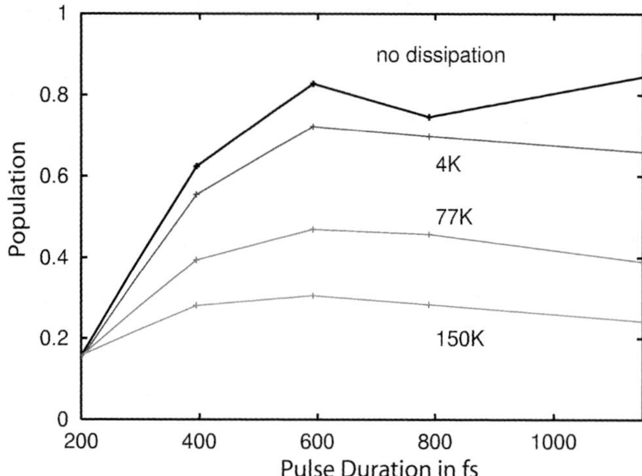

Fig. 9.27. Excitation energy localization in the FMO complex demonstrated by the population P_{tar}, (9.10) of the single BChl Q_y–state $m_{\mathrm{tar}} = 7$ (cf. Fig. 9.25) at t_f. The population is drawn versus the pulse duration $t_f - t_0$ as well as in the absence of dissipation and for different temperatures. All data have been obtained in neglecting two–exciton states as well as disorder and after up to 50 iterations of the coupled density matrix equations for forward and backward propagation.

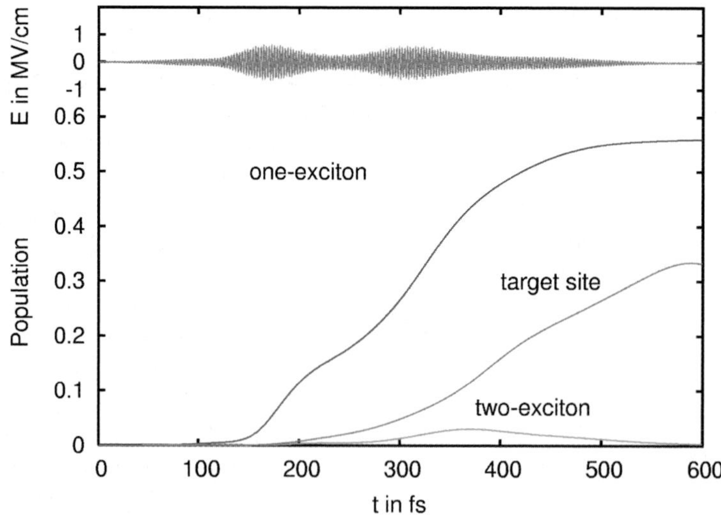

Fig. 9.28. Temporal evolution of excitation energy localization in the FMO complex, including transitions into the two–exciton manifold and a decay via exciton–exciton annihilation (target site $m_{\mathrm{tar}} = 7$, pulse duration $\tau_f - t_0 = 600$ fs, T = 4 K, neglect of disorder) Upper part: time–dependence of the control field, lower panel: overall single– and two–exciton state population as well as the target site population.

equations compared to about 1000 for the inclusion of two–exciton states). In any case the presence of disorder further decreases the control yield as shown in Fig. 9.29 (upper panel). Fortunately, this inefficiency can be compensated when changing to laser pulse control including polarization shaping. The possibility of an independent control of both polarizations of the field increases the control yield considerably (cf. the lower panel of Fig. 9.29). As demonstrated also in Fig. 9.29 the time interval where the two components (x– and y–component) of the field are non–zero becomes longer compared to the control with a linearly polarized field (the different sub pulses of the two components act out of phase indicating strong elliptic polarization of the control field).

9.5 Light induced singlet oxygen generation

I. Corral, L. González, A. Lauer, and K. Heyne

During the last decade drugs with endoperoxide groups have become more and more important. They are synthesized as antimalarial drugs [137, 138] or as photosensitizers for cancer treatment in photodynamic therapy (PDT) [139].

Malaria, along with tuberculosis and the human immunodeficiency virus (HIV), form a disease triad that accounts for almost half of all the infectious disease mortality. Malaria is transmitted by the bite of an infected female *Anopheles* mosquito that transfers to the human host up to four different species of *Plasmodium* parasites. Traditionally, malaria has been treated with quinolines. Unfortunately, most of the parasites responsible for the vast majority of fatal malaria infections, have become resistant to quinolines. Today, the most potent antimalarials available are artemisinins, extracted from sweet wormwood (*Artemisia annua*), rapidly killing all asexual stages of *Plasmodium falciparum* by specifically inhibiting its sarcoendoplasmic reticulum Ca^{2+} ATPase ortholog [140]. The special feature of Artemisinin is an endoperoxide group which can be cleaved or dissociated by interaction with iron ions or light [141]. Therefore, the synthesis of new drugs with endoperoxide groups plays an important role in the development of new prodrug prototypes to treat malaria [138].

Molecules with endoperoxide groups are also interesting for photodynamic therapy (PDT) [139]. In traditional PDT, a photosensitive drug is excited by light in the presence of ground state molecular oxygen, which is abundant in tissues/cells, to produce singlet molecular oxygen, $^1O_2$, as well as other cytotoxic agents. These, and $^1O_2$ cause severe physiological damage in selected cell subunits, such as cell membranes, and lead subsequently to cell death [142]. In the first step of PDT the dye is administered to the patient, where it accumulates in the tumor tissues [143]. Next, the target area is irradiated directly by a light source of a wavelength where the photosensitizer absorbs. Upon

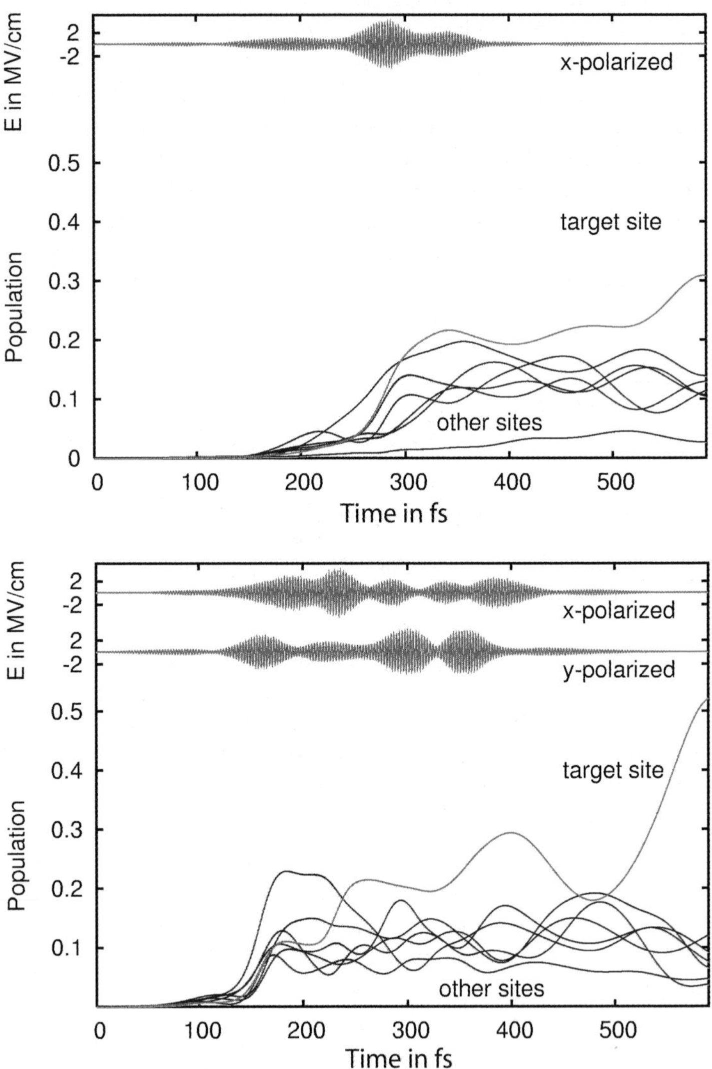

Fig. 9.29. Disorder averaged chromophore populations in the FMO complex at the target site $m_{\mathrm{tar}} = 7$ (and at all other sites). Upper panel: use of a linearly polarized control pulse, lower panel: the optimization covered the independent two polarization directions of the field. An ensemble of 10 complexes with randomly chosen spatial orientation and fluctuating on–site energies E_e ($\sigma = 100\ \mathrm{cm}^{-1}$) has been considered. The temporal evolution of the field components is shown in the upper part of both panels.

light absorption, (see Fig. 9.30), the photosensitizer is excited to the first excited singlet state S_1, from where it can again return to the ground state S_0 or alternatively reach the first excited triplet state T_1 through intersystem crossing (ISC). Once the first triplet excited state is formed, two different types of interaction mechanisms might occur between the photosensitizer and the target substrate [144]. Photodynamic reactions of type I are those where the photodynamic damage is mainly caused by radical photosensitizer intermediates. Reactions of type II, on the other hand, involve energy transfer (ET) from the triplet state of the photosensitizer to ground state triplet oxygen $^3O_2$, resulting in $^1O_2$ and other reactive oxygen species (ROS).

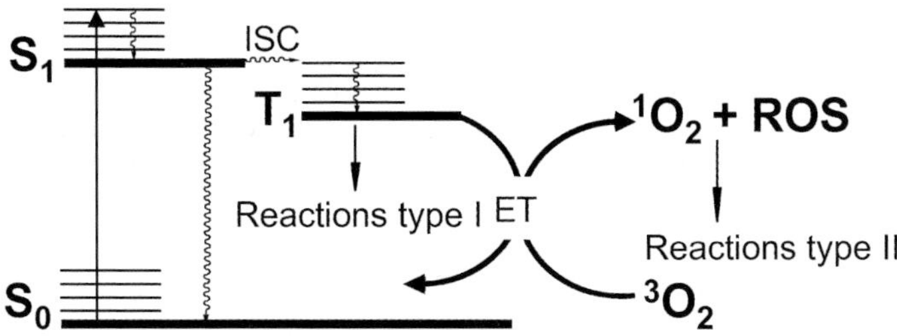

Fig. 9.30. Scheme for singlet oxygen generation [142]. The initial step is the excitation of the photosensitizer ($S_0 \rightarrow S_1$). By intersystem crossing (ISC) the triplet T_1 is populated ($S_1 \rightarrow T_1$). Deactivation of the triplet photosensitizer state is followed by conversion of the oxygen triplet ground state $^3O_2$ into the singlet oxygen state $^1O_2$ and the generation of other reactive oxygen species (ROS).

Necessary properties for drugs to be used as photosensitizers are high and selective accumulation in target tissues, absorption in the phototherapeutic window 600-1200 nm, high yields in generation of the excited triplet state T_1, and of course minimal adverse effects. The use of PDT in biological tissues is limited to a few millimeters depth, due to strong scattering and photon loss $I = I_0 \exp(-\sigma_t x)$. Here, I_0 and I are the light intensities in front of and inside the biological tissue, σ_t is the total damping cross section, which is a sum of the scattering cross section σ_s and the absorption cross section σ_a, and x is the depth in the sample. Typically, the scattering coefficient σ_s decreases from $120\,\mathrm{mm}^{-1}$ at 400 nm to $20\,\mathrm{mm}^{-1}$ at 1800 nm, whereas the absorption coefficient varies between $40\,\mathrm{mm}^{-1}$ and $0.01\,\mathrm{mm}^{-1}$ in the same spectral region. Thus, the transmitted light intensity in 1 mm depth of biological tissue is only 0.027 % at 600 nm and 4.07 % at 1600 nm. Since the intensity of the transmitted light increases for longer wavelengths, an ideal photosensitizer should preferably absorb at long wavelengths. However, the generation of singlet oxygen from abundant ground state oxygen requires an energy of more than 1270 nm, restricting the absorption of photosensitizers to wavelengths shorter than 1270 nm. As a consequence, for single photon excitation the phototherapeutic window ranges from 600 nm to 1200 nm. An alternative approach in PDT is the use of two-photon excitation processes with shaped laser pulses of longer wavelengths, and concomitantly higher light transmission in biological tissues. This could accomplish better excitation and higher selectivity in the presence of molecules with similar one-photon absorption, e.g. the melanin pigment [145].

A special kind of photosensitizers are those which contain an endoperoxide group, so-called oxygen-carrier photosensitizers. These photosensitizers are able to generate $^1O_2$ directly upon irradiation, contrary to conventional ones. One such case is anthracene-9,10-endoperoxide, (APO), which has been chosen in the present study as a model to understand the photochemical behavior of aromatic oxygen-carrier photosensitizers.

Oxygen-carrier photosensitizers, such as APO, can show different photochemical behaviors and reaction mechanisms depending on the excitation wavelength. As shown in Fig. 9.31, two competing photodissociation mechanisms can take place in APO upon ultraviolet (UV) irradiation: a) O-O homolysis, resulting in oxygen biradical rearrangement products, and b) cycloreversion, either concerted or stepwise, producing singlet oxygen and the parent hydrocarbon.

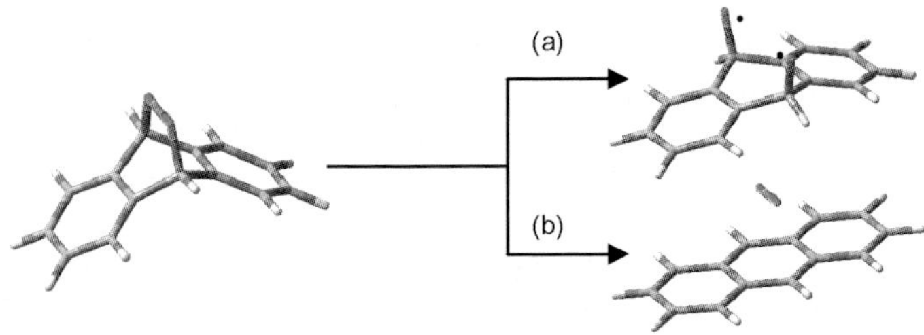

Fig. 9.31. Photodissociation products of APO by: (a) O-O homolysis (b) cycloreversion.

Much has been published on APO and related derivatives; however, the detailed photodissociation mechanism is still unclear. In the following we present new experimental and theoretical data on APO's photochemistry attempting to increase the understanding of the APO photochemical mechanisms.

The first studies dealing with the photochemistry of endoperoxides date from 1969. Through orbital and state correlation diagrams obtained from thermochemical/spectroscopical data and symmetry/spin selection rules, Kearns and Khan [146] postulated that the excitation of small size endoperoxides to their first excited singlet or triplet states leads to the cleavage of the O-O endoperoxide bond, and therefore to the formation of oxygen biradicals, (see Fig. 9.31a), while the excitation to higher-lying excited states, which imply promotion of an electron from a π^*_{OO} orbital to a σ^*_{CO} orbital, induces the endoperoxide photodissociation into the parent hydrocarbon and singlet oxygen, (see Fig. 9.31b). This can be easily rationalized taking into account that the population of antibonding σ^*_{OO} and σ^*_{CO} orbitals leads to a weakening of the O-O and C-O bonds, respectively, and therefore favors the cleavage of these respective bonds.

Some time later, in 1984 Brauer and coworkers [147] recorded experimental absorption spectra of APO in CH_2Cl_2 and studied its UV photolysis. In agreement with Kearns and Khan's predictions [146], they assigned the first excited singlet state S_1 at 384-300 nm to the $\pi^*_{OO} \rightarrow \sigma^*_{OO}$ transition, which they found leads to rearrangement products with origin in the homolytic rupture of the endoperoxide bridge. They also found cycloreversion products, with origin in higher energy excited states, $S_n \geq 2$, at $\lambda \leq 278$ nm, characterized by $\pi_{CC} \rightarrow \pi^*{}_{CC}$ transitions.

Experimental wavelength-dependent photodissociation studies on phenylated and methylated derivatives of anthracene-9,10-endoperoxides, were performed by Eisenthal et al. [148] and Rigaudy et al. [149]. They found the same general photodissociation trends as reported by Brauer and coworkers [147]. The picosecond time-resolved measurements carried out by Eisenthal et al. [148] allowed, in addition, to postulate a nonconcerted mechanism for the loss of molecular oxygen, based on the unusual long rise times (50-75 ps) observed for the cycloreversion process.

More recent Ar-matrix experiments supported by semiempirical CNDO/S and INDO/S calculations carried out by Klein and coworkers [150] suggested, however, that the well-accepted assumption of cycloreversion occurring from the S_2 excited state was wrong. Excitation to the S_1 state at 275 nm led to cycloreversion products and their calculations predicted a $\pi_{CC} \rightarrow \pi^*_{CC}$ character for the S_1 state, concluding that O-O homolysis products may have their origin in the triplet manifold.

These controversial results clearly call for further experiments as well as for high level ab initio quantum chemical calculations, beyond semiempirical and state correlation diagrams. In the following, results obtained by using multiconfigurational ab initio theory and femtosecond polarization resolved UV pump IR probe spectroscopy are presented. This combined theoretical and experimental study contributes to clarifying the mechanisms behind both cycloreversion and O-O homolysis reactions, with the long term goal to understand and optimize drugs with endoperoxide groups.

Computational Details

All the calculations on APO were done at the ab initio (see Chapter 3, Sect. 3.2.1) multiconfigurational Complete Active Space Self Consistent Field [151] (CASSCF) level of theory in combination with the Small Atomic Natural Orbital Basis Set [152] (ANO-S) as implemented in the quantum chemistry package MOLCAS 6.0 [153]. The calculations were performed on the C_{2v}

equilibrium geometry optimized at MP2 level of theory with the 6-311G(d,p) Pople Split Valence Triple Zeta Basis Set, supplemented with d and p diffuse functions on heavy and hydrogen atoms, respectively. In order to include dynamic correlation effects, the CASSCF electronic energies were corrected using Multi-State Complete Active Space Perturbation Theory to the Second Order [154] (MS-CASPT2).

The full active space for the description of the system in its equilibrium geometry consists of 6 pairs of π_{CC} and π^*_{CC} orbitals belonging to the anthracene moiety, together with the π_{OO}, π^*_{OO}, σ_{OO} and σ^*_{OO} orbitals localized in the endoperoxide bridge. This amounts to a total of 18 electrons in 16 orbitals, resulting in 27810640 configurations, impracticable from the computational point of view. Our effective active space for describing the equilibrium structure of APO is therefore reduced to 14 electrons in 12 orbitals, CAS(14,12), disregarding the two lowest and highest energy π_{CC} and π^*_{CC} occupied and virtual orbitals, respectively. To describe the C-O dissociation problem, the σ_{CO} and σ^*_{CO} orbitals are additionally included in the active space.

In order to assess the performance of the chosen theoretical procedure and basis set, other basis sets and active spaces were as well employed. The comparison of MS-CASPT2(14,12)/ANO-S//CASSCF(14,12)/ANO-S vertical excitation energies for the first singlet excited states with those obtained with the Large Atomic Natural Orbital basis set [155] (ANO-L) and with a (16,14) active space, shows differences of 2-3 kcal/mol, that lie within the error inherent of the employed methods, justifying our choice.

Experimental Details

APO was dissolved in deuterated chloroform ($CDCl_3$) at a concentration of 70 mM (sample thickness 0.1 mm, temperature 293 K). The APO absorption spectrum exhibits clear absorption peaks at 271 nm and 279 nm in the UV spectrum as shown in Fig. 9.32 and gives rise to the infrared absorption bands shown in Fig. 9.33. Femtosecond pump and probe pulses at 273 nm and 1170 cm^{-1}, respectively, were generated with a cross correlation time between the pump and probe pulses of 300 fs (FWHM). Measurements were performed with parallel and perpendicular linear polarization of pump and probe pulses. The absorbance change was derived from measurements with parallel $\Delta A_{\|}(t_D) = -\log(T_{\|}(t_D)/T_{\|0})$; ($T_{\|}(t_D)$, $T_{\|0}$: sample transmission with and without excitation (respectively), t_D: delay time), and perpendicular ($\Delta A_{\perp}(t_D)$) polarization. The relative angle of the electronic transition dipole moment to the vibrational transition dipole moment was derived using the formulas: $D = \Delta A_{\|}(t_D)/\Delta A_{\perp}(t_D)$ and $\vartheta = \arccos\sqrt{\frac{2D-1}{D+2}}$. The orientations of the calculated electronic transition dipole moments within the APO molecule are shown in Fig. 9.32. The orientations of the vibrational dipole moments in the electronic ground state were calculated on the geometry given in

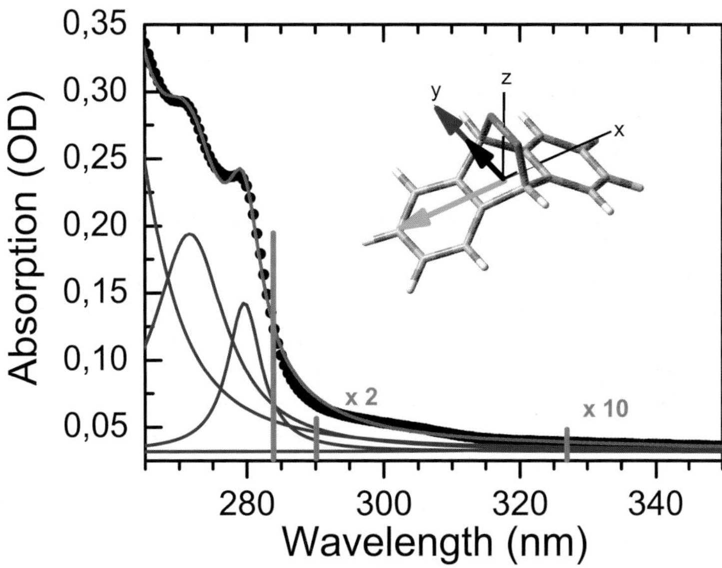

Fig. 9.32. Absorption spectrum of APO in chloroform ($CDCl_3$) (circles); Multiple Lorentzian Fit: green line; individual peaks: blue lines; calculated transitions: red bars. The inset shows the APO optimized geometry used in the calculations. The arrows indicate the orientations of transition dipole moments. $S_0 \rightarrow S_1$ and $S_0 \rightarrow S_2$ transition dipole moment (yellow arrow); vibrational transition dipole moment of the C-H bending vibrations of APO (black arrow); $S_0 \rightarrow S_4$ transition dipole moment (blue arrow).

Fig. 9.32 using Density Functional Theory (DFT) at B3LYP/6-31G* level of theory, as implemented in Gaussian03 [156].

9.5.1 Assignment of the absorption spectrum

In Fig. 9.32 the lowest energy part of the linear absorption spectrum of APO in deuterated chloroform (circles) is plotted together with the gas phase calculated spectrum (red bars). In the experimental absorption spectrum two distinct bands appear at 271 nm and 279 nm. The band at 271 nm has an oscillator strength 4.5 times higher than the band at 279 nm. A third very weak band around 300 nm can only be found at higher concentrations (data not shown). To gain deeper insight into the origin of these absorption bands quantum chemistry calculations were performed on the gas phase optimized geometry given in Fig. 9.32. Gas phase MS-CASPT2 vertical excitation energies are presented in Table 9.1. The first singlet excited state S_1 of APO is found to have B_1 symmetry and is assigned to a $\pi^*_{OO} \rightarrow \sigma^*_{OO}$ transition

Table 9.1. MS-CASPT2/CASSCF vertical singlet excitation energies (ΔE in nm), corresponding assignments, oscillator strengths f, and x,y,z transition dipole moment (TDM) components. Most intense bands are highlighted in color.

State	Symmetry	Assignment	ΔE	f	TDM (x/y/z)
S_0	1^1A1	—	0.00	—	—
S_1	1^1B1	$\pi^*_{OO} \rightarrow \sigma^*_{OO}$	327	0.0003	(0.06/0/0)
S_2	2^1B1	$\pi_{CC} \rightarrow \pi^*_{CC}$	290	0.0021	(-0.14/0/0)
S_3	1^1A2	$\pi^*_{OO} \rightarrow \pi^*_{CC}$	286	0.0000	—
S_4	1^1B2	$\pi^*_{OO} \rightarrow \pi^*_{CC}$	284	0.0220	(0/-0.45/0)
S_5	2^1A2	$\pi_{CC} \rightarrow \pi^*_{CC}$	267	0.0000	(0/0/0.10)
S_6	2^1A1	$\pi_{CC} \rightarrow \pi^*_{CC}$	266	0.0011	—

absorbing at 327 nm with a very weak oscillator strength. The second singlet state S_2 displays also B_1 symmetry and corresponds to a $\pi_{CC} \rightarrow \pi^*_{CC}$ transition. This band is centered at 290 nm and according to the computed oscillator strength it should be about 7 times more intense than the S_1 band. The transition from the ground to the fourth singlet state S_4 with B_2 symmetry is the next band with nonzero oscillator strength. The excitation is assigned to a $\pi^*_{OO} \rightarrow \pi^*_{CC}$ transition absorbing at 284 nm with a 10 times higher oscillator strength than the $S_0 \rightarrow S_2$ transition.

The comparison of theoretically and experimentally determined absorption bands (cf. Fig. 9.32) shows a good agreement of the experiment and the gas phase calculations, if taking into account a small shift due to the solvent. The experimental absorption bands of APO in chloroform are located at 271 nm, 279 nm, and about 305 nm with decreasing oscillator strengths. The calculated ab initio MS-CASPT2 excitation energies (red bars) are located red shifted at 284 nm, 290 nm, and 327 nm, also with decreasing oscillator strengths (see Table 9.1). Taking into account the blue shifting of the experimental bands due to solvent effects with respect to the gas phase theoretical calculations, the following tentative assignment of the experimental spectrum is made: the experimental band centered at 271 nm is assigned to the $\pi^*_{OO} \rightarrow \pi^*_{CC}$ transition, and the band at 279 nm is assigned to the $\pi_{CC} \rightarrow \pi^*_{CC}$ transitions. Both are red-shifted compared to the gas phase theoretical calculations. Finally, the spectral tail lying between 290-320 nm is attributed to the $\pi^*_{OO} \rightarrow \sigma^*_{OO}$ transition, which is theoretically predicted at 327 nm with a very low oscillator strength.

These results are also consistent with the early theoretical work of Kearns and Khan [146]. Furthermore, they are in reasonable agreement with the ex-

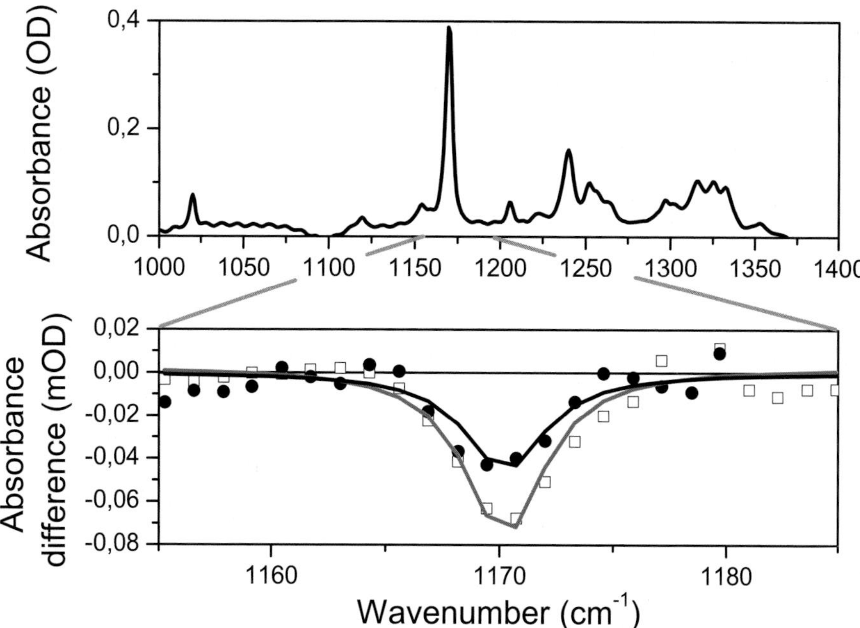

Fig. 9.33. Upper panel: Infrared absorption spectrum of APO in chloroform; Lower panel: Absorbance difference 25 ps after APO excitation with 273 nm. Black line: Lorentzian fit of the bleaching signal (solid circles) with perpendicular polarized light pulses. Blue line: Lorentzian fit of the bleaching signal (open squares) with parallel polarized light pulses. The ratio of the bleach amplitudes correspond to a relative angle of $28 \pm 20°$ between electronic and vibrational TDM.

perimental APO absorption spectrum recorded in solutions of CH_2Cl_2 and CH_3CN, reported by the groups of Brauer [147], Rygaudy [149], and Eisenthal [148], respectively.

On the other hand, our results differ significantly from the semiempirical calculations and experimental work of Gudipati and coworkers [150] who did not find any low-lying transitions involving a significant contribution of endoperoxide orbitals, and reported, instead, anthracene and $^1O_2$ cycloreversion products as the only reaction products observed upon UV irradiation.

In order to confirm the theoretical assignments, the character of the excited states are probed experimentally by determining the orientation of the electronic transition dipole moment (TDM) within the molecule. This is done by measuring the relative angle between the orientations of the electronic TDM and a specific vibrational transition dipole moment, whose orientations are related to the geometry of the molecule.

Comparison of the calculated and measured relative angles between the electronic and vibrational transition dipole moments can help elucidating the character of the excited state transition. In APO, the orientations of the electronic TDMs are the same for $S_0 \rightarrow S_1$ and $S_0 \rightarrow S_2$ transitions: parallel to the x-axis (orange arrow in Fig. 9.32). However, the $S_0 \rightarrow S_4$ transition is perpendicular to these, and parallel to the y-axis (blue arrow in Fig. 9.32). The investigated C-H bending vibration $\delta(CH)$ of APO (black arrow in Fig. 9.32) around 1170 cm^{-1} has a vibrational dipole moment oriented parallel to the y-axis. Excitation at 273 nm results in a bleaching of the $\delta(CH)$ absorption band. The bleaching signal of the vibrational $\delta(CH)$ transition at 1170 cm^{-1} is plotted in Fig. 9.33 for different relative polarization directions of the pump and the probe beam. The bleaching for probing with perpendicular polarized light is weaker than for probing with parallel polarized light indicating a relative angle of less than 54.7. Analysis of the bleaching signal gives an angle of 28 ± 20, indicating an orientation of the electronic TDM excited at 273 nm more parallel to the y-axis than to the x-axis. Thus, the absorption band at 273 nm can be assigned to the $S_0 \rightarrow S_4$ transition, and not to the $S_0 \rightarrow S_1$ or $S_0 \rightarrow S_2$ transitions. These findings support the above assignment of the absorption bands and the calculated character of the transitions. To summarize: on the basis of the new data we assign the absorption band at 271 nm to the $S_0 \rightarrow S_4$ transition, the band at 279 nm to the $S_0 \rightarrow S_2$ transition and the band around 300 nm, which exhibits a very low oscillator strength, both in the experiment as well as in the calculations, to the $S_0 \rightarrow S_1$ transition.

9.5.2 Analysis of Photodissociation Mechanisms

In order to get more insight into the dual photodissociation mechanism of APO, the dissociation curves for its ground and lowest energy excited states along the two relevant coordinates leading to oxygen biradical rearrangement, R_{OO}, and cycloreversion products, R_{CO}, were calculated at the MS-CASPT2/ANO-S//CASSCF/ANO-S level of theory. The potential energy curves for homolytic O-O endoperoxide bond breaking are plotted in Fig. 9.34, while the potentials for the concerted and step-wise photodissociation mechanism associated to the C-O cleavage are shown in Figs. 9.35a and 9.35b, respectively.

From the S_1 energy profile plotted in Fig. 9.34, which shows a $\pi^*_{OO} \rightarrow \sigma^*_{OO}$ character, it can be observed that the electronic energy decreases as the R_{OO} distance is increased, reaching its minimum around 2.1 Å. From there, it starts to increase again, showing thereby a binding character. As already stated before, electron excitation into σ^*_{OO} orbitals would significantly weaken the O-O endoperoxide bridge favoring its homolytic rupture. This way, APO molecules excited into the S_1 state would be "trapped" in this potential. Preliminary wave packet simulations in this one-dimensional model indicate vibrational oscillations of ca. 80 fs (full period); at longer time scales, the vibrational energy is expected to be dissipated into other normal modes, leading to relaxed oxygen diradical molecules that would finally evolve to lower energy rearrangement products. This is consistent with the photolysis results published in the works of Brauer [147] and Eisenthal, [148] who obtain rearrangement

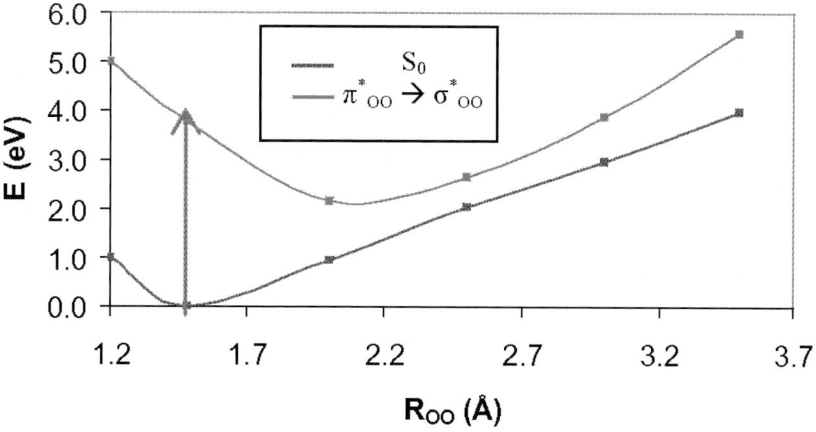

Fig. 9.34. Potential energy curves for the ground and the lowest energy B_1 singlet excited states of APO along the R_{OO} dissociation coordinate. The red arrow indicates the experimental irradiation frequency, which results in the observation of oxygen diradical rearrangement products [147].

products and the parent aromatic hydrocarbon, upon irradiation of APO or 1,4-dimethyl-9,10-diphenyl-9,10-anthracene-endoperoxide at $\lambda \geq 270$nm.

In order to study the mechanism behind cycloreversion processes, the lowest B_1 states along the C-O dissociation coordinate for the concerted and step-wise singlet oxygen loss were also computed. The concerted dissociation of the two C-O bonds in APO is described by considering the stretching of both C-O distances, while keeping the rest of the degrees of freedom frozen. The potential's profile of Fig. 9.35a points to the existence of a conical intersection between the $\pi_{CC} \rightarrow \pi^*_{CC}$ (S_2) excited state, populated upon irradiation, and the $\pi^*_{OO} \rightarrow \sigma^*_{CO}$ dissociative state. The population of σ^*_{CO} orbitals through the conical intersection is consistent with the weakening of the CO bond and the subsequent oxygen loss. However, this process is highly unlikely to occur as the curve crossing is located significantly above the experimental irradiation energy [147]. These findings, therefore, indicate that the possibility that the cycloreversion process takes place through a concerted mechanism should be ruled out a step-wise cycloreversion is more favourable.

In order to model the step-wise mechanism for the molecular singlet oxygen loss, we considered the cleavage of a single C-O bond, keeping the second C-O distance as well as the other reaction coordinates frozen. Figure 9.35b displays schematically the diabatic potential energy curves for the ground electronic state and the lowest energy states of B_1 symmetry of APO. Similar to the concerted cycloreversion mechanism, several singlet/singlet crossings also can be observed for the nonconcerted mechanism. Starting from the $\pi_{CC} \rightarrow \pi^*_{CC}$

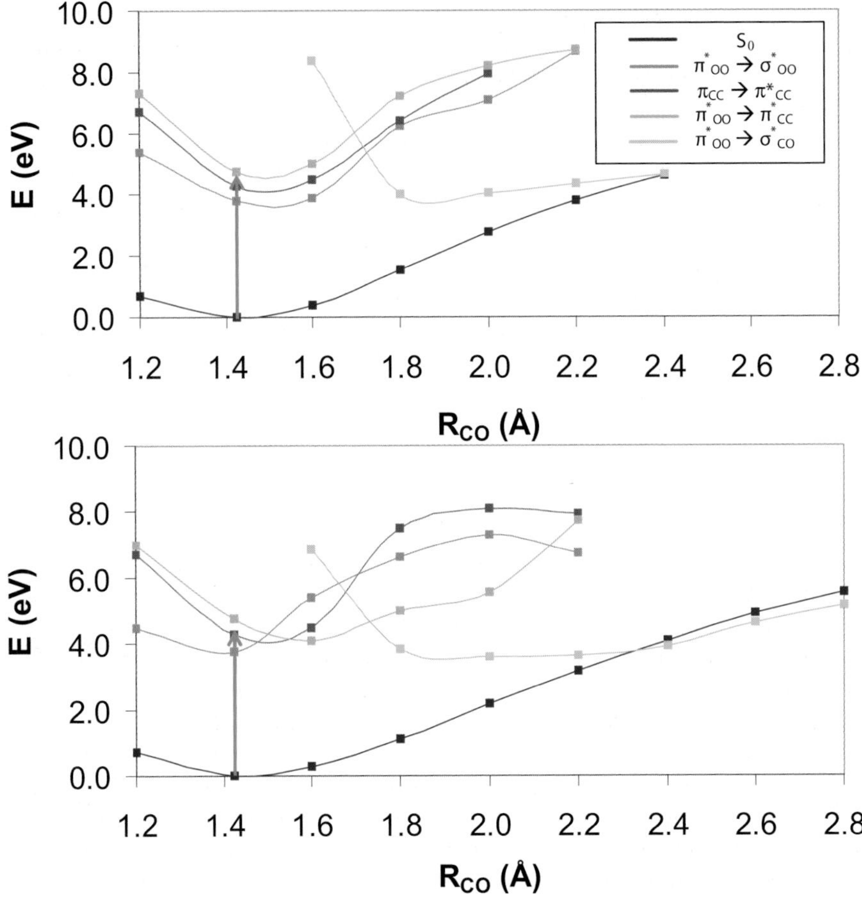

Fig. 9.35. Diabatic potential energy curves for the ground and the lowest energy B_1 singlet excited states of APO along the R_{CO} dissociation coordinate, assuming a concerted (a) and a step-wise mechanism (b) for $^1O_2$. The red arrow represents the experimental irradiation energy that results in the observation of cycloreversion products [147].

S_2 excited state, which is the only accessible state for irradiation at $\lambda \geq$ 279nm, the system would be able to overcome the S_2/S_1 and S_2/S_3 crossings through vibronic couplings, finally reaching the $\pi^*_{OO} \rightarrow \pi^*_{CC}$ S_3 excited state, which in turn presents a crossing around a R_{CO} distance of 1.70 Å with the $\pi^*_{OO} \rightarrow \sigma^*_{CO}$ dissociative state, directly associated with singlet oxygen loss. The main difference to the concerted mechanism is that in this case all the calculated singlet/singlet crossings, including the $\pi^*_{OO} \rightarrow \pi^*_{CC}/\ \pi^*_{OO} \rightarrow \sigma^*_{CO}$ curve crossing lie below the energy stored by the system upon irradiation; therefore, through nonadiabatic dynamics fragmentation can occur. Once more, these results are consistent with the experimental cycloreversion quantum yields ranging from 0.25-0.35, reported in Brauer's [147] and Eisenthal's [148] work when they irradiated at the wavelength region assigned to the S_2 band.

As seen in Fig. 9.35, a second crossing for a R_{CO} distance of 2.35 Å, is also found between the dissociative $\pi^*_{OO} \rightarrow \sigma^*_{CO}$ and APO's ground. This would account for the portion of the quantum yield returning back to the ground state.

Finally, it is worth mentioning that our high-level quantum mechanical description of the photodissociation mechanisms of APO agrees well with the theoretical calculations made by Kearns and Khan, [146] who predicted cycloreversion and O-O homolysis products from $\pi^*_{OO} \rightarrow \sigma^*_{CO}$ and $\pi^*_{OO} \rightarrow \sigma^*_{OO}$ transitions, respectively.

9.5.3 The role of triplet excited states

Until here, only the role of singlet excited states in O-O homolysis and cycloreversion reactions has been discussed. However, some authors [146, 148, 150] have pointed to the possible role of excited triplet states in the homolytic O-O bond rupture. Our MS-CASPT2/CASSCF vertical triplet state energies show that the first triplet excited state, similar to the first singlet excited state, has a $\pi^*_{OO} \rightarrow \sigma^*_{OO}$ character. The next higher energy triplet states exhibit $\pi_{CC} \rightarrow \pi^*_{CC}$ and $\pi^*_{OO} \rightarrow \pi^*_{CC}$ character and show different symmetry. There are three triplet states lying in between the S_0 and S_1 state and four triplet states in between the S_1 and S_2 state. From these facts, one could expect the occurrence of a nonnegligible probability for intersystem crossing. Nevertheless, no definitive conclusions can be drawn without computing potential energy surfaces along the R_{OO} and R_{CO} dissociation coordinates and corresponding spin-orbit couplings.

In summary, anthracene-9,10-endoperoxide has been chosen as a model to understand the photochemical behavior of aromatic oxygen carrier photosensitizers with endoperoxide groups. These are able to directly release singlet oxygen upon light irradiation, which is the main cytotoxic agent used in photodynamic therapy for the treatment of cancers and against viral/bacterial diseases.

Upon UV irradiation, endoperoxides exhibit at least two different photodissociation processes: i) cycloreversion reactions that lead to molecular singlet oxygen and the parent aromatic hydrocarbon, and ii) O-O homolysis reactions that lead to oxygen diradical rearrangement products. Using nonlinear femtosecond polarization resolved spectroscopy and ab initio multiconfigurational MS-CASPT2/CASSCF calculations the linear absorption spectrum and the character of the low energy excited states of APO has been assigned.

9.6 Metastable conformational structure and dynamics: Peptide between gas phase via clusters and aqueous solution

E. Meerbach, C. Schütte, I. Horenko, and B. Schmidt

The dynamics of biomolecular systems is characterized by the existence of biomolecular conformations which can be understood as metastable geometrical large scale structures [157]. On the longest time scales, biomolecular dynamics is dominated by flipping processes between these conformations, while on shorter time scales, the dynamical behavior is governed by flexibility within these conformations, resulting in a rich temporal multiscale structure of time-dependent observables. An approach to characterize biomolecular dynamics is to construct reduced models reflecting the "effective dynamics" of the system.

It is a promising idea to describe the effective dynamics of a biomolecular systems by means of a Markov chain with discrete states $D_1, \ldots, D_m$, representing the metastable conformations, and a transition matrix $P = (p_{kj})$, describing the "flipping dynamics" between these states. Efficient algorithmic identification of the metastable conformations is a challenging problem, which has recently been tackled by set-oriented approaches [158–160]. In the context of the present work Hidden Markov Models (HMM) are used to extract the effective dynamics between hidden metastable molecule conformations from observable time series. e.g. the torsional angles of the backbone of biopolymers obtained by MD simulations [161]. In addition, the flexibility within conformations can be modeled by stochastic differential equations (SDE), thus comprising the HMMSDE model [162–164]. As the description of internal flexibility by SDEs also accounts for relaxation from one metastable conformation to another, this approach narrows the gap between "flipping dynamics" and transition path computation, as described in, e.g., [165].

9.6.1 Metastability and the transfer operator approach

In the following we shortly summarize the algorithmic idea of the transfer operator approach, omitting most of the theoretical background. Instead we concentrate on the question how to set up an effective dynamics from a given time series, e.g., trajectory data. The reader interested in a mathematically more rigorous description is referred to [159, 166–169], readers not familiar with the basic notations of Markov chain theory are referred to [170]. First we explicate the concept of metastability of a Markov chain and the key idea for the identification of metastable states. Note that the transfer operator is an object in continuous state space, while we present the concept on discrete state space. Therefore only the discretized equivalent of the transfer operator, the transition matrix, appears in the following.

Consider a Markov chain $\{X_k\}_{k\in\mathbb{N}}$ on a discrete state space $\mathbf{X} = \{1,2,\ldots,n\}$ specified by a stochastic transition matrix $P = (p_{kj})$, with

$$p_{kj} = \mathbb{P}\,[X_{l+1} = j|X_l = k],$$

denoting the conditional probability to jump from k to j within one time step. Furthermore, assume that the Markov chain is irreducible, aperiodic and reversible, i.e. a unique and strictly positive stationary distribution $\pi = (\pi_k)$ exists with $\pi_k p_{kj} = \pi_j p_{jk}$ for all $k, j \in \mathbf{X}$. A subset $B \subset \mathbf{X}$ is called metastable if

$$\mathbb{P}\,[X_{l+1} \in B|X_l \in B] \approx 1,$$

i.e., if the process is in subset B it is very likely to stay there within the next time step.

A decomposition $\mathrm{d} = \{D_1,\ldots,D_m\}$ of the state space $\mathbf{X}$ is defined as a collection of disjoint subsets $D_k \subset \mathbf{X}$ covering $\mathbf{X}$, i.e. $\cup_{k=1}^m D_k = \mathbf{X}$. The metastability of a decomposition d is defined as the sum of the metastabilities of its subsets, i.e. for each arbitrary decomposition d_m of the state space $\mathbf{X}$ into m sets its metastability measure is defined as

$$M(\mathrm{d}_m) = \sum_{j=1}^{m} \mathbb{P}\,[X_{l+1} \in D_j|X_l \in D_j].$$

For given m, the optimal metastable decomposition into m sets maximizes the functional M. In particular the appropriate number m of metastable subsets must be identified. Both the determination of m and the identification of the metastable subsets can be achieved via spectral analysis of the transition matrix P, as the following holds:

> Due to reversibility, all eigenvalues of the transition matrix P are real. Metastable subsets can be detected via eigenvalues close to the maximal dominant eigenvalue $\lambda = 1$, i.e., the number of metastable subsets in the metastable decomposition is equal to the number of eigenvalues close to 1, including $\lambda = 1$ and accounting for multiplicity, while the rest of the spectrum is separated through a spectral gap from 1. Among other possibilities, the sign structure of the eigenfunctions allows the identification of the metastable subsets [159, 168, 171].

Therefore the road map to determine metastable states on basis of a time series reads as follows:

1. Discretize the state space of the time series and extract a transition matrix by counting transitions between the discrete states.
2. Use the spectral properties of the transition matrix to obtain metastable sets, yielding a coarse-grained description.

There are two remarks to be made on this road map.
First: Discretizing the state space is a nontrivial task, as typical biomolecular systems contain hundreds or thousands of degrees of freedom. Fortunately,

chemical observations reveal that—even for larger biomolecules—the curse of dimensionality can be circumvented by exploiting the hierarchical structure of the dynamical and statistical properties of biomolecular systems: only relatively few essential degrees of freedom may be needed to describe the conformational transitions.

Second: After discretizing the state space there is a choice in the lag time τ used to obtain the transition matrix. If τ^* is the time step between subsequent data points then the lag time τ can be set to $r\tau^*$ by evaluating transitions from every kth sampled step to every $(k+r)$th, $r \geq 1$, sampled step. Taking $r > 1$ corresponds to a coarser discretization of the time domain of the originally continuous dynamics. Different values of r give rise to different transition matrices. Therefore, subsets of the state space are metastable with respect to a certain timescale. By choosing r sufficiently large one can decrease correlations between subsequent time steps and therefore ensure that the Markov description is a proper description.

9.6.2 Illustrative example

We give a short and simplistic example to highlight the procedure outlined above. Consider the one dimensional time series $(Y_t)_{t=t_1,\ldots,t_N}$, with constant sampling time $\tau^* = t_{j+1} - t_j$, shown in Fig. 9.36, which clearly exhibits metastable behavior. We discretize the state space $[-180\ 180]$ into 9 equidistant boxes, the numbering of the boxes randomly chosen. If $N_r(j,k)$ denotes the number of transitions from box j to box k in r steps and $N_r(k)$ the number of data points in box k, we obtain a reversible transition matrix $P = (p_{kj})$, with respect to the time lag $\tau = r * \tau^*$, by setting

$$p_{kj} = \frac{N_r(j,k) + N_r(k,j)}{N_r(j) + N_r(k)}. \tag{9.11}$$

The obtained matrix seems to exhibit no special structure, but computing the spectrum, for $r = 1$, yields

$$\sigma(P) = \{1, 0.98, 0.55, 0.34, \ldots\},$$

indicating two metastable states. The information contained in the eigenvector belonging to the second eigenvalue is used to identify the metastable subsets, i.e. boxes with the same eigenvector sign are assigned to the same metastable state. Permuting the matrix such that boxes belonging to the same metastable set are neighbors, results in a dominantly blockdiagonal structure. Aggregating the states in each metastable set results in a two state "effective dynamics" with transition matrix

$$\begin{pmatrix} 0.989 & 0.011 \\ 0.013 & 0.987 \end{pmatrix},$$

and a stationary distribution $\pi = (0.56\ 0.44)^T$.

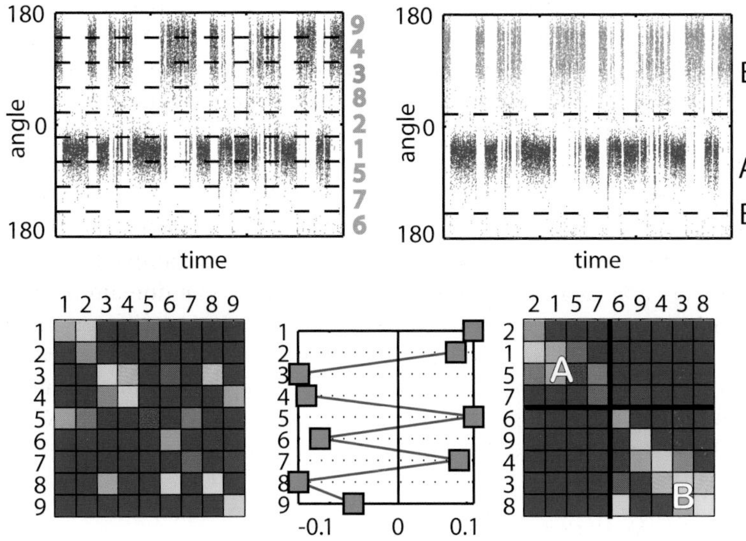

Fig. 9.36. Top left: A time series of circular data. The equidistant space discretization in 9 randomly numbered boxes is indicated with dashed lines. Bottom left: The obtained stochastic transition matrix, blue (dark) color represents entries near zero, while red (light) entries are corresponding to entries close to one. Bottom middle: The sign structure of the second eigenvector allows assignment to metastable states. Bottom right: The matrix permuted according to the eigenvector structure exhibits a block structure. Top right: Aggregating the discretization boxes belonging to the same metastable state yields a two state model.

9.6.3 The hidden Markov model approach

Assume that we extracted a time series Y_t from, e.g., MD-simulation, which do not necessarily completely specify the state of the molecule at time t, but rather some low-dimensional observable, for example, some or all torsion angles or a set of essential degrees of freedom. As the Markov property does not hold for projections of Markov processes in general, we have to be aware that the process on the (torsion angles) subspace might no longer be Markovian. Nevertheless, we assume that there is an unknown metastable decomposition into m sets $D_1, \ldots, D_m$, in the full dimensional system. We then can premise that, at any time t, the system is in one of the metastable states D_{j_t} to which we simply refer by j_t in the following. However, the time series (j_t) is hidden, i.e., neither known in advance nor observed, while the series (Y_t) is called the output series or the observed sequence.

This scenario can be represented by a Hidden Markov Model (HMM). A HMM abstractly consists of two related stochastic processes: a hidden process j_t, that fulfills the Markov property, and an observed process Y_t, that depends on the state of the hidden process j_t at time t. A HMM is fully specified by the initial distribution π, the transition matrix P of the hidden Markov process j_t, a rate matrix in continuous time, and the probability distributions that govern the observable Y_t depending on the respective hidden state j_t.

In the standard versions of HMMs the observables are assumed to be identically and independent distributed (i.i.d.) random variables with stationary distributions that depend on the respective hidden states [161]. Within the scope of molecular dynamics this setting corresponds to the simple case where the (sampling) time lag τ is comparable with the relaxation times within the metastable states, while being are sufficiently smaller than the mean exit times of the metastable states. In other words one expects the process to sample the restricted invariant density before exiting from a metastable state, and the sampling time of the time series is long enough to assume statistical independence between steps. Nevertheless, if this is not the case, only a slight modification of the model structure is required to include the relaxation behavior: Instead of i.i.d. random variables, an Ornstein-Uhlenbeck (OU) process serves as a model for the output behavior in each hidden state. The HMM then takes the form [162]:

$$\dot{Y}_t = -\nabla V^{(j_t)}(Y_t) + \sigma^{(j_t)}\dot{W}_t, \tag{9.12}$$

$$j_t \; : \; \mathbf{R}^1 \to \{1, 2, \ldots, m\}, \tag{9.13}$$

where j_t are the realizations of the hidden Markov process with discrete state space, W_t is standard "white noise", and $\sigma^{(j)}$ the state dependent diffusion matrix. $V^{(j)}$ is assumed of the form

$$V^{(j)}(Y) = \frac{1}{2}(Y - \mu^{(j)})^T D^{(j)}(Y - \mu^{(j)}) + V_0^{(j)}, \tag{9.14}$$

i.e., to be harmonic potentials with $\mu^{(j)}$ and $D^{(j)}$ denoting equilibrium position and Hesse-matrix of the OU process within conformation j. This process is therefore specified by the parameters $\Theta^{(j)} = (\mu^{(j)}, D^{(j)}, \sigma^{(j)})$. Since the output process is given by a stochastic differential equation we will refer to this model modification as HMMSDE. Its entire parameter set is $\Theta = (\Theta^{(1)}, \ldots, \Theta^{(m)}, P)$, where P denotes the transition matrix of the Markov chain in (9.13).

The parameter set of this model can be estimated from a time series via a modified EM (expectation-maximation) algorithm [172], as described in [162, 163, 173]. Once the model parameters are estimated one can use the Viterbi algorithm [174] to compute the most probable path of hidden states, the Viterbi path, given an observation sequence. So both can be obtained, a dynamical model and the assignment of data points to the hidden, not observed, states. In contrast to the transfer operator approach, where the number of metastable states is extracted from the spectral properties of the transition matrix, we have to specify the number of metastable states as an input parameter for the EM algorithm. Since this number is in general unknown, a combination of both algorithms is used: First, guess a sufficiently large number of metastable sets, compute a Viterbi path and, second, reduce the number of states by set up a transition matrix from the (discrete) Viterbi path and cluster with the transfer operator approach.

9.6.4 Conformation analysis of a glycine dipeptide analogue (GLDA)

As an example we investigate the dynamics of glycine dipeptide analogue (CH_3–CO–NH–CH_2–CO–NH–CH_3), which is one of the smallest (artificial) peptides containing two peptide bonds (CO–NH). Thus the essential degrees of freedom are the torsional rotations of the individual peptide units (–CO–NH–) about the backbone of the chain, where Φ and Ψ describe the torsion of the N-terminus (CH_3–CO–NH–) and the C-terminus (–CO–NH–CH_3), respectively, with regard to the central CH_2 group, see Fig. 9.37. The plane spanned by the two angles Φ, Ψ is referred to as Ramachandran plane [175], with values of $(\pm 180^\circ, \pm 180^\circ)$ corresponding to a fully extended conformation of the chain. For longer polypeptide chains these angles serve to characterize typical secondary structural motifs such as helices and sheets.

9.6.4.1 GLDA in the gas phase

We used an empirical force field (Gromos 53a6 [176,177]) to obtain a potential energy surface in the two essential degrees, i.e., varying the Ramachandran angles and minimizing the potential energy wrt. the other degrees of freedom.

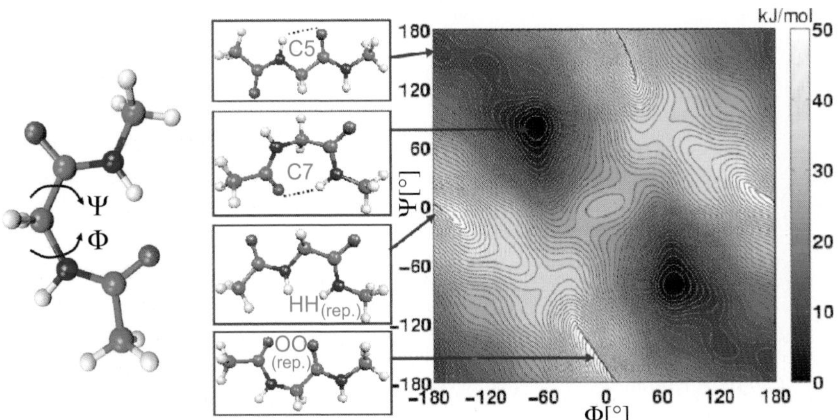

Fig. 9.37. Left: Glycine dipeptide analogue. The two marked torsion angles are the essential degrees of freedom as the peptide units are essentially planar. Right: Potential energy surface with respect to the two torsion angles. The local minima are due to the formation of hydrogen bonds between the peptide groups, the local maxima are due to repulsion.

The potential energy surface shown in Fig. 9.37, reflects the symmetry of the peptide. Local minima correspond to energetically favorable formations of ringlike structures, including seven or five atoms (C5 and C7), closed by (strongly) frustrated intramolecular hydrogen bonds. The maximal regions are corresponding to intramolecular repulsion of *–O O–* and *–H H–*. The accuracy of the potential energy surface is of course limited by the quality of the empirical force field used, but comparison with the potential energy surface computed by quantum chemical calculations yields a qualitatively similar picture [178–182].

Performing a finite temperature MD-simulation at 300K using a Berendsen thermostat of the dipeptide in vacuum, samples the low energy regions of C5 and C7 in the Ramachandran plane, see Fig 9.38E. As these regions are separated by a barrier of approx. 9 kJ/mol one would expect a metastable behavior at a reasonable timescale. To confirm this assumption, we use HMMSDE to extract a Viterbi path, assuming 4 (hidden) states, for each of the Ramachan-

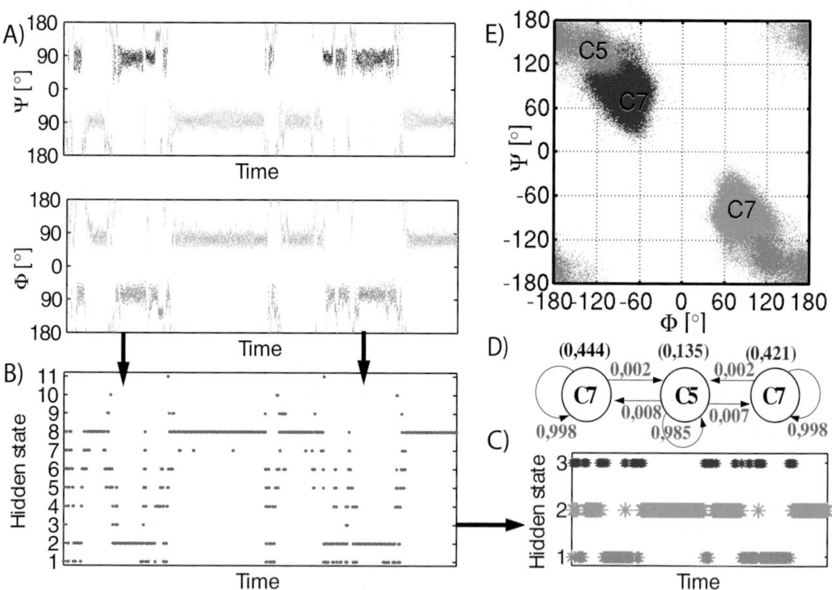

Fig. 9.38. A) A sample of the two dimensional torsion angle time series extracted from a 300K MD-simulation. The data points are colored according to the allocation to hidden states by HMM-SDE. B) Superposition of the two Viterbi paths yields into a joint Viterbi path with 11 hidden states. C) Using the transfer operator approach lumps the 11 hidden states to 3 metastable hidden states. D) A transition network for the 3 hidden states with a time lag of $\tau = 0.1ps$, red numbers denote the conditional transition probabilities, numbers in brackets the weight of each state. E) The data points of the torsion angle time series in the Ramachandran plane colored according to their allocation to metastable states.

dran angles Φ, Ψ. Superposition of these two Viterbi paths yields a Viterbi path with 11 states. As the number of hidden states is only determined by our initial guess, we use the transfer operator approach to further reduce the number of states. Setting up the stochastic transition matrix P with time lag $\tau = 0.1ps$, and computing the first five eigenvalues:

$$\sigma(P) = \{1, 0.9948, 0.9296, 0.8152, 0.6540\ldots\},$$

indicates 3 metastable sets. Using the information coded in the three dominant eigenvectors we aggregate the 11 states to 3 states. In Fig. 9.38E the assignment of the data points to these metastable (hidden) states is shown, the plot reveals, that we identified the (symmetric) C7 conformations and the C5 conformation. Transition probabilities between these hidden states can be obtained by using (9.11), Fig. 9.38D. Thus we have obtained a detailed dynamical picture of the effective finite temperature dynamics of GLDA in vacuum.

9.6.4.2 GLDA in aqueous solution

To compare these results with the dynamics in solution phase we consider GLDA in a $(3.5\text{nm})^3$ box filled with 1405 rigid water molecules. Using a cutoff for electrostatic interactions of 1.1 nm and a Berendsen-temperature coupling to the solvent of 300K, we performed an MD-simulation over 2.5 ns with an integration timestep of 2 fs using again the Gromos 53a6 force field and recording the atom positions every 20 fs. After discretizing the Ramachandran plane in $5^\circ \times 5^\circ$ boxes, the free energy for each box B_i can be calculated by [183, 184]

$$\Delta G(B_i) = -k_B T(\log(\mathbb{P}[B_i]) - \log(\max_i \mathbb{P}[B_i])).$$

This free energy surface, see Fig. 9.39, has, due to intermolecular interactions (water-GLDA), a considerably richer structure than the potential energy surface in gas phase, Fig. 9.37. Analyzing the torsion angle time series with the HMMSDE approach, assuming 24 metastable sets, perfectly distinguishes regions belonging to different local minima in the free energy surface, Fig. 9.39. As these local minima are separated by low energy barriers, compared with thermally available energies, it is not a priori clear that they correspond to metastable states on timescales of e.g. 1 ps. An instructive picture is obtained by setting up the transition matrix, based on the 24 states of the obtained Viterbi path, and plotting the eigenvalues against the time lag used, Fig. 9.40. It can be clearly seen that 4 metastable states are persistent even for larger time lags, as there is an obvious gap after the first 4 eigenvalues.

The cause of the metastable states can be revealed by taking the intermolecular interactions into account. These interactions are mainly due to H-bond bridges between the peptide groups ($-CO-NH-$) and neighboring water molecules. Each peptide group provides a donor pair (NH) and an acceptor (O) for

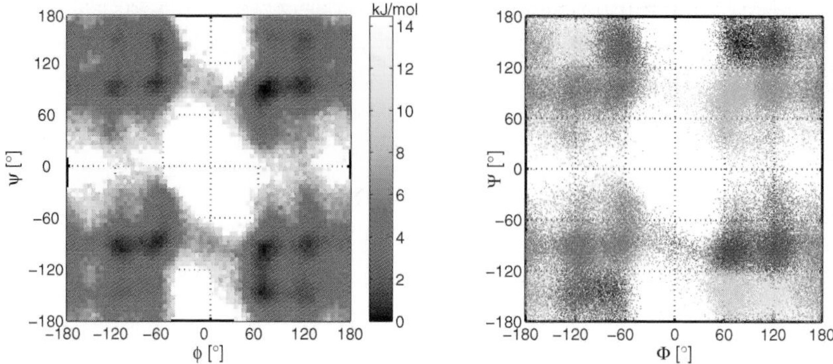

Fig. 9.39. Left: The free energy surface obtained from MD-Simulation with water has a considerably richer structure than the potential energy surface in vacuum. Right: Coloring the data points according to state allocation obtained by HMMSDE perfectly distinguishes the local minima in the free energy surface (for 24 metastable sets assumed).

H-bond bridges attracting solvent molecules. If we restrict to microsolvation structures, i.e. GLDA with 1 or 2 waters, it is clear that ring like structures, as shown in Fig. 9.41, are energetically favorable, as each water molecule can participate in two H-bond bridges [185]. Besides the possible extension of the C7 and C5 structure to C7+2 (atoms), C7+2+2, C5+2 and C5+2+2 structures, H-bond bridges can stabilize structures that do not occur in vacuum, namely the C6 structures shown, where the –O O– and –H H– repulsion is overcome by inserting water molecules to form a ring structure. In the following we denote by C7+X the C7+2+2 and C7+2 structures collectively (with analogous meaning of C6+X and C5+X).

These microsolvation ring structures can also be identified in the fully solvated system. Comparison of the plots in Fig. 9.40 reveals the nature of the four metastable states. They correspond to regions where C7+X/C5+X or C6+X ring structures occur.

The assumption that microsolvation structures cause metastability can be further supported by redoing the analysis based on six metastable states, see Fig. 9.42. Even though there are regions in the Ramachandran plane allowing different microsolvation structures, e.g. regions allowing C7+X or C5+X structures have an overlap, the plots of the data points belonging to a metastable state and the plots of data points with certain microsolvation structures show an obvious similarity. Again, this indicates that the origin of the metastable conformational structures is related to the formation of different microsolvation environments of the solute molecule.

In conclusion, we demonstrated the ability of the HMMSDE approach to reflect structural properties of the complete simulated system by analysis of only two essential degrees of freedom. The effective reduction of dimension-

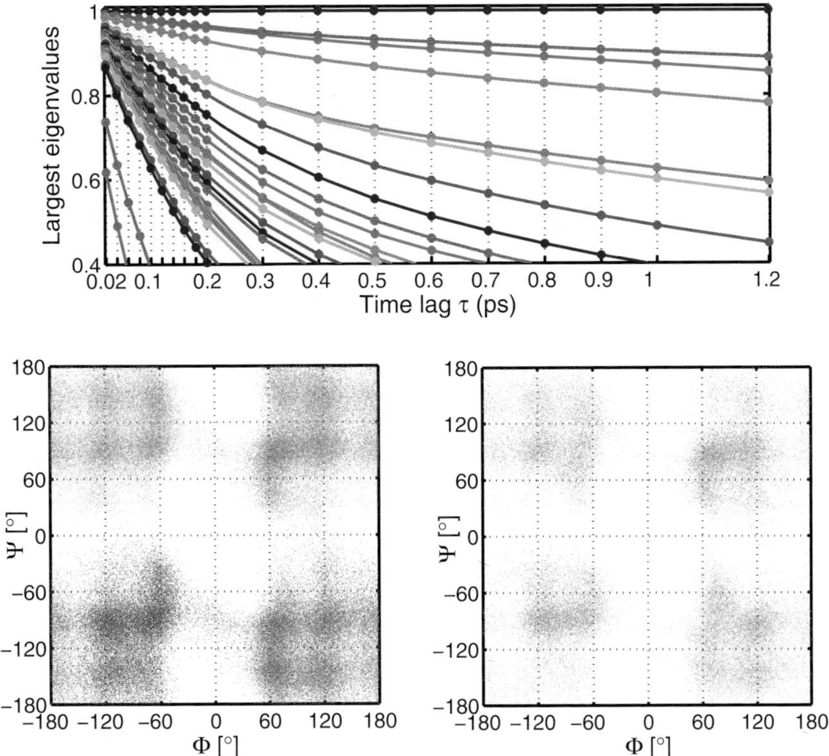

Fig. 9.40. Left: Dependence of the eigenvalues of the transition matrix obtained from the Viterbi path on different time lags. Middle: Data points are colored according to a clustering of the Viterbi path in 4 metastable sets. Right: Data points exhibiting a C6 microsolvation structure (magenta) and data points exhibiting a C7 or C5 microsolvation structure (cyan).

ality achieved for the GLDA example is due to the capability of HMMSDE to distinguish different dynamical behavior in time series. Although the system investigated here is of moderate size, HMMSDE appears to be a promising approach to beat the curse of dimensionality in more complex systems. Currently conformational analysis of DNA fragments containing 15 base pairs has been pursued in our laboratory [173]. Hence, it is believed that this approach is much more general and can be used beyond the context of MD-simulations. Possible applications range, e.g., from transient spectroscopy to the analysis of climate or financial data.

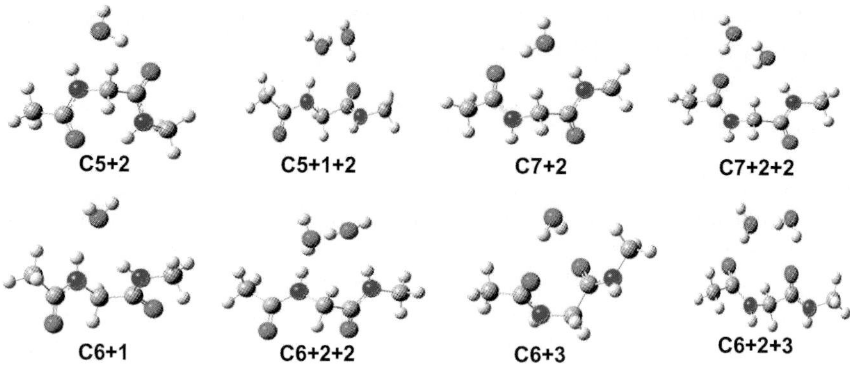

Fig. 9.41. Energetically favorable microsolvation structures with one or two water molecules.

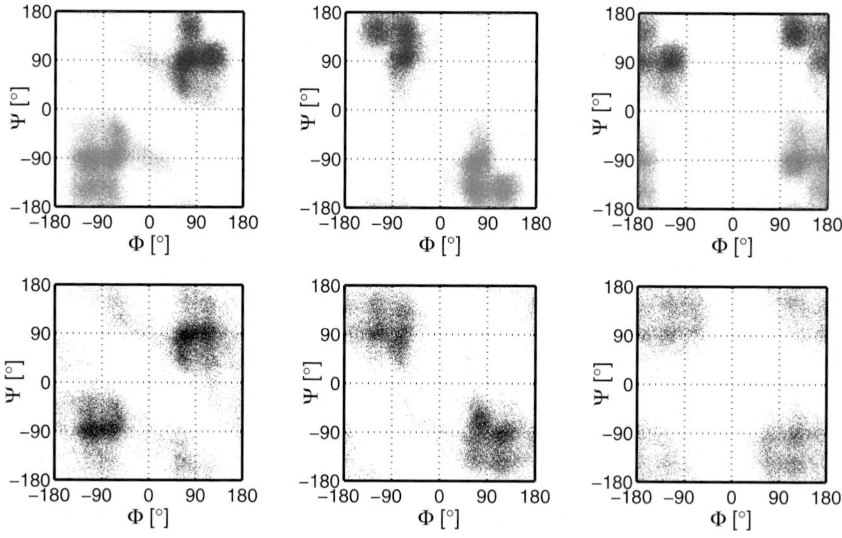

Fig. 9.42. *Top:* Data points belonging to different metastable states after clustering in 6 states, two symmetric equivalent states are shown in one plot. *Bottom:* Data points exhibiting C6+X (left), C7+X (middle) or C5+X (right) ring structures.

9.7 Detection and identification of bacteria in air using femtosecond spectroscopy

J.-P. Wolf

Rapid detection and identification of pathogenic aerosols such as *Bacillus anthracis* (anthrax) and *variola major* (smallpox) from potential bioterrorism release, as well as infectious diseases such as severe acute respiratory syndrome (SARS) or legionellosis, are urgent safety issues. However, in order to efficiently protect populations from bioterrorism in public locations or to prevent nosocomial infections in hospitals (e.g. Methicillin resistant *Staphylococcus aureus*) or epidemic spread, bioaerosol detectors need to be very fast (typ. minutes) and very selective (to discriminate pathogen from nonpathogen particles and minimize false alarm rates). Unfortunately such detectors are today sorely lacking.

One can roughly distinguish two research tracks that were followed these latter years in order to reach this important but difficult objective: [186] biochemical identification procedures, which are selective but slow, and [187] optical devices, which are fast but not specific enough.

The currently available biochemical techniques, such as polymerase chain reaction (PCR) [186–189], fluorescence in situ hybridization (FISH) [186,187, 190], antibiotic resistance determination [191,192], or chip matrix of biochemical microsensors [193,194], can identify the genus and species of the bacteria or virus. However, after the alarm has sounded, these biochemical assay procedures require time (many hours or even days). Even the recently developed ultrafast B cell adaptive immune-based sensor [195] and real-time PCR [188] will require more than an hour [195] for the total assay.

Several groups have developed sophisticated optical systems to distinguish bioaerosols from nonbioaerosols based on fluorescence [196–199] and/or elastic scattering [200,201]. The most advanced experiments address each individual aerosol particle, which is spectrally analyzed [196,202,203]. These instruments can run continuously, in-situ, and in real-time to provide rapid warning/alarm for the existence of a few potentially life threatening bio-aerosols in the midst of a vast number of nonbio-aerosols. The major flaw inherent in these instruments is frequent false alarms because the UV-Visible fluorescence systems are incapable of distinguishing different molecules with similar fluorescence peaks (such as tryptophan and diesel particles or cigarette smoke) [202,204]. Fig 9.43 shows, as an example, the similitude in the fluorescence spectra of diesel fuel, tryptophan and *Bacillus subtilis*, which is a biosimulant for *Bacillus anthracis*. It is even more elusive using LIF to expect identifying different kinds of harmless from pathogen microorganisms. In this latter case, extensive studies based on principle components analysis of the excitation vs fluorescence spectral matrix have been performed to distinguish the spectrum from one type of bacteria from another, without success [205].

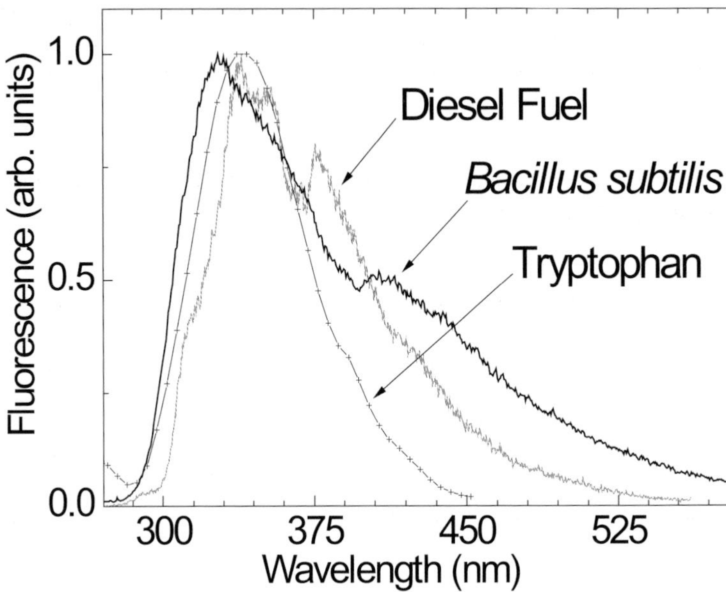

Fig. 9.43. Comparison of the fluorescence spectra of Tryptophan, *B. Subtilis* and Diesel fuel.

However, an attractive capability of optical techniques is remote sensing, and especially Lidar (Light Detection and Ranging). The Lidar technique [206] allows mapping aerosols in 3D over several kilometers, similar to an optical Radar. It might be able to detect the release and spread of potentially harmful plumes (such as bioterrorism release or legionella from industrial cooling towers) at large distance and then provide time to take the necessary measures to protect populations or identify sources. So far, Lidar detection of bioaerosols has been demonstrated either using elastic scattering [206] or LIF [206, 207]. However, the distinction between bioaerosols and nonbio-aerosols was either impossible (elastic scattering only) or very unsatisfactory for LIF-Lidars (for the same reason as point monitors, e.g. interference with pollens and organic particles like traffic related soot or polycyclic aromatic hydrocarbons (PAHs)).

To overcome these difficulties, there is an interest in exciting the fluorescence with ultrafast laser pulses in order to access specific molecular dynamical features. Recent experiments using coherent control and multiphoton ultrafast spectroscopy have shown the ability of discriminating between molecular species that have similar one-photon absorption and emission spectra [208, 209]. Two-photon excited fluorescence (2PEF) and pulse shaping techniques should allow for selective enhancement of the fluorescence of one molecule versus another that has similar spectra. Optimal dynamic discrimination (ODD) [210] of similar molecular agent provides the basis for generating optimal signals for detection.

9.7.1 Multiphoton excited fluorescence (MPEF) and multiphoton ionization (MPI) in aerosol microparticles

Femtosecond laser pulses provide very high pulse intensity at low energy, which allows inducing nonlinear processes in particles without deformation due to electrostrictive and thermal expansion effects.

The most prominent feature of nonlinear processes in aerosol particles is strong localization of the emitting molecules within the particle, and subsequent backward enhancement of the emitted light [211, 212]. This unexpected behavior is extremely attractive for remote detection schemes, such as Lidar applications. Localization is achieved by the nonlinear processes, which typically involve the n-th power of the internal intensity $\mathrm{I}^n(\mathrm{r})$ (r for position inside the particle). The backward enhancement can be explained by the reciprocity (or time reversal) principle: Re-emission from regions with high $\mathrm{I}^n(\mathrm{r})$ tends to return toward the illuminating source by essentially retracing the direction of the incident beam that gave rise to the focal points. This backward enhancement has been observed for both spherical and non spherical [213] microparticles. More precisely, we investigated, both theoretically and experimentally, incoherent multiphoton processes involving $n = 1$ to 5 photons [212]. For $n = 1$, 2, 3, MPEF occurs in bioaerosols because of natural fluorophors such as amino acids (tryptophan, tyrosin), NADH (nicotinamide adenine dinucleotide), and flavins. Figure 9.44 shows the LIF spectra of various bacteria, and the contribution of each fluorophor in the fluorescence spectrum under 266 nm excitation. The strongly anisotropic MPEF emission was demonstrated on individual microdroplets containing tryptophan, riboflavin, or other synthetic fluorophors [211–213]. The experiment was performed such that each individual microparticle was hit by a single laser shot. The aerosol source was based on a piezo driven nozzle, which precisely controlled the time of ejection of the microparticles. Figure 9.45 shows the MPEF angular distribution and the comparison between experimental and theoretical (Lorentz-Mie calculations) results for the one- (400 nm) (Fig. 9.45a), two- (800 nm) (Fig. 9.45b) and three-photon $(1, 2\mu m)$ (Fig. 9.45c) excitation process. They show that the fluorescence emission is maximum in the direction toward the exciting source. The directionality of the emission is dependent on the increase of n, because the excitation process involves the nth power of the intensity $\mathrm{I}^n(\mathrm{r})$. The ratio Rf= $\mathrm{P}(180°)/\mathrm{P}(90°)$ increases from 1.8 to 9 when n changes from 1 to 3 (P is the emitted light power). For 3PEF, fluorescence from aerosol microparticles is therefore mainly backwards emitted, which is ideal for Lidar experiments. For $n = 5$ photons we investigated laser induced breakdown (LIBS) in water microdroplets, initiated by multiphoton ionization. The ionization potential

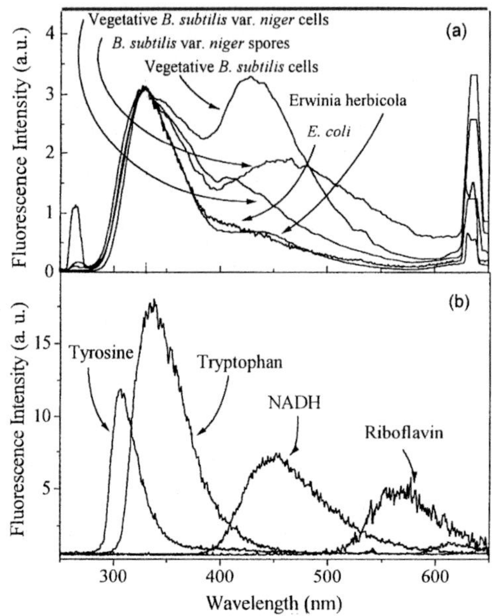

Fig. 9.44. Fluorescence spectra of bacteria and respective contributions of amino acids, NADH and flavins [202].

of water molecules is E_{ion}=6.5 eV, so that 5 photons are required at a laser wavelength of 800 nm to initiate the process of plasma formation. The growth of the plasma is also a nonlinear function of $I^n(r)$. We showed that both localization and backward enhancement strongly increases with the order n of the multiphoton process, exceeding $Rf = U(180°)/U(90°) = 35$ for $n = 5$ [214]. As for MPEF, LIBS has the potential of providing information about the aerosols composition.

9.7.2 MPEF-Lidar detection of biological aerosols

The first multiphoton excited fluorescence Lidar detection of biological aerosols was performed using the Teramobile system. The Teramobile (www.teramobile.org) is the first femtosecond-terawatt laser based Lidar [215], and was developed by a French-German consortium, formed by the Universities of Jena, Berlin, Lyon, and the Ecole Polytechnique (Palaiseau).

The bioaerosol particles, consisting of 1 μm size water droplets containing 0.03 g/l Riboflavin (typical for bacteria), were generated at a distance of 50 m from the Teramobile system. Riboflavin was excited by two photons at 800 nm and emitted a broad fluorescence around 540 nm. This experiment [215, 216] is the first demonstration of the remote detection of bioaerosols using a 2PEF-femtosecond Lidar. The broad fluorescence signature is clearly

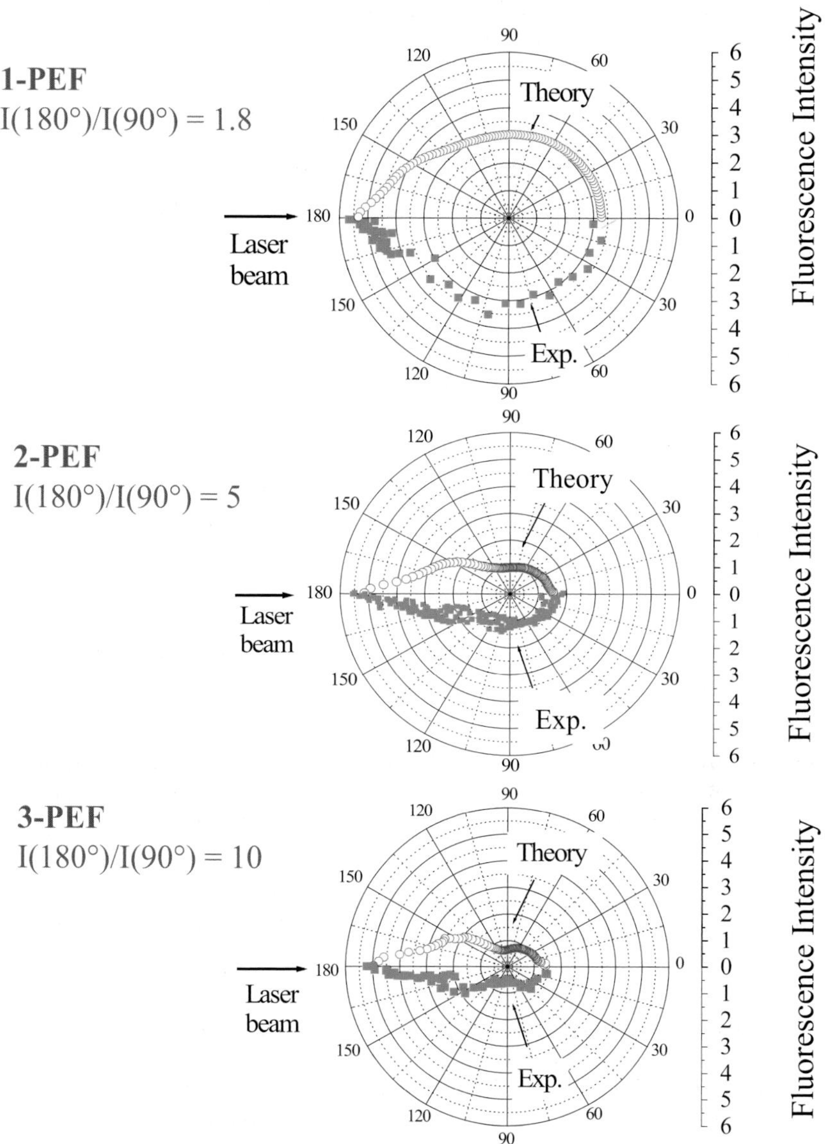

Fig. 9.45. Backward enhancement of the fluorescence from aerosol microparticles in the case of nonlinear excitation (MPEF) [211].

observed from the particle cloud (typ. $10^4 p/cm^3$), with a range resolution of a few meters (Fig. 9.46). As a comparison, droplets of pure water do not exhibit any parasitic fluorescence in this spectral range. However, a background is observed for both types of particles, arising from the scattering of white light generated by filaments in air.

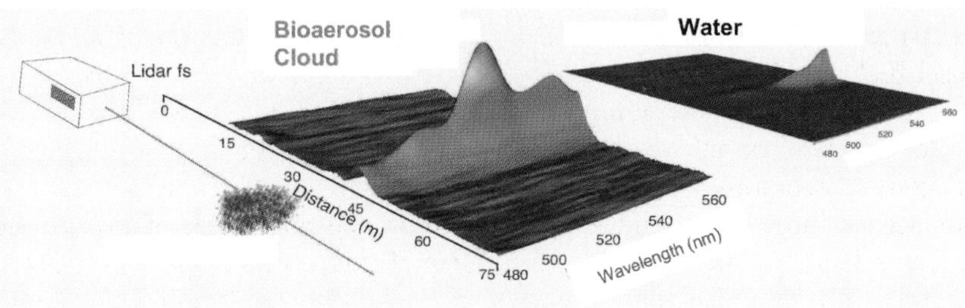

Fig. 9.46. Two-photon excited fluorescence (2PEF)-Lidar detection of bioaerosols [215, 216].

Primarily, MPEF might be advantageous as compared to linear LIF for the following reasons: (1) MPEF is enhanced in the backward direction and (2) the transmission of the atmosphere is much higher for longer wavelengths. For example, if we consider the detection of tryptophan (another typical bio-tracer that can be excited with 3 photons of 810 nm), the transmission of the atmosphere is typically 0.6 km^{-1} at 270 nm, whereas it is $3 \cdot 10^{-3}$ km^{-1} at 810 nm (for a clear atmosphere, depending on the background ozone concentration). This compensates the lower 3-PEF cross-section compared to the 1-PEF cross-section at distances larger than a couple of kilometers. The most attractive feature of MPEF is, however, the possibility of using pump-probe techniques, as described hereafter in order to discriminate bioaerosols from background interferents such as traffic related soot or PAHs.

9.7.3 Pump-probe spectroscopy to distinguish bioaerosols from background organic particles in air

As mentioned before, a major drawback inherent in LIF instruments is the frequency of false identification because UV-Vis fluorescence is incapable of discriminating different molecules with similar absorption and fluorescence signatures. While mineral and carbon black particles do not have strong fluorescence signals, aromatics and polycyclic aromatic hydrocarbons (PAH) from organic particles and Diesel soot strongly interfere with biological fluorophors such as amino acids [202, 204]. The similarity between the spectral signatures of organic and biological molecules under UV-Vis excitation lies in the fact that similar π-electrons from carbonic rings are involved. Therefore, aromatics or PAHs (such as naphthalene) exhibit absorption and emission bands similar to those of amino acids like Tyrosine or Trp. Some shifts are present because of differences in specific bonds and the number of aromatic rings, but the broad featureless nature of the bands renders them almost indistinguishable. Moreover, the different environments of Trp in bacteria (e.g., they are contained in many proteins) and the mixtures of PAHs in transportation generated particles, blur their signatures. As shown before in Fig. 9.43, the fluorescence spectra (resulting from excitation at 270 nm) of Trp, *Bacillus Subtilis* and Diesel fuel are almost identical.

A novel femtosecond pump-probe depletion (PPD) concept was recently developed [217], based on the time-resolved observation of the competition between excited state absorption (ESA) into a higher lying excited state and fluorescence into the ground state. This approach makes use of two physical processes beyond that available in the usual fluorescence spectrum: (1) the dynamics in the intermediate pumped state and (2) the coupling efficiency to a higher lying excited state. More precisely, as shown in Fig. 9.47, a femtosecond pump pulse (τ=120fs) at 270 nm transfers a portion of the ground state S_0 population of Trp to its $S_1(\{v'\})$ excited state (corresponding to a set of vibronic levels, *i.e.*, {v'}). The vibronic excitation relaxes by internal energy redistribution to lower {v'} modes, associated with charge transfer processes (CT), conformational relaxation, and intersystem crossing with repulsive $\pi\sigma*$ states [218]. After vibronic energy redistribution, fluorescence is emitted from $S_1(\{v'\})$ within a lifetime of 2.6 ns. By illuminating the amino acid with a

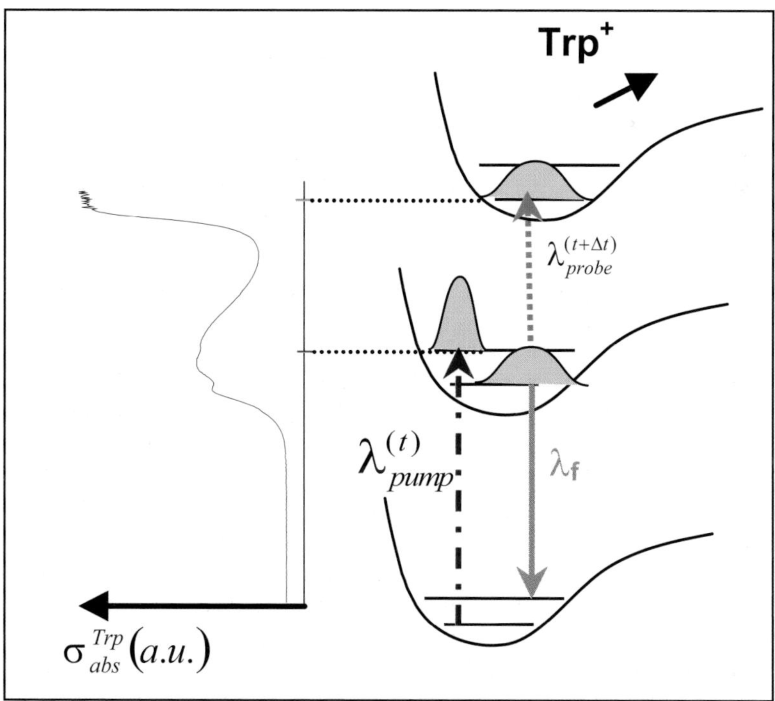

Fig. 9.47. Femtosecond pump-probe depletion (PPD) technique to distinguish biomolecules from organic interferents.

second pulse (probe) at 800 nm after a certain time delay from the pump pulse, the S_1 population is decreased and the fluorescence depleted. The 800 nm femtosecond laser pulse induces a transition from S_1 to an ensemble of higher lying S_n states, which are likely to be autoionizing [219], but also undergo radiationless relaxation into S_0 [220]. By varying the temporal delay between the pump and probe, the dynamics of the internal energy redistribution within the intermediate excited potential surface S_1 is explored. In principle, as different species have distinct S_1 surfaces, discriminating signals can be enhanced in this fashion.

Figure 9.48 shows the pump-probe depletion dynamics of the S_1 state in Trp as compared to a circulation of Diesel fuel and Naphthalene in cyclohexane, one of the most abundant fluorescing PAHs in Diesel. While depletion reaches as much as 50% in Trp for an optimum delay of $\Delta t = 2$ ps, Diesel fuel and Naphthalene appear almost unaffected (within a few percent), at least on these timescales. This remarkable difference allows for efficient discrimination between Trp and organic species, although they exhibit very similar linear

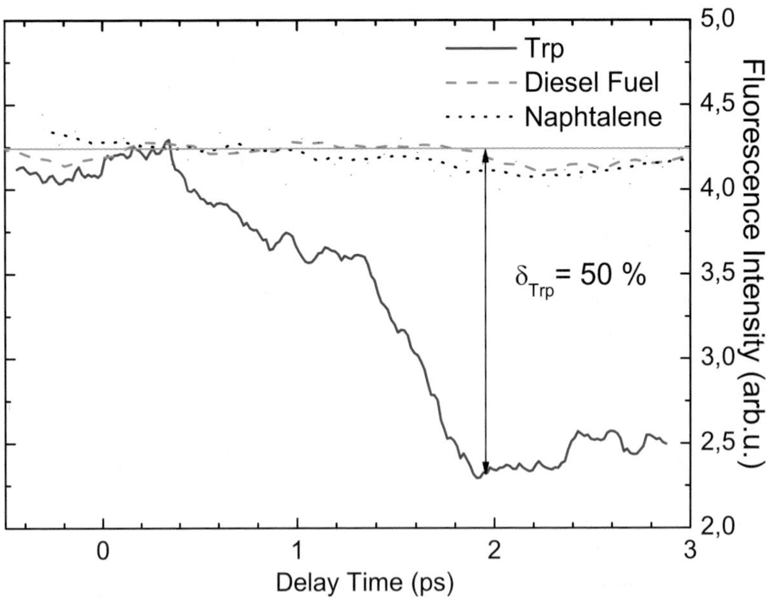

Fig. 9.48. Femtosecond pump-probe depletion PPD applied to the distinction of Trp from traffic related PAH's [217].

excitation/fluorescence spectra (Fig. 9.43). Two reasons might be invoked to understand this difference: (1) the intermediate state dynamics is predominantly influenced by the NH- and CO- groups of the amino acid backbone and (2) the ionization potential and other excited states are higher for the PAHs contained in Diesel than for Trp by about 1 eV so that excitation induced by the probe laser is much less likely in the organic compounds. The particular dynamics of the internal energy redistribution in the S_1 state of Trp (Fig. 9.48), and in particular the time needed to start efficient depletion, is not fully interpreted as yet. Further electronic structure calculations are required to better understand the process, especially on the higher lying S_n potential surfaces.

In order to more closely approach the application of detecting and discriminating bioagents from organic particles in air, we repeated the experiment with bacteria, including *Escherichia coli*, *Enterococcus* and *Bacillus subtilis* (BG). Artefacts due to preparation methods have been avoided by using a variety of samples, i.e. lyophilized cells and spores, suspended either in pure or in biologically buffered water (i.e. typically $10^7 - 10^9$ bacteria per cc). The observed pump-probe depletion results are remarkably robust (Fig. 9.49), with similar depletion values for all the considered bacteria (results for Enterococcus, not shown in the figure, are identical), although the Trp microenvironment within the bacteria proteins is very different from water. These unique features can be used for a novel selective bioaerosol detection technique that avoids interference from background traffic related organic particles in the air: The excitation shall consist of a pump-probe sequence with the optimum delay $\Delta t = 2$ps, and the fluorescence emitted by the mixture will be measured as the probe laser is alternately switched on and off. This pump-probe two-photon differential fluorescence method will be especially attractive for an active remote detection technique such as MPEF-Lidar, where the lack of discrimination between bioaerosols and transportation related organics is currently most acute.

These results, based on very simple pump-probe schemes, are very encouraging and open new perspectives in the discrimination capability of bioaerosols in air. We intend to extend the technique by applying more sophisticated excitation schemes (e.g., optimally shaped pulses), related to coherent control, in order to better distinguish bioaerosols from non-bioaerosols, but also to gain selectivity among the bacteria themselves. Theoretical calculations were recently performed, which show that under some conditions, optimal dynamic discrimination (ODD) can lead to efficient distinction between 3 species that exhibit almost the same spectral characteristics [210].

9.7.4 Toward LIBS identification of bacteria in air

As mentioned above, laser induced plasma line emission is also enhanced in the backward direction when microparticles are excited by a femtosecond laser. Although nanosecond-laser LIBS (nano-LIBS) has already been applied to the

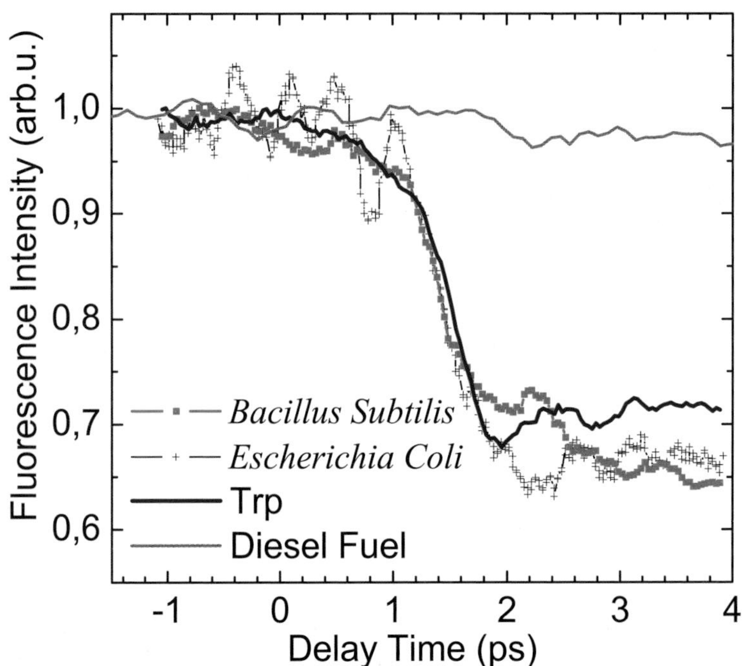

Fig. 9.49. Femtosecond pump-probe depletion of in vivo bacteria [217].

study of bacteria [221, 222], femtosecond lasers open new perspectives in this respect. The plasma temperature is, indeed, much lower in the case of femtosecond excitation, which strongly reduces the blackbody background and interfering lines from excited N_2 and O_2 from the air. This allows performing time gated detection with very short delays, and thus observing much richer and cleaner spectra from the biological sample. This crucial advantage is shown in Fig. 9.50, where the K line emitted by a sample of *Escherichia coli* is clearly detected in femto-LIBS and almost unobservable under ns-laser excitation [223]. Thanks to the low thermal background in fs-LIBS, 20-50 lines are recorded for each bacterial sample considered in the study (*Acinetobacter*, *Escherichia coli*, *Erwinia*, *Shewanella* and *Bacillus subtilis*) . A systematic sorting with sophisticated algorithms is in progress in order to evaluate whether the spectra are sufficiently different to unambiguously identify each species [224]. The results are already promising and we show (Fig. 9.51), as an example, the difference between *Escherichia coli* and *Bacillus subtilis* for the Li line intensity (also observed for the Ca line). This observation can be un-

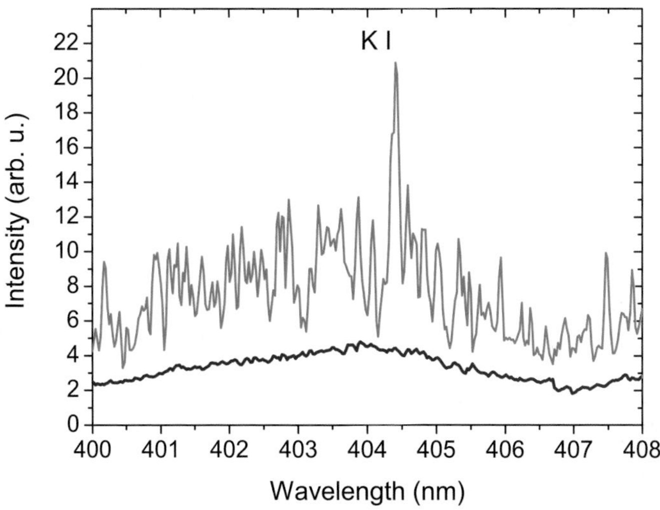

Fig. 9.50. Comparison of ns-LIBS (blue) and fs-LIBS (red) results on the K line (KI) of *Escherichia coli* [224].

derstood by the typical difference of the cell wall structure between a Gram+ and a Gram- bacterium. The ratio of these lines as compared to Na for example constitutes an "all optical Gram test". Low temperature plasma is not the only advantage of fs-LIBS: the ablation process itself seems different. fs-LIBS acts more as a direct bond-breaking and evaporation process instead of thermal vaporization. This particular ablation process could be put into evidence as not only atoms and ions lines were observed but also molecular signatures such as CN or C_2 [225]. It was shown in particular that these molecular species are directly ablated from the sample, and not created by recombination of C atoms or ions with Nitrogen from the air (which occurs for ns excitation). Obtaining molecular signatures in addition to trace elements is a significant improvement of the method. The presence of CN molecules is, for instance, a good indicator for a biological material.

In conclusion, femtosecond spectroscopy opens new ways for the optical detection and identification of bioaerosols in air. Its unique capability of distinguishing molecules that exhibit almost identical absorption and fluorescence signatures is a key feature for identifying bacteria in a background of urban aerosols. The technique can also be applied for the remote detection of the microorganisms, if a nonlinear Lidar based configuration is chosen, as for the Teramobile. A more difficult task will be the distinction of one bacteria species from another, and in particular the identification of pathogen from

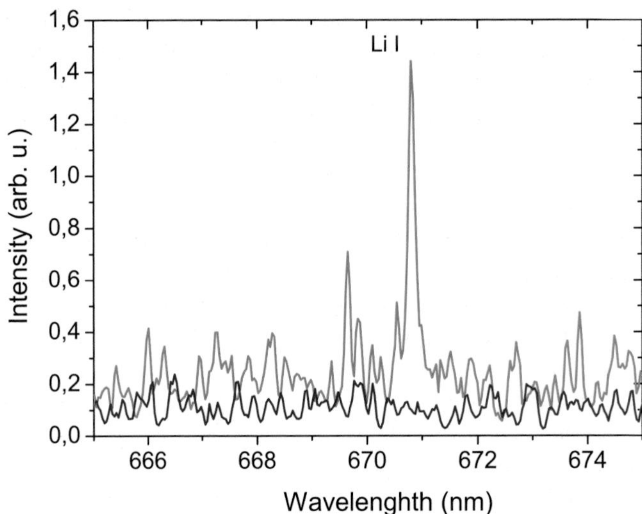

Fig. 9.51. Distinction of *Escherichia coli* (blue) from *Bacillus subtilis* (red) through their different Li-line emission (Li l) [224].

nonpathogen bioaerosols. A possible way to reach this difficult goal might be ODD. Another option could be provided by LIBS, as the wealth of emission lines under femtosecond excitation might allow to target some biological process that is characteristic from one type of bacteria. Moreover, it was recently demonstrated that LIBS could also be performed with a femtosecond-Lidar [226] and thus allow remote identification of particles in air.

9.8 Outlook

The investigation of biological systems with laser spectroscopic techniques remains a vibrant field. In particular, the strength of femtosecond spectroscopy for characterization of key reactive events and relaxation pathways, and identification of important inter- and intramolecular interactions has been illustrated in this chapter. Another avenue of growing importance is the use of femtosecond laser spectroscopies to obtain structure information in real-time on changes induced by optical excitation. For example, the monitoring of time-dependent absorption band shifts of tryptophan residues, induced by excitation of nearby bacteriorhodopsin was shown as a tool to this end. These experiments established the connection between translocation of charge [27] and the skeletal changes of the conjugate chain [23]. In addition, it also stressed the role of the dynamic electric force fields, which drive structural dynamics and govern enzymatic reactions. The capacity of infrared probing for determine the orientations of molecular electronic transition dipole moments within the molecular frame was moreover demonstrated on anthracene-endoperoxide, and in combination with quantum mechanical calculations a definitive assignment of reactive electronic states could be made. For more complex biological systems and for tackling both the fast and the slower dynamics, Molecular Dynamics simulations, such as presented in Sect. 9.6, will prove to be a powerful tool for singling out crucial 3-dimensional conformations that can be connected to key events. Methods such as these, are expected to greatly contribute in the near future to enhancing our knowledge of 3-dimensional structural details in complex biological systems and their connection to the mechanisms that govern essential biological processes.

Insightful demonstrations were also presented of the use of creative multiple pulse manipulation schemes for separating overlapping spectral bands, manipulating excited and ground state populations, and knocking the investigated system out of equilibrium in biological systems of varying complexity, Sect. 9.3. These inherently multiphoton techniques can also be used to investigate excitation interactions in multichromophore systems [59, 60]. These new abilities allow significant new insights into fundamental biological events governing photosynthetic light harvesting [49,59,60,82], biological proton transfer reactions [69] and primary photoreception events [48, 48]. The fact that the control is achieved by manipulating electronic state populations allows the clear and concise interpretations of the experimental data.

On the theoretical front calculations of time-dependent wave packet propagations can provide information on time scales and branching ratios for competing reaction channels. These calculations also provide essential input for simulations based on a quantum dynamical Optimal Control Theory formulation, that can be used to predict the possibilities of experimentally manipulating the quantum yields of different reaction channels, or steering or even suppressing energy transfer and delocalization, by the use of shaped pulses. Input parameters for control possibilities are the amplitude (intensity), the phase, and the polarization of the different applied electromagnetic frequencies. We are only standing at the beginning of the tremendously exciting road of experimentally testing the possibilities optically altering the natural outcome biological processes.

Last but not least, femtosecond spectroscopy has led to new ways for outdoor detection and identification of airborne bioaerosols. The unique capability of distinguishing molecules with almost identical absorption and fluorescence signatures is a key feature for identifying bacteria in a background of urban aerosols, which will lead to clearly outlined applications.

The state-of-the-art achievements in laser technology, together with the continuously increasing computing power, due to the persisting validity of Moore's law, has clearly led to new spectroscopic and theoretical tools that continue to open up the possibilities of bringing the understanding of light-induced biological reactions to a higher levels.

Acknowledgments

I. Corral gratefully acknowledges La Fundación Ramón Areces for Financial Support and L. González the Berliner Förderprogram. The Teramobile project (www.teramobile.org) is funded by the Center National de la Recherche Scientifique (CNRS) and the Deutsche Forschungsgemeinschaft (DFG). J. P. Wolf gratefully acknowledges the members of the groups of R. Sauerbrey at the University Jena, A. Mysyrowicz at the ENSTA (Palaiseau). In particular, we wish to thank J. Kasparian, V. Boutou, F. Courvoisier, J. Yu, M. Baudelet, E. Salmon, D. Mondelain, G. Méjean, M. Rodriguez, H. Wille, S. Frey, R. Bourayou, and Y. B. Andre. The laboratory measurements on aerosols were performed in collaboration with the groups of R.K. Chang (Yale University), S.C. Hill (US Army Research Laboratories) and H. Rabitz (Princeton University). For these studies, RKC and JPW acknowledge NATO support SST-CLG977928.

The authors of Section 9.3 would like to thank colleagues that helped with this work including Luuk van Wilderen, Dr. Dorte Madsen, Dr. Ivo H.M. van Stokkum, Dr. Michael van der Horst. Results presented here were supported by the Netherlands Organization for Scientific Research (NWO) via the Dutch Foundation for Earth and Life Sciences (ALW) and the Stichting voor Fundamenteel Onderzoek der Materie, Netherlands (FOM). D.S.L. is grateful to the Human Frontier Science Program Organization for providing financial support with a long-term fellowship.

References

1. W. Behrens, U. Alexiev, R. Mollaaghababa, H.G. Khorana, M.P. Heyn, Biochem. **37**, 10411 (1998)
2. W. Behrens, (2004). Private communication
3. N. Grigorieff, T.A. Ceska, K.H. Downing, J.M. Baldwin, R. Henderson, J. Mol. Biol. **259**, 393 (1996)
4. P. Hamm, M. Zurek, T. Röschinger, H. Patzelt, D. Oesterhelt, Chem. Phys. Lett. **263**, 613 (1996)
5. L. Song, M.A. El-Sayed, J.K. Lanyi, Science **261**, 891 (1993)
6. K. Heyne, J. Herbst, B. Dominguez-Herradon, U. Alexiev, R. Diller, J. Phys. Chem. B **104**, 6053 (2000)
7. J. Herbst, K. Heyne, R. Diller, Science **297**, 822 (2002)
8. M.K. Johansson, H. Fidder, D. Dick, R.M. Cook, J. Am. Chem. Soc. **124**, 6950 (2002)
9. A. Warshel, Acc. Chem. Res. **14**, 284 (1981)
10. M.F. Perutz, Science **201**, 1187 (1978)
11. I. Gitlin, J.D. Carbeck, G.M. Whitesides, Angew. Chem. Int. Ed. **45**, 3022 (2006)
12. B.E. Cohen, T.B. McAnaney, E.S. Park, Y.N. Jan, S.G. Boxer, L.Y. Jan, Science **296**, 1700 (2002)

13. G.N. Phillips, M.L. Teodoro, T.S. Li, B. Smith, J.S. Olson, J. Phys. Chem. B **103**, 8817 (1999)
14. J.B. Jackson, A.R. Crofts, FEBS Letters **4**, 185 (1969)
15. J.L. Herek, T. Polivka, T. Pullerits, G.J. Fowler, C.N. Hunter, V. Sundström, Biochem. **37**, 7057 (1998)
16. R.A. Mathies, L. Stryer, Proc. Natl. Acad. Sci. **73**, 2169 (1976)
17. J. Huang, Z. Chen, A. Lewis, Phys. Chem. **93**, 3314 (1989)
18. D. Xu, C. Martin, K. Schulten, Biophys. J. **40**, 453 (1996)
19. J.T.M. Kennis, D.S. Larsen, K. Ohta, F.M. T, R.M. Glaeser, G.R. Fleming, J. Phys. Chem. B **106**, 6067 (2002)
20. A. Lewis, Proc. Natl. Acad. Sci. **75**, 549 (1978)
21. G. Wald, Science **162**, 230 (1968)
22. Q. Zhong, S. Ruhman, M. Ottolenghi, M. Sheves, N. Friedman, G.H. Atkinson, J. Delaney, J. Am. Chem. Soc. **118**, 12828 (1996)
23. T. Kobayashi, T. Saito, H. Ohtani, Nature **414**, 531 (2001)
24. D.W. McCamant, P. Kukura, R.A. Mathies, J. Phys. Chem. B **109**, 10449 (2005)
25. J. Dobler, W. Zinth, W. Kaiser, D. Oesterhelt, Chem. Phys. Lett **144**, 215 (1988)
26. S.J. Doig, P.J. Reid, R.A. Mathies, Phys. Chem **95**, 6372 (1991)
27. R. González-Luque, M. Garavelli, F. Bernardi, M. Merchán, M.A. Robb, M. Olivucci, Proc. Natl. Acad. Sci. **97**, 9379 (2000)
28. L. Salem, P. Bruckmann, Nature **258**, 526 (1975)
29. G.I. Groma, J. Hebling, C. Ludwig, J. Kuhl, Biophys. J. **69**, 2060 (1995)
30. J. Xu, A.B. Stickrath, P. Bhattacharya, J. Nees, G. Váró, J.R. Hillebrecht, L. Ren, R.R. Birge, Biophys. J. **85**, 1128 (2003)
31. G.I. Groma, A. Colonna, J.C. Lambry, J.W. Petrich, G. Váró, M. Joffre, M.H. Vos, J.L. Martin, Proc. Natl. Acad. Sci. **101**, 7971 (2004)
32. U. Zadok, A. Khatchatouriants, A. Lewis, M. Ottolenghi, M. Sheves, J. Am. Chem. Soc. **124**, 11844 (2002)
33. Y.R. Shen, *The principles of nonlinear optics* (Wiley-Interscience, Hoboken, N. J., 2003)
34. E. Pebay-Peyroula, G. Rummel, J.P. Rosenbusch, E.M. Landau, Science **277**, 1676 (1997)
35. H. Luecke, B. Schobert, H.T. Richter, J.P. Cartailler, J.K. Lanyi, J. Mol. Biol. **291**, 899 (1999)
36. P.R. Callis, Meth. Enzym. **278**, 113 (1997)
37. S. Schenkl, F. van Mourik, G. van der Zwan, S. Haacke, M. Chergui, Science **309**, 917 (2005)
38. S. Schenkl, F. van Mourik, N. Friedman, M. Sheves, R. Schlesinger, S. Haacke, M. Chergui, Proc. Nat. Acad. Sci. **103**, 4101 (2006)
39. M.R. Eftink, L.A. Selvige, P.R. Callis, A.A. Rehms, J. Phys. Chem. **94**, 3469 (1990)
40. M.L. Horng, J.A. Gardecki, A. Papazyan, M. Maroncelli, J. Phys. Chem. **99**, 17311 (1995)
41. B. Becher, F. Tokunaga, T.G. Ebrey, Biochem. **17**, 2293 (1978)
42. S. Georgakopoulou, R.N. Frese, E. Johnson, C. Koolhaas, R.J. Cogdell, R. van Grondelle, G. van der Zwan, Biophys. J. **82**, 2184 (2002)
43. Z.R. Grabowski, K. Rotkiewicz, W. Rettig, Chem. Rev. **103**, 3899 (2003)

44. W.M. Kwok, C. Ma, P. Matousek, A.W. Parker, D. Phillips, W.T. Toner, M. Towrie, S. Umapathy, J. Phys. Chem. A **105**, 984 (2001)
45. K. Nakanishi, V. Baloghnair, M. Arnaboldi, K. Tsujimoto, B. Honig, J. Am. Chem. Soc. **102**, 7945 (1980)
46. L.H. Andersen, I.B. Nielsen, M.B. Kristensen, M.O.A.E. Ghazaly, S. Haacke, M.B. Nielsen, M.A. Petersen, J. Am. Chem. Soc. **127**, 12347 (2005)
47. R. Jimenez, G.R. Fleming, *Ultrafast spectroscopy of photosynthetic systems, in Biophysical techniques in photosynthesis* (Kluwer Academic Publishers, Dordrecht, 1996)
48. D.S. Larsen, I.H.M. van Stokkum, M. Vengris, M.A. van der Horst, F.L. de Weerd, K.J. Hellingwerf, R. van Grondelle, Biophys. J. **87**, 1858 (2004)
49. D.S. Larsen, E. Papagiannakis, I.H.M. van Stokkum, M. Vengris, J.T.M. Kennis, R. van Grondelle, Chem. Phys. Lett. **381**, 733 (2003)
50. F. Gai, J.C. McDonald, P. Anfinrud, J. Am. Chem. Soc. **119**, 6201 (1997)
51. S. Ruhman, B. Hou, N. Freidman, M. Ottolenghi, M. Sheves, J. Am. Chem. Soc. **124**, 8854 (2002)
52. P. Changenet-Barret, C. Choma, E. Gooding, W. DeGrado, R.M. Hochstrasser, J. Phys. Chem. **104**, 9322 (2000)
53. S.A. Kovalenko, J. Ruthmann, N.P. Ernsting, J. Chem. Phys. **109**, 1894 (1998)
54. S.L. Logunov, V.V. Volkov, M. Braun, M.A. El-Sayed, Proc. Natl. Acad. Sci. **98**, 8475 (2001)
55. A.R. Holzwarth, *Data analysis in time-resolved measurements*, in *Biophysical techniques in photosynthesis* (Kluwer, Dordrecht, The Netherlands, 1996)
56. I.H.M. van Stokkum, D.S. Larsen, R. van Grondelle, Biochim. Biophys. Acta Bioenergetics **1657**, 82 (2004)
57. I.H.M. van Stokkum, D.S. Larsen, R. van Grondelle, Biochim. Biophys. Acta Bioenergetics **1658**, 262 (2004)
58. S.A. Rice, M. Zhao, *Optical control of molecular dynamics* (New York: Wiley Interscience, 2000)
59. E. Papagiannakis, I.H.M. van Stokkum, M. Vengris, R.J. Cogdell, R. van Grondelle, D.S. Larsen, J. Phys. Chem. B **110**, 5727 (2006)
60. E. Papagiannakis, M. Vengris, L. Valkunas, R.J. Cogdell, R. van Grondelle, D.S. Larsen, J. Phys. Chem. B **110**, 5737 (2006)
61. R.Y. Tsien, Ann. Rev. Biochem. **67**, 509 (1998)
62. M. Chattoraj, B.A. King, G.U. Bublitz, S.G. Boxer, Proc. Natl. Acad. Sci. **93**, 8362 (1996)
63. F. Yang, L. G, Moss, G.N. Phillips, Nat. Biotech. **14**, 1246 (1996)
64. T.M.H. Creemers, A.J. Lock, V. Subramaniam, T.M. Jovin, S. Volker, Nat. Struct. Biol. **6**, 557 (1999)
65. K. Brejc, T. Sixma, P. Kitts, S. Kain, R. Tsien, M. Ormo, S. Remington, Proc. Natl. Acad. Sci. **94**, 2306 (1997)
66. G.J. Palm, A. Zdanov, G.A. Gaitanaris, R. Stauber, G.N. Pavlakis, A. Wlodawer, Nat. Struct. Biol. **4**, 361 (1997)
67. H. Lossau, A. Kummer, R. Heinecke, F. Pollinger-Dammer, C. Kompa, G. Bieser, T. Jonsson, C.M. Silva, M.M. Yang, D.C. Youvan, M.E. Michel-Beyerle, Chem. Phys. **213**, 1 (1996)
68. K. Winkler, J.R. Lindner, V. Subramaniam, T.M. Jovin, P. Vöhringer, Phys. Chem. Chem. Phys. **4**, 1072 (2002)
69. J.T.M. Kennis, D.S. Larsen, I.H.M. van Stokkum, M. Vengris, J.J. van Thor, R. van Grondelle, Proc. Natl. Acad. Sci. **101**, 17988 (2004)

70. G. Striker, V. Subramaniam, C.A.M. Seidel, A. Volkmer, J. Phys. Chem. B **103**, 8612 (1999)
71. T. Polivka, V. Sundström, Chem. Rev. **104**, 2021 (2004)
72. F.L. de Weerd, I.H.M. van Stokkum, R. van Grondelle, Chem. Phys. Lett. **354**, 38 (2002)
73. C.C. Grandinaru, J.T.M. Kennis, E. Papagiannakis, I.H.M. van Stokkum, R.J. Cogdell, G.R. Fleming, R.A. Niederman, R. van Grondelle, Proc. Natl. Acad. Sci. **98**, 2364 (2001)
74. D. Zigmantas, T. Polivka, R.G. Hiller, A. Yartsev, V. Sundström, J. Phys. Chem. A **105**, 10296 (2001)
75. J.P. Zhang, T. Inaba, Y. Watanabe, Y. Koyama, Chem. Phys. Lett. **332**, 351 (2000)
76. E. Papagiannakis, J.T.M. Kennis, I.H.M. van Stokkum, R.J. Cogdell, R. van Grondelle, Proc. Natl. Acad. Sci. **99**, 6017 (2002)
77. P. Tavan, K. Schulten, J. Chem. Phys. **85**, 6602 (1986)
78. E. Papagiannakis, S.K. Das, A. Gall, I.H.M. van Stokkum, B. Robert, R. van Grondelle, H.A. Frank, J.T.M. Kennis, J. Phys. Chem. B **107**, 5642 (2003)
79. J.A. Bautista, R.E. Connors, B.B. Raju, R.G. Hiller, F.P. Sharples, D. Gosztola, M.R. Wasielewski, H.A. Frank, J. Phys. Chem. B **103**, 8751 (1999)
80. D. Zigmantas, R.G. Hiller, A. Yartsev, V. Sundström, T. Polivka, J. Phys. Chem. B **107**, 5339 (2003)
81. S. Shima, R.P. Ilagan, N. Gillespie, B.J. Sommer, R.G. Hiller, F.P. Sharpless, H.A. Frank, R.R. Birge, J. Phys. Chem. A **107**, 8052 (2003)
82. E. Papagiannakis, D.S. Larsen, I.H.M. van Stokkum, M. Vengris, R.G. Hiller, R. van Grondelle, Biochem. **43**, 15303 (2004)
83. T.E. Meyer, Biochim. Biophys. Acta **806**, 175 (1985)
84. T.E. Meyer, E. Yakali, M.A. Cusanovich, G. Tollin, Biochem. **26**, 418 (1987)
85. W.W. Sprenger, W.D. Hoff, J.P. Armitage, K.J. Hellingwerf, J. Bacteriol. **175**, 3096 (1993)
86. M. Baca, G.E. Borgstahl, M. Boissinot, P.M. Burke, D.R. Williams, K.A. Slater, E.D. Getzoff, Biochem. **33**, 14369 (1994)
87. W.D. Hoff, P. Dux, K.H.B. Devreese, I.M. Nugteren-Roodzant, W. Crielaard, R. Boelens, R. Kaptein, J. van Beeumen, K.J. Hellingwerf, Biochem. **33**, 13959 (1994)
88. Y. Imamoto, T. Ito, M. Kataoka, F. Tokunaga, FEBS Lett. **374**, 157 (1995)
89. U.K. Genick, G.E. Borgstahl, K. Ng, Z. Ren, C. Pradervand, P.M. Burke, V. Srajer, T.Y. Teng, W. Schildkamp, D.E. McRee, K. Moffat, E.D. Getzoff, Science **275**, 1471 (1997)
90. U.K. Genick, S.M. Soltis, P. Kuhn, I.L. Canestrelli, E.D. Getzoff, Nature **392**, 206 (1998)
91. B. Perman, V. Srajer, Z. Ren, T. Teng, C. Pradervand, T. Ursby, D. Bourgeois, F. Schotte, M. Wulff, R. Kort, K. Hellingwerf, K. Moffat, Science **279**, 1946 (1998)
92. A. Xie, L. Kelemen, J. Hendriks, B.J. White, K.J. Hellingwerf, W.D. Hoff, Biochem. **40**, 1510 (2001)
93. R. Brudler, R. Rammelsberg, T.T. Woo, E.D. Getzoff, K. Gerwert, Nat. Struct. Biol. **8**, 265 (2001)
94. A. Baltuška, I.H.M. van Stokkum, A. Kroon, R. Monshouwer, K.J. Hellingwerf, R. van Grondelle, Chem. Phys. Lett. **270**, 263 (1997)

95. T. Gensch, C.C. Gradinaru, I.H.M. van Stokkum, J. Hendriks, K. Hellingwerf, R. van Grondelle, Chem. Phys. Lett. **356**, 347 (2002)
96. A. Xie, W.D. Hoff, A.R. Kroon, K.J. Hellingwerf, Biochem. **35**, 14671 (1996)
97. Y. Zhou, L. Ujj, T.E. Meyer, M.A. Cusanovich, G.H. Atkinson, J. Phys. Chem. A **105**, 5719 (2001)
98. H. Chosrowjan, N. Mataga, Y. Shibata, Y. Imamoto, F. Tokunaga, J. Phys. Chem. B **102**, 7695 (1998)
99. M.A. Cusanovich, T.E. Meyer, Biochem. **42**, 4759 (2003)
100. K.J. Hellingwerf, J. Hendriks, T. Gensch, J. Phys. Chem. A **107**, 1082 (2003)
101. D.S. Larsen, R. van Grondelle, ChemPhysChem **6**, 828 (2005)
102. L. Ujj, S. Devanathan, T.E. Meyer, M.A. Cusanovich, G. Tollin, G.H. Atkinson, Biophys. J. **75**, 406 (1998)
103. S. Devanathan, A. Pacheco, L. Ujj, M. Cusanovich, G. Tollin, S. Lin, N. Woodbury, Biophys. J. **77**, 1017 (1999)
104. Y. Imamoto, M. Kataoka, F. Tokunaga, T. Asahi, H. Masuhara, Biochem. **40**, 6047 (2001)
105. H. Chosrowjan, N. Mataga, N. Nakashima, I. Yasushi, F. Tokunaga, Chem. Phys. Lett. **270**, 267 (1997)
106. P. Changenet, H. Zhang, M.J. van der Meer, K.J. Hellingwerf, M. Glasbeek, Chem. Phys. Lett. **282**, 276 (1998)
107. M. Vengris, M.A. van der Horst, G. Zgrablic, I.H.M. van Stokkum, S. Haacke, M. Chergui, K.J. Hellingwerf, R. van Grondelle, D.S. Larsen, Biophys. J. **87**, 1848 (2004)
108. S. Devanathan, S. Lin, M.A. Cusanovich, N. Woodbury, G. Tollin, Biophys. J. **79**, 2132 (2000)
109. S. Mukamel, *Principles of nonlinear optical spectroscopy* (Oxford University Press, 1995)
110. V. May, O. Kühn, *Charge and energy transfer dynamics in molecular systems* (Wiley-VCH, Berlin, 2004)
111. H. van Amerongen, L. Valkunas, R. van Grondelle, *Photosynthetic excitons* (World Scientific, Singapore, 2000)
112. T. Renger, V. May, O. Kühn, Phys. Rep. **343**, 137 (2001)
113. J. L. Herek (ed.), *Coherent Control of Photochemical and Photobiological Processes*, special issue, J. Photochem. Photobiol. A Chemistry, **180**, 225 (2006).
114. T. Halfmann, Opt. Comm. **264**, 247 (2006)
115. J.L. Herek, W. Wohlleben, R.J. Cogdell, D. Zeidler, M. Motzkus, Nature **417**, 533 (2002)
116. B. Brüggemann, V. May, J. Phys. Chem. B **108**, 10529 (2004)
117. B. Brüggemann, K. Sznee, V. Novoderezhkin, R. van Grondelle, V. May, J. Phys. Chem. B **108**, 13563 (2004)
118. B. Brüggemann, V. May, Chem. Phys. Lett. **400**, 573 (2004)
119. B. Brüggemann, T. Pullerits, V. May, in [113], p. 322.
120. D. Voronine, D. Abramavicius, S. Mukamel, J. Chem. Phys. **124**, 034104 (2006)
121. T. Brixner, G. Krampert, T. Pfeifer, R. Selle, G. Gerber, M. Wollenhaupt, O. Graefe, C. Horn, D. Liese, T. Baumert, Phys. Rev. Lett. **92**, 208301 (2004)
122. A. Kaiser, V. May, J. Chem. Phys. **121**, 2528 (2004)
123. A. Kaiser, V. May, Chem. Phys. Lett. **405**, 339 (2005)
124. A. Kaiser, V. May, Chem. Phys. **320**, 95 (2006)

125. D.E. Tronrad, B.W. Matthews, in *Photosynthetic reaction centers*, ed. by J. Deisenhofer, J.R. Norris (Academic Press, 1993)
126. M. Wendling, M.A. Przyjalgowski, D. Gülen, S.I.E. Vulto, T.J. Aartsma, R. van Grondelle, H. van Amerongen, Phot. Res. **71**, 99 (2002)
127. B. Brüggemann, V. May, J. Chem. Phys. **118**, 746 (2003)
128. B. Brüggemann, V. May, J. Chem. Phys. **120**, 2325 (2004)
129. B. Brüggemann, D.V. Tsivlin, V. May, *Quantum dynamics in complex molecular systems* (Springer Series in Chemical Physics 83, 2006)
130. V. May, Int. J. Quant. Chem. **106**, 3056 (2006)
131. S.A. Rice, M. Zhao, *Optical control of molecular dynamics* (Wiley, New York, 2000)
132. M. Shapiro, P. Brumer, *Principles of the quantum control of molecular processes* (Wiley, New York, 2003)
133. Y. Ohtsuki, W. Zhu, H. Rabitz, J. Chem. Phys. **110**, 9825 (1999)
134. T. Mančal, V. May, Euro. Phys. J. D **14**, 173 (2001)
135. T. Mančal, U. Kleinekathöfer, V. May, J. Chem. Phys. **117**, 636 (2002)
136. R. Xu, Y.J. Yan, Y. Ohtsuki, Y. Fujimura, H. Rabitz, J. Chem. Phys. **120**, 6600 (2004)
137. S. Pagola, P.W. Stephens, D.S. Bohle, A.D. Kosar, S.K. Madsen, Nature **404**, 307 (2000)
138. P.M. ONeill, P.A. Stocks, M.D. Pugh, N.C. Araujo, K.E.E. Orshin, J.F. Bickley, S.A. Ward, P.G. Bray, E. Paini, J. Davies, V.E. Erissimo, M.D. Bachi, Angew. Chem. Int. Ed. **43**, 4193 (2004)
139. W. Freyer, H. Stiel, M. Hild, K. Teuchner, D. Leupold, Photochem. Photobiol. **66**, 596 (1997)
140. A.C. Uhlemann, A. Cameron, U. Eckstein-Ludwig, J. Fischbarg, P. Iserovich, F.A. Zuniga, M. East, A. Lee, L. Brady, R.K. Haynes, S. Krishna, Nat. Struc. Biol. **12**, 628 (2005)
141. S. Kapetanaki, C. Varotsis, J. Med. Chem. **44**, 3150 (2001)
142. R. Bonnet, Chem. Soc. Rev. **24**, 19 (1995)
143. M.R. Hamblin, E.L. Newman, J. Photochem. Photobiol., B **23**, 3 (1994)
144. I.J. Macdonald, T.J. Dougherty, J. Porphyrins Phthalocyanines **5**, 105 (2001)
145. J.M.D. Cruz, I. Pastirk, M. Comstock, V.V. Lozovoy, M. Dantus, Proc. Natl. Acad. Sci. **101**, 16996 (2004)
146. D. Kearns, A. Khan, Photochem. Photobiol. **10**, 193 (1969)
147. R. Schmidt, K. Schaffner, W. Trost, H.D. Brauer, J. Phys. Chem. **88**, 956 (1984)
148. K.B. Eisenthal, N.J. Turro, C.G. Dupuy, D.A. Hrovat, J. Langan, T.A. Jenny, E.V. Sitzmann, J. Phys. Chem. **90**, 5168 (1986)
149. J. Rigaudy, C. Breliere, P. Scribe, Tetrahedron Lett. **7**, 687 (1978)
150. A. Klein, M. Kalb, M.S. Gudipati, J. Phys. Chem. A **103**, 3843 (1999)
151. B.O. Roos, *Advances in Chemical Physics: The complete active space self-consistent field method and its application in electronic structure calculations*, vol. 69 (John Wiley & Sons Ltd., 1987)
152. K. Pierloot, B. Dumez, P.O. Widmark, B.O. Roos, Theor. Chim. Acta **90**, 87 (1995)
153. G. Karlström, R. Lindh, P.A. Malmqvist, B.O. Roos, U. Ryde, V. Veryazov, P.O. Widmark, M. Cossi, B. Schimmelpfennig, P. Neogrady, L. Seijo, Comput. Mat. Sci. **28**, 222 (2003)

154. K. Anderson, P.A. Malmqvist, B.O. Roos, A.J. Sadlej, K. Wolinski, J. Chem. Phys. **96**, 1218 (1990)
155. P.O. Widmark, P.A. Malmqvist, B.O. Roos, Theor. Chim. Acta **77**, 291 (1990)
156. M.J. Frisch, G.W. Trucks, H.B. Schlegel, G.E. Scuseria, M.A. Robb, J.R. Cheeseman, V.G. Zakrzewski, J.A. Montgomery, R.E. Stratmann, J.C. Burant, S. Dapprich, J.M. Millam, A.D. Daniels, K.N. Kudin, M.C. Strain, O. Farkas, J. Tomasi, V. Barone, M. Cossi, R. Cammi, B. Mennucci, C. Pomelli, C. Adamo, S. Clifford, J. Ochterski, G.A. Petersson, P.Y. Ayala, Q. Cui, K. Morokuma, D.K. Malick, A.D. Rabuck, K. Raghavachari, J.B. Foresman, J. Cioslowski, J.V. Ortiz, B.B. Stefanov, G. Liu, A. Liashenko, P. Piskorz, I. Komaromi, R. Gomperts, R.L. Martin, D.J. Fox, T. Keith, M.A. Al-Laham, C.Y. Peng, A. Nanayakkara, C. Gonzalez, M. Challacombe, P.M.W. Gill, B.G. Johnson, W. Chen, M.W. Wong, J.L. Andres, M. Head-Gordon, E.S. Replogle, J.A. Pople, *Gaussian 98 (Revision A.2)* (Gaussian, Inc., Pittsburgh PA, 1998)
157. P. Deuflhard, C. Schütte, in *Applied Mathematics entering the 21st century. Proc. ICIAM 2003*, ed. by J.M. Hill, R. Moore (2004), pp. 91–119
158. C. Schütte, A. Fischer, W. Huisinga, P. Deuflhard, J. Comput. Phys., Special Issue on Computational Biophysics **151**, 146 (1999)
159. P. Deuflhard, W. Huisinga, A. Fischer, C. Schütte, Lin. Alg. Appl. **315**, 39 (2000)
160. M. Dellnitz, O. Junge, SIAM J. Num. Anal. **36**, 491 (1999)
161. A. Fischer, S. Waldhausen, I. Horenko, E. Meerbach, C. Schütte, J. Chem. Phys., submitted (2004)
162. I. Horenko, E. Dittmer, C. Schütte, Comp. Vis. Sci. **9**, 89 (2005)
163. I. Horenko, E. Dittmer, A. Fischer, C. Schütte, Multiscale modeling and simulation, accepted (2005)
164. E. Meerbach, E. Dittmer, I. Horenko, C. Schütte, in *Computer simulations in condensed matter systems*, *Lecture Notes in Physics*, vol. 703 (2006), *Lecture notes in Physics*, vol. 703, pp. 475–497
165. P. Metzner, C. Schütte, E. Vanden-Eijnden, J. Chem. Phys. **125**, 084110 (2006)
166. F. Cordes, M. Weber, J. Schmidt-Ehrenberg, Metastable conformations via successive Perron cluster analysis of dihedrals (2002). ZIB-Report 02-40, Zuse-Institute-Zentrum, Berlin
167. P. Deuflhard, M. Weber, Lin. Alg. Appl. **398**, 164 (2005)
168. W. Huisinga, B. Schmidt, in *New algorithms for macromolecular simulation*, *Lecture notes in Computational Science and Engineering*, vol. 49, ed. by C. Chipot, R. Elber, A. Laaksonen, B. Leimkuhler, A. Mark, T. Schlick, C. Schütte, R. Skeel (Springer, 2005), *Lecture notes in Computational Science and Engineering*, vol. 49, pp. 167–182
169. C. Schütte, W. Huisinga, in *Handbook of Numerical Analysis*, vol. X, ed. by P.G. Ciaret, J.L. Lions (North-Holland, 2003), pp. 699–744
170. P. Bremaud, *Gibbs fields, Monte Carlo simulation, and queues*, *Texts in applied mathematics*, vol. 31 (Springer, New York, 1999)
171. M. Weber, Improved Perron cluster analysis (2004). ZIB-Report 03-04, Zuse-Institute-Zentrum, Berlin
172. A.P. Dempster, N.M. Laird, D.B. Rubin, J. Roy. Stat. Soc. B **39**, 1 (1977.)
173. I. Horenko, E. Dittmer, F. Lankas, J. Maddocks, P. Metzner, C. Schütte, J. Appl. Dyn. Syst., submitted (2005)
174. A.J. Viterbi, IEEE Trans. Informat. Theory **IT-13**, 260 (1967)

175. G.N. Ramachandran, V. Sasiskharan, Advan. Prot. Chem. **23**, 283 (1968)
176. D. van der Spoel, E. Lindahl, B. Hess, G. Groenhof, A. Mark, H.J. Berendsen, J. Chem. Phys. **26**, 1701 (2005)
177. C. Oostenbrink, A. Villa, A.E. Mark, W.F.V. Gunsteren, J. Comp. Chem. **25**, 1656 (2004)
178. J. Antony, B. Schmidt, C. Schütte, J. Chem. Phys. **122**, 014309 (2005)
179. T. Head-Gordon, M. Head-Gordon, M.J. Frisch, C.L. Brooks III, J.A. Pople, J. Am. Chem. Soc. **113**, 5989 (1991)
180. A. Perczel, Ö. Farkas, I. Jakli, I.A. Topol, I.G. Csizmadia, J. Comp. Chem. **24**, 1026 (2003)
181. L. Schäfer, C.V. Alsenoy, J.N. Scarsdale, J. Chem. Phys. **76**, 1439 (1982)
182. H. Hu, M. Elstner, J. Hermans, Proteins **50**, 451 (2003)
183. Y. Mu, G. Stock, J. Phys. Chem. B **106**, 5294 (2002)
184. Y. Mu, P.H. Nguyen, G. Stock, Proteins **58**, 45 (2004)
185. W.G. Han, K.J. Jalkanen, M. Elstner, S. Suhai, J. Phys. Chem. B **102**, 2587 (1998)
186. F.C. Tenover, J.K.R. eds P R Murray, E.J. Baron, M.A. Pfaller, *Manual of clinical Microbiology* (Amer. Society for Microbiology, 1999)
187. J. Ho, Analy. Chim. Acta **457**, 125 (2002)
188. P. Belgrader, W. Benett, D. Hadley, J. Richards, P. Stratton, R. Mariella, F. Milanovich, Science **284**, 449 (1999)
189. S.I. Makino, H.I. Cheun, M. Wateral, I. Uchida, K. Takeshi, Appl. Microbio. **33**, 237 (2001)
190. B. Beatty, S. Mai, J. Squire (eds.), *FISH: a practical approach* (Oxford University Press, 2002)
191. F. Pourahmadi, M. Taylor, G. Kovacs, K. Lloyd, S. Sakai, T.S.B. Helton, L. Western, S. Zaner, J. Ching, B. McMillan, P. Belgrader, M.A. Northrup, Clin. Chem. **46**, 1151 (2000)
192. B.K. De, S.L. Bragg, G.N. Sanden, K.E. Wilson, L.A. Diem, C.K. Marston, A.R. Hoffmaster, G.A. Barnett, R.S. Weyant, T.G. Abshire, J.W. Ezzell, T. Popovic, Emerging Infection Diseases **8**, 1060 (2002)
193. C. Hagleitner, A. Hierlemann, D. Lange, A. Kummer, N. Kerness, O. Brand, H. Baltes, Nature **414**, 293 (2001)
194. P. Francois, M. Bento, P. Vaudaux, J. Schrenzel, J. Micros. Meth. **55**, 755 (2003)
195. T.H. Rider, M.S. Petrovick, F.E. Nargi, J.D. Harper, E.D. Schwoebel, R.H. Mathews, D.J. Blanchard, L.T. Bortolin, A.M. Young, J.Z. Chen, M.A. Hollis, Science **301**, 213 (2003)
196. Y.L. Pan, J. Hartings, R.G. Pinnick, J.H. S C Hill, R.K. Chang, Aerosol Sci. Tech **37**, 627 (2003)
197. F.L. Reyes, T.H. Jeys, N.R. Newbury, C.A. Primmerman, G.S. Rowe, A. Sanchez, Field Anal Chem. Tech. **3**, 240 (1999)
198. J.D. Eversole, W.K. Cary, C.S. Scotto, R. Pierson, M. Spence, A.J. Campillo, Field Anal. Chem. Tech. **5**, 205 (2001)
199. G.A. Luoma, P.P. Cherrier, L.A. Retfalvi, Field Anal. Chem. Tech. **3**, 260 (1999)
200. Y.L. Pan, K.B. Aptowicz, R.K. Chang, M. Hart, J.D. Eversole, Opt. Lett. **28**, 589 (2003)
201. P. Kaye, E. Hirst, T.J. Wang, Appl.Opt **36**, 6149 (1997)

202. S.C. Hill, R.P. Pinnick, S. Niles, Y.L. Pan, S. Holler, R.K. Chang, J. Bottiger, B.T. Chen, C.S. Orr, G. Feather, Field Anal Chem. Tech. **5**, 221 (1999)
203. Y.L. Pan, P. Cobler, S. Rhodes, A. Potter, T. Chou, S. Holler, R.K. Chang, R.G. Pinnick, J.P. Wolf, Rev.Sci.Inst. **72**, 1831 (2001)
204. R.G. Pinnick, S.C. Hill, Y.L. Pan, R.K. Chang, Atmosph.Environ. **38**, 1657 (2004)
205. Y.S. Cheng, E.B. Barr, B.J. Fan, P.J. Hargis, D.J. Rader, T.J. OHern, T.R. Torczynski, G.C. Tisone, B.L. Preppernau, S.A. Young, R.J. Radloff, Aero.Sci.Tech. **31**, 409 (1999)
206. C. Weitkamp (ed.), *Lidar*, vol. 102 (Springer-Verlag, New York, 2005)
207. F. Immler, D. Engelbart, O. Schrems, Atmos.Chem.Phys **4**, 5831 (2004)
208. T. Brixner, N.H. D, P. Niklaus, G. Gerber, Nature **414**, 57 (2001)
209. T. Brixner, N.H. D, B. Kiefer, G. Gerber, J.Chem. Phys. **118** (2003)
210. B. Li, H. Rabitz, J.P. Wolf, J. Chem. Phys. **122**, 154103 (2005)
211. S.C. Hill, V. Boutou, J. Yu, S. Ramstein, J.P. Wolf, Y. Pan, S. Holler, R.K. Chang, Phys.Rev.Lett. **85**, 54 (2000)
212. V. Boutou, C. Favre, S.C. Hill, Y. Pan, R.K. Chang, J.P. Wolf, App.Phys.B **75**, 145 (2002)
213. Y. Pan, S.C. Hill, J.P. Wolf, S. Holler, R.K. Chang, J.R. Bottiger, Appl. Opt. **41**, 2994 (2002)
214. C. Favre, V. Boutou, S.C. Hill, W. Zimmer, M. Krenz, H. Lambrecht, J. Yu, R.K. Chang, L. Woeste, J.P. Wolf, Phys.Rev.Lett. **89**, 035002 (2002)
215. J. Kasparian, M. Rodriguez, G. Méjean, J. Yu, E. Salmon, H. Wille, R. Bourayou, S. Frey, Y.B. Andre, A. Mysyrowicz, R. Sauerbrey, J.P. Wolf, L. Woeste, Science **301**, 61 (2003)
216. G. Méjean, J. Kasparian, J. Yu, S. Frey, E. Salmon, J.P. Wolf, Appl. Phys. **78**, 535 (2004)
217. F. Courvoisier, V. Boutou, V. Wood, J.P. Wolf, A. Bartelt, M. Roth, H. Rabitz, App. Phys. Lett. **87**, 063901 (2005)
218. A.L. Sobolewski, W. Domcke, C. Dedonder-Lardeux, C. Jouvet, Phys. Chem. Chem. Phys. **4**, 1093 (2002)
219. J. Teraoka, P.A. Harmon, S.A. Asher, J. Am. Chem. Soc. **112**, 2892 (1989)
220. H.B. Steen, J. Chem. Phys. **61**, 3997 (1974)
221. S. Morel, N. Leon, P. Adam, J. Amouroux, Appl. Opt. **42**, 6184 (2005)
222. P.B. Dixon, D.W. Hahn, Anal. Chem. **77**, 631 (2005)
223. M. Baudelet, L. Guyon, J. Yu, J.P. Wolf, T. Amodeo, E. Fréjafon, P. Laloi, to be published in J. App. Physics (2006)
224. M. Baudelet, M. Bossu, J. Jovelet, J. Yu, J.P. Wolf, T. Amodeo, E. Fréjafon, P. Laloi, submitted (2006)
225. M. Baudelet, L. Guyon, J. Yu, J.P. Wolf, T. Amodeo, E. Fréjafon, P. Laloi, App. Phys. Lett. **88**, 053901 (2006)
226. K. Stelmanszczyk, P. Rohwetter, G. Méjean, J. Yu, E. Salmon, J. Kasparian, R. Ackermann, J.P. Wolf, L. Woeste, App. Phys. Lett. **85**, 3977 (2004)

Author Index

Subject Index

Printing: Krips bv, Meppel
Binding: Stürtz, Würzburg